Essentials of Glycobiology

FOURTH EDITION

Full text of this work and associated links and downloadable PowerPoint slides of all of the figures are available at http://www.ncbi.nlm.nih.gov/books/n/glyco4

Essentials of Glycobiology

FOURTH EDITION

EDITED BY

Ajit Varki
Richard D. Cummings
Jeffrey D. Esko
Pamela Stanley
Gerald W. Hart
Markus Aebi
Debra Mohnen
Taroh Kinoshita
Nicolle H. Packer
James H. Prestegard
Ronald L. Schnaar
Peter H. Seeberger

CSH PRESS
COLD SPRING HARBOR LABORATORY PRESS
Cold Spring Harbor, New York • www.cshlpress.org

Essentials of Glycobiology, Fourth Edition

Published by Cold Spring Harbor Laboratory Press, Cold Spring Harbor, New York
Printed in the United States of America

Library of Congress Control Number: 2021950841

Publisher and Acquisition Editor	John Inglis
Senior Project Manager	Inez Sialiano
Permissions Coordinator	Carol Brown
Production Editor	Kathleen Bubbeo
Production Manager	Denise Weiss
Cover Designers	Lorenzo Casalino and Rommie Amaro
Illustrator and Illustrations Coordinator	Richard D. Cummings

Front cover artwork: Molecular representation of the full-length, fully glycosylated, all-atom model of the SARS-CoV-2 spike protein in the open state, embedded in the viral membrane. The model of the spike has been developed by Casalino et al. (*ACS Cent Sci* **6:** 1722–1734 [2020]) based on the cryo-EM structure by Wrapp et al. (*Science* **367:** 1260–1263 [2020]) (PDB ID: 6VSB). N-/O-glycans have been modeled according to Watanabe et al. (*Science* **369:** 330–333 [2020]) and Shajahan et al. (*Glycobiology* **30:** 981–988 [2020]).

On the *left*, the full-length SARS-CoV-2 spike in the open state—that is, with one receptor binding domain (RBD) in the "up" conformation—is shown with a cyan transparent surface overlaid on the cartoon representation of its secondary structure. The conformation of the spike was selected from molecular dynamics simulations performed by Casalino et al. (*ACS Cent Sci* **6:** 1722–1734 [2020]). N-linked and O-linked glycans are depicted using the Symbol Nomenclature for Glycans (SNFG), in which blue filled squares are for *N*-acetyl-D-glucosamine (GlcNAc), green filled circles for D-mannose, yellow filled squares for *N*-acetyl-D-galactosamine (GalNAc), yellow filled circles for D-galactose (Gal), red filled triangles for L-fucose (Fuc), and purple filled diamonds for *N*-acetyl-D-neuraminic acid (Neu5Ac). The lipid bilayer of the viral membrane is depicted with a surface representation, in which the POPC (1-palmitoyl-2-oleoyl-sn-glycero-3-phosphocholine) are colored in pink, POPE (1-palmitoyl-2-oleoyl-sn-glycero-3-phosphoethanolamine) in purple, POPI (1-palmitoyl-2-oleoyl-sn-glycero-3-phosphoinositol) in orange, POPS (1-palmitoyl-2-oleoyl-sn-glycero-3-phosphoserine) in red, and cholesterol in yellow.

On the *right*, the glycan shield (dark-blue bush-like structures) in the SARS-CoV-2 spike protein (cyan transparent surface) is shown by overlaying multiple conformations of the N-linked and O-linked glycans obtained at multiple, interspersed frames along 1 μsec of molecular dynamics simulations (Casalino et al., *ACS Cent Sci* **6:** 1722–1734 [2020]). For each glycan, each conformation sampled along the dynamics is shown with dark-blue sticks. When multiple conformations of each glycan are overlaid, they form a protective bush-like structure providing a visual representation of the extent of protein surface covered over a specific time frame. When the receptor-binding domain (RBD), located in the apical portion of the spike, is in the "up" conformation, it emerges from the glycan shield (as shown in the image with transparent cyan surface) and becomes available for binding to the angiotensin-converting enzyme 2 (ACE2) receptors located on the host cell. The binding event between the RBD and ACE2 initiates infection.

The cover artwork was designed and created by Dr. Lorenzo Casalino in the Amaro Laboratory (University of California San Diego), based on the all-atom model published in *ACS Cent Sci* **6:** 1722–1734 (2020).

Library of Congress Cataloging-in-Publication Data
Identifiers: LCCN 2021950841 | ISBN 978-1-621824-21-3 (hardcover) | ISBN 978-1-621824-22-0 (ePub3)

10 9 8 7 6 5 4 3 2 1

For a complete catalog of all Cold Spring Harbor Laboratory Press publications, visit our website at www.cshlpress.org.

Contents

METHODS AND APPLICATIONS

Foreword

THE GOAL OF BARRIER-FREE ACCESS TO SCHOLARLY information online has preoccupied academic researchers for more than two decades. The debates have focused principally on how to liberate research papers from the constraints of journal subscriptions. But in more recent years, as more and more journals have become open access, the spotlight has shifted to books, including textbooks, and the challenge of making their content freely available to readers while retaining financial viability. Amid the arguments about paths forward, the 20-year success of *Essentials of Glycobiology* in achieving the goal is less well-known than it should be. That success is attributable to a unique partnership between two not-for-profit institutions and a consortium of scientist–editors led by the book's executive editor, the indefatigable Ajit Varki.

The first edition of *Essentials of Glycobiology* was published in 1999, in the same year that the Director of the National Institutes of Health (NIH), Harold Varmus, proposed creating a freely available archive of all new research manuscripts. The radical "Ebiomed" scheme did not gain traction but, pushed by Ajit, then its Editor-in-Chief, *The Journal of Clinical Investigation* was already making all its articles freely available online at publication. This bold gambit had greatly expanded the readership of the journal's content. Inspired by this success, and as an energetic proselyte for the emerging, cross-disciplinary field of glycobiology, Ajit was eager to give his field the widest possible audience. So, on publication of the textbook, discussion began with its publisher, Cold Spring Harbor Laboratory Press, about ways of making its content freely available. The idea of an open-access book was intriguing to an institution with a 100-year history of innovation in the communication of science.

The National Center for Biotechnology Information (NCBI), a division of the National Library of Medicine at NIH, was led by another entrepreneurial scientist interested in breaking barriers to information flow, David Lipman. Among NCBI's innovations was the creation of Bookshelf, a platform for open distribution of book content. Brokered by Ajit and his colleagues, the Consortium of Glycobiology Editors, who collectively owned the copyright of *Essentials in Glycobiology*, an agreement was forged between the Laboratory and NCBI that enabled the full text of the book to be discovered, searched, and read online on the Bookshelf, while the print edition remained available for sale through the Press. The risk, seemingly significant at the time, was that the availability of a free, high-quality rendition of the book's content online would limit print sales and prevent the Press from recovering the investment required to create the book.

The online edition of *Essentials* appeared in 2003 and its usage was substantial, with hundreds of people each week from all over the world continuing to consult it even five years later. The content was discoverable through indexing by major search engines, citations in Wikipedia, listing and linking by libraries, and entries in PubMed. And gratifyingly, interest in the print edition did not appear to waver. So, in 2008, when a second edition of *Essentials* was ready for publication, another bold decision was taken, to release both the print and Bookshelf editions simultaneously, the first time this had been done with a major textbook. And once again, the gambit paid off, as sales of the print book prospered while usage of the online content was brisk. The same approach was adopted for the third edition in 2017. And now, for this fourth edition, the project continues to evolve, as the freely accessible Bookshelf edition is released simultaneously with both a print edition and an e-book downloadable exclusively from the Press website that can be read easily on a variety of portable devices.

Nearly two decades after the editors, NCBI, and the Press embarked on this joint venture, devotees of glycobiology have more choices than ever in how they consult and use the content of the latest *Essentials.* Tastes and information habits have changed greatly during that time—a print edition of such a large book now has much less appeal to readers, despite the elegance of its design and illustrations by Consortium member Rick Cummings and its high production values. But the mission of *Essentials* has not changed. The Consortium of Glycobiology Editors remains substantially intact but has also recruited a new generation of younger authors who can carry the torch forward in the future. With each edition, they collectively strive to create an up-to-date knowledge resource that will satisfy established investigators but also entice new recruits to the study of the important, diverse, and expanding biology of the glycans. CSHL Press and successive leaders at NCBI remain committed to ensuring that the information the editors have curated reaches the widest possible audience, however and wherever those individuals want to use it. On behalf of our organizations, we congratulate The Consortium of Glycobiology Editors on creating another remarkable, important, and pioneering contribution to the literature of Glycoscience.

John Inglis
Executive Director and Publisher
Cold Spring Harbor Laboratory Press

Rita Sarkar
Program Officer
National Heart, Lung, and Blood Institute (NHLBI)
National Institutes of Health

Stephen T. Sherry
Acting Director
National Center for Biotechnology Information (NCBI)
National Library of Medicine

Preface

BECAUSE THIS IS THE FOURTH EDITION OF THIS BOOK, it seems appropriate to look back to the origins of the project and to acknowledge the many individuals and organizations that have contributed. Such memories of the past are not stored in our brains like computer files but must rather be reconstituted each time we recall, making them prone to fade and/or "mutate" over time. But since the 1990s, we have archived e-mails and other electronic records to cross-check if our memories are accurate. Digging through such archives and contacting various people, I was able to reconstitute a reasonably accurate (although likely incomplete) history below.

Until the 1970s, the study of biological systems included equal attention to all major classes of macromolecules: nucleic acids, proteins, lipids, and glycans. But by the end of the 1980s, most popular books, monographs, and manuals about Molecular Biology had a strong emphasis on DNA, RNA, and proteins, with some attention to lipids, but effectively leaving out glycans, which were poorly understood, far more complex, diverse, and difficult to study. This anomaly also affected the popular *Current Protocols in Molecular Biology*, established in 1987 by experts at Harvard, in collaboration with Sarah Greene of Greene Publishing Associates. But one of the editors (the late John Smith, who was biochemically inclined) proposed to the board a chapter about glycans. Apparently, this proposal came after discussions with John Coligan at NIH (later an editor of *Current Protocols in Immunology*). The outcome was a series of Units on Preparation and Analysis of Glycoconjugates, led by Adriana Manzi, Hud Freeze, and myself and coordinated by Kaaren Janssen as the Series Editor.

By the 1990s, the now somewhat segregated minority of scientists working primarily on glycans had embraced Raymond Dwek's term "Glycobiology" and a Society and Journal of the same name followed. At UC San Diego, a small group of afficionados had formed a Glycobiology Program within the UCSD Cancer Center and, along with the Division of Cellular and Molecular Medicine (created by George Palade and Marylin Farquhar), recruited Jamey Marth and then Jeff Esko—while also collaborating with a parallel program at the La Jolla Cancer Research Foundation (now Sanford Burnham Prebys) to organize an annual San Diego Glycobiology Symposium. A graduate-level, elective course entitled "Essentials of Glycobiology" had already been taught at UCSD for many years, and it seemed logical to convert it into a textbook. In a December 23, 1996 e-mail to the Glycobiology Program members, I wrote that "[w]hile I was at the GlycoImmunology meeting in Greece, I had a long talk with Jerry Hart and Rick Cummings about their proposed Glycobiology Book. They were interested in the possibility of combining forces with our group to do a Glycobiology course and produce a book in 1998." Working together, we were able to put together a "Consortium of Glycobiology Editors" (CGE) who proposed a graduate-level textbook called *Essentials of Glycobiology*.

In looking for a publisher, we had the advantage of the prior connection with Kaaren Janssen (mentioned above), who was now at Cold Spring Harbor Laboratory Press (CSHLP), and who introduced us to John Inglis, the Executive Director of the Press. After considering some options, it was an easy choice to make an agreement with the highly prestigious and nonprofit CSHLP. An additional major consideration was John's forward-thinking willingness to consider our rather unusual request: to make the entire book freely available for online searching and reading. This first-of-its-kind example for a major textbook was possible because of yet another coincidence—I had just finished a stint as Editor in Chief of the *Journal of Clinical Investigation* (JCI), during which we took advantage of the online presentation of the journal (a new concept) to also make it free to all readers. This approach was later called "open access."

The success of the JCI experiment generated an invitation to the PubMed Central Advisory Committee from David Lipman, then the new Director of the National Center for Biotechnology Information (NCBI) of the National Library of Medicine—the home of GenBank, PubMed, and many other free online bioinformatic resources. Lipman did not hesitate to accept a proposal from the editors and CSHLP to make the entire book freely available at NCBI Bookshelf for searching and reading by anyone anywhere in the world with an internet connection—eventually including downloadable slides for each figure. Jo McEntyre (followed by Marilu Hoeppner) at NCBI worked closely with the Press and the editors to make this all possible. Since the recent departure of Lipman from NCBI, we were fortunate that the NCBI leaders who followed (Jim Ostell and now Steve Sherry) have continued their support of the online version. Although the Office of the NIH Director later considered supporting the effort, they could not do so without a national competition, and an anonymous donor stepped in to help the Press partially absorb the costs. But the editorial "office" still needed administrative support. Fortunately, a long-standing, glycobiology-focused program project from the National Heart, Lung, and Blood Institute (NHLBI) to members of the UCSD Glycobiology Research and Training Center (GRTC; co-directed by Jeff Esko and myself) could be used to support such efforts, followed by the NHLBI Programs of Excellence in Glycosciences (2011–2019), and then a National Career Development Consortium for Excellence in Glycosciences (K12, 2018–2023), which focused on immersive postdoctoral training in all aspects of glycosciences. All NHLBI support for these programs (and hence for the book) was spearheaded by program officer Rita Sarkar, without whose dedication and relentless support none of this would have happened.

As with the last edition, the editors of the CGE (now registered as a U.S. nonprofit association) have agreed to do this work purely as a service to the scientific community and will not be accepting any personal income from the book. Rather, we will be assigning any remaining income, after covering production costs, to further the impact of the book in the glycoscience community. Editors of the earlier editions have agreed to forward residual income from that edition toward the same goal.

Two other spinoffs that the NCBI supported with the third edition are now independently run by internationally representative committees: the NCBI Glycans Page (led now by Natasha Zachara) and the Symbol Nomenclature for Glycans (SNFG; led now by Sriram Neelamegham). Another special feature of the book has been the outstanding artwork executed largely by Rick Cummings (in this and all previous editions) and the downloadable slides of all figures. Special thanks also to Lorenzo Casalino and Rommie Amaro for the images of the COVID spike protein on the front cover of this edition.

Over the four editions, we have had progressive improvements in gender balance among editors and coauthors from the first to the current edition (from 8% to 39% female), the total number of editor and coauthors (from 13 to 131 participants), and in international participation (from 8% to 56%). Usage statistics from NCBI indicate access by almost 2,036,226 unique IP addresses in 2,213,203 sessions since 2011. Many kudos are due to the CSHLP Team (Denise Weiss, Kathleen Bubbeo, Inez Sialiano, Carol Brown, and Mala Mazzullo), the NCBI Bookshelf team (Stacy Lathrop, Kim Pruitt, Susan Douglas, Diana Jordan, and Jeff Beck), Evan Bolton at PubChem, and last but definitely not least, Amanda Cuervo, the "Glycobook Administrator" at the UCSD GRTC, who coordinated everything.

Most editors of this fourth edition are "Boomers" like me and will likely not be involved in the next one. Thus, the future of the book, and the field, will be determined by a younger generation of afficionados such as those we have recruited as coauthors in the current edition. We hope they will complete the process of reintroducing glycans back into the mainstream of "molecular biology."

Ajit Varki
for the Consortium of Glycobiology Authors

Praise for the Previous Editions of *Essentials of Glycobiology*

FIRST EDITION

"Glycobiology is a field undergoing spectacular advances that progressively reveal the critical role of glycoproteins and proteoglycans in the organization and function of eukaryotic cells, especially in multicellular organisms. Glycobiology is . . . poised for further major advances in the immediate future. This book . . . will greatly help all those ready to enter this challenging, yet promising field."

– **George E. Palade**, *Nobel Laureate in Medicine, 1974*

"Precise laws govern the linear sequences of nucleic acids and proteins, but what about complex carbohydrates? Is there a carbohydrate code? Do carbohydrate sequences dictate the behavior of the proteins and lipids that carry them? This book provides the background necessary to answer these timely questions. In the Era of Functional Genomics, this book will be indispensable to anyone who encounters a new molecule with an unknown sugar modification."

– **Michael S. Brown and Joseph L. Goldstein,**
Nobel Laureates in Medicine, 1985

"As the literature that is available to biological scientists continues to expand at an ever increasing rate we tend to specialize . . . This practice is, of course, fraught with dangers. The most exciting developments are often those that bring together knowledge from one area and show that it is applicable to another . . . Thus, *Essentials of Glycobiology* will be of great value to all of us and will be important for people working in essentially all branches of biology."

– **Edwin G. Krebs**, *Nobel Laureate in Medicine, 1992*

"The prefix glyco- strikes dread into the hearts of many molecular biologists . . . Thank heavens for this new treatise on Glycobiology, which does much to demystify this complex area. The treatment is logical and thorough, yet individual topics are presented with refreshing brevity. Although the inherent biology of glycosylation is poorly understood, what is known is presented with clarity and rigor. This is the textbook I wished for six years ago, when I developed an interest in the area. I recommend it as a knowledgeable and readable source of information . . ."

– **Richard J. Roberts**, *Nobel Laureate in Medicine, 1993*

SECOND EDITION

"My own research career has repeatedly intersected with glycobiology—ganglioside affinity purification of *E. coli* enterotoxin, isolation of red cell Rh blood group antigens, and molecular elucidation of N-glycosylated aquaporin water channels. The biological importance of these carbohydrates was always intriguing but at times confusing. The availability of the second edition of *Essentials of Glycobiology* now provides succinct coverage of this important area of science for non-experts, like me, as well as the cognoscenti. The basic principles of glycobiology are clearly articulated in this volume, and the roles of complex carbohydrates in disease are an important read for all biomedical scientists."

– **Peter Agre,** *Nobel Laureate in Chemistry, 2003*

"My first experience with glycoproteins was the research for my thesis (1957) on the effects of proteins on the physical characteristics of the long-chain sugar hyaluronic acid. In 1990 I began an association with the Glycobiology Institute of Oxford University, a pioneer glycobiology laboratory where the term glycobiology was introduced by its Director, Raymond Dwek. We initiated a program [that] led to the design of novel potential therapies—iminosugars—for HCV and potentially many other viruses that could have a profound effect on the prevention and treatment of viral disease. *Essentials of Glycobiology* is a major resource for understanding these posttranslational biochemical reactions that affect the function and fate of proteins produced by the genes that are profoundly changed by their added sugars."

– **Baruch S. Blumberg,** *Nobel Laureate in Medicine, 1976*

"Paul Valery, the French author, poet, and mathematician, once said: '*Ce qui est simple est faux, ce qui est compliqué est incomprehensible*' (that which is simple is wrong, that which is complicated is incomprehensible). Well, yes, but not this time. The second edition of *Essentials of Glycobiology*, superbly printed and illustrated, develops in simple and absolutely precise terms the complicated intricacies of glycobiology. I would have killed to get this encyclopedic treatise 40 years ago when I was working my way through this field."

– **Edmond H. Fischer,** *Nobel Laureate in Medicine, 1992*

THIRD EDITION

"The field of glycobiology has matured. The comments of Nobel Laureates on the previous editions reflect the long-held belief that central functional roles played by the diversity of glycan chains would be revealed by research in this field. Now, as the result of advances in analytical chemistry and much deeper understanding of genomes, cell and tissue organization, this field has arrived. The third edition of *Essentials of Glycobiology* stands as the authoritative treatise on the subject, covering all aspects of the field and written by the world leaders in current research."

– **James E. Rothman,** *Nobel Laureate in Medicine, 2013*

"Difficult to analyze and synthesize artificially, glycans are often simply ignored. To do so is to avert one's gaze from an important part of life. More than mere decoration, glycans magnify the diversity of the already diverse molecules to which they are attached, affect protein folding and stability, direct traffic within cells, serve as signposts of self vs. non-self, create barriers that protect us, and conversely, defend microbes, making some of them the pathogens they are. It is hard to imagine a world without complex sugars, but if such a world existed, it would be much diminished. The third edition of *Essentials of Glycobiology* may be life changing for scientists who have not yet engaged with glycobiology and will certainly be a treasured resource for those who already have."

– **Bruce Beutler,** *Nobel Laureate in Medicine, 2011*

"The importance of glycans has long been recognized and great advances have been reported on the synthesis and chemical analysis of this class of natural compounds. In my field, structural biology, carbohydrate moieties in glycoproteins and in complex multi-component macromolecular systems have been and continue to be difficult to handle. I greatly welcome the effort made in this multiauthor volume to present results obtained with methods of structural biology in the context of the wealth of currently available chemical and biological data. I recommend the third edition of *Essentials of Glycobiology* as a highly useful reference on the current state of the field."

– **Kurt Wüthrich**, *Nobel Laureate in Chemistry, 2002*

"We think conventionally of the immune system as having evolved to deal with invading pathogens that express "foreign" proteins (and peptides), targeted via specific recognition units, particularly secreted antibodies and cell-bound T lymphocyte receptors. As the molecular revolution has unfolded, such science that relies on a reasonably direct correlation between genotype and phenotype has been relatively straightforward. Much more difficult to assess is the part played by glycosylation profiles in immune recognition and pathogen elimination. Perhaps even more intriguing is the issue of glycan abnormalities and recognition in cancer and many other disease processes. Now, in a third edition of *Essentials of Glycobiology* written by authoritative leaders in the field, we learn how this knowledge has been rapidly advancing, and see possibilities for real breakthroughs in understanding and therapy."

– **Peter C. Doherty**, *Nobel Laureate in Chemistry, 1996*

Contributors

EXECUTIVE EDITOR

Ajit Varki, Distinguished Professor of Medicine and of Cellular & Molecular Medicine, Founding Director, Glycobiology Research and Training Center, University of California, San Diego, La Jolla, California, USA

EDITORS

Richard D. Cummings, Professor of Surgery, Beth Israel Deaconess Medical Center, Director, National Center for Functional Glycomics, Harvard Medical School, Boston, Massachusetts, USA

Jeffrey D. Esko, Distinguished Professor of Cellular & Molecular Medicine, Founding Director, Glycobiology Research and Training Center, University of California, San Diego, La Jolla, California, USA

Pamela Stanley, Horace W. Goldsmith Foundation Chair, Professor of Cell Biology, Albert Einstein College of Medicine, Bronx, New York, USA

Gerald W. Hart, Sr. Eminent Scholar in Drug Discovery, Georgia Research Alliance William Henry Terry Professor of Biochemistry & Molecular Biology, Complex Carbohydrate Research Center, University of Georgia, Athens, Georgia, USA

Markus Aebi, Professor of Mycology, ETH Zürich, Institute of Microbiology, Department of Biology, Zürich, Switzerland

Debra Mohnen, Distinguished Research Professor, Georgia Athletics Association Professor in Complex Carbohydrate Research, Professor of Biochemistry and Molecular Biology, Complex Carbohydrate Research Center, University of Georgia, Athens, Georgia, USA

Taroh Kinoshita, Endowed Chair and Professor, Yabumoto Department of Intractable Disease Research, Research Institute for Microbial Diseases, Immunology Frontier Research Center, Osaka University, Osaka, Japan

Nicolle H. Packer, Distinguished Professor of Molecular Sciences, Macquarie University, Sydney, Australia and Principal Research Leader, Institute for Glycomics, Griffith University, Gold Coast, Australia

James H. Prestegard, Professor and Eminent Scholar Emeritus, Complex Carbohydrate Research Center, University of Georgia, Athens, Georgia, USA

Ronald L. Schnaar, Professor, Departments of Pharmacology and Neuroscience, Johns Hopkins University, Baltimore, Maryland, USA

Peter H. Seeberger, Director of Max-Planck-Institute of Colloids and Interfaces, Potsdam, Germany, Professor, Freie Universität, Berlin, Germany

CO-AUTHORS

Yoshihiro Akimoto, Professor, Kyorin University School of Medicine, Tokyo, Japan

Sonja-Verena Albers, Professor, University of Freiburg, Freiberg im Breisgau, Germany

Takashi Angata, Associate Research Fellow, Institute of Biological Chemistry, Academia Sinica, Taipei, Taiwan

Jesús Angulo, Sr. Distinguished Researcher, University of Seville, Seville, Spain

Kiyoko F. Aoki-Kinoshita, Professor, Glycan and Life Systems Integration Center, Soka University, Tokyo, Japan

Susan L. Bellis, Professor and Endowed Chair, University of Alabama, Birmingham, Alabama, USA

Carolyn R. Bertozzi, Professor, Stanford University, Stanford, California, USA
Michael Boyce, Associate Professor, Duke University, Durham, North Carolina, USA
Thomas Braulke, Professor, University Medical-Center Hamburg-Eppendorf, Hamburg, Germany
Inka Brockhausen, Associate Professor, Queen's University, Kingston, Ontario, Canada
Matthew P. Campbell, Research Fellow, Institute for Glycomics, Griffith University, Queensland, Australia
Xi Chen, Professor, University of California, Davis, Davis, California, USA
Elise Chiffoleau, Researcher, French National Institute of Health and Medical Research, University of Nantes, Nantes, France
Henrik Clausen, Professor, Copenhagen Center for Glycomics, University of Copenhagen, Copenhagen, Denmark
Karen J. Colley, Professor and Dean, University of Illinois Chicago, Chicago, Illinois, USA
John Couchman, Professor Emeritus, University of Copenhagen, Copenhagen, Denmark
Nancy Dahms, Professor, Medical College of Wisconsin, Milwaukee, Wisconsin, USA
Alan G. Darvill, Director and Regents Professor, Complex Carbohydrate Research Center, University of Georgia, Athens, Georgia, USA
Benjamin G. Davis, Professor and Science Director, The Rosalind Franklin Institute and University of Oxford, Oxford, England
Martina Delbianco, Group Leader, Max-Planck-Institute of Colloids and Interfaces, Potsdam, Germany
Matthew P. DeLisa, Professor, Chemical and Biomolecular Engineering, Cornell University, Ithaca, New York, USA
Anne Dell, Professor and Department Head Life Sciences, Imperial College, London, London, England
Tamara L. Doering, Professor, Washington University School of Medicine, St. Louis, Missouri, USA
Kurt Drickamer, Professor, Imperial College London, London, England
Jerry Eichler, Professor, Ben-Gurion University of the Negev, Beer-Sheva, Israel
Marilynn E. Etzler, Professor, University of California, Davis, Davis, California, USA
Michael A.J. Ferguson, Regius Professor, School of Life Sciences, University of Dundee, Dundee, Scotland
Darón I. Freedberg, Senior Scientist, Center for Biologics Evaluation and Research, U.S. Food and Drug Administration, Silver Spring, Maryland, USA
Hudson H. Freeze, Professor, Sanford Burnham Prebys, San Diego, California, USA
Morihisa Fujita, Professor, Jiangnan University, Wuxi, Jiangsu, China
Pascal Gagneux, Professor, University of California, San Diego, La Jolla, California, USA
Kamil Godula, Associate Professor and Associate Director, Glycobiology Research and Training Center, University of San Diego, La Jolla, California, USA
Michael G. Hahn, Professor, Complex Carbohydrate Research Center, University of Georgia, Athens, Georgia, USA
Robert S. Haltiwanger, GRA Eminent Scholar, Complex Carbohydrate Research Center, University of Georgia, Athens, Georgia, USA
Vincent Hascall, Professor, Lerner Research Institute, Cleveland Clinic, Cleveland, Ohio, USA
Stuart M. Haslam, Professor, Imperial College London, London, England
Thierry Hennet, Professor, Institute of Physiology, University of Zurich, Switzerland, Zurich, Switzerland
Bernard Henrissat, Professor, Technical University of Denmark, Kongens Lyngby, Denmark
Karin M. Hoffmeister, Professor, Endowed Chair and Director, Versiti Translational Glycomics Center and Blood Research Institute, Wauwatosa, Wisconsin, USA
Cornelis H. Hokke, Professor, Leiden University Medical Center, Leiden, Netherlands
Anne Imberty, Research Director, CERMAV, Centre National de la Recherche Scientifique, Grenoble, France
Hamed Jafar-Nejad, Professor, Baylor College of Medicine, Houston, Texas, USA
Reiji Kannagi, Distinguished Research Fellow, Institute of Biomedical Science, Academia Sinica, Taipei, Taiwan
Niclas G. Karlsson, Professor, Oslo Metropolitan University, Oslo, Norway
Koichi Kato, Professor, Exploratory Research Center, National Institutes of Natural Sciences, Okazaki, Japan
Kay-Hooi Khoo, Distinguished Research Fellow, Institute of Biological Chemistry, Academia Sinica, Taipei, Taiwan
Jennifer J. Kohler, Associate Professor, UT Southwestern Medical Center, Dallas, Texas, USA

Sneha Sudha Komath, Professor, Jawaharlal Nehru University School of Life Sciences, New Delhi, India
Stuart Kornfeld, Professor, Washington University School of Medicine in St. Louis, St. Louis, Missouri, USA
Gordan Lauc, Professor, University of Zagreb, Croatia
Carlito B. Lebrilla, Distinguished Professor, University of California, Davis, Davis, California, USA
Dirk J. Lefeber, Professor, Radboud University Medical Center, Nijmegen, Netherlands
Nathan E. Lewis, Associate Professor, University of California, San Diego, La Jolla, California, USA
Amanda L. Lewis, Professor and Associate Director, Glycobiology Research and Training Center, University of San Diego, La Jolla, California, USA
Ulf Lindahl, Professor Emeritus, Uppsala University, Uppsala, Sweden
Robert J. Linhardt, Chair, Rensselaer Polytechnic Institute, Troy, New York, USA
Frédérique Lisacek, Group Leader, Swiss Institute of Bioinformatics, University of Geneva, Geneva, Switzerland
Fu-Tong Liu, Vice President, Academia Sinica, Taipei, Taiwan
Jian Liu, Professor, Eshelman School of Pharmacy University of North Carolina, Chapel Hill, North Carolina, USA
Todd L. Lowary, Distinguished Research Fellow, Institute of Biological Chemistry, Academia Sinica, Taipei, Taiwan
Lara K. Mahal, Professor, Canada Excellence Research Chair, University of Alberta, Edmonton, Alberta, Canada
Rodger P. McEver, Vice President of Research, Oklahoma Medical Research Foundation, Oklahoma City, Oklahoma, USA
Catherine L.R. Merry, Professor of Stem Cell Glycobiology, Nottingham University, Nottingham, England
Benjamin H. Meyer, Molecular Enzyme Technology and Biochemistry, University of Duisburg-Essen, Germany
Antonio Molina, Professor and Director, Centre for Plant Biotechnology and Genomics, Technical University of Madrid, Madrid, Spain
Robert J. Moon, Materials Research Engineer, USDA-Forest Service-Forest Products Laboratory, Madison, Wisconsin, USA
Kelley W. Moremen, Professor, Complex Carbohydrate Research Center, University of Georgia, Athens, Georgia, USA
Jenny C. Mortimer, Associate Professor, University of Adelaide, Glen Osmond, South Australia, Australia
Barbara Mulloy, Visiting Professor, Glycosciences Laboratory, Imperial College London, London, England
Hiroshi Nakato, Professor, University of Minnesota, Minneapolis, Minnesota, USA
Sriram Neelamegham, Professor, State University of New York, Buffalo, Buffalo, New York, USA
Malcolm A. O'Neill, Associate Research Scientist, Complex Carbohydrate Research Center, University of Georgia, Athens, Georgia, USA
Tetsuya Okajima, Professor, School of Medicine, Nagoya University, Nagoya, Japan
Hermen S. Overkleeft, Professor, Leiden University, Leiden, Netherlands
Yasuhiro Ozeki, Professor, School of Sciences, Yokohama City University, Yokohama, Japan
Vladislav Panin, Professor, Texas A&M University, College Station, Texas, USA
Armando Parodi, Professor, Fundación Instituto Leloir, Buenos Aires, Argentina
Katharina Paschinger, FWF Fellow, Institute of Biochemistry, University of Natural Resources and Life Sciences, Vienna, Austria
Markus Pauly, Professor, Heinrich-Heine University, Düsseldorf, Germany
Serge Perez, Research Director Emeritus, Centre National de la Recherche Scientifique, Grenoble, France
J. Michael Pierce, Distinguished Research Professor and Mudter Professor of Cancer Research, University of Georgia, Athens, Georgia, USA
Gabriel A. Rabinovich, Professor, Institute of Biology and Experimental Medicine, National Scientific and Research Council, University of Buenos Aires, Buenos Aires, Argentina
T.N.C. Ramya, Principal Scientist, Council of Scientific and Industrial Research, Institute of Microbial Technology, Chandigarh, India
Celso A. Reis, Professor, i3S Institute for Research and Innovation in Health, University of Porto, Porto, Portugal

James M. Rini, Professor, University of Toronto, Toronto, Ontario, Canada
Françoise H. Routier, Professor, Hannover Medical School, Institute for Clinical Biochemistry, Hannover, Germany
Pauline M. Rudd, Visiting Investigator, Bioprocessing Technology Institute A*Star, Singapore and Adjunct Professor, University College, Dublin, Ireland
Robert Sackstein, Senior Vice-President for Health Affairs and Dean, College of Medicine, Florida International University, Miami, Florida, USA
Roger Sandhoff, German Cancer Research Center, Heidelberg, Germany
Liliana Schaefer, Professor, Goethe University, Frankfurt, Germany
Melanie A. Simpson, Professor and Department Head, North Carolina State University, Raleigh, North Carolina, USA
Chad Slawson, Associate Professor, University of Kansas Medical Center, Kansas City, Kansas, USA
Richard Steet, Director of Research, Greenwood Genetic Center, Greenwood, South Carolina, USA
Sean R. Stowell, Associate Professor, Brigham and Women's Hospital, Harvard Medical School, Boston, Massachusetts, USA
Avadhesha Surolia, Professor, Indian Institute of Science, Bangalore, India
Tadashi Suzuki, Chief Scientist, Riken Cluster for Pioneering Research, Wako, Japan
Christine M. Szymanski, Professor, University of Georgia, Athens, Georgia, USA
Naoyuki Taniguchi, Director of Research Center, Osaka International Cancer Institute, and Professor Emeritus Osaka University, Osaka, Japan
Maureen E. Taylor, Reader, Imperial College London, London, England
Kelly G. Ten Hagen, Senior Investigator and Chief, National Institutes of Health, Bethesda, Maryland, USA
Nicolas Terrapon, Associate Professor, Aix-Marseille University, Marseille, France
Morten Thaysen-Andersen, Associate Professor, ARC Future Fellow, Macquarie University, Sydney, Australia
Michael Tiemeyer, Professor, Complex Carbohydrate Research Center, University of Georgia, Athens, Georgia, USA
Breeanna Urbanowicz, Assistant Professor, Complex Carbohydrate Research Center, University of Georgia, Athens, Georgia, USA
Yvette van Kooyk, Professor, Amsterdam University Medical Center, Amsterdam, Netherlands
Gerardo R. Vasta, Professor, Institute of Marine and Environmental Technology, University of Maryland School of Medicine, Baltimore, Maryland, USA
David J. Vocadlo, Professor, Simon Fraser University, Burnaby, British Columbia, Canada
Stephan von Gunten, Professor, Institute of Pharmacology, University of Bern, Bern, Switzerland
Hans H. Wandall, Professor, Copenhagen Center for Glycomics, Faculty of Health and Medical Sciences, University of Copenhagen, Copenhagen, Denmark
Lance Wells, Professor, Complex Carbohydrate Research Center, University of Georgia, Athens, Georgia, USA
Christopher M. West, Professor and Department Head, University of Georgia, Athens, Georgia, USA
Chris Whitfield, Professor, University of Guelph, Guelph, Ontario, Canada
Göran Widmalm, Professor, Stockholm University, Stockholm, Sweden
Iain B.H. Wilson, Associate Professor, University of Natural Resources and Life Sciences, Wien, Austria
Robert J. Woods, Professor, Complex Carbohydrate Research Center, University of Georgia, Athens, Georgia, USA
Manfred Wuhrer, Professor, Center for Proteomics and Metabolomics, Leiden University, Leiden, Netherlands
Ding Xu, Associate Professor, University at Buffalo, Buffalo, New York, USA
William S. York, Emeritus Professor, Complex Carbohydrate Research Center, University of Georgia, Athens, Georgia, USA
Natasha E. Zachara, Associate Professor, Johns Hopkins University School of Medicine, Baltimore, Maryland, USA
Jochen Zimmer, Professor, University of Virginia, Charlottesville, Virginia, USA

Editors of Earlier Editions

The current Consortium of Glycobiology Editors thank the following colleagues for their major contributions as editors of earlier editions of this book:

Carolyn R. Bertozzi, *Anne T. and Robert M. Bass Professor of Chemistry, Professor, Chemical and Systems Biology and Radiation, Stanford University, Stanford, California, USA*

Alan G. Darvill, *Director and Regents Professor, Complex Carbohydrate Research Center, GRA Senior Fellow, University of Georgia, Athens, Georgia, USA*

Marilynn E. Etzler, *Professor Emerita of Biochemistry, University of California, Davis, Davis, California, USA*

Hudson H. Freeze, *Director, Human Genetics Program, Director, Sanford Children's Health Research Center, La Jolla, California, USA*

Jamey Marth, *Director and Professor, Immunity and Pathogenesis Program, Sanford Burnham Prebys, La Jolla, California, USA*

Consultant Reviewers

The Editors would like to thank the following reviewers for their valuable help:

Anabel Gonzalez-Gil, *Johns Hopkins University School of Medicine, USA*

Laura Bacete, *Norwegian University of Science and Technology, Norway*

Felix Broecker, *Idorsia Pharmaceuticals Ltd., Germany*

Dillon Chen, *University of California, San Diego School of Medicine, USA*

Jennifer Groves, *Johns Hopkins University School of Medicine, USA*

Graham Heberlig, *University of California, San Diego School of Medicine, USA*

So-Young Kim, *University of California, San Diego School of Medicine, USA*

Junko Matsuda, *Kawasaki Medical School, Japan*

Satyajit Mayor, *National Center for Biological Sciences, India*

Bobby Ng, *Sanford Burnham Prebys, USA*

Kyriakos Papanicolaou, *Johns Hopkins University School of Medicine, USA*

Ryan Porell, *University of California, San Diego School of Medicine, USA*

Ruth Siew, *University of California, San Diego School of Medicine, USA*

Priya Umapathi, *Johns Hopkins University School of Medicine, USA*

Yan Wang, *Cleveland Clinic Lerner Research Institute, USA*

The Editors and Co-Authors gratefully acknowledge the efforts of Richard D. Cummings in helping to coordinate and prepare the hundreds of illustrations in this book.

Classic Books and Monographs

Previous editions of this book featured a list of books and monograph resources. Given current day Internet searching and frequent review articles, such a listing is no longer as critical. But certain classic books and monographs still have much value for those who want to explore the history of the field. Accordingly, we have retained the following list from the 20th century.

–The Editors

Gottschalk A, ed. 1960. *The chemistry and biology of sialic acids and related substances.* Cambridge University Press, Cambridge.

Stacey M, Barker SA. 1960. *Polysaccharides of micro-organisms.* Oxford University Press, London.

Ginsburg V, Neufeld E, eds. 1966. Complex carbohydrates, part A. *Methods in enzymology,* Vol. 8. Academic, San Diego.

Whistler R, ed. 1968-80. *Methods in carbohydrate chemistry,* Vols. I-VIII. Academic, San Diego.

Hunt S. 1970. *Polysaccharide-protein complexes in invertebrates.* Academic, London.

Ginsburg V, ed. 1972. Complex carbohydrates, part B. *Methods in enzymology,* Vol. 28. Academic, San Diego.

Gottschalk A, ed. 1972. *Glycoproteins: their composition, structure and function.* Elsevier, New York.

Rosenberg A, Schengrund C-L., eds. 1976. *Biological roles of sialic acids.* Plenum, New York.

Ginsburg V, ed. 1978. Complex carbohydrates, part C. *Methods in enzymology,* Vol. 50. Academic, San Diego.

Sweeley CC, ed. 1979. *Cell surface glycolipids.* American Chemical Society, Washington, DC.

Lennarz WJ, ed. 1980. *The biochemistry of glycoproteins and proteoglycans.* Plenum, New York.

Ginsburg V, Robbins P, eds. 1981. *Biology of carbohydrates,* Vol. 1. Wiley, New York.

Ginsburg V, ed. 1982. Complex carbohydrates, part D. *Methods in enzymology,* Vol. 83. Academic, San Diego.

Horowitz M, Pigman W, eds. 1982. *The glycoconjugates.* Academic, New York.

Schauer R, ed. 1982. *Sialic acids, chemistry, metabolism, and function.* Springer-Verlag, New York.

Ginsburg V, Robbins P, eds. 1984. *Biology of carbohydrates,* Vol. 2. Wiley, New York.

Ivatt RJ, ed. 1984. *The biology of glycoproteins.* Plenum, New York.

Beeley JG, ed. 1985. *Glycoprotein and proteoglycan techniques.* Elsevier, Amsterdam.

Evered D, Whelan J, eds. 1986. *Functions of the proteoglycans.* Ciba Foundation Symposium. Wiley, Chichester, UK.

Liener IE, Sharon N, Goldstein IJ, eds. 1986. *The lectins: properties, functions, and applications in biology and medicine.* Academic, Orlando, FL.

Chaplin MF, Kennedy JF, eds. 1987. *Carbohydrate analysis: a practical approach.* IRL Press, Oxford.

Ginsburg V, ed. 1987. Complex carbohydrates, part E. *Methods in enzymology,* Vol. 138. Academic, San Diego.

Wight TN, Mecham RP. 1987. *Biology of proteoglycans.* Academic, London.

Evered D, Whelan J. 1988. *The biology of hyaluronan.* Ciba Foundation Symposium. Wiley, Chichester, UK.

Evered D, Whelan J, eds. 1989. *The biology of hyaluronan.* Ciba Foundation Symposium, Vol. 143. Wiley, New York.

Feizi T. 1989. *Carbohydrate recognition in cellular function.* Ciba Foundation Symposium, Vol. 145. Wiley, New York.

Ginsburg V, ed. 1989. Complex carbohydrates, part F. In *Methods in enzymology,* Vol. 179. Academic, San Diego.

Greiling H, Scott JE, eds. 1989. *Keratan sulphate: chemistry, biology, chemical pathology.* The Biochemical Society, London.

Margolis RU, Margolis RK, eds. 1989. *Neurobiology of glycoconjugates.* Plenum, New York.
Lane DG, Lindahl U, eds. 1990. *Heparin: chemical and biological properties.* CRC Press, Boca Raton, FL.
Ginsburg V, Robbins P, eds. 1991. *Biology of carbohydrates,* Vol. 3. Wiley, New York.
Allen HJ, Kisailus EC, eds. 1992. *Glycoconjugates: composition, structure, and function.* Marcel Dekker, New York.
Fukuda M, ed. 1992. *Cell surface carbohydrates and cell development.* CRC Press, Boca Raton, FL.
Fukuda M, ed. 1992. *Glycobiology: a practical approach.* IRL Press, Oxford.
Roth J, Rutishauser U, Troy F, eds. 1992. *Polysialic acids.* Birkhauser Verlag, Basel, Switzerland.
Roberts DD, Mecham RP, eds. 1993. *Cell surface and extracellular glycoconjugates: structure and function.* Academic, San Diego.
Varki A, Guest Ed. 1993. Analysis of glycoconjugates. In *Current protocols in molecular biology* (ed. Ausubel F, et al.), Chap. 17. Green Publishing/Wiley Interscience, New York.
Bock K, Clausen H, eds. 1994. *Complex carbohydrates in drug research: structural and functional aspects.* Munksgaard, Copenhagen.
Fukuda M, Hindsgaul O, eds. 1994. *Molecular glycobiology.* Oxford University Press, New York.
Lee YC, Lee RT. 1994. Neoglycoconjugates, part B. Biomedical applications. In *Methods in enzymology,* Vol. 247. Academic, London.
Lennarz WJ, Hart GW, eds. 1994. Guide to techniques in glycobiology. In *Methods in enzymology,* Vol. 230. Academic, San Diego.
Alavi A, Axford JS. 1995. *Advances in experimental medicine and biology,* Vol. 376, *Glycoimmunology.* Plenum, New York.
Montreuil J, Vliegenthart JFG, Schachter H, eds. 1995. *Glycoproteins.* Elsevier, New York.
Rosenberg A, ed. 1995. *Biology of the sialic acids.* Plenum, New York.
Verbert A, ed. 1995. *Methods on glycoconjugates: a laboratory manual.* Harwood Academic, Chur, Switzerland.
Montreuil J, Vliegenthart JFG, Schachter H, eds. 1996. *Glycoproteins and disease.* Elsevier, New York.
Brockhausen I, Kuhns W. 1997. *Glycoproteins and human disease.* R.G. Landes, Austin, TX.
Gabius HJ, Gabius S, eds. 1997. *Glycosciences: status and perspectives.* Chapman and Hall, New York.
Montreuil J, Vliegenthart JFG, Schachter H, eds. 1997. *Glycoproteins II.* Elsevier, New York.
Townsend RR, Hotchkiss AT, eds. 1997. *Techniques in glycobiology.* Marcel Dekker, New York.
Conrad HE, ed. 1998. *Heparin-binding proteins.* Academic, San Diego.
Hounsell EF, ed. 1998. *Methods in molecular biology,* Vol. 76, *Glycoanalysis protocols.* Humana, Totowa, NJ.
Laurent TC. 1998. *The chemistry, biology and medical applications of hyaluronan and its derivatives.* Portland Press, London.
Varki A, Esko J, Cummings R, Freeze HH, Hart GW, Marth J, eds. 1999. *Essentials of glycobiology.* Cold Spring Harbor Laboratory Press, Cold Spring Harbor, NY.

Abbreviations

2-AA	2-aminobenzoic acid
AA	amino acid
AAL	*Aleuria aurantia* lectin
2-AB	2-aminobenzamide
ABC	ATP-binding cassette
ACE	affinity co-electrophoresis OR angiotensin-converting enzyme
ACG-IB	achrondrogenesis type IB
AchE	acetylcholinesterase
AD	Alzheimer's disease
ADAMTS	a disintegrin and metalloproteinase with thrombospondin motifs
ADCC	antibody-dependent cellular cytotoxicity
ADPKD	autosomal-dominant polycystic kidney disease
AEP	aminoethylphosphonate
AFM	atomic force microscopy
AG	arabinogalactan
AGA	automated glycan assembly
AGEs	advanced glycation end products
AGP	arabinogalactan protein
AGX	arabinoglucuronoxylan
AH	actinohivin
AHM	anhydromannitol
ALG	Asn-linked glycosylation
α-DG	α-dystroglycan
ALS	amyotrophic lateral sclerosis
AMR	Ashwell–Morell receptor
ANTS	8-aminonaphthalene-1,3,6-trisulfonic acid
AOII	atelosteogenesis type II
AOX	alcohol oxidase
APAP	arabinoxylan pectin arabinogalactan protein
APC	antigen-presenting cell
Api	apiose
API	application programming interface
APLP	amyloid precursor–like protein
APTS	1-aminopyrene-3,6,8-trisulfonic acid
ASGPR	asialoglycoprotein receptor
AST	arylsulfotransferase
AuNP	gold nanoparticle
AVM	arteriovenous malformation
BCR	B-cell receptor
BCSDB	Bacterial Carbohydrate Structure Database
βCD	β-cyclodextrin
BMP	bone morphogenetic protein OR bis(monoacylglycero)phosphate
BoNT	Botulinum neurotoxin
BSA	bovine serum albumin
CAD	collision-activated dissociation
CASPER	Computer Assisted Spectrum Evaluation of Regular Polysaccharides [program]
CAZy	Carbohydrate-Active Enzymes [database]
CBM	carbohydrate-binding molecule
CCP	complement control protein
CCSD	Complex Carbohydrate Structure Database [aka CarbBank database]
CD	cluster of differentiation
CDG-II	congenital disorders of glycosylation type II
CD-MPR	cation-dependent mannose 6-phosphate receptor
CE	capillary electrophoresis
CEBiP	chitin oligosaccharide elicitor-binding protein
CER	ceramide
CESA	cellulose synthase
CFG	Consortium for Functional Glycomics
CFTR	cystic fibrosis transmembrane conductance regulator
CGD	congenital disorders of glycosylation
CHIME	coloboma, heart defects, ichthyosiform dermatitis, mental retardation, and ear anomalies with hearing loss [syndrome]
CHO	Chinese hamster ovary
CID	collision-induced dissociation
CI-MPR	cation-independent mannose 6-phosphate receptor
CK	creatine kinase
CM	cytoplasmic membrane
CMAH	CMP–Neu5Ac hydroxylase
CMD	congenital muscular dystrophy
CMP	cytidine monophosphate
CMS	congenital myasthemic syndrome
CMV	cytomegalovirus

CN cellulose nanomaterial
CNS central nervous system
CNT carbon nanotube
CNTFR ciliary neurotrophic factor receptor
CNX calnexin
COG conserved oligomeric Golgi
ConA concanavalin A
COPD chronic obstructive pulmonary disease
COPI coat protein I
COSY coupling correlated spectroscopy
COVID-19 coronavirus disease 2019
CPS capsular polysaccharide
CRD carbohydrate-recognition domain OR cross-reacting determinant
CREB cAMP response element binding
CREG cellular repressor of E1A-stimulated genes
CRISPR clustered regularly interspaced short palindromic repeats
CRP C-reactive protein
crRNA CRISPR-targeting guide RNA
CRT calreticulin
cryo-EM cryo-electron microscopy
CS chondroitin sulfate OR circumsporozoite
CsA contact site A
CSDB Carbohydrate Structure Database
CSL CBF1/Su(H)/Lag-1
CT cholera toxin
CTD carboxy-terminal domain
CTL C-type lectin
CTLD C-type lectin domain
CTLF C-type lectin fold

DAF decay accelerating factor
DAG diacylglycerol
DAMP damage-associated molecular pattern
DANA 2-deoxy-2,3-di-dehydro-*N*-acetylneuraminic acid
DAP DNAX activation protein
DBL *Dolichos biflorus* lectin
DC dendritic cell
DCAL dendritic cell–associated lectin
DC-SIGN dendritic cell–specific intercellular adhesion molecule-3-grabbing nonintegrin
DDD2 Dowling–Degos disease type 2
DEEP-STD differential epitope mapping saturation transfer difference
DFT density functional theory
DGDG 1,2-diacyl-3-O-(α-D-Gal-(1,6)-O-β-Gal)-sn-glycerol OR digalactosyldiacylglycerol
DIC disseminated intravascular coagulation
DLL Delta-like ligand
DMAP DNA methylase-associated protein
DMB 1,2-diamino-4,5-methylenedioxybenzene
Dol-P dolichol phosphate
Dol-PP dolichol pyrophosphate
DON deoxynorleucine OR 6-diazo-5-oxo-L-norleucine
dp degree of polymerization
D-PDMP D-threo-1-phenyl-2-decanoylamino-3-morpholino-1-propanol
Dpp Decapentaplegic
DS dermatan sulfate
DSA *Datura stramonium* agglutinin
DspB dispersin B
DTD diastrophic dystrophy

eAChE erythrocyte acetylcholinesterase
EAE experimental allergic encephalomyelitis
EBA erythrocyte-binding antigen
EBI European Bioinformatics Institute
EBL erythrocyte-binding-like
EBR1 egg binding receptor 1
EBV Epstein–Barr virus
EC Enzyme Commission
ECD electron capture dissociation
ECM extracellular matrix
EcorL *Erythrina corallodendron* lectin
EDEM ER degradation-enhancing α-mannosidase I–like protein
EET enzyme enhancement therapy
EGF epidermal growth factor
EGT ecdysteroid glucosyltransferase
EI electron impact
eIF2 eukaryotic initiation factor 2
EIS electrochemical impedance spectroscopy
Elf-1 E74-like factor 1
ELISA enzyme-linked immunoabsorbent assay
EMT epithelial–mesenchymal transition
ENDO F2 endoglycosidase F2
ENDO H endoglycosidase H
eNOS endothelial nitric oxide synthase
EOGT EGF-specific O-GlcNAc transferase
EPEC enteropathogenic *E. coli*
EPG endopolygalacturonase
E-PHA erythroagglutining phytohemagglutinin from *Phaseolus vulgaris*
EPO erythropoietin
EPS exopolysaccharide OR extracellular polysaccharide
ER endoplasmic reticulum
ERAD endoplasmic reticulum–associated degradation
ERGIC ER–Golgi intermediate compartment
ERK extracellular mitogen-regulated protein kinase

ERT enzyme replacement therapy
ESI electrospray ionization
ETD electron transfer dissociation
EtNP ethanolamine phosphate
EUL rice *Euonymus*-related lectin

FAB fast atom bombardment
FACE fluorophore-assisted carbohydrate electrophoresis
FAK focal adhesion kinase
FCMD Fukuyama congenital muscular dystrophy
f-CNT functionalized carbon nanotube
FDA Food and Drug Administration
FET field-effect transistor
FGE formylglycine-generating enzyme
FGF fibroblast growth factor
FGFR fibroblast growth factor receptor
FHA filamentous hemagglutinin
FKRP fukutin-related protein
FKTN fukutin
FRC fibroblastic reticular cell
FSH follicle-stimulating hormone
FSP fucose sulfate polymer
FT Fourier transform OR fucosyl-transferase

GABA γ-aminobutyric acid
GAG glycosaminoglycan
Gal galactose
GALE uridine diphosphate galactose-4-epimerase
GALK galactokinase
GALNT polypeptide GalNAc-transferase
GALT galactose-1-phosphate uridyltransferase
GalT galactosyltransferase
GARP glutamic acid/alanine-rich protein
GAS Group A streptococcus
GBP glycan-binding protein
GBS Group B streptococcus OR Guillain–Barré syndrome
GC gas chromatography
GCS glycosylceramide synthase
GDNFR-α glial-cell-[line-]derived neurotrophic factor receptor-α
GFAT glucosamine:fructose aminotransferase
GFP green fluorescent protein
GH glycoside hydrolase [aka glycosidase]
GIPL glycoinositol phospholipid
Glc glucose
GlcA glucuronic acid
GlcNAc N-acetylglucosamine
GLIC Glycome Informatics Consortium
GlycO Glycomics Ontology
GNA *Galanthus nivalis* agglutinin
GNE N-acetylglucosamine-2-epimerase/N-acetylmannosamine kinase
GPCR G protein–coupled receptor
GPI glycosylphosphatidylinositol
GPI-AP GPI-anchored protein
GPP Glycan Pathway Predictor
GRIFIN galectin-related interfiber protein
GRP glycan-recognizing probe OR galectin-related protein OR glycan-recognizing protein
GS glycogen synthase
GSC germline stem cell
GSK3β glycogen synthase kinase-3β
GSL glycosphingolipid
GT glycosyltransferase
GWAS genome-wide association study
GXM glucuronoxylomannan
GXMG glucuronoxylomannogalactan

HA hyaluronan OR hemagglutinin
HAPLN hyaluronan and proteoglycan link protein
HAS hyaluronan synthase
HAT histone acetyltransferase
HBHA heparin-binding hemagglutinin adhesin
HBP hexosamine biosynthetic pathway
HCD high-energy collision dissociation
HCELL hematopoietic cell E-/L-selectin ligand
HDL high-density lipoprotein
HDX hydrogen deuterium exchange
HEMPAS hereditary erythroblastic multinuclearity with positive acidified-serum test
HEV high endothelial venule
HF hydrogen fluoride
HG homogalacturonan
HGF hepatocyte growth factor
Hh Hedgehog
HIBM-II hereditary inclusion body myopathy type II
HIF hypoxia-inducible factor
HILIC hydrophilic interaction liquid chromatography
Hint-1 human intelectin 1
HIT heparin-induced thrombocytopenia
HIV human immunodeficiency virus
HMBC heteronuclear multiple-bond correlation
HME hereditary multiple exocytosis
HMG high mobility group
HMM hidden Markov model
HMO human milk oligosaccharide
HPA *Helix pomatia* agglutinin
HPAEC high-pH anion-exchange chromatography

HPCE	high-performance capillary electrophoresis
HPLC	high-performance liquid chromatography
HPMRS	hyperphosphatasia with mental retardation syndrome
HRGP	hydroxyproline-rich glycoprotein
HRP	horseradish peroxidase
HS	heparan sulfate
HSCT	hematopoietic stem cell therapy
HSQC	heteronuclear single quantum coherence
HSV	herpes simplex virus
HUPO-PSI	Human Proteome Organization-Proteomics Standards Initiative
Hyp	hydroxyproline
IC	intermediate compartment
ICAM	intercellular adhesion molecule
ID	intellectual disability
IDGF	imaginal disc growth factor
IdoA	iduronic acid
IEF	isoelectric focusing
IFN-γ	interferon-γ
IGD	inherited GPI deficiency
IGSF	immunoglobulin superfamily
IL-1	Interleukin-1
IR	insulin receptor OR infrared
IRS	insulin receptor substrate
ITAM	immunoreceptor tyrosine-based activatory motif
ITC	isothermal titration calorimetry
ITI	inter-α-trypsin inhibitor
ITIM	immunoreceptor tyrosine-based inhibitory motif
ITP	idiopathic thrombocytopenic purpura
IUBMB	International Union of Biochemistry and Molecular Biology
IUPAC	International Union of Pure and Applied Chemistry
JAG	Jagged
JCGGDB	Japan Consortium for Glycobiology and Glycotechnology Database
KCF	KEGG Chemical Function
KDN	2-keto-3-deoxy-D-glycero-D-galactonononic acid
KEGG	Kyoto Encyclopedia of Genes and Genomics
KIR	killer cell immunoglobulin-like receptor
KLH	keyhole limpet hemocyanin
KO	KEGG Orthology [database]
KS	keratan sulfate
LAA	*Laburnum alpinum* agglutinin
LAD	leukocyte adhesion deficiency
LAM	lipoarabinomannan
LAR	leukocyte common antigen-related protein
LC	liquid chromatography
LCA	*Lens culinaris* agglutinin
LDL	low-density lipoprotein
LDN	LacdiNAc
LDNF	fucosylated LacdiNAc
LecRK	L-type lectin receptor kinase
LERP	lysosomal enzyme receptor protein
Le^x	Lewis x
Lev	levulinoyl
LFNG	Lunatic fringe
LIF	leukemia inhibitory factor OR laser-induced fluorescence
LLO	lipid-linked oligosaccharide
LM	lipomannan
LMW	low-molecular-weight
LNP	lectin nucleotide phosphohydrolase
LOS	lipooligosaccharide
LPG	lipophosphoglycan
L-PHA	L-phytohemagglutinin
LPPG	lipopeptidophosphoglycan
L-PROSY	looped, projected spectroscopy
LPS	lipopolysaccharide
LSC	leukemic stem cell
LSL	*Lonchocarpus sericeus* agglutinin
LSPR	localized surface plasmon resonance
LTA	lipoteichoic acid OR *Lotus tetragonolobus* agglutinin
LysM	lysozyme motif
mAb	monoclonal antibody
MAG	myelin-associated glycoprotein
MAH	*Maackia amurensis* hemagglutinin
MAL	*Maackia amurensis* lectin
MALDI-MS	matrix-assisted laser desorption/ionization mass spectrometry
MALDI-TOF	matrix-assisted laser desorption/ionization time of flight
Man	mannose
MAP	mitogen-activated protein
MAPK	MAP kinase
MASP	MBL-associated serine protease
MBL	mannose-binding lectin
MBP	mannose-binding protein OR major basic protein
MCD	macular corneal dystrophy
MD	molecular dynamics
***m*-DAP**	*meso* diaminopimelic acid
MDCK	Madin–Darby canine kidney
MDL	myeloid DAP12-associating lectin

MDO	membrane-derived oligosaccharide
MDP	metallodipeptidase
MEB	muscle–eye–brain disease
MFNG	Manic fringe
MGDG	1,2-diacyl-3-O-(β-D-Gal)-sn-glycerol OR monogalactosyldiacylglycerol
MGL	macrophage galactose-binding lectin
MGUS	monoclonal gammopathy of unknown significance
MHC	major histocompatibility complex
MIAPE	Minimum Information for A Proteomics Experiment [initiative]
MICL	myeloid inhibitory CTL-like receptor
MINCLE	macrophage inducible Ca^{++}-dependent lectin
MIP-1β	macrophage inflammatory protein-1β
MIRAGE	Minimum Information Required for A Glycomics Experiment [initiative]
MM	molecular mechanics
MMDB	Molecular Modeling Database
MN	membranous neuropathy
MND	motor neuron disease
MNP	magnetic nanoparticle
mOGT	mitochondrial OGT
MPL	monophosphoryl lipid A
MPR	M6P receptor
MPS	mucopolysaccharidosis
MR	mannose receptor
MRH	M6P receptor homology OR mannose receptor homologous
MRI	magnetic resonance imaging
mRNA	messenger RNA
MS	mass spectrometry
MSD	multiple sulfatase deficiency
MSP	merozoite surface protein
MTD	major tropism determinant
MTG	mind-the-gap [protein]
MTX	mosquitocidal toxin
4-MU	4-methylumbelliferone
MVB	multivesicular body
MWCNT	multiwalled carbon nanotube
MytiLec	*Mytilus galloprovincialis* β-trefoil lectin
NAD	nicotinamide adenine dinucleotide
NADPH	nicotinamide adenine dinucleotide phosphate
NCAM	neural cell adhesion molecule
NCBI	National Center for Biotechnology Information
ncOGT	nucleocytoplasmic OGT
Neu	neuraminic acid
NF-κB	nuclear factor-κB
NHS	*N*-hydroxysuccinimide
NIR	near-infrared
NK	natural killer
NKT	natural killer T
NLR	NOD-like receptor
NLS	nuclear localization sequence
NMJ	neuromuscular junction
NMR	nuclear magnetic resonance
NOD	nucleotide-binding oligomerization domain
NOE	nuclear Overhauser effect
NP	nanoparticle
NSID	nonsyndromic intellectual disability
NulO	nonulosonic acid
ODR-2	odorant response abnormal 2
OGA	O-GlcNAcase
OGT	O-GlcNAc transferase
OMIM	Online Mendelian Inheritance in Man
OPG	osmoregulated periplasmic glucan
O-PS	O-antigen polysaccharide
ORF	open reading frame
OST	oligosaccharyltransferase
OTase	oligosaccharyltransferase
PAD	pulsed amperometric detection
PAH	pulmonary arterial hypertension
PAI-1	plasminogen activator inhibitor-1
PAM	protospacer adjacent motif
PAMP	pathogen-associated molecular pattern
PAP	pulmonary arterial pressure
PAPS	3′-phosphoadenosine-5′-phosphosulfate
PAS	periodic acid–Schiff
PBP	penicillin-binding protein
PC	phosphorylcholine OR polycystin
PcG	Polycomb group
PCLP	podocalyxin-like protein
PCP	polysaccharide copolymerase
PCR	polymerase chain reaction
PCT	pharmacological chaperone therapy
PDB	Protein Data Bank
PDI	protein disulfide isomerase
PDK	phosphatidylinositol-dependent kinase
PDX1	pancreatic and duodenal homeobox 1
PE	phosphoethanolamine
PECAM	platelet endothelial cell adhesion molecule
PEP	phosphoenolpyruvate
PG	phosphoglycan
PGC	porous graphitized carbon
PGIP	polygalacturonase-inhibiting proteins
PGM	phosphoglucomutase
PHAL	phytohemagglutinin L
PHF	paired helical filament
PI	phosphatidylinositol

PI3K phosphatidylinositol 3-kinase
PILR paired immunoglobulin-like receptor
PIM phosphatidylinositol mannoside
PKC protein kinase C
PLAP placental alkaline phosphatase
PLC phospholipase C
PLE protein-losing enteropathy
PLM phospholipomannan
PMAA partially methylated alditol acetate
PMT protein mannosyltransferase
PNA peanut agglutinin
PNAG poly-N-acetylglucosamine
PNGase F peptide-N-glycosidase F
PNH paroxysmal nocturnal hemoglobinuria
PNS peripheral nervous system
POFUT protein O-fucosyltransferase
POGLUT protein O-glucosyltransferase
PPG proteophosphoglycan
PrP prion protein
PRR pattern-recognition receptor
PS phosphatidylserine
PsA prespore antigen
PSA *Pisum sativum* agglutinin OR polysialic acid
PSGL P-selectin glycoprotein ligand
PSL *Pisum sativum* lectin (pea lectin)
PSP promastigote surface protease
PVD pulmonary vascular resistance

QCM quartz crystal microbalance
QD quantum dot
QM quantum mechanics
QTOF quadrupole time of flight

RA rheumatoid arthritis
RAGE receptor for advanced glycation end products
RANKL receptor activator of nuclear factor (NF)-κB ligand
Rb retinoblastoma
RCA *Ricinus communis* agglutinin
RCL reactive center loop
RDF Resource Description Framework
rER rough endoplasmic reticulum
RF radio frequency
RFNG Radical fringe
RG rhamnogalacturonan
RHO ras homolog
RI refractive index
RINGS Resource for Informatics of Glycomes at Soka
RIP ribosome-inactivating protein
RME receptor-mediated endocytosis
RNAi RNA interference
RNA-seq RNA sequencing
ROE rotating-frame Overhauser effect
ROS reactive oxygen species
RPTP receptor protein tyrosine phosphatase
rRNA ribosomal RNA
RRV relative reactivity value
RU repeating unit OR resonance unit

SAG surface antigen
SAM self-assembled monolayer
SAMP self-associated molecular pattern
SAP sphingolipid activator protein OR serum amyloid P
SAX strong anion exchange chromatography
SAXS small-angle X-ray scattering
SBA soybean agglutinin
SCD sickle cell disease
scFv single-chain variable fragment
SDF-1 stromal cell–derived factor-1
SDS-PAGE sodium dodecyl sulfate polyacrylamide gel electrophoresis
SEC size exclusion chromatography
SED spondyloepimetaphyseal dysplasia
SeviL *Mytilisepta virgata* R-type lectin
SHP-1 Src homology region 2 domain–containing phosphatase-1
shRNA short hairpin RNA
Sia sialic acid
SiaT sialyltransferase
SIB Swiss Institute of Bioinformatics
Siglec Sia-binding immunoglobulin-like lectin
siRNA small interfering RNA
SLE systemic lupus erythematosus
SLe^x sialyl-Lewis x
SLRP small leucine-rich proteoglycan
Sn sialoadhesin
SNA *Sambucus nigra* agglutinin
SNFG Symbol Nomenclature for Glycans
SNP single-nucleotide polymorphism
sOGT short OGT
SPA single-particle analysis
SPR surface plasmon resonance
SRT substrate reduction therapy
SSA *Sambucus sieboldiana* lectin
SSEA stage-specific embryonic antigen
ST sialyltransferase
STAT5 signal transducer and activator of transcription 5
STD saturation transfer difference
SVM support vector machine
SWCNT single-walled carbon nanotube

tACE testis angiotensin-converting enzyme
TACE TNF-α-converting enzyme

TAIR The Arabidopsis Information Service
TALE transcription activator–like effector
TALEN transcription activator–like effector nuclease
TBS tertbutyldimethylsilyl
TCGA The Cancer Genome Atlas
TCR T-cell receptor
T-DNA transfer DNA
TEM transmission electron microscopy
TET ten-eleven translocation
TF transcription factor
TGF-β transforming growth factor β
TGN *trans*-Golgi network
Th1 T helper 1
TLC thin-layer chromatography
TLR Toll-like receptor
TM transmembrane domain
TMT tandem mass tag
TNALP tissue-nonspecific alkaline phosphatase
TNF-α tumor necrosis factor-α
TOCSY total correlation spectroscopy
TP transpeptidase
TPA tissue plasminogen activator
TPR tetratricopeptide repeat
Treg regulatory T cell
trNOE transferred nuclear Overhauser effect
TS *trans*-sialidase
TSG TNF-α-stimulated gene
TSP thrombospondin
TSR thrombospondin type-1 repeat
TS-TEM ultrathin section transmission electron microscopy

UCH ubiquitin carboxyhydrolase
UEA *Ulex europaeus* agglutinin
UGT or UGGT UDP-glucose glycoprotein glucosyltransferase
UniLectin3D a database covering the three-dimensional features of lectins
uPA(R) urokinase-type plasminogen activator (receptor)
UTI urinary tract infection
UV-Vis ultraviolet-visible

VEGF vascular endothelial growth factor
VIPL VIP3b-like
VNTR variable number tandem repeat
VSA variant surface antigen
VSG variant surface glycoprotein
VWF von Willebrand factor
VZV Varicella zoster virus

WAX weak anion exchange
WFA *Wisteria floribunda* agglutinin
WGA wheat germ agglutinin
WTA wall teichoic acid
WWS Walker–Warburg syndrome

XGA xylogalacturonan
XLNSID X-linked nonsyndromic intellectual disability
XPS X-ray photoelectron spectroscopy

YY1 Yin Yang 1

1 Historical Background and Overview

Ajit Varki and Stuart Kornfeld

This chapter provides a historical background to the field of glycobiology and an overview of this book. General terms found in the volume are considered, common monosaccharide units of glycoconjugates mentioned, and a uniform symbol nomenclature used for structural depictions presented. The major glycan classes discussed in the book are described, and an overview of the general pathways for their biosynthesis is provided. Topological issues relevant to biosynthesis and functions of glycoconjugates are also considered, and the growing role of these molecules in medicine, biotechnology, nanotechnology, bioenergy, and materials science is mentioned.

WHAT IS GLYCOBIOLOGY?

Defined in the broadest sense, glycobiology is the study of the structure, biosynthesis, biology, and evolution of saccharides (also called carbohydrates, sugar chains, or glycans) that are widely

published by Cold Spring Harbor Laboratory Press; doi: 10.1101/glycobiology.4e.1

distributed in nature and of the proteins that recognize them. How does glycobiology fit into the modern concepts of molecular biology? The central paradigm driving research in molecular biology has been that biological information flows from DNA to RNA to protein. The power of this concept lies in its template-based precision, the ability to manipulate one class of molecules based on knowledge of another, and the patterns of sequence homology and relatedness that predict function and reveal evolutionary relationships. With ongoing sequencing of numerous genomes, spectacular gains in understanding the biology of nucleic acids and proteins have occurred. Thus, many scientists assume that studying just these molecules will explain the makeup of cells, tissues, organs, physiological systems, and intact organisms. In fact, making a cell requires many small molecule metabolites as well as two other major classes of macromolecules—lipids and carbohydrates—which serve as intermediates in generating energy and as signaling effectors, recognition markers, and structural components. Taken together with the fact that they encompass some of the major posttranslational modifications of proteins, lipids, and carbohydrates help explain how the relatively small number of genes in the typical genome can generate the enormous biological complexities inherent in the development, growth, and functioning of diverse organisms.

The biological roles of carbohydrates are particularly prominent in the assembly of complex multicellular organs and organisms, which requires interactions between cells and the surrounding matrix. However, without any known exception, all cells and numerous macromolecules in nature carry an array of covalently attached sugars (monosaccharides) or sugar chains (oligosaccharides), which are generically referred to in this book as "glycans." Sometimes glycans can also be freestanding entities. Being on the outer surface of cellular and secreted macromolecules, many glycans are in a position to modulate or mediate a variety of events in cell–cell, cell–matrix, and cell–molecule interactions critical to the development and function of a complex multicellular organism. They can also mediate interactions between organisms (e.g., between host and a parasite, pathogen, or a symbiont). In addition, simple, rapidly turning over, protein-bound glycans are abundant within the nucleus and cytoplasm, in which they can serve as regulatory switches. A more complete paradigm of molecular biology must therefore include glycans, often in covalent combination with other macromolecules (i.e., glycoconjugates, such as glycoproteins and glycolipids). In analogy to the current situation in Cosmology, glycans can be considered as the "dark matter" of the biological universe: a major and critical component that has yet to be fully incorporated into the "standard model" of biology. However, unlike the situation with dark matter in the Universe, there is already a lot known about glycans.

The chemistry and metabolism of carbohydrates were prominent matters of interest in the first part of the 20th century. Although engendering much attention, they were primarily considered as a source of energy or as structural materials, apparently lacking other biological activities. Furthermore, during the molecular biology revolution of the 1970s, studies of glycans lagged far behind those of other major classes of molecules. This was in part because of their inherent structural complexity, the difficulty in determining their sequences, and the fact that their biosynthesis could not be directly predicted from a DNA template. The development of many new technologies for exploring the structures and functions of glycans has since opened a new frontier of molecular biology called "glycobiology"—a word first coined in the late 1980s to recognize the coming together of the traditional disciplines of carbohydrate chemistry and biochemistry with a modern understanding of the cell and molecular biology of glycans and, in particular, their conjugates with proteins and lipids. Glycobiology is now one of the more rapidly growing fields in the natural sciences, with broad relevance to many areas of basic research, biomedicine, and biotechnology. The field includes the chemistry of carbohydrates, the enzymology of glycan formation and degradation, the recognition of glycans by specific proteins, roles of glycans in complex biological systems, and their analysis or manipulation by various techniques. Research in glycobiology thus requires a foundation not only in the nomenclature, biosynthesis, structure, chemical synthesis, and functions of glycans, but also in the general disciplines of molecular genetics, protein chemistry, cell biology, developmental biology, physiology,

FIGURE 1.1. Nobel laureates in fields related to the history of glycobiology. Listed are the Laureates and their original Nobel citations: Hermann Emil Fischer (Chemistry, 1902), *"in recognition of the extraordinary services he has rendered by his work on sugar and purine syntheses"*; Eduard Buchner (Chemistry, 1907), *"for his discovery of cell-free fermentation"*; Otto Fritz Meyerhof (Physiology or Medicine, 1922), *"for his discovery of the fixed relationship between the consumption of oxygen and the metabolism of lactic acid in the muscle"*; Karl Landsteiner (Physiology or Medicine, 1930), *"for his discovery of human blood groups"*; Walter Norman Haworth (Chemistry, 1937), *"for his investigations on carbohydrates and vitamin C"*; Carl Ferdinand and Gerty Theresa Cori (Physiology or Medicine, 1947), *"for their discovery of the course of the catalytic conversion of glycogen"*; Luis F. Leloir (Chemistry, 1970), *"for his discovery of sugar nucleotides and their role in the biosynthesis of carbohydrates"*; George E. Palade (Physiology or Medicine, 1974), *"for discoveries concerning the structural and functional organization of the cell"*; and James E. Rothman and Randy W. Schekman (Physiology or Medicine 2013), *"for their discoveries of machinery regulating vesicle traffic, a major transport system in our cells."* (C. Cori, G. Cori, L. Leloir, G. Palade, The Nobel Foundation/TT/Sipa USA; J. Rothman, R. Schekman, © Nobel Media AB. Photo: Alexander Mahmoud.)

and medicine. This book provides an overview of the field, with relative emphasis on the glycans of animal systems. It is assumed that the reader has a basic background in advanced undergraduate-level chemistry, biochemistry, and cell biology. Some of the major investigators who influenced the early development of Glycobiology are shown in Figure 1.1, and more are listed in Online Appendix 1A. Many others have made major contributions as well, and a summary of the general principles gained from all this research is presented in Table 1.1.

MONOSACCHARIDES ARE THE BASIC STRUCTURAL UNITS OF GLYCANS

Carbohydrates are defined as polyhydroxyaldehydes, polyhydroxyketones and their simple derivatives, or larger compounds that can be hydrolyzed into such units. A monosaccharide is a carbohydrate that usually cannot be hydrolyzed into a simpler form. It has a potential carbonyl group at the end of the carbon chain (an aldehyde group) or at an inner carbon (a ketone group). These two types of monosaccharides are therefore named aldoses and ketoses, respectively (for examples, see below, and for more details, see Chapter 2). Free monosaccharides can exist in open-chain or ring forms (Figure 1.2). Ring forms of the monosaccharides are the general rule in oligosaccharides, which are linear or branched chains of monosaccharides attached to one another via glycosidic linkages (the term "polysaccharide" is typically used for large glycans composed of repeating oligosaccharide motifs; for examples, see Chapter 3). The ring form of a monosaccharide generates a chiral anomeric center at C-1 for aldo sugars or at C-2 for keto sugars (for details, see Chapter 2). A glycosidic linkage involves the attachment of a

TABLE 1.1. General principles of glycobiology

Occurrence

All cells in nature are covered with a dense and complex array of sugar chains (glycans).

The cell walls of bacteria and Archaea are composed of several classes of glycans and glycoconjugates.

Most secreted proteins of eukaryotes carry large amounts of covalently attached glycans.

In eukaryotes, cell-surface and secreted glycans are mostly assembled via the ER–Golgi pathway.

The extracellular matrix, secretions, and body fluids of eukaryotes are also rich in glycans.

Cytosolic and nuclear glycans are common in eukaryotes.

For topological, evolutionary, and biophysical reasons, there are limited similarities between cell-surface/secreted and nuclear/cytosolic glycans.

Chemistry and structure

Glycosidic linkages can be in α- or β-linkage forms, which are biologically recognized as distinct.

Glycan chains can be linear or branched.

Glycans can be modified by a variety of different substituents, such as acetylation and sulfation.

Complete sequencing of glycans is feasible but usually requires combinatorial or iterative methods and attention to potential loss of labile substitutions.

Modern methods allow in vitro chemoenzymatic synthesis of both simple and complex glycans.

Biosynthesis

The primary units of glycans (monosaccharides) can be synthesized within a cell or salvaged from the environment.

Monosaccharides are activated into nucleotide sugars or lipid-linked sugars before they are used as donors for glycan synthesis.

Whereas lipid-linked sugar donors can be flipped across membranes, nucleotide sugars must be transported into the lumen of the ER–Golgi pathway of eukaryotes.

Each linkage unit of a glycan or glycoconjugate is assembled by one or more unique glycosyltransferases.

Many glycosyltransferases are members of multigene families with related functions.

Most glycosyltransferases recognize only the underlying glycan of their acceptor target, but some are protein- or lipid-specific.

Many biosynthetic enzymes (glycosyltransferases, glycosidases, sulfotransferases, etc.) are expressed in a cell type– and in a tissue-specific, temporally regulated manner.

Diversity

Monosaccharides can generate far greater combinatorial diversity than nucleotides or amino acids.

Further diversity arises from branching or covalent modifications of glycans.

Glycosylation introduces a marked diversity in glycoproteins.

Only a limited subset of the potential diversity is found in a given organism or cell type.

Intrinsic diversity (microheterogeneity) can exist in glycoprotein glycans within a cell type or even at a single glycosylation site.

The total expressed glycan repertoire (glycome) of a given cell type or organism is far more complex than the genome or proteome.

The glycome of a given cell type or organism is also dynamic, changing in response to intrinsic and extrinsic signals.

Glycome differences in cell type, space, and time generate biological diversity and can help to explain why only a limited number of genes are expressed from the typical genome.

Recognition

Glycans are recognized by specific glycan-binding proteins that are intrinsic to an organism.

Glycans are also recognized by many extrinsic glycan-binding proteins of pathogens and symbionts.

Glycan-binding proteins fall in two general categories: those that can usually be grouped by shared evolutionary origins and/or similarity in structural folds (lectins) and those that emerged by convergent evolution from different ancestors (e.g., sulfated GAG-binding proteins).

Glycan-binding proteins often show a high degree of specificity for binding to specific glycan structures, but they typically have relatively low affinities for single-site binding.

Biologically relevant lectin recognition often requires multivalency of both the glycan and glycan-binding protein, to generate high avidity of binding.

Genetics

Naturally occurring genetic defects in glycans appear relatively rare in intact organisms. This apparent rarity may be attributed to lack of survival, or a failure of detection, owing to unpredictable or pleiotropic phenotypes.

Genetic defects in cell-surface/secreted glycans in cultured cells often have few biological consequences, whereas the same mutations in intact multicellular organisms may have major phenotypic effects.

(Continued)

TABLE 1.1. *(Continued)*

Complete elimination of major glycan classes generally causes early developmental lethality.
Organisms bearing tissue-specific alteration of glycans often survive, but with both cell autonomous and distal biological defects.

Biological roles
Biological roles for glycans span the spectrum from nonessential activities to those that are crucial for the development, function, and survival of an organism.
Glycans can have different roles in different tissues or at different times in development.
Terminal sequences, unusual glycans, and modifications are more likely to mediate specific biological roles and may also reflect evolutionary interactions with microorganisms and other noxious agents.
Many of the major roles of cell-surface glycans involve cell–cell and/or extracellular interactions.
Nuclear/cytosolic glycans have more cell-intrinsic roles (e.g., in signaling).
A priori prediction of functions of a specific glycan or its relative importance to the organism is difficult.

Evolution
Relatively little is still known about glycan evolution.
Interspecies and intraspecies variations in glycan structure are common, suggesting rapid evolution.
The dominant mechanisms for glycan evolution include ongoing selection pressure by pathogens that recognize or mimic host glycans combined with the need to preserve critical intrinsic glycan functions.
Interplay between pathogen selection pressure and preservation of intrinsic roles could result in the formation of "junk" glycans from which new intrinsic functions might arise during evolution.

(ER) Endoplasmic reticulum.

monosaccharide to another residue, typically via the hydroxyl group of this anomeric center, generating α-linkages or β-linkages that are defined based on the relationship of the glycosidic oxygen to the anomeric carbon and ring (Chapter 2). These two linkage types confer very different structural properties and biological functions on sequences that are otherwise identical in composition, as classically illustrated by the differences between starch and cellulose (both are homopolymers of glucose, the former largely α1-4-linked and the latter β1-4-linked throughout). A glycoconjugate is a compound in which one or more monosaccharide or oligosaccharide units (the glycone) are covalently linked to a noncarbohydrate moiety (the aglycone). An oligosaccharide that is not attached to an aglycone possesses the reducing power of the aldehyde or ketone in its terminal monosaccharide component, with the exception of oligosaccharides in which the sugars are linked together at their reducing ends, as in derivatives of sucrose or trehalose. The end of a glycan exposing the aldehyde or ketone group is therefore named the reducing terminus or reducing end, terms that tend to be used even when the sugar chain is

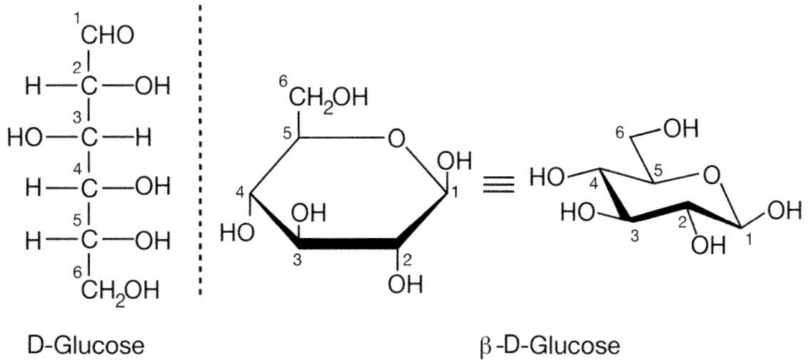

D-Glucose

β-D-Glucose

FIGURE 1.2. Open-chain and ring forms of glucose. Changes in the orientation of hydroxyl groups around specific carbon atoms generate new molecules that have a distinct biology and biochemistry (e.g., galactose is the C-4 epimer of glucose). In the ring form, glucose and other sugars adopt one of two hydroxyl group orientations (α or β) of which the latter is shown here.

attached to an aglycone and has thus lost its reducing power. Correspondingly, the opposite end of the chain tends to be called the nonreducing end (note the analogy to the amino and carboxyl ends of proteins, or the 5′ and 3′ ends of DNA and RNA).

GLYCANS CAN CONSTITUTE A MAJOR PORTION OF THE MASS OF A GLYCOCONJUGATE

In naturally occurring glycoconjugates, the portion of the molecule comprising the glycans can vary greatly in contribution to its overall size. In many cases, the glycans comprise a substantial portion of the mass of glycoconjugates (for a typical example, see Figure 1.3). For this reason, the surfaces of all types of cells in nature (which are heavily decorated with different kinds of glycoconjugates) are effectively covered with a dense array of sugars, the so-called "glycocalyx." This cell-surface structure was observed many years ago by electron microscopists as a negatively charged coat external to the cell surface membrane in bacteria, which could be stained with ruthenium red and in animal cells in which the anionic coat could be decorated with polycationic reagents (Figure 1.4.) Evidence that the "glycocalyx" was enriched in sugars, including protein-bound sialic acid in animal cells, first came from studies of the effect of proteolytic enzymes on the behavior of erythrocytes in cell electrophoresis along with studies of the nature

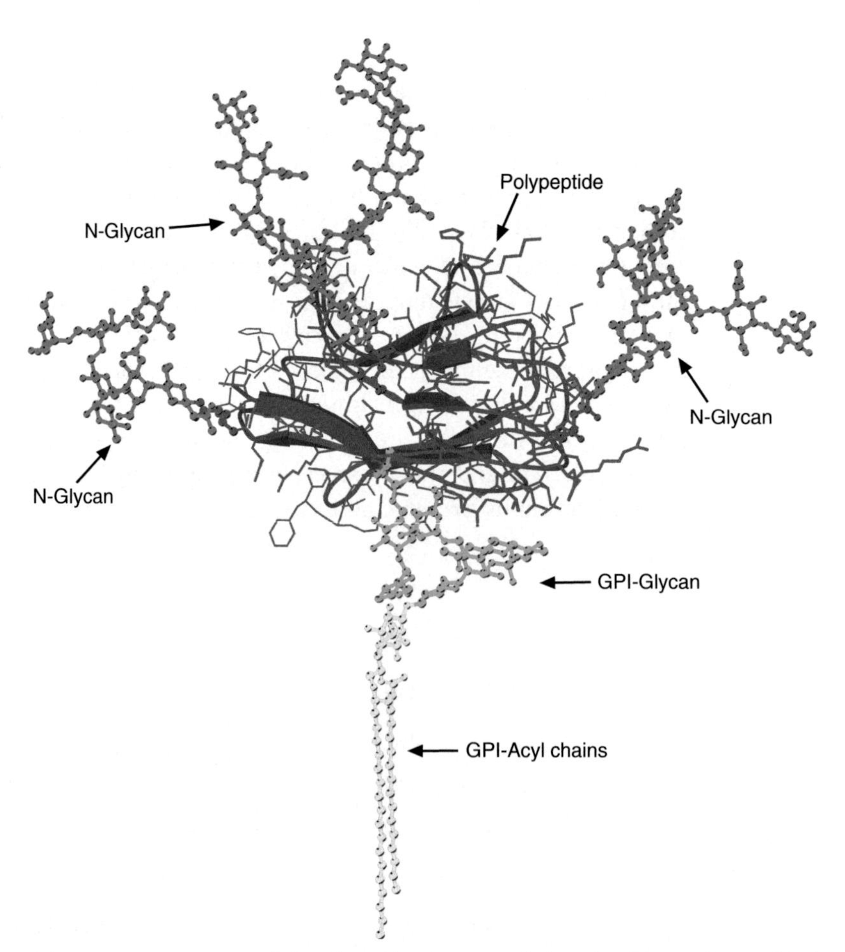

FIGURE 1.3. Schematic representation of the Thy-1 glycoprotein including the three N-glycans (*blue*) and a glycosylphosphatidylinositol (GPI-glycan; *green*) lipid anchor whose acyl chains (*yellow*) would normally be embedded in the membrane bilayer. Note that the polypeptide (*purple*) represents only a relatively small portion of the total mass of the protein. (Original art courtesy of Mark Wormald and Raymond Dwek, Oxford Glycobiology Institute.)

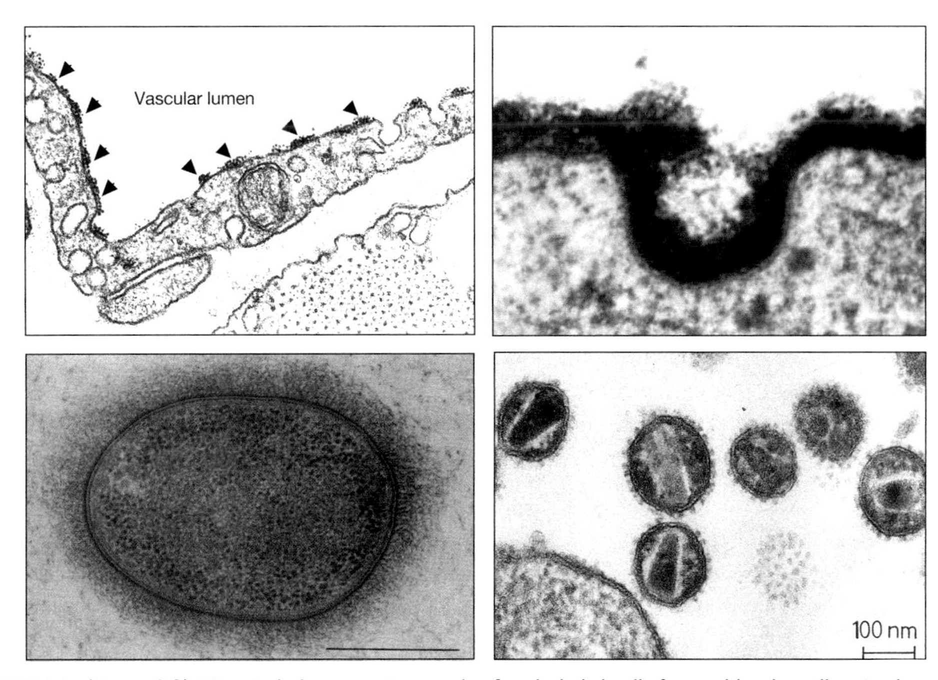

FIGURE 1.4. (*Upper left*) Historical electron micrograph of endothelial cells from a blood capillary in the rat diaphragm muscle, showing the lumenal cell membrane of the cells (facing the blood) decorated with particles of cationized ferritin (*arrowheads*). These particles are binding to acidic residues (sialic acid–containing glycans and sulfated glycosaminoglycans) contained in the cell-surface glycocalyx. Note that the particles are several layers deep, indicating the remarkable thickness of this layer of glycoconjugates. (Courtesy of the late George E. Palade, University of California, San Diego). (*Upper right*) Electron micrograph of fibroblast glycocalyx stained with ruthenium red (120,000×). (From Martinez-Palomo A, et al. 1969. *Cancer Res* **29:** 925–937, with permission from American Association for Cancer Research.) (*Lower left*) Thin section of *Klebsiella pneumonia* showing capsular layers substitution fixed with osmium tetraoxide. Scale bar, 0.5 μm. (From Amako K, et al. 1988. *J Bacteriol* **170:** 4960–4962, with permission from American Society for Microbiology.) (*Lower right*) Morphologically mature human immunodeficiency virus (HIV-1) particles shown by ultrathin section transmission electron microscopy (TS-TEM). Virus-producing H9 cells were treated with 1% tannic acid to render glycoprotein spikes on the virion detectable. The ill-defined aggregate of "spots" represents spikes in a tangential section of a virion. (From Gelderblom HR, et al. 1987. *Virology* **156:** 171–176, with permission from Elsevier.)

of viral and plant lectin binding sites on this cell type. The density of glycans in the glycocalyx can be remarkably high. For example, it has been calculated that the concentration of sialic acids in the glycocalyx of a typical human B lymphocyte may be >100 mM.

MONOSACCHARIDES CAN BE LINKED TOGETHER IN MANY MORE WAYS THAN AMINO ACIDS OR NUCLEOTIDES

Nucleotides and proteins are linear polymers that can each contain only one basic type of linkage between monomers. In contrast, each monosaccharide can theoretically generate either an α- or a β-linkage to any one of several positions on another monosaccharide in a chain or to another type of molecule. Thus, whereas three different nucleotides or amino acids can only generate six trimers, three different hexoses could theoretically produce (depending on which of their forms are considered) anywhere from 1056 to 27,648 unique trisaccharides. This difference in complexity becomes even greater as the number of monosaccharide units found in natural glycan increases (now numbering in the hundreds). Fortunately for the student of glycobiology, naturally occurring biological macromolecules in a given species tend to contain relatively few of the possible monosaccharide units, and they are in a limited number of combinations. Of course, the great majority of glycans in most species have yet to be discovered and structurally defined. Thus, much of the possible diversity may yet exist in nature.

COMMON MONOSACCHARIDE UNITS OF GLYCOCONJUGATES

Although several hundred distinct monosaccharides are known in nature, only a minority of these are commonly found in well-studied glycans. Examples of common monosaccharides in vertebrate cells are listed below, along with their standard abbreviations (for details regarding their structures, see Chapter 2, and embedded links from the symbols in Online Appendix 1B).

- *Pentoses*: five-carbon neutral sugars—D-xylose (Xyl)
- *Hexoses*: six-carbon neutral sugars—for example, D-glucose (Glc)
- *Hexosamines*: hexoses with an amino group at the 2-position, which can be either free or, more commonly, N-acetylated—for example, *N*-acetyl-D-glucosamine (GlcNAc)
- *6-Deoxyhexoses*: for example, L-fucose (Fuc)
- *Uronic acids*: hexoses with a carboxylate at the 6-position—for example, D-glucuronic acid (GlcA)
- *Nonulosonic acids*: family of nine-carbon acidic sugars, of which the most common in animals is the sialic acid *N*-acetylneuraminic acid (Neu5Ac, also sometimes called NeuAc or, historically, NANA) (see Chapter 15)

For simplicity, the symbols D- and L- are omitted from the full names of common monosaccharides from here on unless a less common variant occurs. This limited set of monosaccharides dominates the glycobiology of more recently evolved (so-called "higher") animals, but several others have been found in "lower" animals (e.g., tyvelose [Chapters 25 and 26], bacteria and Archaea [e.g., keto-deoxyoctulosonic acid, rhamnose, L-arabinose, and muramic acid; Chapters 21 and 22], and plants [e.g., arabinose, apiose, and galacturonic acid; Chapter 24]). A variety of modifications of glycans further enhance their diversity in nature and often serve to mediate specific biological functions. Thus, the hydroxyl groups of different monosaccharides can be subject to modifications such as phosphorylation, sulfation, methylation, O-acetylation, or fatty acylation. Although amino groups are commonly N-acetylated, they can be N-sulfated or remain unsubstituted. Carboxyl groups are occasionally subject to lactonization to nearby hydroxyl groups or even lactamization to nearby amino groups.

Details regarding the structural depiction of monosaccharides, linkages, and oligosaccharides are discussed in Chapter 2. Many figures in this volume use a symbolic depiction of sugar chains (see Online Appendix 1B and examples in Figure 1.5). This Symbol Nomenclature for Glycans (SNFG) is expanded from the Second Edition and has been vetted and adopted by many investigators. For reader convenience, a full table of symbols is reproduced on the inside cover of the book, is available for view at the NCBI Books website (Online Appendix 1B), and can be downloaded in drawing and text formats from this website. Detailed notes regarding the logic used for the symbol nomenclature can also be found at the Online Appendix 1B, along with embedded links to online databases featuring details about each monosaccharide, as well as color settings information for artists who wish to use the system.

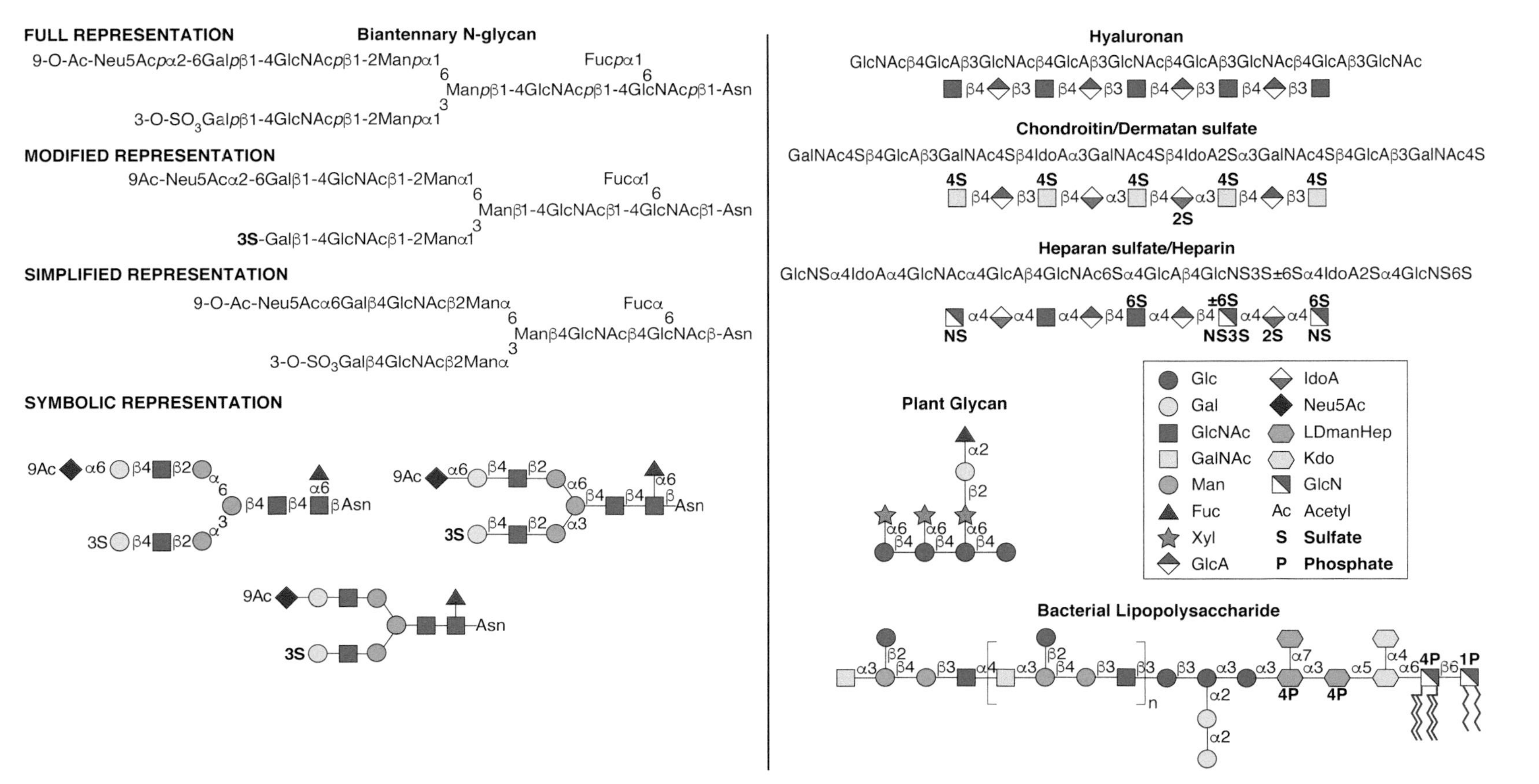

FIGURE 1.5. Examples of symbols and conventions for drawing glycan structures. The monosaccharide symbol set from the Third Edition of *Essentials of Glycobiology* remains intact but has been extended to cover a wider range of monosaccharides found in nature. For reader convenience, a full table of symbols is reproduced on the inside cover of this book and is available as Online Appendix 1B, Table 1 at the NCBI Books website. A few examples of symbols and the use of this system to depict naturally occurring glycans are shown here. (*Left*) Branched "biantennary" N-glycan from a vertebrate, with two types of outer termini, depicted at different levels of structural details. (*Right*) Examples of glycosaminoglycan (GAG) chains from metazoan animals, a plant glycan, and a bacterial lipopolysaccharide.

MAJOR CLASSES OF GLYCOCONJUGATES AND GLYCANS

The common classes of glycans are primarily defined according to the nature of the linkage to the aglycone (protein or lipid) (for common eukaryotic examples, see Figures 1.6 and 1.7). A glycoprotein is a glycoconjugate in which a protein carries one or more glycans covalently attached to a polypeptide backbone, usually via N- or O-linkages. An N-glycan (N-linked oligosaccharide, N-[Asn]-linked oligosaccharide) is a sugar chain covalently linked to an asparagine residue of a polypeptide chain, commonly involving a GlcNAc residue in eukaryotes, and the consensus peptide sequence: Asn-X-Ser/Thr. Animal N-glycans also share a common pentasaccharide core region and can be generally divided into three main classes: oligomannose (or high-mannose) type, complex type, and hybrid type (Chapter 9). An O-glycan (O-linked oligosaccharide) is frequently linked to the polypeptide via *N*-acetylgalactosamine (GalNAc) to a hydroxyl group of a serine or threonine residue and can be extended into different structural core classes (Chapter 10). A mucin is a large glycoprotein that carries many O-glycans that are clustered (closely spaced). Several other types of O-glycans also exist (e.g., O-linked fucose, glucose, or mannose). A proteoglycan is a glycoconjugate that has one or more glycosaminoglycan (GAG) chains (see definition below) attached to a "core protein" through a typical core region that has at its reducing end a xylose residue linked to the hydroxyl group of a serine residue. The distinction between a proteoglycan and a glycoprotein is otherwise arbitrary, because some proteoglycan polypeptides can carry both GAG chains and different O- and N-glycans (Chapter 17). Many cytoplasmic and nuclear proteins have single GlcNAc residue on one or more of their serine or threonine residues (Chapter 19). Figure 1.7 provides a listing of known glycan–protein linkages in nature.

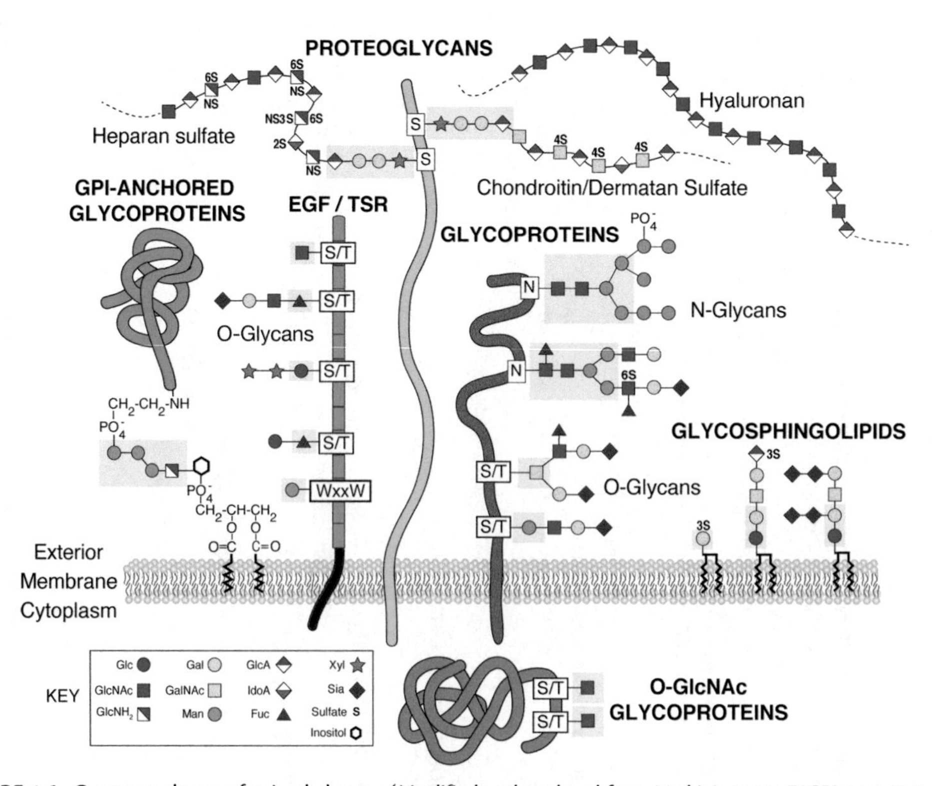

FIGURE 1.6. Common classes of animal glycans. (Modified and updated from Varki A. 1997. *FASEB J* **11:** 248–255; Fuster M, Esko JD. 2005. *Nat Rev Can* **7:** 526–542, with permission from Springer Nature and Stanley P. 2011. *Cold Spring Harb Perspect Biol* **3:** a005199.) (See Online Appendix 1B for full names and designated symbols for monosaccharides.)

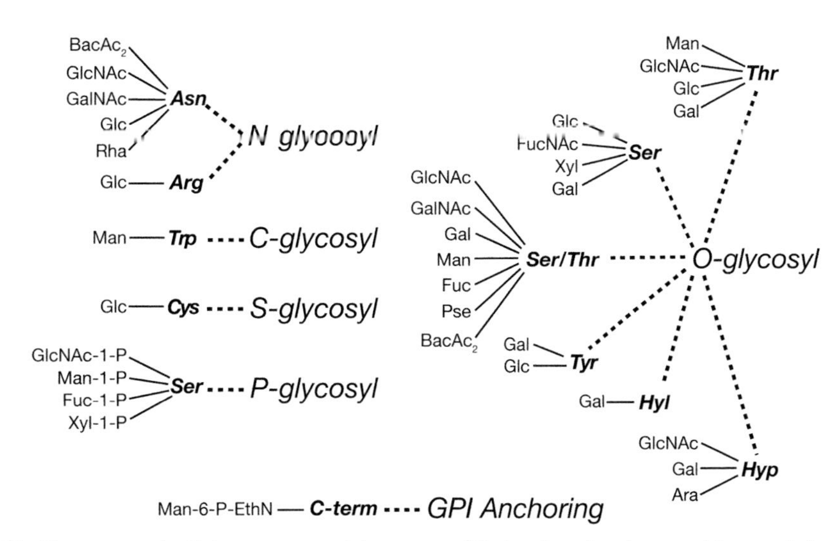

FIGURE 1.7. Glycan–protein linkages reported in nature. (Updated and redrawn, with permission of Oxford University Press, from Spiro RG. 2002. *Glycobiology* **12:** 43R–56R.) Diagrammatic representation of six distinct types of sugar-peptide bonds that have been identified in nature, to date. (See Online Appendix 1B for full names of monosaccharides.) (Hyl) Hydroxylysine; (Hyp) hydroxyproline; (C-term) carboxy-terminal amino acid residue. Glypiation is the process by which a glycosylphosphatidylinositol (GPI) anchor is added to a protein. For other details, including anomeric linkages, please see the Spiro (2002) review cited above.

A glycosylphosphatidylinositol anchor is a glycan bridge between phosphatidylinositol and a phosphoethanolamine that is in amide linkage to the carboxyl terminus of a protein. This structure typically constitutes the only anchor to the lipid bilayer membrane for such proteins (Chapter 12). A glycosphingolipid (often named a glycolipid) consists of a glycan usually attached via glucose or galactose to the terminal primary hydroxyl group of the lipid moiety ceramide, which is composed of a long chain base (sphingosine) and a fatty acid (Chapter 11). Glycolipids can be neutral or anionic. A ganglioside is an anionic glycolipid containing one or more residues of sialic acid. It should be noted that these represent only the most common classes of glycans reported in eukaryotic cells. There are several other less studied glycan types found on one or the other side of the cell membrane in animal cells (Chapters 13, 17, and 18) and many others in plants, algae, and prokaryotes.

Although different glycan classes have unique core regions by which they are distinguished, certain outer structural sequences are often shared among different classes of glycans. For example, animal N- and O-glycans and glycosphingolipids frequently carry the subterminal disaccharide Galβ1-4GlcNAcβ1-(*N*-acetyllactosamine or LacNAc) or, less commonly, GalNAcβ1-4GlcNAcβ1-(LacdiNAc) units. The LacNAc units can sometimes be repeated, giving extended poly-*N*-acetyllactosamines (sometimes incorrectly called "polylactosamines") (Chapter 14). Less commonly, the LacdiNAc motif can also be repeated (termed polyLacdiNAc). Outer LacNAc units can be modified by fucosylation or by branching and are typically capped in vertebrates by sialic acids or, less commonly, by sulfate esters, Fuc, α-Gal, β-GalNAc, or β-GlcA units (Chapters 14 and 15). In contrast, eukaryotic glycosaminoglycans are linear copolymers of acidic disaccharide repeating units, each typically containing a hexosamine (GlcN or GalN) and a hexose (Gal) or hexuronic acid (GlcA or IdoA) (Chapter 17). The type of disaccharide unit defines the glycosaminoglycan as chondroitin or dermatan sulfate (GalNAcβ1-4GlcA/IdoA), heparin or heparan sulfate (GlcNAcα1-4GlcA/IdoA), or keratan sulfate (Galβ1-4GlcNAc). Keratan sulfate is actually a 6-O-sulfated form of poly-*N*-acetyllactosamine attached to an N- or O-glycan core, rather than to a typical Xyl-Ser-containing proteoglycan linkage region. Another type of glycosaminoglycan, hyaluronan (a polymer of GlcNAcβ1-4GlcA), appears to exist primarily as a free glycan unattached to any aglycone (Chapter 16). Some glycosaminoglycans have sulfate esters substituting either amino or hydroxyl groups (i.e., N- or O-sulfate groups). Another anionic polysaccharide that can be extended from LacNAc units is polysialic acid, a homopolymer of sialic acid that is selectively expressed only on a few proteins in vertebrates. Polysialic acids and hyaluronan are also found as the capsular polysaccharides of certain pathogenic bacteria (Chapter 15). For simplicity, this section has focused primarily on vertebrate glycans. Many other classes of glycans exist in other branches of the tree of life (Chapters 21–26).

GLYCAN STRUCTURES ARE NOT ENCODED DIRECTLY IN THE GENOME

Unlike protein sequences, which are primary gene products, glycan structures are not encoded directly in the genome and are secondary gene products. A few percent of known genes in the human genome are dedicated to producing the enzymes and transporters responsible for the biosynthesis and assembly of glycans (Chapter 8), typically as posttranslational modifications of proteins or by glycosylation of core lipids. The glycans themselves represent numerous combinatorial possibilities, generated by a variety of competing and sequentially acting glycosidases and glycosyltransferases (Chapter 6) and the subcompartmentalized "assembly-line" mechanisms of glycan biosynthesis in the Golgi apparatus of eukaryotes (Chapter 4). Thus, even with full knowledge of the expression levels of all relevant gene products, one cannot accurately predict the precise structures of glycans elaborated by a given cell type. Furthermore, small changes in environmental cues can cause dramatic changes in glycans produced by a given cell. It is this variable and dynamic nature of glycosylation that makes it a powerful way to generate and modulate biological diversity and complexity. Of course, it also makes glycans much more difficult to study than nucleic acids and proteins.

SITE-SPECIFIC STRUCTURAL DIVERSITY IN PROTEIN GLYCOSYLATION

One of the most fascinating and yet frustrating aspects of protein glycosylation is the phenomenon of microheterogeneity. Thus, at any particular glycan attachment site on a given protein synthesized by a particular cell type, a range of variations might be found in the structures of the attached glycan (and in some instances, the glycan may be missing). Effectively, a given polypeptide encoded by a single gene can exist in numerous "glycoforms," each constituting a distinct molecular species. For some glycoproteins the microheterogeneity at a particular site may be quite limited, whereas for other sites it may be extensive, even within the same glycoprotein species. Mechanistically, microheterogeneity might be generated by the rapidity with which multiple, sequential, partially competitive glycosylation and deglycosylation reactions take place in the endoplasmic reticulum (ER) and Golgi apparatus, through which a newly synthesized glycoprotein passes, along with the lack of a template for directing the synthesis and the accessibility of glycans at a site to the modifying enzymes (Chapter 4). An alternate possibility is that each individual cell or cell type is in fact exquisitely specific in the glycosylation it produces, but that intercellular variations result in the observed microheterogeneity of samples from natural multicellular sources. Whatever the origin of microheterogeneity, it explains the anomalous behavior of glycoproteins in analytical/separation techniques and makes complete structural analysis of a glycoprotein a difficult task. From a functional point of view, the biological significance of microheterogeneity remains unclear. It is possible that this is a type of diversity generator, intended for diversifying endogenous recognition functions and/or for evading microbes and parasites, each of which can bind with high specificity only to certain glycan structures (Chapters 37 and 42).

CELL BIOLOGY OF GLYCOSYLATION

Most well-characterized pathways for the biosynthesis of major classes of eukaryotic glycans are within ER and Golgi compartments (Chapter 4). Newly synthesized proteins originating in the ER are either cotranslationally or posttranslationally modified with glycans at various stages in their itinerary toward their final destinations. N-glycans are partially assembled on lipid donors on the cytoplasmic face of the ER and then flipped across the membrane where the oligosaccharide assembly is completed and transfer to the nascent protein occurs. This oligosaccharide is then trimmed and extended by the addition of one monosaccharide at a time as the protein passes through the ER and the Golgi. These glycosylation reactions use activated forms of monosaccharides (nucleotide sugars; Chapter 5) as donors for reactions that are catalyzed by glycosyltransferases (for details about their biochemistry, molecular genetics, and cell biology, see Chapters 4, 6, and 8). The nucleotide sugar donors are synthesized within the cytosolic or nuclear compartment from monosaccharide precursors of endogenous or exogenous origin and then actively transported across a membrane bilayer into the lumen of the ER and Golgi compartments (Chapter 5). Notably, the portion of the glycoconjugate that faces the inside of these compartments will ultimately face the outside of the cell or the inside of a secretory granule or lysosome and will be topologically unexposed to the cytosol. The biosynthetic enzymes (glycosyltransferases, sulfotransferases, etc.) responsible for catalyzing these reactions are well studied (Chapter 6), and their location has helped to define various functional compartments of the ER–Golgi pathway. A classical model envisioned that these enzymes are physically lined up along this pathway in the precise sequence in which they actually work. This model appears to be oversimplified, as there is considerable overlap in the distribution of these enzymes, and the actual distribution of a given enzyme depends on the cell type.

All topological considerations mentioned above are reversed with regard to nuclear and cytoplasmic glycosylation, because the active sites of the relevant glycosyltransferases face the cytosol, which is in direct communication with the interior of the nucleus. Until the mid-1980s, the accepted dogma was that glycoconjugates occurred exclusively on the outer surface of cells, on the internal (luminal) surface of intracellular organelles, and on secreted molecules. The cytosol and nucleus were assumed to be devoid of glycosylation capacity. However, it is now clear that certain distinct types of glycoconjugates are synthesized and reside within the cytosol and nucleus (Chapter 18). Indeed, one of them, named O-GlcNAc (Chapter 19), may well be numerically the most common type of glycoconjugate in many cell types. The fact that this major form of glycosylation was missed by so many investigators for so long serves to emphasize the relatively unexplored state of the field of glycobiology.

Like all components of living cells, glycans are constantly being turned over by degradation and the enzymes that catalyze this process cleave glycans either at the outer (nonreducing) terminus (exoglycosidases) or internally (endoglycosidases) (Chapters 4 and 44). Some terminal monosaccharide units such as sialic acids are sometimes removed and new units reattached during endosomal recycling, without degradation of the underlying chain. The final complete degradation of most eukaryotic glycans is generally performed by multiple glycosidases in the lysosome. Once degraded, their individual unit monosaccharides are then typically exported from the lysosome into the cytosol for reuse (Figure 1.8). In contrast to the relatively slow turnover of glycans derived from the ER–Golgi pathway, the O-GlcNAc monosaccharide modifications of the nucleus and cytoplasm appear more dynamic (Chapter 19). In some instances, extracellular or intracellular free glycans can also serve as signaling molecules (Chapter 40).

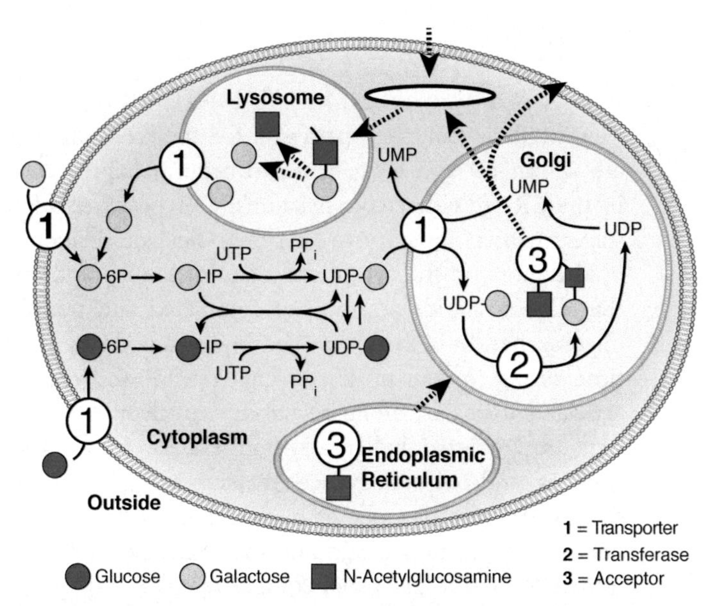

FIGURE 1.8. Biosynthesis, use, and turnover of a common monosaccharide. This schematic shows the biosynthesis, fate, and turnover of galactose, a common monosaccharide constituent of animal glycans. Although small amounts of galactose can be taken up from the outside of the cell, most cellular galactose is either synthesized de novo from glucose or recycled from degradation of glycoconjugates in the lysosome. The figure is a simplified view of the generation of the UDP nucleotide sugar UDP-Gal, its equilibrium state with UDP-glucose, and its uptake and use in the Golgi apparatus for synthesis of new glycans. (*Solid lines*) Biochemical pathways; (*dashed lines*) pathways for the trafficking of membranes and glycans.

TOOLS USED TO STUDY GLYCOSYLATION

Unlike oligonucleotides and proteins, glycans are not commonly found in a linear, unbranched fashion. Even when linear (e.g., GAGs), they often contain a variety of substituents, such as sulfate groups, which are not uniformly distributed. Thus, complete sequencing of glycans is usually impossible by a single method and requires iterative combinations of physical, chemical, and enzymatic approaches that together yield the details of structure (for a discussion of low- and high-resolution separation and analysis, including mass spectrometry and nuclear magnetic resonance [NMR], see Chapter 50). Less detailed information on structure may be sufficient to explore the biology of some glycans and can be obtained by simple techniques, such as the use of enzymes (endoglycosidases and exoglycosidases), lectins, and other glycan-binding proteins (Chapters 48 and 50), chemical modification or cleavage, metabolic radioactive labeling, antibodies, or cloned glycosyltransferases (Chapters 53 and 54). Glycosylation can also be perturbed in various ways, for example, by glycosylation inhibitors and primers (Chapters 55 and 56) and by genetic manipulation of glycosylation in intact cells and organisms (Chapter 49). Directed in vitro synthesis of glycans by chemical and enzymatic methods has also made great strides in recent years, providing many new tools for exploring glycobiology (Chapters 53, 54, and 57). The generation of complex glycan libraries by a variety of routes has further enhanced this interface of chemistry and biology (Chapters 53 and 54), including the generation of glycan microarrays.

GLYCOMICS

Analogous to genomics and proteomics, glycomics represents the systematic methodological elucidation of the "glycome" (the totality of glycan structures) of a given cell type or organism (Chapters 51 and 52). In reality, the glycome is far more complex than the genome or proteome. In addition to the vastly greater structural diversity in glycans, one is faced with the complexities of glycosylation microheterogeneity (see above) and the dynamic changes that occur in the course of development, differentiation, metabolic changes, aging, malignancy, inflammation, or infection. Added diversity arises from intraspecies and interspecies variations in glycosylation. Thus, a given cell type in a given species can manifest a large number of possible glycome states. Glycomic analysis today generally consists of extracting entire cell types, organs, or organisms; releasing all the glycan chains from their linkages; and cataloging them via approaches such as mass spectrometry. In a variation called glycoproteomics, the glycans are analyzed while still attached to protease-generated fragments of glycoproteins. The results obtained represent a spectacular improvement over what was possible a few decades ago, but they are still analogous to cutting down all the trees in a forest and cataloging them, without attention to the layout of the forest and the landscape (Chapter 15 discusses this complex issue from the perspective of just one monosaccharide class, sialic acids; see Figure 15.3).

Glycomic analysis thus needs to be complemented by classical methods such as tissue-section staining or flow cytometry, using lectins or glycan-specific antibodies that aid in understanding the glycome by taking into account the heterogeneity of glycosylation at the level of the different cell types and subcellular domains in the tissue under study. This is even more important because of the common observation that removing cells from their normal milieu and placing them into tissue culture can result in major changes in the glycosylation machinery of the cell. However, such classical approaches suffer from poor quantitation and relative insensitivity to structural details. A combination of the two approaches is now potentially feasible via laser-capture microdissection of specific cell types directly from tissue sections, with the resulting samples being studied by mass spectrometry. New methods for in situ imaging and characterization of glycans in the intact "forest" are clearly needed.

Because most of the genes involved in glycan biosynthetic pathways have been cloned from multiple organisms, it is possible today to obtain an indirect genomic and transcriptomic view of the glycome in a specific cell type (Chapter 8). However, given the relatively poor correlation between mRNA and protein levels, and the complex assembly line and competitive nature of the cellular Golgi glycosylation pathways, even complete knowledge of the mRNA expression patterns of all relevant genes in a given cell cannot allow accurate prediction of the distribution and structures of glycans in that cell type. Thus, there is as yet no reliable indirect route toward elucidating the glycome, other than by actual analysis using an array of methods.

GLYCOSYLATION DEFECTS IN ORGANISMS AND CULTURED CELLS

Many mutant variants of cultured cell lines with altered glycan structures and specific glycan biosynthetic defects have been described, the most common of which are lectin-resistant (Chapter 49). Indeed, with few exceptions, mutants with specific defects at most steps of the major pathways of glycan biosynthesis have been found in cultured animal cells. The use of such cell lines has been of great value in elucidating the details of glycan biosynthetic pathways. Their existence implies that many types of glycans are not crucial to the optimal growth of single cells growing in the sheltered and relatively unchanging environment of the culture dish. Rather, most glycan structures must be more important in mediating cell–cell and cell–matrix interactions in intact multicellular organisms and/or interactions between organisms. In keeping with this supposition, genetic defects completely eliminating major glycan classes in intact animals all cause embryonic lethality (Chapter 45). As might be expected, naturally occurring viable animal mutants of this type tend to have disease phenotypes of intermediate severity and show complex phenotypes involving multiple systems. Less severe genetic alterations of outer chain components of glycans tend to give viable organisms with more specific phenotypes

(Chapter 45). Overall, there is much to be learned by studying the consequences of natural or induced genetic defects in intact multicellular organisms, including humans (Chapter 45).

BIOLOGICAL ROLES OF GLYCANS ARE DIVERSE

A major theme of this volume is the exploration and elucidation of the biological roles of glycans. It is interesting to note that, in the short time since the first edition, we have gone from asking "what is it that glycans do anyway?" to having to explain a large number of complex and sometimes nonviable glycosylation-modified phenotypes in humans, mice, flies, and other organisms. Like any biological system, the optimal approach carefully considers the relationship of structure and biosynthesis to function (Chapter 7). As might be imagined from their ubiquitous and complex nature, the biological roles of glycans are remarkably varied. Indeed, asking what these roles are is akin to asking the same question about proteins. Thus, all of the proposed theories regarding glycan function turn out to be partly correct, and exceptions to each can also be found. Not surprisingly for such a diverse group of molecules, the biological roles of glycans also span the spectrum from those that are subtle to those that are crucial for the development, growth, function, or survival of an organism (Chapter 7). The diverse functions ascribed to glycans can be simply divided into two general categories: (i) structural and modulatory functions (involving the glycans themselves or their modulation of the molecules to which they are attached) and (ii) specific recognition of glycans by glycan-binding proteins. Of course, any given glycan can mediate one or both types of functions. The binding proteins in turn fall into two broad groups: lectins and sulfated GAG-binding proteins (Chapters 27, 28, and 38). Such molecules can be either intrinsic to the organism that synthesized the cognate glycans (e.g., see Chapters 31–36, 38, and 39) or extrinsic (see Chapters 37 and 42 for information concerning microbial proteins that bind to specific glycans on host cells). The atomic details of these glycan–protein interactions have been elucidated in many instances (Chapters 29 and 30). Although there are exceptions to this notion, the following general theme has emerged regarding lectins: monovalent binding tends to be of relatively low affinity, and such systems typically achieve their specificity and function by achieving high avidity, via interactions of multivalent arrays of glycans with cognate lectin-binding sites.

GLYCOSYLATION CHANGES IN DEVELOPMENT, DIFFERENTIATION, AND MALIGNANCY

Whenever a new tool (e.g., an antibody or lectin) specific for detecting a particular glycan is developed and used to probe its expression in intact organisms, it is common to find exquisitely specific temporal and spatial patterns of expression of that glycan in relation to cellular activation, embryonic development, organogenesis, and differentiation (see Chapter 41 for examples). Certain relatively specific changes in expression of glycans are also often found in the course of transformation and progression to malignancy (Chapter 47), as well as other pathological situations such as inflammation (Chapter 46). These spatially and temporally controlled patterns of glycan expression imply the involvement of glycans in many normal and pathological processes, the precise mechanisms of which are understood in only some cases.

EVOLUTIONARY CONSIDERATIONS IN GLYCOBIOLOGY

Remarkably little is still known about the evolution of glycosylation. There are clearly shared and unique features of glycosylation in different kingdoms and taxa. Among animals, there may be a trend toward increasing complexity of N- and O-glycans in more recently evolved ("higher") taxa. Intraspecies and interspecies variations in glycosylation are also relatively common. It has been suggested that the more specific biological roles of glycans are often mediated by uncommon structures, unusual presentations of common structures, or further modifications of the commonly occurring saccharides themselves. Such unusual structures likely result from unique expression patterns of the relevant glycosyltransferases or other glycan-modifying enzymes. On the other hand, such uncommon glycans can be targets for specific recognition by infectious microorganisms and various toxins. Thus, at least some of the diversity in glycan expression in nature must be related to the evolutionary selection pressures generated by interspecies interactions (e.g., of host with pathogen or symbiont). In other words, the two different classes of glycan recognition mentioned above (mediated by intrinsic and extrinsic glycan-binding proteins) are in competition with each other with regard to a particular glycan target. The specialized glycans expressed by parasites and microbes that are of great interest from the biomedical point of view (Chapters 21, 22, 23, and 43) may themselves be subject to evolutionary selection pressures. These issues are further considered in Chapter 20, which also discusses the limited information concerning how various glycan biosynthetic pathways appear to have evolved and diverged in different life-forms.

GLYCANS IN MEDICINE AND BIOTECHNOLOGY

Numerous natural bioactive molecules are glycoconjugates, and the attached glycans can have dramatic effects on the biosynthesis, stability, action, and turnover of these molecules in intact organisms. For example, the sulfated glycosaminoglycan heparin and its derivatives are among the most commonly used drugs in the world. The aminoglycoside antibiotics all have carbohydrate components essential for activity. For this and many other reasons, glycobiology and carbohydrate chemistry have become increasingly important in modern biotechnology. Patenting a glycoprotein drug, obtaining FDA approval for its use, and monitoring its production all require knowledge of the structure of its glycans. Moreover, glycoproteins, which include monoclonal antibodies, enzymes, and hormones, are by now the major products of the biotechnology industry, with sales in the tens of billions of dollars annually, which continues to grow at an increasing rate. In addition, several human disease states involve changes in glycan biosynthesis that can be of diagnostic and/or therapeutic significance. The emerging importance of glycobiology in medicine and biotechnology is discussed in Chapters 56 and 57.

GLYCANS IN NANOTECHNOLOGY, BIOENERGY, AND MATERIALS SCIENCE

Although not traditionally considered part of "Glycobiology," many natural and synthetic glycans are key components of Nanotechnology, Bioenergy, and Material Science. Glyconanomaterials (Chapter 58) have tunable chemical and physical properties and can be built on different scaffolds to probe cellular, tissue, and organismal interactions. Attached glycans can change nanomaterial properties, optimizing solubility and biocompatibility and lowering cytotoxicity. Glyconanomaterials have been used as imaging agents, spectroscopic tools, monitors of cellular systems, and vehicles for vaccination and drug delivery. Plant glycans are used for many purposes: energy sources, building materials, clothes, paper products, animal feed, and food and beverage additives (Chapter 59). Concerns about detrimental environmental effects and diminishing reserves of petroleum and its by-products have greatly renewed interest in using plant glycans for energy production, generation of polymers with improved or new functionalities, and as sources of high-value chemosynthetic precursors (Chapter 59).

ACKNOWLEDGMENTS

The authors acknowledge contributions to the previous version of this chapter from the late Nathan Sharon, as well as helpful comments and suggestions from other editors.

FURTHER READING

Rademacher TW, Parekh RB, Dwek RA. 1988. Glycobiology. *Annu Rev Biochem* **57**: 785–838. doi:10.1146/annurev.bi.57.070188.004033

Varki A. 1993. Biological roles of oligosaccharides: all of the theories are correct. *Glycobiology* **3**: 97–130. doi:10.1093/glycob/3.2.97

Drickamer K, Taylor ME. 1998. Evolving views of protein glycosylation. *Trends Biochem Sci* **23**: 321–324. doi:10.1016/s0968-0004(98)01246-8

Etzler ME. 1998. Oligosaccharide signaling of plant cells. *J Cell Biochem Suppl* **30–31**: 123–128.

Gagneux P, Varki A. 1999. Evolutionary considerations in relating oligosaccharide diversity to biological function. *Glycobiology* **9**: 747–755. doi:10.1093/glycob/9.8.747

Roseman S. 2001. Reflections on glycobiology. *J Biol Chem* **276**: 41527–41542. doi:10.1074/jbc.r100053200

Hakomori S-I. 2002. The glycosynapse. *Proc Natl Acad Sci* **99**: 225–232. doi:10.1073/pnas.012540899

Spiro RG. 2002. Protein glycosylation: nature, distribution, enzymatic formation, and disease implications of glycopeptide bonds. *Glycobiology* **12**: 43R–56R. doi:10.1093/glycob/12.4.43r

Haltiwanger RS, Lowe JB. 2004. Role of glycosylation in development. *Annu Rev Biochem* **73**: 491–537. doi:10.1146/annurev.biochem.73.011303.074043

Sharon N, Lis H. 2004. History of lectins: from hemagglutinins to biological recognition molecules. *Glycobiology* **14**: 53R–62R. doi:10.1093/glycob/cwh122

Drickamer K, Taylor ME. 2006. *Introduction to glycobiology*, Vol. 2. Oxford University Press, Oxford.

Ohtsubo K, Marth JD. 2006. Glycosylation in cellular mechanisms of health and disease. *Cell* **126**: 855–867. doi:10.1016/j.cell.2006.08.019

Patnaik SK, Stanley P. 2006. Lectin-resistant CHO glycosylation mutants. *Methods Enzymol* **416**: 159–182. doi:10.1016/s0076-6879(06)16011-5

Kamerling J, Boons G-J, Lee Y, Suzuki A, Taniguch N, Voragen AGJ. 2007. *Comprehensive glycoscience*, pp. 1–4. Elsevier Science, London.

Freeze HH, Ng BG. 2011. Golgi glycosylation and human inherited diseases. *Cold Spring Harb Perspect Biol* **3**: a005371. doi:10.1101/cshperspect.a005371

Hart GW, Slawson C, Ramirez-Correa G, Lagerlof O. 2011. Cross talk between O-GlcNAcylation and phosphorylation: roles in signaling, transcription, and chronic disease. *Annu Rev Biochem* **80**: 825–858. doi:10.1146/annurev-biochem-060608-102511

Sarrazin S, Lamanna WC, Esko JD. 2011. Heparan sulfate proteoglycans. *Cold Spring Harb Perspect Biol* **3**: a004952. doi:10.1101/cshperspect.a004952

Varki A. 2011. Evolutionary forces shaping the Golgi glycosylation machinery: why cell surface glycans are universal to living cells. *Cold Spring Harb Perspect Biol* **3**: a005462. doi:10.1101/cshperspect.a005462

Aebi M. 2013. N-linked protein glycosylation in the ER. *Biochim Biophys Acta* **1833**: 2430–2437. doi:10.1016/j.bbamcr.2013.04.001

Prasanphanich NS, Mickum ML, Heimburg-Molinaro J, Cummings RD. 2013. Glycoconjugates in host–helminth interactions. *Front Immunol* **4**: 240. doi:10.3389/fimmu.2013.00240

Varki A. 2013. Omics: account for the 'dark matter' of biology. *Nature* **497**: 565. doi:10.1038/497565

Belardi B, Bertozzi CR. 2015. Chemical lectinology: tools for probing the ligands and dynamics of mammalian lectins in vivo. *Chem Biol* **22**: 983–993. doi:10.1016/j.chembiol.2015.07.009

Endo T. 2015. Glycobiology of α-dystroglycan and muscular dystrophy. *J Biochem* **157**: 1–12. doi:10.1093/jb/mvu066

Misra S, Hascall VC, Markwald RR, Ghatak S. 2015. Interactions between hyaluronan and its receptors (CD44, RHAMM) regulate the activities of inflammation and cancer. *Front Immunol* **6**: 201. doi:10.3389/fimmu.2015.00201

Varki A, Cummings RD, Aebi M, Packer NH, Seeberger PH, Esko JD, Stanley P, Hart G, Darvill A, Kinoshita T, et al. 2015. Symbol nomenclature for graphical representations of glycans. *Glycobiology* **25**: 1323–1324. doi:10.1093/glycob/cwv091

Aoki-Kinoshita K, Agravat S, Aoki NP, Arpinar S, Cummings RD, Fujita A, Fujita N, Hart GM, Haslam SM, Kawasaki T, et al. 2016. GlyTouCan 1.0—the international glycan structure repository. *Nucl Acids Res* **44**: D1237–D1242. doi:10.1093/nar/gkv1041

Cook GMW. 2016. Glycobiology of the cell surface: its debt to cell electrophoresis 1940-65. *Electrophoresis* **37**: 1399–1406. doi:10.1002/elps.201500476

Varki A. 2017. Biological roles of glycans. *Glycobiology* **27**: 3–49. doi:10.1093/glycob/cww086

2 Monosaccharide Diversity

Peter H. Seeberger

This chapter covers the basic building blocks of glycans and fundamental considerations regarding glycan structure by introducing chemical concepts. Modes of linking glycans and structural depiction of the same are discussed to provide the groundwork for understanding longer glycans (Chapter 3).

INTRODUCTION TO GLYCAN TERMINOLOGY

In this book, as well as in the earlier editions, the term glycan is used. Still, a host of names are commonly used to refer to sugar polymers in other textbooks and the literature. In the 19th century, sugar-based substances were referred to as carbohydrates, or "hydrates of carbon," that are based on the general formula $C_x(H_2O)_n$ that also possess a carbonyl group, either an aldehyde or a ketone. Monosaccharides are the simplest of these polyhydroxylated carbonyl compounds (saccharide is derived from the Greek word for sugar or sweetness).

Monosaccharides are joined together to give rise to oligosaccharides or polysaccharides. Typically, the term "oligosaccharide" refers to any glycan that contains less than 20 monosaccharide residues connected by glycosidic linkages. The term "polysaccharide" is typically used to denote any linear or branched polymer consisting of monosaccharide residues, such as cellulose (Chapters 14

published by Cold Spring Harbor Laboratory Press; doi: 10.1101/glycobiology.4e.02

and 24). Thus, the relationship of monosaccharides to oligosaccharides or polysaccharides is analogous to that of amino acids and proteins or nucleotides and nucleic acids (polynucleotides).

The term “glycoconjugate” is often used to describe a macromolecule that contains monosaccharides covalently linked to proteins or lipids. The prefix “glyco” and the suffixes “saccharide” and “glycan” indicate the presence of carbohydrate constituents (e.g., glycoproteins, glycolipids, and proteoglycans). Just as is observed with proteins in nature, additional structural diversity can be imparted to glycans by modifying their hydroxyl groups with phosphate, sulfate, or acetyl esters, and/or their amino groups with acetyl or sulfate groups.

A carbohydrate may be termed “complex” if it contains more than one type of monosaccharide building unit. The glucose-based polymer cellulose is an example of a “simple” carbohydrate, whereas a galactomannan polysaccharide, composed of both galactose and mannose, is an example of a complex carbohydrate. However, even so-called simple glycans, such as cellulose and starch, often have very complex molecular structures in three dimensions. The term complex carbohydrates includes glycoconjugates, whereas the term carbohydrates per se would not. Additional nomenclature issues are covered in this chapter and Chapter 3. A more detailed and comprehensive listing of carbohydrate nomenclature rules has been published (see McNaught 1997 and Varki et al. 2015 in Further Reading at the end of this chapter, and Online Appendix 1B).

MONOSACCHARIDES: BASIC STRUCTURES AND STEREOISOMERISM

The classification of monosaccharide structures began in the late 19th century with the pioneering work of Emil Fischer. All simple monosaccharides have the general empirical formula $C_x(H_2O)_n$, where n is an integer ranging from 3 to 9. As mentioned briefly in Chapter 1, all monosaccharides consist of a chain of chiral hydroxymethylene units, which terminates at one end with a hydroxymethyl group and at the other with either an aldehyde group (aldoses) or an α-hydroxy ketone group (ketoses). Glyceraldehyde is the simplest aldose and dihydroxyacetone is the simplest ketose (Figure 2.1). The structures of glyceraldehyde and dihydroxyacetone are distinct in that glyceraldehyde contains an asymmetric (chiral) carbon atom (Figure 2.1), whereas dihydroxyacetone does not. With the exception of dihydroxyacetone, all monosaccharides have at least one asymmetric carbon atom, the total number being equal to the number of internal (CHOH) groups ($n-2$ for aldoses and $n-3$ for ketoses with n carbon atoms). The number of stereoisomers corresponds to 2^k, where k equals the number of asymmetric carbon atoms. For example, an aldohexose with the general formula $C_6H_{12}O_6$ and four asymmetric carbon atoms (i.e., four (CHOH) groups) can be described in 16 possible isomeric forms (Figure 2.1).

The numbering of carbon atoms follows the rules of organic chemistry nomenclature. The aldehyde carbon is referred to as C-1 and the carbonyl group in ketoses is referred to as C-2. The overall configuration (D or L) of each sugar is determined by the absolute configuration of

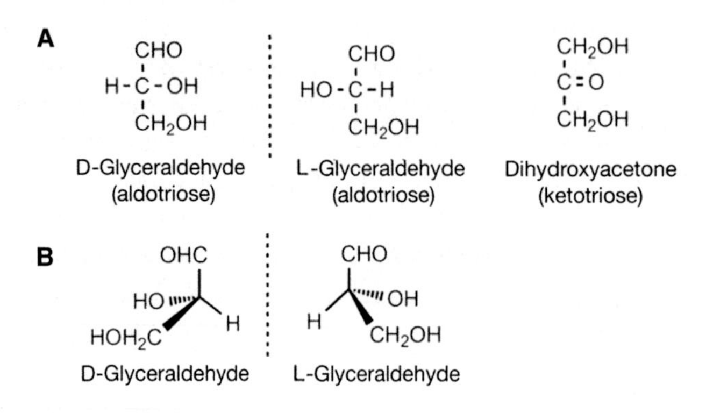

FIGURE 2.1. Structures of glyceraldehyde and dihydroxyacetone. (*A*) Fischer projection. (*B*) D- and L-glyceraldehyde. The chiral central carbon in glyceraldehyde gives rise to two possible configurations of the molecule, termed D and L.

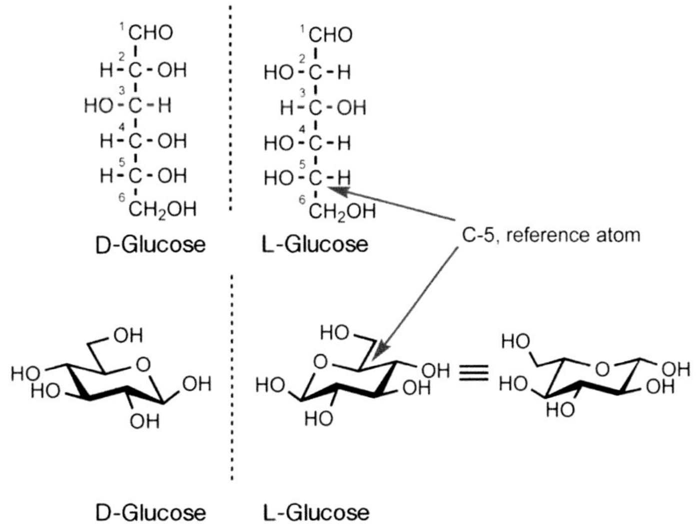

FIGURE 2.2. D- and L-glucopyranose in Fischer projection and chair conformation.

the stereogenic center furthest from the carbonyl group (i.e., with the highest numbered asymmetric carbon atom; this is C-5 in hexoses and C-4 in pentoses). The configuration of a monosaccharide is most easily determined by representing the structure in a Fischer projection. If the OH (or other non-H group) is on the right in the Fischer projection, the overall configuration is D. If the OH (or other non-H group) is on the left, the overall configuration is L (Figure 2.2). This figure also shows D- and L-glucose in the cyclic form (chair conformation) found in solution. Most vertebrate monosaccharides have the D configuration with the exception of fucose and iduronic acid (IdoA) L sugars. The Fischer projections shown in Figure 2.3 illustrate the acyclic structures of all D-aldoses through the aldohexose group.

Any two sugars that differ only in the configuration around a single chiral carbon atom are called epimers. For example, D-mannose is the C-2 epimer of D-glucose, whereas D-galactose is the C-4 epimer of D-glucose (Figure 2.4). Monosaccharide names are frequently abbreviated; most common are three-letter abbreviations for simple monosaccharides (e.g., Gal, Glc, Man, Xyl, Fuc). There are nine common monosaccharides found in vertebrate glycoconjugates (Figure 2.4). Once incorporated into a glycan, these nine monosaccharide building blocks can be further modified to generate additional sugar structures. For example, glucuronic acid (GlcA) can be epimerized at C-5 to generate IdoA. Many more monosaccharides exist in glycoconjugates from other species and as intermediates in metabolism. We use a symbolic notation for the monosaccharides that are most abundant in vertebrate glycoconjugates (see Chapter 1).

MONOSACCHARIDES EXIST PRIMARILY IN CYCLIC FORM

Monosaccharides exist in solution as an equilibrium mixture of acyclic and cyclic forms. The percentage of each form depends on the sugar structure. The cyclic form of a monosaccharide is characterized by a hemiacetal group formed by the reaction of one of the hydroxyl groups with the C-1 aldehyde or ketone. For reasons of chemical stability, five- and six-membered rings are most commonly formed from acyclic monosaccharides. Hexoses (six-carbon aldoses) and hexuloses (six-carbon ketoses) form six-membered rings via a C-1—O—C-5 ring closure; they form five-membered rings through a C-1—O—C-4 ring closure (Figure 2.5). A five-membered cyclic hemiacetal is labeled a "furanose" and a six-membered cyclic hemiacetal is called a "pyranose." Pentoses can form both pyranose and furanose forms.

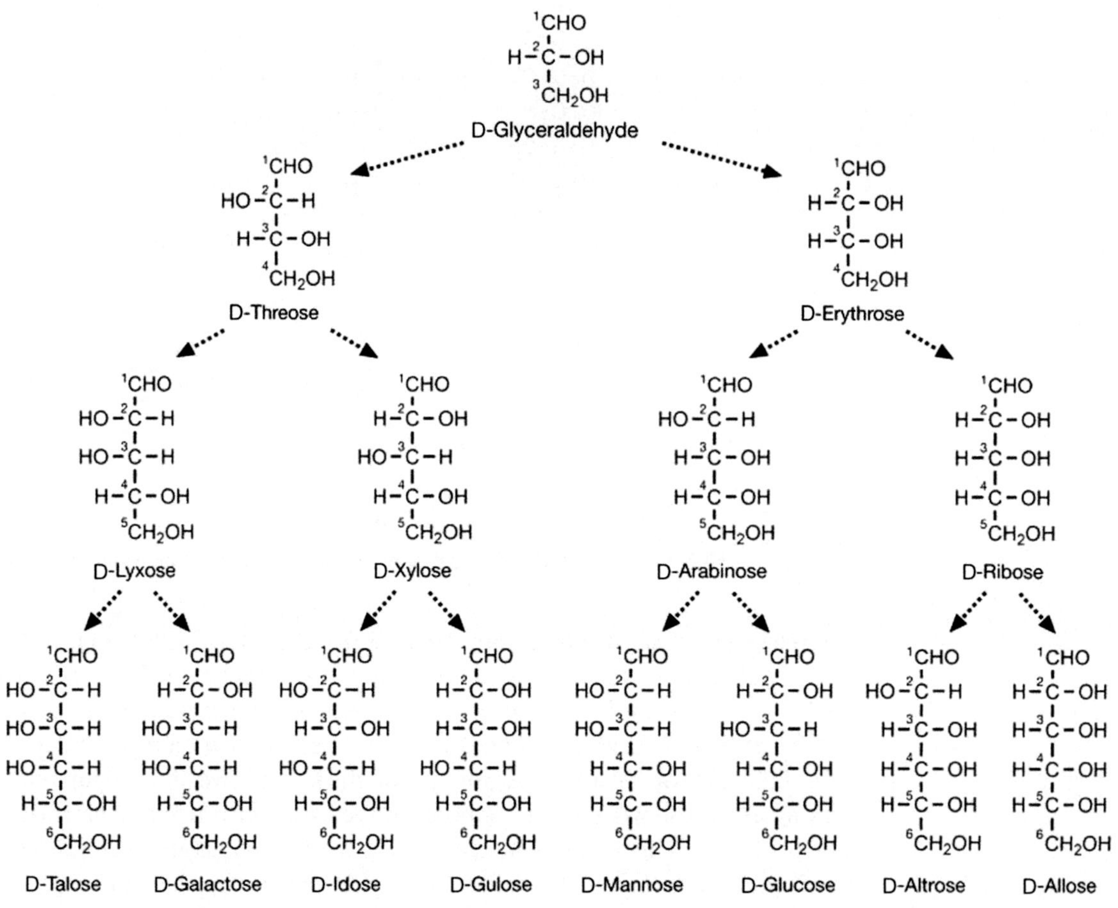

FIGURE 2.3. Fischer projections for the acyclic forms of the D series of aldoses, ranging from triose to hexose.

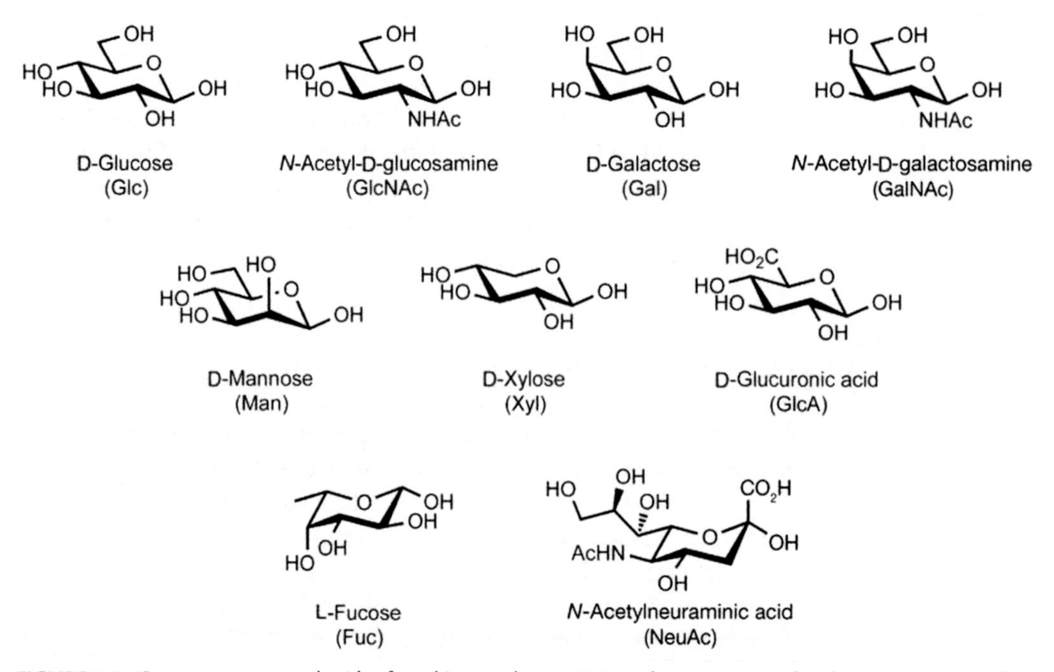

FIGURE 2.4. Common monosaccharides found in vertebrates. *N*-Acetylneuraminic acid is the most common form of sialic acid.

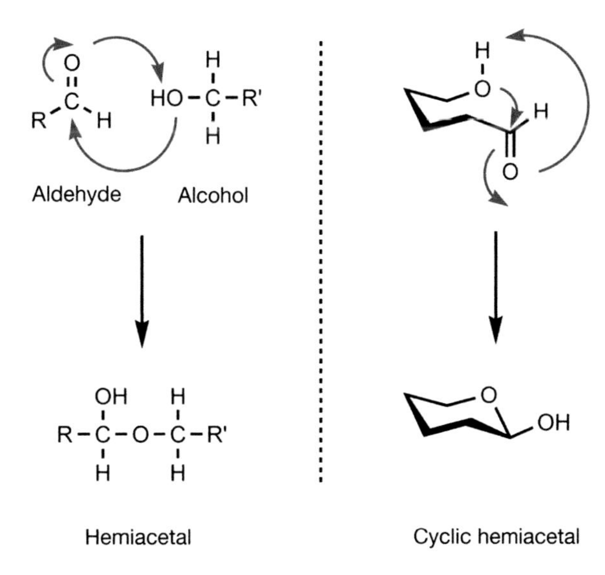

FIGURE 2.5. Cyclization of acyclic D-glucose to form pyranose and furanose structures. The cyclization reaction produces both the α and β anomers (i.e., C-1 epimers).

Formation of Hemiacetals

Monosaccharides can also be represented as Haworth projections in which both five- and six-membered cyclic structures are depicted as planar ring systems, with the hydroxy groups oriented either above or below the plane of the ring (Figure 2.6). Although not truly representative of the three-dimensional structure of a monosaccharide, the Haworth representation has been used since the late 1920s as an easy-to-draw formula that permits a quick evaluation of stereochemistry around the monosaccharide ring. The Haworth representations are preferably drawn with the ring oxygen atom at the top (for furanose) or the top right-hand corner (for pyranose) of the structure; the numbering of the ring carbons increases in a clockwise direction.

For any D sugar, the conversion of a Fischer projection into a Haworth projection proceeds as follows: (1) any groups (atoms) that are directed to the right in the Fischer structure are given a downward orientation in the Haworth structure, (2) any groups (atoms) that are directed to the left in the Fischer structure are given an upward orientation in the Haworth structure, and (3) the terminal —CH_2OH group is given an upward orientation in the Haworth structure. For an L sugar, (1) and (2) are the same, but the terminal —CH_2OH group is projected downward.

The planar Haworth structures are distorted representations of the actual molecules. The preferred conformation of a pyranose ring is the "chair" conformation, similar to the structure of

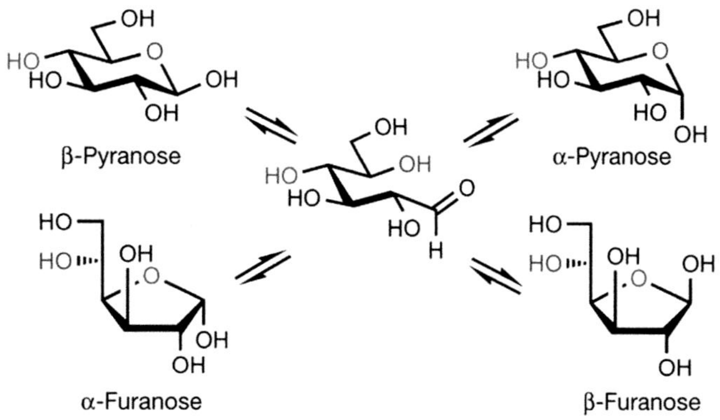

β-Pyranose

α-Pyranose

α-Furanose

β-Furanose

FIGURE 2.6. Conversion from Fischer to Haworth projection. Each hydroxyl group projected to the right in the Fischer projection points down in the Haworth formula.

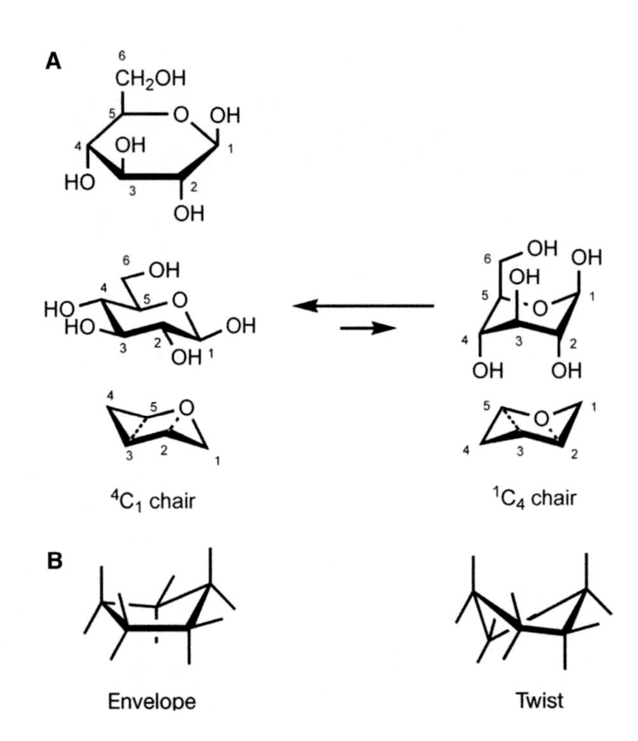

FIGURE 2.7. Chair conformations. (*A*) β-D-Glucose in Haworth projection and in its $^4C_1$ and $^1C_4$ chair conformations; (*B*) envelope and twist conformations for a five-membered ring structure.

cyclohexane. The conversion from Haworth projection to chair conformation leaves the downward or upward orientation of ring substituents unaltered. Two chair conformations can be distinguished and designated as $^4C_1$ and $^1C_4$, respectively (Figure 2.7A), and these conformers can interconvert by a process called the "ring flip." The first numeral in the chair conformer designation (superscript) indicates the number of the ring carbon atom above the "seat of the chair (C)" and the second numeral (subscript) indicates the number of the ring carbon atom below the plane of the seat (spanned by C-2, C-3, C-5, and the ring O). Chair conformations are designated from structures with the ring oxygen atom in the top right-hand corner of the ring "seat," resulting in the clockwise appearance of the ring numbering. To determine the stereochemistry in the chair form as it corresponds to the Fischer projection, one can locate C-6 and then trace along the carbon skeleton of the sugar, bisecting the C—O and C—H bonds formed from each atom. The OH (or OR) and H groups are found on the right (R) or left (L) sides, just as in the Fischer projection (Figure 2.8).

The more structurally accurate chair representations are preferred to Haworth projections for depicting pyranoses. However, Haworth projections are convenient and are commonly used to depict furanoses. The furanose ring is rather flexible and not entirely flat in any of its energetically favored conformations; for example, it has a slight pucker when viewed from the side, as seen in the representations of the so-called envelope and twist (or skew) conformations (Figure

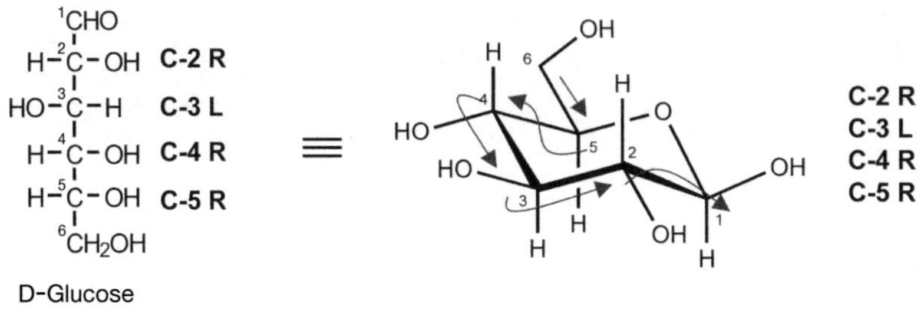

C-2 R
C-3 L
C-4 R
C-5 R
D-Glucose

FIGURE 2.8. Conversion from Fischer to chair projection formula; (R) right; (L) left. *Red arrows* illustrate the path to follow along the sugar backbone when correlating the stereochemistry of the Fischer projection with the chair conformation.

2.7B). Because furanoses can adopt many low-energy conformations, researchers have adopted the Haworth projection as a simple means to avoid this complexity.

CHEMISTRY AT THE ANOMERIC CENTER

Mutarotation

When cyclized into rings, monosaccharides acquire an additional asymmetric center derived from the carbonyl carbon atom (Figure 2.5). The new asymmetric center is termed the "anomeric carbon" (i.e., C-1 in the ring form of glucose). Two stereoisomers are formed by the cyclization reaction because the anomeric hydroxy group can assume two possible orientations. When the configurations are the same at the anomeric carbon and the stereogenic center furthest from the anomeric carbon, the monosaccharide is defined as the α anomer. When the configurations are different, the monosaccharide is defined as the β anomer (Figure 2.9). Unlike the other stereocenters on the monosaccharide ring, which are configurationally stable, the anomeric center can undergo an interconversion of stereoisomers via the process of mutarotation. Catalyzed by dilute acid or base, the reaction proceeds by the reverse of the cyclization reaction. The monosaccharide ring opens up and then recloses to form a ring with the other anomeric configuration (Figure 2.5). The term mutarotation derives from the rapid change in optical rotation (denoted [α] D) that is observed when an anomerically pure form of a monosaccharide is dissolved in water. For example, β-D-glucopyranose shows an initial rotation of +19°, whereas the α anomer shows an initial rotation of +112°. When either anomer is allowed to undergo the mutarotation reaction, an equilibrium mixture containing both anomers is obtained, producing a rotation of +52.5°.

Oxidation and Reduction

Generally, the acyclic (aldehyde or ketone) form of a monosaccharide is only present in minor amounts in an equilibrium mixture ($<$0.01%). Nevertheless, the open-chain aldehydes or ketones can participate in chemical reactions that drive the equilibrium and eventually consume the sugar.

Aldoses and ketoses were historically referred to as "reducing sugars" because they responded positively in a chemical test that effected oxidation of their aldehyde and hydroxyketone functionalities, respectively. The carboxylic acid formed by oxidation of the aldehyde in an aldose is referred to as a glyconic acid (e.g., gluconic acid is the oxidation product of glucose). It is also possible to oxidize the hydroxyl groups of monosaccharides, most notably the terminal OH group (i.e., C-6 of glucose). In this reaction, a glycuronic acid is produced, and if both terminal groups are oxidized, the product is a glycaric acid. The three acids derived from D-glucose are illustrated in Figure 2.10. These compounds have a tendency to undergo

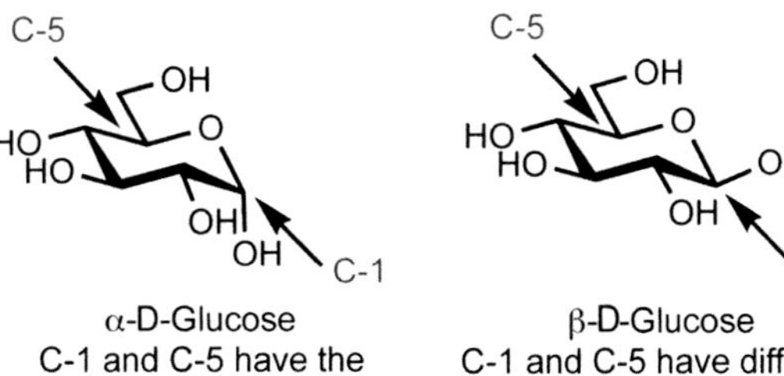

FIGURE 2.9. Determination of configuration at the anomeric center.

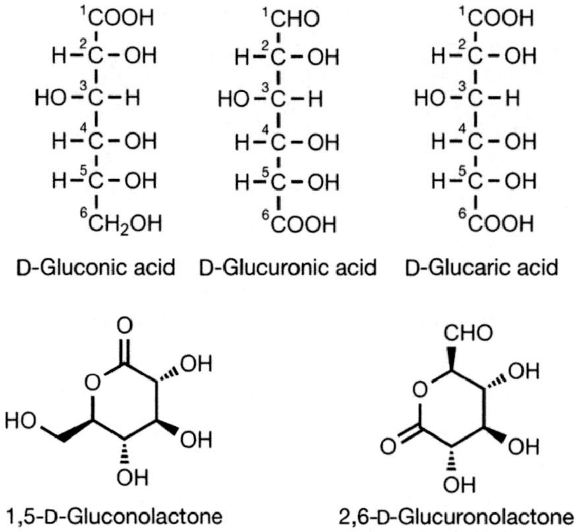

FIGURE 2.10. Oxidized forms of D-glucose.

intramolecular cyclization reactions, preferably yielding six-membered lactones. Two examples of lactonized forms are shown in Figure 2.11. Oxidized forms of monosaccharides can be found in nature. For example, GlcA is an abundant component of many glycosaminoglycans (see Chapter 17).

The carbonyl groups of aldoses and ketoses also can be reduced with sodium borohydride ($NaBH_4$) to form polyhydroxy alcohols, referred to as alditols. This reaction is widely used to introduce a radiolabel at C-1 of the monosaccharide by reduction with NaB^3H_4 (Figure 2.11).

Schiff Base Formation

The aldehyde and ketone groups of monosaccharides can also undergo Schiff base formation with amines or hydrazides, forming imines and hydrazones, respectively (Figure 2.12). This reaction is often used to conjugate the monosaccharide to proteins (via their lysine residues) or to biochemical probes such as biotin hydrazide. It should be noted that the imines formed with amino groups are not stable to water and are typically reduced with sodium cyanoborohydride ($NaCNBH_3$) in a process termed reductive amination.

As aldehydes, reducing sugars can also form Schiff bases with amino groups of the lysine residues in proteins. This nonenzymatic process that links glycans to proteins is termed "glycation" and is distinct from "glycosylation," which involves the formation of a glycosidic bond between the sugar and protein. Glycation products can undergo further reactions that lead to the formation of protein cross-links, and these can have pathogenic consequences (i.e., they are immunogenic and change the properties of the protein). Glycation products of glucose accumulate at higher levels in diabetics than in healthy individuals because of elevated blood glucose levels. These modified proteins are thought to underlie some of the pathologies associated with diabetes.

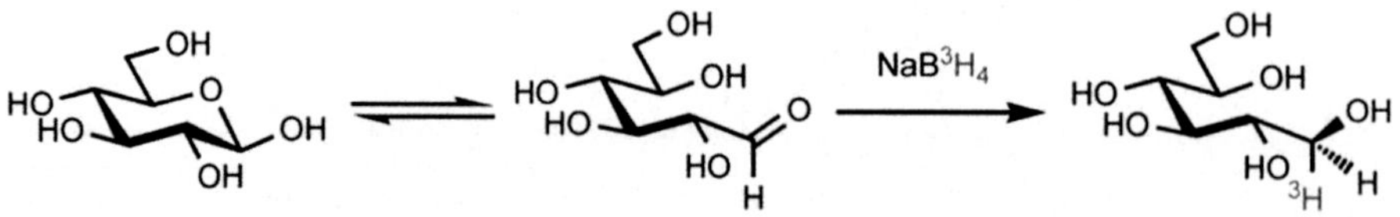

FIGURE 2.11. Conversion of a monosaccharide to a tritium-labeled alditol by reduction with NaB^3H_4.

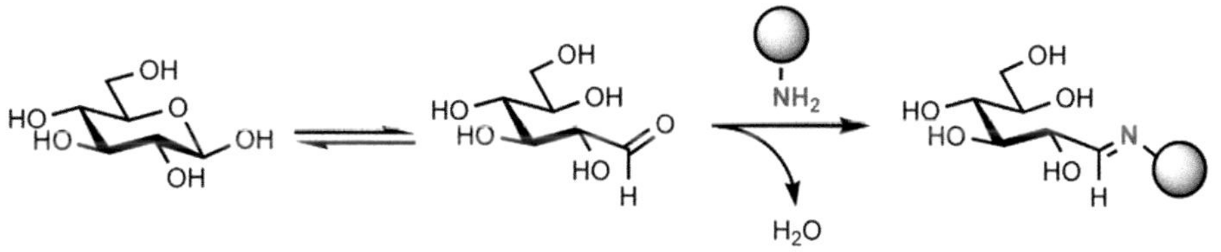

FIGURE 2.12. Conjugation of a monosaccharide to an amino group by formation of an imine. The *filled circle* represents any small molecule or macromolecule containing an amine.

Glycosidic Bond Formation

Two monosaccharide units can be joined together by a glycosidic bond—this is the fundamental linkage among the monosaccharide building blocks found in all oligosaccharides. The glycosidic bond is formed between the anomeric carbon of one monosaccharide and a hydroxyl group of another. In chemical terms, a hemiacetal group reacts with an alcohol group to form an acetal. Glycosidic bonds can be formed with virtually any hydroxylated compound, including simple alcohols such as methanol (Figure 2.13) or hydroxy amino acids such as serine, threonine, and tyrosine. Indeed, glycosidic linkages are formed between sugars and these amino acids within proteins to form glycoproteins (see Chapters 9 and 10). Like the hemiacetal, the acetal or glycosidic linkage can exist in two stereoisomeric forms: α and β. But unlike the hemiacetal, the acetal is configurationally stable under most conditions. Thus, once a glycosidic bond is formed, its configuration is maintained. Like acetals in general, glycosidic bonds can be hydrolyzed in dilute acid, generating the constituent monosaccharides from oligosaccharides.

Glycosidic bond construction is the central challenge of glycan synthesis and immense efforts have been devoted to high-yielding and stereoselective glycosylation reactions. An overview of glycan synthesis strategies is provided in Chapters 53 and 54.

CHEMISTRY OF MONOSACCHARIDE FUNCTIONAL GROUPS

Methylation of Hydroxyl Groups

The hydroxyl groups present in both monosaccharides and oligosaccharides can be chemically modified without affecting the glycosidic linkages. Methylation is used in the structural analysis of glycans (see Chapter 50). Natural products containing partially methylated glycans are known and a number of methyltransferases have been identified.

Esterification of Hydroxyl Groups

A variety of different enzymes can esterify the hydroxyl groups of glycans to transiently vary glycan structure. Esterification is sometimes required for interactions with other biomolecules. The most important types of sugar esters in nature are phosphate esters (including diphosphate

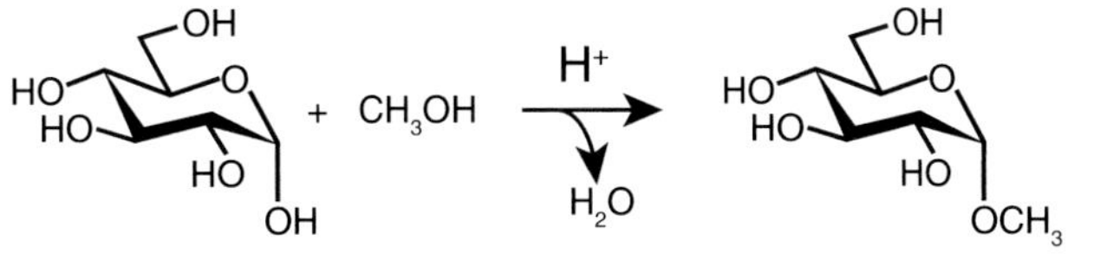

α-D-Glucopyranose + CH_3OH $\xrightarrow{H^+}$ Methyl α-D-Glucopyranose + H_2O

FIGURE 2.13. Glycoside formation. Conversion of a hemiacetal into an acetal.

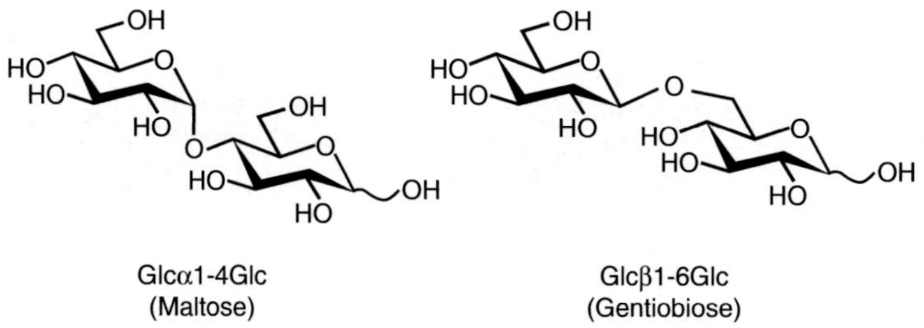

Glcα1-4Glc (Maltose)

Glcβ1-6Glc (Gentiobiose)

FIGURE 2.14. Two isomeric disaccharides.

esters), acyl esters (with acetic acid or fatty acids), and sulfate esters. Acyl esters can sometime migrate to other hydroxyl groups on the same monosaccharide.

Deoxygenation of Hydroxyl Groups

The replacement of monosaccharide hydroxyl groups with hydrogen atoms forms deoxysugars. Nature has evolved enzymes to perform this reaction in a minimum number of steps, whereas chemically multistep procedures are required. Deoxygenation of ribose within a ribonucleotide to form the 2-deoxyribonucleotide is a critical reaction in DNA biosynthesis. Fucose (Fuc), one of the common vertebrate monosaccharides, is deoxygenated at C-6 during its biosynthesis from mannose (Chapter 5).

Amino Groups

Many monosaccharides contain N-acetamido groups, such as GlcNAc, GalNAc, and NeuNAc. Free amino groups, formed by de-N-acetylation of the N-acetamido group, are rare and found in heparan sulfate (HS) (Chapter 17), glycosylphosphatidylinositol (GPI) anchors (Chapter 12), neuraminic acid (Chapter 15), and in many bacterial glycan structures (Chapter 20). Amino groups can be modified with sulfates, similar to hydroxyl groups, as found in HS.

GLYCOSIDIC LINKAGES

A variety of linkages can be formed between two monosaccharides. The glycosidic linkage can give rise to two possible stereoisomers at the anomeric carbon of one sugar (α or β). Also, the many hydroxyl groups of the other sugar permit several possible regioisomers. Two glucose residues, for example, can be joined together in numerous ways, as illustrated by maltose (Glcα4Glc) and gentiobiose (Glcβ6Glc) (Figure 2.14). These isomers have very different three-dimensional structures and biological activities. A monosaccharide can engage in more than two glycosidic linkages, thus serving as a branchpoint. The common occurrence of branched sequences (as opposed to the linear sequences that are found in almost all peptides and oligonucleotides) is unique to glycans and contributes to their structural diversity.

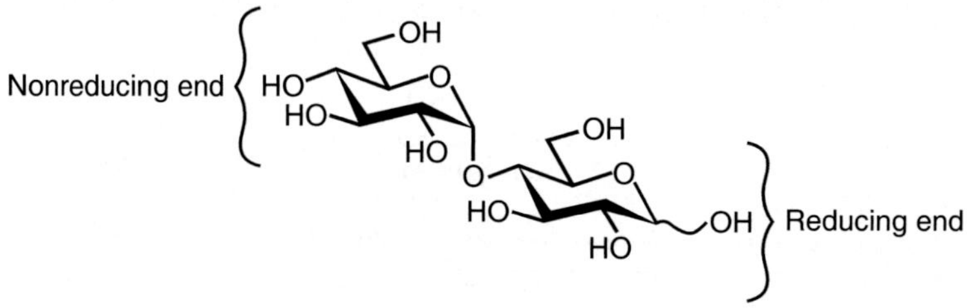

Nonreducing end

Reducing end

FIGURE 2.15. Reducing and nonreducing ends of a disaccharide.

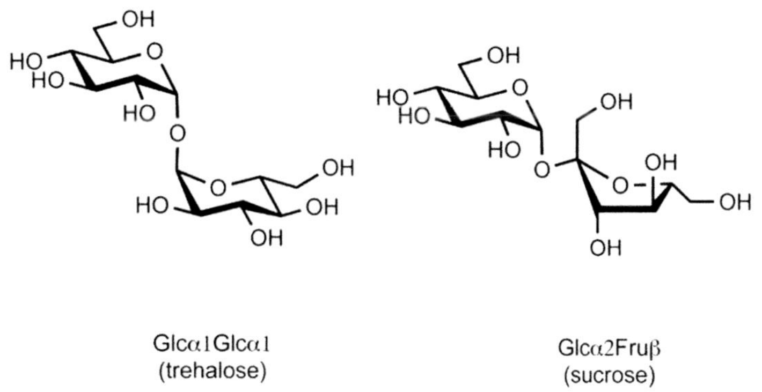

Glcα1Glcα1
(trehalose)

Glcα2Fruβ
(sucrose)

FIGURE 2.16. Nonreducing disaccharides.

The relationship of the glycosidic bond to oligosaccharides is analogous to the relationship of the amide bond to polypeptides and the phosphodiester bond to polynucleotides. However, amino acids and nucleotides are linked in only one fashion during the formation of polypeptides and nucleic acids, respectively; there is no stereochemical or regiochemical diversity in these biopolymers. The number of monomeric residues contained in an oligosaccharide is designated in the nomenclature—disaccharide, trisaccharide, and so on. Just as polypeptides have amino and carboxyl termini and polynucleotides have 5′ and 3′ termini, oligosaccharides have a polarity that is defined by their reducing and nonreducing termini (Figure 2.15). The reducing end of the oligosaccharide bears a free anomeric center that is not engaged in a glycosidic bond and thus retains the chemical reactivity of the aldehyde. However, it continues to be referred to as reducing end even when it is engaged in a linkage (e.g., to the hydroxyl of serine or threonine in glycoproteins). Structures are commonly written from the nonreducing end on the left toward the reducing end on the right. For some structures, there is no reducing end. For example, the common disaccharides sucrose and trehalose have glycosidic linkages between the anomeric centers of two monosaccharide constituents (Figure 2.16).

The glycosidic linkage is the most flexible part of a disaccharide structure. Whereas the chair conformation of the constituent monosaccharides is relatively rigid, the torsion angles around the glycosidic bond (φ, ψ, and ω; Figure 2.17) can vary. Thus, a disaccharide of well-defined

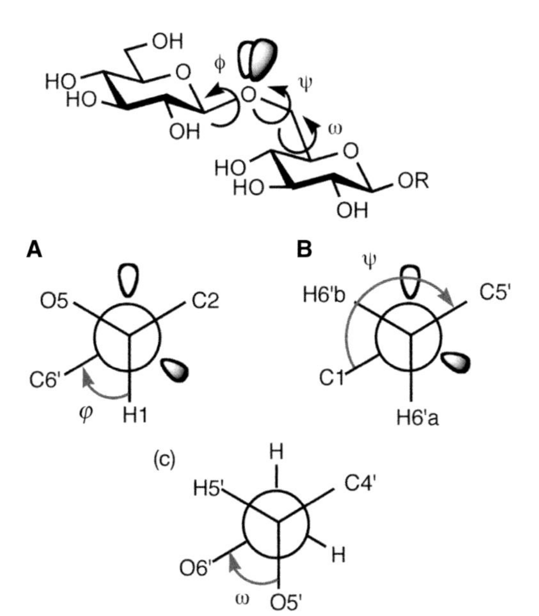

FIGURE 2.17. Torsion angles that define the conformation of the glycosidic linkages φ, ψ, and ω. (*A*) Newman projection along the C1—O1 bond illustrating φ for the 1-6 glycosidic bond. (*B*) Newman projection along the C6′—O1 bond illustrating ψ for a 1 → 6 linkage (*C*) Newman projection along the C5′—C6′ bond illustrating ω for the 1 → 6 linkage. The lobes on the glycosidic oxygen atom represent lone pairs of electrons. The torsion angles depicted are arbitrary and do not necessarily reflect the most stable conformation.

primary structure can adopt multiple conformations in solution that differ in the relative orientation of the two monosaccharides. The combination of structural rigidity and flexibility is typical of complex carbohydrates and essential to their biological functions.

Glycans are linked to other biomolecules, such as lipids or amino acids within polypeptides, through glycosidic linkages to form glycoconjugates (see Chapters 9, 10, 11, and 12). Glycans are often referred to as the glycone of a glycoconjugate and the noncarbohydrate component is named the aglycone. The glycan may be a mono- or an oligosaccharide.

In conclusion, monosaccharide building blocks can be linked to various regio- and stereochemistries, and the resulting oligosaccharides can be assembled on protein or lipid scaffolds (see Chapter 3).

ACKNOWLEDGMENTS

The authors acknowledge contributions to previous versions of this chapter by Carolyn R. Bertozzi and David Rabuka and appreciate helpful comments and suggestions from Dirk Lefeber, Todd Lowary, and Sriram Neelamegham.

FURTHER READING

El Khadem HS. 1988. *Carbohydrate chemistry: monosaccharides and their oligomers.* Academic, San Diego.

Allen HJ, Kisailus EC. 1992. *Glycoconjugates: composition, structure, and function.* Marcel Dekker, New York.

McNaught AD. 1997. Nomenclature of carbohydrates. *Carbohydr Res* **297**: 1–92. doi:10.1016/S0008-6215(97)83449-0

Bill MR, Revers L, Wilson IBH. 1998. *Protein glycosylation.* Kluwer Academic, Boston.

Boons G-J. 1998. *Carbohydrate chemistry.* Blackie Academic and Professional, London.

Stick RV. 2001. *Carbohydrates: the sweet molecules of life.* Academic, New York.

Varki NM, Varki A. 2007. Diversity in cell surface sialic acid presentations: implications for biology and disease. *Lab Invest* **87**: 851–857. doi:10.1038/labinvest.3700656

Varki A, Cummings RD, Aebi M, Packer NH, Seeberger PH, Esko JD, Stanley P, Hart G, Darvill A, Kinoshita T, et al. 2015. Symbol nomenclature for graphical representations of glycans. *Glycobiology* **25**: 1323–1324. doi:10.1093/glycob/cwv091

3 Oligosaccharides and Polysaccharides

Carlito B. Lebrilla, Jian Liu, Göran Widmalm, and James H. Prestegard

This chapter discusses the diversity in structure and properties that results when multiple monosaccharides (Chapter 2) are linked together to form oligosaccharides and polysaccharides (the latter comprising much of the biomass on the planet). Some examples of the more complex polymeric assemblies that occur in nature are presented, and how these remarkable structures are generated is discussed.

GLYCANS IN NATURE ARE OFTEN CONJUGATED

Except in their roles as sources of energy for living organisms, sugars seldom occur in nature as monosaccharides. Instead, they serve as building blocks for more complex molecules. In the most common process, an initial sugar is linked to an aglycone (often a lipid or a protein) and this sugar is further elaborated by covalently joining other sugars through glycosidic linkages (Chapter 2) between the anomeric carbon of the sugar being added and a hydroxyl oxygen of an existing sugar. The resulting glycans are called oligosaccharides (usually less than a dozen monosaccharides) or polysaccharides (usually more than a dozen monosaccharides). The latter are usually built on a core of repeating subunits of linked monosaccharides. The way in which assembly of both oligosaccharides and polysaccharides occurs produces structures of enormous diversity and widely varying properties. This allows glycans to fill roles that vary from cell-surface interactions with proteins important in differentiation, recognition, and proliferation of cells, to interactions with other glycans that generate the mechanical properties of plant and microbial cell walls.

published by Cold Spring Harbor Laboratory Press; doi: 10.1101/glycobiology.4e.03

DIVERSITY FROM OLIGOSACCHARIDE BRANCHING

Diverse structures can be created by simply linking different monosaccharides through glycosidic bonds, to make oligosaccharides or polysaccharides. The diversity arises not only from the choice of sugars but also from the way they are linked. If there were just one way to link monosaccharides, the choice among the dozen or so commonly used sugars would make the resulting polysaccharides more diverse than polynucleotides (four nucleotide choices for DNA and RNA) but less diverse than polypeptides (20 amino acid choices for mammalian proteins). However, the possibility of making glycosidic bonds between the anomeric carbon of one sugar and any one of the unmodified hydroxyl groups in another mono- or oligosaccharide adds to the diversity, by allowing not only more linear products, but also branched products in which more than one hydroxyl group on a given sugar is used to make glycosidic bonds. In addition, each anomeric carbon is a stereogenic center, and therefore each glycosidic linkage can be constructed having either the α- or β-configuration. Building an oligosaccharide, such as a tetrasaccharide (four sugars) with an unlinked reducing end, using only a single sugar in one ring form, such as glucopyranose, the authors could construct 1792 distinct structures. Of course, not all theoretically possible molecules are produced in nature because of the lack of enzymes to build them, but many are made contributing to the wide range of functional properties that allow carbohydrates to play many important roles.

Branching is a prime characteristic of many glycans found on mammalian cell surfaces. Glycans representing two major types of eukaryotic protein glycosylation are shown in Figure 3.1. An N-glycan makes a glycosidic bond with the side-chain nitrogen of an asparagine residue that is a part of a consensus peptide sequence NX(S/T). An O-glycan makes a glycosidic bond with the terminal oxygen of a serine or threonine residue. N-Glycans contain a core composed of three mannose residues and two *N*-acetylglucosamine residues (Manα1-6[Manα1-3]Manβ1-4 GlcNAcβ1-4GlcNAcβ1-N-Asn). The depicted glycan is a biantennary glycan with branches linked at the 3- and 6-positions of the first mannose residue in the glycan chain. However, more complex structures exist with three and four branches. Details of the synthesis and biological importance of these glycans are presented in Chapter 9. The depicted O-glycan contains a typical core structure (one of four common cores) that begins at the reducing end with an *N*-acetylgalactosamine α-linked to a serine or threonine (GlcNAcβ1-6[Galβ1-3]GalNAcα1-O-Ser/Thr). It begins as a biantennary structure but can be further extended toward the nonreducing end to form more

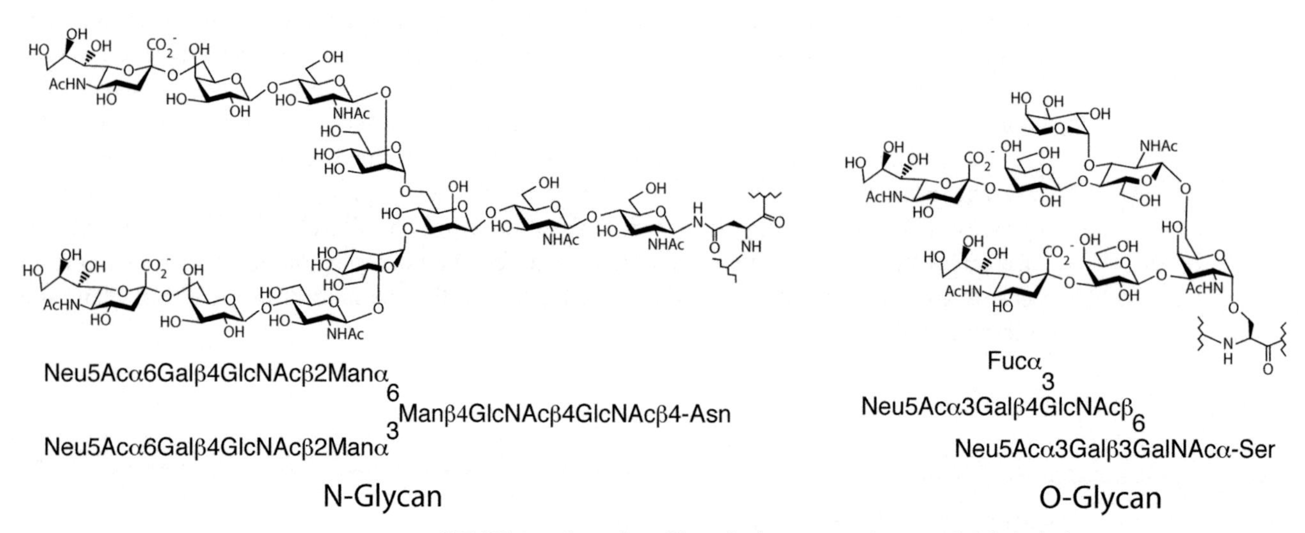

FIGURE 3.1. Examples of branched structures in N- and O-linked glycans.

complex structures. The synthesis and biological importance of O-glycans are described in Chapters 10 (O-GalNAc) and 19 (O-GlcNAc).

Both glycans depicted are terminated with a sialic acid (often Neu5Ac in humans) at their nonreducing end. Sialylation is characteristic of mammalian glycans and important for immune response. In protein–glycan interactions, not only are particular residues recognized, but often their position in a branching structure is recognized as well. An interesting example is the interaction of the enzyme that adds the 2-6-linked sialic acids to the termini of N-glycans. It has a high preference for adding to the 1-3-linked branch (by more than an order of magnitude) despite the fact that the residues on both branches are identical back to the mannose at the branch point (Galβ1-4GlcNAcβ1-2Manα-). This may give some indication of the extent to which branching plays a role in recognition processes.

Unique among mammals are free oligosaccharides present in milk. Human milk oligosaccharides (HMOs) are highly abundant in breast milk. Unlike most glycans that are conjugated to either proteins or lipids, HMOs are unconjugated in their native states. Thus, they contain a reactive aldehyde on the reducing end. HMOs decrease during lactation but are often even more abundant than proteins. Although the precise details of their synthesis are still not well known, HMO structures are more similar to glycans on glycolipids and O-glycans on proteins. The structures are based on a lactose core (Galβ1-4Glc) and extend either in branched or linear forms. Extensions are composed primarily of glucose, galactose, N-acetylglucosamine, fucose, and sialic acids. As molecules, they are generally small, with the majority of the structures in the range of three to six monosaccharides; however, structures with more than 20 monosaccharides have been observed. HMO structures have been characterized by advanced glycomic analytical tools such as liquid chromatography–mass spectrometry (LC-MS) (see Chapter 50). Hundreds of structures have been observed, although one mother typically produces about 100 structures.

There are two HMO phenotypes (from four genotypes) that determine the types of structures in the milk. Mothers who secrete their blood type or the Lewis b epitope are termed secretors. Secretors produce structures that contain α1-2-linked fucose. Nonsecretors produce little or no α1-2-linked fucose but instead produce fucosylated glycans that are primarily (1-3)- or (1-4)-linked. One of the most abundant components in milk of secretor mothers is 2-fucosyllactose, whereas nonsecretor mothers produce very little or none of this compound. Similarly, secretor milk is slightly richer in fucosylated compounds, whereas nonsecretor milk is slightly richer in sialylated compounds.

HMOs have altered our concept of the function of food. Human milk was thought to be primarily nutritive. However, despite the large abundances of HMOs in the milk, there are no human enzymes in the infant gut that can degrade them. Instead, characterization of the gut microbiome of infants have revealed that HMOs are primarily food for the gut bacteria. Indeed, *Bifidobacteria*, found in the microbiome of heathy breastfed infants, have glycosyl hydrolases that specifically catabolize HMOs. Additionally, HMOs may also have a protective role by blocking the binding of pathogens in the gut.

STRUCTURAL AND STORAGE POLYSACCHARIDES

Linkage variation plays an important role in the structural properties of polysaccharides as illustrated for two closely related glucose polymers having repeating units (RUs) of -[4Glcβ1-]$_n$ and -[4Glcα1-]$_n$. The former is the structural polymer, cellulose, that forms the foundation of all plant cell walls and is a major component of materials such as wood and cotton. The latter is starch, an easily digestible material with no significant structural utility. The ability of one polymer to associate in long fibrils having both crystalline and amorphous regions contributing to its structural properties, and the lack of these tendencies in the other polymer, is clearly linked to the stereochemistry at the anomeric carbon (β for cellulose and α for starch) and the preferred torsional angles about the C1′-O4 and O4-C4 bonds of the glycosidic linkages. The torsion angles are referred to as φ and ψ, much the same as primary structural variables in polypeptides; by International Union of Pure and Applied Chemistry (IUPAC) convention they are defined by four connected atoms, O5′-C1′-O4-C4 and C1′-O4-C4-C3, respectively. Because the torsion angle can be monitored directly through couplings between protons at the ends of the glycosidic linkage observed in nuclear magnetic resonance (NMR) spectra, an alternate NMR definition is also in common use—namely, H1′-C1′-O4-C4 for φ and C1′-O4-C4-H4 for ψ. Glycosidic torsion angles differ considerably in crystalline cellulose and starch. Using IUPAC definitions they prefer ~−95° and ~+95° in the former and ~+115° and ~+120° in the latter. Figure 3.2 depicts these differences in the repeating disaccharide units of cellulose and starch as represented by the isolated disaccharides having common names cellobiose and maltose, respectively. These local preferences influence association properties and ultimately structural characteristics. When extended to a long cellulose polymer, cellobiose units generate elongated strands that can pack and interact with other strands through hydrogen bonds to form layers. Layers in turn interact through a combination of forces to form fibrils composed of 18 polymer chains. The more helical strands in starch cannot pack easily and result in a more amorphous material.

Other important polysaccharides, such as pectins found in plant cell walls, help plants accommodate changes during growth. Pectins are some of the most complex polymers in nature. They are based on polymers of α(1-4)-linked galacturonic acid (GalA) or the RU, -2)-α-L-Rha-(1-4)-α-D-GalA-(1-, and may contain additional sugar and nonsugar substituents, including methyl and acetyl esters. The negatively charged carboxyl group at the 6-position of the GalA residues contributes to the water solubility of these polymers and the long-range interactions that give them gelling properties useful in the food industry. Oligo- and polysaccharides of plant origin are discussed in detail in Chapter 24.

Animals also use polysaccharides for various purposes. Glycogen is a storage polymer related to starch in that it is a glucose polymer with primarily α(1-4)-linkages connecting glucose residues, but it is highly branched having additional α(1-6)-linkages to some of the glucose residues. Structural polymers also exist; for example, the repeating polymer of *N*-acetylglucosamine, -[4GlcNAcβ1-]$_n$, is the primary component of chitin, the material that forms the exoskeletons of arachnids, crustaceans, and insects (Chapter 26). Modifications of glucose residues by replacing the hydroxyl group at 2-position with an amine group and subsequent N-acetylation change the structural properties significantly. These changes allow for the formation of composites with proteins and minerals that lead to additional variation in structure and function.

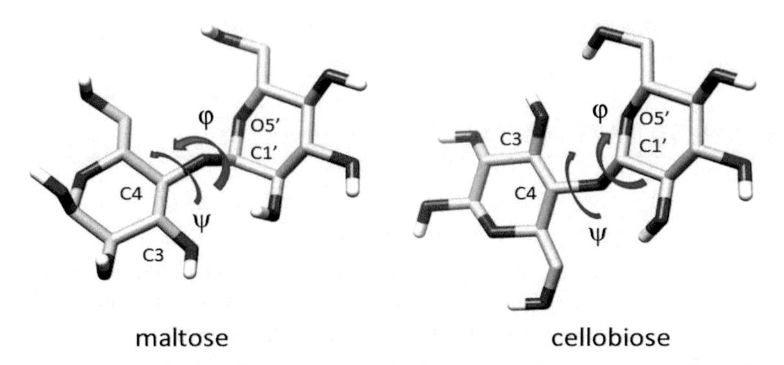

FIGURE 3.2. Repeating units from cellulose and starch showing conformation and glycosidic torsion angles φ and ψ.

CELL-SURFACE POLYSACCHARIDES OF ANIMALS

Most cell-surface polysaccharides found in animals belong to a class of glycans known as glycosaminoglycans (GAGs) (Chapter 17). Abundantly present on the cell surface as well as in the extracellular matrix, GAGs are linear macromolecules with molecular mass of >15,000 Da. The building blocks of most GAGs are composed of an amino-substituted sugar and a hexuronic acid residue. Modifications on the sugar residues—in particular, the sulfation of hydroxyl or the amino groups—are common. The sulfates and hexuronic acid carboxylate groups are negatively charged under physiological conditions. Thus, GAGs are the most anionic molecules present in the animal kingdom. Commonly found GAGs include chondroitin sulfate, dermatan sulfate, heparan sulfate (HS), hyaluronic acid, and keratan sulfate. These GAGs structurally differ in their disaccharide RUs. For example, chondroitin sulfates consist of disaccharide RUs of [4GlcAβ1-3GalNAcβ1-]$_n$, whereas HS consists of disaccharide RUs of [4GlcAβ1-4GlcNAcα1-]$_n$ or [4IdoAα1-4GlcNSα1-]$_n$ (Figure 3.3A). Structural diversity of these polymers is primarily a result of additional sulfation of hydroxyl groups and is discussed below for the diverse HS polymers found in mammals. GAGs found in other animals can be distinct from mammalian GAGs by virtue of further modifications. For example, marine invertebrates can carry particularly unique sulfation patterns (i.e., 3-O-sulfation on GlcA residues) and distinct side-chain modifications (i.e., fucosylation on chondroitin sulfate).

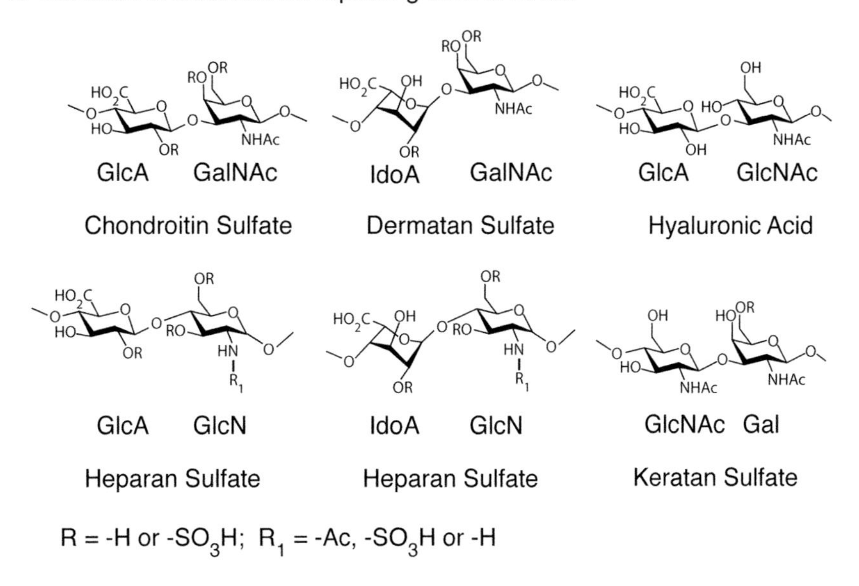

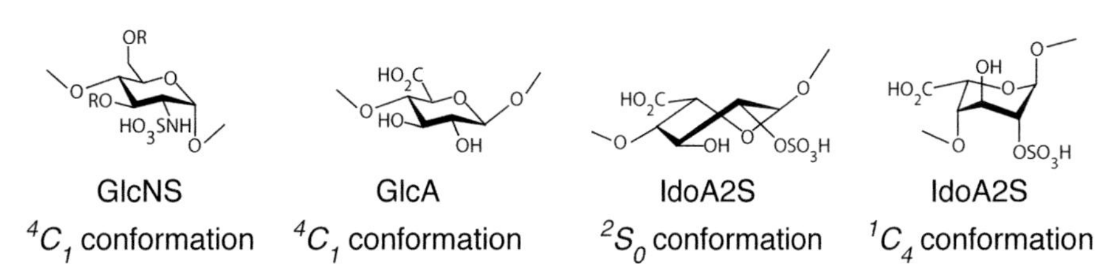

FIGURE 3.3. Structures of (*A*) disaccharide repeating units of different glycosaminoglycans and (*B*) the conformations of monosaccharides from heparan sulfate.

Structure–Function Relationships of Animal Polysaccharides

GAGs show biological function primarily through their interactions with the hundreds of GAG-binding proteins found on cellular surfaces and in extracellular spaces. The structural factors that affect the strength and specificity of binding are key to the elicitation of a proper biological response. HS, the most studied member of the GAG family, provides a good illustration of the wide range of physiological and pathophysiological functions affected. HS, for example, is involved in trimolecular interactions with growth factors and growth factor receptors, and thereby participates in regulating embryonic development. HS interacts with proteases and protease inhibitors in the blood to control the coagulation process and binds to viral envelope proteins as a receptor for viral infections. Also, heparin, a highly sulfated form of HS, is a commonly used anticoagulant drug in the clinic. A more complete discussion of the biological function of HS can be found in Chapters 17 and Chapter 38. This section uses the interaction between HS and proteins to illustrate how structural factors affect binding between GAGs and proteins in general.

One factor that contributes to the structural diversity of HS-protein interactions is the conformational flexibility of L-IdoA and its derivative, L-IdoA2S (2-*O*-sulfo iduronic acid). Present in the pyranose form, the six-membered ring structure of IdoA or IdoA2S can adopt either chair or skew-boat conformations (Figure 3.3B). Until now, only the $^4C_1$ chair conformation has been verified experimentally for GlcA and GlcN in HS, but both the $^1C_4$ chair and the $^2S_0$ conformations have been found for IdoA2S residues in crystal structures containing HS. In solution, NMR studies show IdoA2S and IdoA residues to be present in a mixture of $^1C_4$ and $^2S_0$ conformations. The conformational flexibility in IdoA residues likely allows orientation of the sulfate groups in HS to maximize the binding affinity to proteins. Structural elements that dictate preferences for one form over the other, including the possible effects from the sulfated monosaccharide sequences around the IdoA/IdoA2S residue, are subject to further investigation.

A second factor contributing to structural diversity and selective interactions with certain HS-binding proteins is the size of the sulfated saccharide domains. HS isolated from natural sources shows domain-like structures, with clusters of six to eight saccharides forming the highly sulfated domains known as S-domains. These regions are separated by nonsulfated saccharide residues, consisting of GlcA-GlcNAc RUs, known as NAc-domains. The S-domains contain primarily IdoA residues that likely contribute the flexibility needed to optimize binding to proteins and elicit desired biological activities. The contribution of NAc-domains to the functions of HS has not been fully established. However, a possible role may be to appropriately position S-domains in a single polysaccharide chain for interaction with multiple proteins. One example is the interaction of heparin with antithrombin and thrombin. In this complex, one part of the heparin chain interacts with antithrombin, and another part of the heparin chain interacts with thrombin. Between the antithrombin-binding domain and thrombin-binding domain, a linker of six to seven saccharide residues exists.

Cellular Regulation of the Structures of GAGs

Unlike proteins and nucleic acids, the biosynthesis of GAGs is not regulated by a well-defined template. Instead, each member of the GAG family is synthesized by a unique pathway (Chapter 17). The HS synthesis pathway, for example, involves multiple enzymes, including specific glycosyltransferases (or HS polymerase), an epimerase, and several sulfotransferases. Moreover, HS is biosynthesized as a polymer covalently linked to a proteoglycan that consists of a core protein and polysaccharide side chains; the functions of these proteoglycans are, in fact, dominated by the properties of the HS chains added. Although this is a non-template-driven process, the overall structures of HS generally remain unchanged between generations. There is considerable interest in understanding the mechanism that controls the structure of HS.

The biosynthesis of chondroitin sulfate and dermatan sulfate is less understood than that of HS. Chondroitin sulfate tends to be synthesized on different proteoglycan core proteins and requires a different set of polymerases and sulfotransferases than the enzymes specifically recognizing HS polysaccharides. To synthesize the IdoA residues in dermatan sulfate, a specialized epimerase is required. Hyaluronic acid biosynthesis is quite different: it is not synthesized on a core protein, it does not take place in the endoplasmic reticulum and Golgi, and it only requires one hyaluronic acid synthase (a dual activity glycosyltransferase), as the polysaccharide does not contain sulfate groups or IdoA residues (Chapter 16).

BACTERIAL POLYSACCHARIDES

The interactions of bacteria with their environment provide an excellent example of how polysaccharide properties play an important role in an organism's survival. Bacterial polysaccharides are especially diverse, in that they can include a larger number of distinct sugar residues in their RUs (usually two to six residues), and they can include branching. Many of them are parts of bacterial cell membranes where they serve important structural and protective roles. Because of their location on the outside of the cell, bacterial polysaccharides such as lipopolysaccharides (LPSs), capsular polysaccharides (CPSs), and exopolysaccharides (EPSs) are often potent antigens that elicit a strong immune response in humans. LPSs carry long polysaccharides called O-antigens and are unique to Gram-negative bacteria where they constitute the major component of the outer leaflet of the outer membrane. These types of bacteria may also carry CPSs forming a relatively dense additional layer around the bacterial cell. Cell wall lipoteichoic acids and teichoic acids are unique components of Gram-positive bacteria, which are often surrounded by a CPS or a less dense EPS layer.

The variation in biosynthetic pathways among bacterial species is ultimately responsible for the diversity in bacterial polysaccharides. The biosynthesis of bacterial polysaccharides is discussed further in Chapter 21. Here the authors present a few examples to illustrate how synthesis of backbone structures, branching, and postpolymerization modification lead to a diverse set of polymer structures. In one of the biosynthetic pathways, sugar residues are added sequentially onto an anchor molecule; thus, the polymer is growing from the terminal, nonreducing end until a termination entity/substituent is added that precludes further chain elongation—for example, in the O-antigen polysaccharide of the LPS from *Escherichia coli* O8 (Figure 3.4A). The polymer is linear, and although sugar residues are added step-by-step to form the polysaccharide, RUs can be identified.

Some heteropolysaccharides contain two alternating sequentially added sugar residues and a formal RU can be identified. Often a processive glycosyltransferase is responsible for forming this pattern as is the case in the synthesis of the O-polysaccharide of *Salmonella enterica* O54 (Figure 3.4B). When two sugar residues alternate, branched structures can be formed, as one sugar may give rise to the polymer backbone and one a side chain.

Polysaccharides are not always built from the nonreducing end, and preformed subunits can be used in the assembly. For example, synthesis of the *E. coli* antigens O5ab and O5ac relies on a preformed linear oligosaccharide, with five sugar residues constituting the RU. The oligosaccharide is built on an undecaprenyl pyrophosphoryl glycoside anchor molecule. This oligosaccharide is then added *onto* another oligosaccharide–lipid anchor to grow the polymer from the "reducing end." When polymerization is taking place at the penultimate sugar of the oligosaccharide a branched RU results (cf. O-antigen of *E. coli* O168 [Figure 3.4C]).

Branching can be introduced by the addition of sugars following polymer backbone formation. *Helicobacter pylori* O-antigens often contain human blood group structures as part of their RUs. *N*-acetyl-D-glucosamine and D-galactose are added in a processive manner onto an undecaprenyl pyrophosphoryl carrier, thereby forming a linear polysaccharide composed of Galβ1-

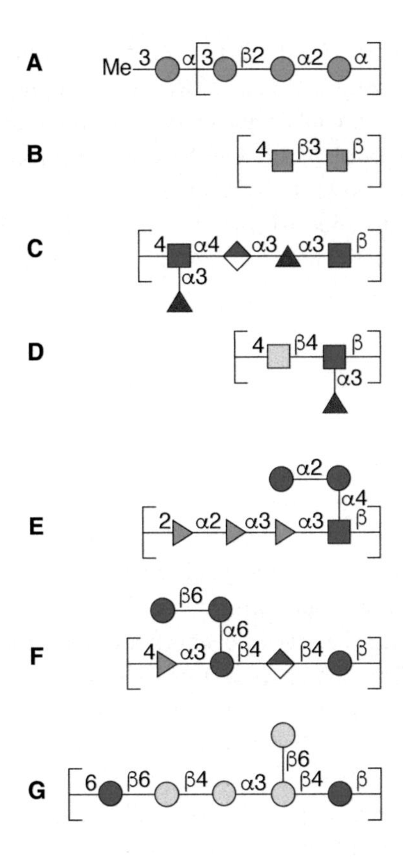

FIGURE 3.4. Schematic representation in SNFG (Symbol Nomenclature for Glycans) format of repeating units of bacterial polysaccharides: (*A*) O-antigen of *E. coli* O8, (*B*) O-polysaccharide of *Salmonella enterica* O54, (*C*) O-antigen of *E. coli* O168, (*D*) O-polysaccharide of *Helicobacter pylori*, (*E*) O-antigen of *Shigella flexneri* serotype 7a, (*F*) exopolysaccharide S-194 (rhamsan) from *Alcaligenes*, and (*G*) exopolysaccharide from *Lactobacillus helveticus*.

4GlcNAc (LacNAc) disaccharide RUs. Subsequently, L-fucosyl residues are added onto the backbone polysaccharide resulting in branched Lewis-type structures (Figure 3.4D).

Bacterial saccharides consist of diverse monosaccharides (Chapter 2) including 6-deoxy-hexoses like L-rhamnose or L-fucose as well as rare sugars that are often found at the terminal position of the oligosaccharide–lipid acceptors. On polymerization, the rare monosaccharides become part of the side chain in branched structures and give rise to structural epitopes that are characteristic for bacterial species. In some cases, these epitopes are the basis for molecular mimicry (Chapter 42). Furthermore, these sugar residues constitute the terminal entity of the polysaccharide chain, which can be recognized by antibodies of the immune system. The polysaccharide may also be decorated by substituents such as amino acids, O-acyl, or phosphodiester groups. It is not uncommon to find O-acetyl groups at the branch-point sugar residue, thereby leading to a highly crowded substitution pattern in which all positions on the sugar residue are either part of a glycosidic linkage or carry a nonsugar substituent. Examples of diversity by substituent addition are the O-antigens of *Shigella flexneri* in which glucosyl, O-acetyl, or phosphoethanolamine groups are added to the backbone (Figure 3.4E).

Branching and substituent addition affect the properties of polysaccharide solutions, such as gelling and high viscosity. Some polymers are known under their commercial names gellan, welan, and rhamsan (S-194) (Figure 3.4F). The differences among these and similar types of polymers are based on acyl substituents such as O-acetyl or O-succinyl groups and side chains consisting of L-Rha/L-Man, di-glucosyl or di-rhamnosyl groups, as well as an L-Rha/L-Man backbone modification.

The size of polysaccharides produced by bacteria can vary widely. Whereas the O-antigens in LPSs have less than 100 RUs, CPSs and EPSs have higher numbers of RUs (10^3–10^5). Branched structures are often present in these polysaccharides (Figure 3.4G) with either more than one sugar residue in the side chain or with two branches within the RU. More information on these complex materials is presented in Chapter 42.

Charge can significantly affect polysaccharide properties. Charged polysaccharides form a common subclass of bacterial polymers and are mostly present in the form of negatively charged sugar residues or as a result of the addition of negatively charged substituents. Uronic acids (e.g., GlcA) and nonulosonic acids (e.g., Neu5Ac, Sia) introduce negative charge into the RU, thereby rendering the polysaccharide a polyanionic polymer, much like the GAGs found in mammalian systems. Substituents on sugar residues such as pyruvate, phosphate and sulfate groups confer polyanionic character. The charged groups can be present both in the polysaccharide backbone and the side chains. Positively charged amines are sometimes present in the RU in conjunction with a negatively charged group. The CPSs from *Neisseria meningitidis* are representative for these types of polysaccharides. Types B and C are homopolymers of Neu5Ac and types W-135 and Y contain disaccharide RUs with Neu5Ac and a hexose.

Flexibility is a significant variable in bacterial polysaccharides (cf. the torsion angles φ and ψ). However, when the linkage is formed via O6, an additional degree of freedom becomes available at the glycosidic linkage (namely, the torsion angle ω) because of the exocyclic hydroxymethyl group in hexopyranoses. A (1-6)-linkage in the backbone of polysaccharides (Figure 3.4G) may result in a less rigid polymer with higher flexibility and random coil character. Likewise, the occurrence of this linkage in the side chains will make them more flexible. When furanose residues are part of the polymer, different ring conformations provide more options for introducing flexibility.

Cross-linking is another way that physical properties of bacterial polysaccharides can be varied. The cell wall of Gram-positive bacteria contains a particularly thick layer of peptidoglycan covalently cross-linked via short peptide sequences. Furthermore, teichoic acid polymers, built from glycerol or ribitol residues joined by phosphodiester linkages, are located within the cell wall. Mono- or disaccharide RUs containing amino sugars can be part of these repetitive structures, thereby giving rise to different cell wall teichoic acids. Polymers having a phosphodiester linkage in the backbone as part of the RU are referred to as "teichoic acid type." CPSs of *Haemophilus influenzae* are built on this theme in which two (serotypes a and b) have RUs consisting of [ribitol-*P*-Hex-]$_n$ and two others (serotypes c and f) are made of [Hex-*P*-Hex-]$_n$.

In summary, bacterial polysaccharides highlight the many ways in which diversity is built into oligo- and polysaccharides. Some diversity results from a larger set of sugar residues, some from branching, and some from modification with a wide variety of substituents, such as phosphate, sulfate, acyl, and amino groups. This diversity gives rise to different physical properties. It allows

TABLE 3.1. Oligosaccharides and polysaccharide repeating units

Common name	Figure	Representative structure	Chapter
N-linked glycan	3.1	Galβ1-4GlcNAcβ1-2Manα1-6[Neu5Acα1-6Galβ1-4GlcNAcβ1-2 Manα1-3]Manβ1-4GlcNAcβ1-4GlcNAcβ1-N-Asn	9
O-linked glycan	3.1	GlcNAcβ1-6[Galβ1-3]GalNAcα1-O-Ser/Thr	10
O-linked glycan		GlcNAcβ1-O-Ser/Thr	19
Cellulose	3.2	-[4Glcβ1-]$_n$	24
Starch	3.2	-[4Glcα1-]$_n$	24
Chitin		-[4GlcNAcβ1-]$_n$	26
Chondroitin sulfate	3.3	-[4GlcAβ1-3GalNAc4/6Sβ1-]$_n$	17
Heparan sulfate	3.3	-[4GlcAβ1-4GlcNAcα1-]$_n$+-[4IdoA2Sα1-4GlcNS6Sα1-]$_m$	17
Capsular polysaccharide		-[6Glcα1-4Neu5Acα2-]$_n$	21
LPS O-antigen	3.4D	-[4GalNAcβ1-4[Fucα1-3]GlcNAcβ1-]$_n$	21
Exopolysaccharide	3.4G	-[6Glcβ1-6Galβ1-4Galα1-3[Galβ1-6]Galβ1-4Glcβ1-]$_n$	21

(LPS) Lipopolysaccharide.

bacteria to mimic their hosts in an attempt to evade detection, and it provides a means of distinguishing self from competitive organisms. Structures discussed in this chapter are depicted in abbreviated IUPAC/IUBMB (International Union of Biochemistry and Molecular Biology) form in Table 3.1 for comparison to the common names, symbolic representations, and actual chemical structures used in the text and figures. This chapter serves as a preview of glycans discussed more thoroughly in the following chapters.

ACKNOWLEDGMENTS

The authors acknowledge contributions to previous versions of this chapter by Carolyn R. Bertozzi and David Rabuka and appreciate helpful comments and suggestions from Jenny Mortimer, Breeanna Urbanowicz, and Darón Freedberg.

FURTHER READING

Laremore TN, Zhang F, Dordick JS, Liu J, Linhardt RJ. 2009. Recent progress and applications in glycosaminoglycan and heparin research. *Curr Opin Chem Biol* **13**: 633–640. doi:10.1016/j.cbpa.2009.08.017

Zivkovic AM, German JB, Lebrilla CB, Mills DA. 2011. Human milk glycobiome and its impact on the infant gastrointestinal microbiota. *Proc Natl Acad Sci* **108**: 4653–4658. doi:10.1073/pnas.1000083107

DeAngelis PL, Liu J, Linhardt RJ. 2013. Chemoenzymatic synthesis of glycosaminoglycans: re-creating, re-modeling, and re-designing nature's longest or most complex carbohydrate chains. *Glycobiology* **23**: 764–777. doi:10.1093/glycob/cwt016

Cosgrove DJ. 2021. Re-constructing our models of cellulose and primary cell wall assembly. *Curr Opin Plant Biol* **22**: 122–131. doi:10.1016/j.pbi.2014.11.001

Schmid J, Sieber V. 2015. Enzymatic transformations involved in the biosynthesis of microbial exo-polysaccharides based on the assembly of repeat units. *Chembiochem* **16**: 1141–1147. doi:10.1002/cbic.201500035

Higel F, Seidl A, Soergel F, Friess W. 2016. N-glycosylation heterogeneity and the influence on structure, function and pharmacokinetics of monoclonal antibodies and Fc fusion proteins. *Eur J Pharm Biopharm* **100**: 94–100. doi:10.1016/j.ejpb.2016.01.005

Moradali MF, Rehm BHA. 2020. Bacterial biopolymers: from pathogenesis to advanced materials. *Nat Rev Microbiol* **18**: 195–210. doi:10.1038/s41579-019-0313-3

Whitfield C, Wear SS, Sande C. 2020. Assembly of bacterial capsular polysaccharides and exopolysaccharides. *Annu Rev Microbiol* **74**: 521–543. doi:10.1146/annurev-micro-011420-075607

4 | Cellular Organization of Glycosylation

Karen J. Colley, Ajit Varki, Robert S. Haltiwanger, and Taroh Kinoshita

This chapter provides an overview of glycosylation from the perspective of a single cell, taking into account the patterns of expression, topology, and other features of the biosynthetic and degradative enzymes that are common to most cell types. The focus is mostly on the organization of glycosylation in eukaryotic cells. Chapters 21 and 22 further address prokaryotic glycosylation mechanisms.

GLYCOSYLATION IS UNIVERSAL IN LIVING ORGANISMS

After more than three billion years of evolution, every free-living cell and every cell type within eukaryotic organisms remains covered with a dense and complex layer of glycans (Chapter 20). Even enveloped viruses that bud from infected cells carry with them the glycosylation patterns of the host. Additionally, most secreted molecules are glycosylated and extracellular matrices of multicellular organisms are rich in glycans and glycoconjugates. Matrices secreted by unicellular organisms when they congregate (e.g., bacterial biofilms [Chapter 21]) also contain glycans. Thus, evolution has repeatedly and consistently selected glycans as being the most diverse and flexible molecules to position at the interface between cells and the extracellular milieu. Possible reasons include their relative hydrophilicity, flexibility, and mobility in aqueous environments and their extreme diversity, allowing facile short-term and long-term adaptations to changing environments and pathogen regimes.

published by Cold Spring Harbor Laboratory Press; doi: 10.1101/glycobiology.4e.04

In bacteria, Archaea, and fungi, glycans serve critical structural roles in the cell wall and in resisting large differences in osmolarity between cytoplasm and environment. In eukaryotes, both secretory proteins and membrane proteins typically pass through an endoplasmic reticulum (ER)–Golgi pathway, the cellular system in which many major glycosylation reactions occur (see below). Most proteins in the blood plasma of animals (with the exception of albumin) are also heavily glycosylated, and the glycosylation of these and other secreted proteins may provide solubility, hydrophilicity, and negative charge, thus reducing unwanted intermolecular interactions and protecting against proteolysis. Cell-surface membrane proteins like receptors, adhesion molecules, and channels are typically glycosylated, and this modification can promote their proper folding, ensure their stability, and impact function.

TOPOLOGICAL ISSUES IN GLYCAN BIOSYNTHESIS

The ER–Golgi Pathway of Eukaryotes

The classic work of George Palade indicated that most cell-surface and secreted proteins in eukaryotic cells are cotranslationally translocated into the ER where they are folded, modified, and subjected to quality control mechanisms. They then make their way via an intermediate compartment (IC) through multiple stacks of the Golgi apparatus, finally being distributed to various destinations from the *trans*-Golgi network (TGN). Secretory pathway proteins can be N-glycosylated, O-glycosylated, and/or modified with glycosylphosphatidylinositol (GPI) anchors, and some called proteoglycans are modified with attached glycosaminoglycan chains. The enzymes involved in each of these modification pathways are distinct. N-linked glycans and GPI anchors are preassembled before being transferred to proteins and then further modified in the ER–Golgi pathway. The stepwise assembly of O-linked glycans and glycosaminoglycans, as well as the glycosylation of lipids, involve reactions in both the ER and Golgi (Chapters 9–13 and 17). Figure 4.1 superficially depicts some steps in the synthesis of the major glycan classes in the ER–Golgi pathway of animal cells.

In the ER–Golgi pathway, some glycan chains are made on the cytoplasmic face of intracellular membranes and flipped across to the other side, but most are added to the growing chain on the *inside* of the ER or the Golgi (Figure 4.1). Regardless, the portion of a molecule that faces the inside of the lumen of the ER or Golgi will ultimately face the *outside* of the cell or the inside of a secretory granule or lysosome. To date, there are no well-documented exceptions to this topological rule. Of course, these topological considerations are reversed for nuclear and cytoplasmic glycosylation (see below), because the active sites of the relevant glycosyltransferases for these reactions face the cytoplasm. Not surprisingly then, the types of glycans found on the two sides of the cell membrane so far appear to be distinct from each other. Some "leaderless secretory proteins" destined for the extracellular space never enter the ER lumen, but are instead transferred directly through the plasma membrane, by as yet poorly defined mechanisms, including cytokines such as IL-1β and IL-18, growth factors such as fibroblast growth factor 2 (FGF2), and galectins. Interestingly, many of these proteins have glycan binding properties, which would presumably cause problems if they instead traversed the glycan rich ER–Golgi pathway.

Donors for Glycosylation Reactions

Regardless of location, most glycosylation reactions use activated forms of monosaccharides (often nucleotide sugars and in some cases lipid-phosphate-linked sugars, such as dolichol phosphate mannose) as donors for glycosyltransferases (see Chapter 5 for a listing of some enzymes, their donors and corresponding transporters, and details about their biochemistry).

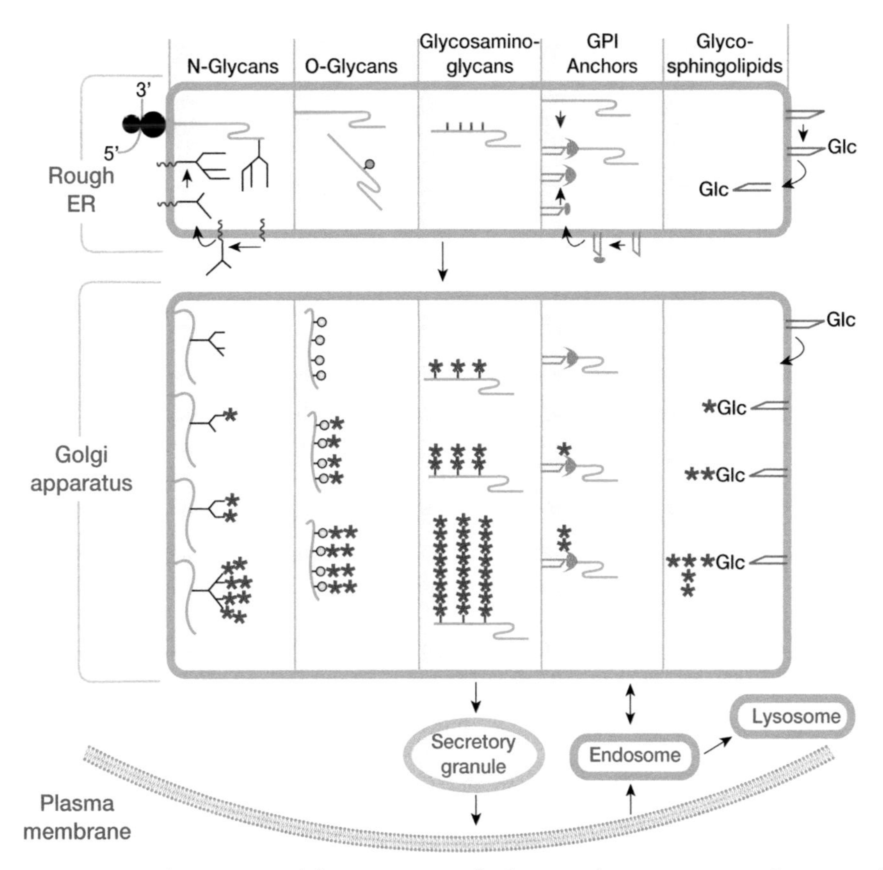

FIGURE 4.1. Initiation and maturation of the major types of eukaryotic glycoconjugates in relation to subcellular trafficking in the ER–Golgi–plasma membrane pathway. This illustration outlines the different mechanisms and topology for initiation, trimming, and elongation of the major glycan classes in animal cells. *Asterisks* represent the addition of outer sugars to glycans in the Golgi apparatus. N-Glycans and glycosylphosphatidylinositol (GPI) anchors are initiated by the en bloc transfer of a large preformed precursor glycan to a newly synthesized glycoprotein. O-glycans and sulfated glycosaminoglycans are initiated by the addition of a single monosaccharide to a serine or threonine in the case of O-glycans or to specific tetrasaccharide linkers in the case of sulfated glycosaminoglycans, followed by extension. The most common glycosphingolipids are initiated by the addition of glucose to ceramide on the outer face of the ER–Golgi compartments, and the glycan is then flipped into the lumen to be extended. For a better understanding of the events depicted in this figure, see details in other chapters of this book: N-glycans (Chapter 9); O-glycans (Chapters 10 and 13); glycosphingolipids (Chapter 11); GPI anchors (Chapter 12); and sulfated glycosaminoglycans (Chapter 17).

A variety of glycan modifications also occur in nature (Chapter 6). Of these, the most common are generated by sulfotransferases, acetyltransferases, and methyltransferases, which use activated forms of sulfate (3′-phosphoadenyl-5′-phosphosulfate; PAPS), acetate (acetyl-CoA), and methyl groups (*S*-adenosylmethionine; AdoMet), respectively. Almost all donors for glycosylation reactions and glycan modifications are synthesized within the cytosolic compartment, from precursors of endogenous origin. In eukaryotes, most of these donors are actively transported across a membrane bilayer by specific multipass transporter proteins, becoming available for reactions within the lumen of the ER–Golgi pathway.

Nuclear and Cytosolic Glycosylation in Eukaryotes

For many years, the nucleus and cytosol (which are topologically semicontinuous via nuclear pores) were assumed to be devoid of glycosylation capacity. It is now established that distinct glycoconjugates are synthesized and reside within these compartments. Indeed, one of them

(O-linked GlcNAc; Chapter 19) may well be quantitatively the most common type of glycoconjugate in many cells. Both the O-GlcNAc transferase (OGT) responsible for synthesizing O-linked GlcNAc on nuclear and cytosolic proteins and the O-GlcNAcase (OGA) that removes this monosaccharide are soluble proteins in these compartments. In addition, select cytosolic proteins can be modified with O-Glc, O-Fuc, or O-Man (Chapters 18 and 19).

Glycosylation Reactions at the Plasma Membrane

Because prokaryotic cells do not have an ER–Golgi pathway, they typically generate their cell-surface glycans at the interface of the cytoplasmic membrane and the cytoplasm or in the periplasm (see below). Other glycoconjugates like hyaluronan in vertebrate cells, chitin in invertebrate cells, and cellulose in plant cells are synthesized on the cytoplasmic face of the plasma membrane and simultaneously extruded across the membrane to the outside (Chapters 16, 24, and 26). The enzymes involved in the synthesis of these glycoconjugates appear to mediate this extrusion. Typical eukaryotic Golgi glycosylation enzymes have also been found at the cell surface or as soluble forms in the extracellular space, as described below. Whether these glycosyltransferases would routinely have adequate supplies of nucleotide sugar donors to modify cell-surface glycans is unclear, but at least one example has been documented (see below). On the other hand, there are examples of remodeling of cell-surface glycans in animal cells—for example, the endosulfatases (Sulf enzymes) that modify heparan sulfate glycosaminoglycan (Chapter 17) and endogenous sialidases that remove cell-surface sialic acids (Chapter 15). Some protozoan parasites, such as trypanosomes, transfer sialic acids from host glycoconjugates to their cell-surface glycans using *trans*-sialidases (Chapter 43).

Glycosylation Pathways in Eubacteria and Archaea

Prokaryotic assembly pathways of polysaccharides and oligosaccharides are very similar to pathways found in the ER and the plasma membrane of eukaryotes. They are assembled in the cytoplasm and then translocated across the plasma membrane. For many biosynthetic pathways, such as O-antigen biosynthesis in Gram-negative bacteria, N-linked protein glycosylation in Archaea, and some Gram-negative bacteria or S-layer biosynthesis in Gram-positive bacteria, oligosaccharides are assembled on lipid carriers at the inner site of the plasma membrane and then flipped to the periplasmic site (Chapters 21 and 22). Synthesis of oligosaccharides can continue in the periplasm, but for these reactions, isoprenoid-linked monosaccharides serve as activated substrates as is the case in cell wall biogenesis of actinobacteria (mycobacteria) (Chapter 21).

GLYCOSYLATION IN THE EUKARYOTIC SECRETORY PATHWAY

Much effort has gone into understanding the mechanisms of glycosylation and glycan modification within the ER and the Golgi apparatus, and it is clear that a variety of interacting and competing factors determine the final outcome of the reactions. The glycosyltransferases and processing glycosidases are well-studied (Chapter 6), and their location has helped to define various functional compartments of the ER–Golgi pathway.

Many Golgi Enzymes Share a Similar Topology

Despite lack of sequence homology among different glycosyltransferase families, most Golgi enzymes share some features. Early studies of vertebrate glycosyltransferases found some of these activities in soluble form in secretions and body fluids; others were identified as membrane-bound activities within cells, and some showed both properties. Subsequent molecular cloning defined the sequences of Golgi glycosyltransferases, revealing that they share a common topology and domain structure that can account for these observations.

The majority of Golgi glycosylation enzymes are type II membrane proteins consisting of three parts: an amino-terminal cytoplasmic tail, followed by a transmembrane (TM) region that also acts as an uncleavable signal sequence, and a large carboxy-terminal region containing a membrane-proximal, proteolytically sensitive stem region as well as a large catalytic domain. The type II topology of these Golgi enzymes places their catalytic sequences in the Golgi lumen, where they participate in the synthesis of the glycan chains on proteins and lipids during their transit through the secretory pathway (Figure 4.2).

Many Golgi enzymes are secreted by cells, sometimes in large quantities, and can be found in cell culture supernatants and various body fluids. These soluble, secreted enzymes are derived from their membrane-associated forms by one or more proteolytic cleavage events that occur within the enzyme's stem region (Figure 4.2). These cleavage events are catalyzed by proteases in the *trans* regions of the Golgi and in post-Golgi compartments. The production of soluble enzymes from cell types such as hepatocytes and endothelium can be dramatically up-regulated under inflammatory conditions. Because circulating and cell-surface localized glycosyltransferases are not expected to have access to adequate concentrations of donor nucleotide sugars (primarily located inside cells), it was thought that they should be functionally incapable of performing transfer reactions in extracellular spaces. However, recent evidence suggests that the release of nucleotide sugar donors by activated platelets may allow soluble, secreted sialyltransferase ST6Gal-I to modify glycans on cell surfaces beyond the original source of the enzyme.

Not all glycosylation enzymes in the secretory pathway are type II membrane proteins. For example, the UDP-GlcNAc:lysosomal enzyme N-acetylglucosamine-1-phosphotransferase (GlcNAc-phosphotransferase) is a multisubunit complex and the GlcNAc-1-phosphodiester α-N-acetylglucosaminidase is a type I membrane protein with its amino terminus in the lumen of the Golgi. These enzymes are involved in the synthesis of the Man-6-P targeting signal of newly synthesized lysosomal hydrolases (see Chapter 33). Some ER glycosylation enzymes are synthesized as soluble proteins. These include the UDP-glucose glycoprotein glucosyltransferase (UGGT) involved in ER quality control (Chapter 39), and enzymes involved in epidermal growth factor (EGF) repeat or thrombospondin type 1 repeat (TSR) glycosylation such as the two protein O-fucosyltransferases, POFUT1 (EGF-like repeats) and 2 (TSR), protein O-glucosyltransferases 1, 2, and 3 (POGLUT1-3; EGF-like repeats), EGF-specific O-GlcNAc transferase (EOGT; EGF-like repeats), and β1-3-glucosyltransferase (B3GLCT; TSR) (Chapter 13). In addition, one of the sulfotransferases involved in heparan sulfate synthesis, GlcNAc 3-O-sulfotransferase 1, is a soluble enzyme in the Golgi.

Localization of Glycosylation Enzymes in Golgi Compartments

All forms of glycosylation in the secretory pathway are highly ordered and sequential processes, typically involving glycosyltransferase reactions. These enzymes, their glycan substrates (attached to protein or lipid), and the appropriate nucleotide sugar donor must be located in the same compartment. Biochemical and ultrastructural studies indicate that glycosyltransferases segregate into distinct overlapping compartments within the secretory pathway. Generally speaking, enzymes acting early in the biosynthetic pathway localize to *cis*- and *medial*-Golgi compartments, whereas those acting later in the pathway tend to localize in the *trans*-Golgi cisternae and the TGN. These observations prompted extensive exploration of mechanisms whereby glycosyltransferases and processing glycosidases achieve this compartmental segregation. Early studies were directed at identifying enzyme sequences required for their retention in the Golgi

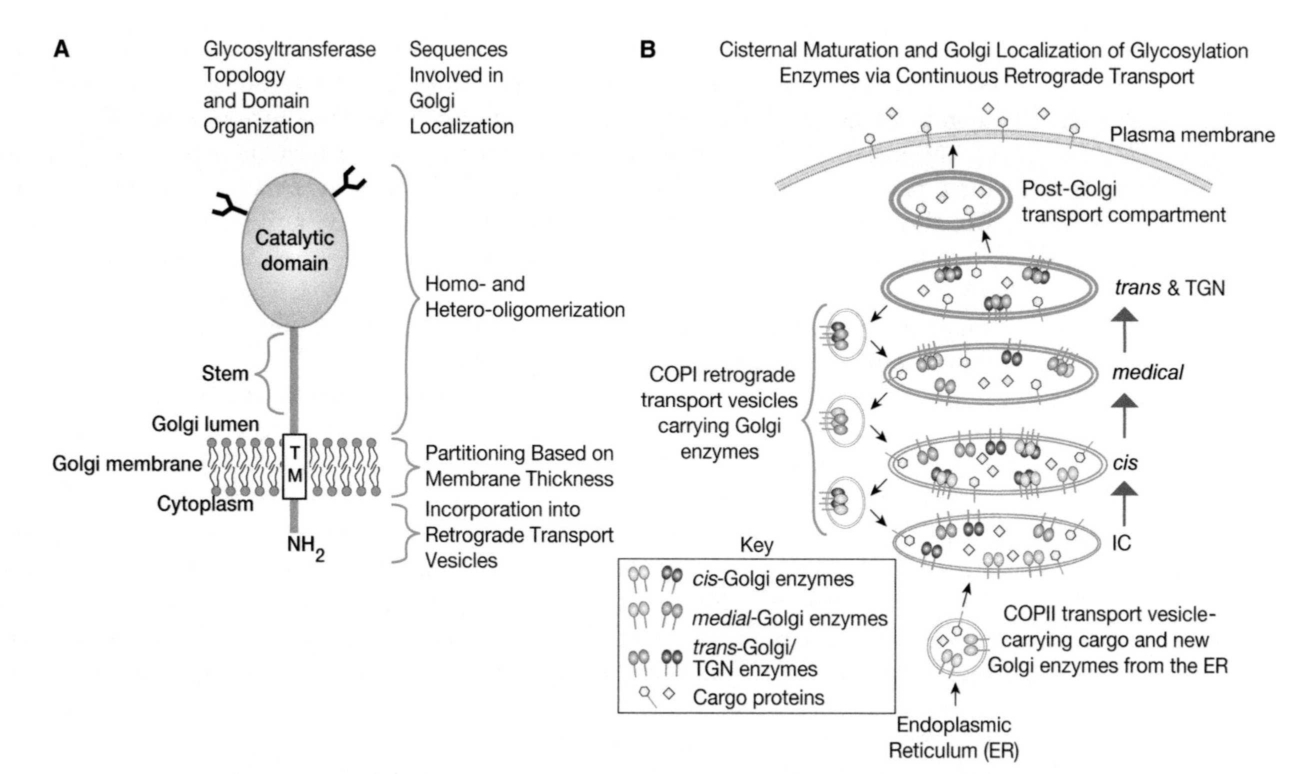

FIGURE 4.2. Topology and localization of Golgi glycosylation enzymes. Golgi glycosyltransferases and glycosidases are type II membrane proteins with their catalytic sequences facing the lumen of the Golgi. According to the cisternal maturation model of *intra*-Golgi transport, these glycosylation enzymes are maintained in the Golgi and segregated into different cisternae via continuous retrograde transport in COPI-coated vesicles. Their incorporation into these vesicles is likely mediated by the interaction between sequences in their cytoplasmic tails and proteins associated with the coated vesicles. For example, two select COPI coatomer subunits have been demonstrated to bind to a specific sequence motif in the amino-terminal cytoplasmic tails of a number of Golgi glycosylation enzymes. Mutations in this motif in the GlcNAc-1-phosphotransferase enzyme that modifies lysosomal enzymes lead to its mislocation, resulting in the lysosomal storage disease, mucolipidosis III (Chapter 33). The selective partitioning of glycosylation enzymes into membranes also plays a role in their Golgi localization. Differences in transmembrane (TM) length and hydrophobicity allows the selective partitioning of these enzymes into different compartments with low cholesterol content (thinner membranes) or high cholesterol content (wider membranes). In general, the relatively short TM regions of these enzymes may prevent their partitioning into post-Golgi transport compartments that share the wider, cholesterol-rich membrane found on the cell surface. The dimerization and hetero-oligomerization of Golgi glycosylation enzymes, mediated by their luminal sequences, can also be important for enzyme localization and efficient glycosylation in some pathways. The Golgi localization of these enzymes is not always absolute, and some are found on the cell surface or are cleaved within their proteolytically sensitive stem region by proteases in late Golgi or post-Golgi compartments and secreted into the extracellular space (not shown). (IC) intermediate compartment; (TGN) *trans*-Golgi network.

cisternae per the vesicular transport model of protein trafficking, whereas more recent studies have included the framework of the cisternal maturation model (see below).

The view of how proteins traverse the Golgi stack and how Golgi enzymes "retain" their relative positions in the Golgi cisternae has evolved substantially and includes the two primary models mentioned above, which are not mutually exclusive and may function together in cells. The vesicular transport model posits that the Golgi is a stable compartment, and that cargo proteins are transported in coated vesicles from the ER to an intermediate compartment and between each Golgi cisterna in a vectorial fashion, during which time these proteins are modified by Golgi glycosylation enzymes retained in each cisterna. More recent data support a cisternal maturation model that can explain the intra-Golgi transport of larger cargo molecules, which cannot fit into small transport vesicles (Figure 4.2). In this model, a new Golgi cisterna is formed on the *cis* face of the stack by the transport of cargo molecules in COPII-coated vesicles from the ER to an intermediate compartment, and the retrograde transport of *cis*-Golgi enzymes in COPI-coated vesicles from an "older" *cis* cisterna into the newly formed compartment that then becomes the *cis* cisterna. The cisterna and its cargo mature as later Golgi glycosylation enzymes are sequentially transported into the "younger" cisternae. The cisternae progress and mature until they effectively dissolve at the TGN stage as membrane and cargo are transported to the plasma membrane for residence or constitutive secretion, to secretory granules for regulated secretion, or to the endosome/lysosome system. The cisternal maturation model is thus distinct from the vesicular transport model in that Golgi enzymes are not retained in stable compartments but continuously transported in a retrograde fashion to "mature" younger cisternae and their cargo.

The role of the cisternal maturation in Golgi enzyme localization is supported by observations that mutations in conserved oligomeric Golgi (COG) complex proteins involved in retrograde vesicular transport impact Golgi enzyme distribution and overall protein glycosylation. The COG complex is a hetero-oligomer of eight subunits that is believed to function as a cytoplasmic tethering complex that links incoming vesicles to their target compartments before vesicle fusion. This complex is thought to cooperate with COPI subunits in retrograde vesicular transport related to intra-Golgi and Golgi-to-ER trafficking. Mutations in COG subunits lead to the instability and/or mislocalization of several Golgi glycosyltransferases across the stack and to corresponding glycosylation defects. The COG complex does not directly interact with Golgi enzymes, but it is critical for retrograde vesicular transport in the Golgi system and in this way impacts overall Golgi structure and thus ensures efficient glycosylation. Notably, several congenital disorders of glycosylation type II (CDG-II) are the result of mutations in COG subunits (Chapter 45).

Studies using mutant and chimeric Golgi enzymes showed that different enzymes have different requirements for their localization. Early work pointed to the TM regions of enzymes such as the GlcNAcT-1 (*medial*-Golgi), GalT-1 (*trans*-Golgi), and ST6Gal-I (*trans*-Golgi and TGN), but later studies revealed that for many enzymes, multiple signals and mechanisms are responsible. The role of both homo- and hetero-oligomerization in the localization of some Golgi enzymes has been established. In addition, substantial evidence supports the role of glycosyltransferase cytoplasmic tails in enzyme retrograde transport and Golgi localization (see below).

The length and hydrophobicity of a membrane protein's TM region determine its ability to partition into membrane microdomains and are now appreciated to be involved in membrane protein trafficking and localization throughout the cell. Both the concentration of cholesterol and the width of the membrane increase throughout the secretory pathway with the widest, most cholesterol-rich membranes found at the cell surface. Experiments using cholesterol-containing model membranes showed that shorter TM peptides partition into thinner membranes, whereas longer TM peptides partition into thicker membranes. It is possible that cholesterol's tendency to "straighten" the lipid acyl chains may make it more energetically difficult to partition TM peptides into membranes with mismatched thicknesses. In support of the idea that membrane thickness may contribute to membrane protein localization in the secretory pathway, it has been noted that ER proteins have shorter TM regions than plasma membrane proteins, and that the TM regions of Golgi enzymes are intermediate between those of ER and plasma membrane proteins. However, among the Golgi enzymes there is not a strict increase in TM length as one moves from the *cis* to the *trans* face of the organelle. One possibility is that the relative impact of TM region length on cisternal localization depends on what other sequences and mechanisms are involved in the localization of a specific enzyme. Nevertheless, at minimum, the shorter TM regions of Golgi enzymes may prevent these proteins from leaving the Golgi system by reducing their ability to partition to the thicker, cholesterol-rich membranes of carriers destined for post-Golgi compartments like the plasma membrane (Figure 4.2).

Another mechanism contributing to Golgi localization of enzymes is their ability to form oligomeric complexes (Figure 4.2). Nearly all enzymes in the N-linked and O-linked glycosylation pathways form homodimers, and many also form heteromeric complexes. In some cases, heteromeric complex formation is pH dependent. Heteromeric complex formation is observed between enzymes that catalyze sequential reactions in the same pathway and that are localized in the same cisternae. For example, in the N-glycosylation pathway, complexes are formed between two N-acetylglucosaminyltransferases (GlcNAcT-I and GlcNAcT-II) in the *medial*-Golgi and between GalT-I and ST6Gal-I in the *trans*-Golgi. Notably, enzymes not in the same pathway (e.g., O-glycosylation and N-glycosylation enzymes) and enzymes in the same pathway, but that catalyze competing or nonsequential events, do not form heteromeric complexes even if they are localized in the same cisterna. Complexes between sequential enzymes in a pathway could increase the efficiency of glycosylation by promoting substrate channeling, wherein one enzyme hands the newly modified substrate off to the next enzyme in the pathway.

Taken together, evidence indicates that glycosylation enzymes use multiple mechanisms to maintain their Golgi localization. The number of signals and mechanisms used by an enzyme could determine how stable its Golgi localization is, whether it is able to move to a later compartment, and whether it is cleaved and secreted into the extracellular space.

GLYCOSYLATION IN UNEXPECTED SUBCELLULAR LOCATIONS

There are scattered reports of glycosylation in unexpected locations—for example, gangliosides in mitochondria and GAGs and N-glycans in the nucleus. Many of these claims are based on incomplete evidence (Chapter 18). One possibility is that there are indeed glycans in these unexpected locations, but that their true structures are novel. Conversely, although structural evidence might be strong, there may be inadequate evidence to be certain about the topology of the claimed structures. Regardless, past experience tells us that the cell biology of glycosylation can hold many surprises, and dogmatic positions about such controversial issues are not warranted.

TURNOVER AND RECYCLING OF GLYCANS

Like all components of living cells, glycans constantly turn over. Some glycoconjugates, such as transmembrane heparan sulfate proteoglycans, turn over by shedding from the cell surface through limited proteolysis. Most glycoconjugate turnover occurs by endocytosis and subsequent degradation in lysosomes (Chapter 44). Endoglycosidases can initially cleave glycans internally, producing substrates for exoglycosidases in the lysosome. Once broken down, individual monosaccharides are then typically exported from the lysosome into the cytoplasm, so that they can be reused (Figure 1.8, Chapter 1). In contrast to the relatively slow turnover of glycans derived from the ER–Golgi pathway, glycans of the nucleus and cytoplasm may be more dynamic and rapidly turned over (Chapters 18 and 19). Glycans in bacterial cells (especially those in the cell wall) also turn over during cell division when the cell wall undergoes cleavage and remodeling.

ACKNOWLEDGMENTS

The authors acknowledge contributions to previous versions of this chapter by Jeffrey D. Esko and appreciate helpful comments and suggestions from James H. Prestegard.

FURTHER READING

Paulson JC, Colley KJ. 1989. Glycosyltransferases. Structure, localization, and control of cell type specific glycosylation. *J Biol Chem* **264**: 17615–17618. doi:10.1016/S0021-9258(19)84610-0

Calo D, Kaminski L, Eichler J. 2010. Protein glycosylation in Archaea: sweet and extreme. *Glycobiology* **20**: 1065–1076. doi:10.1093/glycob/cwq055

Dell A, Galadari A, Sastre F, Hitchen P. 2010. Similarities and differences in the glycosylation mechanisms in prokaryotes and eukaryotes. *Int J Microbiol* **2010**: 148178. doi:10.1155/2010/148178

Nothaft H, Szymanski CM. 2010. Protein glycosylation in bacteria: sweeter than ever. *Nat Rev Microbiol* **8**: 765–778. doi:10.1038/nrmicro2383

Banfield DK. 2011. Mechanisms of protein retention in the Golgi. *Cold Spring Harb Perspect Biol* **3**: a005264. doi:10.1101/cshperspect.a005264

Glick BS, Luini A. 2011. Models for Golgi traffic: a critical assessment. *Cold Spring Harb Perspect Biol* **3**: a005215. doi:10.1101/cshperspect.a005215

Reynders E, Foulquier F, Annaert W, Matthijs G. 2011. How Golgi glycosylation meets and needs trafficking: the case of the COG complex. *Glycobiology* **21**: 853–863. doi:10.1093/glycob/cwq179

Varki A. 2011. Evolutionary forces shaping the Golgi glycosylation machinery: why cell surface glycans are universal to living cells. *Cold Spring Harb Perspect Biol* **3**: a005462. doi:10.1101/cshperspect.a005462

Moremen KW, Tiemeyer M, Nairn AV. 2012. Vertebrate protein glycosylation: diversity, synthesis and function. *Nat Rev Mol Cell Biol* **13**: 448–462. doi:10.1038/nrm3383

5 Glycosylation Precursors

Hudson H. Freeze, Michael Boyce, Natasha E. Zachara, Gerald W. Hart, and Ronald L. Schnaar

In nature, most glycans are synthesized by glycosyltransferases, enzymes that transfer activated forms of monosaccharides from nucleotide sugars and lipid-linked sugar intermediates to acceptors, including proteins, lipids, and growing glycan chains. Monosaccharide precursors are imported into the cell, salvaged from degraded glycans, or created enzymatically from other sugars within the cell. In eukaryotic cells, glycosylation occurs mostly in the endoplasmic reticulum (ER) and Golgi apparatus, whereas monosaccharide activation and interconversions occur mostly in the cytoplasm. Nucleotide sugar–specific transporters carry activated sugar donors into the Golgi and to a lesser extent into the ER. In some cases, nucleotide sugars are used to synthesize activated lipid-linked intermediates before glycan transfer. This chapter describes how cells accomplish these tasks, with an emphasis on animal cells.

GENERAL PRINCIPLES

Glucose and fructose are the major carbon and energy sources for organisms as diverse as yeast and humans. Most organisms can synthesize the other monosaccharides needed for glycan biosynthesis from these sources. Not all of these biosynthetic pathways are equally active in all types of cells. However, there are some general principles. Monosaccharides must be activated to a high-energy donor for use in glycan synthesis. This process requires nucleoside triphosphates

published by Cold Spring Harbor Laboratory Press; doi: 10.1101/glycobiology.4e.05

TABLE 5.1. Activated sugar donors in animal cells

Sugar	Activated form
Glc	UDP-sugar
Gal	
GlcNAc	
GalNAc	
GlcA	
Xyl	
Man	GDP-sugar
Fuc	
Sia	CMP-Sia

(typically UTP or GTP) and a glycosyl-1-P (monosaccharide with a phosphate at the anomeric carbon). They can be activated by a kinase (reaction 1) or generated from a previously synthesized activated nucleotide sugar (reactions 2 and 3):

Reaction 1 Sugar + NTP $\xrightarrow{H_2O \rightarrow NDP}$ Sugar-P $\xrightarrow{NTP \rightarrow PPi}$ Sugar-NDP

Reaction 2 Sugar(A)-NDP $\rightleftharpoons$ Sugar(B)-NDP

Reaction 3 Sugar(A)-NDP + Sugar(B)-1-P $\rightleftharpoons$ Sugar(B)-NDP + Sugar(A)-1-P

The most common nucleotide sugar donors in animal cells are shown in Table 5.1. Sialic acids and their evolutionary ancestors (prokaryotic nonulosonic acids and Kdo) are the only monosaccharides in animals activated as CMP-mononucleotides. Iduronic acid does not have a nucleotide sugar parent because it is formed by epimerization of glucuronic acid after it is incorporated into glycosaminoglycan (GAG) chains. In some instances, one nucleotide sugar can be formed from another either by direct epimerization (reaction 2 above) or by a nucleotide exchange reaction (reaction 3 above). For example, UDP-Gal is made from UDP-Glc by exchange of Gal-1-P for Glc-1-P.

EXTERNAL SUGAR SOURCES AND SUGAR TRANSPORTERS

Three types of sugar transporters carry sugars across the plasma membrane into cells. First are energy-independent facilitated diffusion transporters, such as the glucose transporter (GLUT) family of hexose transporters found in yeast and most mammalian cells. The genes encoding these proteins are named *SLC2A* (solute carriers 2A). Second are energy-dependent transporters—for example, the sodium-dependent glucose transporters (SGLT; gene names *SLC5A*) in intestinal and kidney epithelial cells. The third type includes transporters that couple ATP-dependent phosphorylation with sugar import. These are found in bacteria (Chapter 21) and are not covered in this chapter.

GLUT family transporters were first described in yeast, where at least 18 genes are known. Humans have 14 GLUT homologs. GLUT transporters range in size from ~40 to 70 kDa and have a similar structure, containing 12 membrane-spanning domains, which is typical of many eukaryotic transporters. The transmembrane domains form a barrel with a small pore for sugar passage. Compared with GLUT1, the other family members have a modest (28%–65%) amino acid identity. There are "sugar transporter signatures" consisting of one or a few amino acids in specific positions relative to the membrane-spanning domains, but no major transporter motifs.

Typically, the GLUTs have K_m values for glucose uptake in the 1–20 mM range. In yeast, many transport glucose, but others are specific for galactose, fructose, or disaccharides. Most mammalian GLUT proteins transport glucose or fructose with variable efficiency but without fully characterized specificity in physiological contexts. However, GLUT5 primarily transports fructose, and the GLUT called HMIT is a proton-coupled *myo*-inositol transporter. GLUT2 also efficiently transports glucosamine.

Glucose is transported from the gut lumen by the energy-requiring SGLT-1 and is recovered from the kidney filtrates by a related transporter, SGLT-2. The SGLT-type transporters have K_m values of <1 mM for glucose.

GLUT1–5 have different distributions among different mammalian cells and different K_m values that enable them to respond to the availability of glucose. Although most of the human GLUT members are located on the cell surface, a portion of GLUT4 resides in intracellular vesicles, which are recruited to the cell surface in response to insulin. Following carbohydrate-rich meals, glucose transported by SGLT-1 in the intestine is thought to promote the recruitment of GLUT2 to the apical surface for enhanced glucose uptake.

INTRACELLULAR SOURCES OF MONOSACCHARIDES

Salvage

Monosaccharides can be salvaged from glycans degraded within cells (Chapter 44). Most of the degradation occurs at low pH in lysosomes. Salvage pathways have received relatively little attention, but their contribution to glycosylation may be quite substantial. For example, 80% of the radiolabeled *N*-acetylglucosamine from glycoproteins degraded in liver lysosomes is converted into UDP-GlcNAc, and at least one-third is used to synthesize secreted glycoproteins. Also, fibroblasts endocytose labeled glycans and reuse about 50% of the amino sugars for de novo glycoprotein synthesis. Efficient salvage is not limited to GlcNAc. Much of the sialic acid derived from endocytosed extracellular glycans may be reused.

Monosaccharides released by degradation must exit the lysosome. Different lysosomal carriers exist for neutral hexoses (glucose, mannose, and galactose), N-acetylated amino sugars, and acidic sugars; the neutral sugar carrier has a K_m value of 50–75 mM for hexose substrates, but also transports fucose and xylose. The *N*-acetylhexosamine carrier ($K_m \sim 4$ mM) cannot transport nonacetylated amino sugars. The sialic acid and glucuronic acid carrier ($K_m \sim$ 300–550 μM) is important because its loss leads to an accumulation of these sugars in the lysosome and secretion into the urine, with genetic mutations resulting in a human lysosomal storage disease (Chapter 44). Most monosaccharides that reach the cytoplasm are activated and reused, as described below. However, the uronic acids cannot be reused in animals and are degraded via the pentose phosphate pathway. Mannose released from N-glycan processing or turnover is transported out of the cell by a hexose transporter/exchanger with little or no direct reutilization.

Glycan salvage pathways are clinically exploited to mitigate the impact of certain rare congenital disorders of glycosylation that target sugar biosynthesis or transport (Chapter 45). Experimentally, salvage pathways are used to manipulate nucleotide sugar pools, to incorporate glycans with biorthogonal chemical handles (Chapter 56), and to generate metabolic glycosyltransferase inhibitors (Chapter 55).

Activation and Interconversion of Monosaccharides

Glycogen

Glycogen is an immense molecule that contains up to 100,000 glucose units, arranged in Glcα1–4Glc repeating disaccharides with periodic α1–6Glc branches. It is synthesized on a cytoplasmic protein called glycogenin (Chapter 18). Glycogen is the major storage polysaccharide in animal cells, and its synthesis and degradation (glycogenolysis) are highly regulated for energy use. Glycogen is synthesized by the addition of single glucose units from UDP-Glc, and it is degraded by glycogen phosphorylase. This ATP-independent reaction forms glucose-1-P by phosphorolysis of glycogen. This substrate can be used directly to form UDP-Glc or converted to glucose-6-P for further catabolism via glycolysis or direct oxidation via glucose-6-phosphate dehydrogenase.

Glucose

Glucose is the central monosaccharide in carbohydrate metabolism, and it can be converted directly or indirectly into all other sugars (Figure 5.1). Glucose is first converted to glucose-6-P by hexokinase. In the glycolytic pathway, glucose-6-P is converted to fructose-6-P by phosphoglucose isomerase or into glucose-1-P by phosphoglucomutase. Reaction of glucose-1-P

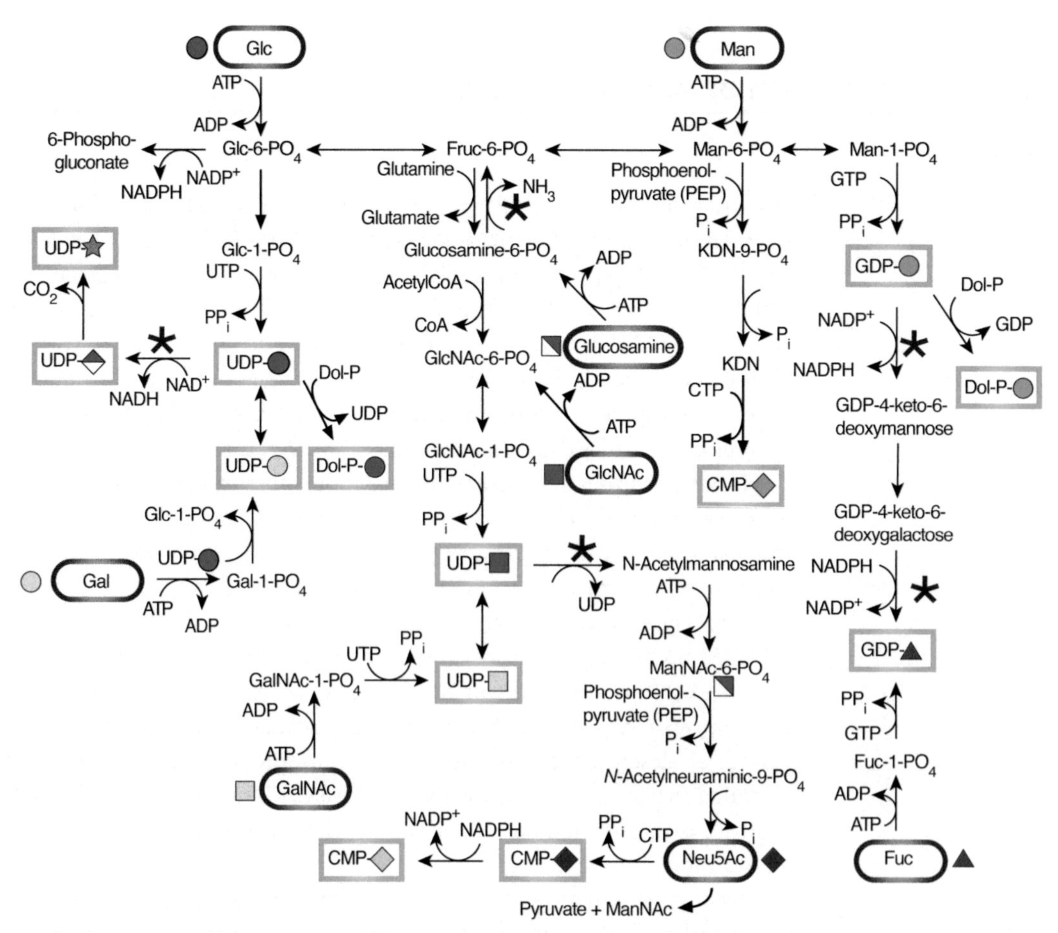

FIGURE 5.1. Biosynthesis and interconversion of monosaccharides. The relative contributions of each pathway under physiological conditions are unknown. (*Rectangles*) donors; (*ovals*) monosaccharides; (*asterisks*) control points; (Kdn) 2-keto-3-deoxy-D-glycero-D-galactonononic acid; (Dol) dolichol. See Online Appendix 1B for full names and designated symbols for monosaccharides.

with UTP by the enzyme UDP-glucose pyrophosphorylase forms the high-energy donor UDP-Glc. The UDP-Glc pool is used to synthesize glycogen and other glucose-containing molecules such as glucosylceramide (Chapter 11) and dolichol-P-glucose, which is used in the N-linked glycan biosynthetic pathway (Chapter 9).

Glucose-6-phosphate is also the substrate for glucose-6-P dehydrogenase, the entry point for its oxidation via the pentose phosphate pathway that subsequently produces 6-phosphogluconate and then ribose-5-phosphate. These reactions generate nicotinamide adenine dinucleotide phosphate (NADPH), which is needed to maintain proper redox status.

Glucuronic Acid

UDP-GlcA (glucuronic acid) is synthesized directly from UDP-Glc by a two-stage reaction requiring two NAD^+-dependent oxidations at C-6. UDP-GlcA is used primarily for GAG biosynthesis (Chapters 16 and 17), but some N- and O-linked glycans and glycosphingolipids contain glucuronic acid as well. In liver cells, the addition of glucuronic acid to bile acids and xenobiotic compounds (such as drugs and toxins) increases their solubility, and a large class of microsomal glucuronosyltransferases is devoted to these reactions.

Iduronic Acid

Iduronic acid is the C-5 epimer of GlcA, and it is found in GAGs dermatan sulfate, heparan sulfate, and heparin. Iduronic acid is not directly synthesized from a nucleotide sugar donor. Instead, it is generated by epimerization of GlcA following its incorporation into the growing GAG chain (Chapter 17).

Xylose

Decarboxylation of UDP-GlcA generates UDP-Xyl, which is used to initiate GAG synthesis (Figure 5.2; Chapter 17) in vertebrates. Xylose is also found on proteins that have O-glucose modifications in epidermal growth factor (EGF) modules (Chapter 13) and on O-mannose-based glycans on α-dystroglycan (Chapters 13 and 45), as well as on plant N-glycans. A type II membrane protein performs the decarboxylation reaction using UDP-GlcA transported into the ER

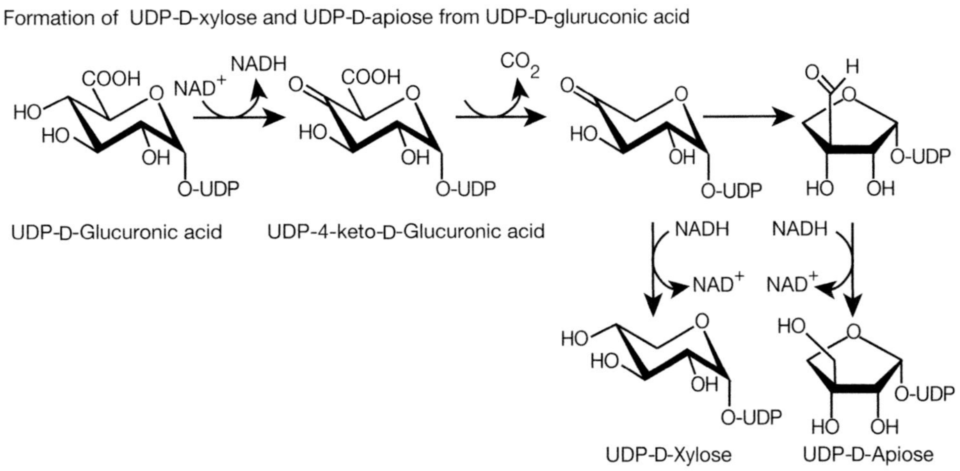

Formation of UDP-D-xylose and UDP-D-apiose from UDP-D-gluruconic acid

FIGURE 5.2. Biosynthesis of UDP-xylose and the branched sugar donor UDP-apiose from UDP-GlcA. Xylose is found in animals and plants, whereas apiose is used for plant polysaccharides, such as apiogalacturonan in *Lemna minor*. Note the similarity and overlap in the synthesis of xylose and apiose. The only differences are that the C-3 is removed by an unknown mechanism, and the reduction of the newly formed aldehyde creates the branched sugar donor.

or Golgi. In *Caenorhabditis elegans*, the decarboxylase is called SQV-1, and it colocalizes with the UDP-GlcA transporter (Chapter 25). In *Arabidopsis*, another UDP-GlcA decarboxylase also occurs in the cytoplasm, but no ortholog has been identified in animals.

Mannose

Mannose is used for multiple types of glycans (Chapters 9, 11, 12, and 13). Guanosine diphosphate mannose (GDP-Man) is the primary activated donor. Its production requires prior synthesis of mannose-6-P and its conversion to mannose-1-P. Two ways to produce mannose-6-P are by direct phosphorylation via hexokinase or conversion of fructose-6-P to mannose-6-P using the enzyme phosphomannose isomerase. In yeast, loss of the latter enzyme is lethal. In humans, loss of this enzyme produces a potentially fatal disease called congenital disorder of glycosylation (CDG) type Ib or MPI-CDG (Chapter 45). Phosphomannose isomerase is important because free exogenous mannose is not common in the diet, and this enzyme is the key link between mannose and glucose. Both yeast and human phosphomannose isomerase deficiencies are rescued by providing exogenous mannose (although excess mannose is toxic). Mice lacking phosphomannose isomerase activity die in utero because of an accumulation of Man-6-P that inhibits glycolysis and depletes ATP. In mammals, mannose-6-P is converted to mannose-1-P using phosphomannomutase. Because mannose-6-P and mannose-1-P are both obligate precursors of GDP-Man, failure to make sufficient amounts of either one reduces the formation of GDP-Man, which is a direct donor for lipid-linked oligosaccharides (see below) and a precursor for dolichol-P-mannose, which serves multiple glycosylation pathways.

Mannose-6-P can also condense with phosphoenolpyruvate to form 2-keto-3-deoxy-D-glycero-D-galactonononic acid (Kdn). This molecule is activated with CTP to produce CMP-Kdn and is abundant in fish (e.g., in trout testis and on their sperm), in which it is thought to be important for sperm–egg adhesion.

Fucose

Guanosine diphosphate fucose (GDP-Fuc) can be derived from GDP-Man by the sequential action of two enzymes involving three steps. In the first step, the C-4 hydroxyl group of GDP-Man is oxidized to a ketone (GDP-4-keto-6-deoxy-mannose) by the enzyme GDP-Man 4,6-dehydratase, along with the reduction of $NADP^+$ to NADPH. The next two reactions are catalyzed by a single polypeptide (GDP-L-Fuc synthase in humans) that has epimerase and reductase activity and is well-conserved from bacteria to mammals. GDP-4-keto-6-deoxymannose is epimerized at C-3 and C-5 to form GDP-4-keto-6-deoxyglucose, which is then reduced with NADPH at C-4 to form GDP-Fuc (Figure 5.3A). The first oxidation step is feedback-inhibited by GDP-Fuc. GDP-Fuc can also be synthesized directly from fucose. The first step uses a kinase to make fucose-1-P, which is then converted to GDP-Fuc. Mutant CHO cells that cannot convert GDP-Man to GDP-Fuc form hypofucosylated proteins, but this can be corrected by providing exogenous fucose in the medium. Also, mice genetically deficient in the GDP-Man to GDP-Fuc conversion can be rescued by providing fucose in their food or drinking water. Plasma membrane transporters for sugars other than glucose have been characterized, and fucose transporters may exist as well, although they have not been fully characterized, so their quantitative contribution is not known. As with many monosaccharides other than glucose, free fucose concentration in the blood is in the very low micromolar range.

Galactose

UDP-Gal can be made in a two-step reaction by ATP-dependent phosphorylation of galactose at C-1 (by galactokinase) to form galactose-1-P, which then displaces glucose-1-P in UDP-Glc to form UDP-Gal by the action of galactose-1-phosphate uridylyltransferase. A deficiency in this

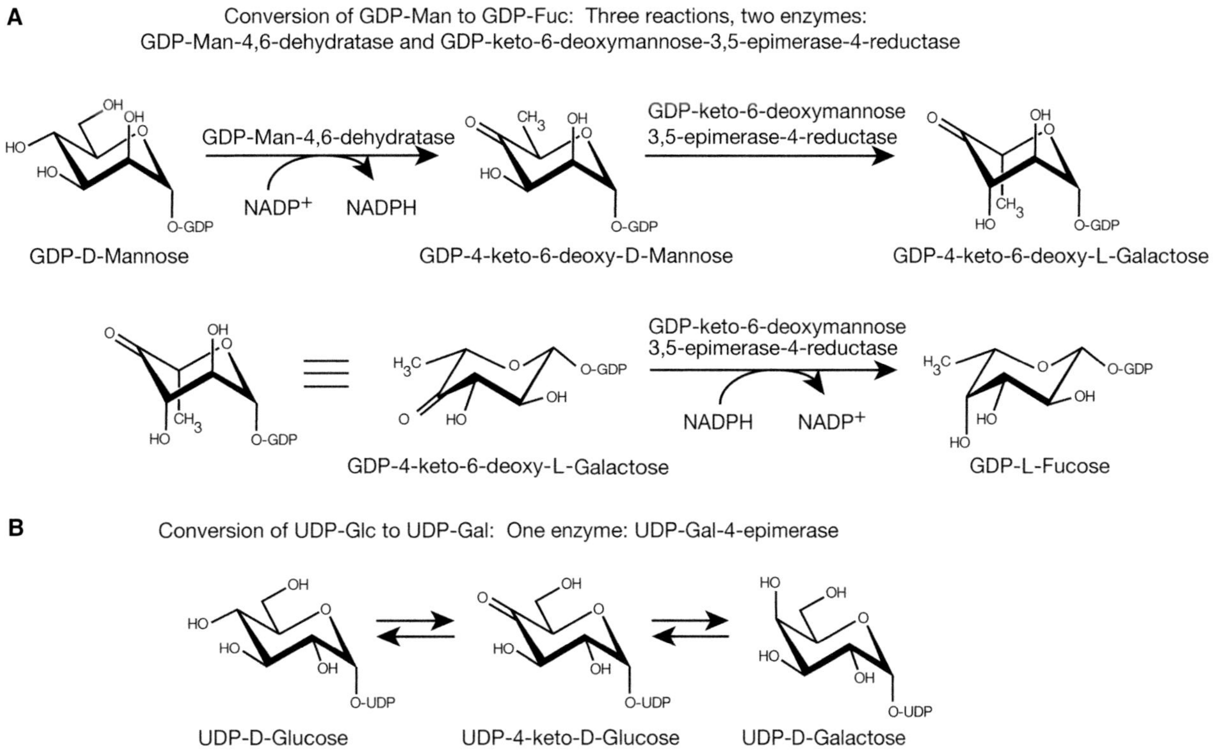

A Conversion of GDP-Man to GDP-Fuc: Three reactions, two enzymes: GDP-Man-4,6-dehydratase and GDP-keto-6-deoxymannose-3,5-epimerase-4-reductase

GDP-D-Mannose → GDP-Man-4,6-dehydratase ($NADP^+$ → NADPH) → GDP-4-keto-6-deoxy-D-Mannose → GDP-keto-6-deoxymannose 3,5-epimerase-4-reductase → GDP-4-keto-6-deoxy-L-Galactose

GDP-4-keto-6-deoxy-L-Galactose → GDP-keto-6-deoxymannose 3,5-epimerase-4-reductase (NADPH → $NADP^+$) → GDP-L-Fucose

B Conversion of UDP-Glc to UDP-Gal: One enzyme: UDP-Gal-4-epimerase

UDP-D-Glucose ⇄ UDP-4-keto-D-Glucose ⇄ UDP-D-Galactose

FIGURE 5.3. Conversion of activated sugar donors. Steps in the synthesis of (*A*) GDP-Fuc from GDP-Man and (*B*) UDP-Gal from UDP-Glc. Details of the various enzymes are given in the text. GDP-Fuc synthesis by this route is not reversible, whereas the interconversion of UDP-Glc and UDP-Gal is easily reversible.

enzyme activity results in a severe human disease called galactosemia, which leads to intellectual disability, liver damage, and eventual death if galactose intake is not controlled (Chapter 45). Finally, UDP-Gal can be formed from UDP-Glc by the NAD-dependent reaction catalyzed by UDP-Gal 4-epimerase. The enzyme first converts the C-4 hydroxyl group to a keto derivative forming NADH from bound NAD^+. In the next step, the keto group is converted back to a hydroxyl group with opposite orientation and NAD^+ reforms (Figure 5.3B). In mammals, the same enzyme interconverts UDP-GalNAc and UDP-GlcNAc.

Galactose usually occurs as a pyranose (*p*) ring in "higher" animals, but bacteria and pathogenic eukaryotes, such as *Leishmania* and *Aspergillus,* incorporate galactofuranose (*f*) into their glycans (Chapter 21). The donor is formed by conversion of UDP-Gal(*p*)→UDP-Gal(*f*) using a flavin adenine dinucleotide–dependent mutase.

N-*Acetylglucosamine*

Enzymatic synthesis of uridine diphosphate *N*-acetylglucosamine (UDP-GlcNAc) begins with the formation of glucosamine-6-P from fructose-6-P by transamination using glutamine as the $-NH_2$ donor by the enzyme glutamine–fructose-6-phosphate transaminase (GFAT). Glucosamine-6-P is then N-acetylated via an acetyl-CoA-mediated reaction to form *N*-acetylglucosamine-6-P and then isomerized to *N*-acetylglucosamine-1-P via a 1,6-bis-phosphate intermediate. Similar to the other activation reactions, *N*-acetylglucosamine-1-P then reacts with UTP to form UDP-GlcNAc and pyrophosphate. Alternatively, GlcNAc can be directly phosphorylated to form *N*-acetylglucosamine-6-P by a kinase that also synthesizes *N*-acetylmannosamine-6-P from *N*-acetylmannosamine. Phospho-*N*-acetylglucosamine mutase then converts *N*-acetylglucosamine-6-P to *N*-acetylglucosamine-1-P. This route may account for the efficient salvage of GlcNAc from lysosomal degradation of glycans. Glucosamine can also be salvaged following sequential phosphorylation and acetylation.

N-Acetylgalactosamine

In animals, UDP-GalNAc can arise from two routes. One is the direct reaction of *N*-acetylgalactosamine-1-P with UTP. *N*-Acetylgalactosamine-1-P is formed by a specific kinase that is distinct from galactose-1-kinase. UDP-GalNAc can also be formed by epimerization of UDP-GlcNAc using the same NAD-dependent epimerase that converts UDP-Glc to UDP-Gal.

Sialic Acids

The sialic acids include three parent compounds—*N*-acetylneuraminic acid (Neu5Ac), *N*-glycolylneuraminic acid (Neu5Gc), and Kdn (Chapter 15)—all of which are converted to CMP nucleotide sugars. Generation of cytidine-5′-monophospho-*N*-acetylneuraminic acid (CMP-Neu5Ac) is more complicated than formation of the other activated sugars. First, UDP-GlcNAc is converted to *N*-acetylmannosamine-6-P by bifunctional UDP-N-acetylglucosamine 2-epimerase/N-acetylmannosamine kinase (GNE), a single enzyme with two catalytic activities. The first activity epimerizes the GlcNAc of UDP-GlcNAc at C-2 and cleaves the UDP to yield *N*-acetylmannosamine (ManNAc). In the next reaction, the kinase activity uses ATP to form ManNAc-6-P. Mutations in this enzyme cause two completely distinct metabolic disorders: sialuria and inclusion body myopathy type 2 (Chapter 45). Knocking out this gene in mice causes early embryonic lethality. In the next step, ManNAc-6-P is condensed with phosphoenolpyruvate to form *N*-acetylneuraminic acid-9-P. Phosphate is then removed by a phosphatase. Activation with CTP yields CMP-Neu5Ac, which is the target in some organisms for a hydroxylase that converts some of it to CMP-Neu5Gc. For unknown reasons, the last steps occur in the nucleus in vertebrate cells, with subsequent export of the activated precursors to the cytoplasm. Other modifications of sialic acid can occur in the Golgi after transfer of the sialic acid to the oligosaccharide acceptor.

Sialic acids can be salvaged from internal glycoprotein turnover or from plasma and activated by phosphorylation and the addition of CMP from CTP. In addition, GlcNAc can be activated to UDP-GlcNAc and reenter the biosynthetic pathway to CMP-Neu5Ac (Chapter 15).

Diverse Monosaccharides in Bacteria and Plants

Fucose is the only deoxyhexose commonly found in animal cell glycans. In contrast, bacterial and plant polysaccharides and glycoproteins frequently contain a variety of deoxysugars, deoxyaminosugars, and branched-chain sugars. These diverse sugars often have potent biological properties. For example, aminoglycoside antibiotics like streptomycin bind to the bacterial ribosome to disrupt protein synthesis. Deoxyhexoses are often immunological determinants of lipopolysaccharides or O-antigens of the *Salmonella* species. Five of the eight possible 3,6-dideoxyhexoses have been found in these organisms at the nonreducing end of the Gram-negative cell wall lipopolysaccharide. Other deoxyhexoses, such as a 4,6-dideoxy-hexose and a 2,3,6-trideoxyhexose, are also biologically significant but so far seem to be uncommon in nature.

Biosynthesis of both deoxysugars and dideoxysugars begins with the oxidation of C-4, similar to the first step of the conversion of GDP-Man to GDP-Fuc. The nucleotide (N) differs for the various sugars, and the individual pathways use different dehydratases. For example, biosynthesis of most 3,6-dideoxyhexoses (except colitose) begins with conversion of CDP-glucose to CDP-4-keto-6-deoxyhexose by NAD^+-dependent CDP-glucose dehydratase. In the biosynthesis of abequose (3,6-dideoxy-D-xylohexose), the product, CDP-6-deoxy-L-threo-D-glycerohexulose, is then converted in two additional steps to CDP-3,6-dideoxy-D-glycero-D-glycero-4-hexulose by a second dehydratase followed by a reductase.

Amino sugars, such as glucosamine, arise from keto sugars by the addition of an amino group from glutamine (Figure 5.1). In addition, bacteria and plants have many 6-deoxy-hexoses with amino groups in the 2, 3, or 4 positions. For example, daunosamine is a 3-amino-6-deoxyhexose that is found in the antibiotic daunomycin. Here, TDP-glucose is dehydrated to 3-keto-6-deoxyglucose and the amino group is added via a transamination reaction, probably involving a vitamin B6-dependent reaction.

Plants and bacteria also contain a number of branched-chain sugars. For instance, apiose is a component of the polysaccharide apiogalacturonan of *L. minor*, and strepose is a component of the antibiotic streptomycin, produced by *Streptomyces griseus*. Apiose (Figure 5.2) is synthesized from UDP-GlcA via a 4-keto intermediate that can yield UDP-Xyl or UDP-apiose. Apiose synthesis removes carbon 3 from the chain to give the branched sugar by an unknown mechanism. Although the syntheses of other branched-chain sugars have not been delineated, they probably follow similar reaction pathways.

NUCLEOTIDE SUGAR TRANSPORTERS

In eukaryotes, nucleotide sugars are synthesized in the cytoplasm or nucleus, whereas most glycosylation occurs inside the ER or Golgi compartments (exceptions being hyaluronan [Chapter 16] and nucleocytoplasmic glycosylation [Chapters 18 and 19]). Therefore, newly synthesized nucleotide sugars are on the "wrong" side of the membrane for most glycosylation reactions and so must be transported into the ER and Golgi. Negative charge prevents these donors from simply diffusing into these compartments. To overcome this hurdle, eukaryotic cells have a set of energy-independent nucleotide sugar antiporters that deliver nucleotide sugars into the lumen of these organelles, with the simultaneous exiting of nucleoside monophosphates, most of which must first be generated from the nucleoside diphosphates by a nucleoside diphosphatase (Figure 5.4). This transport mechanism was determined biochemically in isolated vesicles and genetically in various mutant cell lines. The K_m of the transporters ranges from 1 to 10 μM. Using in vitro systems, the transporters have been shown to increase the concentration of the nucleotide sugars within the Golgi lumen by 10- to 50-fold. This is usually sufficient to reach or exceed the calculated K_m of glycosyltransferases that use these donors.

Most of these antiporters are found in the Golgi, but some are also found in the ER. They are organelle-specific and their location usually corresponds to the location of the relevant glycosyltransferases (Table 5.2; Figure 5.4). Nucleotide sugar import into the Golgi is not energy-dependent nor affected by ionophores. However, the import is competitively inhibited by the corresponding nucleoside monophosphates and diphosphates in the cytosol but not by the monosaccharides. There are also transporters for ATP and PAPS (3′-phosphoadenosine-5′-phosphosulfate), which are used for carbohydrate and protein sulfation.

The glucuronidation of bile and xenobiotic compounds in the ER explains the need for UDP-GlcA transporter in the ER of liver cells. The presence of a Golgi transporter is consistent with the location of the glycosyltransferases that use UDP-GlcA for the formation of GAG chains and other classes of glycans. The observation that reglycosylation of misfolded glycoproteins occurs in the ER (Chapter 39) explains the need for an ER UDP-Glc transporter. Under stressful conditions, synthesis of lumenal uridine diphosphatase increases to accommodate increased transport of UDP-Glc needed for reglycosylation of misfolded glycoproteins. The existence of UDP-GlcNAc, UDP-GalNAc, and UDP-Xyl transporters in the ER may mean that some reactions thought to occur exclusively in the Golgi may also occur in the ER. A good example is the

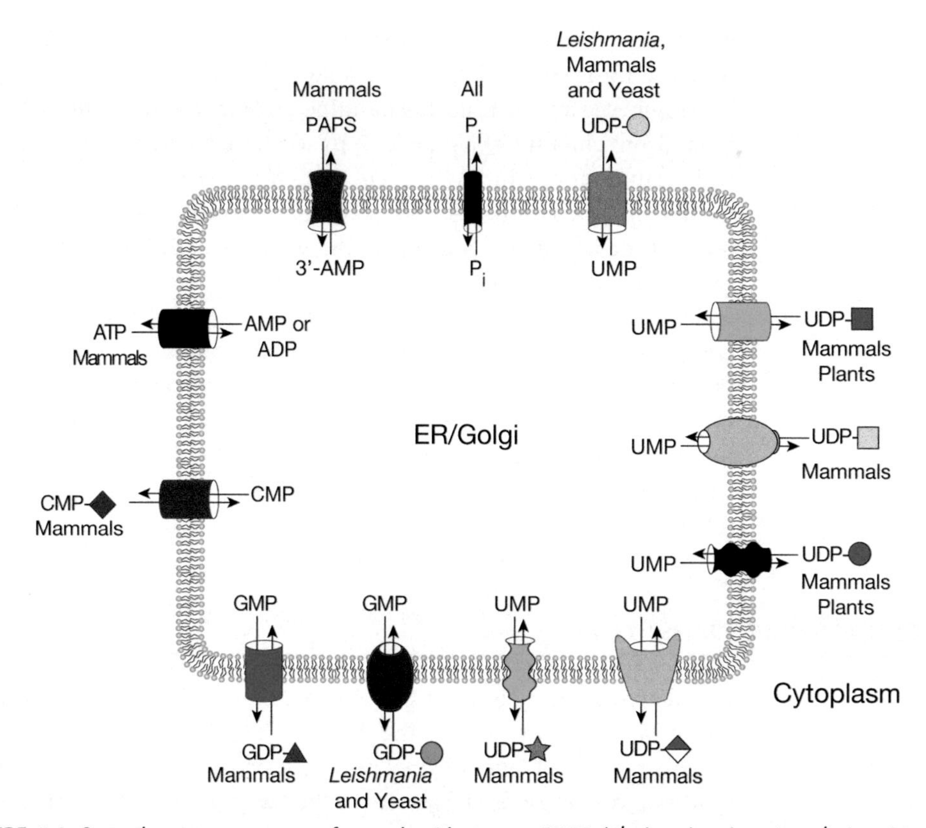

FIGURE 5.4. Some known transporters for nucleotide sugars, PAPS (3′-phosphoadenosine-5′-phosphosulfate), and ATP are located in the Golgi membranes of mammals, yeast, protozoa, and plants. These proteins are antiporters and return the corresponding nucleoside monophosphate back to the cytoplasm when the nucleotide sugar is delivered into the endoplasmic reticulum (ER) or Golgi compartment. Because most glycosylation reactions produce a nucleoside diphosphate, this requires conversion to the nucleoside monophosphate. For PAPS, the corresponding exiting molecule is unknown, and for ATP, it is AMP, ADP, or both. The phosphate (P_i) transporter is hypothetical. See Online Appendix 1B for full names and designated symbols for monosaccharides.

synthesis of O-fucosylated proteins, such as Notch, in the ER versus fucosylation of N- and O-linked chains in the Golgi (Chapter 13). Other, as-yet-undiscovered, glycosylation reactions may also occur in the ER.

Mutations in several nucleotide sugar transporters, UDP-Gal, CMP-Sia, GDP-Fuc, UDP-GlcA/UDP-GalNAc, and UDP-GlcNAc cause human glycosylation disorders (Chapter 45), generating incomplete sugar chains. Mutant mammalian cell lines also lack specific nucleotide sugar transporters (e.g., for UDP-Gal or CMP-Sia; Chapter 49). However, there is some "leakiness" in such mutants. For instance, loss of the UDP-Gal transporter in the Golgi of mutant MDCK (Madin–Darby canine kidney) cells decreases the synthesis of keratan sulfate and galactosylated glycoproteins and glycolipids, but leaves heparan and chondroitin sulfate unaffected. This is probably because the galactosyltransferases that synthesize the core region tetrasaccharide common to GAG chains have lower K_m values for their donors (Chapter 17).

Many putative transporters were identified by homology in the genomes of mammals, *Drosophila melanogaster*, *C. elegans*, plants, and yeast. Like the plasma membrane GLUT transporters discussed above, all are multimembrane-spanning (type III) proteins, but the level of amino acid identity does not give any clue to the substrate specificity. The UDP-GlcNAc transporters from mammalian cells and yeast are 22% identical, whereas mammalian CMP-Sia, UDP-Gal, and UDP-GlcNAc transporters have 40%–50% identity. Clever domain-swapping experiments

TABLE 5.2. Nucleotide transport in Golgi and ER

Nucleotide	ER	Golgi
CMP-Sia	−	+++
GDP-Fuc	+	++++
UDP-Gal	−	++++
PAPS	−	++++
GDP-Man	−	++++
UDP-GlcNAc	++	++++
UDP-GalNAc	++	++++
UDP-Xyl	++	++++
ATP	+++	++++
UDP-GlcA	++++	++++
UDP-Glc	++++	+

The relative distribution of the nucleotide transporters in ER and Golgi is indicated by the number of plus signs (+). A minus sign (−) indicates that the transporter is not found in that compartment.

ER, Endoplasmic reticulum.

show that distinct regions are responsible for functional transport, and engineered chimeric transporters can carry both CMP-Sia and UDP-Gal. Biophysical studies of eukaryotic GDP-mannose and CMP-sialic acid transporters illuminate the structural basis for nucleotide sugar selectivity and membrane lipid interactions in transporter function.

Heterologous expression or rescue of transporter-deficient cell lines can be used to analyze the function of putative transporters. For example, expressing the *C. elegans* gene *SQV-7* in yeast showed that this one protein can transport UDP-GlcA, UDP-GalNAc, and UDP-Gal, whereas mutant alleles cannot transport any of these donors. The human gene *SLC35B4* encodes a bifunctional transporter that recognizes UDP-Xyl and UDP-GlcNAc. Another example is the GDP-Man transporter of *Leishmania,* which can also transport GDP-Fuc and GDP-arabinose. This point illustrates that functional, biochemical analyses are essential; genetic homology is insufficient to infer specificity. Moreover, not all of the potential transport-like genes have been assigned specific physiological substrates.

Theoretically, glycosylation may be controlled in part by regulating the availability of nucleotide sugars within the Golgi, presumably by regulating the transporters. The subcompartmental location (*cis*, *medial*, *trans*) of the transporters in the Golgi is not known nor are the physical relationships of the transporters to the various glycosyltransferases they service. Clearly, a functional Golgi compartment requires both the nucleotide sugar donor and the acceptor with a colocalized transferase. There have been few studies on how the actual glycosylation reactions occur within the Golgi. Is it more like solution chemistry or like solid-state transfers? Are there really "soluble pools" of nucleotide sugars? Dramatic time-lapse videos of green fluorescent protein (GFP)-tagged glycosyltransferases show that the proteins are highly mobile within the Golgi, but there is also physical evidence for multiglycosyltransferase complexes involved in the biosynthesis of N-linked glycans, glycosphingolipids, and heparan sulfate. Many transporters appear to function as homodimers, and the GDP-Man transporter in *Saccharomyces cerevisiae* (VRG4) oligomerizes in the ER and appears to be transported to the Golgi by an active process. Also, synthesis of galactosylceramide occurs in the ER and a portion of the UDP-Gal transporter binds specifically to galactosylceramide synthase and is retained in the ER to provide donor substrate (Chapter 11).

The discovery of protein O-GlcNAcylation, as well as the enzymes that cycle O-GlcNAc, in mitochondria suggests the presence of nucleotide sugar transporters that bridge the outer and inner mitochondrial membranes to the cytosol. Recent data suggest that a pyrimidine nucleotide carrier is an effective transporter of UDP-GlcNAc.

CONTROL OF GLYCOSYLATION PRECURSORS

Biochemical control of the intracellular concentrations of glycosylation precursors (ultimately nucleotide sugars) is a complex and important area of ongoing research. In some cases, key biosynthetic enzymes are inhibited by their final products. A human genetic disorder called sialuria is a clear example. In this condition, massive amounts of sialic acid (several grams each day) are secreted into the urine, along with various intermediates in the CMP-Sia biosynthetic pathway. Sialuria is caused by mutations in the enzyme responsible for the first step in sialic acid biosynthesis, GNE. Sialuria-causing mutations impair the normal feedback inhibition of GNE by CMP-Sia, the end product of the precursor pathway (Chapter 45).

Most glycosylation precursor pools turn over within a matter of minutes. The relevant steady state concentrations of the nucleotide sugars have been difficult to reliably determine. It is especially difficult to measure the concentrations of nucleotide sugar donors at their sites of action, which are often confined to the lumen of the ER or Golgi. In whole-animal studies, the GDP-Fuc pool and fucosylated glycans in the intestine can be regulated by the diet and time of weaning. Considering that resident bacteria in the small intestine participate in the induction of fucosylation pathways in the enterocytes, and that gut bacteria can release and use monosaccharides, dietary manipulation of glycosylation introduces another level of unexplored complexity.

Enzymes within the hexosamine biosynthetic pathway (e.g., GFAT and UDP-N-acetylhexosamine pyrophosphorylase) are often up-regulated in cancer. Altered glycan profiles are well-known markers of tumor progression (Chapter 47) and likely result from changes in both nucleotide sugar pools and glycosyltransferases. One factor that may drive up-regulation of hexosamine synthesis in tumors is activation of the unfolded protein response (UPR), which may increase GFAT and other nucleotide sugar metabolic enzymes, as well as glycosyltransferases. This axis linking the UPR to nucleotide sugar metabolism has also been linked to lifespan regulation and cardiac hypertrophy.

DONORS FOR GLYCAN MODIFICATION

Glycans can be modified postsynthetically, imparting additional complexity and biological information. Sulfation, phosphorylation, methylation, pyruvylation, acetylation, succinylation, and acylation have been found, and the corresponding donors are listed in Table 5.3. In "lower" eukaryotes and bacteria, pyruvic acid is often found as a 1-carboxyethylidene bridge between two hydroxyl groups on a sugar, such as galactose. Because all of these reactions occur in the Golgi in eukaryotes, there must be carriers or transporters that deliver and orient activated donors for efficient synthesis. As additional modifications of sugar chains made in the ER–

TABLE 5.3. Donors for glycan modifications

Modification	Precursor	Transporter
Phosphate	ATP (?)	yes
Sulfate	PAPS	yes
Methyl	S-adenosylmethionine	?
Acetyl	acetyl-CoA	yes
Pyruvate	phosphoenolpyruvate	?
Acyl	acyl-CoA (?)	?
Succinyl	succinyl-CoA (?)	?

PAPS, 3′-Phosphoadenosine-5′-phosphosulfate.

Golgi pathway are uncovered, they will likely turn out to require specific transporters to carry the activated donors into the lumen of these compartments.

SYNTHESIS OF CARRIER LIPIDS

Multiple glycosylation pathways in prokaryotes and eukaryotes require lipid carriers to present monosaccharides and oligosaccharides at the proper location. Undecaprenyl-P (bactoprenol) is the glycosyl carrier for O-antigen, peptidoglycan, capsular polysaccharides, teichoic acid, and mannans in bacteria (Chapter 21). Dolichol-P serves an analogous function in eukaryotic cells (Chapter 9). Dolichol-P-mannose provides all of the mannose for glycophospholipid anchors, C-mannosylated proteins, O-mannose-based chains, and four of the mannose residues of the precursor oligosaccharide used for N-glycan biosynthesis. Dolichol-P-glucose provides glucose for the mature N-linked glycan precursor $Glc_3Man_9GlcNAc_2$, which itself is built on dolichol pyrophosphate (dolichol-PP).

The formation of dolichol-P involves elongation of farnesyl pyrophosphate with multiple *cis*-isopentenyl pyrophosphate units. The total number of isoprene units can vary, from typically 11 in bacteria (making a C_{55} bactoprenol chain) to up to 21 in mammals. In eukaryotes, the double bond nearest the pyrophosphate must be reduced for the carrier to be functional in glycosylation. Studies in yeast, mice, and humans indicate that direct reduction of polyprenol to dolichol is a major pathway, but an alternate pathway must also exist. It is unclear whether the phosphates are removed before or after the reduction step. The evolutionary significance of the different chain lengths and reduction of the double bond is not known. Dolichol is phosphorylated by an ATP-dependent dolichol kinase to generate dolichol-P as needed. Because dolichol, dolichol-P, and dolichol-PP are all generated from a common metabolically stable pool, they must be recycled and interconverted as needed. Dolichol occurs in the ER and Golgi and turns over very slowly.

EMERGING KNOWLEDGE

The finding that disruptions in nucleotide sugar metabolism underpin disease resulted in an influx of data, expanding our knowledge of the signaling pathways controlling nucleotide sugar flux, highlighting noncanonical roles for nucleotide sugars, and identifying new drugs for modulating metabolism. The map of biosynthetic and signaling pathways regulating nucleotide sugars is incomplete, especially with respect to tissue specificity, development, nutrient availability, and stress. New approaches to study metabolic flux and assess or modulate nucleotide sugar levels in an organelle- or cell-specific manner, as well as structural insight into nucleotide sugar transporters and glycosyltransferases, are providing more clarity. Innovations in isotopic labeling of metabolites, mass spectrometry methods, genetic and chemical biosensors of nucleotide sugars, and cryo-electron microscopy hold great promise in this area.

ACKNOWLEDGMENTS

The authors acknowledge contributions to previous versions of this chapter by the late Alan Elbein and appreciate helpful comments and suggestions from Dirk J. Lefeber and Stephan von Gunten.

FURTHER READING

Leloir LF. 1970. Two decades of research on the biosynthesis of saccharides. *Nobel Lecture*. http://www.nobelprize.org/nobel_prizes/chemistry/laureates/1970/leloir-lecture.pdf

Kresge N, Simoni RD, Hill RL. 2005. Luis F. Leloir and the biosynthesis of saccharides. *J Biol Chem* **280**: 158–160. doi:10.1016/s0021-9258(20)67598-6

Park D, Ryu KS, Choi D, Kwak J, Park C. 2007. Characterization and role of fucose mutarotase in mammalian cells. *Glycobiology* **17**: 955–962. doi:10.1093/glycob/cwm066

Holden HM, Cook PD, Thoden JB. 2010. Biosynthetic enzymes of unusual microbial sugars. *Curr Opin Struct Biol* **20**: 543–550. doi:10.1016/j.sbi.2010.08.002

Bar-Peled M, O'Neill MA. 2011. Plant nucleotide sugar formation, interconversion, and salvage by sugar recycling. *Annu Rev Plant Biol* **62**: 127–155. doi:10.1146/annurev-arplant-042110-103918

Freeze HH, Ng BG. 2011. Golgi glycosylation and human inherited diseases. *Cold Spring Harb Perspect Biol* **3**: a005371. doi:10.1101/cshperspect.a005371

Sharma V, Ichikawa M, Freeze HH. 2014. Mannose metabolism: more than meets the eye. *Biochem Biophys Res Commun* **453**: 220–228. doi:10.1016/j.bbrc.2014.06.021

Yonekawa T, Malicdan MC, Cho A, Hayashi YK, Nonaka I, Mine T, Yamamoto T, Nishino I, Noguchi S. 2014. Sialyllactose ameliorates myopathic phenotypes in symptomatic GNE myopathy model mice. *Brain* **137**: 2670–2679. doi:10.1093/brain/awu210

Chen LQ, Cheung LS, Feng L, Tanner W, Frommer WB. 2015. Transport of sugars. *Annu Rev Biochem* **84**: 865–894. doi:10.1146/annurev-biochem-060614-033904

6 Glycosyltransferases and Glycan-Processing Enzymes

James M. Rini, Kelley W. Moremen, Benjamin G. Davis, and Jeffrey D. Esko

Glycosyltransferases and glycosidases are responsible for the assembly, processing, and turnover of glycans. In addition, there are a number of transferases that modify glycans by the addition of acetyl, methyl, phosphate, sulfate, and other groups. This chapter covers the general characteristics of enzymes involved in glycan initiation, assembly, and processing, including aspects of substrate specificity, primary sequence relationships, structures, and enzyme mechanisms.

GENERAL PROPERTIES

The biosynthesis of glycans is primarily determined by glycosyltransferases that assemble monosaccharide moieties into linear and branched glycan chains. As might be expected from the complex array of glycan structures found in nature, glycosyltransferases constitute a very large family of enzymes. In many cases, they catalyze a group-transfer reaction in which the monosaccharide moiety of a simple nucleotide sugar donor (electrophile) substrate (e.g., UDP-Gal, GDP-Fuc, or CMP-Sia; Chapter 5) is transferred to the acceptor (nucleophile) substrate. In some instances, the donor substrates contain a lipid moiety, such as dolichol-phosphate, linked to mannose or glucose. For other glycosyltransferases, the donor substrate is dolichol-pyrophosphate, linked to an oligosaccharide, and in these cases the entire oligosaccharide is

published by Cold Spring Harbor Laboratory Press; doi: 10.1101/glycobiology.4e.06

transferred en bloc to the acceptor substrate (Chapter 9). Similarly, other lipid-linked sugars serve as donor substrates for bacterial glycosyltransferases involved in the assembly of peptidoglycan (e.g., GlcNAc-MurNAc(pentapeptide)-undecaprenyl pyrophosphate), lipopolysaccharide, and capsules (Chapters 21 and 22).

Glycosyltransferases that use monosaccharides, oligosaccharides, proteins, lipids, small organic molecules, and DNA as acceptor substrates have been characterized (activity to RNA has also been suggested), but only glycosyltransferases involved in the biosynthesis of glycoproteins, proteoglycans, and glycolipids are discussed in this chapter. Among these enzymes, the vast majority is responsible for elongating the glycan moieties of these glycoconjugates; the remainder is responsible for the transfer of either a mono- or oligosaccharide directly to the polypeptide or lipid. Generally speaking, the enzymes that elongate glycans act sequentially so that the product of one enzyme yields a preferred acceptor substrate for the subsequent action of another. The end result is a linear and/or branched structure composed of monosaccharides linked to one another. Acceptor recognition by these glycan-elongating glycosyltransferases does not typically involve the polypeptide or lipid moiety of the acceptor substrate when it exists, although there are several notable exceptions as discussed below.

Glycosidases that remove monosaccharides to form intermediates that are then acted on by glycosyltransferases also play a role in the biosynthesis of some glycan types. These are to be contrasted with glycosidases that are involved in the degradation of glycans (e.g., in lysosomes [Chapter 44]). In addition, glycans can be modified by many other enzyme types, including sulfotransferases, phosphotransferases, O-acetyltransferases, O-methyltransferases, pyruvyltransferases, and phosphoethanolamine transferases.

GLYCOSYLTRANSFERASE SPECIFICITY

Most glycosyltransferases show a high degree of specificity for both their donor and acceptor substrates, and this led Saul Roseman and coworkers to advance the "one enzyme–one linkage" hypothesis. The human B blood group α1-3 galactosyltransferase exemplifies this concept. This enzyme catalyzes a glycosylation reaction in which galactose is added in α-linkage to the C-3 hydroxyl group of a galactose residue on the acceptor substrate (Figure 6.1). However, the enzyme only acts on galactose modified by fucose in α1-2-linkage and prior modification by other monosaccharides, such as α2-6-linked sialic acid, yields a glycan that is not a substrate (Figure 6.1).

We now know that there are instances in which more than one glycosyltransferase can use the same acceptor to make the same linkage. The human fucosyltransferases III–VII, for example, all attach fucose in α1-3-linkage to *N*-acetyllactosamine moieties on glycans (Chapter 14). Examples of relaxed acceptor specificity are provided by the α2-3 sialyltransferases and β1-4 galactosyltransferases that act broadly on β-linked galactose and *N*-acetylglucosamine, respectively. In some rare cases, a single enzyme can catalyze more than one reaction. Human fucosyltransferase III can attach fucose in either α1-3- or α1-4-linkage, and an enzyme called EXTL2 can attach either *N*-acetylgalactosamine or *N*-acetylglucosamine in α-linkage to glucuronic acid (Chapter 17). The β1-4 galactosyltransferase involved in *N*-acetyllactosamine formation shows an unusual flexibility in specificity. When β1-4 galactosyltransferase binds α-lactalbumin (the complex is called lactose synthase), it switches its acceptor specificity from *N*-acetylglucosamine to glucose, which enables the synthesis of lactose and other oligosaccharides during milk production (Chapter 14). Finally, some glycosyltransferases have two separate active sites with different substrate specificities. For example, the enzymes that synthesize the backbones of heparan sulfate (EXT1) and hyaluronan (HAS) have one active site that catalyzes the attachment of *N*-acetylglucosamine to glucuronic acid and another that attaches glucuronic acid to *N*-acetylglucosamine (Chapters 16 and 17). However, the examples described above are all

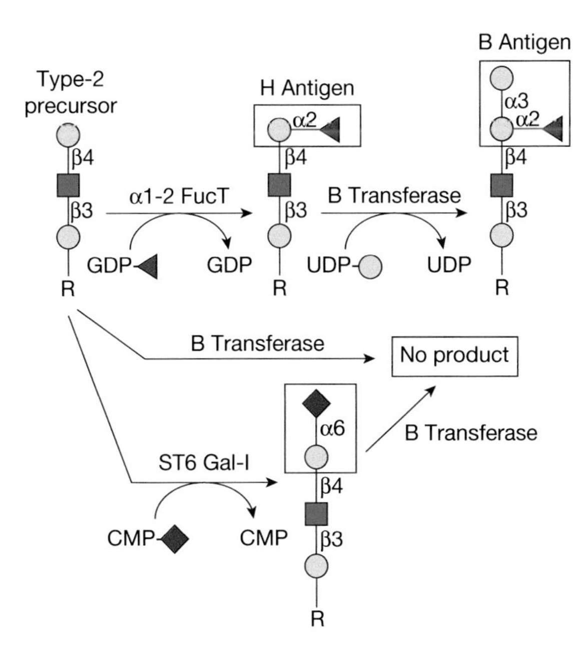

FIGURE 6.1. The strict acceptor substrate specificity of glycosyltransferases is illustrated by the human B blood group α1-3 galactosyltransferase. The B transferase adds galactose in α1-3-linkage to the H antigen (*top middle*) to form the B antigen (*top right*). This enzyme requires the α1-2-linked fucose modification of the H antigen for activity because the B transferase does not add to an unmodified type-2 precursor (*top left*) or precursors modified by sialyl residues (*bottom*) or other monosaccharides (not shown). (For the monosaccharide symbol code, see Figure 1.5.)

exceptions to the generally strict donor, acceptor, and linkage specificity shown by most glycosyltransferases, a property that serves to define and limit the number and type of glycan structures observed in a given cell type or organism.

The glycosyltransferases that transfer monosaccharides or oligosaccharides directly to polypeptide or lipid moieties also show a high degree of substrate specificity and this will be discussed in more detail below for those involving polypeptide. Glycosyltransferases that initiate the synthesis of glycosphingolipids transfer a monosaccharide moiety to what was originally a serine residue in the ceramide lipid precursor of sphingolipids (see Chapter 11). Because different glycolipids have different ceramide moieties, it appears that some glycosyltransferases, such as the sialyltransferases, differentially recognize their substrates based on the nature of the ceramide moiety.

GLYCOSYLTRANSFERASES THAT RECOGNIZE THE PROTEIN MOIETY OF THEIR ACCEPTOR SUBSTRATES

The glycosyltransferases that transfer sugar directly to the polypeptide chain of a protein or glycoprotein recognize their acceptor substrates in a number of different ways. All eukaryotic N-glycans are initiated by oligosaccharyltransferase (OST), generally an ER-resident multisubunit enzyme that transfers en bloc $Glc_3Man_9GlcNAc_2$ to the side chain of asparagine residues in the sequence motif Asn-X-Ser/Thr (where X can be any amino acid except proline; Chapter 9). In contrast, the polypeptide GalNAc transferases, responsible for the initiation of mucin-type O-glycans, act after the protein has been folded and transported to the Golgi (Chapter 10). The polypeptide O-GalNAc transferases do not recognize a specific sequence motif and in general, they transfer *N*-acetyllgalactosamine to the side chain hydroxyl group of serine and threonine residues found in relatively unstructured regions of the folded protein. Some polypeptide O-GalNAc transferases possess a lectin domain that serves to direct the glycosyltransferase to regions of the polypeptide that already possess O-glycan chains. In this way, regions of polypeptide that have a high degree of O-glycan substitution, typical of mucin structures, can be synthesized.

In addition to the O-GalNAc linkage formed by the polypeptide O-GalNAc transferases, a number of other glycosyltransferases can glycosylate the side chain hydroxyl groups of serine and threonine to generate O-GlcNAc, O-Fuc, O-Glc, O-Man, and O-Xyl linkages (Chapters 13, 17, and 19). Specificity for a particular serine or threonine residue is achieved in different ways. The xylosyltransferase that *O*-xylosylates a serine residue in chondroitin and heparan sulfate proteoglycans, for example, has an absolute requirement for a glycine residue carboxy-terminal to the serine and/or more acidic residues in the vicinity of the glycosylation site. In contrast, the O-GlcNAc transferase (OGT), responsible for adding *N*-acetylglucosamine to serine and threonine residues on thousands of nuclear and cytoplasmic proteins (Chapter 19), lacks any obvious consensus sequence associated with acceptor substrate binding specificity. See Table 6.1 for some amino acid–consensus sequences or glycosylation motifs used in the formation of glycopeptide bonds.

The endoplasmic reticulum (ER)-resident O-fucosyltransferases, POFUT1 and POFUT2, that specifically fucosylate epidermal growth factor (EGF)-like domains and thrombospondin type 1 repeats (TSRs), respectively (Chapter 13), differ fundamentally from most other glycosyltransferases. In addition to recognizing a specific sequence motif containing the target serine or threonine residue (Table 6.1), these enzymes only act on EGF-like domains and TSRs that are properly folded and disulfide-bonded. Glycosyltransferases that add O-Glc and O-GlcNAc to other serine and threonine residues in the EGF-like domain also exist and these enzymes also recognize a specific sequence motif and require a folded EGF-like domain.

A number of the glycan-elongating glycosyltransferases that act on glycoproteins also recognize the polypeptide moiety of their acceptor substrates. The glycoprotein hormone GalNAc transferase (a member of the family of β1-4GalNAcTs) provides an interesting example in which modification of an N-glycan is dependent on the presence of the protein sequence motif Pro-X-Arg/Lys positioned several amino acids amino-terminally to the N-glycan being

TABLE 6.1. Amino acid–consensus sequences or glycosylation motifs for the formation of glycopeptide bonds

Glycopeptide bond	Consensus sequence or peptide motif
N-linked:	
GlcNAc-β-Asn	Asn-X-Ser/Thr (X = any amino acid except Pro)
Glc-β-Asn	Asn-X-Ser/Thr
O-linked:	
GalNAc-α-Ser/Thr	repeat domains rich in Ser, Thr, Pro, Gly, Ala with no specific sequence
GlcNAc-α-Thr	Thr-rich domain near Pro residues
GlcNAc-β-Ser/Thr	Ser/Thr-rich domains near Pro, Val, Ala, Gly
	EGF modules (Cys-X-X-G-X-Ser/Thr-G-X-X-Cys)
Man-α-Ser/Thr α	Ser/Thr-rich domains
Fuc-α-Ser/Thr	EGF modules (Cys-X-X-X-X-Ser/Thr-Cys)
	TSR modules (Cys-X-X-Ser/Thr-Cys-X-X-Gly)
Glc-β-Ser	EGF modules (Cys-X-Ser-X-Pro/Ala-Cys)
Xyl-β-Ser	Ser-Gly (in the vicinity of one or more acidic residues)
Glc/GlcNAc-Thr	Rho: Thr-37; Ras, Rac; Cdc42: Thr-35
Gal-Thr	Gly-X-Thr (X = Ala, Arg, Pro, Hyp, Ser) (vent worm)
Gal-β-Hyl	collagen repeats (X-Hyl-Gly)
Ara-α-Hyp	repetitive Hyp-rich domains (e.g., Lys-Pro-Hyp-Hyp-Val)
GlcNAc-Hyp	Skp1: Hyp-143
Glc-α-Tyr	glycogenin: Tyr-194
GlcNAc-α-1-P-Ser	Ser-rich domains (e.g., Ala-Ser-Ser-Ala)
Man-α-1-P-Ser	Ser-rich repeat domains
C-linked:	
Man-α-C-Trp	Trp-X-X-Trp

Modified from Spiro RG. 2002. *Glycobiology* **12:** 43R–56R, by permission of Oxford University Press; see also Table 1.7.

EGF, epidermal growth factor; TSR, thrombospondin repeat; Hyp, hydroxyproline; Hyl, hydroxylysine; GPI, glycosylphosphatidylinositol; Ara, arabinose.

modified. The motif is typically followed closely by additional positively charged residues. The X-ray crystal structure of its acceptor substrate, human chorionic gonadotropin, shows that the Pro-X-Arg/Lys motif is at the beginning of a short surface-exposed helix that also contains the additional positively charged residues (Figure 6.2). The *N*-acetylgalactosamine residue transferred by this enzyme subsequently undergoes a biologically important 4-*O*-sulfation reaction that, in the case of luteinizing hormone and follicle-stimulating hormone, generates a determinant recognized by specific liver clearance receptors that remove them from the blood (Chapter 34). Glycosyltransferases that only elongate the fucose moiety added by POFUT1 or POFUT2 also exist (Chapter 13). The specificity shown by these enzymes stems from their ability to recognize both the fucose moiety and the EGF-like and TSR components of their acceptor substrates. Additional examples of enzymes that elongate glycans on specific glycoprotein substrates include the polysialyltransferases that act on neural cell adhesion molecule (NCAM) and neuropilin-2 (Chapter 15), and the N-acetylglucosaminyltransferase, EXTL3, that adds *N*-acetylglucosamine in α1-4-linkage to glucuronic acid in the first committed step to heparan sulfate biosynthesis on proteoglycans (Chapter 17).

In a variation on this theme, GlcNAc-1-phosphotransferase selectively modifies the N-glycans found on a large family of lysosomal enzymes that differ in three-dimensional structure and that lack any obvious and common protein sequence motif. In this case, modification by this enzyme has been shown to be dependent on lysine residues appropriately spaced and positioned relative to the N-glycan site (Chapter 33). The ER-resident glucosyltransferase, UGGT, is also able to transfer sugar to the N-glycans on a large but unique subset of glycoprotein substrates (Chapter 39). In this case, the enzyme adds a glucose moiety to the N-glycans of misfolded glycoproteins rendering them substrates for the ER-resident lectins calnexin and calreticulin. These lectins in turn recruit folding catalysts that promote correct disulfide bond formation and the *cis–trans* isomerization of proline residues.

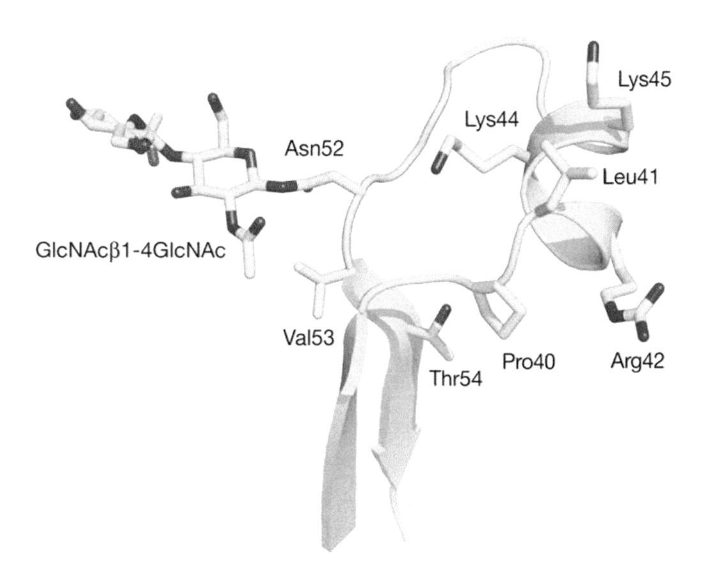

FIGURE 6.2. Human chorionic gonadotropin showing the determinants of recognition used by glycoprotein hormone N-acetylgalactosaminyl (GalNAc) transferase. Ribbon diagram of a fragment (residues 34–58) of human chorionic gonadotropin (PDB [protein data bank] ID 1HRP). The Pro40-Leu41-Arg42 tripeptide and residues Lys44 and Lys45 correspond to residues essential for recognition by glycoprotein hormone GalNAc transferase. Residues Asn52-Val53-Thr54 correspond to the N-glycosylation sequence motif of the N-glycan (at Asn52) modified by the glycoprotein hormone GalNAc transferase. Only the chitobiose core (GlcNAcβ1-4GlcNAc) of the full acceptor N-glycan is shown.

GLYCOSYLTRANSFERASE SEQUENCE FAMILIES AND FOLD TYPES

Approximately 1% of genes in the mammalian genome are involved in the production or modification of glycans. More than 750,000 glycosyltransferase sequences are known across all kingdoms, and the numbers are growing rapidly. Based on sequence analysis, they have been categorized into more than 110 glycosyltransferase families as described in the Carbohydrate-Active Enzymes (CAZy) database (Chapter 8). Although the absence of significant sequence similarity between members of one family and another constitutes the basis for this classification, short sequence motifs common to the members of more than one family have been identified. These sequence elements are typically found among glycosyltransferases with a given donor substrate specificity; the sialyl motifs common to eukaryotic sialyltransferases are a good example (Figure 6.3). Sequence motifs common to galactosyltransferases, fucosyltransferases, and N-acetylglucosaminyltransferases have also been identified. In contrast, the so-called "DXD" motif (Asp–any residue–Asp) is not associated with any particular substrate specificity; this motif is involved in metal ion binding and catalysis as discussed in more detail below.

Despite the large number of sequence families that have been defined, structural analysis has shown that glycosyltransferases possess a limited number of fold types. To date, structures for 262 members representing 59 of the CAZy families have been determined by X-ray crystallography or cryo-electron microscopy (cryo-EM); of these, all but a few possess what have been termed the GT-A, GT-B, GT-C, and lysozyme-type folds (Figure 6.4). The GT-A and GT-B enzymes use nucleotide sugar donor substrates, whereas the GT-C and lysozyme-type enzymes use lipid-linked sugar donors. The GT-A fold glycosyltransferases appear to have evolved from a common ancestor harboring a ~231 amino acid core structure and a conserved set of structural features. The core contains elements of the Rossmann fold, which mediates the interaction with the nucleotide sugar, as well as the generally conserved DXD motif responsible for divalent cation binding (usually Mn^{++} or Mg^{++}) and catalysis. Recent analysis has shown how the insertion of loops into the core structure has facilitated the acquisition, over evolutionary time, of the diverse acceptor specificities shown by the GT-A enzymes. The GT-B enzymes possess two distinct domains, and although the carboxy-terminal domain is primarily responsible for binding the nucleotide sugar donor substrate, both domains possess elements of the Rossmann fold. Acceptor substrates are typically bound in the cleft between the two domains, and unlike the GT-A enzymes, the GT-B glycosyltransferases are metal-ion-independent and do not possess a DXD motif. The GT-C fold enzymes are multi-spanning integral membrane proteins characterized by their use of lipid-linked sugar donors. Recent X-ray and cryo-EM structures are now providing important insights into how these enzymes mediate substrate binding and

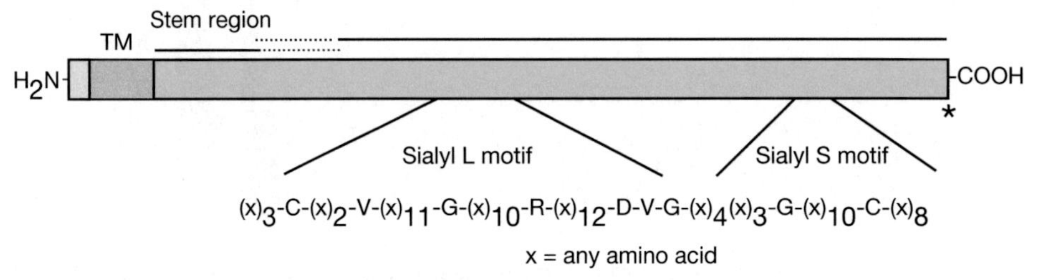

FIGURE 6.3. Sialyl motifs. Domain structure of a typical sialyltransferase, showing the sialyl motifs shared by this family of enzymes. The sialyl L motif of 48–49 amino acids shares significant similarity among members and may be up to 65% identical in amino acid sequence. The sialyl S motif is smaller (~23 amino acids) and diverges more among members of the family, with only two absolutely conserved residues. In both cases, identical residues are indicated, and residues showing similarity are denoted by parentheses. (*Asterisk*) Position of a highly conserved sequence H-X_4-E. Additional conserved motifs may exist as well.

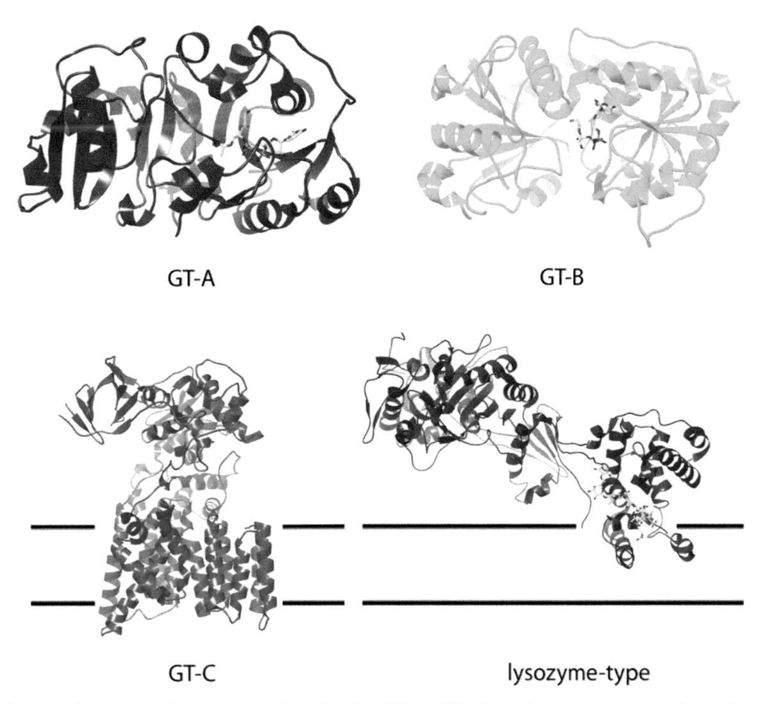

FIGURE 6.4. Ribbon diagrams of representative GT-A, GT-B, GT-C, and lysozyme-type fold glycosyltransferases. The GT-A and GT-B structures correspond to those of rabbit β1-2 N-acetylglucosaminyltransferase I (PDB ID 1FOA) and T4 phage β-glucosyltransferase (PDB ID 1J39), respectively. In both cases, the bound nucleotide sugar donor substrate is shown in stick representation. The GT-C structure is that of *Campylobacter lari* oligosaccharyltransferase, PglB (PDB ID 3RCE), and the lysozyme-type structure is that of the *Staphylococcus aureus* bacterial peptidoglycan glycosyltransferase (PDB ID 2OLV) in complex with moenomycin (stick representation).

catalysis (Figure 6.4). Notable among them are the structures of both prokaryotic (PglB and AglB) and eukaryotic (yeast and human) oligosaccharyltransferases that have provided a model for N-glycosylation, a modification conserved across all three domains of life. Structural comparison among the GT-C glycosyltransferases has identified a conserved core of transmembrane helices involved in binding their polyisoprenoid(pyro)phosphate-sugar donors and a variable number of transmembrane- and nonmembrane-associated segments involved in acceptor interactions. The lysozyme-type glycosyltransferases (CAZy family 51) are involved in the biosynthesis of bacterial peptidoglycan and use lipid II (GlcNAc-MurNAc(pentapeptide)-undecaprenyl pyrophosphate) as donor substrate (Chapter 21). In addition to the lysozyme-type domain, these glycosyltransferases possess the jaw subdomain, which embeds in the extracellular face of the cytoplasmic membrane and provides access to the lipid II substrate. In a very recent advance, it has been shown that the dual activity glycosyltransferase-phosphorylases, important for the turnover of the storage polysaccharide mannogen and virulence in *Leishmania* parasites, possess a β-propeller fold catalytic domain not previously observed for a glycosyltransferase.

GLYCOSIDASES

Glycosidases are a very large group of enzymes with more than 870,000 members falling into more than 170 CAZy database families (Chapter 8). Unlike the glycosyltransferases, members of this family have evolved independently many times, a fact reflected in the diverse array of three-dimensional structures observed for these enzymes. Glycosidases play important roles in the degradation of glycan structures for the uptake and metabolism of sugars and for the turnover of glycoconjugates in various cellular processes. Glycosidases are also involved in the formation of intermediates that are used as substrates for glycosyltransferases in the biosynthesis of glycans. The use of glycosidases in this way is particularly important in the biosynthesis of N-glycan-containing glycoproteins in more evolutionarily advanced eukaryotes and is thought to be associated with the acquisition of complex N-glycans during the evolution of multicellular organisms. In this case, the nascent glycoprotein glycan, $Glc_3Man_9GlcNAc_2$-Asn, is trimmed by glucosidases and mannosidases in the ER and Golgi to generate substrates for the glycosyltransferases that lead to elaborated complex and hybrid-type N-glycans (Chapter 9). Glucosidase II, one of the two ER-resident glucosidases involved, also works in conjunction with UGGT during glycoprotein folding to allow repeated deglucosylation/reglucosylation during what has been termed the calnexin/calreticulin quality control cycle (Chapter 39). An interplay between glycosylation and deglycosylation is also found to occur with the O-GlcNAc moiety that appears in many nuclear and cytoplasmic proteins (transferred by O-GlcNAc transferase; Chapter 19). In this case, removal of the N-acetylglucosamine moiety by the glycosidase, O-GlcNAcase, provides a means of dynamically regulating the extent of O-GlcNAcylation and the diverse processes that it mediates.

CATALYTIC AND KINETIC MECHANISMS

Glycosyltransferases catalyze their reactions with either inversion or retention of stereochemistry at the anomeric carbon atom of the donor substrate (Figure 6.5). For example, β1-4 galactosyltransferase, an inverting glycosyltransferase, transfers galactose from UDP-α-Gal to generate a β1-4-linked galactose-containing product. Inversion of stereochemistry follows from the fact that the enzyme uses an S_N2 (substitution nucleophilic bimolecular) reaction mechanism in which an acceptor hydroxyl group attacks the anomeric carbon atom of UDP-Gal from one side and UDP leaves from the other (Figure 6.5A). Typically, enzymes of this type possess an aspartate, glutamate, or histidine residue whose side chain serves to partially deprotonate the incoming acceptor hydroxyl group (acting as a general base), rendering it a better nucleophile (as shown in Figure 6.6 for the β1-4 galactosyltransferase). Additional structures for GT-B fold inverting glycosyltransferases have been identified in which such a base appears to be lacking. In these cases, a water-mediated proton-shuttle mechanism has been proposed to achieve the required acceptor deprotonation. In addition, these enzymes possess features that help to promote leaving-group departure. In the GT-A enzymes, a metal ion, bound by the DXD motif, is typically positioned to interact with the diphosphate moiety. The positively charged metal ion serves to electrostatically stabilize the additional negative charge that develops on the terminal phosphate moiety of the UDP leaving group during breakage of the sugar-phosphate bond of the donor substrate (Figure 6.6). In the few GT-A enzymes that are not metal-ion-dependent, positively charged side chains stabilize the leaving group, a strategy also used by some of the GT-B-fold enzymes.

Although the mechanisms used by retaining glycosyltransferases remain open, initial catalytic models based on glycosidase mechanisms (see below) proposed a double-displacement mechanism (Figure 6.5B). In this case, a nucleophile (e.g., an aspartic acid or glutamic acid side chain) in the enzyme active site makes the first attack with inversion of anomeric configuration, followed by a second attack (on a glycosyl-enzyme intermediate) and inversion, to give overall retention of stereochemistry. However, no enzyme-associated catalytic nucleophile has yet been identified for a retaining glycosyltransferase and attempts to trap and study glycosyltransferase covalent reaction intermediates have not yet clearly supported this mechanism. Thus, while a "double-displacement" mechanism cannot yet be discounted and/or may be limited to specific families of glycosyltransferases, other mechanisms have been proposed. In several cases,

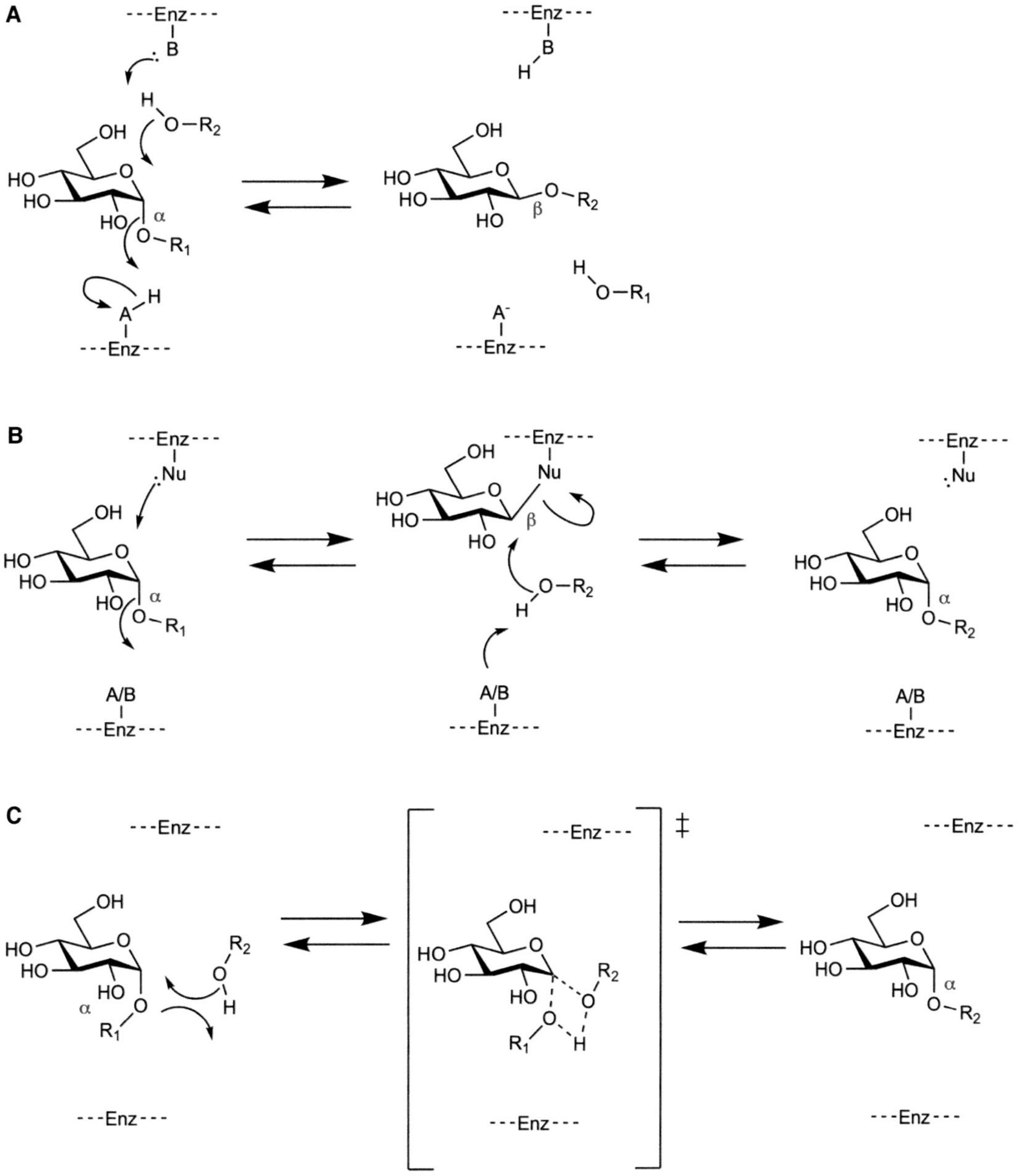

FIGURE 6.5. Schematic representation of inverting and retaining catalytic mechanisms. (*A*) S_N2-like attack of the acceptor leads to inversion of stereochemistry at C1 of the donor sugar (the anomeric carbon). For a glycosidase reaction, R_2 would correspond to a proton (i.e., attack by water as a nucleophile) and R_1 would be the remainder of the glycoside (the so-called aglycone). A and B label general acid and base groups in the catalytic site of the enzyme. For a glycosyltransferase reaction, R_2 would correspond to the remainder of the acceptor substrate and R_1 would typically be the nucleoside monophosphate or diphosphate moiety of the donor substrate. For the latter, a general acid/base residue is typically not required and other mechanisms that assist leaving-group departure operate (see text). (*B*) This mechanism has only been established for glycosidases. The so-called "double-displacement" mechanism involving two successive S_N2-like reactions separated by a glycosyl-enzyme intermediate leads to retention of the configuration at anomeric C1. For a glycosidase reaction, R_1 corresponds to the remainder of the glycoside/aglycone and R_2 to a proton. For a glycosyltransferase, R_1 corresponds to the leaving group and R_2 to the acceptor. Nu labels a residue that could act as a nucleophile for the first S_N2. (*C*) An S_Ni or S_Ni-like dissociative, same/front-side mechanism in which the leaving group "guides" the nucleophilic acceptor to the same face of the anomeric carbon with resulting retention of anomeric configuration. Substrate-assisted acceptor deprotonation by the β-phosphate of the donor leaving group has been proposed. For a glycosyltransferase, R_1 corresponds to the leaving group (and so the oxygen that guides is a phosphate) and R_2 corresponds to the acceptor.

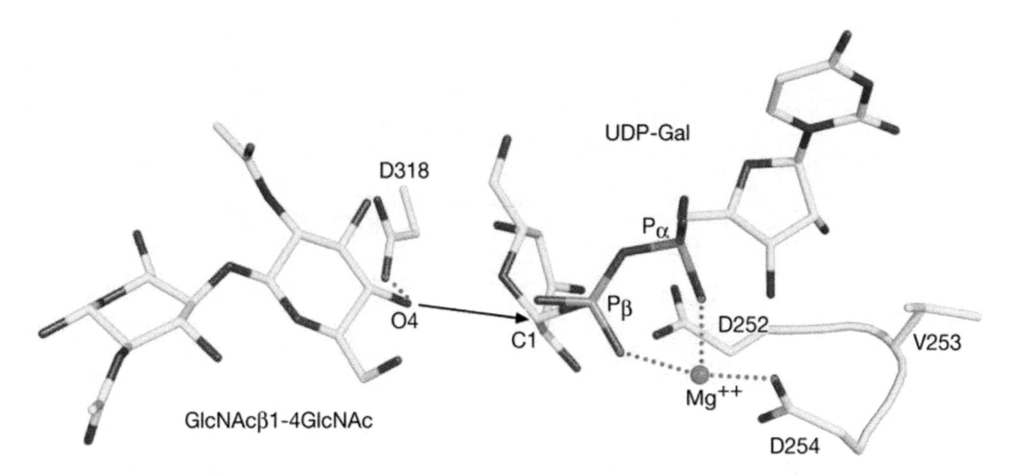

FIGURE 6.6. Catalytic site of bovine β1-4 galactosyltransferase. Composite figure shows selected residues/atoms of the superimposition of the donor complex (PDB ID 1TW1) on the acceptor complex (PDB ID 1TW5). O4 designates the C4 hydroxyl group of the GlcNAcβ1-4GlcNAc acceptor substrate positioned for in-line S_N2 attack (*arrow*) on C1 of the UDP-Gal donor substrate. The carboxylate form of D318 serves to act as a base to partially deprotonate the C4 hydroxyl group rendering it a better nucleophile. The positively charged Mg^{++} ion coordinates the two phosphates of the UDP leaving group, promoting cleavage of the $C1-OP_\beta$ bond by stabilizing the additional negative charge that develops on P_β of the leaving group. D252-V253-D254 corresponds to the DXD motif in bovine β1-4 galactosyltransferase.

retaining glycosyltransferases appear to position the hydroxyl nucleophile of the acceptor adjacent to the β-phosphate group of the donor nucleotide. Substrate-assisted acceptor deprotonation by the donor phosphate has been proposed with a dissociative, same/front-side mechanism in which the leaving group "guides" the nucleophilic acceptor to the same face of the anomeric carbon with resulting retention of anomeric configuration for the resulting glycosidic bond (Figure 6.5C). This S_Ni or S_Ni-like mechanism is supported by computational modeling and, in specific cases, by kinetic analyses and X-ray crystal structures of transition-state analog inhibitor complexes.

On the basis of much structural and enzyme kinetic analysis, it is well established that inverting glycosidases proceed via a single S_N2 displacement mechanism (Figure 6.5A), whereas retaining glycosidases use a double-displacement mechanism involving a covalent glycosyl-enzyme intermediate (Figure 6.5B). In the double-displacement mechanism, an aspartic acid or glutamic acid side chain in the enzyme active site makes the first attack and inversion, followed by a second water-mediated attack (on the glycosyl-enzyme intermediate) and inversion, to give overall retention of stereochemistry. Using mechanism-based inhibitors, the glycosyl-enzyme intermediate has been trapped and studied by X-ray crystallography for a number of glycosidases.

Many glycosyltransferases have been shown to possess a Bi-Bi sequential kinetic mechanism in which the donor substrate binds before the acceptor substrate, and the glycosylated acceptor is released before the nucleoside monophosphate or diphosphate, depending on the reaction. Such kinetics are readily explained by a structural model in which the active site represents a deep pocket, with the nucleotide sugar substrate at the bottom and the acceptor substrate stacked on top. If the acceptor substrate was to bind first, it would sterically preclude donor substrate binding and, as such, catalysis would not occur. Because of the stacked arrangement, it also follows that release of the glycosylated product must precede release of the nucleotide leaving group. Although largely consistent with such a model, the X-ray crystal structures of glycosyltransferase-substrate complexes also show that substrate-dependent ordering of flexible loops is a feature common to glycosyltransferases. Typically, donor substrate binding orders a loop(s) that in turn facilitates acceptor substrate binding. Moreover, loop ordering may serve to exclude bulk water from the active site, a strategy proposed to be used by most enzymes to create an active site environment that reduces the energy of the transition state and promotes catalysis. For glycosyltransferases, this would also serve to remove waters as a competing hydrolytic nucleophile, thereby favoring the acceptor substrate.

SULFATION AND OTHER MODIFICATIONS

The sulfotransferases are a large family of enzymes found in both the cytoplasm and the Golgi. Sulfotransferases play a particularly important role in the production of glycosaminoglycans (Chapter 17) and the formation of L-selectin ligands, glycans, and glycoconjugates required for the trafficking of lymphocytes across high-endothelial venules in lymph nodes (Chapter 34). All sulfotransferases use 3′-phospho-adenosine-5′-phosphosulfate (PAPS) as the sulfate donor (Chapter 5). Although the sequence similarity among sulfotransferases can be very low, they all possess conserved sequence motifs responsible for binding the 5′ and 3′ phosphate groups of PAPS. Moreover, structural analysis has shown that all of the sulfotransferases solved, to date, possess the same basic structure. Mechanistically, the enzyme proceeds via an S_N2-like reaction, with the hydroxyl group of the acceptor making an in-line attack on the sulfate group, although it should be understood that substitution at sulfur may be mechanistically very different to that at carbon. Structural studies and mutagenesis have shown that a histidine residue serves to activate the hydroxyl nucleophile and that a lysine residue assists in stabilizing the PAP leaving group. Interestingly, a conserved serine residue seems to be involved in modulating the activity of these enzymes to prevent PAPS hydrolysis in the absence of acceptor substrate, another example of the need for enzymes to control the role of the competing water.

Phosphorylation of sugar residues by ATP-dependent kinases occurs at C-2 of *O*-xylose in proteoglycans (Chapter 17) and at the C-6 position of *O*-mannose in α-dystroglycan after the mannose has been further glycosylated (Chapter 13). Phosphoglycosylation (or glycophosphorylation), a process in which a sugar phosphoryl is transferred from a nucleotide sugar donor directly to a serine residue on a protein (e.g., GlcNAc-P-serine and Man-P-serine), occurs in *Dictyostelium* and *Leishmania*, respectively. In eukaryotic cells, mannose-6-phosphate on the N-glycans of lysosomal enzymes occurs by a two-step process involving UDP-GlcNAc as the phosphate donor. In the first step, GlcNAc-1-phosphotransferase uses UDP-GlcNAc to generate GlcNAcα1-P-6-Manα1-(N-glycan), and in the second step the *N*-acetylglucosamine moiety is removed by a second enzyme to give P-6-Manα1-(N-glycan) (Chapter 33).

Glycan *O*-acetylation occurs in bacteria and plants and the *O*-acetylation of sialic acids is found in bacteria, parasites, and vertebrates. The structure of a plant acetyltransferase suggests the use of a catalytic triad and a double-displacement mechanism for acetyl group transfer analogous to esterases and serine proteases. CASD1, the sialate O-acetyltransferase (SOAT) in humans is a multi-spanning membrane protein with a globular extramembranous catalytic domain. The enzyme transfers the acetyl group from acetyl-CoA to CMP-Neu5Ac. As such, it is the sialytransferase-mediated delivery of 9-*O*-acetylated sialic to acceptor substrates that leads to the 9-*O*-acetylated glycoconjugates observed. 9-*O*-acetylated sialic acid plays a role in CD22 signaling, a process regulated by the opposing action of a sialate 9-*O*-acetylesterase (SIAE). It also serves as receptor or coreceptor for influenza C and a number of coronaviruses. *N*-Deacetylation of *N*-acetylglucosamine residues to glucosamine occurs during heparin/heparan sulfate formation (Chapter 17), lipopolysaccharide assembly (Chapters 21 and 22), and GPI-anchor synthesis (Chapter 12). The bacterial enzyme is zinc-dependent, but in-depth studies of the vertebrate N-deacetylases have not been performed. *N*-Deacetylation of *N*-acetylneuraminic acid (the most common sialic acid) has also been reported (Chapter 15).

Finally, glycans can be modified in many other ways, including pyruvylation (e.g., in the formation of *N*-acetylmuramic acid; Chapters 21 and 22), the addition of ethanolamine phosphate (e.g., during GPI-anchor synthesis; Chapter 12), and alkylation, deoxygenation, and halogenation in microbial glycans. All of these reactions are catalyzed by unique transferases or oxidoreductases and represent areas of active research.

ACKNOWLEDGMENTS

The authors appreciate helpful comments and suggestions from David Vocadlo, Mike Ferguson, Susan Bellis, and Tetsuya Okajima.

FURTHER READING

Roseman S. 2001. Reflections on glycobiology. *J Biol Chem* **276**: 41527–41542. doi:10.1074/jbc.R100053200

Lairson LL, Henrissat B, Davies GJ, Withers SG. 2008. Glycosyltransferases: structures, functions, and mechanisms. *Annu Rev Biochem* **77**: 521–555. doi:10.1146/annurev.biochem.76.061005.092322

Schwarz F, Aebi M. 2011. Mechanisms and principles of N-linked protein glycosylation. *Curr Opin Struct Biol* **21**: 576–582. doi:10.1016/j.sbi.2011.08.005

Bennett EP, Mandel U, Clausen H, Gerken TA, Fritz TA, Tabak LA. 2012. Control of mucin-type O-glycosylation: a classification of the polypeptide GalNAc-transferase gene family. *Glycobiology* **22**: 736–756. doi:10.1093/glycob/cwr182

Hurtado-Guerrero R, Davies GJ. 2012. Recent structural and mechanistic insights into post-translational enzymatic glycosylation. *Curr Opin Chem Biol* **16**: 479–487. doi:10.1016/j.cbpa.2012.10.013

Albesa-Jové D, Giganti D, Jackson M, Alzari PM, Guerin ME. 2014. Structure–function relationships of membrane-associated GT-B glycosyltransferases. *Glycobiology* **24**: 108–124. doi:10.1093/glycob/cwt101

Janetzko J, Walker S. 2014. The making of a sweet modification: structure and function of O-GlcNAc transferase. *J Biol Chem* **289**: 34424–34432. doi:10.1074/jbc.R114.604405

Speciale G, Thompson AJ, Davies GJ, Williams SJ. 2014. Dissecting conformational contributions to glycosidase catalysis and inhibition. *Curr Opin Struct Biol* **28**: 1–13. doi:10.1016/j.sbi.2014.06.003

Moremen KW, Haltiwanger RS. 2019. Emerging structural insights into glycosyltransferase-mediated synthesis of glycans. *Nat Chem Biol* **15**: 853–864. doi:10.1038/s41589-019-0350-2

Bai L, Li H. 2020. Protein N-glycosylation and O-mannosylation are catalyzed by two evolutionarily related GT-C glycosyltransferases. *Curr Opin Struct Biol* **68**: 66–73. doi:10.1016/j.sbi.2020.12.009

Taujale R, Venkat A, Huang L-C, Yeung W, Rasheed K, Edison AS, Moremen KW, Kannan N. 2020. Deep evolutionary analysis reveals the design principles of fold A glycosyltransferases. *eLife* **9**: e54532. doi:10.7554/eLife.54532

7 Biological Functions of Glycans

Pascal Gagneux, Thierry Hennet, and Ajit Varki

This chapter provides an overview of the biological functions of glycans in three broad categories: structural roles in, on, and outside cells; energy metabolism, including nutrient storage and sequestration; and information carriers, that is, specific recognition—most commonly by glycan-binding proteins of intrinsic or extrinsic origin. The chapter then presents some general principles for understanding and further exploring biological functions of glycans. For details, see the sources cited and other chapters in this book.

GENERAL PRINCIPLES

As with other major classes of macromolecules, the biological functions of glycans span the spectrum from relatively subtle to crucial for development, growth, maintenance, or survival

published by Cold Spring Harbor Laboratory Press; doi: 10.1101/glycobiology.4e.07

of the organism that synthesizes them. For many glycans, specific functions are not yet evident. The same glycan may also have different functions depending on which aglycone (protein or lipid) it is attached to. Over the years, many theories have been advanced regarding biological roles of glycans. Although there is evidence to support all the theories, exceptions to each one can also be found. This should not be surprising, given the abundance and enormous diversity of glycans in nature. Complexities also arise because glycans are frequently bound by microbes and microbial toxins, making them a liability to the organism that synthesizes them.

Biological functions of glycans can be divided into three broad categories: (1) structural contributions (e.g., extracellular scaffolds: cell walls and extracellular matrices, protein folding and function); (2) energy metabolism (e.g., carbohydrates as carbon sources for storage and manipulation of animal behavior: pollination and seed dispersal) ; and (3) information carriers (e.g., molecular patterns recognized by glycan binding proteins [GBPs]) (Figure 7.1). The GBPs can be subdivided into two groups: (1) intrinsic GBPs, which recognize glycans from the same organism and (2) extrinsic GBPs, which recognize glycans from another organism. Intrinsic GBPs typically mediate cell–cell interactions or recognize extracellular molecules, but they can also recognize glycans on the glycocalyx of same cell. Extrinsic GBPs comprise pathogenic microbial adhesins, agglutinins, or toxins, which evolved for host colonization or invasion, but also include proteins that mediate symbiotic relationships or host defense directed at microbial glycans. Intrinsic and extrinsic glycan recognition can also act as opposing selective forces, simultaneously constraining and driving evolutionary change respectively (Chapter 20), likely accounting for the enormous diversity of glycans in nature. Further diversity arises because microbial pathogens also engage in "molecular mimicry," evading immune reactions by decorating themselves with glycans typical of their hosts, and even in "glycan gimmickry,"

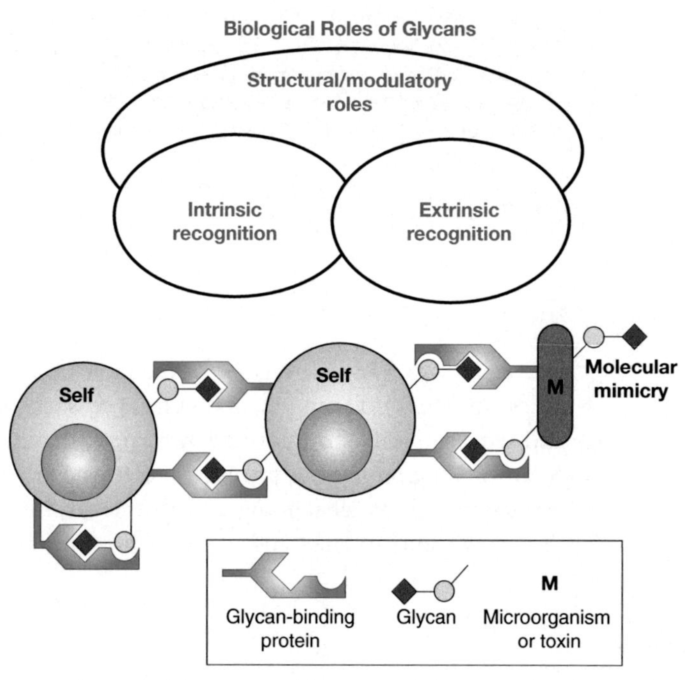

FIGURE 7.1. General classification of the biological functions of glycans. A simplified and broad classification is presented, emphasizing the roles of organism-intrinsic and -extrinsic glycan-binding proteins in recognizing glycans. There is some overlap between the groups (e.g., some structural properties involve specific recognition of glycans). In the *lower* part of the figure, intrinsic recognition is represented by the binding shown on the *left* of the central "self" cell and extrinsic recognition is represented by the binding shown to the *right* of that cell. (Modified and redrawn, with permission of Oxford University Press, from Gagneux P, Varki A. 1999. *Glycobiology* **9:** 747–755.)

modulating host immunity toward increased tolerance. Finally, most microbes are also themselves targets of pathogens that use microbial glycans for attachment and infection (e.g., bacteriophages that invade bacteria).

Other general principles emerge based on existing literature. The biological consequences of experimentally altering glycosylation in various systems seem to be highly variable and unpredictable. Also, a given glycan can have different roles in different tissues, at different times in development (organism-intrinsic functions) or in different environmental contexts (organism-extrinsic functions). As a broad generalization, it can be stated that terminal sequences, unusual structures, and modifications of glycans are more likely to mediate specific biological functions within the organism. However, such glycans or their modifications are also more likely to be targets for pathogens and toxins. Perhaps as a consequence, intra- and interspecies variations in glycosylation are relatively common, and at least some of the diversity of glycans in nature may represent signatures of past or current host–pathogen or symbiotic interactions (Chapter 20). Finally, genetic defects in glycosylation are easily obtained in cultured cells, but often have limited biological consequences in vitro. In contrast, the same defects often have major and even catastrophic consequences in whole organisms. This generalization indicates that many major functions of glycans are operative mainly within an intact, multicellular organism. Some of these principles are briefly discussed below.

BIOLOGICAL CONSEQUENCES OF EXPERIMENTALLY ALTERING GLYCOSYLATION ARE VARIABLE

Experimental approaches to elucidating biological roles of glycans include prevention of initial glycosylation, prevention of glycan chain elongation, alteration of glycan processing, enzymatic or chemical deglycosylation of completed chains, genetic elimination of glycosylation sites, addition of unnatural monosaccharides, and studies of mutant and naturally occurring genetic variants in glycosylation enzymes. The consequences of such manipulations can range from being essentially undetectable, to the complete loss of particular functions, or even to loss of the entire glycoconjugate bearing the altered glycan. Even within a particular class of molecules (e.g., cell-surface receptors), the effects of altering glycosylation are variable and unpredictable. Moreover, the same glycosylation change can have markedly different effects in different cell types or when studied in vivo or in vitro. The effect may depend on the structure of the glycan, the biological context (including interactions with cognate glycan receptors), GBPs, and the specific biological function. Given all of the above considerations, it is difficult to predict the functions that a given glycan on a given glycoconjugate might mediate and its relative importance to the organism.

STRUCTURAL FUNCTIONS OF GLYCANS

Glycans have many protective, stabilizing, organizational, and barrier functions. The glycocalyx that covers all eukaryotic cells and the polysaccharide coats of various prokaryotes represent a substantial physical barrier. Cellulose provides the material for cell walls in plants, and chitin fulfills this role in fungi. Arthropods evolved resistant chitin-rich exoskeletons. Cellulose and chitin represent the two most abundant biopolymers on the planet. In multicellular organisms, glycan constituents of matrix molecules, such as proteoglycans, are important for maintenance of tissue structure, porosity, and integrity. Such molecules can also contain binding sites for other specific glycans, which, in turn, aid overall matrix organization. The external location of glycans on most glycoproteins can provide a general shield, protecting the underlying polypeptide from recognition by proteases (mucins), blocking antibody binding (viral glycoproteins), and even (as in the case of mucins) protecting entire tissue surfaces from microbial attachment.

Another structural role of glycans is their involvement in folding of newly synthesized polypeptides in the endoplasmic reticulum (ER) and/or in the subsequent maintenance of protein solubility and conformation (Chapter 39). Indeed, when some proteins are incorrectly glycosylated, they can fail to fold properly and/or to exit the ER. Such misfolded glycoproteins are translocated to the cytoplasm for degradation in proteasomes. Conversely, there are examples of glycoproteins whose synthesis, folding, trafficking, sensitivity to proteolysis, or immune recognition seem unaffected by altering their glycosylation. Moreover, inhibitors (Chapter 55) or genetic mutations (Chapter 45) that only affect later steps of glycan processing often do not interfere with basic structural functions. Although structural functions of glycans are obviously of great importance to the intact organism, they do not explain the evolution of such a diverse and complex range of molecules.

A further structural function of glycans is to act as a protective storage depot ("sponge") for biologically important molecules. For example, many heparin-binding growth factors (Chapters 17 and 38) bind glycosaminoglycan (GAG) chains of the extracellular matrix, adjacent to cells that need to be stimulated (e.g., in the basement membrane underlying epithelial and endothelial cells). This prevents diffusion of factors away from the site of secretion (sometimes generating morphogenic gradients), protects them from nonspecific proteolysis, prolongs their active lives, and allows them to be released under specific conditions. Likewise, GAG chains in secretory granules can bind and protect protein contents of the granule and modulate their functions. There are several other instances in which glycans act as sinks or depots for biologically important molecules, such as water, ions, and immune regulatory proteins. The asymmetry in localization of most glycans on the outer plasma membrane leaflet can contribute to a signal for the intracellular rupture of vesicles, wherein exposed glycans are detected by cytoplasmic lectins (e.g., galectins) and trigger autophagy.

Polymerized glycans such as starch and cellulose (in plants) and glycogen (in animals) can serve a major role in nutrient storage and sequestration. The relative resistance to digestibility of the internal linkages of such polymers by enzymes of intrinsic or extrinsic origin determines their relative role as structural components or as a mechanism for nutrient sequestration. Flowering plants have evolved important glycan-based strategies for exploiting animal behavior by making available sucrose-rich rewards to pollinators and seed dispersers (nectar and fruit). Many animal species have secondarily evolved strategies to store collected nectar (bees and honey pot ants). Flowering plants have also evolved countless ways of storing starch and other polysaccharide nutrients in the endosperm of their seeds, as a way to give the next generation a head start.

GLYCANS AS INFORMATION CARRIERS: SPECIFIC LIGANDS FOR CELL–CELL INTERACTIONS (INTRINSIC RECOGNITION)

The first intrinsic animal glycan receptors (GBPs) identified in vertebrates were those mediating clearance, turnover, and intracellular trafficking of soluble blood-plasma glycoproteins (Chapters 33 and 34), specifically recognizing certain terminal or subterminal glycans on the circulating glycoprotein. However, even the most precise examples, such as the role of mannose 6-phosphate (Man-6-P) in the intracellular trafficking of lysosomal enzymes to lysosomes (Chapter 33), feature exceptions. For instance, Man-6-phosphorylation is not absolutely required for trafficking of lysosomal enzymes in certain cell types, nor is it operative in some "lower" eukaryotes. There are also endocytic receptors recognizing specific glycan sequences, whose functions have yet to be assigned. Several instances exist wherein free glycans can have hormone-like actions that induce specific responses in a structure-specific manner. Examples include interaction of small glycans (lipochitooligosaccharides) from bacterial and fungal symbionts with GBPs of plant roots (Chapter 40) and bioactive properties of hyaluronan fragments recognized by GBPs in mammalian systems (Chapter 16), both of which induce biological responses in a size- and structure-dependent manner. Likewise, free heparan or dermatan sulfate fragments released by certain cell types can have biological effects in situations such as wound healing. In many of these instances, the putative GBP receptors for these molecules and their precise mechanisms are still being defined.

It is now clear that glycans have many specific information-bearing biological roles in cell–cell recognition and cell–matrix interactions. A well-characterized example is the selectin family of adhesion molecules. These cell-surface proteins recognize glycans on ligands and mediate critical interactions between and amongst blood cells and vascular cells in a variety of normal and pathological situations but are also involved in implantation of embryos on the endometrium (Chapter 34). Glycans and GBPs on cell surfaces can interact specifically with molecules in the matrix or even with glycans on the same cell surface (in *cis*). Some critical recognition sites are actually combinations of glycans and protein. For example, P-selectin recognizes the generic selectin ligand sialyl Lewis x with high affinity only in the context of the amino-terminal 13 amino acids of P-selectin glycoprotein ligand-1 (PSGL-1), which include required sulfated tyrosine residues (Chapter 34). Glycan-binding sites of cell-surface receptors like Siglecs can be masked/ “blocked” by cognate glycans on the same cell surface, making them unavailable for recognition of external ligands (Chapter 35). On the other hand, some glycans can act as “biological masks,” preventing recognition of underlying residues (e.g., sialic acids can mask recognition of underlying β-galactosides by galectins or other GBPs).

Carbohydrate–carbohydrate interactions can also have specific biological roles. A classic example is the species-specific interaction between marine sponges, mediated via homotypic binding of glycans on a large cell-surface glycoprotein, perhaps indicating the critical importance of glycosylation for the evolution of multicellularity. Another example is the compaction of the mouse embryo at the morula stage, which seems to be facilitated by a Lewis x–Lewis x interaction. The single-site affinities of such interactions are often weak and thus difficult to measure. However, if the molecules are present in very high copy numbers, a large number of relatively low-affinity interactions can collaborate to produce a high-avidity “Velcro” effect that is sufficient to mediate biologically relevant interactions.

Glycosylation can also modulate interactions of proteins with one another. Some growth factor receptors acquire their binding abilities in a glycosylation-dependent manner while in transit through the Golgi apparatus. This may limit unwanted early interactions of a newly synthesized receptor with a growth factor that is synthesized in the same cell. Glycosylation of a polypeptide can also mediate an on–off switching effect. For example, when the hormone β-human chorionic gonadotrophin is deglycosylated, it still binds to its receptor with similar affinity, but it fails to stimulate adenylate cyclase. In most instances, the effects of glycosylation are incomplete; that is, glycosylation appears to be “tuning” a primary function of the protein rather than turning it on or off. For example, the activity of some glycosylated growth factors and hormones can be modulated over a wide range by the extent and type of their glycosylation. Otherwise identical antibodies can mediate differential effects depending on the precise structure(s) of the N-glycan in their Fc domain. This becomes particularly evident when recombinant glycoprotein molecules are produced in biotechnology, bearing different glycosylation patterns based on the evolutionary history of the cell expression system in use (bacterial, yeast, insect, nonhuman mammal, or human).

The intricate modulation of Notch–ligand interactions by the Fringe glycosyltransferase, whereby *cis* and *trans* ligand interactions are tuned, is a prime example of the importance of informational content of specific glycan modification for development (Chapter 13). Another striking example is the role of polysialic acid (polySia) chains on the neural cell adhesion molecule (NCAM). This adhesion receptor normally mediates homophilic binding between neuronal cells. In the embryonic state, or in other states of neural “plasticity,” polySia chains tend to be long, thereby interfering with homophilic binding (Chapter 15). There are also instances wherein functions can be tuned by glycans attached to other neighboring structures. For example, polySias of embryonic NCAM can interfere with interactions of other unrelated receptor–ligand pairs, simply by physically separating the cells. Also, tyrosine phosphorylation of the epidermal growth factor (EGF) and the insulin receptors can be modulated by endogenous cell-surface gangliosides, possibly by organizing them into membrane microdomains (Chapter 11). Although the precise mechanisms of the latter effects are uncertain, specificity is implied by the requirement for a defined glycan sequence on the ganglioside. Because most of such "tuning" effects of glycans are partial, their overall importance tends to be questioned. However, the sum total of several such partial effects can result in a dramatic effects on the final biological outcome. Thus, glycosylation appears to be a mechanism for generating important functional diversity from the limited set of basic receptor–ligand interactions that are possible, when using gene products from a typical genome. Of course, as with most other glycan functions, exceptions can be found. There are many receptors whose ligand binding is independent of glycosylation and many peptide ligands whose binding and action are not obviously affected by glycosylation.

GLYCANS AS SPECIFIC LIGANDS FOR CELL–MICROBE INTERACTIONS (EXTRINSIC RECOGNITION)

Many glycans are specifically recognized and engaged by various viruses, phages, bacteria, and parasites and also targeted by many toxins (Chapter 37). Given rapid evolution of pathogens and ongoing selection, there is typically excellent recognition specificity for the sequence of the glycans involved. For example, the hemagglutinins of many viruses specifically recognize the type of host sialic acid, its modifications, and its glycosidic linkage to the underlying sugar chain. Likewise, various toxins bind with great specificity to certain gangliosides but not to the same glycan epitopes presented on proteins (Chapters 11 and 37). There is little doubt about the importance of structural specificity with respect to these functions of glycans. Indeed, some of the microbial binding proteins involved have been harnessed as molecular probes for studying the expression of their cognate glycans. However, providing such signposts to aid the success of pathogenic microorganisms has little obvious value to the organism that synthesized such glycans. To counter such deleterious consequences, some organisms have also evolved the ability to mask or modify glycans recognized by microorganisms or toxins. Meanwhile, glycan sequences on soluble glycoconjugates, such as secreted mucins, act as decoys for microorganisms and parasites. Thus, a pathogenic organism or toxin evolving to bind to mucosal cell membranes may first encounter its cognate ligand attached to a soluble mucin, which can then be washed away, removing the danger to cells underneath. In contrast, instances occur in which symbiosis is mediated by specific glycan recognition, such as some commensal bacteria in the gut lumen of animals and the bacteria involved in forming plant root nodules in nitrogen fixing legumes or mycorrhizal symbioses between trees and fungi (Chapter 40). Many pathogen-associated molecular patterns (PAMPs) often consist of foreign glycans and/or glycan patterns that, on invading microbes, are recognized by innate immune cells and are detected by specific receptors such as Toll-like receptors (TLRs) and C-type lectins (Chapter 34). Last but not least, innate immune mechanisms are kept in check in vivo via recognition of self-associated molecular patterns (SAMPs).

MOLECULAR MIMICRY OF HOST GLYCANS AND GLYCOGIMMICKRY BY PATHOGENS AND HOSTS

Pathogens that invade multicellular animals sometimes decorate themselves with glycan structures that are identical or nearly identical to those on host cell surfaces (Chapters 42 and 43). Such glycans block recognition of underlying antigenic epitopes, restrict immune cell complement system activation, and can also mimic SAMPs of hosts, all of which are successful strategies for evading host immune responses. Perhaps not surprisingly, pathogens have evolved this state of molecular mimicry by making use of "every possible trick in the book," including direct or indirect appropriation of host glycans, convergent evolution toward similar biosynthetic pathways, and even lateral gene transfer. In some instances, the impact of the pathogen is aggravated by autoimmune reactions, resulting from host reactions to these self-like antigens. Pathogens and parasites also engage in glycogimmickry—to cite just a few examples, the gut symbiont *Bacteroides fragilis* induces tolerogenic states of gut immunity using a pentasaccharide called polysaccharide A, mycobacterial lipoarabinomannan (LAM) binding to host dendritic cell–specific intercellular adhesion molecule-3-grabbingcnonintegrin (DC-SIGN) induces expression of the inhibitory cytokine IL-10, and helminthic parasites have evolved many ways to modulate the immune response of their hosts toward prolonged tolerance, using specific immunomodulatory glycans. Secreted and soluble milk oligosaccharides (all extensions of the disaccharide lactose) of mammals provide an example of trans-generational glycan manipulation of microbial symbionts or pathogens in the infant by maternal glycans. These glycans have potent effects by favoring symbionts and blocking pathogenic microbes (bacteria and viruses) of the infant gut.

THE SAME GLYCAN CAN HAVE DIFFERENT FUNCTIONS WITHIN AN ORGANISM

Expression of certain glycans on different glycoconjugates in different tissues at different stages of development implies diverse roles for these glycan structures within the same organism. For example, Man-6-P-containing glycans were first found on lysosomal enzymes and are involved in lysosomal trafficking (Chapter 33). However, such glycans are now known to occur on some apparently unrelated proteins, for different functional roles. Likewise, the sialylated fucosylated lactosamines critical for selectin recognition (Chapter 34) are found in a variety of unrelated cell types in mammals, and the polySia chains that play an important part in embryonic nervous system neural cell adhesion molecule (NCAM) function (Chapter 15) are also found on a G protein–coupled receptor (GPCR) (CCR7) expressed in dendritic cells, where they appear to be important for targeting of these cells to the lymph nodes. Given that glycans are added posttranslationally, these observations should not be surprising. Once a new glycan or modification has been expressed in an organism, several distinct functions could evolve independently in different tissues and at different times in development. If any of these situations mediated a function valuable to survival and reproduction, the genetic mechanisms responsible for expression of the glycan and its expression pattern would remain conserved in evolution. A further example of dual roles is the recognition of fragments of structural glycans as danger or damage-associated molecular patterns (DAMPs) by the immune systems of multicellular organisms. Examples include cell wall fragments in plants and hyaluronan fragments in vertebrates.

INTRASPECIES AND INTERSPECIES VARIATIONS IN GLYCOSYLATION

The core structures of major glycan classes tend to be conserved across many species; for example, the core structure of N-glycans is conserved across all eukaryotes and at least some Archaea (Chapter 9). However, there can be considerable diversity in outer-chain glycosylation, even among relatively similar species. Such interspecies variation in glycan structure indicates that some glycan sequences do not have fundamental and universal roles in all tissues and cell types in which they are expressed. Of course, such diversity could be involved in generating differences in morphology and function between species. Such variations could also reflect the outcome of differing selection pressures by exposure to variant pathogen regimes. Furthermore, significant intraspecies polymorphism in glycan structure can exist without obvious functional value, although they may be involved in the interplay between parasites and host populations (Chapter 20).

POTENTIAL IMPORTANCE OF TERMINAL SEQUENCES, MODIFICATIONS, AND UNUSUAL STRUCTURES

Given all of the above, it is challenging to predict which glycan structures are likely to mediate the more specific or crucial biological roles within an organism. As mentioned above, terminal sugar sequences, unusual structures, or modifications of glycans are more likely to be involved in such specific roles. The predictive value of this observation is reduced by the fact that such entities are also more likely to be involved in interactions with microorganisms and other potentially noxious agents. Further complexity arises from "microheterogeneity" in glycan structure (Chapter 20), wherein the same glycosylation site on the same protein in the same species can carry a variety of related glycan structures or none at all. The challenge then is to predict and sort out which distinct roles can be assigned to a given glycan structure in a given cell type and organism. In the coevolutionary arms race between pathogens and hosts, pathogens will tend to exploit glycans that are difficult for the host to change (those constrained by their intrinsic host function), and because hosts are unable to alter their glycans easily owing to their slower evolutionary rates and functional constraints of existing glycans, the extrinsic and intrinsic functions of these glycans are bound to overlap. Hosts can, however, to the degree possible, be expected to evolve glycans that are difficult for microbes to mimic, such as the vertebrate-specific sialic acid Neu5Gc (Chapter 15) and sulfated GAGs, prokaryotes do not seem to have reinvented (Chapter 17).

ARE THERE "JUNK" GLYCANS?

Because microorganisms and parasites that bind glycans evolve in parallel with their multicellular hosts, they must adapt their glycan-binding "repertoire" to any change in glycan structure presented by the host. At the same time, each of these microorganisms is itself host to its own suite of parasites (phages), known to exploit the cell surface glycans of microbes in order to infect these. Host populations (multicellular organisms infected by microbes or microbes infected by phages) may evolve new modifications of the structures targeted by their parasites, especially if the latter had meanwhile evolved a vital function elsewhere within the organism. This would constrain the preservation of the underlying scaffolding on which the latest modification was placed, while adding yet another layer of complexity to host glycans. Such cycles of evolutionary interaction between (nested sets of) microbes and hosts might explain some of the complex and extended sugar chains found in multicellular organisms, especially in areas of frequent microbial contact, such as in secreted mucins on mucosal surfaces. In this manner, "junk" glycans (i.e., glycans without *current* adaptive value) could accumulate, akin to "junk" DNA. An important distinction is that much of the junk DNA consists of selfish DNA elements capable of replicating, whereas glycans cannot encode additional copies of themselves, given that their synthesis relies on the activity of a coordinated network of gene products. Although such structures may still function as structural scaffolding, they may have no other specific role in a particular cell type at that particular time in evolution. They would, of course, provide fodder for future evolutionary selection, either for new organism-intrinsic functions or selective responses to a new pathogen. Additionally, neutral drift (which is now acknowledged as a major process in evolution) may explain some of the apparent "junk" glycans. The microheterogeneity inherent in glycan synthesis processes may be producing "neutral noise," but this noise may also contribute to evolutionary flexibility.

APPROACHES TO ELUCIDATING SPECIFIC BIOLOGICAL FUNCTIONS OF GLYCANS

Some functions of glycans are discovered serendipitously. In other instances, the investigator who has elucidated complete details of the structure and biosynthesis of a specific glycan is left without knowing its functions. It is necessary to design experiments that can differentiate between trivial and crucial functions mediated by each glycan. Various approaches are discussed below, emphasizing the advantages and disadvantages of each approach (also presented in schematic form in Figure 7.2).

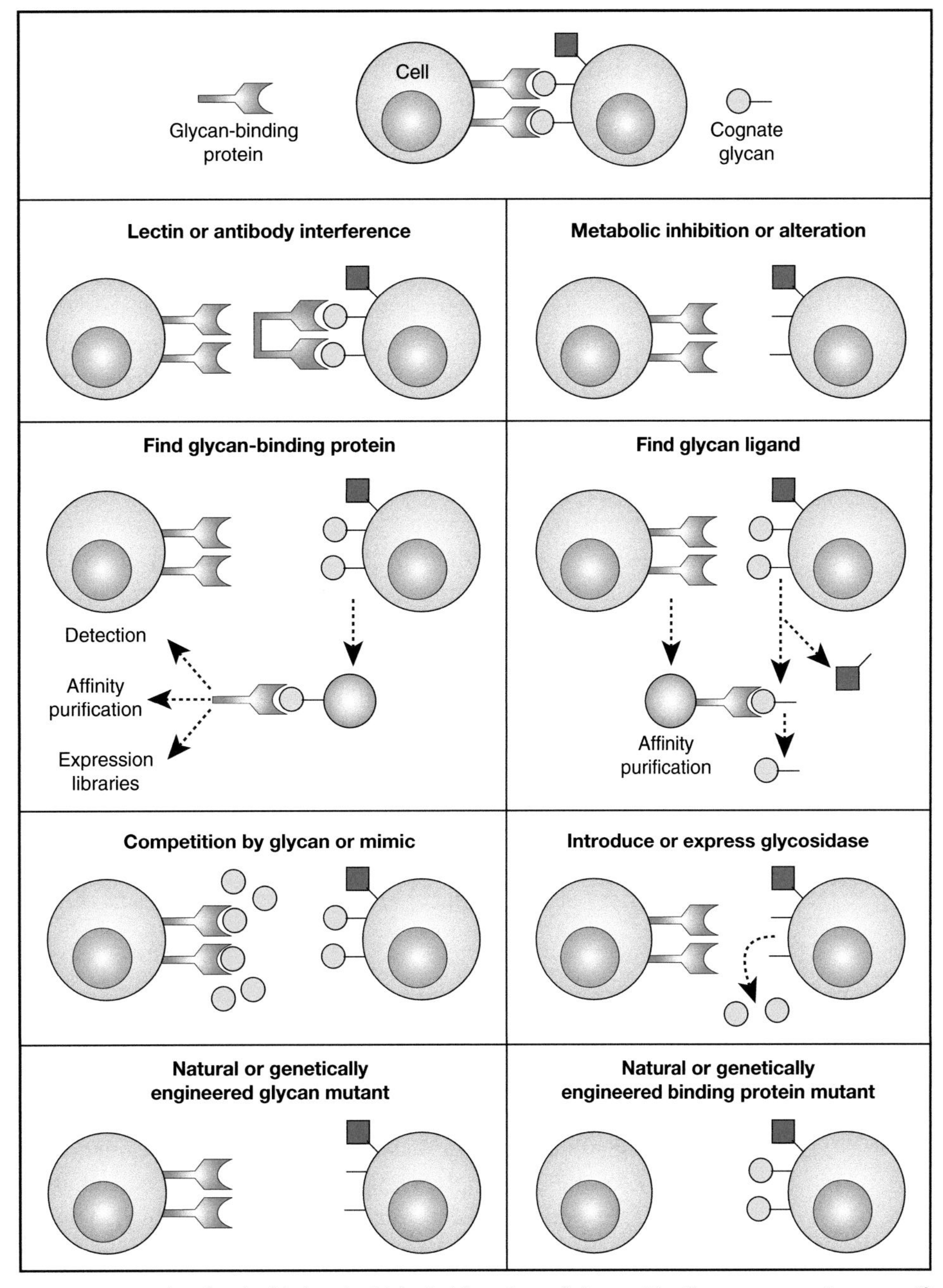

FIGURE 7.2. Approaches for elucidating the biological functions of glycans. The figure assumes that a specific biological role is being mediated by recognition of a certain glycan structure by a specific glycan-binding protein. Clues to this biological role could be obtained by various different approaches (for discussion of each approach, see text).

Localization or Interference with Specific Glycans Using Glycan-Recognizing Probes (GRPs)

Many current approaches to understanding glycan diversity (Chapters 50 and 51) involve extraction and identification of the entire complement of glycans found in a given organ or tissue, without regard to the fact that individual cell types and even basal versus apical sides of the same cell can have widely varying glycan expression patterns. However, cell type–specific localization of glycans can be explored using highly specific GRPs (GBPs or antibodies, see Chapter 48). Once a specific glycan has been localized in an interesting biological context, it is natural to consider introducing the cognate GRP into the intact system, hoping to interfere with a specific function and generate an interpretable phenotype. Such an approach is likely to give confusing results with regard to glycan function. Some GRPs (e.g., antibodies against glycans) tend to have weak affinity and show cross-reactivity. Although some plant lectins seem very specific for animal glycans, they often originate from organisms that typically do not contain the same ligand. Thus, their apparent specificity may not be as reliable when introducing them into complex animal biological systems in which they might bind unknown, cross-reacting glycans. Finally, most GRPs are multivalent, and their cognate ligands (the glycans) tend to be present in multiple copies on multiple glycoconjugates. Thus, introduction of a GRP into a complex biological system may cause nonspecific aggregation of various molecules and cell types, and the effects seen may extend well beyond the biological functions of the glycan in question. It would seem worthwhile to develop recombinant monovalent GRP modules that are derived from the same system being investigated, provided they are of high enough affinity. Effects of introducing such monovalent GRPs as competitors of the native function may yield more interpretable clues.

Metabolic Inhibition or Alteration of Glycosylation

Many pharmacological agents can metabolically inhibit or alter glycosylation in intact cells and animals (Chapter 49). Although such agents are powerful tools to elucidate biosynthetic pathways, they can yield confusing results in complex systems. One concern is that an inhibitor may have effects on other unrelated pathways. For example, the bacterial inhibitor tunicamycin that blocks N-linked glycosylation can also cause ER stress and inhibit UDP-Gal uptake into the Golgi. The second concern is that the inhibitor may cause such global changes in glycan synthesis that the physical properties of the glycoconjugates and/or membranes are altered, making it difficult to interpret the results. Somewhat more useful results can be obtained by introducing low-molecular-weight primers of terminal glycosylation (Chapter 55), which can act as alternate substrates for Golgi enzymes, diverting synthesis away from the endogenous glycoproteins. However, this approach can simultaneously generate incomplete glycans on endogenous glycoconjugates, as well as produce secreted glycan chains, each with their own biological effects.

Finding Natural Glycan Ligands for Specific Receptors

Because specific "carbohydrate-recognition domains" can be identified within a primary amino acid sequence (Chapters 28 and 29), it is now possible to predict whether a newly cloned protein can bind glycans. If a potential GBP can be produced in sufficient quantities, techniques such as hemagglutination, flow cytometry, surface plasmon resonance, and affinity chromatography (Chapters 29 and 28) can then be used to search for specific ligands. However, the monovalent affinity of the putative GBP for its ligand may not be high. Thus, high densities and/or multivalent arrays may be needed to avoid missing a biologically relevant interaction. The question also arises as to where exactly to look for the biologically relevant ligands in a complex multicellular system. Furthermore, because many glycan structures can be expressed in different tissues at different times in development and growth, a recombinant GBP may detect a cognate structure in a location and at a time that it is not of major biological relevance. Careful consideration of the natural occurrence and expression profile of the GBPs should lead to a rational decision as to where to look for its biologically relevant glycan ligands.

Finding Receptors That Recognize Specific Glycans

The converse situation arises when an unusual glycan is found to be expressed in an interesting context and is hypothesized to be a ligand for a specific receptor. It is possible to search for such a receptor by techniques similar to those above, such as hemagglutination, flow cytometry, and affinity chromatography (Chapters 28 and 29). To facilitate the search, it is necessary to have reasonable quantities of the pure defined glycan in question, as well as a variety of closely related structures as negative controls. Because many biologically relevant lectin-like interactions are of low affinity, it is probably advisable to use a multivalent form of the glycan as the probe (bait). Finally, it may not be obvious where to look for the glycan-binding protein. For example, the receptor that recognizes the unusual sulfated N-glycans of pituitary gonadotropin hormones was eventually found not in the pituitary nor in any of the target tissues for these hormones, but in liver endothelial cells, wherein it regulates the circulating half-life of the hormones (Chapter 31). Indeed, the most biologically relevant receptor for a particular glycan might even be found in another organism (a pathogen, a symbiont, or a mate).

Interference with Soluble Glycans or Structural Mimics

The addition of soluble glycans or structural mimics into a system can block the interaction between an endogenous GBP and a specific glycan (Chapters 28 and 29). If a sufficient concentration of the specific inhibitor can be achieved, the resulting phenotypic changes can be instructive. When studying in vitro systems, even monosaccharides can be used in such experiments, as exemplified by the Man-6-P receptor pathway (Chapter 33). However, it is often necessary to use competing glycans at high concentrations to block the relatively low-affinity site interactions between a GBP and its ligand. Intrinsically low site affinity can sometimes be overcome using a multivalent form of the cognate glycan. Finally, especially when studying complex multicellular systems, introduced glycans could be cross-recognized by other known or unknown binding proteins, giving a confusing phenotypic readout.

Eliminating Specific Glycan Structures by Glycosidases

A powerful approach is the use of degradative enzymes known to be highly specific for a particular glycan sequence. Many such specific enzymes can be obtained from microbial pathogens. The advantage of this approach lies in eliminating certain structures selectively after normal synthesis has been completed rather than interfering with the biosynthetic cellular machinery. Thus, for example, sialidase treatment abolished lymphocyte binding to the high endothelial venules of lymph nodes and provided the first indication of endogenous ligands for L-selectin (Chapter 32); injection of endoneuraminidase into the developing retina suggested specific roles for polySias (Chapter 15); and injection of heparanase into developing embryos gave randomization of left–right axis formation (Chapter 17). In all such studies, the purity of the enzyme used is critical and appropriate controls are necessary (preferably including a specific inhibitor of the enzyme or a catalytically inactive version of the enzyme). If the enzyme is of bacterial origin, trace amounts of contaminants such as endotoxin are also of concern. A genetic approach can be used to avoid problems of contamination by expressing a cDNA for the glycan-modifying enzyme in the intact cell or animal. For example, transgenic expression of an influenza-C sialic acid–specific 9-O-acetylesterase in mice resulted in either early or late abnormalities in development, depending on the promoter used. Unfortunately, many such glycosidases may not function well or at all in the context of an intact animal, which can limit the spectrum of glycan structures that may be probed for function with this approach.

Studying Natural or Genetically Engineered Glycan Mutants

This is intuitively a powerful approach for understanding glycan function. It is easiest to study glycosylation mutants in cultured cell lines (Chapter 49). Although genetic or acquired defects in glycosylation are obtained relatively easily in cells, these defects may have limited or not easily discernible biological consequences. This may be because of lack of other factors or cell types that would be present in the intact organism. For example, the cognate receptor for the glycan may not be present in the same cell type. Of course, such mutants can still be used to analyze basic structural functions of the glycans and their relevance to the physiology of a single cell. Furthermore, one can reintroduce external factors or other cell types thought to interact with the modified glycan. Some mutants can also be reintroduced into intact organisms, for example, to study tumorigenicity or metastatic behavior of malignant cells.

Although much useful information can be gained by such approaches, many of the more specific roles of glycans need to be uncovered by studying mutations in the intact multicellular organism. Looking at the various glycosylation mutants that have been recently discovered in flies, worms, mice, and humans, it is clear that glycan changes often affect multiple systems (are pleiotropic) and that the phenotypes are unpredictable and highly variable. This has become apparent, in part, by comparing the genotype–phenotype relationships in naturally occurring human disorders of glycosylation and in experimentally induced glycosylation disorders in mice. In humans, naturally occurring disease-associated mutations in glycosylation pathways often leave some residual enzymatic function intact (Chapters 44 and 45), whereas deletion of the corresponding enzymatic locus in mice often leads to a lethal phenotype during embryogenesis. Regardless, the value of engineering glycosylation mutants in intact animals is evident. Indeed, complete elimination of most major glycan classes of vertebrates has been accomplished in mice, and in every instance has led to embryonic lethality. Given the complex phenotypes and the potential for early developmental lethality, the ability to disrupt glycosylation-related genes in a temporally controlled and cell type–specific manner can be particularly valuable. In addition, genetically engineered tissue models now allow demonstration of context-dependent functions of each class of glycans in a human tissue models, such as those reflecting natural differentiation of the human skin.

Studying Natural or Genetically Engineered Glycan Receptor Mutants

Eliminating a specific glycan receptor can yield a phenotype that may be very instructive with regard to the functions of the glycan. As with genetic modification of the glycan, the results are more likely to be useful if studied in the intact organism. However, the receptor protein may have other functions unrelated to glycan recognition. Conversely, the glycan in question may have other functions not mediated by the receptor. For example, genetic elimination of the CD22/Siglec-2 receptor and the ST6Gal-I enzyme that generates its ligand gave complementary, but not identical, phenotypes (Chapter 35). However, breeding both mutations into the same mouse indicated that there were indeed epistatic interactions. Similar results were obtained by mating mice deficient in making polysialic acid with mice lacking the protein carrier of polysialic acids, NCAM.

ACKNOWLEDGMENTS

The authors appreciate helpful comments and suggestions from Hans Wandall.

FURTHER READING

Berger EG, Buddecke E, Kamerling JP, Kobata A, Paulson JC, Vliegenthart JF. 1982. Structure, biosynthesis and functions of glycoprotein glycans. *Experientia* **38**: 1129–1162. doi:10.1007/bf01959725

Kobata A. 1992. Structures and functions of the sugar chains of glycoproteins. *Eur J Biochem* **209**: 483–501. doi:10.1111/j.1432-1033.1992.tb17313.x

Lis H, Sharon N. 1993. Protein glycosylation. Structural and functional aspects. *Eur J Biochem* **218**: 1–27. doi:10.1111/j.1432-1033.1993.tb18347.x

Varki A. 1993. Biological roles of oligosaccharides: all of the theories are correct. *Glycobiology* **3**: 97–130. doi:10.1093/glycob/3.2.97

Drickamer K, Taylor ME. 1998. Evolving views of protein glycosylation. *Trends Biochem Sci* **23**: 321–324. doi:10.1016/s0968-0004(98)01246-8

Ferguson MA. 1999. The structure, biosynthesis and functions of glycosylphosphatidylinositol anchors, and the contributions of trypanosome research. *J Cell Sci* **112**: 2799–2809. doi:10.1242/jcs.112.17.2799

Gagneux P, Varki A. 1999. Evolutionary considerations in relating oligosaccharide diversity to biological function. *Glycobiology* **9**: 747–755. doi:10.1093/glycob/9.8.747

Spiro RG. 2002. Protein glycosylation: nature, distribution, enzymatic formation, and disease implications of glycopeptide bonds. *Glycobiology* **12**: 43R–56R. doi:10.1093/glycob/12.4.43r

Haltiwanger RS, Lowe JB. 2004. Role of glycosylation in development. *Annu Rev Biochem* **73**: 491–537. doi:10.1146/annurev.biochem.73.011303.074043

Ohtsubo K, Marth JD. 2006. Glycosylation in cellular mechanisms of health and disease. *Cell* **126**: 855–867. doi:10.1016/j.cell.2006.08.019

Bishop JR, Schuksz M, Esko JD. 2007. Heparan sulphate proteoglycans fine-tune mammalian physiology. *Nature* **446**: 1030–1037. doi:10.1038/nature05817

Moremen KW, Tiemeyer M, Nairn AV. 2012. Vertebrate protein glycosylation: diversity, synthesis and function. *Nat Rev Mol Cell Biol* **13**: 448–462. doi:10.1038/nrm3383

Hart GW. 2013. Thematic minireview series on glycobiology and extracellular matrices: glycan functions pervade biology at all levels. *J Biol Chem* **288**: 6903. doi:10.1074/jbc.r113.453977

Hardivillé S, Hart GW. 2014. Nutrient regulation of signaling, transcription, and cell physiology by O-GlcNAcylation. *Cell Metab* **20**: 208–213. doi:10.1016/j.cmet.2014.07.014

Van Breedam W, Pöhlmann S, Favoreel HW, de Groot RJ, Nauwynck HJ. 2014. Bitter-sweet symphony: glycan–lectin interactions in virus biology. *FEMS Microbiol Rev* **38**: 598–632. doi:10.1111/1574-6976.12052

Schnaar RL. 2017. Glycobiology simplified: diverse roles of glycan recognition in inflammation. *J Leukoc Biol* **99**: 825–838. doi:10.1189/jlb.3ri0116-021r

Stanley P. 2017. What have we learned from glycosyltransferase knockouts in mice? *J Mol Biol* **428**: 3166–3182. doi:10.1016/j.jmb.2016.03.025

Varki A. 2017. Biological rules of glycans. *Glycobiology* **27**: 3–49. doi:10.1093/glycob/cww086

Broussard AC, Boyce M. 2019. Life is sweet: the cell biology of glycoconjugates. *Mol Biol Cell* **30**: 525–529. doi:10.1091/mbc.e18-04-0247

Dabelsteen S, Pallesen EMH, Marinova IN, Nielsen MI, Adamopoulou M, Rømer TB, Levann A, Andersen MM, Ye Z, Thein D, et al. 2020. Essential functions of glycans in human epithelia dissected by a CRISPR-Cas9-engineered human organotypic skin model. *Dev Cell* **54**: 669–684. doi:10.1016/j.devcel.2020.06.039

Schjoldager KT, Narimatsu Y, Joshi HJ, Clausen H. 2020. Global view of human protein glycosylation pathways and functions. *Nat Rev Mol Cell Biol* **21**: 729–749. doi:10.1038/s41580-020-00294-x

8 A Genomic View of Glycobiology

Nicolas Terrapon, Bernard Henrissat, Kiyoko F. Aoki-Kinoshita, Avadhesha Surolia, and Pamela Stanley

A multitude of glycosyltransferases (GTs), glycoside hydrolases (GHs), other enzymes, and nucleotide sugar transporters are required to synthesize and metabolize glycans. Also, many genes encode glycan-binding proteins (GBPs), which recognize specific glycan structures. This chapter provides a genomic perspective of the genes that code for GTs, GHs, and GBPs.

THE GLYCOME

The glycome comprises all the glycan structures synthesized by an organism. It is analogous to the genome, the transcriptome, and/or the proteome but even more dynamic, and it has higher structural complexity that has yet to be fully defined. Cells of different types synthesize a subset of the glycome based on their differentiation state and physiological environment. The human and mouse glycomes have many glycan structures in common, but a few are unique or have divergent functional properties. For example, unlike humans, rodents synthesize cytidine monophospho-N-glycolylneuraminic acid (CMP-Neu5Gc), for the transfer of Neu5Gc to N- and

published by Cold Spring Harbor Laboratory Press; doi: 10.1101/glycobiology.4e.08

O-glycans (Chapter 15). Similarly, the gene encoding α-1,3-galactosyltransferase (A3GALT2) is functional in the mouse but not in human (Chapter 20). The human and fly genomes include orthologous genes encoding GTs that catalyze the same reaction, but they also have GTs that are unique. Thus, protein O-fucosyltransferase 1 (POFUT1) in mammals and Ofut1 in flies transfer fucose to Notch receptors and are examples of an evolutionarily conserved GT. In contrast, flies do not make complex N-glycans with four branches, which are common in mammalian glycoproteins (Chapters 9 and 20). Additionally, flies make unique glycolipids absent from mammals that are important for conserved signaling pathways mediated by the epidermal growth factor (EGF) receptor or Notch receptors (Chapter 26).

GENOMICS OF GLYCOSYLATION

The genome encodes all of the enzymes, transporters, and other activities necessary to construct and regulate the glycome of an organism. There were few complete genomes available in 1999 for the first edition of this book, approximately 650 genomes by the second edition (2009), 25,000 by the third edition (2015), and in August 2021 there were 259,000 permanent draft genomes in the Genomes OnLine Database (GOLD) of which more than 22,000 finished genomes are included in the manually curated Carbohydrate-Active enZYmes (CAZy) database. Likewise, the number of GTs with a known three-dimensional structure has grown from one in 1999 to 158 in 2015 and 282 in 2020. The CAZy database is dedicated to (i) coping with the staggering increases in sequences released in GenBank by the NCBI (National Center for Biotechnology Information), (ii) creating new families based on the literature, and (iii) reporting new functions/substrate specificities and 3D structures within existing families.

In the pre-genomic era, the glycobiology of mammals, invertebrates, plants, bacteria, and viruses did not overlap extensively, and progress in one domain did not immediately benefit others. With several genomes being released each day, the evolutionary history of GT and GH sequences has emerged. We now know that GTs from various organisms display the same basic structural folds, and we can harness the relationships between the sequence and the specificity of a GT (Chapter 6). The content of genomes can also be examined from a glycobiology perspective (e.g., by listing candidate GHs or GTs in a genome) and compared across genomes to see which families have expanded or disappeared during evolution. Examination of completely sequenced genomes suggests that a few percent of any genome encodes GTs and GHs. The number of GT genes in different organisms is variable, but varies much less within a taxonomic clade. The number of GTs tends to be greater than the number of GHs, except for organisms that forage on complex glycans as a carbon source. Thus, the genomes of saprophytic fungi and bacteria of the intestinal flora can encode several hundred enzymes for the breakdown of glycans (i.e., five to 10 times more GHs than the number of GTs). The number of genes that encode GBPs is more difficult to define because many have other functional domains and have not been annotated as GBPs. A conservative estimate is that mammalian genomes encode 100 to 200 GBPs.

The genome of each new organism is now routinely searched for genes that show similarity to known GTs, GHs, or GBPs. To annotate a new gene, its sequence is compared to previously annotated genes (pairwise alignments) or gene families (model alignments). Sequence similarity is used as a proxy for homology, to infer the biochemical activity of the predicted protein based on its distance from previously characterized proteins. GTs can be highly versatile, and GHs, despite being more specific in terms of the substrates they recognize, can form large families with various activities. Thus, although it can be difficult to predict the precise reaction catalyzed by a GT or GH on the basis of its sequence relatedness to biochemically characterized enzymes, it is often possible to predict the anomeric linkage of the sugar transferred or hydrolyzed, respectively, or the broad glycan category targeted. Prokaryotic GHs have similarities to those of eukaryotes, and with some exceptions, prokaryotic and eukaryotic GTs have one of three folds, GT-A GT-B or GT-C, indicating three ancestors for essentially all GTs (Chapter 6). Additional information can also be obtained from the sequence of a new gene by searching for conserved motifs such as signal peptides, transmembrane domains, glycosylphosphatidylinositol (GPI) anchors (Chapter 12), or carboxy-terminal retrieval sequences.

GENE FAMILIES

Glycosyltransferase (GT) and Glycoside Hydrolase (GH) Families

Carbohydrate-active enzymes can be classified according to several criteria. Substrate specificity forms the simplest basis to assign an Enzyme Commission (EC) number by the International Union of Biochemistry and Molecular Biology (IUBMB). GHs are given the code EC 3.2.1.x, where x represents substrate specificity. Similarly, GTs are described as EC 2.4.y.z, where y defines the sugar transferred and z describes the precise donor and acceptor. The EC number is given only after experimental determination of enzymatic specificity. Currently there are 462 EC numbers describing the known activities of GTs and 213 for GHs.

The intrinsic problem with a classification system based on substrate (or product) specificity is that it does not appropriately accommodate enzymes that act on several substrates. It also fails to reflect the sequence or the structural features of these enzymes in relation to evolution. To circumvent these problems, a novel system was introduced for classifying GHs and GTs based on the relationship between amino acid sequence and folding similarities. Regardless of activity and substrate specificity, sequences that display similarity are grouped in the same family. For GTs, such classification was initiated in 1997 (approximately 500 sequences; 27 families). This classification is continuously updated in the CAZy database, which listed approximately 850,000 GT sequences in approximately 110 families in August 2021. CAZy gives access to the various families of GTs and GHs, but also polysaccharide lyases and their carbohydrate-binding modules, as well as several families of auxiliary activities, such as the recently described lytic polysaccharide monooxygenases. Each CAZy family is annotated with known enzyme activities and often includes catalytic and structural features. This summary is followed by a list of the proteins and open reading frames (ORFs) belonging to the family, with links to sequence and structural information in public databases. CAZy also features summary pages for approximately 22,000 publicly available genomes (~95% of bacterial origin) in August 2021.

The earliest observation from the sequence-based families was that many are "polyspecific" and contain enzymes of different substrate specificity. Polyspecific families indicate that (1) the acquisition of new specificities by GHs and GTs is a common evolutionary event, (2) their substrate specificities can be engineered for experimental or applied purposes, and (3) their substrate (or product) specificities are governed by fine details of three-dimensional structure, not by global fold. Human GTs with experimentally determined activities are compiled in several excellent resources at Kyoto Encyclopedia of Genes and Genomes (KEGG), UniProt, and the GlycoGene Database (GGDB) in ACGG-DB (Chapter 52). In contrast, assignments for other organisms, like microbes, are often erroneous, because of misassignment of EC numbers based on too-distant sequence relatedness. The challenge of the postgenomic era is to characterize this ever-growing list of ORFs whose encoded proteins are candidate GTs of unknown donor, acceptor, and product specificities.

Glycan-Binding Proteins (GBPs)

The information presented by the wide variety of glycans on glycoconjugates is deciphered by an equally versatile number of GBPs that recognize specific sugars, glycans, or glycopeptides (Chapters 28–38). To understand the biology behind protein–glycan interactions, it is imperative to identify all GBPs and their glycan ligands.

GBPs were identified in the past by systematic biochemical studies that determined their glycan-recognition properties. However, the recent explosion of sequenced genomes makes it possible to identify genes likely to encode GBPs by sequence similarity. For example, the mannose-binding lectins (MBLs) can easily be identified because they show motifs found in C-type lectins (Chapter 34), as well as a collagen-like domain that promotes their oligomerization and is necessary for host defense through complement activation. A variant allele with changes in both the promoter and structural regions of the human *MBL2* gene influences the stability and serum concentration of the protein. Epidemiological studies have suggested that genetically determined variations in MBL serum concentration influence the susceptibility to, and the course of, different types of infections, autoimmune reactions, and metabolic and cardiovascular diseases. The fact that genetic variations in MBL are frequent indicates a dual role for MBL in host defense and highlights the power of genomics to aid our understanding of human disease.

Most of the studies on GBPs have been restricted to mammalian proteins (e.g., C-type lectins, galectins, and Siglecs). Their counterparts in plants and other "lower" organisms are underexplored. An extended classification of GBPs is proposed in Chapter 28, where they are contrasted with binding proteins that recognize sulfated glycosaminoglycans (which seem to have emerged independently of each other, by convergent evolution).

Knowledge of the ligand specificity of a GBP is required to assign in vivo functions. In cases in which there is a dearth of information, rational predictions based on the framework and sequence of existing carbohydrate-recognition domains (CRDs) have been found useful. Legume lectins represent a class of GBPs identified decades ago that continues to provide perhaps the best model for protein–glycan recognition. Moreover, discovery of the legume lectin fold (jelly-roll motif) in mammalian lectins, such as galectins (Chapter 36), calnexin, and calreticulin (Chapter 34), highlights the preeminence of this fold in carbohydrate recognition across phylogeny. Earlier work in the identification of monosaccharide-binding specificities of legume lectins provided the framework for finding their relatives in all forms of life (Chapter 32). This approach led to the assignment of glycan specificities for proteins involved in the sorting of vesicular compartments and in glycoprotein folding in the endoplasmic reticulum (ER) and Golgi compartments of mammalian cells (Chapter 39). Of similar importance is the discovery of new galectins in the galectin-10 family and galectin-like proteins in genome databases using similarity searches (Chapter 36). Likewise, Siglecs, a family of sialic acid–binding I-type lectins involved in regulating multiple biological responses (Chapter 35), show signature sequence motifs, and 17 members have been identified in primates to date. It is important to note that the mere presence of a CRD does not necessarily translate into functional glycan-recognizing activity. This is because sequence motifs used to identify CRDs are often found in a functionally inactive, lectin-like, CRD fold (Chapter 34).

Glycan microarrays provide a high-throughput means of detecting the interactions of GBPs with the diverse oligosaccharide sequences of glycoproteins, glycolipids, and polysaccharides (Chapter 30). The use of glass slides, microarray printing technology, and surface patterning of engineered glycophages displaying unique carbohydrate epitopes allows the production of glycan microarrays with the potential to examine binding of all types of GBPs (lectins, antiglycan monoclonal or serum antibodies, and glycan-binding cytokines or chemokines) to several thousand unique glycans, simultaneously. Binding is assessed by fluorescent or spectrometric techniques. Glycan microarray data are provided by resources such as the Consortium for Functional

Glycomics (http://www.functionalglycomics.org/glycomics/publicdata/primaryscreen.jsp) and the Imperial College Microarray Data Online Portal (https://glycosciences.med.ic.ac.uk/data.html), and several analysis software packages, including GLAD (https://glycotoolkit.com/GLAD/), MotifFinder (https://haablab.vai.org/tools/), MCAW (https://mcawdb.glycoinfo.org/) and CCARL (https://github.com/andrewguy/CCARL). Note that arrays that use different linkers and/or different attachment chemistries can give quite variable results, and the results need to be evaluated in the context of natural binding phenomena.

THE GLYCOME IN DIFFERENT ORGANISMS

Viruses

It has long been known that many viruses use host glycans as specific binding receptors for entering the cell (Chapter 37). Similarly, several viruses encode lytic enzymes that break down host cell surface glycans to release viral particles after viral replication. Genome sequencing reveals that many double-stranded DNA viruses also take advantage by adding sugars to host glycoproteins through the use of viral GTs (Chapter 42). Although biological roles of viral GTs are poorly understood, some functions have been identified. For example, the T4 bacteriophage encodes nucleases that degrade host cell DNA. To protect its own genome, the phage modifies its DNA by replacing cytosine with 5-hydroxymethylcytosine and subsequently transferring glucose (Glc) to the 5-hydroxymethylcytosine using a specific UDP-Glc:DNA Glc-transferase. The baculovirus enzyme ecdysteroid glucosyltransferase (EGT) disrupts the hormonal balance of the insect host by catalyzing the conjugation of ecdysteroid hormones with Glc or galactose (Gal). Expression of the EGT gene allows the virus to block molting and pupation of infected insect larvae. Similarly, Chloroviruses have enzymes in CAZy family GT4 for the glycosylation of their structural proteins. Serotype conversion in *Shigella flexneri* is mediated by temperate bacteriophages, which encode GTs that mediate O-antigen conversion by the addition of Glc to O-antigen units. Finally, giant viruses such as *Acanthamoeba polyphaga* mimivirus encode 12 putative GTs for the synthesis of complex O-glycans.

Bacteria

Bacterial GTs play a major role in their symbiosis and virulence. Some bacteria such as *Campylobacter* are able to N-glycosylate their proteins, but the most universal role for bacterial glycosylation is in the synthesis of cell-wall peptidoglycan, simple glycolipids, lipopolysaccharides, and complex exopolysaccharides (Chapter 21). The GTs involved in peptidoglycan biosynthesis are GT28 MurG, which adds N-acetylglucosamine (GlcNAc) to undecaprenyl diphospho-N-MurNAc, and GT51 MtgA, which polymerizes undecaprenyl diphospho-MurNAc-GlcNAc. *Mycobacterium tuberculosis* produces an extremely complex envelope that includes all of the above. In bacteria, the role of these glycans is to provide a barrier that affords mechanical, chemical, and biological protection to the cell. Some pathogenic or commensal bacteria produce an outer glycan layer that mimics that of their hosts, in order to evade host immune surveillance (Chapters 15 and 42). *Pasteurella multocida* produces a thick hyaluronan capsule. Oral streptococci produce two GTs for adhesion and virulence, and the EPax GT enables colonization of the gut by *Enterococcus faecalis.* Other pathogens such as *Escherichia coli* K1 and *Neisseria meningitidis* produce a poly α2–8 sialic acid capsule. Mammalian gut bacteria produce capsular polysaccharides thought to help the maturation of the host immune system.

Archaea

Archaea devote ~1% of their genes to GTs but, on average, devote only 0.25% to GHs, and there is almost no correlation with the number of GH genes and the overall number of genes. Surprisingly, ~20% of sequenced archaeal genomes appear to be completely devoid of GHs. The most striking example is *Methanosphaera stadtmanae*, whose genome encodes at least 43 GTs but apparently no GHs. This is not due to sequence divergence, because GHs are readily detected in some Archaea. These observations suggest that (1) horizontal transfer is likely the determining factor behind archaeal GH repertoires, and (2) the Archaea in question do not recycle glycosidic bonds elaborated by their own GTs. Although they do not make peptidoglycans like bacteria, Archaea use nucleotide-activated oligosaccharides to produce a variety of extracellular polysaccharides such as the heteropolysaccharide "methanochondroitin" made by *Methanosarcina barkeri*, which resembles eukaryotic chondroitin sulfate (Chapter 17). Archaea also make glycophospholipids and one relevant GT is GDP-Glc:glucosyl-3-phosphoglycerate synthase from family GT81 in *Methanococcoides burtonii.* In CAZy family GT55, there are several archaeal GDP-Man:mannosyl-3-phosphoglycerate synthases. A number of Archaea have GTs related to bacterial and eukaryotic oligosaccharyltransferases (CAZy family GT66), which is consistent with the fact that Archaea use oligosaccharyldiphospholipids as sugar donors. It has been shown that the archaeon *Methanococcus voltae* uses this strategy to transfer N-glycans to flagellin and S-layer proteins. Like bacteria and eukaryotes, evolution toward an obligate symbiont lifestyle was also accompanied by gene loss in Archaea. For example, the tiny genome of *Nanoarchaeum equitans* appears to encode only three GTs and no GH.

EUKARYOTES

With their large genomes and complex body plans that require regulated gene expression in different tissues and/or at different developmental stages, genomes of eukaryotes encode many more GTs and GHs than those of individual bacteria and Archaeal species. But, overall, prokaryotes appear to use a greater diversity of the monosaccharides that exist in nature (Chapters 20–23). Several eukaryotes have also undergone genome reduction and lost most of their GT genes. Thus, *Plasmodium falciparum* and *Encephalitozoon cuniculi* have only nine and eight GTs, respectively. Overall, the abundance of GTs in free-living eukaryotes correlates with evolution to multicellularity. Free-living fungi and the unicellular marine green alga *Ostreococcus tauri* have a number of GTs similar to certain bacteria.

Plants

The genomes of "higher" plants encode more GTs than any other organism, with approximately 560 in *Arabidopsis*, approximately 800 in poplar, and approximately 1200 in *Arachis hypogaea*! "Higher" plants have huge genomes resulting from several rounds of complete genome duplication. The massive number of GTs in plants is due to the expansion of several extremely populated GT families. For example, *Arabidopsis*, poplar, and *A. hypogaea* have about 120, 280, and 400 GT1 genes, respectively. "Higher" plants are characterized by extremely complex cell walls made of various polysaccharides that can be rather simple like cellulose, more complex as in hemicelluloses (e.g., xylans, glucuronoxylans, galactomannans, xyloglucans), or extremely complex like the "hairy" regions of pectin (Chapter 24). Biosynthesis of pectin alone requires the action of dozens of GTs. Differential expression in various tissues is probably one of the driving forces behind the accumulation of hundreds of genes encoding GTs in plants. Likewise, a diverse array of GHs is involved in the remodeling of the plant cell wall during plant growth. Thus, *Arabidopsis*, poplar, and *A. hypogaea* genomes encode about 420, 620, and 950 GHs, respectively.

Vertebrates

Vertebrates are characterized by a large diversity of GT genes. Human GTs fall into 46 CAZy families, a number similar to that of plants. Families present only in vertebrates are GT6, GT12, and most GT29 family members. Vertebrates usually have many different GT29 sialyltransferases belonging to several distinct subfamilies whereas invertebrates have only one member of a particular sialyltransferase subfamily. However, there are no GT families that are unique to humans or primates. The completion of the first animal genomes also revealed a relative paucity in the number of encoded GHs. Thus, the human genome codes for only 93 GHs, with a dozen devoted to the digestion of only three glycans: sucrose, lactose, and a portion of starch. The digestion of the immense majority of the plant cell-wall polysaccharides in the diet is "outsourced" to the multitude of different microorganisms that colonize the human gut. The genetic material of this flora, the "microbiome," greatly enlarges our limited genome. For instance, a single species of our gut bacteria such as *Bacteroides cellulosilyticus* WH2 encodes four times more GHs (408) than our own genome.

Invertebrates

One of the initial surprises that came with the completion of the first genomes was that the human genome encodes fewer GTs (242) than that of the nematode *Caenorhabditis elegans* (273). Interestingly, *Drosophila melanogaster* has only 155 GT genes. These gross numbers, however, mask important biological differences. The comparative abundance of GTs in *C. elegans* compared with humans is essentially due to four GT families more highly represented in the nematode: GT1 glucuronyltransferases (79 in *C. elegans*, 35 in human), GT11 fucosyltransferases (26 in *C. elegans*, three in human), GT14 β-xylosyltransferases and β1-6 GlcNAc-transferases (20 in *C. elegans*, 11 in human), and GT92 galactosyltransferases (27 in *C. elegans*, none in human). For most other GT families, *C. elegans* appears to have the same number, or fewer, GT genes than humans. With more than 415 GT genes, the bdelloid rotifer *Adineta vaga* is the animal with the largest known number of GTs. This large number is probably due to the ameiotic reproduction mode of this animal, which is accompanied by a large rate of horizontal gene acquisition from other organisms. *C. elegans* has 114 GHs, whereas *D. melanogaster* has 104 and humans have 93. GH18 chitinase is highly represented in *C. elegans* and *D. melanogaster* with 43 and 22 members, respectively.

MODULAR GLYCOSYLTRANSFERASES AND GLYCOSIDE HYDROLASES

In addition to catalytic specificity, the amino acid sequence of some GTs and GHs can also contain one or more additional domains that modulate the activity of the GT or GH. The most striking example is the two-domain mammalian heparan synthases that have evolved for the addition of alternating sugars to form a polysaccharide (Chapters 16 and 17). The amino-terminal domain, which adds β1-4 glucuronic acid (GlcA) residues, belongs to GT47, whereas the carboxy-terminal domain, which adds α1-4 GlcNAc residues, belongs to GT64. Some strains of bacteria have a heparan synthase, which also consists of two catalytic modules from families GT2 and GT45 (Figure 8.1), thereby providing a beautiful example of convergent evolution. A similar example of convergent evolution is found among chondroitin synthases, in which human enzymes are made of GT31 and GT7 catalytic domains, whereas the bacterial equivalent is made of tandem GT2 catalytic domains. Human LARGE is another bifunctional glycosyltransferase made of two domains. The amino-terminal domain, which adds α1-3 xylose (Xyl) residues, belongs to GT8, whereas the carboxy-terminal domain, which adds β1-3 GlcA residues, belongs to GT49.

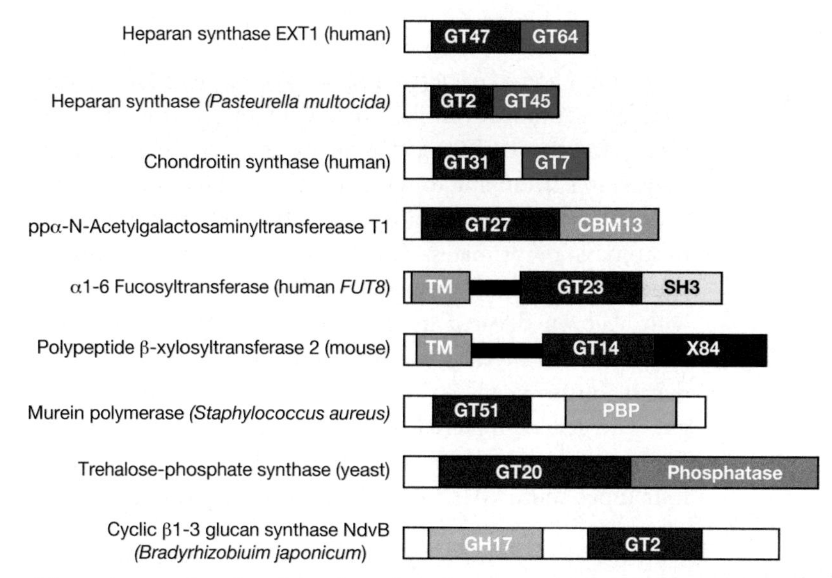

FIGURE 8.1. Schematic examples of modular glycosyltransferases. GT family modules are shown in *red* and *blue*. Other modules in various colors are CBM13, ricin-like carbohydrate-binding module; SH3, src homology domain 3; X84, putative glycan-binding module; PBP, penicillin-binding protein; TM, transmembrane domain; phosphatase, trehalose 6-phosphate phosphatase; GH17, glycosidase family 17 module. Regions without labels have unknown function(s).

Other modular GTs can feature an appended GBP domain. The best-known examples are polypeptide N-acetylgalactosaminyltransferases (ppGalNAcTs; GALNT) that transfer *N*-acetylgalactosamine (GalNAc) to Ser or Thr residues (Chapters 6 and 10). In these enzymes, a GT27 catalytic domain is linked to a GBP domain related to ricin and classified as CBM13 in the CAZy database. The GBP domain binds to the GalNAc residue transferred to protein by the GT27 catalytic domain and tethers the enzyme to the substrate. Another example is mouse polypeptide β-xylosyltransferase 2, in which a GT14 catalytic domain is linked to a carboxy-terminal domain that is thought to act as a GBP.

The GHs can also be modular, with the catalytic domain appended to one or more other modules whose role is to bind polysaccharides. Although human GHs are infrequently modular, those of microbes involved in plant cell-wall degradation can have more than five different modules assembled in a single polypeptide. Human acidic chitinase is an example of a mammalian modular GH having a CBM14 domain appended at the carboxyl terminus of the GH18 catalytic domain. The most intricate architecture of GHs is found in certain bacteria, such as *Clostridium thermocellum*, which elaborate a macromolecular complex called a "cellulosome" in which a large variety of modular plant cell-wall hydrolases are assembled together on a scaffolding protein. This strategy enables the assembly of dozens of catalytic modules simultaneously targeting the various polysaccharides that make up the plant cell wall.

RELATIONSHIPS OF GENOMICS TO GLYCOMICS

In summary, the genome, comprising the DNA of an organism, includes all the genes that produce the glycome, which comprises all of the glycans made by an organism. Although, within an organism, almost every cell that contains a nucleus and mitochondria has an identical genome; cells typically differ in the portion of the genome and, therefore the glycome, they express. Thus, the glycan complement of a cell depends on which genes are actively transcribed and which transcripts are translated and stably expressed. Transcription, splicing, translation, and posttranslational processing may vary depending on the state of differentiation and the physiological environment of a cell. Therefore, during development and differentiation, and under different environmental conditions, the glycan repertoire of a cell represents a subset of all the glycans that an organism is capable of making. To describe this variation, it is common to qualify the term glycome when referring to the glycans made by a particular tissue or cell type (e.g., T-cell glycome, hepatocyte glycome, or serum glycome), and to note the particular stage of development (e.g., fetal liver glycome, breast cancer serum glycome).

REGULATION OF THE GLYCOME

The glycome of a given cell in an organism can undergo substantial changes in response to environmental stimuli ranging from pH and ionic strength to hormonal stimulation or inflammation. Combined with the "assembly-line" nature of the Golgi apparatus (Chapter 4) and potential remodeling by GHs, full knowledge of the GT and GH transcriptome is only a rough predictor of the actual glycome of a given cell type, albeit a very helpful one. Apart from transcriptional control, the glycome is regulated by posttranscriptional control by microRNAs (miRs). For example, *GALNT7* is a target of miR-30d, a microRNA that is known to promote melanoma metastasis in patients and mouse models. The down-regulation of *GALNT7* phenocopied the expression of miR-30d. Subsequently, miRNAs have emerged as key regulators of the glycome owing to their ability to regulate multiple GT and GH mRNA targets. Nearly 80 GT and GH genes have been identified as targets of miRNAs to date.

ACKNOWLEDGMENTS

The authors appreciate helpful comments and suggestions from Kelley Moremen and Tadashi Suzuki.

FURTHER READING

El Kaoutari A, Armougom F, Gordon JI, Raoult D, Henrissat B. 2013. The abundance and variety of carbohydrate-active enzymes in the human gut microbiota. *Nature Rev Microbiol* **11:** 497–504. doi:10.1038/nrmicro3050

Lombard V, Golaconda Ramulu H, Drula E, Coutinho PM, Henrissat B. 2014. The Carbohydrate-Active enZYmes database (CAZy) in 2013. *Nucleic Acids Res* **42:** D490–D495. doi:10.1093/nar/gkt1178

Kohler A, Kuo A, Nagy LG, Morin E, Barry KW, Buscot F, Canback B, Choi C, Cichocki N, Clum A, et al. 2015. Convergent losses of decay mechanisms and rapid turnover of symbiosis genes in mycorrhizal mutualists. *Nat Genet* **47**: 410–415. doi:10.1038/ng.3223

Kremkow BG, Lee KH. 2018. Glyco-Mapper: a Chinese hamster ovary (CHO) genome-specific glycosylation prediction tool. *Metab Eng* **47**: 134–142. doi:10.1016/j.ymben.2018.03.002

Moremen KW, Haltiwanger RS. 2019. Emerging structural insights into glycosyltransferase-mediated synthesis of glycans. *Nat Chem Biol* **15**: 853–864. doi:10.1038/s41589-019-0350-2

Jayaprakash NG, Singh A, Vivek R, Yadav S, Pathak S, Trivedi J, Jayaraman N, Nandi D, Mitra D, Surolia A. 2020. The barley lectin, horcolin, binds high-mannose glycans in a multivalent fashion, enabling high-affinity, specific inhibition of cellular HIV infection. *J Biol Chem* **295**: 12111–12129. doi:10.1074/jbc.ra120.013100

Thu CT, Mahal LK. 2020. Sweet control: microRNA regulation of the glycome. *Biochemistry* **59**: 3098–3110. doi:10.1021/acs.biochem.9b00784

Huang Y-F, Aoki K, Akase S, Ishihara M, Liu Y-S, Yang G, Kizuka Y, Mizumoto S, Tiemeyer M, Gao X-D, et al. 2021. Global mapping of glycosylation pathways in human-derived cells. *Dev Cell* **56**: 1195–1209.e7. doi:10.1016/j.devcell.2021.02.023

9 N-Glycans

Pamela Stanley, Kelley W. Moremen, Nathan E. Lewis,
Naoyuki Taniguchi, and Markus Aebi

N-Glycans are covalently attached to protein at asparagine (Asn) residues by an N-glycosidic bond. Although diverse sugars are attached to Asn in prokaryotes (Chapters 21 and 22), all eukaryotic N-glycans begin with GlcNAcβ1–Asn and are the focus of this chapter. The biosynthesis of N-glycans is most complex in mammals and is described here in detail. Terminal sugars that largely determine the diversity of N-glycans are described in Chapter 14. Glycosylation-mediated quality control of protein folding by N-glycans is presented in Chapter 39, and the mannose 6-phosphate recognition determinant on N-glycans, necessary for targeting lysosomal hydrolases to lysosomes, is described in Chapter 33. Human congenital disorders of glycosylation arising from defects in N-glycan synthesis are discussed in Chapter 45.

DISCOVERY AND BACKGROUND

The GlcNAcβ1–Asn linkage was discovered by biochemical analyses of ovalbumin. The minimal amino acid sequence to receive an N-glycan is Asn-X-Ser/Thr in which "X" is any amino acid except Pro. However, not all Asn residues in this sequon are N-glycosylated, as discussed

published by Cold Spring Harbor Laboratory Press; doi: 10.1101/glycobiology.4e.09

below. Other linkages to Asn include Glc to Asn in laminin of mammals, the S-layer in Archaea and adhesins in some Gram-negative bacteria, GalNAc to Asn in Archaea, and rhamnose or bacillosamine to Asn in bacteria. In a sweet corn glycoprotein, Arg is found in N-linkage to Glc.

N-Glycan synthesis begins on a lipid-like polyisoprenoid molecule termed dolichol-phosphate (Dol-P) in eukaryotes. Following synthesis of an oligosaccharide that contains as many as 14 sugars, the N-glycan is transferred "en bloc" to protein. This synthetic pathway is conserved in all metazoa, plants, and yeast. Bacteria use related mechanisms to synthesize cell wall (Chapter 21). N-Glycans affect many properties of glycoproteins including their conformation, solubility, antigenicity, activity, and recognition by glycan-binding proteins (GBPs). Introduction of an N-glycan site (Asn-X-Ser/Thr) is used as a method to localize or orient a glycoprotein in the secretory pathway, or to follow its trafficking through the cell. Defects in N-glycan synthesis lead to a variety of human diseases (Chapter 45).

MAJOR CLASSES AND NOMENCLATURE IN EUKARYOTES

All eukaryotic N-glycans share a common core sequence, Manα1-3(Manα1-6)Manβ1-4GlcNAcβ1–4GlcNAcβ1–Asn-X-Ser/Thr, and are classified into three types: (1) oligomannose, in which only Man residues extend the core; (2) complex, in which "antennae" initiated by GlcNAc extend the core; and (3) hybrid, in which Man extends the Manα1-6 arm of the core and one or two GlcNAc-initiated antennae extend the Manα1-3 arm (Figure 9.1).

PREDICTING SITES OF N-GLYCOSYLATION IN EUKARYOTES

N-Glycans are added to secreted and membrane-bound glycoproteins at Asn-X-Ser/Thr sequons. About 70% of proteins contain this sequon, and ~70% of sequons carry an N-glycan. Mapping of the murine N-glycoproteome revealed more than 10,000 different N-glycosylation sites. Occasionally, N-glycans occur at Asn-X-Cys or, rarely, with a different amino acid in the third position. The transfer of an N-glycan to Asn-X-Ser/Thr occurs on the lumenal side of the

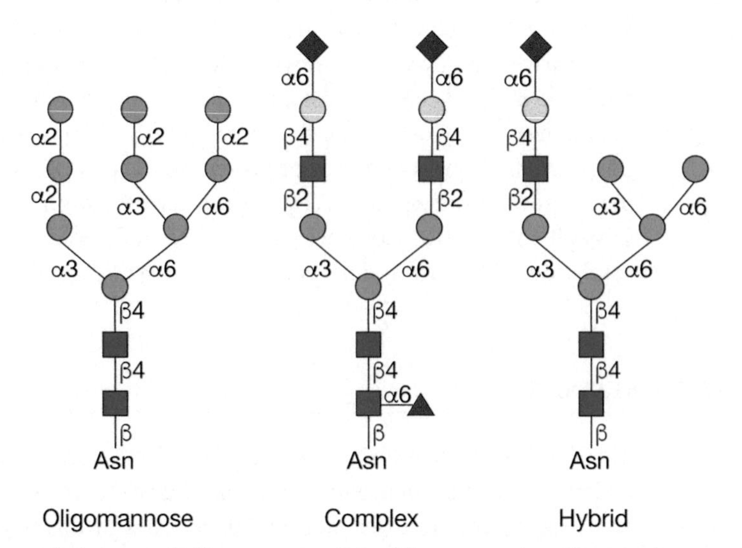

FIGURE 9.1. Types of N-glycans. N-Glycans at Asn-X-Ser/Thr sequons in eukaryote glycoproteins are of three general types: oligomannose, complex, and hybrid. Each N-glycan contains the common core $Man_3GlcNAc_2Asn$. Complex N-glycans can have up to five branches initiated by GlcNAc and elongated with Galβ1-4GlcNAc (LacNAc) repeats (see Figure 9.6).

endoplasmic reticulum (ER) membrane, during or after the translocation of the protein substrate. There is no definitive evidence that N-glycans occur on cytoplasmic or nuclear proteins nor on the cytoplasmic portions of membrane proteins. However, N-glycans covalently attached to RNA and expressed at the cell surface have recently been discovered. In proteins, only Asn-X-Ser/Thr sequons accessible to the ER lumen are known to receive an N-glycan. "X" may not be a Pro or may reduce the efficiency of glycosylation, such as when "X" is acidic (Asp or Glu), or enhance the efficiency, such as when Phe is in an adjacent reverse turn. However, although the presence of Asn-X-Ser/Thr is necessary for the receipt of an N-glycan, transfer does not always occur, because of conformational or other constraints during glycoprotein folding. Thus, Asn-X-Ser/Thr sequons encoded by an mRNA are referred to as potential N-glycan sites. Proof that an N-glycan is actually present requires experimental evidence, as described later in this chapter.

ISOLATION, PURIFICATION, AND ANALYSIS

N-Glycans of eukaryotes may be released from Asn using the bacterial enzyme peptide-N-glycosidase F (PNGase F). This enzyme will remove oligomannose, hybrid, and complex N-glycans attached to Asn unless the N-glycan core has certain modifications found in slime molds, plants, insects, and parasites. Another enzyme termed PNGase A (from almonds) will remove all N-glycans. Both enzymes are amidases that release N-glycans attached to the nitrogen of Asn, thereby converting Asn to Asp. Therefore, sites of glycosylation can be deduced by amino acid sequence analysis performed before and after PNGase F or A treatment. Other bacterial enzymes cleave between the two GlcNAc residues of the N-glycan core, leaving one GlcNAc attached to Asn. Endoglycosidase H releases oligomannose and hybrid N-glycans, but not complex N-glycans. Endoglycosidase F1 is similar to endoglycosidase H, whereas endoglycosidase F2 releases primarily biantennary N-glycans, and endoglycosidase F3 releases bi- and triantennary N-glycans, with a preference for those with a Fuc residue in the core. N-Glycans may also be released by hydrazinolysis or by exhaustive digestion with a protease that removes all amino acids except for the Asn. Released N-glycans may be purified by conventional ion-exchange and size-exclusion chromatography, high-performance liquid chromatography (HPLC) methods, and affinity chromatography on GBPs such as lectins. Lectins for glycan analysis are usually obtained from plants (Chapter 48). Release of N-glycans using chemical and enzymic methods, purification, and analysis are described in Chapters 50 and 51.

SYNTHESIS OF N-GLYCANS IN EUKARYOTES

N-Glycan biosynthesis occurs in two phases and in two compartments of eukaryotic cells, the ER and the Golgi (Chapter 4). The first phase is a highly conserved pathway that proceeds at the ER membrane on the lipid carrier dolichol-phosphate (Dol-P). An oligosaccharide assembled on Dol-P is transferred to Asn in select Asn-X-Ser/Thr sequons of secretory and membrane proteins during their translocation into the ER. The second phase begins with processing of N-glycans by glycosidases and glycosyltransferases in the lumen of the ER and continues in the Golgi in a species-, cell type–, protein-, and even site-specific manner. Many of the glycosidases and glycosyltransferases are differentially expressed and their activity is exquisitely sensitive to the physiological state of the cell. All glycosyltransferases use activated sugars (nucleotide sugars, dolichol sugars) as substrates (Chapter 5). Thus, a mature glycoprotein carries N-glycans that depend on the complement of expressed glycosylation genes in the cell type in which the glycoprotein is made and on the physiological state of that cell that may affect the localization and activity of glycosylation enzymes and nucleotide sugar transporters.

Dolichol phosphate

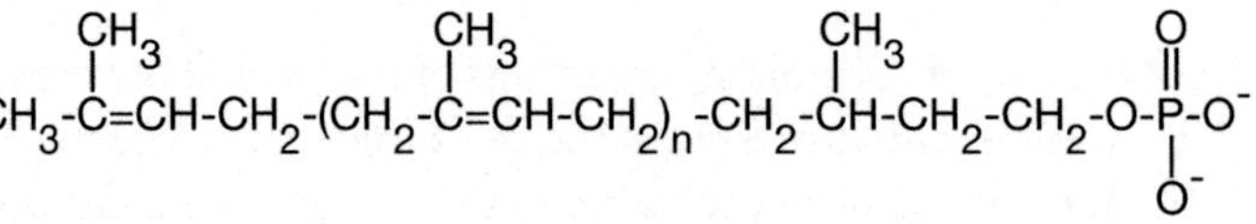

$$CH_3\text{-}\overset{\overset{CH_3}{|}}{C}\text{=CH-}CH_2\text{-}(CH_2\text{-}\overset{\overset{CH_3}{|}}{C}\text{=CH-}CH_2)_n\text{-}CH_2\text{-}\overset{\overset{CH_3}{|}}{CH}\text{-}CH_2\text{-}CH_2\text{-O-}\overset{\overset{O}{||}}{\underset{\underset{O^-}{|}}{P}}\text{-}O^-$$

FIGURE 9.2. Dolichol phosphate (Dol-P). N-Glycan synthesis begins with the transfer of GlcNAc-1-P from UDP-GlcNAc to Dol-P to generate dolichol pyrophosphate N-acetylglucosamine (Dol-P-P-GlcNAc). This reaction is inhibited by tunicamycin, which blocks N-glycan synthesis.

Synthesis of the Dolichol-Linked Precursor

Dolichol is a polyisoprenoid comprised of five-carbon isoprene units (Figure 9.2). The most common yeast dolichol has 14 isoprene units, whereas dolichols from other eukaryotes, including mammals, may have up to 19 isoprene units. The structure of the mature N-glycan precursor synthesized on Dol-P is shown in Figure 9.3. Genetic studies in *Saccharomyces cerevisiae* have identified conserved *ALG* (Asn-linked glycosylation) loci that encode the biosynthetic machinery for the assembly of the lipid-linked oligosaccharide in eukaryotes (Figure 9.3).

The first step is catalyzed by ALG7 (DPAGT1 in mammals), a GlcNAc-1-phosphotransferase that transfers GlcNAc-1-P from UDP-GlcNAc to form Dol-P-P-GlcNAc. Tunicamycin, an inhibitor of this enzyme, is used to inhibit N-glycosylation in cells. A second GlcNAc and five Man residues are subsequently transferred from UDP-GlcNAc and GDP-Man, respectively, to generate $Man_5GlcNAc_2$-P-P-Dol on the cytoplasmic side of the ER membrane (Figure 9.3). All of these enzymes transfer only the sugar portion of the nucleotide sugar. The $Man_5GlcNAc_2$-P-P-Dol precursor translocates across the ER membrane bilayer via a "flippase" encoded by the *RFT1* locus in yeast. $Man_5GlcNAc_2$-P-P-Dol is extended by the addition of four Man and three Glc residues transferred from Dol-P-Man and Dol-P-Glc, respectively. Dol-P-Man and Dol-P-Glc donors are formed on the cytoplasmic side of the ER membrane from GDP-Man and UDP-Glc. Dol-P-Man and Dol-P-Glc must also be flipped across the ER membrane. Mammalian MPDU1 is an ER membrane protein necessary for the utilization of Dol-P-Man and Dol-P-Glc in the ER lumen in the synthesis of the mature N-glycan precursor $Glc_3Man_9GlcNAc_2$-P-P-Dol (Figure 9.3). This 14-sugar glycan is transferred by oligosaccharyltransferase (OST) to Asn in receptive Asn-X-Ser/Thr sequons in protein regions that have translocated across the ER membrane.

Transfer of the Dolichol-Linked Precursor to Nascent Proteins

OST is a multisubunit protein complex in the ER membrane, except in the case of the kinetoplastids (Chapter 43). OST catalyzes the transfer of the oligosaccharide from Dol-P-P to Asn-X-Ser/Thr in newly synthesized regions of proteins during passage through the translocon into the ER. OST has a high specificity for the completely assembled oligosaccharide, which is $Glc_3Man_9GlcNAc_2$ in most eukaryotes. When incomplete oligosaccharides are assembled, transfer efficiency is reduced, resulting in hypoglycosylation of glycoproteins that mature with empty N-glycan sites. All OST subunits are transmembrane proteins with between one and 13 transmembrane domains. The OST complex cleaves the high-energy GlcNAc-P bond, releasing Dol-P-P in the process (Figure 9.3). Yeast OST is comprised of eight different subunits Stt3p, Ost1p, Wbp1p, Swp1p, Ost2p, Ost4p, Ost5p, and Ost3p or Ost6p. Stt3p is the catalytic subunit of the enzyme. The two OST complexes (containing either of the thioredoxin-subunit Ost3p or Ost6p) have a different protein–substrate specificity. The complexity of OST increases in multicellular organisms. In mammals, there are two different catalytic STT3 subunits that both associate with ribophorins I and II, OST48, OST4, and DAD1 proteins (homologs of the yeast

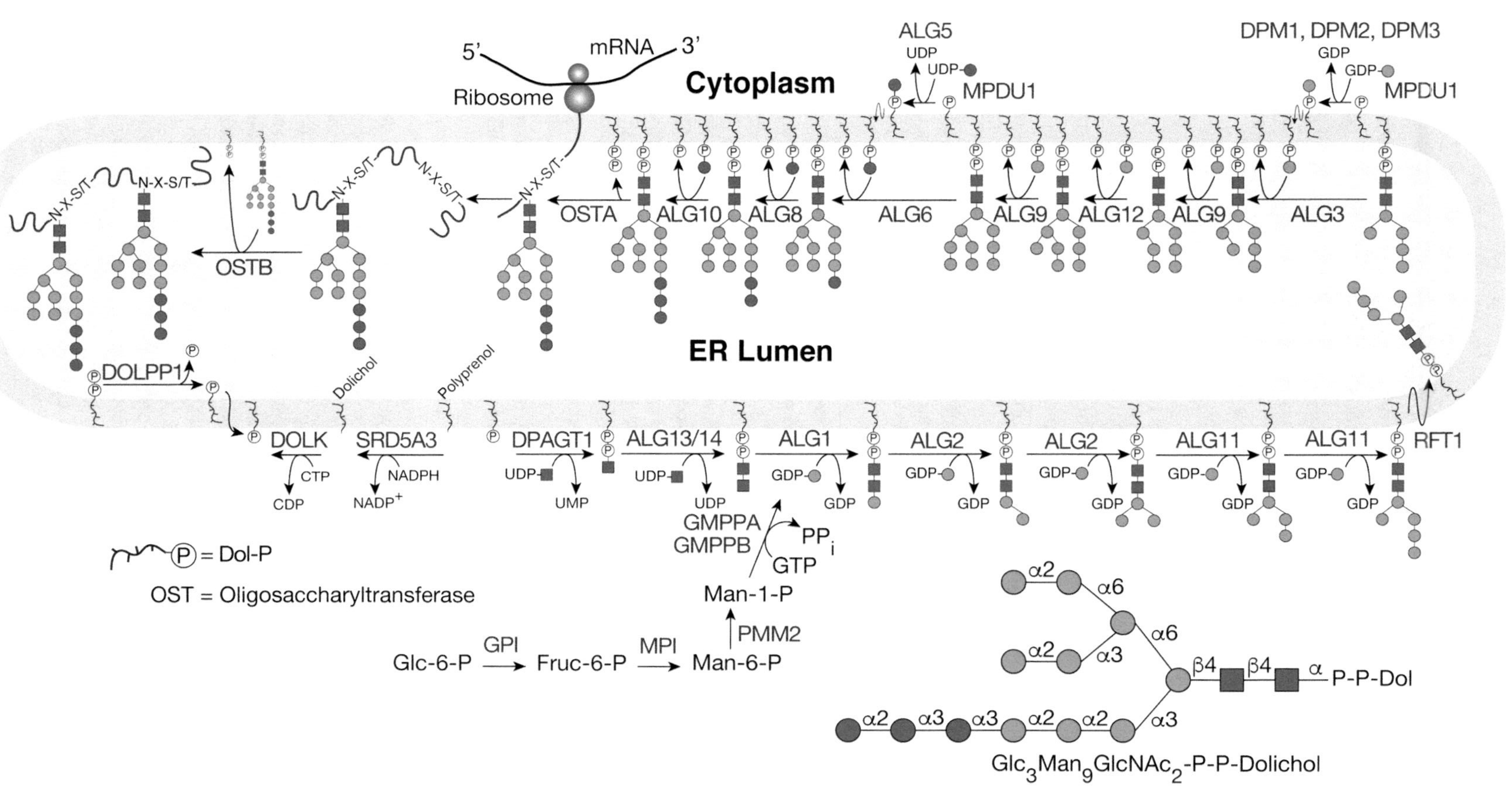

FIGURE 9.3. Synthesis of Dolichol-P-P-GlcNAc$_2$Man$_9$Glc$_3$. Dolichol (*red squiggle*) phosphate (Dol-P) located on the cytoplasmic face of the endoplasmic reticulum (ER) membrane receives GlcNAc-1-P from UDP-GlcNAc in the cytoplasm to generate Dol-P-P-GlcNAc. Dol-P-P-GlcNAc is extended to Dol-P-P-GlcNAc$_2$Man$_5$ before being "flipped" across the ER membrane to the lumenal side. Subsequently, four Man residues are added from Dol-P-Man and three Glc residues from Dol-P-Glc. Dol-P-Man and Dol-P-Glc are also made on the cytoplasmic side of the ER membrane and "flipped" onto the lumenal side. Yeast mutants defective in an *alg* (asparagine-linked glycosylation) gene were used to identify the ALG enzyme responsible for many reactions. The 14-sugar oligosaccharide is transferred to the Asn side chain within the sequon N-X-S/T by an oligosaccharyltransferase complex (OST) of 8-9 transmembrane proteins. In mammalian cells, the OSTA complex is associated with the translocon in the ER membrane and preferentially glycosylates nascent polypeptides traversing the translocon, whereas the OSTB complex modifies proteins that have left the translocon and are in the ER lumen. Enzyme names are from the Human Genome Nomenclature Committee (HGNC).

Ost1p, Swp1p, Wbp1p, Ost4p, and Ost2p, respectively). The STT3A complex (OSTA), closely associated with the translocon, contains the KCP2 and DC2 subunits, whereas the STT3B complex (OSTB) that has either MAGT1 or TUSC3 (homologs of yeast Ost3p/Ost6p) glycosylates polypeptides posttranslationally after their translocation into the ER. On binding to the catalytic STT3 subunit, the client peptide adopts a 180° turn, making polypeptide folding a competing reaction for N-glycosylation. Indeed, the thioredoxin subunits of the OST complex (Ost3p/Ost6p; MAGT1/TUSC1) modulate the oxidative folding of the client polypeptide, thereby extending the polypeptide substrate range of OST. As of February 2021, the UniProt database reports 1911 N-glycosylation sites in yeast and 13,648 in murine glycoproteins.

Early Processing Steps: $Glc_3Man_9GlcNAc_2Asn$ to $Man_5GlcNAc_2Asn$

Following the covalent attachment of the 14-sugar glycan to Asn-X-Ser/Thr in a protein, processing reactions trim the N-glycan in the ER. The initial steps have key roles in regulating glycoprotein folding via interactions with ER chaperones that recognize specific features of the trimmed N-glycan (Chapter 39). Processing of $Glc_3Man_9GlcNAc_2Asn$ begins with the sequential removal of Glc residues by α-glucosidases I (MOGS) and II (GANAB) (Figure 9.4). Both glucosidases function in the lumen of the ER, with α-glucosidase I acting specifically on the terminal α1–2Glc and α-glucosidase II sequentially removing the two inner α1–3Glc residues.

FIGURE 9.4. Processing and maturation of an N-glycan. The mature glycan attached to Dol-P-P (Figure 9.3) is usually transferred to Asn-X-Ser/Thr sequons during protein synthesis as proteins are being translocated into the endoplasmic reticulum (ER). Some transfer also occurs after translocation is complete. Following transfer of $Glc_3Man_9GlcNAc_2$ to protein, glucosidases in the ER remove three Glc residues, and ER mannosidase removes a Man residue. These reactions are intimately associated with the folding of the glycoprotein assisted by the lectins calnexin and calreticulin and the UDP-glucose:glycoprotein glucosyltransferase (UGGT), which together determine whether the glycoprotein with $Man_9GlcNAc_2$ continues to the Golgi or is degraded (Chapter 39). For misfolded proteins in the ER, mannose trimming by ER degradation-enhancing α-mannosidase I–like proteins (EDEMs) leads to formation of $Man_7GlcNAc_2$ N-glycans, which are recognized by the lectin OS9, which escorts them via retrotranslocation into the cytoplasm for degradation (Chapter 39). The removal of the first Glc (and therefore all Glc) can be experimentally blocked by castanospermine, leaving $Glc_3Man_9GlcNAc_2$, which may subsequently have terminal Man residues removed during passage through the Golgi. For most glycoproteins, additional Man residues are removed in the *cis* compartment of the Golgi by a family of mannosidases (MAN1A1, MAN1A2, MAN1C1) until $Man_5GlcNAc_2$ is generated. The mannosidase inhibitors deoxymannojirimycin and kifunensine block the removal of these Man residues, leaving $Man_9GlcNAc_2$, which is not further processed in the Golgi. The action of MGAT1 on $Man_5GlcNAc_2$ in the *medial*-Golgi initiates the first branch of a complex or hybrid N-glycan. This reaction is blocked in Lec1 CHO mutants in which MGAT1 is inactive, leaving $Man_5GlcNAc_2$, which is not further processed. α-Mannosidase II activity, catalyzed by MAN2A1 or MAN2A2, removes two outer Man residues in a reaction that may be blocked by the inhibitor swainsonine. The action of α-mannosidase II generates the substrate for MGAT2. The resulting biantennary N-glycan is extended by the addition of Fuc, Gal, and Sia to generate a complex N-glycan with two branches. The addition of Gal does not occur in Lec8 CHO mutants, which have an inactive UDP-Gal transporter. Thus, in Lec8 mutants, complex N-glycans terminate in GlcNAc. The addition of Sia does not occur in Lec2 CHO mutants, which have an inactive CMP-Sia transporter. In Lec2 mutants, complex N-glycans terminate in Gal. Complex N-glycans can have many more sugars than shown in this figure, including additional residues attached to the core, additional branches, branches extended with poly-LacNAc units, and different "capping" epitopes (Chapter 14). Also shown is the special case of lysosomal hydrolases that acquire a GlcNAc-1-P at C-6 of Man residues on oligomannose N-glycans in the *cis*-Golgi by N-acetylglucosamine-1-phosphotransferase (heterodimer of GNPTAB and GNPTG). The GlcNAc is removed in the *trans*-Golgi by a phosphodiester N-acetylglucosamindase (NAGPA), thereby exposing Man-6-P that is recognized by a Man-6-P receptor (M6PR) and routed to an acidified, prelysosomal compartment (Chapter 33). Chemical inhibitors of N-glycan processing are described in Chapter 55, and CHO mutants blocked in N-glycan synthesis are described in Chapter 49. (Adapted, with permission of the *Annual Review of Biochemistry*, from Kornfeld R, Kornfeld S. 1985. *Annu Rev Biochem* 54: 631–634.)

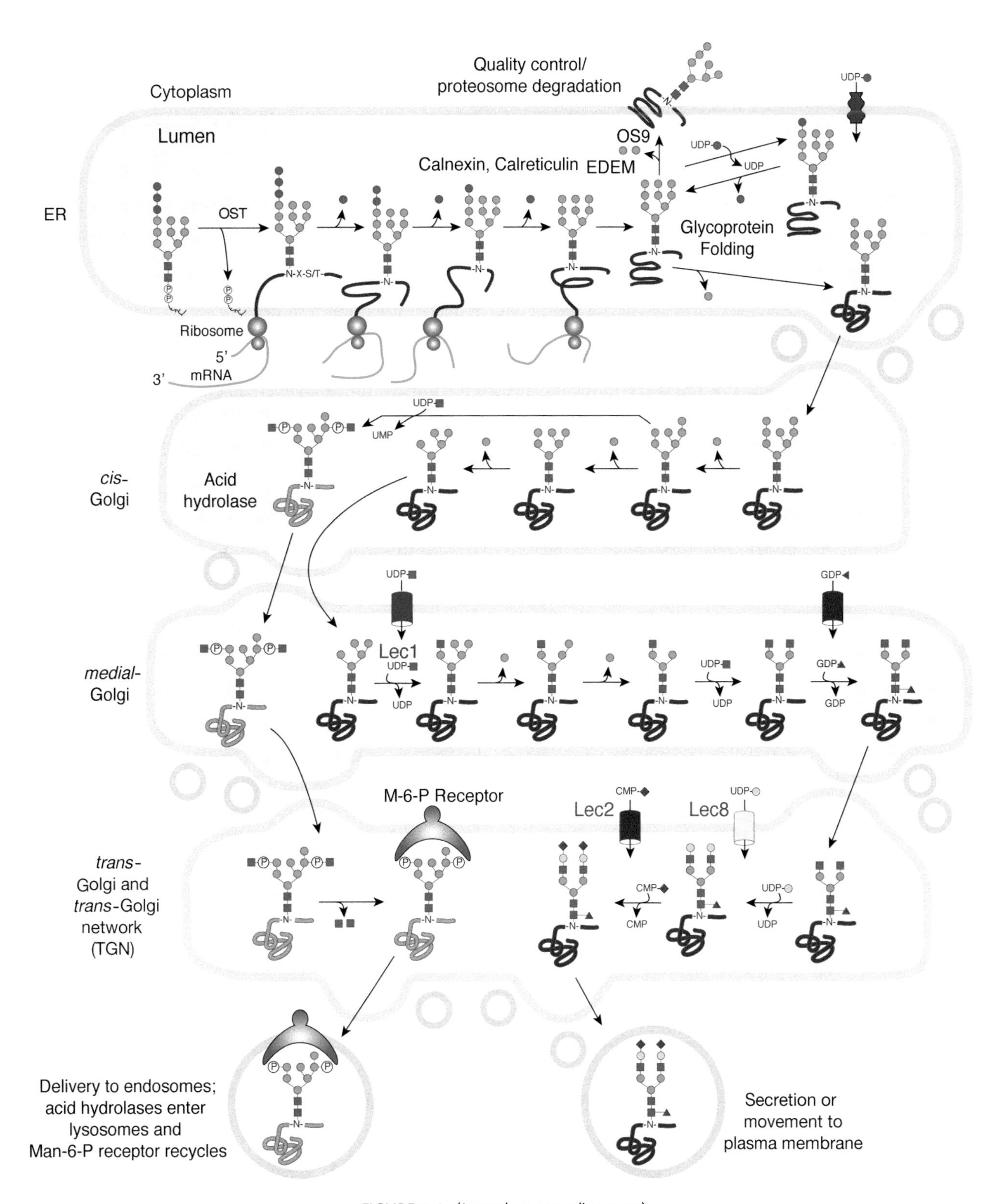

FIGURE 9.4. (*Legend on preceding page.*)

Removal of Glc residues, and the transient re-addition of the innermost α1–3Glc during protein folding, contribute to ER retention time. The removal of Glc may be prevented experimentally by the use of glucosidase I inhibitors such as castanospermine or deoxynojirimycin (Chapter 55). Following inhibition, N-glycans retain the three Glc residues and usually lose one or two Man residues as they pass through the ER and Golgi, resulting in Glc_3Man_{7-9} $GlcNAc_2$ N-glycans on mature glycoproteins. Before exiting the ER, ER α-mannosidase I (MAN1B1), removes the terminal α1-2Man from the central arm of $Man_9GlcNAc_2$ to yield a $Man_8GlcNAc_2$ isomer (Figure 9.4). Slow cleavage of terminal α1-2Man residues on misfolded glycoproteins by EDEM mannosidases leads to recognition of the trimmed glycans by OS9 and targets them for ER degradation (Chapter 39). The majority of glycoproteins exiting the ER to the Golgi carry N-glycans with either eight or nine Man residues.

Some N-glycans in the *cis*-Golgi retain a Glc residue because of incomplete processing in the ER. In this case, Golgi endo-α-mannosidase (MANEA) cleaves internally between the two Man residues of the Glcα1-3Manα1-2Manα1-2 moiety, thereby generating a $Man_8GlcNAc_2$ isomer different from that produced by ER α-mannosidase I (MAN1B1). Trimming of α1-2Man residues continues with the action of α1-2 mannosidases IA, IB, and IC (MAN1A1, MAN1A2, MAN1C1) in the *cis*-Golgi to give $Man_5GlcNAc_2$ (Figure 9.4) to yield $Man_5GlcNAc_2$, a key intermediate in the pathway to hybrid and complex N-glycans (Figure 9.1). Some $Man_{5-9}GlcNAc_2$ glycans may also escape further modification. In these cases, a mature membrane or secreted glycoprotein will carry $Man_{5-9}GlcNAc_2$ N-glycans. In addition, the action of MAN1B1, MAN1A1, MAN1A2, and MAN1C1 can all be blocked experimentally by the inhibitors, deoxymannojirimycin or kifunensine, resulting in $Man_{8-9}GlcNAc_2$ on mature glycoproteins. Most mature glycoproteins have some oligomannose N-glycans that are not processed in the *cis*-Golgi.

Late Processing Steps: $Man_5GlcNAc_2Asn$ to Hybrid and Complex N-Glycans

Biosynthesis of hybrid and complex N-glycans (Figure 9.1) is initiated in the *medial*-Golgi by the action of an N-acetylglucosaminyltransferase called GlcNAc-TI (MGAT1), which adds a GlcNAc residue to the C-2 of the α1-3Man in the core of $Man_5GlcNAc_2$ (Figure 9.4). Subsequently, the majority of N-glycans are trimmed by α-mannosidase II enzymes (MAN2A1 or MAN2A2) in the *medial*-Golgi, which remove the terminal α1-3Man and α1-6Man residues from $GlcNAcMan_5GlcNAc_2$ to form $GlcNAcMan_3GlcNAc_2$. It is important to note that α-mannosidase II cannot trim $Man_5GlcNAc_2$ unless it has been acted on by MGAT1. Once both Man residues are removed, a second GlcNAc is added to the C-2 of the α1-6Man in the N-glycan core by the action of GlcNAc-TII (MGAT2) to yield the precursor for all biantennary, complex N-glycans. Hybrid N-glycans are formed if the $GlcNAcMan_5GlcNAc_2$ glycan produced by MGAT1 is not acted on by α-mannosidase II. Incomplete action of α-mannosidase II can result in $GlcNAcMan_4GlcNAc_2$ hybrids. Small oligomannose N-glycans have been found in relatively large amounts in invertebrates and plants. These $Man_{3-4}GlcNAc_2$ N-glycans (paucimannose N-glycans) are formed from $GlcNAcMan_{3-4}GlcNAc_2$ following removal of the peripheral GlcNAc by a Golgi hexosaminidase that acts after α-mannosidase II (Chapters 24 and 26). Paucimannose glycans are also found in mammals, and can be elevated in cancer, inflammation, and stem cell development.

The complex N-glycan shown in the *medial*-Golgi of Figure 9.4 has two antennae or branches initiated by the addition of two GlcNAc residues. Additional branches can be initiated at C-4 of the core α1-3Man by GlcNAc-TIV (MGAT4A, MGAT4B, MGAT4C) and at C-6 of the core α1-6Man by GlcNAc-TV (MGAT5) to yield tri- and tetra-antennary N-glycans (Figure 9.5). MGAT5B or GlcNAc-TIX catalyzes the same reaction but preferentially on O-mannose glycans in brain. Another branch, found in birds and fish, can be initiated at C-4 of the core α1-6Man by GlcNAc-TVI (MGAT6; Figure 9.5). Genes related to GlcNAc-TVI exist in

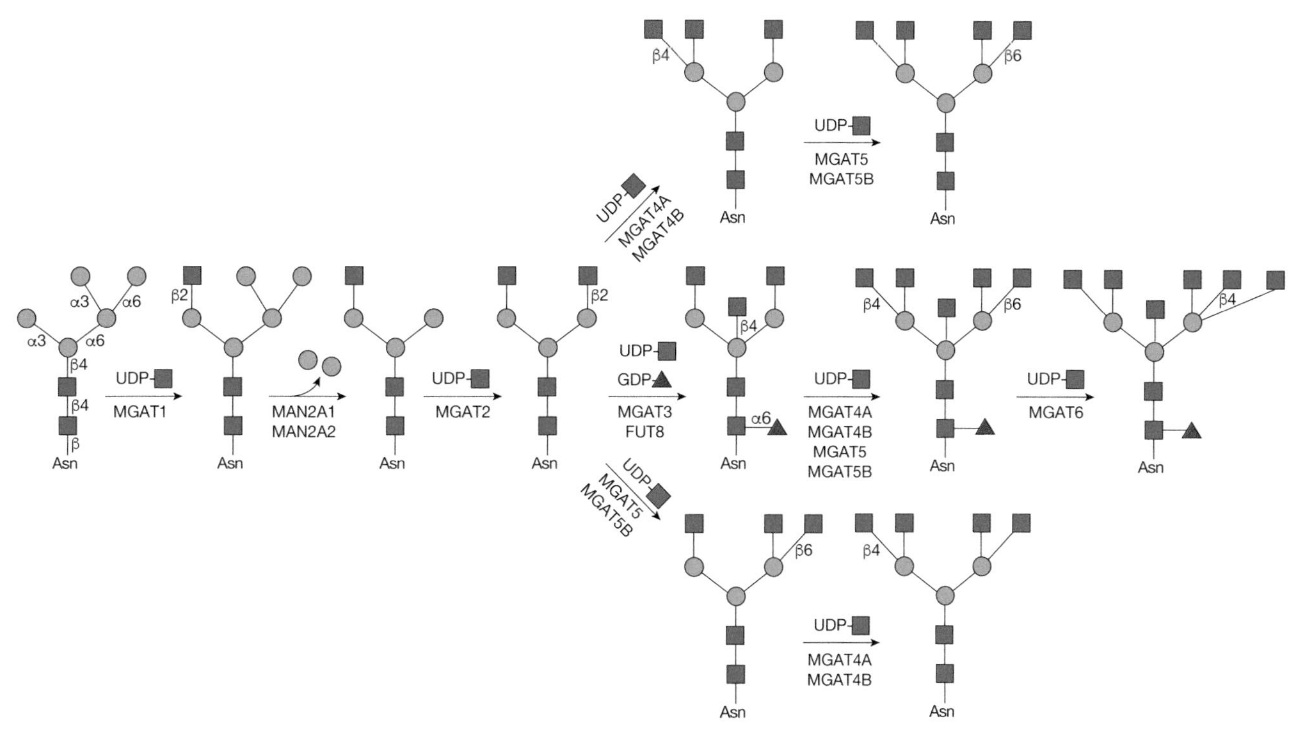

FIGURE 9.5. Branching and core modification of complex N-glycans. The hybrid and mature, biantennary, complex N-glycans shown in Figure 9.4 may contain more branches because of GlcNAc-transferases in the Golgi that act only after MGAT1 has acted. If the α-mannosidase II enzymes, MAN2A1 or MAN2A2, do not act, a hybrid N-glycan results. When the mannosidases act, a biantennary N-glycan is generated by MGAT2. This substrate may accept the bisecting GlcNAc from MGAT3, or a Fuc from FUT8, or a GlcNAc from the branching enzymes MGAT4A, MGAT4B, MGAT5, or MGAT5B in mammals. MGAT6 is present in birds and fish and potentially mammals. Each GlcNAc branch may be elongated with Gal, GlcNAc, Sia, and Fuc (Figure 9.6 and Chapter 14). The bisecting GlcNAc is not usually elongated unless MGAT2 is missing. The core Fuc is not elongated in mammals. The linkage of the sugar transferred at each step is shown.

mammalian genomes (MGAT4C). Complex and hybrid N-glycans may also carry a "bisecting" GlcNAc residue that is attached to the β-Man of the core by GlcNAc-TIII (MGAT3) (Figure 9.5). A bisecting GlcNAc on a biantennary N-glycan is shown in Figure 9.5, and it may be present in all of the more highly branched N-glycans. However, it is not usually extended.

Maturation of N-Glycans

Further sugar additions convert the limited repertoire of hybrid and branched N-glycans into an extensive array of mature, complex N-glycans comprising (1) sugar additions to the N-glycan core, (2) elongation of branching GlcNAc residues by sugar additions, and (3) "capping" or "decoration" of elongated branches.

The major core modification in vertebrate N-glycans is the addition of α1-6Fuc to the Asn-linked GlcNAc in the N-glycan core (Figure 9.5). The α1-6 fucosyltransferase (FUT8) usually requires the prior action of MGAT1 (Figure 9.4). In invertebrate glycoproteins, both core GlcNAc residues may receive a Fuc in an α1–3 and/or α1-6 linkage (Chapters 25 and 26). In plants, Fuc is transferred to the Asn-linked GlcNAc only in α1-3 linkage (Chapter 24). Also, in plant and helminth glycoproteins, the addition of β1-2Xyl to the β-Man of the core

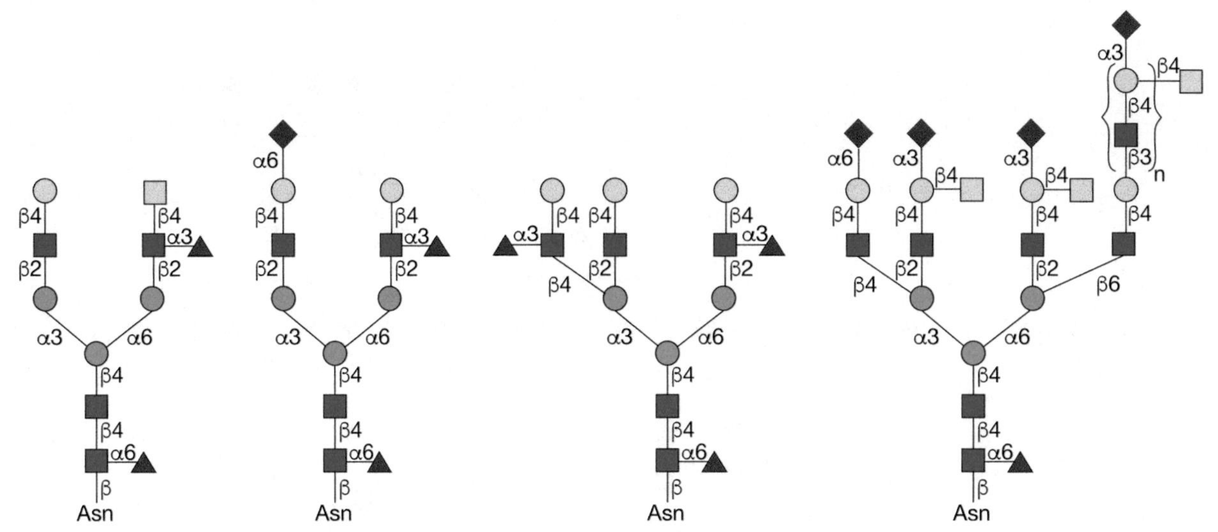

FIGURE 9.6. Typical complex N-glycans found on mature glycoproteins. A LacNAc unit (*bracketed*) on any branch may be repeated many times.

is common. This xylosyltransferase also requires the prior action of MGAT1. Xylose has not been detected in vertebrate N-glycans.

The majority of complex and hybrid N-glycans have extended branches that are made by the addition of Gal to the initiating GlcNAc to produce the ubiquitous building block Galβ1-4GlcNAc, referred to as a type-2 N-acetyllactosamine or "LacNAc" sequence (Figure 9.6). The sequential addition of LacNAc disaccharides gives tandem repeats termed poly-LacNAc. In some glycoproteins, β-linked GalNAc is added to GlcNAc instead of Gal, yielding antennae with a GalNAcβ1–4GlcNAc (LacdiNAc) extension (Chapter 14).

The most important "capping" reactions involve the addition of Sia, Fuc, Gal, GlcNAc, and sulfate to complex N-glycan branches. Capping sugars are most commonly α-linked and therefore protrude away from the β-linked poly-LacNAc branches, thus facilitating the presentation of terminal sugars to lectins and antibodies. Many of these structures are shared by N- and O-glycans and by glycolipids (Chapter 14). Terminal sialic acids can have further chemical modifications (e.g., O-acetylation) (Chapter 15).

The various reactions described above potentially yield a myriad of complex N-glycans that differ in branch number, composition, length, capping arrangements, and core modifications. Some examples to illustrate this diversity are shown in Figure 9.6.

THE PHOSPHORYLATED N-GLYCANS ON LYSOSOMAL HYDROLASES

Lysosomal hydrolases degrade proteins, lipids, and glycans in the lysosome. Many of these enzymes are targeted to lysosomes by a specialized trafficking pathway that requires phosphorylated oligomannose N-glycans. The phosphorylation step involves the transfer of GlcNAc-1-P to C-6 of Man residues of oligomannose N-glycans on lysosomal hydrolases in the *cis*-Golgi by N-acetylglucosamine-1-phosphotransferase (heterodimer of GNPTAB and GNPTG) (Figure 9.4). A phosphodiester N-acetylglucosamindase (NAGPA) in the *trans*-Golgi removes the GlcNAc to generate Man-6-P recognized by a Man-6-P receptor (M6PR). M6PRs transport lysosomal hydrolases to an acidified compartment which ultimately fuses with the lysosome. The details of this trafficking pathway are presented in Chapter 33.

TRANSFERASES AND TRANSPORTERS IN N-GLYCAN SYNTHESIS

The glycosyltransferases in the ER are mainly multitransmembrane proteins woven into the ER membrane. In contrast, the glycosyltransferases in Golgi compartments are generally type II membrane proteins with a small cytoplasmic amino-terminal domain, a single transmembrane domain, and a large lumenal domain that includes an elongated stem region extending from the membrane, and a globular catalytic domain (Chapter 6). The stem region is often cleaved by signal peptide peptidase-like proteases, particularly SPPL-3, releasing the catalytic domain into the lumen of the Golgi and allowing its secretion. Thus, extracellular soluble forms of many glycosyltransferases exist in tissues and sera and may function as transferases if nucleotide sugars are present extracellularly. Nucleotide sugars are synthesized in the cytoplasm, except for CMP-sialic acids, which are synthesized in the nucleus (Chapter 5). They are subsequently concentrated in the appropriate compartment following transport across the membrane by specialized nucleotide sugar transporters that translocate CMP-Sia, UDP-Gal, UDP-Glc, UDP-GlcNAc, GDP-Fuc, and other nucleotide sugars. A few of these transporters can transport more than one nucleotide sugar. Each transporter is a multitransmembrane protein that usually contains 10 membrane-spanning domains. Some CMP-sialic acids can be further modified within the lumen of the Golgi by O-acetyl groups before transfer to a glycan acceptor (Chapter 15).

GLYCOPROTEINS COMPRISE MANY GLYCOFORMS

Glycoproteins usually have a range of different N-glycans on a particular Asn-X-Ser/Thr N-glycosylation sequon, leading to glycan heterogeneity at each site. Furthermore, when there is more than one Asn-X-Ser/Thr sequon per molecule, different molecules in a population may have different subsets of N-glycans on different sequons, leading to glycoprotein microheterogeneity. Glycoproteins that differ only in their N-glycan complement are termed glycoforms. The variation in N-glycans of a glycoprotein may be due to protein conformation affecting substrate availability for Golgi glycosidases or glycosyltransferases, nucleotide sugar metabolism, transport rate of the glycoprotein through the lumen of the ER and Golgi, and the proximity of an Asn-X-Ser/Thr sequon to a transmembrane domain. Also, localization of glycosyltransferases within subcompartments of the Golgi can determine which enzymes encounter N-glycan acceptors. It is important to note that glycosylation enzymes often compete for the same acceptor and that most glycosyltransferases and glycosidases require the prior actions of other glycosyltransferases and glycosidases before they can act (Figure 9.4).

FUNCTIONS OF N-GLYCANS

Determining the functions of N-glycans may be accomplished using inhibitors including tunicamycin, which blocks the first step of N-glycosylation, or castanospermine, deoxynojirimycin, deoxymannojirimycin, kifunensine, and swainsonine, which block N-glycan processing; or glycosylation mutants of model organisms such as yeast, cultured mammalian cells, *Drosophila melanogaster*, *Caenorhabditis elegans*, zebrafish, and mouse. The various chemical inhibitors of N-glycan synthesis are discussed in Chapter 55. Many yeast mutants in the synthesis and initial processing of N-glycans are identified in Figure 9.3, and three mutants of cultured cells with altered glycosylation are identified in Figure 9.4 and described in detail in Chapter 49. Mutant cells or organisms with an altered N-glycosylation ability provide enormous insights into the biological functions of N-glycans, and their contributions to the biochemical properties of a glycoprotein in terms of structure, activity, susceptibility to proteases, and antigenicity. In addition, mutant cells and organisms allow glycosylation pathways that operate in vivo to be defined. A cell or organism with a loss-of-function mutation usually accumulates the biosynthetic intermediate that is the substrate of the activity lost in the mutant. Gain-of-function mutations reveal alternative pathways or glycosylation reactions that may occur depending on cell type. N-glycan functions have also been determined from the features of human congenital disorders of glycosylation (CDG) (Chapter 45).

Mouse mutants, in particular, have provided enormous insights into the functions of individual sugars present in N-glycans, as well as the functions of whole classes of N-glycans. Thus, deletion of the *Mgat1* gene that encodes MGAT1 prevents the synthesis of complex and hybrid N-glycans so that $Man_5GlcNAc_2$ is found at all complex and hybrid N-glycan sites (Figure 9.4). Whereas the absence of MGAT1 does not affect the viability or growth of Lec1 cultured cells under normal conditions, elimination of MGAT1 in the mouse results in death during embryonic development (Chapter 41). The complex N-glycans are important in retaining growth factor and cytokine receptors at the cell surface, probably through interactions with GBPs such as galectins, or cytokines such as transforming growth factor-β. Deletion of genes encoding sialyltransferases, fucosyltransferases, or branching N-acetylglucosaminyltranferases other than MGAT1 has generally produced viable mice with defects in immunity or neuronal cell migration, emphysema, or inflammation. N-Glycans may carry the sugar determinants recognized by selectins that mediate cell–cell interactions important for leukocyte extravasation from the bloodstream and regulation of lymphocyte homing to lymph nodes (Chapter 34). N-Glycans are known to become more branched when cells become cancerous, and this change facilitates cancer progression (Chapter 47). Tumors formed in mice lacking MGAT5 are retarded in their progression. Thus, certain glycosyltransferases may be appropriate targets for the design of cancer therapeutics.

ACKNOWLEDGMENTS

The authors acknowledge the contributions of Harry Schachter to previous versions of this chapter and appreciate helpful comments and suggestions from Thierry Hennet to this version.

FURTHER READING

Waechter CJ, Lennarz WJ. 1976. The role of polyprenol-linked sugars in glycoprotein synthesis. *Annu Rev Biochem* **45**: 95–112. doi:10.1146/annurev.bi.45.070176.000523

Snider MD, Sultzman LA, Robbins PW. 1980. Transmembrane location of oligosaccharide-lipid synthesis in microsomal vesicles. *Cell* **21**: 385–392. doi:10.1016/0092-8674(80)90475-4

Kornfeld R, Kornfeld S. 1985. Assembly of asparagine-linked oligosaccharides. *Annu Rev Biochem* **54**: 631–664. doi:10.1146/annurev.bi.54.070185.003215

Herscovics A. 1999. Importance of glycosidases in mammalian glycoprotein biosynthesis. *Biochim Biophys Acta* **1473**: 96–107. doi:10.1016/s0304-4165(99)00171-3

Berninsone PM, Hirschberg CB. 2000. Nucleotide sugar transporters of the Golgi apparatus. *Curr Opin Struct Biol* **10**: 542–547. doi:10.1016/s0959-440x(00)00128-7

Schenk B, Fernandez F, Waechter CJ. 2001. The ins(ide) and out(side) of dolichyl phosphate biosynthesis and recycling in the endoplasmic reticulum. *Glycobiology* **11**: 61R–70R. doi:10.1093/glycob/11.5.61r

Spiro RG. 2002. Protein glycosylation: nature, distribution, enzymatic formation, and disease implications of glycopeptide bonds. *Glycobiology* **12**: 43R–56R. doi:10.1093/glycob/12.4.43r

Patnaik SK, Stanley P. 2006. Lectin-resistant CHO glycosylation mutants. *Methods Enzymol* **416**: 159–182. doi:10.1016/s0076-6879(06)16011-5

Zielinska DF, Gnad F, Wisniewski JR, Mann M. 2010. Precision mapping of an in vivo N-glycoproteome reveals rigid topological and sequence constraints. *Cell* **141**: 897–907. doi:10.1016/j.cell.2010.04.012

Moremen KW, Tiemeyer M, Nairn AV. 2012. Vertebrate protein glycosylation: diversity, synthesis, and function. *Nat Rev Cell Mol Biol* **13**: 448–462. doi:10.1038/nrm3383

Aebi M. 2013. N-linked protein glycosylation in the ER. *Biochim Biophys Acta* **1833**: 2430–2437. doi:10.1016/j.bbamcr.2013.04.001

Shrimal S, Cherepanova NA, Gilmore R. 2015. Cotranslational and posttranslocational N-glycosylation of proteins in the endoplasmic reticulum. *Semin Cell Dev Biol* **41**: 71–78. doi:10.1016/j.semcdb.2014.11.005

Taniguchi N, Kizuka Y. 2015. Glycans and cancer: role of *N*-glycans in cancer biomarker, progression and metastasis, and therapeutics. *Adv Cancer Res* **126**: 11–15. doi:10.1016/bs.acr.2014.11.001

Harada Y, Ohkawa Y, Kizuka Y, Taniguchi N. 2019. Oligosaccharyltransferase: a gatekeeper of health and tumor progression. *Int J Mol Sci* **20**: 6074–6087. doi:10.3390/ijms20236074

Zheng L, Liu Z, Wang Y, Yang F, Wang J, Huang W, Qin J, Tian M, Cai X, Liu X, et al. 2021. Cryo-EM structures of human GMPPA–GMPPB complex reveal how cells maintain GDP-mannose homeostasis. *Nat Struct Mol Biol* **28**: 1–12. doi:10.1038/s41594-021-00591-9

10 O-GalNAc Glycans

Inka Brockhausen, Hans H. Wandall, Kelly G. Ten Hagen, and Pamela Stanley

Many glycoproteins carry glycans initiated by GalNAc attached to the hydroxyl of Ser or Thr residues. Mucins are the class of glycoproteins carrying the greatest number of O-GalNAc glycans (also called mucin-type O-glycans), but this posttranslational modification is common among many glycoproteins. The sugars found in O-GalNAc glycans include GalNAc, Gal, GlcNAc, Fuc, and Sia, whereas Man, Glc, or Xyl residues are not represented. Sialic acids may be modified by O-acetylation, and Gal and GlcNAc by sulfation. The length of O-GalNAc glycans may vary from a single GalNAc to more than 20 sugar residues and can include blood group and other glycan epitopes. This chapter describes the structures, biosynthesis, and functions of O-GalNAc glycans in mammals.

MUCIN GLYCOPROTEINS

About 150 years ago, E. Eichwald and E. Hoppe-Seyler noted that highly glycosylated proteins that contain hundreds of O-GalNAc glycans, which they termed mucins, are found throughout the body (Figure 10.1). Since then we have learned that O-GalNAc glycans are not only found as dense clusters on mucins, but also at single sites on most secreted and membrane-bound proteins. O-GalNAc glycans are involved in almost every aspect of biology, including cell–cell communication, cell adhesion, signal transduction, immune surveillance, epithelial cell protection, and host–pathogen interactions.

GalNAc O-linked to Ser/Thr is the initiating sugar of O-GalNAc glycans and is usually extended to form one of four common core structures (Table 10.1; Figure 10.1). Each core can subsequently be extended to give a mature linear or branched O-GalNAc glycan.

published by Cold Spring Harbor Laboratory Press; doi: 10.1101/glycobiology.4e.10

TABLE 10.1. O-GalNAc glycan cores and antigenic epitopes of mucins

Core	
Tn antigen	GalNAcαSer/Thr
Sialyl-Tn antigen	Siaα2-6GalNAcαSer/Thr
Core 1 or T antigen	Galβ1-3GalNAcαSer/Thr
Core 2	GlcNAcβ1-6(Galβ1-3)GalNAcαSer/Thr
Core 3	GlcNAcβ1-3GalNAcαSer/Thr
Core 4	GlcNAcβ1-6(GlcNAcβ1-3)GalNAcαSer/Thr
Epitope	
Blood group H	Fucα1-2Gal-
Blood group A	GalNAcα1-3(Fucα1-2)Gal-
Blood group B	Galα1-3(Fucα1-2)Gal-
Linear blood group B	Galα1-3Gal-
Blood group i	Galβ1-4GlcNAcβ1-3Gal-
Blood group I	Galβ1-4GlcNAcβ1-6(Galβ1-4GlcNAcβ1-3)Gal-
Blood group Sd(a), Cad	GalNAcβ1-4(Siaα2-3)Gal-
Blood group Lewis a	Galβ1-3(Fucα1-4)GlcNAc-
Blood group Lewis x	Galβ1-4(Fucα1-3)GlcNAc-
Blood group sialyl-Lewis x	Siaα2-3Galβ1-4(Fucα1-3)GlcNAc-
Blood group Lewis y	Fucα1-2Galβ1-4(Fucα1-3)GlcNAc-

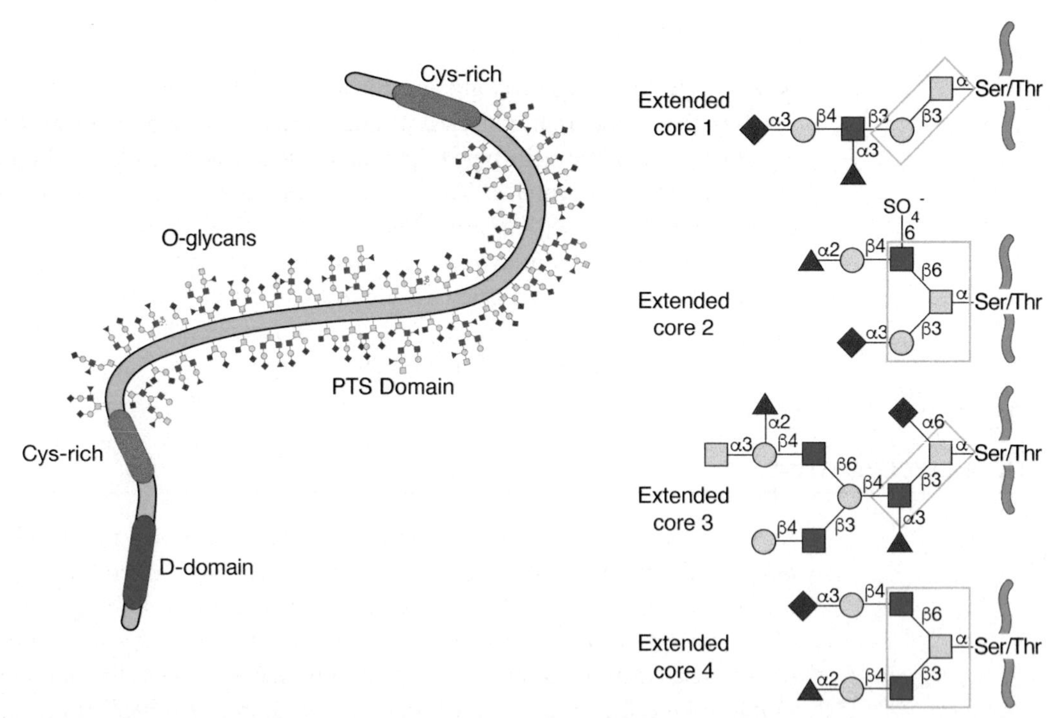

FIGURE 10.1. A simplified model of a large secreted mucin. The PTS domain, rich in Pro, Thr, and Ser, is highly O-glycosylated on Ser/Thr. The mucin assumes an extended "bottle brush" conformation. The cysteine-rich regions form disulfide bonds generating large polymers of several million Daltons. D domains have similarity to von Willebrand factor and are also involved in polymerization. Attached to the mucin are complex O-GalNAc glycans with different cores boxed in gray. Extended core 1, 2, 3, or 4 O-GalNAc glycans are from human respiratory mucins, and the extended core 3 O-GalNAc glycan is also from human colonic mucins. All four core structures (boxed) may be extended, branched, and terminated with Fuc, Sia, or blood group antigenic determinants (Table 10.1; Chapter 14). Core 1 and 3 O-GalNAc glycans may also carry α2-6Sia linked to the core GalNAc. *Green lines* are protein.

The great variety of O-GalNAc glycans often makes it very difficult to assign functions to individual O-GalNAc glycans at particular attachment sites, and most functions have historically been ascribed to the densely glycosylated mucins. In mucins, O-GalNAc glycans control chemical, physical, and biological properties. Because O-GalNAc glycans are hydrophilic, and usually negatively charged, they promote binding of water and salts and are major contributors to the viscosity and adhesiveness of mucins and the mucus they form. Mucins line the epithelial surfaces of the body, including the gastrointestinal, genitourinary, and respiratory tracts, where they shield epithelial cells against physical and chemical damage and protect against infection. Mucins can be antiadhesive and repel cell-surface interactions or, alternatively, promote adhesion by mediating recognition of glycan-binding proteins via their O-GalNAc glycans. A number of diseases are associated with abnormal mucin gene expression and abnormal mucin O-GalNAc glycans. These include cancer (Chapter 47), inflammatory bowel disease, congenital disorders of glycosylation (Chapter 45), and hypersecretory bronchial and lung diseases.

Humans have about 20 different mucin genes encoding both secreted and membrane-bound mucins, which vary considerably in their primary sequence and tissue-specific expression. Mucins are characterized by having densely glycosylated regions, previously termed "variable number tandem repeat" (VNTR) regions, but now called PTS domains (for the abundance of proline, threonine, and serine), which carry the majority of the O-GalNAc glycans on Ser and Thr residues (50%–80% of their molecular weight) (Figure 10.1). The O-GalNAc glycans expressed on a mucin are the result of the spectrum of glycosyltransferases active in the cell type producing the mucin. The first mucin polypeptide gene to be cloned, *MUC1*, encodes a transmembrane mucin that is ubiquitous in epithelium. The levels of MUC1 are high in certain tumors, in which its glycosylation is often abnormal. The dense and elongated O-glycans on membrane-bound mucins or other glycoproteins are thought to confer extended "bottle brush" conformations, lifting them above the surface of the cell (Figure 10.1), as well as providing protection from proteolytic cleavage. Additionally, dense O-GalNAc glycans on the large, secreted, gel-forming, intestinal mucin MUC2 allow it to adopt a hydrated, membrane-like structure that acts as the first line of defense to protect the underlying epithelium and mediate proper interactions with the microbiome. The importance of this mucin in intestinal health is illustrated by mice lacking *Muc2*, which spontaneously develop colorectal cancer.

O-GalNAc GLYCAN CORE STRUCTURES

The O-GalNAc glycans of mucins have four major core structures (cores 1–4; Table 10.1). Each core can be extended (Figure 10.1) by a variety of sugar residues to give linear or branched chains that resemble those on N-glycans (Chapter 9) and glycolipids (Chapter 11). Blood group determinants are commonly found in mucins at nonreducing termini of O-GalNAc glycans (Chapter 14). The extension of O-GalNAc with a β1-3Gal forms core 1, the most common O-GalNAc glycan. Core 2 is formed by the addition of β1-6GlcNAc to the GalNAc of core 1. Less common cores are core 3, in which β1-3GlcNAc is added to O-GalNAc, and core 4, in which core 3 is branched by the addition of β1-6GlcNAc (Table 10.1). Core 1 and 2 O-GalNAc glycans are found in glycoproteins and mucins produced in many different cell types. However, core 3 and 4 O-GalNAc glycans are more restricted to mucins and glycoproteins in gastrointestinal and bronchial tissues.

A single GalNAc residue attached to Ser/Thr forms the Tn antigen. As mentioned above, the core 1 O-GalNAc glycan (Galβ1-3GalNAc on Ser/Thr) forms the Thomsen–Friedenreich (TF or T antigen). Although Tn and T antigens are usually cryptic because they are extended by other sugars, they are found at increased levels in mucins from cancer cells. They can also carry Sia and form sialyl-Tn or sialyl-T antigens.

O-GalNAc cores are often extended to form complex O-GalNAc glycans that may include the ABO and Lewis blood group determinants (Chapter 14), polysialic acid, the linear i antigen (Galβ1-4GlcNAcβ1-3Gal), and the GlcNAc β1-6-branched I antigens (Table 10.1). Extensions by Type 1 (Galβ1-3GlcNAc) or Type 2 (Galβ1-4GlcNAc) units can be repeated and provide scaffolds for the attachment of additional sugars or functional groups. The termini of O-GalNAc glycans may contain Fuc and Sia in α-linkages, and Gal, GalNAc, and GlcNAc in both α- and β-linkages, and sulfate. Many of these terminal sugars are antigenic or recognized by lectins. In particular, the sialylated and sulfated Lewis antigens are ligands for selectins (Chapter 34), and Gal-terminating structures are ligands for galectins (Chapter 36). Some sugar residues, or their modifications, may mask underlying antigens or receptors. For example, O-acetyl groups on the Sia of the sialyl-Tn antigen prevent recognition by anti-sialyl-Tn antibodies. Gut bacteria may actively remove this mask. Dense O-glycosylation of mucin domains provides almost complete protection from protease degradation.

ISOLATION, PURIFICATION, AND ANALYSIS OF MUCIN O-GalNAc GLYCANS

The O-linkage between GalNAc and Ser/Thr residues is labile under alkaline conditions. Thus, O-GalNAc glycans can be released by a reaction termed β-elimination (i.e., treatment with 0.1 M sodium hydroxide). The hemiacetal GalNAc produced will undergo rapid alkali-catalyzed degradation under these conditions (called peeling), but can be reduced with sodium borohydride to yield stable *N*-acetylgalactosaminitol at the reducing end of the released O-glycan. β-Elimination is the method of choice to release O-glycans from glycoproteins that also have N-glycans, because the latter are not susceptible to cleavage under mild conditions. O-GalNAc glycans, as well as other Ser/Thr-linked glycans (Chapter 13), are released as alditols by β-elimination, but with losses of labile O-acetyl or sulfate esters. An alternative method that preserves the reducing end of O-GalNAc uses ammonia followed by boric acid. O-GalNAc that is not substituted with another sugar can be enzymatically released by a N-acetylgalactosaminidase. Another glycosidase, termed O-glycanase, releases core 1 (Galβ1-3GalNAc-) from Ser/Thr, provided the disaccharide is not further substituted. Thus, sialidase treatment followed by O-glycanase releases most simple, core 1 O-GalNAc glycans. Terminal Sia residues can also be easily removed with mild acid treatment. There are no known enzymes that can release more complex and extended whole O-GalNAc glycans, but mixtures of exoglycosidases can be used to sequentially remove sugars from O-GalNAc glycans on a glycoprotein. Glycoproteins with clusters of sialylated O-GalNAc glycans may be digested by an O-sialoglycoprotein endopeptidase.

Released, intact O-GalNAc glycans may be separated by different chromatographic methods, including high-performance liquid chromatography (HPLC). Chemical derivatization of GalNAc at the reducing end helps in the separation and subsequent analysis of sugar composition and linkages by gas chromatography and mass spectrometry (MS) (Chapter 50). Another tool to isolate O-glycans with specific epitopes is affinity chromatography using lectins. For example, *Helix pomatia* agglutinin binds to terminal GalNAc, whereas peanut lectin binds to unsubstituted core 1 (Table 10.1).

The structures of O-GalNAc glycans released from mucins and other glycoproteins may be determined by a combination of liquid or gas chromatography, MS, and nuclear magnetic resonance (NMR) spectroscopy. The anomeric linkage of each sugar can be determined using specific glycosidases that distinguish between α- or β-linked sugars, and by one- and two-dimensional NMR methods (Chapter 50).

The sites of O-GalNAc glycan modification in mucins are difficult to determine directly, but this has been achieved by sensitive MS methods and new enzymatic tools. A big step forward in understanding the extent of the O-GalNAc glycoproteome came with the proteome-wide mapping of O-GalNAc glycosylation sites. Several MS-based strategies have been applied including modifications of the endogenous glycan structures through chemical labeling, lectin enrichment of glycans derived from native or glycoengineered cells, or O-GalNAc-specific endo-peptidase treatment. Additionally, the use of mucin-type O-proteases (OpeRATOR and StcE) that specifically cleave N-terminal to an O-glycosylated Ser/Thr have aided in mapping sites of glycosylation. Based on these methodologies, we now know that >80% of the proteins passing through the secretory pathway are modified with O-GalNAc glycans, although the occupancy and the nature of an O-GalNAc glycan associated with a glycosylated site remains elusive.

BIOSYNTHESIS OF O-GalNAc GLYCANS

O-GalNAc glycans are added to Ser/Thr residues in proteins in the Golgi apparatus. The biosynthetic glycosyltransferases are type II transmembrane proteins with a short cytoplasmic tail at the amino terminus, a transmembrane domain, a stem region, and a catalytic domain in the lumen of the Golgi. The arrangement within Golgi membranes appears to be similar to an "assembly line" with early reactions occurring in the *cis*-Golgi and late reactions in the *trans*-Golgi (Chapter 4). Many of the enzymes, however, are diffusely distributed in Golgi compartments.

The subcellular localization, activity levels, and substrate specificities of glycosyltransferases involved in the assembly of O-GalNAc glycans play a critical role in determining the range of O-glycans synthesized by a cell (Table 10.2 and Figures 10.2 and 10.3). The glycosyltransferases that are involved in the assembly of O-GalNAc glycans are listed in Table 10.2. However, other enzymes that contribute to the synthesis of N-glycans and glycolipids also act on O-glycans, and some of these prefer O-glycans as acceptor substrates (Chapter 14). In vitro assays have shown that the activities of glycosyltransferases are controlled by factors such as metal ions and pH.

Polypeptide-N-Acetylgalactosaminyltransferases

The first and essential step of O-GalNAc glycosylation is the addition of GalNAc in α-linkage to Ser or Thr by a polypeptide GalNAc-transferase (ppGalNAcT; GALNT) (Table 10.2; Figure 10.2). Humans have 20 genes encoding GALNTs. The large number of GALNTs provides redundancy and also reflects differences in substrate specificity. Studies in the fly indicate that certain GALNTs (PGANTs in the fly) are required for normal development (Chapter 26). Deletion

TABLE 10.2. Glycosyltransferases that synthesize O-GalNAc glycans

Enzyme name	Short form	HGNC name
Polypeptide N-acetylgalactosaminyltransferase	ppGalNAcT-1 to -20	GALNT1-20
Core 1 β1-3 galactosyltransferase 1	C1GalT-1 or T synthase	C1GALT1
Essential chaperone for T synthase	COSMC	C1GALT1C1
Core 2 β1-6 N-acetylglucosaminyltransferase	C2GnT-1, C2GnT-3	GCNT1, GCNT4
Core 3 β1-3 N-acetylglucosaminyltransferase 6	C3GnT-6	B3GNT6
Core 2/4 β1-6 N-acetylglucosaminyltransferase 2	C2GnT-2	GCNT3
Elongation β1-3 N-acetylglucosaminyltransferase	elongation β3GnT	B3GNT3
I branching β1-6 N-acetylglucosaminyltransferase	I GnT	GCNT2
β1-3 galactosyltransferase	β3GalT5	B3GALT5
Core 1 α2-3 sialyltransferase	ST3Gal I	ST3GAL1
α2-6 sialyltransferase	ST6GalNAc I, II, III, IV	ST6GALNAC1-4
Core 1 3-O-sulfotransferase	Gal3ST4	GAL3ST4
α1-2 fucosyltransferase	FucT-I, FucT-II	FUT1, FUT2

HGNC, Human Genome Nomenclature Committee.

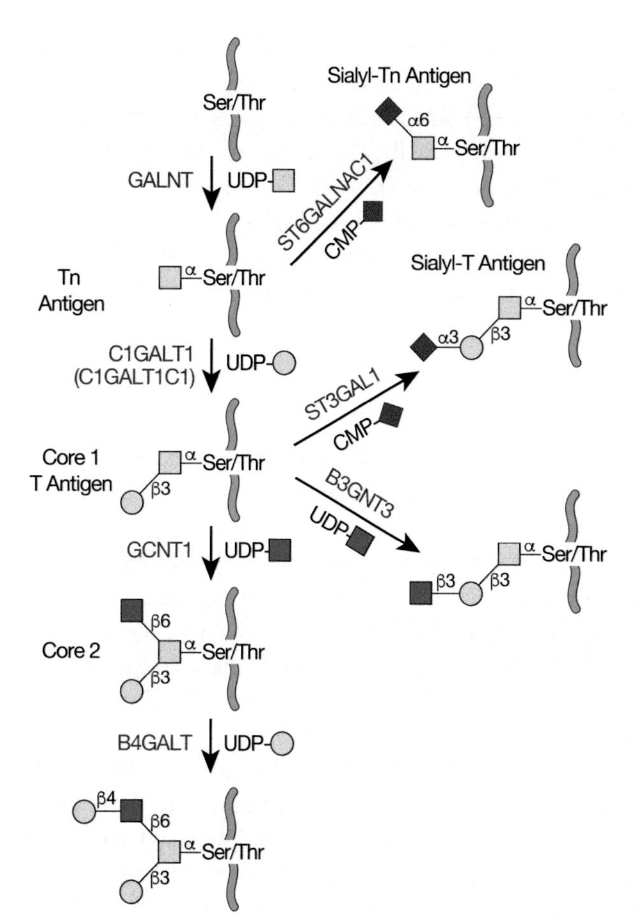

FIGURE 10.2. Biosynthesis of core 1 and 2 O-GalNAc glycans as described in the text. *Green lines* are protein.

of single GALNTs in mammals results in organ and cell differentiation defects. The GALNTs are found throughout the animal kingdom but not in bacteria, yeast, or plants. All GALNTs are classified in the GT27 CaZy family with a GT-A fold (Chapter 8), and most have a lectin (ricin-like) domain at the carboxyl terminus, which is unique among glycosyltransferases. GALNTs coordinate the transfer of GalNAc from the donor substrate (UDP-GalNAc) to the hydroxyl group of Ser/Thr on acceptor substrates and fall into two general categories: those that require the presence of an extant GalNAc on a peptide or protein before they will add additional GalNAcs (glycopeptide-preferring transferases), and those that will transfer GalNAc to a modified or unmodified protein (peptide transferases). Crystal structures of a number of GALNTs have revealed key details in the mechanisms of action of these enzymes and their unique substrate specificities. The conserved DXH motif found in all GALNTs coordinates Mn^{++} and UDP-GalNAc binding. Acceptor substrate preferences are dictated by unique amino acids present within the catalytic domain of each GALNT. For example, some GALNTs have "proline pockets" within the catalytic domain, thus conferring a strong preference for Pro residues near the site of O-glycosylation. The lectin domain recognizes extant O-GalNAc residues on previously glycosylated substrates to position the catalytic domain for further GalNAc addition. More recent work has demonstrated that charged residues within the lectin domain can also influence the glycosylation of charged protein substrates (in the absence of prior glycosylation). Additionally, the flexible linker in between the catalytic and lectin domains (which varies in length and sequence among the GALNTs) has been shown to influence sites of GalNAc addition. Databases that estimate the likelihood of O-glycosylation at a specific site are based on known sequences around O-glycosylation sites, as well as in vitro determinations of sequence preferences for certain GALNTs (e.g., NetOGlyc and ISOGlyP). However, these predictions

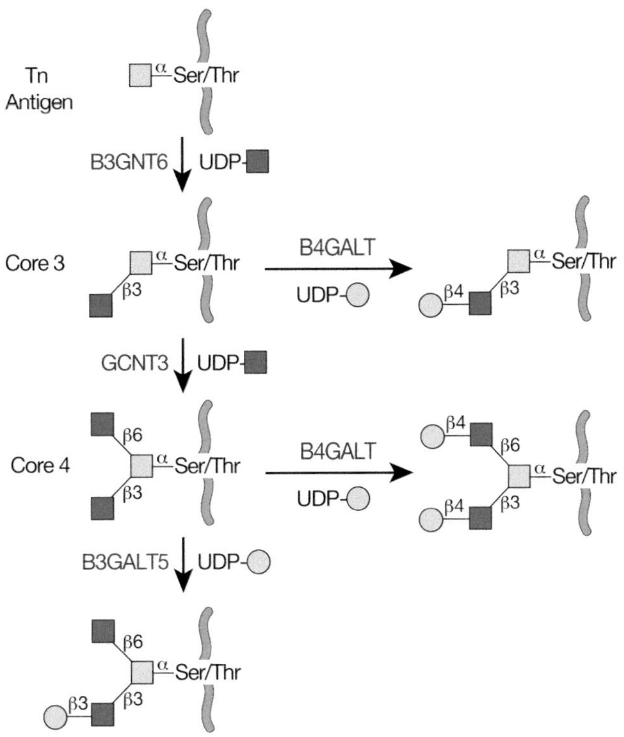

FIGURE 10.3. Biosynthesis of core 3 and 4 O-GalNAc glycans as described in the text. *Green lines* are protein.

do not account for the different GALNTs expressed in the cell type producing a particular glycoprotein and do not apply to mucins. Because of overlapping localization of GALNTs and the extension enzymes, it is likely that a heterogeneous mixture of different O-GalNAc glycans is present on all mature glycoproteins. Also, the presence of O-GalNAc glycans on proteins is likely often missed because of this limited predictive value of surrounding amino acid sequence.

Synthesis of O-GalNAc Glycan Cores

As mentioned above, O-GalNAc glycan synthesis begins with the transfer of GalNAc from UDP-GalNAc to Ser/Thr catalyzed by a GALNT. Although a single, unextended GalNAc linked to Ser/Thr (the Tn antigen) is uncommon in normal mucins, it is often found at increased levels in tumor mucins. This suggests that the extension of O-GalNAc glycans beyond the first sugar is blocked in some cancer cells. Sia added to GalNAc by the α2-6 sialyltransferase ST6GALNAC1 generates the sialyl-Tn antigen, which is common in advanced tumors. Other sugars are not known to be added to this O-glycan, but it can be O-acetylated by Sia O-acetyltransferase, which prevents detection by anti-sialyl-Tn antibodies.

The addition of one or two neutral sugars to O-GalNAc generates one of the different cores of O-GalNAc glycans (Table 10.1; Figure 10.2). Core 1 (Galβ1-3GalNAc-O-Ser/Thr) is generated by C1GALT1. This activity is present in most cell types but absolutely requires the molecular chaperone C1GALT1C1 (COSMC) during synthesis in the endoplasmic reticulum (ER) in mammalian cells to ensure proper folding and activity in the Golgi. Lack of core 1 synthesis in certain cell types can be due to either defective C1GALT1 or the absence of functional C1GALT1C1 and is reflected in high expression of Tn and sialyl-Tn antigens. For example, Jurkat T cells and colon cancer leukemic stem cells (LSCs) lack the C1GALT1C1 chaperone, and thus C1GALT1 activity, and have high expression of Tn and sialyl-Tn antigens.

A GlcNAc β1-6 branch added to the GalNAc residue of core 1 forms core 2 O-GalNAc glycans (Tables 10.1 and 10.2; Figure 10.2). Core 2 O-GalNAc glycans are more cell type–specific than the essentially ubiquitous core 1 O-GalNAc glycans, and their expression is highly regulated during activation of lymphocytes, cytokine stimulation, and embryonic development. Leukemia and cancer cells and other diseased tissues have abnormal amounts of core 2 O-GalNAc glycans. The enzymes responsible for core 2 synthesis are core 2 β1-6 N-acetylglucosaminyltransferases 1, 2, and 3 (GCNT1, GCNT3 and GCNT4). These glycosyltransferases do not require divalent cations as cofactors, and X-ray crystallography shows that positively charged amino acids replace the function of divalent metal ions. There are two different types of core 2 β1-6 N-acetylglucosaminyltransferases. One type synthesizes only core 2 O-GalNAc glycans (GCNT1, the leukocyte or L-type enzyme, and GCNT4), whereas the M-type enzyme (mucin type, GCNT3) synthesizes core 2 and 4 O-GalNAc glycans (Table 10.1; Figure 10.3). The L-type enzyme is active in many tissues and cell types, but the M type is found only in mucin-secreting tissues such as the intestine, stomach, and respiratory tissues. The expression and activity of both the L and M types are altered in certain tumors. The synthesis of core 2 O-GalNAc glycans has been correlated with tumor metastasis, possibly because of selectin ligands that are preferentially assembled on core 2 O-GalNAc glycans and that facilitate egress from the circulation (Chapter 34).

The synthesis of core 3 O-GalNAc glycans appears to be restricted mostly to mucus epithelia from the gastrointestinal and respiratory tracts and to the salivary glands. The enzyme responsible is core 3 β1-3 N-acetylglucosaminyltransferase 6 (B3GNT6) (Table 10.2, Figure 10.3). The enzyme has low in vitro activity but must be highly efficient in vivo because colonic mucins are rich in core 3 O-GalNAc glycans. The expression and activity of B3GNT6 is especially low in colonic tumors and virtually absent from tumor cells in culture. Overexpression of the enzyme in colon cancer cells decreases their ability to metastasize. Mice deficient in B3GNT6 show increased susceptibility to colitis and tumor development. The synthesis of core 4 by the M-type GCNT3 requires the prior synthesis of a core 3 O-GalNAc glycan (Figure 10.3). Transfection of colon cancer cells HCT116 with GCNT3 suppresses cell growth and invasive properties. In a xenograft model in nude mice, transfection with GCNT3 also suppresses tumor growth. Thus, both core 3 and 4 O-GalNAc glycans repress tumor progression, although the mechanisms of repression are not clear.

Synthesis of Complex O-GalNAc Glycans

The elongation of O-GalNAc glycans is catalyzed by families of β1-3 GlcNAc-transferases and β1-3 and β1-4 Gal-transferases to form repeated type 1 and 2 poly-N-acetyllactosamines (Table 10.1). Although most of the extension enzymes act on a number of different glycans, core 3 is a preferred acceptor for β3GalT5 (B3GALT5). In addition, the Galβ1-3 residues of core 1 and 2 O-GalNAc glycans are preferred substrates for the elongation enzyme β1-3 N-acetylglucosaminyltransferase 3 (B3GNT3). Less common elongation reactions are the formation of GalNAcβ1-4GlcNAc (LacdiNAc) and Galβ1-3GlcNAc- sequences. Linear poly-N-acetyllactosamine units can be branched by members of the β1-6 N-acetylglucosaminyltransferase family (e.g., GCNT2), resulting in the I antigen (Table 10.2). Most of these elongation and branching reactions also occur on O- and N-glycans and glycolipids.

Some sialyltransferases and sulfotransferases prefer O-GalNAc glycans as substrates but many of these enzymes have an overlapping specificity and also act on N-glycans. A family of α2-6 sialyltransferases (ST6GALNAC1–ST6GALNAC4) with distinct specificities synthesizes sialyl-Tn and sialylated core 1 O-GalNAc glycans. A family of α2-3 sialyltransferases is responsible for the synthesis of sialylated O-GalNAc glycans, with ST3GAL1 being mainly involved in the sialylation of the Galβ1-3 residue of core 1 and 2 O-GalNAc glycans. Sialylation blocks the further linear extension of O-glycan chains.

Sulfotransferases are localized in the Golgi and cap O-GalNAc glycans with a sulfate ester linked to the 3-position of Gal or the 6-position of GlcNAc. The sulfate group is transferred from 3′-phosphoadenosine-5′-phosphosulfate (PAPS). This adds a negative charge to O-GalNAc glycans of lung, intestinal, and other mucins that has a considerable effect on the chemical and metal ion binding properties of these glycans. GAL3ST4 is the major sulfotransferase acting on the Gal residue of core 1 O-glycans. Skeletal type keratan sulfate (KS) is also an O-GalNAc-linked highly sulfated polysaccharide (Chapter 17). O-acetyltransferases that add O-acetyl esters to one or more hydroxyl groups of Sia residues remain poorly characterized. Some evidence suggests that the esters can be added to CMP-Sias before transfer.

The α1-2 fucosyltransferases FUT1 and FUT2 synthesize the blood group H determinant of O-GalNAc glycans which can be converted by an α1-3 Gal-transferase to blood group B or by an α1-3 GalNAc-transferase to blood group A (Table 10.1). In addition, α1-3 and α1-3/4 fucosyltransferases synthesize the Lewis antigens (Table 10.1). A number of uncommon and antigenic sugars are also found on O-GalNAc glycans. For example, neuropilin-2 in the nervous system has core 1 and 2 O-GalNAc glycans that carry polysialic acid residues synthesized by polysialytransferase IV (ST8SIA4). These highly charged glycans play a critical role in the negative regulation of cell adhesion during maturation of the nervous system. An α1-4 GlcNAc-transferase in gastric tissue adds GlcNAc in α1-4 linkage to β1-4 Gal in core 1 and 2 O-GalNAc glycans. The α1-4 GlcNAc-containing glycans appear to inhibit colonization by *Helicobacter pylori.*

FUNCTIONS OF O-GalNAc GLYCANS

The functions of O-GalNAc glycans are many and varied, depending on their structure and density, as well on as the protein to which they are attached. As mentioned above, in densely glycosylated proteins such as mucins, O-glycans aid in hydration, structural support, interaction with the microbiome, and protection from proteolysis. In contrast, the function of single sites of O-GalNAc glycosylation vary widely and are still being determined. For example, O-GalNAc glycosylation at one or a few sites in certain proteins has been shown to regulate proprotein convertase cleavage, shedding of ectodomains, ligand binding, and cell–cell and cell–matrix interactions, influencing tissue formation and differentiation. Moreover, O-glycans, either in defined sites or clusters, may serve as carriers of terminal ligands such as blood type antigens like sialyl-Lewis x, which are important for host–pathogen interactions, and the circulation and homing of immune cells. The ligands for certain selectin-mediated interactions between endothelial cells and leukocytes require sialyl-Lewis x that is commonly attached to core 2 O-GalNAc glycans (Table 10.1). Finally, O-glycan terminal motifs on some immune cell subtypes can engage specific Siglecs and induce tolerance to self-antigens.

Methods to determine the function of O-GalNAc glycans are diverse and include the use of biosynthetic inhibitors, the construction of cell lines that lack or overexpress specific enzymes in O-glycosylation pathways, the use of O-GalNAc glycan-specific lectins or antibodies, removal of specific sugar residues by glycosidases, and deletion/mutation of specific GALNTs in model organisms. For example, a critical role of O-GalNAc glycans in selectin-mediated cell adhesion was revealed by treating cells with GalNAc-O-benzyl, which is a competitive inhibitor of core 1 and core 2 synthesis. GalNAc-O-benzyl acts as a decoy substrate for C1GALT1 and thereby reduces the synthesis of core 1 and core 2 O-GalNAc glycans on glycoproteins. Inhibitor-treated cancer cells lose the ability to bind to E-selectin and endothelial cells in vitro. Cancer cells often express sialyl-Lewis x and may thus use the selectin-binding properties of sialyl-Lewis x as a mechanism to invade tissues. Small molecule inhibitors that inactivate GALNTs are being developed to block the initiation of all O-GalNAc glycans by each GALNT.

Cell lines engineered to express altered O-GalNAc glycans, as well as mice with targeted mutations, are excellent models to identify roles for O-GalNAc glycosylation. Cells lacking C1GALT1C1 have been developed to identify a partial O-GalNAc proteome (i.e., the specific locations of core 1 or core 2 O-GalNAc glycans in the glycoproteome). The same approach has identified roles for core 1 and core 2 O-GalNAc glycans in cell transformation and cancer cell progression. Mice lacking C1GALT1, and thus lacking core 1 and core 2 O-GalNAc glycans, die during embryogenesis because of defective angiogenesis and hemorrhaging (Chapter 41). Moreover, a variety of phenotypes have been observed in mice with conditional deficiencies of core 1 O-glycans in specific tissues including spontaneous colitis, thrombocytopenia, defective lymphocyte homing, defects in podocyte function, and blood/lymphatic misconnections. Mice that lack core 1 and core 2 O-glycans within intestinal epithelial cells develop spontaneous duodenal tumors, illustrating the role of O-glycans in intestinal protection and homeostasis. Similarly, the influence of individual GALNTs has also been examined in model organisms and human tissue models. Loss or down-regulation of individual GALNTs have shown the importance for epithelial differentiation in human models. In *Drosophila melanogaster*, loss of GALNTs results in loss of viability, and defects in secretion and secretory vesicle formation, leading to epithelial cell damage and loss of cell–cell adhesion (Chapter 26). Mice deficient for GALNT1 display defects in cardiac development, hemostasis, immune cell homing, and ECM composition. Humans and animal models lacking GALNT2 result in a complex phenotype that includes dyslipidemia, hypercholesterolemia, and neurodevelopmental disorders. In humans, mutations in GALNT3 result in the disease hyperphosphatemic familial tumoral calcinosis, characterized by high blood phosphate levels and the development of calcified tumors. This disease results from the loss of GALNT3-mediated glycosylation that protects the phosphate regulating hormone FGF23, from inactivating cleavage. Mouse models of this disease display similar phenotypes, along with disruption of the oral microbiome. Finally, mice deficient for GALNT11 exhibit low-molecular-weight proteinuria, because of glycosylation changes in the proximal tubule endocytic receptor megalin that alter its ability to bind ligands.

Given the many potential functions of O-GalNAc glycans, there are surprisingly few direct medical and biotherapeutic applications. One of the most prominent examples includes the targeting of PSGL-1 in inflammatory conditions, either by antibodies or small molecule mimetics, which most recently has been successfully applied to children with sickle cell crisis. Cancer is another area in which disease-specific changes in O-GalNAc glycans can be exploited. Because of the association of increased levels of Tn, sialyl-Tn, and T antigens with cancer, several vaccine candidates based on these cancer-associated O-GalNAc-glycans are being investigated. An alternative and promising strategy is the use of high-affinity antibodies that are selective for cancer-associated O-GalNAc-glycans. Such antibodies may be especially effective with the new potent antibody-based therapeutic strategies, including Bi-specific T-cell engagers (BiTEs) and chimeric antigen receptors inserted in cytotoxic T cells (CAR-Ts). Indeed, CAR-Ts directed to glycopeptide epitopes in MUC1 have shown antitumor effects in preclinical animal models.

ACKNOWLEDGMENTS

The authors acknowledge the contributions of Harry Schachter to previous versions of this chapter.

FURTHER READING

Brockhausen I. 2010. Biosynthesis of complex mucin-type O-glycans. In *Comprehensive natural products. II: chemistry and biology* (ed. L Mander, H-W Lui, PG Wang), Vol. 6, pp. 315–350. Elsevier, Oxford. doi:10.1016/b978-008045382-8.00643-2

Jensen PH, Kolarich D, Packer NH. 2010. Mucin-type O-glycosylation—putting the pieces together. *FEBS J* **277**: 81–94. doi:10.1111/j.1742-4658.2009.07429.x

Bennett EP, Mandel U, Clausen H, Gerken TA, Fritz TA, Tabak LA. 2012. Control of mucin-type O-glycosylation: a classification of the polypeptide GalNAc-transferase gene family. *Glycobiology* **22**: 736–756. doi:10.1093/glycob/cwr182

Brockhausen I, Gao Y. 2012. Structural glycobiology: applications in cancer research. In *Structural glycobiology* (ed. E Yuriev, PA Ramsland), pp. 177–213. CRC Press, Boca Raton, FL. doi:10.1201/b12965

Steentoft C, Vakhrushev SY, Joshi HJ, Kong Y, Vester-Christensen MB, Schjoldager KT, Lavrsen K, Dabelsteen S, Pedersen NB, Marcos-Silva L, et al. 2013. Precision mapping of the human O-GalNAc glycoproteome through SimpleCell technology. *EMBO J* **32**: 1478–1488. doi:10.1038/emboj.2013.79

de Las Rivas M, Lira-Navarrete E, Gerken TA, Hurtado-Guerrero R. 2018. Polypeptide GalNAc-Ts: from redundancy to specificity. *Curr Opin Struct Biol* **56**: 87–96. doi:10.1016/j.sbi.2018.12.007

Khoo KH. 2019. Advances toward mapping the full extent of protein site-specific O-GalNAc glycosylation that better reflects underlying glycomic complexity. *Curr Opin Struct Biol* **56**: 146–154. doi:10.1016/j.sbi.2019.02.007

Tian E, Wang S, Zhang L, Zhang Y, Malicdan MC, Mao Y, Christoffersen C, Tabak LA, Schjoldager KT, Ten Hagen KG. 2019. Galnt11 regulates kidney function by glycosylating the endocytosis receptor megalin to modulate ligand binding. *Proc Natl Acad Sci* **116**: 25196–25202. doi:10.1073/pnas.1909573116

van Tol W, Wessels H, Lefeber DJ. 2019. O-glycosylation disorders pave the road for understanding the complex human O-glycosylation machinery. *Curr Opin Struct Biol* **56**: 107–118. doi:10.1016/j.sbi.2018.12.006

Bagdonaite I, Pallesen EM, Ye Z, Vakhrushev SY, Marinova IN, Nielsen MI, Kramer SH, Pedersen SF, Joshi HJ, Bennett EP, Dabelsteen S, Wandall HH. 2020. O-glycan initiation directs distinct biological pathways and controls epithelial differentiation. *EMBO Rep* **21**: e48885. doi:10.15252/embr.201948885

May C, Ji S, Syed ZA, Revoredo L, Paul Daniel EJ, Gerken TA, Tabak LA, Samara NL, Ten Hagen KG. 2020. Differential splicing of the lectin domain of an *O*-glycosyltransferase modulates both peptide and glycopeptide preferences. *J Biol Chem* **295**: 12525–12536. doi:10.1074/jbc.ra120.014700

Wandall HH, Nielsen MAI, King-Smith S, de Haan N, Bagdonaite I. 2021. Global functions of O-glycosylation: promises and challenges in O-glycobiology. *FEBS J* doi:10.1111/febs.16148

11 Glycosphingolipids

Ronald L. Schnaar, Roger Sandhoff, Michael Tiemeyer, and Taroh Kinoshita

Glycosphingolipids (GSLs), a subclass of glycolipids found in the cell membranes of organisms from bacteria to humans, are the major glycolipids of animals. The emphasis of this chapter is on vertebrate glycosphingolipids. Information on glycolipids of fungi, plants, and invertebrates is covered elsewhere (Chapters 20 and 23–26), as are glycosylphosphatidylinositols (GPIs), glycolipids attached to proteins as membrane anchors (Chapter 12). This chapter describes the characteristic features of GSLs, pathways for their biosynthesis, and insights into their biological roles in membrane structure, host–pathogen interactions, cell–cell recognition, and modulation of membrane protein function.

DISCOVERY AND BACKGROUND

The first GSL to be characterized was galactosylceramide (GalCer). Among the simplest of glycolipids, it is also one of the most abundant molecules in the vertebrate brain. It consists of a single galactose residue in β-glycosidic linkage to the C-1 hydroxyl group of a lipid moiety called ceramide (Figure 11.1). The structure of the lipid, a long-chain amino alcohol in amide linkage to a fatty acid, was so difficult to determine that the amino alcohol was dubbed "sphingosine" after the enigmatic Egyptian Sphinx. Animal cells synthesize various sphingosines and related long-chain amino alcohols, together referred to as sphingoid bases. Nearly all glycolipids in vertebrates are GSLs, which, in turn, are part of the larger family of sphingolipids (lipids built on sphingoid bases) that includes the major membrane phospholipid, sphingomyelin, and the second messenger sphingosine 1-phosphate that regulates angiogenesis and immune cell

published by Cold Spring Harbor Laboratory Press; doi: 10.1101/glycobiology.4e.11

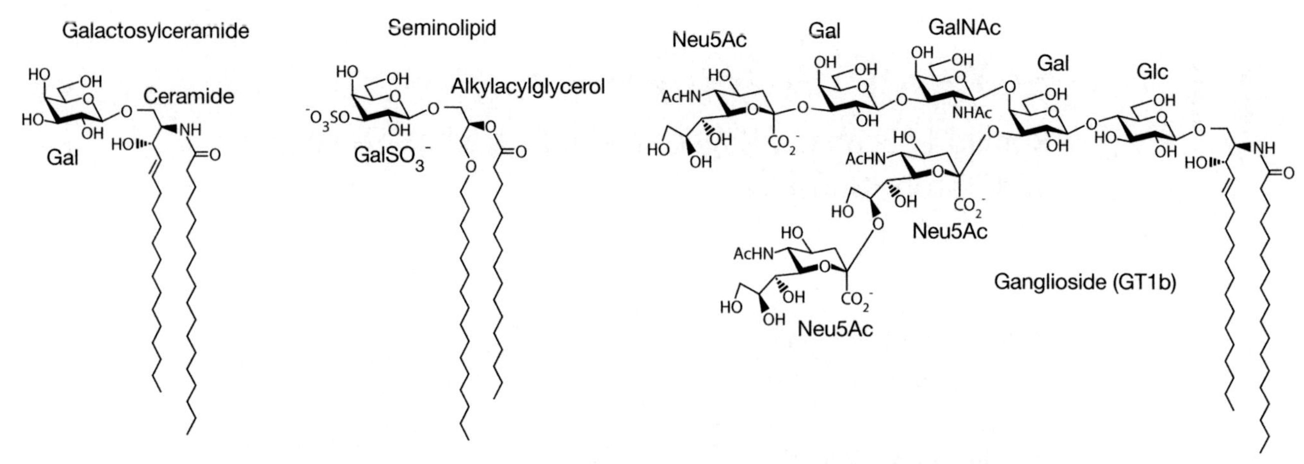

FIGURE 11.1. Structures of representative glycosphingolipids (GSLs) and glycoglycerolipids. GSLs, such as GalCer, are built on a ceramide lipid moiety that consists of a long-chain amino alcohol (sphingosine) in amide linkage to a fatty acid. In comparison, glycoglycerolipids, such as seminolipid, are built on a diacyl or alkyl(acyl)glycerol lipid moiety. Most animal glycolipids are GSLs, which have a large and diverse family of glycans attached to ceramide. Shown is one example of a complex sialylated GSL, GT1b ($IV^3Neu5AcII^3[Neu5Ac]_2Gg_4Cer$).

trafficking. Other GSLs were later identified because they accumulate to pathological levels in tissues of patients with lysosomal storage diseases, genetic disorders in which glycan degrading enzymes are faulty or missing (Chapter 44). For example, a sialic acid–containing GSL (GM2) was first isolated from the brain of a victim of Tay–Sachs disease, in which it accumulates, and was named "ganglioside" based on its location in nerve clusters or "ganglia" in the brain. Likewise, glucosylceramide (GlcCer) was first isolated from the spleen of a Gaucher disease patient, where it accumulates. As purification, separation, and analytical techniques improved, GSLs were found in all vertebrate tissues. Hundreds of unique GSL structures were found that vary in glycan structures alone, each of which are presented on several distinct ceramides.

Glycoglycerolipids are distinguished from GSLs by their lipid, having glycans linked to the C-3 hydroxyl of diacylglycerol or alkyl(acyl)glycerol (Figure 11.1). Very minor constituents of most animal tissues (other than the testes), glycoglycerolipids are widely distributed in microbes and plants. Glycolipids of fungi, plants, and invertebrates are covered in Chapters 20 and 23–26. Glycosylphosphatidylinositols (GPIs), a different family of glycolipids, are often attached to proteins as membrane, and may also exist as free glycolipids (discussed in Chapter 12). The lipopolysaccharides of Gram-negative bacteria are discussed in Chapter 21.

MAJOR CLASSES AND NOMENCLATURE

The ceramide lipid of GSLs consists of a sphingoid base with a fatty acid amide at the C-2 amine. Sphingosine is the most common sphingoid base in mammals, with hydroxyls at the C-1 and C-3 carbons and a *trans* double bond between C-4 and C-5 (Figure 11.1). Sphinganine is the same structure without the double bond, and phytosphingosine lacks the double bond and has an additional hydroxyl on C-4. Ceramides containing sphinganine and phytosphingosine are less abundant in animals, whereas phytosphingosine is prominent in the glycosphingolipids of plants and fungi. The fatty acid components of ceramides vary widely, with lengths ranging from C14 to C30 or greater. Although often saturated, they may be variably unsaturated or have α-hydroxyl groups. Ceramide structures modulate membrane associations and functions of GSLs.

Although ceramide variations add significant diversity to GSL structures, major structural and functional classifications are based on the glycans. The first sugars linked to ceramide in vertebrates are typically β-linked galactose (GalCer) or glucose (GlcCer). GalCer and its analog sulfatide, with sulfate at the C-3 hydroxyl of galactose, are the major glycans in the kidney and the brain. In the brain they have essential roles in the structure and function of myelin, the insulator that allows for rapid nerve conduction. Interestingly, the related sulfogalactoglycerolipid, seminolipid (Figure 11.1), is abundant only in the male reproductive tract, where it is essential for spermatogenesis. Sialylated GalCer (Neu5Acα2-3GalβCer; GM4) is also found in myelin. These galactolipids are seldom extended with larger saccharide chains; rather, most other members of the large and diverse family of GSLs in animals are built on GlcCer (Figure 11.2). GlcCer itself is abundant in certain tissues. In skin, it is a precursor of special ceramides that are required for creating the essential surface water barrier when embedded into the stratum corneum of the epidermis (see below).

The vast majority of GSL structures are classified based on seven common tetrasaccharide neutral sugar core sequences (Figure 11.2). GSLs sharing the same neutral core sequence are said to be of that "series"; the quantitatively major series in vertebrates are ganglio-, globo-, and neolacto-series, whereas in invertebrates the mollu- and arthro-series predominate.

GSL series are expressed in tissue-specific patterns. In mammals, ganglio-series GSLs, although broadly distributed, predominate in the brain, whereas neolacto-series glycolipids are common on certain hematopoietic cells including leukocytes. In contrast, lacto-series glycolipids are prominent in secretory organs and globo-series glycolipids are the most abundant in erythrocytes. This diversity presumably reflects important differences in GSL functions.

GSLs are further subclassified as neutral (no charged sugars or ionic groups), sialylated (having one or more sialic acid residues), or sulfated. Traditionally, all sialylated GSLs are known as "gangliosides" regardless of whether they are based on the ganglio-series neutral sugar core. In the official nomenclature, saccharide and other substituents that extend or branch from the neutral core structures are indicated by a roman numeral designating which of the neutral core sugars carries the substituent (counting the sugar closest to the ceramide as "I"), and a superscript designating which hydroxyl on that sugar is modified (Figure 11.2). This nomenclature is too complex for daily use, so the most common GSLs are usually referred to by unofficial names. For example, in the widely used nomenclature of Svennerholm, the ganglioside Galβ1-3GalNAcβ1-4(Neu5Acα2-3)Galβ1-4GlcβCer is designated "GM1" (Figure 11.3). In this nomenclature, G refers to ganglioside series, the second letter refers to the number of sialic acid residues (mono, di, tri, quattro [or tetra], penta, etc.), and the number (1, 2, 3, etc.) refers to the order of migration of the ganglioside on thin-layer chromatography (TLC) (e.g., GM3 > GM2 > GM1) with respect to the starting point. Therefore, the neutral core oligosaccharide shrinks in size (and moves farther by TLC) as the number increases.

series	symbol	IV–III	III–II	II–I
ganglio	Gg	β3	β4	β4
lacto	Lc	β3	β3	β4
neolacto	nLc	β4	β3	β4
globo	Gb	β3	α4	β4
isoglobo	iGb	β3	α3	β4
mollu	Mu	β2	α3	β4
arthro	At	β4	β3	β4

FIGURE 11.2. Glycosphingolipid (GSL) neutral cores and their designation based on IUPAC (International Union of Pure and Applied Chemistry) Nomenclature. In the official nomenclature, saccharide and other substituents that extend or branch from the neutral core structures are indicated by a roman numeral designating which of the neutral core sugars (counting the sugar closest to the ceramide as "I") carries the substituent, and a superscript designating which hydroxyl on that sugar is modified. For example, see ganglioside GT1b ($IV^3Neu5AcII^3[Neu5Ac]_2Gg_4Cer$) in Figure 11.1. See Online Appendix 1B for full names and designated symbols for monosaccharides.

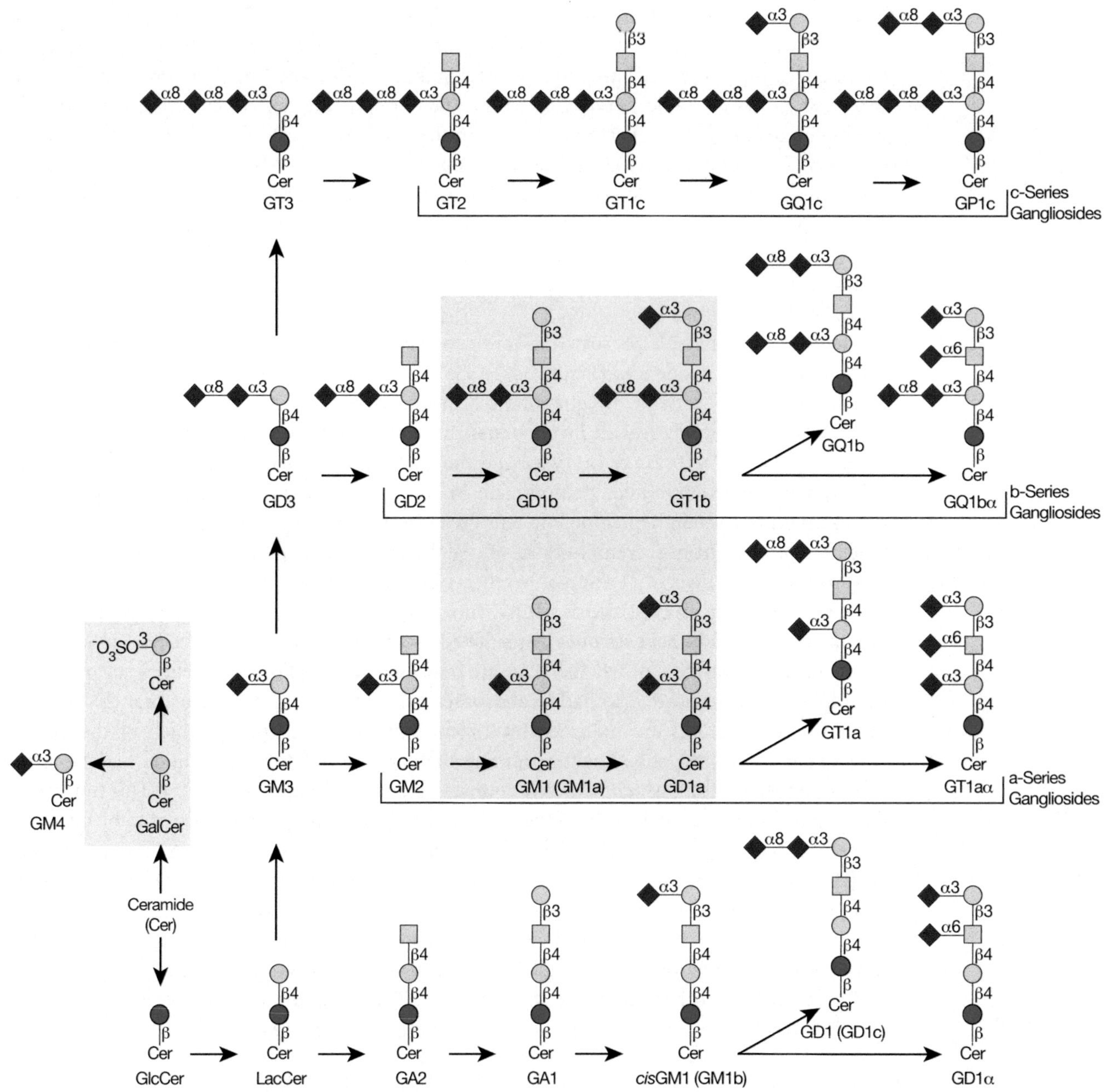

FIGURE 11.3. Glycosphingolipids (GSLs) are synthesized by the stepwise addition of sugars first to ceramide and then to the growing glycans. Shown as examples are major brain GSLs. Ceramide (Cer) is the acceptor for UDP-Gal:ceramide β-galactosyltransferase or UDP-Glc:ceramide β-glucosyltransferase in the major pathways to biosynthesis of the quantitatively major GSLs in oligodendrocytes (shaded *pink*) and nerve cells (shaded *blue*), respectively. GalCer is the acceptor for GalCer sulfotransferase, which adds a sulfate group to the C-3 of galactose to form sulfatide. Extension of GlcCer to the major brain gangliosides occurs by the action of UDP-Gal:GlcCer β1-4 galactosyltransferase to make lactosylceramide (LacCer) and then CMP-Neu5Ac:lactosylceramide α2-3 sialyltransferase to make the simple ganglioside GM3. GM3 is a branch point and acts as the acceptor for UDP-GalNAc:GM3/GD3 β1-4 N-acetylgalactosaminyltransferase to generate a-series gangliosides and for CMP-Neu5Ac:GM3 α2-8 sialyltransferase to generate GD3 and the b-series gangliosides. Similarly, the action of an α2-8 sialyltransferase on GD3 gives rise to GT3 and the c-series gangliosides. Enzymes for subsequent elongation are common to the a-, b-, and c-series gangliosides.

ISOLATION, PURIFICATION, AND ANALYSIS

Organic solvents are used to extract glycolipids from tissues and cells, where they are found primarily in the external leaflet of plasma membranes. Extraction procedures, most often using defined chloroform–methanol–water mixtures, are optimized to precipitate and remove proteins and nucleic acids while maximizing solubilization of GSLs (along with other lipids). Because GSLs aggregate with one another and other lipids in aqueous solution, organic solvents are used throughout subsequent purification steps, which typically involve solvent partition, ion-exchange, and adsorption chromatographies.

Because of their amphipathic nature, glycolipids are well-suited for TLC analysis, which is useful for monitoring their purification, qualitative and quantitative determination of their expression in normal and diseased tissues, partial structural analysis, and detecting biological activities including immunoreactivity and binding activity toward toxins, viruses, bacteria, and eukaryotic cells. After separation by TLC, picomole to nanomole quantities of glycolipids can be chemically detected with orcinol reagent for hexoses and with resorcinol-HCl reagent for sialic acid. Normal phase TLC robustly resolves GSLs based on oligosaccharide diversity but is less effective at separating differences due to ceramides. To study glycolipid receptors, TLC-resolved glycolipids are overlaid with potential binding proteins or organisms (e.g., antibodies, lectins, toxins, viruses, bacteria, or cells). After washing, glycolipid species that bind can be identified by detection of the bound material at precise positions on the TLC plate. Two-dimensional TLC with exposure to ammonia vapor after the first dimension can be useful in identifying the presence of O-acetyl modifications on sialic acids, which are especially common in the brain.

Complete structural analyses of glycolipids require a combination of techniques to determine the composition, sequence, linkage positions, and anomeric configurations of the glycan moiety and the fatty acid and long-chain base of the ceramide moiety. Glycan composition is determined by hydrolysis and analysis of the released monosaccharides. Mass spectrometry (MS) of underivatized GSLs or of their permethylated derivatives is a powerful tool for sequence determination and ceramide identification and can sometimes be performed directly from the TLC plate. However, the methylation process causes destruction of labile substitutions like O-acetyl esters. Linkage determination is most rigorously performed by methylation analysis, and anomeric configurations can be obtained by nuclear magnetic resonance (NMR) spectroscopy. Information on glycan sequence, linkage, and anomeric configuration can also be obtained by combining enzymatic hydrolysis using specific glycosidases with TLC. Intact glycans are released from many GSLs enzymatically using ceramide glycanases or endoglycoceramidases. The resulting oligosaccharides can be analyzed using the methods described in Chapter 50. Additionally, there are many monoclonal antibodies that detect specific glycosphingolipid structures.

BIOSYNTHESIS, TRAFFICKING, AND DEGRADATION

Glycosphingolipid biosynthesis occurs in a stepwise fashion, with an individual sugar added first to ceramide and then subsequent sugars transferred by glycosyltransferases from nucleotide sugar donors. Ceramide is synthesized on the cytoplasmic face of the endoplasmic reticulum (ER), equilibrates to the luminal face, and then traffics to the Golgi apparatus. GlcCer is synthesized on the cytoplasmic face of the early Golgi apparatus. It is flipped to the luminal side of the Golgi (or after retrograde transport to the ER), and then is elongated by a series of Golgi glycosyltransferases. In contrast, GalCer is synthesized on the luminal face of the ER and then traffics through the Golgi, where it may be sulfated to form sulfatide. In both cases, the final orientation of GSLs during biosynthesis is consistent with their nearly exclusive appearance on the outer leaflet of the plasma membrane, facing the extracellular milieu. Specific exceptions occur, such as the high concentration of lactosylceramide in neutrophil granules. Although ceramide resides on intracellular organelles, such as mitochondria, GSLs beyond GlcCer are not typically found on the membrane leaflet facing the cytoplasm, although they appear in the nucleus.

The biosynthesis of GSLs in the brain provides an example of how competing biosynthetic pathways can lead to glycan structural diversity (Figure 11.3). Stepwise biosynthesis of GalCer and sulfatide occurs in oligodendrocytes, the cells that elaborate myelin. Gangliosides, in contrast, are synthesized by all cells, with concentrations of the different forms varying according to cell type. Expression patterns of GSLs are determined by the expression and intracellular distribution of the enzymes required for their biosynthesis. In some cases, multiple glycosyltransferases compete for the same GSL precursor (Figure 11.3). For example, the ganglioside GM3 may be acted on by N-acetylgalactosaminyltransferase, thereby forming GM2 and going down the "a-series" gangliosides, or by sialyltransferase, thereby forming GD3, the simplest of the "b-series" gangliosides. Each branch is a committed pathway, because sialyltransferases cannot directly convert a-series gangliosides (beyond GM3) to their corresponding b-series gangliosides. Because of this branch exclusivity, competition between two enzymes at a key branch point determines the relative expression levels of the final GSL products. The transfer of *N*-acetylgalactosamine to a-, b-, and c-series gangliosides, transforming GM3 into GM2, GD3 into GD2, or GT3 into GT2, is catalyzed by the same N-acetylgalactosaminyltransferase. Likewise, the transfer of galactose to GM2 to form GM1, to GD2 to form GD1b, or to GT2 to form GT1c is accomplished by a single galactosyltransferase. In addition, the levels of nucleotide sugar donors used by glycosyltransferases in the Golgi lumen (including UDP-Gal, UDP-Glc, UDP-GlcNAc, UDP-GalNAc, and CMP-Neu5Ac) ultimately affect the final structure of glycans and are regulated by synthetic enzymes in the cytoplasm or nucleus and by the activity of nucleotide sugar transporters in the Golgi membrane (Chapter 5). The sialic acids of human gangliosides are exclusively in the form of *N*-acetylneuraminic acid (Neu5Ac) and its O-acetylated derivatives, but those of many other mammals contain both Neu5Ac and *N*-glycolylneuraminic acid (Neu5Gc) (Chapter 15). Even in animals with predominantly Neu5Gc in the gangliosides of nonneural tissues, their brain gangliosides have nearly exclusively Neu5Ac. Sialic acids on gangliosides may be further modified by O-acetylation or removal of the N-acyl group, to generate a free amino group (Chapter 15).

An additional level of regulation occurs via stable association of different GSL glycosyltransferases into "multiglycosyltransferase" complexes. The multiple enzymes are thought to act concertedly on the growing GSL without releasing intermediate structures, ensuring progression to the preferred end product.

Although the enzymes that catalyze the initial steps in GSL biosynthesis are specific and used only for GSL biosynthesis, outer sugars, such as the outermost sialic acid, fucose, or glucuronic acid residues, are sometimes added by glycosyltransferases that also act on glycoproteins, resulting in terminal structures being shared by GSL and glycoprotein glycans (Chapter 14). One example is the blood group ABO antigen system. The α1-3 N-acetylgalactosaminyltransferase encoded by the blood group A gene produces the blood group A determinant GalNAcα1-3(Fucα1-2)Galβ on glycoproteins and glycolipids. Correspondingly, the α1-3 galactosyltransferase encoded by the allelic blood group B gene transfers galactose to both glycoproteins and glycolipids.

Major brain ganglioside structures are highly conserved among all mammals and birds. In contrast, differences in blood cell glycolipids are well known even among humans, as in the case of the ABO, Lewis, and P blood group antigens (Chapter 14). Species differences among glycolipids also occur, one example being the expression of Forssman antigen, GalNAcα1-3GalNAcβ1-3Galα1-4Galβ1-4GlcβCer. This molecule is a good immunogen in Forssman antigen–negative species, such as rabbit, rat, and human, which have a mutated α1-3 N-acetylgalactosaminyltransferase that cannot transfer *N*-acetylgalactosamine to the precursor Gb4Cer. In contrast, guinea pig, mouse, sheep, and goat are Forssman antigen–positive.

The breakdown of GSLs occurs stepwise by the action of lysosomal hydrolases. GSLs on the outer surface of the plasma membrane are internalized, along with other membrane components, in invaginated vesicles that then fuse with endosomes, resulting in the GSL glycan facing the endosome lumen. GSL-enriched areas of the endosomal membrane may then invaginate once again to form intraluminal vesicles within the endosome. When endosomes fuse with primary lysosomes, GSLs become exposed to lysosomal hydrolases. Absence of any one of these glycosidases results in lysosomal storage diseases (Chapter 44).

As GSLs are successively cleaved to smaller structures, the remaining "core" monosaccharides become inaccessible to the water-soluble lysosomal hydrolases and require assistance from activator proteins that are referred to as "liftases." These include GM2-activator protein and four structurally related saposins, all of which are derived from a single polypeptide precursor by proteolytic cleavage. Saposins are thought to bind to their glycolipid substrate, disrupt its interaction with the local membrane environment, and facilitate access of the glycans to the water-soluble hydrolytic enzymes. In certain lysosomal storage diseases (Chapter 44), mutations in activator proteins result in pathological accumulation of glycolipids, even though there is an abundance of the hydrolase responsible for degradation, thus demonstrating the essential role of activator proteins in GSL catabolism in vivo. The final breakdown products of GSL catabolism—monosaccharides, fatty acids, and free sphingoid bases—are then available for reuse by salvage pathways.

BIOLOGICAL AND PATHOLOGICAL ROLES

Glycosphingolipid-Enriched Membrane Microdomains

GSLs comprise from $<5\%$ (erythrocytes) to $>20\%$ (myelin) of the total membrane lipids in the plasma membranes of vertebrate cells. However, they are not uniformly distributed in the plane of the membrane but cluster in "lipid rafts," small lateral microdomains of self-associating membrane molecules. Although the precise structure and makeup of lipid rafts is a matter of ongoing debate, their outer leaflets are believed to be enriched in sphingolipids, including GSLs and sphingomyelin (the phosphocholine derivative of ceramide). The self-association of sphingolipids is driven through the unique biophysical properties imparted by their long saturated carbon chains (Figure 11.1). Besides sphingolipids, lipid rafts are enriched in cholesterol and selected proteins, including GPI-anchored proteins and some transmembrane signaling proteins such as receptor tyrosine kinases. On the cytoplasmic side, acylated proteins, such as Src family protein tyrosine kinases and Gα subunits of G proteins, associate with lipid rafts.

Lipid rafts are apparently dynamic, short-lived (msec), and small (10–50 nm in diameter), each containing perhaps hundreds of lipid molecules along with a few protein molecules. It has been argued that external clustering of lipid rafts into larger structures might bring signaling molecules such as kinases and their substrates together to enhance intracellular signaling. Thus, GSLs may act as intermediaries in the flow of information from the outside to the inside of cells. This idea is supported by the observation that antibody-induced GSL clustering activates lipid-raft-associated signaling and has led to the concept of plasma membrane "glycosignaling domains" or "glycosynapse." Interactions between glycans and glycan-binding proteins, as well as glycan–glycan interactions, are also very much influenced by the density of glycans in terms of binding affinity. Multiple glycans clustered in a limited area can increase the avidity of cognate binding proteins compared with a single molecule of a glycan (Chapter 29). Natural multivalency adds unique functional properties to plasma membrane glycolipids. Indeed, several growth factor receptors including the epidermal growth factor (EGF) receptor, insulin receptor, and the nerve growth factor receptor are localized in membrane microdomains, and structural studies identified binding sites occupied by specific GSLs that modulate receptor signaling functions.

Physiological Functions of Glycosphingolipids

When expressed in the outer leaflet of the plasma membrane of cells, with their glycans facing the external milieu, GSL functions fall into two major categories: mediating cell–cell interactions via binding to complementary molecules on apposing plasma membranes (*trans* recognition) and modulating activities of proteins in the same plasma membrane (*cis* regulation).

At the single-cell level, GSLs are not essential for life. Using specific chemical inhibitors and genetic ablation of biosynthetic genes, cells without GSLs survive, proliferate, and even differentiate. However, GSLs are required for development at the whole-animal level. Mice engineered to lack the gene for GlcCer synthesis fail to develop, with arrest occurring just past the gastrula stage because of extensive apoptosis in the embryo. These and other recent observations lead to a basic principle: GSLs mediate and modulate intercellular coordination in multicellular organisms. Sometimes this occurs in quite subtle ways, as exemplified by the role of GalCer and sulfatide in myelination.

GalCer and its 3-O-sulfated derivative, sulfatide, are predominant glycans in the brain, where they constitute >50% of the total glycoconjugates. In the brain, they are expressed by oligodendrocytes, which elaborate myelin, the multilayered membrane insulation that ensheathes nerve axons. GalCer and sulfatide constitute >20% of myelin lipids and were widely believed to be essential to myelin structure. This turned out to be true, but in a much subtler way than anticipated. Mice engineered to lack the enzyme responsible for GalCer synthesis (UDP-Gal:ceramide β-galactosyltransferase) do not make any GalCer or sulfatide. However, they myelinate axons, and the myelin appears grossly normal. Nevertheless, the mice show all the signs of failed myelination, including tremor, ataxia, slow nerve conduction, and early death. In both normal and mutant mice, myelination occurs in short stretches along axons, with intermittent gaps called "nodes of Ranvier" where concentrated ion channels pass nerve impulses along to the next gap. At the edge of the node, myelin membranes normally curve downward and attach to the axon to seal off the node. In animals lacking GalCer and sulfatide, these myelin "end feet" fail to attach to the axon, instead turning upward, away from the axon. The result is a faulty node of Ranvier, with ion channels and adhesion molecules in disarray. Without the proper structure at the node of Ranvier, rapid nerve conduction is disrupted. A similar phenotype is shared by mice lacking the enzyme that adds the sulfate group to GalCer to make sulfatide. The conclusion is that sulfatide is essential for myelin–axon interactions, and its absence results in severe neurological deficits.

A key function of GlcCer has been learned from studies on its catabolism in postnatal animals. Ceramide is a key component of the outer layer of the skin and is responsible for the epidermal permeability barrier—a key defense against dehydration. Infants with severe Gaucher disease, in which β-glucocerebrosidase activity is nearly absent and GlcCer is not catabolized, are prone to dehydration because of high skin permeability. The relationship among GlcCer, ceramide, and skin permeability was confirmed in mice engineered with the same mutation in β-glucocerebrosidase as found in these infants. The mice unable to catabolize GlcCer died within days of birth by dehydration through the skin. This established the role of GlcCer as the obligate precursor to the ceramide required to build the outermost protective layer (stratum corneum) of the skin. GlcCer is synthesized, transported to the stratum corneum, and then enzymatically hydrolyzed, resulting in ceramide deposition.

More complex GSLs function both in cell–cell recognition and in the regulation of signal transduction. As with sulfatide, these functions are sometimes subtle, as exemplified by the effects of blocking ganglioside biosynthesis on nervous system physiology. Given the complexity of complex ganglioside biosynthesis (Figure 11.3), it was surprising to discover that major alterations in ganglioside expression resulted in only modest phenotypic changes in mice. When the N-acetylgalactosaminyltransferase responsible for ganglioside elongation (GM2/GD2 synthase) was inactivated in mice, none of the major complex gangliosides (GM1, GD1a, GD1b, or GT1b) were expressed, and a comparable concentration of the simple gangliosides GM3 and GD3 were found in the adult brain. The resulting mice were grossly normal, but as they aged, mice without the normal spectrum of brain gangliosides displayed signs of axon degeneration and demyelination, hallmarks of a problem in myelin–axon cell–cell communication. These deficits arise from loss of ganglioside binding to a protein on the myelin membrane, myelin-associated glycoprotein (MAG), a member of the Siglec family of sialic acid–dependent carbohydrate-binding proteins (Chapter 35). MAG is expressed on the innermost myelin wrap, directly across from the axon surface. Mice engineered to lack MAG have many of the same phenotypic changes as mice lacking GM2/GD2 synthase, and biochemical and cell-biological studies showed that the major brain gangliosides GD1a and GT1b are excellent ligands for MAG. These results support the conclusion that MAG on the innermost myelin membrane binds to GD1a and GT1b on the axon cell surface to stabilize myelin–axon interactions. Disruption of the same ganglioside biosynthetic gene in humans leads to a similar phenotype: hereditary spastic paraplegia.

A second *trans* recognition role for GSLs is evident in the interaction of leukocytes with the blood vessel wall during the process of inflammation, the body's protection against bacterial infection. As discussed in Chapter 34, the first step in inflammation is the binding of white blood cells (leukocytes) to the endothelial cells lining the blood vessel near sites of infection (activated endothelium). This cell–cell interaction is initiated when glycan-binding proteins of the selectin family, expressed on the activated endothelium, bind to complementary glycans on the surface of passing leukocytes. One of the selectins, E-selectin, binds to glycoconjugates found on human leukocytes. The glycan receptor(s) for E-selectin are resistant to protease treatment, indicating that they may be GSLs. A candidate class of GSLs, myeloglycans, has been identified in leukocytes. The candidate GSLs have long sugar chains consisting of a neutral core with Galβ1-4GlcNAcβ1-3 repeats (Chapter 14), substituted with a terminal sialic acid and fucose residues on one or more of the *N*-acetylglucosamine residues.

Another role for GSLs in mediating inflammatory responses involves lipid-specific antigen presentation. Natural killer T (NKT) cells, which carry both T- and NKT-cell receptors, are involved in the suppression of autoimmune reactions, cancer metastasis, and the graft rejection response. The MHC class I-related molecule (CD1d) of dendritic cells presents glycolipid antigens via T-cell receptor recognition to activate NKT cells. NKT cells can be activated by non-mammalian Galα-ceramide, expressed by some bacteria in the large intestine of mice, as well as the isoglobo-series GSL iGb_3Cer (Galα1-3Galβ1-4GlcβCer), which has been proposed as an endogenous NKT activator.

In addition to their action as *trans* recognition molecules, GSLs also interact laterally with proteins in the same membrane to modulate their activities (*cis* regulation). Notable among these *cis*-regulatory interactions are those between gangliosides and members of the receptor tyrosine kinase family. Gangliosides regulate the activity of the EGF receptor, platelet-derived growth factor receptor, fibroblast growth factor receptor, TrkA neurotrophin receptor, and insulin receptor. Ganglioside GM3, for example, down-regulates insulin receptor responsiveness to insulin. Mice that lack the enzyme responsible for the biosynthesis of GM3 display increased insulin receptor phosphorylation, enhanced glucose tolerance, and enhanced insulin sensitivity and are less susceptible to induced insulin resistance. *Cis*-regulatory interactions such as GM3-mediated insulin sensitivity may arise from direct interactions with signaling receptors or from the contribution of GSLs to the biophysical characteristics of membrane microdomains.

Glycosphingolipids in Human Pathology

Mutations in glycosphingolipid biosynthetic genes are exceedingly rare in humans, perhaps because of their devastating effects (Chapter 45). A single case of UDP-glucose ceramide glucosyltransferase (Ugcg) deficiency has been described, resulting in a lethal form of ichthyosis. This mutation evidently left some residual activity, because complete loss is associated with embryonic lethality in mice. Mutations in a gene required for biosynthesis of ganglioside GM3, *ST3GAL5*, result in severe infantile seizures accompanied by profound motor and intellectual deficits as well as blindness and deafness. Mutations in another ganglioside-specific biosynthetic gene, *B4GALNT1*, responsible for biosynthesis of GM2 and GD2, are less severe, resulting in hereditary spastic paraplegia accompanied by intellectual disability. Loss of globo-series GSLs does not cause obvious disease, but increases spontaneous abortions associated with the rare p blood group resulting from α4-galactosyltransferase deficiency. Mutations in GSL degradation genes, which are also rare, cause GSL storage diseases that lead to the accumulation of GSLs in lysosomes. They typically result from mutations in glycosidases, and less frequently from mutations in activator proteins (Chapter 44). The symptoms depend on the tissues in which the unhydrolyzed GSL accumulates and on the extent of loss of enzyme activity. The most common GSL storage disease is Gaucher disease, which is caused by mutations in the enzyme β-glucocerebrosidase, resulting primarily in accumulation of GlcCer in the liver and spleen (and other tissues in more severe cases). Enzyme replacement therapy has been successful in treating Gaucher disease, and drugs to block GlcCer synthesis ("substrate reduction therapy") are in clinical use (Chapter 55). Another example, Tay–Sachs disease, is caused by mutations in a β-hexosaminidase and results in the buildup of GM2, culminating in irreversible fatal deterioration of brain function. Unfortunately, enzyme replacement delivery to the brain has not yet been successfully developed, which is also a problem for long-term treatment of Gaucher disease. Glycosphingolipid storage and related diseases are considered more extensively in Chapter 44.

Anti-GSL antibodies are involved in certain autoimmune diseases (Chapter 46). Some forms of Guillain–Barré syndrome, the most common form of paralytic disease worldwide, clearly involve autoantibodies against gangliosides. One form of Guillain–Barré syndrome occurs subsequent to infection with particular strains of the common diarrheal bacterial agent *Campylobacter jejuni.* These bacteria produce near-exact replicas of brain ganglioside glycans (such as GD1a) attached to their lipooligosaccharide cores. Following infection and immune clearance of the bacteria, the antiglycan antibodies produced to fight the bacteria go on to attack the patient's own nerves, causing paralysis. In some patients with multiple myeloma (a malignancy of antibody-producing plasma cells), the tumor cells secrete monoclonal antibodies against glycolipids, such as the rare sulfoglucuronyl epitope of nervous system GSLs termed HNK-1 (IV^3GlcA[3-sulfate]-nLc_4Cer). These patients suffer severe peripheral neuropathy.

Several bacterial toxins take advantage of GSLs to gain access to cells (Chapter 37). Cholera toxin and the structurally related *Escherichia coli* heat-labile enterotoxins are produced in the intestinal tract of infected individuals, bind to intestinal epithelial cell surfaces, and insert their toxic polypeptide "payload" through the cell membrane, where it disrupts ion fluxes, causing severe diarrhea. The toxins behave as docking modules with gangliosides acting as the site of attachment. Five identical polypeptide B subunits in a ring each bind to ganglioside GM1 on the cell surface, and a sixth A subunit (the "payload") is then inserted through the membrane. A similar mechanism is used by *Shiga* toxin (also called verotoxin), which binds to the glycolipid Gb_3Cer (globotriaosylceramide, Galα1-4Galβ1-4GlcβCer) via five subunits in a ring, each with three GSL-binding sites. In contrast, tetanus and related botulinum toxins are multidomain single polypeptides. One domain binds b-series gangliosides on nerve cells, whereas the other domains translocate the toxin into cells and disrupt proteins essential for synaptic transmission. Custom-designed multivalent glycans and glycoconjugates are being evaluated as high-affinity blockers of certain bacterial toxins. In addition to soluble toxins, certain intact bacteria also bind to specific GSLs via bacterial surface proteins called adhesins. This adherence is essential for successful colonization and symbiosis. Microbial adhesins are addressed in more detail in Chapter 37.

Malignant transformation in cancer progression is often associated with changes in the glycan structures of glycoproteins and glycolipids. The changes result mainly from altered levels of glycosyltransferase activities involved in glycolipid biosynthesis. The increase of GD3 or GM2 in melanoma, and of sialyl-Lewis a antigen (Neu5Acα2-3Galβ1-3[Fucα1-4]GlcNAcβ1-3Galβ1-4GlcβCer) in gastrointestinal cancers, and of GD2 in neuroblastoma are typical examples (Chapter 47). Certain cancers also produce and shed gangliosides that have immunosuppressive effects.

ACKNOWLEDGMENTS

The authors acknowledge contributions to previous versions of this chapter by Akemi Suzuki and appreciate helpful comments and suggestions from Anabel Gonzalez-Gil, Tetsuya Okajima, and Ryan N. Porell.

FURTHER READING

Hakomori S. 1981. Glycosphingolipids in cellular interaction, differentiation, and oncogenesis. *Ann Rev Biochem* **50**: 733–764. doi:10.1146/annurev.bi.50.070181.003505

Todeschini AR, Hakomori S-I. 2008. Functional role of glycosphingolipids and gangliosides in control of cell adhesion, motility, and growth, through glycosynaptic microdomains. *Biochim Biophys Acta* **1780**: 421–433. doi:10.1016/j.bbagen.2007.10.008

Simons K, Gerl MJ. 2010. Revitalizing membrane rafts: new tools and insights. *Nat Rev Mol Cell Biol* **11**: 688–699. doi:10.1038/nrm2977

Merrill AH Jr. 2011. Sphingolipid and glycosphingolipid metabolic pathways in the era of sphingolipidomics. *Chem Rev* **111**: 6387–6422. doi:10.1021/cr2002917

D'Angelo G, Capasso S, Sticco L, Russo D. 2013. Glycosphingolipids: synthesis and functions. *FEBS J* **280**: 6338–6353. doi:10.1111/febs.12559

Jennemann R, Grone HJ. 2013. Cell-specific in vivo functions of glycosphingolipids: lessons from genetic deletions of enzymes involved in glycosphingolipid synthesis. *Prog Lipid Res* **52**: 231–248. doi:10.1016/j.plipres.2013.02.001

Julien S, Bobowski M, Steenackers A, Le Bourhis X, Delannoy P. 2013. How do gangliosides regulate RTKs signaling? *Cells* **2**: 751–767. doi:10.3390/cells2040751

Sandhoff K, Harzer K. 2013. Gangliosides and gangliosidoses: principles of molecular and metabolic pathogenesis. *J Neurosci* **33**: 10195–10208. doi:10.1523/jneurosci.0822-13.2013

Feingold KR, Elias PM. 2014. Role of lipids in the formation and maintenance of the cutaneous permeability barrier. *Biochim Biophys Acta* **1841**: 280–294. doi:10.1016/j.bbalip.2013.11.007

Platt FM. 2014. Sphingolipid lysosomal storage disorders. *Nature* **510**: 68–75. doi:10.1038/nature13476

Schnaar RL, Gerardy-Schahn R, Hildebrandt H. 2014. Sialic acids in the brain: gangliosides and polysialic acid in nervous system development, stability, disease and regeneration. *Physiol Rev* **94**: 461–518. doi:10.1152/physrev.00033.2013

Sandhoff R, Schulze H, Sandhoff K. 2018. Ganglioside metabolism in health and disease. *Prog Mol Biol Transl Sci* **156**: 1–62. doi:10.1016/bs.pmbts.2018.01.002

Dunn TM, Tifft CJ, Proia RL. 2019. A perilous path: the inborn errors of sphingolipid metabolism. *J Lipid Res* **60**: 475–483. doi:10.1194/jlr.s091827

Yu J, Hung J-T, Wang S-H, Cheng J-Y, Yu AL. 2020. Targeting glycosphingolipids for cancer immunotherapy. *FEBS Lett* **594**: 3602–3618. doi:10.1002/1873-3468.13917

12 Glycosylphosphatidylinositol Anchors

Sneha Sudha Komath, Morihisa Fujita, Gerald W. Hart, Michael A.J. Ferguson, and Taroh Kinoshita

Plasma membrane proteins are either peripheral proteins or integral membrane proteins. The latter include proteins that span the lipid bilayer once or several times, and a second class that are covalently attached to lipids. Proteins attached to glycosylphosphatidylinositol (GPI) via their carboxyl termini are generally found in the outer leaflet of the lipid bilayer facing the extracellular environment. The GPI membrane anchor may be conveniently thought of as an alternative to the single transmembrane domain of type-I integral membrane proteins. This chapter reviews the discovery, distribution, structure, biosynthesis, properties, and suggested functions of GPI anchors and related molecules, as well as their roles in diseases.

BACKGROUND AND DISCOVERY

The first tentative evidence for the existence of protein-phospholipid anchors appeared in 1963 with the finding that crude bacterial phospholipase C (PLC) selectively releases alkaline phosphatase from mammalian cells. Phosphatidylinositol (PI)-protein anchors were first postulated in the mid-1970s when highly purified bacterial PI-specific PLCs were observed to release proteins, such as alkaline phosphatase and 5′-nucleotidase, from mammalian plasma membranes. By 1985, these predictions were confirmed by compositional and structural data from studies on *Torpedo* acetylcholinesterase, human and bovine erythrocyte acetylcholinesterase, rat Thy-1, and

published by Cold Spring Harbor Laboratory Press; doi: 10.1101/glycobiology.4e.12

the sleeping sickness parasite *Trypanosoma brucei* variant surface glycoprotein (VSG). The first complete GPI structures, which were for *T. brucei* VSG and rat Thy-1, were solved in 1988 (Chapter 1, Figure 1.3).

DIVERSITY OF PROTEINS WITH GPI ANCHORS

To date, hundreds of GPI-anchored proteins (GPI-APs) have been identified in many eukaryotes, ranging from protozoa and fungi to plants and humans (Online Appendix 12A). The range of described GPI-APs and the distribution of putative GPI biosynthesis genes suggests that (1) GPI anchors are almost ubiquitous among eukaryotes; (2) GPI-APs are functionally diverse and include hydrolytic enzymes, adhesion molecules, complement regulatory proteins, receptors, protozoan coat proteins, and prion proteins; and (3) in mammals, alternative messenger RNA (mRNA) splicing may lead to the expression of transmembrane and/or soluble and GPI-anchored forms of the same gene product. These variants may be developmentally regulated. For example, neural cell adhesion molecule (NCAM) exists in GPI-anchored and soluble forms when expressed in muscle and in GPI-anchored and two transmembrane forms when expressed in brain.

STRUCTURE OF GPI ANCHORS

Protein-Linked GPI Structures

The substructure Manα1-4GlcNα1-6*myo*-inositol-1-*P*-lipid is a universal hallmark of GPI anchors and related structures. All but one protein-linked GPI anchor share a larger common core structure (Figure 12.1; Online Appendix 12B, panels I–IV). The protein–carbohydrate association in GPI-APs is unique in that the reducing terminus of the GPI oligosaccharide is not attached to the protein but to the D-*myo*-inositol head group of a PI moiety by an α1-6 linkage. A distal, nonreducing mannose (Man) is attached to the protein via an ethanolamine phosphate (EtNP) bridge between its C-6 hydroxyl and the α-carboxyl group of the carboxy-terminal amino acid. GPIs are one of the rare instances in which GlcN is found without either an N-acetyl or N-sulfate (as in proteoglycans) moiety (Chapter 17).

Beyond the common core, the structures of mature GPI anchors are quite diverse, depending on both the protein to which they are attached and the organism in which they are synthesized (Figure 12.1; Online Appendix 12B). Modifications to the core include additional EtNP and a wide variety of linear and branched glycosyl substituents of largely unknown function.

There is considerable variation in the PI moiety. Indeed, GPI is a rather loose term because, strictly speaking, PI refers specifically to D-*myo*-inositol-1-*P*-3(*sn*-1,2-diacylglycerol) (i.e., diacyl-PI), whereas many GPIs contain other types of inositolphospholipids, such as *lyso*acyl-PI, alkylacyl-PI, alkenylacyl-PI, and inositolphosphoceramide (Online Appendix 12B). Another variation is the attachment of an ester-linked fatty acid at the C-2 hydroxyl of the inositol residue, which makes the anchor inherently resistant to bacterial PI-PLCs. The available structural data suggest that (1) inositolphosphoceramide-based protein-linked GPIs are mainly found in "lower" eukaryotes, such as *Saccharomyces cerevisiae*, *Aspergillus fumigatus*, *Aspergillus niger*, *Dictyostelium discoideum*, and *Trypanosoma cruzi*; (2) the lipid structures of GPIs generally do not reflect those of the general cellular PI or inositolphosphoceramide pool; and (3) the lipid structures of some (e.g., trypanosome) GPI-APs are under developmental control.

The factors controlling the synthesis of a mature protein-linked GPI appear to be similar to those for other posttranslational modifications such as N- and O-glycosylation. Thus, primary control occurs at the cellular level, whereby the levels of specific biosynthetic and processing enzymes dictate

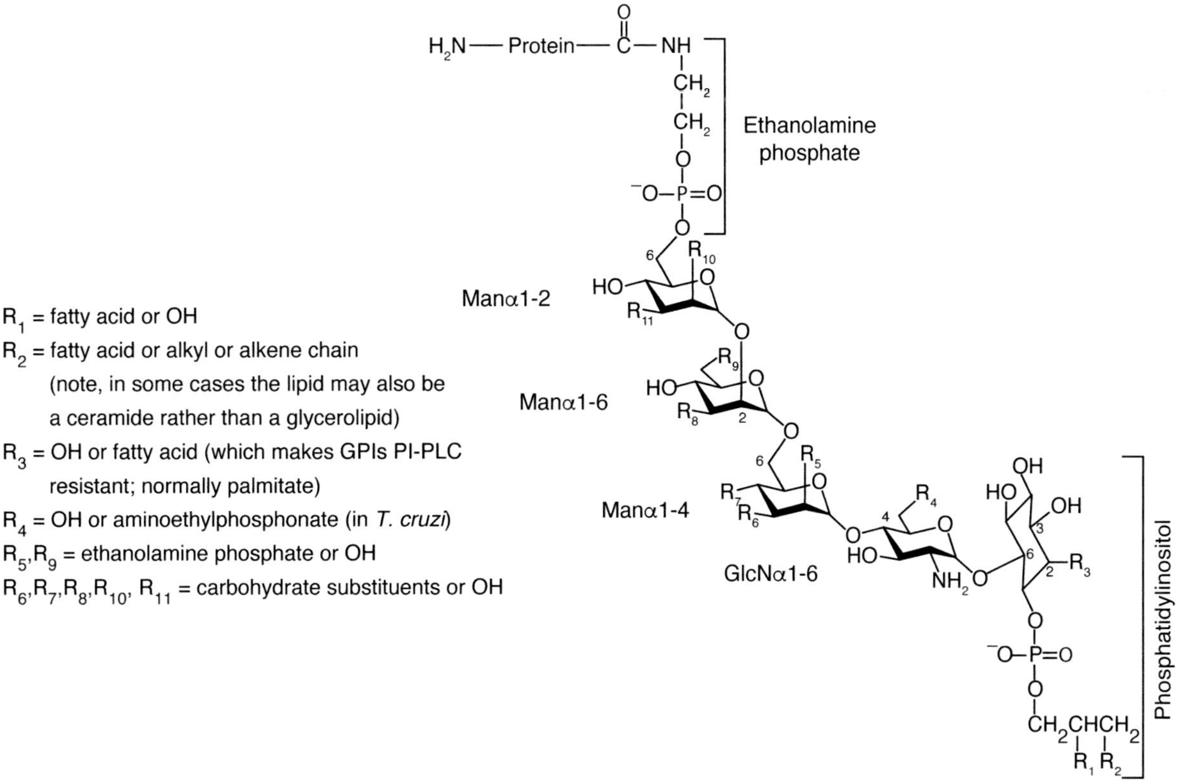
H2N— Protein— C—NH
Ethanolamine phosphate
Manα1-2
Manα1-6
Manα1-4
GlcNα1-6
Phosphatidylinositol
R1 = fatty acid or OH
R2 = fatty acid or alkyl or alkene chain
(note, in some cases the lipid may also be a ceramide rather than a glycerolipid)
R3 = OH or fatty acid (which makes GPIs PI-PLC resistant; normally palmitate)
R4 = OH or aminoethylphosphonate (in T. cruzi)
R5,R9 = ethanolamine phosphate or OH
R6,R7,R8,R10, R11 = carbohydrate substituents or OH

FIGURE 12.1. General structure of glycosylphosphatidylinositol (GPI) anchors attached to proteins. All characterized GPI anchors share a common core consisting of ethanolamine-PO_4-6Manα1-2Manα1-6Manα1-4GlcNα1-6*myo*-inositol-1-PO_4-lipid. Heterogeneity in GPI anchors is derived from various substitutions of this core structure and is represented here as R groups (see Online Appendix 12B). (Adapted, with permission, from Cole RN, Hart GW. 1997. In *Glycoproteins II* [ed. Montreuil J, et al.], pp. 69–88. Elsevier, Amsterdam, © Elsevier.)

the final repertoire of structures. Secondary control occurs at the level of the tertiary/quaternary structures of the GPI-APs, which affect accessibility to processing enzymes. Examples of primary control include (1) differences in GPI glycan side chains in human versus porcine membrane dipeptidase and brain versus thymocyte rat Thy-1 and (2) differences in glycan side chains and the lipid structure of *T. brucei* VSG in the bloodstream versus insect life-cycle stages of the parasite. An example of secondary control is the difference in VSG glycan side chains when VSGs with different carboxy-terminal sequences are expressed in the same trypanosome clone.

Non-Protein-Linked GPI Structures

In mammalian cells, some free GPIs (mature and biosynthetic intermediates) are found at the cell surface, although their functional significance remains unknown. On the other hand, several protozoa (particularly trypanosomatids) express high numbers ($>10^7$ copies per cell) of free GPIs on their cell surface as metabolic end products. These include the so-called glycoinositol phospholipids (GIPLs) and lipophosphoglycans (LPGs) of *Leishmania*. Some protozoan (type-1) GIPLs conform to the Manα1-6Manα1-4GlcNα1-6PI sequence common to protein-linked GPIs, whereas others contain a (type-2) Manα1-3Manα1-4GlcNα1-6PI motif, and still others possess hybrid structures containing the branched motif (Manα1-6)Manα1-3Manα1-4GlcNα1-6PI. The only GPI-linked fungal polysaccharide reported so far is *A. fumigatus* galactomannan-Manα1–2Manα1–2Manα1–6Manα1–4GlcN-inositolphosphoceramide, in which the glycan is attached directly to the anchor without the bridging EtNP (see Online Appendix 12B, panel II).

THE CHEMISTRY OF GPI ANCHORS

The GPI lends itself to selective cleavage by several chemical and enzymatic reagents (Online Appendix 12C, panel I). These were originally used to determine GPI structures and are now applied to confirm the presence, and/or obtain partial structural information, of GPIs. A key reaction is nitrous acid deamination of the GlcN residue, which selectively cleaves the GlcN-inositol glycosidic bond, thereby liberating the PI moiety that can be isolated by solvent partition and analyzed by mass spectrometry. The free reducing terminus, generated on the GPI glycan in the form of 2,5-anhydromannose, can be reduced to [1-^{3}H]2,5-anhydromannitol (AHM) by sodium borotritide to introduce a radiolabel, or can be attached to a fluorophore such as 2-aminobenzamide (2-AB) by reductive amination. Once labeled and dephosphorylated with aqueous HF, the GPI glycan can be conveniently sequenced using exoglycosidases. Partial structural information can also be obtained by tandem mass spectrometry of (i) GPI-peptides generated by proteolytic digestion of GPI-APs or (ii) GPI glycans released by aqueous HF dephosphorylation and permethylated before analysis. Other indirect methods for inferring the presence of GPI-APs are described in Online Appendix 12C, panel II.

GPI BIOSYNTHESIS AND TRAFFICKING

The biosynthesis of GPI anchors occurs in three stages: (1) preassembly of a GPI precursor in the endoplasmic reticulum (ER) membrane, (2) attachment of the GPI to a newly synthesized protein in the ER lumen with concomitant cleavage of a carboxy-terminal GPI-addition signal peptide, and (3) lipid remodeling and/or carbohydrate side-chain modifications in the ER and after transport to the Golgi.

Analysis of biosynthesis of the GPI was made possible by the development of a cell-free system in *T. brucei*. The sequence of events underlying GPI biosynthesis has also been studied in *T. cruzi*, *Toxoplasma gondii*, *Plasmodium falciparum*, *Leishmania major*, Paramecium spp., *S. cerevisiae*, *Candida albicans*, *Cryptococcus neoformans*, and mammalian cells. The emphasis on eukaryotic microbes reflects the abundance of GPI-APs in these organisms and the potential of GPI inhibition for chemotherapeutic intervention. This notion has been genetically validated in the bloodstream form of *T. brucei*, as well as in *P. falciparum*, *S. cerevisiae*, and *C. albicans*.

The essential events in GPI biosynthesis are highly conserved. There are, however, variations on the theme. The *T. brucei*, *S. cerevisiae*, and mammalian cell GPI pathways are used here to showcase these differences (Figure 12.2). In all cases, GPI biosynthesis begins with the transfer of *N*-acetylglucosamine (GlcNAc) from UDP-GlcNAc to PI to produce GlcNAc-PI, which is de-

FIGURE 12.2. Glycosylphosphatidylinositol (GPI)-biosynthetic pathways of *Trypanosoma brucei*, *Saccharomyces cerevisiae*, and mammals. These examples show that, despite the highly conserved core structure of GPI anchors, some diversity in GPI-biosynthesis exists. In particular, yeast and mammalian GPI intermediates are inositol-acylated at the level of GlcN-PI and are not deacylated until after transfer to protein, whereas *T. brucei* GPI intermediates are acylated at the level of Man-GlcN-PI and undergo rounds of deacylation and reacylation throughout the pathway. Mammalian GPI intermediates undergo a first round of lipid remodeling at the level of GlcN-aPI and then again after transport to the Golgi. *S. cerevisiae* GPI intermediates have a compulsory Man-4 added unlike in mammals and lipid remodeling of their GPI-APs occur in two consecutive steps within the endoplasmic reticulum (ER). The protein components of the pathway that catalyze these steps in mammals and *S. cerevisiae* are described in Table 12.1. Some of the *T. brucei* homologs are yet to be identified. The "?" refers to an unknown process. The *double-headed arrows* represent flipping of the GPI intermediate by an unknown mechanism. *Dashed arrows* denote probable alternative pathways or steps that are not compulsory. (*See figure on following page.*)

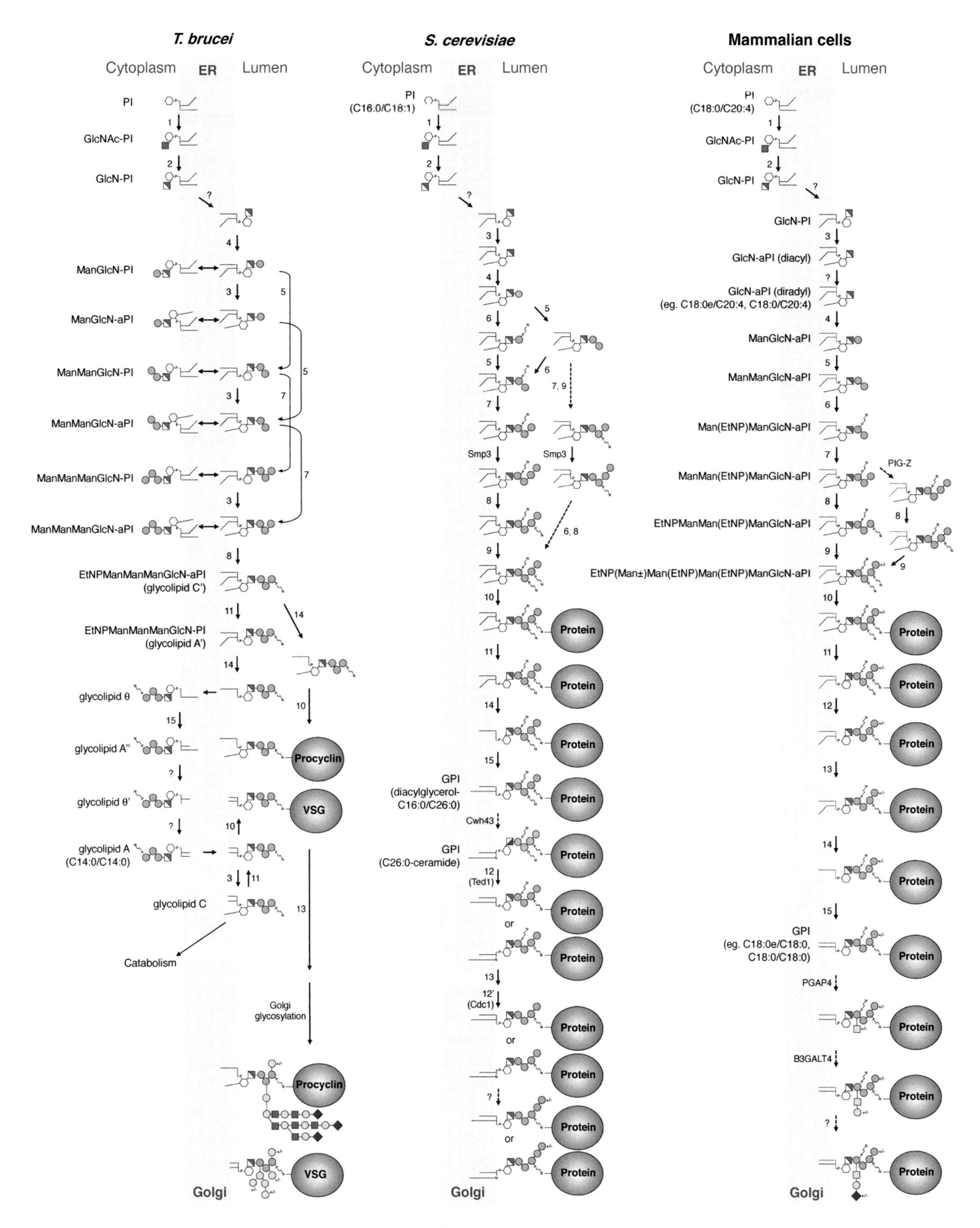

FIGURE 12.2. (*See legend on preceding page.*)

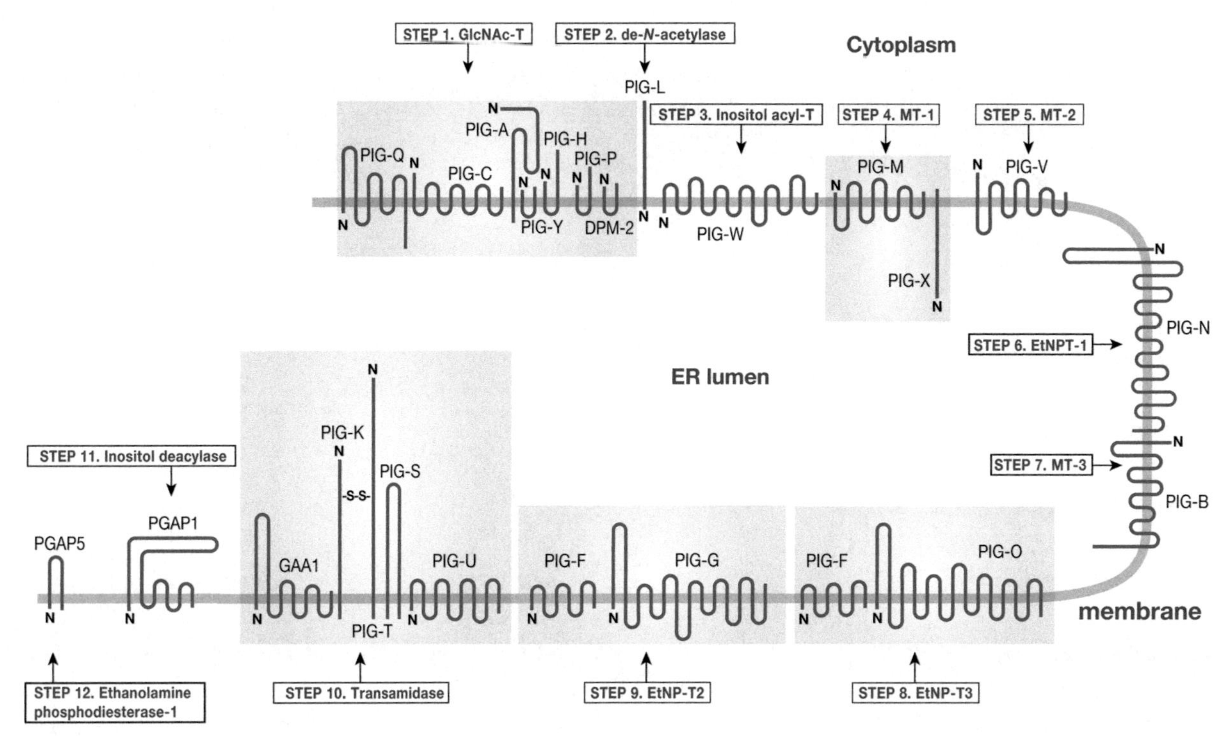

FIGURE 12.3. Predicted topologies of the ER-resident components of glycosylphosphatidylinositol (GPI) biosynthesis in mammalian cells. Components within boxes belong to multisubunit complexes. The step numbers refer to those in Figure 12.2 and Table 12.1. The topologies of the "multispanning" membrane proteins are mostly those predicted by bioinformatics methods. N, amino terminus; ER, endoplasmic reticulum.

N-acetylated in the next step to produce GlcN-PI on the cytoplasmic face of the ER (Figure 12.3; Table 12.1). Notable differences between the *T. brucei* and yeast or mammalian GPI-biosynthetic pathways occur from GlcN-PI onwards. Inositol acylation of GlcN-PI (at the C-2 hydroxyl of the D-*myo*-inositol; labeled as GlcN-aPI), strictly follows the first Man addition in *T. brucei*, whereas these steps are temporally reversed in yeast and mammalian cells. In the yeast and mammalian pathways, inositol acylation and deacylation are discrete steps occurring only at the beginning and end of the pathway, respectively, whereas in *T. brucei* these reactions occur on multiple GPI intermediates. Furthermore, in some mammalian cells such as human erythroblasts, inositol-deacylation never occurs and the mature GPI protein retains three hydrocarbon chains (Online Appendix 12B, panel III).

Fatty-acid remodeling in GPI of bloodstream from *T. brucei* occurs at the end of the pathway, but before transfer to protein, and involves exchanging the *sn*-2 fatty acids (a mixture of C18–C22 species) and the *sn*-1 fatty acid (C18:0) exclusively for C14:0 myristate. In yeast, this occurs in the ER, after transfer to proteins, and involves two distinct but consecutive processes. First, the unsaturated *sn*-2 fatty acid (C18:1) is exchanged for a C26:0 chain. Next, the diacylglycerol is exchanged for ceramide on many, but not all, GPI-APs. Lipid remodeling in mammalian cells is more complex. Many protein-linked GPIs contain *sn*-1-alkyl-2-acyl-PI with two saturated fatty chains, whereas the major cellular PI is *sn*-1-stearoyl-2-arachidonoyl-PI (i.e., with C18:0 and C20:4 fatty acids and few, if any, alkyl or alkenyl species). Two processes are involved in these structural changes. First, remodeling from the diacyl-PI to diradyl-PI (mixtures of 1-alkyl-2-acyl-PI and diacyl-PI), having unsaturated fatty acid at the *sn*-2-position, occurs early in the pathway, within the ER, to produce GlcN-aPI with the remodeled lipid tail. The reaction responsible for this is not known, but alkyl phospholipids synthesized in the peroxisome serve as the alkyl donors. Second, fatty-acid remodeling, involving exchange of the unsaturated *sn*-2 fatty acid

TABLE 12.1. Components of the core mammalian and *Saccharomyces cerevisiae* glycosylphosphatidylinositol (GPI)-biosynthetic machinery

Step 1. UDP-GlcNAc:PI α1-6 N-acetylglucosaminyltransferase (GlcNAc-T) transfers GlcNAc from UDP-GlcNAc to PI to produce GlcNAc-PI

Components

PIG-A[a,b] (Gpi3), 484 (452) AA, 1 (1) TM, catalytic subunit
PIG-H[b] (Gpi15), 188 (229) AA, 2 (2) TM, associated with PIG-A
PIG-C[b] (Gpi2), 292 (280) AA, 8 (6) TM, associated with PIG-Q (with Gpi1)
PIG-Q[b] (Gpi1), 581 (609) AA, 6 (6) TM, associated with PIG-C
PIG-P[b] (Gpi19), 134 (140) AA, 2 (2) TM, associated with PIG-A, PIG-Q (with Gpi2)
DPM2 (Yil102c-A), 84 (75) AA, 2 (2) TM, subunit of Dol-P-Man synthase
PIG-Y[b] (Eri1), 71 (68) AA, 2 (2) TM, associated with PIG-A

Note: PIG-A/PIG-H complex is associated with the PIG-C/PIG-Q complex. Active site faces the cytoplasm. In yeast, but not in mammals, Ras physically associates with GlcNAc-T. Ras2 inhibits GlcNAc-T in *S. cerevisiae*. Ras1 activates GlcNAc-T in *C. albicans*.

Step 2. GlcNAc-PI de-N-acetylase de-*N*-acetylates GlcNAc-PI to GlcN-PI

Components

PIG-L[b] (Gpi12), 252 (304) AA, 1 (1) TM

Note: The active site of PIG-L (Gpi12) faces the cytoplasm. It is generally stimulated by divalent cations but species-specific differences in metal ion preferences seem to exist.

Step 3. Inositol acyltransferase (inositol acyl-T) acylates C2 hydroxyl of myo-inositol, using acyl-CoA as donor, to produce GlcN-aPI

Components

PIG-W[b] (Gwt1), 504 (490) AA, 13 (12) TM

Note: The active site of PIG-W (Gwt1) faces the ER lumen, which requires the GPI intermediate to be flipped inwards from the cytoplasmic face. A flippase for this process has not yet been identified. In mammals, the first stage of lipid remodeling to replace the diacyl form with the diradyl form also occurs at this step. The mechanistic details of the process are unclear.

Step 4. Mannosyltransferase-1 (MT-1) transfers Man-1 from Dol-P-Man to produce Man-GlcN-aPI

Components

PIG-M[b] (Gpi14), 423 (403) AA, 8 (8) TM, catalytic subunit
PIG-X (Pbn1), 252 (416) AA, 1 (1) TM, associated with PIG-M

Note: The active site of PIG-M (Gpi14) faces the ER lumen. A conserved DXD motif is critical for activity. PIG-M and PIG-X must be co-expressed to rescue a *gpi14* mutant strain. Yeast Arv1 is probably involved in flipping of the GPI intermediate and/ or its delivery to MT-1. Human ARV1[b] (271 AA, 3 TM) complements yeast Arv1 (321 AA, 3 TM); its endogenous role in the GPI pathway is unknown.

Step 5. Mannosyltransferase-2 (MT-2) transfers Man-2 from Dol-P-Man to produce Man-Man-GlcN-aPI

Components

PIG-V[b] (Gpi18 with Pga1), 493 (433, 198) AA, 8 (8+2) TM

Note: The active site of PIG-V (Gpi18) faces the ER lumen. PIG-V complements both Gpi18 and Pga1.

Step 6. Ethanolamine phosphate transferase-1 (EtNPT-1) transfers EtNP-1 from phosphatidylethanolamine (PE) to produce Man-(EtNP)Man-GlcN-aPI

Components

PIG-N[b] (Mcd4), 931 (919) AA, 15 (14) TM

Note: The active site of PIG-N (Mcd4) faces the ER lumen. PIG-N, PIG-O and PIG-G possess a conserved phosphatase motif.

Step 7. Mannosyltransferase-3 (MT-3) transfers Man-3 from Dol-P-Man to produce Man-Man-(EtNP)Man-GlcN-aPI

Components

PIG-B[a,b] (Gpi10), 554 (616) AA, 9 (9) TM

Note: The active site of PIG-B faces the ER lumen. Dol-P-Man is the donor of Man. MT-3 in some protozoan parasites may also add Man-4.

Step 8. Ethanolamine phosphate transferase -3 (EtNPT-3) transfers EtNP-3 from PE to produce (EtNP)Man-Man-(EtNP)Man-GlcN-aPI

Components

PIG-O[b] (Gpi13), 880 (1017) AA, 14 (13) TM, catalytic subunit

(*Continued*)

TABLE 12.1. *(Continued)*

PIG-F[b] (Gpi11), 219 (219) AA, 6 (4) TM, associated with PIG-O

Note: The active site of PIG-O faces the ER lumen. Gpi11 is not a major participant at this step in yeast.

Step 9. Ethanolamine phosphate transferase-2 (EtNPT-2) transfers EtNP-2 from PE to produce (EtNP)Man-(EtNP)Man-(EtNP)Man-GlcN-aPI

Components

PIG-G[b] (Gpi7), 983 (830) AA, 13 (9) TM, catalytic subunit

PIG-F[b] (Gpi11), 219 (219) AA, 6 (4) TM, associated with PIG-G

Note: The active site of PIG-G faces the ER lumen. In yeast, Gpi11 is not critical for Gpi7 activity, but prior addition of Man-4 by MT-4 (Smp3, 516 AA, 7 TM) is crucial. In humans, the action of a MT-4, PIG-Z (579 AA, 4 TM), is tissue-specific.

Step 10. GPI transamidase transfers the complete GPI precursor to proteins

Components

GPAA1[b] (Gaa1), 621 (614) AA, 7 (6) TM

PIG-K[b] (Gpi8), 367 (411) AA, 1 (1) TM, catalytic subunit

PIG-T[a,b] (Gpi16), 557 (610) AA, 1 (1) TM, disulfide bonded to PIG-K

PIG-S[b] (Gpi17), 555 (534) AA, 2 (2) TM

PIG-U[b] (Gab1), 435 (394) AA, 8 (8) TM

Note: PIG-K/Gpi8, a cysteine protease-like endopeptidase, cleaves GPIsp. GPAA1/Gaa1, a metallo-protease/synthase, is predicted to catalyze amide bond formation between the GPI anchor and protein. Active sites of both subunits face the ER lumen. Gpi8 is not disulfide linked to Gpi16 in yeast.

Step 11. Inositol deacylase deacylates from inositol of GPI-APs

Components

PGAP1[b] (Bst1), 922 (1029) AA, 6 (8) TM

Note: The active site of PGAP1 faces the ER lumen. A conserved Ser in the active site is essential for activity.

Step 12. Ethanolamine phosphoesterase-1 removes EtNP from Man-2 of GPI-APs

Components

PGAP5 (Ted1), 396 (473) AA, 2 (1) TM

Note: The active site of PGAP5 faces the ER lumen. It is a Mn^{++}-dependent phosphoesterase. In yeast, Cdc1 (491 AA, 3 TM), a paralog of Ted1, is the ethanolamine phosphoesterase-2 that removes EtNP from Man-1 and enables cell wall localization of the GPI-AP. Recent reports suggest that Cdc1 localizes to the Golgi (see Figure 12.2).

Step 13. ER to Golgi transport cargo receptor transports GPI-APs to the Golgi after packaging them into COPII-coated vesicles

Components

TMED9 (Erp1), 235 (219) AA, 1 (1) TM

TMED2 (Emp24), 201 (203) AA, 1 (1) TM

TMED5 (Erp2), 229 (215) AA, 1 (1) TM, GPI-binding subunit

TMED10 (Erv25), 219 (211) AA, 1 (1) TM

Step14. Fatty-acid remodeling step 1 involves removal of the *sn*-2 fatty acid to generate a lyso-PI in the Golgi

Components

PGAP3[b] (Per1), 320 (357) AA, 7 (6) TM, GPI-specific phospholipase A2

Note: In yeast, this reaction is mediated by Per1 in the ER lumen.

Step15. Fatty-acid remodeling step 2 involves reacylation at *sn*-2 with a saturated fatty acid

Components

PGAP2[b] (Gup1), 254 (560) AA, 5 (10) TM.

Note: PGAP2 is a Golgi protein. Gup1 is an ER membrane-bound acyltransferase and not a structural homolog of PGAP2. The diacylglycerol on many GPI-APs of *S. cerevisiae*, may ultimately be replaced with ceramide in the ER by Cwh43 (Fatty-acid remodeling step 3; 953 AA, multiple TM) before Step 13 (above).

The step numbers refer to Figure 12.2. The size of the predicted protein in amino acids (AA) and the number of predicted transmembrane (TM) domains are given alongside the protein names. The names and numbers in brackets correspond to the yeast homologs. The numbers of transmembrane domains in proteins with more than five transmembrane domains are predicted from their amino acid sequences rather than being experimentally determined in most cases.

ER, endoplasmic reticulum.

[a]Mutations in this human gene may produce paroxysmal nocturnal hemoglobinuria (PNH) (Chapter 46).

[b]Mutations in this human gene cause inherited GPI deficiencies (IGDs) (Chapter 45).

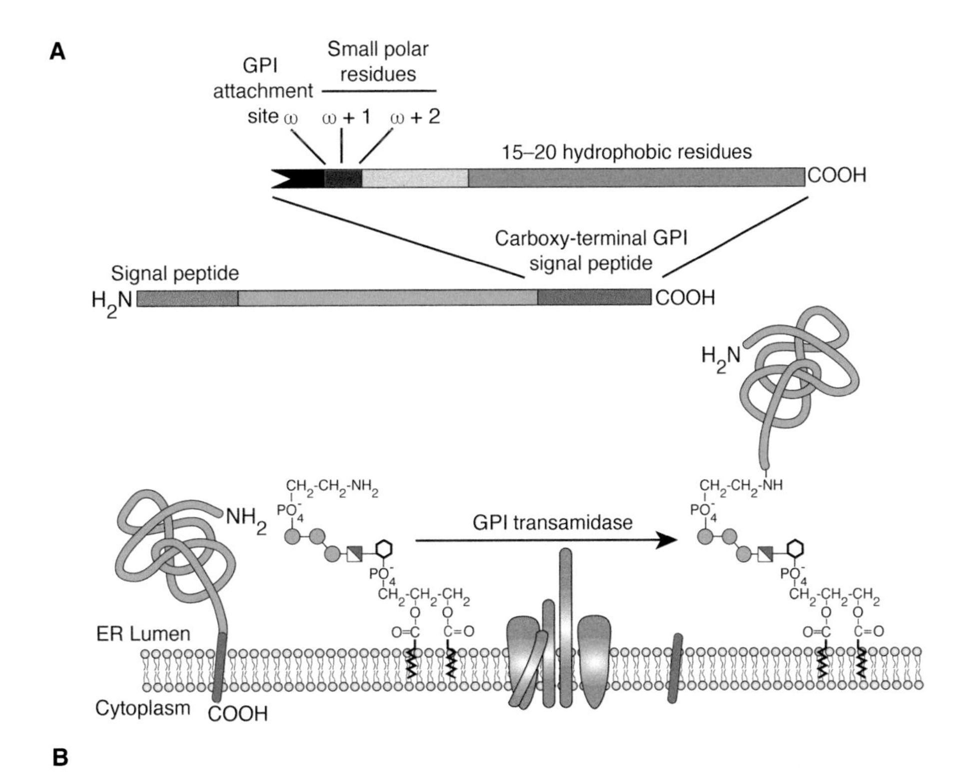

B

Protein	GPI signal sequence	
Acetylcholinesterase *(Torpedo)*	NQFLPKLLNATA**C**	DGELSSSGTSSSKGIIFYVLFSILYLIFY
Alkaline phosphatase *(placenta)*	TACDLAPPAGTT**D**	AAHPGRSVVPALLPLLAGTLLLLETATAP
Decay accelerating factor	HETTPNKGSGTT**S**	GTTRLLSGHTCFTLTGLLGTLVTMGLLT
Thy-1 *(rat)*	KTINVIRDKLVK**C**	GGISLLVQNTSWLLLLLLSLSFLQATDFISL
Prion protein *(hamster)*	QKESQAYYDGRR**S**	SAVLFSSPPVILLISFLIFLMVG
Variant surface glycoprotein *(T. brucei)*	ESNCKWENNACK**D**	SSILVTKKFALTVVSAAFVALLF
PARP *(T. brucei)*	EPEPEPEPEPEP**G**	AATLKSVALPFAIAAAALVAAF
Gas1 *(S. cerevisiae)*	SASSSSSSKK**N**	AATNVKANLAQVVFTSIISLSIAAGVGFALV

Boldfaced amino acid is the site of attachment of the GPI. Sequence to the right of the space is cleaved from the protein by the transamidase upon anchor addition.

FIGURE 12.4. (*A*) Features of glycosylphosphatidylinositol (GPI)-anchored proteins and their processing by GPI transamidase. GPI-anchored proteins have an amino-terminal signal peptide for translocation into the ER and a carboxy-terminal GPI-addition signal peptide (*top*) that is removed and directly replaced by a GPI precursor (*bottom*). (*B*) Examples of carboxy-terminal sequences signaling the addition of GPI anchors.

for a saturated fatty acid, mainly stearate (C18:0), occurs after the GPI-APs are transported to the Golgi.

The GPI pathway genes were identified principally by expression cloning using GPI-deficient mutants of mammalian cells and temperature-sensitive yeast mutants. More recently, epitope tagging/pull-down/proteomic approaches have been used to identify GPI pathway–associated components. Some basic details of the known mammalian and yeast components and their topologies in the ER membrane are described in Table 12.1. Predicted topologies for the mammalian enzymes are shown in Figure 12.3.

The GPI precursor is transferred to proteins via a multisubunit transamidase complex. The reaction involves two complex substrates, the preassembled GPI precursor and the carboxyl terminus of a partially folded nascent protein (Figure 12.4). The carboxy-terminal GPI-addition signal peptide (GPIsp) contains three domains: (1) three relatively small amino acids (Ala, Asn, Asp, Cys, Gly, or Ser) located at ω, ω+1, and ω+2, where ω is the amino acid attached to the GPI anchor and ω+1 and ω+2 are the first two residues of the cleaved peptide; (2) a relatively polar domain of typically five to 10 residues; and (3) a hydrophobic domain typically comprising 15–20 hydrophobic amino acids. These GPIsp sequences have no strict consensus, but are easily identified by eye and by automated algorithms. The final hydrophobic stretch of amino acids often resembles a transmembrane domain, but the absence of positively charged and polar residues immediately downstream makes a GPIsp easy to spot. Like N-glycosylation sequons, a GPIsp will only be functional if the protein is translocated into the ER. Hence, all GPI-APs are synthesized with an amino-terminal ER localization signal peptide as well.

GPIs present microheterogeneity in their glycan side-chains depending on the species, cell types, and proteins (Online Appendix 12B, panels I–IV). A small number of enzymes involved in these reactions in mammals, yeast and *T. brucei* are known (Online Appendix 12B, panel V). Yeast and some mammalian GPIs possess a fourth α1-2Man (Man-4) residue added by Smp3/PIG-Z during GPI precursor assembly in the ER. Yeast GPI-APs have a fifth αMan (Man-5) residue added in the Golgi. A proportion of mammalian GPI-APs contain Man-1 modified by a β1-4*N*-acetylgalactosamine (GalNAc), and which may be further modified with a β1-4galactose (Gal), in the Golgi. An α2-3 sialic acid (Sia) could be added to the Gal. In *T. brucei* bloodstream forms, GPI-APs present glycan side-chains consisting of Gal, whereas the procyclic GPIs carry sialylated poly-*N*-acetyllactosamine and poly-lacto-*N*-biose structures.

The transport of GPI-APs from the ER to the Golgi is mediated by coat protein II (COPII)-coated transport vesicles. Their packaging into the vesicles requires a transmembrane cargo receptor, a complex of four p24 family proteins, that links luminally oriented GPI to COPII components on the cytoplasmic side of the ER membrane.

MEMBRANE PROPERTIES OF GPI-APs

GPI-APs with two long hydrocarbon chains (i.e., those containing diacylglycerol, alkylacylglycerol, alkenylacylglycerol, or ceramide) are stably associated with the lipid bilayer. It follows that inositol-acylated GPI proteins with three fatty-acid chains should be more stably associated. On the other hand, *Leishmania* LPG, with a single C24:0 alkyl chain, has a half-life of only minutes at the cell surface and is secreted intact into the medium.

The thermodynamics of bilayer interactions also depend on the length and degree of saturation of the fatty-acid chains. In this regard, the saturated nature of most (but not all) mammalian GPI anchors (Online Appendix 12B, panels III and IV) is thought to explain why GPI-APs associate with "lipid rafts." The current model for mammalian lipid rafts is that of transient liquid-ordered nanoclusters of membrane components. These are dependent on dynamic cortical actin asters that, through adaptor proteins, cluster phosphatidylserine (PS) on the inner face of the plasma membrane bilayer. This coupling effect occurs because the long, saturated lipid chains from inner leaflet PS overlap and interact with those of GPI-APs and glycosphingolipids in the middle of the bilayer, thereby generating functional membrane domains. A compelling aspect of this model is that, although dynamic cortical actin can organize molecules across the leaflet into nanoclusters, the clustering of GPI-APs or glycosphingolipids at the outer leaflet can also, conversely, organize the inner leaflet and recruit molecules that may favor liquid-ordered domains. This provides a possible explanation for the perplexing, yet well-characterized, ability of GPI-APs to transduce signals across the plasma membrane.

There are many examples of transmembrane signaling via the cross-linking of GPI-APs with antibodies and clustering with a second antibody on various cells, particularly leukocytes. Cellular responses include an increase in intracellular Ca^{++}, tyrosine phosphorylation, proliferation, cytokine induction, and oxidative burst. These signaling events are dependent on the presence of a GPI anchor and might be mostly caused by the induction and coalescence of lipid raft nanoclusters, although their participation in ligand binding in conjunction with signal receptors (such as receptor-like-kinases in *Arabidopsis*) to transduce extracellular signals cannot be ruled out. GPI-APs involved in transmembrane signaling, such as the glial-cell-(line-)derived neurotrophic factor receptor-α (GDNFR-α), need to be associated with transmembrane β co-receptors to transmit their signals. Similarly, GPI-anchored CD14 (the LPS/LPS-binding protein receptor) requires to partner transmembrane Toll-like receptor-4, and can function equally well with a GPI anchor or with a spliced transmembrane domain.

GPI ANCHORS AS TOOLS IN CELL BIOLOGY

The replacement of carboxy-terminal transmembrane domains of type-I integral membrane proteins by GPIsps allows their expression as GPI-APs on the plasma membrane of transfected mammalian cells. This offers a useful method for producing soluble forms of membrane proteins. For example, the T-cell receptor was rendered soluble after it was expressed in the GPI-anchored form (by the action of bacterial PI-PLC), and not by simply expressing the transmembrane-deleted domain. In addition, purified GPI-APs can be used to coat hydrophobic surface plasmon resonance chips, thus providing a convenient method for orienting and presenting proteins for binding studies. It is clear that purified GPI-APs will spontaneously insert into lipid bilayers. The physiological significance of direct GPI-protein exchange between membranes is still uncertain, particularly because all mammals express potent GPI-PLD activity in serum that can remove the lipid (phosphatidic acid) component of the anchor and, thus, prevent GPI-protein re-insertion. However, the re-insertion property of GPI-APs has been exploited experimentally to "paint" exogenous proteins onto cell surfaces.

BIOLOGICAL FUNCTIONS OF GPI ANCHORS

GPI anchors are essential for life in some, but not all, eukaryotic microbes. In *S. cerevisiae*, and probably most fungi, the presence of a GPI anchor targets certain mannoproteins for covalent incorporation into the β-glucan layer of the cell wall via a transglycosylation reaction, whereby Man-1 within the GPI-anchor core is transferred to the β-glucan polymer, probably by the actions of Dfg5 and Dcw1. Defects in cell wall biosynthesis are detrimental to yeast and this may be one of the reasons why GPI biosynthesis is essential for this organism. Certain unique phenotypes of *C. albicans* GPI biosynthetic mutants are strictly associated with a single gene defect, suggestive of highly specific cross-talk of the related GPI biosynthetic step with other cellular pathways. GPI biosynthesis is essential for the bloodstream form of *T. brucei*, even in tissue culture. This may be due to nutritional stress, given that this parasite uses an essential GPI-anchored transferrin receptor. On the other hand, surprisingly, GPI biosynthesis and/or transfer to protein are not essential for the insect-dwelling forms of *T. brucei* or *L. major*. In *Arabidopsis thaliana*, GPI biosynthesis is required for cell wall synthesis, morphogenesis, and pollen tube development. The availability of GPI-deficient mammalian cell lines shows that GPI-APs are not essential at the cellular level. However, mouse knockouts and tissue-specific conditional knockouts of the *PIGA* gene clearly show that GPI-APs are essential for early embryo and tissue development. GPI anchors impart to their attached proteins the ability to be shed from the cell surface in soluble form through the action of cellular or serum GPI-cleaving enzymes. Mammalian sperms acquire the ability to fuse with oocytes after GPI-anchored TEX101 and LY6K are released by a sperm-associated GPI-cleaving enzyme tACE (testis form angiotensin converting enzyme). The cleavage of GPI-APs is required for maturation of ADAM3, a metalloprotease on the sperm surface. Certain proliferating motor neurons initiate differentiation after the GPI-anchored proteinase inhibitor, RECK, is released by the GPI-cleaving enzyme, GDE2. This relieves the inhibition of ADAM10 metalloproteinase, allowing it to degrade the Notch-ligand to terminate Notch signaling. Thus, the cell switches from proliferation to differentiation. GPI-anchored CRIPTO acts as a coreceptor for Nodal signaling. Shedding of CRIPTO by a GPI-phospholipase A2, PGAP6, followed by a PLD-like enzyme, regulates Nodal signaling. In "lower" eukaryotes, GPI anchors may be useful for assembling particularly dense cell-surface protein coats, such as the VSG coat of *T. brucei*. In this case, each parasite expresses five million VSG dimers on the cell surface to protect it against complement-mediated lysis. If each VSG monomer had a single transmembrane domain instead of a GPI anchor, there would be little room for other integral membrane proteins such as hexose and nucleoside transporters. Generally, GPI-APs do recycle through intracellular compartments but, compared to typical transmembrane proteins, they reside in higher proportions on the cell surface and have longer half-lives. There are several examples of the exchange of GPI-APs from one cell surface to another. Some GPI-APs are incorporated into exosomes, suggesting the possibility of exosome-mediated cell-to-cell transfer. Sperm acquires some GPI-APs such as CD52, from epididymis, most likely mediated by exosomes.

GPI ANCHORS AND DISEASE

Paroxysmal nocturnal hemoglobinuria (PNH) is a human disease in which patients suffer from hemolytic anemia. The condition arises from loss of expression of several GPI-APs that protect the blood cells from lysis by the complement system (e.g., decay accelerating factor and CD59). The defect in PNH cells is a somatic mutation in the X-linked *PIGA* gene and appears to occur in a bone marrow stem cell. Unlike other enzymes in the pathway, which are encoded by autosomal genes, PNH caused by *PIGA* mutations is thought to arise at a higher frequency because of X inactivation. In both male and female stem cells, somatic mutation in the one active allele of *PIGA* results in complete loss of GlcNAc-T function (Chapter 46). Atypical PNH caused by *PIGT* or *PIGB* gene mutations has been reported; the patients show autoinflammatory features such as aseptic meningitis, in addition to typical PNH symptoms.

Inherited GPI deficiencies (IGDs) are caused by germline mutations in genes involved in GPI biosynthesis, protein transfer, and remodeling. Because complete GPI deficiency causes embryonic lethality, mutations in IGDs are hypomorphic, causing partial deficiency. Mutations in genes involved in GPI remodeling, such as in PGAP1, can be null and cause GPI-APs with abnormal structure. Patients with IGDs caused by mutations in 23 genes of the GPI-biosynthetic pathway have been reported (Table 12.1). Most of these mutations were identified by whole exome sequencing of patients' cells. Major symptoms of IGDs are neurological problems, such as developmental delay/intellectual disability, seizures, cerebral and/or cerebellar atrophy, hearing loss, and visual impairment. Other symptoms include hyperphosphatasia; brachytelepharangy; abnormal facial features such as hypertelorism and tented mouth; cleft palate; anorectal, renal, and heart anomalies; and Hirschsprung disease (Chapter 45).

GPI biosynthesis and transfer to proteins are essential for yeast, for pathogenic fungi, and for the African sleeping sickness parasite *T. brucei*, as mentioned above. Several key surface molecules of the apicomplexan parasites, *P. falciparum* (malaria), *Toxoplasma*, and *Cryptosporidium* are GPI-anchored, and it is thought that the GPI pathway is likely to be essential in these pathogens. There is evidence that some parasite GPI anchors also play a direct role in modulating the host immune response to infection. Hence, pathogen-specific GPI pathway inhibitors are being actively sought as potential drugs (Online Appendix 12D). Indeed, Gwt1 inhibitors are currently being optimized as leads for antimalarial drugs. Gwt1 is also viewed as an important antifungal drug target. Of particular interest in this context is fosmanogepix, which is presently in phase-II clinical trials. Fungal GPI-APs shield cell wall β-(1,3)-glucans from detection by macrophages. Thus, inhibitors of GPI biosynthesis should support better clearance of the pathogen by the host immune system.

Like other glycoconjugates, GPI-APs can be exploited by pathogens. For example, the GPI anchors themselves are receptors for hemolytic pore-forming toxins, such as aerolysin from *Aeromonas hydrophilia*, which causes gastroenteritis, deep wound infections, and septicemia in humans. In addition, the GPI-AP CD55/DAF is the principal cell-surface ligand for enterovirus and several echoviruses. Finally, the conformational changes that the endogenous prion protein undergoes to become the aberrant spongiform-encephalopathy ("mad cow disease" or scrapie in sheep)-causing form may be associated with a clathrin-independent endocytic pathway that is followed by this GPI-anchored protein in neurons.

ACKNOWLEDGMENTS

The authors appreciate helpful comments and suggestions from Satyajit Mayor, Hiroshi Nakato, and Jerry Eichler.

FURTHER READING

Ferguson MA, Williams AF. 1988. Cell-surface anchoring of proteins via glycosyl-phosphatidylinositol structures. *Annu Rev Biochem* **57:** 285–320. doi:10.1146/annurev.bi.57.070188.001441

Ferguson MA, Homans SW, Dwek RA, Rademacher TW. 1988. Glycosyl-phosphatidylinositol moiety that anchors *Trypanosoma brucei* variant surface glycoprotein to the membrane. *Science* **239:** 753–759. doi:10.1126/science.3340856

Guha-Niyogi A, Sullivan DR, Turco SJ. 2001. Glycoconjugate structures of parasitic protozoa. *Glycobiology* **11:** p45R–p59R. doi:10.1093/glycob/11.4.45r

de Macedo CS, Shams-Eldin H, Smith TK, Schwarz RT, Azzouz N. 2003. Inhibitors of glycosylphosphatidylinositol anchor biosynthesis. *Biochimie* **85:** 465–472. doi:10.1016/s0300-9084(03)00065-8

Maeda Y, Ashida H, Kinoshita T. 2006. CHO glycosylation mutants: GPI anchor. *Methods Enzymol* **416:** 182–205. doi:10.1016/s0076-6879(06)16012-7

Levental I, Grzybek M, Simons K. 2010. Greasing their way: lipid modifications determine protein association with membrane rafts. *Biochemistry* **49:** 6305–6316. doi:10.1021/bi100882y

Nikolaev AV, Al-Maharik N. 2011. Synthetic glycosylphosphatidylinositol (GPI) anchors: how these complex molecules have been made. *Nat Prod Rep* **28:** 970–1020. doi:10.1039/c0np00064g

Tsai YH, Liu X, Seeberger PH. 2012. Chemical biology of glycosylphosphatidylinositol anchors. *Angew Chem Int Ed Engl* **51:** 11438–11456. doi:10.1002/anie.201203912

Guo Z. 2013. Synthetic studies of glycosylphosphatidylinositol (GPI) anchors and GPI-anchored peptides, glycopeptides, and proteins. *Curr Org Synth* **10:** 366–383. doi:10.2174/1570179411310030003

Raghupathy R, Anilkumar AA, Polley A, Singh PP, Yadav M, Johnson C, Suryawanshi S, Saikam V, Sawant SD, Panda A, et al. 2015. Transbilayer lipid interactions mediate nanoclustering of lipid-anchored proteins. *Cell* **161:** 581–594. doi:10.1016/j.cell.2015.03.048

Kinoshita T, Fujita M. 2016. Biosynthesis of GPI-anchored proteins: special emphasis on GPI lipid remodeling. *J Lipid Res* **57:** 6–24. doi:10.1194/jlr.r063313

Muñiz M, Riezman H. 2016. Trafficking of glycosylphosphatidylinositol anchored proteins from the endoplasmic reticulum to the cell surface. *J Lipid Res* **57:** 352–360. doi:10.1194/jlr.r062760

Komath SS, Singh SL, Pratyusha VA, Sah SK. 2018. Generating anchors only to lose them: the unusual story of glycosylphosphatidylinositol anchor biosynthesis and remodeling in yeast and fungi. *IUBMB Life* **70:** 355–383. doi:10.1002/iub.1734

Kinoshita T. 2020. Biosynthesis and biology of mammalian GPI-anchored proteins. *Open Biol* **10:** 190290. doi:10.1098/rsob.190290

13 Other Classes of Eukaryotic Glycans

Robert S. Haltiwanger, Lance Wells, Hudson H. Freeze,
Hamed Jafar-Nejad, Tetsuya Okajima, and Pamela Stanley

This chapter focuses on less-easily categorized types of glycan linkages that occur on certain proteins or domains. O-linked sugars on epidermal growth factor (EGF)-like repeats (O-fucose, O-glucose, and O-GlcNAc) regulate Notch signaling and the functions of several other proteins. O-Fucosylation of thrombospondin type-1 repeats (TSRs) is required for folding of these domains in a number of secreted matricellular proteins. O-mannosylation of α-dystroglycan is essential for interactions with several extracellular matrix (ECM) proteins. Defects in the glycosyltransferases that add these glycans (O-fucose, O-glucose, O-GlcNAc, and O-mannose) result in human diseases. C-Mannosylation is a unique form of glycosylation in which mannose is linked through a carbon–carbon bond to tryptophan. O-Linked Glc-Gal disaccharides are added to hydroxylysine residues and play an important role in collagen fibril formation. Although the glycans described here are found on relatively few glycoproteins, they play specific and important roles in biology.

DISCOVERY OF NOVEL TYPES OF GLYCOSYLATION

In glycoproteins, the linkage between the first sugar of a glycan and the protein defines its glycosylation class (Chapter 1, Figure 1.7) including the common GlcNAc-N-Asn, GalNAc-O-Ser/Thr, and Xyl-O-Ser linkages present in glycoproteins and proteoglycans. The novel, nonclassical glycan Glcβ1–3Fucα-O-Thr was first discovered in human urine but garnered little interest at the time. However, finding O-fucose directly linked to various clotting proteins and signaling

published by Cold Spring Harbor Laboratory Press; doi: 10.1101/glycobiology.4e.13

TABLE 13.1. Other classes of eukaryotic glycoprotein glycosylation

Modification	Examples of proteins modified	Motif
O-α-Fuc	Notch, Delta, Serrate, Jagged, factor IX, urokinase, t-PA, factor XII, factor VII, Cripto	EGF repeat
O-α-Fuc	thrombospondin 1, properdin, F-spondin, ADAMTS13, ADAMTS-like 1	TSR
O-β-Glc	Notch, Delta, Jagged, Crumbs-2, Eyes shut, factor VII, factor IX, protein Z	EGF repeat
O-β-GlcNAc	Notch, Delta, Serrate, Dumpy	EGF repeat
O-β-Gal	collagens, surfactant protein, complement factor C1q, mannan-binding proteins	collagen repeat
O-α-Man	α-dystroglycan, cadherins, plexins	mucin-like domain
C-α-Man	RNase 2, thrombospondin1, properdin	WXXW motif

t-PA, tissue plasminogen activator; EGF, epidermal growth factor; TSR, thrombospondin type-1 repeat.

receptors, such as Notch, sparked considerable interest. Monoclonal antibodies that detect glycans on specific proteins such as α-dystroglycan provided tools to identify other novel glycans. In addition, mass spectrometry revealed unusual protein modifications, such as mannose linked to protein as a C-glycoside. Table 13.1 describes many of the less common linkages synthesized in the endoplasmic reticulum (ER)–Golgi secretory pathway. Chapters 18 and 19 describe the very few known glycosylation linkages synthesized in the nucleus and cytoplasm.

O-LINKED MODIFICATIONS IN EGF REPEATS

EGF repeats, also known as EGF domains, are small protein domains (~40 amino acids) defined by six conserved Cys residues, which form three disulfide bonds (Figure 13.1A). They are found in a few hundred cell-surface and secreted proteins in metazoans and, depending on their sequence, may be modified with O-glycans as described in Table 13.1. Proteins with EGF repeats harboring these O-glycans include several involved in blood clot formation and dissolution, and the Notch family of receptors and canonical Notch ligands (Delta and Serrate/Jagged) involved in cell fate decisions. These glycan modifications are important because they regulate signal transduction during embryonic development and adult organ maintenance, cell differentiation, and the growth of several cancers. Moreover, mutations in several enzymes involved in the addition or elongation of these glycans have been found in human diseases (Chapter 45).

The signaling pathway best known to be regulated by O-glycans on EGF repeats is the Notch signaling pathway. Notch was originally identified in Drosophila, and homologs have been found in all metazoans, with four Notch receptors in mammals. Activation of Notch signal transduction is controlled at numerous levels, and dysregulation of Notch signaling results in a number of human diseases, including several types of cancer and a variety of developmental disorders. Two classes of canonical ligands bind to and activate Notch signaling in Drosophila: Delta and Serrate. Mammals have three Delta-like homologs (DLL1, DLL3, and DLL4), and two Serrate homologs called Jagged 1 and 2 (JAG1 and JAG2). These ligands are single-pass transmembrane glycoproteins that bind to and *trans*-activate Notch receptors on an adjacent cell. Binding of Notch and ligands in the same cell often results in *cis*-inhibition of the pathway. Work in recent years has shown that the O-glycans on Notch receptors affect both *trans*-activatory and *cis*-inhibitory interactions between Notch receptors and their ligands. The extracellular domain of Notch contains up to 36 tandem EGF repeats (Figure 13.2), many of which contain consensus sites for one or more types of O-glycan (Figure 13.1A). Notch EGF repeats can harbor O-fucose, O-GlcNAc and/or O-glucose (Figure 13.2). As discussed below, these sugar modifications regulate various aspects of Notch signal transduction, sometimes in a partially redundant manner and sometimes via distinct mechanisms.

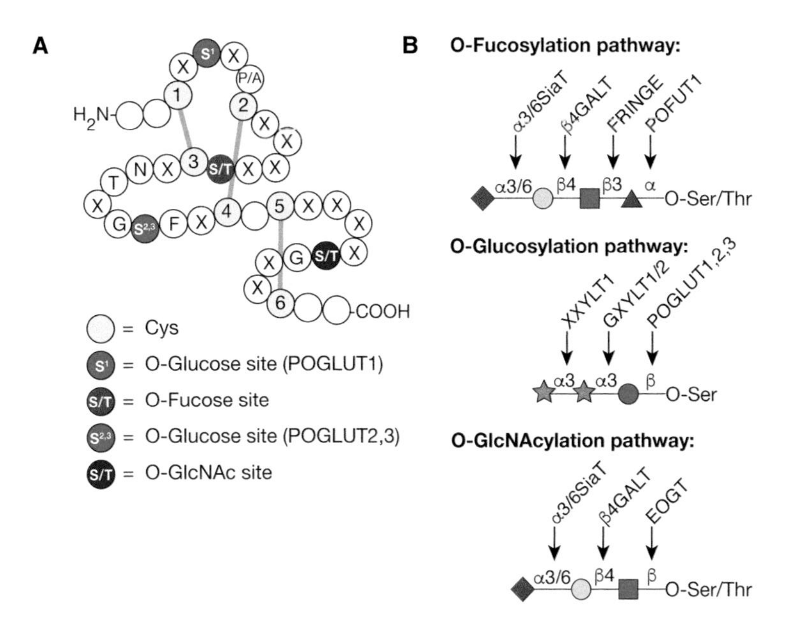

FIGURE 13.1. Modifications of epidermal growth factor (EGF) repeats. (*A*) EGF repeats can be modified by O-fucose, O-glucose, and O-GlcNAc. A schematic representation of an EGF repeat is shown: (*yellow*) conserved Cys residues; (*gray lines between the cysteines*) disulfide-bonding pattern. The modification sites for O-glucose (*blue*), O-fucose (*red*), and O-GlcNAc (*dark blue*) are shown in the context of the consensus sequences for each. S, serine; S^1, POGLUT1 modification site; $S^{2,3}$, POGLUT2/3 modification site; T, Thr; P, Pro; A, Ala; G, Gly; N, Asn; F, Phe; X, any amino acid. (Modified, with permission, from Haltiwanger RS. 2004. In *Encyclopedia of Biological Chemistry*, Vol. 2, pp. 277–282, © Elsevier.) (*B*) EGF repeat O-glycan synthesis in mammals. The largest known structure for O-fucose, O-glucose, and O-GlcNAc glycans on EGF repeats is shown, and the enzyme responsible for the addition of each sugar is indicated.

O-Fucose Glycans

The α-linked O-fucose modification immediately precedes the third conserved Cys of certain EGF repeats (Figure 13.1A). The consensus motif for O-fucosylation is $C^2X_4(S/T)C^3$ (C^2 and C^3 are the second and third conserved Cys of the EGF repeat). Nearly 100 proteins in mouse or human databases contain this sequence. O-Fucose can be elongated to a tetrasaccharide (Siaα2–3/6Galβ1-4GlcNAcβ1-3Fucα-O-Ser/Thr) in certain contexts (e.g., EGF1 from human clotting factor IX and EGF12 from mouse NOTCH1) (Table 13.1), but not in others (e.g., EGF1 from human clotting factor VII and EGF5 from mouse NOTCH1).

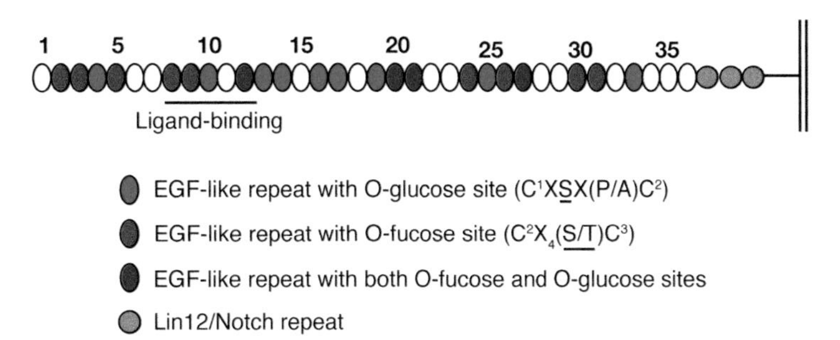

FIGURE 13.2. Extracellular domain of a generic Notch receptor showing the numerous sites for POFUT1 and POGLUT1 modifications that are evolutionarily conserved in Drosophila Notch, mouse NOTCH1 and NOTCH2, and human NOTCH1 and NOTCH2. Epidermal growth factor (EGF) repeats directly involved in ligand binding are indicated by the bar. (Adapted, with permission of the American Society for Biochemistry and Molecular Biology, from Shao L, Moloney DJ, and Haltiwanger RS. 2003. *J Biol Chem* **278:** 7775–7782.)

Protein O-fucosyltransferase 1 (POFUT1) transfers fucose from GDP-Fuc to a properly folded EGF repeat containing the appropriate consensus sequence (Figure 13.1B). O-Fucose can be elongated by a β1-3 N-acetylglucosaminyltransferase (β3GlcNAcT) specific for O-fucose residues in a properly folded EGF repeat. The gene encoding the O-fucose-EGF β3GlcNAcT was originally identified in Drosophila as a modifier of Notch signaling called *fringe*. There are three mammalian homologs: Manic fringe (MFNG), Lunatic fringe (LFNG), and Radical fringe (RFNG). Each of the Fringe proteins catalyzes the transfer of GlcNAc from UDP-GlcNAc to O-fucose on an EGF repeat. The GlcNAc can be further elongated by a β4GalT (e.g., B4GALT1) and capped with a sialic acid, probably using the same sialyltransferases (α2-3/6SiaTs) as those that act on N- or other O-glycans (Chapter 14). Elongation past the disaccharide has not been observed to date on Drosophila Notch.

The O-fucose glycans play important roles in the Notch signal transduction pathway. Elimination of POFUT1 in Drosophila or mice results in embryonic lethality with developmental defects reminiscent of loss of Notch signaling. Moreover, cell-based assays and analysis of knock-in alleles of Drosophila *Notch* and mouse *Notch1* lacking specific O-fucosylation sites indicate that POFUT1 and its downstream enzymes primarily, if not exclusively, regulate Notch signaling by adding O-fucose glycans to Notch receptors. Heterozygous mutations in the human *POFUT1* gene cause a rare autosomal dominant skin discoloration disease called Dowling–Degos disease type 2 (DDD2; OMIM 615327). A homozygous *POFUT1* variant causes a developmental disease termed POFUT1-CDG. POFUT1 is localized to the ER, in which it only modifies properly folded EGF repeats, suggesting that O-fucosylation may have some role in quality control. The Drosophila homolog of POFUT1 has chaperone activity and is proposed to play a role in the proper folding and trafficking of Notch. Loss of POFUT1 also affects the trafficking of Notch receptors to the surface of mammalian cells to varying extents. Structural studies reveal that the O-fucose on EGF12 directly participates in canonical Notch ligand binding. Of note, mutating this O-fucosylation site in Drosophila Notch and in mouse NOTCH1 results in embryonic lethality, with phenotypes milder than loss of Notch/NOTCH1 in each system, suggesting that other Notch O-fucosylation sites also contribute to the effects of POFUT1 on Notch signaling.

The O-fucose on Notch EGF repeats can be elongated by the activity of Fringe enzymes, which transfer a GlcNAc residue to O-fucose on Notch receptors (Figure 13.1B). Elimination of Fringe in Drosophila causes defects in wing, eye, and leg development, and ablation of *Lfng* in mice causes a severe defect in somite formation. Notably, *Pofut1* loss-of-function phenotypes are more severe and broader than *Fringe* loss-of-function phenotypes in both mouse and fly embryos, indicating that elongation of O-fucose by Fringe proteins only regulates a subset of Notch-dependent developmental processes. Mutations in human *LFNG* result in malformed vertebrae and ribs (spondylocostal dysostosis type 3; OMIM 609813). Surprisingly, Fringe reduces Notch activation by Serrate/Jagged ligands, but potentiates Notch activation by Delta ligands (Figure 13.3). For instance, Fringe-mediated elongation of O-fucose on EGF8 and EGF12 of the ligand-binding domain clearly enhances the affinity between Notch and Delta ligands, resulting in enhanced Notch activation. The fucose on NOTCH1 EGF12 functions like a surrogate amino acid as it directly participates in ligand binding. Regulation of Notch signaling by O-fucose glycans provides one of the clearest examples of how a signal transduction pathway may be regulated by altering the glycosylation of a receptor.

O-Glucose Glycans

The first β-linked O-glucose modification of EGF repeats described occurs between the first and second conserved Cys at the consensus sequence $C^1XSX(P/A)C^2$ (Figure 13.1A), which is found in approximately 50 proteins in mouse or human databases. This O-glucose glycan typically exists as the trisaccharide Xylα1–3Xylα1–3Glcβ-O-Ser, although mono- and disaccharide

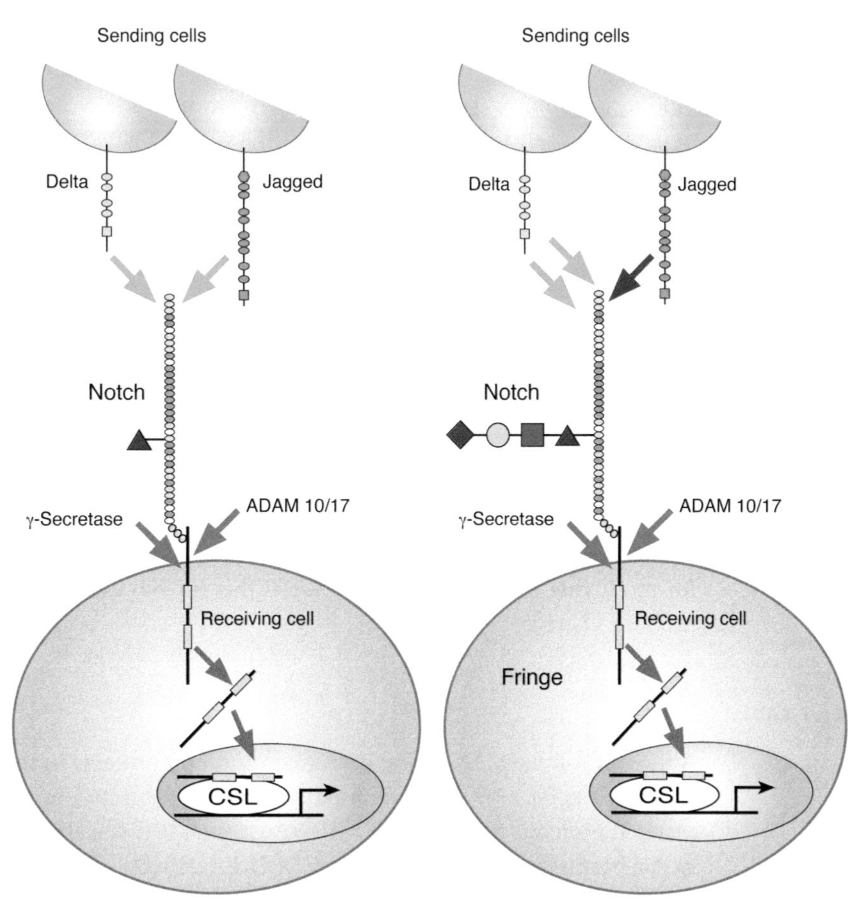

FIGURE 13.3. Notch signaling pathway. Notch exists on the cell surface as a heterodimer in which the extracellular domain is tethered to the transmembrane and intracellular domain by noncovalent, calcium-dependent interactions. Notch is activated by ligand (members of the Delta or Jagged families in mammals) expressed on the surface of a "sending cell." Both ligands are shown, but typically the sending cell would predominantly express one or the other. Binding of ligand induces a conformational change in the Notch receptor, exposing a protease site for a cell-surface protease (either ADAM10 or ADAM17). This extracellular cleavage is followed by a second cleavage, catalyzed by γ-secretase, which results in release of the Notch intracellular domain as a soluble protein in the cytoplasm of the "receiving cell." The Notch intracellular domain translocates to the nucleus where it interacts with members of the CSL (CBF1, Su(H), Lag-1) family of transcriptional regulators and activates transcription of a number of downstream genes. Notch expressed in receiving cells without Fringe (*left*) is compared with cells that express Fringe (*right*). The presence of Fringe results in elongation of O-fucose and alters the ability of Notch to respond to ligands on the sending cell. In this example, Fringe causes an increase in Delta signaling and a decrease in Jagged signaling. (Modified, with permission of the *Annual Review of Biochemistry*, from Haltiwanger RS, Lowe JB. 2004. *Annu Rev Biochem* **73:** 491–537, © Annual Reviews.)

forms are also seen. The human gene encoding this protein O-glucosyltransferase is *POGLUT1* (*rumi* in flies). Mutations in *Poglut1/rumi* cause developmental defects similar to *Notch1/Notch* mutations in mice and flies, although in both organisms, O-glucosylation is important in other glycoproteins as well. Similar to *POFUT1*, heterozygous mutations in human *POGLUT1* cause an autosomal dominant skin pigmentation disease termed Dowling–Degos disease type 4 (DDD4; OMIM 615696). Moreover, recessive mutations in *POGLUT1* have been identified in a new form of limb-girdle muscular dystrophy lacking any other phenotypes in the skin or other organ systems known to depend on Notch signaling (LGMDR21; OMIM 617232). Like POFUT1, POGLUT1 is localized to the ER and requires a properly folded EGF repeat as substrate. Although loss of *Poglut1/rumi* does not affect the surface expression of Notch in fly tissues and of NOTCH1 in several mammalian cell lines, it does affect NOTCH1 surface expression in some other mammalian cell lines. O-Glucosylation does not appear to affect Notch binding to canonical ligands and is likely required for Notch cleavage.

The genes encoding the xylosyltransferases are glucoside α1-3 xylosyltransferase (*GXYLT1* and *GXYLT2*, *shams* in flies) and xyloside α1-3 xylosyltransferase (*XXYLT1*, *Xxylt* in flies) (Figure 13.1B). Genetic and cell-based studies on Notch xylosylation in flies have shown that (1) although addition of O-glucose to Notch promotes Notch signaling, addition of xylose residues to O-glucose on EGF16-20 of Drosophila Notch inhibits Notch signaling; (2) xylose residues on EGF16-20 selectively reduce the binding of Notch to *trans*-Delta, without affecting the binding of Notch to *trans*-Serrate or to *cis*-ligands; and (3) although loss of one copy of the fly *GXYLT*/*shams* does not have any phenotype by itself, it can affect Notch signaling when Notch is haploinsufficient or overexpressed.

Recently, a second β-linked O-glucose modification site has been identified between the third and the fourth Cys of the EGF repeats at the putative consensus sequence $C^3XNTXGSFXC^4$ (Figure 13.1A), which is found in more than 30 proteins in mouse and human databases. Homologs of POGLUT1, termed POGLUT2 and POGLUT3, are responsible for modifying this site. The O-glucose at this site has not been reported to be elongated past the monosaccharide.

O-GlcNAc Glycans

Although β-linked O-GlcNAc is very common on proteins in the nuclear and cytosolic compartments (Chapter 19), the same modification can be found in comparatively few membrane and secreted proteins. Addition of O-GlcNAc occurs between the fifth and sixth conserved Cys of an EGF repeat at the putative consensus $C^5XXXX(S/T)GX_{2\text{-}3}C^6$ (Figure 13.1A) in mouse and Drosophila Notch. However, sites have been mapped on only a few proteins. The O-GlcNAc can be elongated by β1,4Gal followed by a sialic acid to form Siaα2-3/6Galβ1-4GlcNAc. The EGF-specific O-GlcNAc transferase, EOGT, is distinct from OGT, the GlcNAc-transferase that modifies nuclear and cytoplasmic proteins (Chapter 19). Inactivation of Drosophila *Eogt* is pupal lethal and causes a wing-blistering defect if deleted in the wing. The Drosophila phenotype can be explained by an EOGT substrate Dumpy and no obvious Notch-deficient phenotypes have been observed. However, genetic interactions between *eogt* and Notch pathway gene mutations occur in the fly and mutations in human *EOGT* result in skin and skeletal problems in AOS (Adams–Oliver syndrome type 4; OMIM 615297), a disorder with links to the Notch pathway. Consistently, EOGT in mice regulates vascular development and integrity by potentiating Delta-like ligand-mediated Notch signaling.

O-FUCOSE MODIFICATION OF THROMBOSPONDIN TYPE 1 REPEATS (TSRs)

O-Fucose is also found on TSRs. Similar to EGF repeats, TSRs are small protein domains (50–60 amino acids) with six conserved Cys that form three disulfide bonds (Figure 13.4). Like EGF repeats, TSRs are found in a subset of cell-surface and secreted proteins in metazoans, and they appear to function in protein–protein interactions. The O-fucose site occurs between the first and second conserved Cys residues at the putative consensus sequence $C^1X_2(S/T)C^2$ in TSRs in about 50 proteins (Table 13.1). The O-fucose on TSRs can be elongated to form the disaccharide Glcβ1-3Fucα-O-Ser/Thr, first observed in 1975 in glycosides from human urine. The enzyme that fucosylates TSRs is POFUT2. It is expressed in many cells and tissues and is distinct from POFUT1. Elimination of *Pofut2* in mice results in early embryonic lethality with defects in gastrulation. The gene encoding the β3-glucosyltransferase that elongates O-fucose on TSRs is *B3GLCT*. Mutations in *B3GLCT* in humans result in a multifaceted severe developmental

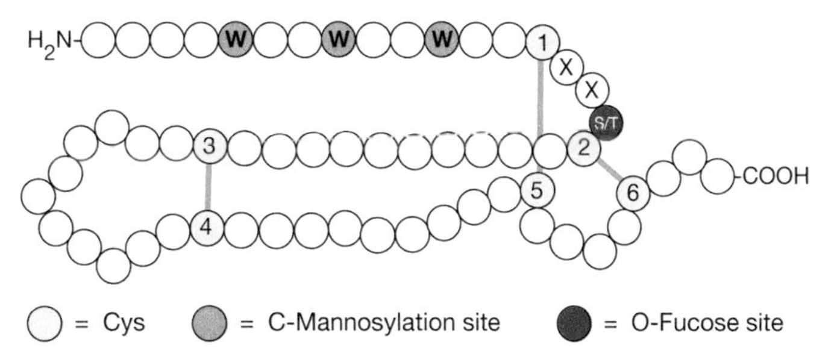

FIGURE 13.4. Modifications of thrombospondin type-1 repeats (TSRs). These repeats have six conserved Cys (*yellow*), three disulfide bonds (*gray lines*), and consensus sites for O-fucose (*red*) and C-mannose (*pink*) addition. W, tryptophan; S, serine; T, threonine; X, any amino acid. (Modified, with permission, from Haltiwanger RS. 2004. In *Encyclopedia of Biological Chemistry* [ed. Lennarz WI, Lane MD], Vol. 2, pp. 277–282, © Elsevier.)

disorder known as Peters-plus syndrome (OMIM 261540). Elimination of *B3glct* in mice phenocopies many of the defects seen in Peters-plus syndrome. TSRs are also often frequently modified with C-mannose (Figure 13.4), although the relationship between the two modifications is not yet clear. A number of biologically interesting proteins are known or predicted to contain this modification, including thrombospondin-1 and -2, all ADAMTS (a disintegrin and metalloproteinase with thrombospondin motifs; ADAMTS 1–20) and ADAMTS-like proteins, properdin, and F-spondin (Table 13.1). O-Fucosylation of TSRs in several proteins including ADAMTS 13 and ADAMTS-like 1 is needed for their secretion, suggesting that O-fucosylation of TSRs may be involved in quality control or folding, similar to the predicted role of O-fucose on EGF repeats in Drosophila. Like POFUT1, POFUT2 is an ER-localized enzyme that only fucosylates properly folded TSRs. POFUT2 and B3GLCT appear to function in a novel ER quality control pathway required for proper folding of TSRs. In thrombospondin-1, the region of the TSR modified with O-fucose mediates interactions between TSRs and several cell-surface receptors. Thus, the presence of O-fucose in this region could influence interactions between TSRs and adjacent cells, much as O-fucose on EGF repeats regulates Notch ligand binding. TSRs allow proteins like the thrombospondins to bind to cells and/or to other components in the ECM, an essential step in the antiangiogenic functions of these proteins.

O-MANNOSE GLYCANS

O-Mannose α-linked to protein was first identified in yeast in the 1950s (Chapter 23). Fungi elongate this mannose with a battery of mannosyltransferases to form extensions similar to those on their N-glycans. O-Mannose was first identified in mammals in 1979, linked to a rat brain proteoglycan. O-Mannose glycans account for up to one-third of all O-glycans in some mammalian tissues, including brain. O-Mannose glycans in mammals are quite varied and belong to three core classes (Figure 13.5). Defects in O-mannosylation of α-dystroglycan, which is the major extracellular glycoprotein component of the dystrophin–glycoprotein complex, result in multiple forms of congenital muscular dystrophy (CMD). These diseases are known as the secondary dystroglycanopathies and include Walker–Warburg syndrome, muscle–eye–brain disease, Fukuyama congenital muscular dystrophy, and various forms of limb-girdle muscular dystrophy (Chapter 45). Some of the O-mannose glycans on α-dystroglycan serve as binding sites for ECM proteins that contain laminin globular domains. These same O-mannose glycans function as receptors for certain arenaviruses, and the loss of these glycans is also associated with metastasis in multiple cancers.

The initiating O-mannose for α-dystroglycan is transferred by O-mannosyltransferases POMT1 and POMT2 that function together (Figure 13.5). However, there are multiple protein O-mannosyltransferases that catalyze the addition of O-mannose to Ser/Thr residues on a host of

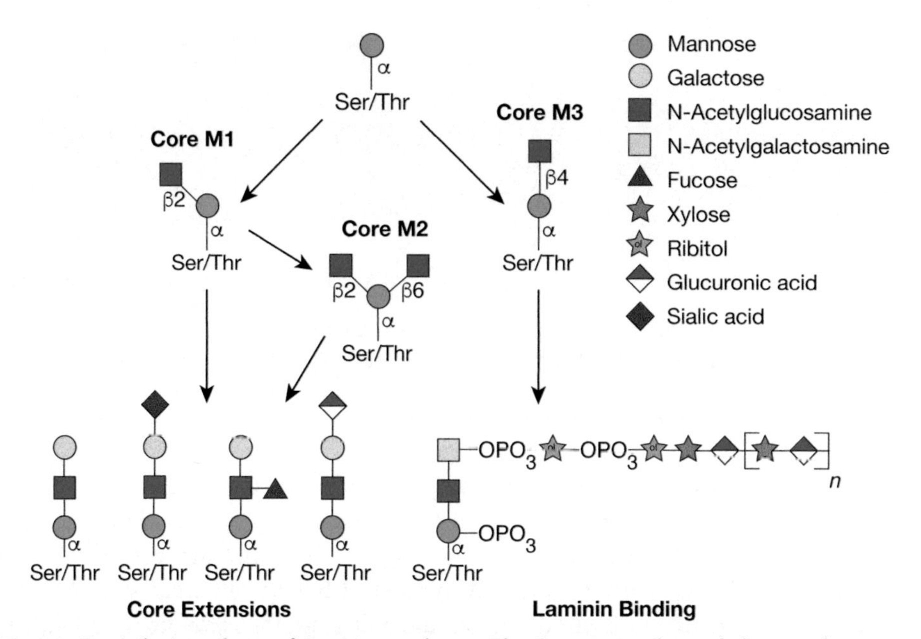

FIGURE 13.5. Biosynthetic pathway of O-mannose glycans. The O-mannose glycans belong to three core classes depending on extension from protein-linked mannose by a sole β1-2GlcNAc (core M1), by two GlcNAcs in β1-2 and β1-6 linkages (core M2), or by β1-4GlcNAc (core M3). Further elaboration of all three core classes can occur by extensions from the mannose as shown. Core M3 glycans may be specific for α-dystroglycan in mammals, and usually carry the repeating disaccharide matriglycan transferred by LARGE that is required for laminin binding.

proteins in the ER using dolichol-P-mannose as the donor. Core M1 O-mannose glycans (Figure 13.5) are then generated by the action of a Golgi-localized N-acetylglucosaminyltransferase (POMGNT1) that adds β1-2GlcNAc to O-mannose on glycoproteins. Core M1 glycans can be further extended by enzymes that also act in other pathways, including galactosyltransferases, fucosyltransferases, glucuronyltransferases, and sialyltransferases. Core M1 glycans can also be converted to Core M2 O-mannose glycans by the addition of a β1-6GlcNAc to the O-mannose, catalyzed by the N-acetylglucosaminyltransferase MGAT5B. This new branch on core M2 glycans can also be extended by the glycosyltransferases that act on core M1 (Figure 13.5 core extensions). Core M1 and M2 O-mannose glycans have been identified on multiple mammalian proteins besides α-dystroglycan.

The core M3 glycans (Figure 13.5) possess key functional properties and have to date been identified only on the mammalian protein α-dystroglycan. Core 3 O-mannose is extended in the ER by an N-acetylglucosaminyltransferase (POMGNT2) that adds a β1-4GlcNAc. This disaccharide is a substrate for an N-acetylgalactosaminyltransferase (B3GALNT2) that adds β1-3GalNAc to the GlcNAc. This trisaccharide is a substrate for a secretory pathway kinase (POMK) that adds a 6-linked phosphate to the core mannose. The immediate next steps are the addition of a ribitol-5-P in a phosphodiester linkage to the GalNAc catalyzed by fukutin (FKTN), followed by the addition of another ribitol-5-P in a phosphodiester linkage to the first ribitol by fukutin-related protein (FKRP). A D-ribitol-5-phosphate cytidylytransferase (CRPPA) generates the sugar nucleotide CDP-ribitol that FKTN and FKRP use. RXYLT1 and B4GAT1 glycosyltransferases add a xylose and a glucuronic acid, respectively, to the outermost ribitol. This action serves to prime the core M3 glycan for the addition of a repeating disaccharide (Xyl-α3-GlcA-β3-)$_n$ that is catalyzed by the dual activity LARGE1 (or LARGE2) glycosyltransferase. This repeating disaccharide, termed matriglycan, serves as the binding site for ECM proteins containing laminin globular domains, including laminins, agrin, pikachurin, and Eyes shut homolog, as well as the spike protein of certain arenaviruses. Mutations in the genes encoding all of the enzymes involved in the synthesis of the M3 glycan have been identified in CMD patients (Chapter 45).

O-GLYCANS IN COLLAGENS

Proteins with collagen domains are modified by a disaccharide, Glcα1-2Galβ-, which is assembled on hydroxylysine (Hyl) residues. First recognized on collagen in the 1960s, this O-glycan has been observed in other proteins that have collagen-like modules, including adiponectin and pulmonary surfactant proteins. Primitive forms of collagen in sponges and sea anemones also have disaccharide-modified Hyl. The first step in the pathway is the hydroxylation of lysine to Hyl via three lysyl hydroxylases. The two-step glycosylation rapidly follows on most of the available acceptor sites. Galactosytransferases COLGALT1 and COLGALT2 initiate collagen glycosylation in the ER. Autosomal recessive mutations in *COLGALT1* have been identified in two unrelated patients with a brain vessel disease associated with fragility of cerebral vessels (OMIM 618360). Importantly, similar phenotypes are observed in some patients with collagen type IV mutations, highlighting the critical role of collagen glycosylation in vascular integrity. Interestingly, although the hydroxylation and addition of galactose are conserved in all animal species examined and occur by the action of two separate enzymes, giant viruses have one bifunctional enzyme capable of hydroxylating lysine and adding glucose instead of galactose to collagen. It is worth noting that the enzyme that adds glucose to the galactose on collagens remains to be identified, although human lysyl hydroxylase 3 (LH3) has been suggested to have this activity besides hydroxylation of lysine. Mice lacking LH3 fail to glycosylate collagen IV properly, which causes embryonic lethality and deposition of misfolded collagen in the ER. Glycosylation is thought to proceed until the protein folds and assembles into the well-known collagen triple helix. Some studies suggest that the extent of glycosylation controls or influences the rate of triple-helix formation and, in turn, the size of collagen fibrils. Other studies suggest that excessive glycosylation of hydroxylysines makes them poor substrates for extracellular enzymatic deamination, the first step in cross-linking collagen.

C-MANNOSYLATION

Another type of glycosylation, first detected in human urine, is called C-mannosylation. The C-1 atom of a single mannose residue is added in α-linkage to the C-2 atom of the indole moiety of tryptophan (Trp) in target proteins (Figure 13.6). Note that this is a rare glycosidic linkage, as it is a C–C bond rather than a C–O or C–N bond. Structural analysis, biosynthetic studies in mammalian cell lines, and a specific antibody against the modification suggest that C-mannosylation is widespread, but it appears to be absent in bacteria and yeast. Conserved C-mannosylation sites are mainly found in proteins with TSRs and in type I cytokine receptors. The critical glycosylation sequence is WXXW/C. In TSRs, the C-mannosylation motif can have up to three Trp residues (WXXWXXWXXC), all three of which can be C-mannosylated (Figure 13.4). Synthetic peptides with this motif can be C-mannosylated in vitro, and more than 300 mammalian proteins have this sequence. Several complement proteins contain TSRs with C-mannose. For instance, in properdin (which is a positive regulator of complement), 14 of 17 consensus sites are fully modified. The biosynthetic donor for C-mannosylation in the ER is Dol-P-Man, which is produced from GDP-Man (Chapter 5). A gene encoding a C-mannosyltransferase has been identified in *Caenorhabditis elegans* (DPY-19) and mutants in DPY-19 cause a neuroblast migration defect. Mammalian genomes encode four DPY-19 homologs, two of which have been demonstrated to have C-mannosyltransferase activity. Protein C-mannosylation is thought to occur before or during protein folding in the ER. Moreover, loss of C-mannosylation results in the ER accumulation of some DPY-19 target proteins. Therefore, C-mannosylation is likely to aid in protein folding in the ER. Given the widespread presence of C-mannosylation on proteins and recent

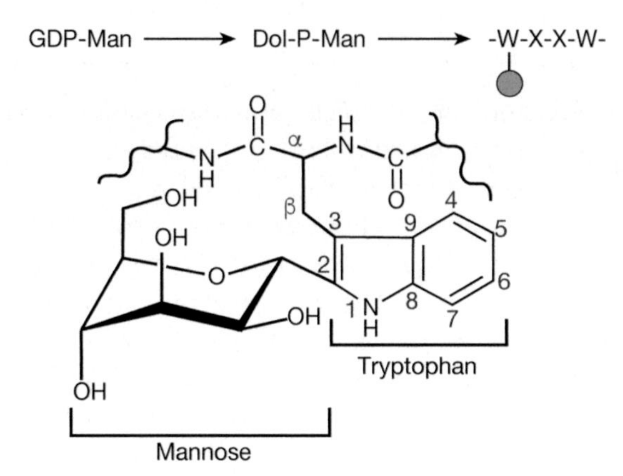

FIGURE 13.6. Biosynthetic pathway for C-mannosylation and structural details of a C-mannosylated tryptophan. The α-linked mannose is shown in the $^1C_4$ conformation.

identification of the enzymes responsible for this posttranslational modification, biological functions should be revealed soon.

ACKNOWLEDGMENTS

The authors appreciate helpful comments and suggestions from Thierry Hennet, Tadashi Suzuki, and Jon Agirre.

FURTHER READING

Ogawa M, Furukawa K, Okajima T. 2014. Extracellular O-linked β-*N*-acetylglucosamine: its biology and relationship to human disease. *World J Biol Chem* **5**: 224–230. doi:10.4331/wjbc.v5.i2.224

Praissman JL, Wells L. 2014. Mammalian O-mannosylation pathway: glycan structures, enzymes, and protein substrates. *Biochemistry* **53**: 3066–3078. doi:10.1021/bi500153y

Yoshida-Moriguchi T, Campbell KP. 2015. Matriglycan: a novel polysaccharide that links dystroglycan to the basement membrane. *Glycobiology* **25**: 702–713. doi:10.1093/glycob/cwv021

Sheikh MO, Halmo SM, Wells L. 2017. Recent advancements in understanding mammalian O-mannosylation. *Glycobiology* **27**: 806–819. doi:10.1093/glycob/cwx062

Hennet T. 2019. Collagen glycosylation. *Curr Opin Struct Biol* **56**: 131–138. doi:10.1016/j.sbi.2019.01.015

Hohenester E. 2019. Laminin G-like domains: dystroglycan-specific lectins. *Curr Opin Struct Biol* **56**: 56–63. doi:10.1016/j.sbi.2018.11.007

Holdener BC, Haltiwanger RS. 2019. Protein *O*-fucosylation: structure and function. *Curr Opin Struct Biol* **56**: 78–86. doi:10.1016/j.sbi.2018.12.005

Shcherbakova A, Preller M, Taft MH, Pujols J, Ventura S, Tiemann B, Buettner FF, Bakker H. 2019. C-mannosylation supports folding and enhances stability of thrombospondin repeats. *eLife* **8**: e52978. doi:10.7554/elife.52978

Varshney S, Stanley P. 2019. Multiple roles for O-glycans in Notch signaling. *FEBS Lett* **592**: 3819–3834. doi:10.1002/1873-3468.13251

Yu H, Takeuchi H. 2019. Protein *O*-glucosylation: another essential role of glucose in biology. *Curr Opin Struct Biol* **56**: 64–71. doi:10.1016/j.sbi.2018.12.001

John A, Järvå MA, Shah S, Mao R, Chappaz S, Birkinshaw RW, Czabotar PE, Lo AW, Scott NE, Goddard-Borger ED. 2021. Yeast- and antibody-based tools for studying tryptophan C-mannosylation. *Nat Chem Biol* **17**: 428–437. doi:10.1038/s41589-020-00727-w

Pandey A, Niknejad N, Jafar-Nejad H. 2021. Multifaceted regulation of Notch signaling by glycosylation. *Glycobiology* **31**: 8–28. doi:10.1093/glycob/cwaa049

14 Structures Common to Different Glycans

Pamela Stanley, Manfred Wuhrer, Gordon Lauc, Sean R. Stowell, and Richard D. Cummings

This chapter describes the variable components of N-glycans, O-glycans, and glycolipids attached to the core of each glycan class and presented in Chapters 9, 10, and 11. The glycan extensions of these cores form the mature glycan and may include human blood group determinants. The terminal sugars of the mature glycan often regulate the function(s) or recognition properties of a glycoconjugate. Also discussed are milk oligosaccharides, that carry many of the same extensions on a lactose core.

REGULATED GLYCOSYLATION OF GLYCAN EXTENSIONS

Many glycan extensions are regulated during embryogenesis and in the postnatal period as part of the normal developmental program (Chapter 41). Changes in terminal glycan structure are

published by Cold Spring Harbor Laboratory Press; doi: 10.1101/glycobiology.4e.14

also often associated with malignant transformation in cancer (Chapter 47). Tissue- and/or lineage-specific regulation of glycan extension biosynthesis is largely due to the regulated expression of the relevant glycosyltransferase and related glycosylation genes. Biological consequences of such changes in glycan extensions are discussed throughout this volume. However, the majority of regulated terminal glycosylations observed probably have many different functions.

Type-2 Glycan Units (LacNAc)

The core structures in Figure 14.1 have a terminal GlcNAc and may receive β1-4Gal to generate a Type-2 unit composed of Galβ1-4GlcNAc, also called N-acetyllactosamine (LacNAc) (Figure 14.2). The terminal Gal so generated can receive a β1-3GlcNAc, which in turn can receive a β1-4Gal, thus forming two LacNAc units. These reactions may recur to form poly-N-acetyllactosamine $[\text{-3Gal}\beta\text{1-4GlcNAc}\beta\text{1-}]_n$ (poly-LacNAc). Poly-LacNAc chains occur in glycans from most cell types. An alternative is a chain composed of LacdiNAc glycan units $[\text{-3GalNAc}\beta\text{1-4GlcNAc}\beta\text{1-}]_n$ generated by the action of a β1-4GalNAc-transferase. LacdiNAc (GalNAcβ1-4GlcNAc) termini occur on N-glycans in bovine milk, rat prolactin, and kidney epithelial cells, as well as in invertebrates such as snails and worms (Chapter 25). These residues are frequently α2-6-sialylated in vertebrates.

Type-1 Glycan Units

Terminal GlcNAc residues in N-glycans, O-glycans, and glycolipids may alternatively be modified by β1-3Gal (Figure 14.2) to generate a Type-1 unit composed of Galβ1-3GlcNAc. In humans, expression of Type-1 units is relatively high in O-glycans of glycoproteins and glycolipids in the epithelia of the gastrointestinal or reproductive tracts.

Type-1 and Type-2 units may be further modified by glycosyltransferases that transfer sugars to terminal Gal or subterminal GlcNAc, generating sialylated, fucosylated, or sulfated structures or distinct blood group determinants. Type-1- or Type-2-based blood group determinants can be distinguished immunologically and may be important to consider when crossing allogenic barriers in the setting of transplantation and transfusion.

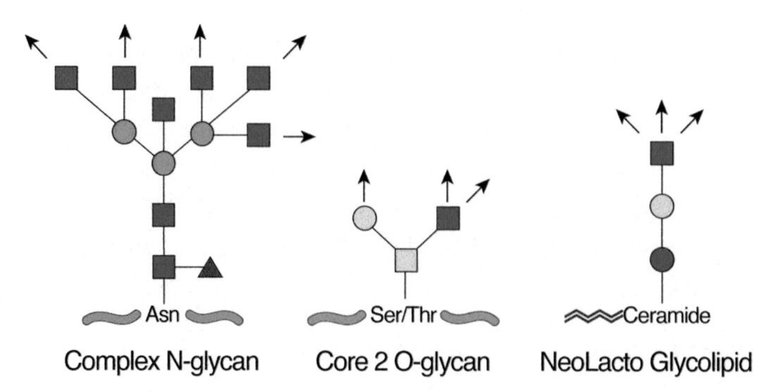

FIGURE 14.1. N-Glycan synthesis (Chapter 9) leads to complex N-glycans with branching GlcNAc residues that are generally extended (*arrows*) in glycosylation reactions that may be tissue-specific, developmentally regulated, or even protein-specific. The GlcNAc linked to the core mannose is termed the bisecting GlcNAc, and it is not usually modified. O-GalNAc glycan synthesis (Chapter 10) includes a core 2 structure with GlcNAc that may be modified subsequently by many of the same enzymes that act on N-glycans. Glycolipid core structures (Chapter 11) and O-fucose and O-mannose glycans (Chapter 13) with a terminal GlcNAc are also modified by many of the same enzymes that act on N- and O-glycans.

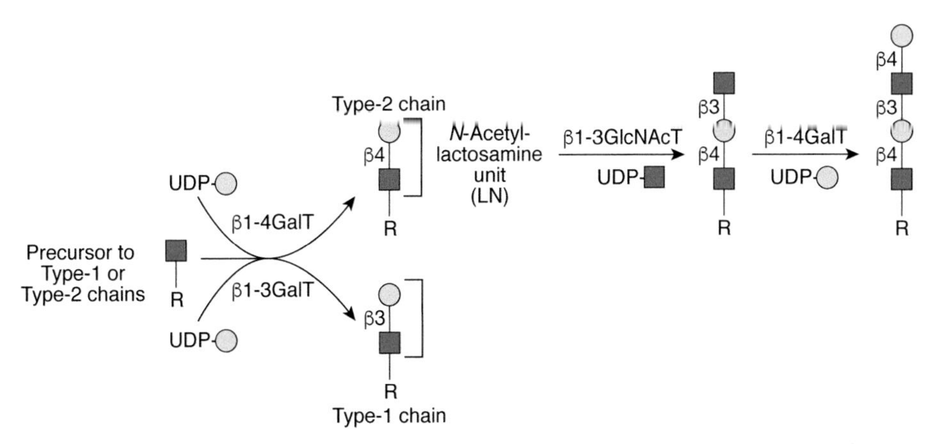

FIGURE 14.2. Terminal GlcNAc residues are usually galactosylated. Modification by β1-4Gal (*top*) occurs in all mammalian tissues. This reaction is catalyzed by β1-4 galactosyltransferases (B4GALT1 to B4GALT6) and yields the Galβ1-4GlcNAc (N-acetyllactosamine) unit termed Type-2. Transfer of β1-3Gal residues (*bottom*) is restricted to certain tissues. This reaction is catalyzed by β1-3 galactosyltransferase (B3GALT1, 2, 4, and 5) and yields the Galβ1-3GlcNAc unit termed Type-1. R indicates N- and O-glycans or glycolipids. Type-2 and -1 units can be further modified by subsequent glycosylation reactions. Poly-N-acetyllactosamine chain initiation is catalyzed by β1-3 N-acetylglucosaminyltransferases (B3GNT2 to B3GNT8).

Poly-*N*-Acetyllactosamines and i and I Human Blood Groups

Some glycoproteins and glycolipids preferentially carry poly-LacNAc. This implies that the glycosyltransferases responsible can discriminate between glycan acceptors with terminal GlcNAc or Gal. For example, poly-LacNAc extensions preferentially occur on multiantennary N-glycans, particularly on the β1-6GlcNAc branch initiated by N-acetylglucosaminyltransferase V (GlcNAc-TV, MGAT5) (Chapter 9). Similarly, poly-LacNAc extensions on O-GalNAc glycans associated with mucin glycoproteins often preferentially occur on the β1-6GlcNAc transferred by a core 2 β1-6GlcNAcT (CGNT1, 3, and 4; Chapter 10). N-Glycans generally have longer poly-LacNAc extensions than O-glycans, and both may receive sialic acid or Fuc residues or sulfate. Thus, poly-LacNAc chains may serve as linear, extended scaffolds for the presentation of specific terminal glycans, whose functions require them to be presented at a certain distance from the plasma membrane. This may be especially important when considering that among the many LacNAc-terminating glycans on the cell surface, LacNAc binding galectins, for example, appear to preferentially engage particular cell-surface glycans. More specifically, galectins display a strong preference for poly-LacNAc chains, especially in the context of the cell surface. These interactions occur through internal as well as terminal Gal engagement (Chapter 36). In the context of cell-surface glycan presentation in a complex glycocalyx, extension of glycans into linear poly-LacNAc structures may allow galectins, and perhaps other lectins, to specifically bind distinct glycoproteins among many potential LacNAc-bearing glycoconjugates. Interestingly, poly-LacNAc is also the backbone of keratan sulfate (KS, see below), in which the Gal and GlcNAc residues are 6-O-sulfated, and such sulfated poly-LacNAc is not recognized by galectins.

Poly-LacNAc chains may also become branched by the addition of β1-6GlcNAc to internal Gal residues. Branched and nonbranched poly-LacNAc chains correspond to the "I" and "i" blood group antigens, respectively (Figure 14.3). These antigens were originally discovered during the analysis of a cold-dependent agglutinating antibody (cold agglutinin) in a patient with acquired hemolytic anemia (cold agglutinin disease [CAD]) (Chapter 46). Cold agglutinin antibodies interact with red blood cells (erythrocytes) that express the I blood group (the "I" antigen). The development of antibodies against the I antigen is often associated with a *Mycoplasma pneumoniae* infection, whereas anti-i antibodies occur most commonly in the setting of

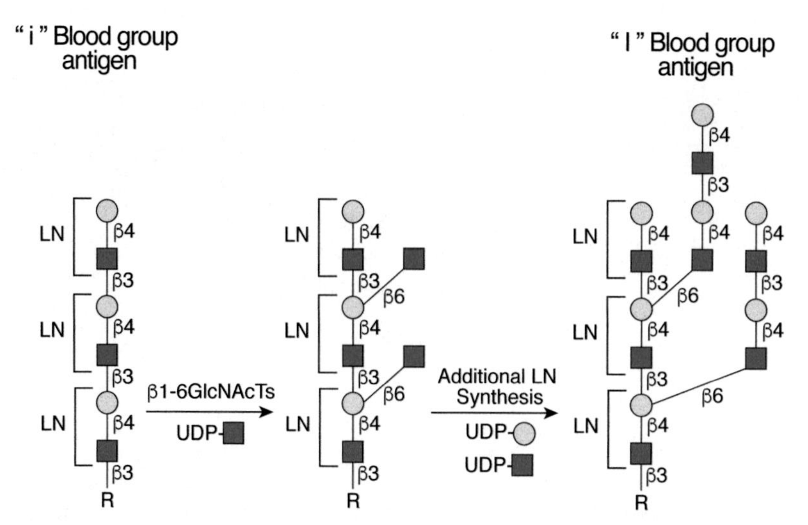

FIGURE 14.3. Blood group i and I antigen synthesis. Linear poly-N-acetyllactosamine chains (i antigen) synthesized on N- and O-glycans or glycolipids (R) may be modified by a β1-6 N-acetylglucosaminyltransferase (GCNT2 or GCNT3). These enzymes transfer GlcNAc in β1-6 linkage to internal Gal residues. The newly added β1-6 N-acetylglucosamine branch (I antigen) may serve as substrate for subsequent poly-N-acetyllactosamine biosynthesis (Figure 14.2). (LN) N-acetyllactosamine unit.

infectious mononucleosis. Distinct β1-6GlcNAcTs yield different types of β1-6-branched glycans (Figure 14.3). The i antigen is abundantly expressed on the surface of embryonic erythrocytes and on erythrocytes during times of altered erythropoiesis. Such cells are relatively deficient in the expression of the I antigen. However, during the first 18 months of life, I antigen reactivity on erythrocytes reaches adult levels, and i antigen reactivity declines to very low levels. This developmental regulation is presumed to be due to the regulated expression of GCNT2 or GCNT3, which both encode a β1-6GlcNAcT. Given the development regulation of I antigen expression, in practice, cord blood is used as a target source of the i antigen when testing whether an individual possesses anti-i antibodies. Rare individuals, who have hereditary persistence of the i antigen, never express the I antigen on erythrocytes and maintain embryonic levels of erythrocyte i antigen expression as adults. There is no obvious pathophysiology associated as yet with the absence of the I blood group in humans, although this phenotype can create challenges when testing patients for CAD.

THE A, B, AND H HUMAN BLOOD GROUPS

The ABO blood group antigens were discovered early in the 20th century by Karl Landsteiner and colleagues. They showed that humans could be divided into different groups according to the presence or absence of serum factors that would agglutinate red blood cells isolated from other humans. We now know that these serum factors are antibodies and that the corresponding antigens are glycan epitopes determined by the inheritance of genes that, for the most part, encode glycosyltransferases.

The A, B, and H blood group antigens are glycans presented on Type-1 or Type-2 structures (Figure 14.2), on O-GalNAc glycans (Type-3), or on glycolipids (Type-4) (Figure 14.4). The blood group antigens are formed by the sequential action of glycosyltransferases encoded by the *ABO*, *H*, and *Se* genes, now termed the *ABO*, *FUT1*, and *FUT2* loci (Figure 14.5). Blood group antigen synthesis begins with modification of Type-1 or Type-2 structures by the transfer of α1-2Fuc to Gal to form the blood group H determinant. The *H* allele encodes an α1-2FucT

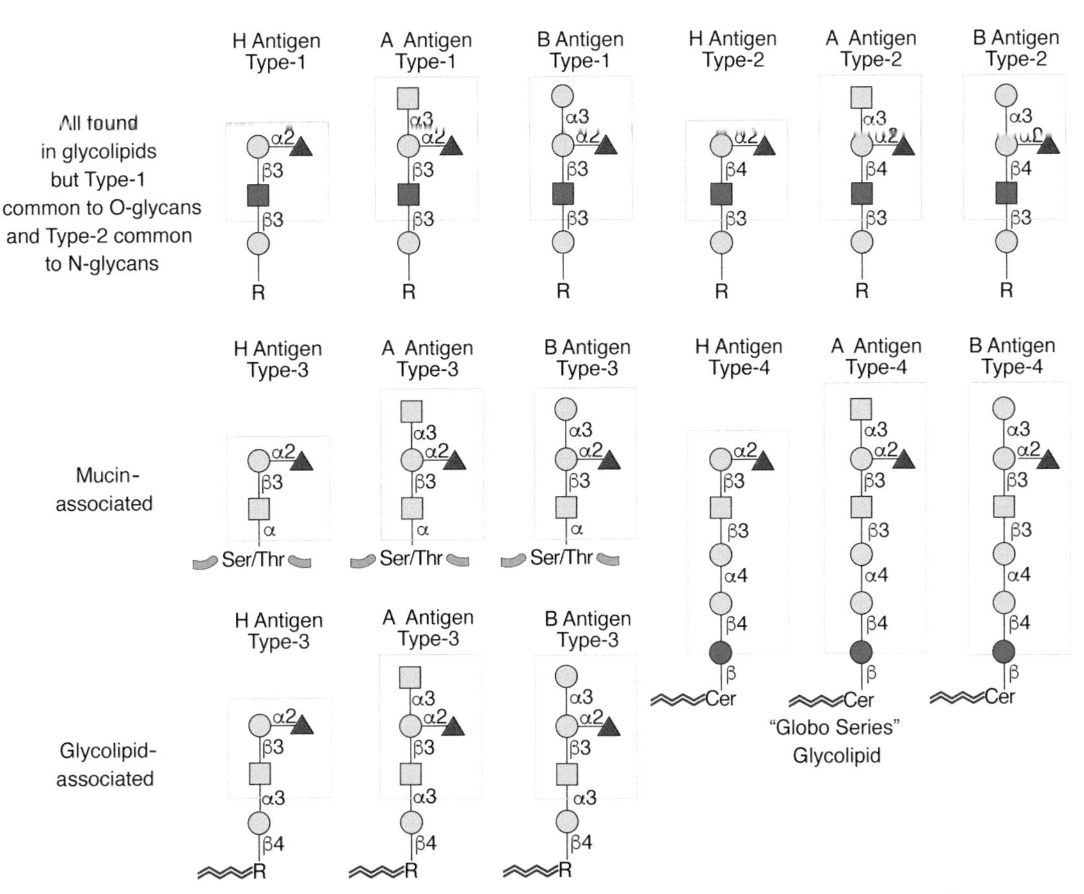

FIGURE 14.4. Type-1, -2, and -3 H, A, and B antigens that form the O (H), A, and B blood group determinants on N- and O-glycans. Type-4 H, A, and B antigens form the O (H), A, and B blood group determinants on glycolipids.

(FUT1) expressed in erythrocyte precursors and transfers Fuc to Type-2 and Type-4 glycan units to form the H antigen on erythrocytes (Figure 14.4). The *Se* allele encodes another α1-2FucT (FUT2) expressed in epithelial cells and uses Type-1 and Type-3 LacNAc to form the H antigen in epithelia lining the lumen of the gastrointestinal, respiratory, and reproductive tracts and in salivary glands (Figure 14.4), as well as modifying milk oligosaccharides.

A or B blood group determinants are subsequently formed from H Type-1, -2, -3, or -4 determinants by glycosyltransferases encoded by the *ABO* locus. The *A* allele encodes the α1-3GalNAcT (A3GALNT) that generates the A glycan epitope forming the A blood group (Figure 14.5). The *B* allele of the *ABO* locus encodes the α1-3GalT (A3GALT1) that forms the B glycan determinant and generates the B blood group (Figure 14.5). *O* alleles at the *ABO* locus encode a functionally inactive A/B glycosyltransferase. Individuals who synthesize exclusively A determinants are blood group A and have the genotype *AA* or *AO*, blood group B individuals are *BB* or *BO*, and individuals that express one *A* and one *B* allele have the genotype *AB*. Blood group O individuals expressing inactive A/B glycosyltransferase have the genotype *OO*. They express only the H antigen. Blood type designations are the same as the blood group genotypes above. In terms of nomenclature, the O blood group includes the H antigen and occasionally the term ABO(H) is used. Additional variants of α1-3GalNAcT and α1-3GalT exist that are enzymatically active, but at a much lower level. These variants are largely responsible for individuals who are initially typed as blood group O, but actually express very low levels of the A or B blood group antigen.

The ABO antigens are expressed on membrane glycoproteins and glycolipids on the surface of erythrocytes and many epithelial or endothelial cells in tissues. Some tissues also synthesize

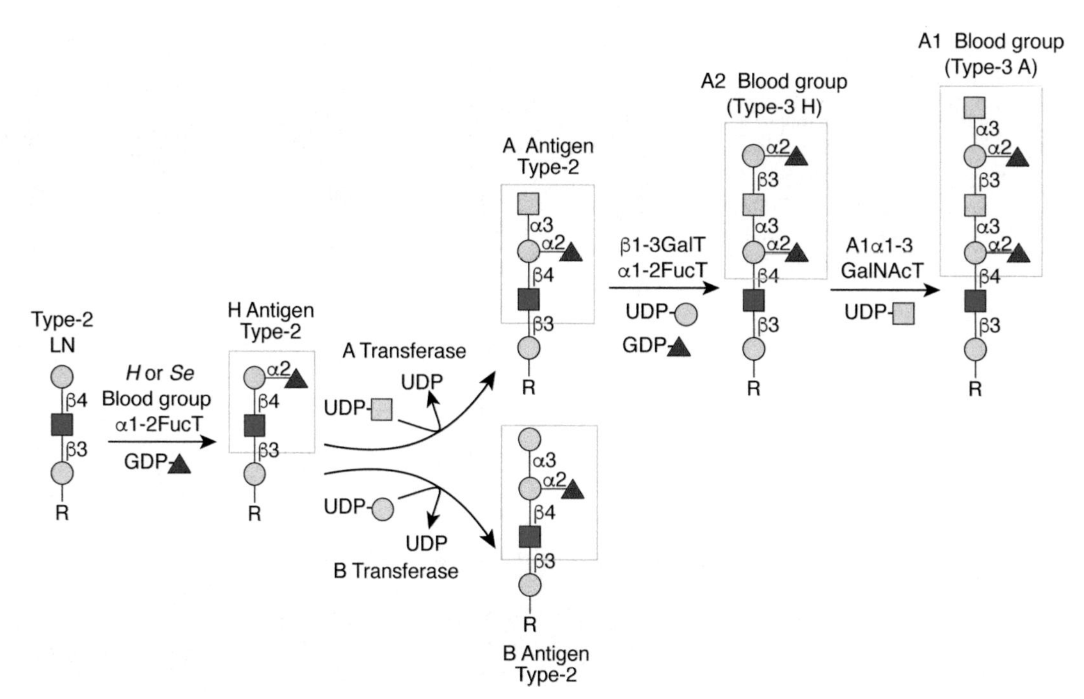

FIGURE 14.5. Synthesis of H (O), A, and B blood group determinants. For details, see text. LN designates N-acetyllactosamine unit.

soluble forms on secreted glycoproteins, glycolipids, and free glycans. As discussed below, the ability to secrete soluble molecules carrying ABO(H) blood group antigens is a genetically determined function of alleles at the *Se* (*FUT2*) locus. On each human red blood cell, ~80% of the 1–2 million ABO(H) determinants are attached to the anion transport protein Band 3, and ~15% are carried by the erythrocyte glucose transport protein Band 4.5. Both of these integral membrane proteins carry ABO(H) antigens on a single, branched N-glycan with poly-LacNAc. Each erythrocyte has other glycoproteins and approximately half a million glycolipids with ABO(H) determinants. Many of these glycolipids have A, B, and H determinants on poly-LacNAc chains and have been termed polyglycosylceramides or macroglycolipids. A, B, and H determinants based on Type-4 chains (Figure 14.4) are also present in human erythrocyte glycolipids.

A, B, and H determinants of the epidermis are primarily on Type-2 units, whereas mucins of the gastric mucosa and in ovarian cyst fluid carry A, B, and H antigens on Type-3 units (Figure 14.4). Epithelial cells lining the digestive, respiratory, urinary, and reproductive tracts, and epithelia of some salivary and exocrine glands, synthesize soluble forms of the ABO(H) determinants, largely carried on Type-1 units (Figure 14.4). Expression of the A, B, and H determinants in secretory tissues is a function of the α1-2FucT encoded by the *Se* gene (*FUT2*), because the *H* gene (*FUT1*) is not expressed there. Humans with an inactive *FUT2* gene do not express soluble forms of the A, B, or H determinants in saliva or milk oligosaccharides or in other tissues and are termed "nonsecretors."

Serology is used to characterize erythrocytes for transfusion and has identified variants of the A and B blood group determinants that typically yield weak reactivity with blood typing reagents. Interestingly, plant lectins were used historically to aid in typing blood. For example, the lectin from *Dolichos biflorus* agglutinates erythrocytes from most blood group A individuals (termed A1 individuals), but it does not agglutinate erythrocytes from individuals of the A2 subgroup, and it is still used currently to distinguish individuals with these two different blood types. The A1 and A2 subgroup antigens are distinct (Figure 14.5) reflecting the different A transferases encoded by the A1 versus A2 allele at the *ABO* locus.

Heritable erythrocyte antigenic polymorphisms determined by the *ABO* locus have important medical implications. Early in the postnatal period, the immune system generates IgM antibodies against ABO antigen(s), even though they are absent from erythrocytes. This is because glycan antigens similar or identical to the A and B blood group determinants are carried by colonizing bacteria and fungi. Thus, type-O individuals do not synthesize A or B determinants but show relatively high titers of circulating IgM antibodies (termed isoagglutinins) against A and B blood group determinants. Similarly, blood group B individuals show circulating IgM anti-A isoagglutinins, but they do not make isoagglutinins against the blood group B determinant, a "self" antigen. Conversely, serum from blood group A individuals contains anti-B but not anti-A antibodies. Finally, people with the AB blood group do not make either anti-A or anti-B IgM isoagglutinins, because both are "self" antigens. Anti-H(O) antibodies are not made in most people because a substantial fraction are converted to A or B determinants or they are a "self" antigen.

IgM isoagglutinins efficiently trigger the complement cascade and circulate in human plasma at titers sufficient to cause complement-dependent lysis of transfused erythrocytes that display the corresponding blood group antigens. Such rapid erythrocyte lysis causes an immediate, acute transfusion reaction, which can lead to hypotension, shock, acute renal failure, and death from circulatory collapse. This problem is avoided by ensuring that the ABO type of transfused erythrocytes is compatible with the recipient's ABO type. Thus, an A recipient may receive erythrocytes from A or O persons but not from a person of type B or AB. Blood banks perform typing and cross-matching assays. First, units of erythrocyte products typed for the A and B antigens are chosen to match the patient's ABO type. To ensure that these are truly "compatible," the patient's serum is cross-matched by mixing with a small aliquot of each prospective erythrocyte unit. Erythrocytes of compatible units do not agglutinate (form an erythrocyte clump), whereas incompatibility is indicated by agglutinated erythrocytes formed by antibodies in the patient's serum. Blood typing is used to ensure compatibility not only for red blood cell transfusions but also for transfusion with platelets and plasma. Similar ABO compatibility concerns are important in heart, kidney, liver, and bone marrow transplantation procedures. The "type and cross" procedures have virtually eliminated ABO blood group transfusion reactions in the developed world. Attempts are being made to enzymatically modify A or B erythrocytes using glycosidases to remove GalNAc (A) and/or Gal (B), in order to convert them to "O", the "universal donor" type. The few individuals with AB type are "universal acceptors." In addition to accurately typing AB blood types, A1 and A2 blood group distinctions are important clinically as A2 individuals can generate anti-A1 antibodies, although these antibodies are often not clinically meaningful. Furthermore, when crossing ABO(H) barriers during solid organ transplantation, A2 donor organs have better overall survival when compared to A1 following transplantation into a blood group O or B recipient; this is presumably because of the lower expression of the A antigen in A2 individuals.

In addition to transfusion and transplantation, ABO(H) alloantigens are some of the most frequent allogeneic barriers crossed during fetal development. IgG anti-A,B antibodies present in blood group O mothers, which react with the A and B antigen, can cross the placenta and engage A or B antigens on the surface of fetal red blood cells (RBCs). However, unlike IgG antibodies against other alloantigens, such as RhD, IgG anti-A,B antibodies rarely cause life-threatening anemia in a fetus or neonate, presumably because of the relatively low level of A and B antigen expression present on fetal RBCs. However, despite the lower level of A and B expression on fetal RBCs, anti-A and anti-B antibodies present in maternal circulation can rapidly remove A or B positive fetal RBCs following exposure during parturition, reducing the probability of alloimmunization to other alloantigens. This was the first example of antibody-mediated immunosuppression clinically and formed the basis of highly successful anti-RhD formularies designed to prevent Rh alloimmunization and therefore hemolytic disease of the fetus and newborn.

Cross-matching procedures helped to identify a rare ABO blood group phenotype termed the Bombay phenotype, so named because the first identified individual lived in that city (now Mumbai). Affected persons have erythrocytes and tissue cells lacking A, B, and H determinants because they have inactive *FUT1* and *FUT2* genes, and therefore no α1-2FucT enzyme. Bombay sera contain IgM antibodies that react with erythrocytes from virtually all donors, including O erythrocytes (H antigen-positive, A and B antigen-negative). They show robust titers of anti-H, anti-A, and anti-B IgM antibodies and cannot receive erythrocytes from any donor except those of the same Bombay blood type. A related phenotype, termed para-Bombay, occurs in people with an inactive *FUT1* gene, but at least one functional *Se* (*FUT2*) allele (secretor-positive). The fact that Bombay individuals appear generally healthy implies that developmental or physiological functions for the A, B, and H antigens, if they ever existed, are no longer relevant. However, a variety of associations have been made between the ABO blood group phenotype and relative risk for infection by some pathogens and the acquisition of a spectrum of diseases. For example, people with blood group O who are also Lewis-antigen positive (see below) are the most susceptible to infection by *Helicobacter pylori*. This is because *H. pylori* binds well to glycans with terminal Fuc, such as the H and Lewis antigens. The AB blood group is associated with infection by Brucella (Brucellosis) and noroviruses that cause gastroenteritis, while blood group O individuals appear to be at an increased risk for cholera. Levels of von Willebrand factor (VWF) correlate with ABO(H) blood group inheritance, with blood group A individuals on average having the highest VWF levels. Differences in VWF among individuals of different blood group status may in part account for associations observed between blood group A individuals and the likelihood of thromboembolic events and other forms of cardiovascular disease. Blood group status has also been associated with risk of stomach and pancreatic cancers. ABO status may also be protective. Thus, enveloped viruses carry the ABO(H) glycans of their hosts and are susceptible to lysis following infection of another individual with an ABO-incompatible type. Finally, differences in susceptibility to severe complications of malaria appear to be affected by ABO blood groups. Protective mechanisms relevant to these roles for blood group determinants may explain why the ABO system has survived more than 50 million years of primate evolution.

LEWIS BLOOD GROUPS

The Lewis blood group antigens are a related set of glycans that carry α1-3/α1-4 Fuc residues (Figure 14.6). The term "Lewis" derives from a family who suffered from a RBC incompatibility. The Lewis a antigen (Le^a) is synthesized by an α1-3/α1-4FucT (FUT3) encoded by the *Lewis* (*LE* or *FUT3*) blood group locus. The Lewis b antigen (Le^b) is synthesized by the concerted actions of FUT3 and FUT2. Secretor-positive individuals express FUT2 and convert Type-1 units to Type-1 H determinants that may be acted on by FUT3 to form the Le^b determinant (Figure 14.6). Nonsecretors who do not synthesize Type-1 H determinants in secretory epithelia express the Le^a determinant via FUT3 (Figure 14.6). Individuals with an inactive *FUT3* locus (~10%–20% of the population), are termed Lewis-negative. Lewis-negative secretors express Type-I H determinants that cannot be converted to Le^a or Le^b determinants. Lewis-negative nonsecretors express Type-1 units that are devoid of Fuc.

Expression of Le^a and Le^b glycans and FUT3 is largely restricted to the same epithelia that express FUT2. Thus, soluble forms of these antigens are released into secretions and body fluids. Le^a and Le^b antigens are also detectable on erythrocytes. However, the precursors of erythrocytes do not synthesize these determinants. Instead, Lewis antigens are acquired by the erythrocyte membrane through passive adsorption of Lewis-positive glycolipids that circulate in plasma in lipoprotein complexes and aqueous dispersions. Antibodies against the Le^a antigens have been implicated in occasional transfusion reactions. However, these reactions are rare,

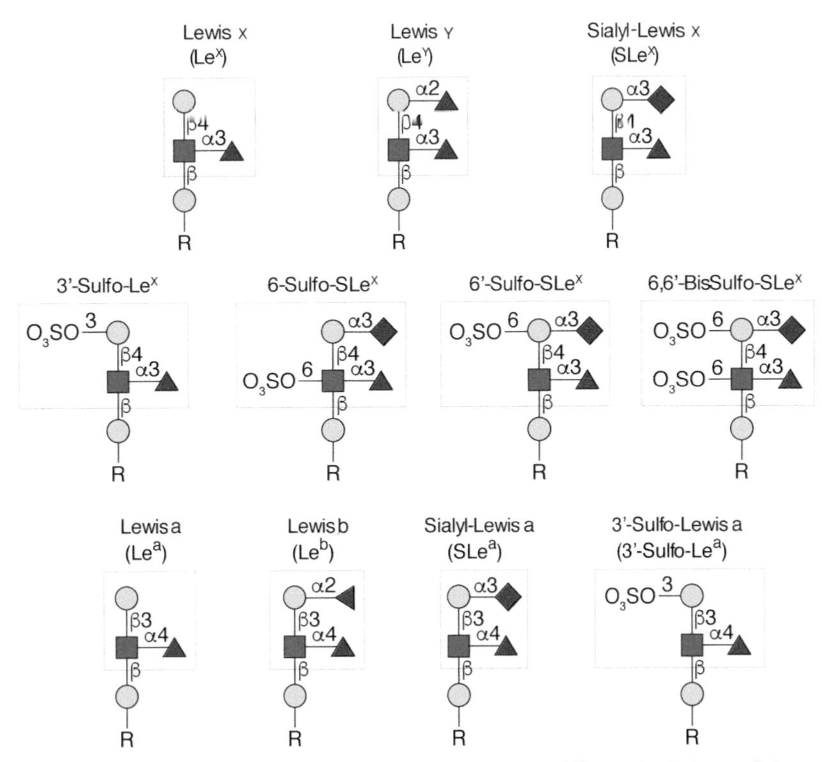

FIGURE 14.6. Type-1 and -2 Lewis determinants. Type-1 and -2 units differ in the linkage of the outermost galactose (β1-3 or β1-4, respectively), and thus in the linkage of fucose to the internal GlcNAc (α1-4 or α1-3, respectively). R represents N- or O-glycan or glycolipid. Lewis x determinants may contain sulfate at C3 or C6 of Gal and at C6 of GlcNAc. The 6-O-sulfated GlcNAc is necessary for lymphocyte homing to peripheral lymph nodes. The Type-1 Lewis blood group determinants on glycoproteins and glycolipids (R) are characterized by the presence or absence of the fucosyltransferases FUT2 and FUT3.

presumably because of the ability of soluble Lewis glycolipids to bind anti-Lewis antibodies and therefore prevent these antibodies from engaging transfused red blood cells and causing a hemolytic transfusion reaction.

Because of structural similarities, the term "Lewis antigen" was applied to other glycan sequences and includes the Lewis x (Le^x) and Lewis y (Le^y) determinants and forms of the Le^a and Le^x determinants that are sialylated and/or sulfated (Figure 14.6). These glycan epitopes are formed through the actions of FUT3, FUT4, FUT5, FUT6, FUT7, and/or FUT9. Some Lewis blood group antigens have important functions in selectin-dependent leukocyte extravasation and tumor cell metastasis. Most strongly implicated are the sialylated and/or sulfated determinants represented by sialyl Le^x and its sulfated variants (Figure 14.6), which function as selectin ligands on glycoproteins and glycolipids of leukocytes and tumor cells (Chapters 34 and 47). The Lewis blood group antigens have also been proposed to function in the pathogenesis of *H. pylori*, the causative agent in chronic active gastritis associated with hypertrophic gastropathy, duodenal ulcer, gastric adenocarcinoma, and gastrointestinal lymphoma (Chapter 37). Also, Lewis antigens may be expressed on glycoproteins in plants (Chapter 24).

P BLOOD GROUPS

The P1PK blood group includes the P1 and P^k antigens. Their synthesis involves two pathways, each beginning with lactosylceramide (Figure 14.7). The P^k antigen is synthesized by an α1-4GalT (P^k transferase; A4GALT) and may be modified by a β1-3GalNAcT (P transferase;

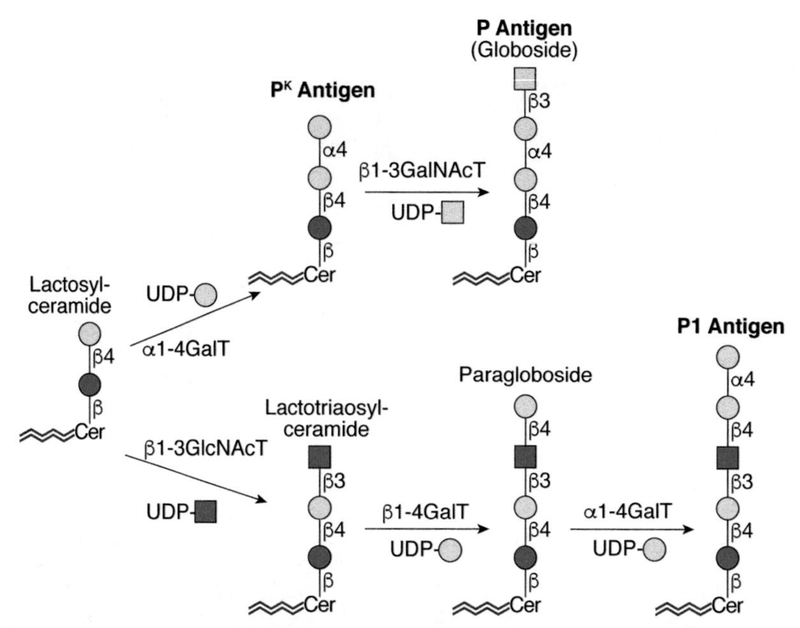

FIGURE 14.7. Biosynthesis of antigens of the P1PK blood group system: P^k, P, and P1.

B3GALNT1) to form the P antigen. In the second pathway, P1-antigen biosynthesis begins with two reactions leading to paragloboside, which is acted on by A4GALT to form the P1 antigen, the most common P blood group. Individuals with this blood group possess both pathways, and their erythrocytes express P and P1 antigens and small amounts of P^k not converted into P determinants. Individuals with low P1 transferase are quite common and express the P2 blood group. Their erythrocytes show normal levels of P and P^k antigens but are deficient in P1 determinants caused by low expression of A4GALT. Antibodies against P, P1, and P^k determinants have been implicated in transfusion reactions. Complement-fixing, cold-reactive anti-P antibodies known as "Donath–Landsteiner" antibodies cause intravascular hemolysis observed in a syndrome called paroxysmal cold hemoglobinuria (see Chapter 46). This syndrome is distinct from CAD in that the antibodies are typically IgG.

Various uropathogenic strains of *Escherichia coli* express adhesins that bind to the terminal Galα1-4Gal moiety of the P^k and P1 antigens (Chapter 37). The P1 determinant is expressed on the urothelium and probably facilitates infection by mediating attachment of bacteria. P1 individuals have a higher relative risk for urinary tract infections and pyelonephritis. The adhesion of a pyelonephritic strain of *E. coli* to renal tissue is mediated by a bacterial adhesin specific for the Galα1-4Gal epitope, and deficiency of the adhesin severely attenuates the pyelonephritic activity of the organism. The P blood group antigens may also have a role as receptors for human parvovirus B19. This virus causes erythema infectiosum and leads to congenital anemia and hydrops fetalis following infection in utero. It is also associated with transient aplastic crisis in patients with hemolytic anemia and with cases of pure erythrocyte aplasia and chronic anemia in immunocompromised individuals. Parvovirus B19 replication is restricted to erythroid progenitor cells owing to an interaction between the virion and glycolipids with P antigen.

MILK OLIGOSACCHARIDES

Mammals make a series of oligosaccharides with a core of lactose (Galβ1-4Glc) and secrete them into the milk. Human milk comprises many hundreds of different glycans with small amounts of glycans containing more than 15 sugars. The distribution of structures and amount of each oligosaccharide vary between individuals and during the lactation cycle. There is a relatively high concentration of complex, sialylated, and fucosylated oligosaccharides in milk, and these glycans appear to protect infants against enteric pathogens. Interestingly, in individuals that are "nonsecretors" and Lewis-negative, fucosylated glycans are not present beyond 2′-fucosyllactose, and the overall repertoire of oligosaccharides is reduced. Most other mammals synthesize lactose but also express species-specific repertoires of glycans that largely differ from humans—often being much simpler—but may have similar functions. Beside lactose, which is nutritionally important, the larger human milk oligosaccharides are thought to be important in infant immunoprotection and to have prebiotic activity by contributing to the development of healthy microbiota. Surprisingly, there is essentially no information on the hormone-regulated biosynthesis of these abundant glycans. Lactose is generated by a β1-4 galactosyltransferase (termed lactose synthase) only in the lactating mammary gland, because of the lactation-specific expression of the modifier protein α-lactalbumin, which causes the enzyme to transfer Gal from UDP-Gal to Glc rather than to GlcNAc. Even though this process has been demonstrated to occur in intact Golgi, the precise mechanisms of how lactose is modified by addition of other sugars by specific glycosyltransferases during lactation are unknown. It is assumed that the same enzymes involved in making termini of other glycan classes are responsible.

THE Galα1-3Gal TERMINUS

The Galα1-3Gal epitope (often called "alpha-Gal") is synthesized on Type-2 units on glycolipids and glycoproteins by a specific α1-3GalT (Figure 14.8). This epitope and the α1-3GalT that synthesizes it are expressed by New World primates and many nonprimate mammals, but the α1-3GalT gene (*GGTA1*) is inactivated in humans and Old World primates. Mice engineered to lack the α1-3GalT develop cataracts. Species that do not express the Galα1-3Gal epitope, including humans, carry anti-Galα1-3Gal antibodies, likely because of immunization through exposure to the Galα1-3Gal epitope on microbes and food. Anti-Galα1-3Gal antibodies present a major barrier to the use of porcine and other nonprimate organs for xenotransplantation in humans, because they bind to Galα1-3Gal epitopes on the vascular endothelium of xenotransplants and cause hyperacute graft rejection through complement-mediated endothelial cell cytotoxicity. Efforts are in progress to overcome this barrier by using animal organ donors that have been genetically modified. Approaches include transgenic expression of enzymes, such as FUT1, that diminishes Galα1-3Gal expression by diverting Type-2 units toward H antigen synthesis. Unfortunately, pig tissues lacking Galα1-3Gal elicit a graft rejection reaction to other pig antigens. Anti-Galα1-3Gal antibodies have also been shown to significantly diminish the infective efficiency of recombinant retroviruses. The problem has been solved through the generation of packaging cell lines that are deficient in α1-3GalT. Allergic reactions toward therapeutic

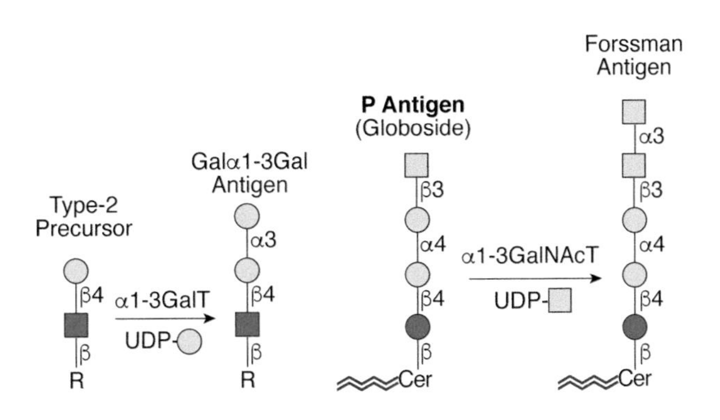

FIGURE 14.8. Structure and synthesis of the Galα1-3Gal antigen. The α1-3GalT uses unsubstituted Type-2 units on glycoproteins or glycolipids (R) to form the Galα1-3Gal terminal epitope. The glycolipid globoside serves as the substrate for the Forssman α1-3GalNAcT (GBGT1) that forms globopentosylceramide, also termed the Forssman glycolipid.

monoclonal antibodies expressing Galα1-3Gal first illustrated the importance of preparing recombinant glycoproteins for therapeutic use in humans in cells that do not express α1-3GalT. Severe allergies to red meat consumption can occur when high titer IgE antibodies against this epitope appear in adult humans, claimed to be the consequence of a prior bite by the Lone Star tick, which may be expressing the same epitope in its saliva.

THE FORSSMAN ANTIGEN

The Forssman antigen (also known as globopentosylceramide) is a glycolipid that contains terminal α1-3GalNAc linked to the terminal GalNAc of globoside transferred by the α1-3 GalNAcT GBGT1, which is related in sequence to ABO transferases (Figure 14.8). The Forssman antigen, first discovered in sheep erythrocytes by John Frederick Forssman, is expressed during embryonic and adult stages in many mammals. Humans have a mutated GBGT1 and cannot synthesize the Forssman antigen but carry anti-Forssman antibodies in their serum. Rare individuals have a reversion mutation that restores activity to GBGT1, and synthesis of the Forssman antigen. Anti-Forssman antibodies may contribute to the pathogenesis of Guillain–Barré syndrome by binding to cross-reactive glycolipid components of peripheral nerve myelin. It is interesting that anti-Forssman antibodies can disrupt tight junction formation, apical–basal polarization, and cell adhesion.

SULFATED GalNAc: PITUITARY GLYCOPROTEIN HORMONES

Glycans with sulfated terminal β-linked GalNAc are found on the pituitary glycoprotein hormones lutropin (LH) and thyrotropin (TSH) but not on follicle-stimulating hormone (FSH), although it is made in the same cells. These heterodimeric glycoproteins contain a common α-subunit and a unique β-subunit, each with biantennary N-glycans. The N-glycans of TSH and LH have an unusual 4-O–sulfated GalNAc attached to GlcNAc residues (Figure 14.9). This contrasts with the N-glycans on FSH (and most N-glycans), in which GlcNAc residues are substituted with β1-4Gal, often extended by α2-3 or α2-6 sialic acid residues (Chapter 9). A free α-subunit common to LH, TSH, and FSH is present in pituitary cells and it also carries this determinant, as do other glycoproteins synthesized by the pituitary and elsewhere (e.g., on the O-glycans of pro-opiomelanocortin). Synthesis of the sulfated GalNAc determinant is controlled by a β1-4GalNAcT, either B4GALNT3 or B4GALNT4 (Figure 14.9). The terminal β1-4GalNAc is then sulfated by a sulfotransferase (CHST8 or CHST9), also expressed in pituitary cells. In some tissues, including the pituitary, the β1-4GalNAc is substituted by an α2-6 sialic acid residue. Both β1-4GalNAcT and β1-4GalT enzymes are expressed in pituitary cells, but the N-glycans on LH and TSH carry the uncommon β1-4GalNAc, whereas the N-glycans on FSH carry the common β1-4Gal residue. This protein-specific glycosylation is a consequence of interactions between B4GALNT3 or B4GALNT4 and a specific peptide motif present on the combined αβ-subunits of LH and TSH. This interaction causes an increase in the catalytic efficiency of the β1-4GalNAcT that modifies biantennary N-glycans on LH and TSH at the expense of the competing β1-4GalT. Importantly, the peptide motif recognized by the β1-4GalNAcT is not present in the β-subunit of FSH, and the recognition motif on the α-subunit of FSH is not accessible to the enzyme. Consequently, the biantennary N-glycans on FSH are modified exclusively by a β1-4GalT.

These differential glycosylation events have profound consequences for the ovulatory cycle in vertebrates. Circulating LH levels increase and decrease in a highly pulsatile manner. This assures maximal stimulation of the ovarian LH receptor at the preovulatory surge, because sustained high LH levels would lead to LH receptor desensitization. The increase and decrease in

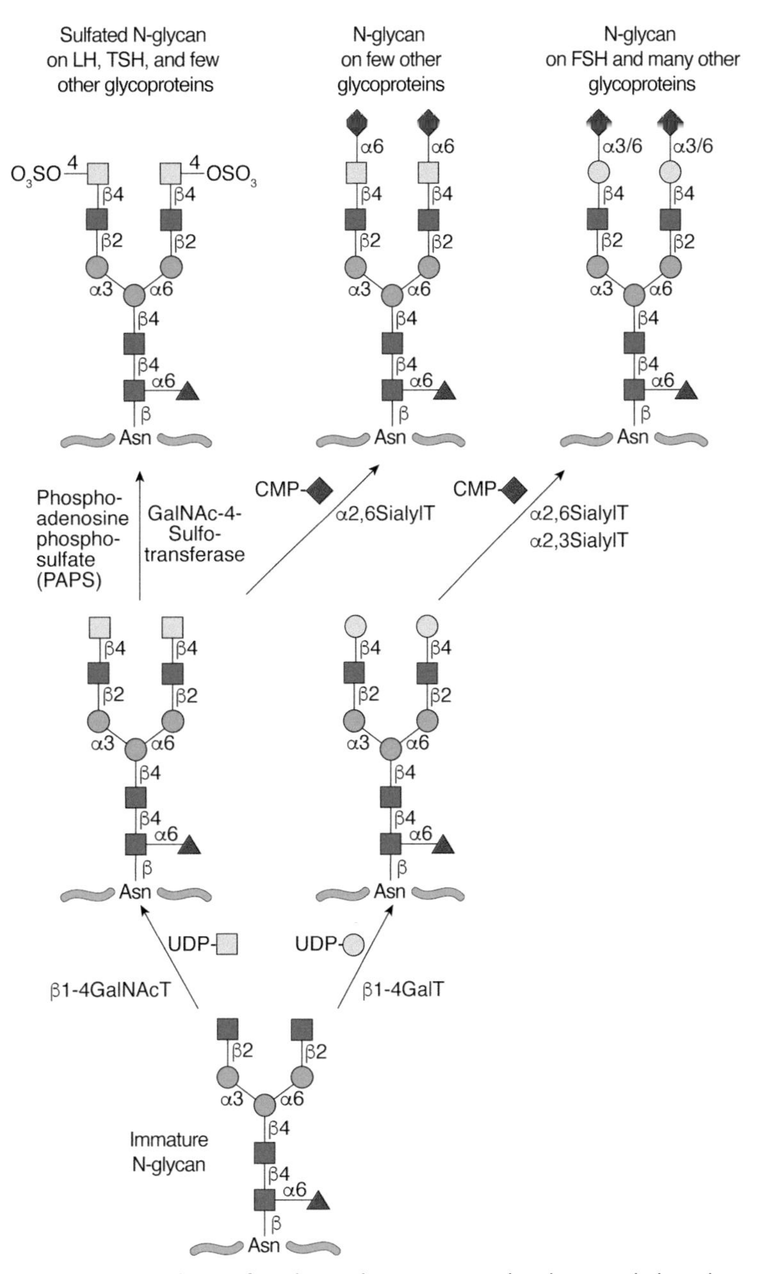

FIGURE 14.9. Structure and synthesis of N-glycans bearing terminal GalNAc, including those with sulfated-GalNAc found on the pituitary hormones lutropin (LH) and thyrotropin (TSH), but not on follicle-stimulating hormone (FSH).

LH levels is due, in part, to pulsatile release of the hormone by the pituitary. However, the peaks and troughs are accentuated markedly by the rapid clearance of LH from the circulation mediated via recognition of its terminal sulfated-GalNAcβ1-4GlcNAc determinant(s) by the "mannose receptor" MRC1, which is expressed by hepatic endothelial cells and by Kupffer cells in liver. LH binding is followed by internalization and lysosomal degradation. MRC1 is also expressed in macrophages. In liver, MRC1 recognizes sulfated-GalNAc via an R-type lectin domain, whereas in macrophages, the same receptor recognizes mannose via an L-type lectin domain (Chapters 31 and 32).

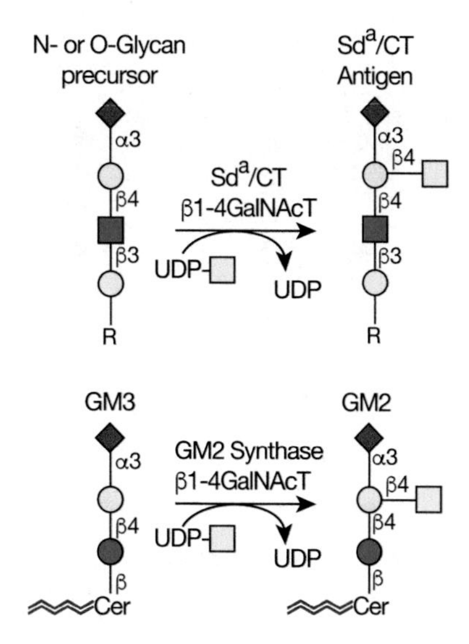

FIGURE 14.10. Synthesis of the human Sd[a] or mouse CT antigen and the glycolipid GM2.

TERMINAL β-LINKED GalNAc: THE Sd[a] BLOOD GROUP

The addition of GalNAc to Gal already substituted with α2-3Sia may also occur on glycoproteins and glycolipids (Figure 14.10). On glycoproteins, this structure forms the human Sd[a] blood group, which is expressed by most individuals. In mice, the Sd[a] antigen was first described on cytotoxic T lymphocytes (CTLs) and was termed the CT antigen. On glycolipids, the same trisaccharide terminus is shared by the ganglioside GM2 (Chapter 11). A related structure to the Sd[a] antigen GalNAcβ1-4(Neu5Acα2-3)Galβ1-3(Neu5Acα2-6)GalNAc is found in O-glycans and glycolipids. The human Sd[a] antigen was first sequenced in N-glycans of Tamm–Horsfall glycoprotein from human urine. Both human and mouse β1-4GalNAcT (B4GALNT2) transfers GalNAc to N- and O-glycans on glycoproteins, but not to the glycolipid GM3 (Siaα2-3Galβ1-4Glc-Cer), even though both can efficiently use 3-sialyllactose (Siaα2-3Galβ1-4Glc) as a substrate in vitro. In the mouse, Sd[a] antigens are recognized by IgM monoclonal antibodies termed CT1 and CT2, which were isolated for their ability to block lysis of cellular targets by a murine CTL clone. Rare humans lack the determinant and form naturally occurring antibodies against it, but they show no apparent pathophysiology. Mice with a dominantly inherited form of von Willebrand's disease express B4GALNT2 aberrantly in vascular endothelium. The presence of the β1-4GalNAcT in this abnormal location generates von Willebrand factor (VWF) carrying the Sd[a] determinant. This VWF glycoform is rapidly cleared from the circulation by the asialoglycoprotein receptor (ASGR) in liver, leading to VWF deficiency and hemorrhagic disease.

The glycolipid equivalent of the Sd[a] determinant, termed GM2, which is synthesized by GM2 synthase (B4GALNT1) (Figure 14.10), is widely expressed in the central and peripheral nervous systems and in the adrenal gland. Mice homozygous for a null mutation in *B4galnt1* show modest conduction defects in the peripheral nervous system and male sterility (Chapter 41).

α2-3-SIALYLATED GLYCANS

Sialic acids in α2-3 linkage are found on N-glycans, O-glycans, and glycolipids generated by six different α2-3 sialyltransferases (ST3GAL1 to ST3GAL6). ST3GAL3 and ST3GAL4 are broadly expressed in mammals. In mouse, *St3gal1* transcripts are most abundant in the spleen, liver, bone marrow, thymus, and salivary glands. *St3gal2* expression is most abundant in brain, where α2-3-sialylated glycolipids are common. *St3gal5* is expressed well in brain, skeletal muscle, adrenals, and liver, and *St3gal6* is most expressed in testes. In vertebrates, α2-3 sialic acid residues are found on terminal Gal residues. The addition of α2-3 sialic acid to Gal inhibits the action of other enzymes, including α1-2FucTs, α1-3GalT, GlcNAcTs, and GalNAcTs, which compete with terminal α2-3 sialyltransferases. Although most α2-3 sialic acid on glycoproteins is found on complex N-glycans (Chapter 9) and O-GalNAc glycans (Chapter 10), sialylation also occurs on O-fucose and O-mannose glycans found on a limited subset of glycoproteins (Chapter 13). As discussed above and in Chapter 34, selectin ligands are α2-3-sialylated glycans.

Glycans bearing α2-3 sialic acid contribute to the circulating half-life of plasma glycoproteins by "masking" terminal Gal residues that contribute to the removal of glycoproteins from serum by the asialoglycoprotein receptor (Chapter 34). ST3GAL1 generates Siaα2-3Galβ1-3GalNAcα-Ser/Thr that is important for the viability of peripheral $CD8^+$ T cells. Mice lacking ST3GAL1 show decreased cytotoxic T-cell responses with an increase in the apoptotic death of naïve $CD8^+$ T cells (Chapter 36).

Sialic acid recognition is important for binding by viruses and bacteria. Binding to sialic acid and subsequent release by neuraminidase are important for infection by influenza virus. Sialic acid residues in α2-3 linkage are recognized by the hemagglutinin (HA) in the envelope of influenza viruses from birds and pigs. Human influenza viruses bind more commonly to sialic acid residues in α2-6 linkage. Mutations in the HA gene of influenza viruses from birds may lead to a human influenza pandemic partly because of their enhanced ability to infect human cells through HA recognition of α2-6Neu5Ac (Chapter 42). Glycans bearing α2-3 sialic acid residues have also been implicated in bacterial pathogenesis. Glycans terminating in α2-3 sialic acid residues support the adhesion of *H. pylori*, which causes gastritis, gastric ulcers, and stomach cancer. The ganglioside GM1 (Galβ1-3GlcNAcβ1-4[Siaα2-3]Galβ1-4GlcβCer) is a receptor for cholera toxin produced by *Vibrio cholerae* and the heat-labile enterotoxin (LT-1) produced by enterotoxigenic *E. coli* (Chapter 42); recent data suggest that LT-1 may also recognize the H antigen, possibly providing an explanation for the worse prognosis observed in blood group O individuals. Glycan-based inhibitors are currently under evaluation in humans for their ability to diminish the symptoms and progression of cholera. A variety of other pathogens and toxins bind to sialylated termini bearing one of many possible modified sialic acids (Chapter 15).

α2-6-SIALYLATED GLYCANS

Sialic acid in α2-6 linkage is found on N-glycans, O-glycans, and glycolipids. Two α2-6 sialyltransferases, ST6GAL1 and ST6GAL2, transfer to Gal, whereas ST6GALNAC1–ST6GALNAC6 transfer to GalNAc (ST6GALNAC5 and ST6GALNAC6 also transfer to GlcNAc) (Figure 14.11). In vertebrates, α2-6 sialic acid is found on terminal Gal, on terminal or subterminal GalNAc, or, in the case of reactions catalyzed by ST6GALNAC3 and ST6GALNAC4, on a core GalNAc. α2-6 sialic acid is less common than α2-3 sialic acid in tissue glycoproteins but more common in plasma glycoproteins. Glycans with terminal α2-6 sialic acid are generally not modified further. In mouse, *St6gal1* is expressed at a relatively high level in hepatocytes and lymphocytes and is responsible for α2-6 sialylation of serum glycoproteins and glycoproteins of the antigen receptor complex in lymphocytes. *St6gal2* expression is mainly restricted to the embryonic and adult brain, and its functions are currently unknown.

The α2-6-sialylated glycans from ST6GALNAC1 and ST6GALNAC2 are restricted to O-glycans. ST6GALNAC3 and ST6GALNAC4 are responsible for transferring α2-6 sialic acid to the GalNAcα-Ser/Thr core of O-GalNAc glycans and to GalNAc in glycolipids. ST6GALNAC5 and ST6GALNAC6 appear to use glycolipids as preferred acceptors. Many strains of influenza infectious for humans bind terminal α2-6 sialic acid residues (Chapter 37), and glycoproteins

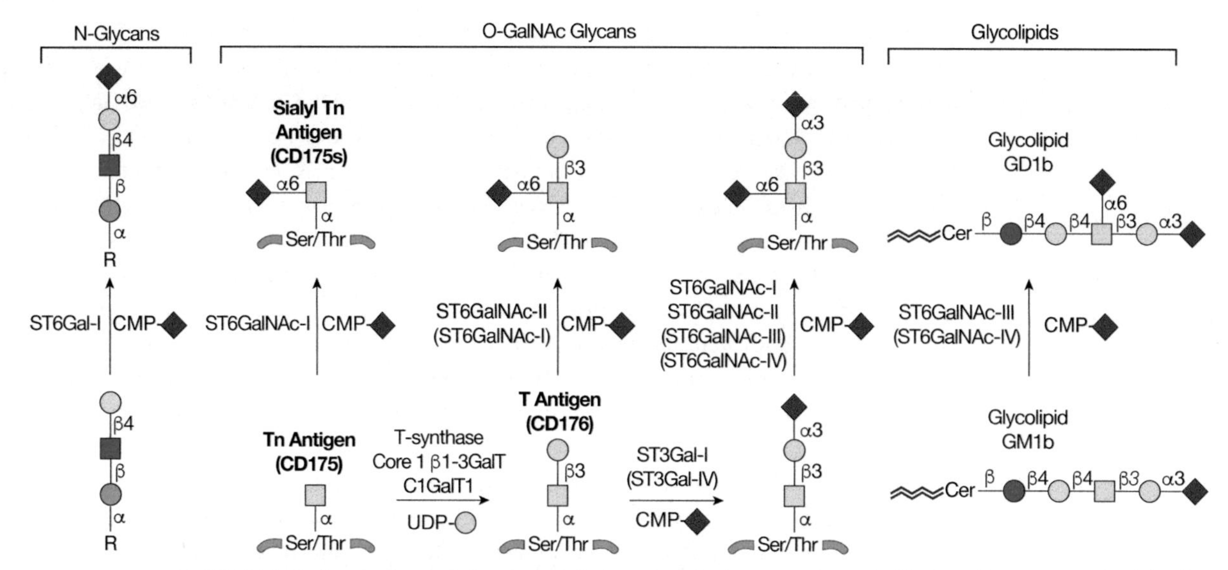

FIGURE 14.11. Synthesis of α2-6 and α2-3 sialic acid on O-glycans and glycolipids (see Chapters 9 and 10) by the ST3Gal and ST6GalNAc families of sialyltransferases. Enzymes in parentheses contribute at relatively low levels in vitro to the reactions indicated.

bearing α2-6 sialylation can be cleared from the circulation by the asialoglycoprotein receptor (Chapter 31).

Mice lacking ST6GAL1 show diminished antibody responses to T-lymphocyte-dependent and -independent antigens, reduced B-lymphocyte proliferation response, reduced B-cell surface IgM and CD22 levels, ~65% reduction in serum IgM levels, and reduced B-cell receptor (BCR) signaling (Chapter 35). The extracellular domain of CD22 on B lymphocytes specifically recognizes Siaα2-6Galβ1-4GlcNAc-. In the absence of α2-6 sialic acid on glycans, CD22 shows increased clustering with the BCR, and BCR signaling is down-regulated. The addition of α2-3 and α2-6 sialic acid may not only provide ligands for siglecs, but also may prevent terminal glycan recognition by several galectin family members, and perhaps other lectins. In this way, terminal sialylation can be viewed as an analogous regulatory pathway to intracellular phosphorylation, in which the addition and linkage of sialic acid can control the susceptibility of a wide variety of cell-surface glycans toward glycan binding protein-induced changes in cellular activity.

α2-8-SIALYLATED GLYCANS

Glycans modified by α2-8 polysialylation occur in vertebrates mainly in developing brain and are carried primarily on the neural cell adhesion molecule NCAM. α2-8-sialylated glycans are also expressed on a few glycoproteins in nonneuronal cells and on tumor cells. There are six α2-8 sialyltransferases, ST8SIA1 through ST8SIA6, that transfer sialic acid in α2-8 linkage to a terminal α2-3- or α2-6-linked sialic acid, generally on an N-glycan (Figure 14.12). ST8SIA2 (also called STX) and ST8SIA4 (also called PST) catalyze the synthesis of linear polymers of up to 400 α2-8 sialic acid residues to give polysialic acid (PolySia or PSA) on NCAM. Both ST8SIA2 and ST8SIA4 are autocatalytic and synthesize polySia on their own N-glycans, although polysialylation is not a prerequisite for their sialyltransferase activity. Thus, some cultured cells that do not express a known substrate of these sialytransferases may express surface polySia when transfected with ST8SIA2 or ST8SIA4 because of their autosialylation. Glycoproteins with N- or O-glycans carrying only one α2-8 sialic acid or two (disialic acid) or up to seven α2-8 sialic

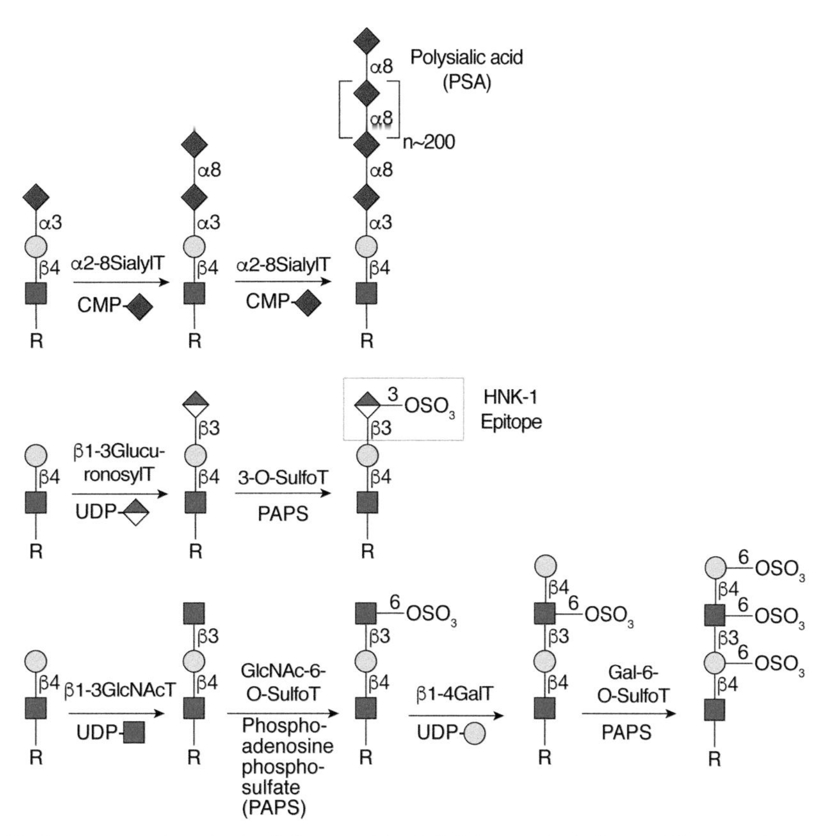

FIGURE 14.12. Structure and synthesis of glycans with α2-8 sialic acids including polySia on N-glycans by ST8SIA1 to ST8SIA6. Synthesis and structure of the HNK-1 epitope. (GlucuronosylT) Glucuronosyltransferase, (SulfoT) sulfotransferase.

acids (oligosialic acid) have been described and may be synthesized by ST8SIA3 and/or ST8SIA6. However, functional studies have focused on polySia on NCAM (Chapter 15).

PolySia is highly negatively charged, highly hydrated, and contributes up to one third of the molecular mass of NCAM. The embryonic form of NCAM is extensively modified by polySia, which exerts an antiadhesive effect and reduces homotypic interactions. PolySia can also diminish interactions promoted by other adhesion molecules, including L1-dependent attachment to laminin or collagen, and also binds extracellular signaling molecules like BDNF and FGF2. Mice lacking ST8SIA4 show reduced PSA in certain brain regions and have altered neuronal responses in the hippocampal CA1 region. Mice lacking ST8SIA2 have a distinct neuronal phenotype because of misguided migration of a subset of hippocampal neurons and ectopic synapses. When both ST8SIA2 and ST8SIA4 are inactivated, mice have severe neuronal and other problems and die precociously. However, this phenotype is rescued by also removing NCAM in a triple-knockout strain. This shows that the presence of NCAM lacking PSA is a major cause of the severe defects in double-knockout mice (Chapter 41).

Certain glycolipids also carry α2-8 sialic acid linkages, which are constructed by three α2-8 sialyltransferases termed ST8SIA1 (also known as GD3 synthase), ST8SIA3, and ST8SIA5 (Chapter 11). They generate single or oligomeric α2-8 sialic acid residues but not polymeric PSA. The three enzymes are generally thought to act primarily on glycolipid substrates, but studies suggest that ST8SIA3 can also use N- and O-glycans to generate oligosialic acids. These α2-8 sialyltransferases are expressed in the brain, where each shows a distinct developmentally regulated expression pattern. ST8SIA1 is also present in kidney and thymus. In vitro experiments imply that certain α2-8-sialylated glycolipids may participate in signal transduction processes in neuronal cell types. Inactivation of the ST8SIA1 gene in the mouse causes alterations in sensory neuron responses to pain.

SULFATED AND PHOSPHORYLATED GLYCANS

In principle, any free hydroxyl group on a monosaccharide could be modified by sulfation or phosphorylation. However, in vertebrates, glycan sulfation is restricted to Gal, GlcNAc, GlcA, and GalNAc at internal or terminal positions, and phosphorylation has been observed only on Man, GlcNAc, and Xyl to date. The internally sulfated glycans in heparin, heparan sulfate (HS), and chondroitin sulfate (CS) proteoglycans and the sulfated N-glycans on KS are discussed in Chapter 17. This chapter describes sulfated glycans recognized by L-selectin, the HNK-1 epitope, and the pituitary glycoprotein hormones mentioned above. Chapter 17 also mentions transient phosphorylation of the Xyl that initiates the proteoglycan core. Mannose phosphorylation is described in Chapters 9, 13, and 33. Parasites, fungi, and bacteria have a wide variety of phosphorylated glycans (Chapters 21, 22, 23, and 43). However, glycan sulfation is remarkably rare in prokaryotes.

In vertebrates, L-selectin on lymphocytes binds to the high endothelial venules (HEVs) in lymph nodes through recognition of L-selectin ligands present on O-GalNAc glycans of HEV glycoproteins. Sulfated forms of the sialyl Lewis x determinant (Figure 14.6) provide an essential contribution to L-selectin recognition of these glycoproteins. Sulfation occurs at C-6 of Gal by CHST1 and C-6 of GlcNAc by CHST2 and CHST4, both of which contribute to L-selectin ligand activity. Mice lacking the two sulfotransferases show almost no homing of lymphocytes to HEV. The biosynthesis of sulfated L-selectin ligands and the enzymes that participate in this process are discussed in Chapter 34. Sulfated forms of the sialyl Lewis x determinant are also thought to contribute to Siglec recognition, as mentioned in Chapter 35.

The HNK-1 antigen is a terminally sulfated glycan that was first described on human natural killer cells, and is also called CD57. The HNK-1 epitope is expressed in the vertebrate nervous system, and expression patterns change during neural development (Chapter 41). The HNK-1 determinant comprises 3-O-sulfated GlcA attached in β1-3 linkage to a terminal Gal (Figure 14.12) of N-glycans, O-glycans, proteoglycans, and glycolipids. Two different GlcA-transferases participate in HNK-1 GlcA addition: GlcAT-P (B3GAT1) and GlcAT-S (B3GAT2). They have very different activities for glycoprotein or glycolipid substrates in vitro and may generate functionally different HNK-1 epitopes in vivo. Glucuronylation is followed by 3-O sulfation of the GlcA by a sulfotransferase (CHST10). The HNK-1 epitope is present on a variety of neuronal cell glycoproteins, including NCAM, contactin, myelin-associated glycoprotein, telencephalin, L1, and P0 (the major glycoprotein of peripheral nerve myelin). There is evidence that HNK-1 can function as a ligand for laminin, L-selectin, P-selectin, and a cerebellar adhesion protein termed amphoterin. HNK-1 has also been shown to mediate homotypic adhesive interactions involving P0. HNK-1-dependent adhesive interactions have been implicated in cell migration processes involving cell–cell and cell–matrix interactions and are proposed to participate in reinnervation of muscles by motor neurons.

Phosphorylation of sugars is also important in recognition events. In mammals, the phosphorylation of Man on oligomannose N-glycans at the C-6 position of lysosomal hydrolases occurs by a phospho-GlcNAc transferase to create a GlcNAc-phospho-6-mannose diester (Chapters 9 and 33). Subsequent removal of the GlcNAc exposes monophosphoester Man-6-P for recognition of lysosomal hydrolases by the Man-6-P receptors. Interestingly, mannose-1-6-phosphate-mannose is a common modification in yeast mannans on their cell walls. Phosphorylation of the C-2 of the Xyl that initiates proteoglycan core linker synthesis is mediated by a Golgi kinase FAM20B and is essential for addition of the second Gal in the core linker

GlcAβ1-3Galβ1-3Galβ1-4Xylβ1-Ser/Thr core. The phosphate must be removed by a phosphatase, PXYLP1, before GlcA can be added to the core linker, generating the substrate for HS, CS, or DS addition. Interestingly, a GlcNAc may be added to the phosphorylated trisaccharide to block extension of the core linker glycan. This regulation by Xyl phosphorylation is essential for physiological homeostasis of proteoglycans (Chapters 17 and 41). In another example of regulation by phosphorylation, the Golgi kinase POMK phosphorylates the C-6 position of the O-Man that initiates O-mannose glycans on α-dystroglycan, and this has been shown to be essential for the subsequent action of the glycosyltransferase LARGE, which adds a polymer of -3GlcAβ1-3Xylα1- disaccharide repeats to the O-Man core (Chapters 13 and 45).

ACKNOWLEDGMENTS

The authors appreciate helpful comments and suggestions from Takashi Angata and Nicolle H. Packer.

FURTHER READING

Yamamoto F. 2004. Review: ABO blood group system—ABH oligosaccharide antigens, anti-A and anti-B, A and B glycosyltransferases, and ABO genes. *Immunohematology* **20:** 3–22. doi:10.21307/immunohematology-2019-418

Audry M, Jeanneau C, Imberty A, Harduin-Lepers A, Delannoy P, Breton C. 2011. Current trends in the structure–activity relationships of sialyltransferases. *Glycobiology* **21:** 716–726. doi:10.1093/glycob/cwq189

Fiete D, Beranek M, Baenziger JU. 2012. Molecular basis for protein-specific transfer of *N*-acetylgalactosamine to N-linked glycans by the glycosyltransferases β1,4-*N*-acetylgalactosaminyl transferase 3 (β4GalNAc-T3) and β4GalNAc-T4. *J Biol Chem* **287:** 29194–29203. doi:10.1074/jbc.m112.371567

Patnaik SK, Helmberg W, Blumenfeld OO. 2014. BGMUT database of allelic variants of genes encoding human blood group antigens. *Transfus Med Hemother* **41:** 346–351. doi:10.1159/000366108

Yamamoto F, Cid E, Yamamoto M, Saitou N, Bertranpetit J, Blancher A. 2014. An integrative evolution theory of histo-blood group ABO and related genes. *Sci Rep* **4:** 6601. doi:10.1038/srep06601

Bode L. 2015. The functional biology of human milk oligosaccharides. *Early Hum Dev* **91:** 619–622. doi:10.1016/j.earlhumdev.2015.09.001

Jost T, Lacroix C, Braegger C, Chassard C. 2015. Impact of human milk bacteria and oligosaccharides on neonatal gut microbiota establishment and gut health. *Nutr Rev* **73:** 426–437. doi:10.1093/nutrit/nuu016

Quraishy N, Sapatnekar S. 2017. Advances in blood typing. *Adv Clin Chem* **77:** 221–269. doi:10.1016/bs.acc.2016.06.006

Stowell CP, Stowell SR. 2019. Biologic roles of the ABH and Lewis histo-blood group antigens. Part I: infection and immunity. *Vox Sang* **114:** 426–442. doi:10.1111/vox.12787

Stowell SR, Stowell CP. 2019. Biologic roles of the ABH and Lewis histo-blood group antigens. Part II: thrombosis, cardiovascular disease and metabolism. *Vox Sang* **114:** 535–552. doi:10.1111/vox.12786

Rahfeld P, Withers SG. 2020. Toward universal donor blood: enzymatic conversion of A and B to O type. *J Biol Chem* **295:** 325–334. doi:10.1074/jbc.rev119.008164

15 Sialic Acids and Other Nonulosonic Acids

Amanda L. Lewis, Xi Chen, Ronald L. Schnaar, and Ajit Varki

Sialic acids (Sias) are abundant on vertebrate glycoproteins, glycolipids, and milk oligosaccharides, as well as on some microbial surface glycans, mediating diverse functional roles. Originally discovered within the Deuterostome lineage of animals and associated microbes, they are actually a subset of a more ancient family of α-keto acid monosaccharides with a 9-carbon backbone called nonulosonic acids (NulOs), which are also found in some Eubacteria and Archaea. Biosynthesis of all NulO-glycans requires the activation of NulO to a CMP-sugar, before the NulO is transferred to glycan acceptors. NulOs are remarkable for the number and the type of functional groups on one monosaccharide. Further complexity arises from various epimers,

published by Cold Spring Harbor Laboratory Press; doi: 10.1101/glycobiology.4e.15

modifications, and diverse linkages to other glycans, making these molecules well-suited to carry information for glycan–protein, cell–cell, and pathogen–host recognition. NulOs are among the most rapidly evolving classes of monosaccharides in nature and exist in tremendous variety, particularly in the microbial world. Given their high density and widespread location on vertebrate cells, Sias also exert many functions via electronegative charge, such as repulsion of cell–cell interactions, protein stabilization, ion binding, and ion transport.

DISCOVERY AND GENERAL CLASSIFICATION

Early nomenclature of these molecules was tied to their discovery, being first isolated by Gunnar Blix from salivary mucins in 1936 and independently by Ernst Klenk from brain glycolipids in 1941. Blix named his substance "sialic acid" after the Greek word for saliva (σίαλον), and Klenk named his "neuraminic acid" for neurons in the brain. By the time the relationship of these substances became evident, both names were already in use and have persisted. Although 5-*N*-acetylneuraminic acid (Neu5Ac, sometimes called "NANA") is the most common sialic acid (Sia) in humans, the Sia family is comprised of related structures (Figure 15.1) that vary at the C-5 carbon, including 5-*N*-glycolylneuraminic acid (Neu5Gc) and 3-deoxy-D-glycero-D-galacto-non-2-ulosonic acid (also known as 3-deoxy-nonulosonic acid, 2-keto-3-deoxy-nononic acid, or Kdn) with a hydroxyl group at C-5. Similar 9-carbon backbone 2-keto acid monosaccharides were later discovered in some bacterial lipopolysaccharides and initially called "bacterial sialic acids." The resulting confusion was resolved by suggesting that the term "sialic acid" (Sia) be limited to its original use in describing neuraminic acid (Neu), Kdn, and their derivatives in deuterostomes and their pathogens, and that the term "nonulosonic acid" (NulO) be used to encompass the entire group of 9-carbon backbone non-2-ulosonic acids (Figure 15.1). Base structures representing the larger family of NulO molecules are shown in Figure 15.1 for comparison. This chapter will largely deal with the biology, metabolism, and functions of Sias in mammals. The evolution, distribution, and functions of the larger family of NulOs are briefly described toward the end of this chapter.

THE SIALIC ACID FAMILY

The most common Sia form in humans is Neu5Ac, a relatively strong acid (pKa=2.6) with the electron-withdrawing C-1 carboxylate attached to the C-2 anomeric carbon. The exocyclic glycerol-like side chain (C-7, C-8, and C-9, each carrying a hydroxyl group) provide opportunities for hydrogen bonding. The *N*-acetyl group facilitates hydrophobic interactions, changing to hydrophilic properties when it is replaced with an *N*-glycolyl group. Each of these moieties can contribute to binding specificities of Sia-binding proteins and functions of Sia-containing glycans.

SIALOGLYCAN DIVERSITY

Many publications assume that Neu5Ac is the Sia present in a given biological sample, and it is indeed common in vertebrates. But in addition to Neu5Gc, a second level of diversity arises from various natural modifications (e.g., mono- or multiple-*O*-acetylation at any hydroxyl groups, 8-*O*-methylation, 8-*O*-sulfation, and 9-*O*-lactylation), some of which are also common. The carboxylate group at C-1 can condense with hydroxyl groups of adjacent sugars to form an uncharged lactone, or with a free amino group at C-5 to form an uncharged lactam. Such modifications can determine or modify recognition by Sia-binding proteins and direct Sia function.

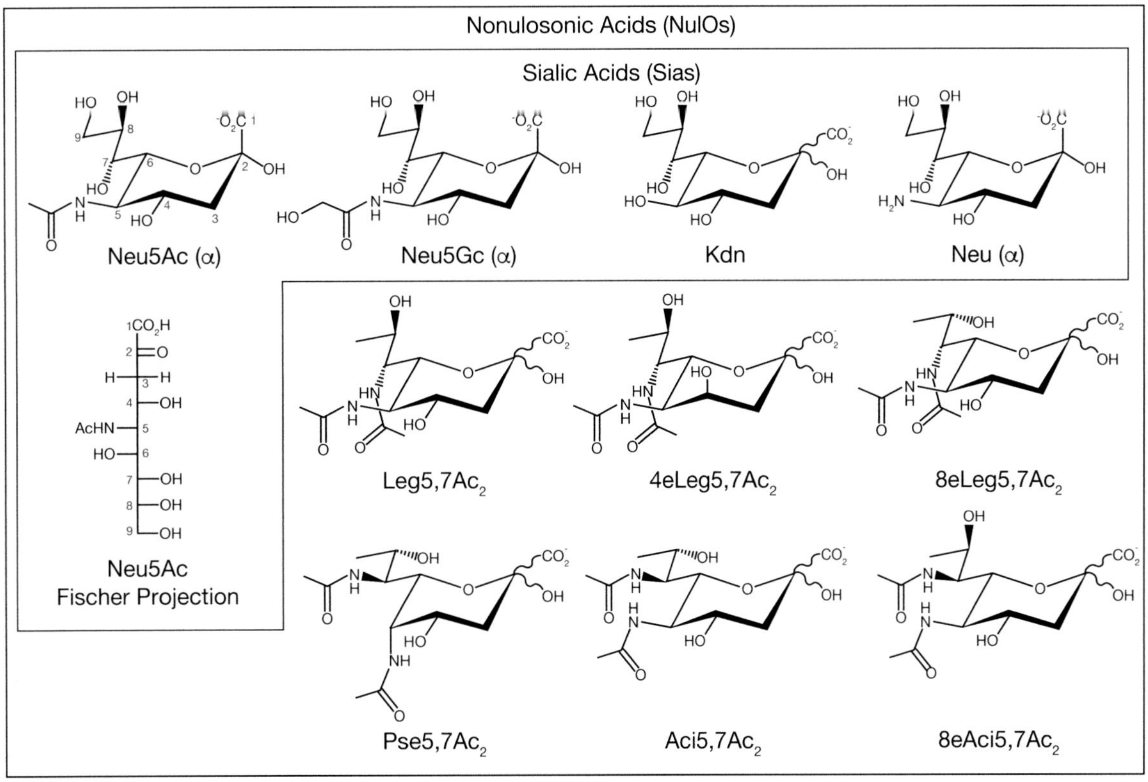

FIGURE 15.1. Sialic acids (Sias) and other nonulosonic acids (NulOs). As shown in the figure, Sias are a subset of NulOs. All NulOs carry a negatively charged carboxylate (C-1) and a 3-carbon exocyclic glycerol-like side-chain (C-7 to C-9). The anomeric center (C-2) is drawn with the carboxyl group in the axial orientation (α-anomer for Sias), the configuration that typically occurs in glycosidically bound states of Sias, but the β-anomer is more abundant in the unbound state in solution and is the norm in CMP-Sias. There are examples of most of the prokaryotic NulOs in α- and β- configurations. The *inset* box shows the structures of the core Sias. These molecules vary widely in the substituent attached to C-5. For Sias, the *N*-acetyl (Neu5Ac, *N*-acetylneuraminic acid) is the most common form, and the *N*-glycolyl (Neu5Gc, *N*-glycolylneuraminic acid) form is common in many nonhuman vertebrates and "higher" invertebrates. Depending on the species, 3-deoxy-D-glycero-D-galacto-non-2-ulosonic acid (also known as 2-keto-3-deoxy-nononic acid or Kdn, with a hydroxyl at the C-5), appears less abundant and is not found in the bound form in mammals, whereas the free amine form (Neu, neuraminic acid) is rare in nature and found only in the glycosidically linked form. Non-Sia NulOs reported to date differ from Neu5Ac in having no hydroxyl group at C-9, an additional often substituted amino group (e.g., acetyl, formyl, acetimidoyl) at C-7, and various different stereochemistries at C-5 and C-7 as well as other carbon positions (C-4 or C-8). Consensus among the experts on nomenclature has not been reached (e.g., both Leg and Leg5,7Ac_2 have been used to represent di-*N*-acetyllegionaminic acid). So far, 9-deoxy NulOs with free amino groups at C-5 and C-7 have not been reported in nature. Here, structures representing six types of prokaryotic NulOs—Leg5,7Ac_2 (di-*N*-acetyllegionaminic acid), 8eLeg5,7Ac_2 (8-epi-di-*N*-acetyllegionaminic acid), 4eLeg5,7Ac_2 (4-epi-di-*N*-acetyllegionaminic acid), Pse5,7Ac_2 (di-*N*-acetylpseudaminic acid), Aci5,7Ac_2 (di-*N*-acetylacinetaminic acid), and 8eAci5,7Ac_2 (8-epi-di-*N*-acetylacinetaminic acid)—are shown.

Despite this complexity, it may be sufficient in some biological studies to simply know that a generic Sia residue is present at the terminal position of a glycan under study (a red diamond in the symbol nomenclature system) (Figure 15.2). Dehydro-Sias also exist in nature. For example, Neu2en5Ac (2,3-dehydro-2-deoxy-*N*-acetylneuraminic acid, sometimes called DANA) has natural inhibitory properties toward sialidases and was the starting point for the design of more potent inhibitors such as the anti-influenza drug Relenza (Chapter 57).

The sum total of diverse sialoglycans on a cell is dubbed its "sialome." As mentioned, Sias are often found as the terminal (outermost) saccharide on branches of N-glycans, O-glycans,

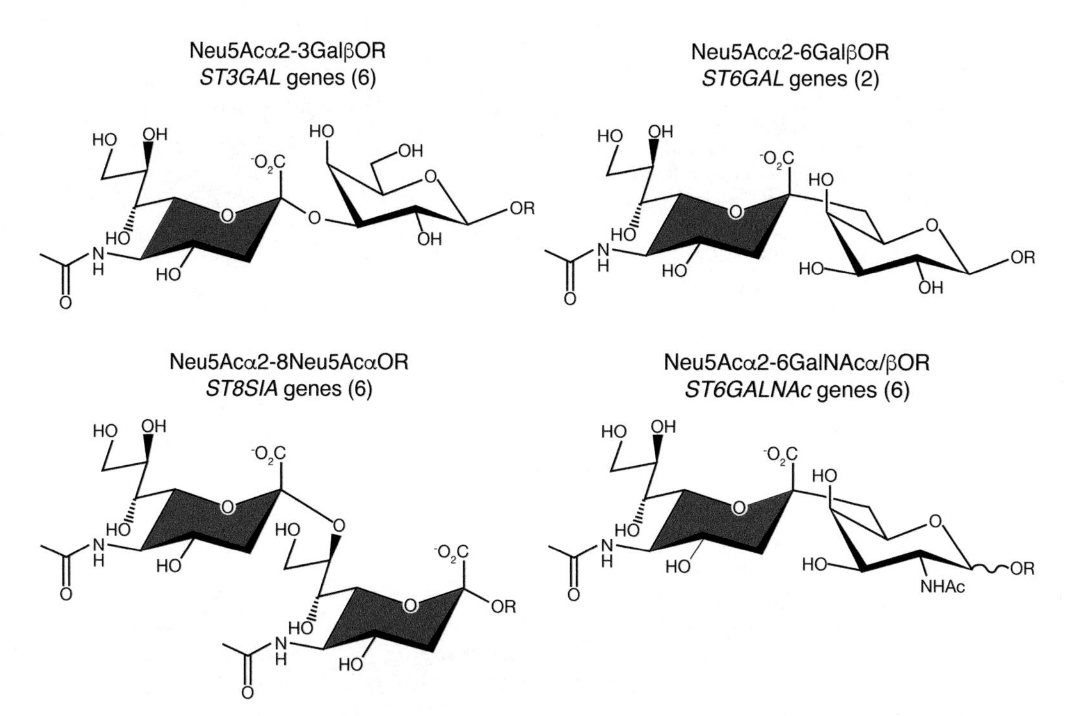

FIGURE 15.2. Diversity in sialic acid linkages. Sialic acids in vertebrates are commonly (but not exclusively) found in α-glycosidic linkage to the C-3 or C-6 hydroxyl of galactose, the C-6 hydroxyl of *N*-acetylgalactosamine, the C-8 hydroxyl of another sialic acid, or less commonly the C-6 hydroxyl of *N*-acetylglucosamine. There are multiple sialyltransferase genes responsible for creating each type of linkage. In humans and mice (and other mammals), there are six genes coding for enzymes that transfer sialic acid α2-3 to Gal, two genes for α2-6 to Gal, six genes for α2-6 to GalNAc, and six genes for α2-8 to Sia. (Modified, with permission, from Schnaar RL et al. 2014. *Physiol Rev* **94:** 461–518.)

glycosphingolipids, milk oligosaccharides, and occasionally capping the side chains of glycosylphosphatidylinositol (GPI) anchors (Chapter 12). A useful conceptualization of the sialome is as a "forest canopy" of the cell-surface glycan "forest" (the whole glycome) (Figure 15.3). At the outermost level are different Sia structures (leaves and flowers), which are in glycosidic linkage to underlying linear and branched oligosaccharides (stems and branches) (see Figures 15.2 and 15.3), which in turn are components of glycoproteins and glycolipids (the tree trunks). At the cell surface, glycolipids and glycoproteins organize into lateral domains (forests). Just as the forest varies greatly from place to place, the sialome varies among cell types and among domains on a given cell surface, variations that serve specific functions. As with the glycome, each cell type in each species expresses its own distinct sialome. For example, although mammalian liver sialoglycans are most abundant on glycoproteins, brain sialoglycans are much more abundant as sialoglycolipids (Chapter 11), so much so that they are termed "gangliosides" for the nerve ganglia on which they were discovered. The sialome also varies markedly in response to environmental cues.

Sias can also exist as internally linked structures rather than on termini of glycans. In vertebrates, the terminal Sia of a glycan chain is rarely extended further, except with another Sia (Figure 15.3). Among bacteria, Sias and the related NulOs can also be found in diverse cellular structures, including capsular polysaccharides, lipopolysaccharide O-antigens or as part of flagella, pili, or structures such as the cell wall or S-layer (Chapter 21). Bacteria with close relationships to vertebrate hosts often mimic the terminal placement of Sias on capsules and O-antigens. However, many NulOs of bacteria, including Sias, can occur as internal glycan components within repeating units of polysaccharides, which have also been reported in some

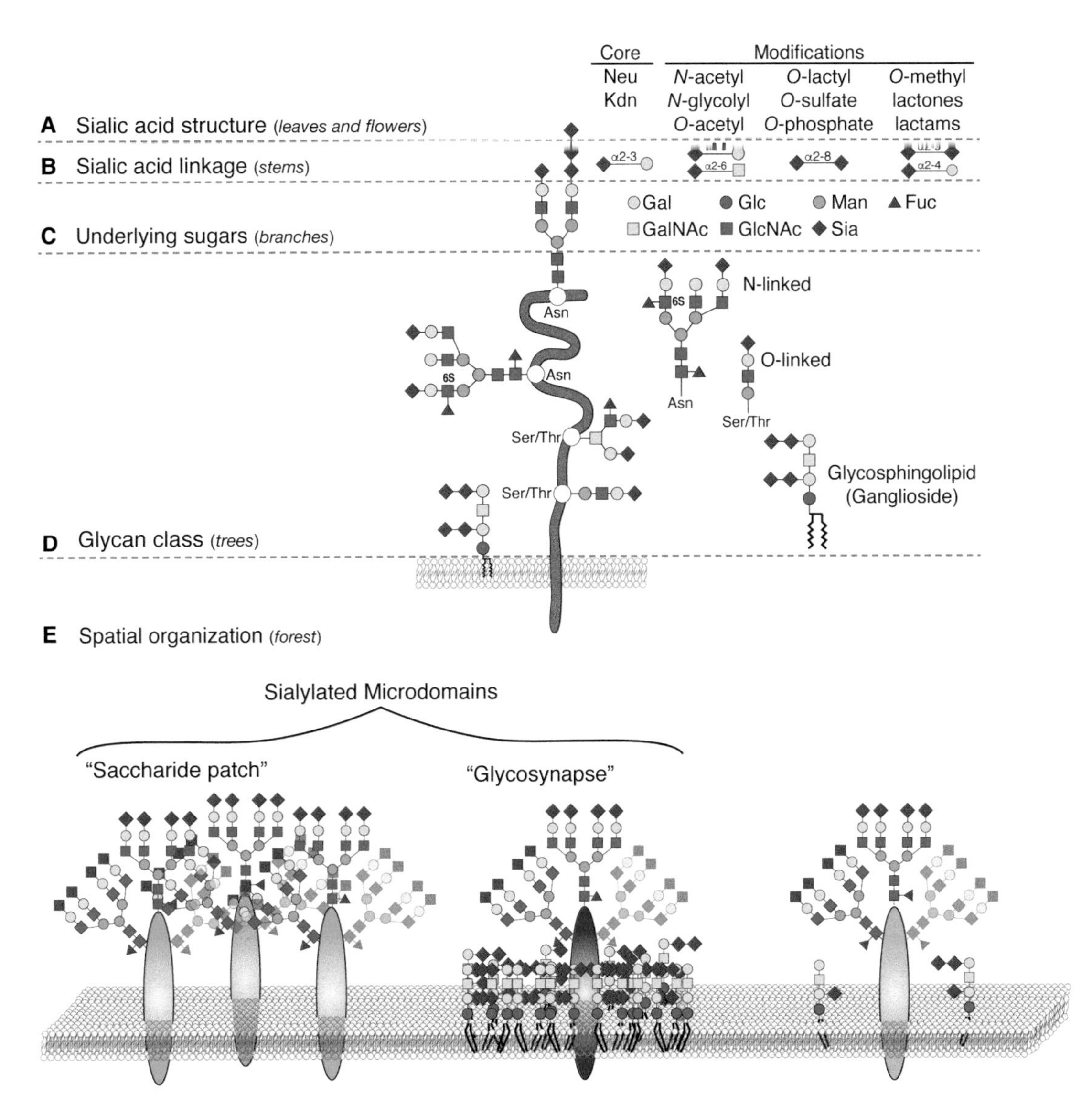

FIGURE 15.3. Hierarchical levels of sialome complexity. The sialome can vary at the following complexity levels: (*A*) sialic acid core and core modification: esterification (with various groups), *O*-methylation, lactonization, or lactamization yielding more than 80 different structures; (*B*) linkage to the underlying sugar (four major and many minor linkages); (*C*) identity and arrangement of the underlying sugar that can also be further modified by fucosylation or sulfation; (*D*) glycan class (N-linked or O-linked glycoproteins, glycosphingolipids); and (*E*) spatial organization of the Sia in sialylated microdomains, which have been referred to as "clustered saccharide patches" or "the glycosynapse." (Reproduced, with permission, from Cohen M, Varki A. 2010. *OMICS* **14:** 455–464.)

invertebrates such as echinoderms. In the latter, the C-4 hydroxyl of Neu5Ac is sometimes glycosylated with other monosaccharides such as fucose, galactose, and glucose. Overall, it appears that we do not yet know the full extent of sialoglycan diversity in nature.

POLYSIALIC ACIDS

Polysialic acid (polySia, previously called PSA) is a linear homopolymer of Sias that sometimes reaches lengths of more than 100 residues. Shorter chains of two to three residues, called oligosialic acids, are common components of gangliosides (Figure 15.3 and Chapter 11) and are found occasionally as terminal structures on glycoprotein glycans. PolySia is a prominent structural feature of a highly select group of acceptor proteins, the best studied of which is the neural cell adhesion molecule (NCAM). Beginning early in vertebrate brain development, long polySia chains (Neu5Acα2-8)$_n$ are added to NCAM (often then called PSA-NCAM), on the termini of two specific N-glycans. The large hydration shell formed by polySia markedly increases the hydrodynamic volume of its protein carrier and interferes with NCAM's natural cell–cell adhesion function. PolySia thus converts an adhesive protein into a repelling one. During brain development, a high level of polySia likely ensures that neural precursors can migrate to their final anatomical sites. Once a destination is reached, polySia is down-regulated and firm adhesion proceeds, keeping the cells in place. When the sialyltransferases (STs) responsible for polySia formation are genetically ablated in mice, some nerve progenitor cells adhere prematurely, become stuck in place and fail to reach their proper destination. In contrast to the above-mentioned "repulsive field effects," polySia can also act an "attractive field," binding and locally concentrating neurotrophic factors like BDNF and FGF2. Notably, human genetic and molecular evidence indicate that variations in polysialylation may be a factor in human psychiatric diseases like schizophrenia.

PolySia is found less abundantly on a few other proteins, including another cell adhesion molecule (SynCAM 1), a peptide receptor (neuropilin-2), and on O-glycans of fish egg glycoproteins. It is also a key component of the capsular polysaccharides of certain pathogenic bacteria (e.g., *Escherichia coli* K1, K92; *Neisseria meningitidis* serogroups B and C). The linkages between Sia units in polySia chains vary in bacteria, including α2-8, α2-9, and alternating α2-8/α2-9. The α2-8-polySia in *E. coli* K1 is also called colominic acid, which can be O-acetylated at Sia C-7 or C-9. The α2-9-polySia in *N. meningitidis* serogroup C can also be O-acetylated at Sia C-7 or C-8. A bacteriophage that attacks polySia-expressing bacteria produces a highly specific endosialidase that only clips chains that are ≥8 residues long. This enzyme (Endo-N) and its inactivated form (which binds polySia) are powerful tools to study polySia functions.

HUMAN LOSS OF *N*-GLYCOLYLNEURAMINIC ACID

Neu5Gc is common in mammals including human's closest evolutionary relatives (bonobos and chimpanzees). However, it is not synthesized by humans, because of a fixed single-exon deletion mutation in the *CMAH* gene encoding a hydroxylase that converts CMP-Neu5Ac to CMP-Neu5Gc. Why did the human lineage lose Neu5Gc? Many pathogens use Sias to bind and infect vertebrates (Chapter 42), and some specifically target Neu5Gc. An organism recognizing Neu5Gc may have been the cause of initial selection of the CMAH null state. Subsequent development of anti-Neu5Gc antibodies in CMAH-null females may have then reduced fertility with CMAH-positive males by killing their sperm, possibly serving to speciate the null population. Other lineage-specific losses of Neu5Gc expression may have occurred in a similar fashion. This is part of a broader consideration of Sias as both a pathogen target and a physiological regulator, a combination that apparently contributed to rapid evolution of both Sia expression patterns and Sia-binding proteins, especially of the Siglecs (Chapter 35).

Despite *CMAH* inactivation in humans, traces of Neu5Gc are found in normal human tissues. This, as well as the presence of Neu5Gc in human tumor cells and tissues, apparently represents metabolic incorporation of Neu5Gc ingested from foods, particularly "red meats." Most healthy humans have circulating anti-Neu5Gc antibodies, raising the possibility that their subsequent interaction with metabolically incorporated Neu5Gc contributes to inflammation in diseases that correlate with red meat consumption, such as atherosclerosis and epithelial cancers, pathologies uncommon in Neu5Gc-positive primates. Other possible consequences of human Neu5Gc loss include resistance to some Neu5Gc-binding animal pathogens such as *E. coli* K99, and the emergence of Neu5Ac-preferring pathogens exclusive to human hosts such as the malarial parasite *Plasmodium falciparum* and the toxin from *Salmonella* Typhi.

An as-yet-unexplained observation is the scarcity of Neu5Gc in the brains of all vertebrates studied to date, including those that have high Neu5Gc levels in other tissues. The evolutionary advantage of excluding Neu5Gc from this vital organ is unknown but may relate to selection pressure by unknown pathogens and/or selective roles of Neu5Ac in optimal brain development and function.

METABOLISM

Synthesis of Sialoglycans

Metabolic pathways of Neu5Ac in vertebrates are shown in Figure 15.4. Neu5Ac is derived by reaction of ManNAc-6-P with phosphoenolpyruvate (PEP). The ManNAc-6-P is produced by a bifunctional enzyme, UDP-GlcNAc-2-epimerase/N-acetylmannosamine kinase, coded by the *GNE* gene. Missense, recessive mutations in *GNE* cause hereditary inclusion body myopathy (HIBM) in humans (Chapter 45), and gene inactivation causes embryonic lethality in mice. Reaction of ManNAc-6-P with PEP yields Neu5Ac-9-P, which is then dephosphorylated by a specific phosphatase to release free Neu5Ac in the cytoplasm. The same pathway can use Man-6-P instead of ManNAc-6-P to generate Kdn. In contrast, Neu5Ac biosynthesis in prokaryotes involves direct reaction of ManNAc with PEP, a pathway that proceeds without a Neu5Ac-9-P intermediate. Many other bacteria synthesize 9-deoxy forms of NulOs using pathways bearing ancient homology to the Neu5Ac synthetic pathways. Notably, synthetic unnatural mannosamine derivatives can enter the Sia biosynthetic pathway in cells, allowing metabolic engineering of the cell surface with modified Sias (Chapter 56).

The β-anomers of free Sias are converted to CMP-Sias by reactions with CTP, which occur in the nuclear compartment, for unknown reasons. The CMP-Sia products then return to the cytosol and transported into Golgi compartments by a CMP antiporter, which concentrates them in the Golgi lumen (Chapter 5). In contrast, prokaryotic CMP-Sias are synthesized in the cytoplasm and used without compartmentalization. In eukaryotes, cytosolic CMP-Sia feeds back to inhibit GNE. In a rare human disease called sialuria, dominant mutations in the allosteric site of *GNE* result in failed feedback regulation and overproduction of Sia.

The linkage-specific STs mentioned earlier are type II membrane proteins in animals, with signals leading to their Golgi localization. Animal STs share amino acid sequence motifs (sialylmotifs) that include shared sites for CMP-Sia recognition (Chapters 6 and 8). In contrast, prokaryotic STs emerged independently by convergent evolution, do not have sialylmotifs, and some do not even share homology. In addition to the Sia linkage, several eukaryotic STs show strong preferences for glycolipids versus glycoproteins and/or the terminal saccharides and sequences. Modified Sias, such as Neu5Gc, O-acetylated species, and unnatural synthetic Sias can be used to form activated CMP-Sias, which can serve as donors for STs. Some mammalian STs transfer both Neu5Ac and Kdn, but others transfer only one or the other. All result in α-sialyl linkages. Furthermore, "*trans*-sialidases" of some pathogenic trypanosome species and some bacteria transfer Sias from one sialoside to form another (Chapter 43). Although *trans*-sialidases tend to be specific regarding the sialyl linkage they generate, they can be promiscuous toward the donor and acceptor glycans.

Modifications of Sialic Acids

Once CMP-Neu5Ac is converted to CMP-Neu5Gc in the cytosol, there is no known way to reverse the reaction, perhaps explaining the accumulation of diet-derived Neu5Gc in human tissues. Further structural diversity in Sias is generated by enzymatic modification of Neu5Ac, Neu5Gc, and Kdn via reactions that occur in the lumen of the Golgi and related organelles,

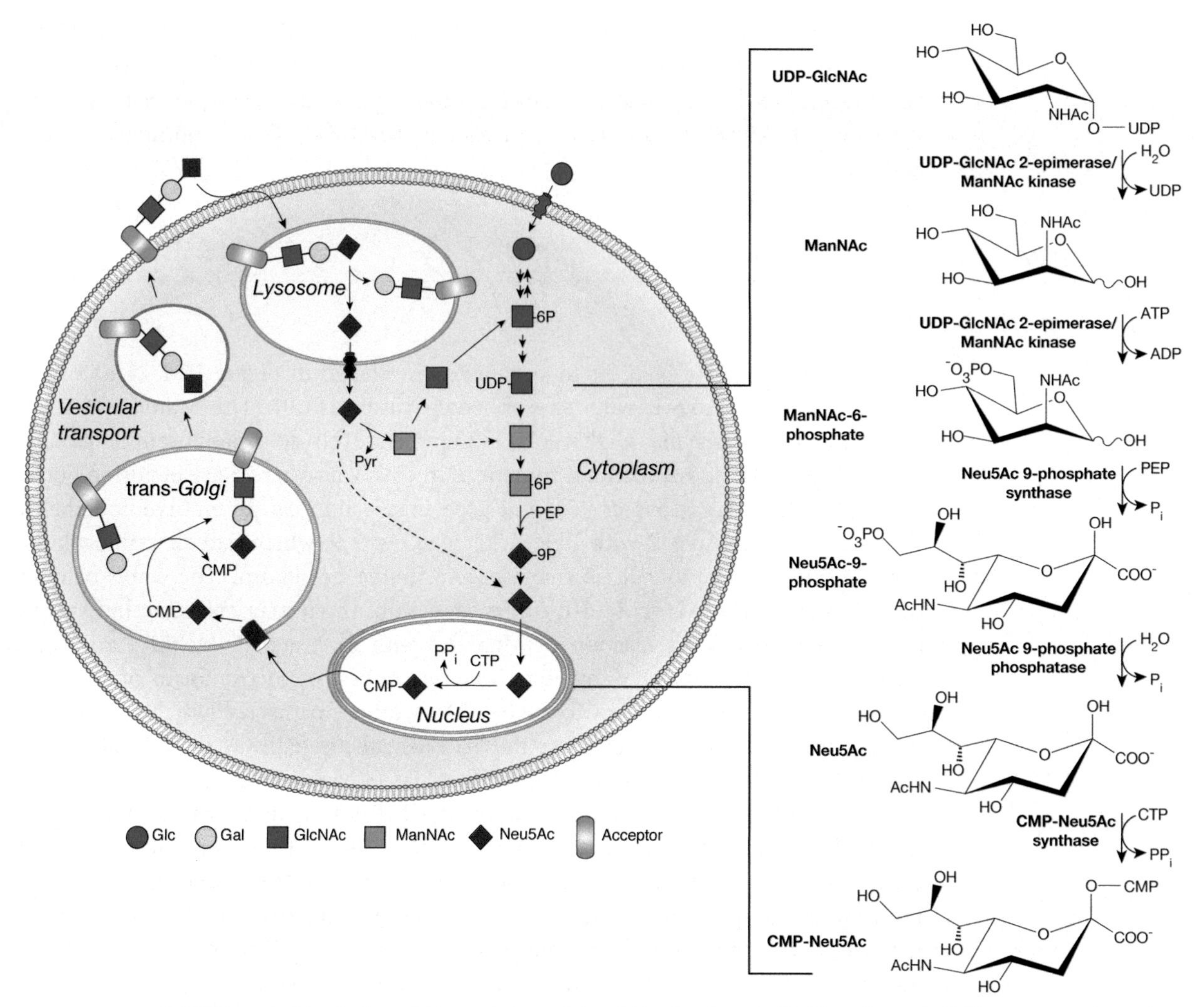

FIGURE 15.4. Metabolism of *N*-acetylneuraminic acid in vertebrate cells. The schematic (*left*) represents the pathways and compartments of biosynthesis of Neu5Ac in vertebrates: biosynthesized in the cytosolic compartment, converted to activated CMP-Neu5Ac in the nucleus, transferred to lipid and protein glycans in the Golgi, transported to the cell surface, eventually secreted or taken up into lysosomes, degraded, and metabolites reused. Details of the enzymatic steps leading to CMP-Neu5Ac biosynthesis in vertebrates are represented on the *right*. Not shown are the additional steps involving modifications such as hydroxylation of the N-acetyl group of CMP-Neu5Ac to generate CMP-Neu5Gc (occurs in the cytosol) and addition of various modifications such as O-acetylation. Methylation and sulfation likely occur in the Golgi apparatus. (Modified, with permission Bentham Science Publishers, from Münster-Kühnel AK, Hinderlich S. 2013. In *Sialobiology: Structure, Biosynthesis and Function* (ed, Tiralongo J, Martinez-Duncker I), pp. 76–114.)

and on mature sialoglycoconjugates, as well as at the CMP-Sia level. Different enzymes may O-acetylate Sia at C-7/9 (most common) and C-4 (less common). O-Acetyltransferases (some likely yet to be discovered) appear to have specificity for Sias of different glycosidic linkages or carried on different classes of mature glycoconjugates, or on CMP-Neu5Ac. For example, a human O-acetyltransferase (CASD1) acetylates CMP-Neu5Ac using acetyl-CoA as the acetyl donor and the resulting CMP-Sia is then used to transfer O-acetylated Neu5Ac to proteins. In bacteria, Neu5Ac O-acetylation can occur by at least two routes. In group B *Streptococcus*, *O*-acetylation occurs intracellularly, prior to CMP-activation via the action of the NeuD enzyme. *E. coli* K1 also has a NeuD homolog, but it appears to de-O-acetylate prior to capsule assembly, then NeuO O-acetylates the assembled polySia. Other substitutions of the hydroxyl groups arise from the use of appropriate donors (*S*-adenosylmethionine for methylating Sias or 3′-phosphoadenosine 5′-phosphosulfate for sulfating Sias). With the exception of Neu5Gc, other modified Sias do not appear to be very effective substrates for reactivation by vertebrate CMP-Sia synthetases, so their catabolism also involves enzymes like Sia O-acetylesterases.

Enzymatic Release of Sialic Acids

Sias are released from glycoconjugates by sialidases, also historically called "neuraminidases." The term "sialidase" is now preferred, but influenza virus sialidases are still called neuraminidases because of the use of "N" in their strain designation (e.g., H1N1 influenza), in which N1 refers to its neuraminidase/sialidase. In eukaryotes, sialidases are encoded by *NEU* genes, of which there are four in the human genome. Although NEU1 and NEU4 are thought to act primarily intracellularly in endosomal/lysosomal compartments to recycle sialoglycans, NEU1 can also be recruited to the cell surface where it may regulate sialylation and modulate receptor-mediated signaling. In contrast, NEU3 appears primarily on the cell-surface and has a preference for cleaving gangliosides, whereas NEU2 exists in the cytosol, presumably to recycle sialoglycans that enter that compartment following autophagy or phagocytosis. Mammalian sialidases have also been implicated in the disordered inflammation seen in colitis and sepsis.

Many microorganisms, including many pathogens, express sialidases. Unlike the case with STs, bacterial, fungal, and invertebrate sialidases are evolutionarily related to their vertebrate counterparts. In contrast, viral sialidases that are essential for their virulence represent distinct families. The 3D structures of pathogen sialidases are used for drug development (Chapter 42). Sialidases vary in their substrate specificity from highly selective (e.g., the α2-3-specific sialidase SpNanB and SpNanC from *Streptococcus pneumoniae*) to broadly acting (e.g., the α2-3, α2-6, α2-8 sialidase from *Arthrobacter ureafaciens* and SpNanA from *S. pneumoniae*). The products of sialidases also differ with free Sias as the most common ones (e.g., released by SpNanA) whereas 2,7-anhydro-Sias (e.g., released by SpNanB) and 2,3-dehydro-2-deoxy-sialic acids (Sia2ens) (e.g., released by SpNanC) are less common. Interestingly, some pathogen sialidases have additional lectin (sugar-binding) domains that appear to direct the action of the enzymes to particular sites.

Some Sia linkages resist release even by sialidases with broad specificity. A branched α2-3Sia on the brain ganglioside GM2 requires a special helper protein (GM2-activator) to facilitate its cleavage, and mutation of GM2-activator protein results in GM2 buildup in the brain (Chapter 44). O-Methylation and O-acetylation of Sias also can hinder or even prevent hydrolysis by sialidases, as is the case for 4-O-acetyl group. These properties are likely biologically significant, but have not been well investigated.

Microbial sialidases and *trans*-sialidases (see also Chapter 43) can be powerful virulence factors that may promote cellular invasion, unmask potential binding sites, and modulate the immune system. Sias liberated by sialidases at mucosal sites also provide nutrients for some bacteria or provide access to underlying carbohydrates. Viral neuraminidases facilitate release and spreading of newly formed viruses, and specific viral neuraminidase inhibitors are widely used as the antiviral drugs such as anti-influenza A virus drugs Relenza and Tamiflu (Chapter 42).

Sialic Acid Recycling and Degradation

Once a free Sia molecule is released in the lysosome of a vertebrate cell, it is returned to the cytosol by a specific exporter (SLC17A5, also called Sialin), in which it is either reused or degraded. Human mutations in Sialin cause Salla disease and infantile sialic acid storage disease, resulting in accumulation of Sia in lysosomes and excretion of excess Sia in the urine. Some pathogens scavenge Sias from the extracellular space, using high-efficiency transporters to concentrate them from trace levels. In contrast, there is no evidence for plasma membrane Sia transporters in eukaryotic cells. Nevertheless, free Sias can be taken up into mammalian cells via fluid-phase macropinocytosis, eventually arriving in the lysosomes, from which they are exported into the cytoplasm by Sialin. Intact sialoglycans can also be taken up and transported to lysosomes, where sialidases release Sias for delivery to the cytoplasm and then to the nucleus to be reused by the cellular CMP-Sia synthetase. However, at the whole-body level, free Sias in the bloodstream derived from cellular sources or digestive processes are rapidly excreted in the urine. Thus, uptake of sialoglycoproteins (followed by lysosomal release by NEU1 and export by Sialin) may better account for the incorporation of dietary Neu5Gc in human tissues.

In vertebrates, O-acetylated Sias are de-O-acetylated by Sia-specific O-acetylesterases, including the human enzyme coded by the *SIAE* gene, which may regulate B-cell development and immune tolerance by modulating recognition by Siglecs (Chapter 35). Enzymes with Sia 9-O-acetylesterase activity are also found in some bacteria and viruses. Influenza C virus and some coronavirus hemagglutinins bind specifically to 9-O-acetyl Sia for host infection, and their O-acetylesterases help in their further dissemination after newly budded virions are released from host cells. Notably, all known O-acetylesterases specific for C-9 ester are incapable of releasing O-acetyl esters from C-7. However, a 7-O-acetyl group can migrate to C-9 under physiological conditions and thus become a substrate for these enzymes. This process can even result in stepwise de-esterification of 7/8/9-tri-O-acetyl Sias, which do exist in nature. Some vertebrate and viral esterases are specific for 4-O-acetyl groups, but relatively little is known about their functions. Among bacteria, Sia-specific esterases appear to participate in normal processes of mucus turnover in the mammalian gut, but when in excess, they have also been linked to disease states such as colorectal cancer.

If not reused or excreted, Sias are degraded by a cytosolic Sia-specific pyruvate lyase (encoded by *NPL*) that cleaves the molecule into *N*-acyl-mannosamine and pyruvate. Such *N*-acyl-mannosamines may either reenter the Sia biosynthetic pathway after phosphorylation by kinases in eukaryotes, or be further metabolized via conversion to *N*-acyl-glucosamines and their corresponding phosphates, with eventual deacylation (Figure 15.4). Notably, GlcNGc derived from Neu5Gc can either be de-acylated or converted to UDP-GlcNGc or even to UDP-GalNGc, which can provide donors for glycosyltransferases to incorporate the monosaccharide into glycans, such as chondroitin sulfate with GalNGc instead of GalNAc. Sia-specific pyruvate lyases also exist in various microorganisms, allowing the use of host Sias as a nutrient source. In some bacteria, pathways for Sia biosynthesis and breakdown appear in the same genomes, suggesting that recycling of Sias may also occur in some bacteria. In other bacterial taxa, there is either anabolic or catabolic machinery.

METHODS FOR STUDYING SIALIC ACIDS

Sias are amenable to analysis as components of isolated glycans, as released monosaccharides, and also in situ on cell surfaces. Among animal monosaccharides, Sias have a distinct molecular mass that makes them readily identifiable by mass spectrometry (MS), currently among the most sensitive methods for glycan analysis (Chapters 50 and 51). The glycerol-like side chain of Sias is uniquely sensitive to periodate oxidation under very mild conditions, providing a method to selectively generate aldehyde groups only on sialoglycans (remarkably, this chemical reaction can even show this specificity when applied to some intact cell surfaces). Subsequent addition of a tagged nucleophile (e.g., biotin hydrazide) results in a tag (e.g., biotin) covalently bound to Sia that was originally cleaved by periodate. Selective release of most Sias from intact cells or isolated sialoglycans is accomplished using microbial sialidases. Use of α2-3 or α2-8-specific sialidases can provide some insight on Sia linkages as well. Sia-glycosides are also more susceptible to acid hydrolysis than most other sugars, being released using 0.1 M HCl or even with weak acids like 2 M acetic acid or propionic acid at 80°C (albeit still with some loss or migration of labile modifications). This allows selective release of Sias from complex glycans for subsequent analysis by chromatographic methods. Free Sias react with 1,2-diamino-4,5-methylenedioxybenzene (DMB) to generate fluorescent compounds amenable to qualitative and quantitative high-pressure liquid chromatography. There are still some major limitations to methods available for Sia analysis. For example, some mass spectrometric methods result in selective loss of Sias before mass detection. Some Sia-linkages are partly or completely resistant to sialidases and some are even relatively resistant to acid release. In terms of identifying Sia modifications, the most accurate analysis requires complete release and purification with the modifications intact. The stability of each modification to analytical conditions must be considered—for example, O-acetyl groups are labile to common methods that use alkaline conditions (permethylation and beta-elimination).

Sia-binding lectins are also useful for detecting sialoglycans and their linkages in situ and on isolated sialoglycans. *Sambucus nigra* (elderberry) lectin binds selectively to α2-6-linked Sias and *Maackia amurensis* lectins bind selectively to α2-3-linked Sias. Because plants do not express Sias, these lectins may perhaps deter animal ingestion. Alternatively, the natural ligands of these lectins may not be Sias. Improved probes are now being developed, taking advantage of the fact that many microbes have spent millions of years optimizing their binding to vertebrate sialomes (e.g., the B subunits of various toxins). An interesting tool for detecting a specific Sia modification is the hemagglutinin of influenza C or bovine coronavirus, which bind specifically to 9-O-acetyl Sias on isolated sialoglycans, cells, and tissues, if the esterase function is inactivated.

FUNCTIONS OF SIALIC ACIDS

Sias are endowed with a rich diversity of chemical modalities within a single monosaccharide, which evolution uses to modulate the biophysical environment, to mask underlying glycans, and for specific recognition by complementary Sia-binding proteins that mediate biological processes. These functions are not mutually exclusive—for example, highly sialylated molecules with major biophysical roles (such as mucins) may also express specific sialoglycans that serve as ligands in recognition processes. For simplicity, different roles of Sias are addressed separately below.

Biophysical Roles of Sialic Acids

Mucins are heavily glycosylated O-GalNAc-modified sialoglycoproteins secreted or membrane-bound at epithelial surfaces in airways and in the gastrointestinal and urogenital tracts of animals (Chapter 10). Mucins can be remarkably large (several in the range of 5 to >20 MDa). Their dense anionic charge and their propensity to bind water make them efficient as a hydrating and protective barrier at tissue surfaces in contact with the environment. On mucins, as well as other glycoproteins, closely spaced O-GalNAc-linked sialoglycans generate extended polypeptide configurations that would collapse without glycosylation. Clustered sialoglycans along the polypeptide chain can also protect the underlying protein from proteases. Membrane-bound sialoglycans on glycoproteins and glycolipids, which are often quite dense, provide cell surface anionic charge that acts as a barrier and regulates cell surface functions. An extreme example of these biophysical effects is polySia, whose functional impacts are detailed above.

Sialic Acid Recognition by Pathogens and Toxins

Given Sia abundance on cell surfaces (see Figure 15.3), it is not surprising that numerous animal pathogens evolved to target these molecules. Indeed, highly specific Sia-binding proteins were first discovered on human pathogens and their toxins. In the search for host cell receptors for influenza viruses, it was discovered that isolated viruses cross-linked red blood cells, resulting in hemagglutination. Over time, the red cells dispersed and could no longer be agglutinated by fresh viruses. This led to the concept of a viral "hemagglutinin" (a term still used in virus typing; e.g., the "H" in H1N1) that binds to a receptor on the red blood cell, which is also susceptible to a "receptor destroying enzyme" from the virus. The component released by receptor-destroying enzyme was identified as Sia, the hemagglutinin as a Sia-binding protein, and the receptor-destroying enzyme as a sialidase (viral neuraminidase, the "N" in H1N1). Interestingly, the Sia-linkage specificity of the hemagglutinin defines host restrictions; with bird influenza virus preferentially binding to Sias in α2-3-linkage and human virus to Sias in α2-6-linkage. Scientists monitoring influenza virus outbreaks test α2-3- versus α2-6-specificity to detect potential emerging human influenza strains. Molecules designed to block the viral sialidases are useful anti-influenza drugs (Chapter 57).

Sia-specific binding proteins are widespread in pathogens and include numerous viral hemagglutinins, bacterial adhesins, bacterial toxins, and parasite-binding proteins (see Table 15.1 for a few examples). *Helicobacter pylori* causes stomach ulcers and cancers. It expresses a Sia-specific adhesin called SabA, which contributes to its chronic colonization of the stomach lining. The merozoite stage of the malarial parasite *P. falciparum* can bind to erythrocyte

TABLE 15.1. A few classic examples of sialic acid–binding proteins in nature

Vertebrate
C-type: E-, P-, and L-selectin
I-type: Siglecs
Unclassified: Complement factor H
Invertebrates[a]
Crab lectin: Limulin (*Limulus polyphemus*)
Lobster lectin: L-agglutinin (*Homarus americanus*)
Scorpion lectin: (*Mastigoproctus giganteus*)
Insect lectin: Allo A-II (*Allomyrina dichotoma*)
Slug lectin: LFA, *Limax flavus*
Oyster lectin: Pacific oyster lectin (*Crassostrea gigas*)
Snail lectin: Achatinin-H (*Achatina fulica*)
Protozoa
Parasite lectins: Merozoite erythrocyte-binding antigens (EBAs) (*Plasmodium falciparum*)
Plants[a]
Elderberry bark lectin (*Sambucus nigra*)
Maackia agglutinin (*Maackia amurensis*)
Wheat-germ agglutinin (*Triticum vulgaris*)
Bacteria
Bacterial adhesins: S-adhesin (*Escherichia coli* K99), SabA, and SabB (*Helicobacter pylori*)
Bacterial toxins: Cholera toxin (*Vibrio cholerae*), tetanus toxin (*Clostridium tetani*), botulinum toxin (*Clostridium botulinum*), and pertussis toxin (*Bordetella pertussis*)
Mycoplasma lectins: *Mycoplasma pneumoniae* hemagglutinin
Viruses
Influenza hemagglutinins
Reovirus σ1
Adenovirus fiber
Rotavirus VP4
Coronavirus hemagglutinins

[a]The natural ligands for some of these lectins may well be other nonulosonic acids.

sialoglycans to initiate cell entry. Several bacterial toxins also target sialoglycans. Cholera toxin and the structurally related *E. coli* heat-labile enterotoxin both bind to Sias on ganglioside GM1 on the intestinal epithelium, whereas tetanus and related botulinum toxins bind to more complex gangliosides (Chapter 37). The toxin from the human pathogen *Salmonella* Typhi binds specifically to Neu5Ac on glycans but not Neu5Gc; influenza C virus expresses a hemagglutinin that binds only to 9-O-acetylated Sias. The complexities of Sia diversification are thought to be the outcome of an ongoing evolutionary "arms race" between animals and microbial pathogens (Chapter 20). In this regard, it is notable that O-acetyl and N-glycolyl groups on Sias can limit action of bacterial sialidases and block the binding of some pathogens. Alternately, the same modifications may facilitate binding of pathogens that have adapted to them. Sialoglycans at environmental surfaces (like mucins) or free in biological fluids might provide protection by virtue of their "decoy" inhibition of microbial adherence and/or sialidases. The evolutionary persistence of sialoglycans despite their role in virulent diseases suggests critical physiological roles, some of which are addressed below.

Sialic Acid Recognition within Vertebrates

Sias in vertebrates can act as "biological masks," preventing recognition of underlying glycans (especially β-linked Gal residues) by intrinsic and extrinsic glycan-binding proteins. They are also essential components of recognition molecules, with roles that vary according to cell type, tissue type, and species. The first mammalian Sia-binding protein found was the complement regulatory molecule factor H, a soluble serum factor that binds cell-surface Sias and protects cells from autoimmune attack, providing recognition of "self-associated molecular patterns" (SAMPs), which are also recognized by CD33-related Siglecs (Chapter 35). Human mutations in the Sia-binding site of factor H result in atypical hemolytic uremic syndrome in humans, a disease caused by excessive complement activation, because of the failure of factor H function. Other variations of the factor H polyanion binding region increase the risk of age-dependent macular degeneration, a common cause of blindness resulting from inflammation in the eye retina.

Sias are also involved in interactions of white blood cells with the endothelial lining of blood vessels. Circulating neutrophils must bind to and move across the blood vessel wall to clear bacterial infections in tissues. Endothelial cells respond to nearby bacterial infection and quickly send Sia-binding lectins to their surface. E- and P-selectins (Chapter 34) connect with passing neutrophils, which have the complementary sialoglycans on their surface lipids or proteins. Initially rolling, and finally snagged, neutrophils then respond to other signals (including cytokines presented by GAGs; Chapter 17) and work their way into the tissue to fight bacteria. Without Sia-dependent binding by E- and P-selectins, inflammation is compromised and persistent tissue infections ensue. A related lectin, L-selectin, is also involved in trafficking of lymphocytes from blood to lymph.

Siglecs (Sia-binding immunoglobulin-like lectins) constitute a Sia-binding family within the I-type lectins (Chapter 35). In humans, there are 14 Siglecs with sialic acid–dependent functions (13 and 17 are missing in humans and 12 does not bind sialic acid), all but one of which are found on the surface of different blood cell types. In several cases, binding of Siglecs to their sialoglycan targets on other cells or their own cell surface modulates ongoing immune responses, either damping them to protect from hyperimmune reactions or activating them. The one Siglec not found in the immune system, Siglec-4 (myelin-associated glycoprotein), is in the nervous system, where it aids cell–cell interactions between nerve cells and myelin, the protective and insulating membrane sheath essential to rapid nerve conduction.

Notably, all the above examples of vertebrate Sia-recognizing proteins were discovered serendipitously, via the unexpected abrogating effect of sialidases on a biological process. To date, there has been no systematic effort to search for other examples, but other likely vertebrate Sia-binding lectins are PECAM-1, PILRs, L1-CAM, HMGB-1, and an as-yet-unidentified uterine agglutinin.

The many functions of Sias in vertebrates make these molecules particularly attractive targets for microbes that co-exist with the host, whether harmoniously or not. As discussed elsewhere, there are several independent mechanisms of Sia mimicry, which promote bacterial interactions with host factor H and Siglecs to enable survival in the host.

Lectins in Organisms without Sialic Acids

Some Sia-binding lectins are found in organisms that do not themselves appear to express Sias (see Table 15.1 for examples). In these cases, Sia binding may defend against sialylated pathogens, such as the protein limulin in the hemolymph (blood-like fluid) of the horseshoe crab, which can trigger foreign cell lysis. Sia-binding lectins in plants, such as elderberry shrubs, may suppress consumption by animals. Of course, some of these Sia-binding properties might be incidental, with the true lectin ligands being related to prokaryotic NulOs such as Leg or Pse, and/or Kdo, which is also found in some plants.

SIALIC ACIDS IN DEVELOPMENT AND MALIGNANCY

Although cell lines deficient in Sias grow in culture, disruption of Sia synthesis in mice results in lethality at embryonic day 8.5, with poor differentiation of nerve cells as well as cardiac and skeletal muscle cells. In contrast, disruption of some ST genes in mice is relatively well tolerated, in some cases because of complementarity among related genes. For example, both *St8sia2* and *St8sia4* must be deleted in mice to fully eliminate polySia, resulting in severe developmental deficits in the brain. Likewise, both *St3sia2* and *St3sia3* must be deleted to fully block terminal sialylation of brain gangliosides, resulting in early motor and behavioral deficits. Deleting both *St3gal4* and *St3gal6*, but not either one separately, results in loss of leukocyte rolling on selectins. One benefit of having multiple genes and enzymes that make a Sia linkage (see Figure 15.2) is that the results of single gene mutations are tempered. In contrast, human mutations in *ST3GAL5* alone result in severe early seizures with profound deficits in postnatal cognitive and motor development, mutations in the *ST3GAL3* gene result in intellectual impairment, and mutation of human *ST8SIA2* is associated with mental illness. Mutations of other STs in mice cause defects in immune system development and function and/or fertility. In Drosophila, mutation of the single ST results in severe phenotypes.

Cellular and environmental regulation of Sia fine-structure implies roles for Sia modifications. Certain classes of lymphocytes have O-acetylated Sias, and mutations in a sialic acid esterase results in altered immune function. Expression of polySia and O-acetylation of brain gangliosides varies with developmental stage and location. Differences in O-acetylation of brain gangliosides are reported between cold- and warm-blooded species and between awake and hibernating animals. Developmental regulation of Neu5Gc expression and O-acetylation expression in the gut mucosa occurs in response to microbial colonization and has been suggested to protect against microorganisms. Similarly, although adult bovine submandibular glands produce large amounts of highly O-acetylated mucins, this modification is scarcely present in the corresponding fetal tissue. The types and linkages of endothelial, plasma protein, and erythrocyte Sias can undergo marked changes in responses to inflammatory stimuli. Abnormalities have also been reported in mice depleted in 9-O-acetyl Sias (transgenic for a viral 9-O-acetylesterase). The physiological roles of these Sia modifications are not well-established, and there is much yet to be learned.

Certain changes in Sias are also characteristic of cancer. In general, total Sia increases and there are quantitative changes in linkages, with α2-6Gal(NAc) linkages becoming particularly prominent. O-Acetylation of Sia at C-9 can disappear (colon carcinomas) in concert with increased levels of O-acetylesterase activity in fecal specimens. Alternatively, 9-O-acetyl-GD3 becomes more prominent in melanomas and basal cell carcinomas. Increased expression of polySia may also facilitate cell migration by some cancers. The precise mechanisms by which Sia changes enhance tumorigenesis and/or invasive behavior remain uncertain, although it has been proposed that tumor sialoglycans enhance metastatic dissemination by their ability to alter functions of galectins (Chapter 36) and by enhancing the binding of tumor cells to endothelial, leukocyte, and platelet selectins (Chapter 34). Increased sialylation may also mask antigenic sites on tumor cells, which also become more like "self" and evade immune surveillance by engaging factor H and/or Siglecs. Regardless of the mechanisms involved, certain sialylated molecules are also specific markers for some cancers and potential ligands for targeted therapies (Chapter 47).

SIALIC ACIDS IN PHARMACOLOGY

Each role of Sias in physiology and pathology provides opportunities for therapeutic development. Best known are competitive inhibitors of viral sialidases (Relenza and Tamiflu), which hinder budding and spreading of susceptible strains of influenza A and B viruses (Chapters 42 and 57). Another strategy to block Sia-binding pathogens has been to engineer therapeutic sialidases to remove pathogen-binding sites on human tissues. Efforts to use sialoglycans themselves as anti-infectives have not been as successful, although sialylated milk oligosaccharides have been claimed to have some value. Sia-mimetic drugs designed to bind to and block selectins (e.g., rivipansel) have shown promise in human trials as anti-inflammatory drugs. Strategies are also under study for inducing immune attack of cancer cells with sialoglycoconjugates and for suppressing antibody responses by engaging Siglecs on B cells.

Many sialoglycoproteins have been developed as drugs, including erythropoietin to enhance blood cell production and monoclonal antibodies to target various diseases including cancers (Chapter 57). In these cases, appropriate sialylation is a desired goal, not only to optimize serum half-life, stability, and receptor binding, but also to assure FDA approval. Chemical, recombinant, and other methods are thus in use to maintain appropriate sialylation on biological drugs, and efforts are underway to engineer "humanized" glycosylation (including sialylation) in yeast, insect, or plant cells (that do not typically express Sias but are more economical sources of expressed glycoproteins) (Chapter 56). Notably, differences in sialylation of endogenous and exogenous erythropoietin are used by the World Anti-Doping Agency to detect illicit administration, and have resulted in rescinding of many Olympic and Tour de France medals.

The presence of Neu5Gc is of potential importance in biotherapeutics, because humans have circulating anti-Neu5Gc antibodies, with marked variations in levels and isotypes. The use of animal-derived proteins in biopharmaceutical development can lead to Neu5Gc incorporation, with unwanted results. Similarly, cell therapies under development often use animal sera or support cells, which can provide Neu5Gc to the cell products. Finally, Neu5Gc on animal organs likely contributes to the failure of xenotransplantation (use of animal organs in humans), which led to engineering of *CMAH* knockout pigs.

EVOLUTIONARY DISTRIBUTION OF SIALIC ACIDS

Early studies suggested species specificity in the occurrence of specific Sia variants. With analytical improvements, it became evident that Sia variations are more widely expressed across species but occur at differing levels. Sias are prominent in the deuterostome lineage (Chapter 27), which comprises vertebrates and some “higher” invertebrates (such as echinoderms). Indeed, with exceptions, Sias are not generally reported in plants or invertebrates. This may change as molecular techniques improve. For example, insects were believed to be free of Sias until their discovery in cicada Malpighian tubules and Drosophila brains, and later identification of a single Drosophila gene (*SiaT*) with sequence similarities to mammalian STs. When *SiaT* is inactivated, flies suffer locomotor abnormalities and defects in neuromuscular junctions. Although of very low abundance, Sias are both present and crucial to the survival of Drosophila. Sias have also been reported in the nervous systems of behaviorally complex protostomes like octopus and squid. On the other hand, no Sias or ST genes were found in the well-studied round worm, *Caenorhabditis elegans* (Chapter 25).

It appears that Sias (other than Kdn, which probably appeared earlier) were “invented” in a common ancestor of protostomes and deuterostomes and then became essential in deuterostomes, while being partially or completely discarded in some protostome lineages. Interestingly, there is wide variation in Sia expression and complexity within deuterostomes, with the sialome of echinoderms appearing very complex and that of humans being among the simplest. Although expression of Neu5Gc and 9-O-acetylated Sias is very common in deuterostomes, exceptions exist, such as independent loss of Neu5Gc biosynthesis in humans, New World monkeys, Sauropsids (birds and reptiles, the descendants of dinosaurs), the Pinnipedia and Musteloidia members of the Carnivora, and likely in Monotremes.

Sias in capsular polysaccharides and lipopolysaccharides can have beneficial effects for microbes in their hosts, protecting them from complement activation and/or antibody recognition of underlying glycans, also sometimes engaging the Siglec family receptors (Chapter 35) to dampen innate immune cell reactivity. The bacterial enzymes involved in synthesizing and metabolizing Sias do not appear to be the result of lateral transfer of genes from animals, but evolved independently, apparently being “reinvented” at least twice from a more ancient prokaryotic NulO pathway. In fact, Sias that are synthesized by some present-day vertebrate pathogens share closer phylogenetic relationships to biosynthetic pathways for legionaminic acids than to Sia pathways of vertebrates. Other microbes “steal” Sias from their host using various mechanisms to achieve “molecular mimicry.” *Neisseria gonorrheae* even has a remarkably efficient surface ST that can scavenge trace amounts of CMP-Neu5Ac from body fluids of its exclusive host, humans.

PROKARYOTIC NONULOSONIC ACIDS

A diverse array of Sia-like molecules (see Figure 15.1) with ancient evolutionary roots has been described in Bacteria and archaea and the relevant biosynthetic pathways are predicted to occur in ~20% of prokaryotes. These molecules differ from Sias in several ways. First, they lack a hydroxyl group at C9 and nearly all described forms have amino functions at C7 that are often decorated with substituents.

Di-*N*-acetyllegionaminic acid (Leg5,7Ac_2) is a common well-characterized microbial NulO initially discovered in the LPS of *Legionella pneumophila*. Like Neu5Ac, it has a D-glycero-D-galacto stereochemistry in the backbone, but with an amino function instead of a hydroxyl at C-7 and no hydroxyl at C-9. Di-*N*-acetylpseudaminic acid (Pse5,7Ac_2), first described in *Pseudomonas aeruginosa*, has a similar structure to Leg5,7Ac_2 but with opposite stereochemical configurations at C-5, C-7, and C-8 (Figure 15.1). Later discoveries of NulOs include 4epi-di-*N*-acetyllegionaminic acid (4eLeg5,7Ac_2), 8epi-di-*N*-acetyllegionaminic acid (8eLeg5,7Ac_2), di-*N*-acetylacinetaminic acid (Aci5,7Ac_2), 8epi-di-*N*-acetylacinetaminic acid (Figure 15.1). Following the earlier tradition of naming NulO classes by the organism in which they were first

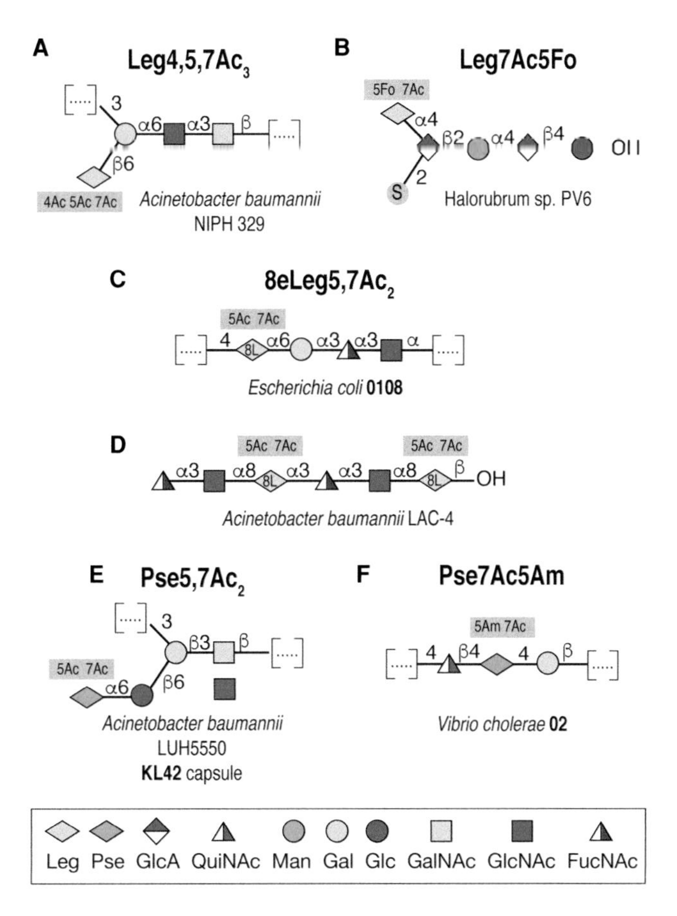

FIGURE 15.5. Examples of legionaminic and pseudaminic acid–containing glycans in bacteria and Archaea, selected from the Bacterial Carbohydrate Structure Database (BCSDB). The sialic acid–like monosaccharides legionaminic acid (Leg) (*A,B*), 8-epi-legionaminic acid (8eLeg) (*C,D*), and pseudaminic acid (Pse) (*E,F*) are displayed in terminal as well as internal positions of polysaccharides from bacteria and Archaea. The BCSDB identifiers of these structures are 1047, 25389, 30524, 30859, and 115488. Inside the Leg symbol, "8L" refers to the 8-epimer. Ac, acetyl; Fo, formyl; and Am, acetamidino.

reported, acinetaminic acid was first described in *Acinetobacter baumannii*. Recently, another novel NulO was reported in *Fusobacterium nucleatum*. Although the configuration is still tentatively defined, "fusaminic acid" was coined to refer to the new NulO—a 4-epimer of pseudaminic acid with a hydroxyl instead of an amino function at C7. NulOs are rather unusual among monosaccharides in being activated as CMP-linked sugars. Interestingly, the 8-carbon α-keto acid monosaccharide 2-keto-3-deoxyoctonic acid (Kdo; see Chapter 21) is also activated in a CMP-linked form and the enzymes involved in activation are homologous. Although the distribution of Kdo is limited to Gram-negative bacteria and plants, NulOs are more widespread among bacterial taxa and archaea. Together, the evidence suggests that NulO biosynthetic pathways are an ancient invention by the common ancestor of all lifeforms, but persisted only in some taxa. Sias were likely "invented" early in life on earth. They appear in bacteria from diverse taxa with varied lifestyles in the environment or with animals, including "lower" animals often in internally linked positions that do not mimic known structures in "higher" animals. Sias became prominent in animals when the Deuterostome lineage emerged at the Cambrian Explosion ~530 million years ago. Sequence homology suggests that ancient prokaryotic NulO biosynthetic pathways later became an evolutionary template that allowed some microbes to reinvent Sias of "higher" animals by convergent evolution. Many important commensals and pathogens use Sia mimicry to evade the vertebrate immune system by one or more of several established mechanisms. Some of the prokaryotic NulOs that share the same stereochemistry as Sias can engage in mechanisms of Sia mimicry, at times conveying pathogenic properties in unintended hosts following environmental exposure or zoonosis. Some examples of bacterial NulO-containing structures are provided (Figure 15.5).

ABBREVIATED NAMES OF NONULOSONIC ACIDS

Complete chemical names of NulOs are rather cumbersome for routine use. For example, the name of the closed ring Neu5Ac is 5-acetamido-3,5-dideoxy-D-*glycero*-D-*galacto*-non-2-ulopyranosonic acid. Beyond the currently known core units (Figure 15.1), additional substitutions are designated by letter codes (Ac, acetyl; Gc, glycolyl; Me, methyl; Lt, lactyl; and S, sulfate), and these are listed along with numbers indicating their location relative to the carbons (e.g., 9-*O*-acetyl-8-*O*-methyl-*N*-acetylneuraminic acid is Neu5,9$Ac_2$8Me). Among NulOs as a whole, there are now at least 138 known structural variants. If one is uncertain of the exact type of molecule present, the generic abbreviations Sia or NulO should be used. Partial structural information can also be incorporated. For example, a Sia of otherwise unknown type with an *O*-acetyl substitution at the C-9 position could be written as Sia9Ac. With regard to the bacterial NulOs, the names Leg, Pse, 4eLeg, 8eLeg, and Aci are often used synonymously with the 5,7-di-N-acetylated forms of these molecules (Figure 15.1). To be consistent with Sia nomenclature (as with Neu) and to avoid confusion, it is recommended that the substituted forms be specified whenever possible (e.g., Leg5,7Ac_2). Free amino forms of the prokaryotic NulOs have not to our knowledge been reported in nature. Instead, many different types of substitutions of the C-5 and C-7 positions have been reported for the exclusively prokaryotic NulOs, including acetyl (Ac), formyl (Fo), (*R*)-hydroxybutyryl (3_RHb), (*S*)-3-hydroxybutyryl (3_SHb), 4-hydroxybutyryl (4Hb), 3,4-dihydroxybutyryl (3,4Hb), acetimidoyl (Am), *N*-methyl-acetimidoyl (AmMe), methyl (Me), D-alanyl (Ala), *N*-acetyl-D-alanyl (AlaNAc), *N*-methyl-5-glutamyl (GluNMe), L-glyceryl (Gr), and/or 2,3-di-*O*-methyl-glyceryl (Me2Gr) groups. Additional variety derives from acetyl groups at O4 and O8, as well as acetyl, *N*-acetyl-glutaminyl (GlnNAc), and glycyl (Gly) substitutions at O8.

ACKNOWLEDGMENTS

The authors acknowledge contributions to previous versions of this chapter by the late Roland Schauer and appreciate helpful comments and suggestions from Hamed Jafar-Nejad.

FURTHER READING

Blix FG, Gottschalk A, Klenk E. 1957. Proposed nomenclature in the field of neuraminic and sialic acids. *Nature* **179:** 1088. doi:10.1038/1791088b0

Schauer R. 1982. Chemistry, metabolism, and biological functions of sialic acids. *Adv Carbohydr Chem Biochem* **40:** 131–234. doi:10.1016/s0065-2318(08)60109-2

Taylor G. 1996. Sialidases: structures, biological significance and therapeutic potential. *Curr Opin Struct Biol* **6:** 830–837. doi:10.1016/s0959-440x(96)80014-5

Kelm S, Schauer R. 1997. Sialic acids in molecular and cellular interactions. *Int Rev Cytol* **175:** 137–240. doi:10.1016/s0074-7696(08)62127-0

Knirel YA, Shashkov AS, Tsvetkov YE, Jansson PE, Zähringer U. 2003. 5,7-Diamino-3,5,7,9-tetradeoxynon-2-ulosonic acids in bacterial glycopolymers: chemistry and biochemistry. *Adv Carbohydr Chem Biochem* **58:** 371–417. doi:10.1016/s0065-2318(03)58007-6

Toukach P, Joshi H, Ranzinger R, Knirel Y, von der Leith C. 2007. Sharing of worldwide distributed carbohydrate-related digital resources: online connection of the Bacterial Carbohydrate Structure Database and GLYCOSCIENCES.de. *Nucleic Acids Res* **35:** D280–D286. doi:10.1093/nar/gkl883

Lewis AL, Desa N, Hansen EE, Knirel YA, Gordon JI, Gagneux P, Nizet V, Varki A. 2009. Innovations in host and microbial sialic acid biosynthesis revealed by phylogenomic prediction of nonulosonic acid structure. *Proc Natl Acad Sci* **106:** 13552–13557. doi:10.1073/pnas.0902431106

Schauer R. 2009. Sialic acids as regulators of molecular and cellular interactions. *Curr Opin Struct Biol* **19:** 507–514. doi:10.1016/j.sbi.2009.06.003

Chen X, Varki A. 2010. Advances in the biology and chemistry of sialic acids. *ACS Chem Biol* **5:** 163–176. doi:10.1021/cb900266r

Xu G, Kiefel MJ, Wilson JC, Andrew PW, Oggioni MR, Taylor GL. 2011. Three *Streptococcus pneumoniae* sialidases: three different products. *J Am Chem Soc* **133:** 1718–1721. doi:10.1021/ja110733q

Li Y, Chen X. 2012. Sialic acid metabolism and sialyltransferases, natural functions and applications. *Appl Microbiol Biotechnol* **94:** 887–905. doi:10.1007/s00253-012-4040-1

Miyagi T, Yamaguchi K. 2012. Mammalian sialidases: physiological and pathological roles in cellular functions. *Glycobiology* **22:** 880–896. doi:10.1093/glycob/cws057

Varki A, Gagneux P. 2012. Multifarious roles of sialic acids in immunity. *Ann NY Acad Sci* **1253:** 16–36. doi:10.1111/j.1749-6632.2012.06517.x

Petit D, Teppa RE, Petit JM, Harduin-Lepers A. 2013. A practical approach to reconstruct evolutionary history of animal sialyltransferases and gain insights into the sequence–function relationships of Golgi-glycosyltransferases. *Methods Mol Biol* **1022:** 73–97. doi:10.1007/978-1-62703-465-4_7

Tiralongo J, Martinez-Duncker I. 2013. *Sialobiology: structure, biosynthesis and function*. Bentham, Oak Park, IL. doi:10.2174/97816080538651130101

Schnaar RL, Gerardy-Schahn R, Hildebrandt H. 2014. Sialic acids in the brain: gangliosides and polysialic acid in nervous system development, stability, disease, and regeneration. *Physiol Rev* **94:** 461–518. doi:10.1152/physrev.00033.2013

Stencel-Baerenwald JE, Reiss K, Reiter DM, Stehle T, Dermody TS. 2014. The sweet spot: defining virus–sialic acid interactions. *Nat Rev Microbiol* **12:** 739–749. doi:10.1038/nrmicro3346

Chen X. 2015. Human milk oligosaccharides (HMOS): structure, function, and enzyme-catalyzed synthesis. *Adv Carbohydr Chem Biochem* **72:** 113–190. doi:10.1016/bs.accb.2015.08.002

Lundblad A. 2015. Gunnar Blix and his discovery of sialic acids. Fascinating molecules in glycobiology. *Ups J Med Sci* **120:** 104–112. doi:10.3109/03009734.2015.1027429

Toukach PV, Egoroca KS. 2016. Carbohydrate structure database merged from bacterial, archaeal, plant and fungal parts. *Nucleic Acids Res* **44:** D1229–D1236. doi:10.1093/nar/gkv840

Pearce OM, Läubli H. 2017. Sialic acids in cancer biology and immunity. *Glycobiology* **26:** 111–128. doi:10.1093/glycob/cwv097

Wasik BR, Barnard KN, Parrish CR. 2017. Effects of sialic acid modifications on virus binding and infection. *Trends Microbiol* **24:** 991–1001. doi:10.1016/j.tim.2016.07.005

Sato C, Kitajima K. 2020. Polysialylation and disease. *Mol Aspects Med* **27:** 100892. doi:10.1016/j.mam.2020.100892

16 Hyaluronan

Melanie Simpson, Liliana Schaefer, Vincent Hascall, and Jeffrey D. Esko

Animal cells and some bacteria produce hyaluronan, a high-molecular-weight, nonsulfated glycosaminoglycan synthesized at the cell surface and extruded into the extracellular environment. This chapter describes the structure and metabolism of hyaluronan, its chemical and physical attributes, and its highly diverse and versatile biological functions.

HISTORICAL AND EVOLUTIONARY PERSPECTIVES

Sulfated glycosaminoglycans were first isolated in the late 1800s, and the isolation of hyaluronic acid (now called hyaluronan) followed in the early 1930s. In their classic paper, Karl Meyer and John Palmer named the "polysaccharide acid of high molecular weight" that they purified from bovine vitreous humor as "*hyaluronic acid*" (from hyaloid, meaning vitreous), and they showed that it contained "uronic acid (and) an amino sugar." It took almost 20 years to determine the actual structure of the repeating disaccharide motif (GlcNAcβ4GlcAβ3) of hyaluronan (Figure 16.1). In contrast to the other classes of glycosaminoglycans, hyaluronan is not further modified by sulfation or by epimerization of the glucuronic acid moiety to iduronic acid (Chapter 17). Thus, the chemical structure shown in Figure 16.1 is faithfully reproduced by any cell that synthesizes hyaluronan, including animal cells and bacteria.

The simplicity of hyaluronan might suggest that it arose early in evolution relative to other glycosaminoglycans. However, this is not the case, because *Drosophila melanogaster* and

published by Cold Spring Harbor Laboratory Press; doi: 10.1101/glycobiology.4e.16

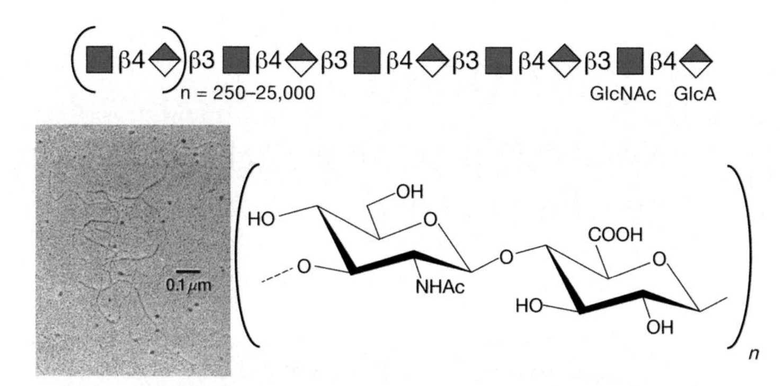

FIGURE 16.1. Hyaluronan consists of repeating disaccharides composed of *N*-acetylglucosamine (GlcNAc) and glucuronic acid (GlcA). It is the largest polysaccharide found in vertebrates, and it forms hydrated matrices. (Electron micrograph provided by Dr. Richard Mayne and Dr. Randolph Brewton, University of Alabama at Birmingham.)

Caenorhabditis elegans do not contain the necessary synthases for its assembly (Chapters 25 and 26). Instead, it appears that hyaluronan arose during the evolution of the notochord shortly before or concurrent with the advent of cartilage and appendicular skeletons apparently as a paralog of the more ancient cell-surface enzymes producing other β-linked polymers like cellulose and chitin (Chapters 24 and 26). Virtually all cells from vertebrate species can produce hyaluronan, and its expression correlates with tissue expansion and cell motility. As discussed below, hyaluronan has essential roles in development, tissue architecture, cell proliferation, signaling reactions across the plasma membrane, inflammation, and microbial virulence.

STRUCTURE AND BIOPHYSICAL PROPERTIES

Hyaluronan has an indefinite and very high degree of polymerization, typically in the range of 10^4 disaccharides, with an end-to-end length of ~10 μm (~1 nm/disaccharide). Thus, a single molecule of hyaluronan could stretch about halfway around the circumference of a typical mammalian cell. The carboxyl groups on the glucuronic acid residues (pK_a 4–5) are negatively charged at physiological pH and ionic strength, making hyaluronan polyanionic. The anionic nature of hyaluronan together with spatial restrictions around the glycosidic bonds confer a relatively stiff, random coil structure to individual hyaluronan molecules in most biological settings. Hyaluronan chains occupy a large hydrodynamic volume such that in a solution containing 3–5 mg/mL hyaluronan, individual molecules occupy essentially all of the solvent. This arrangement creates a size-selective barrier in which small molecules can diffuse freely, whereas larger molecules are partially or completely excluded. Additionally, this solution shows high viscosity with viscoelastic properties, conditions found in the vitreous humor and in synovial fluid of joints. Hyaluronan in synovial fluids of articular joints is essential for distributing load during joint motion and for protecting the cartilaginous surfaces. Thus, in both eye and joint tissues, the physical properties of hyaluronan relate directly to tissue function.

BIOSYNTHESIS

Hyaluronan biosynthesis is catalyzed by hyaluronan synthases (HASs) (Figure 16.2). The first bona fide HAS gene (*spHAS*) was cloned from *Streptococcus*, and the protein expressed in *Escherichia coli* was shown to synthesize high-molecular-weight hyaluronan from the UDP-sugar substrates. The gene shows homology with a *Xenopus* gene, *DG42* (now known as *xlHAS1*; Chapter 27). The homology was instrumental in the subsequent identification of the three members of the

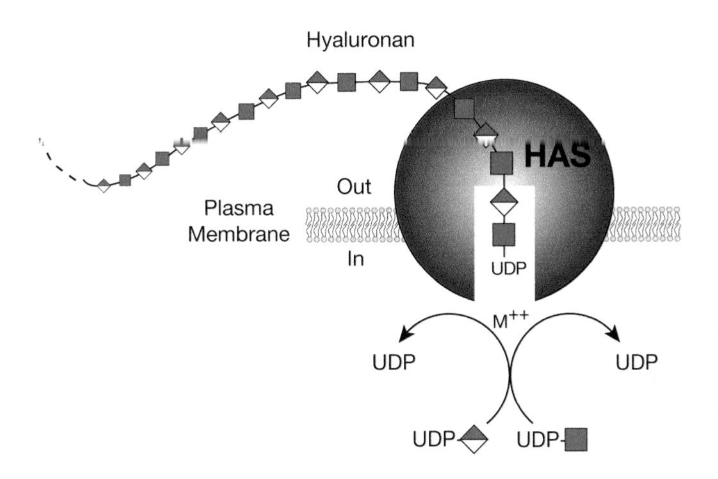

FIGURE 16.2. Hyaluronan biosynthesis by hyaluronan synthase (HAS) occurs by addition of UDP-sugars (UDP-GlcNAc and UDP-GlcA) to the reducing end of the polymer with release of the anchoring UDP. M^{++} refers to a metal ion cofactor.

mammalian *HAS* gene family, *HAS1–3*. These genes code for homologous proteins predicted to contain five to six membrane-spanning segments and a central cytoplasmic domain.

As described in Chapter 17, cells synthesize sulfated glycosaminoglycans (heparan sulfate, chondroitin sulfate, and keratan sulfate) on core proteins of proteoglycans as they transit through the Golgi, and elongation of the chains occurs at their nonreducing ends. In contrast, hyaluronan synthesis normally occurs at the inner surface of the plasma membrane in eukaryotic cells and at the cytoplasmic membrane of bacteria that produce hyaluronan capsules. The synthases use the cytosolic substrates UDP-GlcA and UDP-GlcNAc and extrude the growing polymer through the membrane to form extracellular matrices (Figure 16.2). According to the model, the reducing end of the growing chain would have a UDP moiety that is displaced when the next nucleotide sugar is added. In keeping with its evolutionary history, it appears HA synthesis is initiated by production of a chitin oligosaccharide $(\text{GlcNAc}\beta4\text{GlcNAc})_n$.

In typical cultured mammalian cells, division under conditions of hyperglycemia (two to four times normal glucose level) results in hyaluronan synthesis in the endoplasmic reticulum (ER), Golgi, and transport vesicles. Under these conditions, the elongating hyaluronan chains are inserted inappropriately into these compartments, inducing abnormalities in cellular functions (e.g., kidney nephropathy and proteinurea). The activity of the HAS enzymes also can be regulated by phosphorylation and the addition of O-GlcNAc (Chapter 19).

Hyaluronan biosynthesis in bacteria involves the expression of multiple enzymes, usually as an operon. For example, in *Streptococcus*, *hasC* encodes an enzyme that makes UDP-Glc from UTP and glucose-1-P; *hasB* encodes the dehydrogenase that converts UDP-Glc to UDP-GlcA; *hasD* generates UDP-GlcNAc from glucosamine-1-P, acetyl CoA, and UTP; and *hasA* (*spHas*) encodes the hyaluronan synthase. The *Streptococcus hasA* gene encodes a bifunctional protein that contains both transferase activities and assembles the polysaccharide from the reducing end. The synthase spans the membrane multiple times, presumably forming a pore for hyaluronan extrusion during capsule formation. In contrast, *Pasteurella* synthesizes hyaluronan by an enzyme that is unrelated to *hasA* and the mammalian *Has* gene family. In this case, the enzyme has two separable domains with independent glycosyltransferase activities—one for UDP-GlcNAc and the other for UDP-GlcA, and the elongation is on the nonreducing end.

THE HYALURONIDASES AND HYALURONAN TURNOVER

Animal cells express a set of catabolic enzymes that degrade hyaluronan. The human hyaluronidase gene (*HYAL*) family is complex, with two sets of three contiguous genes located on two chromosomes, a pattern that suggests two ancient gene duplications followed by a block duplication. In humans, the cluster on chromosome 3p21.3 (*HYAL1*, *2*, and *3*) appears to have major roles in somatic tissues. *HYAL4* in the cluster on chromosome 7q31.3 codes for a protein that appears to have chondroitinase, but not hyaluronidase, activity; *PHYAL1* is a pseudogene; and *SPAM1* (sperm adhesion molecule 1, PH-20) is restricted to testes. The role of SPAM1 in fertilization is discussed below. Two additional enzymes with hyaluronan depolymerizing activity at neutral pH have been subsequently identified in multiple tissues and termed TMEM2 (transmembrane protein 2) and KIAA1199 (CEMIP, cell migration–inducing protein).

The turnover of hyaluronan in most tissues is rapid (e.g., a half-life of ~1 day in epidermal tissues), but its residence time in some tissues can be quite long and dependent on location (e.g., in cartilage). It has been estimated that an adult human contains ~15 g of hyaluronan and that about one-third turns over daily. Turnover appears to occur by receptor-mediated endocytosis and lysosomal degradation either locally or after transport by lymph to lymph nodes or by blood to liver. The endothelial cells of the lymph node and liver sinusoids remove hyaluronan via specific receptors, such as LYVE-1 (a homolog of CD44) and HARE (hyaluronan receptor for endocytosis). HARE appears to be the major clearance receptor for hyaluronan delivered systemically by lymph and blood. The current understanding of this catabolic process is that hyaluronidases at the cell surface and in the lysosome cooperate to degrade the chains. Large hyaluronan polymers in the extracellular space can be internalized following interaction with cell-surface receptors that trigger endocytosis, which may be facilitated by the membrane-associated hyaluronidases, HYAL2 and/or TMEM2, or through association with CEMIP. The hyaluronan fragments may be returned to the extracellular space via recycling endosomes or enter a pathway to lysosomes for complete degradation to monosaccharides, probably involving HYAL1 and the two exoglycosidases β-glucuronidase and β-N-acetylglucosaminidase. The importance of the turnover process for growth and survival is demonstrated by the respective embryonic lethal phenotypes in animal models. *Hyal2*-null mice and Tmem2-deficient zebrafish both suffer embryonic lethality as a result of cardiac developmental defects. CEMIP is important for clearing hyaluronan accumulation during bacterial infection in mice. In humans, a lysosomal storage disorder was found to result from mutation of *HYAL1* (Chapter 44).

Hyaluronan fragments have been suggested to act as an endogenous signal of injury, or during infection by Group A *Streptococcus*, which contains a hyaluronan capsule. The signaling activity of hyaluronan fragments is mediated through binding of cell-surface receptors, such as CD44, which in turn modulates response through Toll-like receptors. Signaling through these and other receptors is affected by the size of the hyaluronan fragments, but the mechanism underlying the dependence of activity on the size of the fragments remains an area of active research.

HYALURONAN FUNCTION IN THE EXTRACELLULAR MATRIX

Hyaluronan has multiple roles in early development, tissue organization, and cell proliferation. The *Has2*-null mouse shows an embryonic lethal phenotype at the time of heart formation, whereas *Has1*-, *Has3*-null, and *Has1/3* compound mutant mice show no obvious developmental phenotype. Interestingly, explanted cells from the *Has2*-null embryonic heart do not synthesize hyaluronan or undergo epithelial–mesenchymal transformation and migration unless small amounts of hyaluronan are added to the culture medium. This finding indicates that the production of hyaluronan at key points may be essential for many tissue morphogenetic transformations—in this case, formation of the tricuspid and mitral valves.

Many of the activities of hyaluronan depend on binding proteins present on the cell surface and/or secreted into the extracellular matrix. A class of proteins that bind selectively to hyaluronan was first discovered in cartilage. This class is now referred to as the link module family of hyaladherins (Figure 16.3). Proteoglycans were efficiently extracted from this tissue with denaturing solvents and were shown to reaggregate when restored to renaturing conditions. An essential protein, referred to as the link protein (HAPLN-1), was shown to be necessary for stabilizing the proteoglycan aggregates, and subsequently, the structure of the aggregate was defined (Figure 16.4). The link protein contains two homologous repeats of a sequence motif, now named the link module. Proteins that have a link module, including link proteins (HAPLN-1 through HAPLN-4 in humans), several proteoglycans, and other extracellular matrix proteins, can interact specifically with hyaluronan. The major cartilage proteoglycan, now named aggrecan (Chapter 17), also contains a globular domain, the G1 domain, with two homologous link modules that interact with hyaluronan. An additional domain in HAPLN-1 cooperatively interacts with a homologous domain in G1, which locks the proteoglycan on the hyaluronan chain. In the absence of the HAPLN-1, aggrecan fails to anchor to hyaluronan. Mice deficient in HAPLN-1 show defects in cartilage development and delayed bone formation (short limbs and craniofacial anomalies). Most mutant mice die shortly after birth as a result of respiratory failure, and the few survivors develop progressive skeletal deformities.

Interestingly, there are four proteoglycan genes with homologous G1 domains that interact with hyaluronan (versican, neurocan, brevican, and aggrecan) (Figure 16.3). Versican is a major component of many soft tissues and is especially important in vascular biology. Neurocan and brevican are expressed predominantly in brain tissue. Versican and aggrecan are anchored to hyaluronan in tissues by similar link protein-dependent mechanisms, and it is likely that

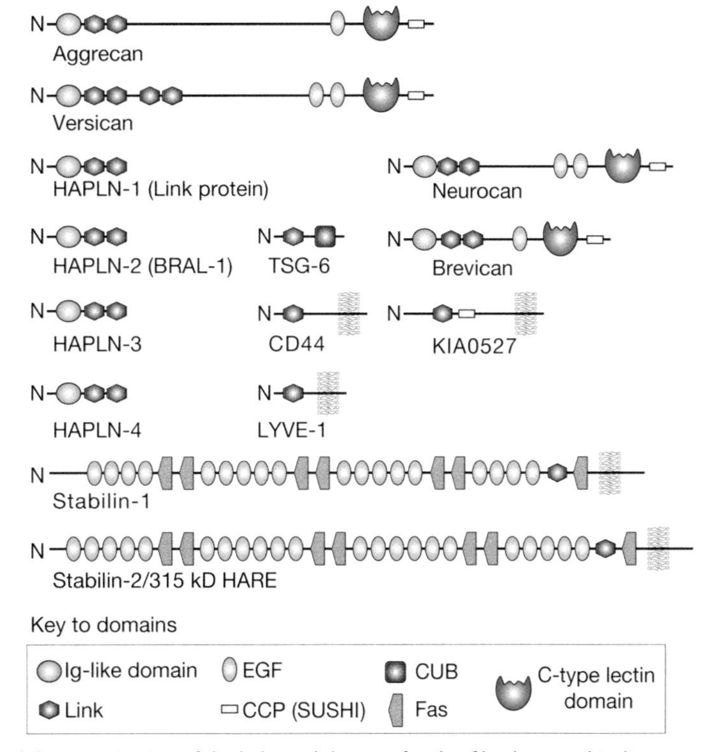

FIGURE 16.3. Modular organization of the link module superfamily of hyaluronan-binding proteins. These proteins contain one or two link modules that bind to hyaluronan. Like many extracellular matrix proteins, the link module superfamily members contain various subdomains, depicted by the following annotated symbols: IG, immunoglobulin-like domain; EGF, epidermal growth factor–like domain; CCP, complement control protein module; Link, hyaluronan-binding module; CUB, domain found in some complement proteins, peptidases, and bone morphogenetic protein; FAS, domain found in fasciclin I family of proteins. (See the SMART database at EMBL for additional information on these domains: http://smart.embl-heidelberg.de/.) (Redrawn, with permission, from Blundell CD. 2004. In *Chemistry and biology of hyaluronan* [ed. Garg HG, Hales CA], pp. 189–204, © Elsevier.)

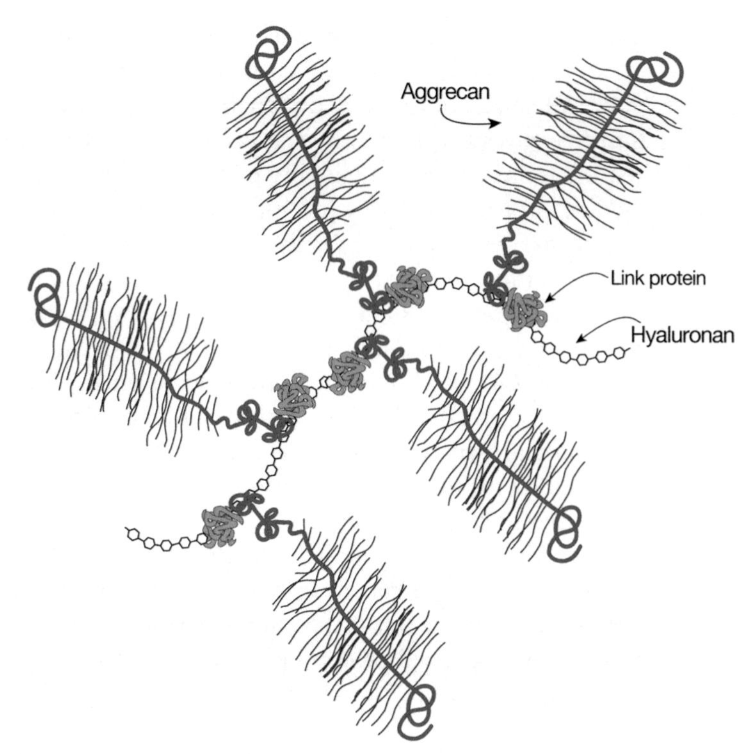

FIGURE 16.4. The large cartilage chondroitin sulfate (CS) proteoglycan (aggrecan) forms an aggregate with hyaluronan and link protein.

neurocan and brevican are organized similarly. In the brain, these complexes form perineuronal nets. Thus, hyaluronan acts as a scaffold on which to build proteoglycan aggregate structures adapted to diverse tissue functions.

An impressive example of the requirement for a hyaluronan-based matrix occurs during the process of cumulus oophorus expansion in the mammalian preovulatory follicle. At the beginning of this process, the oocyte is surrounded by about 1000 cumulus cells tightly compacted and in gap-junction contact with the oocyte. In response to hormonal stimuli, the cumulus cells up-regulate HAS2 and tumor necrosis factor (TNF)-α-stimulated gene 6 (TSG-6; see the next section). The expression of these proteins initiates production of hyaluronan and its organization into an expanding matrix around the cumulus cells. Concurrently, the follicle becomes permeable to serum, which introduces an unusual molecule called inter-α-trypsin inhibitor (ITI), composed of the trypsin inhibitor bikunin and two heavy chains all covalently bound to a chondroitin sulfate chain. TSG-6 catalyzes the transfer of heavy chains that are covalently linked to chondroitin sulfate onto the newly synthesized hyaluronan. In the absence of either TSG-6 or ITI, the matrix does not form, and the phenotype of mice null for either of these molecules is female infertility. At the time of ovulation, hyaluronan synthesis ceases, and ovulation of the expanded cumulus cell–oocyte complex occurs. Before fertilization, individual sperm undergo capacitation enabling them to penetrate and fertilize an ovum. During this process, SPAM1, a GPI-anchored hyaluronidase, redistributes and accumulates in the sperm head. SPAM1 binds hyaluronan in the cumulus, causing an increase in Ca^{++} flux and sperm motility. It also helps dissolve the cumulus matrix as the sperm moves through the hyaluronan vestment. A soluble form of SPAM1 is secreted during the acrosome reaction. The release of acrosomal hyaluronidase and proteases renders the sperm capable of fusing with the egg and eventually destroys the entire matrix to allow the fertilized oocyte to implant and develop.

HYALURONAN-BINDING PROTEINS WITH LINK MODULES

There are several hyaluronan-binding proteins with homologous link modules (Figure 16.3). The four homologous link proteins belong to a subfamily called the "hyaluronan and proteoglycan link proteins" (HAPLNs); these are expressed in many tissues. Four cell-surface receptors have extracellular domains with one link module: CD44, LYVE-1 (lymphatic vessel endothelial hyaluronan receptor), HARE/Stabilin-2 (hepatic hyaluronan clearance receptor), and Stabilin-1, which are expressed on discontinuous endothelial cells and some activated macrophages. Of these, all functionally bind hyaluronan through the link module except Stabilin-1. Other hyaluronan-binding proteins are secreted and include the chondroitin sulfate proteoglycans that comprise the aggrecan superfamily and TSG-6, which has one link module.

The three-dimensional structure of the link module fold in TSG-6 has been determined by nuclear magnetic resonance and defines a consensus fold of the two α-helices and two triple-stranded antiparallel β-sheets (Figure 16.5). The fold consists of about 100 amino acids and contains four cysteines disulfide-bonded in the pattern Cys1-Cys4 and Cys2-Cys3. This fold has only been found in vertebrates, consistent with the fact that hyaluronan is a relatively recent evolutionary invention. The link module fold is related to that found in the C-type lectins, but it lacks the Ca^{++} binding motif (Chapter 34). In the case of TSG-6, the interaction of hyaluronan with the protein involves ionic interactions between positively charged amino acid residues and the carboxyl groups of the uronic acids, and hydrophobic interactions between the acetamido side chains of two *N*-acetylglucosamine residues and hydrophobic pockets on either side of adjacent tyrosines (Figure 16.5). Many of these features are conserved in other members of the hyaluronan-binding proteins. Subgroups, however, differ in the preferred size and length of hyaluronan for binding (e.g., hexasaccharides to decasaccharides).

Some hyaluronan-binding proteins do not contain a link module (RHAMM, ITI, SPACR, SPACRCAN, CD38, CDC37, HABP1/P-32, Siglec-9, and IHABP4), and most of these are

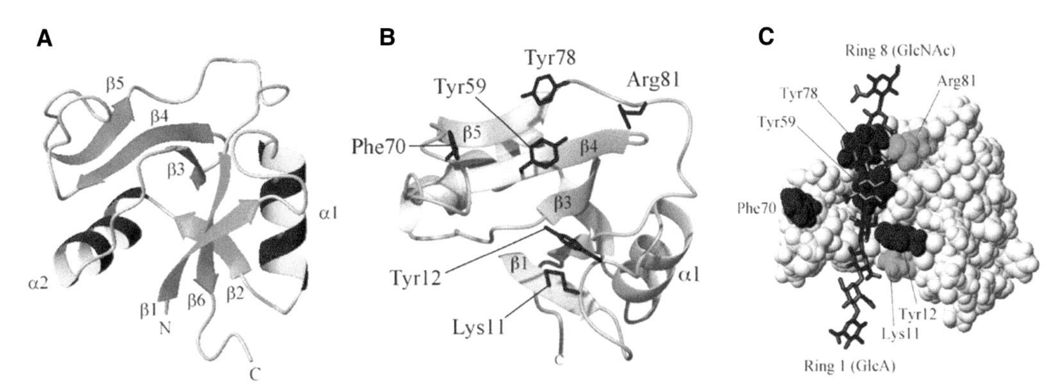

FIGURE 16.5. Structure of the link module. (*A*) TSG-6 contains a prototypical link module defined by two α-helices (α1 and α2) and two triple-stranded antiparallel β-sheets (β1,2,6 and β3–5). (Redrawn, with permission, from Blundell CD, et al. 2003. *J Biol Chem* **278:** 49261–49270, © American Society for Biochemistry and Molecular Biology.) (*B*) The hyaluronan-bound conformation of the protein, with a view showing key amino acids. (Redrawn, with permission, from Blundell CD, et al. 2004. In *Chemistry and biology of hyaluronan* [eds. Garg HC, Hales CA], pp. 189–204, © Elsevier.) (*C*) A model of the TSG-6 link module/hyaluronan complex. Binding of the hyaluronan to the protein is mediated through ionic interactions between positively charged amino acid residues (*green*) and the carboxylate groups of the uronic acids and by a combination of aromatic ring stacking and H-bond interactions between the polysaccharide and aromatic amino acids (*red*). In addition, hydrophobic pockets (on either side of Tyr59) can accommodate the methyl groups of two *N*-acetylglucosamine side chains, which is likely to be a major determinant in specificity of link modules for hyaluronan. (Redrawn, with permission, from Blundell CD, et al. 2005. *J Biol Chem* **280:** 18189–18201, © American Society for Biochemistry and Molecular Biology.)

unrelated to one another by primary sequence. Some of these proteins contain sequences of nine amino acids, referred to as BX_7B motifs (where B is either lysine or arginine and X can be any amino acid other than acidic residues), but the actual hyaluronan docking site of the chain with this motif has not been established. Thus, the presence of the BX_7B motif should not be taken as proof that the protein interacts with hyaluronan.

HYALURONAN AND CELL SIGNALING

Hyaluronan production has long been implicated in enhanced cell adhesion and locomotion because it is present abundantly during morphogenesis and in both physiological and pathological invasive processes. Five different types of cell signaling proteins have been found to mediate the complex breadth of functions associated with hyaluronan (Figure 16.6).

Toll-like receptors TLR2 and TLR4, which are components of the innate immune system responsive to bacterial pro-inflammatory stimuli such as lipopolysaccharide (LPS), have been found to signal differentially in response to the generation of hyaluronan oligosaccharides by hyaluronidase. Depending on cell type, this may occur through direct binding of the hyaluronan oligosaccharides by TLR2/4. Alternatively, because hyaluronan encapsulation can obstruct access of TLR2/4 by other known ligands, hyaluronidase-mediated capsule degradation can also facilitate TLR2/4 activation by allowing other ligands to bind.

HARE is expressed on the cell surface of a variety of sinusoidal endothelial cell types, where it is capable of binding hyaluronan and other glycosaminoglycan ligands, subsequently triggering receptor-mediated endocytosis and thereby clearing these components from circulation. Signaling

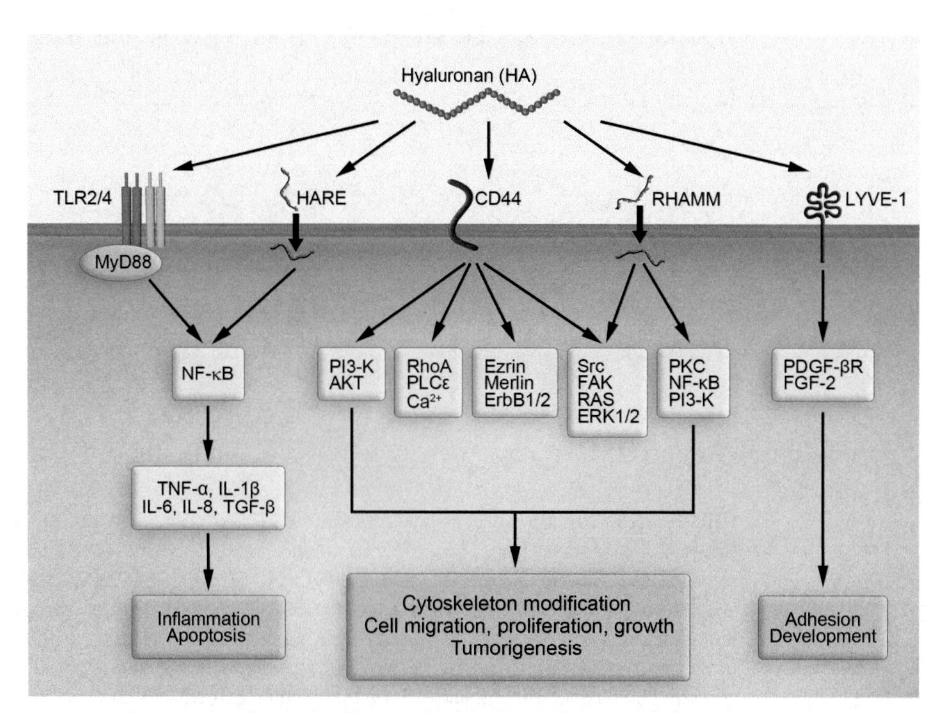

FIGURE 16.6. Hyaluronan signaling in health and disease. CD44, cluster of differentiation antigen, a type I transmembrane receptor; FAK, focal adhesion kinase; FGF, fibroblast growth factor; HARE, hyaluronan receptor for endocytosis; LYVE-1, lymphatic vessel endothelial receptor 1; MyD88, myeloid differentiation primary response gene 88; NF-κB, nuclear factor kappa-light-chain enhancer of activated B cells; PDGF, platelet-derived growth factor; PKC, protein kinase C; RHAMM, receptor for hyaluronan-mediated motility; TGF, transforming growth factor; TLR, Toll-like receptor; TNF, tumor necrosis factor.

occurs optimally during uptake of 50- to 250-kDa hyaluronan; smaller or larger hyaluronan does not activate signaling and blocks signaling by 50- to 250-kDa hyaluronan. Signaling through both TLR2/4 and HARE activate NF-κB and promote the secretion of pro-inflammatory and pro-apoptotic cytokines.

CD44 is a transmembrane receptor expressed by many cell types, and it varies markedly in glycosylation, oligomerization, and protein sequence because of differential mRNA splicing. CD44 contains a cytoplasmic domain, a transmembrane segment, and an ectodomain with a single link module that can bind hyaluronan. When hyaluronan binds to CD44, the cytoplasmic tail can interact with regulatory and adaptor molecules, such as SRC kinases, RHO (ras homolog) GTPases, VAV2 (a human proto-oncogene), GAB1 (a GRB2-associated binding protein), and ankyrin and ezrin (which regulate cytoskeletal assembly/disassembly and cell migration). Binding of hyaluronan to CD44H (the isoform expressed by hematopoietic cells) can mediate leukocyte rolling and extravasation in some tissues. Interaction of hyaluronan with CD44 can also regulate ErbB-family (including epithelial growth factor receptor, HER2) signaling, thereby activating the PI3K (phosphatidylinositol 3-kinase)–PKB (protein kinase B/AKT) signaling pathway and phosphorylation of FAK (focal adhesion kinase) and BAD (BCL2-antagonist of cell death), which promote cell survival. Changes in CD44 expression, notably expression of CD44 variants, are associated with a wide variety of tumors and the metastatic spread of cancer. LYVE-1 is a lymphatic vessel endothelial hyaluronan receptor that is a homolog of CD44. Importantly, LYVE-1 signals uniquely through growth factor stimuli in lymphatic vasculature and lymph nodes to promote circulating leukocyte adhesion and support lymphatic development.

Many cells also express RHAMM, which is also known as HMMR (homeostasis, mitosis, and meiosis regulator) and is involved in cell motility and cell division. The RHAMM pathway is thought to induce focal adhesions to signal the cytoskeletal changes required for elevated cell motility seen in tumor progression, invasion, and metastasis. RHAMM splice variants exist, some of which may be intracellular. RHAMM possesses a BX_7B sequence that promotes electrostatic interaction with hyaluronan but is also part of a basic leucine zipper motif that is capable of centrosome and microtubule interactions. Thus, RHAMM is also a nonmotor spindle assembly factor. Hyaluronan binding to RHAMM activates SRC, FAK, ERK (extracellular mitogen-regulated protein kinase), and PKC (protein tyrosine kinase C) (Chapter 40). These pathways are relevant to tumor cell survival and invasion; their inhibition by hyaluronan oligomers and soluble hyaluronan-binding proteins suggests novel therapeutic approaches for treating cancer (Chapter 47).

HYALURONAN CAPSULES IN BACTERIA

Some pathogenic bacteria (e.g., certain strains of *Streptococcus* and *Pasteurella*) produce hyaluronan as an extracellular capsule; also see Chapter 21). Capsular hyaluronan, like other capsular polysaccharides, increases virulence by helping to shield the microbe from host defenses. For example, the capsule blocks phagocytosis and protects against complement-mediated killing. Because bacterial hyaluronan is identical in structure to host hyaluronan, the capsule can also prevent the formation of protective antibodies. Thus, the formation of hyaluronan capsules by bacteria is a form of molecular mimicry. The capsule also can aid in bacterial adhesion to host tissue, facilitating colonization (Chapter 37). Finally, the production of hyaluronan by invading bacteria can also induce a number of signaling events through hyaluronan-binding proteins that modulate the host physiology (i.e., cytokine production; Chapter 42).

In addition to bacteria, an algal virus (*Chlorella*) encodes a hyaluronan synthase. The functional significance of viral hyaluronan production is unknown, but could be related to prevention of secondary viral infection, increase in host capacity to produce virus, or viral burst size. The origin of viral HAS is unknown, but based on sequence homology it most likely arose from a vertebrate.

HYALURONAN AS A THERAPEUTIC AGENT

Hyaluronan has been used therapeutically for a number of years. Patients with osteoarthritis obtain short-term relief by direct injection of high-molecular-weight hyaluronan into the synovial space of an affected joint. The mechanism of action is complex and probably involves both the viscoelastic properties of the polymer as well as effects on the growth of synovial cells in the joint capsule. Hyaluronan suppresses cartilage degeneration, acts as a lubricant (thereby protecting the surface of articular cartilage), and reduces pain perception.

The application of hyaluronan in ophthalmology is widespread. During surgery for lens replacement due to cataracts, a high potential for injury of fragile intraocular tissues exists, especially for the endothelial layer of the cornea. High-molecular-weight hyaluronan is injected to maintain operative space and structure and to protect the endothelial layer from physical damage. Hyaluronan also has been approved for cosmetic and surgical uses. For example, hyaluronan can be subdermally injected to fill wrinkles or pockets under the skin. As a component of topical dressings, hyaluronan facilitates skin wound healing, and as a bioresorbable component of anti-adhesive gels applied during surgical incision closure, hyaluronan prevents adhesions that can occur following multiple standard procedures.

The fact that the small organic molecule 4-methylumbelliferone (4-MU) acts as a sink for consumption of UDP-GlcA has been taken advantage of to reduce hyaluronan synthesis. A low-dose formulation of 4-MU is already clinically approved for treatment of biliary dyskinesia, presumably by altering the glucuronidation of bile acids. However, depletion of UDP-GlcA could affect many other biochemical pathways, and these high levels of 4-MU could well have other effects.

Low-molecular-weight hyaluronan oligosaccharides ($\sim 10^3 - 10^4$ Da) also have potent biological activities by altering selective signaling pathways (see Chapter 40). In cancer cells, hyaluronan oligosaccharides induce apoptosis and inhibit tumor growth in vivo. Thus, short hyaluronan chains may prove useful for preventing cancer metastasis by boosting certain immune responses or altering new blood vessel growth. Recombinant forms of the *Pasteurella* synthase (pmHas) have been engineered to produce hyaluronan oligosaccharides of defined size. This strategy has great promise for exploring the relationship of hyaluronan size to function, which may in turn yield new therapeutic agents with selective activities.

ACKNOWLEDGMENTS

The authors appreciate helpful comments and suggestions from Yan Wang, Ding Xu, and Ulf Lindahl.

FURTHER READING

Simoni RD, Hill RL, Vaughan M, Hascall V. 2002. The discovery of hyaluronan by Karl Meyer. *J Biol Chem* **277**: e27. doi:10.1016/S0021-9258(18)36679-1

Toole BP. 2004. Hyaluronan: from extracellular glue to pericellular cue. *Nat Rev Cancer* **4**: 528–539. doi:10.1038/nrc1391

Hascall VC, Wang A, Tammi M, Oikari S, Tammi R, Passi A, Vigetti D, Hanson RW, Hart GW. 2014. The dynamic metabolism of hyaluronan regulates the cytosolic concentration of UDP-GlcNAc. *Matrix Biol* **35**: 14–17. doi:10.1016%2Fj.matbio.2014.01.014

McAtee CO, Barycki JJ, Simpson MA. 2014. Emerging roles for hyaluronidase in cancer metastasis and therapy. *Adv Cancer Res* **123**: 1–34. doi:10.1016/B978-0-12-800092-2.00001-0

Vigetti D, Karousou E, Viola M, Deleonibus S, De Luca G, Passi A. 2014. Hyaluronan: biosynthesis and signaling. *Biochim Biophys Acta* **1840**: 2452–2459. doi:10.1016/j.bbagen.2014.02.001

Liang J, Jiang D, Noble PW. 2017. Hyaluronan as a therapeutic target in human diseases. *Adv Drug Deliv Rev* **97**: 186–203. doi:10.1016/j.addr.2015.10.017

Weigel PH, Baggenstoss BA, Washburn JL. 2017. Hyaluronan synthase assembles hyaluronan on a [GlcNAcβ1,4]n-GlcNAcα-UDP primer and hyaluronan retains this residual chitin oligomer as a cap at the nonreducing end. *Glycobiology* **27**: 536–554. doi:10.1093/glycob/cwx012

Jackson DG. 2019. Hyaluronan in the lymphatics: the key role of the hyaluronan receptor LYVE-1 in leucocyte trafficking. *Matrix Biol* **78–79**: 219–235. doi:10.1016/j.matbio.2018.02.001

Yamaguchi Y, Yamamoto H, Tobisawa Y, Irie F. 2019. TMEM2: a missing link in hyaluronan catabolism identified? *Matrix Biol* **78–79**: 139–146. doi:10.1016/j.matbio.2018.03.020

Dokoshi T, Zhang L-J, Li F, Nakatsuji T, Butcher A, Yoshida H, Shimoda M, Okada Y, Gallo RL. 2020. Hyaluronan degradation by Cemip regulates host defense against *Staphylococcus aureus* skin infection. *Cell Rep* **30**: 61–68.e4. doi:10.1016/j.celrep.2019.12.001

He Z, Mei L, Connell M, Maxwell CA. 2020. Hyaluronan mediated motility receptor (HMMR) encodes an evolutionarily conserved homeostasis, mitosis, and meiosis regulator rather than a hyaluronan receptor. *Cells* **9**: 819. doi:10.3390/cells9040819

Roedig H, Damiescu R, Zeng-Brouwers J, Kutija I, Trebicka J, Wygrecka M, Schaefer L. 2020. Danger matrix molecules orchestrate CD14/CD44 signaling in cancer development. *Semin Cancer Biol* **62**: 31–47. doi:10.1016/j.semcancer.2019.07.026

17 Proteoglycans and Sulfated Glycosaminoglycans

Catherine L.R. Merry, Ulf Lindahl, John Couchman, and Jeffrey D. Esko

This chapter focuses on the structure, biosynthesis, and general biology of proteoglycans. Topics include a description of the major families of proteoglycans, their characteristic polysaccharide chains (glycosaminoglycans), biosynthetic pathways, and general concepts about proteoglycan function. Proteoglycans, like other glycoconjugates, have many essential roles in biology.

HISTORICAL PERSPECTIVE

The study of proteoglycans dates back to the beginning of the 20th century with investigations of "chondromucoid" from cartilage and anticoagulant preparations from liver (heparin). From 1930 to 1960, great strides were made in analyzing the chemistry of the polysaccharides of these preparations (also known as "mucopolysaccharides"), yielding the structure of hyaluronan (Chapter 16), dermatan sulfate (DS), keratan sulfate (KS), different isomeric forms of chondroitin sulfate (CS), heparin, and heparan sulfate (HS). Together, these polysaccharides came to be known as glycosaminoglycans (sometimes abbreviated as GAGs) to indicate the presence of

published by Cold Spring Harbor Laboratory Press; doi: 10.1101/glycobiology.4e.17

amino sugars and other sugars in a polymeric form. Subsequent studies provided insights into the linkage of the chains to proteoglycan core proteins. These structural studies paved the way for biosynthetic studies.

The 1970s marked a turning point in the field, when improved isolation and chromatographic procedures were developed for the purification and analysis of tissue proteoglycans and glycosaminoglycans. Density-gradient ultracentrifugation allowed separation of the large aggregating proteoglycans from cartilage, revealing a complex of proteoglycan, hyaluronan, and link protein. Also during this period, it was realized that the production of proteoglycans was a general property of animal cells and that proteoglycans and glycosaminoglycans were present on the cell surface, inside the cell, and in the extracellular matrix (ECM). This observation led to a rapid expansion of the field and the eventual appreciation of proteoglycan function in cell adhesion, signaling, and other biological activities (Chapter 38). Today, studies with animal cell mutants (Chapter 49) as well as experiments using gene knockout and silencing techniques in a variety of model organisms are aimed at extending our understanding of the role of proteoglycans in development and physiology (Chapters 25–27) and human diseases (Chapters 41–47). The application of a variety of newly developed analytical tools, including mass spectrometry (Chapter 50) and glycan arrays (Chapter 48), are leading to a better understanding of proteoglycan structure and function.

PROTEOGLYCAN AND GLYCOSAMINOGLYCAN COMPOSITION

Proteoglycans consist of a "core" protein and one or more covalently attached glycosaminoglycan chains (Figure 17.1). Glycosaminoglycans are linear polysaccharides that can be broken down chemically or enzymatically into disaccharides, each consisting of an amino sugar (glucosamine [GlcN] that is N-acetylated [GlcNAc] or N-sulfated [GlcNS] or *N*-acetylgalactosamine [GalNAc]) and a uronic acid (glucuronic acid [GlcA] or iduronic acid [IdoA]) or galactose (Gal). Figure 17.2 depicts schematic short segments of glycosaminoglycans and their characteristic features. Hyaluronan does not occur covalently linked to a protein core but instead interacts noncovalently with some proteoglycans via hyaluronan-binding motifs (Chapter 16). Generally, invertebrates produce the same types of glycosaminoglycans as vertebrates, except that hyaluronan is not present and the chondroitin chains are predominantly, although not exclusively, nonsulfated. Most proteoglycans also contain N- and O-glycans typically found on glycoproteins (see Chapters 9 and 10). The glycosaminoglycan chains are much larger than these other types of glycans (e.g., a 20-kDa glycosaminoglycan chain contains approximately 80 sugar residues, whereas a typical biantennary N-glycan contains 10 to 12 residues). Keratan sulfate is a sulfated poly-*N*-acetyllactosamine chain present on a limited number of proteins as an N-linked or O-linked chain. The composition of the glycosaminoglycan chains, the structure of the protein cores, and the distribution of the proteoglycan all determine the biological activities associated with proteoglycans.

PROTEOGLYCANS ARE DIVERSE IN STRUCTURE AND FUNCTION

Virtually all mammalian cells produce proteoglycans and secrete them into the ECM, insert them into the plasma membrane, or store them in secretory granules. The ECM, an essential component of all multicellular animals, determines the physical characteristics of tissues and many of the biological properties of the cells embedded in it. The major components of the ECM are fibrillar proteins that provide tensile strength and elasticity (e.g., various collagens and elastins), adhesive glycoproteins (e.g., fibronectin, laminins, and tenascins), and proteoglycans that interact with other ECM components to promote ECM assembly, govern its physical

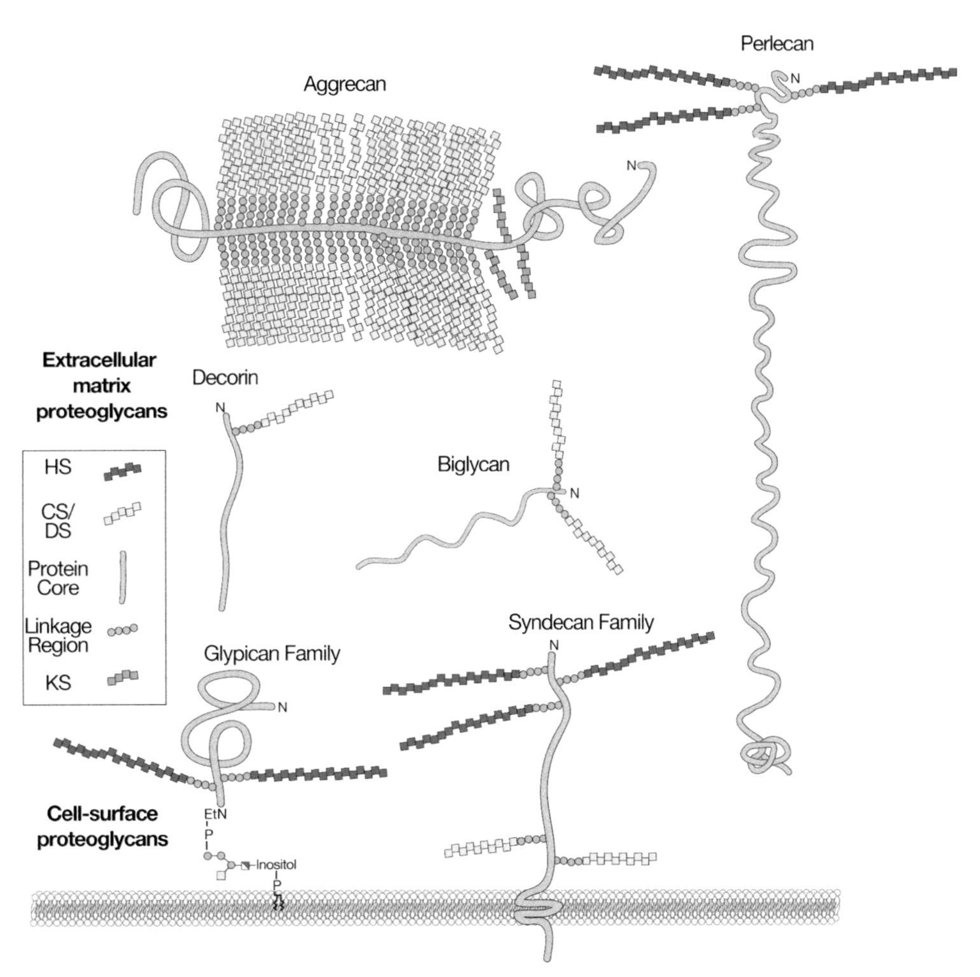

FIGURE 17.1. Proteoglycans consist of a protein core (*brown*) and one or more covalently attached glycosaminoglycan chains (*dark blue*, HS, heparan sulfate; *yellow*, CS/DS, chondroitin sulfate/dermatan sulfate; *light blue*, KS, keratan sulfate). Membrane proteoglycans either span the plasma membrane (type I membrane proteins) or are linked by a glycosylphosphatidylinositol (GPI) anchor. Extracellular matrix proteoglycans are usually secreted, but some proteoglycans can be proteolytically cleaved and shed from the cell surface (not shown).

properties, and serve as a reservoir of biologically active small proteins such as growth factors. A single cell type can express multiple proteoglycans. Vascular endothelial cells, for example, synthesize several different cell-surface proteoglycans, secretory granule proteoglycans, as well as several ECM proteoglycans.

Compared with the hundreds, perhaps thousands, of glycoproteins that carry N- and O-linked glycans, to date, relatively few proteins have been identified that carry glycosaminoglycans (less than 50), but the application of new glycoproteomic approaches has already led to the discovery of several new proteoglycans. Tremendous structural variation of proteoglycans exists due to a number of factors. First, many proteoglycans can be substituted with one or more types of glycosaminoglycan chain; for example, glypicans contain heparan sulfate, whereas syndecan-1 contains both heparan sulfate and chondroitin sulfate chains. Some proteoglycans contain only one glycosaminoglycan chain (e.g., decorin), whereas others have more than 100 chains (e.g., aggrecan). Another source of variability lies in the stoichiometry of glycosaminoglycan chain substitution. For example, syndecan-1 has five attachment sites for glycosaminoglycans, but not all of the sites are used equally. Other proteoglycans can be "part-time"—that is, they may exist with or without a glycosaminoglycan chain or with only a truncated oligosaccharide. A given proteoglycan present in different cell types often shows differences in the

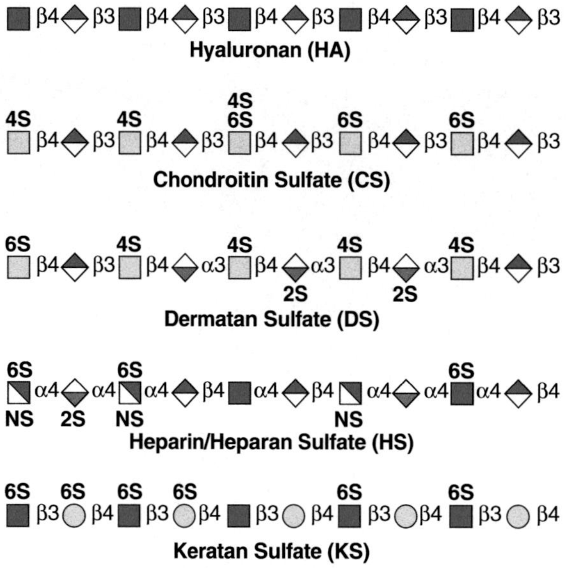

FIGURE 17.2. Glycosaminoglycans consist of alternating N-acetylated (GlcNAc or GalNAc) or N-sulfated (GlcNS) glucosamine and either a uronic acid (GlcA or IdoA) or galactose (Gal). Hyaluronan lacks sulfate groups, but the rest of the glycosaminoglycans contain sulfates at various positions. Dermatan sulfate is distinguished from chondroitin sulfate by the presence of IdoA. Heparan sulfate and heparin are the only glycosaminoglycans that contain an N-sulfated glucosamine. Keratan sulfates lack uronic acids and instead consist of sulfated galactose and *N*-acetylglucosamine residues. Reducing termini are to the *right* in all sequences.

number of glycosaminoglycan chains, their lengths, and the arrangement of sulfated residues along the chains. Thus, a preparation of any one proteoglycan (defined by its core protein) represents a diverse population of molecules, each potentially representing a unique structural entity. These characteristics, typical of all proteoglycans, create enormous diversity and potential biological variation in activity.

Mammalian Proteoglycans—Form and Function

The major classes of proteoglycans can be classified by their distribution, homologies, and function. Table 17.1 provides an overview of many of the well-known and characterized proteoglycans.

The aggrecan family of ECM proteoglycans (also known as lecticans) consists of aggrecan, versican, brevican, and neurocan. In all four members, the protein moiety contains an amino-terminal domain capable of binding hyaluronan, a central region that contains covalently bound chondroitin sulfate chains, and a carboxy-terminal domain containing a C-type lectin domain (Chapter 34). Aggrecan is the best-studied member of this family, because it represents the major proteoglycan in cartilage where it forms a stable matrix capable of withstanding compressive forces by water desorption and resorption. Versican, which is produced predominantly by connective tissue cells, undergoes alternative splicing events that generate a family of proteins. Neurocan is expressed in the late embryonic central nervous system (CNS) and can inhibit neurite outgrowth. Brevican is expressed in the terminally differentiated CNS, particularly in perineuronal nets.

The small leucine-rich proteoglycans (SLRPs) contain leucine-rich repeats flanked by cysteines in their central domain. At least nine members of this family are known and some carry chondroitin sulfate, dermatan sulfate, or keratan sulfate chains. These proteoglycans help to stabilize and organize collagen fibers but have other roles in innate immunity and regulation of growth factor signaling.

The SLRPs and aggrecan family of proteoglycans appear to be unique to vertebrates. *Caenorhabditis elegans* and *Drosophila melanogaster* express other proteoglycans, suggesting that the core proteins have undergone enormous diversification during evolution, presumably to accommodate different needs of the organism. In contrast, the biosynthetic machinery for glycosaminoglycan assembly has been evolutionarily conserved, demonstrating conservation of function for the glycosaminoglycan chains.

TABLE 17.1. Diversity among known vertebrate proteoglycans

Proteoglycan	Core protein (kDa)	Number and type of glycosaminoglycan chains	Tissue distribution	Proposed functions	Human disease associations
Secreted proteoglycans—Aggrecan (lectican) family					
Aggrecan	208–220	~100 CS, ~20 KS II (human, bovine)	cartilage, brain	with hyaluronan forms hydrated ECM to resist compressive forces	excessive degradation/loss in arthritic disease
Versican/PG-M	265 (50–450 kDa splice forms)	0–15 CS/DS	pericellular and interstitial ECM; blood vessels; brain, leukocytes	multiple ECM interactions, regulation of inflammation, cell adhesion and migration	supports progression of atherosclerosis
Neurocan	145	1–2 CS/DS	brain	regulates neurite outgrowth	schizophrenia, bipolar disorder
Brevican	96	0–4 CS/DS	brain-perineuronal nets	regulates synaptic plasticity	promotes glioma invasiveness
Secreted proteoglycans—SLRPs					
Decorin	36	1 CS/DS	connective tissue cells	regulates interstitial collagen fibrillogenesis, inhibition of TGF-β signaling	overexpression in systemic sclerosis, mutations in congenital stroma corneal dystrophy
Biglycan	38	0–2 CS/DS	connective tissue cells, macrophages	collagen matrix assembly, soluble forms activate innate immune system	
Lumican	37	3–4 KS I	widely distributed	collagen matrix assembly	
Keratocan	37	3–4 KS I	widely distributed but sulfated only in cornea	collagen matrix assembly, roles in corneal transparency	mutations associated with cornea plana
Fibromodulin	59	2–4 KS I	widely distributed	collagen matrix assembly	intronic variations and SNPs associate with high-degree myopia
Osteoglycin/Mimecan	25	2–3 KS I	widely distributed but sulfated only in cornea	collagen matrix assembly, roles in bone formation, corneal transparency	potential biomarker of ischemic heart failure
Other secreted proteoglycans					
Perlecan	400	1–3 HS, 0–2 CS	basement membranes, stem cell niche, other ECM, cartilage	ECM assembly, regulated cell migration through integrin interactions, sequestration of growth factors (e.g., FGFs)	rare mutations causing severe skeletal malformations
Agrin	200	1–3 HS	basement membranes, brain and neuromuscular junctions	neuromuscular junction maturation, ligand for integrin and α-dystroglycan	
Collagen type IX, α2 chain	68	1 CS/DS	cartilage, vitreous humor	stabilization of cartilage collagen II/XI fibrils	mutated in some forms of multiple epiphyseal dysplasia
Collagen type XVIII	147	2–3 HS	basement membranes, longest isoform more widespread	basement membrane stability; monomeric carboxy-terminal domain (endostatin) is antiangiogenic	mutations give rise to Knobloch syndrome, with multiple ocular and neural tube closure defects

(Continued)

TABLE 17.1. *(Continued)*

Proteoglycan	Core protein (kDa)	Number and type of glycosaminoglycan chains	Tissue distribution	Proposed functions	Human disease associations
Membrane-bound proteoglycans					
Syndecans 1–4	31–45	1–3 HS, 0–2 CS	most nucleated cells	regulator of cell adhesion, migration and actin cytoskeletal organization, control of ligand clearance from cell surface, co-receptor role in protein signaling	dysregulation in several cancers (e.g., myeloma, mammary carcinoma)
Betaglycan	110	0–1 HS, 0–1 CS	fibroblasts	coreceptor regulating ligand binding (e.g., inhibins) and signaling through TGF-β receptors	tumor suppressor; commonly lost in ovarian cancer
Glypicans 1–6	~60	1–3 HS	epithelial and mesenchymal cells, brain	coreceptors regulating signaling through associated (e.g., tyrosine kinase) receptors	Simpson–Golabi–Behmel overgrowth syndrome, hepatocellular carcinoma progression (GPC3)
Phosphacan/PTPζ	175	2–5 CS/DS	brain	cortical neuronal migration	schizophrenia
Thrombomodulin	58	1 CS/DS	endothelial cells, but also a variety of nonvascular cells	anticoagulant through protein C activation	mutations—bleeding disorders; therapeutic potential for soluble recombinant form in disseminated intravascular coagulation
CD44	37	0–2 CS/DS/HS	widely distributed including lymphocytes, used as a marker of mesenchymal stem cells	hyaluronan and growth factor receptor	used as a marker of breast cancer stem cells
NG2/CSPG4	251	2–3 CS/DS	some stem cells, glial progenitors, vascular mural cells, melanocytes	cell-ECM adhesion, growth factor interactions, integrin activation	supports melanoma invasion and survival
Invariant chain/CD74	31	1 CS	antigen-processing cells	main MHC class II chaperone; CS bearing forms enhance T-cell activation	humanized monoclonal anti-CD74 antibody, milatuzumab, under study for treatment of some lymphomas
SV2	80	1–3 KS I	synaptic vesicles	regulation of presynaptic transmitter exocytosis	target for some epilepsy syndromes
Intracellular granule proteoglycans					
Serglycin	10–19	10–15 heparin/CS	mast cells, other leukocytes, endothelial cells	packaging of granule contents, maintenance of protease activity, regulation after secretion of coagulation, host defense, and wound repair	inflammation, cancer progression

HS, heparan sulfate; CS, chondroitin sulfate; DS, dermatan sulfate, KS, keratan sulfate; ECM, extracellular matrix; SLRP, small leucine-rich proteoglycan; TGF-β, transforming growth factor-β; SNP, single-nucleotide polymorphism; FGF, fibroblast growth factor; MHC, major histocompatibility complex.

Basement membranes are highly specialized thin layers of the ECM that lie flush against epithelial cells and surround muscle and fat cells. Major components are laminins, nidogens, and collagens, as well as three unrelated basement membrane proteoglycans—perlecan, agrin, and type XVIII collagen. These proteoglycans interact with other basement membrane components and cell-surface adhesion receptors but can also be important reservoirs of heparan sulfate–binding signaling factors.

The membrane-bound proteoglycans are diverse. The syndecan family consists of four members, each with a short hydrophobic domain that spans the membrane, linking the larger extracellular domain containing the glycosaminoglycan attachment sites to a smaller intracellular cytoplasmic domain. The syndecans are expressed in a tissue-specific manner and facilitate cellular interactions with a wide range of extracellular ligands, such as growth factors and matrix molecules. Because of their membrane-spanning properties, the syndecans can transmit signals from the extracellular environment to the intracellular cytoskeleton via their cytoplasmic tails. Syndecans are sensitive to proteolytic cleavage by matrix metalloproteases, resulting in shedding of the ectodomains bearing the glycosaminoglycan chains that retain potent biological activity (Chapter 38). *C. elegans* and *D. melanogaster* express only one syndecan (Chapters 25 and 26).

Glypicans carry only HS chains, which can bind a wide array of factors essential for development and morphogenesis. Six glypican family members exist in mammals, and only two are expressed in *D. melanogaster* and *C. elegans*. Each member of the glypican family of cell-surface proteoglycans has a glycosylphosphatidylinositol anchor attached at the carboxyl terminus, which embeds them in the outer leaflet of the plasma membrane (Chapter 12). The amino-terminal portion of the protein has multiple cysteine residues and a globular shape that distinguishes the glypicans from the syndecan ectodomains, which tend to be extended structures (Figure 17.1).

A number of other membrane proteoglycans are expressed on the surface of many different cell types including the widespread CD44, NG2 (also known as CSPG4), phosphacan (PTPζ), thrombomodulin, and invariant chain of the major histocompatibility complex (MHC) class II system. Serglycin is the major cytoplasmic secretory granule proteoglycan that is present in endothelial, endocrine, and hematopoietic cells. Depending on the species, it has a variable number of glycosaminoglycan attachment sites that can carry chondroitin sulfate or heparin chains. In fact, many proteoglycans show variation in the degree of substitution by GAG chains, giving rise in some cases to so-called "part-time" proteoglycans.

To a large extent, the biological functions of proteoglycans depend on the interaction of the glycosaminoglycan chains with different protein ligands. However, the protein core determines when and where expression takes place and can itself interact with other components of the extracellular environment and the cytoskeleton. Table 38.1 lists examples of proteins known to interact with glycosaminoglycans. Proteins that bind to the sulfated glycosaminoglycan chains appear to have evolved by convergent evolution (i.e., they do not contain a specific fold present in all glycosaminoglycan-binding proteins, in contrast to other groups of glycan-binding proteins). These interactions have profound physiological effects and are discussed further in Chapter 38.

LINKAGES OF GLYCOSAMINOGLYCANS TO PROTEINS

Different subtypes of sulfated glycosaminoglycans are attached to their core proteins by unique linkages. There are two types of keratan sulfate, distinguished by the nature of their linkage to protein (Figure 17.3). KS I, originally detected in cornea, is found on an N-glycan linked to protein through an asparagine residue (Chapter 9). KS II (skeletal keratan sulfate) is found on an O-glycan core 2 structure and is thus linked through *N*-acetylgalactosamine to serine or threonine (Chapter 10). The structural features in control of keratan sulfate substitution remain unclear, as the underlying poly-*N*-acetyllactosamine backbone can be found on many other

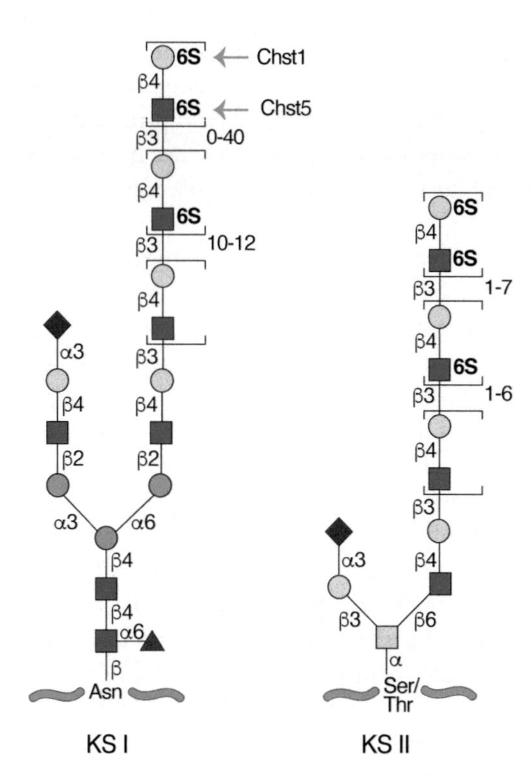

FIGURE 17.3. Keratan sulfates (KS) contain a sulfated poly-*N*-acetyllactosamine chain, linked to either asparagine or serine/threonine residues. Chst1 and 5 can add sulfate groups at the indicated positions. The actual order of the various sulfated and nonsulfated disaccharides occurs somewhat randomly along the chain.

glycoproteins. Notably, in humans and bovine, the large chondroitin sulfate proteoglycan found in cartilage (aggrecan) contains a segment of four to 23 hexapeptide repeats (E-E/L-P-F-P-S) where the keratan sulfate chains are located, whereas aggrecan in rats and other rodents lacks this motif and does not contain keratan sulfate.

Two classes of glycosaminoglycan chains—chondroitin sulfate/dermatan sulfate and heparan sulfate/heparin—are linked to serine residues in proteins by way of xylose (Figure 17.4). Xylosyltransferase initiates the process using UDP-xylose as donor. Two isoforms of the enzyme are known in vertebrates (XYLT1 and XYLT2), but only one isozyme exists in *C. elegans* and *D. melanogaster*. A glycine residue invariably lies to the carboxy-terminal side of the serine attachment site, but a perfect consensus sequence for xylosylation does not exist. At least two acidic amino acid residues are usually present, and they can be located on one or both sides of the serine, usually within a few residues. Several proteoglycans contain clustered glycosaminoglycan attachment sites, raising the possibility that xylosyltransferase could act in a processive manner. Xylosylation is an incomplete process in some proteoglycans, which may explain why proteoglycans with multiple potential attachment sites contain different numbers of chains in different cells.

After xylose addition, a linkage tetrasaccharide assembles by the transfer of two galactose residues catalyzed by unique members of the β4 galactosyl-, β3 galactosyl-, and β3 glucuronosyltransferase families of enzymes (Figure 17.4). This intermediate can undergo phosphorylation at the C-2 position of xylose and in the case of chondroitin sulfate, sulfation of the galactose residues. In general, phosphorylation and sulfation occur substoichiometrically, but phosphorylation may be transient. Phosphorylation occurs early in the assembly process and creates the preferred substrate for B4GALT7; a phosphatase removes the phosphate at a later stage of biosynthesis. The function of galactose sulfation in chondroitin sulfate remains unclear.

The linkage tetrasaccharide lies at a bifurcation in the biosynthetic pathway: addition of β4-linked *N*-acetylgalactosamine, which initiates chondroitin sulfate assembly, or addition of α4-linked *N*-acetylglucosamine, which initiates heparan sulfate assembly (Figure 17.4). Genetic evidence from studies of *C. elegans* suggests that *N*-acetylgalactosamine addition during

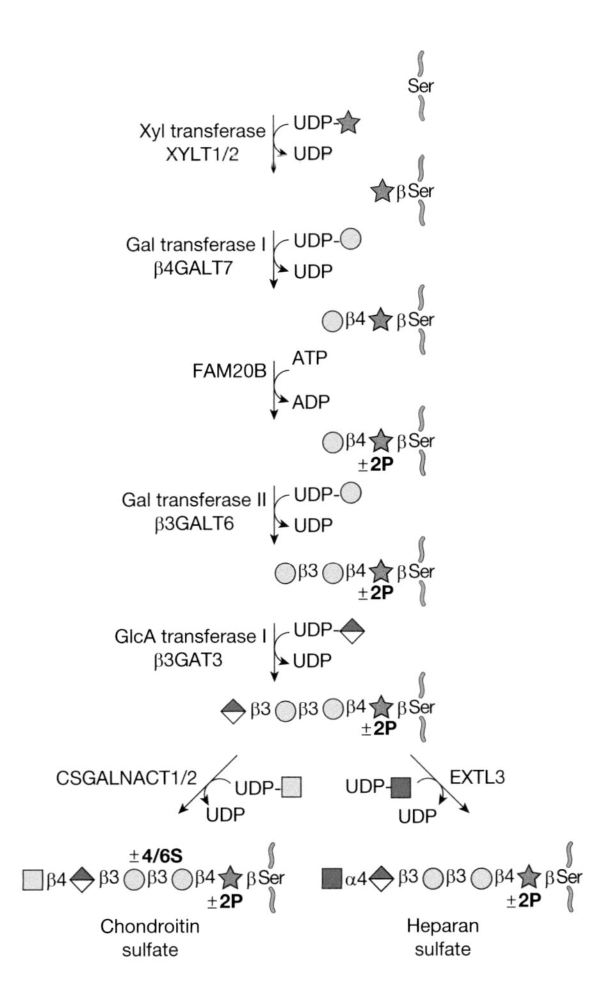

FIGURE 17.4. The biosynthesis of chondroitin sulfate (*left chain*) and heparan sulfate (*right chain*) is initiated by the formation of a linkage region tetrasaccharide (*green circles* in Fig. 17.1). Addition of the first hexosamine commits the intermediate to either chondroitin sulfate (GalNAc) or heparan sulfate (GlcNAc).

chondroitin assembly is mediated by the same enzyme that is involved in chain polymerization (Sqv5), but biochemical evidence suggests that more than one enzyme may exist in vertebrates. In heparin/heparan sulfate formation, the addition of the first *N*-acetylglucosamine residue is catalyzed by an enzyme called EXTL3, which differs from the transferases involved in heparan polymerization (called EXT1 and EXT2). These enzymes are important control points because they ultimately regulate the type of glycosaminoglycan chain that will assemble. Control of the addition of β4GalNAc or α4GlcNAc appears to be manifested at the level of enzyme recognition of the polypeptide substrate.

GLYCOSAMINOGLYCAN BIOSYNTHESIS

Keratan Sulfate

Keratan sulfate chains contain a mixture of nonsulfated (Galβ4GlcNAcβ3), monosulfated (Galβ4GlcNAc6Sβ3), and disulfated (Gal6Sβ4GlcNAc6Sβ3) disaccharide units (Figure 17.2). The biosynthesis of the poly-*N*-acetyllactosamine backbone is described in Chapter 14. At least two classes of sulfotransferases, one or more N-acetylglucosaminyl 6-O-sulfotransferases (e.g., CHST4 or CHST6), and one or two galactosyl 6-O-sulfotransferases (CHST1 and CHST3) catalyze the sulfation reactions. These enzymes, like other sulfotransferases, use activated sulfate (PAPS [3′-phosphoadenyl-5′-phosphosulfate]) as a high-energy donor (Chapter 5). 6-*O*-Sulfation of *N*-acetylglucosamine occurs on the nonreducing terminal residue, encouraging further chain elongation, whereas sulfation of galactose residues takes place on nonreducing terminal and internal galactose residues, with a preference for galactose units adjacent to a sulfated *N*-acetylglucosamine residue. Sulfation of a nonreducing terminal galactose residue blocks further elongation of the chain, providing a potential mechanism for controlling chain length. The poly-*N*-acetyllactosamine chains of KS I are generally longer than those of KS II and may contain up to 50 disaccharide units (20–25 kDa). The chains can be fucosylated and sialylated as well (Chapter 14).

Chondroitin Sulfate

Vertebrate chondroitin sulfate consists of repeating sulfate-substituted GalNAcβ4GlcAβ3 disaccharides polymerized into long chains (Figure 17.2). In contrast, invertebrates such as *C. elegans* and *D. melanogaster* make either nonsulfated or low sulfated chains. The assembly process for the backbone appears to be highly conserved, based on the presence of homologous genes for all of the reactions (Chapters 25 and 26). As described above, the assembly process is initiated by the transfer of GalNAcβ3 to the linkage tetrasaccharide (Figure 17.4). In both vertebrates and invertebrates, the polymerization step is catalyzed by one or more bifunctional enzymes (chondroitin synthases) that have both β3 glucuronosyltransferase and β4 N-acetylgalactosaminyltransferase activities. Vertebrates also express homologs that can transfer individual sugars to the chain. Chondroitin polymerization also requires the action of the chondroitin polymerizing factor (CHPF), a protein that lacks independent activity but collaborates with the polymerases to enhance the formation of polymers. Sulfation of chondroitin in vertebrates is a complex process, with multiple sulfotransferases involved in 4-*O*-sulfation and 6-*O*-sulfation of *N*-acetylgalactosamine residues (Figure 17.5).

Additional enzymes exist for epimerization of D-glucuronic acid to L-iduronic acid in dermatan sulfate (DSE1-2), sulfation at the C-2 position of the uronic acids, and other patterns of sulfation found in unusual species of chondroitin (Table 17.2). The location of sulfate groups is easily assessed using bacterial chondroitinases (ABC, B, and ACII) that cleave the chains into disaccharides. Many chains are hybrid structures containing more than one type of chondroitin disaccharide unit. For example, dermatan sulfate is defined as having one or more iduronic acid–containing disaccharide units (chondroitin sulfate B) as well as glucuronic acid–containing disaccharides (chondroitin sulfate A and C). Animal cells also degrade chondroitin sulfate in lysosomes using a series of exolytic activities (Chapter 44).

Heparan Sulfate

Heparan sulfate assembles as a copolymer of GlcNAcα4GlcAβ4 (Figure 17.5), which then undergoes extensive modification reactions, catalyzed by at least four families of sulfotransferases and one epimerase. *N*-acetylglucosamine N-deacetylase/N-sulfotransferases (NDST1-4) act on a subset of *N*-acetylglucosamine residues to generate N-sulfated glucosamine ($GlcNSO_3$) units, many of which occur in clusters along the chain. Generally, the enzyme deacetylates *N*-acetylglucosamine and rapidly adds sulfate to the free amino group to form $GlcNSO_3$, but a small number of glucosamine residues with unsubstituted amino groups may arise from incomplete N-sulfation. An epimerase (GLCE), different from the one involved in dermatan sulfate synthesis, then acts on some glucuronic acid residues, followed by 2-*O*-sulfation of some of the iduronic acid units (catalyzed by HS2ST). Some glucuronic acid units also undergo 2--*O*-sulfation by the same enzyme. The addition of 2-*O*-sulfate groups to glucuronic or iduronic acid prevents the reversible epimerization reaction. Next, 6-*O*-sulfotransferases (HS6ST1-3) add sulfate groups to selected glucosamine residues. Finally, certain subsequences of sulfated sugar residues and uronic acid epimers provide targets for 3-O-sulfotransferases (HS3ST1-6).

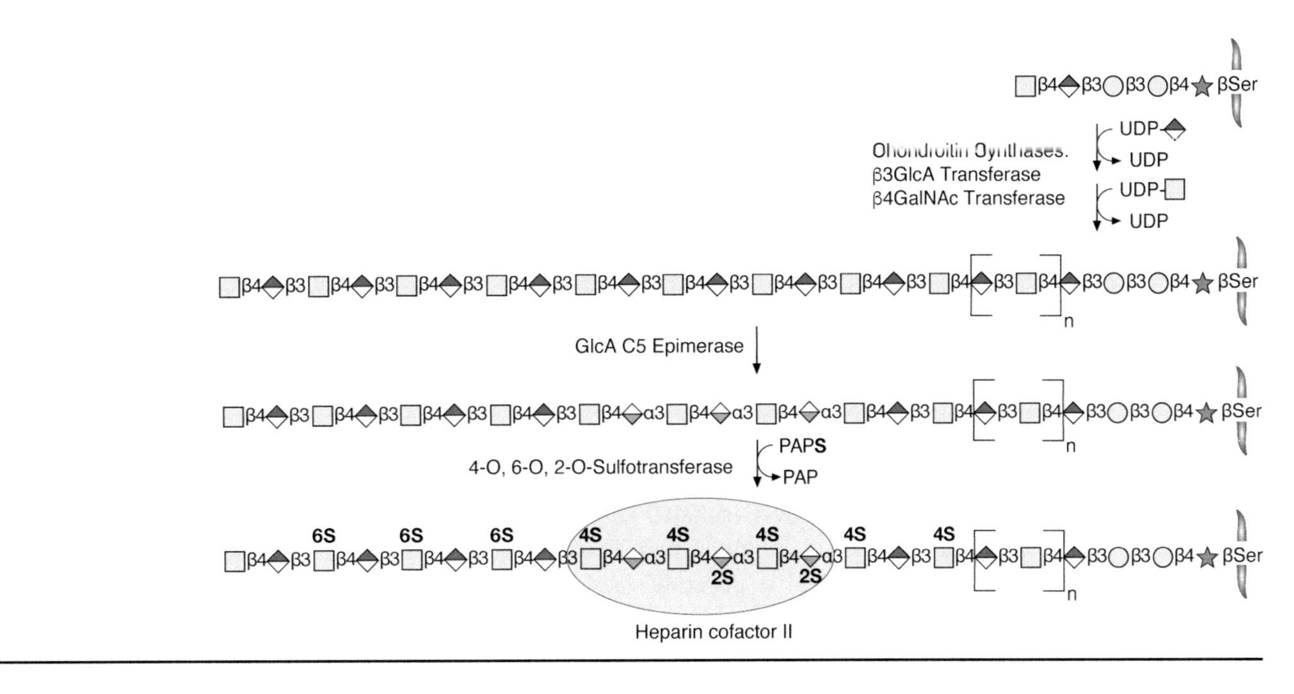

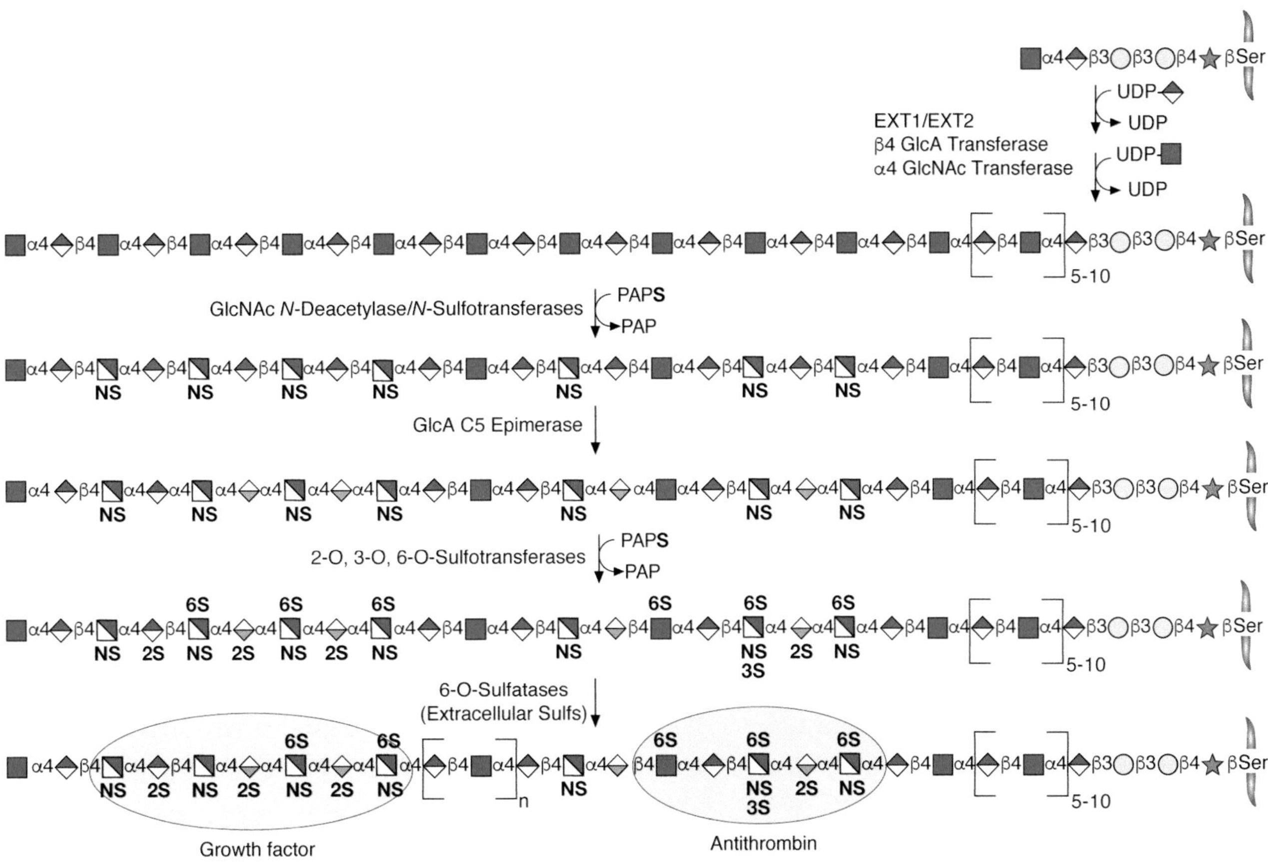

FIGURE 17.5. Biosynthesis of chondroitin sulfate/dermatan sulfate involves the polymerization of *N*-acetylgalactosamine and glucuronic acid units and a series of modification reactions including O-sulfation and epimerization of glucuronic acid to iduronic acid. Heparan sulfate biosynthesis involves copolymerization of *N*-acetylglucosamine and glucuronic acid residues. Additional modification by uronic acid epimerization and N- and O-sulfation occur in an incomplete and interdependent manner throughout the chain, with clustered sulfation forming sulfated domains separated by regions of low or no sulfation.

TABLE 17.2. Types of chondroitin sulfates

Chondroitin sulfate type	Disaccharide type	Source
A	GlcAβ1-3GalNAc4S	cartilage and other tissues
B	IdoAα1-3GalNAc4S	skin; tendon
C	GlcAβ1-3GalNAc6S	cartilage and other tissues
D	GlcA2Sβ1-3GalNAc6S	shark cartilage; brain
E	GlcAβ1-3GalNAc4,6diS	squid; secretory granules

This list is not meant to be exhaustive because many types of chondroitins exist with unusual modifications. For example, dermatan sulfate disaccharide can also contain sulfate at the C-2 position of IdoA and sulfate at C-6 instead of C-4, and 2-O-sulfated and 3-O-sulfated GlcA has been described in some cartilage chondroitin sulfate.

Unlike other glycosaminoglycans, heparan sulfate is further modified once displayed at the plasma membrane. A family of plasma membrane endosulfatases (SULFs) can remove sulfate groups from internal 6-O-sulfated glucosamine residues in heparan sulfate and a heparanase can cleave the chains at limited sites. This post-assembly processing of the chains at the cell surface results in altered response of cells to growth factors and morphogens. Intriguingly, the mammalian genome contains other sulfatases of unknown function, raising the possibility that other post-assembly processing reactions of glycosaminoglycans may occur.

In contrast to chondroitin chains, which tend to have long tracts of fully modified disaccharides, the modification reactions in heparan sulfate biosynthesis occur in clusters along the chain, with regions devoid of sulfate separating the modified domains. In general, the reactions proceed in the order indicated, with evidence for functional interdependence within and between the biosynthetic enzyme families, but they often fail to go to completion, resulting in tremendous chemical heterogeneity.

Readout and Regulation of Glycosaminoglycan Assembly

The disaccharide composition of glycosaminoglycan chains can be readily assessed using bacterial lyases or chemical degradation methods (which are more useful for differentiating glucuronic and iduronic acids). Direct sequencing of the chains has proved difficult because of their heterogeneity, however the application of specific lysosomal exoenzymes involved in glycosaminoglycan degradation and new mass spectrometry methods are making significant inroads into sequencing of glycosaminoglycans (Chapter 50). The specific arrangement of sulfated residues and uronic acid epimers in heparin/heparan sulfate and dermatan sulfate gives rise to binding sequences for proteins. The three examples shown in Figure 17.5 show sequences that can interact with antithrombin, heparin cofactor II, and other potential proteins. More modified sequences can interact as well, and the binding of many ligands is dependent on the presence of correctly orientated charged clusters rather than the specific presence of individual sulfated groups. Binding of glycosaminoglycans to proteins is described in greater detail in Chapter 38. A major question remains regarding how the enzymes and biosynthetic pathways are regulated to achieve tissue-specific expression of protein-binding sequences.

During the last decade, most if not all of the enzymes involved in glycosaminoglycan synthesis have been purified and molecularly cloned from mammals and model organisms. Several important features have emerged from these studies, which may shed light on how different protein-binding sequences arise.

- Several of the enzymes appear to have dual catalytic activities. Thus, a single protein bearing two catalytic domains catalyzes N-deacetylation of *N*-acetylglucosamine residues and subsequent N-sulfation (NDSTs) in heparan sulfate formation. The same is true of the co-polymerases, which transfer *N*-acetylglucosamine and glucuronic acid (heparan sulfate) and *N*-acetylgalactosamine and glucuronic acid (chondroitin sulfate) from the corresponding UDP sugars to the growing polymer. In contrast, the epimerases and O-sulfotransferase activities appear to be unique properties of independent enzymes.

- In several cases, multiple isozymes exist that can catalyze either a single or a pair of reactions. Thus, four N-deacetylase/N-sulfotransferases, three 6-O-sulfotransferases, seven 3-O-sulfotransferases and two endo-sulfatases (SULFs) have been identified in heparan sulfate biosynthesis. Their tissue distribution varies and differences exist in substrate preference, which may cause differences in the pattern of sulfation. However, some overlap in expression and in substrate utilization occurs as well. Multiple isozymes of 4-O- and 6-O-sulfotransferases also can participate in chondroitin sulfate formation.
- The polymerization and polymer modification reactions probably colocalize in the same stacks of the Golgi complex. Thus, the enzymes may form supramolecular complexes that coordinate these reactions. The composition of these complexes may play a part in regulating the fine structure of the chains.
- In general, the composition of heparan sulfate, and likely chondroitin and dermatan sulfate, on a given proteoglycan varies more between cell types than on different core proteins expressed in the same cell. This observation suggests that each cell type may express a unique array of enzymes and potential regulatory factors. The mechanisms behind the generation of apparently cell-specific glycosaminoglycan chains through regulated, yet partly stochastic modification reactions are yet to be fully elucidated. However, recently identified transcriptional regulation of key heparan sulfate biosynthetic enzymes suggests we may be close to this understanding.
- Recombinant enzymes and new synthetic schemes are increasingly used to generate defined glycosaminoglycan oligosaccharides, which can be used to probe ligand-binding affinities and specificities. Recent innovations include re-engineering biosynthetic enzymes to introduce tailored specificities, allowing the creation of glycosylation patterns not seen in nature and the exploration of their biological impact.

Arrays of synthetic glycosaminoglycans used to probe the importance of chain length, sulfation pattern, and chain density are now in common use and enable the rapid screening of glycosaminoglycan–protein interactions. As highlighted above, care needs to be taken when inferring a biological effect from these interactions. However, recently developed chemical strategies for creating artificial proteoglycans at the cell surface together with new opportunities for altering cell surface display of glycosaminoglycans using gene editing, or by application of soluble inhibitors/activators to influence components of the biosynthetic pathway, offer controlled environments with which to test glycosaminoglycan–protein interactions at the cellular level.

HEPARIN VERSUS HEPARAN SULFATE

Considerable confusion exists regarding the definition of heparin and heparan sulfate (even the spelling is often in error!). Heparin is produced by a limited number of cells, notably connective tissue-type mast cells and bipotential glial progenitor cells, whereas heparan sulfate is made by virtually all types of cells. During biosynthesis, heparin undergoes more extensive sulfation and uronic acid epimerization, such that $>80\%$ of the *N*-acetylglucosamine residues are N-deacetylated and N-sulfated and $>70\%$ of the glucuronic acids undergo epimerization to iduronic acid. Heparin derived from porcine entrails and bovine lung is prepared commercially by selective precipitation and is sold by pharmaceutical companies as an anticoagulant because of its high capacity to bind to antithrombin (Chapters 38 and 57). The active sequence is a pentasaccharide shown in Figure 17.5, which is now sold as a purely synthetic anticoagulant (Arixtra). Low-molecular-weight heparins are derived from commercial unfractionated heparin by chemical or enzymatic cleavage, depending on the brand. Selectively desulfated forms of heparin and heparin oligosaccharides are also available commercially, some of which lack anticoagulant activity, but still retain other potentially useful properties. Heparan sulfate also can contain anticoagulant activity, but typical preparations from cells or tissues are much less active than heparin. Care should be taken in extrapolating data obtained with heparin (e.g., binding of a protein to heparin-Sepharose) versus binding to heparan sulfate and heparan sulfate proteoglycans; binding to heparin can occur owing to the high charge content of the polysaccharide, whereas the same factor might bind to heparan sulfate with lower affinity or not at all. On the other hand, specific protein-binding motifs expressed in subspecies of heparan sulfate may occur also in heparin, although concealed by additional, redundant, sulfate residues.

PROTEOGLYCAN PROCESSING AND TURNOVER

Cells secrete matrix proteoglycans directly into the extracellular environment (e.g., members of the aggrecan family, the basement membrane proteoglycans, SLRPs, and serglycin). However, others are shed from the cell surface through proteolytic cleavage of the core protein through matrix metalloproteases (e.g., the syndecans).

Extracellular heparanase, an endo-β-glucuronidase, cleaves heparan sulfate at restricted sites resulting in release of growth factors or chemokines immobilized on heparan sulfate proteoglycans at cell surfaces or in the ECM. The activity of shedding extracellular domains can contrast with that of the intact transmembrane proteoglycan and digestion by heparanase can similarly generate glycosaminoglycan fragments with distinct and contrasting activity. This is a particular feature of invading cells in which secreted heparanase can act in concert with matrix metalloproteases to remodel the ECM. Because heparanase has been implicated in the progression of various cancers, there is considerable interest in the development of inhibitors. This has led to clinical trials involving inhibitors either alone or in combination with other therapeutic agents (Chapter 47).

Cells also internalize a large fraction of cell-surface heparan sulfate proteoglycans by endocytosis. These internalized proteoglycans first encounter proteases and heparanase and the core protein and the heparan sulfate chains are cleaved endolytically. The resulting smaller heparan sulfate fragments eventually appear in the lysosome and undergo complete degradation by way of a series of exoglycosidases and sulfatases (Chapter 44). Chondroitin sulfate and dermatan sulfate proteoglycans follow a similar endocytic route. One of the human hyaluronidases (HYAL-4) has been found to be involved in endolytic degradation of chondroitin sulfate.

ACKNOWLEDGMENTS

The authors appreciate helpful comments and suggestions from Barbara Mulloy.

FURTHER READING

Thacker BE, Xu D, Lawrence R, Esko JD. 2014. Heparan sulfate 3-*O*-sulfation: a rare modification in search of a function. *Matrix Biol* **35**: 60–72. doi:10.1016/j.matbio.2013.12.001

Gallagher J. 2015. Fell-Muir Lecture: heparan sulphate and the art of cell regulation: a polymer chain conducts the protein orchestra. *Int J Exp Pathol* **96**: 203–231. doi:10.1111/iep.12135

Mizumoto S, Yamada S, Sugahara K. 2015. Molecular interactions between chondroitin-dermatan sulfate and growth factors/receptors/matrix proteins. *Curr Opin Struct Biol* **34**: 35–42. doi:10.1016/j.sbi.2015.06.004

Neill T, Schaefer L, Iozzo RV. 2015. Decoding the matrix: instructive roles of proteoglycan receptors. *Biochemistry* **54**: 4583–4598. doi:10.1021/acs.biochem.5b00653

Mitsou I, Multhaupt HM, Couchman JR. 2017. Proteoglycans, ion channels and cell-matrix adhesion. *Biochem J* **474**: 1965–1979. doi:10.1042/BCJ20160747

Caterson B, Melrose J. 2018. Keratan sulfate, a complex glycosaminoglycan with unique functional capability. *Glycobiology* **28**: 182–206. doi:10.1093/glycob/cwy003

Kjellén L, Lindahl U. 2018. Specificity of glycosaminoglycan–protein interactions. *Curr Opin Struc Biol* **50**: 101 – 108. doi:10.1016/j.sbi.2017.12.011

Vlodavsky I, Ilan N, Sanderson R. 2020. Forty years of basic and translational heparanase research. *Adv Exp Med Biol* **1221**: 3–59. doi:10.1007/978-3-030-34521-1_1

Annaval T, Wild R, Crétinon Y, Sadir R, Vivès RR, Lortat-Jacob H. 2020. Heparan sulfate proteoglycans biosynthesis and post synthesis mechanisms combine few enzymes and few core proteins to generate extensive structural and functional diversity. *Molecules* **25**: 4215. doi:10.3390/molecules25184215

Merry CLR. 2021. Exciting new developments and emerging themes in glycosaminoglycan research. *J Histochem Cytochem* **69**: 9–11. doi:10.1369/0022155420974361

18 Nucleocytoplasmic Glycosylation

Christopher M. West, Chad Slawson, Natasha E. Zachara, and Gerald W. Hart

The glycosylation of macromolecules is a highly compartmentalized process. Most of the enzyme donor precursors are synthesized in the cytoplasm or nucleoplasm and transferred into the secretory pathway. There they are incorporated into the glycoproteins, glycolipids, or polysaccharides that are fated for delivery to the extracellular environment or organelles such as the lysosome. In some cases, the donor sugars are transferred to intermediate lipids presented at the cytoplasmic surface of secretory pathway membranes, and subsequently flipped into the secretory pathway for final incorporation into glycoconjugates. In other cases, polysaccharides are synthesized at the cytoplasmic surface of the plasma membrane and simultaneously translocated to the cell surface. Thus the cytoplasm has a key role in the assembly of glycans that are, however, destined to function outside of the cytoplasm. In addition, entirely different glycosyltransferases exist in the cytoplasm or nucleus to glycosylate proteins and lipids that remain to function in the cytoplasm or nucleoplasm. There is also evidence for cytoplasmic and nuclear glycoconjugates that have acquired secretory pathway–type glycans by unexplained

published by Cold Spring Harbor Laboratory Press; doi: 10.1101/glycobiology.4e.18

mechanisms. The origin and role of these nucleocytoplasmic glycoconjugates, mainly glycoproteins, are the focus of this chapter. We begin with examples of monoglycosylation, including both endogenous processes and those associated with parasitism, and note that one form, O-linked β-*N*-acetylglucosamine (O-β-GlcNAc), is so prevalent that an entire chapter is devoted to it (Chapter 19). We then continue to complex glycans and glycosylation associated with mitochondria and chloroplasts and conclude with an assessment of nucleocytoplasmic carbohydrate binding proteins that may serve as glycan readers.

MONOGLYCOSYLATION OF NUCLEOCYTOPLASMIC PROTEINS

Eukaryotic Monoglycosylation

Monoglycosylation, the addition of a single monosaccharide to the hydroxyl or amide moieties of amino acid side-chains, was first discovered in the form O-β-GlcNAc and is now known as a widespread modification of thousands of nuclear mitochondrial and cytoplasmic proteins throughout the animal and plant kingdoms and some protists, fungi, and bacteria (see Chapter 19). The addition of O-GlcNAc is linked to the side-chains of Ser or Thr and is mediated by a highly conserved lineage of O-GlcNAc transferases (OGT), a nonmembrane bound enzyme that resides in the cytoplasm and nucleus. A splicing variant is also found in animal mitochondria in which O-GlcNAcylation is found on respiratory enzymes. A highly related enzyme lineage, termed O-fucosyltransferase (OFT), mediates the addition of O-fucose (O-Fuc) to dozens of nucleocytoplasmic proteins in plants, numerous protists, and possibly bacteria. The OFT was first discovered genetically in plants as Spy, in which it controls transcriptional processes. In the protist pathogen *Toxoplasma gondii*, OFT modifies numerous proteins involved in transcription, mRNA biogenesis, nuclear transport, and cell signaling. Recently, Greb1, derived from another lineage of enzymes, was shown to O-β-GlcNAcylate estrogen receptor-α and potentially other nuclear proteins, which if verified is an interesting example of convergent evolution. Paralogs of these enzymes have not been described in the well-studied *Saccharomyces cerevisiae*, but biochemical data suggests that this yeast modifies intracellular proteins with mannose (Man) and even di- and tri-Man. Furthermore, several reports suggest that select intracellular proteins of mammalian cells are modified by *N*-acetylgalactosamine (GalNAc). The significance of these intriguing latter examples awaits identification of the responsible glycosyltransferases and evidence for their roles in cells. Examples in which the glycosyltransferases are known are represented schematically in Figure 18.1A and summarized in Table 18.1.

Prokaryotic Monoglycosylation

Highly conserved homologs of OGT and OFT are found in numerous bacterial genomes, although studies of their functional roles are essentially nonexistent. An example of bacterial monoglycosylation that leads to a known function comes from the EF-P protein, the bacterial homolog of eukaryotic elongation initiation factor 5a. In some bacteria, EF-P suppresses translational stalling by a mechanism that involves oxidation of a critical lysyl residue, and its eukaryotic ortholog is regulated by hypusylation. A recent phylogenetic analysis of EF-P sequences revealed a subset of enzymes with an Arg in place of Lys, and a coevolving gene that was subsequently identified by biochemistry and mass spectrometry as an argininyl rhamnosyltransferase. Assembly of the Rha-Arg linkage activates EF-P and is required for pathogenicity of *Pseudomonas*, a Gram-negative, opportunistic human pathogen, and a number of other bacteria. The discovery of this linkage by a phylogenetic approach suggests that more noncanonical examples of glycosylation are yet to be discovered.

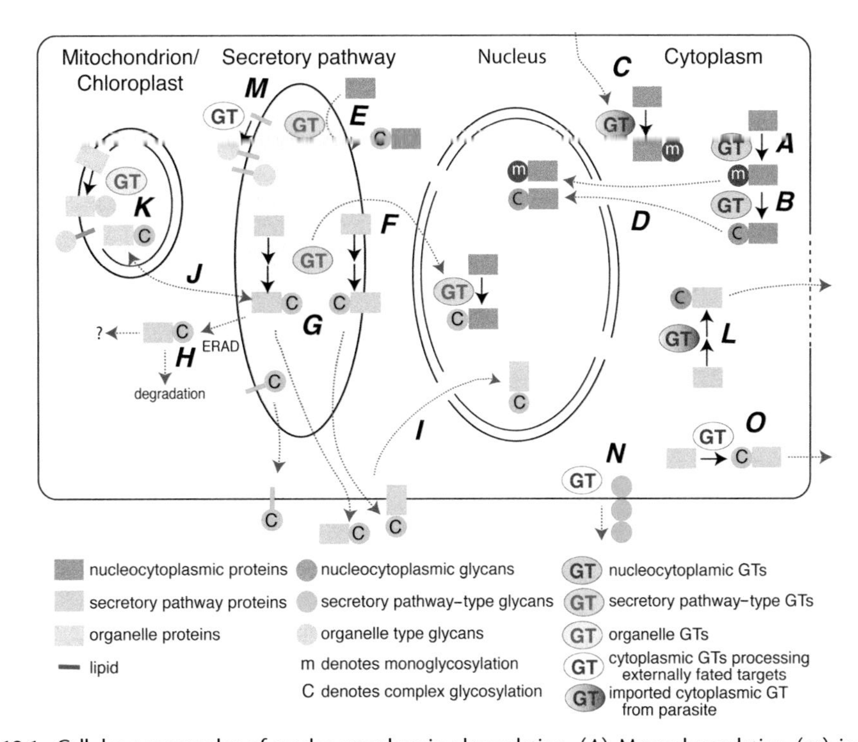

FIGURE 18.1. Cellular topography of nucleocytoplasmic glycosylation. (*A*) Monoglycosylation (m) in the cytoplasm (e.g., O-GlcNAc and O-Fuc). (*B*) Complex (c) glycosylation in the cytoplasm by O-GlcNAc transferases (GTs) that extend the first sugar. (*C*) Cytoplasmic monoglycosylation mediated by bacterial toxins introduced from the outside. (*D*) The responsible GTs and glycoproteins may enter the nucleus via nuclear pores. (*E*) Potential mechanism by which nucleocytoplasmic proteins acquire secretory pathway–type glycans, involving transient entry into the rough endoplasmic reticulum (rER) or Golgi. (*F*) Potential mechanism by which nucleocytoplasmic proteins acquire secretory pathway–type glycans, involving movement of GTs out of the secretory pathway, shown into the nucleus, by an unknown mechanism. (*G*) Conventional glycosylation of soluble and membrane bound proteins (N- or O-) and lipids in the secretory pathway. Glycoproteins and glycolipids are typically fated for secretion, or transport to the plasma membrane– or organelle-like lysosomes (not shown). (*H*) Glycoproteins generated in the secretory pathway might become nucleocytoplasmic by endoplasmic reticulum–associated degradation (ERAD)-associated retrograde transport, but diversion from subsequent degradation. (*I*) A potential source of external or cell surface glycoproteins that may accumulate in the nucleus, by an unknown mechanism that might involve endocytosis and retrograde transport via the secretory pathway. (*J*) A pathway by which mitochondrial or chloroplast proteins receive secretory pathway type glycans, via ill-defined trafficking with the rER. (*K*) Glycosylation of mitochondrial or chloroplast proteins by intrinsic glycosyltransferases (topology relative to membranes not specified). (*L*) Example of a virally encoded protein modified by virally encoded cytoplasmic GTs, but fated to the exterior following cell lysis. (*M*) The synthesis of certain secretory pathway–type glycans (e.g., LLO for N-glycosylation) is initiated by cytoplasmic GTs on specific membrane lipids before they are flipped inside. (*N*) Polysaccharides (e.g., cellulose, hyaluronic acid) are synthesized by cytoplasmically oriented, membrane embedded GTs but coordinately translocated to the exterior. (*O*) In prokaryotes, many proteins are glycosylated before translocation across the membrane, rather than after as is usually the case in eukaryotes.

A directed proteomics search for glycoproteins in the Gram-positive intestinal bacterium *Lactobacillus plantarum*, biased only because potential glycopeptides were affinity enriched with the lectin wheat germ agglutinin (WGA), identified several glycoproteins with cytoplasmic localization and functions: the molecular chaperone DnaK, the PdhC E2 subunit of the pyruvate dehydrogenase complex, the DNA translocase FtsK1, the signal recognition particle receptor FtsY, FtsZ involved in cell division, and others of unknown function. Peptides from these proteins were modified with single HexNAc residues attached to Ser residues. Some of the sites were variably modified, hinting at a novel regulatory process, but nothing is known about the mechanism of this glycosylation.

TABLE 18.1. Examples of nuclear or cytoplasmic glycosylation events

Target	GT/CAZy family	GT source	Linkage formed	GT donor	Organism(s)	Function (or compartment)
Acceptor target is retained within the nucleocytoplasmic compartment						
Self	glycogenin/GT8	same[a]	Glcα-Tyr	Glc-UDP	animals, yeast	primes glycogen synthesis
Self	glycogenin/GT8	same	Glcα1,4Glc	Glc-UDP	animals, yeast	primes glycogen synthesis
Skp1	Gnt1/GT60	same	GlcNAcα-Hyp	GlcNAc-UDP	protists	E3-Ub ligase assembly
Skp1	PgtA/GT2	same	Galβ1,3GlcNAc	Gal-UDP	protists	E3-Ub ligase assembly
Skp1	PgtA/GT74	same	Fucα1,2Gal	Fuc-GDP	protists	E3-Ub ligase assembly
Skp1	AgtA/GT77	same	Galα1,3Fuc Galα1,3Gal	Gal-UDP	*Dictyostelium*	E3-Ub ligase assembly
Skp1	AgtA/GT77	same	Galα1,3Gal	Gal-UDP	*Dictyostelium*	E3-Ub ligase assembly
Skp1	Glt1/GT32	same	Glcα1,3Fuc	Glc-UDP	*Toxoplasma*/*Pythium*	E3-Ub ligase assembly
Skp1	Gat1/GT8	same	Galα1,3Glc	Gal-UDP	*Toxoplasma*/*Pythium*	E3-Ub ligase assembly
Many proteins	OGT/GT41[b]	same	GlcNAcβ-Ser/Thr	GlcNAc-UDP	numerous	numerous
Many proteins	OFT/GT41	same	Fucα-Ser/Thr	Fuc-GDP	numerous	numerous
ERα	Greb1/GT2	same	GlcNAcβ-Ser/Thr	GlcNAc-UDP	animals	stability
Many proteins	unknown	same	$Man\alpha_{(1-3)}$-Ser/Thr	unknown	yeast	numerous?
self	MDR1/GT8	same	Glcα-Glc	Glc-UDP	numerous	mitochon. fission
unknown	Fut1	same	Fucα1,2Gal	Fuc-GDP	trypanosomes	mitochondrial
EF-P	EarP/GT104	same	Rhaα-Arg	Rha-dTDP	bacteria[c]	protein translation editing
HMW1	HMW1C/NGT	same	Hex or Hex-Hex (Gal/Glc)-Asn	Glc/Gal-UDP	*Haemophilus influenzae*	adhesiveness
DnaK	unassigned	same	HexNAc-Ser	unknown	*Lactobacillus*	molecular chaperone
PdhC E2	unassigned	same	HexNAc-Ser	unknown	*Lactobacillus*	pyruvate dehydrogenase
FtsK1	unassigned	same	HexNAc-Ser	unknown	*Lactobacillus*	DNA translocase
FtsY	unassigned	same	HexNAc-Ser	unknown	*Lactobacillus*	SRP receptor
FtsZ	unassigned	same	HexNAc-Ser	unknown	*Lactobacillus*	septal ring in cell division
UPTG	UPTG/GT75	same	Glc-Arg	Glc-UDP	potato	polysaccharide primer?
DNA	JGT/GT2	same	Glcβ-OMeUra	Glc-UDP	trypanosomes	transcriptional termination
GTPases	α-toxin(TcdA)/GT44	*Clostridia*	GlcNAcα-Thr	GlcNAc-UDP	animals	cytoskeletal remodeling
GTPases	α-toxin/GT44	*Clostridia*	Glcα-Thr	Glc-UDP	animals	cytoskeletal remodeling
EF1A	Lgt1/GT88	*Legionella*	Glcα-Ser	Glc-UDP	vertebrate	protein translation
Unknown	SetA/GT44	*Legionella*	unknown	unknown	vertebrate	vesicle trafficking
GTPases	Afp18/GT8	*Yersinia*	GlcNAcα-Tyr	GlcNAc-UDP	vertebrate	cytoskeletal remodeling
Death dom.[d]	NleB	*Escherichia coli*	GlcNAcα-Arg	Gal-UDP	animals	inhibits death signaling
DNA	αGlcT/GT72	phages	Glcα-OMeCyt	Glc-UDP	bacteria	restriction
DNA	βGlcT/GT63	phages	Glcβ-OMeCyt	Glc-UDP	bacteria	restriction
DNA	βGlcT	phages	Glcβ-Glc	Glc-UDP	bacteria	restriction
Acceptor protein is exported from cell						
Adhesin	NGT/?	same	Glcβ-Asn	Glc-UDP	*Actinobacillus*	adhesion
Adhesin	α6GalT/GT4	same	Glcα1-6Glc	Glc-UDP	*Actinobacillus*	adhesion
Autolysin	unassigned	same	GlcNAc-Ser	GlcNAc-UDP	*Lactobacillus*	peptidoglycan remodeling
MCP/VP54	A075L/GT114	PBCV-1[e]	Xylβ1-4Fuc	Xyl-NDP	*Chlorella* alga	viral capsid
MCP	A064R-D1	PBCV-1	Rhaβ1-4Xyl	β-L-Rha-UDP	*Chlorella* alga	viral capsid
MCP	A064R-D2	PBCV-1	Rhaα1-2Rha	β-L-Rha-UDP	*Chlorella* alga	viral capsid
MCP	A071R	PBCV-1	D-Rhaα1-3Fuc	D-Rha-NDP	*Chlorella* alga	viral capsid
MCP	unassigned[f]	PBCV-1	about 6 others[f]	sugar-NDP	*Chlorella* alga	viral capsid
Plastid glycoconjugates						
Carbonic anhydrase-1 (CAH-1)			N-glycan	imported	*Arabidopsis*	chloroplast
Glx, Syn, MutS			N-glycan	imported	diatom[g]	complex plastid

(*Continued*)

TABLE 18.1. *(Continued)*

Target	GT/CAZy family	GT source	Linkage formed	GT donor	Organism(s)	Function (or compartment)
α-amylase, phosphodiesterase			N-glycan	imported	rice	chloroplast
Electron transport chain proteins, others[b]			GlcNAcβ-Ser/Thr	GlcNAc-UDP	eukaryotes	mitochondrion
45-kDa inner membrane protein			N-glycan	imported	rat liver	mitochondrion
Mitochondrion dividing ring-1 (MDR1), plastid dividing ring-1 (PDR1)			α-glucan	Glc-NDP	algae	mitochondrion, chloroplast
Diacyl-glycerol			Galβ-, Galα1,6Galβ-	Gal-UDP	plants	chloroplast, other membranes (in stress)

This list highlights glycoproteins that function in the cytoplasm or nucleus, with the corresponding glycosyltransferases responsible for their glycosylation (top section). DNA glycosyltransferases are included in this group. The list also includes residential cytoplasmic glycosyltransferases whose product glycoproteins are exported from the cell and function externally (middle section). Finally, examples of plastid-associated glycosylation of proteins and cytoplasmically oriented lipids are given in the bottom section. Glycosyltransferases that modify lipids or assemble polysaccharides on the cytoplasmic face of membranes, but whose products are flipped or externalized, are not listed.

[a] "same" refers to endogenous enzyme, not derived from an external bacterium or virus.

[b] The O-GlcNAc modification is described in Chapter 19.

[c] For example, *Shewanella.*

[d] Found in TRADD, FADD, RIPK1, and TNFR1; also GAPDH.

[e] *Paramecium bursaria* Chlorella virus-1.

[f] The PBCV-1 genome encodes greater than or equal to seven predicted soluble glycosyltransferase genes, some with multiple domains, to assemble the described decasaccharide. Some are assigned to CAZy GT2.

[g] *Phaeodactylum.*

Bacterial Glycosyltransferase Toxins

Pathogenic bacteria have developed a wide array of toxin and effector virulence factors that also monoglycosylate host-cell proteins (Figure 18.1C). By virtue of their glycosyltransferase domains, these impair the host's cytoplasmic or nuclear machinery and disrupt host-cell immune response. For example, small cytoplasmic G-proteins (GTP-binding proteins) of the Rho family are involved in regulating the cytoskeleton. Certain toxins from anaerobic bacteria were found to contain retaining glycosyltransferase activities that inhibit G-proteins by attaching a glycosyl moiety to a threonine residue (Thr-37) in their GTP-binding sites. These secreted toxins show the remarkable ability of translocating across the surface membrane into the cytoplasm of mammalian target cells. The enterotoxins from *Clostridium difficile* (ToxA) and *Clostridium sordellii* are α-glucosyltransferases from CAZy family GT44 that use UDP-glucose (UDP-Glc) as the donor. In contrast, a similar toxin from *Clostridium novyi* is an O-αGlcNAc transferase that uses UDP-GlcNAc as the donor. The *C. novyi* toxin has no primary sequence relationship to host glycosyltransferases. A distantly related toxin from *Legionella pneumophila*, also delivered by a type IV secretion system, installs an αGlc residue on elongation factor 1A in a broad range of eukaryotic host cells. The target residue, Ser53, is located in the G domain near the switch-1 region of the GTPase, and glucosylation inhibits its activity in vitro and in vivo. The recent discovery of a new effector glucosyltransferase, LtpM, shows that this is an expanding field of glycosyltransferase discovery. Interestingly, an unrelated toxin from *Yersinia*, which enters host cells via a phage tail-derived translocation system, was found to be an αGlcNAc-transferase that modifies Tyr-34 of RhoA. Typically, the sugar–amino acid linkage is not recognized by endogenous glycosidases, allowing the pathogen to irreversibly alter the function of the target—typically a GTPase domain.

Other Gram-negative bacteria use a type 3 secretion system to inject virulence factors that mediate a different form of monoglycosylation: GlcNAcα-Arg. This novel reaction is catalyzed by NleB-related glycosyltransferases that promote host-cell longevity during infection. NleB and NleB orthologs target arginine side-chains within death domains of death receptors and associated proteins TRADD, FADD, RIPK1, and TNFR1. Importantly, these factors all work to attenuate NF-κB signaling to dampen host-cell antibacterial and inflammatory responses. Furthermore, various components of glucose metabolism such as GAPDH (glyceraldehyde 3-phosphate dehydrogenase) and HIF-1α are also targeted to reduced NF-κB signaling. Although these examples involve monoglycosylation mediated by exogenous glycosyltransferases (see Table 18.1), their existence reinforces the potential impact of glycosylation as a regulatory mechanism, and the diversity of glycosylation events that probably remain to be discovered.

Glycosylation of DNA

Although the focus of this chapter is on nuclear and cytoplasmic glycoproteins, it should be noted that specific bases in DNA have long been known to be a target of bacteriophage-encoded glycosyltransferases. Modification of T4-phage DNA hydroxymethylcytosine residues by βGlcTs or αGlcTs (Table 18.1) aids in the distinction of native and foreign DNA, rendering them resistant to host restriction enzyme digestion and impairing a subset of CRISPR/Cas systems. The use of Glc disaccharide, arabinose, and other sugar modifications of bases suggests that the full diversity of sugar modifications remains to be described. A related modification occurs in a protist, *Trypanosoma brucei*, in the form of base J. Using a mechanism resembling Skp1 glycosylation (*vide infra*), DNA thymidine residues are initially hydroxylated by an O_2-dependent nonheme dioxygenase generating hydroxymethyl moieties suitable for glycosylation by a novel nuclear β-glucosyltransferase, JGT, a CAZy GT2 glycosyltransferase (Table 18.1). Base J influences Pol II termination by recruiting a complex containing protein phosphatase 1 to dephosphorylate RNA polymerase II. Disruption in complex formation affects polycistronic transcription termination in the wider group of kinetoplastids.

COMPLEX GLYCOSYLATION OF NUCLEOCYTOPLASMIC PROTEINS

Complex glycans, consisting of more than a single sugar, have also been reported in the nucleocytoplasmic compartment (Figure 18.1B). The better understood examples are described here, together with evidence for glycosylation associated with two organelles that reside in the cytoplasm: mitochondria and chloroplasts. Less well described candidates are addressed in a subsequent section.

Hydroxyproline-Linked Skp1 Glycans

The Skp1 glycoprotein, which bears a single linear pentasaccharide, was discovered in the nucleus and cytoplasm of a free-living soil amoeba, the cellular slime mold *Dictyostelium discoideum.* Skp1 is an adaptor in SCF (Skp1/Cullin1/F-box protein)-type E3-ubiquitin ligase complexes that function in the cytoplasm and nucleus of all eukaryotes. There they mediate the polyubiquitination and eventual proteasomal degradation of hundreds of proteins involved in cell cycle regulation, transcription, and signal transduction. Skp1 is subject to O_2-dependent hydroxylation at Pro143, which generates the substrate for five glycosyltransferase reactions. Based on mass spectrometric, sequential exoglycosidase treatments, and nuclear magnetic resonance (NMR) studies (Chapter 50), the glycan consists of a core trisaccharide equivalent to the type 1 blood group H structure, Fucα1-2Galβ1-3GlcNAc1α-, substituted at the 3-position of Fuc by a Galα1,3Galα- disaccharide (Figure 18.2B). In most other protists that modify Skp1, including the apicomplexan parasite *T. gondii*, a similar core trisaccharide is capped by a Galα1-3Glcα1- disaccharide (Figure 18.2A). The linkage of this glycan to protein, its localization in the cytoplasm, and the structures themselves are currently unique. Because the Pro residue is conserved in the Skp1 genes of plants, invertebrates, and unicellular eukaryotes, and gene

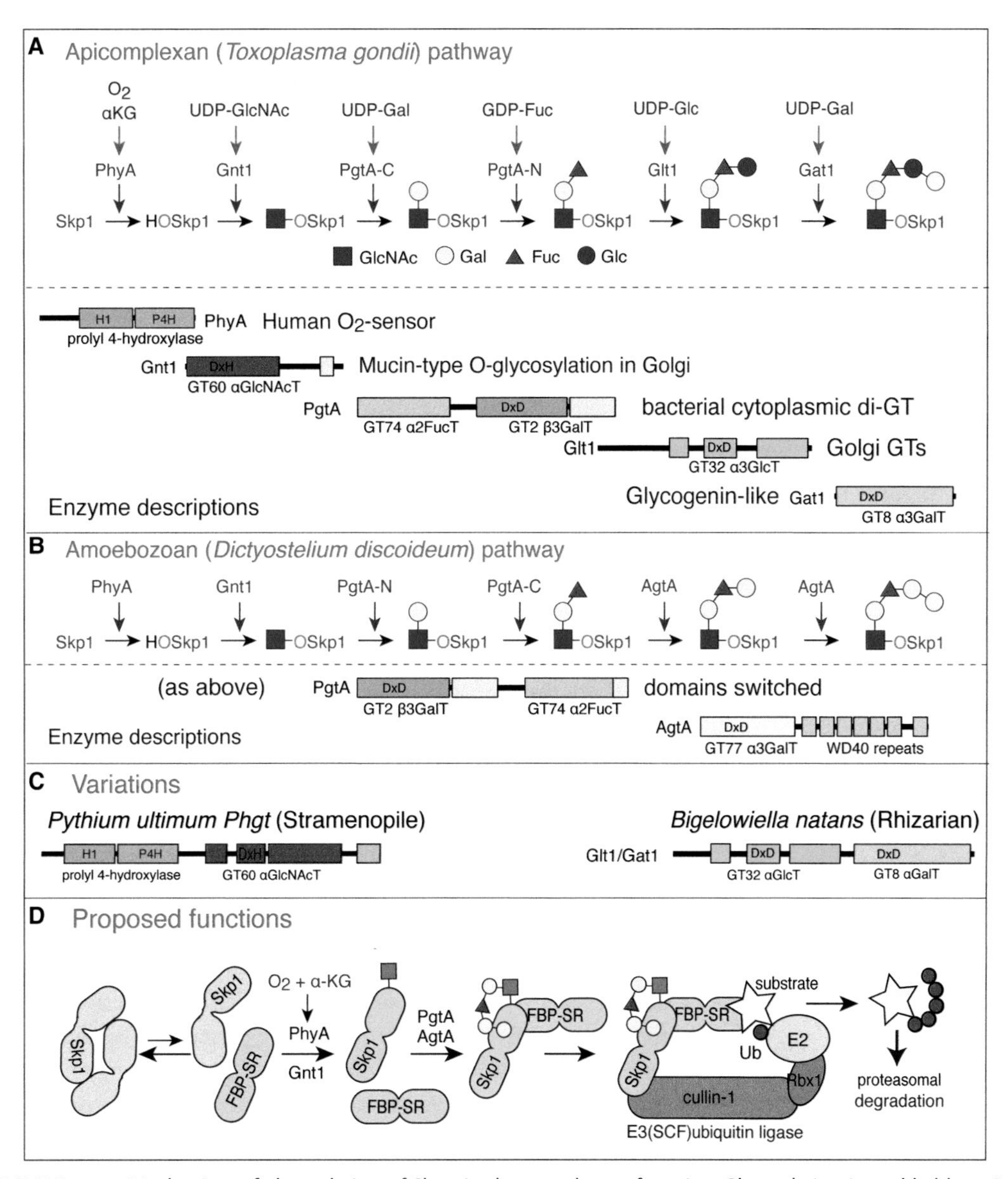

FIGURE 18.2. Mechanism of glycosylation of Skp1 in the cytoplasm of protists. Glycosylation is enabled by prior hydroxylation of a single Pro-residue by the action of a cytoplasmic, O_2-dependent prolyl 4(*trans*)-hydroxylase that is homologous to the HIFα prolyl hydroxylases of animals that regulate transcriptional responses to hypoxia. (*A*) In the parasite *Toxoplasma gondii*, Skp1 hydroxyproline-154 is sequentially modified by a series of five soluble, sugar nucleotide–dependent glycosyltransferase activities expressed by four proteins. Below, domain diagrams of these enzymes, and the CAZy family designations of the glycosyltransferases, are illustrated. Notably, they are all cytoplasmic proteins lacking rough endoplasmic reticulum (rER) or nuclear targeting sequences. (*B*) In another protist, *Dictyostelium*, the equivalent Pro-residue is similarly modified, but the addition of the final two sugars is mediated by a different, dual function glycosyltransferase, suggesting convergent evolution to generate a similar glycan. (*C*) Schematic diagrams of enzyme homologs from other protists, depicting their expression as fusion proteins. Related glycosyltransferase genes are found in select representatives of all of the major branches of protist evolution and some pathogenic fungi, but not "higher" plants or animals. (*D*) Model for the role of hydroxylation and αGlcNAcylation in inhibiting Skp1 dimerization and affecting its conformation, the role of glycan extension on promoting interactions with F-box proteins, and predicted effects on the turnover of polyubiquitin ligase substrates.

sequences related to those of known Skp1 modification enzymes (see below) are present in the genomes of many protists, this modification appears to be widespread in aerobic unicellular eukaryotes and pathogenic fungi.

The glycosyltransferases are all NDP-sugar-dependent and colocalize with Skp1 in the cytoplasm or nucleus (Figure 18.2A-C; Table 18.1). The Skp1 αGlcNAc-transferase (EC 2.4.1.229), which adds the first sugar, is related to the Thr/Ser polypeptide αGalNAc-transferases that initiate mucin-type O-glycosylation in the Golgi of animals, and therefore forms an anomeric linkage opposite to that of the aforementioned OGT (Chapter 19). However, unlike known polypeptide αGalNAc-transferases (Chapters 6 and 10), the Skp1 αGlcNAc-transferase lacks an amino-terminal signal anchor, consistent with its biochemical fractionation as a soluble cytoplasmic protein. Thus, the mechanism of initiation of Skp1 glycosylation is like that of mucin-type domains of secretory proteins, except that a different *N*-acetylhexosamine (GlcNAc vs. GalNAc) is attached to a distinct hydroxyamino acid (Hyp vs. Thr or Ser) in the cytoplasm versus the Golgi lumen. Additions of the second (βGal) and the third (α-L-Fuc) sugars are catalyzed by separate domains of the same soluble protein, PgtA. The β3-galactosyltransferase catalytic domain is most like bacterial lipopolysaccharide and capsular glycosyltransferases, which are also cytoplasmic, suggesting that cytoplasmic glycosylation in *Dictyostelium* has its evolutionary origins in bacterial glycolipid synthesis. The β-galactosyltransferase catalytic domain belongs to the inverting CAZy GT2 family, which includes numerous glycosyltransferases with catalytic domains exposed to the cytoplasm. In *Dictyostelium*, the fourth and fifth sugars are added by the two-domain glycosyltransferase AgtA, in which the α3-galactosyltransferase domain is fused to a β-propeller-like domain, which has a second function that may involve constitutive Skp1 sequestration activity. The catalytic domain is related to a large number of plant Golgi glycosyltransferases implicated in pectin biosynthesis and here catalyzes two successive additions. In contrast, most protists employ separate glycosyltransferases with independent evolutionary origins to catalyze the completion of the glycan (Figure 18.2A). In *Toxoplasma* and *Pythium ultimum*, a crop plant pathogen and agent for human pythiosus, addition of the αGlc residue is carried out by a CAZy GT31 enzyme most related to Golgi glycosyltransferases that extend mannose chains in yeast, and the terminal αGal residue is added by Gat1, a CAZy GT8 enzyme that is most related to the enzyme discussed below, glycogenin. Finally, in other protists, gene fusions consolidate the first two or final two enzymes as separate domains of the same protein (Figure 18.2C), which may facilitate efficient extension of the glycan to its final length.

Dictyostelium cells aggregate when starved and the resulting multicellular slug differentiates to form a fruiting body with stress-tolerant spores. This process is sensitive to the level of O_2, which is thought to help the cells determine whether they are below or above ground in their native soil environment. Considerable biochemical, genetic, and physiological evidence supports the model that O_2 levels regulate the rate of Skp1 hydroxylation, which in turn controls the rate of glycosylation. Glycosylation regulates the relative representation of different F-box protein substrate receptors in the Skp1 interactome (Figure 18.2D), by a mechanism that most likely controls the conformational profile of Skp1 rather than recognition by a glycan receptor. This may differentially regulates the lifetime of proteins that critically control or execute developmental progression in response to starvation.

Glycogenin

A second well-studied example is glycogen, a large branched homopolysaccharide used as a short-term storage form of glucose in bacteria, yeast, and animals. Glucose is added to and removed from the nonreducing termini per the availability of the sugar donor UDP-Glc and the nutritional needs of the cell. Glycogen stores in the liver are important in maintaining glucose homeostasis in the blood. Glycogen is assembled initially from a linear oligoglucoside that is extended by glycogen synthase and rearranged by the so-called branching enzyme. The initial oligoglucose is assembled by another enzyme, glycogenin, which attaches the first sugar to itself in an uncommon Glcα1-Tyr linkage, at Tyr-195 in human glycogenin-1. Thus, every glycogen molecule, which can contain up to 10^5 Glc residues and 12 generations of branchpoints, is thought to have a single glycogenin protein molecule at its nonreducing end. Therefore, glycogen is a glycoprotein whose glycosylation is initiated by glycogenin, extended by glycogen synthase, and modified by branching enzyme. Free glucose-1-PO_4 is released, as needed, by the actions of debranching enzyme and glycogen phosphorylase.

Glycogenin naturally assembles as a homodimer under cellular conditions. Evidence indicates that the first Glc is transferred to the acceptor hydroxyl on Tyr across subunits, and extended up to a chain length of ~10 by mixed inter- and intrasubunit additions of αGlc residues to the 4-position of underlying Glc residues. Interestingly, glycogenin shows triphasic kinetics in which there is a priming step typically with UDP-Glc; however, UDP-Gal can form this initial bond. After priming there is a short extension step again with some flexibility with substrate sugars, and finally a refining step catalyzed by the adjacent glycogenin that shows strong substrate specificity for UDG-Glc. Despite the in vitro evidence, aberrant levels of glycogen are still formed in mice lacking glycogenin-1 and its paralog glycogenin-2. Mutations in glycogenin-1 leading to decreased enzymatic function or loss of the protein cause glycogen storage disease type XV characterized by the presence of polyglucosan bodies of abnormally structured glycogen that cannot be digested by amylase. Interestingly, these polyglucosan bodies are derived from aberrant glycogen formation by glycogen synthase. These mutations contribute to respiratory distress, limb-girdle muscular dystrophy–like symptoms, and progressive weakness. However, patients do not show hypoglycemia, hepatomegaly, or hyperlipidemia as with other glycogen storage diseases. Crystallographic and mutational evidence for formation of a complex between glycogenin and glycogen synthase that affects glycogen synthase activity in vitro and in vivo offers a potential explanation for the mutational effects. Thus, levels of glycogenin may operate together with the better understood hormonally controlled mechanisms of glycogen formation that involve enzyme phosphorylation/dephosphorylation and that regulate glycogen elongation.

Glycogenin-like proteins are found in a wide range of plants, animals, and free-living single-celled eukaryotes. The *Escherichia coli* homolog is incapable of autoglucosylation. The most closely related sequence homolog in alveolates and stramenopile protists, which lack glycogenin, modifies Skp1 (*vide supra*). Glycogenin is a GT-A superfamily member from CAZy family GT8 (Carbohydrate-Active enZYmes sequence database) (Chapter 52), and more distantly related glycosyltransferases include bacterial lipopolysaccharide (LPS) glucosyl and galactosyl transferases and galactinol synthases—all cytoplasmic as for glycogenin—as well as other glycosyltransferases in the eukaryotic Golgi. The eukaryotic parasite Leishmania assembles mannogen, a polymer of β2-linked mannose polymers, instead of glucose polymers, but a protein primer is not involved for the cytoplasmically localized mannogen synthase. Interestingly, a novel Glc-Arg linkage has been described in a plant protein potentially associated with the synthesis of starch, which is related to glycogen, but this protein has no apparent sequence similarity to glycogenin. The function of this and the glycogenin-like proteins deserve further investigation for their potential to mediate other cytoplasmic glycosylation events.

Mitochondrial and Chloroplast Glycosylation

Mitochondria and chloroplasts are independently replicating organelles that reside in the cytoplasmic compartment. Because of their evolutionary origin from bacteria, which are now known to have substantial glycosylation capacity, these organelles can be expected to possess glycosylation machinery. Indeed, there is evidence for peptidoglycan-like networks between the inner and outer membranes of chloroplasts of certain algae. In other algae, glycosyltransferase MDR1 (mitochondrion dividing-ring 1), an integral membrane protein with a CAZy GT8 glycosyltransferase domain, assembles the mitochondrial dividing ring using α-linked polyglucose nanofilaments that have been visualized between the two membranes. A similar process occurs in chloroplasts, and the conservation of these genes suggests that this might be a heretofore unrecognized example of organelle glycosylation for many eukaryotes.

More recent cell biological studies emphasize intracellular connections between mitochondria and rough endoplasmic reticulum (rER) elements of the secretory pathway. This might explain a pioneering observation that two nuclear-encoded mitochondrial glycoproteins appear to be conventionally N-glycosylated in the rER based on pulse-chase labeling studies and susceptibility to *N*-glycanase (Figure 18.1J). Indeed, lectin-binding studies suggest that mitochondria contain complex glycoconjugates. In Ewing's sarcomas, the plasma membrane protein MRP-1 (multidrug resistance-associated protein 1) is glycosylated and localized to the mitochondrial outer membrane and could play a role in drug resistance. Several plant chloroplast proteins, including carbonic anhydrase-1, α-amylase, and pyrophosphatase/phosphodiesterase in "higher" plants and Glx, Syn, MutS in diatoms, might be transported from the rER as N-glycosylated proteins (Table 18.1). The presence of N-glycans highlights their potential biosynthetic origin in the conventional secretory pathway, and how these glycoproteins transit to these organelles warrants further investigation.

As noted above, mammalian mitochondria import OGT that O-GlcNAcylates some of their interior proteins. In addition, mono- and digalactolipids (MGDG: 1,2-diacyl-3-*O*-[β-D-Gal]-sn-glycerol and DGDG: 1,2-diacyl-3-O-[α-D-Gal-(1,6)-O-β-D-Gal]-sn-glycerol) are prominent lipids in chloroplast thylakoid membranes, and MGDG is essential for photosystem function (Chapter 24). Their synthesis is controlled by UDP-Gal-dependent galactosyltransferases residing in the inner and outer layer membranes of the chloroplast (Figure 18.1K). Under conditions of phosphate deprivation, the synthesis of these glycolipids is dramatically increased, becoming up to 70 mol% of glycerolipids. The glycolipids replace conventional phospholipids such as phosphatidylcholine in a variety of other cellular membranes including vacuoles, mitochondria, and the plasma membrane (Table 18.1). DGDG almost exclusively resides in the cytoplasmic leaflet of these membranes and oriented toward the cytoplasm, which is opposite of the conventional orientation of glycolipids as known from animal and yeast cells. Variants containing sulfoquinovose and GlcUA, which conserve the negative charge, are also produced. Little is known about the trafficking of these glycolipids. Given the functional importance of DGDG for cells based on genetic studies, it will be interesting to understand how the asymmetric distribution contributes to cytoplasmic functionality of the cellular membranes.

POSSIBILITIES OF "CONVENTIONAL" SECRETORY-TYPE GLYCANS ON NUCLEOCYTOPLASMIC GLYCOPROTEINS

N-Glycan Types

The above examples describe well-documented examples from both prokaryotes and eukaryote of bona fide glycoproteins that function in the nucleus or cytoplasm and are glycosylated by residential nucleocytoplasmic glucosyltransferases (GTs). This section addresses other potential sources of nucleocytoplasmic glycoproteins. As suggested by the examples below, these proteins might be glycosylated by glycosyltransferases normally associated with the rER or Golgi, either by transient occurrence of the glycoprotein in the secretory pathway, or translocation of the secretory pathway glycosyltransferases into the nucleocytoplasmic compartment.

A sizable fraction of nascent proteins fail quality control of folding and assembly in the rER (Chapter 39) and are retrotranslocated to the cytoplasm as a part of ERAD (ER-associated degradation), as depicted in Figure 18.1H. Most of these are N-glycosylated and normally processed by a cytoplasmic *N*-glycanase (*NGLY1*) or an endo-β-N-acetylglucosaminidase (ENGase) to remove their N-glycans before their degradation by the 26S-proteasome or an autophagic vacuole. There are also N-glycan-dependent E3 ubiquitin ligases that assemble polyubiquitin chains that serve as a signal for proteasomal degradation. Interestingly, *NGLY1* is essential for mitochondrial turnover via mitophagy, although evidence suggests it functions via a role other than deglycosylating mitochondrial proteins. If these processes do not operate on a given protein, because the N-glycan is modified in such a way that it is not recognized or is sequestered in a protein complex or the nucleus, then it may accumulate in the cytoplasm and potentially execute a novel function that might depend on its ER-derived glycan. In addition, there is evidence that endocytosed proteins, such as cholera toxin, can access the rER via retrograde transport and then enter the cytoplasm via retrotranslocation or a related process.

Other studies suggest that glycoproteins conventionally secreted or transported to the plasma membrane have the potential to later move to the nucleus (Figure 18.1I). This includes cytokines, growth factors, and sometimes their transmembrane receptors, some of which have been reported to exert direct effects on transcription. For example, fragments of the neural cell adhesion molecule (NCAM), which is modified by polysialic acid, appear to be internalized into cells in which they differentially impact transcriptional programs according to their polysialic acid content. These findings have been considered controversial because the mechanism of such translocation remains mysterious.

Different approaches support the occurrence of nucleus-associated glycoproteins. The sialic acid–specific lectin, *Sambucus nigra* agglutinin (SNA), was shown to bind to several proteins on the cytoplasmic face of the nuclear envelope, including two major nucleoporins: p62 and p180. Prior sialidase treatment blocked binding of SNA to these nuclear pore proteins and, based on peptide-*N*-glycosidase sensitivity, the sialic acids on p180 appeared to be on N-linked glycans. Furthermore, SNA blocked nuclear protein import in neuroblastoma cells, suggesting that the sialic acids might have functional importance on the nuclear pore proteins. Other notable reports of nuclear glycoproteins include a heat-shock-like nuclear chaperone protein with GlcNAc-binding activity, CBP70, which has been suggested to be N-glycosylated. A subpopulation of nuclear prion proteins may be N-glycosylated and possibly associated with the GlcNAc-binding lectin.

N-glycosylation of cytoplasmic proteins or protein domains has also been proposed. For example, the α-subunit of the dog kidney sodium pump (Na^+, K^+-ATPase), a transmembrane protein, was reported to contain traditional N-linked glycans with terminal GlcNAc residues in its cytosolic domain. This conclusion was based on the enzymatic attachment of radioactive Gal to GlcNAc residues by galactosyltransferase labeling of permeabilized right-side-out membrane vesicles. Peptide-*N*-glycosidase sensitivity of the radiolabeled products suggested that the acceptors are N-linked glycans (Chapters 9 and 50). However, this and other claims will be considered provocative until they are confirmed by direct site mapping of the putative glycans. Based on conventional principles of the separate cellular compartmentalization of the N-glycosylation pathway and cytoplasmic and nuclear proteins (not within the nuclear envelope, which is an extension of the rER) (Figure 18.1) and known pathways of translocation of proteins across membranes, there is not a simple explanation for how these modifications are acquired.

Finally, findings of cytoplasmic glycosylation in pathological tissues, such as tau N-glycosylation during neurodegeneration, raise the possibility that compartmental breakdowns result in the exposure of cytoplasmic proteins to normally latent Golgi glycosyltransferases.

Secretory Pathway O-Glycan Types

Glycosylation that is more reminiscent of secretory pathway–type O-glycans has also been reported. In plants, a nuclear pore–associated protein is recognized by the lectin WGA and could be labeled using tritiated UDP-Gal and β4-galactosyltransferase, which is specific for nonreducing terminal GlcNAc. The glycan, with an approximate size of five sugars, was released by mild alkaline degradation consistent with β-elimination from a Thr or Ser residue. The identity of the glycosyltransferases is unknown. More recently, evidence was presented for animal nuclear proteins, including lamins, ribonucleoproteins, and estrogen receptor-α, that are O-glycosylated by αGalNAc (Tn antigen) or core 1 O-glycans Galβ1,3-GalNAc-. Further data show that several intracellular proteins including p53 are enriched by the lectin VVA indicative of modification with O-GalNAc. Another potential example of a cytoplasmic glycan comes from a study on purified mammalian cytokeratin, which was reported to contain GalNAc and bind lectins that recognize α1,3-linked GalNAc in an *N*-acetylgalactosaminidase-sensitive fashion.

A distinct example of glycosylation has been described on parafusin, a cytoplasmic protein from the phosphoglucomutase family, in several eukaryotes. Evidence suggested that this protein possesses a Glcα1-PO_4-Man- linkage, and data were provided for glucose phosphotransferase and Glc-1-phosphate phosphodiesterase activities in cytosolic extracts. However, little is known about the enzymes that catalyze these modifications or the specific structures themselves.

Secretory Pathway Glycosyltransferases in the Nucleus?

A possible mechanism for the occurrence of nuclear O-glycans in the nucleus is if the enzymes that are normally in the secretory pathway gain access to the nucleus or cytoplasm (Figure 18.1F). For example, several reports describe glycosyltransferase activities in highly purified preparations of rat liver nuclei judged to be >99% pure by marker enzyme analysis. These studies documented that incorporation of GlcNAc was inhibited by low concentrations of the antibiotic tunicamycin (an inhibitor of formation of the N-glycan precursor), and later studies showed the direct transfer of chitobiose (GlcNAcβ1-4GlcNAc) from chitobiosyl-dolichol to endogenous nuclear acceptors by these mammalian nuclear preparations, suggesting a novel pathway of N-glycosylation. The products of these in vitro reactions were validated by their sensitivity to peptide *N*-glycosidase F and hydrazinolysis but insensitivity to alkali-induced β-elimination (Chapter 50). Supporting these observations, recent data suggests that both the core GalNAcT1 and GalNAcT2 can localize to the nucleus. Similar studies have documented the presence of nuclear mannosyltransferases. The finding that CMP (cytidine 5′-monophosphate)-sialic acid is synthesized in the mammalian nucleus offers further support for the concept of nuclear glycosyltransferases. Although these studies are provocative, they must also be interpreted with caution. The rER, which is the widely accepted site of N-glycosylation (Chapter 9), is functionally contiguous with the outer nuclear envelope, which also can fold into the interior. Even a minor contamination of nuclear envelope could lead to misinterpretation of enzyme assays. Also, it is very difficult to purify nuclei such that other cellular components do not nonspecifically adhere to the otherwise "pure" nuclei during their preparation. Given these potential problems, widespread acceptance of the existence of these nuclear glycosyltransferases must await independent confirmation by alternative methods.

Glycosaminoglycans

Another class of glycans traditionally thought to reside at the cell surface and the extracellular matrix are the glycosaminoglycans (or GAGs) (Chapter 17). Early studies provided evidence of GAGs in purified nuclei, and soon thereafter studies on sea urchin embryos indicated developmentally regulated effects of stage-specific heparan sulfate (HS) on transcription that were not observed for chondroitin sulfates or hyaluronan (Chapter 16). Given recent advances in the tools to study proteoglycans, these potentially exciting observations warrant a critical reexamination. Years later, subcellular fractionation of rat hepatocytes radiolabeled with $^{35}SO_4$ revealed substantial enrichment of specific fragments of HS (after cleavage) in their nuclear fractions. Given the unique nature of these structures, it is hard to conceive how they could be derived by contamination from cell surface HS molecules, which lack these structures. Subsequent pulse-chase studies suggested that exogenous or cell surface HS is taken into the nucleus and modified to these unique nuclear molecular species. In comparison, others found dermatan sulfates, but not HSs, associated with the nucleus of a different cell type. Given the ability of HSs to influence gene transcription in vitro, these studies could eventually prove highly significant. However, all of the known enzymes involved in the biosynthesis of glycosaminoglycans have their active sites in luminal compartments (Chapter 17), and pathways for nuclear uptake of such large and negatively charged molecules have not been described.

Using another approach, immunocytochemical studies probing for a GPI-anchored HS proteoglycan (PG), glypican, have provided strong evidence for the nuclear accumulation of this PG in addition to its traditionally accepted presence at the cell surface. Although the mechanism of nuclear compartmentalization is not known, the glypican polypeptide harbors a nuclear localization sequence that is functional when grafted onto a neutral carrier expressed in the cytoplasm. In a separate study, overexpression of a chondroitin sulfate proteoglycan core protein in cultured cells resulted in accumulation of the core polypeptide in the cytoplasm and nucleus in addition to the secretory pathway, indicating, as has been suggested for other proteins, the possibility of dual compartmentalization. However, although these studies provide an explanation for how proteoglycan core proteins might accumulate in the nucleus, it is difficult to imagine how they have been modified by the glycosaminoglycan synthases residing in the Golgi lumen.

Cytochemical studies using naturally occurring proteins with high-affinity binding to hyaluronic acid have been used to document accumulation of hyaluronic acid in the nuclei of cultured cells. In response to inflammatory conditions, such as hypoglycemia, cells appear to synthesize hyaluronan in intracellular compartments. After cell division, the intracellular hyaluronan is extruded outside cells in cables that recruit inflammatory cells initiating a stress response. Mechanistically, this is likely caused by a redistribution of the hyaluronan synthases to intracellular membranes from the plasma membrane, across which hyaluronic acid is normally extruded (Figure 18.1N). Notably, hyaluronic acid–binding proteins naturally occur within the nucleus raising the possibility of a physiological role for nuclear hyaluronan via interaction with these hyaluronic acid–binding proteins.

VALIDATED CYTOPLASMIC GLYCOSYLTRANSFERASES WHOSE TARGETS ARE EXPORTED FROM THE CELL

A striking example of the assembly of complex glycans in the cytoplasm comes from the PBCV-1 viruses that infect the alga *Paramecium*. Its major capsid protein, MCP or VP54, is synthesized, posttranslationally modified, and incorporated into viral structures within the cytoplasm of the host. X-ray diffraction analysis shows that VP54 possesses four sites for unconventional Asn-linked-glycosylation, which do not match the canonical N-glycosylation sequon used for N-glycosylation in the eukaryotic secretory pathway, and the linkage sugar is βGlc. A major glycoform consists of a branched, partially methylated decasaccharide containing seven different sugars, including D-Glc, D-Gal, D-Man, D-Xyl, L-Fuc, L-Ara, D-Rha, L-Rha, and dimethylated L-Rha. The glycan is assembled stepwise by virally encoded, sugar nucleotide-dependent, cytoplasmic glycosyltransferases (Table 18.1). Glycosylation of MCP is at least partially virally encoded, and indeed the genome of PBCV-1 contains seven sequences predicted to encode cytoplasmically localized glycosyltransferases that lack targeting sequences for the secretory pathway. Some of these enzymes appear to have multiple enzymatic domains, as observed in the Skp1 glycosylation pathway. The structure is polymorphic between viral lineages and, because not all linkages have been assigned to predictable glycosyltransferase sequences, it is possible that new classes of glycosyltransferase genes are yet to be discovered. Because PBCV-1 eventually lyses its host, the product glycoprotein is expected to function outside of the alga (Figure 18.1L). The MCP pathway elegantly demonstrates the potential of the cytoplasmic compartment to mediate complex glycan assembly that, however, does not ultimately appear to contribute a cytoplasmic or nuclear function.

Protein glycosylation is becoming increasingly recognized as a common modification in bacteria with up to 50% of surface proteins predicted to bear glycans through N-, O-, S-, and C-linkages (Chapter 21). In comparison to eukaryotes, protein glycans appear to vary more among bacterial species. A substantial fraction of protein glycosylation occurs directly on the cytoplasmic surface of the cell membrane prior to export to the surface (Figure 18.1O). In Gram-positive bacteria, the serine-rich repeat protein (SRRP), a bacterial adhesin, is modified with GlcNAc residues by the combined action of GftA/B. Depending on the strain, this modification can be extended into longer structures containing GlcNAc or glucose/galactose residues. These glycoproteins are exported by an accessory secretion (SecA2/Y2) system and the glycan modification appears to play a role in modulating the adhesive properties of the bacterium. *Haemophilus influenzae* also modifies high molecular weight adhesions 1 (HMW1) prior to export. HMW1 is modified at some 31 asparagine residues, predominantly in the N-X-S/T amino acid motif, with mono- or dihexose modifications containing glucose and galactose. Glycosylation is initiated and extended by HMW1C (NGT), an enzyme with homology to the O-GlcNAc transferase. Of note, HMW1C uses UDP-Glucose and UDP-Galactose to initiate and extend glycans in a stepwise manner. Glycosylation of HMW1 protects against premature degradation and promotes tethering to the bacterial cell surface. Recent studies suggest that closely related proteins in *Actinobacillus pleuropneumoniae*, *Kingella kingae*, and *Aggregatibacter aphrophilus* are also glycosyltransferases. Among several other known examples, flagellin filament proteins of *Caulobacter crescentus* are glycosylated intracellularly with pseudaminic acid by the hitherto unknown FlmG glycosyltransferase. This modification is important for successful cell surface flagellation.

INTERMEDIATES IN ASSEMBLY OF EXPORTED GLYCOCONJUGATES OR POLYSACCHARIDES

The biosynthesis of several polysaccharides and precursors of secretory pathway glycans are mediated by membrane-associated glycosyltransferases whose catalytic domains are situated at the cytoplasm-facing surface. This includes early steps in the synthesis of glucoceramides, GPI anchors, and dolichyl-linked N-glycosylation precursors (Figure 18.1M). However, these precursors are ultimately "flipped" to the other side of the membrane where they are extended and, in the case of the GPI-anchor and N-glycan precursors, are transferred to proteins within the lumen of the ER in the secretory pathway (Chapter 9). Cytosolic-oriented glycosyltransferases also include transmembrane proteins that polymerize hyaluronic acid (Chapter 16), cellulose, chitin, and lipopolysaccharides, in which the products are directly translocated across the plasma membrane (Figure 18.1N). The glycan products of these membrane-associated glycosyltransferases normally exit the cytoplasmic space but, if this did not occur, they could serve as the origin of novel, cytoplasmic, nonprotein-linked glycoconjugates.

NUCLEAR AND CYTOPLASMIC LECTINS AND ENZYMES

The occurrence of cytoplasmic or nuclear glycoproteins or glycolipids (oriented toward the cytoplasm) underlies hypotheses for the parallel existence of carbohydrate-binding proteins (or lectins) within these same compatments. O-β-GlcNAc is a prominent nucleocytoplasmic modification, and the heat-shock chaperone protein CBP70 reportedly recognizes O-β-GlcNAc-modified nuclear, cytoplasmic, and mitochondrial proteins. Furthermore, recent evidence shows that the 14-3-3 family of proteins are O-GlcNAc lectins. Early experiments demonstrating that BSA-based neoglycoproteins derivatized with L-Rha, D-GlcNAc, D-Glc, lactose, Man-6-PO_4, and L-Fuc all bind to nuclei at threefold higher affinity than underivatized BSA (see below) reinforced the possible existence of cytoplasmic and nuclear lectins. However, little is known about the molecular nature of these putative carbohydrate-binding proteins.

The galectin family of lectins in animals, and the discoidin family of cytoplasmic lectins in the amoebozoan *Dictyostelium*, are high-abundance proteins that have carbohydrate-binding specificity generally directed toward β-linked Gal and/or GalNAc (Chapter 36). Large pools of these proteins are soluble in the cytoplasm, but cell biological studies show these proteins also reside at the cell surface and in the pericellular matrix. These proteins lack typical amino-terminal signal peptides and exit the cytoplasm by a poorly characterized, posttranslational, unconventional, secretory mechanism that may avoid premature association with glycoproteins in the secretory pathway. Biochemical and genetic studies have uncovered primarily extracellular functions for these lectins (Chapter 36); strikingly, the soluble galectins appear to act as "cytoplasmic sentinels" regulating cellular response to endomembrane damage. On lysosomal damage, galectins-3 and -8 recognize exposed cytoplasmic glycoconjugates. These galectins recruit autophagy and ESCRT proteins to either repair or remove the damaged lysosome. Concurrently, galectin-9 will interact with and activate AMP-dependent protein kinase, and galectin-8 acts to inhibit mTOR activity.

Galectin-3 (CBP35) and possibly galectin-1 are present in the nucleus as part of the hnRNP complex, where they appear to be required for normal mRNA splicing in extracts. However, at present there is only limited evidence for a functional role of the carbohydrate-binding activity of galectin-3 in mRNA splicing. In addition, a cytoplasmic function in regulating apoptosis has been implicated for galectin-1, although a role for the carbohydrate-binding activity has not been established. Galectin-3 is a regulator of mitotic spindle by interacting with the O-GlcNAcylated form NuMA1 protein to stabilize the microtubule-organizing center. Thus, these findings demonstrate that galectins have numerous important cellular functions and potentially other functions waiting to be discovered.

Several filamentous fungi express cytoplasmic lectins with glycan-binding specificities that do not seem to be expressed in the fungi themselves (Chapter 23). These lectins are toxic to predators like roundworms, mosquitoes, and amoebas, and represent a form of innate immunity. In a potentially related observation, several galectins show selective reactivity to glyco-epitopes on pathogenic bacteria. It is interesting to speculate that the cytoplasm represents a "safe harbor" to stow carbohydrate reactive proteins as a defense for other cells in the community should damage cause their release from individual members. Thus, the occurrence of a lectin does not necessarily imply the existence of a cognate glycan in the same compartment.

Numerous other soluble cytoplasmic proteins have been assigned sugar-binding activity that point to the potential significance of cytoplasmic glycoconjugates. In many cases, these must be identified based on biochemical or genetic studies because of the many paths by which these activities seem to have evolved. In plants, cytoplasmic mannose- and GlcNAc-binding lectins are induced by various kinds of biotic and abiotic effectors. Further studies are warranted to evaluate the functional significance of the glycan-binding activities of these interesting proteins.

Indirect evidence for the existence of intracellular lectins comes from evidence that sugars may also serve as nuclear localization signals. Molecules larger than ~40 kDa do not diffuse freely through nuclear pores and must be specifically and actively transported into and out of the nucleus. The synthetically generated neoglycoproteins BSA-Glc, BSA-Fuc, and BSA-Man are rapidly transported into the nucleus of permeabilized or microinjected living HeLa cells, whereas bovine serum albumin (BSA, ~66 kDa) itself is not. Like the classical basic peptide–mediated NLS pathway, the sugar-mediated nuclear transport requires energy and is blocked by the lectin WGA, which binds O-GlcNAc at nuclear pores. However, unlike the basic peptide system, the sugar-mediated pathway does not require cytosolic factors and is not blocked by sulfhydryl-reactive chemicals. Additional evidence shows that BSA-GlcNAcβ1-4GlcNAc is rapidly localized to purified nuclei in vitro by a pathway distinct from the classically defined NLS systems. Validation of these fascinating results awaits characterization of the components involved, and identification of natural counterparts to the neoglycoproteins.

A final line of evidence for nucleocytoplasmic glycans is the colocalization of glycosidases. In addition to the lysosomal hexosaminidases Hex A and Hex B, mammals possess two neutral hexosaminidases Hex C and Hex D. Hex C is the O-GlcNAcase (OGA) that catalyzes the removal of O-βGlcNAc. Like OGA, HexD has a neutral pH optimum and is localized to the nucleus and cytoplasm; however, HexD has a preference for galactosaminide substrates. Isoforms of the Neu4 sialidase have been localized at the outer member of the mitochondrion. As described above, processing of products of ERAD have been ascribed to resident cytoplasmic NGLY1 and ENGase, but if the translocated proteins are not degraded, ENGase provides a mechanism to generate mono-β-GlcNAcylation at Asn residues (Figure 18.1H), and the cytoplasmically localized α-mannosidase MAN2C1 might process high-mannose N-glycans to simplified core structures.

CONCLUSION

Overall, there are many tantalizing clues for the existence and importance of glycoconjugates with simple or complex glycans within the nucleus and cytoplasm, in addition to the ubiquitous O-βGlcNAc (Chapter 19) and the analogous O-fucose outside of animals. Well-characterized examples involve novel linkages to the protein via tyrosine in animal and yeast glycogenin, via hydroxyproline in protist Skp1, and via arginine in bacteria (Table 18.1). The known glycosyltransferases that mediate these modifications are traditional cytoplasmically localized proteins that evolved from the same evolutionary lineages that generated the enzymes of the secretory pathway. At present, these modifications appear to be directed to specific protein targets.

Cytoplasmic glycosylation is also a strategy for pathogens to control host-cell responses, and these mechanisms also frequently involve novel linkages and target single proteins. These examples clearly establish the importance of cytoplasmic and nuclear glycosylation in protein-specific regulation, which contrasts with the relatively broad distribution and heterogeneity of glycans on cell surface, extracellular matrix, and blood proteins. However, much remains to be explored to establish the generality of this concept. Significantly, there exists substantial indirect evidence for much more extensive complex cytoplasmic glycosylation in animal cells, and cell biologists continue to illustrate novel mechanisms of protein compartmentalization that could allow a protein that is glycosylated in one place to be transferred to another, as more commonly occurs in prokaryotes. Nevertheless, proof of the implied prevalence of familiar or unfamiliar glycans will require detailed structural evidence in conjunction with supporting biosynthetic, cell biological, and functional studies. Given that much of what is known about cytoplasmic glycosylation has only recently emerged, it is indeed likely that much remains to be discovered in this realm, and that these pathways are much more common in both eukaryotes and prokaryotes than is currently appreciated. This promises to be an exciting and important area of research in the future.

ACKNOWLEDGMENTS

The authors appreciate helpful comments from Priya Umapathi.

FURTHER READING

Hart GW, Haltiwanger RS, Holt GD, Kelly WG. 1989. Glycosylation in the nucleus and cytoplasm. *Annu Rev Biochem* **58**: 841–874. doi:10.1146/annurev.bi.58.070189.004205

Chandra NC, Spiro MJ, Spiro RG. 1998. Identification of a glycoprotein from rat liver mitochondrial inner membrane and demonstration of its origin in the endoplasmic reticulum. *J Biol Chem* **273**: 19715–19721. doi:10.1074/jbc.273.31.19715

Hascall VC, Majors AK, De La Motte CA, Evanko SP, Wang A, Drazba JA, Strong SA, Wight TN. 2004. Intracellular hyaluronan: a new frontier for inflammation? *Biochim Biophys Acta* **1673**: 3–12. doi:10.1016/j.bbagen.2004.02.013

Monsigny M, Rondanino C, Duverger E, Fajac I, Roche AC. 2004. Glyco-dependent nuclear import of glycoproteins, glycoplexes and glycosylated plasmids. *Biochim Biophys Acta* **1673**: 94–103. doi:10.1016/j.bbagen.2004.03.015

Funakoshi Y, Suzuki T. 2009. Glycobiology in the cytosol: the bitter side of a sweet world. *Biochim Biophys Acta* **1790**: 81–94. doi:10.1016/j.bbagen.2009.03.024

Haudek KC, Spronk KJ, Voss PG, Patterson RJ, Wang JL, Arnoys EJ. 2010. Dynamics of galectin-3 in the nucleus and cytoplasm. *Biochim Biophys Acta* **1800**: 181–189. doi:10.1016/j.bbagen.2009.07.005

Lannoo N, Van Damme EJ. 2010. Nucleocytoplasmic plant lectins. *Biochim Biophys Acta* **1800**: 190–201. doi:10.1016/j.bbagen.2009.07.021

Fredriksen L, Moen A, Adzhubei AA, Mathiesen G, Eijsink VG, Egge-Jacobsen W. 2013. *Lactobacillus plantarum* WCFS1 O-linked protein glycosylation: an extended spectrum of target proteins and modification sites detected by mass spectrometry. *Glycobiology* **23**: 1439–1451. doi:10.1093/glycob/cwt071

Peschke M, Hempel F. 2013. Glycoprotein import: a common feature of complex plastids? *Plant Signal Behav* **8**: e26050. doi:10.4161/psb.26050

Boudière L, Michaud M, Petroutsos D, Rébeillé F, Falconet D, Bastien O, Roy S, Finazzi G, Rolland N, Jouhet J, et al. 2014. Glycerolipids in photosynthesis: composition, synthesis and trafficking. *Biochim Biophys Acta* **1837**: 470–480. doi:10.1016/j.bbabio.2013.09.007

Bullard W, Lopes da Rosa-Spiegler J, Liu S, Wang Y, Sabatini R. 2014. Identification of the glucosyltransferase that converts hydroxymethyluracil to base J in the trypanosomatid genome. *J Biol Chem* **289**: 20273–20282. doi:10.1074/jbc.m114.579821

Naegeli A, Michaud G, Schubert M, Lin CW, Lizak C, Darbre T, Reymond JL, Aebi M. 2014. Substrate specificity of cytoplasmic *N*-glycosyltransferase. *J Biol Chem* **289**: 24521–24532. doi:10.1074/jbc.m114.579326

Jank T, Belyi Y, Aktories K. 2015. Bacterial glycosyltransferase toxins. *Cell Microbiol* **17**: 1752–1765. doi:10.1111/cmi.12533

Lassak J, Keilhauer EC, Fürst M, Wuichet K, Gödeke J, Starosta AL, Chen JM, Søgaard-Andersen L, Rohr J, Wilson DN, et al. 2015. Arginine-rhamnosylation as new strategy to activate translation elongation factor P. *Nat Chem Biol* **11**: 266–270. doi:10.1038/nchembio.1751

West CM, Blader IJ. 2015. Oxygen sensing by protozoans: how they catch their breath. *Curr Opin Microbiol* **26**: 41–47. doi:10.1016/j.mib.2015.04.006

Westphal N, Theis T, Loers G, Schachner M, Kleene R. 2017. Nuclear fragments of the neural cell adhesion molecule NCAM with or without polysialic acid differentially regulate gene expression. *Sci Rep* **7**: 13631. doi:10.1038/s41598-017-14056-x

Yoshida Y, Kuroiwa H, Shimada T, Yoshida M, Ohnuma M, Fujiwara T, Imoto Y, Yagisawa F, Nishida K, Hirooka S, et al. 2017. Glycosyltransferase MDR1 assembles a dividing ring for mitochondrial proliferation comprising polyglucan nanofilaments. *Proc Natl Acad Sci* **114**: 13284–13289. doi:10.1073/pnas.1715008114

Johannes L, Jacob R, Leffler H. 2018. Galectins at a glance. *J Cell Sci* **131**: jcs208884. doi:10.1242/jcs.208884

Curtino JA, Aon MA. 2019. From the seminal discovery of proteoglycogen and glycogenin to emerging knowledge and research on glycogen biology. *Biochem J* **476**: 3109–3124. doi:10.1042/bcj20190441

Sernee MF, Ralton JE, Nero TL, Sobala LF, Kloehn J, Vieira-Lara MA, Cobbold SA, Stanton L, Pires DEV, Hanssen E, et al. 2019. A family of dual-activity glycosyltransferase-phosphorylases mediates mannogen turnover and virulence in *Leishmania* parasites. *Cell Host Microbe* **26**: 385–399. doi:10.1016/j.chom.2019.08.009

Speciale I, Duncan GA, Unione L, Agarkova IV, Garozzo D, Jiménez-Barbero J, Lin S, Lowary TL, Molinaro A, Noel E, et al. 2019. The *N*-glycan structures of the antigenic variants of chlorovirus PBCV-1 major capsid protein help to identify the virus-encoded glycosyltransferases. *J Biol Chem* **294**: 5688–5699. doi:10.1074/jbc.ra118.007182

Yoshida Y, Mizushima T, Tanaka K. 2019. Sugar-recognizing ubiquitin ligases: action mechanisms and physiology. *Front Physiol* **10**: 104. doi:10.3389/fphys.2019.00104

Jia J, Claude-Taupin A, Gu Y, Choi SW, Peters R, Bissa B, Mudd MH, Allers L, Pallikkuth S, Lidke KA, et al. 2020. Galectin-3 coordinates a cellular system for lysosomal repair and removal. *Dev Cell* **52**: 69–87. doi:10.1016/j.devcel.2019.10.025

Koh E, Cho HS. 2021. NleB/SseKs ortholog effectors as a general bacterial monoglycosyltransferase for eukaryotic proteins. *Curr Opin Struct Biol* **68**: 215–223. doi:10.1016/j.sbi.2021.02.004

Sun TP. 2021. Novel nucleocytoplasmic protein *O*-fucosylation by SPINDLY regulates diverse developmental processes in plants. *Curr Opin Struct Biol* **68**: 113–121. doi:10.1016/j.sbi.2020.12.013

West CM, Malzl D, Hykollari A, Wilson IBH. 2021. Glycomics, glycoproteomics, and glycogenomics: an inter-taxa evolutionary perspective. *Mol Cell Proteomics* **20**: 100024. doi:10.1074/mcp.r120.002263

19 The O-GlcNAc Modification

Natasha E. Zachara, Yoshihiro Akimoto, Michael Boyce, and Gerald W. Hart

This chapter presents an overview of the dynamic modification of serine or threonine hydroxyl moieties on nuclear, mitochondrial, and cytoplasmic proteins by O-linked β-*N*-acetylglucosamine, termed O-β-GlcNAc or simply O-GlcNAc. This seemingly simple carbohydrate modification plays key roles in cellular physiology and disease progression. Underpinning these observations are the thousands of O-GlcNAc-modified (O-GlcNAcylated) proteins that regulate cellular processes, such as epigenetics, gene expression, translation, protein degradation, signal transduction, mitochondrial bioenergetics, the cell cycle, and protein localization.

HISTORICAL BACKGROUND

O-GlcNAc was unexpectedly discovered in 1983, when purified bovine milk galactosyltransferase and its radiolabeled donor substrate (UDP-[^{3}H]galactose) were used to probe for GlcNAc-terminating glycoconjugates on the surface of living murine thymocytes, splenic B- and T-lymphocytes, and macrophages. Galactosyltransferase is a Golgi glycosyltransferase that attaches galactose in a

published by Cold Spring Harbor Laboratory Press; doi: 10.1101/glycobiology.4e.19

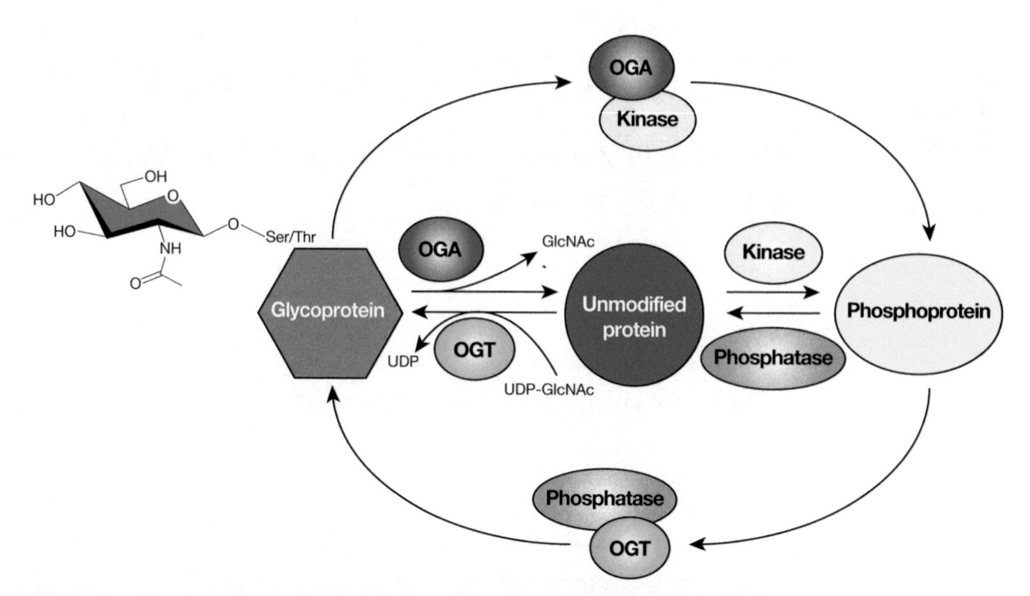

FIGURE 19.1. Many nuclear, mitochondrial, and cytoplasmic proteins are modified by monosaccharides of O-linked β-*N*-acetylglucosamine (O-GlcNAc). In mammals, O-GlcNAc is dynamically added and removed from proteins by just two enzymes, the O-GlcNAc transferase (OGT) and the O-GlcNAcase (OGA). On a subset of proteins, O-GlcNAc competes with phosphorylation for Ser/Thr residues. Evidence suggests that some proteins are rapidly cycled between phosphorylated and glycosylated states by protein complexes composed of OGT and protein phosphatase 1 β, γ, or OGA and protein kinases.

β1-4 linkage to almost any terminal *N*-acetylglucosamine residues, which are common on extracellular glycoconjugates (Chapter 6). Contrary to expectations, product analyses, including release from protein by alkali-induced β-elimination and resistance to cleavage by peptide-N-glycosidase F (PNGase F; Chapter 50), demonstrated most of the labeled galactosylated glycans existed as single O-linked GlcNAc moieties (Figure 19.1). Studies of the subcellular localization of O-GlcNAc in rat liver established that O-GlcNAc is abundant within chromatin, concentrated on the nuclear pore complex, and present within the cytoplasm (Figure 19.2; Table 19.1). Recently established databases show that more than five thousand proteins are O-GlcNAcylated and these glycoproteins fall in almost all protein classes (O-GlcNAcAtlas; The O-GlcNAc Database).

O-GlcNAc is distinct from other common forms of protein glycosylation in several respects: (1) It occurs mostly within the nuclear, mitochondrial, and cytoplasmic compartments of the cell; (2) the GlcNAc moiety is generally not elongated or modified to form more complex structures; and (3) O-GlcNAc is attached and removed multiple times in the life of a protein, often cycling rapidly on a substrate. Of note, this modification is distinct in origin from the recently described extracellular O-GlcNAc found on proteins containing epidermal growth factor (EGF) repeats (Chapter 13). The enzyme that catalyzes the addition of O-GlcNAc to EGF repeats (EOGT) is located in the endoplasmic reticulum (ER) and is closely related to the enzymes that catalyze mucin-like glycosylation, but not the O-GlcNAc transferase (OGT; catalyzes the addition of O-GlcNAc). Unlike intracellular O-GlcNAcylation, modification of EGF repeats by O-GlcNAc is static and can be extended by the addition of galactose.

WHY DID O-GlcNAc REMAIN UNDETECTED FOR SO LONG?

Recent studies show that O-GlcNAc is found on thousands of nuclear, mitochondrial, and cytoplasmic proteins, including many heavily studied proteins such as RNA polymerase II, histones, and ribosomal proteins (Figure 19.2; Table 19.1). Yet, although phosphorylation was discovered

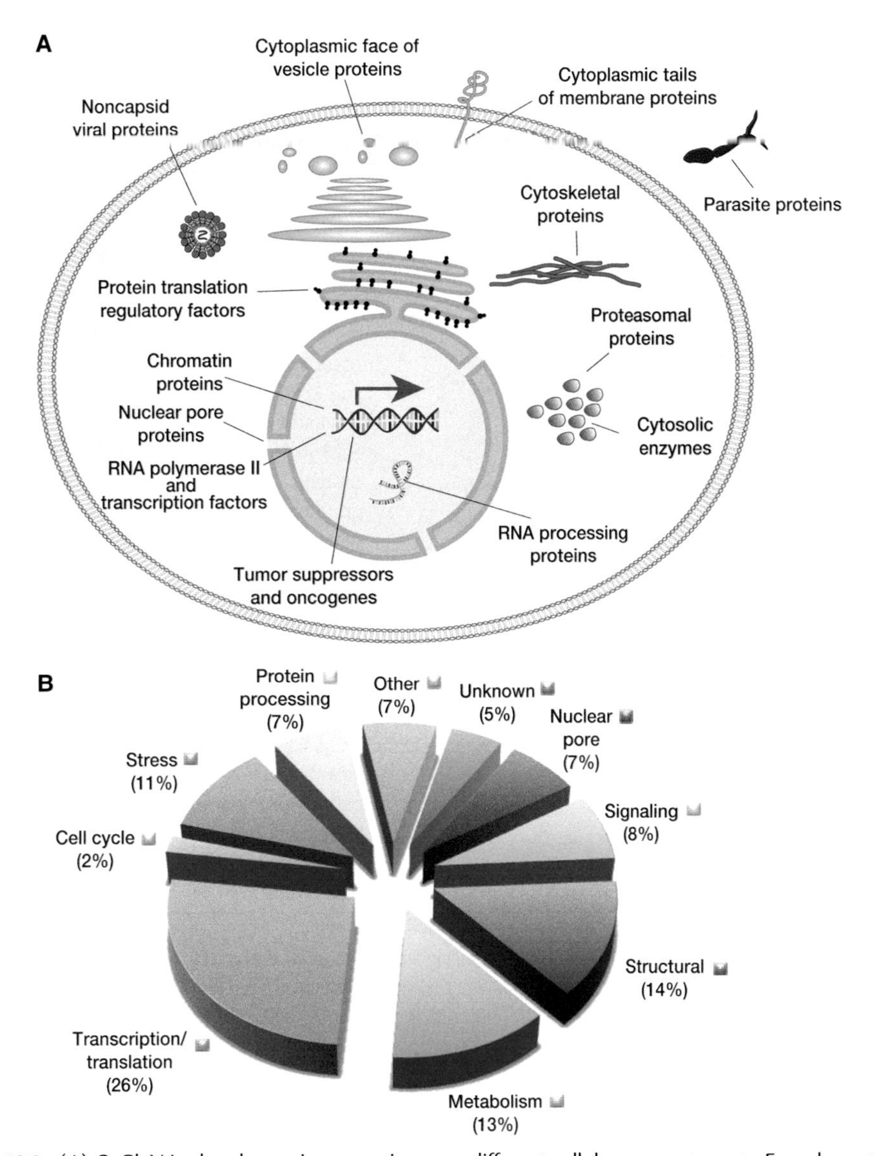

FIGURE 19.2. (*A*) O-GlcNAcylated proteins occur in many different cellular compartments. Found mostly within the nucleus, they are also in the cytoplasm and mitochondria. The subcellular distribution of O-GlcNAc is similar to that of Ser/Thr-O-phosphorylation. (Redrawn, with permission, from Comer FI, Hart GW. 2000. *J Biol Chem* **275:** 29179–29182.) (*B*) O-GlcNAcylated proteins belong to many different functional classes. The pie chart illustrates the functional distribution of O-GlcNAcylated proteins identified thus far. (Redrawn from Love DC, Hanover JA. 2005. *Science STKE* **312:** re13.)

in 1954, O-GlcNAc was not discovered until 1983. So why did O-GlcNAc remain undetected for so long? First, the long-standing dogma was that glycosylation (other than glycogen storage) did not occur with the nucleocytosolic compartment. Second, unlike charged modifications (e.g., phosphate), addition and removal of O-GlcNAc generally does not affect the migration of polypeptides on SDS-PAGE (sodium dodecyl sulfate polyacrylamide gel electrophoresis). A small shift in protein migration might be seen if the O-GlcNAc residues are highly clustered or if the protein is extensively O-GlcNAcylated at multiple sites (e.g., p62 nuclear pore protein). Third, all cells contain high levels of hydrolases, including abundant lysosomal hexosaminidases and nucleocytoplasmic β-N-acetylglucosaminidases, that rapidly remove O-GlcNAc from intracellular proteins when the cell is damaged or lysed. Thus, O-GlcNAc is often lost during the isolation of a protein. Finally, O-GlcNAc is particularly difficult to detect by physical techniques,

TABLE 19.1. Selected O-GlcNAcylated proteins

Protein class	Some examples
Nuclear pore proteins	p62, Nup54, Nup155, Nup180, Nup153, Nup214, Nup358
Epigenetics and transcription	Sp1, c-fos, c-jun, CTF, HNF1, v-ErbA, pancreas-specific TF, serum response factor, c-Myc, p53, estrogen receptors, β-catenin, NF-κB, elf-1, pax-6, enhancer factor D, human C1, Oct1, plakoglobin, YY1, PDX-1, CREB, Rb, p107, RNA polymerase II, ATF-2, HCF-1, SRC-1, TLE-4, CCR4-NOT4, histone H2A, H2B, H3, H4, TET proteins.
RNA-binding proteins	HnRNP-G, Ewing sarcoma RNA-binding protein, EF4A1, EF-1α, RNA-binding motif protein 14, 15 ribosomal proteins
Phosphatases, kinases, adapter proteins	nuclear tyrosine phosphatase p65, casein kinase II, insulin receptor substrates 1 and 2, GSK3β, p85 and p110 PI3-kinase subunits, PKCs, AKT and CamKIIγ
Cytoskeletal proteins	cytokeratins 8, 13, and 18, neurofilaments H, M, and L, band 4.1, talin, vinculin, ankyrin, synapsin 1, myosin, E-cadherin, cofilin, tau, microtubule-associated proteins 1B, 2 and 4, dynein, α-tubulin, AP3, AP180, β-amyloid precursor protein, β-synuclein, piccolo, spectrin β-chain, WNK-1, PDZ-GEF, synaptopodin, vimentin, β-actin
Chaperones	heat-shock protein 27, α-crystallin, HSC70, HSP70, HSP90, HSP60
Metabolic enzymes	eNOS, enolase, glyceraldehyde-3-phosphate dehydrogenase, phosphoglycerate kinase, pyruvate kinase, UDP-glucose pyrophosphorylase, glycogen synthase, phosphofructokinase, Glucose 6 phosphate dehydrogenase, glycolytic enzymes
Other regulatory proteins	eukaryotic peptide chain initiation factor 2, p67, O-GlcNAc transferase, CRMP-2, UCH, GLUT1, annexin 1, nucleophosmin, many proteasome subunits including Rpt2 ATPase, Q04323 UCH homolog, Sec24, Sec23, Sec31, Ran, peptidylprolyl isomerase, Rho GDP-dissociation inhibitor, GABA receptor interacting protein, Milton
Viral proteins	adenovirus fiber protein, SV40 large T antigen, cytomegalovirus basic phosphoprotein, NS26 rotaviral protein, baculoviral tegument protein

(ATF-2) activation transcription factor 2, (CREB) cAMP response element binding, (CRMP-2) collapsin response mediator protein 2, (CTF) CCAAT box–binding transcription factor, (Elf-1) E74-like factor 1, (eNOS) endothelial nitric oxide synthase, (GABA) γ-aminobutyric acid; nitric oxide synthase, (HCF-1) host-cell factor 1, (HNF1) hepatocyte nuclear factor-1, (hnRNP-G) heterogeneous nuclear ribonucleoprotein G, (GSK3β) glycogen synthase kinase-3β, (Nup) nuclear pore protein, (PDX-1) pancreatic and duodenal homeobox 1, (PDZ-GEF) PDZ-domain-containing guanine nucleotide exchange factor, (PKC) protein kinase C, (Rb) retinoblastoma, (SRC-1) steroid receptor coactivator 1, (TLE-4) transducin-like enhancer 4, (UCH) ubiquitin carboxyhydrolase, (NF-κB) nuclear factor-κB, (TET) Ten-eleven translocation, (TF) transcription factor, (YY1) Yin Yang 1.

such as mass spectrometry, because it often occurs at substoichiometric amounts on a protein and is readily lost during the ionization process in a mass spectrometer. In both electrospray ionization (ESI) mass spectrometry and matrix-assisted laser desorption time of flight (MALDI-TOF) mass spectrometry analyses of a mixture of unmodified and O-GlcNAc-modified peptides, not only is the O-GlcNAc lost during ionization but the signal from the O-GlcNAc-modified peptides that remain is suppressed by the presence of the unmodified peptide. Our ability to detect O-GlcNAc has been improved in recent years by the development of monoclonal antibodies, lectins and mutant glycoside hydrolases that bind O-GlcNAc, the use of "click chemistry" compatible sugars (Chapters 51 and 56), and the invention of more sophisticated mass spectrometric techniques, such as electron transfer dissociation (ETD) mass spectrometry.

ENZYMES CONTROLLING O-GlcNAc CYCLING

In mammals, the dynamic cycling of O-GlcNAc on proteins is regulated by the concerted actions of enzymes encoded by just two genes: OGT and a neutral β-hexosaminidase known as O-GlcNAcase (OGA) (Figure 19.1). The regulation of the transcription start site and

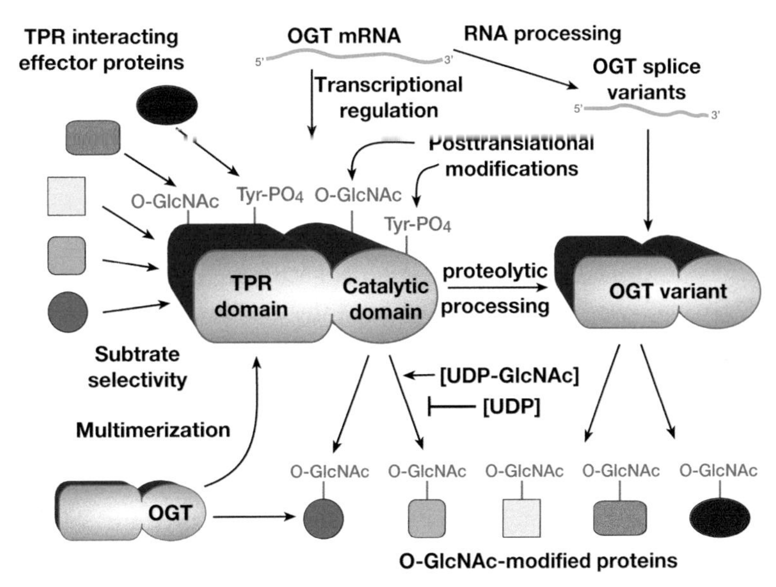

FIGURE 19.3. O-GlcNAc transferase (OGT) is regulated by multiple complex mechanisms, including transcriptional regulation of its expression, differential mRNA splicing, proteolytic processing, posttranslational modification, and multimerization with itself and other proteins. The target specificity of OGT is regulated by a large number of proteins that bind to the tetratricopeptide repeats (TPRs). However, the most important regulator of O-GlcNAc levels is the availability of its donor substrate UDP-GlcNAc. mRNA, Messenger RNA. (Redrawn, with permission, from Comer FI, Hart GW. 2000. *J Biol Chem* **275:** 29179–29182.)

alternative splicing results in the synthesis of at least three isoforms of OGT and two isoforms of OGA. Nonetheless, the question remains: "How can so few enzymes specifically glycosylate and deglycosylate so many substrates?" Although both OGT and OGA each have only a single catalytic subunit, they actually exist within the cell as a multitude of different holoenzymes in which their catalytic subunits are noncovalently bound to a myriad of accessory proteins that appear to control their targeting (Figure 19.3). In addition, both OGT and OGA are modified by both tyrosine and serine/threonine phosphorylation, nitrosylation, ubiquitination, and O-GlcNAcylation. Functions for some of these posttranslational modifications have been identified, including substrate targeting, altering protein localization, and altering specific activity. Notably, cellular O-GlcNAc levels control the maturation of OGT and OGA mRNA and, in turn, the abundance of each enzyme. Thus, when O-GlcNAc levels are elevated, OGT abundance is depressed and OGA abundance augmented. In contrast, in response to low levels of O-GlcNAc, OGA abundance is depressed, and OGT levels are augmented.

O-GlcNAc Transferase

OGT (uridine diphospho-*N*-acetylglucosamine: polypeptide β-N-acetylglucosaminyltransferase; EC 2.4.1.255) catalyzes the addition of *N*-acetylglucosamine from UDP-GlcNAc to specific serine or threonine residues to form a β-O-glycosidic linkage. The enzyme was first identified and purified from rat liver and subsequently cloned from rat, human, *Caenorhabditis elegans*, and other organisms. *OGT* resides on the X chromosome (Xq13 in humans) near the centromere and is among the most highly conserved proteins from worms to man. Mutations in OGT, both in its tetratricopeptide repeats (TPRs) and in its catalytic domain, cause X-linked intellectual disability in humans. To date, three isoforms of OGT have been well characterized: (1) the nucleocytoplasmic or full-length variant (ncOGT), which is 110 kDa; (2) a short isoform of OGT (sOGT), which is 78 kDa; and (3) a variant of OGT that is targeted to the mitochondria (mOGT; ~90 kDa). In the nucleus and cytoplasm, OGT appears to form multimers, consisting of one or more 110-kDa subunits and 78-kDa subunits. Several useful small molecule inhibitors of OGT have been developed, including OSMI-4b, Ac45SGlcNAc, and 5SGlcNHex, with the latter effective in animal models. However, some off-target effects have been reported for these inhibitors, likely because they impact other enzymes that use UDP-GlcNAc.

ncOGT has two distinct domains separated by a putative nuclear localization sequence. The amino terminus of each OGT subunit contains TPRs, which can number up to 13 (species-dependent). The major variation between the three (above) variants is in the number of the TPRs. The crystal structure of the human TPR domain of ncOGT shows that the repeats occur as stacked α-helical domains, forming a "tube-like" structure with a remarkable structural similarity to the armadillo repeat domain of the nuclear transport protein importin-α. The TPR domain mediates multimerization of OGT subunits and acts as a scaffold for protein–protein interactions, including protein substrates (Figure 19.3).

OGT's glycosyltransfer reaction occurs by an ordered bi–bi mechanism with UDP-GlcNAc binding first. The TPR domains are required for substrate binding, enzymatic activity, and stability. Recent studies have demonstrated that residues in the TPR domain are critical for interactions with substrates. Such findings likely underpin the observation that there is no strict consensus motif that dictates glycosylation by OGT, although valine and alanine are preferred at the −3 and +2 positions, respectively, and there is a preference for aromatic residues at −4 position. In addition to catalyzing O-GlcNAcylation, OGT also has the remarkable ability to proteolytically cleave HCF-1, a transcription factor regulating the cell cycle. HCF-1 binds to the TPRs positioning the cleavage site at the active site of OGT. Cleavage occurs between cysteine and glutamate residues, yielding a pyroglutamate. The cleavage site glutamate is converted to serine, which is then O-GlcNAcylated. The cleavage reaction requires UDP-GlcNAc at the active site, which participates in the mechanism.

The regulation of OGT is quite complex and still not well-understood. OGT is itself O-GlcNAcylated and also tyrosine-phosphorylated (Figure 19.3). Tyrosine phosphorylation appears to activate the enzyme, but the role of O-GlcNAc on OGT is not yet clear. Recent studies suggest that OGT is also targeted and regulated by Ser/Thr phosphorylation; AMPK, CAMKIV, CAMKII, and GSK3β have been reported to modify OGT, altering either the localization or activity of the enzyme. Purified or recombinant OGT modifies small synthetic peptides based on known sites from O-GlcNAcylated proteins, but OGT appears to require accessory proteins to modify many full-length protein substrates efficiently.

The high-energy nucleotide sugar used by OGT is UDP-GlcNAc, which is synthesized by the hexosamine biosynthetic pathway (HBP). Increased flux through the HBP, a result of increased glucose levels or certain upstream stimuli, such as the unfolded protein response, has been shown to modulate O-GlcNAc levels (Figure 19.4). On transfer of GlcNAc to proteins, UDP is released, which is a potent feedback inhibitor of OGT. Under conditions in which UDP is rapidly removed (as occurs within the cell), OGT activity is dependent on the UDP-GlcNAc level over a remarkable range of concentrations (from the low nM range to >50 mM). Of note, the substrate specificity of OGT appears to change at different UDP-GlcNAc concentrations, suggesting that OGT regulates cellular processes in a manner dependent on nutritional status.

O-GlcNAcase

Nucleocytoplasmic OGA (EC 3.2.1.169) was first identified as a neutral cytosolic hexosaminidase (referred to as "hexosaminidase C"). OGA was purified from rat kidney and bovine brain, and the human gene was cloned using peptide sequence information. The *OGA* gene was found to be identical to *MGEA5*, a putative hyaluronidase genetically identified because of its

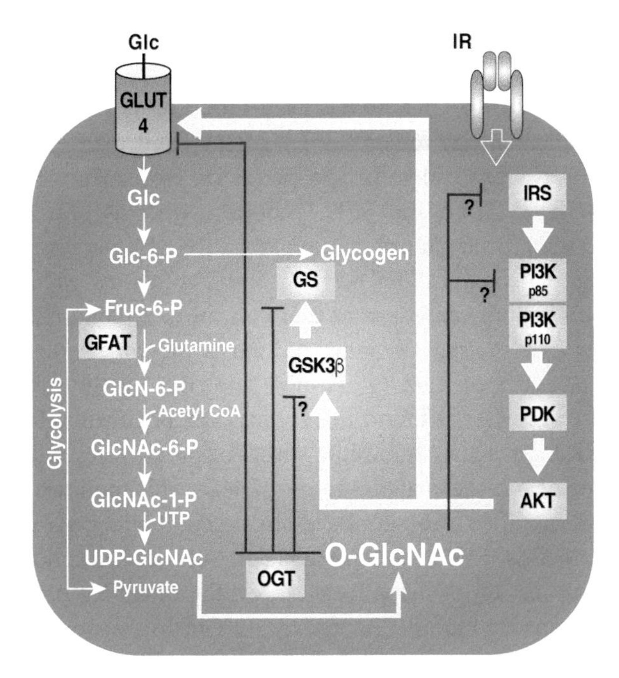

FIGURE 19.4. Elevating O-GlcNAc blocks insulin signaling at many points. Glucose flux via glucose transporters (e.g., GLUT4 in insulin-sensitive cells) through the hexosamine biosynthetic pathway (HBP; which accounts for 2%–5% of total glucose usage) leads to the production of UDP-GlcNAc, the donor for O-GlcNAcylation. Artificially elevating O-GlcNAcylation through increased activity of O-GlcNAc transferase, increased flux through the HBP, or decreased activity of O-GlcNAcase leads to many of the hallmarks of type 2 diabetes. Of note, elevated levels of O-GlcNAc block insulin signaling at several early steps, blocks insulin-stimulated glucose uptake, and prevents insulin from activating glycogen synthase. (IR) Insulin receptor, (PI3K) phosphatidylinositol 3-kinase. (Redrawn, from Slawson C, Housley MP, Hart GW. 2006. *J Cell Biochem* **97:** 71–83, with permission from John Wiley and Sons.)

association with meningiomas. There are two well-characterized isoforms of OGA (short and full-length), which appear to arise from alternative splicing. Short OGA is identical to full-length OGA (916 amino acids) for the first 662 amino acids but possesses an alternative carboxy-terminal sequence 15 amino acids in length. Sequence and structural analyses reveal that OGA contains an amino-terminal hexosaminidase domain and a carboxy-terminal domain, with homology to GCN5 histone acetyltransferases (HAT), separated by an intervening sequence. OGA forms an unusual arm-in-arm homodimer that is mediated by a helix within the intervening sequence. Dimerization is critical for activity and the formation of a substrate binding cleft. Notably, the HAT domain lacks the motif required for binding acetyl-CoA, indicating it lacks HAT enzymatic activity. Like OGT, OGA is thought to be regulated by its protein interactors, posttranslational modifications, and localization. OGA can be efficiently inhibited in vitro and in vivo by a panel of highly effective inhibitors. Although many inhibitors display some cross-reactivity toward the lysosomal hexosaminidases, GlcNAcstatin B and Thiamet-G are selective for OGA.

O-GlcNAc IS A HIGHLY DYNAMIC MODIFICATION

Unlike the relatively static nature of mature N- and O-glycans on glycoproteins, O-GlcNAc cycles rapidly on and off most substrates. Early studies showed that mitogen or antigen activation of lymphocytes rapidly decreased O-GlcNAcylation of many cytoplasmic proteins, but concomitantly increased O-GlcNAcylation of many nuclear proteins. Likewise, neutrophils were shown to rapidly modulate O-GlcNAcylation of several proteins in response to chemotactic agents. Recently, changes in O-GlcNAc cycling have been shown in response to cellular stress, the cell cycle, developmental stages, neuron depolarization, nutrient sensing, and insulin signaling. Pulse-chase analyses have revealed that O-GlcNAc residues on small heat-shock protein in the lens (α-crystallin) and on intermediate filament proteins (cytokeratins) turn over more rapidly than the polypeptide chains to which they are attached. These observations suggest that O-GlcNAc is a regulatory posttranslational modification analogous to phosphorylation.

O-GlcNAc IS UBIQUITOUS AND ESSENTIAL IN METAZOANS

To date, nucleocytoplasmic O-β-GlcNAc has been found in all multicellular organisms investigated, ranging from filamentous fungi, worms, insects, and plants to humans. However, to date no OGA ortholog has been identified in plants or protozoans. Plants encode two OGT paralogs: *SECRET AGENT (SEC)* and *SPY*. Curiously, whereas SEC encodes an O-GlcNAc transferase, *SPY* encodes a fucosyltransferase. Notably, the SPY paralog in *Dictyostelium discoideum, Giardia lamblia, Cryptosporidium parvum,* and *Toxoplasma gondii* also fucosylates proteins. To date, O-GlcNAc and the enzymes that control its cycling do not appear to exist in yeast, such as *Saccharomyces cerevisiae* or *Schizosaccharomyces pombe*. However, recently, it was discovered that *S. cerevisiae* have O-mannose moieties on the same proteins and at the same sites on which O-GlcNAc is found in multicellular organisms. Recent data suggest that the O-GlcNAc modification is also found in a subset of prokaryotes, such as *Listeria monocytogenes*. It is currently unclear if the enzymes responsible for this modification are related to OGT or EOGT. O-GlcNAc-modified proteins have also been found on many viruses that infect metazoans (e.g., adenovirus, SV40, cytomegalovirus, rotavirus, baculovirus, plum pox, HIV, and others). In viruses, instead of being localized on the outside capsid of the virus in which "classical" N- or O-glycans are found (Chapter 42), O-GlcNAc is found deep within the viruses, on tegument and other regulatory proteins, close to the nucleic acid components.

Genetic and subsequent biochemical studies in *Arabidopsis thaliana* showed that the gene *SEC* regulates growth hormone (gibberellic acid) signaling, and later studies established that *SEC* encodes an O-β-GlcNAc transferase. Mutations in *SEC* cause severe growth phenotypes, but they are not lethal. Interestingly, simultaneous mutation of *SEC* and the paralogous *SPY* gene, which encodes an OGT-related fucosyltransferase, is lethal. Studies in rice, potatoes, and other plants have also indicated that O-GlcNAcylation is important for growth regulation.

Mammals and insects appear to have only a single gene encoding the catalytic subunit of the *OGT*. Using *Cre-LoxP* conditional gene disruption in mice, *OGT* was shown to be required for the viability of embryonic stem cells. Tissue-targeted disruption in mice and disruption of *OGT* expression in cell culture have established that O-GlcNAcylation is essential for viability at the single-cell level in many mammalian cell types. Disruption of *OGT* in the worm *C. elegans* causes defective carbohydrate metabolism, abnormalities in dauer formation, and reduced life span. In Drosophila, homozygous mutants in the OGT ortholog super sex combs survive for >5 days until the late larval stage, but subsequently die in the pupal cases. Targeted, inducible deletion of *OGT* in αCAMKII-positive (excitatory) neurons in brains of adult mice results in morbidly obese mice because of defects in satiety, further supporting the roles of O-GlcNAcylation as a nutrient sensor. The highly homologous and conserved nature of *OGT* among metazoans allows a transgene encoding the human sequence of OGT to be used to rescue *OGT* null *Drosophila melanogaster*. Recently, OGA has been disrupted in both *C. elegans* and mammalian models. Like the OGT null, the *C. elegans* OGA null has defective carbohydrate metabolism and abnormalities in dauer formation. In contrast to the OGT null, in *C. elegans* in which OGA has been disrupted, the worms have an extended life span. In murine models, deletion of OGA leads to perinatal lethality, genomic instability and severe epigenetic and metabolic phenotypes. The OGA heterozygous mouse shows widespread changes in transcription and metabolism, and an enhanced dependence on glucose metabolism.

O-GlcNAc Is Found on a Wide Range of Proteins

Studies from numerous laboratories have described O-GlcNAcylated proteins from virtually all cellular compartments, representing nearly all functional classes of protein (Figure 19.2;

Table 19.1). Indeed, recent databases contain more than 6,000 O-GlcNAcylated proteins with more than 7,000 sites mapped (O-GlcNAcAtlas; The O-GlcNAc Database). O-GlcNAc is particularly abundant within the nucleus, in which it occurs on the transcriptional regulatory machinery, including the carboxy-terminal domain (CTD) of the RNA polymerase II catalytic subunit and basal, as well as other, transcription factors, histones, chromatin-modifying proteins, mRNA biogenesis proteins, and DNA methyltransferases. O-GlcNAcylation also appears to be particularly abundant on proteins involved in signaling, stress responses, and energy metabolism. Nearly 90 proteins within the mitochondria are dynamically O-GlcNAcylated, with the highest concentration found within the electron transport chain. O-GlcNAc also occurs on many cytoskeletal regulatory proteins, such as those regulating actin assembly (e.g., vinculin, talin, vimentin, and ankyrin) and tubulin assembly (e.g., MAPs, dynein, and tau). Even α-tubulin itself is dynamically modified by O-GlcNAc, but the stoichiometry appears to be low. Intermediate filaments, such as cytokeratins and neurofilaments in brain, are also heavily modified by O-GlcNAc.

O-GlcNAc HAS A COMPLEX DYNAMIC INTERPLAY WITH O-PHOSPHATE

Site-mapping studies have shown that protein kinases and OGT can often modify the same serine and threonine residues, suggesting that a complex interplay exists between these two posttranslational modifications that fine-tunes signal transduction networks. In fact, major enzymes that remove O-phosphate, protein phosphatase 1 β and γ, are in a dynamic complex with OGT, suggesting that in many cases the same enzyme complex both removes O-phosphate and concomitantly attaches O-GlcNAc (Figure 19.1). One example of a protein in which phosphorylation and O-GlcNAc appear competitive is the CTD repeat domain (YSPTSPS) of RNA polymerase II, which can contain as many as three O-GlcNAc residues per repeat. After the initiation step of the transcription cycle, O-GlcNAc on the CTD is removed and replaced with O-phosphate, initiating the elongation phase of transcription. In vitro studies have shown that synthetic peptides with up to ten CTD repeats (comprising 70 amino acids) cannot be phosphorylated by CTD kinases if they contain even a single O-GlcNAc moiety. Conversely, these CTD peptides cannot be O-GlcNAcylated if even one of the repeats contains a single O-phosphate residue.

On some proteins (e.g., casein kinase II), O-GlcNAc and O-phosphate occur at separate but adjacent sites, yet they still appear to be reciprocal with each other. However, on other proteins, the relationship between O-GlcNAc and O-phosphate dynamics remains unclear. For example, on cytokeratins, O-GlcNAcylation and O-phosphorylation appear to be independently regulated but may occur mutually exclusively on completely different subsets of the same polypeptides. Complicating this relationship further, recent data has shown that many kinases are O-GlcNAc-modified, and that glycosylation can alter the activity and their association with substrates.

BIOLOGICAL FUNCTIONS OF O-GlcNAc

Like other posttranslational modifications, the specific functions of O-GlcNAc depend on the protein and sites to which the moiety is attached. Nonetheless, O-GlcNAc acts like a rheostat to modulate protein's activities, localizations, and interactions in response to environmental, nutritional, and developmental cues. The glycan regulates nearly all cellular processes, including transcription, translation, and mitochondrial function.

O-GlcNAc Regulates Epigenetics and Transcription

Genome-wide studies have shown that OGT, OGA, and O-GlcNAc are found on thousands of promoters in *C. elegans*. Deletion of either OGT or OGA has a profound, but complicated, effect on transcription. Consistent with key roles for O-GlcNAc, deletion of OGT from *C. elegans* (L1 stage) results in the up-regulation of 299 transcripts and the suppression of 389 transcripts, whereas deletion of OGA results in the up-regulation of 218 transcripts and the suppression of 291 transcripts. These complicated effects likely result from the observation that RNA polymerase II and the basal transcription complex are O-GlcNAc-modified. Moreover, O-GlcNAc directly regulates the activities of a variety of transcription factors, including Sp1, estrogen receptors, STAT5 (signal transducer and activator of transcription 5), NF-κB (nuclear factor-κB), p53, YY1 (Yin Yang 1), Elf-1 (E74-like factor 1), c-Myc, Rb (retinoblastoma), PDX-1 (pancreatic and duodenal homeobox 1), CREB (cAMP response element binding), forkhead, and others.

In addition to modulating transcription directly, O-GlcNAc has been strongly implicated in mediating epigenetics. For instance, recent studies have shown that histones (H2A, H2B, H3, and H4) are modified by O-GlcNAc. Moreover, many epigenetic regulators are themselves O-GlcNAc-modified or associate with OGT/OGA. OGT associates with the ten-eleven translocation (TET) proteins, mSin3A/HDAC complexes, and the Polycomb group (PcG) proteins that regulate DNA methylation. In Drosophila, OGT plays a role in Polycomb repression. The association of OGT with the Polycomb repressive complex 2 (PRC2) is critical in maintaining methylation of DNA and suppressing transcription.

O-GlcNAc Regulates Protein Translation, Stability, and Turnover

O-GlcNAc appears to modulate various stages of protein expression, stability, and turnover. First, at least 34 well-characterized ribosomal proteins and several associated translation factors are O-GlcNAcylated. Second, the activity of eukaryotic initiation factor 2 (eIF2) is regulated by its binding to p67. O-GlcNAcylation of p67 protein prevents eIF2 phosphorylation, thus promoting translation. Third, O-GlcNAcylation of eukaryotic translation initiation factor 4E (eIF4E)-binding protein 1 (4E-BP1) promotes binding to eIF4E, thus promoting 7-methyl-GTP-cap-independent translation. Fourth, O-GlcNAc has been reported to prevent protein aggregation, and this is important in models of neurodegenerative disease and during injury. Fifth, O-GlcNAcylation of proteins has been reported to increase their half-life. Reports suggest that this is caused by increased stability of recently translated proteins, suppression of proteasome activity, and reduced targeting of proteins for degradation. Last, it appears that the 26S proteasome itself is O-GlcNAc-modified. Recent proteomic analyses of the 26S proteasome have shown that five of 19 and nine of 14 proteins of the catalytic and regulatory cores, respectively, are modified by O-GlcNAc. Increased O-GlcNAcylation of the Rpt2 ATPase, a component of the 19S cap of the proteasome, blocks its ATPase activity, reducing proteasome-catalyzed degradation. It has been suggested that O-GlcNAcylation of the proteasome allows the cell to respond to metabolic needs by controlling the availability of amino acids and altering the half-lives of key regulatory proteins.

O-GlcNAc Is Involved in Neurodegenerative Disease

Impaired glucose metabolism has been linked to the onset of several neurodegenerative diseases, and one feature is reduced O-GlcNAcylation of key proteins. OGT and OGA map to loci linked to Parkinson's dystonia and late-onset Alzheimer's disease (AD), suggesting that changes in the expression and activity of these enzymes may contribute to the onset of neurodegenerative diseases. Consistent with a model in which decreased O-GlcNAcylation exacerbates the side effects of AD, frontotemporal dementia and parkinsonism, increasing O-GlcNAc levels artificially reduces plaque formation and improves cognition in murine models.

AD is characterized by the production and oligomerization of amyloid peptide β1-42, which is derived from proteolytic processing of amyloid-β precursor protein. The appearance of amyloid peptide in cerebrospinal fluid is concomitant with a reduction in glucose metabolism and hyperphosphorylation and oligomerization of tau. Hyperphosphorylated tau is ultimately secreted into the cerebrospinal fluid, in which it aggregates to form toxic neurofibrillary tangles (PHF [paired helical filament]-tau). These events precede neurodegeneration and brain atrophy. At a molecular level, O-GlcNAc is thought to counteract the effects of AD at several points. First, O-GlcNAcylation of γ-secretase suppresses its activity, reducing the production of amyloid peptide β1-42. Second, on tau, O-GlcNAcylation and phosphorylation appear reciprocal. Recent studies on tau have established that O-GlcNAcylation negatively regulates its O-phosphorylation in a site-specific manner both in vitro and in vivo. These data suggest that O-GlcNAcylation can suppress phosphorylation of tau, and thus reduce the formation of the toxic PHF-tau. Last, O-GlcNAcylated tau appears less likely to aggregate in vitro when compared with unmodified tau. These data suggest that not only does O-GlcNAc prevent toxic hyperphosphorylation of tau, but that O-GlcNAc stabilizes tau protein structure. Drugs that inhibit O-GlcNAcase to raise O-GlcNAc have shown efficacy in murine AD models and are in human clinical trials.

O-GlcNAcylation has been reported to regulate other proteins involved in neurodegenerative disease. Like tau, studies using expressed protein ligation have demonstrated that O-GlcNAcylation reduces aggregation of α-synuclein, which in turn reduces molecular pathologies that contribute to Parkinson's disease. Neurofilaments appear to be hypo-O-GlcNAcylated in neurons from patients with Lou Gehrig's disease (ALS [amyotrophic lateral sclerosis] and MND [motor neuron disease]). In giant axonal neuropathy, neurofilament protein accumulation and aggregation impairs neuronal function and viability. Recently, O-GlcNAcylation of gigaxonin has been demonstrated to promote neurofilament turnover, suggesting another mechanism by which defects in glucose metabolism and O-GlcNAcylation can impact proteostasis and thus disease progression. Clathrin-assembly proteins AP-3 and AP-180 are both modified by O-GlcNAc, and these modifications decline in AD, suggesting that reduced O-GlcNAc is associated with the loss of synaptic vesicle recycling. Altogether, the current data point to potentially significant roles of the O-GlcNAc modification in normal neuronal function and in the molecular mechanisms underlying the pathology of neurodegenerative disease.

Elevated O-GlcNAc Underlies Diabetes Mellitus and Glucose Toxicity

Perhaps the best-understood function of O-GlcNAc is its role in the regulation of insulin signaling and as a mediator of glucose toxicity (see Figure 19.4). The HBP is in a unique position to sense nutrients, coordinating cellular metabolism in response to nucleotide levels, acetyl-CoA, nitrogen metabolism (via glutamine), and glucose levels. Flux through the HBP and subsequent changes in O-GlcNAc levels are thought to mediate signaling pathways, inducing an appropriate response from cells given their nutritional state. For instance, increased O-GlcNAcylation of PDX-1, a pancreatic β-cell transcription factor that controls insulin transcription, increases its affinity for DNA and results in increased proinsulin transcription.

The first studies to link glucosamine metabolism directly with the toxicity of glucose in diabetes showed that in cultured adipocytes, glucosamine is more potent than glucose in inducing insulin resistance, one hallmark of type 2 diabetes (formerly known as adult-onset diabetes or noninsulin-dependent diabetes mellitus). These studies also demonstrated that the ability of glucose to induce insulin resistance could be blocked by deoxynorleucine (DON), a drug that inhibits glutamine:fructose amidotransferase (GFAT), the enzyme that converts fructose-6-P to glucosamine-6-P, and that this blockage could be bypassed by adding glucosamine to the culture media (Chapter 5). In 2002, two seminal studies suggested that aberrant O-GlcNAcylation, a result of elevated UDP-GlcNAc levels, is one molecular mechanism by which glucose and glucosamine metabolism led to insulin resistance. Collectively, these studies showed that artificially increasing O-GlcNAcylation in adipocytes or muscle blocks insulin signaling at several points, and overexpression of *OGT* in muscle or adipose tissue in transgenic mice causes insulin resistance and hyperleptinemia. Recent data shows that overexpression of OGT in the liver also induces insulin resistance and dyslipidemia. Consistent with these data, overexpression of OGA in the liver rescues circulating glucose levels in diabetic mice. Aberrant O-GlcNAcylation is also associated with many of the complications arising from type 2 diabetes. For instance, inappropriate O-GlcNAcylation is associated with mitochondrial dysfunction and contractile defects in the hearts of diabetic mice. Based on these and other studies, elevated levels of O-GlcNAc have been characterized in several models of diabetes and are being investigated as a marker of prediabetes.

O-GlcNAcylation and Cancer

The Warburg effect describes a common metabolic phenotype of cancer cells in which anaerobic glycolysis is used in preference to oxidative phosphorylation. One consequence of this phenotype is increased glucose transport and subsequently increased flux of metabolites through the HBP. Recent studies suggest that one outcome of the Warburg effect is increased O-GlcNAcylation (prostate, breast, lung, colon, and liver) and that these changes in glycosylation modulate signaling pathways, metabolism, and transcriptional profiles. Collectively, changes in O-GlcNAcylation are thought to promote resistance to cell death stimuli and augment angiogenesis, invasion, metastasis, and proliferation, thus enhancing the cancer cell phenotype. Supporting these data is the fact that suppressing the activity of OGT reduces proliferation and migration of cancer cells. Numerous mechanisms have been reported, and these changes in O-GlcNAcylation target changes in transcription, metabolism, and signaling. For example, O-GlcNAcylation of phosphofructokinase 1 (PFK1) shifts metabolites into the pentose phosphate pathway, augmenting glutathione levels and thus enhancing the ability of cancer cells to withstand oxidative stress. Together, these data suggest that detection of key O-GlcNAcylated proteins and expression of OGT/OGA may provide novel biomarkers for the early detection of human cancer. Given the dependence of cancer cells on O-GlcNAc, targeting OGT and other components of the HBP may reduce the aggressiveness of cancer cells while sensitizing them to chemotherapeutic drugs.

O-GlcNAc and Cellular Stress Survival

In every mammalian cell type examined to date, cellular stress initiates a signal that results in a rapid and global increase in O-GlcNAcylation on a multitude of proteins. Levels of O-GlcNAc increase rapidly in response to cellular injury in both in vitro (heat shock, ethanol, UV, hypoxia/reoxygenation, reductive, oxidative, and osmotic stress) and in vivo (ischemia preconditioning and remote ischemic preconditioning) models. Dynamic changes in the O-GlcNAc modification appear to result from changes in the activity and expression of OGT and OGA, as well as flux through the HBP. Several lines of evidence suggest that stress-induced elevations in O-GlcNAcylation promote a survival signaling program in cells and tissues: (1) suppressing O-GlcNAc levels, pharmacologically and genetically, sensitizes cells to oxidative, osmotic, and heat stress; and (2) elevating O-GlcNAc levels, pharmacologically and genetically, promotes survival in models of heat stress, hypoxia reoxygenation, oxidative stress, trauma hemorrhage, and ischemic reperfusion injury of the heart. Consistent with these observations, artificial modulation of O-GlcNAc levels appears to target pathways known to modulate cellular survival. For instance, in models of heat stress, dynamic O-GlcNAcylation promotes the induction of heat shock proteins, chaperones that promote survival by refolding proteins and inhibiting pro-apoptotic pathways. In models of myocardial infarction (heart attack), elevating O-GlcNAcylation has been shown to suppress all of the hallmarks of ischemia reperfusion injury, and include mitochondrial dysfunction, ER stress, increased reactive oxygen species, opening of the mitochondrial permeability transition pore, and calcium overload. Despite these compelling data, our understanding of the molecular events and OGT/OGA substrates that underlie these observations remains unclear. Elucidating how increasing O-GlcNAcylation helps a cell to survive stressful conditions should provide novel targets for the development of therapeutics for treating conditions such as stroke and myocardial infarction.

FUTURE DIRECTIONS

During the past three decades, O-GlcNAcylation has been shown to be a sensor of cellular state (nutrition, stress, cell cycle phase), regulating or modulating nearly every cellular process, including signaling, transcription, translation, cytoskeletal function, and cell division. O-GlcNAc plays critical roles in chronic diseases of aging, including diabetes, cancer, neurodegeneration, and cardiomyopathies. However, the global importance of O-GlcNAc remains unappreciated by most researchers. The modification is difficult to detect by standard biochemical methods, is labile in mass spectrometers, and lacks facile tools to study its biological functions. Improved methods for quantitatively site-mapping O-GlcNAc, methods for altering its stoichiometry at single sites on proteins, and site-specific antibodies for hundreds of key proteins will be required if we are to understand the biological significance of this essential and ubiquitous protein modification. Elucidation of O-GlcNAc's roles in glucose toxicity in diabetes and in mechanisms of neoplasia and its functions in neurons are key future areas of investigation. After almost 40 years of investigation, we are still only beginning to understand the remarkable significance of O-GlcNAcylation in all aspects of eukaryotic biology.

ACKNOWLEDGMENTS

The authors appreciate helpful comments and suggestions from Jarrod W. Barnes, Anabel Gonzalez-Gil, Albert Lee, and Krithika Vaidyanathan.

FURTHER READING

Torres C-R, Hart GW. 1984. Topography and polypeptide distribution of terminal *N*-acetylglucosamine residues on the surfaces of intact lymphocytes. Evidence for O-linked GlcNAc. *J Biol Chem* **259:** 3308–3317. doi:10.1016/s0021-9258(17)43295-9

Hart GW. 1997. Dynamic O-linked glycosylation of nuclear and cytoskeletal proteins. *Annu Rev Biochem* **66:** 315–335. doi:10.1146/annurev.biochem.66.1.315

Comer FI, Hart GW. 2000. O-Glycosylation of nuclear and cytosolic proteins. Dynamic interplay between O-GlcNAc and O-phosphate. *J Biol Chem* **275:** 29179–29182. doi:10.1074/jbc.r000010200

Wells L, Vosseller K, Hart GW. 2001. Glycosylation of nucleocytoplasmic proteins: signal transduction and O-GlcNAc. *Science* **291:** 2376–2378. doi:10.1126/science.1058714

Vocadlo DJ, Hang HC, Kim EJ, Hanover JA, Bertozzi CR. 2003. A chemical approach for identifying O-GlcNAc-modified proteins in cells. *Proc Natl Acad Sci* **100:** 9116–9121. doi:10.1073/pnas.1632821100

Liu F, Iqbal K, Grundke-Iqbal I, Hart GW, Gong CX. 2004. O-GlcNAcylation regulates phosphorylation of tau: a mechanism involved in Alzheimer's disease. *Proc Natl Acad Sci* **101:** 10804–10809. doi:10.1073/pnas.0400348101

Love DC, Hanover JA. 2005. The hexosamine signaling pathway: deciphering the "O-GlcNAc code". *Science STKE* **312:** re13. doi:10.1126/stke.3122005re13

Slawson C, Housley MP, Hart GW. 2006. O-GlcNAc cycling: how a single sugar posttranslational modification is changing the way we think about signaling networks. *J Cell Biochem* **97:** 71–83. doi:10.1002/jcb.20676

Groves JA, Lee A, Yildirir G, Zachara NE. 2013. Dynamic O-GlcNAcylation and its roles in the cellular stress response and homeostasis. *Cell Stress Chaperones* **18:** 535–558. doi:10.1007/s12192-013-0426-y

Dassanayaka S, Jones SP. 2014. O-GlcNAc and the cardiovascular system. *Pharmaco Ther* **142:** 62–71. doi:10.1016/j.pharmthera.2013.11.005

Hardivillé S, Hart GW. 2014. Nutrient regulation of transcription, signaling, and cell physiology by O-GlcNAcylation. *Cell Metab* **20:** 208–213. doi:10.1016/j.cmet.2014.07.014

Lewis BA, Hanover JA. 2014. O-GlcNAc and the epigenetic regulation of gene expression. *J Biol Chem* **289:** 34440–34448. doi:10.1074/jbc.r114.595439

Ma Z, Vosseller K. 2014. Cancer metabolism and elevated O-GlcNAc in oncogenic signaling. *J Biol Chem* **289:** 34457–34465. doi:10.1074/jbc.r114.577718

Vaidyanathan K, Wells L. 2014. Multiple tissue-specific roles for the O-GlcNAc posttranslational modification in the induction of and complications arising from type II diabetes. *J Biol Chem* **289:** 34466–34471. doi:10.1074/jbc.r114.591560

Zhu Y, Shan X, Yuzwa SA, Vocadlo DJ. 2014. The emerging link between O-GlcNAc and Alzheimer disease. *J Biol Chem* **289:** 34472–34481. doi:10.1074/jbc.r114.601351

Halim A, Larsen IS, Neubert P, Joshi HJ, Petersen BL, Vakhrushev SY, Strahl S, Clausen H. 2015. Discovery of a nucleocytoplasmic O-mannose glycoproteome in yeast. *Proc Natl Acad Sci* **112:** 15648–15653. doi:10.1073/pnas.1511743112

Ma J, Liu T, Wei AC, Banerjee P, O'Rourke B, Hart GW. 2015. Protein O-GlcNAcylation regulates cardiac mitochondrial function. *J Biol Chem* **290:** 29141–29153. doi:10.1074/jbc.m115.691741

Lagerlöf O, Blackshaw S, Hart GW, Huganir RL. 2017. The nutrient sensor OGT regulates feeding in αCaMKII-positive neurons of the PVN. *Science* **351:** 1293–1296. doi:10.1126/science.aad5494

Hart GW. 2019. Nutrient regulation of signaling and transcription. *J Biol Chem* **294:** 2211–2231. doi:10.1074/jbc.AW119.003226

Ma J, Li Y, Hou C, Wu C. 2021. O-GlcNAcAtlas: a database of experimentally identified O-GlcNAc sites and proteins. *Glycobiology* **31:** 719–723. doi:10.1093/glycob/cwab003

Wulff-Fuentes E, Berendt RR, Massman L, Danner L, Malard F, Vora J, Kahsay R, Olivier-Van Stichelen S. 2021. The human O-GlcNAcome database and meta-analysis. *Scientific Data* **8:** 25. doi:10.1038/s41597-021-00810-4

20 Evolution of Glycan Diversity

Pascal Gagneux, Vladislav Panin, Thierry Hennet, Markus Aebi, and Ajit Varki

This chapter provides an overview of glycosylation patterns across biological taxa and discusses glycan complexity and diversity from an evolutionary perspective. As much of the currently available information concerns vertebrates, this chapter emphasizes comparisons between vertebrate glycans and those of other taxa. Evolutionary processes that likely determine generation of glycan diversity are briefly considered, including intrinsic host glycan-binding protein functions and interactions of hosts with extrinsic pathogens or symbionts.

RELATIVELY LITTLE IS KNOWN ABOUT GLYCAN DIVERSITY IN NATURE

The genetic code is shared by all known organisms, and core functions such as gene transcription and energy generation are conserved across taxa. Complex glycans are found in all organisms in nature, and some have argued that polysaccharides were the original macromolecules contributing to the origin of life itself. Regardless of their origins, they vary immensely in

published by Cold Spring Harbor Laboratory Press; doi: 10.1101/glycobiology.4e.20

structure and expression (abundance and patterns of distribution in and on cells, secretions, and extracellular matrices) both within and between evolutionary lineages. Partly because of inherent difficulties in elucidating their structures, our knowledge about this diversity remains limited, and there are few comprehensive data sets. For many taxa, there is a lack of any information on glycan profiles. Sufficient data are available to indicate that even though all living cells require a glycocalyx (a dense and complex array of cell-surface glycans), there is no evidence for a universal "glycan code," akin to the genetic code.

Importantly, glycan structures are not directly encoded in the genome. They are synthesized and modified by a network of enzymes in a template-independent manner, and glycophenotypes represent the outcome of co-expressed gene networks and nutrients. Glycans expressed by most free-living Bacteria (Eubacteria) and Archaea have relatively little in common with those of eukaryotes. They contain a much larger number of monosaccharide types and include many glycans exclusive to such microbes. In contrast, most major glycan classes in animal cells seem to be represented in some related form among other eukaryotes, and sometimes in Archaea. Figure 20.1 shows a circular depiction of the phylogeny of cellular life on earth. The rich glycan diversity encountered in the best-studied vertebrate species suggests similar diversity in other groups of organisms, and existing information points to complicated patterns. On the one hand, glycan patterns can form "trends" and characterize entire phylogenetic lineages, wherein one encounters further biochemical variation with subsets unique to certain sublineages. Conversely, many glycans show discontinuous distribution across phyla and distantly related organisms can produce surprisingly similar glycans, using either shared, ancient pathways or convergently (independently) evolved mechanisms.

EVOLUTIONARY VARIATIONS IN GLYCANS

N-Glycans

The broadest base of evolutionary information concerns asparagine–N-linked glycans, a "general" glycosylation system found in all domains of life (Chapter 9). In prokaryotes, protein N-glycosylation takes place in the periplasm at the plasma membrane, whereas in eukaryotes, covalent attachment of an oligosaccharide takes place intracellularly at the endoplasmic reticulum (ER) membrane, with the protein-bound glycan being further processed in the ER and Golgi. Detailed studies of N-glycosylation systems in model organisms reveal that the covalent modification of proteins at asparagine side chains within the N-X-S/T sequon is a homologous process in all taxa, characterized by certain common properties: Nucleotide-activated monosaccharides serve as building blocks for assembly of an oligosaccharide on an isoprenoid lipid carrier in the cytoplasm. The lipid-linked oligosaccharide is translocated across the ER membrane (eukaryotes) or the plasma membrane (prokaryotes). In most eukaryotes, the lipid-linked oligosaccharide is extended further before transfer to protein. Translocated proteins with N-X-S/T consensus sequences can serve as acceptors for the oligosaccharyltransferase (OST), which catalyzes the en bloc transfer from the lipid-linked precursor to the amido group of asparagine. The structural diversity of the initially transferred oligosaccharide is highest in the archaeal domain and lowest in eukaryotes. However, the protein-bound glycan is further processed in the ER and in the Golgi compartments of eukaryotes, generating greater structural diversity of glycans exposed at the cell surface.

Analysis of N-glycosylation in model organisms from all three domains of life offers the opportunity to visualize evolutionary trends and to propose selective forces at work. Structural diversity of surface-exposed N-linked glycans between populations and species is common, likely driven by an evolutionary "arms race" caused by exploitation of host glycans by parasites

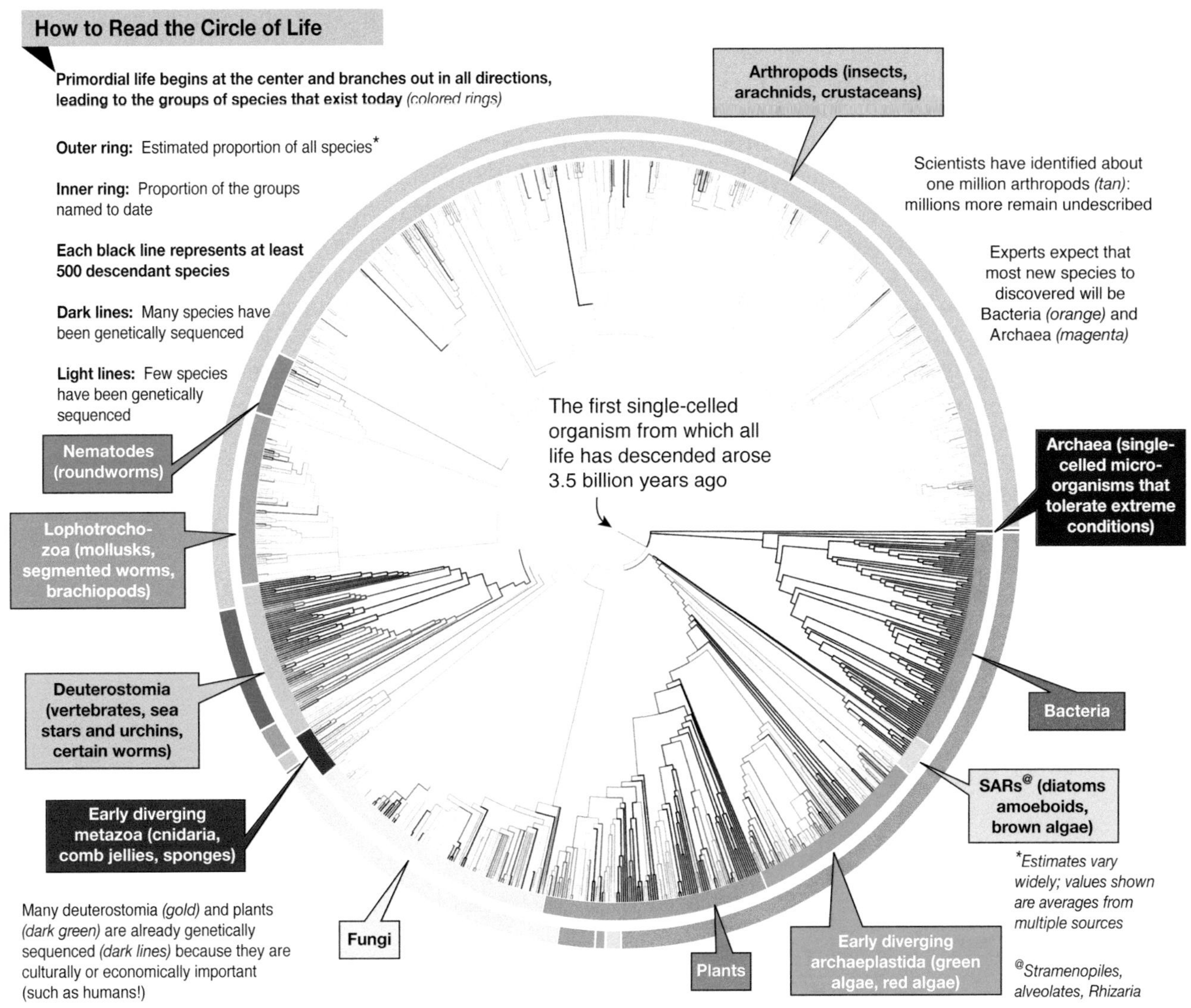

FIGURE 20.1. Circular depiction of phylogeny of cellular forms of life on earth. The lines inside the circle represent all 2.3 million species that have been named. However, biologists have genomic sequences for only ~5% of them; as more sequences become available, the relationships within and across groups of species may change. (*Outer ring*) Estimated proportion of all species (estimates vary widely; values shown are averages from multiple sources). (*Inner ring*) Proportions of the groups named to date. (*Black lines*) Each line represents at least 500 descendant species. (*Dark lines*) Many species with available genomic sequences. (*Light lines*) Few species with available genomic sequences. (Reproduced, with permission, from *Scientific American*, March 2016, p. 76 [Artist: Stephen Smith]. Source: Hinchliff et al. 2015. *Proc Natl Acad Sci* **112:** 12764–12769.)

and pathogens. The intracellular attachment of the oligosaccharide to proteins that are in the process of folding is the basis for the role of highly defined glycan structures in the modulation and the quality control of protein folding in the ER (Chapter 39). This intracellular function of N-linked glycans is reflected in the high degree of conservation of the ER pathway in eukaryotes: only some phylogenetically ancient protists are known to transfer truncated forms of otherwise similar lipid-linked oligosaccharides.

In addition to structural diversity, significant quantitative evolution of N-glycosylation is apparent. Because of the short N-X-S/T sequon in polypeptides directing the OST substrate, a general modification system has evolved that affects many proteins, some with multiple N-glycosylation sites. Analysis of the N-glycome reveals a correlation: There are far fewer

N-glycoproteins in unicellular than multicellular organisms. Thus, N-glycan-mediated intrinsic cell–cell interaction likely forms a selective force that leads to increased N-glycoprotein diversity with multicellularity.

Newer analytical techniques (Chapter 50) reveal ever-increasing structural diversity of N-glycans in eukaryotes, arising from remodeling in the ER and Golgi via variation in trimming, extension, and branching by different building blocks. In addition, N-glycans can be modified by phosphorylation, methylation, etc. There are evolutionary trends of N-glycan processing in eukaryotes (Figure 20.2). In fungi, trimming is restricted to the quality control process of glycoprotein folding, and the diversity of building blocks used for extension is limited. In contrast, trimming to $Man_3GlcNAc_2$ is a characteristic of plants and animals. Interestingly, each unit of this pentasaccharide can serve as a substrate for branching and extension, but only animals seem to have branching from terminal mannose residues, resulting in multiantennary complex-type N-glycans. In contrast, modification of the β-linked mannose by xylose is a characteristic of plants.

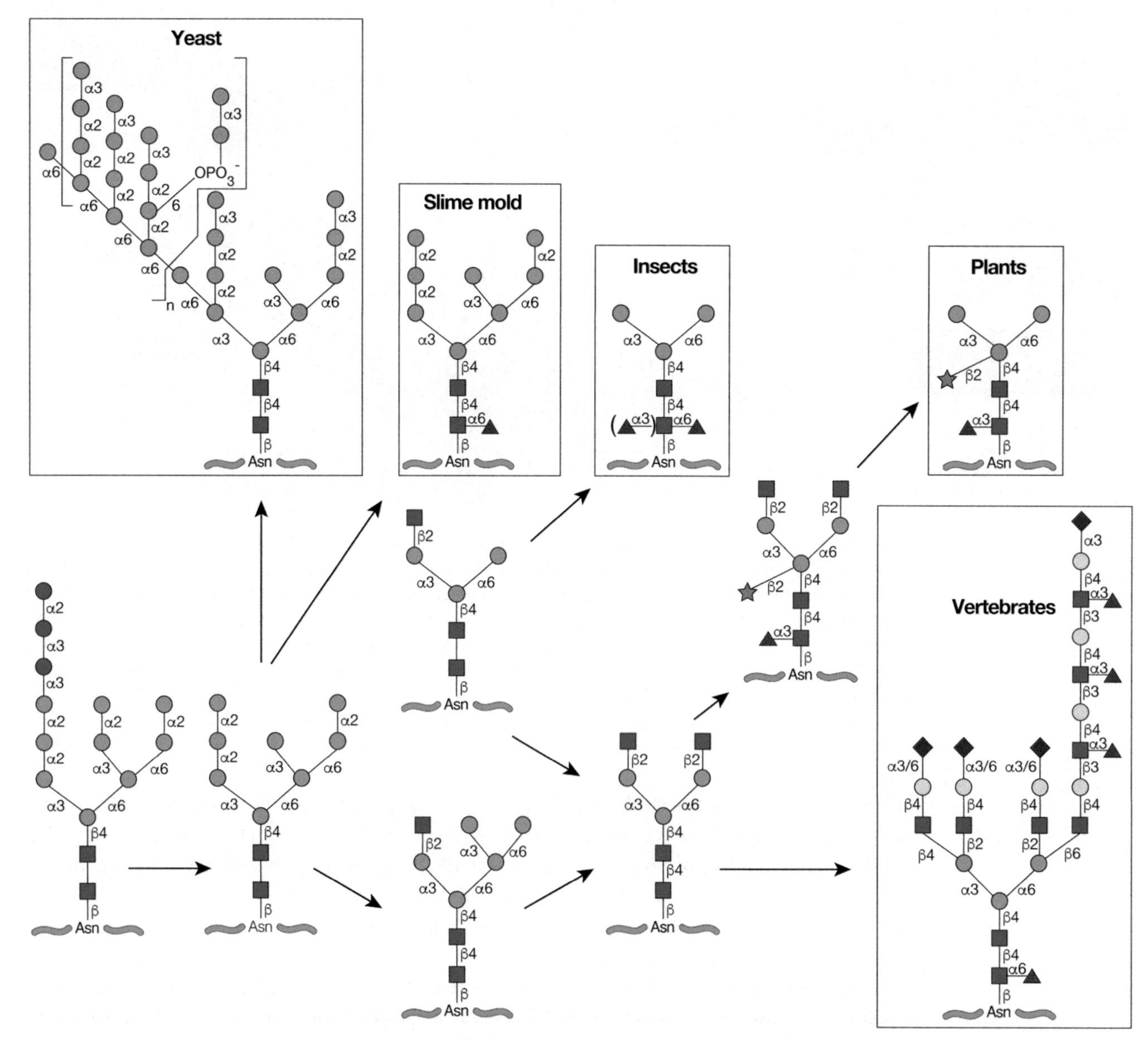

FIGURE 20.2. Characteristic pathways of N-glycan processing among different eukaryotic taxa. See text for discussion.

Lineage-specific N-glycosylation pathways are characterized by the presence or absence of functional glycosyltransferases resulting in defined glycan structures (e.g., LacdiNAc structures) and/or building blocks (e.g., sialic acids). At the organismal level, N-glycan structures can be organ-, cell type–, or sex-specific, because of differential expression of processing enzymes. Their structures can also vary at different developmental stages, during senescence, and in response to diseases and pathogens. The common outer-chain Galβ1-4GlcNAcβ1-(*N*-acetyllactosamine or "LacNAc") is a prime example of phylogenetic variation of glycosyltransferase machinery. The structure, common in invertebrates (Chapter 26), is also found in plants. Some plants even add outer-chain Fucα1-3 residues to the GlcNAc residues of LacNAc units, generating Lewis x–like structures identical to those in animal cells (Chapter 24). In some taxa, such as mollusks, an outer GalNAcβ1-4GlcNAcβ1-structure (the so-called "LacDiNAc" or LDN) tends to dominate, in place of the typical LacNAc structure more commonly seen in vertebrates. The SO_4-4-GalNAcβ1-4GlcNAcβ1-terminal units of pituitary glycoprotein hormones (Chapter 14) are conserved throughout vertebrate evolution, suggesting importance for biological activity.

With respect to monosaccharide building blocks, some appear restricted to certain evolutionary lineages. Arabinose, rhamnose, and xylose are typical in plants, but helminths share some of these monosaccharides. Many bacteria produce monosaccharides bearing unique modifications absent from animals, which in turn secrete defensive lectins (e.g., intelectin-1 or RegIIIa) specific for these microbial glycans. Interestingly, mammalian glycans are assembled from a small number of different monosaccharide units, whereas microbial glycans consist of more than 700 different monosaccharide building blocks. Although more data are needed to be certain, there appears to be a general trend toward reduction of monosaccharide complexity in more recently evolved multicellular taxa with multiple internal organ systems and even more so in complex multicellular organisms that have evolved adaptive immune systems—perhaps because the evolutionary pressure to change the cell surface as a response to viruses is reduced. On the other hand, the much deeper evolutionary history of microbes may contribute to this difference as well. In this regard, the evolution of N-glycosylation apparently took a different route in giant viruses, such as Chloroviruses. They have evolved to use atypical sequons and rely on their own unique glycosylation machinery rather than hijacking host glycosylation. However, the evolutionary forces that brought about these differences remain unclear.

Sialic Acids

Sialic acids are prominent at the outer termini of N-glycans, O-glycans, and glycosphingolipids of deuterostomes (Chapter 15). They were once thought to represent an evolutionary innovation unique to this lineage, which originated during the Cambrian Expansion (500 myr), and other scattered reports of sialic acids in a few other taxa were thought to reflect lateral gene transfer and/or convergent evolution (i.e., independent evolution of sialic acid synthesis in these taxa). However, although lateral transfer mechanisms exist and may explain the presence of sialic acids in some bacterial taxa, sialic acids are also reported in some fungi and mollusks. Genome sequences reveal the presence of a set of genes for sialic acid production and addition in some protostomes (e.g., in insects such as *Drosophila* or mollusks such as *Octopus*). Drosophila *sialyltransferase* and *CMP-sialic acid synthase* genes were found to express functional enzymes structurally and functionally similar to vertebrate counterparts, clearly indicating an earlier evolutionary origin of sialylation. *Caenorhabditis elegans*, the free-living nematode, on the other hand, does not contain genes for synthesizing or metabolizing sialic acid. Earlier claims for sialic acids in plants are probably due to environmental contamination and/or incorrect identification of the chemically related sugar Kdo (2-keto-3-deoxy-octulosonic acid). However, sialic acid biosynthetic genes of some bacterial species share homology with those of vertebrates. It is also now evident that sialic acids are an invention derived from genes of more ancient pathways for nonulosonic acid (NulO) synthesis (Chapter 15). In this scenario, NulOs were differentially exploited during evolution, becoming prominent as sialic acids only in the deuterostome lineage, although being abandoned or substantially reduced in complexity and/or biological importance in other animal and fungal taxa. Sialic acids also appear to contribute to self-associated molecular patterns (SAMPs) in vertebrates, which have immune-modulating intrinsic sialic acid–binding lectins (like Siglecs) and/or recruit blood plasma factor H and thereby dampen complement activation. Meanwhile, a variety of bacteria evade immunity by synthesizing sialic acid–like molecules via convergent evolution of the ancestral NulO biosynthetic pathway (Chapter 15). The ability to produce sialylated glycans is under positive selection in pathogenic microbes that commonly decorate their surface with sialic acids to evade the vertebrate host's immune responses. It is curious that invertebrates, such as echinoderms (sea urchins and starfish) have the highest sialic acid diversity in deuterostomes and that the simplest profiles are found in humans. Thus, sialic acids seem to have evolved in many possible directions, disappearing altogether, or undergoing respective diversification or simplification of their structures. Although there is a tendency for some types of sialic acids to be dominant in certain mammalian species (e.g., *N*-glycolylneuraminic acid in pigs and 4-O-acetylated sialic acids in horses), careful investigation reveals lower quantities of such sialic acids in many other species. Notably, humans are "knockout" primates for CMP–Neu5Ac hydroxylase (CMAH). The *CMAH* gene was inactivated in the hominid lineage ~3 myr. Thus, unlike the closely related great apes, humans cannot synthesize the sialic acid Neu5Gc, which, however, can still be incorporated in human glycans from dietary sources (Chapter 15). Independent loss of function of *CMAH* has also occurred in other mammalian lineages, including New World primates, ferrets and other *Mustelids*, seals and sea lions (*Pinnipeds*), and two lineages of microbats. CMAH inactivation happened at least eight times during mammalian evolution, with some events as early as 200 myr (such as inactivation in platypus). Sauropsids (birds and reptiles, descendants of dinosaurs) appear to represent another lineage that lost Neu5Gc.

O-Glycans

O-glycans encompasses a large class of glycoconjugates, which comprise the monosaccharide core structures O-GalNAc, O-Man, O-Fuc, O-Glc, O-Gal, and O-GlcNAc. Homologs of UDP-GalNAc:polypeptide N-acetylgalactosaminyltransferases (ppGalNAcTs) initiating synthesis of mucin-type O-GalNAc, the most common O-glycan class in vertebrates, have been found throughout the animal kingdom (Chapter 10). Multiple ppGalNAcTs with different polypeptide substrate specificity work in metazoans. The common Core-1 Galβ1-3GalNAcα1-O-Ser/Thr structure of vertebrates exists in insects, where it also forms part of a mucin-like protective layer in the gut. Heavily O-glycosylated gel-forming mucins have been recruited and diversified in metazoans to mediate lubrication and protection of hydrated epithelia in direct contact with the environment. O-Glycosylation of mucin-like domains is also common among protists, in which the glycans are usually initiated by ppGlcNAcTs—evolutionary precursors of the ppGalNAcTs. O-Linked mannose represents another example of Ser/Thr glycosylation that is conserved in prokaryotes and eukaryotes, from yeast to mammals. O-Mannosylation plays diverse functions, supporting cell structural stability, facilitating protein folding, and affecting cell adhesion. In vertebrates, O-mannosylation is mediated by several families of glycosyltransferases, including distantly related protein O-mannosyltransferases (POMT1-2) and transmembrane and tetratricopeptide repeat containing proteins (TMTC1-4). These enzymes evolved to recognize distinct substrates, including unstructured regions (POMTs) and specific protein folds, such as cadherin domains

(TMTCs). On α-dystroglycan, POMTs initiate the biosynthesis of the matriglycan, one of the most elaborate and highly specialized vertebrate glycans that connects muscles with the extracellular matrix (Chapters 13, 27, 45). In contrast, plants do not appear to have O-GalNAc and O-mannose. Instead, they express arabinose O-linked to hydroxyproline and galactose O-linked to serine and threonine (Chapter 24). Far less is known about O-glycosylation in prokaryotes, although O-glycans, such as O-mannose and novel Galβ1-O-Tyr modifications, can be found in bacteria (Chapter 21).

Glycosphingolipids

Glucosylceramide is found in both plants and animals (Chapter 11). However, the most common core structure of vertebrate glycosphingolipids (Galβ1-4Glc-Cer) is varied in other organisms (e.g., Manβ1-4Glc-Cer and GlcNAcβ1-4Glc-Cer in certain invertebrates). Another variation is inositol-1-*O*-phosphorylceramide (e.g., mannosyldiinositolphosphorylceramide, the most abundant sphingolipid of yeast) and GlcNAcα1-4GlcAα1-2-*myo*-inositol-1-*O*-phosphorylceramide, found in tobacco leaves. Galactosylceramide and its derivatives seem to be limited to the nervous system of the deuterostome lineage of animals. In contrast, protostome nerves contain mainly glucosylceramides. An evolutionary trend is suggested: A transition from gluco- to galactoceramides corresponds with changes in the nervous system from loosely structured to highly structured myelin. Regarding complex gangliosides of the deuterostome nervous system, some general trends are seen in comparing reptiles and fish to mammals: an increase in sialic acid content, a decrease in complexity, and a decrease in "alkali-labile" gangliosides (bearing O-acetylated sialic acids). The lower the temperature, the more polar the composition of brain gangliosides; poikilothermic (cold-blooded) animals tend to express many polysialylated gangliosides in the brain.

Glycosaminoglycans

Heparan and chondroitin sulfate are found in many animal taxa, including insects (Chapter 26) and in mollusks. The most widely distributed and evolutionarily ancient class appears to be chondroitin chains, which are not always sulfated (e.g., in *C. elegans*) (Chapter 25). The more highly sulfated and epimerized forms of heparin and dermatan sulfate tend to be found primarily in more recently evolved animal species of the deuterostome lineage. The same is true of hyaluronan, a secreted free glycosaminoglycan synthesized at the cell membrane and likely evolved from its precursor chitin to facilitate the movement of metazoan cells and shaping of organs during development and normal physiology. Echinoderms such as the sea cucumber make typical chondroitin chains, but some glucuronic acids have branches containing fucose sulfate. Simpler multicellular animals such as sponges can have unusual glycosaminoglycans that include uronic acids but do not have the typical repeat units of chondroitin and heparan sulfate. Plants do not have typical animal glycosaminoglycans. Instead, they have secreted acidic pectin polysaccharides not attached to protein cores (only hyaluronan is unattached among animal glycosaminoglycans), characterized by galacturonic acid and its methyl ester derivative (Chapter 24). Bacteria have mostly completely distinct polysaccharides (Chapter 21), although certain pathogenic strains can mimic mammalian glycosaminoglycan chains (see below).

Nuclear and Cytoplasmic Glycans

The O-β-GlcNAc modification common on cytoplasmic, mitochondrial, and nuclear proteins (Chapter 19) is widely expressed in "higher" animals and in plants, and its occurrence on histone tails implicates this modification in epigenetic regulation mechanisms. The connections between nutrient state, UDP-GlcNAc levels, and O-GlcNAcylation mean that gene regulation has an important metabolic dimension. Conserved homologs of the responsible O-GlcNAc transferase have been found in many eukaryotic taxa and in a wide range of bacteria. An independent clade of enzymes evidently originating in bacteria mediates O-fucosylation instead of O-GlcNAcylation in plants and numerous protists (Chapter 18). Most interestingly, in some animal pathogenic members of Pasteurellaceae and Enterobacteriaceae, homologs of the O-GlcNAc transferase catalyze a cytoplasmic N-glycosylation within N-X-S/T sequons of adhesins. Molecular mimicry of eukaryotic N-glycoproteins might be the driving force for this convergent evolution of an N-glycosylation system. Surprisingly, O-GlcNAcylation is apparently absent in yeast. However, nucleocytoplasmic yeast proteins can carry O-mannose in the same conserved regions that are modified with O-GlcNAc in mammals, suggesting that a cytoplasmic O-mannosylation evolved in yeast by convergent evolution to function similarly to O-GlcNAcylation.

Structural Glycans

The most abundant biopolymers in nature include secreted polysaccharides such as cellulose, hemicellulose, chitin, and glycosaminoglycans. These large, chemically stable molecules provide crucial structural support for capsules, cell walls, exoskeletons, and extracellular matrices of countless organisms. The repeated β1-4 glycosidic linkages of cellulose and chitin also represent chemically extremely resistant sequestrations of accumulated glycans, given the inability of most organisms to hydrolyze these robust linkages. Compared to other types of glycans, structural polysaccharides show relatively little variation in evolution, which may indicate a more stringent selection for chemical and mechanical properties required for their functions. Interestingly, some of these structures, such as glycosaminoglycans, can also play nonstructural roles in other biological contexts, including cell signaling, which adds additional selection forces driving their evolution.

VIRUSES HIJACK HOST GLYCOSYLATION

Viruses typically have minimalist genomes and use host-cell machinery for replication. Thus, glycosylation of enveloped viruses reflects that of the host. However, there are exceptions. Giant viruses such as the algae-targeting chlorella virus and members of the amoeba-targeting *Mimiviridae* family have genomes large enough that they can express their own glycosylation machineries. On a smaller scale, specific viruses express glycosyltransferases as virulence factors. For instance, a baculovirus-encoded glucosyltransferase glycosylates insect host ecdysteroid hormones to block molting, and bacteriophage-derived glucosyltransferases modify 5-hydroxymethyl cytosine bases in the phage DNA to protect it from bacterial restriction enzymes. Host-derived glycosylation in enveloped viruses is typically extensive and the resulting "glycan shield" protects the virus from immune reactions against the underlying polypeptide. In this regard, it has been suggested that the high frequency of heterozygous states for human congenital disorders of glycosylation (Chapter 44) may reflect selection for genomes that limit glycosylation of invading viruses. Host lectins may also be "hijacked" by glycans on viral surface glycoproteins. For example, Sialoadhesin (Siglec-1; Chapter 15) is used by the heavily sialylated porcine reproductive and respiratory syndrome virus (PRRSV) to gain entry into macrophages. Viral protein sequence can also influence host glycosylation (e.g., by structural constraints on access for glycosyltransferases or hydrolases), in a manner that favors viral antigenicity, such as the high-mannose N-glycans on HIV-1 envelope gp120 trimers.

VAST DIVERSITY IN BACTERIAL AND ARCHAEAL GLYCOSYLATION

Despite the enormous potential for structural diversity, a rather limited subset of all possible monosaccharides and their possible linkages and modifications are found in eukaryotic cells. Why one encounters only such a limited subset of the possible glycan structures is a puzzling question. Could there be trade-offs between the variety of glycan structures and the additional resources, such as energy and biosynthetic pathways, that cells/organisms need in order to maintain that variety? Or maybe the variety of structures is limited by a combination of positive and negative selection due to both, its intrinsic functions and interactions with symbionts and pathogens? Regardless, this limited subset has allowed better elucidation of structures of eukaryotic glycans. In contrast, Bacteria and Archaea have had several billion additional years to respond to selective pressure from pathogens; in particular, phages. These organisms also have short generation times and can exchange genetic material across vast phylogenetic distances via plasmid-mediated horizontal gene flow. Glycosylation in Bacteria and Archaea is far more diverse, both in terms of the range of monosaccharides used or synthesized, and in the types of linkages and modifications (Chapters 21 and 22). In addition, prokaryotic cell–cell interactions both within and between species are often mediated by glycans. However, most work to date has focused on the glycans of pathogens, and we may have barely scratched the surface of prokaryotic glycan diversity.

MOLECULAR MIMICRY OF HOST GLYCANS BY PATHOGENS

Despite great differences between pathways generating glycan structures of bacteria and those of vertebrates, occasional microbial surface structures are strikingly similar to those of mammalian cells. Interestingly, most such examples of "molecular mimicry" occur in pathogenic/symbiotic microorganisms, apparently adapting them for better survival in the host by avoiding, reducing, or manipulating host immunity. A few examples include *Escherichia coli* K1 and Meningococcus group B (polysialic acid), *E. coli* K5 (heparosan, heparan sulfate backbone), Group A Streptococcus (hyaluronan), Group B Streptococcus (sialylated *N*-acetyllactosamines), and *Campylobacter jejuni* (ganglioside-like glycans). Initially, it was thought that the responsible microbial genes arose via lateral gene transfer from eukaryotes. However, in all instances in which genetic information is available, evidence points toward convergent evolution rather than gene transfer. For example, genes synthesizing sialic acids in bacteria seem to have been derived from preexisting prokaryotic pathways for nonulosonic acids, an ancestral family of monosaccharides with a structural resemblance. In contrast, bacterial sialyltransferases bear little resemblance to those of eukaryotes, and the vast sequence differences between different bacterial sialyltransferases indicate that these have even been reinvented on several separate occasions. Of course, lateral gene transfer is common among Bacteria and Archaea, facilitating rapid phylogenetic dissemination of such enzymatic "inventions."

INTERSPECIES AND INTRASPECIES DIFFERENCES IN GLYCOSYLATION

Why do closely related species differ with regard to the presence or absence of certain glycans? Does the same glycoprotein have the same type of glycosylation in distinct but related species? Relatively little data exist regarding these issues, but examples of both, extreme conservation and diversification are found. A reasonable explanation is that conservation of glycan structure reflects specific functional constraints for the glycans in question. In other instances, considerable evolutionary drift in the details of glycan structure might be tolerated, as long as the underlying protein is able to carry out its primary functions (changes with no consequences for survival or reproduction, i.e., those that are selectively neutral). Even in the absence of important endogenous functions, glycans can have key roles in mediating interactions with symbionts and pathogens. The evolution of diversity and microheterogeneity (across tissues and cell types) in glycosylation could well be of value to the organisms in providing additional obstacles to pathogens that use host glycans for attachment and entry. Free glycans (e.g., milk oligosaccharides) can also have important roles in attracting and feeding symbiont microbial communities needed for internal functions (e.g., immune maturation) and in accommodating or restricting these to particular areas of the host.

There can even be significant variation in glycosylation among members of the same species, particularly in terminal glycan sequences. The classic example is the ABH(O) histo-blood group system (Chapter 14), a polymorphism found in all human populations, which has also persisted for tens of millions of years of primate evolution and has even been independently rederived in some instances. Despite its clinical importance for blood transfusion, this polymorphism appears to cause no major differences in the intrinsic biology of individuals of the species (Chapter 14) beyond conferring a variable susceptibility to viruses, such as noroviruses, that use ABO glycans as receptors. Like other blood groups, the ABO polymorphism is accompanied by production of natural antibodies against the variants absent from an individual. These antibodies may be protective by causing complement-mediated lysis of enveloped viruses generated within other individuals who express the target structure. Thus, an enveloped virus generated in a B blood group individual might be susceptible to complement-mediated lysis on contact with an A or O blood group individual who has circulating anti-B antibodies. The glycosyltransferase gene ABO responsible for synthesizing ABO antigens is one of the few loci in the human genome shown to be under balancing (frequency-dependent) selection.

This latter mechanism may also provide an explanation for interspecies diversity via selection exerted by pathogens recognizing glycans as targets for attachment and entry into cells. This mechanism is likely operative in generating the extreme diversity of sialic acid types and linkages (Chapter 15). Recent analyses have tried to combine the two mechanisms: the antibody-mediated protection from intracellular but enveloped viruses and possible frequency-dependent protection from glycan-exploiting extracellular pathogens, such as Noro- and Rotaviruses and *Plasmodium falciparum* (the causative agent of malignant malaria). Modeling approaches have successfully generated observed frequencies of ABO by incorporating these two simultaneous selection pressures. The evolutionary persistence of the ABO system still needs further explanation.

Another unexplained phenomenon is genetic inactivation in Old World primates of the ability to synthesize the otherwise very common terminal Galα1-3Galβ1-4GlcNAc-R structure (Chapter 14). This variation is also associated with spontaneously appearing and persistently circulating antibodies against the missing glycan determinant, thus forming a kind of "interspecies blood group." This glycan difference may also be protective for the primate lineage which lost the "αGal" structure and has a high-titer circulating antibody, as it is now better protected against infection by enveloped viruses emanating from other mammals. Independent losses of the vertebrate-specific sialic acid Neu5Gc in humans and some other mammalian clades is a further example of glycan evolution by loss of function (Chapter 15). In the process, these lineages have lost a potent signal of self, resulting in protection from microbes that synthesize this glycan.

Regardless of the mechanisms maintaining these types of polymorphisms, such intra- and interspecies diversity might also provide for "herd immunity," a phenomenon whereby one glycan variant–resistant individual can indirectly protect other susceptible individuals by restricting the spread of a pathogen through the population. Such proposed protective functions of glycan diversity are only apparent at the level of populations. This complicates their study in model organisms in which the focus is classically on the individual. It is important to point out here that evolution itself is a process that occurs at the level of populations.

Future studies will have to test precisely how much of the extant inter- and intraspecies glycan variation is directly driven by host–pathogen interactions. Although glycan variation forms an important determinant of host susceptibility, variation on target tissues as well as defensive secretions (mucins) must be considered when trying to understand disease, especially epidemics or zoonotics involving different host species and their interactions (e.g., influenza A) (Chapter 15). Finally, recent evidence suggests that antibodies against glycans absent in a subset of females in an extant species might aid in speciation via the killing of sperm from the remaining males in the population still bearing the glycan.

USE OF MODEL ORGANISMS TO STUDY GLYCAN DIVERSITY

For obvious reasons, the most detailed information about glycan structures is available for various popular model organisms as well as certain well-studied pathogens, and useful comparative knowledge can be gleaned from the relevant chapters that follow in this part of the book. But we must be careful about extrapolating data from organisms long maintained under optimal laboratory conditions to the overall taxa that they represent. The realization that rodents are the closest evolutionary cousins to primates has provided added justification for their use to understand human disease. However, the late Nobel laureate Sydney Brenner suggested there is now enough information about humans to consider ourselves to be the "next model organism." Indeed, the studies that combine tractable questions about the pathobiology of naturally occurring mutations affecting glycans in humans with mechanistic studies in suitable model organisms tend to provide deeper insights into glycan functions.

WHY DO WIDELY EXPRESSED GLYCOSYLTRANSFERASES SOMETIMES HAVE LIMITED INTRINSIC FUNCTIONS?

It was once popular to suggest that every glycan on every cell type must have a critical intrinsic host function. Analysis of data on glycosyltransferase-deficient mice suggests that this is not the case. For example, ST6Gal-I α2-6 sialyltransferase is the main enzyme that produces Siaα2-6Galβ1-4GlcNAcβ1- termini on vertebrate glycans. Although this glycan serves as a specific ligand for the B-cell regulatory molecule CD22 (Siglec-2; Chapter 35), it is also found on many other cell types, as well as on many soluble secreted glycoproteins. Furthermore, ST6Gal-I mRNA varies markedly among cell types, and *ST6Gal-I* transcription is regulated by several cell type–specific promoters, which are in turn modulated by hormones and cytokines. Despite these data, the prominent consequences of eliminating its expression in mice so far seem to be restricted to the immune and hematopoietic systems and some cell adhesion, apoptosis, and oncogenic pathways. If the specific intrinsic functions of the ST6Gal-I glycan product are in fact restricted to some systems, why is it expressed in so many other locations? And why up-regulate its expression so markedly in the liver and endothelium during a so-called "acute phase" inflammatory response? Besides feedback on the immune system, could it be that scattered expression of this structure in other locations functions as a "smoke screen" or temporary "firewall," restricting intra-organismal spread of an invading pathogen? Could it also be that heavily glycosylated nonnucleated cells like mammalian erythrocytes act as "decoy traps" for viral pathogens that require nucleated cells for replication? Answers to these questions must take into account the evolutionary selection pressures (both intrinsic and extrinsic recognition phenomena such as host–pathogen interactions and innate immune contributions) on glycosyltransferase products. Many effects may also not be apparent in inbred genetically modified mice living in hygienic vivaria, requiring studies in a natural, pathogen-rich environment. It is also possible that other gene products are masking phenotypes in these model systems, compensating for the loss. Furthermore, it is likely we have not looked hard enough at such genetically modified mice nor applied the relevant environmental pressures to elicit phenotypes.

EVOLUTIONARY FORCES DRIVING GLYCAN DIVERSIFICATION IN NATURE

Based on available data, it is reasonable to suggest that the evolution of glycan diversification in complex multicellular organisms has been driven by selection pressures of both intrinsic and extrinsic origin relative to the organism under study (Chapter 7). Glycans are particularly susceptible to the "Red Queen" effect, in which host glycans must keep changing to stay ahead of the pathogens, which have more rapid evolutionary rates because of their short generation times, high mutation rates, and rampant horizontal gene transfer. Glycan evolution is expected to be under selective constraints—that is, opposing selective forces influencing their course of evolution. Given their important role in definition of the molecular frontier of cells and organisms, the same glycan can come under opposing selective pressures at different times in the life of a cell or organism. Glycans favoring cell motility (e.g., polysialic acid) will be favored for development but become detrimental when accidentally exploited by malignant cells. Glycans on reproductive tract secretions favoring survival of sperm (e.g., Glycodelin S) might be counter-selected in females who benefit from a different glycan form (Glycodelin A), which challenges male gametes as part of female quality control. Unique glycans evolved as reliable SAMPs can become a liability if exploited by pathogens through molecular mimicry. Given the rapid evolution of extrinsic pathogens and their frequent use of glycans as targets for host recognition, it seems likely that a significant portion of the overall diversity in vertebrate cell-surface glycan structure reflects such pathogen-mediated selection processes. Meanwhile, even one critical intrinsic role of a glycan could disallow its elimination as a mechanism to evade pathogens. Thus, glycan expression patterns may represent trade-offs between evading pathogens (or accommodating symbionts) and preserving intrinsic functions.

More gene disruption studies in intact animals could help differentiate intrinsic and extrinsic glycan functions. More systematic comparative glycobiology could also contribute, making predictions about intrinsic glycan function—that is, the consistent (conserved) expression of the same structure in the same cell type across several taxa would imply a critical intrinsic role. Such work might also help define the rate of glycan diversification during evolution, better define the relative roles of the intrinsic and extrinsic selective forces, and eventually lead to a better understanding of the functional significance of glycan diversification during evolution. The possibility that pathogen-driven glycan diversification might even favor the process of sympatric speciation (via reproductive isolation) also needs to be further explored.

ACKNOWLEDGMENTS

The authors appreciate helpful comments and suggestions from Cristina De Castro and Christopher Mark West.

FURTHER READING

Warren L. 1963. The distribution of sialic acids in nature. *Comp Biochem Physiol* **10:** 153–171. doi:10.1016/0010-406x(63)90238-x

Kishimoto Y. 1986. Phylogenetic development of myelin glycosphingolipids. *Chem Phys Lipids* **42:** 117–128. doi:10.1016/0009-3084(86)90047-2

Galili U. 1993. Evolution and pathophysiology of the human natural anti-α-galactosyl IgG (anti-Gal) antibody. *Springer Semin Immunopathol* **15:** 155–171. doi:10.1007/bf00201098

Kappel T, Hilbig R, Rahmann H. 1993. Variability in brain ganglioside content and composition of endothermic mammals, heterothermic hibernators and ectothermic fishes. *Neurochem Int* **22:** 555–566. doi:10.1016/0197-0186(93)90030-9

Martinko JM, Vincek V, Klein D, Klein J. 1993. Primate ABO glycosyltransferases: evidence for trans-species evolution. *Immunogenetics* **37:** 274–278. doi:10.1007/bf00187453

Dairaku K, Spiro RG. 1997. Phylogenetic survey of endomannosidase indicates late evolutionary appearance of this N-linked oligosaccharide processing enzyme. *Glycobiology* **7:** 579–586. doi:10.1093/glycob/7.4.579

Drickamer K, Taylor ME. 1998. Evolving views of protein glycosylation. *Trends Biochem Sci* **23:** 321–324. doi:10.1016/s0968-0004(98)01246-8

Gagneux P, Varki A. 1999. Evolutionary considerations in relating oligosaccharide diversity to biological function. *Glycobiology* **9:** 747–755. doi:10.1093/glycob/9.8.747

Freeze HH. 2001. The pathology of N-glycosylation—stay the middle, avoid the risks. *Glycobiology* **11:** 37G–38G.

Angata T, Varki A. 2002. Chemical diversity in the sialic acids and related α-keto acids: an evolutionary perspective. *Chem Rev* **102:** 439–469. doi:10.1021/cr000407m

Varki A. 2006. Nothing in glycobiology makes sense, except in the light of evolution. *Cell* **126:** 841–845. doi:10.1016/j.cell.2006.08.022

Stern R, Jedrzejas MJ. 2008. Carbohydrate polymers at the center of life's origins: the importance of molecular processivity. *Chem Rev* **108:** 5061–5085. doi:10.1021/cr078240l

van Die I, Cummings RD. 2010. Glycan gimmickry by parasitic helminths: a strategy for modulating the host immune response? *Glycobiology* **20:** 2–12. doi:10.1093/glycob/cwp140

Springer SA, Gagneux P. 2013. Glycan evolution in response to collaboration, conflict, and constraint. *J Biol Chem* **288:** 6904–6911. doi:10.1074/jbc.r112.424523

Clark GF. 2014. The role of glycans in immune evasion: the human fetoembryonic defence system hypothesis revisited. *Mol Hum Reprod* **20:** 185–199. doi:10.1093/molehr/gat064

Le Pendu J, Nyström K, Ruvoën-Clouet N. 2014. Host–pathogen co-evolution and glycan interactions. *Curr Opin Virol* **7:** 88–94. doi:10.1016/j.coviro.2014.06.001

Corfield AP, Berry M. 2015. Glycan variation and evolution in the eukaryotes. *Trends Biochem Sci* **40:** 351–359. doi:10.1016/j.tibs.2015.04.004

Hinchliff CE, Smith SA, Allman JF, Burleigh JG, Chaudhary R, Coghill LM, Crandall KA, Deng J, Drew BT, Gazis R, et al. 2015. Synthesis of phylogeny and taxonomy into a comprehensive tree of life. *Proc Natl Acad Sci* **112:** 12764–12769. doi:10.1073/pnas.1423041112

Springer SA, Gagneux P. 2017. Glycomics: revealing the dynamic ecology and evolution of sugar molecules. *J Proteom* **135:** 90–100. doi:10.1016/j.jprot.2015.11.022

Van Etten JL, Agarkova I, Dunigan DD, Tonetti M, De Castro C, Duncan GA. 2017. Chloroviruses have a sweet tooth. *Viruses* **9:** 88. doi:10.3390/v9040088

Varki A. 2017. Biological roles of glycans. *Glycobiology* **27:** 3–49. doi:10.1093/glycob/cww086

Joshi HJ, Narimatsu Y, Schjoldager KT, Tytgat HLP, Aebi M, Clausen H, Halim A. 2018. SnapShot: O-glycosylation pathways across kingdoms. *Cell* **172:** 632. doi:10.1016/j.cell.2018.01.016

Suzuki N. 2018. Glycan diversity in the course of vertebrate evolution. *Glycobiology* **29:** 625–644. doi:10.1093/glycob/cwz038

West CM, Malzl D, Hykollari A, Wilson IBH. 2021. Glycomics, glycoproteomics, and glycogenomics: an inter-taxa evolutionary perspective. *Mol Cell Proteomics* **20:** 100024. doi:10.1074/mcp.r120.002263

21 Eubacteria

Chris Whitfield, Christine M. Szymanski, Amanda L. Lewis, and Markus Aebi

Glycoconjugates are integral components of the cell surfaces of bacteria and are often the immediate point of contact with the environment. Surface glycoconjugates contribute to the essential permeability barrier properties of the cell envelope, influence both the susceptibility and resistance of bacteria to antibiotics and other harmful compounds, participate in the formation and dispersion of biofilms, act as receptors for bacteriophages, and play pivotal roles in pathogenic and symbiotic host–microbe interactions. Reflecting these many functions, surface glycoconjugates are remarkably diverse, enabled by the propensity for bacterial genetic recombination and lateral gene transfer and shaped by environmental interactions that impart niche-specific selective pressures. Surface glycoconjugates drive a variety of important interactions with host innate and adaptive immune defenses. Some are recognized as pathogen-associated molecular patterns (PAMPs), for example, via Toll-like receptor (TLR)-mediated pathways. Others are natural targets of adaptive immunity and have been exploited in successful vaccine strategies. Because of their importance in cell viability, surface glycoconjugates are also frequent targets of antimicrobial strategies. This chapter will provide an overview of the structure and biosynthesis of glycoconjugates, together with some examples of their functions.

published by Cold Spring Harbor Laboratory Press; doi: 10.1101/glycobiology.4e.21

AN OVERVIEW OF CELL ENVELOPE ARCHITECTURES

Bacteria were historically divided into two major groups based on their response in the Gram-staining procedure (i.e., Gram-positive and Gram-negative organisms), reflecting the organization of the cell wall. The amount and location of peptidoglycan in the cell wall is an important contributor to the outcome of Gram staining. Peptidoglycan is essential for the viability of most bacteria. It consists of polysaccharide strands covalently cross-linked by short peptides, creating a three-dimensional structure that confers shape and rigidity to the cell. In Gram-negative bacteria, such as *Escherichia coli*, the cell wall consists of two membranes separated by a cellular compartment termed the periplasm in which thin layers of peptidoglycan reside (Figure 21.1). In Gram-positive bacteria, substantially thicker layers of peptidoglycan surround a single membrane and provide an attachment point for other glycan structures. Another class of bacteria, the Negativicutes, are related to the Gram-positive Firmicutes but nevertheless have a two-membrane architecture and stain as Gram-negative. Other bacteria produce variable responses in the Gram-staining reaction, largely because of the presence of other glycans and cell wall lipids (see below).

The periplasm of Gram-negative bacteria contains proteins associated with cell-surface assembly and nutrient uptake but may also contain free oligosaccharides (fOS) that protect against osmotic stress. The outer membrane is an asymmetric lipid bilayer, with an outer leaflet composed mainly of a unique glycolipid called lipopolysaccharide (LPS) that is essential for the integrity of the permeability barrier imposed by the outer membrane. Many Gram-negative bacteria are covered in a surface-bound polysaccharide layer known as a capsule, and, in some cases, this capsular polysaccharide (CPS) is released from the cell in large amounts as free exopolysaccharide (EPS). Bacteria producing these products are often readily identified by their highly mucoid colonies. Structural variations in the LPS and capsules of different bacterial species are diverse and influence many types of interactions between bacteria and their environmental and host niches.

Gram-positive bacteria lack the outer membrane (Figure 21.1) and depend on a much thicker multilayered peptidoglycan layer for viability. The Gram-positive cell wall is modified with additional specialized cell wall glycan polymers covalently linked to peptidoglycan (e.g., wall teichoic acids [WTAs]), whereas glycolipids, such as lipoteichoic acids (LTAs), are anchored in the cell membrance. Mycobacteria and related Actinobacteria are not considered to be Gram-positive or Gram-negative, and are instead classified as acid-fast bacteria because they possess distinctive cell walls that confer an aberrant response to the classical Gram-staining procedure. Their complex cell walls have a high content of remarkable glycan and glycolipid structures (Figure 21.1). The unique arabinogalactan (AG) component that characterizes these organisms is covalently attached to peptidoglycan and provides an attachment point for characteristic long-chain (C_{60}–C_{90}) α-alkyl-β-hydroxymycolic acids. The resulting mycolyl-AG-peptidoglycan complex underpins a permeability barrier conferring resistance to antibiotics and other harmful molecules. Mycolic acids help form an envelope layer referred to as an "outer membrane," although its structure shares no similarity with its namesake in Gram-negative bacteria.

PEPTIDOGLYCAN, A DYNAMIC STRESS-BEARING LAYER

Structure, Arrangement, and Function

Peptidoglycan (also known as murein) forms a rigid sacculus enveloping the cytoplasmic membrane, conferring cell shape, and contributing to the ability of bacteria to resist the effects of internal osmotic (turgor) pressure. Although most bacteria possess peptidoglycan, it is absent in some obligate intracellular pathogens, such as *Mycoplasma sp.*, and can be conditionally expendable in certain bacteria that produce L-forms. Given its distribution, peptidoglycan is recognized by host

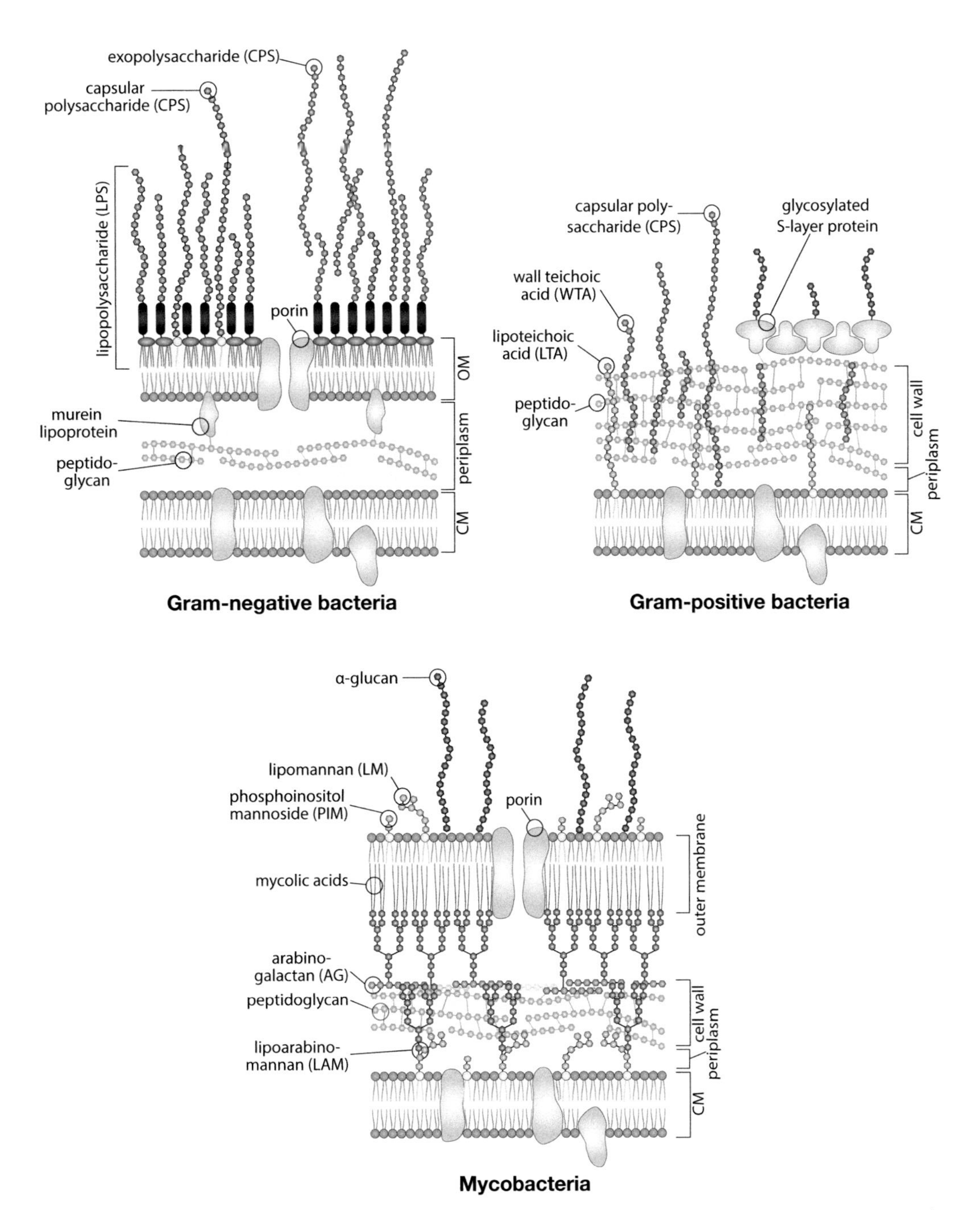

FIGURE 21.1. Conceptual organization of the cell envelopes of Gram-negative bacteria, Gram-positive bacteria, and mycobacteria. The schematic comparison of the cell walls (or cell envelopes) of Gram-negative bacteria, Gram-positive bacteria, and mycobacteria illustrates the major classes of structural glycoconjugates and their locations. Each envelope type possesses peptidoglycan located outside the cytoplasmic membrane (CM) as a major component conferring shape and integrity to the cell wall. In Gram-negative bacteria, a selectively permeable asymmetric outer membrane (OM) usually contains lipopolysaccharides. In Gram-positive bacteria, the much thicker peptidoglycan layer is augmented by covalently linked wall teichoic acids (WTAs) and a variety of lipoglycans (including lipoteichoic acids [LTAs]) embedded in the cell membrane. Mycobacteria also possess an "outer membrane," but it is much different from the Gram-negative OM. The major lipid components are long-chain mycolic acids, linked via a branched arabinogalactan structure to the peptidoglycan. Mycobacterial walls possess a wide diversity of lipoglycans in the cytoplasmic and OMs. All three wall types may be covered with capsular and extracellular polysaccharides.

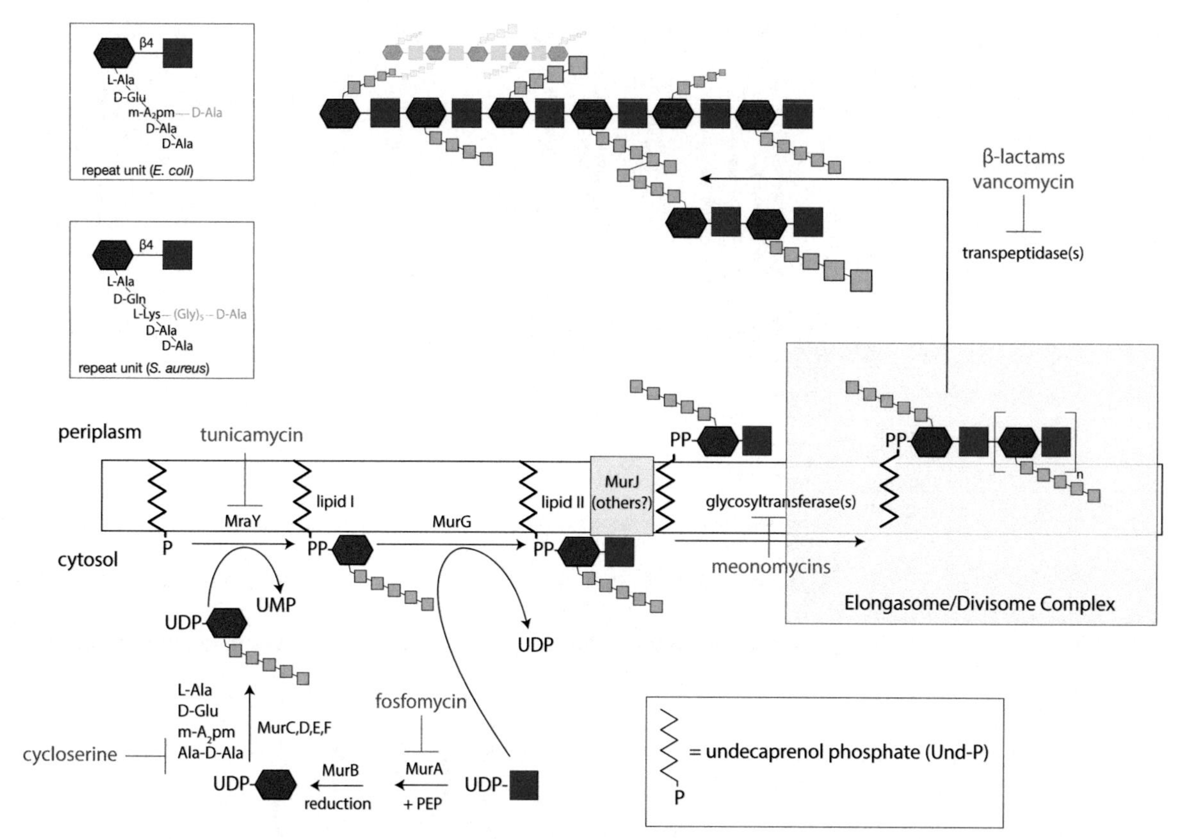

FIGURE 21.2. Structure, biosynthesis, and inhibition of peptidoglycan assembly. Peptidoglycan creates a meshwork composed of glycan chains with a disaccharide unit containing β1-4-linked *N*-acetylglucosamine and *N*-acetylmuramic acid (MurNAc, *purple*; GlcNAc, *blue*). The chains are synthesized with pentapeptide stems that are used to create periodic cross-links between adjacent glycan strands in three dimensions. The pentapeptides may differ in amino acid residues but always contain a dibasic residue (e.g., *meso*-diaminopimelic acid [*m*-DAP] or L-Lys) at position 3 to facilitate cross-linking and they typically end with D-Ala-D-Ala. The cross-links may be direct (e.g., *E. coli*) or involve a short oligopeptide (e.g., *Staphylococcus aureus*). Peptidoglycan assembly begins at the interface of the cytosol and the cytoplasmic membrane with formation of an undecaprenol diphosphate-linked muropeptide subunit (lipid II). After being flipped to the periplasmic face of the membrane, these subunits are incorporated into longer chains and then inserted into the growing cell wall by a process known as transglycosylation. Cross-linking of adjacent peptide stems by transpeptidation completes the process. The placement and coordination of these activities are complex processes and the precise content of the assembly enzyme complexes depends on the context (e.g., for cell elongation or division). Peptidoglycan has been a rich source of targets for antibiotics and prominent examples are identified in *red*. Fosfomycin and cycloserine are structural analogs of phosphoenolpyruvate (PEP) and D-Ala, respectively, and block the completion of lipid I. Moenomycins are phosphoglycolipids that bind to the active sites of peptidoglycan GTs to inhibit the transglycosylation process. Finally, β-lactams and vancomycin both prevent transpeptidation by different strategies. β-Lactams mimic D-Ala-D-Ala and bind to penicillin-binding proteins to directly affect the peptidase step by creating a slowly turned over substrate that occupies the enzyme nonproductively. In contrast, vancomycin sterically inhibits the transpeptidases by binding to terminal D-Ala-D-Ala substrates.

defenses as a marker of infection. Released peptidoglycan is a PAMP recognized by TLR2 and pattern recognition proteins, such as nucleotide-binding oligomerization domain (NOD)-like receptors (NLRs), to activate an inflammatory response as part of innate immunity.

Peptidoglycan makes up ~10% of the dry weight of the cell wall in Gram-negative bacteria, in which it exists in a structure one to three layers thick. In contrast, Gram-positive bacteria have a thicker cell wall containing 10 to 20 layers of peptidoglycan, representing as much as 20%–25% of the dry weight of the cell. The overall chemical structure of peptidoglycan is similar across bacterial genera (Figure 21.2) and consists of parallel strands of polysaccharides composed of a β1-4-linked disaccharide of *N*-acetylglucosamine (GlcNAc) and *N*-acetylmuramic acid (MurNAc). The average chain length of the glycan strands in *E. coli* is 25–35 disaccharide units but may increase several-fold in Gram-positive bacteria. Adjacent strands are connected to one another via cross-links between short peptide strands attached to the MurNAc residues (Figure 21.2). Individual glycan strands are thought to be arranged parallel to the membrane and the helical conformation of the glycan strands enables cross-linking in three dimensions, within and between layers, to create the functional higher-order structure. Although the amino acid composition varies in different bacteria, the peptides are typically composed of L-Ala, unusual D-amino acids (e.g., D-Glu and D-Ala) and a dibasic amino acid, such as *meso*-diaminopimelic acid (*m*-A_2pm or *m*-DAP) or L-Lys, to facilitate the cross-linking. Cross-links may be direct peptide bonds or (in some Gram-positive bacteria) include several amino acids, such as Gly_5. This, and the frequency of cross-linking, will have spatial implications for the final meshwork. Variations in peptidoglycan structure may occur within the same organism at different stages of the cell cycle, during developmental processes, or at different sites in the same cell.

A characteristic property of many Gram-positive cell walls is the attachment of secondary cell wall polymers to MurNAc residues in the glycan backbone. In addition, the dibasic amino acid provides a potential site for protein attachment. Sortase-family enzymes link a wide range of proteins (including pili, surface-layer [S-layer] subunits, and metabolite-binding proteins) to the cell wall. In Gram-negative bacteria, outer membrane lipoproteins are attached to this site in peptidoglycan.

Despite its role in cell integrity, peptidoglycan is a dynamic structure, allowing insertion of new material during rapid cell growth and division, as well as accommodating the assembly of macromolecular structures (e.g., flagella and protein secretion systems), which pass through the cell wall. As a result, peptidoglycan undergoes ~50% turnover every generation, catalyzed by a collection of peptidoglycan hydrolases (glycosidases, peptidases, and amidases). An imbalance in the metabolism of peptidoglycan can lead to a rapid loss of cell wall integrity, followed by osmotic swelling and cell lysis. Such effects underpin the actions of cell wall–active antibiotics (described below) and the effect of lysozyme, an enzyme secreted by leukocytes as part of innate immunity.

Assembling the Cell Wall

Peptidoglycan synthesis involves a complex series of well-conserved enzymatic reactions localized to three cellular compartments (Figure 21.2). In the initial stage, activated nucleotide precursors are synthesized in the cytoplasm. UDP-MurNAc is generated through the condensation of phosphoenolpyruvate (PEP) with UDP-GlcNAc and subsequent reduction. In some mycobacteria UDP-MurNAc is hydroxylated into UDP-MurNGc for unknown reasons, generating the only other known N-glycolyl group in nature (besides *N*-glycolylneuraminic acid, which is exclusive to deuterostomes) (Chapter 15). Sequential addition of the amino acids by specific ATP-dependent amino acid ligases leads to the formation of the UDP-MurNAc-pentapeptide. The disaccharide repeat unit is built on a C_{55} polyprenol carrier (undecaprenol or bactoprenol phosphate) in the cytoplasmic membrane in a process resembling the early stage of eukaryotic N-linked glycan formation (Chapter 9). The resulting lipid-linked disaccharide subunit (lipid II) is flipped across the membrane mainly by MurJ, although other flippases may also participate in some circumstances. Lipid II is polymerized at the periplasmic face of the membrane and new glycans are inserted into the growing cell wall. This requires glycosyltransferases (GTs) to extend the glycan backbones and transpeptidases (TPs) to catalyze cross-linking. Each organism possesses multiple GT and TP activities, existing as monofunctional enzymes or as bifunctional GT-TPs, and structures of key enzymes have provided insight into their mechanisms. GTs mediate stepwise assembly of the glycan using a molecule of lipid II as the acceptor and releasing one undecaprenyl diphosphate in each cycle.

The TP and GT reactions are context-dependent; active complexes containing different enzyme combinations are coupled with peptidoglycan remodeling enzymes that are guided by cytoskeletal elements (actin and tubulin homologs) to achieve wall elongation or cell division. This process is an essential driver of cell shape determination. Division is particularly complex, with a multiprotein "divisome" that includes regulatory proteins dedicated to fidelity in the timing and placement of the division site. Among the remodeling enzymes are lytic transglycosylases, which cleave glycan chains to create a characteristic terminal 1,6-anhydroMurNAc residue. This reaction may determine the size of the glycan chains (effectively marking the end of biosynthesis) but is also required for the insertion of *trans*–cell wall structures, signaling in the regulation of some β-lactamases, and the release of PAMPs.

Antibacterials Targeting Peptidoglycan Assembly

Peptidoglycan synthesis has offered a classic target for antibiotics and continues to provide new avenues for drug discovery. Penicillins and other β-lactam antibiotics mimic the D-Ala-D-Ala motif and create a slow-turnover acyl-enzyme complex with TPs (hence, proteins with TP modules are also referred to as penicillin-binding proteins, or PBPs). The action of these drugs requires active bacterial replication. Resistance to these compounds involves their exclusion from the cell, replacement of the target enzyme, or production of β-lactamases. Vancomycin and other glycopeptides also affect the cross-linking step by binding to the terminal D-Ala-D-Ala to prevent its turnover. Members of the moenomycin family of compounds are phosphoglycolipids that are thought to inhibit transglycosylase GTs by mimicking the growing glycan chain and binding the donor substrate site in the enzyme. The points of action of these and other antibacterial compounds are shown in Figure 21.2.

GRAM-POSITIVE BACTERIA PRODUCE ADDITIONAL CELL WALL GLYCOPOLYMERS

Gram-positive cell walls contain large amounts of additional polymers that provide further functionality and structure to the cell wall. The best known examples are WTAs, which often consist of polyribitol or polyglycerol chains (40 to 60 repeats) variably substituted with monosaccharides and/or D-Ala. These chains are linked by a phosphodiester bond to the C6 position of MurNAc through a conserved linkage region (Figure 21.3). WTAs extend through and beyond the peptidoglycan layers and are exposed at the cell surface. They are expendable for viability in the laboratory, yet are required for pathogenicity and potentially for growth in other environments. A wide range of functions are (directly or indirectly) attributed to WTAs, including regulation of peptidoglycan remodeling, cell wall growth and morphology, binding divalent cations, adhesion, biofilm formation, and resistance to innate immunity (lysozyme and antimicrobial peptides), as well as resistance to antibiotics and other environmental stresses. WTAs also provide a means of tethering some proteins, including S-layer proteins, to the cell surface. Under conditions of phosphate starvation, some bacteria replace the phosphate-rich WTAs by teichuronic acids, where carboxyl groups create their anionic character (Figure 21.3).

WTAs are synthesized from nucleotide-activated precursors on undecaprenyl carriers and, once complete, are exported across the cytoplasmic membrane by an ATP-binding cassette (ABC) transporter (like LPS O-PS, see below) and attached to MurNAc acceptors. The timing (during peptidoglycan assembly) and location of WTA insertion is not fully resolved.

LTAs are also commonly found in Gram-positive cell envelopes. Some LTAs may possess similar carbohydrate structures to WTAs (Figure 21.3), whereas others are substantially more complex. They are attached to membrane anchor glycolipids that vary in sugar composition but typically contain terminal diacylglycerol (DAG). The assembly pathways of WTAs and LTAs are quite

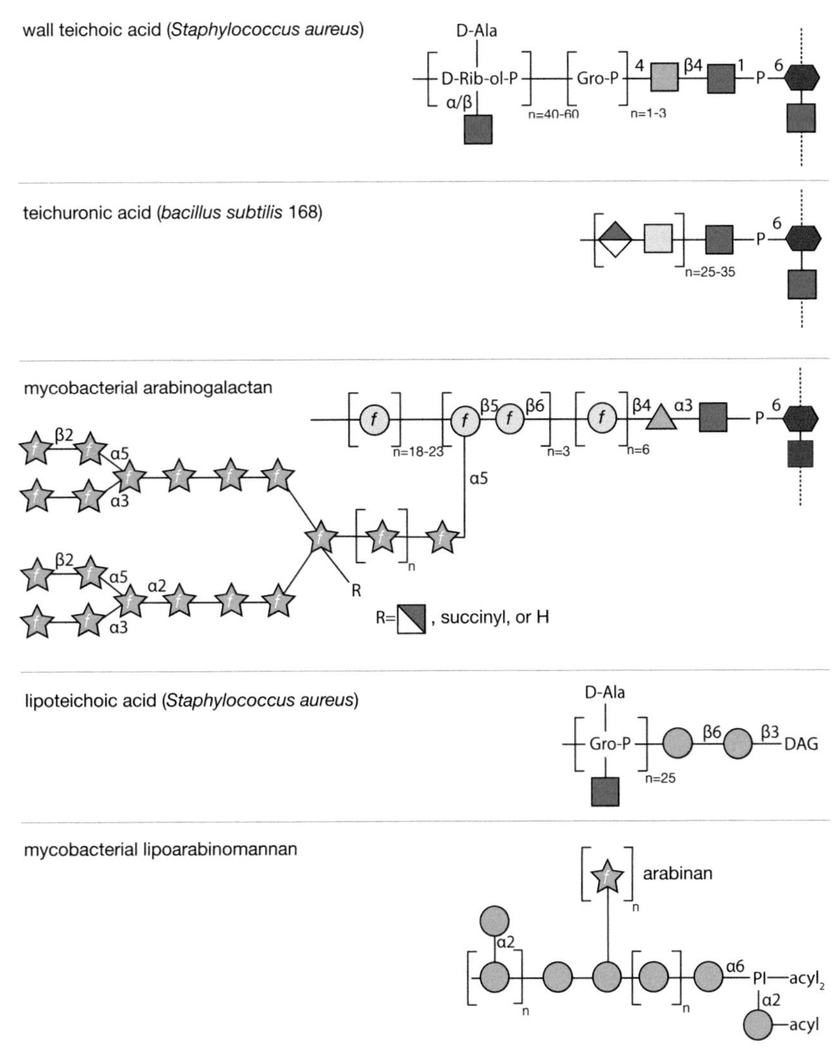

FIGURE 21.3. Structures of additional cell wall polymers in classical Gram-positive bacteria and mycobacteria. A range of additional (secondary) wall polymers extend the structure and functionality of peptidoglycan. Wall teichoic acids (WTAs) typically contain a polyol-phosphate backbone, such as that containing a polyribitol (Rib-ol) structure from *S. aureus*. WTAs are attached via a linkage unit, which is phosphodiester-linked to the C6 hydroxyl of 1 in 9 MurNAc residues in peptidoglycan. The conserved linkage unit is composed of a disaccharide and 1-2 glycerol phosphate (Gro-P) residues. Some hydroxyl groups in the polyribitol chain may be modified nonstoichiometrically with D-Ala and GlcNAc residues. Teichuronic acid production is environmentally regulated; under conditions of phosphate limitation, these molecules can replace the more prevalent (and phosphate-rich) polyol-phosphate-based WTAs. Teichuronic acids are also attached to peptidoglycan via phosphodiester linkage to a terminal GlcNAc (like WTAs), although the fine details of the linkage region and polymer have not been resolved. In mycobacteria, complex arabinogalactan structures are linked to peptidoglycan in a similar manner to WTAs, comprising a backbone of 30–35 Gal*f* residues with alternating β1-5 and β1-6 linkages attached to MurNAc via a disaccharide linker. Three antennary arabinan oligosaccharides with 22 or 31 Ara*f* monomers (depending on the species) are attached to the galactan backbone. The terminal β1-2-linked Ara*f* residues provide the attachment points for mycolic acids. Gram-positive cell walls also contain a wide variety of lipoglycans. Some are attached to the cell membrane by diacylglycerol (DAG) and, in the case of lipoteichoic acids (LTAs), the glycan chain may share some similarities with WTAs. One of the major classes of lipoglycans in mycobacteria is the phosphatidyl-*myo*-inositol mannosides (PIMs). These molecules are composed of a phosphatidylinositol anchor carrying one to six mannose residues and bearing up to four acyl chains. These can be further elaborated into lipoarabinomannan (LAM) by elongation and branching of the mannan chain and addition of an arabinan structure. These structures are more heterogeneous than arabinogalactan, and the figure shows a conceptual composite that incorporates different structural elements seen in variants.

different. For example, the prevalent polyglycerol-based LTAs require phosphatidylglycerol as the glycerol donor, rather than a nucleotide-activated precursor seen in WTA synthesis, and LTAs are assembled on the DAG-containing glycolipid without the participation of undecaprenyl phosphate. LTAs may coexist with WTAs, and some of their proposed functions overlap, although LTAs may be of particular importance in those bacteria that do not contain teichoic acids. Mutations eliminating both LTA and WTA synthesis show synthetic lethality. LTAs have also been implicated in inflammatory responses and are proposed as ligands for a range of host receptors, although information concerning their status as PAMPs recognized by TLR2 is contradictory.

MYCOBACTERIA POSSESS UNUSUALLY COMPLEX CELL ENVELOPE GLYCOCONJUGATES

Pathogenic mycobacteria, including the organisms that cause tuberculosis and leprosy, are intracellular parasites that replicate within modified phagosomes of macrophages. Mycobacteria are also able to enter a dormant (metabolically inactive) state, aiding their persistence. These properties are critically dependent on a remarkable cell envelope architecture based on a wide array of unusual cell wall constituents (Figure 21.1). AG is crucial to the integrity of the cell wall and the organism's resistance to many antibiotics. AG is anchored to approximately one in every 10 disaccharides in a peptidoglycan strand, in a similar manner as WTAs (Figure 21.3). The main carbohydrate poly-Gal*f* backbone of AG is synthesized on a decaprenyl-P carrier in a process resembling some LPS O-antigens (see below), before its export to the periplasm, which is proposed to involve a transporter. Elaboration of the arabinofuran is performed in the periplasm, using decaprenyl-P-Ara*f* as the immediate donor. After ligation to peptidoglycan, the reducing terminal Ara*f* residues provide attachment sites for the characteristic mycolic acids. The lipids form a major part of the mycobacterial outer membrane (Figure 21.1). This unusual outer membrane offers a formidable barrier against antibiotics and harmful compounds. The essential role of the mycolyl-AG-peptidoglycan complex in mycobacterial physiology makes it an attractive target for drug discovery. Indeed, the frontline drug ethambutol is a structural analog inhibitor of arabinosyl GTs.

The cell envelope also contains a wide range of bioactive glycolipids. Abundant examples are based on a core structure of phosphatidylinositol mannosides (PIMs), with varying acylation and glycan lengths. PIMs and PI can account for $>50\%$ of the total envelope phospholipid content and are important for cell wall integrity and division. After synthesis and export, these lipids are extended into lipomannans (LMs) by additions from decaprenyl-P-Man donors. As is the case for AG, periplasmic extension of LMs is catalyzed by specific type C GTs that further elaborate into lipoarabinomannans (LAMs) using decaprenyl-P-Ara*f*, creating a covalently linked arabinan resembling the structure in AG, in a LAM molecule containing more than 100 sugars (Figure 21.3). The terminal Ara*f* residues can be further capped by Man residues in ManLAM. These compounds are all potentially recognized by an adaptive immune response, the mannose receptor on macrophages (facilitating phagocytosis), the lectin pathway of complement activation, and by DC-SIGN (dendritic cell–specific intercellular adhesion molecule-3-grabbing nonintegrin). ManLAM and (particularly) LM are potent ligands for TLR2 but differ in their immunomodulatory effects. Indeed, their exact role(s) in the complex pathogenesis process remain unresolved.

LIPOPOLYSACCHARIDE (ENDOTOXIN)—A KEY COMPONENT IN MOST GRAM-NEGATIVE BACTERIA

The outer membrane of Gram-negative bacteria (Figure 21.1) is an asymmetric bilayer, with the outer leaflet that is mostly composed of the unique glycolipid known as LPS (see Figure 21.5). LPS is absent in only a few Gram-negative bacteria, such as the endosymbionts *Borrelia* and

Wolbachia and in *Sphingomonas*, in which LPS is functionally replaced by sphingolipids. In *Salmonella*, there are ~10^6 molecules of LPS per cell (vs. 10^7 total phospholipids), covering ~75% of the cell surface. LPS is usually essential for outer membrane integrity in producing organisms. However, a few species (e.g., *Neisseria meningitidis* and *Acinetobacter baumannii*) survive in the face of mutations that prevent LPS synthesis, although this imparts a cost on fitness and virulence. Much of our knowledge of the structure, synthesis, and functions of LPS arises from work in *E. coli* and *Salmonella*, but the general principles are widely applicable. LPS consists of three structural domains: lipid A, core oligosaccharide, and the O-antigen polysaccharide (O-PS) (Figure 21.4). Some mucosal pathogens, such as *Neisseria* and *Campylobacter*, produce a truncated (nonrepeating) glycan that may be subject to phase variation. These LPS forms are often referred to as lipooligosaccharide (LOS) (Figure 21.4).

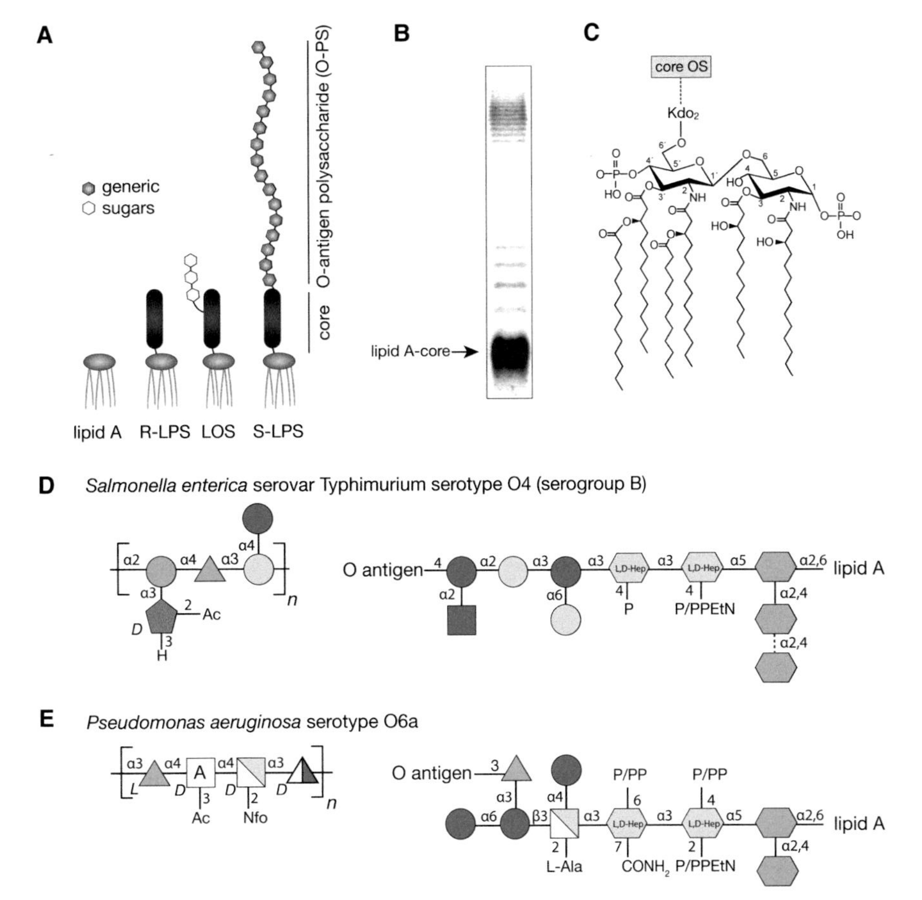

FIGURE 21.4. Structural organization of lipopolysaccharides (LPSs). LPSs are characteristic components of Gram-negative outer membranes, and the cartoon in *A* illustrates different formats of LPS structures. LPS composed of only the lipid moiety (lipid A) is rare in nature but can occur in *E. coli* LPS-assembly mutants, providing additional compensatory mutations are present to accommodate the defect. "Rough LPS" (R-LPS) consists of lipid A plus core oligosaccharide, whereas in "smooth LPS" (S-LPS), the molecule is capped with a long O-antigen polysaccharide chain. Some bacteria extend the core with a short oligosaccharide and are often called lipooligosaccharide (LOS). The LPS species produced by a given isolate are heterogeneous, as illustrated when the molecules from an isolate with S-LPS are separated by SDS-PAGE (*B*). Fast migrating molecules are composed of lipid A plus varying lengths of core oligosaccharide, whereas the characteristic ladder of larger molecules reflects the substitution of complete lipid A-core with O-antigen glycans of varying lengths. Diversity in LPS structures increases in the parts distal to the cell surface. The general composition and features of lipid A are highly conserved; the key biosynthetic intermediate hexacyl lipid A-Kdo_2 is shown in *C*. Within a given species, one to a few core structures are found among different isolates, whereas the structures of O-antigens are highly variable within most species (*D*,*E*). Ⓐ, GalNAcA3Ac(3-*O*-acetyl-2-acetamido-2-deoxy-D-galacturonic acid).

Structure and Function

Lipid A is a glycolipid containing a variable number of fatty acyl chains covalently attached to a disaccharide backbone that anchors LPS in the outer membrane. The most common lipid-A structure consists of a β1-6-linked diglucosamine backbone. In *E. coli*, the reducing terminal sugar contains phosphate at C-1 and ester- and amide-linked β-hydroxymyristic acid residues at C-3 and C-2, respectively (Figure 21.4). The second glucosamine residue also contains phosphate at C-4′ and two β-hydroxymyristic acids, in ester linkage at C-3′ and amide linkage at C-2′, which carry additional lauroyl groups on their β-hydroxyls. LPS was first discovered in the late 1800s as a heat-stable toxin associated with bacteria, and chemical synthesis of *E. coli* lipid A in 1985 confirmed that this moiety is responsible for the biological (endotoxic) properties of LPS in mammals, resulting in fever, septic shock, and other deleterious physiological effects (Chapter 42). When released from the cell surface, lipid A is a PAMP recognized by the TLR4/myeloid differentiation factor 2 (MD2) receptor, which dimerizes to trigger secretion of proinflammatory mediators. This can result in beneficial stimulation of the adaptive immune response and has been used as a vaccine adjuvant (i.e., monophosphoryl lipid A). However, LPS-mediated inflammation can also cause morbidity and mortality and some bacteria (e.g., commensals of the gut microbiome) have adapted LPS chemistries that minimize inflammation. In some Gram-negative species, these different outcomes are dictated by variable phosphorylation and acylation of the lipid-A structure, which are subject to environmental regulation and result in altered signaling downstream from TLR4/MD2.

Free lipid A occurs in a few rare exceptions (e.g., *Francisella novicida*) because, in most bacteria, lipid A is modified by sugars forming the core oligosaccharide. This branched oligosaccharide is conceptually separated into a more variable outer core that provides an attachment point for the O-antigen, whereas the inner core is more conserved. All lipid-A molecules contain one to four units of the unusual eight-carbon sugar, Kdo (3-deoxy-D-*manno*-oct-2-ulosonic acid [or a derivative thereof]), located at the linkage region between lipid A and core, and many inner cores contain L (or D)-glycero-*manno*-heptose residues. Negatively-charged components in the inner core, such as phosphates and uronic acids, provide further binding sites for the divalent cations that stabilize the outer membrane. The outer core is more variable; for example, five different structures are found in different isolates of *E. coli*.

The outermost portion of LPS is the hypervariable O-PS, which forms the O-antigen used in serotyping clinical isolates (e.g., *E. coli* O157). There are more than 180 structurally and serologically distinct O-antigens in *E. coli* alone and these arise from recombination and horizontal gene transfer events and selection pressures, including host immune response and O-PS specific bacteriophages. This diversity reflects a remarkable range of sugar residues, including free and amidated uronic acids, amino sugars, methylated and deoxygenated derivatives, acetylated sugars, and nonsugar substituents (e.g., amino acids and phosphate). O-PSs are repetitive polysaccharides with one to eight sugars (and noncarbohydrate substituents) per repeat unit. Each chain can contain one to hundreds of repeats, although there is a preferred (modal) cluster of chain lengths in LPS isolated from an individual strain. This is evident in the characteristic cluster of bands in the sodium dodecyl sulfate polyacrylamide gel electrophoresis (SDS-PAGE) profile (Figure 21.4), where each higher band reflects an LPS molecule containing an O-PS with one additional repeat unit. In the mammalian gut, LPSs are targets of multivalent IgA that aggregates growing bacteria, leading to entrapped growth and prevention of (pathogen) spread. In turn, the high variability and phase-variation of LPS structures or phage-mediated LPS remodeling may be outcomes of selection in the face of adaptive immunity. In some mammalian pathogens, O-PS chain length is critical for resistance to complement-mediated killing, but it can also influence interactions of bacteria with macrophages. Other mammalian pathogens that synthesize (nonpolymeric) LOS incorporate structural variations that enable resistance to one or more arms of the complement system. The exposure of

O-PSs makes them prime targets as receptors for bacteriophages, and this may be a driving force in the evolution of their structural diversity.

The Complex Path for Assembly and Export of LPS

Substantial effort has been invested in studying LPS biosynthesis and in developing approaches to target the enzymes involved in LPS formation. Lipid A biosynthesis involves a highly conserved nine-step pathway (the Raetz pathway), named after Christian Raetz, whose research group was largely responsible for its discovery and characterization. In the *E. coli* prototype, it begins with the precursor UDP-GlcNAc and culminates with the product lipid A-Kdo_2. The early steps in this pathway occur in the cytosol, whereas the acylation steps are membrane-associated. The two β-hydroxyl-linked fatty acids are added only after the Kdo units. As such, mutations (or inhibitory compounds) affecting the precursor CMP-Kdo block lipid A synthesis at the stage of the tetraacyl (lipid IV_A) intermediate. The formation of Kdo_2–lipid A is normally essential for survival of *E. coli*, but this does not hold true in a few bacteria (including *N. meningitidis* and *A. baumannii*). Once complete, lipid A-Kdo_2 molecules may enter the LPS export pathway or provide an acceptor for addition of the core oligosaccharide by the sequential actions of nucleotide sugar-dependent GTs and kinases before export. Lipid A (or lipid A-core) are substrates for MsbA, an ABC transporter that flips these molecules to the periplasm.

At the periplasmic face, lipid A-core molecules may be modified by the addition of O-PS or translocated directly to the cell surface (Figure 21.5), contributing to heterogeneity in the final

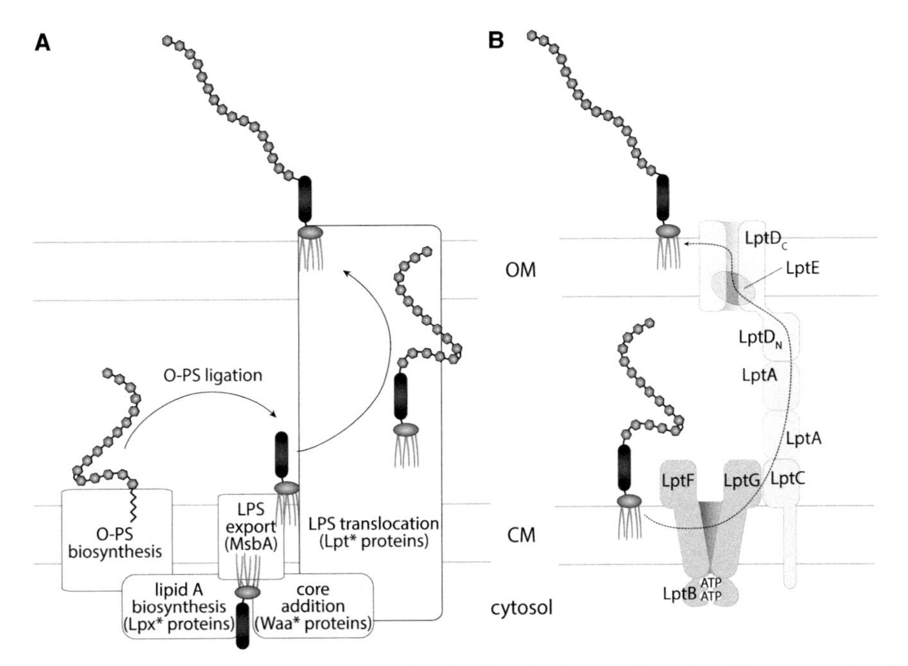

FIGURE 21.5. Assembly and export of lipopolysaccharides (LPSs). The complexity in LPS structures is reflected in the assembly pathways (*A*). Lipid A is assembled at the cytosol cytoplasmic membrane (CM) interface by proteins designated Lpx* (where the asterisk marks a different letter for the particular enzyme). Waa* enzymes extend the core region of lipid A. The completed lipid core is then flipped to the periplasm by the ATP-binding cassette (ABC) transporter MsbA. O-antigens (O-PS) are synthesized by one of three fundamentally different assembly mechanisms. All are *trans*-CM processes, require an undecaprenyl diphosphate carrier, and end with the polyprenol-linked intermediates located at the periplasmic face of the CM. There, a ligase enzyme joins the two parts of the molecule and the completed product enters the export pathway mediated by Lpt* proteins (*B*). The Lpt system extracts completed LPS from the CM, transfers it to a *trans*-periplasm scaffold, and integrates it into the outer membrane. OM, outer membrane. (Adapted, with permission, from Whitfield C, Trent MS 2014. *Annu Rev Biochem* **83:** 99–128, © Annual Reviews.)

LPS species that is reflected in the SDS-PAGE profile (Figure 21.4). The assembly of O-PS occurs independently of the lipid A-core, with most bacterial species using one of two main synthetic strategies. Both strategies involve the formation of undecaprenyl diphosphate-linked intermediates at the cytoplasmic face of the inner membrane, using nucleotide sugars as activated donors. In one pathway, individual lipid-linked repeating units are synthesized and flipped to the periplasmic face of the membrane by a transporter (Wzx), related to MurJ from peptidoglycan assembly (see above). In the periplasm, the polymerase (Wzy) transfers a nascent O-PS chain from its undecaprenol carrier to the nonreducing terminus of the newly exported lipid-linked repeating unit, extending the chain one repeat unit at a time. An additional protein, Wzz (a member of the polysaccharide copolymerase [PCP] family), regulates the polymerization reaction to yield O-PS chains within a particular (modal) size range that is appropriate for their function(s). This generates the distribution seen in SDS-PAGE profiles (Figure 21.4). In the alternative pathway, the full-length O-PS is completed in the cytoplasm by sequential sugar transfer to the nonreducing terminus of the lipid-linked glycan. The final polymerized product is then exported by a dedicated ABC transporter, which shares some structural (and presumably functional) similarity with the WTA transporter. Depending on the bacterial species, the length of the O-PS is determined either by competition between the export and chain-elongation enzymes, or by a chain-capping enzyme. Regardless of which pathway is used, completed undecaprenol diphosphate-linked O-PSs are available in the periplasm, where O-PS ligase (WaaL) completes the glycosylation of the lipid A-core acceptor (Figure 21.5). This enzyme is an integral membrane GT whose catalytic site resides in the periplasm.

Final translocation of LPS molecules to the cell surface is mediated by a conserved LPS translocation pathway comprising seven principle protein components (Figure 21.5). This ATP-driven molecular machine exports an estimated 70,000 molecules/min in growing *E. coli.* An ABC protein complex (LptBFG) extracts the LPS from the inner membrane and transfers it into the periplasmic part of the pathway. In *E. coli*, a scaffold comprised of LptA is flanked by domains from LptC and LptD to bridge the two membranes. These domains share the same fold with LptA and are thought to create a unidirectional groove that sequesters the acyl chains in the trafficking LPS molecules. At the outer membrane, the LPS molecules are proposed to enter the lumen of LptD and escape into the membrane via a lateral opening. Ongoing ATP hydrolysis drives the continuing unidirectional export.

PROTEIN GLYCOSYLATION—AN EXPANDING FACET OF BACTERIAL GLYCOBIOLOGY

Protein glycosylation was once thought to be a property confined to eukaryotes. However, it is now apparent that N- and O-linked protein glycosylation systems are also present in Archaea and many genera of bacteria.

Extensive advances have been made in understanding the biosynthesis of bacterial glycoproteins and four general pathways have been identified. There are the traditional pathways for N- and O-linked glycosylation resembling their eukaryotic counterparts and nontraditional pathways that are unique to bacteria. Traditional N-glycosylation systems have only been described in Gram-negative bacteria, with *Campylobacter jejuni* as the model. Nucleotide-activated sugars in the cytoplasm are assembled onto undecaprenylphosphate in steps resembling the early stages of Wzx/Wzy-dependent O-PS biosynthesis. The completed oligosaccharide is then flipped *en bloc* across the inner membrane into the periplasm, the topological equivalents of the eukaryotic endoplasmic reticulum membrane and lumen, respectively. Subsequently, the bacterial ortholog of STT3 from the oligosaccharyltransferase (OTase) complex transfers the oligosaccharide posttranslationally onto asparagine residues of at least 80 different proteins by recognizing the bacterial sequon: D/E-X1-N-X2-S/T (where X1 and X2 cannot be proline). The sole OTase protein, PglB, also hydrolyzes the oligosaccharide from the lipid and releases this glycan as fOS into the periplasm. In *C. jejuni*, fOS concentrations can comprise 2.5% of the dry cell weight and are influenced by osmotic stress. In *C. jejuni*, a cluster of 10 *pgl* genes encodes enzymes responsible for the synthesis and transfer of the heptasaccharide: GalNAcα1-4GalNAcα1-4[Glcβ1-3]GalNAcα1-4 GalNAcα1-4GalNAcα1-3-diNAcBac-β1 (diNAcBac is 2,4-diacetamido-2,4,6-trideoxyglucopyranose). The N-glycans of *Helicobacter pullorum*, *Helicobacter winghamensis*, *Wolinella succinogenes*, and *Desulfovibrio gigas* have also been examined. An active full-length PglB OTase from *Campylobacter lari* was cocrystallized with an acceptor peptide, providing the first structure of any OTase. The crystal structure of the *C. jejuni* ABC transporter (i.e., PglK flippase) was also solved and showed that the heptasaccharide and pyrophosphate linker are shielded from the inner membrane lipid bilayer, whereas the undecaprenyl anchor interacts with the lipid bilayer and is responsible for triggering translocation.

γ-Proteobacteria, such as *Haemophilus influenzae*, *Yersinia enterocolitica*, and *Actinobacillus pleuropneumoniae*, possess a nontraditional cytoplasmic pathway for N-glycosylation that does not involve assembly of oligosaccharides on a lipid-linked intermediate, but instead transfers single glycan moieties to proteins from nucleotide-activated precursors. The soluble GTs (NGTs) required for N-glycosylation in these organisms are also completely unrelated to the conventional STT3-orthologous OTases, yet prefer the eukaryotic sequon, N-X-S/T. *H. influenzae* N-glycosylates its autotransporter adhesion HMW1 with single glucose and galactose residues. In contrast, *Shewanella oneidensis*, *Pseudomonas aeruginosa*, and *N. meningitidis* add rhamnose to Arg32 of their polyproline specific elongation factor to rescue stalled ribosomes.

In addition to N-glycosylation, bacteria possess both canonical and noncanonical pathways for O-glycosylation. For the O-glycosylation system that mirrors the process in eukaryotes, both Gram-positive and Gram-negative bacteria use nucleotide-activated sugars and soluble GTs in the cytoplasm to modify serine and threonine residues on specific proteins, particularly surface structures, such as flagella, pili, and adhesins. Pseudaminic acids, legionaminic acids, and related nonulosonic sugars are common monosaccharides attached by *C. jejuni*, *Helicobacter pylori*, and *Aeromonas caviae* to their flagella and these modifications are essential for filament assembly and bacterial motility.

The noncanonical pathway for O-glycosylation in Gram-negative bacteria involves glycosyltransferase assembly of nucleotide-activated sugars onto undecaprenylphosphate, flipping of the completed oligosaccharide *en bloc* across the inner membrane into the periplasm, and addition of glycans onto S/T residues by a general OTase. This alternate pathway has been thoroughly characterized in *Neisseria* species, in which multiple proteins (including pilin) are glycosylated with an oligosaccharide containing diNAcBac (or variants such as glyceramidoacetamidotrideoxyhexose or diacetamidodideoxyglucopyranose) at the reducing end, the same carbohydrate attached to Asn residues by the *C. jejuni* N-glycosylation pathway. Gram-positive bacteria lack surface LPS, but many instead possess a regular array of glycoproteins known as S-layer proteins. The oligosaccharides on these S-layer proteins are also first assembled onto an undecaprenylphosphate carrier and subsequently added onto S/T residues of these proteins, but can also be coupled to hydroxyl groups of tyrosine residues.

OSMOREGULATED PERIPLASMIC GLUCANS

Bacteria encounter extreme differences in osmolarity in the environment (and must withstand up to 6 atmospheres of turgor pressure!) and have thus evolved both physical and chemical mechanisms to resist disruption. Peptidoglycan, as discussed earlier, provides a structural barrier to osmotic swelling. In many α-, β-, and γ-proteobacteria, a chemical mechanism also exists to protect the inner membrane. In these Gram-negative organisms, osmoregulated periplasmic glucans (OPGs; previously known as membrane-derived oligosaccharides [MDOs]) contribute to an osmotic buffer. These compounds constitute up to 5% of the dry weight of *E. coli*, and their synthesis is induced by low osmotic conditions. These characteristics are similar to the production of fOS derived from the N-glycosylation pathways of *Campylobacter* species, which lack normal OPGs.

OPGs were first discovered during analysis of *Agrobacterium tumefaciens* culture supernatants and rediscovered in studies of phospholipid turnover in *E. coli*, which showed that the polar head groups of phosphatidylglycerol were transferred to low-molecular-weight, water-soluble oligosaccharides (and thus named MDOs). Other organisms, including *Pseudomonas*, *Rhizobia*, *Brucella*, and *Salmonella*, also make these compounds. However, because not all oligosaccharides contain membrane phospholipid head groups, they were renamed OPGs. Although the precise structure of OPGs varies, they all consist of D-glucose units with a β-linked backbone. The OPGs comprise four families based on their structure. Family I members, including *E. coli*, join 5–12 linear glucose units in β1-2 linkage with α1-6 glucose branches. Family II members, including *Agrobacterium*, synthesize cyclic glucans with 17–25 glucose units joined by β1-2 linkages. Family III/IV members, including *Bradyrhizobium*, also synthesize cyclic glucans with 10–28 glucose units joined by both α- and β-linkages. In addition, the oligosaccharides can contain phosphoethanolamine, phosphoglycerol, and phosphocholine, as well as acetyl, succinyl, methylmalonyl, and phosphoryl groups, which could add charge to the neutral species.

In *E. coli*, one inner membrane glucosyltransferase (OpgH), requiring UDP-Glc as a donor and acyl-carrier protein as a cofactor, is involved in the biosynthesis and transport of OPGs across the inner membrane. A second periplasmic glucosyltransferase (OpgG) is involved in glucose branch addition and size regulation of the OPG molecules. Interestingly, OpgH was recently shown to play another role as a nutrient-dependent regulator of *E. coli* cell size. It is, therefore, not surprising that OPGs influence several biological processes in addition to their involvement in osmoregulation.

EXTRACELLULAR POLYSACCHARIDES

Structure and Function of Capsules and Exopolysaccharides

Bacteria produce long-chain extracellular polysaccharides with extensive structural diversity (Figure 21.6). Extreme examples of this diversity are *E. coli*, with >80 capsule types, and *Streptococcus pneumoniae*, with >90 capsule types. There are two types of extracellular polysaccharides. CPSs maintain an association with the cell surface, encapsulating the bacterium in a hydrophilic coating; in contrast, secreted EPSs have limited association with the cell surface postexport.

The basis for the retention of CPSs at the cell surface is not always known. In *Streptococcus*, this is sometimes achieved by covalent linkage to peptidoglycan, similar to other wall-associated polymers, such as teichoic acids (see above). However, streptococcal CPS can also be attached to an unknown membrane lipid. Some *E. coli* CPSs possess a novel terminal glycolipid composed of an oligosaccharide of several β-linked Kdo residues attached to phosphatidylglycerol, creating a glycolipid with three domains that is conceptually similar to LPS. This terminal glycolipid is conserved in some other Gram-negative mucosal pathogens. In other cases, noncovalent charge-based interactions with other components of the cell surface may be more important.

Extracellular polysaccharides have functions as varied as the environmental niches occupied by the producing bacteria. Given their highly hydrated nature, these polymers are frequently linked to protection against desiccation. Some, such as bacterial cellulose and PNAG (Figure

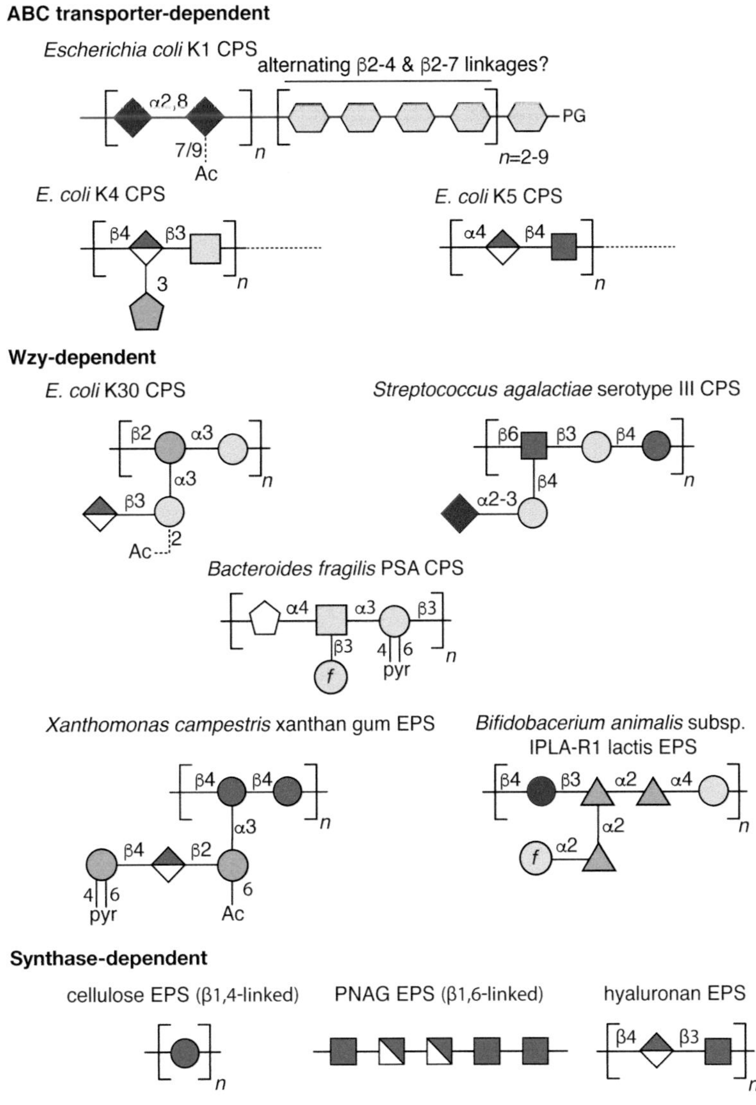

FIGURE 21.6. Structures of exopolysaccharides and capsular polysaccharides (CPSs and EPSs). Like O-antigenic polysaccharides, the structures of CPS and EPS are highly diverse, and the main structural features are covered by the examples shown. The structures are divided according to the three main assembly mechanisms. One group of CPSs is formed by a pathway involving an ATP-binding cassette (ABC) transporter and is (so far) confined to Gram-negative bacteria. The hallmark of this system is the presence of a conserved glycolipid anchor containing an oligosaccharide of β-linked Kdo. Most CPSs with complex branched structures in Gram-negative and Gram-positive bacteria use a conserved mechanism that requires a characteristic polymerase protein, designated Wzy. Relatively simple structures involved in biofilm formation (e.g., cellulose and poly-*N*-acetylglucosamine or PNAG) and some glycosaminoglycans use a pathway involving a dual polymerase-exporter protein referred to as a synthase. D-AAT, 2-acetamido-4-amino-2,4,6-trideoxygalactose.

21.6), are pivotal in biofilm formation. PNAG shows a particularly broad distribution, spanning Gram-positive and -negative species. In many cases, they mask underlying receptors for bacteriophages, although CPSs themselves provide additional receptors and this may help drive the structural diversity. For a bacteriophage to infect encapsulated bacteria, it must penetrate the polysaccharide layer; for this reason, bacteriophages provide a wide range of glycosidases with unique specificities. CPSs are prevalent in pathogens, in which they prevent phagocytosis and (in some cases) complement-mediated killing. Antibodies directed to CPSs overcome these limitations, with the effective pneumococcal vaccines (e.g., Prevnar and Pneumovax) providing excellent examples of the potential of CPSs as vaccine candidates. However, some CPSs in important pathogens are poorly immunogenic because of their mimicry of host cell components. Examples include production of hylauronan by group A streptococci (Chapter 16), α2-3sialyllactosamines by group B streptococci (Chapter 37), α2-8-linked polysialic acid, which mimics embryonic NCAMs, made by *E. coli* K1 and *N. meningitidis* (meningococcus) serogroup B, among others, and production of mimics of the glycosaminoglycans (Chapter 17) heparosan and chondroitin by isolates of *E. coli* and *Pasteurella multocida* (Figure 21.6). These are remarkable examples of convergent evolution by bacteria, generating molecular mimics of "self-associated molecular patterns" (SAMPs) of eukaryotes.

EPSs may also aid in health promotion. CPS production is vital for the ability of *Bacteroides* species to colonize the gut, and are the most abundant members of the gut microbiota. In *Bacteroides fragilis*, a phase-variation mechanism controls the expression of multiple loci (generating multiple structures), but only one is needed for colonization. However, certain CPS structures (e.g., polysaccharide A [PSA], not to be confused with polysialic acids; Figure 21.6), show zwitterionic properties and exert a powerful role in immunomodulation and maturation of the immune system. The capacity of other members of the microbiota and "probiotic" bacteria (e.g., Gram-positive Bifidobacteria) to produce EPSs (Figure 21.6) with wide-ranging physicochemical properties may have additional positive impacts on human health.

EPSs are prominent in a variety of plant pathogens and symbionts. For example, the EPS produced by the Gram-negative phytopathogen *Xanthomonas campestris* is critical for its detrimental effects on feed crop brassicas. The polymer (known as xanthan gum; Figure 21.6) blocks the flow of water, leading to wilting and further damage. The unusual physical properties of xanthan gum have led to a variety of commercial applications, including as food additives (thickeners). Biofilms are particularly dependent on EPSs. Biofilms are dynamic and structurally complex bacterial communities enclosed within a matrix composed of polysaccharides, proteins, and nucleic acids. They can be found on tissue surfaces and teeth (where they form plaque), as well as on nonbiological surfaces, such as pipelines, ship hulls, and medical implants (including catheters and prostheses). Within the biofilm, a combination of the protected environment and altered cell physiology can help protect bacteria from some antimicrobial compounds. Some bacteria may produce multiple EPSs that contribute to biofilms, and their precise structures may be critical for creation of a proper biofilm architecture, as in *Pseudomonas*. Certain glycan structures appear well adapted to roles in biofilm communities and are, therefore, more widely distributed. Many biofilm-forming bacteria produce PNAG, also known as polysaccharide intercellular adhesin, with this being targeted by small-molecule inhibitor and vaccine strategies to eradicate biofilm bacteria (Figure 21.6).

Assembly and Export of Capsules and Exopolysaccharides

Assembly systems for CPSs and EPSs are broadly divided into three mechanisms that are distinguished (in part) by the nature of the acceptor on which they are built (Figure 21.6). A large variety of CPSs and EPSs use undecaprenol diphosphate-linked intermediates and follow a synthetic pathway indistinguishable from that used by Wzx/Wzy-dependent O-antigens, described above. In fact, some bacteria produce O-antigens and CPSs/EPSs with the same structures. The CPS shares the same early steps in the pathway but instead of ligating the product to lipid A-core, a CPS-dedicated export machinery spanning the periplasm and outer membrane is used. The features that distinguish CPSs and EPSs in this type of assembly process are unknown. In Gram-positive bacteria with Wzx/Wzy-dependent processes, the nascent polysaccharide is then either released as an EPS or transferred from undecaprenol and attached to peptidoglycam, often via a phosphodiester linkage to MurNAc residues. In Streptococci, this may be catalyzed by an enzyme related to those that link WTAs to the same site in peptidoglycan. Several Gram-negative bacteria assemble CPSs using a conserved phosphatidylglycerol-$(Kdo)_n$ acceptor. Although prototypes are found in *E. coli* and *N. meningitidis*, the system is conserved across a range of mucosal pathogens. Here, polymerization of the lipid-linked glycan is completed in the cytoplasm, after which it is exported via an ABC transporter. The transporter is linked to additional proteins that span the periplasm and outer membrane in a configuration thought to resemble certain drug efflux pumps in bacteria. The final assembly pathway seems to operate independently of a lipid carrier and requires a processive polymerase called a synthase. Examples are found in Gram-positive and Gram-negative bacteria, but seem to be confined to EPSs. The prototypical hyaluronan and cellulose synthases from Streptococci and Gram-negative bacteria show different architectures, yet both are sufficient for both synthesis and export across the cytoplasmic membrane. The molecular details underlying these processes have been resolved by analysis of the crystal structure of the cellulose synthase. In Gram-negative bacteria, a periplasmic scaffold links the synthase to an outer membrane protein channel.

ADDITIONAL FACETS OF BACTERIAL GLYCOBIOLOGY

Although the focus of this chapter is on the glycobiology of bacterial cell envelopes, there are several additional (and equally important) facets of bacterial glycobiology. Three of these are briefly mentioned here. First, in addition to synthesizing surface structures, bacteria also produce intracellular glycans and cytoplasmic glycosylated proteins and glycolipids. For example, glycogen and trehalose act as storage compounds within some bacteria. Mycobacteria also contain a growing list of unusual glycosylated molecules, which may be related to their ability to withstand a variety of stresses. Second, bacteria produce a wide range of glycan-binding proteins. These include adhesins that facilitate bacterial colonization, exotoxins that bind to host membrane glycans, and single-sugar-binding proteins involved in metabolism (see Chapter 42). Third, bacterial lifestyles are often defined by their capacity to digest and/or metabolize glycans. Complex nanomachines have been described for the degradation of complex carbohydrate structures. Assembling the various enzymes into a highly organized complex affords synergy and efficiency. For example, the roles of glycosyl hydrolases in biomass conversion are well documented (Chapter 59). The cellulose degradation machinery in *Clostridioides sp.* and other organisms comprises a suite of glycosidases and carbohydrate-binding modules, all assembled into a single surface complex (the cellulosome) via noncatalytic cohesin and dockerin structural modules. Within the gut microbiome, Bacteroidetes encode a remarkable array of carbohydrate-active enzymes in polysaccharide utilization loci, allowing them to adapt to dietary changes by sensing, binding, and degrading different complex substrates, followed by import of the products for catabolism. Indeed, bacteria within the many biomes of the human body derive benefit from foraging, not only on plant polysaccharides, but also on host-derived mucosal glycans. We continue to learn more about the intricate pathways for the digestion, capture, and catabolism of environmental and host carbohydrates.

ACKNOWLEDGMENTS

The authors acknowledge contributions to previous versions of this chapter by Jeffrey D. Esko, Tamara L. Doering, and the late Christian Raetz and appreciate helpful comments and suggestions from Graham Heberling and Jerry Eichler.

FURTHER READING

Raetz CR, Whitfield C. 2002. Lipopolysaccharide endotoxins. *Annu Rev Biochem* **71**: 635–700. doi: 10.1146/annurev.biochem.71.110601.135414

Nothaft H, Szymanski CM. 2010. Protein glycosylation in bacteria: sweeter than ever. *Nat Rev Microbiol* **8**: 765–778. doi:10.1038/nrmicro2383

Silhavy TJ, Kahne D, Walker S. 2010. The bacterial cell envelope. *Cold Spring Harb Perspect Biol* **2**: 1–16.

El Kaoutari A, Armougom F, Gordon JI, Raoult D, Henrissat B. 2013. The abundance and variety of carbohydrate-active enzymes in the human gut microbiota. *Nat Rev Microbiol* **11**: 497–504. doi:10.1038/nrmicro3050

Nothaft H, Szymanski CM. 2013. Bacterial protein N-glycosylation: new perspectives and applications. *J Biol Chem* **288**: 6912–6920. doi:10.1074/jbc.r112.417857

Angala SK, Belardinelli JM, Huc-Claustre E, Wheat WH, Jackson M. 2014. The cell envelope glycoconjugates of *Mycobacterium tuberculosis*. *Crit Rev Biochem Mol Biol* **49**: 361–399. doi:10.3109/10409238.2014.925420

Percy MG, Gründling A. 2014. Lipoteichoic acid synthesis and function in Gram-positive bacteria. *Annu Rev Microbiol* **68**: 81–100. doi:10.1146/annurev-micro-091213-112949

Whitfield C, Trent MS. 2014. Biosynthesis and export of bacterial lipopolysaccharides. *Annu Rev Biochem* **83**: 99–128. doi:10.1146/annurev-biochem-060713-035600

Artzi L, Bayer EA, Moraïs S. 2017. Cellulosomes: bacterial nanomachines for dismantling plant polysaccharides. *Nat Rev Microbiol* **15**: 83–95. doi:10.1038/nrmicro.2016.164

Bontemps-Gallo S, Bohin J-P, Lacroix J-M. 2017. Osmoregulated periplasmic glucans. *EcoSal Plus* **7**: doi:10.1128/ecosalplus.esp-0001-2017

Okuda S, Sherman DJ, Silhavy TJ, Ruiz N, Kahne D. 2017. Lipopolysaccharide transport and assembly at the outer membrane: the PEZ model. *Nat Rev Microbiol* **14**: 337–345. doi:10.1038/nrmicro.2016.25

Low KE, Howell PL. 2018. Gram-negative synthase-dependent exopolysaccharide biosynthetic machines. *Curr Opin Struct Biol* **53**: 32–44. doi:10.1016/j.sbi.2018.05.001

Mostowy RJ, Holt KE. 2018. Diversity-generating machines: genetics of bacterial sugar-coating. *Trends Microbiol* **26**: 1008–1021. doi:10.1016/j.tim.2018.06.006

Powers MJ, Trent MS. 2018. Expanding the paradigm for the outer membrane: *Acinetobacter baumannii* in the absence of endotoxin. *Molecular Microbiology* **107**: 47–56. doi:10.1111/mmi.13872

Simpson BW, Trent MS. 2019. Pushing the envelope: LPS modifications and their consequences. *Nat Rev Microbiol* **17**: 403–416. doi:10.1038/s41579-019-0201-x

Dulberger CL, Rubin EJ, Boutte CC. 2020. The mycobacterial cell envelope—a moving target. *Nat Microbiol* **18**: 47–59. doi:10.1038/s41579-019-0273-7

Whitfield C, Wear SS, Sande C. 2020. Assembly of bacterial capsular polysaccharides and exopolysaccharides. *Annu Rev Microbiol* **74**: 1–23. doi:10.1146/annurev-micro-011420-075607

Whitfield C, Williams DM, Kelly S. 2020. Lipopolysaccharide O-antigens—bacterial glycans made to measure. *J Biol Chem* **295**: 10593–10609. doi:10.1074/jbc.rev120.009402

22 Archaea

Benjamin H. Meyer, Sonja-Verena Albers, Jerry Eichler, and Markus Aebi

This chapter describes the current knowledge of archaeal glycobiology. As in bacteria and eukaryotes, the archaeal cell surface is covered with glycans, which serve as an essential part of the cell wall polysaccharides and as a modification of lipids or surface proteins, as well as the major component of the extracellular matrix. Recent discoveries shed light on a tremendous structural and functional diversity of carbohydrates in this domain of life. In particular, the pathways of N-linked protein glycosylation, homologous to the eukaryotic N-glycosylation machinery, generate a wide variety of N-linked glycans in different archaeal species.

BACKGROUND

Based on Carl Woese's pioneering use of 16S ribosomal (r)RNA analysis, the Archaea were first recognized as a separate domain of life, distinct from either Bacteria or Eukarya. As the first Archaea identified were isolated from some of the most physically challenging environments on the planet, such as those defined by extremes in salinity, pH, or temperature, it was assumed that all Archaea were extremophiles. However, the discovery of Archaea in a variety of "normal" as well as "extreme" biological niches revealed that Archaea represent a major portion of the microbial population and that they play crucial roles in geochemical cycles on Earth. Furthermore, the discoveries of new archaeal lineages in the last decade, predominantly based on metagenomics studies, have led to extensive expansion and reconstruction of the archaeal phylogenetic tree. At present, attempts to cultivate these newly identified Archaea have rarely been successful. Nevertheless, the study of cultivated Archaea has led to many important discoveries.

published by Cold Spring Harbor Laboratory Press; doi: 10.1101/glycobiology.4e.22

Since Neuberger's discovery of protein glycosylation in the late 1930s (Chapter 1) and his description of the *N*-acetylglucosamine (GlcNAc)-β-asparagine linkage of a glycan to the modified protein, it became generally accepted that protein glycosylation was a process limited to eukaryotes. This long-held belief was challenged in 1976, when Mescher and Strominger showed that the surface (S)-layer glycoprotein of the archaeon *Halobacterium salinarum* was subject to both N- and O-glycosylation, thus offering the first example of a noneukaryotic glycoprotein. Today, it is clear that protein glycosylation is an almost universal trait of the Archaea.

The discovery that Archaea do not contain murein in their cell wall was one of the main arguments used to distinguish this group of prokaryotes from Bacteria. Indeed, at the time, cell wall composition was considered to be "the only useful phylogenetic criterion, other than direct molecular phylogenetic measurement" to distinguish between the two prokaryotic domains. Some methanogenic Archaea were, however, shown to include a distinct polymer termed pseudomurein (or pseudopeptidoglycan) in their cell wall, whereas other archaeal species were found to assemble cell walls based on different sugar-based polymers. Today, as more and more archaeal species are cultivated, it is becoming clear that Archaea present numerous variations in the composition of the cell surface. For instance, although many species seem to mainly rely on a cell envelope in which the cytoplasmic membrane is enclosed by a two-dimensional crystalline proteinaceous layer called the surface (S)-layer, strains surrounded by two membranes have been identified. Figure 22.1 summarizes current knowledge about archaeal cell surfaces.

THE ARCHAEAL CELL WALL

Similar to the Bacteria, there is no cell wall structure unique to all Archaea. However, like in Bacteria, there are building blocks that are found in different archaeal clades (Figure 22.1). Some of these cell wall components are very similar in structure to their bacterial counterparts, yet seem to be the product of convergent evolution. Other cell wall–generating processes seem to be homologous to pathways used for eukaryotic extracellular matrix assembly. The biophysical properties of the building blocks of the archaeal cell wall provide the basis for the ability of many archaeal species to thrive in extreme habitats.

Archaeal Cell Wall Polysaccharides

Pseudomurein (Pseudopeptidoglycan)

Although pseudomurein was identified as a component of the cell wall early in the study of Archaea, it subsequently became clear that in terms of distribution, the occurrence of this structure was relatively limited. Pseudomurein shares structural similarities with bacterial murein yet presents significant differences (Figure 22.2). It usually consists of *N*-acetyl-L-talosaminuronic acid linked via a β1-3 linkage to *N*-acetyl-D-glucosamine, unlike murein, which contains alternating *N*–acetylmuramic acids linked via β1-4 linkage to GlcNAc. Moreover, the glycan strands of pseudomurein are cross-linked by peptides composed of L-amino acids (glutamic acid, alanine, and lysine), in contrast to the D-amino acids used in murein. Pseudomurein surrounds cells of all species belonging to the genus *Methanopyrus* and the order Methanobacteriales, which can, as in the case of *Methanothermus fervidus*, be bordered by an outer S-layer. Homologs of bacterial murein biosynthesis proteins (e.g., MurG or MraY) have been identified in these methanogens, although the functions of these enzymes have not yet been studied.

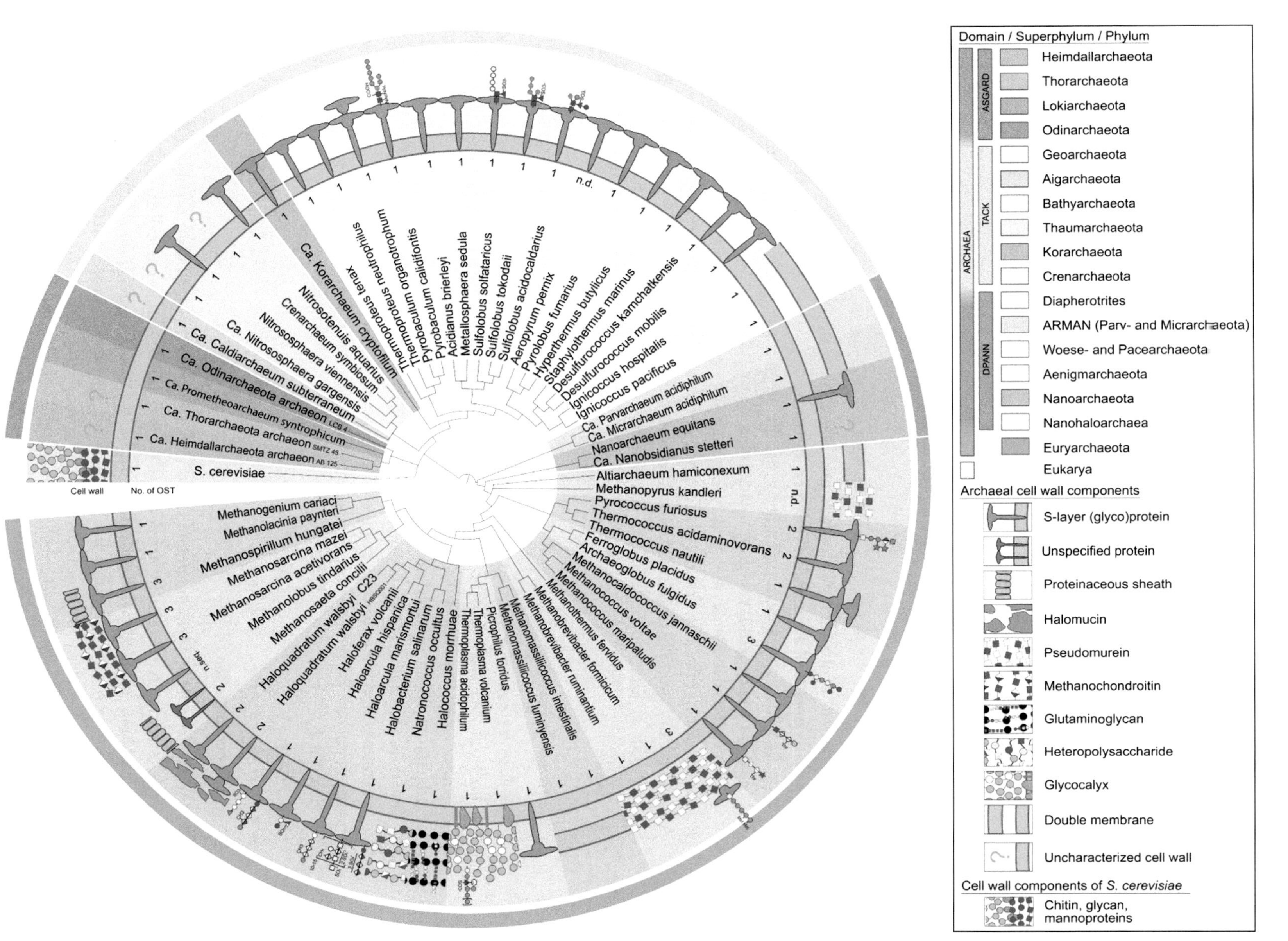

FIGURE 22.1. Diversity of archaeal cell wall structures. The different cell wall components are shown on the *right*. In the *inner circle*, the number of loci encoding putative oligosaccharyltransferases in the respective genomes is shown.

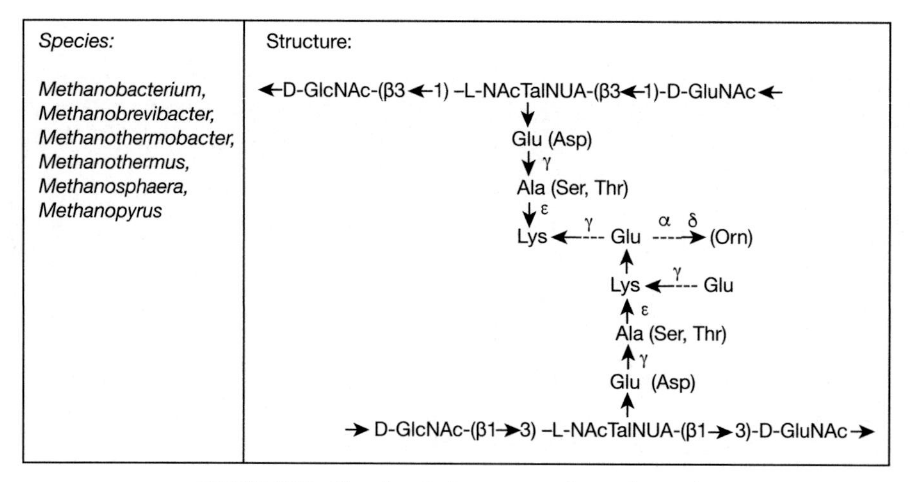

FIGURE 22.2. The chemical structure of pseudomurein.

Glutaminylglycan

The cell wall of the highly halophilic and alkaliphilic genus *Natronococcus* (3.5 M salt and pH 9.5–10) consists of a glutamine polymer. In contrast to poly-γ-D-glutamyl polymers in the bacteria *Bacillus*, *Sporosarcina*, or *Planococcus*, the archaeal polymer is formed from L-glutamines linked via the γ-carboxylic group, yielding a chain of about 60 monomers. Also, in contrast to the bacterial polymer, the poly-γ-L-glutamine chain is glycosylated, containing two types of oligosaccharide. The first oligosaccharide consists of a GlcNAc pentasaccharide at the reducing end and multiple GalA residues at the nonreducing end. The second presents a GalNAc disaccharide at the reducing end and two Glc units at the nonreducing end.

Heteropolysaccharides

Halococcus morrhuae is an extreme halophile surrounded by an electron-dense 50- to 60-nm-thick cell wall composed of a complex, highly sulfated heterosaccharide consisting of glucosamine, galactosamine, gulosaminuronic acid, glucose, galactose, mannose, glucuronic acid, galacturonic acid, N-acetylated amino sugars, and sulfated subunits. Different heteropolysaccharides are thought to be connected via glycine bridges between the amino groups of the glucosamines and the carboxyl groups of the uronic residues. Although the building blocks of heteropolysaccharide have been suggested, biosynthesis of this cell wall structure has yet to be described.

Methanochondroitin

Individual cells of *Methanosarcina* rely on an S-layer as their cell wall. A cubic aggregate of four cells (sarcina) is covered by an additional rigid fibrillar polymer called methanochondroitin. Degradation of methanochondroitin results in the disaggregation of the cells, underlining that the matrix is responsible for maintenance of the aggregate. Methanochondroitin, which is similar to eukaryotic connective tissue chondroitin, is composed of a repeating trimer of uronic acid and two GalNAc residues. Yet unlike chondroitin, methanochondroitin is not sulfated. A pathway of methanochondroitin biosynthesis has been proposed based on activated precursors in *Methanosarcina barkeri* cell extracts. Methanosarcina species can further modify the methanochondroitin condition largely through the addition of glucose and galactose acids.

Lipoglycan

Members of the thermoacidophilic order Thermoplasmatales (pH 1–2 and 60°C), such as *Ferroplasma acidophilum* and *Thermoplasma acidophilum*, lack a rigid cell envelope. Such organisms thus display a pleomorphic shape, similar to mycoplasma. Stabilization of the cell is most likely realized by the oligosaccharide portions of lipoglycans and membrane-associated glycoproteins. The outwardly oriented glycan chains form a protective slime coat called the glycocalyx. A recent study of different *T. acidophilum* cell-surface glycoproteins identified an N-linked branched octosaccharide described below.

Archaeal Surface (S)-Layer Glycoproteins

The majority of characterized Archaea rely on a proteinaceous cell wall, the S-layer, comprising a regularly structured two-dimensional array based on a single protein species, the S-layer glycoprotein, or a limited number of proteins (Figure 22.1). In some archaeal species, the S-layer can be further supported by polysaccharides, by a second S-layer sheet, or by additional surface glycoproteins. For example, the S-layer of *Methanospirillum hungatei* is further enclosed by a tubular proteinaceous sheath. These sheaths form a paracrystalline structure based on a simple p2 lattice, which is distinct from that of the S-layer. Depending on the species, the proteinaceous sheaths can be glycosylated. Another example is the extremely halophilic *Haloquadratum walsbyi*. The unique square-shaped cells, with a thickness of only 0.1–0.5 μm, are surrounded by one or two S-layer sheets. An extremely large glycoprotein, termed halomucin, highly similar to mammalian mucin, is loosely connected to the S-layer. Halomucin is heavily glycosylated, containing more than 280 potential N-glycosylation sites (on average one site every 32 residues). This cell envelope is further enforced by analogues of halomucin, termed Hmu2 and Hmu3, and most likely by a poly-γ-glutamate capsule.

PROTEIN GLYCOSYLATION IN ARCHAEA

The Diversity of N-Linked Glycans in Archaea

To date, glycoproteins from Archaea isolated from a wide range of habitats have been studied to various degrees of detail. Possibly reflecting the varied niches occupied by these organisms, their S-layer glycoproteins and other glycoproteins, such as archaellins and pilins, bear N-linked glycans that present wider diversity in terms of size; degree of branching; the identity of the linking sugar; modification of sugar components by amino acids, sulfate, and methyl groups; and the presence of unique sugars than reported to date in Bacteria or Eukarya. Currently defined archaeal N-linked glycans are depicted in Figure 22.3.

Delineated Pathways of Archaeal N-Glycosylation

The first reported archaeal N-glycosylated protein, the *Halobacterium salinarum* S-layer glycoprotein, was shown to be modified by two different N-linked oligosaccharides, a repeating sulfated pentasaccharide linked via N-glycosylamine to Asn-2 and a sulfated glycan linked by a glucose residue to 10 other Asn residues. The latter glycan is also N-linked to archaellins in this haloarchaeon. Efforts undertaken at the time aimed at deciphering the pathways responsible for the synthesis of these glycans relied solely on biochemical approaches because neither suitable genetic tools nor a genome sequence were available.

Despite these biochemical advances, delineation of archaeal N-glycosylation pathways had to wait until the genome age and the development of tools for the genetic manipulation of various

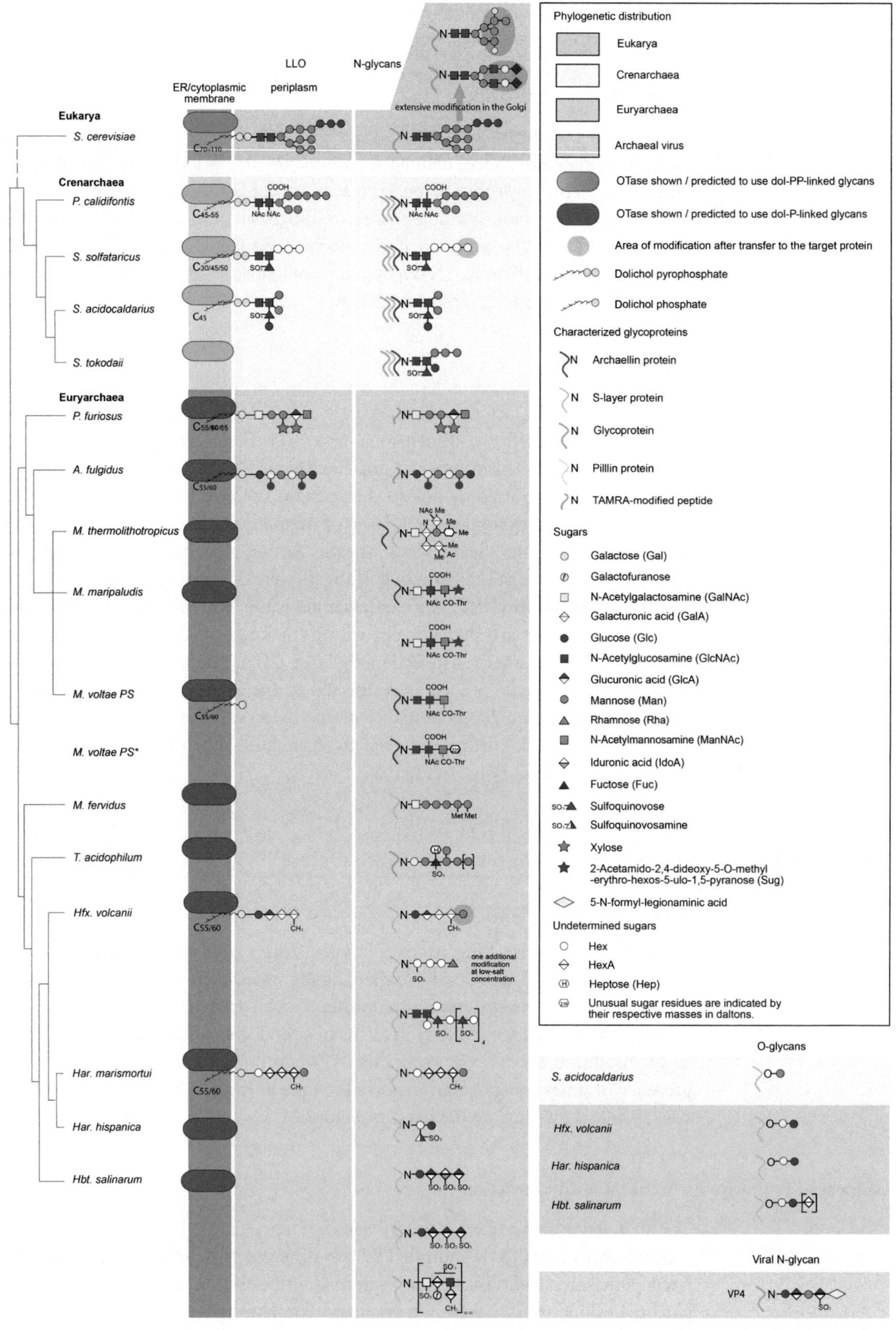

FIGURE 22.3. The structural diversity of N- and O-linked glycans in Archaea. The structures of N-linked glycans found in the Euryarchaeota and Crenarchaeota are shown. The OTase (STT3 or AglB) within the endoplasmic reticulum (ER) or cytoplasmic membrane (*right*), the lipid-linked oligosaccharide (LLO) translocated across the membrane (*middle*) and transferred N-glycan linked to a protein (*right*) are shown. A red background highlights the extensions and modifications of the N-glycan after the transfer of the N-glycan to the target protein. For comparison, the eukaryotic LLO and N-glycan are displayed (*top*). The structure of archaeal O-linked glycans and the N-glycan from one archaeal virus are displayed (*bottom right*).

species. Through the subsequent identification of homologs of eukaryotic and/or bacterial N-glycosylation pathway components, genome scanning for additional components, the generation of deletion strains, and characterization of reporter glycoproteins, *agl* (archaeal glycosylation) genes comprising archaeal N-glycosylation pathways have been identified in several Archaea, including halophilic, methanogenic, and thermophilic species.

Halophilic Euryarchaeota

In the last decade, progress in defining pathways of N-glycosylation has relied on *Haloferax volcanii* as a model organism. In *Hfx. volcanii*, a series of Agl proteins mediate the assembly and attachment of a pentasaccharide with the structure mannose-1,2-[methyl-*O*-4-]glucuronic acid-β1-4-galacturonic acid-α1-4-glucuronic acid-β1-4-glucose-β-Asn to the S-layer glycoprotein and archaellins (Figure 22.4). Acting at the cytoplasmic face of the plasma membrane, the glycosyltransferases AglJ, AglG, AglI, and AglE sequentially add the first four pentasaccharide residues onto a common Dol-P carrier, whereas AglD adds the final pentasaccharide residue, mannose, to a distinct Dol-P (Figure 22.3). Assembly of the Dol-P-linked tetrasaccharide also involves AglF, a glucose-1-phosphate uridyltransferase, AglM, a UDP-glucose dehydrogenase, AglP, a methyltransferase, and AglQ, a predicted isomerase. AglF and AglM have been shown to act in a sequential and coordinated manner in vitro, transforming glucose-1-phosphate into UDP-glucuronic acid. AglB, the archaeal oligosaccharyltransferase, transfers the lipid-linked tetrasaccharide to select Asn residues of target proteins. The final mannose residue is subsequently transferred from its Dol-P carrier to the protein-bound tetrasaccharide in a reaction requiring AglR, a protein that either serves as the Dol-P-mannose flippase or contributes to such activity, and AglS, a Dol-P-mannose mannosyltransferase. Interestingly, N-glycosylation of the S-layer protein is altered in response to a change in environmental conditions. When grown in low-salt medium, *Hfx. volcanii* alters the N-glycan structure in a site-specific manner.

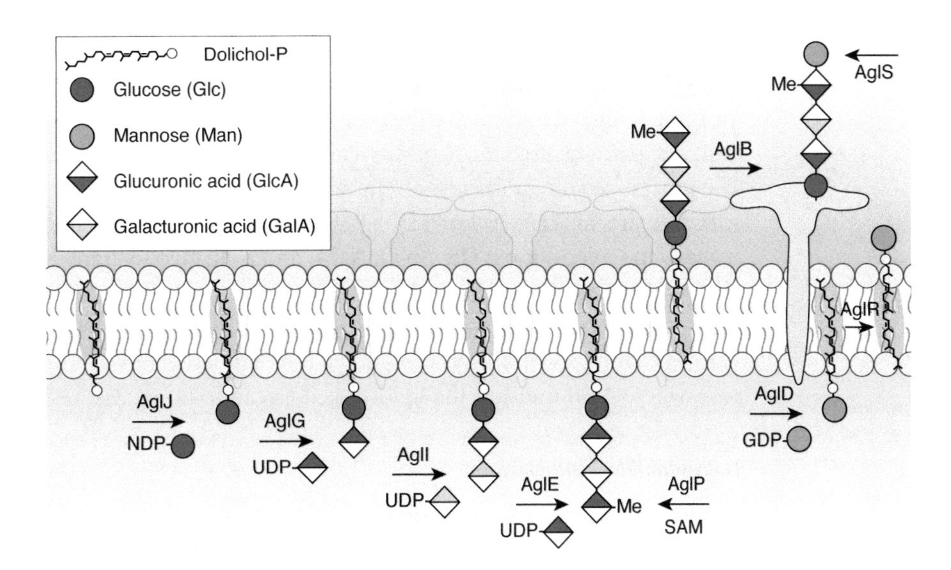

FIGURE 22.4. The pathway of N-glycosylation in *Haloferax volcanii*. The oligosaccharide is assembled on a dolichol phosphate lipid carrier, translocated across the plasma membrane, and transferred to a target protein by the AglB oligosaccharyltransferase. The N-linked glycan is further modified by a mannose residue, originating from dolichol-phosphomannose. The enzymes catalyzing the different reactions are indicated.

Methanogenic Euryarchaeota

Mass spectrometry studies elucidated the glycan N-linked to archaellins of *Methanococcus voltae* strain PS. GlcNAc, the linking sugar, is connected to a diacetylated glucuronic acid, in turn linked to an acetylated mannuronic acid modified by a threonine at the C-6 position (β-Man-pNAcA6Thr-(1-4)-β-Glc-pNAc3NAcA-(1-3)-β-GlcpNAc), although archaellins from other versions of *M. voltae* strain PS presented an N-glycan bearing an additional mass of either 220 or 260 Da at the nonreducing end, likely representing an additional sugar. As with *Hfx. volcanii*, the identification of *M. voltae* N-glycosylation pathway components initially relied on gene deletion and subsequent analysis of the N-linked glycans generated in mutant strain. As such, the oligosaccharyltransferase AglB and the glycosyltransferase AglA, responsible for transfer of the third sugar of the glycan, were discovered. The same strategy was later used to identify AglC and AglK, glycosyltransferases proposed to be involved in the biosynthesis or transfer of the second sugar. A genetic approach also assigned AglH responsibility for adding the linking sugar GlcNAc to the lipid carrier on which the N-linked glycan is assembled. Although *aglH* could not be deleted in *M. voltae*, it was able to complement a conditional lethal mutation in the *alg7* gene of *Saccharomyces cerevisiae*. Alg7, sharing 25% identity with *M. voltae* AglH, catalyzes the conversion of UDP-GlcNAc and Dol-P to UMP and Dol-PP-GlcNAc in the eukaryotic N-glycosylation process. Additional insight into *M. voltae* N-glycosylation has come from in vitro studies. In contrast to the earlier genetics-based studies showing AglH to be the first glycosyltransferase of the pathway, a bacterially expressed version of the enzyme was not able to add GlcNAc to Dol-P. On the other hand, purified AglK catalyzed the formation of Dol-P-GlcNAc from Dol-P and UDP-GlcNAc. The seeming disagreement between the genetics and biochemical results concerning AglH and AglK functions remains to be solved.

Methanococcus maripaludis has become an important model for genetic and structural research on N-glycosylation in the methanogens. In *M. maripaludis*, archaellins are modified by an N-linked tetrasaccharide similar to its *M. voltae* counterpart. In the *M. maripaludis* glycan, the linking sugar is GalNAc and not the GlcNAc used by *M. voltae*. The second sugar in the *M. maripaludis* glycan is a diacetylated glucuronic acid, as in *M. voltae*. Although the third sugar is a modified mannuronic acid with a threonine attached at the C-6 position in both organisms, there is an additional acetamidino group added at position C-3 of the *M. maripaludis* glycan. The fourth and terminal sugar of the *M. maripaludis* glycan is a novel sugar, (5S)-2-acetamido-2,4-dideoxy-5-*O*-methyl-α-L-erythro-hexos-5-ulo-1,5-pyranose. It was later reported that the major *M. maripaludis* pilin is modified by the same N-linked tetrasaccharide bearing an extra hexose branching from the linking GalNAc subunit. The pathway used for N-glycosylation of *M. mariplaudis* archaellin has been largely delineated. The process seemingly starts with the addition of UDP-GalNAc to Dol-P by an unidentified glycosyltransferase. Like *Hfx. volcanii* Dol-P, *M. mariplaudis* Dol-P includes two saturated isoprenes, likely at the α and ω positions. The AglO, AglA, and AglL glycosyltransferases add the next three nucleotide-activated sugars, respectively. AglU adds the threonine moiety to sugar three, apparently only following the addition of the fourth sugar by the glycosyltransferase AglL. AglV then methylates sugar four. The Dol-P-bound tetrasaccharide is then flipped across the membrane by an unidentified flippase, where AglB transfers the lipid-linked glycan to target Asn residues.

Thermophilic Crenarchaeota

Studies on the N-glycosylation process in Crenarchaeota have focused on the thermoacidophilic archaeon *Sulfolobus acidocaldarius*. In *S. acidocaldarius*, the S-layer glycoprotein, archaellin, and cytochrome $b_{558/566}$ are modified by an N-linked hexasaccharide with the structure (Man-α1-6)(Man-α1-4)(Glc-β1-4-Qui6S-β1-3)GlcNAcβ1-4GlcNAc-β-Asn. This glycan is unusual as it contains the typical eukaryotic chitobiose core and a sulfoquinovose, a sugar generally found only in photosynthetic membranes of plants and phototrophic bacteria.

Biosynthesis of the N-linked glycan begins with the transfer of GlcNAc-phosphate, derived from a nucleotide-activated precursor, onto an unusually short and highly saturated Dol-PP lipid carrier by AglH, a UDP-GlcNAc-1-P: Dol-P-GlcNAc-1-P transferase. Information concerning addition of the second and third sugars is lacking. However, Agl3 converts UDP-glucose and sodium sulphite into UDP-sulfoquinovose, which is subsequently added to Dol-PP-bound trisaccharide by an unknown glycosyltransferase. In the final steps of N-linked glycan assembly, the terminal mannose and glucose moieties are added, with Agl16, a soluble glycosyltransferase, adding the final glucose. A so-far unidentified flippase translocates the Dol-PP-bound hexasaccharide across the membrane, where AglB transfers the glycan to target protein Asn residues. In contrast to *Hfx. volcanii, M. voltae*, and *M. maripaludis*, *aglB* is essential in *S. acidocaldarius*.

The Diversity of Archaeal O-Linked Glycans

In comparison to N-glycan biosynthesis, relatively little is known of how archaeal O-glycans are assembled. The O-glycans of four archaeal species have been characterized to a limited extent (Figure 22.3). The only published report on archaeal O-glycan biosynthesis revealed that *Haloarcula hispanica* requires Dol-P-Glc as a sugar donor for the assembly of the N-linked glucose-α-(1,2)-[sulfoquinovosamine-β-(1,6)-]galactose trisaccharide and the O-linked glucose-α-(1,4)-galactose disaccharide.

PHYSIOLOGICAL ROLES OF ARCHAEAL GLYCOSYLATION

N-glycosylation has been considered as assisting Archaea to cope with the challenges of the extreme environments they often occupy. For instance, enhanced surface charge in the face of hypersaline conditions and hence increased solubility was offered as an explanation for the high sulfated sugar content of N-linked glycans decorating the *Hbt. salinarum* S-layer glycoprotein relative to its *Hfx. volcanii* counterpart, given the higher salinity of the locale in which the former lives. In other instances, it is not clear how a given N-glycosylation profile contributes to life in harsh surroundings. In *Hfx. volcanii*, however, N-glycosylation may provide cells with the ability to respond to changes in the surrounding salinity. As noted above, the N-glycosylation profile of the S-layer glycoprotein differs in cells grown in 3.4 or 1.75 M NaCl-containing medium (Figure 22.3). Modified glycosylation in response to environmental conditions has also been reported in the case of *M. hungatei*, in which archaellins are only modified in a low phosphate-containing medium. Furthermore, a fully assembled N-glycan has been shown to be required for a broad range of biological functions, including motility and species-specific cell-cell recognition.

Extracellular Polysaccharides

Archaeal biofilms have been identified in a large variety of habitats (e.g., at low and high temperatures), as well as under acidic, alkaline, and high-salt conditions. It is proposed that Archaea, especially those that require interaction with other species, synthesize biofilms to support these cell–cell interactions. Detailed analyses of the structure and composition of extracellular polymeric substances (EPSs) of archaeal biofilms are limited. The sugar composition of a few EPSs has been analyzed by lectin binding assays. Such studies revealed that in *Sulfolobales* species, for instance, the EPS mainly consists of glucose, galactose, mannose, and *N*-acetyl-D-glucosamine.

ACKNOWLEDGMENTS

The authors acknowledge contributions to previous versions of this chapter by Jeffrey D. Esko, Tamara L. Doering, and the late Christian R.H. Raetz and appreciate helpful comments and suggestions from Ramya Chakravarthy and Debra Mohnen.

FURTHER READING

Sumper M. 1987. Halobacterial glycoprotein biosynthesis. *Biochim Biophys Acta* **906**: 69–79. doi:10.1016/0304-4157(87)90005-0

Lechner J, Wieland F. 1989. Structure and biosynthesis of prokaryotic glycoproteins. *Annu Rev Biochem* **58**: 173–194. doi:10.1146/annurev.bi.58.070189.001133

Kandler O, Konig H. 1998. Cell wall polymers in Archaea (Archaebacteria). *Cell Mol Life Sci* **54**: 305–308. doi:10.1007/s000180050156

Schäffer C, Messner P. 2001. Glycobiology of surface layer proteins. *Biochimie* **83**: 591–599. doi:10.1016/s0300-9084(01)01299-8

Albers SV, Meyer BH. 2011. The archaeal cell envelope. *Nat Rev Microbiol* **9**: 414–426. doi:10.1038/nrmicro2576

Visweswaran GR, Dijkstra BW, Kok J. 2011. Murein and pseudomurein cell wall binding domains of Bacteria and Archaea—a comparative view. *Appl Microbiol Biotechnol* **92**: 921–928. doi:10.1007/s00253-011-3637-0

Eichler J. 2013. Extreme sweetness: protein glycosylation in Archaea. *Nat Rev Microbiol* **11**: 151–156. doi:10.1038/nrmicro2957

Larkin A, Chang MM, Whitworth GE, Imperiali B. 2013. Biochemical evidence for an alternate pathway in N-linked glycoprotein biosynthesis. *Nat Chem Biol* **9**: 367–373. doi:10.1038/nchembio.1249

Jarrell KF, Ding Y, Meyer BH, Albers SV, Kaminski L, Eichler J. 2014. N-linked glycosylation in Archaea: a structural, functional, and genetic analysis. *Microbiol Mol Biol Rev* **78**: 304–341. doi:10.1128/mmbr.00052-13

Klingl A. 2014. S-layer and cytoplasmic membrane—exceptions from the typical archaeal cell wall with a focus on double membranes. *Front Microbiol* **5**: 624. doi:10.3389/fmicb.2014.00624

van Wolferen M, Orell A, Albers SV. 2018. Archaeal biofilm formation. *Nat Rev Microbiol* **16**: 699–713. doi:10.1038/s41579-018-0058-4

23 Fungi

Françoise H. Routier, Tamara L. Doering, Richard D. Cummings, and Markus Aebi

Fungi are a fascinating group of predominantly multicellular organisms. Fungal species, such as *Saccharomyces cerevisiae*, have been instrumental in defining the fundamental processes of glycosylation, but their glycobiology is significantly different from animal or plant systems. This chapter describes the glycan structures that compose the fungal cell wall, offers some insights into novel glycobiology revealed through studying fungal systems, addresses the use of fungi as experimental and synthetic systems, and delineates the relationships of several important glycoconjugates to fungal biology and pathogenesis.

FUNGAL DIVERSITY

More than 70,000 species of fungi have been described, and it is estimated that more than 5,000,000 fungal species exist. The fungal phyla are the Chytridiomycota (zoosporic fungi),

published by Cold Spring Harbor Laboratory Press; doi: 10.1101/glycobiology.4e.23

the Opisthosporidia, the Neocallimastigomycota, the Blastocladiomycota, the Zoopagomycota, the Mucoromycota, the Glomerulomycota (abuscular mycorrhizal fungi), the Ascomycota (sac fungi, e.g., *Saccharomyces*, *Candida*, *Aspergillus*, *Neurospora*, and morel mushrooms), and the Basidiomycota (e.g., mushrooms, rot fungi, and puffballs). The vast majority of species belong to the Ascomycota phylum, which together with the Basidiomycota form the subkingdom Dikarya and comprise the most studied species; this chapter focuses on model organisms in these two clades. Most fungi are primarily made of hyphae (branching filaments) that form the mycelium and multicellular structures such as fruiting bodies, whereas the alternative fungal life-form is growth as unicellular yeast. The extracellular matrix of all fungi, the cell wall, comprises complex polysaccharides including mannans, galactans, glucans, and chitin and represents a major target of fungicides.

FUNGI AS MODEL SYSTEMS FOR GENETICS, BIOCHEMISTRY, AND GLYCOBIOLOGY

Historical Perspective

More than 100 years ago, Louis Pasteur discovered that fermentation requires a viable organism; since then yeast have been used as a model system to study cellular metabolism. In fact, Pasteur coined the word "ferment" during his work on alcohol production by *S. cerevisiae*, or baker's yeast. This organism has been a tremendous resource for biologists and glycobiologists, especially because many of the fundamental enzymes in aerobic and anaerobic metabolism (terms also invented by Pasteur) are shared between yeast and animals. Breakthroughs in enzymology occurred following the 1897 discovery by the Buchner brothers that extracts of yeast could make ethanol and carbon dioxide from glucose, just like intact cells. Mannose is a major component of the yeast cell wall; it was discovered by Emil Fischer in 1888, and the mannose-rich glycans in yeast, historically called yeast gum, have been known since the 1890s. The discovery that the yeast cell wall was composed of D-mannose and work elucidating the chemical structures of other carbohydrates (and vitamin C) led to Sir Walter Norman Haworth's 1937 Nobel Prize in Chemistry. Luis Leloir subsequently discovered the activated precursors required for carbohydrate synthesis, identifying UDP-glucose, GDP-mannose, and other nucleotide sugars from yeast extracts. He was awarded the 1970 Nobel Prize in Chemistry for this work. The discovery of heterothallic yeast strains and the subsequent development of the field of yeast genetics led to multiple groundbreaking discoveries. For example, genetic studies initiated by the laboratory of Phil Robbins led to the molecular characterization of the conserved N- and O-glycosylation pathway in the endoplasmic reticulum (ER) and the biosynthesis of glycosylphosphatidylinositol (GPI)-anchored proteins. Yeast secretory (*sec*) mutants helped define the protein secretory pathway, by which polypeptides travel from the ER through the Golgi apparatus to the cell surface or surrounding milieu, becoming glycosylated en route. This foundational work in cell biology was recognized by the Nobel Prize in Physiology or Medicine awarded to Randy Schekman in 2013.

The Fungal Cell Wall

The fungal cell wall, like the plant cell wall, is composed of highly cross-linked glycan polymers (Figure 23.1), which adapt to growth conditions in a dynamic and flexible way and provide high mechanical stability. In contrast to the plant cell wall, it is directly connected to the plasma membrane and specific cell-wall polysaccharides of fungi differ from those of plants. Fungal cell walls are composed of glycoproteins and complex polysaccharides such as chitin, glucans, mannans, galactomannans, glucomannans, rhamnomannans, and phosphomannans. The nature and relative abundance of cell wall polymers varies between fungal species.

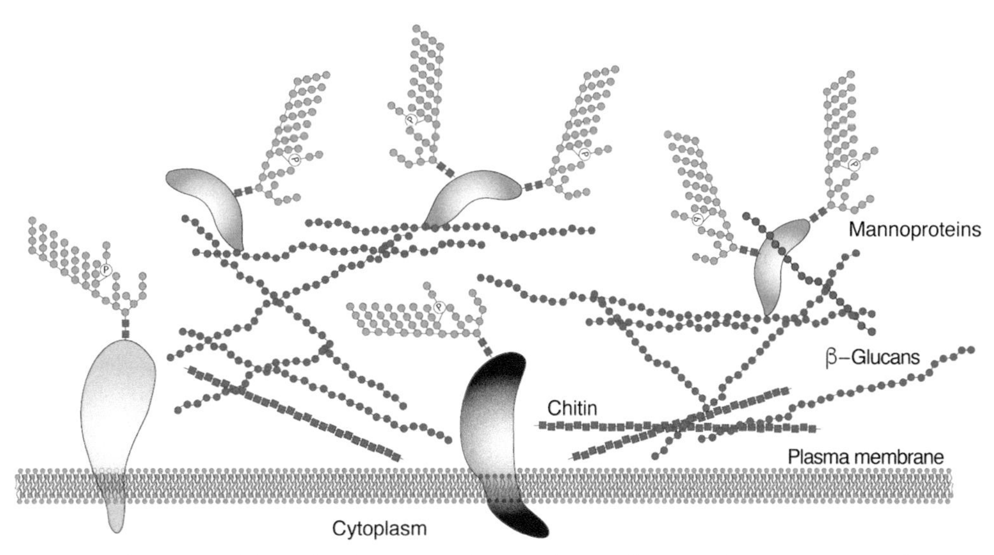

FIGURE 23.1. Illustration of the cell wall of yeasts, showing glycan polymers and mannoproteins. The presence and abundance of different glucans and chitin varies between different fungal species.

Chitin is a polymer of β1-4-linked GlcNAc, which occurs in chains that typically exceed 1000 residues. These chains self-associate to form microfibrils and are deposited primarily at the bud neck of yeast or at septa in filamentous fungi. Chitin is produced by the regulated and coordinated action of multiple chitin synthases, which ensure the timely deposition at specific sites required for normal cell growth and division. Chitin may also be deacetylated to form the cationic polymer, chitosan.

β1-3 Glucan, synthesized from UDP-glucose at the plasma membrane, is a major polysaccharide of fungal cell walls, where it is cross-linked to chitin. A variety of glucans with other linkages, including β1-6, mixed β1-3/β1-4, α1-3, and α1-4, also occur in fungi. β1-6 glucan is a major cell wall component of the yeasts *S. cerevisiae*, *Candida albicans*, and *Cryptococcus neoformans*. In contrast, the filamentous fungi *Aspergillus fumigatus* and *Neurospora crassa* lack β1-6 glucans but synthesize a mixed β1-3/β1-4-glucan. α1-3 Glucan is also a common component of the cell wall of Ascomycota and some Basidiomycota, although it is absent from *S. cerevisiae.*

β-glucan chains act as attachment points for an external glycoprotein layer. The majority of these cell wall proteins are GPI-dependent and carry N- and O-linked glycans. In yeast cell wall proteins, the conserved N-glycan core structure is further elaborated with an extensive repeating α1-6-linked mannose chain. This chain is typically branched by short α1-2- and α1-3-linked mannose structures; some of these may be in phosphodiester linkage (Figure 23.2). These N-glycans are highly heterogeneous in length and branching. In contrast, filamentous fungi and Basidiomycota synthesize small oligomannose *N*-glycans, which may carry substituents such as galactofuranose, N-acetylglucosamine, xylose or fucose. In addition, fungal cell wall proteins bear Ser/Thr-linked O-mannose glycans (Figure 23.3A). During cell wall assembly, cell wall proteins are linked to the β-glucan via a GPI-remnant and/or via their *N*- and *O*-glycans. Cell wall construction is temporally and spatially controlled during the cell cycle, determining cellular shape (hyphae vs. yeast) and function.

Protein Glycosylation

Glycoproteins are a major component of the fungal cell wall, often bearing N- and O-glycans, as well as GPI anchors. As outlined in Chapter 9, N-glycan synthesis begins with synthesis of the conserved lipid-linked core glycan donor, $Glc_3Man_9GlcNAc_2$-P-P-Dol, which is transferred to

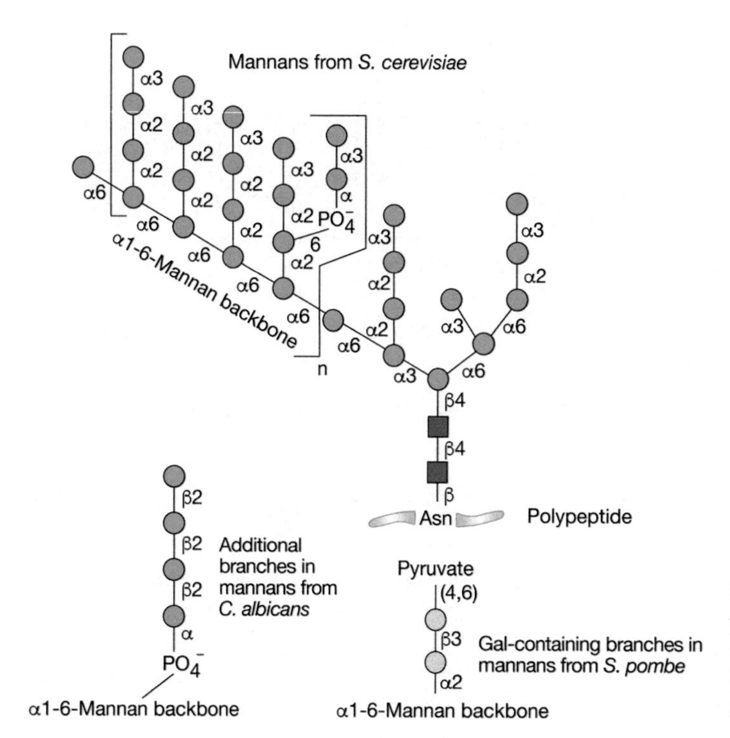

FIGURE 23.2. Structures of selected yeast mannans. Note that a single pyruvate is (R) 4,6 acetyl-(ketal)-linked to the terminal galactose residue in the pyruvylated structure.

nascent polypeptides in the ER. Following core N-glycosylation, the $Glc_3Man_9GlcNAc_2$Asn-R is processed with removal of glucose residues by α-glucosidases I and II to generate $Man_9GlcNAc_2$Asn-R. In mammals and *S. cerevisiae*, the $Man_9GlcNAc_2$Asn-R is further trimmed to $Man_8GlcNAc_2$-Asn-R by an ER-mannosidase. *Schizosaccharomyces pombe*, however, lacks this enzyme and stops processing at $Man_9GlcNAc_2$Asn-R. $Man_8GlcNAc_2$Asn-R in mammals and $Man_9GlcNAc_2$Asn-R in *S. pombe* are then substrates for the UDP-Glc: glycoprotein glucosyltransferase (UGT) that generates $Glc_1Man_8GlcNAc_2$Asn and $Glc_1Man_9GlcNAc_2$Asn in mammals and *S. pombe*, respectively. This reglucosylation, which is part of the quality control system for protein folding in the ER, is absent in *S. cerevisiae* (Chapter 39). The monoglucosylated structure is a ligand for the chaperone lectins calnexin and calreticulin in mammalian cells. Most fungi express calnexin but lack a calreticulin homolog. Specific trimming of the N-linked glycan regulates ER-associated degradation (ERAD) of improperly folded proteins or unassembled protein complex units. In *S. cerevisiae*, trimming of the N-linked glycan to $Man_7GlcNAc_2$ by the mannosidase Htm1p generates a signal that is recognized by the lectin Yos9p and leads to export of the glycoprotein to the cytoplasm and subsequent degradation. Notably, quality control and ERAD processes differ between fungal species; in some cases, components of the complete pathway are absent.

Yeast N-glycans are extended in the Golgi apparatus by mannosyltransferases that use GDP-Man as the donor, with one or more specific glycosyltransferases acting to catalyze synthesis of each linkage and branch.

Fungal proteins are rich in O-linked mannose. This protein modification is initiated by ER protein mannosyltransferases (PMTs) that use Dol-P-Man as a mannose donor. There are several hetero- or homodimeric PMTs in fungi and each may have a different substrate specificity and glycoprotein preference. The Dol-P-Man for the fungal PMTs is synthesized in the cytosol and then flipped into the secretory organelle lumen (Figure 23.4, top right) to be used for both N- and O-glycosylation. Subsequent additions of mannose residues to growing chains occur in the Golgi apparatus, where GDP-Man serves as the donor for reactions catalyzed by Mn^{++}-dependent mannosyltransferases.

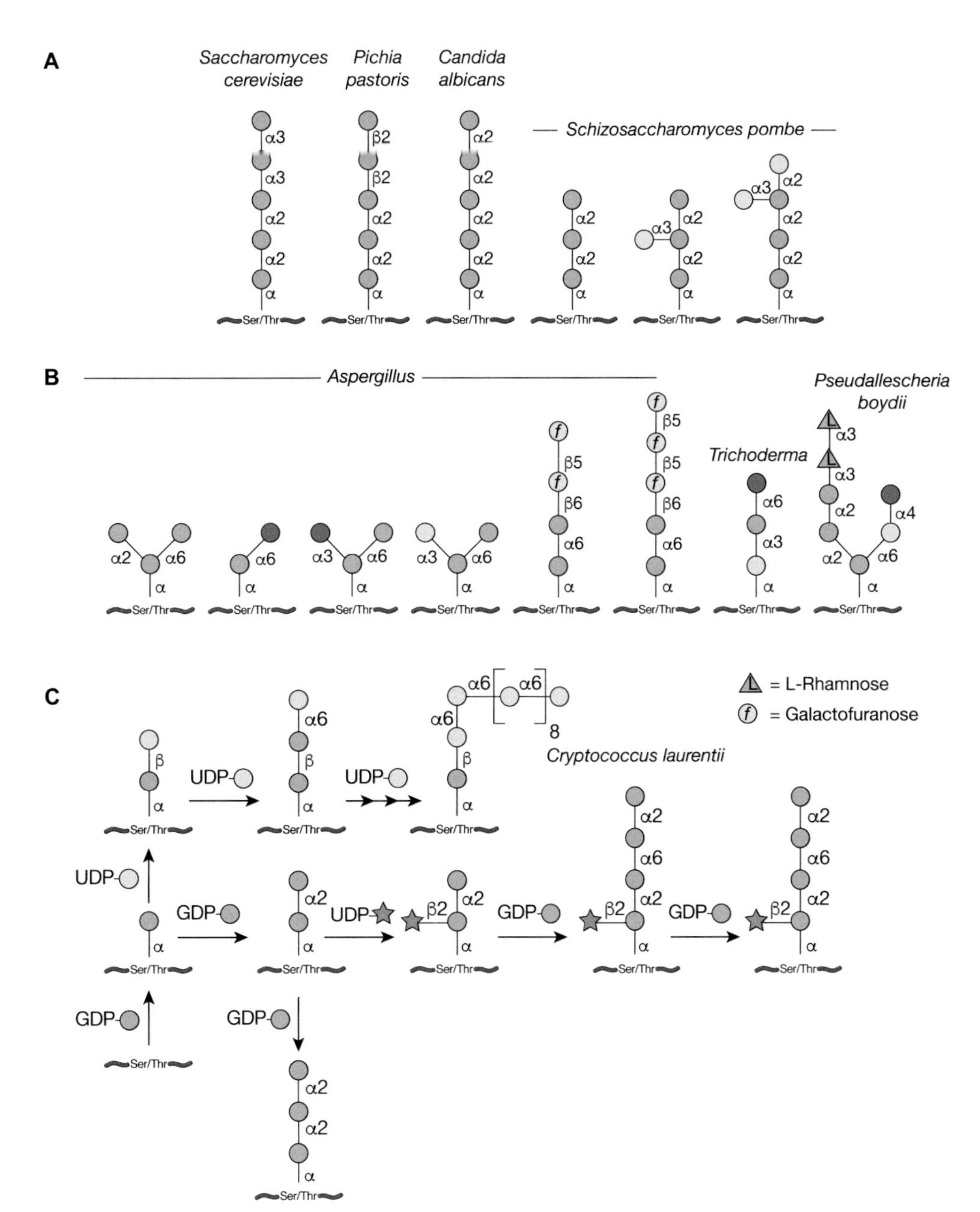

FIGURE 23.3. Structures of selected O-linked glycans in fungi: (*A*) yeast, (*B*) filamentous fungi, and (*C*) *Cryptococcus*.

Fungi express abundant GPI-anchored glycoproteins. As in other systems (Chapter 12), GPI-anchored proteins are synthesized in the ER from a GPI-precursor and a protein precursor having a carboxy-terminal GPI-addition signal peptide. A GPI-transamidase cleaves the signal peptide and replaces it by the GPI precursor. In fungi, addition of an α1-2 mannose residue to the conserved GPI core Manα1-2Manα1-6Manα1-4GlcNα1-6Inositolphospholipid by the enzyme Smp3 is a prerequisite for addition of the ethanolamine phosphate bridge that later carries the protein, although this step is dispensable in mammals. Consequently, fungal GPI-anchors have an extended core with four mannose residues (Figure 23.5). In an interesting divergence from "higher" eukaryotes, yeast GPI anchors may serve as substrates for transglycosylation reactions in cell wall assembly, leading to covalent linkage of glycoproteins to the glucan matrix of the cell

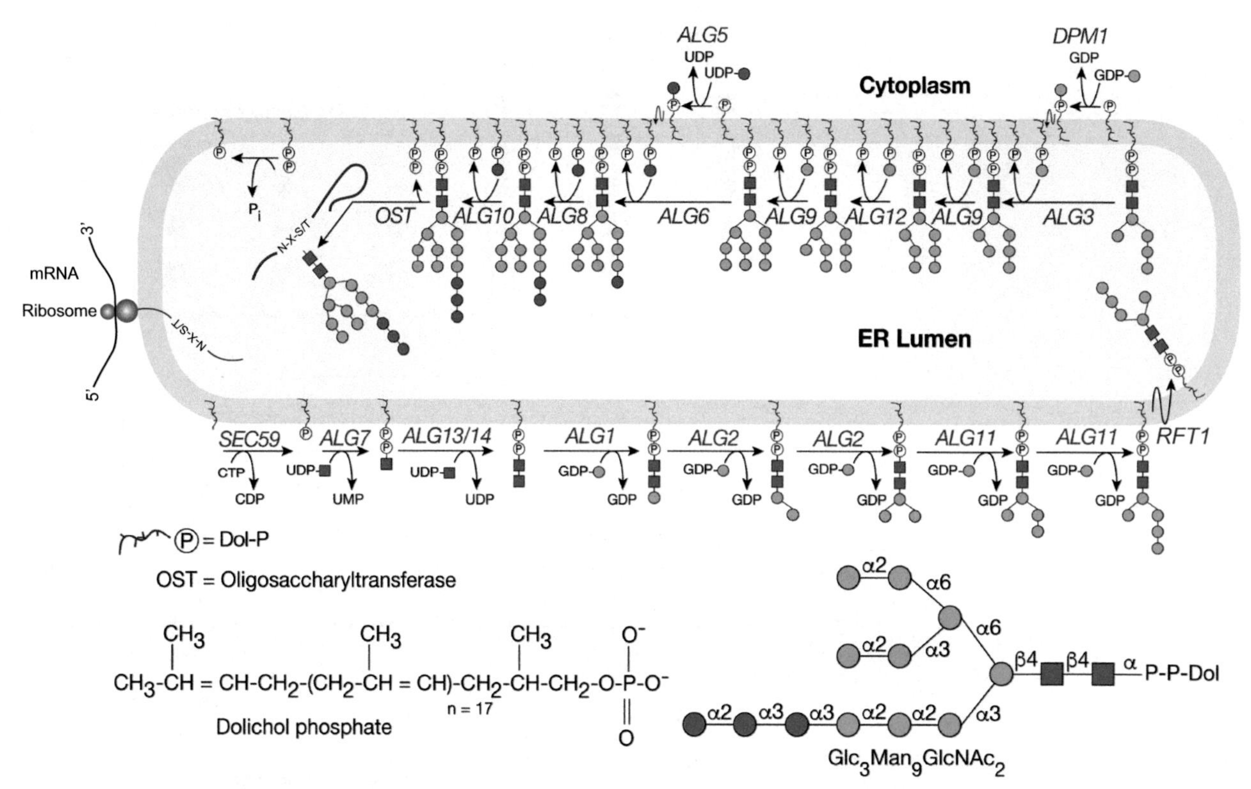

FIGURE 23.4. Biosynthesis of N-glycans and their transfer to -Asn-X-Ser/Thr- sequons of newly synthesized glycoproteins in the fungal endoplasmic reticulum (ER). Individual steps in the biosynthesis pathway from dolichol phosphate (simplest structure at *top right*) and the structure of dolichol (*lower left*) are shown. Biosynthetic steps identified by mutants in yeast are shown as genes designated as *ALG* or *SEC*. The structure of the intermediate $Glc_3Man_9GlcNAc_2$-P-P-Dol, which is common to yeast and mammals, is shown (*lower right*).

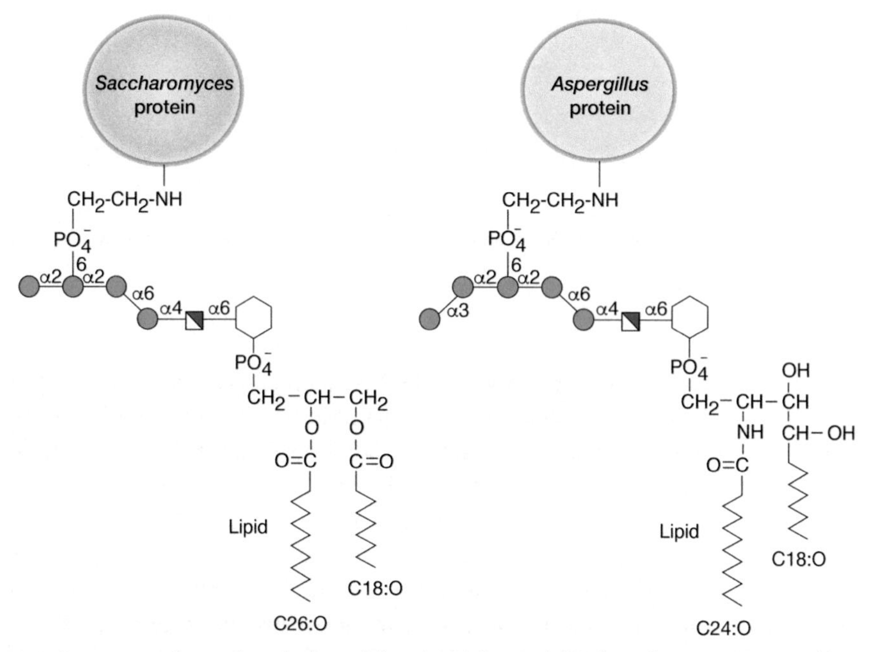

FIGURE 23.5. Structures of two fungal glycosylphosphatidylinositol (GPI) anchors. Hexagon indicates *myo*-D-inositol.

wall. Polysaccharides linked to GPI anchors are also found in the extracellular matrix of filamentous ascomycetes.

Glycolipids

Yeast express a relatively simple array of glycolipids, although *C. albicans* is notable for its large lipid-linked mannans. Many fungi make short-chain glycolipids, commonly containing *myo*-inositol phosphate linked to mannose that may be modified by galactofuranose (as in *Histoplasma capsulatum* or *A. fumigatus*) or additional mannose residues. *S. cerevisiae* generates forms with a single residue of mannose, whereas some longer galactose- and mannose-containing glycolipids are found in Aspergilli. Short-chain glycosylceramides such as Glc-Cer and Gal-Cer are also found in the fungi *Schizophyllum commune* and *A. fumigatus*.

MODEL FUNGI

Saccharomyces cerevisiae as an Experimental System

Yeasts have been valued for baking and brewing for thousands of years, but in the last century or so scientific attention has particularly focused on *S. cerevisiae*, an oval budding yeast 5–10 μm across. The rapid growth of this simple eukaryote, combined with its inexpensive culture and genetic tractability, has made it a powerful and popular model system. Studies of *S. cerevisiae* have enormously influenced the fields of eukaryotic cell biology and genetics, in addition to their impact on basic metabolism and enzymology noted above.

S. cerevisiae contributed to defining the enzymology of GPI lipid precursor biosynthesis (Chapter 12). This complex process, involving more than 20 genes, presented a significant biochemical challenge to researchers in the field. However, as many of the steps are conserved from yeast to mammals, analysis of *S. cerevisiae* mutants offered a complementary and powerful approach to its dissection. Mutants have also been useful for dissecting yeast-specific processes such as mannan synthesis; this was elucidated by identifying *mnn* mutants, which displayed aberrant antibody or dye binding.

Despite the tremendous value of *S. cerevisiae* as a model, it does have certain limitations. These cells do not synthesize complex N-glycans, mucins or mucin-type O-glycans, O-linked *N*-acetylglucosamine (O-GlcNAc), sialic acids, or glycosaminoglycans (GAGs) of the types found in vertebrates. However, *S. cerevisiae* cells use O-mannose on their nucleocytoplasmic proteins in a manner analogous to O-GlcNAc in plants and animals. Like most other fungi, *S. cerevisiae* also lacks long-chain glycolipids (apart from those participating in the GPI synthesis) and does not synthesize complex glycosphingolipids or gangliosides like those found in mammals (although it is still valuable for studying sphingosine and sphingolipid metabolism). *S. cerevisiae* expresses limited glycan diversity even compared with other fungi, with no galactose, xylose, or glucuronic acid reported in its glycans. This must be kept in mind when generalizing from this model to other organisms.

Schizosaccharomyces pombe, a Model for Ultrastructure

S. pombe is a rod-shaped yeast, ~3–4 μm in diameter and 7–14 μm long. Rather than budding, this organism grows by elongation and fission to yield equal-sized daughter cells. Like *S. cerevisiae*, it has a relatively small genome of ~14 million base pairs. *S. pombe* has been useful as a genetically manipulatable model organism for studying the cell cycle. Because it has well-defined organelle structures compared with other yeast, it is a popular choice for studies of intracellular structure. *S. pombe* also synthesizes mannoproteins and mannans, some containing galactose (Figure 23.3A); this may occur as α1-2-linked caps that are also sometimes pyruvylated. The galactose residues are important for lectin recognition in nonsexual flocculation (clumping) of *S. pombe*, as evidenced by inhibition of this process by free galactose. In contrast, flocculation in *S. cerevisiae* is mannose-dependent and inhibited by free mannose. Another difference between *S. pombe* and the more common *S. cerevisiae* model is that its newly-synthesized N-glycans ($Man_9GlcNAc_2Asn$; Figure 23.4) are not trimmed to $Man_8GlcNAc_2Asn$ in the ER.

HARNESSING YEAST FOR PRODUCTION

Pichia pastoris and Its Advantages for Expression

Pichia pastoris is a methylotrophic, nonpathogenic organism that was discovered in 1969 in a screen for yeast capable of using methanol. In this yeast, methanol is oxidized to formaldehyde and hydrogen peroxide by alcohol oxidase (AOX) in the peroxisome. The formaldehyde exits the peroxisome and is oxidized to formate and carbon dioxide in the cytoplasm for energy production. Any remaining formaldehyde is assimilated into glyceraldehyde-3-phosphate and dihydroxyacetone by condensation with xylulose-5-monophosphate, in a reaction catalyzed by the peroxisomal enzyme dihydroxyacetone synthase. *P. pastoris* has become popular as a model system for making recombinant proteins because it is easy to manipulate genetically and can be grown to very high densities. The promoter for AOX is methanol inducible, and transcripts driven by this promoter may comprise up to 5% of the total poly$(A)^+$ RNA in induced cells.

P. pastoris has several advantages over *Escherichia coli* as an expression system, because it does not produce inclusion bodies and it promotes the correct folding of eukaryotic proteins. It also has certain advantages over typical model yeast. First, although its basic N-glycosylation pathway is similar to that of *S. cerevisiae* and yields glycoproteins with oligomannose-type N-glycans, these structures in *P. pastoris* have only five to 15 mannose residues (typically $Man_9GlcNAc_2Asn$ and $Man_8GlcNAc_2Asn$) compared with the 50 to 150 mannose residues found in glycoproteins from *S. cerevisiae* (see Figure 23.2). Hyperglycosylation in *S. cerevisiae* can interfere with protein folding, requiring use of *mnn* mutants to limit hyperglycosylation and avoid this problem. Also, *P. pastoris* does not add outer α1-3-linked mannose residues to its N-glycans. These structures are highly antigenic to humans, making proteins expressed in *S. cerevisiae* unsuitable for human pharmaceutical use. *P. pastoris* synthesizes O-glycans with an O-linked mannose core attached to Ser/Thr residues; most of these are short α1-2-linked mannose structures (see Figure 23.3A). Genetic tools have been used to manipulate the machinery for glycoprotein assembly of both *S. cerevisiae* and *P. pastoris*. Deletion of ER- and Golgi-specific functionalities in combination with the introduction of heterologous hydrolases and glycosyltransferases has generated "humanized" yeasts and filamentous fungi for the production of therapeutic glycoproteins (Chapter 56). *P. pastoris* has emerged as a potent orthogonal host to produce exogenous polysaccharides such as those occurring in the plant cell wall.

Kluyveromyces lactis in Industry

Kluyveromyces lactis metabolizes lactose to lactic acid and, along with *Aspergillus niger* and *E. coli*, is grown to produce rennet for making cheese and other products. *K. lactis* is also a rich source of β-galactosidase, which hydrolyzes lactose. *K. lactis* synthesizes mannans similar to those in *S. cerevisiae*, but they lack mannose phosphate modifications and some side chains are capped with a residue of *N*-acetylglucosamine. A *K. lactis* mutant that lacks these *N*-acetylglucosamine residues because of a deficiency of the Golgi UDP-GlcNAc nucleotide sugar transporter has been productively exploited in studies of heterologous transporters.

BASIDIOMYCETE DIVERSITY

To convey the enormous diversity of fungi we will consider the basidiomycete phylum, which encompasses fungi that produce spores from a pedestal like structure called the basidium. Basidiomycetes range from fungi with gills or pores, such as common mushrooms and bracket fungi, to budding yeast that are deadly human pathogens.

Lifestyle and Polysaccharides

The basidiomycete *Dictyonema glabratum* illustrates a distinct fungal lifestyle. It lives in symbiosis with cyanobacteria Scytonema sp. forming a lichen, which is notable for its many unusual glycans. For example, although the β-glucans of most lichens are linear, in *D. glabratum* they are branched with β1-3 and β1-6 linkages. The mannans of *D. glabratum* also have an α1-3-linked backbone, rather than the typical α1-6 linkages found in other lichens, along with branches at the 2 and 4 positions. Finally, the xylans of this organism are linear β1-4-linked polymers of xylose, more typical of those found in "higher" plants and algae than in fungi. *D. glabratum* also synthesizes several unusual short glycolipids, including glycosyldiacylglycerolipids, which are similar to plant glycolipids and contain monosaccharides, disaccharides, and linear trisaccharides of α1-6-linked galactopyranose. This fungus thus highlights the extensive glycan diversity of the fungal kingdom.

O-Glycans

Another example of glycan diversity is offered by *Cryptococcus laurentii*, which has the unusual property of producing toxins that kill a pathogenic yeast, *C. albicans* (see below). The O-glycans of *C. laurentii* are unusual in that they contain mannose, xylose, and galactose (Figure 23.3C); these glycans are synthesized by a unique set of mannosyl-, xylosyl-, and galactosyltransferases that are not homologous to human enzymes. This diversity would not have been predicted by studies of model yeast alone, emphasizing the importance of examining glycans in a wide range of fungal species.

N-Glycan Diversity

Coprinopsis cinerea is a mushroom species that has gained prominence as a model for studies of diverse topics including mating, sexual development, meiosis, and the evolution of multicellularity. Its N-glycans typically are high mannose type with five to nine mannoses, but may also have a bisecting α1-4GlcNAc at the β-mannose.

PATHOGENIC FUNGI

Pathogenic fungi are a significant cause of plant and animal disease, responsible for devastation of crops, decimation of animal populations (e.g., certain bat, amphibian, and bee species), and serious human diseases that kill an estimated one million people each year. These organisms display diverse glycans, which typically differ from those of the host and have been implicated in multiple pathogenic processes and host–pathogen interactions. For example, during plant infections plant glycosylhydrolases may partially digest wall glucans of invading fungi. Some of the released oligosaccharides, termed oligosaccharins, can then act as signals to promote plant antifungal defenses.

Candida albicans Glycans Are Central in Host Interactions

C. albicans (an ascomycete) is a normal commensal organism that can cause illness ranging from irritations of mucosal surfaces to life-threatening systemic infections. The *C. albicans* cell wall contains β1-3- and β1-6-linked glucans and chitin, similar to the *S. cerevisiae* wall, and immunogenic mannans that are termed phosphopeptidomannans. It also produces unusual short β1-2-linked mannose chains (Figure 23.2) that are highly antigenic and are also expressed on phospholipomannan (PLM) antigens. PLM antigens contain phytoceramide derivatives of *myo*-inositol phosphate. The β1-2 mannosides are linked via an α-mannosylphosphate to the common glycosphingolipid Manα-1,2 inositolphosphoceramide. The abundant GPI-anchored proteins of *C. albicans* have been implicated in fungal adherence to host tissues.

The O-glycans of *C. albicans* are short chains of α1-2-linked mannose (Figure 23.3A), which lack the α1-3-linked mannose caps found in *S. cerevisiae*. As in *S. cerevisiae*, deficiencies of O-mannose addition generated through genetic deletions are lethal, indicating that O-mannosylation is essential in this yeast. *C. albicans* mannans are also important in its interactions with host cells, including macrophages and dendritic cells. In particular, these structures are recognized by the mannose receptor and by dectin-2. These are C-type lectins expressed by immune cells that are important in both innate and adaptive immune responses (see Chapter 34). PLM antigens may be shed by *C. albicans* and, through interactions with Toll-like receptors (TLR-2), they can induce nuclear factor-κB (NF-κB) activation and cytokine responses such as tumor necrosis factor-α (TNF-α) secretion. Galectin-3, a ubiquitous member of the galectin family of lectins that is highly expressed in macrophages, also appears to recognize *C. albicans* expressing β1-2-linked mannose residues, resulting in opsonization of the yeast.

Aspergillus fumigatus

A. fumigatus is an environmental mold that spreads by airborne particles. It causes serious invasive disease in immunocompromised people that is difficult to treat, leading to high mortality rates. As with other fungal pathogens, the surface glycans of *A. fumigatus* are critical for interactions with the host. The cell wall of infectious forms of this fungus is covered with specific proteins and melanin, presumably to alter surface properties and mask these structures from recognition by host immune receptors. The hyphal wall has a core of branched β1-3-glucan covalently linked to other glucan components, chitin, and galactomannan, which consists of a mannose backbone with short galactofuranose side chains. Interestingly, galactomannan also occurs anchored to the plasma membrane by a GPI. This polysaccharide is assembled in the Golgi apparatus and is probably transferred to the cell wall by transglycosidases, in the same way as GPI-anchored proteins. *A. fumigatus* also produces an extracellular matrix composed of monosaccharides, α1-3-glucan, galactomannan, and a galactosaminogalactan composed of variable galactopyranose repeats linked to *N*-acetylgalactosamine, which are partially de-acetylated. This structure has been implicated in adhesion and fungal virulence. The extracellular matrix also plays an important role in concealing the immunogenic β1-3 glucan layer from the immune system.

Cryptococcus neoformans and Its Capsule

C. neoformans is a ubiquitous environmental basidiomycete yeast that causes severe disease in immunocompromised individuals, leading to roughly half a million deaths per year worldwide. It is unique among pathogenic fungi in having an extensive polysaccharide capsule that is required for virulence (Figure 23.6). The capsule is a dynamic structure that changes in thickness and composition depending on the environment and growth conditions. It is

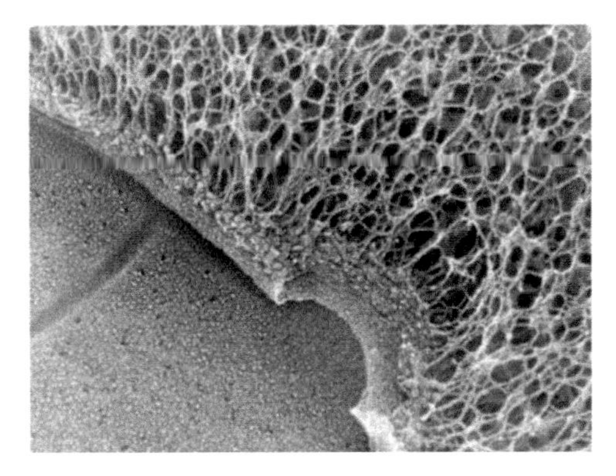

FIGURE 23.6. A quick-freeze deep-etch image of the edge of a *Cryptococcus neoformans* cell. The polysaccharide capsule (open meshwork at *right*) is linked to the cell wall (central structure dividing the image from *upper left* to *lower right*) via α1-3 glucan. The region at *lower left* is the cell membrane, and the arc in the cell wall represents formation of a new bud. (Image by John Heuser and Tamara Doering. Reprinted, from Reese AJ, et al. 2007. *Mol Microbiol* **63:** 1385–1398, with permission of John Wiley and Sons.)

particularly large in the context of mammalian infection, in which it impedes host immune responses. It is composed of two large (millions of Da) polysaccharides named for their monosaccharide components: glucuronoxylomannan (GXM) and glucuronoxylomannogalactan (GXMGal). GXM is an extended α1-3 mannan substituted with β1-2Xyl, β1-4Xyl, and β1-2GlcA (Figure 23.7); a subset of the mannose residues are 6-O-acetylated (not shown). The second polymer, GXMGal, is based on α1-6 galactan, with side chains of galactose, glucuronic acid, mannose, and xylose (Figure 23.7); the backbone is also modified with small amounts of β1-2-linked galactofuranose (not shown). Association of the capsule with the cell surface relies on a cell-wall component, α1-3 glucan. Although α1-3 glucan is not present in the cell walls of *S. cerevisiae* or *C. albicans*, it is common in other fungi. N-glycans of *C. neoformans* are generally high mannose with modest outer chain extensions and may include xylose β1-2-linked to the trimannosyl core. Both N- and O-glycans may also be modified with xylose and xylose phosphate.

Fungal Glycans as Drug Targets

In the context of fungal diseases, the similarity of fungi to their eukaryote hosts becomes a liability, because it is challenging to develop antifungal drugs that are not hampered by toxicity. The unique features of fungal glycans may suggest drug targets to help improve this picture, and decrease the roughly one million deaths each year caused by fungal infections.

The major success story using this approach is the development of echinocandin drugs. These antifungal lipopeptides inhibit β1-3-glucan synthesis in fungi including *Candida* and *Aspergillus*, leading to cell wall compromise, and are used clinically to treat invasive fungal infections, although they are not effective against all fungal pathogens. GPI-synthesis inhibitors also

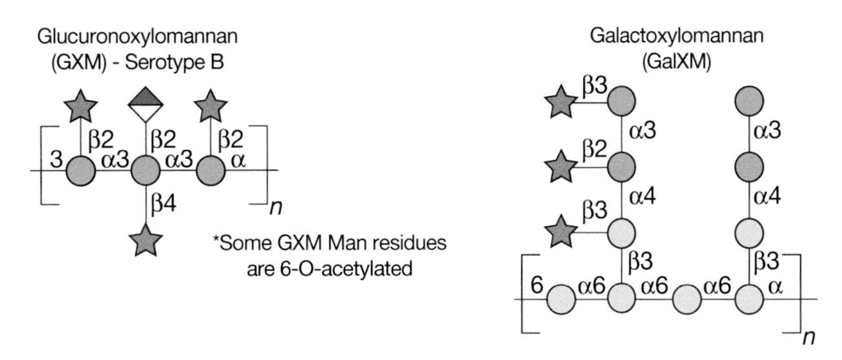

FIGURE 23.7. Structures of capsular polysaccharides in *Cryptococcus neoformans.*

offer promise for treatment of fungal pathogens; compounds that selectively inhibit yeast inositol acylation are currently in clinical trials. Continued efforts to target novel aspects of fungal glycobiology may advance the ongoing search for new therapies.

ACKNOWLEDGMENTS

The authors appreciate helpful comments and suggestions from Anne Imberty and Markus Pauly.

FURTHER READING

Ballou CE, Lipke PN, Raschke WC. 1974. Structure and immunochemistry of the cell wall mannans from *Saccharomyces chevalieri*, *Saccharomyces italicus*, *Saccharomyces diastaticus*, and *Saccharomyces carlsbergensis*. *J Bacteriol* **117**: 461–467. doi:10.1128/jb.117.2.461-467.1974

Huffaker TC, Robbins PW. 1983. Yeast mutants deficient in protein glycosylation. *Proc Natl Acad Sci* **80**: 7466–7470. doi:10.1073/pnas.80.24.7466

Dickson RC, Lester RL. 1999. Yeast sphingolipids. *Biochim Biophys Acta* **1426**: 347–357. doi:10.1016/s0304-4165(98)00135-4

Poulain D, Jouault T. 2004. *Candida albicans* cell wall glycans, host receptors and responses: elements for a decisive crosstalk. *Curr Opin Microbiol* **7**: 342–349. doi:10.1016/j.mib.2004.06.011

Daly R, Hearn MT. 2005. Expression of heterologous proteins in *Pichia pastoris*: a useful experimental tool in protein engineering and production. *J Mol Recognit* **18**: 119–138. doi:10.1002/jmr.687

Klis FM, Ram AF, De Groot PW. 2007. A molecular and genomic view of the fungal cell wall. In *Biology of the fungal cell* (ed. RJ Howard, NAR Gow), 2nd ed, *The Mycota VIII*, pp. 97–120. Springer-Verlag, Berlin. doi:10.1007/978-3-540-70618-2_4

Deshpande N, Wilkins MR, Packer N, Nevalainen H. 2008. Protein glycosylation pathways in filamentous fungi. *Glycobiology* **18**: 626–637. doi:10.1093/glycob/cwn044

De Pourcq K, De Schutter K, Callewaert N. 2010. Engineering of glycosylation in yeast and other fungi: current state and perspectives. *Appl Microbiol Biotechnol* **87**: 1617–1631. doi:10.1007/s00253-010-2721-1

Everest-Dass AV, Jin D, Thaysen-Andersen M, Nevalainen H, Kolarich D, Packer NH. 2012. Comparative structural analysis of the glycosylation of salivary and buccal cell proteins: innate protection against infection by *Candida albicans*. *Glycobiology* **22**: 1465–1479. doi:10.1093/glycob/cws112

Loza LC, Doering TL. 2021. Glycans of the pathogenic yeast *Cryptococcus neoformans* and related opportunities for therapeutic advances. In *Comprehensive glycoscience* (ed. Barch Jr J), 2nd ed., Vol. I, pp. 479–506. Elsevier, Amsterdam. doi:10.1016/B978-0-12-819475-1.00079-1

24 Viridiplantae and Algae

Malcolm A. O'Neill, Alan G. Darvill, Marilynn E. Etzler,
Debra Mohnen, Serge Perez, Jenny C. Mortimer, and Markus Pauly

Viridiplantae (green plants) are a clade of photosynthetic organisms that contain chlorophylls a and b, produce and store their photosynthetic products inside a double-membrane-bounded chloroplast, and have cell walls that typically contain cellulose. As photoautotrophic organisms, green plants are capable of converting carbon dioxide to carbohydrates. Thus, carbohydrates are not limiting and their utilization throughout a plant's life cycle has expanded enormously both in functionality and structural diversity.

The Viridiplantae comprise two clades—the Chlorophyta and the Streptophyta. The Chlorophyta contain most of the organisms typically referred to as "green algae." The term "algae" is also used for several other groups of photosynthetic eukaryotes, including diatoms and the red, brown, golden, and yellow-green algae. The Streptophyta comprise several other lineages that are also referred to as "green algae" and the land plants. Land plants include the liverworts, mosses, hornworts, lycopods, ferns, gymnosperms, and flowering plants. In this chapter, we

published by Cold Spring Harbor Laboratory Press; doi: 10.1101/glycobiology.4e.24

provide an overview of the current knowledge of green plant glycan structures with an emphasis on the features that are unique to land plants.

PLANT GLYCAN DIVERSITY

Green plants synthesize diverse glycans and glycoconjugates that vary in structural complexity and molecular size. Soluble low-molecular-weight compounds encompass the carbon and energy transport disaccharide sucrose as well as glycoconjugates that contain aromatic (e.g., phenolic glycosides) or aliphatic (glycolipids) aglycones. Many of these compounds function in plant protection or defense (e.g., to ward off herbivores). Plant polysaccharides are linear or branched polymers composed of the same or different monosaccharides. Examples of homopolymers made entirely out of glucose include the storage polymer starch and the structural polymer cellulose (Figure 24.1). An example of a structurally complex polysaccharide is the plant cell wall pectic polysaccharide, referred to as rhamnogalacturonan II (RG-II), which contains 12 different monosaccharides linked together by up to 21 distinct glycosidic linkages (Figure 24.2). Plant proteoglycans are structurally diverse glycoconjugates in which carbohydrate, generally O-linked to the protein via hydroxyamino acids, accounts for up to 90% of the molecule (Figure 24.3). Plant glycoproteins typically contain 15% or less carbohydrate in the form of N-linked oligomannose, complex, hybrid, and paucimannose oligosaccharides (see Figure 24.4). Land plants also form O-GlcNAc-modified nuclear and cytosolic proteins (Chapter 19).

NUCLEOTIDE SUGARS—THE BUILDING BLOCKS

Nucleotide sugars are the donors used for the synthesis of glycans, glycoconjugates, and glycosylated secondary metabolites (Chapter 5). In plants, the majority of these activated monosaccharides exist as their nucleotide-diphosphates (e.g., UDP-Glc*p* or GDP-Man*p*), although at least one monosaccharide, Kdo, exists as its cytidine monophosphate derivative (CMP-Kdo*p*). Nucleotide sugars are formed from the carbohydrate generated by photosynthesis, from the monosaccharides released by hydrolysis of sucrose and storage carbohydrates, and by recycling monosaccharides from glycans and the cell wall. Nucleotide sugars are also formed by interconverting preexisting activated monosaccharides. To date, 30 different nucleotide sugars and at least 100 genes encoding proteins involved in their formation and interconversion have been identified in plants.

PLANT GLYCOSYLTRANSFERASES AND GLYCAN-MODIFYING ENZYMES

As plants are carbohydrate-rich organisms, it is perhaps not surprising that their genomes contain a large number of genes encoding proteins involved in the synthesis, metabolism, and modification of glycans and glycoconjugates. These proteins are spread across many enzyme classes in the Carbohydrate-Active enZYmes (CAZy) database (Table 24.1). Many of these proteins may be involved in the formation and modification of the polysaccharide-rich cell wall. Indeed, the unicellular alga *Ostreococcus tauri*, which is one of the few plants that does not form a cell wall, has a much smaller number of genes predicted to be involved in glycan metabolism.

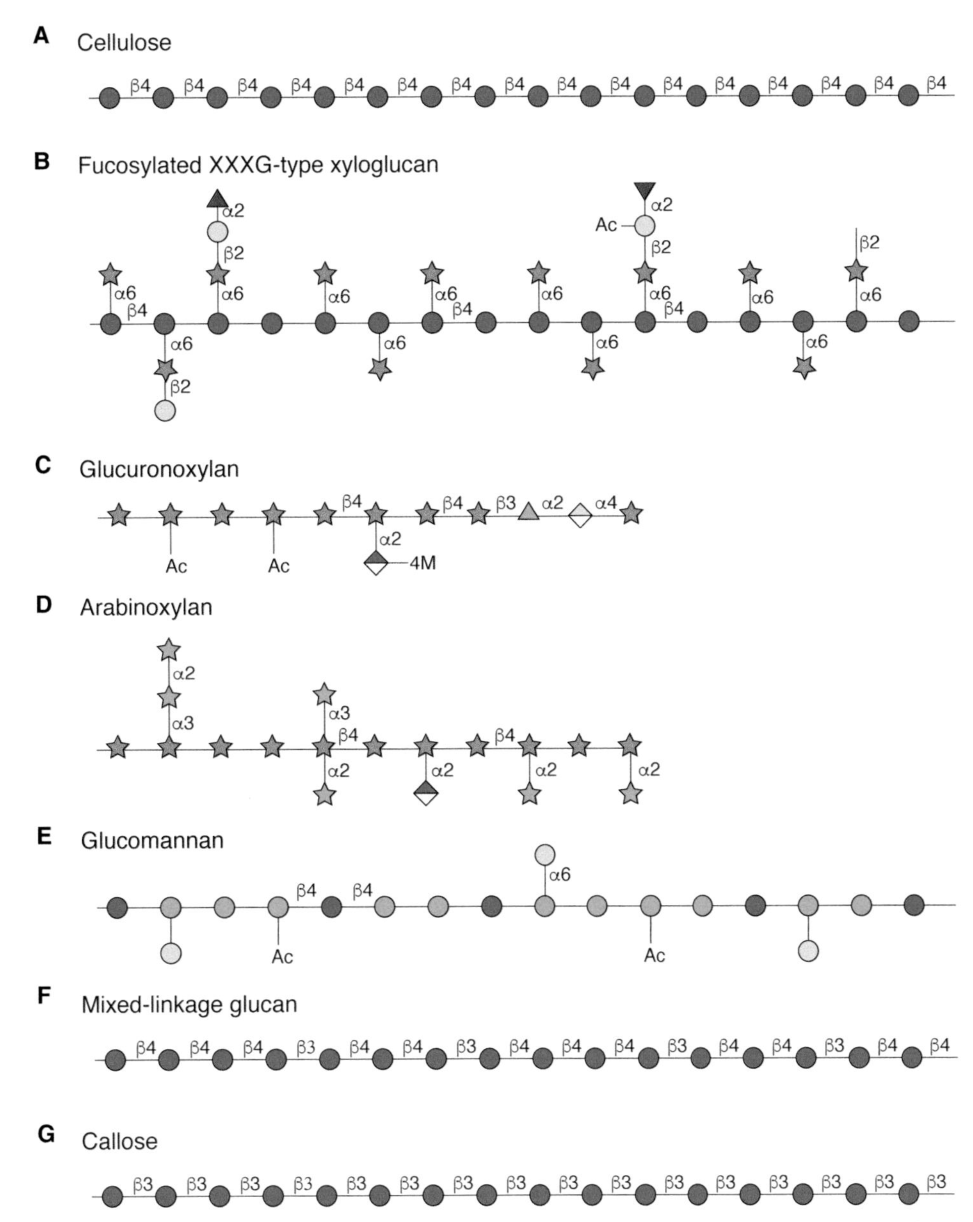

FIGURE 24.1 Glycosyl sequences of (*A*) cellulose, (*B–E*) selected hemicelluloses, (*F*) mixed-linkage glucan, and (*G*) callose.

PLANT MUTANTS PROVIDE CLUES TO GLYCAN FUNCTION

The availability of plant lines carrying mutations in specific genes has yielded considerable insight into glycan biosynthesis and function. *Arabidopsis thaliana* has been widely used as a model dicot as it is easy to grow, has a short life cycle, and its relatively small genome has been sequenced and extensively annotated leading to the generation of chemically induced or transfer DNA (T-DNA) insertion mutant collections. These mutants have been vital in studying both the function and substrate specificity of diverse proteins involved in glycan synthesis and for demonstrating the function or redundancy of glycans and/or glycan substituents in the life-cycle of a plant. With the advent of low-cost whole-genome sequencing, the genomes of nearly 100 plant species including rice, maize, barley, poplar, and potato have been assembled

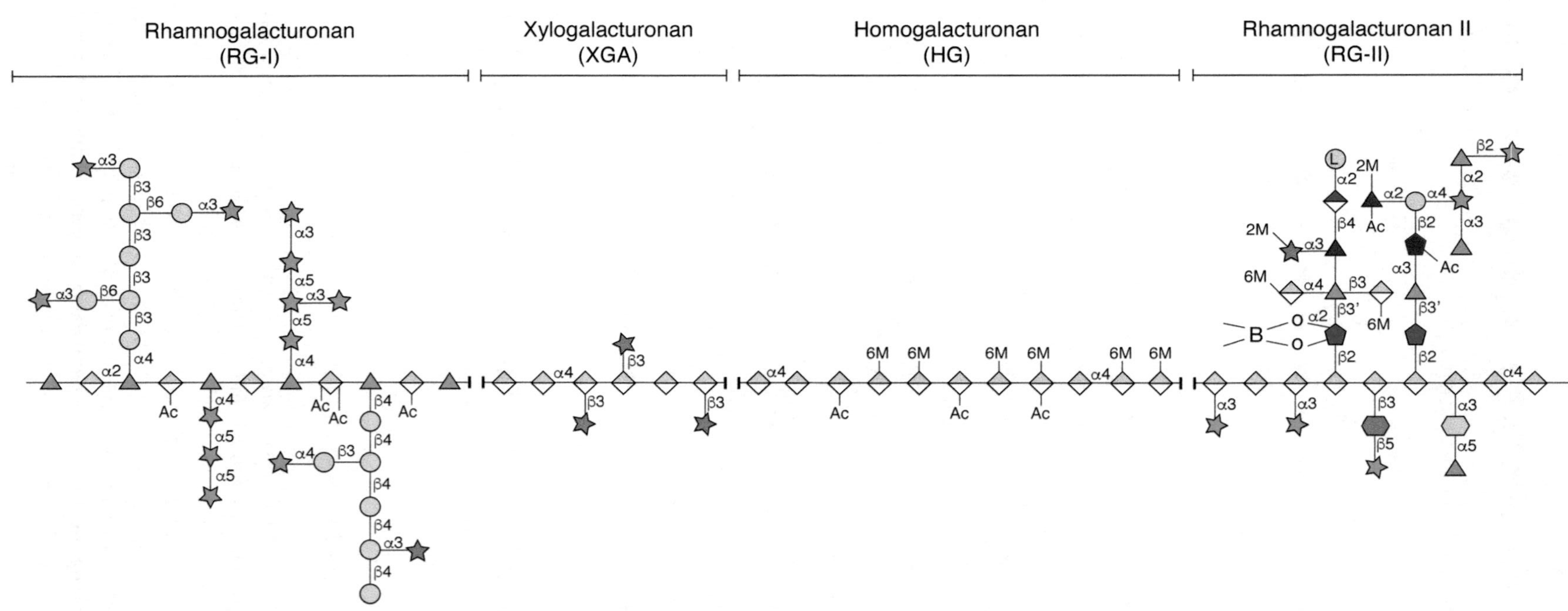

FIGURE 24.2 Schematic structure of pectin. The three main pectic polysaccharides are shown: (RG-I) rhamnogalacturonan I; (HG) homogalacturonan; and (RG-II) rhamnogalacturonan II. The borate ester is formed between the apiosyl residue in side chain A of each RG-II monomer. A region of substituted galacturonan, referred to as xylogalacturonan (XGA), is also shown. XGA is not present in most cell wall pectins. The relative abundance of each pectin domain (HG, RG-I, and RG-II) is dependent on the plant species. The extent of the connection between these pectic polysaccharides is still under investigation.

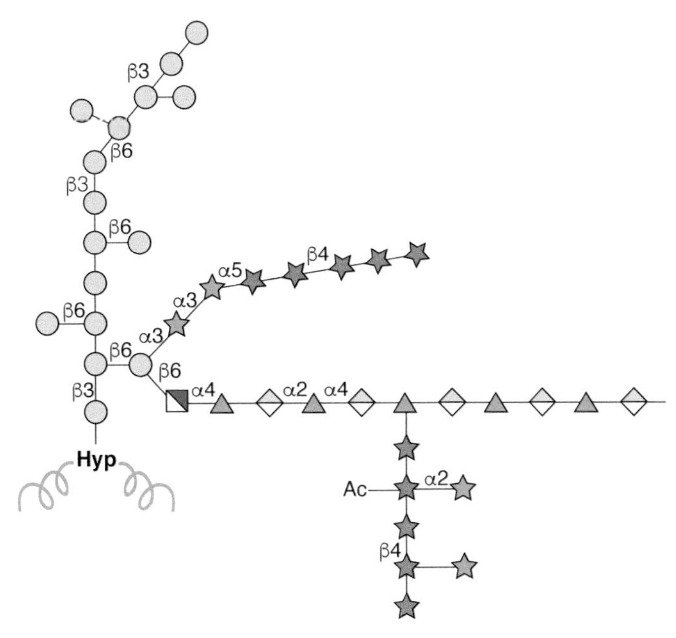

FIGURE 24.3 Schematic structure of the proteoglycan referred to as arabinoxylan pectin arabinogalactan protein1 (APAP1).

(PlantGDB https://www.plantgdb.org). A large amount of transcriptomic data covering most green plant lineages is also available (https://sites.google.com/a/ualberta.ca/onekp/). With this sequence and annotation information, combined with gene editing techniques such as CRISPR/Cas9, plant mutants can be generated to test the functionality of genes related to glycans uniquely present in those plant species.

A forward genetic approach, in which *Arabidopsis* seeds are randomly chemically mutagenized and the resulting plants screened for structural changes in their glycans, has led to the discovery of multiple genes involved in nucleotide sugar interconversion pathways (Chapter 5) as well as plant cell wall–related glycosyltransferases (GTs) (Chapter 6). Reverse genetic approaches, with plants carrying loss-of-function mutations in known genes, have enabled the discovery of GTs involved in the synthesis of primary and secondary cell wall heteroxylans, pectins, N- and O-linked

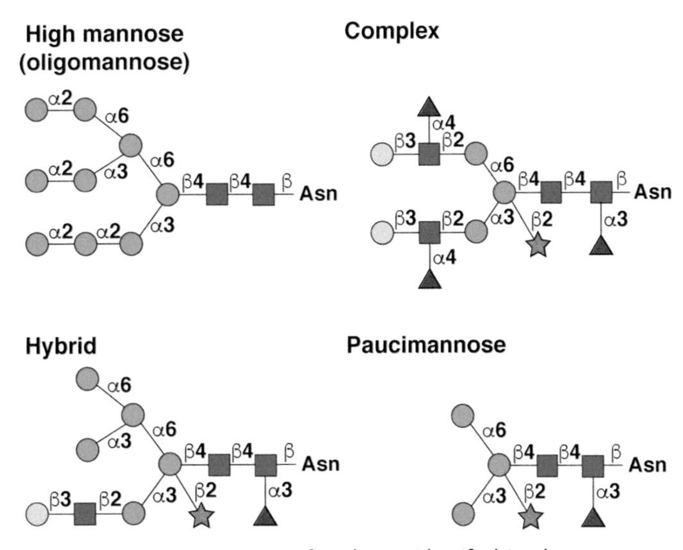

FIGURE 24.4 Types of N-glycans identified in plants.

TABLE 24.1. Estimated number of genes encoding proteins involved in the synthesis and modification of glycans in plants and humans

Organism	GTs	Hydrolases	Lyases	Esterases	CBMs	AA
Arabidopsis	564 (42)	426 (37)	34 (2)	85 (4)	124 (10)	49 (3)
Rice	699 (35)	479 (38)	16 (2)	61 (4)	162 (22)	53 (3)
Human	243 (46)	94 (28)	ni	1 (1)	39 (6)	
Ostreococcus	67 (25)	29 (12)	ni	2 (1)	14 (5)	

Values in parentheses are the number of CAZy classes.
GTs, Glycosyltransferases; CBMs, carbohydrate-binding modules; AA, auxiliary activities; ni, none identified to date.

proteoglycans, glycolipids, and glycosylated metabolites. In addition, glycan-modifying enzymes including O-acetyl- and methyltransferases have been identified and characterized.

PLANT METABOLIC CARBOHYDRATES

The initial product of photosynthesis, a triose phosphate, is used by plants in multiple ways. Triose phosphate can be converted to sucrose, a disaccharide of glucose and fructose (α-D-Glc*p*-1-2-β-D-Fru*f*) that is the most abundant soluble carbohydrate in green plants. Sucrose is the dominant transport carbohydrate for distributing the energy obtained by photosynthesis throughout the plant, in particular to nonphotosynthetic organs such as roots. Other water-soluble carbohydrates that are almost ubiquitous in the plant kingdom include raffinose oligosaccharides (raffinose, stachyose, and verbascose). These oligosaccharides are derivatives of sucrose that contain one or more α-D-Gal*p* residues.

Starch is an abundant branched polysaccharide that is the dominant carbohydrate energy storage form in green plants. Starch in grains, tubers, and fruit accounts for most of the calories that humans consume directly as food or indirectly from livestock-fed plants. Starch consists entirely of glucose and exists as amylopectin and amylose. Amylopectin is a branched polymer comprising α-1-4 and 1-4-6-linked D-Glc*p*, whereas amylose is a linear polymer composed of only α-1-4-linked Glc*p*. Starch polymers are arranged in insoluble, semicrystalline granules in chloroplasts or specialized plastids (amyloplasts).

Plants also produce fructose-containing polysaccharides termed fructans. The simplest fructan is inulin, a linear glycan of sucrose with 1-2-linked β-D-Fru*f*. Other inulins contain two fructan chains on a sucrose core molecule. Levan type fructans in grasses contain linear 2-6-β-D-Fru*f* polymers. Fructans are often utilized by green plants as alternative or additional storage glycans but are stored in the cell's vacuole. It is believed that fructans are also involved in plant protection—in particular, abiotic stresses including drought, salt, or cold stress.

PLANT CELL WALLS

A substantial portion of the carbohydrate formed by photosynthesis is used to produce the polysaccharide-rich walls that surround plant cells. Primary and secondary cell walls are distinguished by their composition, architecture, and functions. A primary wall surrounds growing and dividing plant cells and nongrowing cells in the soft tissues of fruits and leaves. These walls are capable of controlled extension to allow the cell to grow and expand but are sufficiently strong to resist the cell's internal turgor pressure. A much thicker and stronger secondary wall is often formed once a cell has ceased to grow. This secondary wall is deposited between the plasma membrane and the primary wall and is composed of layers that differ in the orientation of their cellulose microfibrils, the type of hemicellulose, the amount of pectin, and the frequent incorporation of the hydrophobic, noncarbohydrate, polyphenolic polymer lignin. For example, the secondary walls of vascular tissues involved in the movement of water and nutrients through the plant are further strengthened by the incorporation of lignin. The ability to form conducting tissues with lignified and rigid secondary walls was an indispensable event in the evolution of vascular land plants, as it facilitated the transport of water and nutrients and allowed extensive upright growth and thus a competitive advantage for the exposure to and capture of sunlight. Secondary cell walls account for most of the carbohydrate in plant biomass, such as straw for animal feed and wood for the production of paper and lumber for construction. Because of its abundance, plant biomass is also considered as a renewable, carbon-neutral feedstock for the production of biofuels, biomaterials, and other value-added commodity chemicals (Chapter 59).

PRIMARY CELL WALL GLYCANS

Primary cell walls are composites that resemble fiber-reinforced porous, aqueous gels. The complex structures and functions of these walls result from the assembly and interactions of a limited number of structurally defined polysaccharides and proteoglycans. Nevertheless, wall structure and organization differ between plant species and in different tissue and cell types within a plant. Moreover, during cell division and differentiation, and in response to biotic and abiotic challenges, a cell often responds by the differential synthesis and modification of the noncellulosic components or by the addition of new components.

Primary walls of land plants contain cellulose, hemicellulose, and pectin, in different proportions. They also contain structural proteins/proteoglycans, enzymes, low-molecular-weight phenolics, and minerals. Pectin and hemicellulose are present in approximately equal amounts in the so-called type I primary walls of gymnosperms, dicots, and nongraminaceous monocots, whereas hemicellulose is far more abundant than pectin in the type II walls of the grasses. Recent studies of walls from diverse land plants and cell type–specific characterization of wall glycans indicate that the breadth of wall composition and structural diversity is enormous and thus one should think of walls as a structural continuum rather than specific types.

Cellulose

Cellulose, the most abundant biopolymer in nature, is a linear glycan composed of 1-4-linked β-D-Glc*p* residues (Figure 24.1A). Several of these chains are hydrogen bonded to one another to form paracrystalline microfibrils. Each microfibril is predicted to contain between 18 and 24 glucan chains. The glucan chain is synthesized by a cellulose synthase complex at the cell's plasma membrane. Three cellulose synthases, encoded by three different genes, are believed to interact to form a trimeric complex, which in turn assembles into a hexameric rosette at the plasma membrane. The catalytic site of each cellulose synthase is located in the cytosol and transfers glucose from UDP-Glc*p* onto the elongating glucan chain. The mechanisms involved in the formation of a microfibril from individual glucan chains are not well understood, although it may involve an assembly process that is facilitated by specific proteins. The newly formed microfibrils are deposited in the wall of a growing cell with an orientation that is transverse to the axis of elongation. This orientation may be guided in part by protein-mediated interactions between cellulose synthase proteins and cortical microtubules.

Many properties of native cellulose depend on interactions that occur at the surface of the microfibrils. The surface chains are accessible and reactive, whereas the hydroxy groups of the internal chains in the crystal participate in extensive intra- and intermolecular hydrogen bonding. Cellulose is insoluble in water and somewhat resistant to hydrolysis by endo- and exoglucanases because of this highly packed arrangement of the glucan chains.

A number of organisms including fungi and bacteria are specialized in depolymerizing cellulose. This involves several types of enzymes including endoglucanases, cellobiohydrolases, and β-glucosidases. Many of these enzymes have a catalytic domain connected to a cellulose-binding module. This module facilitates binding of the enzyme to the insoluble substrate. Some microorganisms also produce copper-dependent oxidases that render crystalline cellulose more susceptible to hydrolysis. Cellulases and other enzymes involved in cellulose hydrolysis often exist as macromolecular complexes referred to as cellulosomes. Improving the effectiveness of cellulosomes is an area of active research, to increase the conversion of plant biomass to fermentable sugar (Chapter 59).

Hemicelluloses

Hemicelluloses are branched polysaccharides with a backbone composed of 1-4-linked β-D-pyranosyl residues with an equatorial O-4 (Glc*p*, Man*p*, and Xyl*p*). Xyloglucan, glucurono/arabinoxylan, and glucomannan (Figure 24.1B–E) are included under this chemical definition of hemicelluloses. Hemicelluloses and cellulose have structural and conformational similarities that allow them to form strong, noncovalent associations with one another in the cell wall, although the biological significance of these interactions is still a subject of debate.

Xyloglucan and cellulose both have a backbone of 1-4-linked β-D-Glc*p* residues, but unlike cellulose, xyloglucan contains sidechain substituents. The xyloglucan backbone is highly substituted with α-Xyl*p* substituents at O-6 and in some cases with O-acetyl-substituents (Figure 24.1B). Each Xyl*p* residue may itself be extended by the addition of one or more monosaccharides including β-D-Gal*p*, α-L-Fuc*p*, α-L-Ara*f*, α-L-Ara*p*, β-D-Xyl*p*, β-D-Gal*p*A, and O-acetyl substituents. Twenty-three structurally unique side chains have been identified to date, although only a subset of these are synthesized by a single plant species.

Early models of dicot primary walls predicted that xyloglucans acted as tethers between cellulose microfibrils and that controlled enzymatic cleavage or reorganization of xyloglucan by proteins facilitated wall expansion and thus plant cell growth. However, this notion has been challenged by genetic engineering of xyloglucan structures in *Arabidopsis* plants, which revealed that eliminating xyloglucan entirely from the primary wall has remarkably little effect on overall plant growth and development. In contrast, removing only selected sidechain substituents of xyloglucan is detrimental to plant growth and replacing the sidechains with various glycosyl moieties restores plant growth, independent of the glycosyl moiety added. These results have led to the suggestion that xyloglucan acts as a spacer molecule to keep cellulose microfibrils apart and that pectin has a more important role in controlling wall expansion than previously believed.

Arabinoxylan (Figure 24.1D) is the predominant noncellulosic polysaccharide in the type II walls of the grasses, with only small amounts present in dicot primary walls. Its backbone is composed of 1-4-linked β-D-Xyl*p* residues, many of which are substituted at O-3 with α-L-Ara*f* residues. These Ara*f* residues may be further substituted at O-2 with an α-L-Ara*f* or a β-D-Xyl*p* residue. A small number of the backbone residues are substituted at O-2 with α-D-Glc*p*A and its 4-O-methylated counterpart (MeGlc*p*A).

The presence of 1-3, 1-4-linked β-glucans (also referred to as mixed-linkage glucans) (Figure 24.1F) in the walls of grasses was once considered to be a unique feature of these plants. However, structurally related mixed-linkage β-glucans have also been identified in the walls of *Selaginella* (lycopod) and *Equisetum* (horsetails), although the evolutionary relationship between these β-glucans is not known. In grasses, mixed linkage glucan is present mainly in young tissues, which has led to the suggestion that it is a carbohydrate storage molecule as also indicated by its metabolism during plant development.

Callose, a polysaccharide composed of 1-3-linked β glucosyl residues, is another β-glucan produced by plants (Figure 24.1G). It is used to form a temporary cell wall at the cell plate during cytokinesis and is involved in regulating the permeability of plasmodesmata. It is often formed in response to abiotic and biotic stresses or damage.

Pectins

Pectins are structurally complex polysaccharides that contain 1-4-linked α-D-Gal*p*A. Three structurally distinct pectins—homogalacturonan, substituted galacturonan, and rhamnogalacturonan—have been identified in plant cell walls (Figure 24.2). Homogalacturonan, which may account for up to 65% of the pectin in a primary wall, is composed of 1-4-linked α-D-Gal*p*A. The carboxyl group may be methyl-esterified, thus neutralizing the negative charge of this glycosyl moiety, and the uronic acid itself may be acetylated at O-2 or O-3. The extent of methyl-esterification is controlled by pectin methylesterases and pectin methylesterase inhibitors present in the wall. The degree of methylesterification affects the ability of homogalacturonan-containing glycans to form ionic calcium cross-links with themselves and with other pectic polymers. Such interactions alter the mechanical properties of the wall and may influence plant growth and development.

Rhamnogalacturonan-I (RG-I) is a family of polysaccharides with a backbone composed of a repeating disaccharide 4-α-D-Gal*p*A-1-2-α-L-Rha*p*-1. Many of the Gal*p*As are acetylated at O-2 and/or O-3. Depending on the plant, between 20% and 80% of the Rha*p* residues may be substituted at O-4 with linear or branched side chains composed predominantly of Ara*f* and Gal*p*, together with smaller amounts of Fuc*p* and Glc*p*A (Figure 24.2). Little is known about the functions of these side chains and their contribution to the properties of the primary wall.

Substituted galacturonans have a backbone composed of 1-4-linked α-D-Gal*p*A acid residues that are substituted to varying degrees with mono-, di-, or oligosaccharides. For example, xylogalacturonans contain single β-D-Xyl*p* residues linked to O-3 of some of the backbone residues (Figure 24.2), whereas apiogalacturonans have β-D-apiose (Api*f*) and apiobiose linked to O-2 of some of the backbone residues. Apiogalacturonans have only been identified in the walls of duckweeds and seagrasses.

The substituted galacturonan referred to as RG-II, which accounts for between 2% and 5% of the primary cell wall, is the most structurally complex polysaccharide yet identified in nature. It is composed of 12 different monosaccharides linked together by up to 21 distinct glycosidic linkages (Figure 24.2). Four structurally different side chains and one or two Ara*f* substituents are attached to the galacturonan backbone. Two structurally conserved disaccharides (side chains C and D) are linked to O-3 of the backbone. The A and B side chains, which contain between 7 and 9 monosaccharides, are linked to O-2 of the backbone. Several of the monosaccharides in RG-II are O-methylated and/or O-acetylated.

Virtually all of the RG-II exists in the primary wall as a dimer cross-linked by a borate ester. The ester is formed between the Api*f* residue in side chain A of each RG-II monomer (Figure 24.2). The dimer forms rapidly in vitro when the RG-II monomer is reacted with boric acid and a divalent cation. However, the mechanism and site of dimer formation in planta has not been determined. Borate cross-linking of RG-II is likely to have substantial effects on the properties of pectin and the primary wall as RG-II is itself linked to homogalacturonan (Figure 24.2). Indeed, mutations that affect RG-II structure and cross-linking result in plants with abnormal walls and severe growth defects. Swollen primary walls and abnormal growth together with reduced RG-II cross-linking are also a characteristic of boron deficient plants. RG-II was believed to be largely resistant to fragmentation by microbial enzymes. However, recent studies have shown that bacteria present in the human gut produce glycanases capable of hydrolyzing all but one of the glycosidic bonds in RG-II.

Pectin is believed to exist in the cell wall as a macromolecular complex comprised of structural domains—homogalacturonan, rhamnogalacturonan, and substituted galacturonan—that are covalently and noncovalently linked to one another. However, there is only a limited understanding of how these structural domains are organized (Figure 24.2). Molecular modeling of a pectin (~50 kDa) containing homogalacturonan (degree of polymerization ~100) and rhamnogalacturonan with arabinogalactan side chains, together with modeling of RG-II conformation have begun to provide insights into the conformations and relative dimensions of each pectin structural domain.

The homogalacturonan region has a persistence length of approximately 20 GalA*p* residues, which is likely to be sufficient to stabilize junction zones formed with Ca^{++}. In vitro studies suggest that the maximum stability of such junction zones is obtained with oligomers containing approximately 15 nonesterified Gal*p*A residues. Thus, controlling the distribution of methyl-ester groups along the homogalacturonan backbone provides a mechanism to regulate the physical properties of pectin, including its ability to form gels. For a gel to form and not to be brittle, other features including sequences that interrupt interchain associations in the pectin macromolecule may be important. For example, the structural diversity and the conformational flexibility of the oligosaccharide side chains of the rhamnogalacturonan domain will limit or prevent interchain pairing. The presence of 1-2-linked Rha*p* residues does not introduce kinks into the backbone geometry of rhamnogalacturonan and thereby limit interchain associations. Rather, it is the side chains linked to these residues that are responsible for preventing or limiting interchain associations.

The conformations of the four side chains attached to the homogalacturonan backbone may lead RG-II to adopt a "disk-like" shape. Well-defined tertiary structures are predicted for the RG-II monomer and the dimer. In the dimer, borate-ester cross-linking and Ca^{++} interchain pairing further stabilizes the two disks. The apparent resistance of RG-II to wall-modifying enzymes together with the formation of a cation-stabilized RG-II dimer likely results in a structure that resists temporal changes in the plant. In contrast, homogalacturonan is continually modified by the action of wall enzymes and its contribution to wall architecture is therefore time dependent.

Increased knowledge of the physical properties of primary wall polysaccharides and proteoglycans is required to understand how modulating the amounts and structural features of a few common polysaccharides and glycan domains lead to primary walls with diverse properties and functionalities. Further research is also needed to determine if wall structure and function results from the noncovalent interactions of polysaccharides and proteoglycans or from the formation of glycan-containing architectural units with specific structural and functional roles. The latter scenario is analogous to the organization of proteoglycans and O-linked mucins in the extracellular matrix of animal cells (Chapters 10, 16, and 17).

PLANT SECONDARY CELL WALL GLYCANS

The secondary walls of woody tissue and grasses are composed predominantly of cellulose, hemicellulose, and the polyphenol lignin. The inclusion of lignin results in a hydrophobic composite that is a major contributor to the structural characteristics of secondary walls.

Heteroxylans are the major hemicellulosic polysaccharide present in the secondary (lignified) cell walls of seed-producing plants. These heteroxylans are classified according to the type and abundance of the substituents on the 1-4-linked β-D-Xyl*p* residues of the polysaccharide backbone. Glucuronoxylans, which are major components in the secondary walls of woody and herbaceous eudicots, have an α-D-Glc*p*A or MeGlc*p*A substituent at O-2 (Figure 24.1C). Gymnosperm secondary walls contain arabinoglucuronoxylans (AGXs), which in addition to MeGlc*p*A substituents, have Ara*f* residues attached to O-3 of some of the backbone residues.

The glucuronoarabinoxylans in the secondary walls of grasses typically contain fewer Ara*f* residues than their primary wall counterpart (Figure 24.1D). Ferulic or coumaric acids are often esterified to the Ara*f* residues of xylan in grass primary and secondary cells walls. There is some evidence that the lignin polymer is covalently linked to the secondary wall hemicelluloses.

Eudicot and gymnosperm secondary wall xylans have a well-defined glycosyl sequence 1-4-β-D-Xyl*p*-1-3-α-L-Rha*p*-1-2-α-D-Gal*p*A-1-4-D-Xyl*p* at their reducing end (Figure 24.1C). This sequence is required for normal xylan synthesis during secondary cell-wall formation and may have a role in regulating the polymers chain length. This sequence is present at the reducing end of heteroxylans of all monocots except the grasses.

HEMICELLULOSE AND PECTIN BIOSYNTHESIS

Genes that encode polysaccharide biosynthetic enzymes, including many of those required for xyloglucan, glucuronoxylan, arabinoxylan, and cellulose synthesis and some of those required for pectin synthesis, have been identified. This information, together with improved methods to demonstrate enzymatic activity of recombinant plant GTs in vitro and wall structural analyses of the corresponding plant mutants is providing a framework for an increased understanding of how plant cell wall polysaccharides are synthesized. Most pectins and hemicellulose are synthesized in the Golgi apparatus and then secreted into the apoplast via vesicles. This is in contrast to cellulose, which is synthesized at the plasma membrane where the glucan chains are extruded into the apoplast to form cellulose microfibrils. Lignin is polymerized in the apoplast by nonenzymatic radical reactions. Despite advances in identifying and understanding the GTs involved in polysaccharide synthesis, we still do not know how many of the wall polymers are synthesized by Golgi-localized multienzyme complexes or if they are assembled by GTs localized in different regions of the Golgi apparatus. We also do not understand how the newly synthesized polymers are assembled into a functional cell wall in the apoplast.

PLANTS PRODUCE PROTEOGLYCANS CONTAINING O-LINKED OLIGOSACCHARIDES AND O-LINKED POLYSACCHARIDES

Plants produce glycoproteins and proteoglycans that contain oligo- or polysaccharides that are linked to hydroxyproline (Hyp) and serine (Ser). The protein component of these is present in relatively low abundance in the wall. Hyp is formed posttranslationally by endoplasmic reticulum (ER)-localized prolyl hydroxylases and is O-glycosylated in the ER and in the Golgi apparatus. The degree and type of Hyp glycosylation is determined to a large extent by the protein's primary sequence and the arrangement of Hyp residues. Hyp glycosylation is initiated by the addition of an Ara*f* or a Gal*p* residue. Contiguous Hyp residues are arabinosylated, whereas clustered but noncontiguous Hyp residues are galactosylated. Ser residues and occasionally threonine residues may also be O-glycosylated in these proteins.

Three classes of structurally distinct plant proteoglycans containing glycosylated Hyp and Ser—the extensins, proline/hydroxyproline-rich proteoglycans, and arabinogalactan proteins—have been identified. Extensins are hydroxyproline-rich proteoglycans with Ser(Hyp)$_4$ repeat sequences and contain between 50% and 60% (w/w) glycan. Most of the carbohydrate exists as oligosaccharides containing one to four Ara*f* residues linked to Hyp together with a small number of single Gal*p* residues α-linked to Ser. The genetic modification of extensin glycosylation levels resulted in the blockage of polarized cell growth in root hairs, underpinning the importance of extensin glycosylation in plant growth and development. The proline/hydroxyproline-rich proteoglycans, which contain from 3% to 70% (w/w) carbohydrate, are distinguished from the extensins by amino acid sequence. Both of these families of hydroxyproline-rich glycoproteins (HRGPs) likely have a structural role in the cell wall. The expression of genes involved in their synthesis is developmentally regulated and is often induced by wounding and fungal attack of plant tissues. Various plant glycopeptide signaling molecules including clavata contain arabinosylated hydroxyproline and have numerous roles in plant growth and development.

Arabinogalactan proteins (AGPs) have a glycan content of up to 90% (w/w). Chains of between 30 and 150 monosaccharides are linked to the protein by Gal*p*-O-Ser and Gal*p*-O-Hyp linkages. These chains have a 1-3-linked β-Gal backbone that is extensively substituted at O-6 with side chains of 1-6-linked β-Gal*p*. These side chains are terminated with Ara*f*, GlcA*p*, and Fuc*p* residues. Some AGPs may contain homogalacturonan, RG-I, and xylan covalently linked to the arabinogalactan (Figure 24.3), thereby forming a protein–hemicellulose–pectin complex referred to as APAP1. The location of this complex in the plant and its biological function remains to be determined.

Several AGPs are secreted into the cell wall, whereas others are linked to the plasma membrane by a glycosylphosphatidylinositol (GPI) anchor. Plant GPI anchors contain a phosphoceramide core. The glycan portion of the GPI anchor of pear cell AGP has the sequence α-D-Man*p*1-2α-D-Man*p*-1-6-α-D-Man*p*-1-4-Glc*p*N-inositol. At least 50% of the Man*p* attached to the Glc*p*N (see Chapter 12) is itself substituted at O-4 with a β-Gal*p*, a feature that may be unique to plants. Many functions have been proposed for the AGPs including their participation in signaling, development, cell expansion, cell proliferation, and somatic embryogenesis.

THE N-LINKED GLYCANS OF PLANT GLYCOPROTEINS HAVE UNIQUE STRUCTURES

Many of the proteins that have passed through the plant secretory system contain N-linked oligomannose, complex, hybrid, or paucimannose-type glycans (Figure 24.4). The initial stages of the synthesis of these N-glycans, including the transfer of the oligosaccharide precursor from its dolichol derivative and the control of protein folding in the ER, are comparable in plants and animals (Chapter 9). However, two modifications of N-glycans during passage through the Golgi are unique to plants.

Oligomannose-type N-glycans are often trimmed in the *cis*-Golgi and then modified in the *medial*-Golgi by N-GlcNAc transferase I (GnT-I) catalyzed addition of GlcNAc to the distal Man of the core. In reactions that are typical to plants, a β-Xyl*p* is often added to O-2 of the core Man*p*. In the *trans*-Golgi, α-Fuc*f* may be added to O-3 of the Glc*p*NAc residue that is itself linked to asparagine (Figure 24.5). The XylT and FucT that catalyze these reactions act independently of one another but do require at least one terminal Glc*p*NAc residue for activity. The FucT is related to the Lewis FucT family, whereas the XylT is unrelated to other known GTs.

The xylosylated and fucosylated N-glycans are often trimmed by α-mannosidase II. A second GlcNAc may then be added by GnT-II. Some plant N-glycans do not undergo further mannose trimming and proceed through the Golgi as hybrid-type N-glycans. Complex and hybrid-type N-glycans may be further modified by the addition of Gal*p* and Fuc*p* in the *trans*-Golgi. Plant glycoproteins are either secreted from the cell or transported to the vacuoles. Many of the glycoproteins present in the vacuoles contain paucimannose type glycans, suggesting that they are trimmed by vacuolar glycosidases (Figure 24.5).

The presence of sialic acid in the N-glycans of plant glycoproteins was claimed but likely represented environmental contamination. Plants do have genes that encode proteins containing sequences similar to sialyltransferase motifs, but their functions have not been established.

Plant-specific modifications of N-glycans result in glycoproteins that are often highly immunogenic and cause allergic responses in humans. The demonstration that complex N-glycans are not essential for plant growth initiated studies to engineer plant N-glycosylation pathways to

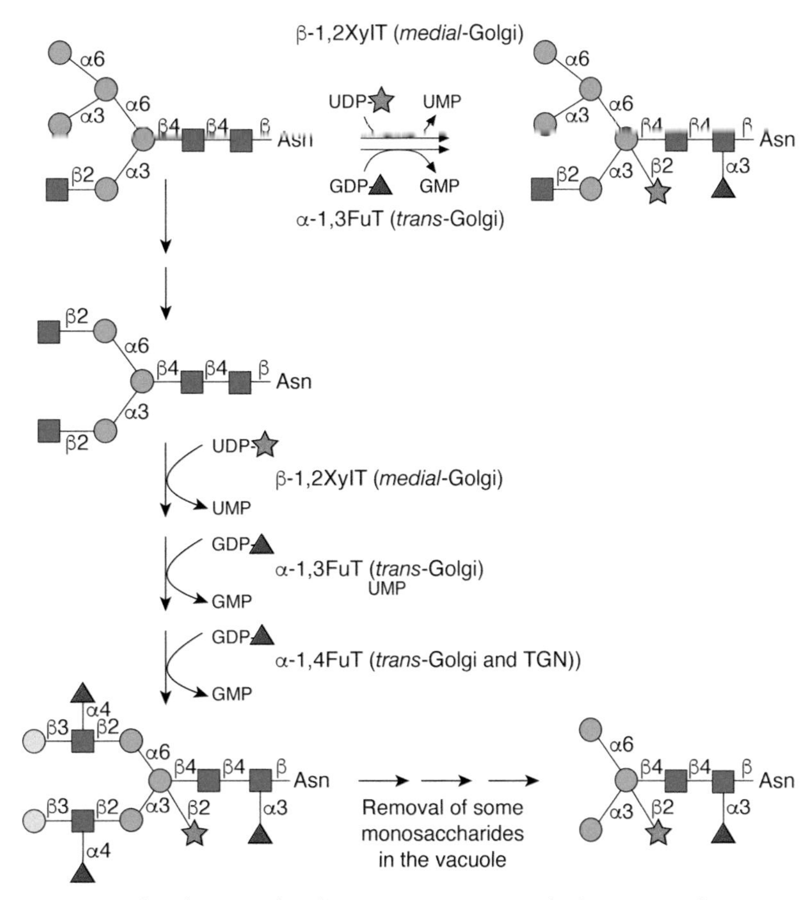

FIGURE 24.5 Processing of N-glycans in the plant secretory system. Only those events that are unique to plants are shown in detail.

produce glycoproteins that do not activate the mammalian immune system. Plants lacking the GTs that add Xyl*p* and Fuc*p* to N-linked glycans produce glycoproteins lacking immunogenic glyco-epitopes. Other glycosylation pathways involved in the addition of sialic acid and Gal*p* must be introduced to fully "humanize" the glycoproteins if plants are to be used to produce recombinant therapeutic glycoproteins.

ALGAL GLYCANS

Only a few glycans of green algae in the Chlorophyta and Streptophyta clades have been studied in detail. However, understanding their diverse polysaccharide structures could have important implications for the evolution of the more complex structures present in land plants.

The Viridiplantae are believed to have diverged into the Chlorophyta and Streptophyta, between 800 and 1200 million years ago. The Chlorophyta, which include diverse marine, freshwater, and terrestrial green algae, often have cell walls that are quite distinct from the walls of the Streptophytes. For example, the extracellular matrix of the chlorophyte *Chlamydomonas reinhardtii* is a crystalline lattice formed from HRGPs, whereas this matrix is rich in Kdo and Dha in the prasinophytes.

The Streptophyta comprise land plants and the charophyte algae including the late diverging Zygnematophyceae, which are currently believed to include the closest living relatives of land plants. Indeed, the cell wall of *Penium margaritaceum*, a unicellular Zygnematophyte, contains cellulose, pectic, and hemicellulose-like glycans.

With the emergence of whole-genome sequencing and extensive transcriptomic analyses, detailed phylogenetic trees of glycan synthesis-related GTs are being established including GTs from green algae and land plants. Such data are required to develop hypotheses predicting when and how different structural forms of diverse cell wall polysaccharides evolved across the Chlorophyta and Streptophyta. Detailed structural and architectural analyses of algal and land plant cell wall glycans and the substrate and acceptor specificities of the corresponding GTs will be required to resolve many of the evolutionary transitions within the Viridiplantae.

Numerous specialized algal polysaccharides are used by humans. For example, several of the polysaccharides produced by red and brown algae are used in the food industry as gelling agents, stabilizers, thickeners, and emulsifiers. They are also used in paints, in cosmetics, in paper, and as reagents for scientific research. These polysaccharides include agarose (agar) and carrageenan, which are sulfated galactans obtained from red seaweeds. These polysaccharides are composed of the repeating disaccharide 3-β-D-Gal*p*-1-4-3,6-anhydro-α-L-Gal*p*-1 unit. Some of the D-Gal*p* and L-Gal*p* units are O-methylated. Pyruvate and sulfate groups may also be present in small quantities. Alginate, a linear polysaccharide composed of 1-4-linked β-D-Man*p*A and its C-5 epimer 1-4-linked α-L-Gul*p*A, produced by various species of brown seaweed is another example of a commercially important polysaccharide. These monosaccharides are typically arranged in blocks of either Man*p*A or Gul*p*A separated by regions comprised of 4-Man*p*A-1-4-Gul*p*A-1 sequences. Brown seaweeds produce polysaccharides that have potential in the treatment of diseases. Laminaran is a linear storage polysaccharide composed of 1-3- and 1-6-linked β-D-Glc*p* residues. There are reports that laminaran has antiapoptotic and antitumor activities. Fucoidans are a group of sulfated polysaccharides isolated from several brown algae that have been reported to have anticoagulant, antitumor, antithrombosis, antiinflammatory, and antiviral properties. Fucoidans have a backbone of 1-3-linked α-Fuc*p* that is substituted at O-2 with fucose and at O-4 with sulfate or fucose. Other fucoidans have backbones of alternating 1-3- and 1-4-linked α-Fuc*p* residues.

PLANT GLYCOLIPIDS

Glycoglycerol lipids are the most abundant glycolipids in plants. Mono- and digalactosyldiacylglycerol have been identified in all plants, whereas tri- and tetragalactosyldiacylglycerol have a more restricted taxonomic distribution (Figure 24.6). The synthesis of these galactolipids is initiated by the formation of diacylglycerol in the ER membrane and the chloroplast membrane. Galactolipids formed in the ER membrane contain predominantly C16 fatty acids at the *sn*-2 position and C18 fatty acids at the *sn*-3 position. The chloroplast pathway produces C18 fatty acids at both positions. Each of these fatty acids is then desaturated to 16:3 or 18:3 acyl groups. Monogalactosyldiacylglycerol (MGDG) is synthesized by the transfer of Gal from UDP-Gal*p* to diacylglycerol by an MGDG synthase. Digalactosyldiacylglycerol (DGDG) is formed from MGDG by the transfer of Gal*p* from UDP-Gal*p* by a DGDG synthase. These reactions occur primarily in the outer chloroplast membrane. The products are then transported to the inner membrane and the thylakoid membranes of the chloroplast. The presence and abundance of MGDG in the chloroplast thylakoid membrane are important for normal photosynthesis to occur. Sulfoquinovosyldiacylglycerol, which is formed from diacylglycerol, is also abundant in the thylakoid membrane and may also have a role in photosynthesis.

Small amounts of MGDG and DGDG are present in the cytosolic leaflet of the plasma membrane, although the mechanism of galactolipid exchange among the membranes is not understood. The outer leaflet of the plasma membrane is instead dominated by glycosylinositolphosphorylceramides (GIPCs), lipids that are absent in animals. GIPCs were first described in the 1950s, but have remained poorly characterized partly because of their insolubility using standard membrane extraction protocols. GIPCs are estimated to comprise up to 40% of the plasma

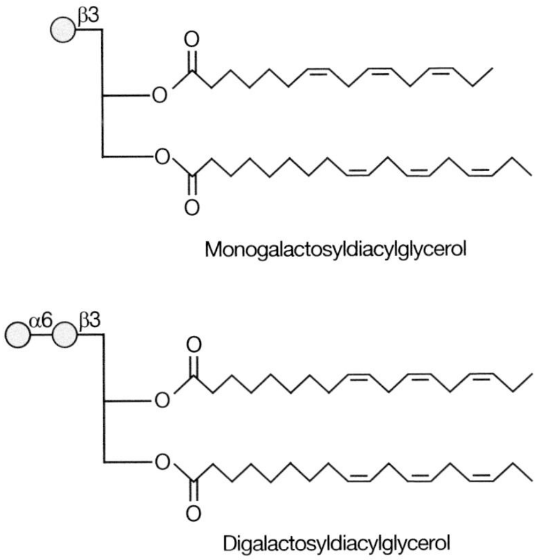

Monogalactosyldiacylglycerol

Digalactosyldiacylglycerol

FIGURE 24.6 The most abundant plant galactolipids.

membrane, with more than 200 species described. Ceramide is synthesized in the ER and is then transported to the Golgi where an inositol phosphate and several glycosyl residues are added. The first sugar added in flowering plants is Glc*p*A, but after that the identity of the sugar, the number of sugars, and their degree of branching depend on tissue and species. GIPCs containing from two (Series A) to seven (Series F) sugars, including Man*p*, Glc*p*, Glc*p*N, Glc*p*NAc, Gal*p*, and Ara*f*, have been consistently identified, although there have been reports of GIPCs containing up to 20 sugars. Further improvements to the isolation and characterization methods are required so that the full array of glycan structures can be described. No gangliosides have been identified in plants. Reports that Kdo*p*-containing lipids with homology to bacterial lipid A are present in plant organelle membranes remain to be confirmed.

GIPC glycosylation mutants can be lethal or have severe developmental phenotypes. Altering the glycan structure can induce severe constitutive defense responses, alter plant–microbe interactions (both pathogens and beneficial microbes), and reduce the cellulose content of the cell wall. Only a limited number of GIPC GTs have yet been described, and much work is required to fully describe their function.

OTHER PLANT GLYCOCONJUGATES

Plants produce numerous phenolics, terpenes, steroids, and alkaloids that are collectively referred to as secondary metabolites. Many of these compounds are O-glycosylated or contain sugars linked via N, S, or C atoms. Glycosylated secondary metabolites often have important roles in a plant's response to biotic and abiotic challenges and may also have value as pharmaceuticals.

In general, the addition of a single sugar or an oligosaccharide may increase water solubility, enhance chemical stability, or alter both chemical and biological activity. For example, the activity of several plant hormones may be regulated by converting them to their glucose esters or their glucosides. Digoxin and oleandrin are potent cardiac glycosides isolated from foxglove and oleander, respectively. Myrosinase-catalyzed cleavage of S-linked Glc*p* from glucosinolates leads to the formation of pungent mustard oils when mustard and horseradish are damaged. The steviol glycosides, which are far sweeter than sucrose, are used as natural sugar substitutes. The bitter taste of citrus fruits is due to naringin, a glycosylated flavonoid.

ACKNOWLEDGMENTS

The authors appreciate helpful comments and suggestions from Todd Lowary, Katharina Paschinger, and Iain B.H. Wilson and thank Bernard Henrissat for help with the CAZy numbers in Table 24.1.

FURTHER READING

Painter T. 1983. Algal polysaccharides. *In The polysaccharides* (ed. Aspinall G), pp. 195–285. Academic, New York.

Pérez S, Mazeau K, du Penhoat CH. 2000. The three-dimensional structures of the pectic polysaccharides. *Plant Physiol Biochem* **38:** 37–55. doi:10.1016/s0981-9428(00)00169-8

Gachon CM, Langlois-Meurinne M, Saindrenan P. 2005. Plant secondary metabolism glycosyltransferases: the emerging functional analysis. *Trends Plant Sci* **10:** 542–549. doi:10.1016/j.tplants.2005.09.007

Hölzl G, Dörmann P. 2007. Structure and function of glycoglycerolipids in plants and bacteria. *Prog Lipid Res* **46:** 225–243. doi:10.1016/j.plipres.2007.05.001

Albersheim P, Darvill A, Roberts K, Sederoff R, Staehelin A. 2010. *Plant cell walls. From chemistry to biology*. Garland Science, New York. doi:10.1201/9780203833476

Gomord V, Fitchette A-C, Menu-Bouaouiche L, Saint-Jore Dupas C, Plasson C, Michaud D, Faye L. 2010. Plant-specific glycosylation patterns in the context of therapeutic protein production. *Plant Biotechnol J* **8:** 564–587. doi:10.1111/j.1467-7652.2009.00497.x

Bar-Peled M, O'Neill MA. 2011. Plant nucleotide sugar formation, interconversion, and salvage by sugar recycling. *Annu Rev Plant Biol* **62:** 127–155. doi:10.1146/annurev-arplant-042110-103918

Kieliszewski MJ, Lamport D, Tan L, Cannon M. 2011. Hydroxyproline-rich glycoproteins: form and function. *Annu Plant Rev* **41:** 321–342. doi:10.1002/9781119312994.apr0442

Popper ZA, Michel G, Hervé C, Domozych DS, Willats WGT, Tuohy MG, Kloareg B, Stengel DB. 2011. Evolution and diversity of plant cell walls: from algae to flowering plants. *Annu Rev Plant Biol* **62:** 567–590. doi:10.1146/annurev-arplant-042110-103809

Atmodjo MA, Hao Z, Mohnen D. 2013. Evolving views of pectin biosynthesis. *Annu Rev Plant Biol* **64:** 747–779. doi:10.1146/annurev-arplant-042811-105534

Pauly M, Gille S, Liu L, Mansoori N, de Souza A, Schultink A, Xiong G. 2013. Hemicellulose biosynthesis. *Planta* **238:** 627–642. doi:10.1007/s00425-013-1921-1

Cosgrove DJ. 2014. Re-constructing our models of cellulose and primary cell wall assembly. *Curr Opin Plant Biol* **22:** 122–131. doi:10.1016/j.pbi.2014.11.001

Knoch E, Dilokpimol A, Geshi N. 2014. Arabinogalactan proteins: focus on carbohydrate active enzymes. *Frontiers Plant Sci* **5:** 198. doi:10.3389/fpls.2014.00198

Lombard V, Golaconda Ramulu H, Drula E, Coutinho PM, Henrissat B. 2014. The Carbohydrate-Active enZYmes database (CAZy) in 2013. *Nucleic Acids Res* **42:** D490–D495. doi:10.1093/nar/gkt1178

Matsubayashi Y. 2014. Posttranslationally modified small-peptide signals in plants. *Annu Rev Plant Biol* **65:** 385–413. doi:10.1146/annurev-arplant-050312-120122

McNamara JT, Morgan JL, Zimmer J. 2015. A molecular description of cellulose biosynthesis. *Annu Rev Biochem* **84:** 895–921. doi:10.1146/annurev-biochem-060614-033930

Höfte H, Voxeur A. 2017. Plant cell walls. *Curr Biol* **27:** R865–R870. doi:10.1016/j.cub.2017.05.025

Ndeh D, Rogowski A, Cartmell A, Luis AS, Baslé A, Gray J, Venditto I, Briggs J, Zhang X, Labourel A, et al. 2017. Complex pectin metabolism by gut bacteria reveals novel catalytic functions. *Nature* **544:** 65–70. doi:10.1038/nature21725

Jiao C, Sørensen I, Sun X, Sun H, Behar H, Alseekh S, Philippe G, Palacio Lopez K, Sun L, Reed R, et al. 2020. The *Penium margaritaceum* genome: hallmarks of the origins of land plants. *Cell* **181:** 1097–1111. doi:10.1016/j.cell.2020.04.019

Mortimer JC, Scheller HV. 2020. Synthesis and function of complex sphingolipid glycosylation. *Trends Plant Sci* **25:** 522–524. doi:10.1016/j.tplants.2020.03.007

25 Nematoda

Iain B.H. Wilson, Katharina Paschinger, Richard D. Cummings, and Markus Aebi

This chapter focuses on the nematode (roundworm) *Caenorhabditis elegans* as an example of the phylum Nematoda. *C. elegans* provides a powerful genetic system for studying glycans during embryological development and in primitive organ systems.

DEVELOPMENTAL BIOLOGY OF *C. ELEGANS*

C. elegans is transparent, and individual cells can be easily visualized in the living organism through all stages of development. Basically, the worm is a tube within a tube (Figure 25.1). A cuticle composed of a collagenous, multilayered, protective exoskeleton surrounds the worm. The "mouth" at the anterior end connects to a tubular intestinal system, which is composed of a muscular pharynx and an intestine. The gonad occupies most of the body cavity. In the hermaphrodite, the gonad is bilobed, with each lobe connecting via an oviduct and spermatheca to a shared midventral vulva and uterus. The worm exists as two sexes, hermaphrodite or male. Eggs pass through the spermatheca where fertilization takes place by stored sperm, and the eggs begin to develop inside the uterus. During sexual reproduction, males fertilize hermaphrodites. The male sperm is also stored in the spermatheca and is preferentially used during fertilization.

Gastrulation starts before egg laying; at this stage, the embryo contains about 30 cells (Figure 25.2). Proliferation results in an embryo of 558 relatively undifferentiated cells. Then organogenesis/morphogenesis begins, terminal differentiation occurs, and the embryo hatches. The animal normally passes through four larval stages, termed L1, L2, L3, and L4 (Figure 25.2).

published by Cold Spring Harbor Laboratory Press; doi: 10.1101/glycobiology.4e.25

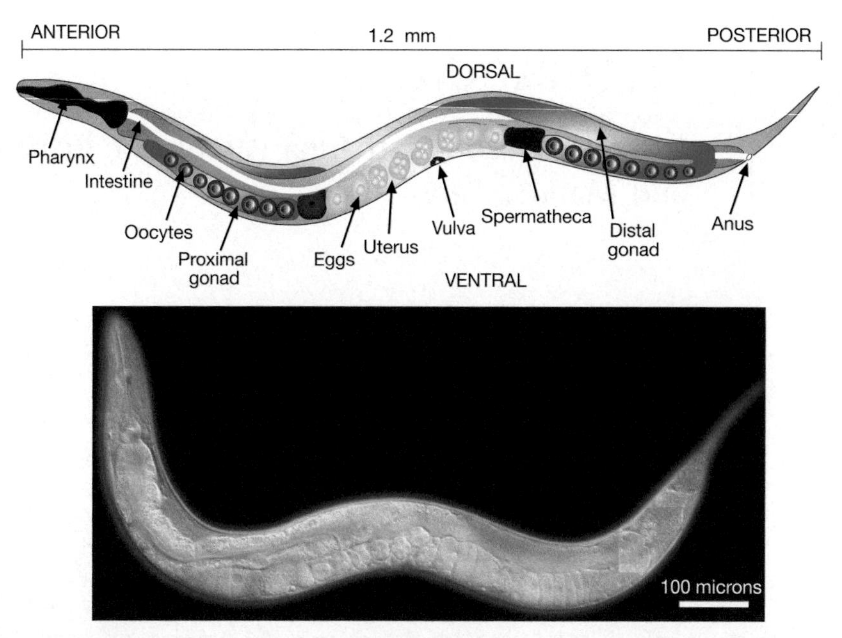

FIGURE 25.1. *Caenorhabditis elegans.* A composite diagram (*upper* panel) and photograph (*lower* panel) of the adult hermaphrodite with labeled body parts. (Photograph kindly provided by Dr. Ian D. Chin-Sang at Queen's University, Kingston, Ontario.) For additional details on the biology of *C. elegans,* see the WormAtlas.

The end of each larval stage is marked by molting, when the cuticle is shed. In L1 larvae, the nervous system, the reproductive system, and the digestive tract begin to develop, and this is completed by the L4 stage. Mature adults hatch after ~45–50 h, whereby hermaphrodites contain 959 somatic cells, including 302 neurons, and 95 body-wall muscle cells. At this time, the hermaphrodite can lay its first eggs, thus completing the 3.5-d life cycle; production of oocytes continues for ~4 d, resulting in about 300 progeny. Afterwards, the animal lives for another 10–15 d. Overcrowding or starvation leads to formation of dauer larvae, a dormant stage, which is easily distinguished from other developmental stages by morphology and behavior.

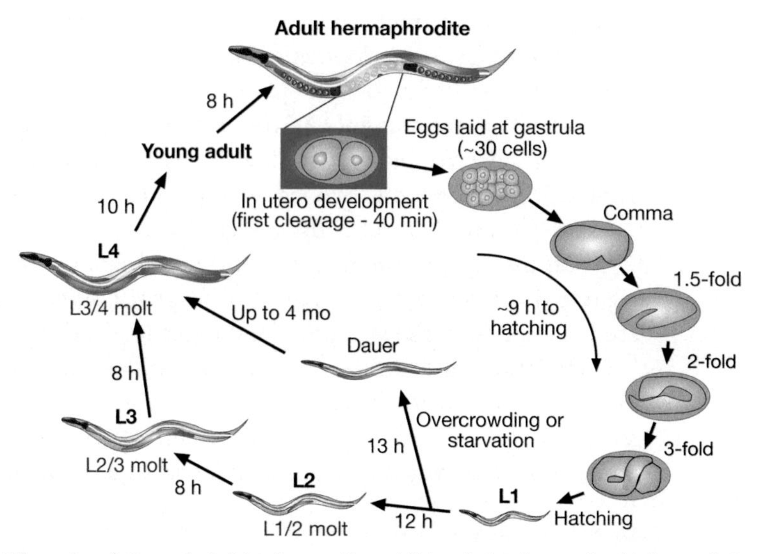

FIGURE 25.2. Life cycle of *Caenorhabditis elegans.* For additional details on the biology of *C. elegans,* see the WormAtlas.

GLYCANS IN *C. ELEGANS*

Considering the large number of glycomic studies on *C. elegans* since 2001, it is probably not an exaggeration to state that this anatomically simple worm has one of the most varied and unusual glycomes of any invertebrate organism studied to date. Although there are a number of conserved elements in its glycans, there are many notable differences between the types of glycans made by *C. elegans* and those in "higher" animals. There are, for example, no sialic acids or other anionic moieties on its N-glycans, but a wide range of fucose modifications on truncated "paucimannosidic" glycans have been found. Moreover, worm O-glycans and glycolipids, which also lack sialic acids, have different core structures. Although some predictions regarding the types of oligosaccharides in the worm can be made from the range of glycosylation-relevant genes in its genome, glycan analyses continue to surprise as to the glycomic potential of this organism.

As in most eukaryotes, the N-glycans of *C. elegans* are derived from a 14-sugar dolichol-linked precursor $Glc_3Man_9GlcNAc_2$, which is transferred to asparagine residues within the -Asn-X-Ser/Thr- sequons of nascent polypeptides in the endoplasmic reticulum (ER) (Chapter 9). Subsequent trimming is akin to that in plants and other animals, but the $GlcNAc_1Man_5GlcNAc_2$-Asn generated in the Golgi is then further processed in a unique way (Figure 25.3). On one hand, *C. elegans* can generate various complex structures with up to three antennae with either single *N*-acetylglucosamine residues or $HexNAc_{2\text{-}3}$ motifs, some of which are modified with phosphorylcholine as in many other nematode glycoproteins. On the other hand, a high degree of modification of the core region is also a feature of this organism. The action of various Golgi β-hexosaminidases and α-mannosidase results in truncated so-called "paucimannosidic" structures; in addition to core α1,6-fucosylation as in "higher" animals, two types of α1,3-fucose substitution of the core chitobiose are known, all of which can be galactosylated. Particularly curious is the addition of a bisecting galactose, which can also be α1,2-fucosylated, to the core β-mannose in a significant population of the N-glycans (Figure 25.3); maximally, five fucose residues as well as multiple galactose and methyl substitutions of the core region are possible.

Many forms of O-glycans familiar from mammals and insects are either structurally defined or genetically predicted in *C. elegans*. These include "mucin-type" O-glycans, many of which are based on the core 1 O-glycan structure common to vertebrates (Chapter 10), but which can be extended by β-glucose, glucuronic acid, and α1-2 fucose residues (Figure 25.4). In terms of glycosaminoglycans (GAGs), chondroitin and heparan sulfate (CS and HS) have been detected in *C. elegans*, but neither keratan sulfate (KS), dermatan sulfate (DS), nor hyaluronan (Chapter 16); whereas the CS chains contain a low amount of sulfate, the overall structure of HS in *C. elegans* is similar to the chains elaborated by vertebrates (see Figure 25.5). Enzymatic and genomic data indicate that there are various domain-specific forms of glycosylation in *C. elegans*, including O-fucose (Fucα1-Ser/Thr) on epidermal growth factor (EGF)-like domains and thrombospondin type-1 repeats (TSRs), in the context of precise consensus sequences (Chapter 13), C-mannosylation of TSRs, and TMTC-dependent O-mannosylation of cadherins. However, POMT-mediated O-mannosylation of dystroglycan is absent. Cytoplasmic and nuclear proteins in *C. elegans* can be modified with O-GlcNAc as in other animals.

Glycosphingolipids (GSLs) (Chapter 11) in *C. elegans* have a core consisting of GlcNAcβ1-3Manβ1-4Glcβ1-Cer, which is based, as in insects, on the arthro-series Manβ1-4Glcβ1-Cer core rather than the common Galβ1-4Glcβ1-Cer core found in vertebrates; some of the glycolipids carry fucose or phosphorylcholine residues (see Figure 25.7). In addition, *C. elegans* has genes encoding the enzymes for synthesis of glycosylphosphatidylinositol (GPI)-anchored glycoproteins, but the GPI anchor structures are not yet defined (Chapter 12).

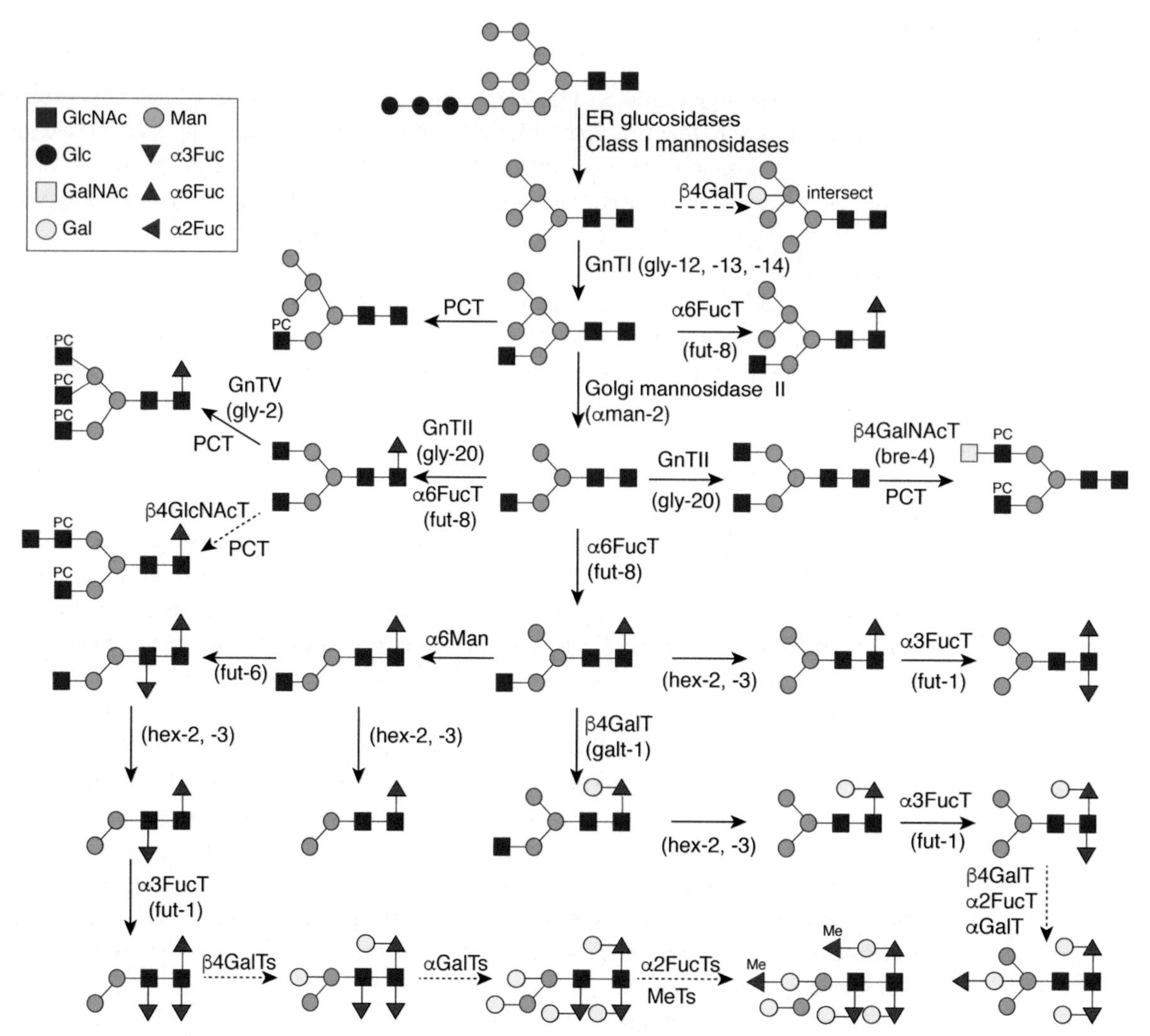

FIGURE 25.3. Biosynthesis of paucimannosidic and core fucosylated N-glycans in *Caenorhabditis elegans*. *Solid* and *dashed lines* refer, respectively, to proven and proposed biosynthetic steps.

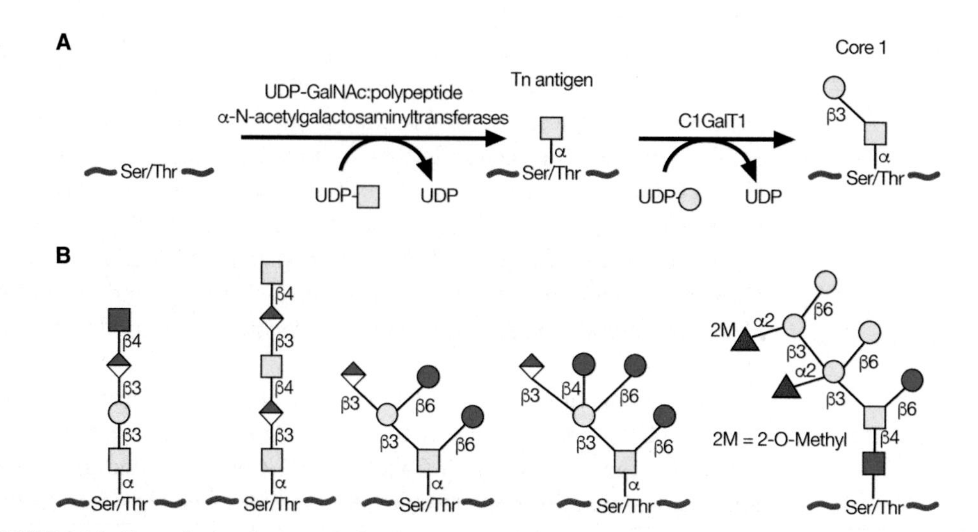

FIGURE 25.4. Biosynthesis of core-1 O-glycans in *Caenorhabditis elegans* (*A*) and some O-glycans proposed to occur in adult worms (*B*).

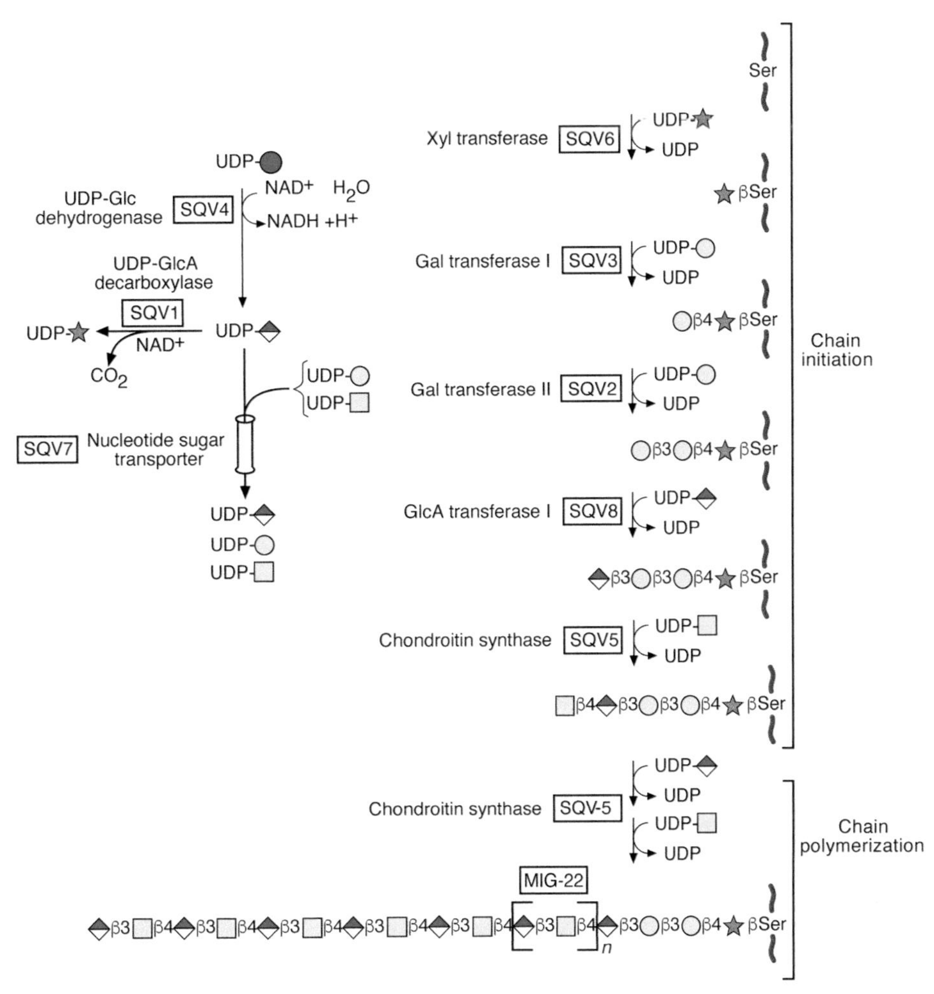

FIGURE 25.5. Biosynthesis of chondroitin in *Caenorhabditis elegans*. Mutations in individual steps were identified as squashed vulva mutants, as described in the text.

GLYCOSYLTRANSFERASE GENES IN *C. ELEGANS*

The *C. elegans* genome encodes homologs of many of the enzymes for glycoconjugate biosynthesis that are found in "higher" animals and humans, including enzymes for the synthesis of O-GalNAc (mucin-type) glycans, GAGs, O-GlcNAc, N-glycans, GSLs, and GPI anchors. In contrast to vertebrates (Chapter 15), the worm lacks sialic acids and any enzymes associated with sialic acid biosynthesis or utilization. Additionally, on the basis of the glycoconjugate structures, *C. elegans* is predicted to express a wide assortment of enzymes involved in their metabolism. Indeed, the genome of *C. elegans* appears to encode about 300 carbohydrate-active enzymes (CAZy database; http://www.cazy.org/), including glycosyltransferases, glycosidases, epimerases, polysaccharide lyases, and carbohydrate esterases (Chapter 52). To date, most of these putative enzymes are not characterized, and very little is known overall about their expression or function in *C. elegans*. Some information has arisen from phenotypes generated by mutagenesis, as discussed below. Together, these studies have identified some interesting differences and similarities between *C. elegans* and vertebrates.

The *C. elegans* genome is predicted to encode some 240 glycosyltransferases and some GT families are highly represented, including 11 putative ppGalNAcTs, 32 α-fucosyltransferase homologs (five α1-3FucTs, 26 α1-2FucTs, and one α1-6FucT), and more than 20 β-N-acetylglucosaminyltransferase homologs. Relatively few glycan-modifying enzymes from *C. elegans* have been shown to be functional; in only a few cases has their acceptor specificity been characterized. Indeed, there are still many gaps in our knowledge as glycosyltransferases required for many glycosidic bonds are still to be identified or there are glycosyltransferase homologs for which no corresponding activity is known.

In terms of Golgi modifications of N-glycans, *C. elegans* contrasts with humans and most other vertebrates in that it has three genes (*gly-12*, *gly-13*, and *gly-14*), rather than just one, encoding the β1-2 N-acetylglucosaminyltransferase (GlcNAcT-1) that catalyzes formation of GlcNAcβ1-2$Man_5GlcNAc_2$-Asn (see Figure 25.3). Triple-knockout worms defective in *gly-12*, *gly-13*, and *gly-14* generate very low amounts of paucimannosidic $Man_{2\text{-}3}GlcNAc_2$ N-glycans (with and without core α1-6-fucose), but $Man_5GlcNAc_2$-Asn is the major structure. On the other hand, core α1-3-fucosylation is not abolished in this mutant.

The GlcNAcβ1-2$Man_5GlcNAc_2$-Asn biosynthetic intermediate is then acted on by a single α-mannosidase II to generate GlcNAcβ1-2$Man_3GlcNAc_2$-Asn. *C. elegans* contains three α-mannosidase activities: the lysosomal enzyme AMAN-1, the α-mannosidase II/IIx-like AMAN-2 involved in N-glycan processing, and the Co^{++}-dependent AMAN-3. A mutant harboring a large deletion in the *aman-2 (F58H1.1)* gene generates largely $Man_5GlcNAc_2$-Asn and $GlcNAc_1Man_5GlcNAc_2$-Asn-R as well as fucosylated and phosphorylcholine-modified variants thereof, but lacks the paucimannosidic structures. In addition, the mutant has reduced levels of the core α1-3 fucose antigen associated with reactivity to antibodies to horseradish peroxidase.

As in many invertebrates, the antennal GlcNAc is normally absent from the final glycan products and indeed *C. elegans* has two relevant β-hexosaminidases (HEX-2 and HEX-3), which cleave GlcNAcβ1-2$Man_3GlcNAc_2$-Asn to generate the paucimannose structure $Man_3GlcNAc_2$-Asn, a reaction not found in vertebrates. Furthermore, there is no evidence for galactosylation of antennal GlcNAc and the closest *C. elegans* homolog to the human β1-4 galactosyltransferase is actually a β1-4 N-acetylgalactosaminyltransferase generating the LacdiNAc sequence (GalNAcβ1-4GlcNAc-R) on N-glycan and glycolipid antennae. Three core fucosyltransferases have been characterized and the glycomes of the corresponding mutants analyzed. Although the *fut-8* gene encodes an enzyme with the same specificity as the mammalian *FUT8* α1-6 fucosyltransferases, two α1-3 fucosyltransferases (FUT-1 and FUT-6) with unusual substrate preferences transfer fucose to the proximal and distal core GlcNAc residues; whereas FUT-1, unlike plant and insect core α1-3 fucosyltransferases, cannot transfer to glycans with a β1-2GlcNAc on the α1-3-mannose, the action of FUT-6 is blocked by the presence of α1-6-mannose. Activity of recombinant forms of two α1-2 fucosyltransferases and a fucose-modifying β1-4-galactosyltransferase (GALT-1) have also been reported.

In terms of peptide-modifying glycosyltransferases, 11 of the nematode UDP-GalNAc polypeptide α-N-acetylgalactosaminyltransferases (ppGalNAcTs) that modify Ser/Thr residues of mucin core polypeptides have been prepared as recombinant proteins, but only five of them are active toward mammalian peptide acceptors. Additionally, the *C. elegans* β1-3 galactosyltransferase (T-synthase), which generates the core 1 O-glycan structure common to vertebrates (Chapter 10), has been characterized. The single *C. elegans* cytosolic or nuclear O-GlcNAc transferase (OGT) and the GAG-initiating O-xylosyltransferase homologs as well as its POFUT1 and 2 O-fucosyltransferases have been characterized at the biochemical and genetic levels. *C. elegans* can also synthesize the canonical core tetrasaccharide of GAGs, GlcAβ1-3Galβ1-3Galβ1-4Xylβ1-Ser, and can extend this to generate CS and HS. In the case of HS biosynthesis, there are numerous sulfotransferases and an HS epimerase, but fewer chondroitin-modifying enzymes, in *C. elegans*. The genes required for initiation of the unusual arthroseries glycolipid core structure (Manβ1-4Glcβ1-Cer) are also known, but *C. elegans* lacks the β1-4 galactosyltransferase that would synthesize the more familiar Galβ1-4Glcβ1-Cer glycolipid core.

Expression of the known glycosyltransferases in *C. elegans* has not been systematically mapped at the cellular level during development or in the adult organism. Many of the relevant studies are based on promoter analyses using GFP as a reporter. Therefore, the promoter region for a gene of interest (usually 0.5–1.5 kb upstream of the gene and sometimes including the upstream elements and the first few exons of the gene) is ligated to the cDNA encoding GFP. Transgenic animals are then produced by direct injection of DNA into the hermaphrodite gonad. Newly developing animals take up this DNA and become transgenic. Thus, one can observe promoter utilization in the different stages of development (see Figure 25.2). Using this approach, it has been found that some glycosyltransferase genes are widely expressed, including the T-synthase, which generates core 1 O-glycans (Figure 25.4), the SQV-2 involved in GAG biosynthesis (Figure 25.5), and the two protein O-fucosyltransferases POFUT1 and POFUT2. In contrast, expression of the six individual glycosyltransferases related to core-2 N-acetylglucosaminyltransferases in vertebrates (see Chapter 10) is tissue-specific; one such gene, *gly-15*, is expressed only in two gland cells. Relevant to N-glycan biosynthesis, promoter analyses show that *gly-12* and *gly-13* are expressed in all cells beginning in embryogenesis, whereas *gly-14* is expressed only in intestinal cells from L1 to adults; the *aman-2* gene is expressed in most cells of the organism. Among the 26 α1-2FucTs that are predicted in *C. elegans*, one of them (*CE2FT-1;* FUT-2) is expressed in a single cell in embryos and exclusively in 20 intestinal cells of larval stages L1–L4 and adult worms. Thus, in large gene families, individual members may be expressed in a localized fashion and have unique activities toward certain substrates, whereas single gene families appear to be expressed in all cells.

FUNCTIONAL ANALYSIS OF GLYCOCONJUGATES

Different methods exist to genetically manipulate *C. elegans* and many of these approaches have yielded important information about the functions of glycosyltransferases, their glycan products, and lectin-binding proteins. Several dozen genes involved in glycosylation pathways have been shown to be developmentally important in *C. elegans* or important in resistance or susceptibility to pathogens in the innate immunity of the worm.

N-Glycans and O-Glycans on Glycoproteins

Glycoproteins in *C. elegans* have both N- and O-glycans, as discussed above. In vertebrates, interference of the early steps in N-glycosylation or O-glycosylation causes embryonic lethality or results in severe developmental phenotypes (Chapter 45). As in vertebrates, interference of the later steps in N- and O-glycan biosynthesis in *C. elegans* does not cause developmental problems, but the effects on the glycome can be very profound. For instance, ablation in *C. elegans* of the three genes (*gly-12*, *gly-13*, and *gly-14*) that encode the three GlcNAcT-I isoforms or of the single mannosidase II gene (*aman-2*) results in drastically altered N-glycomes without affecting development under laboratory conditions, which is in contrast to the effects in mammals; however, altered susceptibility to bacterial infection has been reported for GlcNAcT-I mutant worms. Also, deletion of the OGT transferase required for the modification of cytoplasmic and nuclear O-GlcNAc proteins, although fatal to vertebrate cells (Chapter 19), is not lethal to *C. elegans*, but is accompanied by a phenotype that resembles human insulin resistance. A related phenotype is also produced by deletion of the "antagonistic" OGA O-GlcNAcase.

In terms of O-glycans, the redundancy of ppGalNAcT and α1,2 and α1,3 fucosyltransferase homologues may be a reason for a lack of developmental phenotypes associated with the loss of any members of these families. However, RNAi studies on the worm's POFUT2 (encoded by the *pad-2* gene) suggest that O-fucose–modifications of TSRs (Chapter 13) are required for normal morphogenesis. The biosynthesis of all fucosylated ligands requires the precursor GDP-fucose and its transport into the Golgi apparatus by nucleotide sugar transporters (Chapter 5). One human disease, leukocyte-deficiency type II (LAD II), is caused by a defect in the transport of GDP-fucose (Chapter 34). A screen of candidate *C. elegans* nucleotide sugar transporter genes led to the identification of one which indeed complemented the transport and fucosylation defect in LAD II fibroblasts and thereby to pinpointing of the genetic defect in these patients. A defect in GDP-fucose biosynthesis caused by a mutation in the *bre-1* gene results in resistance to some fungal toxic lectins; furthermore, mutations in genes encoding the FUT-1, -6, and -8 core fucosyltransferases, the GALT-1 galactosyltransferase, the HEX-2 and HEX-3 hexosaminidases, or the SAMT-1 transporter (the latter required for glycan methylation) are associated with altered susceptibility to proteins expressed by nematoxic fungi.

C. elegans is also an interesting model system to study infection and innate immunity. The organism may be colonized by different bacterial pathogens, including *Pseudomonas aeruginosa*, *Yersinia pestis*, and *Yersinia pseudotuberculosis*. The two *Yersinia* species generate a sticky biofilm (an exopolysaccharide matrix encasing a community of bacteria) on the exterior of the worm's head that impairs viability. *P. aeruginosa*, in contrast, colonizes the intestinal tissues. Another bacterium, *Microbacterium nematophilum*, sticks to the anus of the animals and induces an irritation in the underlying hypodermal tissue. *Bacillus thuringiensis* (Bt) infection leads to destruction of the intestine, which is discussed in more detail below in regard to glycolipids. Mutations in the worm, some of them affecting glycosylation processes, have been found that affect colonization by these bacteria. Examples of such mutants include those with defects in *bus* genes encoding putative glycosyltransferases.

An especially interesting set of mutations are the *srf* mutants (altered surface antigenicity mutants). Some of the *srf* mutants were identified by altered antibody or lectin binding to the cuticle, indicating that loss of cuticle components exposed new antigens. *srf-3* mutants are resistant to infection by *M. nematophilum* and lack some O-linked glycoconjugates containing glucuronic acid and galactose as well as reduced levels of some N-glycans. *srf-3* encodes a nucleotide sugar transporter that can transport both UDP-galactose and UDP-*N*-acetylglucosamine, suggesting that altered sugar composition of the cuticle resulting from mutations in this transporter confer resistance to *M. nematophilum*. Interestingly, there are 18 putative nucleotide sugar transporters (Chapter 5) in the genome of *C. elegans*, which is a considerably larger number than the known nucleotide sugars (UDP-galactose, UDP-glucose, UDP-*N*-acetylglucosamine, UDP-*N*-acetylgalactosamine, UDP-xylose, GDP-mannose, and GDP-fucose), suggesting possible functional overlap in these transporters.

Proteoglycans and Glycosaminoglycans

During egg laying, fertilized eggs must pass through the vulva, which is a simple tubular structure that links the gonads with the external cuticle. During postembryonic development, vulva morphogenesis arises through the invagination of a single layer of epithelial cells. Using mutagenesis, several mutations that perturb invagination of the vulva were identified (designated *sqv* or squashed vulva). In the original screen, 25 mutations were identified in eight genes named *sqv-1* through *sqv-8*. All of the mutations produced a similar phenotype: that is, partial collapse of vulval invagination, elongation of the central vulval cells, hermaphrodite sterility associated with maternal-effect lethality, and cytokinesis defects in the early embryo. All eight *sqv* genes show homology with vertebrate enzymes that are involved in the biosynthesis of GAGs (Figure 25.5). *sqv-1, sqv-4,* and *sqv-7* encode proteins that have roles in nucleotide sugar metabolism and transport. The SQV-7 nucleotide transporter was the first example of a carrier that could import more than one nucleotide sugar into the Golgi (Chapter 5). SQV-4 and SQV-1 proteins represent sequential enzymes involved in the formation of UDP-glucuronic acid and UDP-xylose, respectively, showing that the *sqv* mutations most likely affect GAG synthesis. Biochemical analysis of *sqv-6*, *sqv-3*, *sqv-2*, and *sqv-8* showed that they encode worm orthologs of the vertebrate transferases required for the assembly of the linkage region tetrasaccharide common to HS and chondroitin. Finally, characterization of *sqv-5* showed that it encodes the chondroitin synthase, which acts in concert with the chondroitin polymerizing factor encoded by the *mig-22* gene. Thus, the various phenotypes (failed invagination of the epithelial layer that forms the vulva, maternal-effect lethality, and cytokinesis defects) result from defective chondroitin formation.

The requirement for chondroitin assembly in seemingly disparate systems may result from biophysical changes in the lumen of the vulva or between the eggshell and the embryo. One idea is that the high negative charge imparted by the glucuronic acids in chondroitin attracts counterions that raise the local osmolarity, causing a swelling pressure. Another possibility is that the chondroitin acts as a physical scaffold bound to the cell membrane or eggshell. Interestingly, the *sqv* screen did not detect mutations affecting genes that encode proteoglycan core proteins on which the chondroitin chains assemble, but proteomic analysis has led to the identification of nine novel chondroitin sulfate proteoglycan (CPG) core proteins modified with chondroitin chains (Figure 25.6). Two of these (CPG-1 and CPG-2) contain chitin-binding domains that presumably allow the proteoglycans to interact with chitin in the eggshell, thus positioning the proteoglycans between the eggshell and the plasma membrane of the embryo, in which they could serve as spacers or osmotic regulators. Silencing *cpg-1* and *cpg-2* expression by RNAi recapitulates the cytokinesis defect observed in *sqv* mutants, suggesting that these are the relevant proteoglycans. The proteoglycans involved in epithelial invagination have not yet been determined.

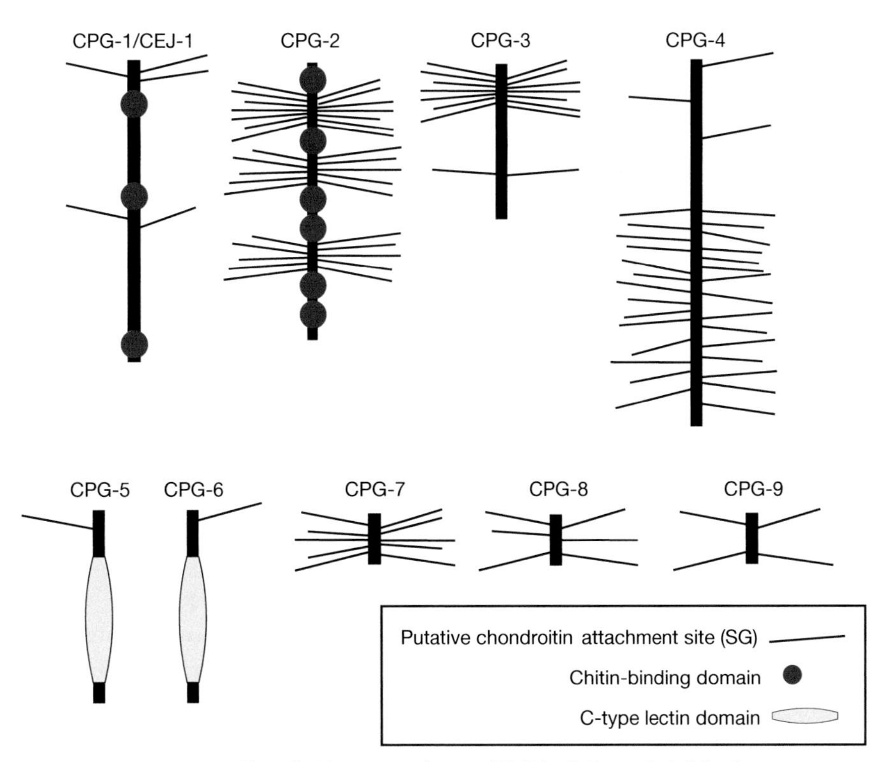

FIGURE 25.6. Chondroitin proteoglycans (CPGs) of *Caenorhabditis elegans*.

HS biosynthesis in *C. elegans* follows the same pattern observed in vertebrate systems (Chapter 17). Mutations in the pathway for HS biosynthesis are lethal in *C. elegans*. Two of the key genes involved in this pathway are *rib-1* and *rib-2*, homologs of the vertebrate genes *Ext2* and *Ext1*, respectively, which catalyze the polymerization of the backbone of HS chains (GlcAβ1-4GlcNAcα1-4) (Chapter 17). Mutants in *rib-2*, the worm homolog of *Ext1*, have defects in development and egg laying. The worm genome also contains a single gene for glucuronic acid C-5 epimerase (*hse-5*) and five genes for sulfotransferase activities (GlcNAc N-deacetylase/N-sulfotransferase [Ndst], *hst-1*; uronyl 2-O-sulfotransferase, *hst-2*; 3-O-sulfotransferases, *hst-3.1 and hst-3.2*; and 6-O-sulfotransferase, *hst-6*), all of which are homologs of vertebrate genes involved in HS synthesis. In contrast, vertebrates contain four Ndsts, three 6-O-sulfotransferases, and seven 3-O-sulfotransferases. Although mutations in the epimerase (*hse-5*) and the sulfotransferases (*hst-6*, *hst-2*) do not affect viability, they cause defects in specific cell migration, axonal outgrowth, and/or neurite branching. Consistent with this finding, inactivation of the cell surface HS proteoglycan syndecan (sdn-1) affects neural migration and axonal guidance. *C. elegans* also produces two GPI-anchored HS proteoglycans. LON-2, a member of the glypican family, negatively regulates a bone morphogenetic protein-like signaling pathway that controls body length in *C. elegans*. Worms also contain a homolog of the vertebrate basement membrane proteoglycan perlecan (encoded by *unc-52*). At least three major classes of UNC-52 isoforms are produced through alternative splicing, and distinct spatial and temporal expression patterns occur throughout development. In keeping with the "uncoordinated" phenotype, *unc-52* mutants affect myofilament assembly in body-wall muscle during embryonic development. Thus, as in vertebrates, HS proteoglycans mediate many fundamental processes during development and in the adult animal.

Glycolipids

The arthro-series glycolipids in *C. elegans* include structures modified with either phosphorylcholine or fucose residues (Figure 25.7A). Recent studies on the resistance of *C. elegans* to bacterial toxins (Chapter 37) have led to interesting insights into the structures and functions of glycolipids. Bt crystal toxins are used in both transgenic and organic farming because of their ability to kill insect pests. Bt is toxic to *C. elegans*, but following mutagenesis, strains resistant to one crystal toxin (Cry5B) were identified and classified as *bre-1* through *bre-5*—which encode either an enzyme required for GDP-fucose biosynthesis or glycolipid-specific glycosyltransferases. None of the strains showed altered development, but they were highly resistant to Bt, whereas the Cry5B-resistant mutants had truncated glycolipids lacking terminal fucose. Therefore, the Cry5B ligands in the intestinal epithelium are concluded to be the largest *C. elegans* glycolipids (see also Figure 25.7 for the relevant enzymatic steps defective in *bre* mutants).

GLYCAN-BINDING PROTEINS IN *C. ELEGANS*

Although the *C. elegans* genome encodes a number of predicted glycan-binding proteins (GBPs), only a few of them have been characterized biochemically or explored by genetic manipulation. The first GBP found in *C. elegans* was a galectin (Chapter 36), which was isolated by affinity-chromatography and sequenced in 1992. This was surprising because until this observation, galectins were thought to be expressed only in vertebrates. Amazingly, the *C. elegans* genome encodes 28 putative galectins, nearly twice as many as in humans. Only two of these proteins have been studied in detail: a tandem-repeat 32-kDa galectin (LEC-6) and a prototypical 16-kDa galectin (LEC-1). Both galectins can bind to galactose-containing ligands.

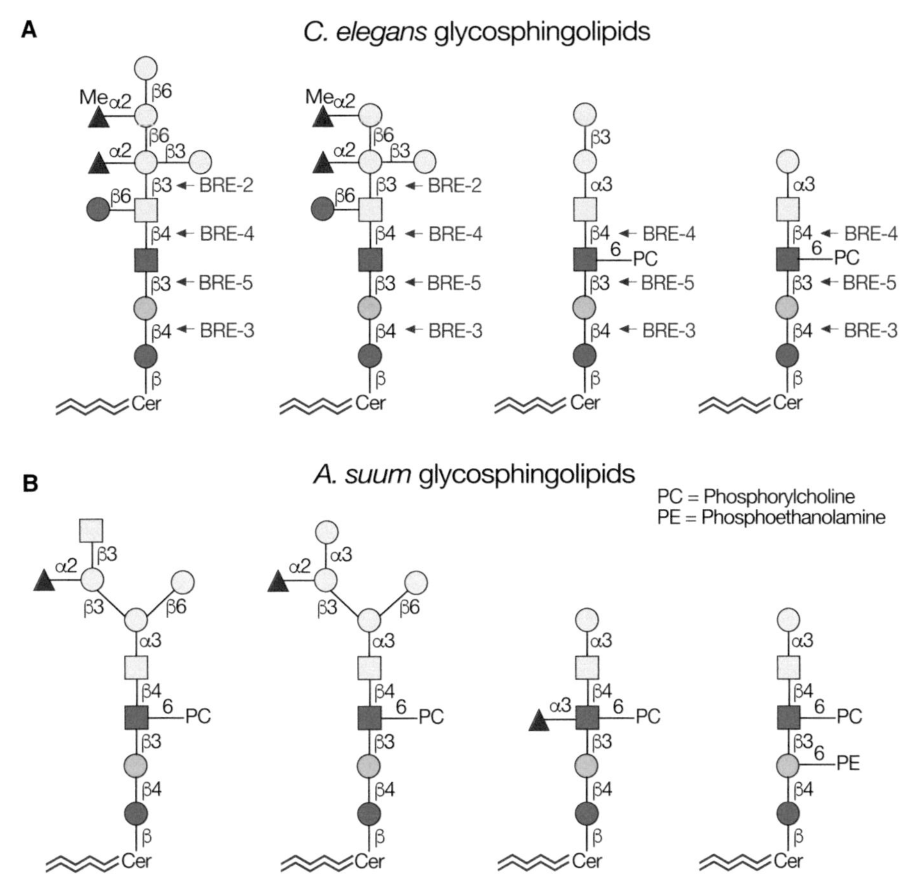

FIGURE 25.7. Examples of nematode glycolipids. (*A*) Structures of glycolipids from *Caenorhabditis elegans.* Mutations that result in truncated glycolipids have been identified as *bre* mutants, as described in the text; the steps absent in these mutants are catalyzed in wild-type worms by the indicated BRE enzymes. (*B*) Structures of glycolipids from *Ascaris suum.*

Some 283 *clec* genes have been identified in the *C. elegans* genome that encode proteins with C-type lectin domains (CTLDs) (Chapter 34), whereby some proteins have multiple CTLDs. Only 19 of these CTLDs have the sequence hallmarks predicting carbohydrate recognition. In contrast to the CTLD-containing proteins in vertebrates, most of the proteins with CTLDs in *C. elegans* have signal sequences and no transmembrane domains, indicating that they are secreted proteins, but their functional roles have not yet been studied in detail in this organism. However, as expression of several of these CTLDs is up-regulated on challenge of the animal with nematocidal Bt strains and other pathogenic bacteria; these lectins may well have roles in the innate immune system of *C. elegans.*

GLYCOBIOLOGY OF OTHER NEMATODES

Studies in the nonparasitic nematode *C. elegans* have been incredibly rewarding because of the ease of genetic manipulation and culture. Much less is known about parasitic nematodes, which cause tremendous death and suffering in animals and people throughout the world. It might be expected that *C. elegans* and other nematodes share much in common in terms of glycoconjugate structures and biosynthesis, but each nematode has differences in

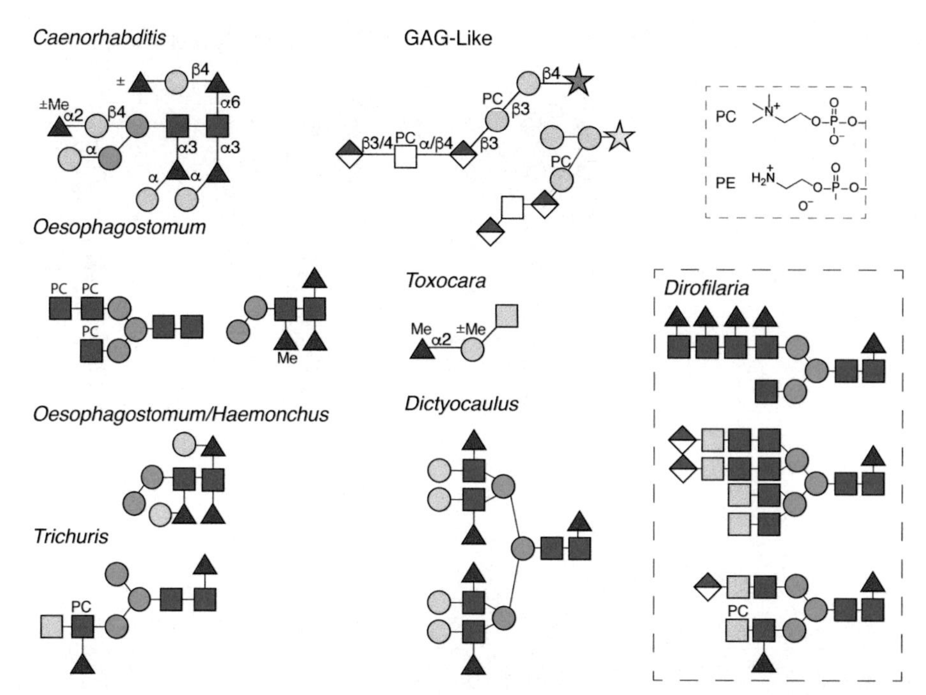

FIGURE 25.8. Examples of nematode glycans. Caenorhabditis core modified N-glycan; *Oesophagostomum* and *Haemonchus* N-glycans; *Trichuris* N-glycan with fucose/PC-modified LacdiNAc; PC-modified glycosaminoglycans from *Oesophagostomum*; *Toxocara* O-glycan; *Dictyocaulus* Lewis-x-modified N-glycan; selected fucosylated and glucuronylated N-glycans from *Dirofilaria*. (For a key to symbols used for monosaccharides, use the SNFG link.)

glycans compared with *C. elegans.* These may pertain to their virulence and parasitic requirements.

Some of the major parasitic nematodes that have been studied in terms of glycoconjugates include *Acanthocheilonema viteae, Ascaris suum, Dictyocaulus viviparous, Dirofilaria immitis, Haemonchus contortus, Oesphagostomum dentatum, Onchocerca volvulus, Trichuris suis, Trichinella spiralis, Toxocara canis,* and *Toxocara cati* (Chapter 43). *A. suum* is a parasitic intestinal nematode of pigs. Like *C. elegans,* the *A. suum* N-glycans are paucimannose-rich and contain phosphorylcholine and core fucose residues. In contrast, the N-glycans of the parasitic nematode of deer *Parelaphostrongylus tenuis* are extensively terminally modified with galactose and carry the terminal structure Galα1-3Galβ1-4GlcNAc-R. The cattle parasite *D. viviparous* has N-glycans with Lewis antigens (see Chapter 14), including Lewis x. The sheep parasite *H. contortus* synthesizes N-glycans containing the fucosylated LacdiNAc antigen GalNAcβ1-4(Fucα1-3)GlcNAc-R (Chapter 43), whereas the dog heartworm *D. immitis* has N-glycans with extending chito-motifs and/or with terminal glucuronic acid. Most nematode N-glycomes reported to date contain phosphorylcholine-modified structures. There are few data on O-glycans of nematode parasites other than short O-glycans in Toxocara or phosphorylcholine-modified GAG-like structures in *O. dentatum* (Figure 25.8).

Nematodes also make unusual glycolipids, and it is likely that each type of nematode synthesizes different glycolipid structures. One of the best-studied nematodes in terms of glycolipids is *A. suum.* Many of its glycolipids, which also contain the arthroseries core, have galactose and fucose modifications, in addition to phosphorylcholine and phosphorylethanolamine (Figure 25.7B). Virtually nothing is known about the genetics regulating glycosylation in these parasitic nematodes, as most studies have focused on the analysis of glycan structures; on the other hand, the application of glycan microarray technology is starting to reveal ligands for proteins of mammalian innate immune systems.

ACKNOWLEDGMENTS

The authors acknowledge contributions to the previous version of this chapter from Jeffrey D. Esko and helpful comments and suggestions from Todd Lowary and Hiroshi Nakato.

FURTHER READING

Brenner S. 1974. The genetics of *Caenorhabditis elegans*. *Genetics* **77**: 71–94. doi:10.1093/genetics/77.1.71

Drickamer K, Dodd RB. 1999. C-Type lectin-like domains in *Caenorhabditis elegans*: predictions from the complete genome sequence. *Glycobiology* **9**: 1357–1369. doi:10.1093/glycob/9.12.1357

Oriol R, Mollicone R, Cailleau A, Balanzino L, Breton C. 1999. Divergent evolution of fucosyl-transferase genes from vertebrates, invertebrates, and bacteria. *Glycobiology* **9**: 323–334. doi:10.1093/glycob/9.4.323

Dodd RB, Drickamer K. 2001. Lectin-like proteins in model organisms: implications for evolution of carbohydrate-binding activity. *Glycobiology* **11**: 71R–79R. doi:10.1093/glycob/11.5.71r

Hirabayashi J, Arata Y, Kasai K. 2001. Glycome project: concept, strategy and preliminary application to *Caenorhabditis elegans*. *Proteomics* **1**: 295–303. doi:10.1002/1615-9861(200102)1:2<295::aid-prot295>3.0.co;2-c

Schachter H. 2004. Protein glycosylation lessons from *Caenorhabditis elegans*. *Curr Opin Struct Biol* **14**: 607–616. doi:10.1016/j.sbi.2004.09.005

Olson SK, Bishop JR, Yates JR, Oegema K, Esko JD. 2006. Identification of novel chondroitin proteoglycans in *C. elegans*: embryonic cell division depends on CPG-1 and CPG-2. *J Cell Biol* **173**: 985–994. doi:10.1083/jcb.200603003

Shi H, Tan J, Schachter H. 2006. N-glycans are involved in the response of *Caenorhabditis elegans* to bacterial pathogens. *Methods Enzymol* **417**: 359–389. doi:10.1016/s0076-6879(06)17022-6

Antoshechkin I, Sternberg PW. 2007. The versatile worm: genetic and genomic resources for *Caenorhabditis elegans* research. *Nat Rev Genet* **8**: 518–532. doi:10.1038/nrg2105

Barrows BD, Haslam SM, Bischof LJ, Morris HR, Dell A, Aroian RV. 2007. Resistance to *Bacillus thuringiensis* toxin in *Caenorhabditis elegans* from loss of fucose. *J Biol Chem* **282**: 3302–3311. doi:10.1074/jbc.m606621200

Laughlin ST, Bertozzi CR. 2009. In vivo imaging of *Caenorhabditis elegans* glycans. *ACS Chem Biol* **4**: 1068–1072. doi:10.1021/cb900254y

Gravato-Nobre MJ, Stroud D, O'Rourke D, Darby C, Hodgkin J. 2011. Glycosylation genes expressed in seam cells determine complex surface properties and bacterial adhesion to the cuticle of *Caenorhabditis elegans*. *Genetics* **187**: 141–155. doi:10.1534/genetics.110.122002

Wohlschlager T, Butschi A, Grassi P, Sutov G, Gauss R, Hauck D, Schmieder SS, Knobel M, Titz A, Dell A, et al. 2014. Methylated glycans as conserved targets of animal and fungal innate defense. *Proc Natl Acad Sci* **111**: E2787–E2796. doi:10.1073/pnas.1401176111

Yan S, Brecker L, Jin C, Titz A, Dragosits M, Karlsson NG, Jantsch V, Wilson IB, Paschinger K. 2015. Bisecting galactose as a feature of N-glycans of wild-type and mutant *Caenorhabditis elegans*. *Mol Cell Proteomics* **14**: 2111–2125. doi:10.1074/mcp.m115.049817

Jiménez-Castells C, Vanbeselaere J, Kohlhuber S, Ruttkowski B, Joachim A, Paschinger K. 2017. Gender and developmental specific N-glycomes of the porcine parasite *Oesophagostomum dentatum*. *Biochim Biophys Acta* **1861**: 418–430. doi:10.1016/j.bbagen.2016.10.011

Martini F, Eckmair B, Štefanić S, Jin C, Garg M, Yan S, Jiménez-Castells C, Hykollari A, Neupert C, Venco L, et al. 2019. Highly modified and immunoactive N-glycans of the canine heartworm. *Nat Commun.* **10**: 75. doi:10.1038/s41467-018-07948-7

26 Arthropoda

Kelly G. Ten Hagen, Hiroshi Nakato, Michael Tiemeyer, and Jeffrey D. Esko

This chapter describes glycosylation in the Arthropoda, focusing primarily on *Drosophila melanogaster*. The major glycan classes are similar to those described in vertebrates, with interesting differences. The powerful genetic systems available for studying gene function in *D. melanogaster* have proved to be effective means for understanding glycan function in early development and have provided some of the first examples of how glycans affect growth factor signaling, morphogen gradients, protein secretion, and neural function in vivo.

HISTORICAL PERSPECTIVE

Arthropods are among the most successful species on Earth and are found in all types of environments. One of their characteristic features is an exoskeleton composed of chitin, which provides support and physical protection. The best-studied example is the fruit fly, *D. melanogaster*. In 1910, T.H. Morgan published the first paper about the genetics of *D. melanogaster*, which showed that white eye color was a sex-linked trait. Since then, this organism has been the predominant model organism for genetic analysis in animals. Its advantages include an easily studied developmental program, a sequenced and actively annotated genome, a relatively

published by Cold Spring Harbor Laboratory Press; doi: 10.1101/glycobiology.4e.26

complex neural system, and the ability to discern literally thousands of different phenotypes in morphology, development, and behavior.

In pursuit of genes that regulate development, many Drosophila geneticists ran head-on into glycans. New analytical techniques have expanded the appreciation for the glycan synthetic capacity of the organism and have helped establish links between interesting phenotypes and altered glycan expression. Some of these associations have proven to be common across species and others are unique to *D. melanogaster*. Given the complexity of glycosylation in Drosophila, it is impossible to cover all aspects of glycans and glycan-binding proteins in this organism. Instead, we provide an overview of the major classes of arthropod glycans and examples of how studying glycans in Drosophila can lead to new discoveries that impact vertebrate as well as invertebrate biology.

INSECT GLYCOPROTEINS

N-Linked Glycan Assembly and Diversity

Although it was once thought that arthropod glycoproteins were exclusively of the high-mannose or paucimannose type (Chapter 9), annotation of the *D. melanogaster* genome predicted the existence of the enzymatic machinery needed to generate hybrid and complex glycans. Moreover, improved analytic techniques allowed the detection of very minor glycans. The commercial and experimental demand for eukaryotic expression systems led to the characterization of the glycosylation pathways of cells derived from the moth *Spodoptera frugiperda* (Sf9 cells) and from *D. melanogaster* (S2 cells). It is now clear that high-mannose and paucimannose glycans account for >90% of the total N-linked glycan diversity in Drosophila and other insects throughout their life cycles. However, hybrid and complex glycans, including sialylated, sulfated, glucuronylated, and zwitterionic structures, are also present, albeit as minor components (Figure 26.1).

Drosophila adds fucose (Fuc) in both α1-3 and α1-6–linkages to the reducing terminal *N*-acetylglucosamine, whereas vertebrates restrict this linkage to α1-6. Fucα1-3GlcNAc is immunogenic in humans and rabbits, resulting in the production of antibodies against the so-called horseradish peroxidase (HRP) epitope. Anti-HRP antibodies show that the Fucα1-3GlcNAc epitope is restricted primarily to neural tissue in a broad range of arthropods. Drosophila does not extend its core fucose residues with additional capping monosaccharides nor modify its N-linked glycans by O-methylation, both of which occur extensively in *Caenorhabditis elegans* (Chapter 25). The demonstration of fucosylated, sialylated, sulfated, hybrid, biantennary complex, and triantennary complex glycans in all stages of the *D. melanogaster* life cycle makes the diversity of the arthropod N-glycans generally comparable to that of mammals (Chapter 27), except for the very limited use of sialic acids (Sias).

The relative paucity of complex N-linked glycans in Drosophila has been attributed at least partially to the presence of a hexosaminidase in the secretory pathway. Encoded by the *fused lobes* (*fdl*) gene, the enzyme is capable of efficiently removing *N*-acetylglucosamine residues that are added by N-acetylglucosaminyltransferase I (GlcNAcT-I encoded by *Mgat1*) to the nonreducing terminal Manα1-3 arm of the $Man_5GlcNAc_2$ core glycan. GlcNAcT-I catalyzes the first committed step toward the generation of hybrid or complex glycans (Chapter 9). Therefore, removal of this *N*-acetylglucosamine residue effectively blocks subsequent extension reactions, generating the observed predominance of high-mannose or paucimannose glycans on glycoproteins.

The presence of hybrid and complex glycans in Drosophila predicts the existence of specific enzymes that act on acceptor substrates that have escaped trimming by Fdl. For instance, α1-6 fucosyltransferase, GlcNAcT-I, -II, -III, -IV, galactosyltransferase (GalT), and sialyltransferase (SiaT) activities are required to generate structures more complex than paucimannose glycans.

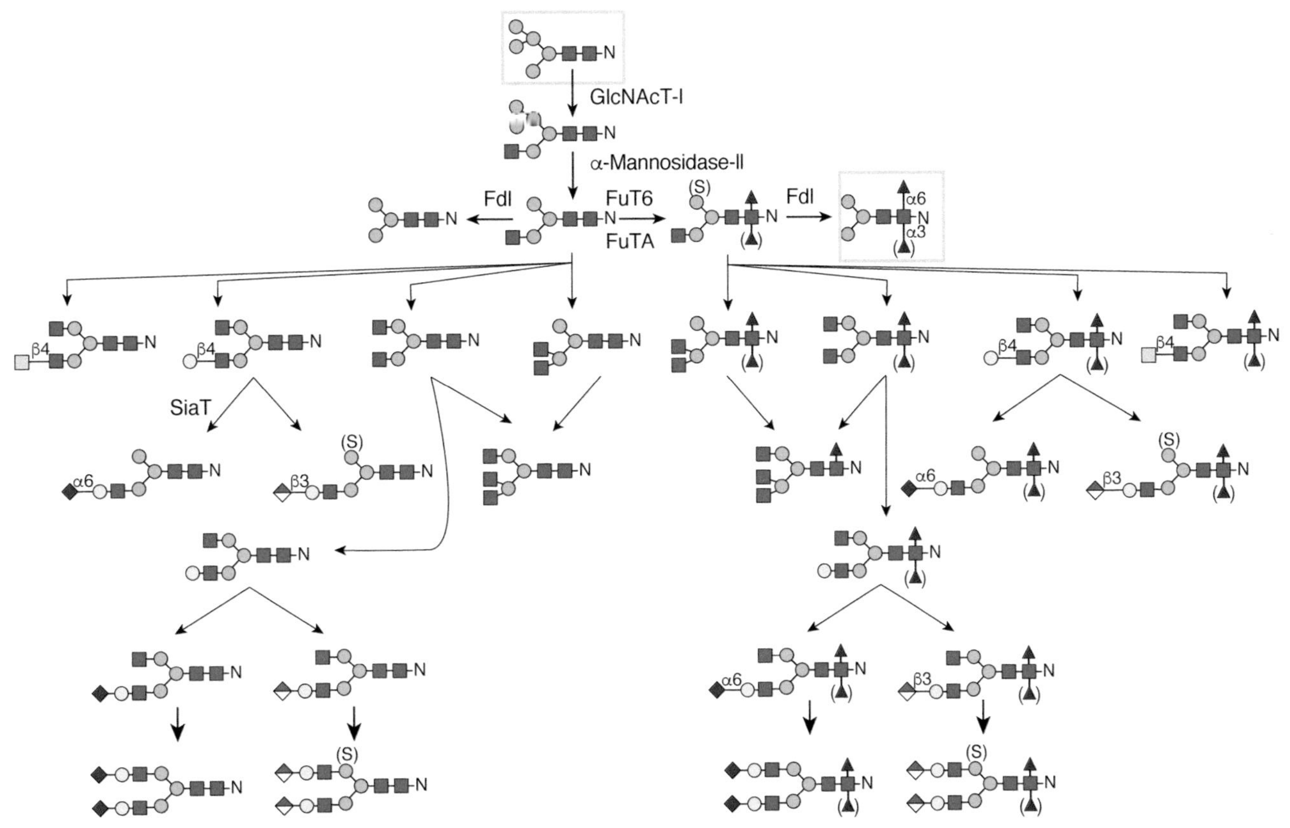

FIGURE 26.1. N-Linked glycan diversity in Drosophila and other insects. N-Linked glycan processing after endoplasmic reticulum (ER) mannosidase trimming to the $Man_5GlcNAc_2$ structure is shown. (*Gray boxes*) The predominant N-linked glycans, $Man_5GlcNAc_2$ and $Man_3GlcNAc_2Fuc$, are found in all stages of Drosophila. N-linked glycan complexity is limited by the Fdl hexosaminidase and expanded by expression of uncharacterized branching N-acetylglucosaminyltransferase activities, undefined galactosyl-, *N*-acetylgalactosaminyl-, and glucuronosyltransferase activities, and a single identified sialyltransferase activity. Core fucosylation occurs at C-6 and/or at C-3 of the reducing terminal *N*-acetylglucosamine residue, catalyzed by FucT6 or FucTA, respectively. Fucosylation at C-3 generates the neural-specific horseradish peroxidase (HRP) epitope. Sulfation of the α1-6-linked core mannose (Man) residue has been detected on all N-linked glycan classes except sialylated glycans in Drosophila and other insects.

Of these enzymes, only GlcNAcT-I and the single SiaT are reasonably well-characterized in Drosophila. Genome annotations predict the existence of the other enzymes, although their activity and expression have not been characterized.

Phenotypes have been described for mutations in key N-linked glycan processing and synthetic enzymes (Table 26.1). Loss of the Golgi-trimming enzyme, α1-2-mannosidase (Golgi mannosidase I, *mas1*) has little or no impact on the processing of high-mannose glycans, which led to the identification of an alternative mannosidase activity that effectively bypasses loss of mannosidase I. However, the embryonic peripheral nervous system, the wing, and the adult eye show mild alterations in the Drosophila *mas1* mutant, suggesting that the bypass is not complete in all tissues, similar to the tissue-specific bypass identified in mouse following targeted disruption of mouse α-mannosidase (*mGMII*, Golgi mannosidase II) (Chapter 9). To date, relatively few mutations in genes affecting core N-linked glycosylation have been reported in Drosophila. One exception is the glucosyltransferase Alg5, which adds glucose (Glc) to the Dol-P-$GlcNAc_2Man_9$ precursor. Mutations in the Drosophila Alg5 gene, known as *wollknäuel* (*wol*), show deficient protein secretion, extracellular matrix deposition, and cuticle elaboration, phenotypes that reflect inefficient protein glycosylation and altered endoplasmic reticulum (ER) function.

TABLE 26.1. Genes that affect the synthesis or function of Drosophila glycans

Gene	Protein	Function	Mutant phenotype
N-linked or O-linked pathways			
PMM2	*phosphomannomutase 2*	generation of Man-6-P for GDP-Man biosynthesis	decreased viability, locomotion, NMJ defects
wollknäuel (wol)	Drosophila *alg-5*, glucosyltransferase	glucosylation of dolichol phosphate to generate Dol-P-Glc	decreased protein secretion, altered extracellular matrix deposition
tumorous imaginal discs (tid)	Drosophila *alg-3*, mannosyltransferase	addition of α1-3 Man to $Man_5GlcNAc_2$-P-P-Dol precursor	neoplastic imaginal disc overgrowth
xiantuan	Drosophila *alg-8*, glucosyltransferase	addition of α1-3 Glc to $Glc_1Man_9GlcNAc_2$-P-P-Dol precursor	defective gastrulation
mas1	α1-2-mannosidase I	removal of mannose to generate $Man_5GlcNAc_2$-N	peripheral nervous system, wing, and eye morphology
fused lobes (fdl)	glycoprotein glycan processing β-hexosaminidase	removal of GlcNAc to generate $Man3GlcNAc_2$-N	larval and adult brain lobe morphology
Mgat1	GlcNAcT-I	generation of $GlcNAc_1Man_5GlcNAc_2$-N	reduced viability and locomotor activity
Tollo/toll-8	Toll-like receptor 8	decreased α1-3 core fucosylation (HRP-epitope)	altered glycosylation in embryonic central nervous system
neurally altered carbohydrate/GFR (nac/GFR)	GDP-Fuc transporter in the Golgi	involved in formation of fucosylated glycans (HRP-epitope)	altered glycosylation in larval, pupal, and adult central nervous system
FucTA	α3-fucosyltransferase	decreased α1-3 core fucosylation (HRP-epitope)	altered glycosylation in larval, pupal, and adult central nervous system
sugar-free frosting (sff)	Sff/SAD serine/threonine protein kinase	decreased α1-3 core fucosylation (HRP-epitope) and increased glycan complexity	altered glycosylation in embryonic nervous system, adult locomotor defects, and Golgi organization defects
SiaT	α6-sialyltransferase	N-linked glycan sialylation	adult locomotor defects, temperature-sensitive seizures, altered neuronal membrane excitability
Csas	CMP-sialic acid synthase	CMP-Sia production	adult locomotor defects, temperature-sensitive seizures, NMJ defects
pgant3	polypeptide GalNAcT	formation of GalNAc linked to Ser/Thr	extracellular matrix secretion and cell adhesion deficits
pgant4	polypeptide GalNAcT	formation of GalNAc linked to Ser/Thr	lethality; disrupted secretory granule formation, Golgi structure, and secretion
pgant5	polypeptide GalNAcT	formation of GalNAc linked to Ser/Thr	lethality; digestive system defects
pgant7	polypeptide GalNAcT	formation of GalNAc linked to Ser/Thr	lethality
pgant9	polypeptide GalNAcT	formation of GalNAc linked to Ser/Thr	altered secretory granule morphology
pgant35A	polypeptide GalNAcT	formation of GalNAc linked to Ser/Thr	lethality; altered trachea formation
C1GalTA	galactosyltransferase	addition of Gal to GalNAc-Ser/Thr	nervous system/NMJ defects
GlcAT-P	glucuronic acid transferase	formation of GlcA linked to Gal	nervous system/NMJ defects
rotated abdomen (rt)	protein-O-mannosyltransferase I	formation of Man linked to Ser/Thr	abnormal abdominal morphology

(*Continued*)

TABLE 26.1. (*Continued*)

Gene	Protein	Function	Mutant phenotype
twisted (tw)	protein-O-mannosyltransferase II	formation of Man linked to Ser/Thr	abnormal abdominal morphology
O-fut1	O-fucosyltransferase I	formation of Fuc linked to Ser/Thr	*Notch*-like defects in cellular differentiation
Efr	GDP-Fuc transporter in the ER, UDP-Xyl and UDP-GlcNAc transport also reported	required for protein O-fucosylation	*Notch*-like defects in cellular differentiation
fringe (fng)	Fuc-specific β3GlcNAcT	addition of *N*-acetylglucosamine to O-linked Fuc	pattern formation defects resulting from altered Notch activation
rumi	protein O-glucosyl and O-xylosyltransferase	addition of O-Glc and O-Xyl to Notch EGF repeats	*Notch*-like defects in cellular differentiation
shams	glucoside xylosyltransferase	addition of Xyl to Glc-O	Increased Notch signaling in some contexts
xxylt	xyloside xylosyltransferase	addition of Xyl to Xyl-Glc-O	Increased Notch signaling in sensitized genetic backgrounds
super sex combs (sxc)	O-GlcNAc transferase	addition of O-GlcNAc to nuclear and cytoplasmic proteins	defects in gene silencing
Eogt	O-GlcNAc transferase	addition of O-GlcNAc to extracellular proteins	larval lethality
Cog7	Cog7, member of COG retrograde trafficking complex	N-glycan processing	larval and adult locomotion defects, abnormal NMJ morphology
GAG/PG pathways			
sugarless	UDP glucose dehydrogenase	involved in chondroitin and heparan sulfate synthesis	Wg, Hh, Dpp signaling defects
slalom	PAPS transporter	involved in synthesis of all sulfated glycans	Wg, Hh signaling defects
fringe connection (frc)	UDP-sugar transporter	involved in synthesis of GAGs and O-linked glycans	Wg, Hh, FGF, Notch signaling defects
sauron (also called *rotini*)	Drosophila GOLPH3	involved in heparan sulfate, perhaps other glycans as well	mislocalization of Ttv, Botv, Sotv resulting in altered HS synthesis and Hh signaling
tout-velu (ttv)	heparan sulfate polymerase (Ext1)	synthesis of heparan sulfate	Wg, Hh, Dpp signaling defects
brother of tout-velu (botv)	N-acetyl-glucosaminyltransferase	initiation of heparan sulfate synthesis	Wg, Hh, Dpp signaling defects
sister of tout-velu (sotv)	heparan sulfate polymerase (Ext2)	synthesis of heparan sulfate	Wg, Hh, Dpp signaling defects
sulfateless (sfl)	N-deacetylase–N-sulfotransferase	sulfation of heparan sulfate	Wg, Hh, Dpp, FGF signaling defects
Hs2st	heparan sulfate 2-O-sulfotransferase	synthesis of heparan sulfate	FGF signaling defects
Hs6st	heparan sulfate 6-O-sulfotransferase	synthesis of heparan sulfate	FGF signaling defects
Hs3st-A	heparan sulfate 3-O-sulfotransferase	synthesis of heparan sulfate	impaired adult midgut homeostasis
Hs3st-B	heparan sulfate 3-O-sulfotransferase	synthesis of heparan sulfate	impaired adult midgut homeostasis
Sulf1/Sulfated	heparan sulfate 6-O-endosulfatase	synthesis of heparan sulfate	Wg, Hh, EGFR signaling defects
dally	glypican-related proteoglycan	GAG core protein involved in signaling	Dpp, Wg, Upd signaling defects
dally-like protein (dlp)	glypican-related proteoglycan	GAG core protein involved in signaling	Wg, Hh, D-LAR, Upd signaling defects
Syndecan	syndecan ortholog	GAG core protein involved in signaling	Slit-Robo, D-LAR signaling defects
terribly reduced optic lobes (trol)	perlecan ortholog	GAG core protein involved in signaling	Hh, FGF signaling defects

(*Continued*)

TABLE 26.1. *(Continued)*

Gene	Protein	Function	Mutant phenotype
Carrier of Wingless (Cow)	testican ortholog	GAG core protein involved in signaling	Wg signaling defects
kon-tiki/perdido (kon/perd)	neuron-glial antigen 2, CSPG4 ortholog	GAG core protein involved in extracellular matrix organization	defects in extracellular matrix organization and glial proliferation
Multiplexin (Mp)	collagen XV/XVIII ortholog	GAG core protein involved in signaling	defects in heart morphogenesis, motor axon pathfinding, and Slit/Robo, Wg signaling
windpipe (wdp)	transmembrane protein with Leucine-rich repeat (LRR) motifs	GAG core protein involved in signaling	Hh, Upd signaling defects
Chitin pathways			
cystic/mummy (cyst/mmy)	UDP-N-acetylglucosamine diphosphorylase	involved in chitin synthesis	tracheal morphology and axon guidance defects
krotzkopf verkehrt (kkv)	chitin synthase	chitin synthesis	tracheal morphology defects
serpentine (serp) and *vermiform (verm)*	multidomain proteins possessing chitin N-deacetylase domains	chitin synthesis	tracheal morphology defects
Glycosphingolipid pathways			
egghead (egh)	GlcCer-specific β4ManT	formation of Manβ1,4Glcβ1-Cer	*Notch*-like defects in cellular differentiation
brainiac (brn)	mactosylceramide-specific β3GlcNAcT	formation of GlcNAcβ3Manβ4Glcβ-Cer	*Notch*-like defects in cellular differentiation
β4GalNAcTA	β4GalNAcT, homology with vertebrate β4GalT	addition of GalNAc to arthroseries triaosyl glycosphingolipid	locomotor behavioral deficits
α4GT1	α4-N-acetyl-galactosaminyltransferase 1	addition of GalNAc to arthroseries tetraosyl glycosphingolipid	regulates Notch signaling by interaction with Serrate
Lectins			
Hemolectin (Hml)	chitin binding domain	chitin binding	hemolymph clotting
furrowed (fw)	furrowed, a C-type lectin	unknown binding preference	bristle and eye morphology defects, planar polarity defects
gliolectin (glec)	gliolectin, an unclassified carbohydrate-binding protein	binding preference for GlcNAc-terminated glycosphingolipids	axon pathfinding abnormalities, *Notch*-like phenotypes in wing
Imaginal disc growth factor 1-5 (Idgf1-5)	chitinase homology lacking key catalytic residues	unknown binding preference	cell proliferation and migration defects in imaginal tissues, impaired wound healing
mind the gap (mtg)	MTG, an unclassified carbohydrate-binding protein	binding preference for GlcNAc	disorganization of larval synaptomatrix at neuromuscular junction

Mutations in genes responsible for subsequent processing steps have begun to reveal the importance of glycan complexity in Drosophila. Reduction in the activity of the Fdl hexosaminidase results in altered brain structure. Brain lobes, normally separated in wild-type adults, are fused together through a continuous stalk at the midline in *fdl* mutants, hence, the original name of the mutation, *fused lobes* (Figure 26.2). In the wild-type adult, the separated lobes form portions of the mushroom body, a brain structure whose function has been implicated in Drosophila learning and memory. In an interesting convergence, a null mutation in *Mgat1* generates an apparently identical fused-lobes phenotype, although the glycan expression profile shifts from more complex in *fdl* mutants to high mannose and paucimannose in *Mgat1* mutants. It is still unclear how loss of complex glycans in one case (*Mgat1*) or enrichment of complex glycans in another case (*fdl*) generates the same neural phenotype. *Mgat1* mutants also show reduced locomotor activity and decreased life span. The decreased life span is rescued by

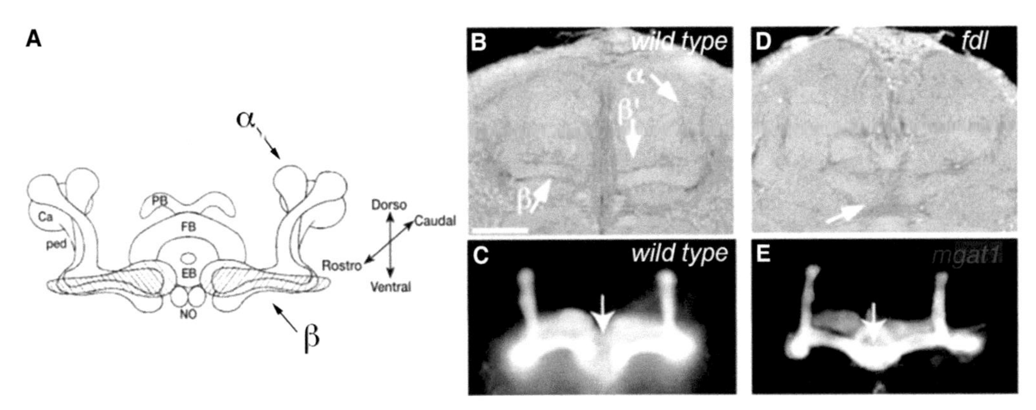

FIGURE 26.2. Mutations in enzymes that process complex N-linked glycans alter adult brain morphology in *D. melanogaster*. (*A*) The major lobes of the adult brain are shown in cross section. (*Hatched areas*) γ-lobes; (α,β) α- and β-lobes; (ped) peduncle; (Ca) calyx; (EB) ellipsoid body; (FB) fan-shaped body; (NO) noduli; (PB) protocerebral bridge. (*B,C*) Symmetrical sets of α- and β-lobes (*B*) are distinctly separated at the midline (*C, arrow*) in the wild-type adult. (*D*) A mutation in the *fused lobes* (*fdl*) gene, which encodes for a hexosaminidase that removes the *N*-acetylglucosamine added by GlcNAcT-I, yields increased expression of complex glycans and midline fusion of the β-lobes (*arrow*). (*E*) A loss-of-function mutation in GlcNAcT-I (*Mgat1*) decreases complex N-linked glycan expression, but it also fuses the β-lobes (*arrow*). Despite driving glycosylation in opposite directions, these two mutations yield convergent brain phenotypes. (*A,B,D*, Reprinted, with permission of Wiley-Liss, Inc., from Boquet I, et al. 2000. *J Neurobiol* **42:** 33–48; *C,E*, reprinted, with permission of the American Society for Biochemistry and Molecular Biology, from Sarkar M, et al. 2006. *J Biol Chem* **281:** 12776–12785.)

re-expression of the enzyme in neural tissue alone, indicating an essential role for complex glycosylation in neural functions that impact whole organism physiology.

A terminal synthetic step in the production of N-linked glycans in mammals is the addition of sialic acid residues. Genomic annotation and biochemical studies have shown that Drosophila possesses some, but not all, of the biosynthetic machinery necessary for the production of CMP-NeuAc. A single SiaT identified in Drosophila shows a two-fold preference for transferring sialic acid to a LacdiNAc (GalNAcβ1-4GlcNAc) acceptor compared to a type II LacNAc (Galβ1-4GlcNAc) acceptor. However, the only identified sialylated N-linked glycans in Drosophila possess subterminal LacNAc instead of LacdiNAc, despite the presence of LacdiNAc terminated N-linked glycans in the organism. SiaT is detected in a very small number of neurons beginning in late embryonic stages, but expands to a larger number of central nervous system (CNS) neurons in larval and adult stages. The relatively restricted expression of the SiaT enzyme reinforces proposals that minor glycans may be restricted to small subsets of cells in specific developmental or adult stages (Chapter 7). In fact, loss of glycoprotein sialylation results in explicit neurologic defects in adult flies, including behavioral abnormalities and temperature-induced seizures, indicating the importance of this minor modification of complex glycans for normal neurophysiological function.

O-Linked Glycan Assembly and Diversity

Insects add glycans in O-linkage to serine and threonine residues on secreted, cell-surface, and intracellular proteins. Structural complexity ranges from single monosaccharides (*N*-acetylgalactosamine, *N*-acetylglucosamine, mannose, glucose, or fucose) to extensively modified glycosaminoglycan chains) (see below). The core-1 structure (Galβ1-3GalNAcα-Thr/Ser), which is found extensively on vertebrate mucin-like proteins (Chapter 10), has also been described in insects (Figure 26.3). As much as 40% of the total mass of these proteins is contributed by glycans. In Drosophila tissues, peanut agglutinin lectin and mucin-specific antibodies reveal developmentally regulated and spatially restricted expression of Galβ1-3GalNAcα-*O*-protein moieties.

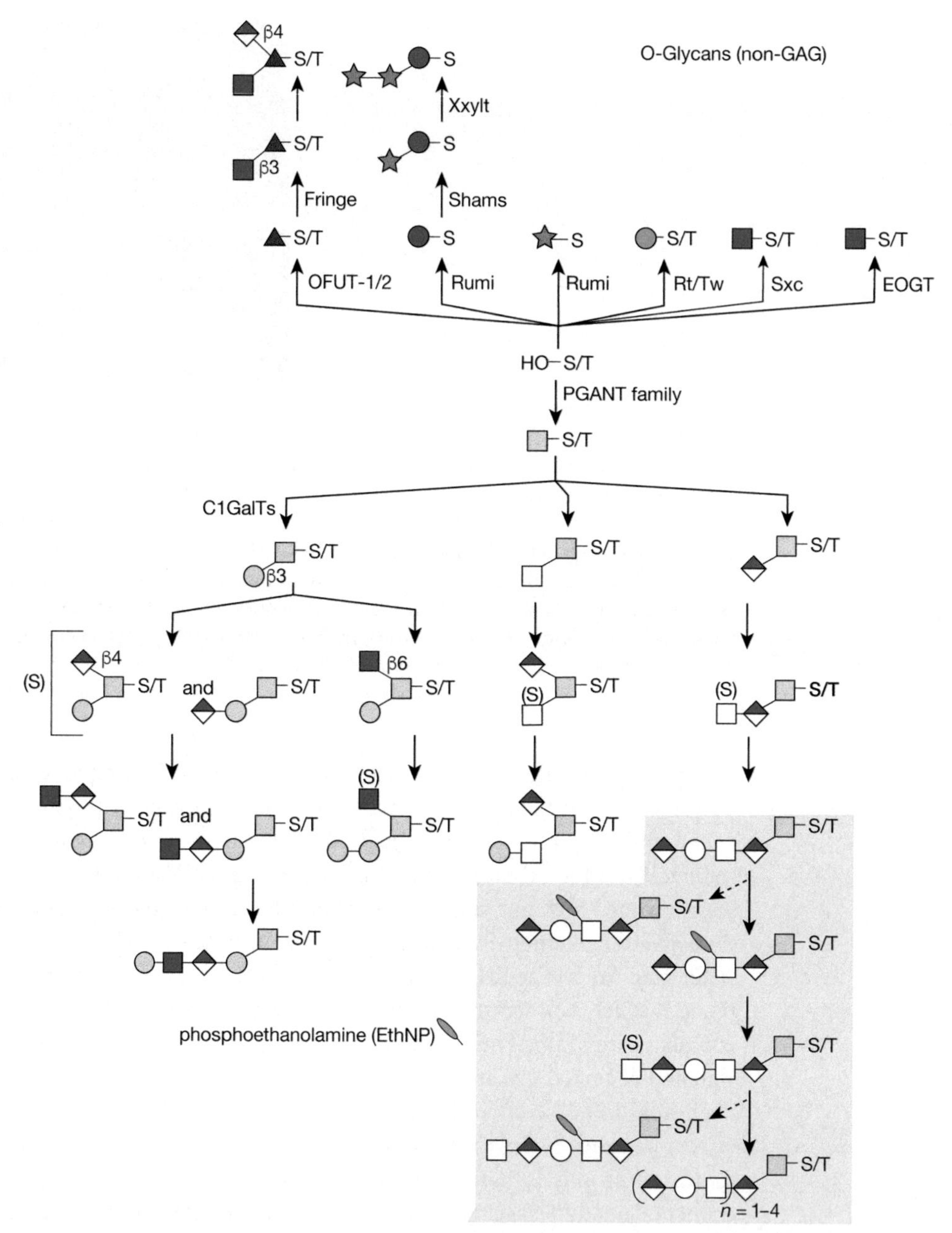

FIGURE 26.3. O-Linked glycan diversity in Drosophila and other insects. Genetic and biochemical analysis of Drosophila mutants were instrumental in defining developmental roles for domain-specific, simple mono-, di- or trisaccharide modifications on serine/threonine residues, including O-Fuc, O-Glc, O-Xyl, O-Man, and O-GlcNAc. A large family of ppGalNAcTs (*pgants*) in Drosophila indicates that arthropods, like vertebrates, express mucin-like glycans through the coordinated activity of an ensemble of partially redundant enzymes with yet-to-be-determined fine specificities. Core-1 mucin-type O-glycans predominate but core-2 structures are also present in Drosophila. Two additional cores define the known complexity of arthropod mucin-type O-glycans, HexNAc-GalNAc-O-Ser/Thr and GlcA-GalNAc-O-Ser/Thr. All detected mucin-type O-glycan cores are characterized by incorporation of glucuronic acid (GlcA), as capping, branching, or internal chain residues. (*Gray box*) Mosquito O-linked glycans of the GlcA-GalNAc-O-Ser/Thr core type are extended by GlcA-Hex-HexNAc repeats that can be modified with phosphoethanolamine (EthNP; *orange ovals*) on the first *N*-acetylhexosamine residue. Sulfation of *N*-acetylhexosamine residues is detected across multiple core types in Drosophila and other arthropod species, although the position of the sulfate moiety on the Drosophila glucuronylated core 1 disaccharide has not yet been assigned.

Drosophila possesses a family of 10 polypeptide N-acetylglucosaminyltransferases (ppGalNAcT/PGANTs encoded by *pgant* genes; Chapter 10). Systematic mutational and RNAi-mediated knockdown analysis of individual *pgant* genes showed that *pgant4*, *5*, *7*, and *35A* are essential for viability. Loss of *pgant3* impacts extracellular matrix secretion and integrin-mediated cell adhesion during development, and *pgant4* modulates proteolytic cleavage of substrates essential for secretory granule formation. Loss of *pgant4* also results in loss of the secreted peritrophic membrane of the digestive tract, leading to disruption of the progenitor cell niche. *pgant9* affects glycosylation of certain mucins and the morphology of secretory granules. Thus, mucin-type O-glycosylation in Drosophila regulates essential developmental programs and modulates trafficking through the secretory pathway, providing blueprints for studying similar functions in vertebrate systems.

Once initiated by ppGalNAcT/PGANT enzymes, the extension of mucin type glycans in Drosophila results in the production of three different core structures. Core 1 structures predominate, but core 2 glycans are also present, as well as an incompletely defined HexNAc-HexNAc core. A family of core 1 galactosyltransferases (C1GalTs) exists in Drosophila and the loss of certain members disrupts the nervous system and neuromuscular junctions (NMJ). Each core type can be modified by the addition of glucuronic acid (GlcA), either as a terminal or as a branching residue. Loss of the glucuronic acid transferase GlcAT-P results in the loss of glucuronic acid on core 1 structures and results in nervous system defects and alterations in the NMJs. Glucuronic acid is also detected as an internal residue within an extended chain, reminiscent of a glycosaminoglycan disaccharide structural unit (Chapter 17) or of the matriglycan repeat on mammalian dystroglycan (Chapter 13). In Drosophila and mosquito, sulfated mucin type glycans are also detected. Species-specific expression also exists with regard to the relative abundance of different core structures and the existence of specific post-synthetic modifications. For example, O-glycans can be decorated with phosphoethanolamine linked to HexNAc residues in mosquito and wasp species, but not in Drosophila.

Other types of O-linked glycan modifications are also present in insects. The discovery that O-GlcNAc decorates protein components of the Drosophila polytene chromosome added weight to initial demonstrations of the existence of nucleocytoplasmic glycosylation in animal cells (Chapters 18 and 19). Subsequently, this finding was reinforced by the observation that the gene encoding the O-GlcNAc transferase (OGT), was affected in super sex combs (*sxc*) mutants. The *sxc* locus is one of several polycomb group (PcG) genes that function as homeotic regulators of gene expression along the antero-posterior axis during Drosophila development. In a dramatic convergence of independent studies, major sites of PcG binding along the Drosophila genome were also shown to be major sites of genomic O-GlcNAc modification, indicating that addition of O-GlcNAc to DNA-binding proteins regulates their genomic binding specificity. Additionally, O-GlcNAc is also added to secreted and membrane proteins by the extracellular O-GlcNAc transferase EOGT. O-GlcNAc has been found on the extracellular matrix protein Dumpy, as well as Notch and Notch ligands. Loss of *Eogt* results in larval lethality.

O-linked mannose, a modification that contributes to the pathophysiology of some human muscular dystrophies (Chapters 13 and 45), is also present in Drosophila and is added to proteins by two transferase genes. Mutations in either of these two genes, *rotated abdomen* (*rt*) or *twisted* (*tw*), cause clockwise helical rotations of abdominal morphology in adult flies. The two genes interact such that the abdominal rotation observed in compound hypomorphic mutants is more severe than that of either single mutant. However, their alleles also show mutually epistatic interactions and their strong loss-of-function alleles are phenotypically indistinguishable, suggesting that Rt and Tw proteins function in a complex. Rt and Tw proteins must be expressed together in the same cell to achieve transfer of mannose onto dystroglycan, a physiologic acceptor for O-mannosylation.

Addition of a simple glycan, O-linked fucose, modulates complex developmental signals. The *Notch* gene was first identified and characterized in Drosophila as encoding a large, multimodular receptor protein involved in cell-fate determination. One characteristic of Notch is the presence of epidermal growth factor (EGF)-like modules. The EGF-like domain modules in the human Notch homolog have been shown to contain three domain-specific types of O-linked glycosylation (O-Glc, O-Fuc, and O-GlcNAc), which have been implicated in various steps in Notch signaling, including Notch folding, cleavage and interactions with its ligands, Delta and Serrate/Jagged (Chapter 13). As in mouse and humans, Drosophila possesses a protein O-glucosyltransferase, named Rumi, and two protein O-fucosyltransferases, O-fut1 and O-fut2. As in vertebrates, Drosophila O-fut1 transfers fucose to EGF-like domains and O-fut2 transfers fucose to thrombospondin-type repeats on different sets of proteins. Loss of O-fut1 in Drosophila yields *Notch* mutant phenotypes, indicating that fucosylation of Notch is essential for ligand-induced activation (Figure 26.4A–C). Loss of Rumi also produces Notch-like phenotypes, but in a temperature-sensitive manner, leading to the hypothesis that O-glucosylation is important for cleavage of Notch by appropriate processing enzymes. In Drosophila and vertebrate organisms, O-linked glucose residues are extended by the addition of xylose (Xyl) residues by the xylosyltransferases Shams and Xxylt (Figure 26.3). In addition, Rumi possesses dual nucleotide sugar donor specificity, allowing it to transfer either glucose or xylose to serine (but not threonine) residues in specific sequence contexts.

Fucosylated Notch is a substrate for an N-acetylglucosaminyltransferase encoded by the *fringe* gene. Elongation of O-linked fucose with *N*-acetylglucosamine yields Notch protein that is more efficiently activated by Delta than by Serrate/Jagged (Chapter 13). Therefore, O-linked glycosylation functions as a switch that activates the Notch receptor and alters its ligand preference. Differential Notch activation by cell-specific expression of O-fut1 and Fringe generates distinct cell fate choices in embryos, larvae, and pupae. The ability of the Drosophila O-fut1 and Fringe proteins to rescue or modify Notch protein or *Notch* mutant phenotypes is well-documented in cultured cells and in whole embryos, although the demonstration of O-Fuc or the GlcNAc-extended disaccharide on Notch protein extracted from Drosophila tissue has not yet been achieved. In mice and humans, but apparently not in Drosophila, the disaccharide can be extended by the addition of galactose (Gal) and then capped with sialic acid (Chapter 13). Drosophila embryos, larvae, and imaginal tissues also contain GlcNAcβ-3Fuc core structures branched by addition of a glucuronic acid to fucose (Figure 26.3). The functional importance of the glucuronyl branching in Drosophila is not yet known.

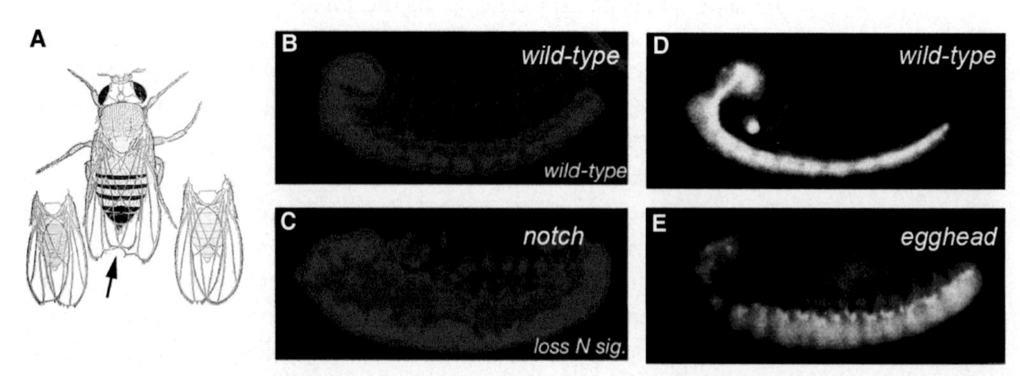

FIGURE 26.4. Cell fate choices dependent on Notch require appropriate glycan expression. (*A*) *Notch* mutations were originally identified based on aberrant wing morphology. Changes in cell fate generate wings that are notched at their margins (*arrow*). Wing notches arise from insufficient numbers of nonneural cells in the developing wing. In the embryo, loss of Notch signaling expands neural tissue at the expense of nonneural ectodermal cell fates. (*B,C*) Staining with a neuron-specific antibody reveals increased neural cell numbers in a *Notch* mutant (*C*) compared with *wild-type* (*B*). In comparison to *wild-type* (*D*), loss of the *egghead* or *brainiac* genes (*E*), which are essential for glycosphingolipid synthesis, results in neurogenic phenotypes similar to loss of Notch signaling. (*B,C*, Reprinted, with permission of the Company of Biologists, from Lai EC. 2004. *Development* **131:** 965–973; *D,E*, reprinted, with permission of the Company of Biologists, from Goode S, et al. 1992. *Development* **116:** 177–192.)

INSECT PROTEOGLYCANS AND GLYCOSAMINOGLYCANS

Glycosaminoglycan (GAG) Structure

Drosophila has the complete set of heparan sulfate (HS) biosynthetic and modifying enzymes found in mammalian species, and produces complex HS structures but has only one gene for each class of these enzymes. Disaccharide profiling of GAGs from Drosophila showed the presence of HS and chondroitin sulfate (CS) remarkably similar in structure to that of GAGs found in vertebrates (Chapter 17). The principal disaccharide species of HS-derived units include N-, 2-O, and 6-O sulfated forms, and mono-, di-, and trisulfated disaccharides. CS detected from whole embryos or larvae is largely unsulfated or 4-O-sulfated, but 6-O-sulfated disaccharides have been detected as well. The covalent attachment of GAG chains to proteins via the canonical tetrasaccharide linker (GlcA-Gal-Gal-Xyl) has also been shown for Drosophila. CS and HS have been documented in many other arthropod species, establishing the conservation of these macromolecules throughout this phylum. In contrast, Drosophila and presumably other arthropods do not produce hyaluronan (Chapter 16). Many of the HS-bearing proteoglycan core proteins known from vertebrate systems are also represented in Drosophila. Drosophila has a single Syndecan (*Sdc*) gene, two glypicans, division abnormally delayed (*dally*) and dally-like protein (*dlp*), testican (*Cow*), and perlecan, which is encoded by the terribly reduced optic lobe (*trol*) gene. Compared to HSPGs, CSPG core proteins are less well-conserved between species. Therefore, identifying new CSPGs cannot rely on the sequence homology to previously identified vertebrate CSPGs. Recently, glycoproteomic approaches have been successfully used to identify new CSPGs in Drosophila and *C. elegans.*

Mutations affecting GAG biosynthesis and modification have provided important insights concerning proteoglycan biosynthesis and development. For example, loss of the single N-deacetylase–N-sulfotransferase gene in Drosophila (*sfl*) results in essentially an unsulfated polymer, N-acetylheparosan, which causes defective patterning decisions orchestrated by several growth factors. Both *Ext1*- and *Ext2*-related genes, *ttv* and *sotv*, are required for HS polymerization in Drosophila. Mutations affecting the single *Hs2st* or *Hs6st* genes have particularly interesting effects. Loss of *Hs2st* eliminates 2-O sulfate groups as expected, but results in compensatory increases in 6-O-sulfation. The converse is also true: loss of *Hs6st* produces a polymer lacking 6-O-sulfate groups but 2-O-sulfation and N-sulfation increase, retaining the overall sulfation state of the polymer. Compensation between 2-O and 6-O modifications has also been observed in CHO cell mutants and mouse embryonic fibroblasts derived from *Hs2st* mutants. HS and CS biosynthesis are also linked in a compensatory fashion; mutations that reduce, but do not eliminate, Ext activity required for HS polymerization increase the net amount of chondroitin polymer formed, similar to observations made in vertebrate systems. These data suggest an important conserved mechanism for retaining the activity of proteoglycans in vivo.

Morphogen Signaling and Organ Size Control

Genetic studies in Drosophila showed that GAGs are critical for signaling mediated by a number of growth factors during development. HS proteoglycans serve as coreceptors for many growth factors, including Wingless (Wg; a Wnt ortholog), Decapentaplegic (Dpp; a BMP4 ortholog), fibroblast growth factor (FGF), Hedgehog (Hh), and Unpaired (Upd; a ligand of the Drosophila Jak/Stat pathway), and affect their distribution and signaling (Table 26.1). For example, patches of cells that express Dally at high levels showed markedly elevated levels of Dpp signaling in a cell-autonomous manner. Overexpression of Dally may increase Dpp protein levels on the cell surface by disrupting receptor-mediated internalization and degradation.

The ability of HS proteoglycans to alter the distribution and levels of growth factors in the matrix has biological significance because many of these secreted proteins are morphogens, secreted protein factors that show graded distributions across tissues and provide an essential mechanism for generating cell diversity during tissue assembly. Morphogen activity has been shown in vivo for Wg, Dpp, and Hh and Upd in Drosophila. HS proteoglycans not only affect their levels in tissues, but are also integrated into regulatory circuits of the morphogen systems. For example, both expression of a Dpp receptor (Thickveins) and Dally, a coreceptor for Dpp, are negatively regulated by Dpp signaling. This feedback system provides a mechanism to adjust cellular responses in the face of inappropriate reductions in Dpp signaling that might arise from genetic or environmental perturbations.

Recent studies have highlighted the role of the glypicans in organ size control in the context of evolutionary developmental biology. Halteres, which contribute to body balance during flight, are characteristic of two-winged Dipteran species, like Drosophila, and have evolved from the hindwings of four-winged ancestral species. This knob-like structure is much smaller than the ancestral wing from which it evolved. The size difference between the two appendages is controlled by a Hox gene, *Ultrabithorax* (*Ubx*), which regulates haltere identity. In the haltere disc, Ubx directly represses expression of the *dally* gene, limiting the bone morphogenetic protein (BMP) signaling and, thereby, organ size. Hox control of morphogen signaling through HS proteoglycan gene expression is one of the general mechanisms used across animal evolution to modify homologous organ structures.

Proteoglycans in the Stem Cell Niche

HS proteoglycans are an evolutionarily conserved, universal component of stem cell niches. In the ovarian germline stem cell (GSC) niche, Dally regulates stem cell number by acting as a niche size determinant. The cells that constitute the niche produce Dpp, which acts on a GSC that is in direct contact with a niche cell. One of the daughter cells arising from the division of a GSC will retain this contact and maintain its stem cell character. The other daughter cell, which loses this contact, will differentiate. Dally expressed in niche cells acts as a "trans" coreceptor, stabilizing Dpp and presenting Dpp to GSCs that are in direct contact with the niche cell (Figure 26.5). In *dally* mutants, Dpp signaling in GSCs is impaired and GSCs are lost to differentiation. In contrast, *dally* overexpression in a somatic cell population outside the niche results in ectopic activation of Dpp signaling in germ cells in this region, leading to an expansion of GSC-like cells. Thus, contact-dependent signaling by HS proteoglycans provides a mechanism to define the physical space of the niche and to control stem cell number.

Proteoglycans in Neural Development

The essential requirement for proteoglycans in neural development has also been illuminated by genetic analysis in Drosophila. Syndecan (Sdc) regulates axon guidance by modulating Slit-Robo signaling in the embryonic CNS. Dally-like (Dlp) also influences nervous system development but has distinct functions from Sdc. Sdc affects axon guidance decisions near the midline of the nervous system, whereas Dlp is required for fascicle formation at a distance from the midline. Trol regulates embryonic motor axon guidance by facilitating transmembrane signaling mediated by Semaphorin–Plexin interactions. HS proteoglycans also regulate synaptogenesis at the neuromuscular junction (NMJ). Both Sdc and Dlp bind to LAR, a receptor tyrosine phosphatase, which controls NMJ growth and active zone morphogenesis. As in axon guidance, these two proteoglycans have distinct functions: Sdc promotes the growth of presynaptic terminals, whereas Dlp regulates active zone formation. Trol regulates the formation of both pre- and postsynaptic structures by localizing Wg protein.

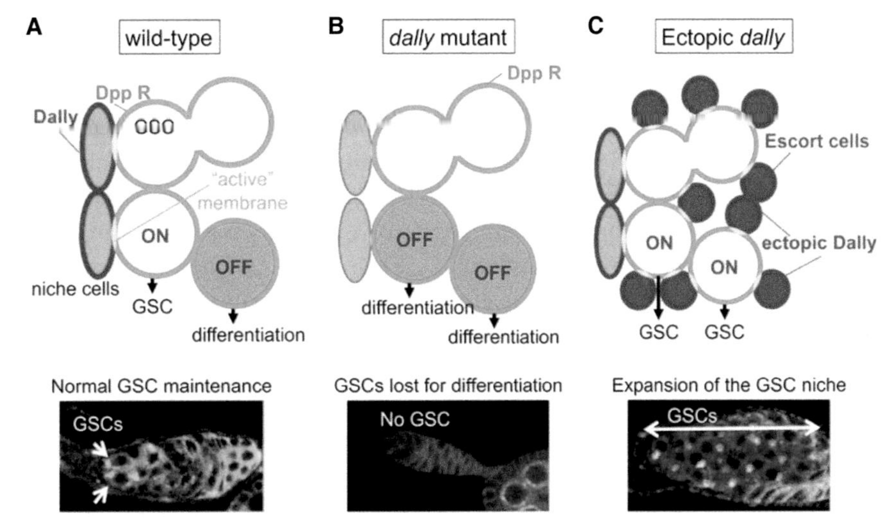

FIGURE 26.5. Glycosaminoglycans regulate the contact-dependent maintenance of germline stem cells (GSCs). (A) The niche cells (*blue*) express Dally on the cell surface (*red*) and germ cells express Dpp receptors (*green*). A dividing GSC is shown at the *top* and two daughter cells are shown at the *bottom*. Dpp signal transduction occurs (ON) where these molecules meet (*yellow*). Dpp signaling is not activated in GSCs that do not directly contact the niche cells (OFF), leading to differentiation. (B) In *dally* mutants, Dpp signaling is reduced and GSCs are lost to differentiation. This results in the loss of germ cells in a germarium. (C) When *dally* is ectopically expressed in a population of somatic cells outside the niche (*magenta*) adjacent to germ cells, Dpp signaling is activated at ectopic sites (*yellow*), resulting in expansion of the GSC niche and loss of differentiating cells. (Modified from Hayashi Y, et al. 2009. *J Cell Biol* **187:** 473–480.)

CHITIN

Chitin, a polymer of $(GlcNAc\beta4)_n$, is one of the most abundant biopolymers on Earth, second only to cellulose (which is interestingly a $(Glc\beta4)_n$ polymer synthesized in a similar manner by extrusion at the cell surface). Chitin is a major component of the rigid, cuticular exoskeleton of all Arthropoda, and therefore these animals expend considerable resources toward its assembly. Chitin fibrils also form a more subtle, protective layer at the apical surface of gut epithelial cells (the peritrophic membrane) and in the lumen of the forming tracheal system. Elegant genetic screens have shown that mutations in genes necessary for chitin polymerization, modification, and disassembly affect tracheal morphology (Table 26.1). The potential for chitin or for small chitin-based oligosaccharides to modulate cell signaling and morphogenesis remains unexplored in Drosophila.

INSECT GLYCOSPHINGOLIPIDS

Insects were once classified as "animals without gangliosides" in reference to sialylated glycosphingolipids (GSLs) found broadly distributed in animal families other than the arthropods (Chapter 11). However, arthropods possess their own family of GSLs, designated as the arthroseries. Sialylated GSLs (gangliosides) have not been found in Drosophila, but the arthroseries glycolipids contain glucuronic acid rather than sialic acid, and like many of the ganglioside sialic acids, the glucuronic acid is linked to a terminal galactose residue (Figure 26.6).

Vertebrates generally build GSLs on a lactosylceramide core (Galβ1-4Glcβ-ceramide). In contrast, the arthropods add a mannose residue to Glcβ-Cer to generate a core (Manβ1-4Glcβ-ceramide), called "mactosylceramide." Addition of the next two monosaccharides, *N*-acetylglucosamine in β1-3 linkage to the underlying mannose and a terminal *N*-acetylgalactosamine in β1-4 linkage, produces a substrate for the addition of ethanolamine-phosphate (EthNP) to C-6 of *N*-acetylglucosamine (EthNP-6GlcNAcβ1-3Manβ1-4Glcβ-Cer). Therefore, the neutral core of most arthroseries glycolipids is more correctly described as zwitterionic rather than neutral. Arthroseries cores with more than one *N*-acetylglucosamine residue are found with zero, one, or two EthNP groups. The functional significance of EthNP modification is currently unknown.

Additional species-specific core diversity has also been identified. For example, the arthroseries triosylceramide in Drosophila can be alternatively extended with galactose rather than *N*-acetylgalactosamine, followed by addition of glucuronic acid (Figure 26.6). Other arthropod species add galactose to mactosylceramide or to the arthroseries tetraosylceramide with subsequent capping and EthNP modification (Figure 26.6). The expanding diversity of alternative core extensions detected in arthropods provides a larger pool of structural variants from which these organisms can tailor GSL expression for specific developmental or tissue-restricted functions, similar to core switching in vertebrate species (Chapter 11).

Mutations that affect the first steps in Drosophila GSL synthesis were originally identified as modulators of cell fate. Biochemical analysis showed that genes called *egghead* (*egh*) and *brainiac*

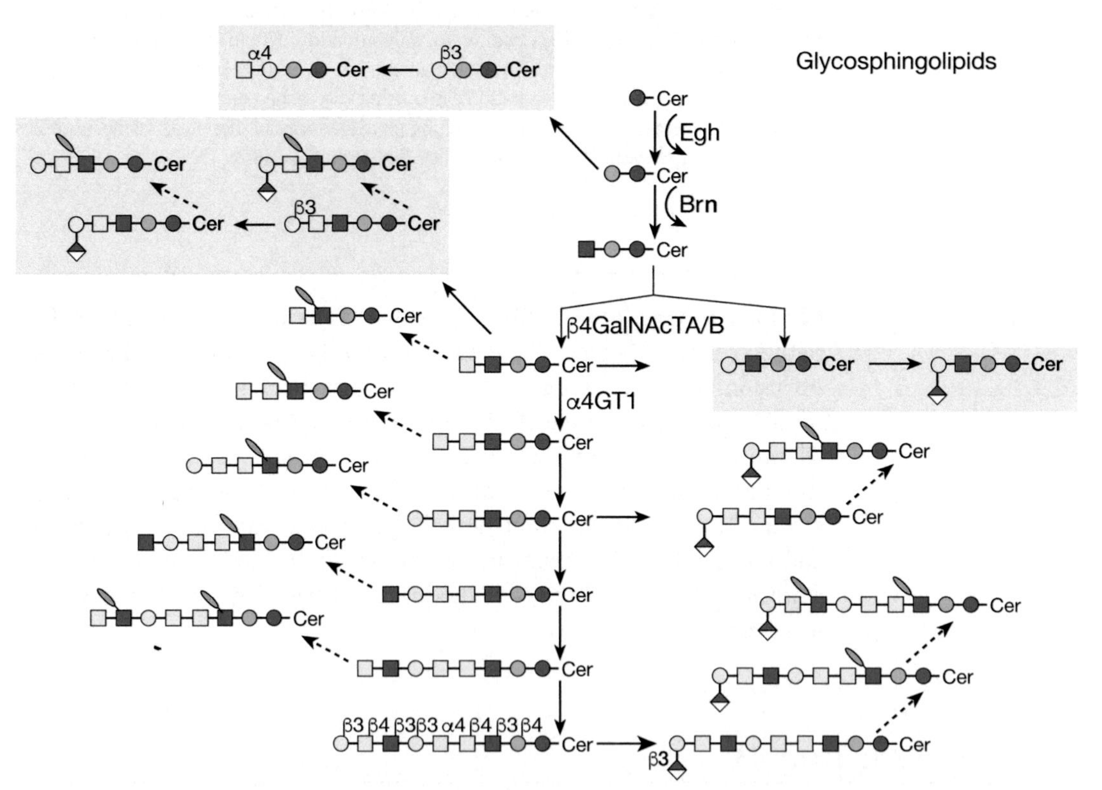

FIGURE 26.6. Glycosphingolipid glycan diversity. The arthroseries glycosphingolipids (GSLs) are built by extension from a mactosylceramide core (Manβ1-4Glcβ-ceramide). Neutral arthroseries glycans are frequently modified by the addition of phosphoethanolamine (EthNP; *orange ovals*) on *N*-acetylglucosamine residues, forming a zwitterionic structure. Acidic charge is imparted by the addition of glucuronic acid to nonreducing terminal galactose residues on neutral or zwitterionic cores. (*Light gray box*) Species-specific core diversity is evident in the alternative extension of the arthroseries triaosylceramide by galactose rather than *N*-acetylgalactosamine in Drosophila. (*Dark gray boxes*) Other arthropods supplement their arthroseries GSLs by distinctive modifications. Moth cells (Sf9) extend the mactosylceramide core with galactose and then *N*-acetylgalactosamine. In two dipteran species, the larvae of the greenbottle fly (*Lucilia caesar*) and the pupae of the blowfly (*Calliphora vicina*), but not in embryos of Drosophila, the arthroseries tetraosylceramide is extended with galactose, capped with glucuronic acid, and modified with EthNP.

(*brn*), respectively, encode the mannosyltransferase and N-acetylglucosaminyltransferase that add the second and third monosaccharide residues (Figure 26.6). Phenotypes associated with loss of maternal and zygotic *egh* and *brn* include overproliferation of neural cells at the expense of epithelial lineages. These *egh*/*brn* phenotypes closely resemble the embryonic phenotype of mutations in the *Notch* gene (Figure 26.4). An additional phenotype of *egh* mutants is the formation of enlarged peripheral nerves due to glial cell overproliferation with immune cell infiltration, reminiscent of the human disorder neurofibromatosis type 1. As in this human disorder, overgrowth results from enhanced phosphatidylinositol-3-kinase activity, likely because of reduced Ras signaling. The neurofibromatosis-like phenotypes indicate the importance of the membrane glycolipid environment for propagating appropriate transmembrane signals. Mutations affecting the two partially redundant N-acetylgalactosaminyltransferases that generate the tetraosylceramide of the arthroseries result in subtle behavioral phenotypes in larvae and altered EGF receptor signaling in ovarian follicle cells, but otherwise produce viable, fertile adults. It remains to be determined whether arthroseries glycans elongated beyond the tetraosylceramide possess structural information that imparts specific function.

INSECT LECTINS

Annotation of the Drosophila genome indicates the presence of representatives for each of the known classes of animal lectins (Chapter 28). Intracellular carbohydrate-binding proteins, which in vertebrates are associated with protein folding and quality control in the ER or with trafficking through the early compartments of the secretory pathway (calnexin, calreticulin, VIP, ERGIC-53), are also found in Drosophila (Chapter 39). Further downstream, lysosomal targeting is mediated by LERP, a sorting protein functionally similar to the vertebrate mannose-6-phosphate receptor (Chapter 33). Although LERP possesses domain architecture and sequence similarity to the P-type lectin sequences of the mannose-6-phosphate receptor, its binding does not appear to be glycan-dependent, suggesting evolutionary divergence in the mechanisms that control the biogenesis of lysosomes. The binding specificities and functions of insect family members of the Galectin (Chapter 36), C-type (Chapter 34), and I-type lectin families (Chapter 35) are not well-characterized. For example, the protein encoded by the *furrowed* gene, a putative C-type lectin with significant homology with vertebrate selectins, has been implicated in planar cell polarity dynamics during eye development, but this activity is independent of its carbohydrate recognition domain (CRD).

Current classification schemes for animal lectins reflect the long history of studying glycan-binding proteins in vertebrates (Chapter 28). However, a handful of lectin activities, first identified in Drosophila, challenge the comprehensiveness of current animal lectin designations. One of these was identified in a search for secreted proteins that stimulate the proliferation and motility of imaginal disc cells. The imaginal disc growth factor (IDGF) family members are structurally related to chitinases but lack amino acid residues essential for catalytic activity. It has been suggested that the IDGF family has evolved away from hydrolysis while maintaining a glycan-binding activity that facilitates mitogenic and trophic support. Another currently unclassified lectin, called "gliolectin," was identified in a screen of Drosophila cDNAs for proteins that mediate cell adhesion to immobilized glycans. Gliolectin is expressed in a subset of embryonic glial cells found at the midline of the developing CNS and in cells at the dorsal/ventral boundary of the wing imaginal disc. Mutants that lack gliolectin show defects in axonal pathfinding, consistent with a role for glycan-mediated cell adhesion in facilitating the transmission of signals between cells. They also show *Notch*-like phenotypes in wing development, indicating a broader role for gliolectin in establishing tissue boundaries. Finally, the *mind-the-gap* gene encodes a protein (MTG) that possesses domain folds consistent with its ability to bind

N-acetylglucosamine, although a definitive structural or functional homolog among well-defined lectin families is yet to be determined. In Drosophila larvae, MTG participates in organizing the glycoprotein matrix at the larval neuromuscular junction.

INSECT NUCLEOTIDE SUGAR TRANSPORTERS

Genetic and biochemical approaches have identified multiple nucleotide sugar transport activities in Drosophila. For several of these genes, glycomic, and other phenotypic consequences associated with knockdown or loss-of-function have been characterized (Table 26.1). For others, further analysis is needed to define their specificity and function. The first nucleotide sugar transporters identified in Drosophila were recovered as mutants from genetic screens that targeted phenotypes associated with altered morphogen or growth factor signaling and, therefore, impacted GAG expression. Fringe connection (*frc*) and Slalom (*sll*) are responsible for the transport of UDP-GlcA, UDP-GlcNAc, UDP-Xyl, and the sulfation donor PAPS, each of which is essential for GAG biosynthesis. Other transport activities that function in important developmental pathways include GFR/Nac and Efr, which transport GDP-fucose into the Golgi and ER, respectively. Loss of these transport activities impacts Notch signaling and neural-specific glycan expression. Other transporter genes are also associated with developmental phenotypes but their transport specificities require further analysis (Table 26.1).

ACKNOWLEDGMENTS

The authors acknowledge helpful contributions from Hamed Jafar-Najad and Iain B.H. Wilson.

FURTHER READING

Wiegandt H. 1992. Insect glycolipids. *Biochim Biophys Acta* **1123**: 117–126. doi:10.1016/0005-2760(92)90101-z

Aoki K, Perlman M, Lim JM, Cantu R, Wells L, Tiemeyer M. 2007. Dynamic developmental elaboration of N-linked glycan complexity in the *Drosophila melanogaster* embryo. *J Biol Chem* **282**: 9127–9142. doi:10.1074/jbc.M606711200

Aoki K, Porterfield M, Lee SS, Dong B, Nguyen K, McGlamry KH, Tiemeyer M. 2008. The diversity of O-linked glycans expressed during *Drosophila melanogaster* development reflects stage- and tissue-specific requirements for cell signaling. *J Biol Chem* **283**: 30385–30400. doi:10.1074/jbc.M804925200

Ten Hagen KG, Zhang L, Tian E, Zhang Y. 2009. Glycobiology on the fly: developmental and mechanistic insights from *Drosophila. Glycobiology* **19**: 102–111. doi:10.1093/glycob/cwn096

Yan D, Lin X. 2009. Shaping morphogen gradients by proteoglycans. *Cold Spring Harb Perspect Biol* **1**: a002493. doi:10.1101/cshperspect.a002493

Crickmore M, Mann RS. 2010. A new chisel for sculpting Darwin's endless forms. *Nat Cell Biol* **12**: 528–529. doi:10.1038/ncb0610-528

Tran DT, Zhang L, Zhang Y, Tian E, Earl LA, Ten Hagen KG. 2012. Multiple members of the UDP-GalNAc: polypeptide N-acetylgalactosaminyltransferase family are essential for viability in *Drosophila. J Biol Chem* **287**: 5243–5252. doi:10.1074/jbc.M111.306159

Nakato H. 2015. Heparan sulfate proteoglycans in the *Drosophila* ovarian germline stem cell niche. In *Glycoscience: biology and medicine* (ed. M Taniguchi, et al.), pp. 825–832. Springer, Tokyo.

Zhang L, Ten Hagen KG. 2019. O-Linked glycosylation in *Drosophila melanogaster. Curr Opin Struct Biol* **56**: 139–145. doi:10.1016/j.sbi.2019.01.014

Nishihara S. 2020. Functional analysis of glycosylation using *Drosophila melanogaster. Glycoconj J* **37**: 1–14. doi:10.1007/s10719-019-09892-0

27 Deuterostomes

Michael Pierce, Iain B.H. Wilson, Katharina Paschinger, and Pamela Stanley

This chapter discusses some general features of glycosylation and glycan-binding interactions in a few examples of species that belong to the deuterostome lineage, with particular emphasis on sea urchins, frogs, zebrafish, and mice. These organisms provide excellent models for studying the functions of glycans in development and physiology.

EVOLUTIONARY BACKGROUND

Animal evolution split into two major lineages ~500–600 million years ago—the deuterostomes and the protostomes (Chapter 20). The superphylum Deuterostomia contains the major phyla of Echinodermata (sea urchins and starfish), Hemichordata (acorn worms), Urochordata (ascidians), Cephalochordata (amphioxus), and Vertebrata (fish, amphibia, reptiles, birds, and mammals). In deuterostomes, cell divisions of the zygote occur by radial cleavage, and cell fates are not precisely determined. Another characteristic feature is that the first opening to form in the blastula becomes the anus and the second opening becomes the mouth. The other superphylum, the Protostomia, contains the major phyla of Porifera (corals and sponges), Cnidara (anemones and hydra), Annelida (segmented worms), Mollusca (clams, oysters, snails, and slugs), and Arthropoda (insects, spiders, and crustacea). In protostomes, unlike deuterostomes,

published by Cold Spring Harbor Laboratory Press; doi: 10.1101/glycobiology.4e.27

cell division during early development occurs in a highly organized manner and cell fate is precisely determined. Examples of well-studied protostome model organisms include the nematode *Caenorhabditis elegans* (Chapter 25) and the arthropod *Drosophila melanogaster* (Chapter 26).

Deuterostomes have been extensively studied because of their relatedness to humans and because they provide excellent model systems for understanding vertebrate biology. Sea urchins, frogs, zebrafish, and mice have received the most attention and each provides certain advantages for studying functions of glycans in reproduction, early development, or adult physiology. In numerous instances, these models have revealed aspects of glycobiology that were later confirmed in more evolutionarily distant organisms, and they often serve as excellent models for understanding human disease. Challenges exist, however, in describing the glycomes of the more ancient deuterostomes, mainly because of poor genome annotation, ignorance of individual glycosyltransferase substrates, and the presence of glycans with unusual structures. Each of these models is briefly described below with references to other sections of the book for additional details.

SEA URCHINS

Sea urchin glycans involved in fertilization have been studied extensively. In fact, much of what we know about the biochemistry of fertilization was first discovered in this organism (Figure 27.1), and the information subsequently applied to mammalian sperm–egg interactions. One of the advantages of studying fertilization in sea urchins is that eggs and sperm are easy to obtain in large quantities. Because fertilization occurs outside the adult body, it is also easy to experimentally manipulate sperm and eggs.

Egg Glycans and Fertilization

Sea urchin eggs are covered by a hydrated jelly coat. About 80% of the weight of egg jelly is a high-molecular-weight linear fucose (Fuc) sulfate polymer (FSP) with a molecular mass of $>10^6$ Da. Receptor proteins on sperm bind to FSP, triggering the opening of two pharmacologically distinct calcium channels that induce the exocytosis of the sperm's acrosome vesicle (the "acrosome reaction"). The ionic mechanisms that trigger the acrosome reaction are conserved in mammals, but the nature of the sperm surface receptors varies. FSP is a species-selective inducer of the sea urchin sperm acrosomal reaction and most FSPs are sulfated α1-3Fuc-based linear polymers made of tri- or tetrasaccharide repeats. The number of fucoses per repeat, the linkage, and sulfation patterns all help ensure species selectivity for induction of the acrosome reaction. About 20% of the egg jelly mass is a large glycoprotein containing a unique polymer of sialic acid (Sia), which can be released from crude sea urchin egg jelly by treatment with mild base (β-elimination). The sialoglycan has a novel structure—[Neu5Gcα2-5-O-glycolylNeu5Gc]$_n$. However, its receptor on the sperm membrane is unknown. When sea urchin sperm undergo the acrosome reaction, a protein named bindin is released from the acrosomal vesicle. Bindin cements the sperm to a large egg surface glycoprotein named egg bindin receptor 1 (EBR1). Bindin can agglutinate mammalian red blood cells much like plant lectins (Chapter 32), and glycopeptide fragments of unfertilized eggs can block bindin-induced red cell agglutination. Bindin is thought to recognize glycans on EBR1. EBR1 itself has lectin-like domains, but its glycan ligands are unknown.

Most studies of glycosylation in sea urchins have identified glycosyltransferase activities and the glycans they synthesize (e.g., including N-glycans with antennal β1-3-linked Gal, β1-4GalNAc, Neu5Gc, and sulphate residues as well as traces of the "invertebrate" feature of core α1-3/α1-6-difucosylation) (Figure 27.2). The functions of glycans and glycan-binding proteins (GBPs) can be investigated using genetic strategies in sea urchins, antisense morpholinos or short hairpin RNAs (shRNAs) to knock down gene expression have been the methods of choice, but these will likely be superseded by

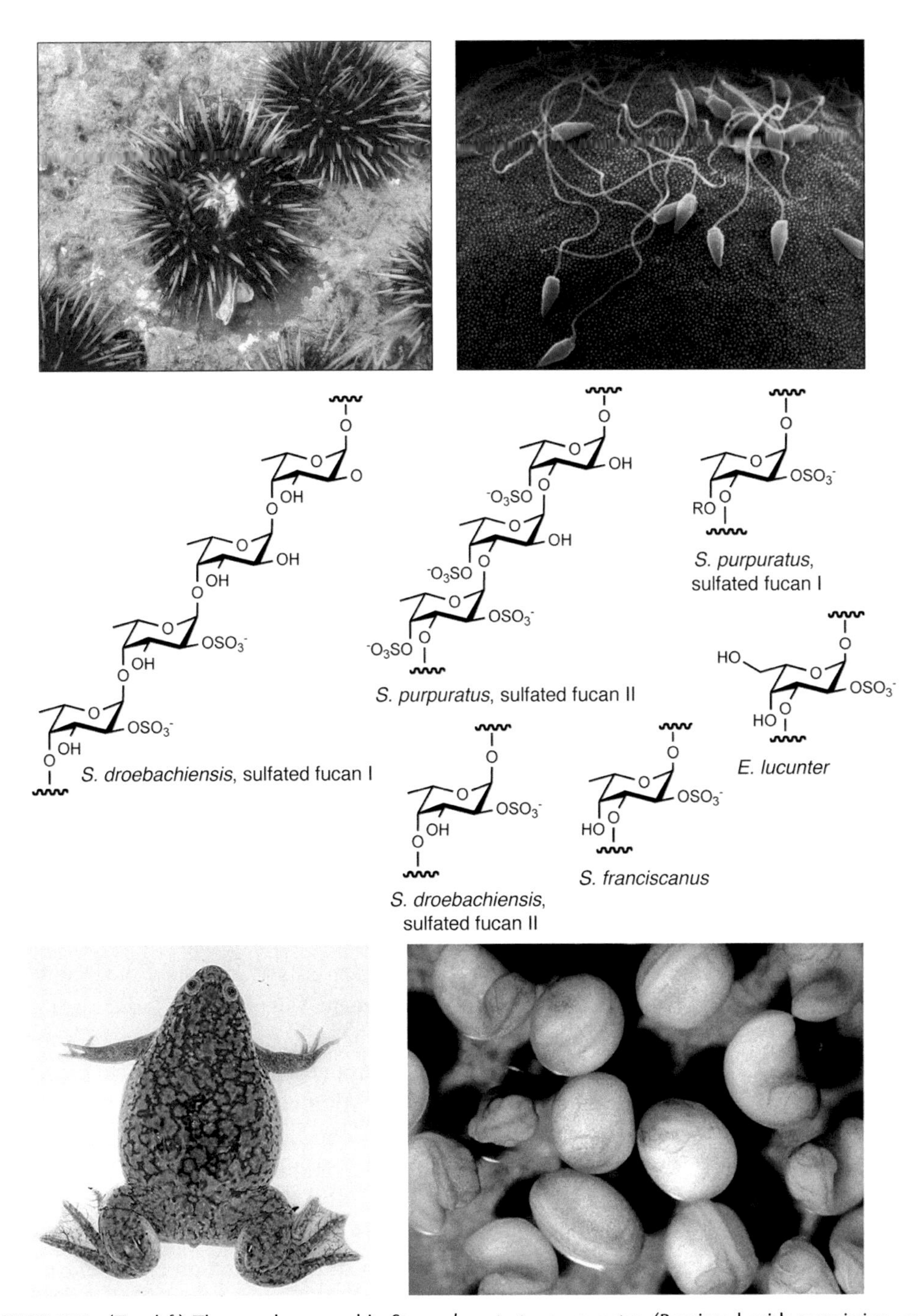

FIGURE 27.1. (*Top left*) The purple sea urchin *Strongylocentrotus purpuratus*. (Reprinted, with permission, courtesy of Charles Hollahan, Santa Barbara Marine Biologicals.) (*Top right*) Sperm binding to a sea urchin egg. (Reprinted, with permission, courtesy of M. Tegner and D. Epel.) Many of the key discoveries regarding glycans and their role in determining species-specific sperm–egg binding and blocks to polyspermy were made in sea urchins. (*Middle*) Structures of sulfated α-fucans and a sulfated α-galactan (marked as *Echinometra lucunter*) from sea urchin egg jelly are shown. The specific pattern of sulfation, the position of the glycosidic linkage, and the constituent monosaccharide vary among sulfated polysaccharides from different species. These structures were deduced from analysis by nuclear magnetic resonance (NMR) spectroscopy. (Redrawn, with permission, from Vilela-Silva AC, et al. 2002. *J Biol Chem* **277:** 379–387, © American Society for Biochemistry and Molecular Biology. Original art has been adapted and redrawn by R.D. Cummings.) (*Bottom left*) *Xenopus laevis*. An adult specimen of the African clawed frog. (Image courtesy of Bruce Blumberg, University of California, Irvine.) (*Bottom right*) Embryonic development from the neurula stage until just before hatching, ~24 hr postfertilization. (Reprinted, courtesy of Thierry Brassac, Exploratorium at www.exploratorium.edu.)

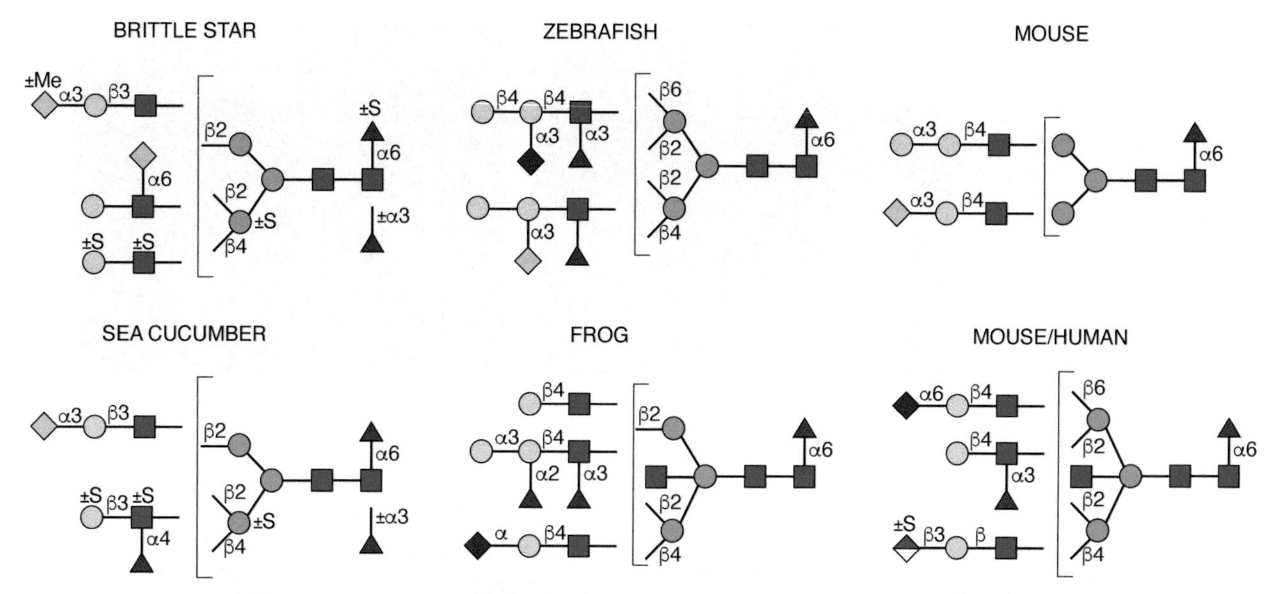

FIGURE 27.2. N-glycan diversity in deuterostomes. Example N-glycan antennae from a brittle star and a sea cucumber (both echinoderms), zebrafish, the African clawed frog, and the mouse are shown as well as those shared between mouse and humans. ±S signifies with or without sulfate. ± Me signifies with or without a methyl group CH_3.

more precise gene editing techniques such as the clustered regularly interspaced short palindromic repeats/Cas9 (CRISPR/Cas9) strategy or transcription activator–like effector (TALE) nucleases (TALENs). A few echinoderm genomes have been sequenced (Echinobase), but as with most organisms, relating glycosylation activities to genes is not straightforward (Chapter 8).

FROGS

The African clawed frog, *Xenopus laevis*, is a well-established model organism for studying fertilization and embryonic development (Figure 27.1). Adults are easy to maintain and can be induced to lay eggs up to three times a year by injection of human chorionic gonadotropin. The very large eggs allow easy microinjection experiments to express exogenous cDNAs encoding RNAs and proteins of interest. Although frog embryos develop quickly, the generation time for adults is long (1–2 yr), and the frog genome is tetraploid, making genetic studies difficult. Nevertheless, as for sea urchins, inhibition of gene expression has been accomplished by injection of antisense morpholinos or shRNAs into early embryos. These techniques suffer from off-target effects and the future will see a shift to newer gene editing strategies including CRISPR/Cas9 and TALENs. In addition, in situ hybridization and overexpression techniques have been used in *Xenopus* embryos. For example, the functions of two polypeptide N-acetylgalactosaminyltransferases in transforming growth factor-β/bone morphogenic protein signaling were distinguished, showing that this organism is useful for discovering glycan biosynthetic steps and mechanisms by which glycans regulate development. Indeed, developmental changes in the *X. laevis* N-glycome (Figure 27.2) have been reported, whereas the O-glycans of frog egg mucins can be extremely complicated in a species-dependent manner.

Lectin Family First Discovered in *Xenopus*

X. laevis oocyte cortical granules contain a lectin called XL35, which is released at fertilization and binds to glycoproteins expressing mucin-type O-glycans (Galα1-3GalNAc) in the jelly surrounding the egg. XL35 is an oligomer made up of monomers, each of which has a molecular weight of 35 kDa. Light scattering experiments showed that the oligomer exists primarily as a 12-mer, allowing it to cross-link its glycoprotein ligands in the egg jelly to form a relatively rigid layer on the fertilization envelope. This molecular complex functions to block polyspermy and also serves as a barrier against microbial infection. Homologs of XL35, known as "X-lectins," exist in some of the most primitive deuterostomes, such as the sea squirt, but also in more complex vertebrates such as zebrafish, mice, and humans. Mice and humans have two orthologs termed intelectins-1 and -2 (Chapter 30) because of their initial discovery in the intestines. It is interesting to note that the sea squirt homolog (*Halocynthia roretzi*) shows a similar monomeric molecular weight to human intelectin 1 (Hint-1) and is 34% identical at the amino acid level. Hint-1 exists as a trimer when secreted from goblet cells in intestinal and respiratory epithelia. IL-13, a cytokine involved in the type TII innate immune response, induces a remarkable increase in Hint-1 transcripts. Constitutively expressed transcripts for Hint-1 are also found in heart and lung tissue. Hint-2, in contrast, is only expressed in Paneth cells that function in immune surveillance in the crypts of the small intestine. Glycan microarray results show that both Hints bind to the glycans of a unique set of human pathogenic bacteria that have in common the presence of particular *cis*-diols on their outer surface polysaccharides and include both Gram-positive and Gram-negative bacteria. Crystal structures of Hint-1 and the *Xenopus* embryonic epidermal lectin show a conserved mechanism of glycan recognition. But clearly, the carbohydrate binding determinants for each intelectin studied are unique. It is becoming clear that the X-lectin family plays key roles in deuterostome innate immune responses against pathogens.

ZEBRAFISH

Zebrafish (*Danio rerio*) provides an excellent model organism for understanding the functions of genes in early vertebrate development (Figure 27.3). Females lay several hundred eggs weekly, which undergo external fertilization. Embryos develop rapidly (in hours), and their translucence allows visualization of early development. Zebrafish are relatively easy to maintain compared with other vertebrate species. Furthermore, they can be mutagenized and outbred to produce mutant strains or manipulated using morpholinos, shRNAs, or CRISPR/Cas9 and TALEN methodologies to silence or delete genes permanently or transiently during early development (Figure 27.3).

Zebrafish live in a nonterrestrial environment and widespread gene duplications and diversification occurred during teleost evolution. All major glycan classes present in mammals have been described in zebrafish and many of them have been manipulated genetically, yielding new insights into glycan functions. Indeed, the N-, O-, and lipid-linked glycomes of zebrafish organs have a range of typical vertebrate features such as β1-4-galactosylation and sialylation, but also a unique β1-4-galactose extension of the "sialyl-Lewis x" epitope (Figure 27.2).

Investigation of Glycosylation-Related Phenotypes

Sophisticated screening methods for the identification of novel mutants that affect specific physiological systems may be applied to zebrafish. For example, high-throughput screenings have identified groups of genes required for fin development and regeneration, which serve as models for cartilage and bone development in mammals. Mutants altered in vascular development and hemostasis have been identified as well. Moreover, the ability to identify novel genes in the system and manipulate glycan expression via gene silencing or mutation is making this model organism a focus of studies in vertebrate glycobiology. Functions of particular protein- and lipid-bound glycans have been uncovered; for example, the *slytherin* mutant that contains a

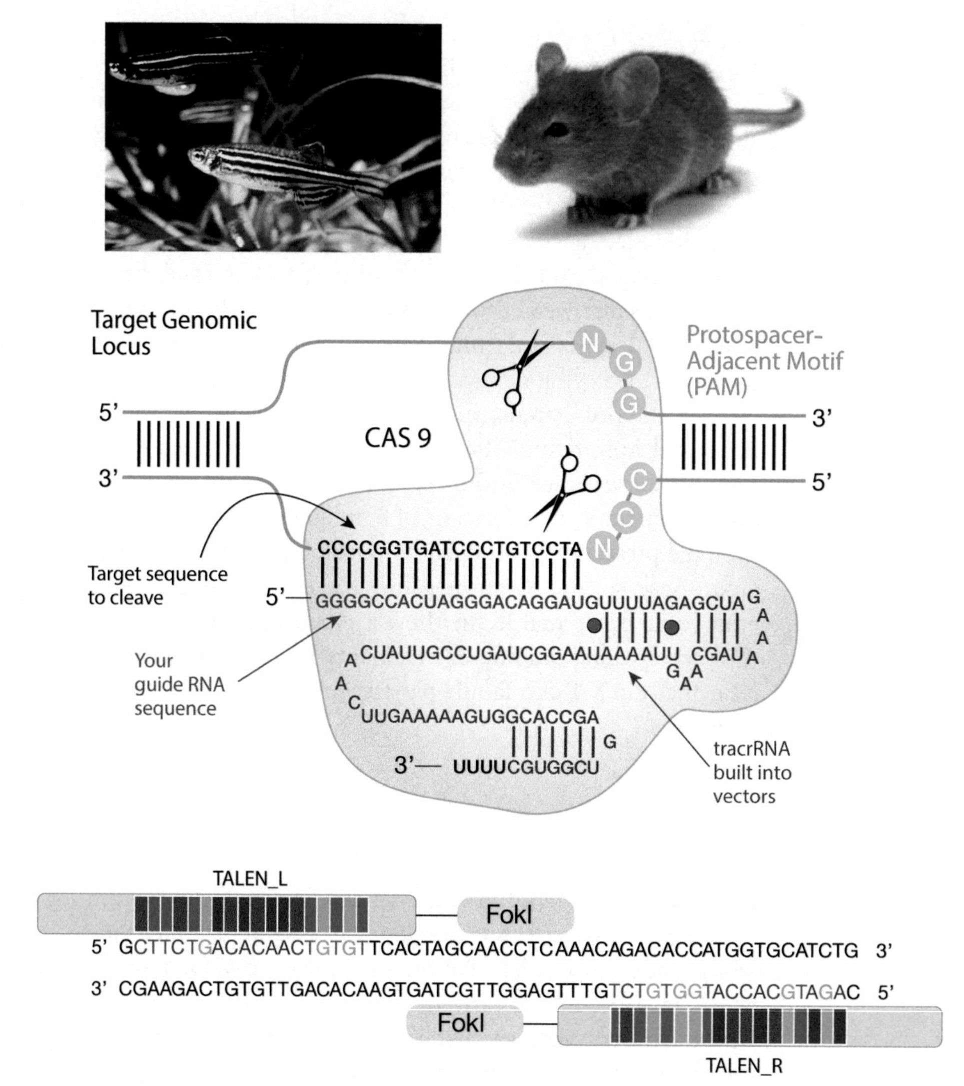

FIGURE 27.3. (*Top left*) Zebrafish (*Danio rerio*). (Photograph from the Zebrafish Information Network [ZFIN], © University of Oregon; see also Sprague J, et al. 2006. *Nucl Acids Res* **34:** D581–D585.) (*Top right*) Adult mouse (*Mus musculus*). (*Middle*) Gene editing using CRISPR/Cas9. A gene-specific guide RNA with a *trans*-activating CRISPR-targeting guide RNA (crRNA) that recruits the Cas9 nuclease for sequence-specific cleavage of genomic DNA. NGG, a protospacer-adjacent motif (PAM), must be located just 3′ of the genomic DNA target sequence. (Reprinted from System Biosciences Inc.; https://www.systembio.com/crispr-cas9-plasmids.) (*Bottom*) Gene editing using a transcription activator–like (TALE) nuclease (TALEN). TALEN vectors encode proteins designed to specifically bind genomic DNA flanking the desired cleavage site linked to the Fok1 nuclease that catalyzes the site-specific cleavage. (Reprinted from Ramalingam S. 2013. *Genome Biol* **14:** 107–110.)

missense mutation affecting GDP-fucose biosynthesis causes several phenotypes impacting neuronal differentiation. Knockdown of genes that regulate polysialylation show the critical function of polysialic acid for neuronal migration and plasticity, including those that may involve motor neurons in somite formation and differentiation. Significant contributions have also been made in the understanding of glycosaminoglycan (GAG) biosynthesis (Chapter 17) and functions during development. Furthermore, the modeling of several human congenital disorders of glycosylation in zebrafish has provided insights into the phenotypes associated with these disorders, including defects in the synthesis of O-mannose glycans (Chapter 13) that give rise to forms of muscular dystrophy (Chapter 45).

MICE

Laboratory mice (*Mus musculus*) are the most widely studied deuterostomes because they are the mammals that are most closely related to primates, reproduce quickly, and their genes can be readily manipulated. Inbred mouse strains are widespread, and spontaneous mutations occur that have specific morphological or pathophysiological phenotypes. Examples of the latter include brachymorphic mice, which are partially deficient in synthesis of PAPS, the universal sulfate donor (Chapter 5), resulting in shortened limbs because of defective chondroitin sulfate (CS) synthesis during endochondral ossification (Chapter 17). Another example is a mutation causing muscular dystrophy, which was traced to $Large^{myd}$, a glycosyltransferase that when defective in humans also causes muscular dystrophy (see Chapters 13 and 45). Much information about mouse glycobiology comes from the use of homologous recombination methods to knock out a specific gene systemically, in specific tissues or at selected times in development (Online Appendix 27A). In numerous cases, human pathological mutations have been engineered into conserved murine homologous genes to generate models for disease (Chapter 41). Notably, humans lack two glycan biosynthetic enzymes that are found in mice: the galactosyltransferase that synthesizes Galα1-3Gal that can terminate N- or O-glycans (Figure 27.2; Chapter 14), and the CMP-Neu5Ac hydroxylase that synthesizes CMP-N-glycolylneuraminic acid (Neu5Gc) from CMP-Neu5Ac (Chapter 15). Mice lack the ABH(O) blood group glycans of humans (Chapter 14).

Mutations That Affect Entire Pathways

Because gene deletion has been used to inactivate numerous genes involved in glycosylation, only a few mutants are mentioned here. Mutant lines have been created in early genes involved in the assembly of all the major subclasses of vertebrate glycans. In most cases, these mutations prevent initiation of the biosynthetic pathways and show embryonic lethality. These include the O-GlcNAc transferase OGT (Chapter 19), whose deletion is lethal even at the single-cell stage. Loss of the entire N-glycan pathway caused by ablation of phosphomannomutase 2 (which generates mannose-1-P and all mannose-containing glycans), and loss of GlcNAc-1-P transferase (which initiates the lipid-linked oligosaccharide precursor; Chapter 9) are lethal within a few days of fertilization. Loss of heparan sulfate (HS) biosynthesis due to deletion of *Ext1* or *Ext2* is also lethal at the E6–7 stage because of a failure to differentiate mesoderm. Systemic deletion of MGAT1 (Chapter 9) is lethal around midgestation (E9–10) showing that early development through gastrulation can proceed in the complete absence of hybrid and complex N-glycans. CS and hyaluronan synthesis are essential for embryonic development as well, and developmental arrest occurs during limb generation and organogenesis. Because there are more than 20 partially redundant polypeptide N-acetylgalactosaminyltransferases in mice, knocking out the O-GalNAc glycan pathway (Chapter 10) completely has not been possible. However, deletion of the single copy core-1 galactosyltransferase C1GALT1, which generates Galβ1-3GalNAcα-O-Ser/Thr, is lethal by stage E14 because of abnormal angiogenesis. Knocking out the protein O-fucosyltransferase POFUT1 that modifies epidermal growth factor (EGF) modules in Notch is lethal and causes developmental defects similar to those seen in global Notch pathway knockout mice. Absence of glycosylphosphatidylinositol (GPI) anchors (Chapter 12) or O-mannose glycans (Chapter 13) also result in embryonic lethality.

Mutations That Affect Specific Glycan Extensions or Modifications

Mutations in genes that encode glycosyltransferases that extend core glycans in each class of glycan usually have less severe phenotypes than those deleting entire pathways. For example, some glycosyltransferase gene knockouts affect the immune system, including leukocyte rolling, lymphocyte homing, and T-cell homeostasis. Knocking out sequential biosynthetic steps or multiple, seemingly redundant genes frequently reveals another layer of glycan-dependent functions, novel structures, and unrecognized biosynthetic pathways. Many mouse mutants serve as models of human genetic disorders (Chapter 41).

Cell Type–Specific Mutations

To study the function of glycans in specific tissues, tissue-specific lesions in a gene can be created by crossing strains bearing conditional null alleles to strains that selectively delete the gene only in those tissues of interest. The most common method is to flank a portion of the gene with *loxP* recombination sites and then mate these mice to a strain that produces bacteriophage Cre-recombinase under the control of a tissue-specific or inducible promoter (Figure 27.4). When activated, the Cre-recombinase recognizes the *loxP* sites and excises the DNA segment contained between them. These techniques can be combined with an environmental stress, dietary modification, or inflammatory challenge to delete genes in both a cell type–specific and temporally regulated fashion.

Sometimes, altering a gene can have a highly specific effect on a given tissue without affecting other tissues because of the unforeseen presence of related isozymes. For example, Golgi α-mannosidase II (MAN2A1) was thought to have the dominant role in N-glycan processing (see Chapter 9), but mice lacking MAN2A1 survive and have a dyserythropoietic anemia. These mice produce many unusual glycans from the abnormally high levels of incomplete precursors that accumulate and drive the synthesis of atypical products. The mutant mice eventually develop a progressive autoimmune nephropathy, accumulating immune complexes in the kidney tubules, probably in response to the abnormal glycan antigens. Their erythrocytes lack complex N-glycans, but other tissues continue to produce them using a different isozyme called α-mannosidase IIx (MAN2A2). Mice lacking MAN2A2 are mostly unaffected, but male mutant mice are sterile because spermatocytes fail to make a glycan required for their binding to Sertoli cells in the testes, blocking spermatogenesis. Deletion of both α-mannosidases prevents synthesis of all complex N-glycans (Chapter 9), and most embryos die between E15 and E18, or shortly after birth from respiratory failure.

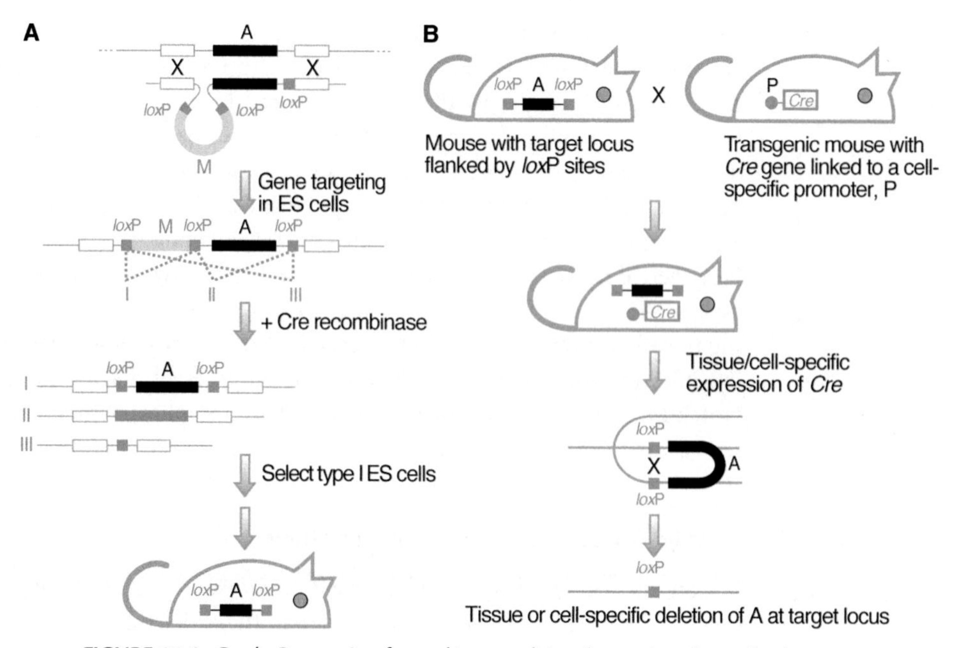

FIGURE 27.4. Cre-*loxP* targeting for making conditional gene knockouts in the mouse.

Strain Differences and Genetic Modifiers

Determining the effect of a mutant gene is usually done in highly inbred strains that are genetically identical except for the gene in question. Sometimes, the phenotype of a mutation changes according to the genetic background (mouse strain) in which the mutation is expressed. For instance, mice lacking N-acetylglucosaminyltransferase II (MGAT2), an enzyme required for complex N-glycan synthesis (Chapter 9), die after birth in one genetic background, but in another, rare survivors occur, with severe gastrointestinal, hematological, and osteogenic abnormalities. Interestingly, these mutants phenocopy the human disorder, CDG-IIa, in which patients lack the same enzyme (Chapter 45). As another example, lines deficient in global fucosylation because of mutations in the enzyme that converts GDP-mannose to GDP-fucose (Chapter 5) die as embryos, but when bred into a different background, they survive until birth and can then be maintained on a fucose-supplemented diet. A third example is the formation of bony outgrowths (exostoses) in the growth plates of endochondral bones. In one background, heterozygous mutations in the Ext-1 copolymerase involved in HS biosynthesis cause infrequent exostoses, whereas in another background, the incidence of tumors increases severalfold. These experiments suggest the existence of modifiers that alter the expression level of an enzyme or the assembly of substrate, that salvage sugars from the diet, or that alter glycan turnover.

Environmentally Driven Phenotypes

In some mutants, phenotypes are only evident under environmental challenge. For example, deletion of both enzymes responsible for polysialic acid synthesis produces "no-fear" mice that tend to be more aggressive and ignore normally stressful or anxiety-producing situations. Deletion of syndecan-1, a major HS core protein (Chapter 17), gives no obvious changes under normal laboratory conditions, but the mice are more resistant to bacterial lung infections. Apparently, bacteria exploit the shed syndecan in the lung to enhance their virulence and modulate host defenses.

Mice deficient in the synthesis of ganglioside GM3 appear to be normal except for enhanced insulin sensitivity, as well as deafness due to inner ear hair cell defects. These mice shunt glycosphingolipid synthesis toward more complex gangliosides that substitute for GM3 loss (Chapter 11). On the other hand, double-mutant mice that synthesize only ganglioside GM3 show tonic–clonic (epileptic) seizures, and 90% die because of seizures in response to sharp sounds. These examples show the importance of having the correct balance of glycans. They also show how phenotypes depend to a large extent on other genetic factors, diet, and environmental cues. Another major factor affecting the phenotype is now thought to be the microbiome, especially in the gut. Thus, mutations that appear silent in a controlled laboratory setting may elicit a strong phenotype in a more natural environment.

HUMANS AND OTHER PRIMATES

Studies of humans and other primates have been limited by cost, ethical controversies, and (in the case of humans) genetically outbred populations. However, there are increasing examples in which findings (especially therapeutic ones) in mice fail to translate to the human conditions. Even chimpanzees (our closest relatives) seem to have markedly different disease profiles, some explained by alterations in sialic acid biology. Although ethically and practically feasible studies on other primates should continue, the late Nobel laureate Sydney Brenner has suggested (Chapter 7) that we now have enough information about humans to consider ourselves to be a viable "model organism" for in-depth study.

ACKNOWLEDGMENTS

The authors acknowledge contributions to previous versions of this chapter by Victor Vacquier and appreciate helpful comments and suggestions from Ajit Varki and Vlad Panin.

FURTHER READING

Lee JK, Baum LG, Moremen K, Pierce M. 2004. The X-lectins: a new family with homology to the *Xenopus laevis* oocyte lectin XL-35. *Glycoconj J* **21**: 443–450. doi:10.1007/s10719-004-5534-6

Ohtsubo K, Marth JD. 2006. Glycosylation in cellular mechanisms of health and disease. *Cell* **126**: 855–867. doi:10.1016/j.cell.2006.08.019

Bishop JR, Schuksz M, Esko JD. 2007. Heparan sulphate proteoglycans fine-tune mammalian physiology. *Nature* **446**: 1030–1037. doi:10.1038/nature05817

Freeze HH, Sharma V. 2010. Metabolic manipulation of glycosylation disorders in humans and animal models. *Semin Cell Dev Biol* **21**: 655–662. doi:10.1016/j.semcdb.2010.03.011

Varki NM, Strobert E, Dick EJ Jr, Benirschke K, Varki A. 2011. Biomedical differences between human and nonhuman hominids: potential roles for uniquely human aspects of sialic acid biology. *Annu Rev Pathol* **6**: 365–393. doi:10.1146/annurev-pathol-011110-130315

Vacquier VD. 2012. The quest for the sea urchin egg receptor for sperm. *Biochem Biophys Res Commun* **425**: 583–587. doi:10.1016/j.bbrc.2012.07.132

Flanagan-Steet HR, Steet R. 2013. "Casting" light on the role of glycosylation during embryonic development: insights from zebrafish. *Glycoconj J* **30**: 33–40. doi:10.1007/s10719-012-9390-5

Bammens R, Mehta N, Race V, Foulquier F, Jaeken J, Tiemeyer M, Steet R, Matthijs G, Flanagan-Steet H. 2015. Abnormal cartilage development and altered N-glycosylation in Tmem165-deficient zebrafish mirrors the phenotypes associated with TMEM165-CDG. *Glycobiology* **25**: 669–682. doi:10.1093/glycob/cwv009

Voglmeir J, Laurent N, Flitsch SL, Oelgeschlager M, Wilson IB. 2015. Biological and biochemical properties of two *Xenopus laevis N*-acetylgalactosaminyltransferases with contrasting roles in embryogenesis. *Comp Biochem Physiol B Biochem Mol Biol* **180**: 40–47. doi:10.1016/j.cbpb.2014.10.003

Wesener DA, Wangkanont K, McBride R, Song X, Kraft MB, Hodges HL, Zarling LC, Splain RA, Smith DF, Cummings RD, et al. 2015. Recognition of microbial glycans by human intelectin-1. *Nat Struct Mol Biol* **22**: 603–610. doi:10.1038/nsmb.3053

Wangkanont K, Wesener DA, Vidani JA, Kiessling LL, Forest KT. 2016. Structures of *Xenopus* embryonic epidermal lectin reveal of conserved mechanism of microbial glycan recognition. *J Biol Chem* **291**: 5596–5610. doi:10.1074/jbc.m115.709212

Stanley P. 2017. What have we learned from glycosyltransferase knockouts in mice? *J Mol Biol* **428**: 3166–3182. doi:10.1016/j.jmb.2016.03.025

Yamakawa N, Vanbeselaere J, Chang LY, Yu SY, Ducrocq L, Harduin-Lepers A, Kurata J, Aoki-Kinoshita KF, Sato C, Khoo KH, et al. 2018. Systems glycomics of adult zebrafish identifies organ-specific sialylation and glycosylation patterns. *Nat Commun* **9**: 4647. doi:10.1038/s41467-018-06950-3

Sigal M, Del Mar Reinés M, Müllerké S, Fischer C, Kapalczynska M, Berger H, Bakker ERM, Mollenkopf H-J, Rothenberg ME, Wiedenmann B, et al. 2019. R-spondin-3 induces secretory, antimicrobial Lgr5$^+$ cells in the stomach. *Nat Cell Biol* **21**: 812–823. doi:10.1038/s41556-019-0339-9

Eckmair B, Jin C, Karlsson NG, Abed-Navandi D, Wilson IBH, Paschinger K. 2020. Glycosylation at an evolutionary nexus: the brittle star *Ophiactis savignyi* expresses both vertebrate and invertebrate *N*-glycomic features. *J Biol Chem* **295**: 3173–3188. doi:10.1074/jbc.ra119.011703

28 Discovery and Classification of Glycan-Binding Proteins

Maureen E. Taylor, Kurt Drickamer, Anne Imberty, Yvette van Kooyk, Ronald L. Schnaar, Marilynn E. Etzler, and Ajit Varki

This chapter provides an overview of naturally occurring glycan-binding proteins (GBPs), the history of their discovery, some of their biological functions, ways in which GBPs are identified, and challenges in defining their biologically relevant ligands. The chapters that follow describe the analysis of glycan–protein interactions (Chapter 29), the physical principles involved (Chapter 30), and the structures and biological functions of important subclasses of GBPs (Chapters 31–38).

TWO DISTINCT CLASSES OF GBPs

GBPs (which exclude glycan-specific antibodies) are found in all living organisms and fall into two overarching groups—lectins and sulfated glycosaminoglycan (GAG)-binding proteins (compared in Online Appendix 28A). Lectins are further classified into evolutionarily related families identified by carbohydrate-recognition domains (CRDs) based on primary and/or three-dimensional structural similarities (Figure 28.1). CRDs can exist as stand-alone proteins or as domains within larger multidomain proteins. They typically recognize terminal groups

published by Cold Spring Harbor Laboratory Press; doi: 10.1101/glycobiology.4e.28

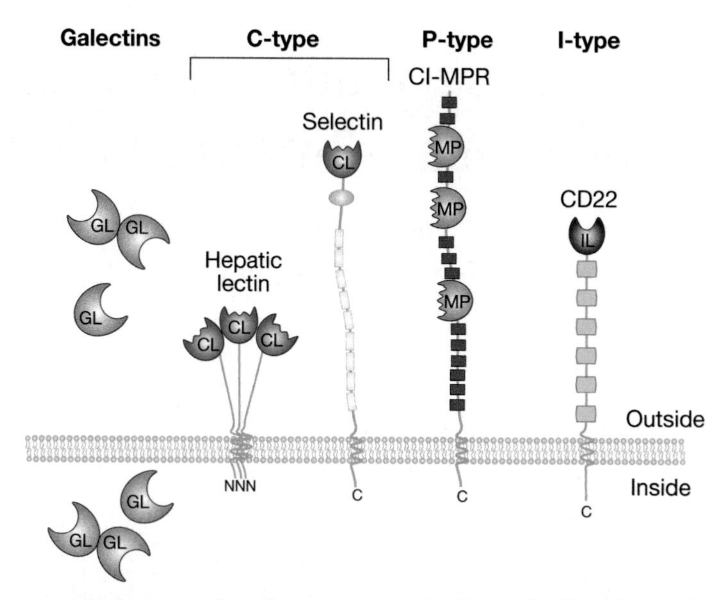

FIGURE 28.1. Representative structures from four common animal lectin families. The emphasis is on the extracellular domain structure and topology. The following are the defined carbohydrate-binding domains (CRDs) shown: (CL) C-type lectin; (GL) galectin; (MP) P-type lectin; (IL) I-type lectin. Other domains may include epithelial growth factor (EGF)-like domains; immunoglobulin C2-set domains; transmembrane domains; and complement regulatory repeats, depending on the family and specific family member. The number of domains accompanying the CRD varies among family members.

on glycans, which fit into shallow but well-defined binding pockets (Chapters 29 and 30). In contrast, proteins that bind to sulfated GAGs (heparan, chondroitin, dermatan, and keratan sulfates; Chapter 17) do so via clusters of positively charged amino acids that bind specific arrangements of carboxylic acid and sulfate groups along GAG chains. Most of these proteins are evolutionarily unrelated. GBPs that bind to the nonsulfated GAG hyaluronic acid (hyaladherins) share an evolutionarily conserved fold that facilitates recognition of short segments of the invariant hyaluronan repeating disaccharide (Chapter 16), so they are best classified as lectins rather than grouped with sulfated GAG-binding proteins. The rest of this chapter emphasizes lectins, different families of which are detailed in Chapters 31–37. Sulfated GAG-binding proteins are also discussed, and are detailed further in Chapter 38.

DISCOVERY AND HISTORY OF LECTINS

Lectins were discovered in plants in 1888 when extracts of castor bean seeds were found to agglutinate animal red blood cells. Subsequently, seeds of many plants were found to contain such "agglutinins," later renamed lectins (Latin for "select") when they were found to distinguish human ABO blood groups (Chapter 14), important for blood transfusions. Lectins are particularly common in the seeds of leguminous plants and these "L-type" lectins, including concanavalin A and phytohemagglutinin, have been extensively studied. Although their specific glycan-binding activities make such plant lectins extremely useful scientific tools, their biological functions in plants remain mostly unknown.

The first animal lectin discovered was the asialoglycoprotein receptor (ASGPR) identified by Anatol Morell and Gilbert Ashwell in the late 1960s during investigations of the turnover of a serum glycoprotein, ceruloplasmin. Like most glycoproteins circulating in blood, ceruloplasmin has complex N-glycans with sialic acid termini. To prepare radiolabeled ceruloplasmin, the terminal sialic acids were removed, leaving an exposed galactose. Surprisingly, asialoceruloplasmin had a circulation half-life (in rabbits) of minutes, whereas intact ceruloplasmin remained in the blood for hours. Glycoproteins with exposed Gal residues were rapidly cleared into liver cells via an endocytic cell-surface receptor that specifically bound to terminal β-linked Gal or GalNAc. ASGPR was purified by affinity chromatography using a column of immobilized asialoglycoprotein.

Other glycan-specific receptors involved in glycoprotein clearance and targeting were subsequently discovered, including mannose 6-phosphate receptors for targeting lysosomal enzymes to the lysosomes (Chapter 33) and mannose receptors that clear glycoproteins with terminal mannose or GlcNAc residues from the blood. Small soluble lectins specific for β-linked galactose (now called "galectins"; Chapter 36) were isolated by affinity chromatography of extracts from many biological sources ranging from the slime mold *Dictyostelium discoideum* to mammalian tissues. By the 1980s, the concept of vertebrate lectins that recognize specific glycans was well established. Although the first animal lectins identified were specific for endogenous glycans, many lectins specific for exogenous glycans of microorganisms were later found. Lectins recognizing exogenous glycans include soluble proteins that circulate in the blood of many species as well as membrane-bound receptors on cells of the immune system.

Lectins and sulfated GAG-binding proteins are also widespread in microorganisms, although they tend to be called by other names such as hemagglutinins and adhesins. The influenza virus hemagglutinin, which binds to sialic acid on host cells (Chapter 15), was the first GBP isolated from a microorganism. The viral hemagglutinins, like many plant lectins, can agglutinate red blood cells. Many bacterial lectins have also been described. They fall into two general classes: adhesins on bacterial surfaces that recognize glycans on host cell-surface glycolipids, glycoproteins, or GAGs to facilitate bacterial adhesion and colonization, and secreted bacterial toxins that bind to host cell-surface glycolipids or glycoproteins (Chapter 37).

DISCOVERY OF SULFATED GAG-BINDING PROTEINS

A large group of GBPs that defy classification based on sequence or structure recognizes sulfated GAGs (Chapter 38). The best-studied example is the interaction of heparin with antithrombin. Heparin was discovered in 1916 by Jay McLean, as a medical student, but it was not until 1939 that heparin was shown to be an anticoagulant in the presence of "heparin cofactor," which was then identified as antithrombin in the 1950s. Many other sulfated GAG-binding proteins were later discovered by affinity chromatography using columns of immobilized heparin. Growth factors and cytokines bearing clusters of positively charged amino acids along their protein surface interact with sulfated GAGs in a looser fashion—that is, they do not always show the high specificity seen with antithrombin. However, in some cases, specific GAG sequences mediate the formation of higher-order complexes, acting as a template for oligomerization or positioning of proteins such as fibroblast growth factor (FGF) and its cell-surface receptor.

MAJOR BIOLOGICAL FUNCTIONS OF GBPs

GBPs function in communication between cells in multicellular organisms and in interactions between microbes and hosts and can also be involved in binding growth factors, chemokines and cytokines. These interactions can take various forms, resulting in movement of molecules, cells, and information.

Trafficking, Targeting, and Clearance of Proteins

Directing movement of glycoproteins within and between cells is a common function for lectins in many organisms. In eukaryotic cells, including yeast as well as "higher" eukaryotes, several groups of lectins are important in glycoprotein biosynthesis and intracellular movement

(Chapter 39). In the endoplasmic reticulum (ER), two lectins—calnexin and calreticulin—bind monoglucosylated high-mannose glycans present on newly synthesized glycoproteins, forming part of a quality control system for protein folding. This binding keeps proteins in the ER until they are correctly folded. Other groups of lectins in the ER, including M-type lectins and proteins containing mannose 6-phosphate receptor homology domains, take part in the process of ER-associated glycoprotein degradation (ERAD), binding partially processed high-mannose glycans on terminally misfolded glycoproteins, causing them to be retrotranslocated into the cytoplasm for deglycosylation, followed by degradation in the proteasome. One of the best characterized functions of GBPs is in delivery of newly synthesized lysosomal enzymes from the *trans*-Golgi to lysosomes. P-type lectins (Chapter 33) recognize mannose 6-phosphate residues that have been added to N-glycans on lysosomal enzymes in the Golgi apparatus, targeting them to endosomes for fusion with lysosomes.

Once released from cells, glycoproteins can also be taken up for degradation in lysosomes. As noted above, the ASGPR on mammalian hepatocytes controls turnover of many serum glycoproteins by recognition of terminal Gal or GalNAc residues. Similarly, the mannose receptor on macrophages and sinusoidal cells of the liver binds and clears glycoproteins with oligomannose N-glycans that are released from cells during inflammation and tissue damage.

Not all lectin-mediated targeting leads to degradation. Glycan-binding subunits of secreted bacterial and plant toxins typically target them to glycolipids on cell surfaces and facilitate entry of the toxins into cells (Chapter 37). Many enzymes contain glycan-binding domains that bring another domain with enzyme activity into close proximity with its substrates. One notable group includes bacterial cellulases in which cellulose-binding modules position the enzymatic domain for optimal degradation of cellulose fibers. Following a similar principle, GalNAc-binding domains in polypeptide-N-acetylgalactosaminyltransferases that initiate O-linked glycosylation in animals position these enzymes to add further GalNAc residues to regions of polypeptides that already bear O-glycans (Chapter 10).

Cell Adhesion

Distinctive glycans on the surfaces of different eukaryotic and prokaryotic cells make them targets for GBPs. Binding of glycans on the surface of one cell by GBPs on an adjacent cell can induce recognition and adhesion, whereas cross-linking glycans on different cells by multivalent soluble lectins provides an alternative mechanism. Such interactions are exploited in specialized situations exemplified by transient contacts between moving cells. The selectins—three receptors that function in interactions between white blood cells, platelets, and endothelia—provide some of the best characterized examples of lectin–glycan interactions in cell–cell adhesion (Chapter 34). For example, L-selectin on lymphocytes binds glycans on specialized endothelial cells of lymph nodes to induce lymphocyte homing, wherein circulating lymphocytes leave the bloodstream and enter the lymph node. Other mammalian GBPs that mediate binding of cells to each other or that recognize ligands on the same cell surface include Siglecs (Chapter 35) and galectins (Chapter 36). Lectins in multicellular organisms also mediate interactions between cells and the extracellular matrix and support the organization of matrix components. For example, proteins containing link modules that bind specifically to hyaluronan in cartilage (and other tissues) are essential for structuring the extracellular matrix (Chapter 38), and other extracellular proteins bind to sulfated GAGs to organize cell–cell and cell–matrix interactions (Chapter 38).

Many bacteria also use lectins to adhere to glycans on host cells, often keeping them from getting washed away. These adhesins are usually present at the ends of long structures called pili or fimbriae that project from the surface of the bacteria (Chapter 37). Adhesion can be part of the infection process. For example, a mannose-specific adhesin on pathogenic strains of *Escherichia coli* that cause urinary infections binds to epithelial cells of the urinary tract. Other glycan–protein interactions between host cells and bacteria provide a mechanism for coexistence. Several bacterial species that are part of the normal gut flora, including nonpathogenic *E. coli,* use adhesins to bind to glycolipids present on cells lining the large intestine.

Immunity and Infection

Many lectins are involved in immune responses in invertebrates as well as in "lower" vertebrates and mammals. Differences between glycans on host and microbial cell surfaces are commonly the basis for innate immune responses. Phagocytosis is a common outcome of the binding of macrophage lectins to nonhost glycans on bacteria, parasites, and fungi, but many of these macrophage lectins, like DC-SIGN, also recognize host glycans on viruses for phagocytosis (Chapters 42 and 43). Other lectins circulating in the blood, such as serum mannose-binding protein and ficolins, bind to pathogen cell surfaces and activate the complement cascade, leading to complement-mediated killing.

Binding of glycans to lectins on immune cells can also trigger intracellular signaling that activates or suppresses cellular responses. Receptors that recognize self-glycans such as sialic acid, as well as several that are specific for glycans characteristic of microorganisms, can initiate such signaling. For example, binding of α2-6 linked Sia to CD22, a member of the Siglec family of vertebrate lectins found on B-lymphocytes, initiates signaling that inhibits activation to prevent self-reactivity (Chapter 35). In contrast, binding of trehalose dimycolate, a glycolipid found in the cell wall of *Mycobacterium tuberculosis* by the macrophage C-type lectin Mincle, induces a signaling pathway that causes the macrophage to secrete proinflammatory cytokines.

Finally, viruses often use their own GBPs to attach to host cells during infection (Chapter 37). Proteins on virus surfaces, including those on influenza virus, reovirus, Sendai virus, and polyomavirus, bind to sialic acids. In addition to bringing the virus into contact with their cell targets, these hemagglutinins typically induce membrane fusion, facilitating virus entry and delivery of nucleic acids into the cytosol. Glycan-binding receptors on viruses are often highly specific for a particular linkage; human influenza viruses preferentially bind to sialic acids α2-6-linked to Gal, whereas bird influenza viruses prefer α2-3-linked sialic acid. Other viruses, such as herpes simplex virus, have adhesins that bind to heparan sulfate proteoglycans on cell surfaces.

ORGANIZATION OF LECTINS

An important concept in identifying, defining, and classifying lectins is that glycan-binding activity is embodied in discrete protein modules or domains, referred to as CRDs. CRDs are typically independently folding segments of proteins; often one can separate the glycan-binding activity from other activities of the protein by expressing its CRD in isolation. In some cases, the CRDs constitute the entire GBP (Figure 28.2).

When a lectin is comprised simply of its CRD, its functions often are dependent on multivalency, which endows lectins with the ability to cross-link glycan-containing structures. This arrangement explains the ability of many plant lectins to agglutinate cells and to cluster glycoproteins on cell surfaces, which can induce mitogenesis as well as other signaling pathways. Other GBPs that function this way include the galectins, which can bridge glycans on one cell surface or between cells. Sometimes other activities are encoded within the structure of the same domain that binds glycans; some cytokines comprised of a single folded domain may have distinct sites for binding glycans and other target receptors. More commonly, other activities of lectins reside in separate modules in multidomain proteins (Figure 28.2). Such arrangements are widespread. The domains associated with CRDs perform many different functions, including binding other types of ligands, performing enzymatic reactions, anchoring

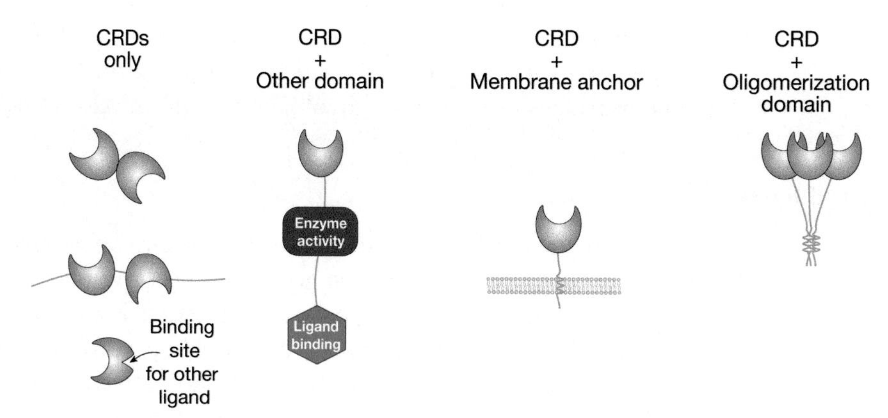

FIGURE 28.2. Arrangements of carbohydrate-recognition domains (CRDs) in lectins. Proteins containing just CRDs or CRDs associated with other types of functional domains, with membrane anchors or with oligomerization domains, are depicted schematically. A single lectin can contain all of these additional domains.

proteins to membranes, and directing oligomerization. GBPs often contain multiple modules, combining several functions in one protein.

Membrane anchors in lectins take multiple forms, but they often span the membrane, linking extracellular CRDs with cytoplasmic domains. This arrangement facilitates the flow of information between glycan-binding sites on the extracellular surface and the cytoplasm. Simple sequence motifs in the cytoplasmic domains of transmembrane lectins often control trafficking of receptors and their bound glycan ligands. Common functions of such intracellular movements are internalization of cell-surface receptors, directing bound ligands to endosomes and lysosomes, and movement through intracellular compartments such as the ER and Golgi apparatus to the cell surface. Flow of information in the opposite direction can lead to stimulation of signaling complexes on the cytoplasmic side of the membrane in response to glycan binding at the cell surface.

Clustering of glycan-binding sites (multivalency) is often critical to both recognition and biological functions and is achieved in different ways: by formation of simple oligomers of CRDs, as a result of the presence of multiple CRDs in a single receptor polypeptide, and through association of CRD-containing polypeptides via independent oligomerization domains. Some oligomers are stable, whereas others, such as those formed by some galectins, are in equilibrium with monomers. These arrangements facilitate multivalent binding to increase avidity and direct the geometrical arrangement of binding sites. Multiple CRDs may face in the same direction for surface recognition or in opposite directions to facilitate cross-linking. Multivalent CRDs may have fixed spacing or flexible spacing to accommodate different target glycans. In some cases, oligomerization domains also form structural features, serving as stalks that project CRDs from the cell surface. Oligomerization domains can also embody other functions, such as the protease-binding sites in the collagen-like domains of mannose-binding protein.

CLASSIFICATION OF LECTINS BASED ON STRUCTURAL SIMILARITIES

It is convenient to classify lectins based on the structures of the CRDs that they contain (Figure 28.3). CRDs are found in a large number of different structural categories, indicating that many different protein folds can accommodate glycan binding (Chapter 30). Based on this observation, glycan recognition must have evolved independently many times and the diversity of CRD structures must have arisen to address a diversity of functions.

GBPs appear across all domains of life, but the types of lectins in each kingdom vary considerably. Several families appear in both prokaryotes and eukaryotes, but their distributions

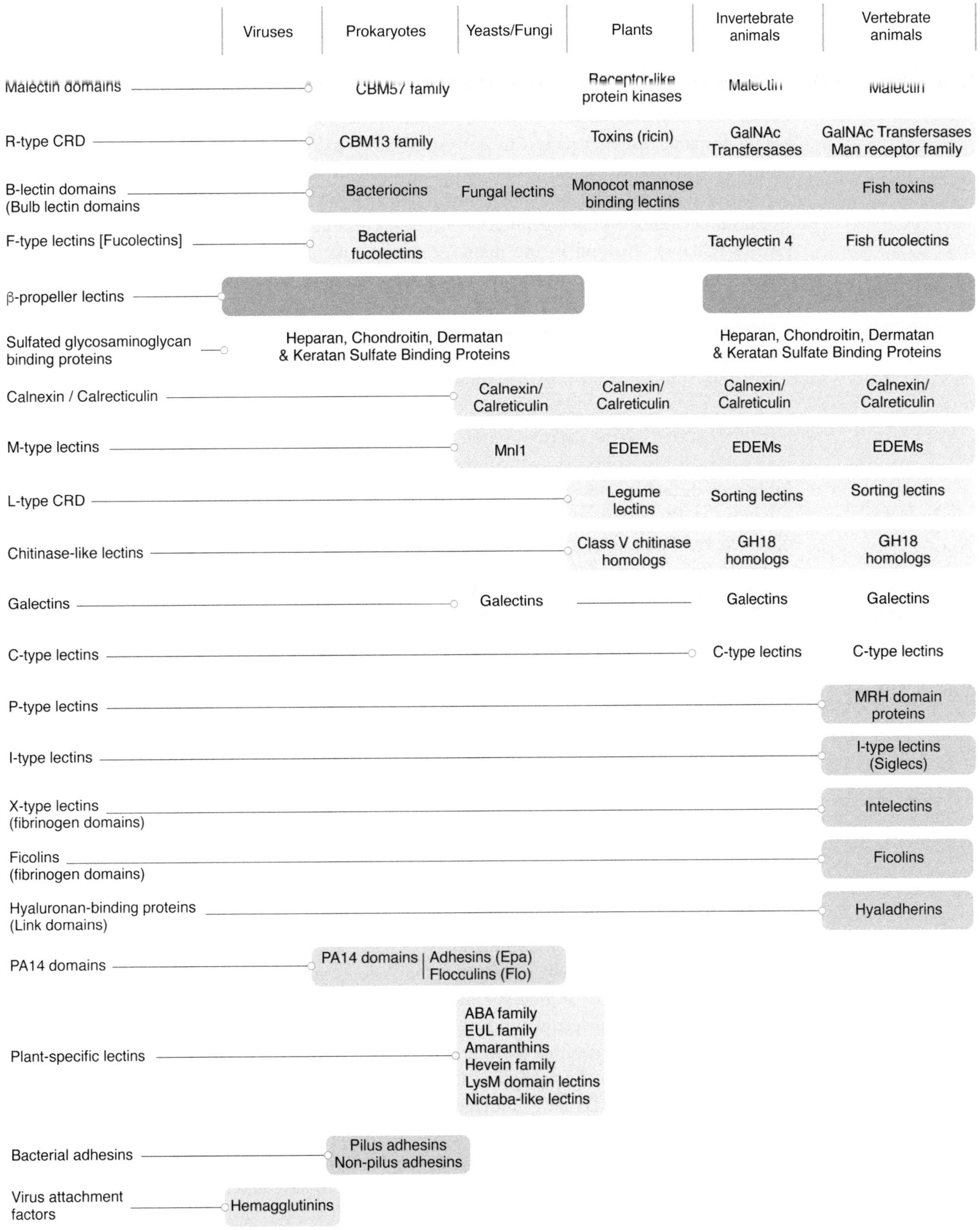

	Viruses	Prokaryotes	Yeasts/Fungi	Plants	Invertebrate animals	Vertebrate animals
Malectin domains		CBM57 family		Receptor-like protein kinases	Malectin	Malectin
R-type CRD		CBM13 family		Toxins (ricin)	GalNAc Transfersases	GalNAc Transfersases Man receptor family
B-lectin domains (Bulb lectin domains		Bacteriocins	Fungal lectins	Monocot mannose binding lectins		Fish toxins
F-type lectins [Fucolectins]		Bacterial fucolectins			Tachylectin 4	Fish fucolectins
β-propeller lectins						
Sulfated glycosaminoglycan binding proteins		Heparan, Chondroitin, Dermatan & Keratan Sulfate Binding Proteins			Heparan, Chondroitin, Dermatan & Keratan Sulfate Binding Proteins	
Calnexin / Calrecticulin			Calnexin/ Calreticulin	Calnexin/ Calreticulin	Calnexin/ Calreticulin	Calnexin/ Calreticulin
M-type lectins			Mnl1	EDEMs	EDEMs	EDEMs
L-type CRD				Legume lectins	Sorting lectins	Sorting lectins
Chitinase-like lectins				Class V chitinase homologs	GH18 homologs	GH18 homologs
Galectins			Galectins		Galectins	Galectins
C-type lectins					C-type lectins	C-type lectins
P-type lectins						MRH domain proteins
I-type lectins						I-type lectins (Siglecs)
X-type lectins (fibrinogen domains)						Intelectins
Ficolins (fibrinogen domains)						Ficolins
Hyaluronan-binding proteins (Link domains)						Hyaladherins
PA14 domains		PA14 domains	Adhesins (Epa) Flocculins (Flo)			
Plant-specific lectins				ABA family EUL family Amaranthins Hevein family LysM domain lectins Nictaba-like lectins		
Bacterial adhesins		Pilus adhesins Non-pilus adhesins				
Virus attachment factors	Hemagglutinins					

FIGURE 28.3. Several major structural families of glycan-binding proteins (GBPs) and their biological distributions. (CBM) carbohydrate-binding module; (EUL) *Euonymus europaeus* lectin; (ABA) *Agaricus bisporus* agglutinin; (EDEM) ER degradation-enhancing α-mannosidase-like proteins; (GH) glycohydrolase; (MRH) mannose receptor homology.

suggest different evolutionary histories. The malectin domain, although conserved in structure and widely distributed in prokaryotes, plants, and animals, is found in proteins with distinct domain organization and different functions in the three groups. Animal malectin is a membrane-anchored CRD of the ER that binds N-linked glycans during glycoprotein biosynthesis. In plants, the malectin CRD is expressed at the cell surface and is linked to a cytoplasmic kinase domain. Bacterial malectins consist of CRDs associated with glycohydrolase domains. Similarly, R-type CRDs (Chapter 31) in plants form the cell-surface-binding component of toxins such as ricin and are linked to glycohydrolase genes in bacteria, but in animals they appear in two distinct contexts: in polypeptide-N-acetylgalactosaminyltransferases that initiate O-GalNAc glycans (Chapter 10) and in the mannose receptor family. Although these CRDs have been adapted to serve different functions in different kingdoms, a glycan-binding function appears to have evolved early and been preserved in subsequent lineages.

In contrast to CRDs with broad evolutionary distribution, two other groups of lectins have sporadic distributions. B-Lectin domains, which are broadly distributed in bacteria in association with hydrolase domains, are found as isolated or tandem CRDs in monocot plants but not in other plants, in bony fishes but not in other animals, and in some fungi. The more recently identified F-type lectins appear in a few bacterial species and several "lower" vertebrates, but have not been found in mammals. In these instances, the presence of related domains in evolutionarily distant species may reflect lateral gene transfer rather than the presence of a precursor lectin in the distant common ancestor that they share. A different pattern of evolution is observed for PA14 domains, another type of CRD found in both bacteria and eukaryotes. Although the PA14 fold is relatively widespread, suggesting that it originated early and was retained across species, only a subset have been shown to have glycan-binding activity: CRDs associated with bacterial glycohydrolases and in adhesins and flocculation factors on the surface of yeast. Similarly, β-propeller lectins have evolved in multiple contexts because of their high avidity for glycans present on cell surfaces, and are present in bacteria, fungi, and animals.

The intracellular sorting lectins mentioned earlier—such as calnexin, calreticulin, and M-type lectins—are the most broadly distributed lectins that evolved from a common eukaryotic ancestor. Their distribution and the conservation of their functions probably reflect an ancient and conserved role in intracellular trafficking of glycoproteins in eukaryotes. Two other groups of CRDs appear to be found in metazoans but not simpler eukaryotes. The L-type CRDs have diverged in function between animals, in which they function in intracellular glycoprotein sorting and trafficking, and plants, in which they serve a protective function (Chapter 32). Chitinase-like glycan-binding domains across a range of species retain the ability to bind polymers of GlcNAc, but their biological functions are not well-understood, so it is unclear if they have shared roles in plants and animals.

In addition to the widely distributed families, certain CRD families are evolutionarily restricted. In addition to animal-specific and vertebrate-specific lectin groups, there are also groups such as the I-type lectins found only in mammals (Chapter 35). The pattern of evolution of animal-specific lectins varies. Galectins seem to be similar in organization in vertebrates and invertebrates, and it may be possible to identify orthologs in quite diverse species (Chapter 36). In contrast, C-type CRDs have undergone independent radiation in vertebrates and invertebrates, and identifying orthologs even between mouse and human proteins in some cases is difficult (Chapter 34). Of the 12 different protein folds found in plant lectins, six appear to be unique to plants. It is also noteworthy that viruses seem to have developed their own approaches to binding glycans rather than borrowing from hosts (Chapter 37).

In addition to families of proteins that share evolutionarily related CRDs, there are individual proteins that bind glycans through domains that are not related to CRDs in other proteins. Examples include proteins with dedicated glycan-binding domains, such as some laminin G domains that recognize glycans on α-dystroglycan (Chapter 45), pentraxins that bind modified and phosphorylated glycans, and macrophage $\alpha_M\beta_2$ integrin that binds fungal glucans and exposed GlcNAc residues on glycoproteins. Other proteins bind to glycans through domains that also have other ligands: Annexin V binds bisecting GlcNAc residues as well as phospholipids, and several cytokines have been reported to bind glycans as well as nonglycan receptors. Sulfated-GAG-binding proteins have also largely evolved by convergent evolution.

IDENTIFYING GBPs BY BIOLOGICAL AND BIOCHEMICAL FUNCTION AND STRUCTURAL SIMILARITY

There are multiple ways in which glycan recognition can be implicated in specific biological processes. One common approach is to show the ability of simple monosaccharides or small glycans to compete with a process. Information can often also be gained by modifying glycans on cells and glycoproteins with enzymes that add or remove glycans, by genetic manipulation, and by chemical inhibitors of glycan metabolism. These strategies have provided information about the glycans involved—for example, those needed for virus or toxin binding or those required for endocytosis of glycoproteins. Based on this information, it is then possible to look for GBPs that target these particular glycans, which can then be linked to the biological process.

The ability to bind specific glycans, assessed in various biochemical assays, has often been the basis for direct identification of novel GBPs without reference to a particular biological function. For example, most galectins (Chapter 36) share a binding preference for β-galactosides and F-type lectins for fucosyl residues. In addition to forming a basis for binding and competition assays, the binding activity is commonly used as a means of isolating these proteins by using affinity chromatography with appropriate immobilized glycan ligands. A wide variety of methods for coupling monosaccharides and complex glycans to create affinity resins have been developed. As mentioned above, many sulfated-GAG-binding proteins have been discovered by affinity chromatography using immobilized GAG chains. A limitation of these approaches is that binding activity does not directly indicate a biological function and the roles of many well-characterized GBPs have not been fully determined.

The observation that many lectins fall into structurally distinct families provides an alternative way to identify novel GBPs through analysis of protein sequences. Sequence motifs characteristic of CRDs are routinely used to screen sequences from whole genome sequencing. These motifs can also be used to screen specific cDNA and gene sequences of interest because of their association with biological functions. Detection of an appropriate motif suggests the presence of a CRD, and structural knowledge of known glycan-binding sites can suggest whether a novel protein is likely to retain glycan-binding activity. In some cases, it can even suggest potential ligands. Such predictions often motivate testing for glycan-binding activity, either by specifically examining binding to predicted ligands or by screening more generally using glycan arrays.

Although structure-based predictions do not directly yield information about biological function, the organization of CRDs and their association with other domains often provide information about potential functions. This type of top–down analysis is limited to discovery of GBPs that contain domains resembling known CRDs. As glycan array screening becomes more widely accessible, more broad-based screening can be envisioned.

GLYCAN LIGANDS FOR LECTINS

Monosaccharides or small oligosaccharides in isolation tend to be low-affinity ligands for GBPs, often with dissociation constants in the millimolar range. These intrinsic affinities are enhanced in several ways (Figure 28.4). At the level of individual glycans, affinity can be enhanced by linking the glycan to other types of structures. Conjugation of glycans to proteins and lipids can lead

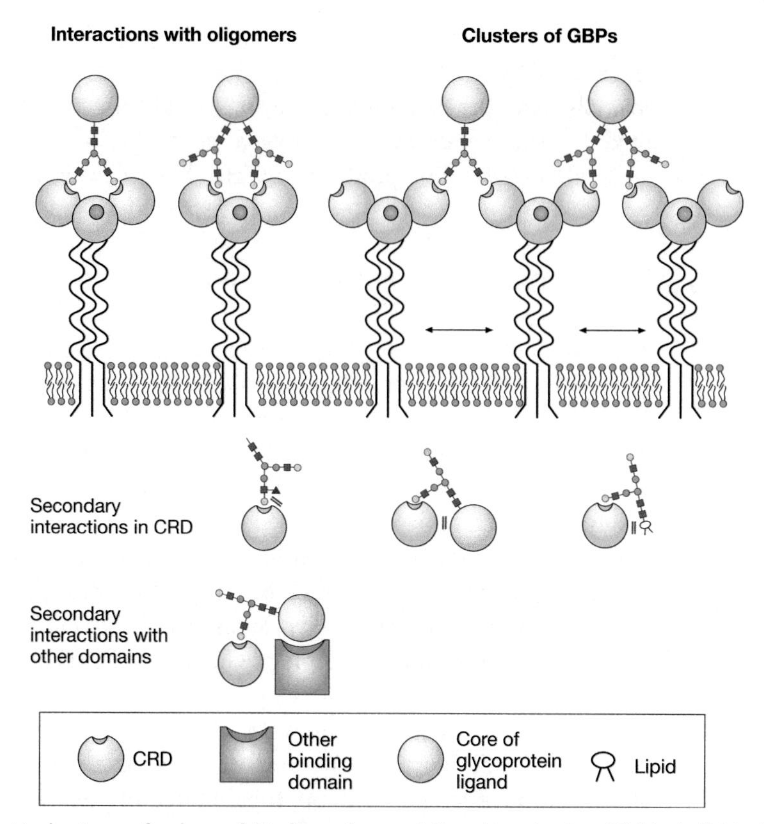

FIGURE 28.4. Mechanisms of enhanced binding of natural ligands to lectins. Within individual carbohydrate-binding domains (CRDs), secondary interactions beyond the primary binding site can be with glycan, protein, or lipid portions of glycoconjugate ligands. Multivalent interactions can reflect interaction of single branched oligosaccharides or multiple oligosaccharides attached to a glycoprotein with multiple CRDs brought together within receptor oligomers or in glycan-binding protein (GBP) clusters on the cell surface.

to enhanced CRD binding. For example, some GBPs such as the macrophage receptor Mincle bind to glycolipids with much higher affinity than they bind to free oligosaccharides. In this case, enhanced affinity can result from the presence of an extended or accessory binding site in a CRD adjacent to the glycan-binding site, which is able to accommodate the hydrophobic tail of the lipid. Other GBPs bind selectively to a particular glycan conjugated to a specific polypeptide motif. Optimal binding of P-selectin to the ligand PSGL-1 requires an O-linked glycan bearing a sialyl Lewis x structure on a peptide with adjacent acidic residues and sulfated tyrosines (Chapter 34). In yet other instances, glycan recognition is combined with other binding domains on a protein. The mannose receptor contains C-type CRDs that bind high-mannose oligosaccharides and a fibronectin type II repeat that binds to triple helical polypeptides. Together, these two modalities facilitate binding to fragments of collagen released at sites of inflammation.

A major determinant of binding to natural lectin ligands is often the interaction of multivalent glycans with clustered CRDs, resulting in high-avidity binding. Ligand clustering can result from multiple binding epitopes in a single oligosaccharide or polysaccharide, multiple glycans attached to a single protein scaffold, or the presence of adjacent glycoproteins or glycolipids in a cell membrane. Similarly, CRD clustering can reflect multiple CRDs in a single polypeptide, formation of polypeptide oligomers that each contains a single CRD, or from clustering of CRD-containing proteins in the cell membrane. Each of these levels of CRD organization has the potential to place geometrical constraints on the optimal arrangement of ligands, depending on the degree to which CRDs are held in a fixed arrangement or are flexibly linked. Clustering of glycans attached to a single polypeptide, particularly in heavily O-glycosylated proteins such as mucins, can also affect their ability to take on different conformations. Because lectins typically interact with a single conformation, there is an entropic penalty associated with binding any one of these which may be reduced when the glycan has fewer potential conformations. In vitro biochemical assays, including glycan arrays, reflect only some of these types of clustering of CRDs and ligands, so they must be interpreted with some caution. In some cases, binding of a CRD to isolated glycans may be essentially undetectable even though binding of the intact CRD-containing protein to its endogenous glycoconjugate may be highly selective and quite strong. Care must also be exercised in use of the term "ligand," to distinguish the glycan part of a ligand from the entire natural glycoconjugate or even cell surface.

TERMINOLOGY FOR SPECIFIC GBP LIGANDS

Based on the above considerations, GBPs may bind optimally to a glycan only when it is conjugated to a particular protein or lipid. In such instances, the GBP ligand is neither the glycan by itself nor the carrier by itself. Examples include P-selectin, which binds to sialyl Lewis x adjacent to sulfated tyrosines on the PSGL-1 protein of leukocytes (see above) and E-selectin, which binds to the same glycan on a variant form of the protein CD44 on hematopoietic stem cells (designated HCELL). Although the concept that GBPs bind glycans in the context of their carriers is well-established, there is no standard terminology to designate a particular glycoform as the ligand for a GBP. For example, stating that a protein (e.g., PSGL-1 or CD44) is "the ligand" for a GBP (P-selectin and E-selectin, respectively) is inaccurate, in that the protein without the specific glycan is not actually a ligand. On the other hand, giving a ligand a completely different name (e.g., HCELL) does not identify the polypeptide carrier. A suggested option is to use a superscript "L" (ligand) to designate these molecules as PSGL-1^{PSL} and $CD44^{ESL}$, respectively. This suggested terminology also has the advantage of distinguishing different glycoforms of the same polypeptide as ligands for different GBPs. For example, subsets of the glycoprotein CD24 recognized by P-selectin and Siglec-10 can be designated as $CD24^{PSL}$ and $CD24^{S10L}$, respectively. Regardless of the nomenclature used, direct proof of a functional interaction in vivo is needed to definitively assign a particular glycoprotein or glycolipid as a physiological GBP ligand.

ACKNOWLEDGMENTS

The authors acknowledge contributions to previous versions of this chapter by Richard Cummings and Jeffrey Esko and appreciate helpful comments and suggestions from T.N.C. Ramya and Gabriel Rabinovich.

FURTHER READING

Stillmark H. 1888. "Ueber Ricin, ein giftiges Ferment aus den Samen von Ricinus comm. L. und einigen anderen Euphoribiaceen." Inaugural dissertation. University of Dorpat, Dorpat (now Tartu), Estonia.

Goldstein IJ, Hughes RC, Monsigny M, Osawa T, Sharon N. 1980. What should be called a lectin. *Nature* **285**: 66. doi:10.1038/285066b0

Ashwell G, Harford J. 1982. Carbohydrate-specific receptors of the liver. *Annu Rev Biochem* **51**: 531–554. doi:10.1146/annurev.bi.51.070182.002531

Drickamer K. 1988. Two distinct classes of carbohydrate-recognition domains in animal lectins. *J Biol Chem* **263**: 9557–9560. doi:10.1016/s0021-9258(19)81549-1

Powell LD, Varki A. 1995. I-type lectins. *J Biol Chem* **270**: 14243–14246. doi:10.1074/jbc.270.24.14243

Lee RT, Lee YC. 2000. Affinity enhancement by multivalent lectin–carbohydrate interaction. *Glycoconj J* **17**: 543–551. doi:10.1023/a:1011070425430

Casu B, Lindahl U. 2001. Structure and biological interactions of heparin and heparan sulfate. *Adv Carbohydr Chem Biochem* **57**: 159–206. doi:10.1016/s0065-2318(01)57017-1

Esko JD, Selleck SB. 2002. Order out of chaos: assembly of ligand binding sites in heparan sulfate. *Annu Rev Biochem* **71**: 435–471. doi:10.1146/annurev.biochem.71.110601.135458

Drickamer K, Taylor ME. 2003. Identification of lectins from genomic sequence data. *Methods Enzymol* **362**: 560–567. doi:10.1016/S0076-6879(03)01037-1

Lee JK, Baum LG, Moremen K, Pierce M. 2004. The X-lectins: a new family with homology to the *Xenopus laevis* oocyte lectin XL-35. *Glycoconj J* **21**: 443–450. doi:10.1007/s10719-004-5534-6

Sharon N, Lis H. 2004. History of lectins: from hemagglutinins to biological recognition molecules. *Glycobiology* **14**: 53R–62R. doi:10.1093/glycob/cwh122

Blundell CD, Almond A, Mahoney DJ, DeAngelis PL, Campbell ID, Day AJ. 2005. Towards a structure for a TSG-6.hyaluronan complex by modeling and NMR spectroscopy: insights into other members of the link module superfamily. *J Biol Chem* **280**: 18189–18201. doi:10.1074/jbc.m414343200

Varki A, Angata T. 2006. Siglecs—the major subfamily of I-type lectins. *Glycobiology* **16**: 1R–27R. doi:10.1093/glycob/cwj008

Van Damme EJM, Lannoo N, Peumans WJ. 2008. Plant lectins. *Adv Bot Res* **48**: 107–209. doi:10.1016/S0065-2296(08)00403-5

Dam TK, Gerken TA, Brewer CF. 2009. Thermodynamics of multivalent carbohydrate-lectin cross-linking interactions: importance of entropy in the bind and jump mechanism. *Biochemistry* **48**: 3822–3827. doi:10.1021/bi9002919

Taylor ME, Drickamer K. 2009. Structural insights into what glycan arrays tell us about how glycan-binding proteins interact with their ligands. *Glycobiology* **19**: 1155–1162. doi:10.1093/glycob/cwp076

Adrangi S, Faramarzi MA. 2013. From bacteria to human: a journey into the world of chitinases. *Biotechnol Adv* **31**: 1786–1795. doi:10.1016/j.biotechadv.2013.09.012

Gilbert HJ, Knox JP, Boraston AB. 2013. Advances in understanding the molecular basis of plant cell wall polysaccharide recognition by carbohydrate-binding modules. *Curr Opin Struct Biol* **23**: 669–677. doi:10.1016/j.sbi.2013.05.005

Cohen M, Varki A. 2014. Modulation of glycan recognition by clustered saccharide patches. *Int Rev Cell Mol Biol* **308**: 75–125. doi:10.1016/b978-0-12-800097-7.00003-8

Nagae M, Yamaguchi Y. 2014. Three-dimensional structural aspects of protein–polysaccharide interactions. *Int J Mol Sci* **15**: 3768–3783. doi:10.3390/ijms15033768

Taylor ME, Drickamer K. 2014. Convergent and divergent mechanisms of sugar recognition across kingdoms. *Curr Opin Struct Biol* **28C**: 14–22. doi:10.1016/j.sbi.2014.07.003

Bishnoi R, Khatri I, Subramanian S, Ramya TN. 2015. Prevalence of the F-type lectin domain. *Glycobiology* **25**: 888–901. doi:10.1093/glycob/cwv029

Drickamer K, Taylor ME. 2015. Recent insights into structures and functions of C-type lectins in the immune system. *Curr Opin Struct Biol* **34**: 26–34. doi:10.1016/j.sbi.2015.06.003

Pees B, Yang W, Zárate-Potes A, Schulenburg H, Dierking K. 2017. High innate immune specificity through diversified C-type lectin-like domain proteins in invertebrates. *J Innate Immun* **8**: 129–142. doi:10.1159/000441475

Bonnardel F, Mariethoz J, Salentin S, Robin X, Schroeder M, Pérez S, Lisacek F, Imberty A. 2019. UniLectin3D, a database of carbohydrate binding proteins with curated information on 3D structures and interacting ligands. *Nucleic Acid Res* **47**: D1236–D1244. doi:10.1093/nar/gky832

29 Principles of Glycan Recognition

Richard D. Cummings, Ronald L. Schnaar, Jeffrey D. Esko,
Robert J. Woods, Kurt Drickamer, and Maureen E. Taylor

Glycans interact with many types of proteins including enzymes, antibodies, and lectins. Protein recognition and interactions with glycans represents a major way in which the information contained in glycan structures is deciphered and promotes biological activities. This chapter describes approaches to study the kinetics and thermodynamics of interactions between glycans and glycan-binding proteins (GBPs).

PROTEIN–GLYCAN RECOGNITION

A tremendous variety of GBPs are known and many are discussed in Chapter 28 and in many other chapters in this book. GBPs differ in the types of glycans they recognize and in their binding affinity and kinetics. The underlying structural basis by which a GBP binds with specificity and high affinity to a very limited number of glycans (or even a single glycan) among the many thousands that are produced by a cell is discussed in Chapter 30. A wide variety of physical techniques are used to identify and quantify protein–glycan interactions. Differential affinities of glycans for different GBPs revealed by these approaches provide insight into the biological roles of glycans and their cognate GBPs. Characterization of protein–glycan recognition using such techniques, in combination with structural studies by nuclear magnetic resonance (NMR) and crystallography, is useful to identify novel antagonists or inhibitors of GBPs. Such approaches are being used, for example, to develop inhibitors of neuraminidases to treat influenza virus infections (Chapter 42) and to screen for high-affinity inhibitors of selectins for the treatment of inflammatory disorders (Chapter 34).

published by Cold Spring Harbor Laboratory Press; doi: 10.1101/glycobiology.4e.29

HISTORICAL BACKGROUND

Much of the initial work on understanding protein–glycan interactions arose from studies on the combining sites of plant lectins and antibodies against specific blood group antigens. These studies led to the development of quantitative assays using glycans to inhibit binding interactions detected by cell agglutination or precipitation of targets, which provided early evidence for the importance of specific sugar structures in biological recognition events. Studies of protein–glycan interactions were instrumental in the development of techniques such as equilibrium dialysis and isothermal titration calorimetry, which are now widely used to analyze protein binding to a variety of types of ligands. On the other hand, methods used to study other types of protein–ligand interactions often need to be adapted to accommodate the specific properties of glycans and the proteins that interact with them.

Valency of GBP Interactions

Because many GBPs are oligomeric, with each subunit typically having a single carbohydrate-binding domain (carbohydrate-recognition domain [CRD]), many GBPs exhibit multivalent interactions with glycan ligands. Thus, although the CRD within a GBP may have a particular affinity for a ligand, the multivalent feature enhances binding through increased avidity and allows ligand cross-linking. Although the term "affinity" in this case, measured at equilibrium as a dissociation constant or affinity K_d, refers to the direct interaction of a single CRD with a monovalent ligand, "avidity," measured as an avidity K_d, refers to the overall strength of multivalent interactions (Figure 29.1). Some researchers use the term "apparent" K_d to denote the nonequilibrium nature of the measurements. Examples of oligomeric and multivalent GBPs include plant lectins, galectins, which are soluble GBPs that typically associate into dimers and higher oligomers, and soluble C-type lectins, such as serum collectins, which are generally oligomeric. In fact, some GBPs, such as the mannose receptor, the mannose 6-phosphate receptors, and some galectins (Chapters 33, 34, and 36), have multiple CRDs within a single

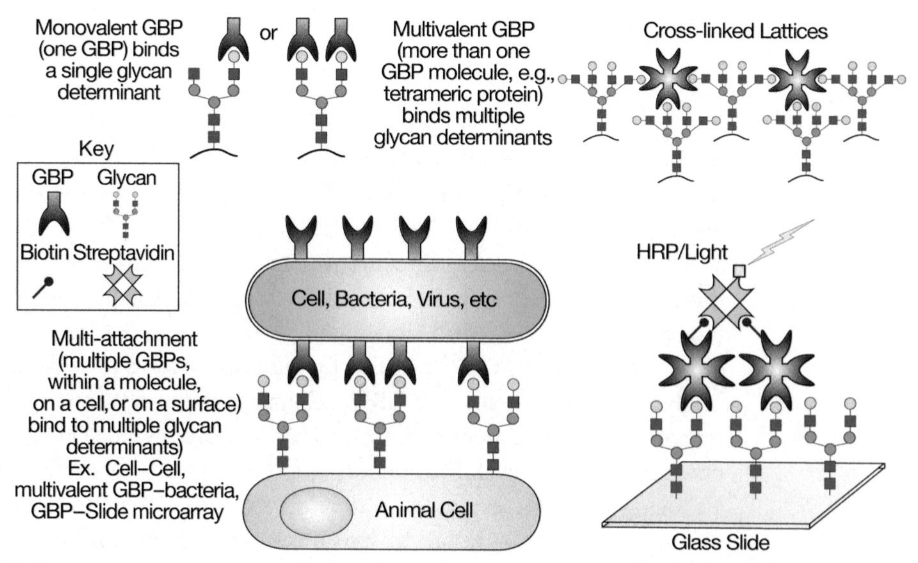

FIGURE 29.1. Monovalent and multivalent interactions of a glycan-binding protein (GBP) with monovalent or multivalent glycan ligands. A variety of interactions is possible, and these affect the equilibrium constant and contribute to the affinity K_d and the avidity K_d, as discussed in the text. An example is shown at the bottom of a microbe interacting with an animal cell in a multivalent manner, as well as a GBP binding in a multivalent manner to glycans on a printed slide or microarray.

polypeptide and can bind multiple ligands. In such cases, the single protein–glycan interaction may be weak (mM–μM K_d), as for the mannose receptor, which binds α-linked mannose with affinity in the low millimolar range, but which can bind with high avidity to the surface of fungi or microbes that may be rich in mannose-containing ligands.

Membrane receptors can also function as oligomeric complexes. For example, the C-type lectin P-selectin, which is dimeric, and membrane Siglecs (Sia-binding immunoglobulin-like lectins) both cluster on the cell surface in the presence of glycan receptors on opposing cells. Similarly, the influenza virus hemagglutinin is trimeric and is present in multiple copies on a virion, whereas cholera toxin is a soluble protein that is an AB_5 complex, consisting of a pentamer of glycan-binding B subunits associated with the catalytic and toxic subunit A. The glycan of ganglioside GM1 binds strongly to each B subunit of cholera toxin with high affinity because of specific and multiple interactions within each CRD (affinity K_d of ~40 nM). As a pentamer in the AB_5 form, cholera toxin has extremely high avidity (avidity K_d of ~40 pM) for cells expressing GM1, which can occur in clusters at the cell surface.

Most plant lectins are dimers or tetramers and are thus multivalent. Of course, the density of glycans on glycoproteins can also affect the affinity of binding, as some glycoproteins carry multiple multiantennary N-linked chains, each of which may interact with a CRD of the GBP (Figure 29.1). In addition, some multivalent lectins GBPs, as seen for galectins, may "bind and jump" within a molecule from glycan to glycan, a type of internal diffusion that can alter the entropy of interactions and promote higher avidity.

THERMODYNAMICS OF BINDING

The interaction of glycans with GBPs can be described thermodynamically and kinetically. Consider the simplest case in which a lectin (L) binds with a single site to a glycan (G) with a single binding determinant in a monovalent interaction. The interaction is governed by Equation 1 (Figure 29.2). At equilibrium the affinity constant, K, is defined as an association constant (or K_a) (Equation 2) and is equal to k_{on}/k_{off}, and the inverse of the K_a is the K_d or dissociation constant (Equation 3). Like any equilibrium constant, K is related to the standard free-energy change of the binding reaction at pH 7 (ΔG^o) in kcal per mole (Equation 4), in which R is the gas constant (0.00198 kcal/mol-K) and T is the absolute temperature (in °K). The affinity constant K is related to the thermodynamic parameters ΔG^o, ΔH^o, and ΔS^o (Equation 4), which represent the changes in free energy, enthalpy, and entropy of binding, respectively.

The on-rate k_{on} is expressed in units of $M^{-1}sec^{-1}$ or $M^{-1}min^{-1}$, whereas k_{off} is expressed in units of sec^{-1} or min^{-1}. Although it is important to define K_a, k_{on}, k_{off}, ΔG^o, ΔH^o, and ΔS^o for each binding phenomenon under consideration, investigators often discuss data in terms of the K_d (Equation 3), because the units are in concentration (millimolar, micromolar, nanomolar, etc.). It is also important to note that an equilibrium constant implies that the experimental

$$\text{Lectin (L) + Glycan (G)} \underset{k_{off}}{\overset{k_{on}}{\rightleftharpoons}} \text{LG} \qquad \textbf{Eq. 1}$$

$$K_a = \frac{[LG]}{[L][G]} = \frac{k_{on}}{k_{off}} \qquad \textbf{Eq. 2}$$

$$K_d = \frac{[L][G]}{[LG]} = \frac{1}{K_a} = \frac{k_{off}}{k_{on}} \qquad \textbf{Eq. 3}$$

$$\Delta G^o = -RT \ln K_a = RT \ln K_d = \Delta H^o - T\Delta S^o \qquad \textbf{Eq. 4}$$

FIGURE 29.2. Equations governing the interactions of a glycan-binding protein or lectin (L) with a glycan ligand (G). The terms are defined in the text.

setup for measuring it includes proof of actual equilibrium. In addition, the equilibrium constants are affected by temperature (increasing temperature decreases the equilibrium constant), buffer conditions, and potential cofactors, such as metals.

Whereas the binding of a monovalent GBP to a monovalent ligand is easily defined by the equilibrium kinetics described above, binding between multivalent ligands or GBPs involves multiple affinities, and the binding equilibria are more complex and more accurately described by a set of equilibrium constants. Typically, for multivalent ligands and GBPs, the values reported for affinity are apparent affinity constants and usually measure avidity.

TECHNIQUES TO STUDY PROTEIN–GLYCAN INTERACTIONS

There are many different ways to study the binding of glycans to proteins, and each approach has its advantages and disadvantages/limitations in terms of thermodynamic rigor, amounts of protein and glycan needed, and speed of analysis. Below is a discussion of some of the major ways in which the binding between a glycan and protein can be studied. Much of the available information about protein–glycan interactions derives from studies of relatively small glycan ligands interacting with a protein. In examining these interactions, two broad categories of techniques have been applied: (1) kinetic and near-equilibrium methods, such as equilibrium dialysis and titration calorimetry; and (2) nonequilibrium methods, such as glycan microarray screening, hapten inhibition, enzyme-linked immunoabsorbent assay (ELISA)-based approaches, and agglutination. In all of these approaches, the concepts of affinity versus avidity must be considered, and because of the multivalency of many GBPs and their ligands, it is difficult to precisely define the kinetic parameters, although the apparent affinity and avidity remain very useful measurements.

Kinetics and Near-Equilibrium Methods

Equilibrium Dialysis for Measuring K_d Values and Interaction Valency

Equilibrium dialysis is one of the earliest and simplest methods to study the binding of a GBP to a glycan. Although the technique is not used that often currently, understanding the principles of equilibrium dialysis helps to clarify the concept of measuring equilibrium constants. A solution of a GBP, such as a lectin or an antibody, is placed in a dialysis chamber of defined volume; the chamber must be permeable to glycans or other small molecules but not to the GBP. The chamber with the GBP inside is then placed in a larger known volume of buffer that contains the glycan in the concentration range of the expected K_d. After equilibrium is achieved, defined as no further changes in concentrations of the glycan either inside or outside the chamber, the total concentration of glycan in the chamber [In] is measured. The value [In] is a combination of bound glycan (associated with the GBP) plus free glycan versus the concentration of glycan outside the chamber [Out], which is the free glycan only. This difference in glycan [In] and [Out] will depend on the amount and affinity of the GBP. From this information, both the K_a and the valence n can be determined from the relationship

$$r/c = K_a n - K_a r, \tag{5}$$

in which r is the molar ratio of glycan bound to GBP, and c is the concentration of unbound glycan [Out]. The concentration of bound glycan is determined by subtracting [Out] from [In].

A plot of r/c versus r for different hapten concentrations approximates a straight line with a slope of $-K_a$. The valence of binding (number of binding sites per mole) is defined by the r intercept at an infinite hapten concentration. If such an analysis were performed with cholera toxin, for example, one would obtain five binding sites per mole of AB_5 complex or 1 mol per mole of B subunit. Nonlinear curve fitting is used to determine equilibrium constants, because older methods using linear conversions have inherent deficiencies and may distort experimental error.

As in any technique for determining binding constants, a number of important assumptions are made and their validity must be considered. These include demonstrating that the protein and glycan(s) are stable and active during the course of the experiment, the glycan is freely diffusible, the complex is at equilibrium, and structurally unrelated small molecules—not expected to bind—show no apparent binding in the experimental setup. If such binding is observed, this may be considered nonspecific and may be subtracted from the specific binding.

There are several advantages to equilibrium dialysis: (1) the approach is relatively easy, and highly sophisticated equipment is not needed; (2) if the affinity is high, then relatively small amounts of protein are needed (typically a few milligrams); (3) if the affinity is high, only small amounts of glycan may be required; (4) if the protein and glycan are very stable, they may be recovered and reused; (5) radioactive or fluorescent-tagged glycans may be used; and (6) reliable equilibrium measurements can be made. Some drawbacks of the approach are that (1) it provides the K_a but not the rate constants (k_{on} or k_{off}); (2) if the affinity of the GBP for the glycan is low, then relatively large amounts of GBP and glycan may be required; (3) the technique is not very adaptable to high-throughput analyses with multiple samples, and many individual measurements with different ligands may be tedious; and (4) many different measurements must be made if the range of affinity is unknown, and this may require many days or weeks to complete.

A variation of this technique is illustrated by the equilibrium gel-filtration method developed by Hummel and Dreyer. In the Hummel–Dreyer method, a GBP is applied to a gel-filtration column that has been preequilibrated with a glycan of interest that is easily detectable (e.g., by radioactive or fluorescent tagging). As the protein binds to the ligand, a complex is formed that emerges from the column as a "peak" above the baseline of ligand alone, followed by a "trough" (where the concentration of ligand is decreased below the baseline) that extends to the included or salt volume of the column. The amount of complex formed is easily determined by the known specific activity of the ligand. Because the amount of complex formed is directly proportional to the amount of protein (or ligand) applied, it is easy to calculate a binding curve from several different Hummel–Dreyer column profiles at different concentrations of either protein or ligand. This binding curve allows the calculation of the equilibrium constant of the interaction. The advantages and drawbacks of this technique are generally the same as for equilibrium dialysis, except that Hummel–Dreyer analyses are often quicker to perform and can be used with ligands of many different sizes. Such an approach has been invaluable in defining the equilibrium binding of selectins to their ligands.

Affinity Chromatography to Assess the Specificity of GBP Binding

Affinity chromatography is a technique typically used to identify interacting partners, but under some variations can be used to measure both affinity and specificity. In this affinity chromatography, a GBP is immobilized to an affinity support, such as Affi-Gel, CNBr-activated Sepharose, Ultralink, or some other activated support. If a glycan or a glycosylated macromolecule binds tightly to an immobilized GBP, a buffer containing a known glycan ligand may be added to force dissociation of the complex. For example, oligomannose-type and hybrid-type N-glycans will bind avidly to an agarose column containing the plant lectin concanavalin A (ConA-agarose) and 10–100 mM α-methyl mannoside is required to elute the bound material efficiently. In contrast, many highly branched complex-type N-glycans will not bind. Biantennary complex-type N-glycans bind to ConA-agarose, but they do not bind as tightly as high-mannose-type N-glycans and their elution can be achieved using 10 mM α-methyl glucoside. In this manner, one can assess the binding specificity of a GBP. In practice, this approach is rather crude,

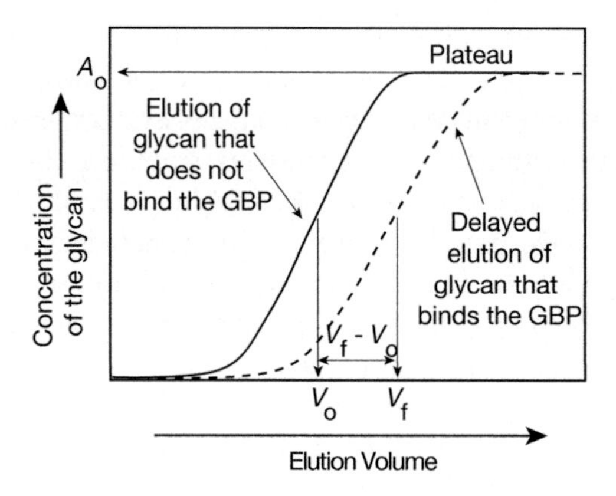

FIGURE 29.3. Example of frontal affinity chromatography, in which different concentrations of a glycan are applied to a column of immobilized GBP. The profile depicts the elution of one glycan that binds the GBP and the elution of another glycan that does not bind the GBP. The V_0 (void volume) and V_f are determined as shown from the elution volume. High V_f values indicate higher affinity.

and although it gives valuable practical information about the capacity of an immobilized lectin to bind specific glycans, it does not provide quantitative affinity measurements. A variant of this method is to immobilize the glycan ligand through covalent linkage or by capturing a biotinylated glycan on a streptavidin-linked surface and then measuring GBP binding.

A more sophisticated version of this approach, termed frontal affinity chromatography, can provide quantitative measurements of the equilibrium binding constants. In this technique, a solution containing a glycan of known concentration is continuously applied to a column of immobilized GBP, and the elution front of the glycan from the column is monitored. Eventually, enough ligand is added through continuous addition that its concentration in the eluant equals that in the starting material. If the glycan has no affinity for the GBP, it will elute in the void volume V_0; if, however, the glycan interacts with the GBP, it will elute after the V_0 and at a volume V_f (Figure 29.3).

The advantages of frontal affinity chromatography are similar to those discussed for equilibrium dialysis: (1) the approach is easy and inexpensive; (2) if the affinity is high, then relatively small amounts of protein are needed (typically a few milligrams), and only a single column is required; (3) correspondingly, small amounts of glycan may be used if the K_d is in the range of 10 nM to 10 mM; (4) if the glycans are stable, they may be recovered and reused; (5) radioactive glycans may be used; and (6) reliable equilibrium measurements can be made. There are some limitations to this approach, including (1) only the K_d can be derived, not k_{on} or k_{off}; (2) the conjugation of the GBP to the matrix must be stable and the protein must retain reasonable activity for many different column runs; (3) the amount of GBP conjugated and active must be defined; (4) many different column runs must be made with a single glycan; and (5) if the K_d is high (>1 mM), this approach is typically not feasible. Overall, frontal affinity chromatography is quite useful and is automated.

Another variation is generally termed a "pull-down assay," akin to a type of immunoprecipitation. In this approach a solution containing potential ligands is incubated with a GBP that may be immobilized on a surface (e.g., a bead). Afterward the bead-GBP is subjected to several steps (e.g., magnetic separation or centrifugation) to remove the unbound material. The material bound to the bead-GBP may then be eluted for measurement and further analyses. Using this pull-down setup, one can perform a concentration-dependent binding assay to obtain an apparent K_a of the ligand for the immobilized GBP.

Isothermal Titration Calorimetry to Measure K_d and Binding Enthalpy

Isothermal titration calorimetry (ITC) is one of the most rigorous means of defining the equilibrium binding constant between a glycan and a GBP or indeed any protein and its ligand. The binding of a glycan to the GBP is measured as a change in enthalpy using a commercial

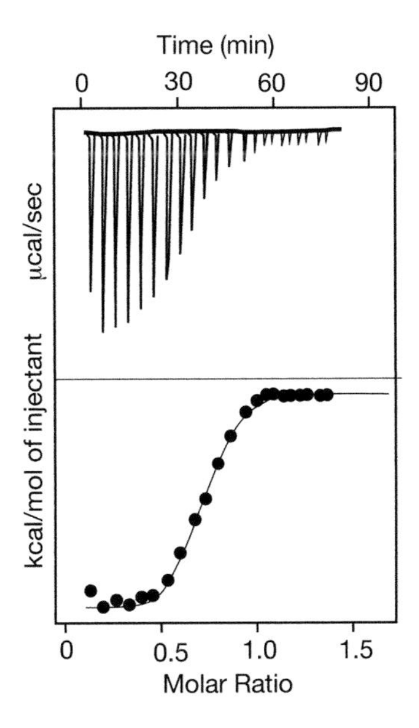

FIGURE 29.4. Example of isothermal titration calorimetry (ITC). (*Top*) Increasing amounts of a glycan are injected to a fixed amount of glycan-binding protein (GBP) in a cell, and the heat produced upon binding is measured as μcal/sec. (*Bottom*) The total kcal/mol of injected glycan relative to the molar ratio is plotted. These data can be used to directly define the thermodynamic parameters of binding and calculate the K_d of interaction between the glycan and the GBP.

microcalorimeter. In this technique, a solution containing a glycan of interest is added in increments into a solution containing a fixed concentration of GBP. The glycan is added at many intervals and the heat evolved from binding is measured relative to a reference cell. Over the course of the experiment, the concentration of glycan is increased in the mixing cell over a glycan-to-GBP molar ratio of 0–10. The heat absorbed or evolved during binding is determined and the data are replotted as kcal/mole of injectant versus the molar ratio (Figure 29.4). These data are then analyzed by replotting data to obtain the K_d. The heat change is directly related to the enthalpy of reaction ΔH^o. From knowledge of the K_d and ΔH^o, and using Equation 4, it is possible to define the binding entropy ΔS^o.

The major advantage of this approach is that it can provide thermodynamic information about the binding of a glycan to a GBP and is thus superior to equilibrium dialysis and affinity chromatography. The limitations of this approach are that (1) relatively large amounts of protein may be required to conduct multiple experiments (>10 mg); (2) relatively large amounts of glycans may be required; and (3) because of the above-mentioned problem, it is not typical for such analyses to use a wide range of different glycans. Nevertheless, this approach is rigorous and if the titration cell dimensions could be decreased in the future, then lower amounts of materials would be required.

Surface Plasmon Resonance to Measure the Kinetics of Binding and the K_d

Surface plasmon resonance (SPR) is used to measure association and dissociation kinetics of ligands (analytes) with a receptor. In SPR, the association of the analyte and receptor, with one or the other immobilized on a sensor chip, which incorporates a critical metal sensing surface, induces a change in total surface plasmon waves resulting in a change in the refractive index of the layer in contact with a gold film (Figure 29.5). This change is recorded as the SPR signal or resonance units (RUs). Binding is measured in real time, and information about the association and dissociation kinetics can thus be obtained, which in turn can be used to obtain K_a and K_d from Equations 2 and 3. There are several instruments available from various companies based on the principle of SPR.

A variety of chemistries are available for coupling ligand or receptor to the surface of the chip, including reaction with amines, thiols, or aldehydes and noncovalent biotin capture. In some approaches, a glycoprotein ligand for a GBP is immobilized and binding of the GBP is measured

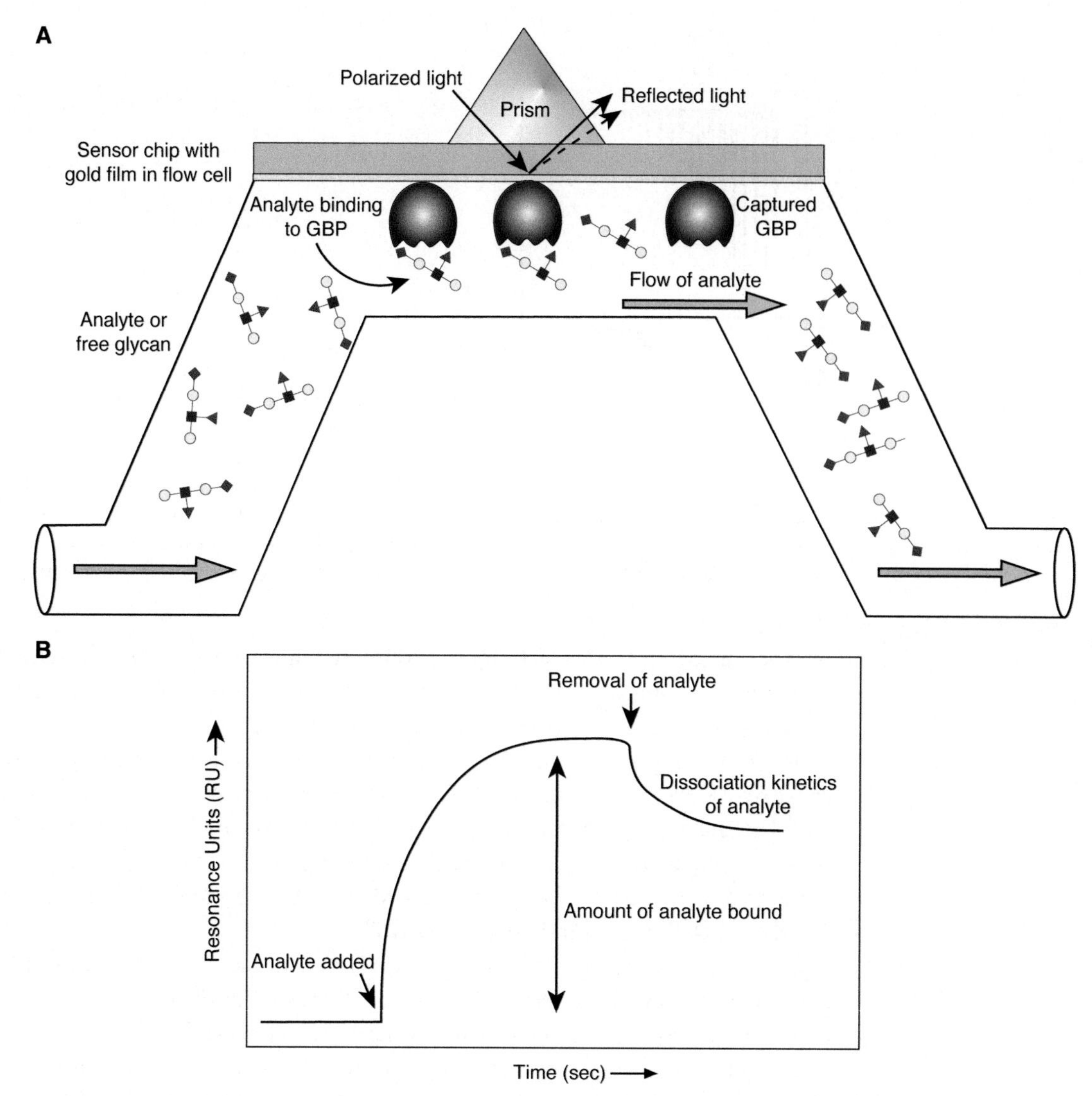

FIGURE 29.5. Example of surface plasmon resonance (SPR). (*A*) In SPR, the reflected light is measured and is altered in response to binding of the analyte in the flow cell to the immobilized GBP. (*B*) An example of a sensorgram showing the binding of the analyte to the ligand and the kinetics of binding and dissociation. RU indicates resonance units.

directly. It is also possible to degrade the immobilized glycoprotein ligand on the chip sequentially by passing over solutions containing exoglycosidases and reexamining at each step the binding to different GBPs, thereby obtaining structural information about the ligand. The immobilized ligand is usually quite stable and can be used repeatedly for hundreds of runs during a period of months.

The advantages of this approach are that (1) affinities in the range from millimolar to picomolar can be measured; (2) complete measurements of k_{on} and k_{off} are routine (see Equations 2 and 3), making calculations of K_d straightforward; (3) for immobilization of a molecule using amine coupling, only 1–5 μg is normally sufficient; (4) typically, the concentration range of the analyte is $0.1-100 \times K_d$ and the typical volumes needed are in the range of 50–150 μL; and (5) measurements are extremely rapid and complete experimental results can be obtained within a few days. The drawbacks of this approach are that (1) analytes must have sufficient mass to cause a significant change in SPR on binding (thus, the glycan is usually immobilized instead of the protein); (2) coupling of free glycans to the chip surface may be inefficient, and thus neoglycoproteins or some other type of large conjugate may be needed; and (3) there may be inhomogeneity in conditions on the SPR instrument because of mass transport effects, which could affect the dissociation rate and thus provide an inaccurate K_d measurement.

A related technology is biolayer interferometry (BLI), in which an optical biosensor analyzes the interference pattern of white light reflected from two surface. One surface carries the GBP immobilized in a biosensor tip, and the other surface is used as a reference. Binding of molecules to the biosensor tip changes the interference pattern in real time, thus allowing measurement of the kinetics of binding and dissociation. A clear advantage of this technique is that it does not require labels, such as fluorescent tags for detection (i.e., it is label-free). Additionally, BLI is generally performed in a 96- or 384-well format enabling it to be high-throughput. It is not quite as sensitive, however, as SPR, and thus small ligands, such as small oligosaccharides, may not provide robust signals. Like SPR, one of the interacting partners (e.g., the GBP) must be immobilized; with multiple tips it is difficult to ensure equal levels of immobilization. Nevertheless, this technique is easy to use and less costly compared to SPR, and when employed with the Cheng–Prusoff model in assays to determine oligosaccharide inhibition, BLI has been shown to generate solution K_d values that agree with those from NMR assays.

Fluorescence Polarization for Measuring K_d

Fluorescence polarization is an established technique but only relatively recently has it been applied to measure the binding constants of glycans to GBPs. This approach is based on the reduced rotational motion of a relatively small glycan when it is bound to a relatively large protein compared with the rotation of the free glycan in solution. Light absorbed by the fluorophore is emitted as fluorescence, but the angle of the emission relative to the incident light is depolarized by rotation of the molecule in solution, as measured through a filter to select molecules oriented close to the plane of incident polarized light. In practice, a fluorescently labeled glycan is incubated with increasing concentrations of a GBP and the fluorescence depolarization is measured. In the absence of the GBP, the fluorescently labeled glycan tumbles randomly and the degree of polarization is low. However, if the fluorescently labeled glycan binds to the GBP, its rate of rotation is diminished and the polarization remains high. By this approach, one can measure directly the K_d of the interaction as a function of the GBP concentration. The advantages of this technique are that (1) it is a homogeneous assay and provides direct measurements of the K_d in solution without derivatization of the GBP; (2) it is relatively simple and can provide rapid measurements of many compounds using microtiter plate–based approaches; (3) it uses relatively small amounts of glycan; (4) the concentrations of all the molecules are known; (5) it avoids complications of multivalent interactions because the glycans are monovalent and free in solution; and (6) it is amenable to inhibition by competitive agents and can be used to determine relative potency of compounds as inhibitors of GBPs. In the latter approach, a single fluorescently labeled glycan is mixed with the GBP, increasing concentrations of inhibitor glycans (which are not fluorescently labeled) are added, and the inhibition of binding is measured. Because the interactions are simple single-site competition, it is possible to use the concentration that causes 50% inhibition (IC_{50}) to derive the approximate K_d for the inhibitor. Some of the disadvantages are that (1) the technique is limited to small glycans ($\leq$2000 Da); (2) it requires fluorescence derivatization of the glycan (the fluorophore may alter the properties of the glycan); and (3) preparation of the glycan and chemical derivatization may be tedious and require large amounts of glycans. However, once a fluorescently labeled glycan has been generated, there is usually enough for many assays.

Additional Methods to Define Protein–Glycan Interactions

More complex approaches to measure noncovalent complexes between proteins and glycan ligands include mass spectrometry (MS) and NMR. Soft ionization methods are used to detect protein–glycan complexes in electrospray ionization MS and in matrix-assisted laser desorption/ionization time-of-flight (MALDI-TOF) MS. Although such methods have several advantages, including sensitivity and the ability to measure binding to multiple ligands in a single experiment, they may also be complicated to interpret, because the noncovalent complex may dissociate during the measurement, the detection may not be quantitative, and the ionization efficiency of the complex versus the free protein or ligand may differ. Another MS approach is hydrogen deuterium exchange (HDX) mass spectrometry, in which the bound ligand alters the deuterium uptake kinetics on the protein at the binding site of the ligand. Thus, the increased association of ligand decreases deuterium uptake, which can be measured on appropriate peptides proteolytically prepared from the protein after the exchange, and this can be plotted to obtain a binding isotherm. This method has the advantages of sensitivity and ease of use, but it also has drawbacks, including the expense of the approach and instrument time, the problem that relatively small glycans may have limited impact on deuterium exchange, and the problem that conformational changes can affect exchange rates in regions outside the CRD.

Several types of NMR spectroscopy measurements are useful for measuring protein–glycan interactions. Interactions can be measured in real time in solution and there is no need to separate the protein–glycan complex from the unassociated GBP or glycan. When weakly bound ligands are in fast exchange with free ligand, both line broadening and changes in chemical shifts can be used; however, the protein–glycan complex has to have clear and detectable chemical shifts distinct from the unbound materials. Saturation transfer difference (STD) measurements can also be used. Here, ^{1}H-NMR is used to detect glycans which are in large excess over GBP. Often these techniques work best for relatively low affinity interactions.

Another spectroscopic approach to measuring glycan binding to a protein relies on changes in intrinsic fluorescence on interacting with a glycan. Finally, atomic force microscopy (AFM) can also be used to define protein–glycan or glycan–glycan interactions. In this approach, the force required to separate a glycan-coated bead from an immobilized GBP can be measured in highly quantitative methods. Remarkably, single-molecule force measurements can be made by AFM. This method not only gives information about the strength of binding, but also provides insight into the molecular nature of noncovalent bond formation between the protein and its ligand.

Nonequilibrium Methods

ELISA to Measure Specificity and Relative Binding Affinity of Ligands

The conventional ELISA has been adapted for studying glycans and GBPs in a variety of formats. Of course, many glycans are antigens and antibodies to them can be analyzed in the conventional ELISA format. Some of the earliest ELISA-type approaches used biotinylated bacterial polysaccharides captured on streptavidin-coated microtiter plates to measure interactions of antibodies to the polysaccharides. In most types of ELISAs used in glycobiology, either an antibody or a GBP of interest is immobilized and the binding of a soluble glycan to the protein is measured, or the reagents are reversed and the glycan or glycoconjugate is immobilized. In either approach, the glycans are modified in some way, such as addition of biotin or a fluorophore, to allow their detection, or they may be coupled to another protein with an attached reporter group (e.g., a fluorescent moiety or an enzyme such as peroxidase).

Competition ELISA-type assays are also used to probe the binding site of a GBP. In this approach, a glycan is coupled to a carrier protein (the target), and its binding to an immobilized GBP is detected directly. Competitive glycans are added to the wells and their competition for the GBP is measured as a function of concentration to obtain an IC_{50}. Under appropriate conditions and concentrations of the ligand, the K_i values and K_d values are similar. The major advantages of this approach are that (1) it is relatively easy; (2) it has high-throughput capability and can be used in an automated fashion by robotic handling; (3) it can provide relative K_d values if the GBP concentration is varied appropriately over a large range and binding is saturable; and (4) it has the capacity to define the relative binding activity of a panel of glycoconjugates. The major limitations of this approach are that (1) it does not provide direct information about affinity constants or other thermodynamic parameters; (2) it can require relatively high amounts of GBP and glycans if used as a general screening array; and (3) it usually requires chemical derivatization of glycans or GBPs.

Glycan Microarrays to Assess Specificity

Glycan microarrays are an extension of both ELISA-type formats and modern DNA and protein microarray technology. In a glycan microarray, glycans are linked, usually covalently, to a solid surface through reaction with *N*-hydroxysuccinimide (NHS)-esters or epoxide-containing supports on a glass slide. Glycans are prepared to contain reactive primary amines at their reducing termini, but other chemical coupling methods are available. In addition, noncovalent immobilization can be used in which lipid-derivatized glycans or glycolipids are deposited onto nitrocellulose-coated slides. Also, neoglycoconjugates (e.g., glycans covalently attached to protein carriers as in neoglycoproteins) can be generated and then coupled to a surface to present glycans for interactions with GBPs. The glycans or glycan-containing materials are printed, much like DNA is printed for DNA microarrays, using contact printers or piezoelectric (noncontact) printing (Figure 29.6). Usually, a few nanoliters of solutions containing glycans in concentrations of $1-100$ μM are deposited by a robotic printer on the glass surface in ~100-μm-diameter spots. Slides are incubated for several hours to allow the chemical reactions to covalently fix the samples on the slides. The slides are then blocked to prevent nonspecific binding of reagents, and these microarrays overlaid with a buffer containing the GBP and incubated for several hours to allow equilibrium binding to occur. The slides are washed to remove unbound GBP and then analyzed. Analyses involve fluorescence detection, which means that either the GBP has to be directly fluorescent-labeled or a fluorescent-tagged antibody to the GBP must be used.

The chief features of the successful microarrays are the variety of glycans they contain. However, the clustered and relatively high amount of glycans densely packed that can bind detectable amounts of GBPs, may also promote binding of even relatively low affinity multivalent GBPs. Thus, the density of the ligand should be taken into account when interpreting the results. The type of linker used and the state of the monosaccharide at the reducing end (i.e., open ring, open or closed ring form attached to the linker) can also affect binding.

Binding is visualized or imaged as intensely fluorescent spots against a dark background. The data are visually imaged on a scanner and then graphically represented. In a typical successful analysis a GBP may bind to several glycans that share structural features, often termed glycan-binding determinants or motifs. If desired, the GBP can then be tested for its binding to the identified candidates by other methods such as titration microcalorimetry or fluorescence polarization, to define the K_d as discussed above. The use of microarrays in characterizing GBPs is a central component of functional glycomics (Chapter 51). Publicly available glycan microarray binding data repositories are increasingly used in the field.

A variation of glycan microarrays is a bead-based Luminex-type assay. In this high-throughput approach glycans are covalently immobilized on Luminex beads of different fluorophore

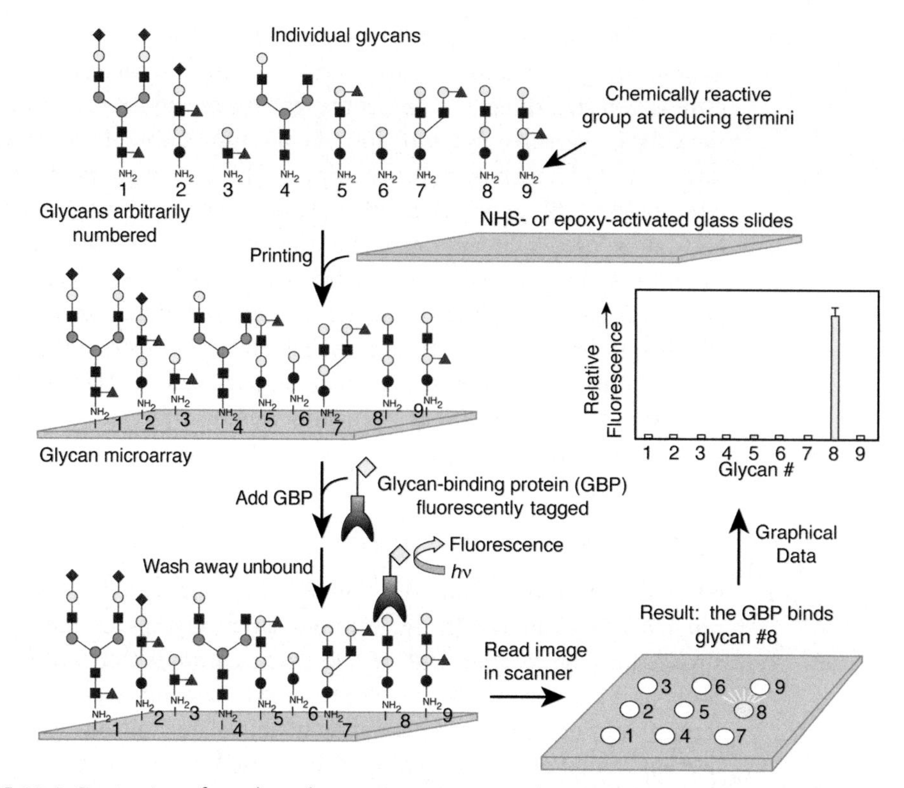

FIGURE 29.6. Preparation of covalent glycan microarrays printed on *N*-hydroxysuccinimide (NHS)- or epoxide-activated glass slides. In this example, the glycans have a free amine at the reducing end and are coupled to the glass slide. After washing to remove unattached materials, and blocking the reactive moieties on the side, the slide is "interrogated" with a glycan-binding protein (GBP). After washing away the unbound GBP, binding is detected by fluorescence. The GBP may be directly fluorescently labeled or detected using fluorescently labeled secondary reagents. The slides are read in a fluorescence scanner. An alternative method uses nitrocellulose-based microarrays, whereby glycolipids, neoglycolipids, glycoproteins, and other molecules can be printed noncovalently and blocked appropriately to reduce nonspecific binding. Otherwise, the assay formats are similar to those in the covalent glycan microarrays.

properties. Thus, a multiplexed approach is taken where a GBP is mixed with a set of conjugated beads each identifiable by its fluorescent properties. The GBP may be biotinylated, allowing detection by fluorescent streptavidin, or the GBP may be detected by binding to specific antibody or another reagent if the GBP is epitope-tagged. The degree of binding of the GBP is a measure of its affinity and specificity for a particular glycan. The advantages of this approach is that it is high-throughput, allowing thousands of assays to be performed automatically, as well as utilizing small amounts of both glycan and GBP. The collection of data and its analysis are also automated. A limitation of the approach is that each glycan to be derivatized must be activated by having a primary amine or other chemistry for conjugation, and each conjugation of a glycan needs to be independently validated.

Agglutination

Multivalent GBPs interacting with multivalent ligands, as expressed on cells, can cause the cells to agglutinate. This can be exploited by allowing one to measure the ability of a soluble glycan to block the agglutination activity of the GBP. The concentration of the soluble glycan that provides 50% inhibition of agglutination is taken as the inhibitory concentration (IC_{50}). Such approaches have been used for many years in studies on lectin-induced agglutination of cells and were useful in elucidating the nature of the human blood group substances. If a sufficiently large panel of soluble glycans is used, then the relative efficacies of each of these can be measured to help define the specificity of the GBP. A major advantage of this technique is that it does not require tagging of the glycans. Furthermore, polystyrene or dextran beads modified with discrete glycans can be used in lieu of cells. In this case, the glycans on the agglutinating particle are better defined. Usually, the IC_{50} does not relate directly to the binding affinity, because inhibition is being measured. The actual binding affinity must be defined by other techniques described earlier in this chapter.

Precipitation

The interaction of a multivalent GBP or antibody with a multivalent ligand allows formation of cross-linked complexes in solution. In many cases, these complexes become insoluble and precipitate. Precipitation may be highly specific and reflects the affinity constant of the ligand for the receptor. To quantify this interaction a fixed amount of GBP or antibody is titrated against a glycoprotein or a glycan to which it binds and a precipitate will form at a precise ratio of ligand to receptor. The amount of protein or ligand in the precipitate can be measured directly by chemical means, using assays for glycans or proteins. The technique of precipitation is useful for studying potentially multivalent ligands, and it has been used recently to show that each branch of terminally galactosylated complex-type di-, tri-, and tetra-antennary N-glycans is independently recognized by galactose-binding lectins. Another precipitation approach takes advantage of the fact that a complex between a GBP and a glycan can be "salted out" or precipitated by ammonium sulfate. A variation of this approach was used in early studies on the characterization of the hepatocyte asialoglycoprotein receptor (AMR), in which the ligand ^{125}I-labeled asialoorosomucoid was incubated with a preparation of receptor. The sample was treated with sufficient ammonium sulfate to precipitate the complex but not the unbound ligand. The precipitated complex was captured on a filter and the amount of ligand in the complex was determined by γ-counting.

Electrophoresis

In this approach, a glycoprotein (or ligand) is mixed with a GBP or antibody and the mixture is electrophoretically separated in polyacrylamide. For glycosaminoglycans, this technique is termed affinity co-electrophoresis (ACE). This method is particularly useful in defining the apparent K_d of the interaction and allows identification of subpopulations of glycosaminoglycans that differentially interact with the GBP. In another method, termed crossed affinity immunoelectrophoresis, a second step of electrophoresis is conducted in the second perpendicular dimension across an agarose gel that contains precipitating monospecific antibody to the glycoprotein or ligand. The gel is then stained with Coomassie Brilliant Blue and an immunoelectrophoretogram is obtained. Glycoproteins not interacting with the GBP or antibody have faster mobility than the complex. The amount of glycoprotein or ligand is determined by the area under the curves obtained in the second dimensional analysis. This method is useful for studying glycoforms of proteins and has been particularly valuable in analyzing glycoforms of α1-acid glycoprotein (an acute-phase glycoprotein) in serum and changes in its α1-3 fucosylation.

Expression of Copy DNAs for Ligands and Receptors

An indirect approach to studying protein–glycan interactions is to express a copy DNA (cDNA) encoding a glycosyltransferase in an animal or bacterial cell (Chapter 56). The adhesion of the modified cell (either transiently or stably transfected) to a GBP or antibody is then measured and taken to reflect the binding of the GBP or antibody to the new glycans (neoglycans) on the cell surface. Conversely, a cDNA encoding a GBP is expressed in cells and their ability to bind glycan ligands is tested. The expression of selectins, Siglecs, and other GBPs on the cell surface of transfected cells has been helpful in evaluating the roles of GBPs in cell adhesion under physiological flow conditions.

FURTHER READING

Cheng Y and Prusoff WH. 1973. Relationship between the inhibition constant (K_I) and the concentration of inhibitor which causes 50 per cent inhibition (I_{50}) of an enzymatic reaction. *Biochem Pharmacol* **22**: 3099–3108. doi:10.1016/0006-2952(73)90196-2

Wu X, Linhardt RJ. 1998. Capillary affinity chromatography and affinity capillary electrophoresis of heparin binding proteins. *Electrophoresis* **19**: 2650–2653. doi:10.1002/elps.1150191514

Fukui S, Feizi T, Galustian C, Lawson AM, Chai W. 2002. Oligosaccharide microarrays for high-throughput detection and specificity assignments of carbohydrate–protein interactions. *Nat Biotechnol* **20**: 1011–1017. doi:10.1038/nbt735

Leppänen A, Penttilä L, Renkonen O, McEver RP, Cummings RD. 2002. Glycosulfopeptides with *O*-glycans containing sialylated and polyfucosylated polylactosamine bind with low affinity to P-selectin. *J Biol Chem* **277**: 39749–39759. doi:10.1074/jbc.m206281200

Berger G, Girault G. 2003. Macromolecule-ligand binding studied by the Hummel and Dryer method: current state of the methodology. *J Chromatog* **797**: 51–61. doi:10.1016/s1570-0232(03)00482-3

Duverger E, Frison N, Roche AC, Monsigny M. 2003. Carbohydrate–lectin interactions assessed by surface plasmon resonance. *Biochimie* **85**: 167–179. doi:10.1016/s0300-9084(03)00060-9

Hirabayashi J. Oligosaccharide microarrays for glycomics. 2003. *Trends Biotechnol* **21**: 141–143. doi:10.1016/s0167-7799(03)00002-7

Sorme P, Kahl-Knutson B, Wellmar U, Nilsson UJ, Leffler H. 2003. Fluorescence polarization to study galectin–ligand interactions. *Methods Enzymol* **362**: 504–512. doi:10.1016/s0076-6879(03)01033-4

Bucior I, Burger MM. 2004. Carbohydrate–carbohydrate interactions in cell recognition. *Curr Opin Struct Biol* **14**: 631–637. doi:10.1016/j.sbi.2004.08.006

Homans SW. 2005. Probing the binding entropy of ligand–protein interactions by NMR. *Chembiochem* **6**: 1585–1591. doi:10.1002/cbic.200500010

Nakamura-Tsuruta S, Uchiyama N, Hirabayashi J. 2006. High-throughput analysis of lectin–oligosaccharide interactions by automated frontal affinity chromatography. *Methods Enzymol* **415**: 311–325. doi:10.1016/s0076-6879(06)15019-3

Paulson JC, Blixt O, Collins BE. 2006. Sweet spots in functional glycomics. *Nat Chem Biol* **2**: 238–248. doi:10.1038/nchembio785

Dam TK, Gerken TA, Brewer CF. 2009. Thermodynamics of multivalent carbohydrate–lectin cross-linking interactions: importance of entropy in the bind and jump mechanism. *Biochemistry* **48**: 3822–3827. doi:10.1021/bi9002919

Rillahan CD, Paulson JC. 2011. Glycan microarrays for decoding the glycome. *Annu Rev Biochem* **80**: 797–823. doi:10.1146/annurev-biochem-061809-152236

de Paz JL, Seeberger PH. 2012. Recent advances and future challenges in glycan microarray technology. *Methods Mol Biol* **808**: 1–12.

Smith DF, Cummings RD. 2013. Application of microarrays for deciphering the structure and function of the human glycome. *Mol Cell Proteomics* **12**: 902–914. doi:10.1074/mcp.r112.027110

Hatakeyama T. 2014. Equilibrium dialysis using chromophoric sugar derivatives. *Methods Mol Biol* **1200**: 165–171. doi:10.1007/978-1-4939-1292-6_15

Dam TK, Brewer CF 2015. Probing lectin – mucin interactions by isothermal titration microcalorimetry. *Methods Mol Biol* **1207**: 75–90. doi:10.1007/978-1-4939-1396-1_5

Palma AS, Feizi T, Childs RA, Chai W, Liu Y 2015. The neoglycolipid (NGL)-based oligosaccharide microarray system poised to decipher the meta-glycome. *Curr Opin Chem Biol* **18**: 87–94. doi:10.1016/j.cbpa.2014.06.009

Xia L, Gildersleeve JC. 2015. The glycan array platform as a tool to identify carbohydrate antigens. *Methods Mol Biol* **1331**: 27–40. doi:10.1007/978-1-4939-2874-3_3

Cockburn D, Wilkens C, Dilokpimol A, Nakai H, Lewinska A, Abou Hachem M, Svensson B. 2016. Using carbohydrate interaction assays to reveal novel binding sites in carbohydrate active enzymes. *PLoS ONE* **11**: e0160112. doi:10.1371/journal.pone.0160112

Dupin L, Noël M, Sonnet S, Meyer A, Géhin T, Bastide L, Randriantsoa M, Souteyrand E, Cottin C, Vergoten G, et al. 2018. Screening of a library of oligosaccharides targeting lectin LecB of *Pseudomonas aeruginosa* and synthesis of high affinity oligoglycoclusters. *Molecules* **23**: 3073. doi:10.3390/molecules23123073

Ji Y, Woods RJ. 2018. Quantifying weak glycan-protein interactions using a biolayer interferometry competition assay: applications to ECL lectin and X-31 influenza hemagglutinin. *Adv Exp Med Biol* **1104**: 2590273. doi:10.1007/978-981-13-2158-0_13

Purohit S, Li T, Guan W, Song X, Song J, Tian Y, Li L, Sharma A, Dun B, Mysona D, et al. 2018. Multiplex glycan bead array for high throughput and high content analyses of glycan binding proteins. *Nat Commun* **9**: 258. doi:10.1038/s41467-017-02747-y

Sood A, Gerlits OO, Ji Y, Bovin NV, Coates L, Woods RJ. 2018. Defining the specificity of carbohydrate-protein interactions by quantifying functional group contributions. *J Chem Inf Model* **58**: 1889–1901. doi:10.1021/acs.jcim.8b00120

30 Structural Biology of Glycan Recognition

Jesús Angulo, Jochen Zimmer, Anne Imberty, and James H. Prestegard

The biological effects that glycans elicit are frequently dependent on recognition of specific glycan features by the proteins with which they interact. In this chapter, some of the key structural features underlying glycan–protein interactions, as well as the primary experimental methods that have led to an understanding of these features, are discussed, specifically X-ray crystallography, nuclear magnetic resonance (NMR), cryo-electron microscopy, and computational modeling.

BACKGROUND

As emphasized in previous chapters, the numbers of distinct glycans produced by various organisms are enormous, but at the same time, glycans lack the diversity in functional groups displayed by other molecules. To achieve specificity in glycan recognition, proteins rely as much on the stereospecific placement of glycan hydroxyl groups at chiral centers, use of different linkage sites, and extensive branching as they rely on specific modifications of hydroxyl groups by processes such as sulfation, phosphorylation, and esterification. This puts placement of various residues and functional groups in three dimensions at a premium. Building

published by Cold Spring Harbor Laboratory Press; doi: 10.1101/glycobiology.4e.30

a three-dimensional picture of how recognition of glycans by proteins occurs is therefore essential if we are to understand how glycans are synthesized and recognized in the many physiological and pathological processes they control. It is also essential if we are to use knowledge of glycan recognition as a basis for the production of therapeutic agents that can control these processes in the event of disease. Building a structure depicting glycan recognition is not without its challenges. Most glycans are highly dynamic in solution, sampling many conformations. Often, a single or a small subset of conformations is selected when a complex forms. This works against the formation of stable complexes for structural studies and the direct use of solution conformational data in defining conformations of bound glycans.

The search for a structural basis of glycan recognition by proteins is not new. The concept of glycans fitting into pockets on protein surfaces dates back to Emil Fischer, who used the phrase "lock and key" to refer to enzymes that recognize specific glycan substrates. Lysozyme was the first "carbohydrate-binding protein" to be crystallized and have its three-dimensional structure determined. Subsequent work in the late 1960s and early 1970s led to a structure complexed with a tetrasaccharide that confirmed the existence of specific interactions occurring between sugars and proteins, and the ability of proteins to select the appropriate "key" from numerous possibilities.

Today, protein crystallography has reached a very high degree of sophistication and is responsible for the vast majority of the more than 170,000 structures deposited in the Protein Data Bank (PDB); however, producing a structure with ligands in place is still challenging. The structures that exist tend to have ligands that are relatively small and interact with particularly high binding constants. Glycan recognition frequently involves contacts with multiple residues to achieve specificity. So, native glycan ligands are often larger than other types of ligands. Often, high avidity is achieved through multivalent interactions, in which case the affinity for an isolated ligand–protein interaction is small. Nevertheless, there are a significant number of crystal structures for glycan–protein complexes, and these have contributed greatly to our understanding of the types of interactions that make glycan recognition possible.

Structural information on bound glycan ligands that is complementary to that from X-ray crystallography is increasingly coming from NMR methods. This is particularly valuable in that it is applicable to ligands with a broader range of affinities, including many that have the lower affinities amplified in multivalent interactions. It is also applicable in solution under near-physiological conditions in which concerns about the effects of crystal lattice contacts and occlusion of some interaction sites are absent. It is even possible to conduct some experiments on assemblies that mimic a membrane surface environment, an environment in which many protein–glycan interactions occur.

It is important to note that structural methodology is continually evolving, with additional information coming from techniques like small-angle X-ray scattering (SAXS) and cryo-electron microscopy (cryo-EM). Recent advances in cryo-EM provide many exciting opportunities to study protein–glycan interactions, which will also be discussed.

The fundamental understanding of glycan–protein interactions, as enriched by experimental studies of all types, has now been encoded in powerful molecular simulation programs that provide a computational approach to generating three dimensional pictures of glycan–protein complexes. These are important because it is difficult to produce complex glycan ligands in the amounts and purity required for most experimental approaches. These methods, although still evolving toward increased confidence in outcomes, provide models for experimentally inaccessible systems that can be tested with a variety of nonstructural approaches. They can also be leveraged with sparse structural data that alone could not provide detailed structural information.

CRYSTALLOGRAPHY

X-ray crystallography is a very powerful method for obtaining details of protein–ligand interactions. It excels in terms of the size range of molecules that can be studied (from small compounds to large multiprotein complexes) and in efficiency of data collection when high-energy X-ray beams at synchrotron sources are used. One of the limitations is still the crystallization step. Crystals of protein–carbohydrate complexes can be obtained by co-crystallizing the two partners or by soaking the carbohydrate ligand into an existing protein crystal. Because the quality of the crystal defines the limit of the diffraction pattern, and therefore the resolution of the structure, flexible oligosaccharide ligands may create structural heterogeneity and therefore limit the quality of the crystal. High-quality crystals of lectins are generally obtained with glycan ligands ranging from mono- to trisaccharides; glycosaminoglycan (GAG)-binding proteins or antibodies, which can bind much larger ligands, are more rarely crystallized in complex with carbohydrate ligands.

Diffraction data are now typically collected at very low temperatures, to protect molecules from radiation damage on high-energy synchrotron beam lines. Because freezing may damage the crystals owing to ice formation, glycerol is often used as cryoprotectant. Glycerol, with its carbohydrate-like hydroxylated carbons, is therefore frequently observed in glycan-binding sites, providing information about the amino acids involved in binding but sometimes competing with the carbohydrate ligand. Often, collaborative efforts with synthetic carbohydrate chemists are necessary to design, for example, nonhydrolyzable carbohydrate derivatives to obtain substrate and product-bound enzyme structures. These efforts can be combined with incorporating heavy atoms into the ligands, which in turn allow localizing them based on specific scattering characteristics.

Databases of Crystal Structures

Crystal structures of protein–carbohydrate complexes can be retrieved from different sources, including the PDB, but also from more specialized databases. The Carbohydrate-Active enZYmes (CAZY) database provides links to the PDB page for all crystal structures of glycosylhydrolases, glycosyltransferases, and their associated carbohydrate-binding modules. UniLectin3D is a database covering the three-dimensional features of lectins, and includes more than 2200 lectin three-dimensional structures, of which >60% are complexed with a carbohydrate ligand. The UniLectin3D classification of more than 500 distinct lectins results in 35 lectin domain folds, 109 classes, and 350 families sharing ~20% and ~70% sequence similarity, respectively. For each structure, links for coordinates, references, and taxonomy are provided, as well as glycan array data when available at the Consortium for Functional Glycomics. Mining for structural data is therefore possible, and structures can be analyzed at different levels revealing not only atomic details of the binding sites but also protein folds and oligomeric states. Examples are given below that illustrate how convergent evolution has built robust systems for efficient recognition of glycans by lectins.

Interactions in Carbohydrate-Binding Sites

The interactions between carbohydrates and amino acids include hydrogen bonds, van der Waals contacts, ionic bonds, and a number of more specialized interactions. CH-π interactions, for example, are associated with the frequent occurrence of aromatic amino acids in carbohydrate-binding sites. Water molecules are often observed that bridge between carbohydrate hydroxyl groups and amino acids. Interestingly, a significant number of enzymes and lectins use divalent ions that directly coordinate to the hydroxyl groups of carbohydrates and to

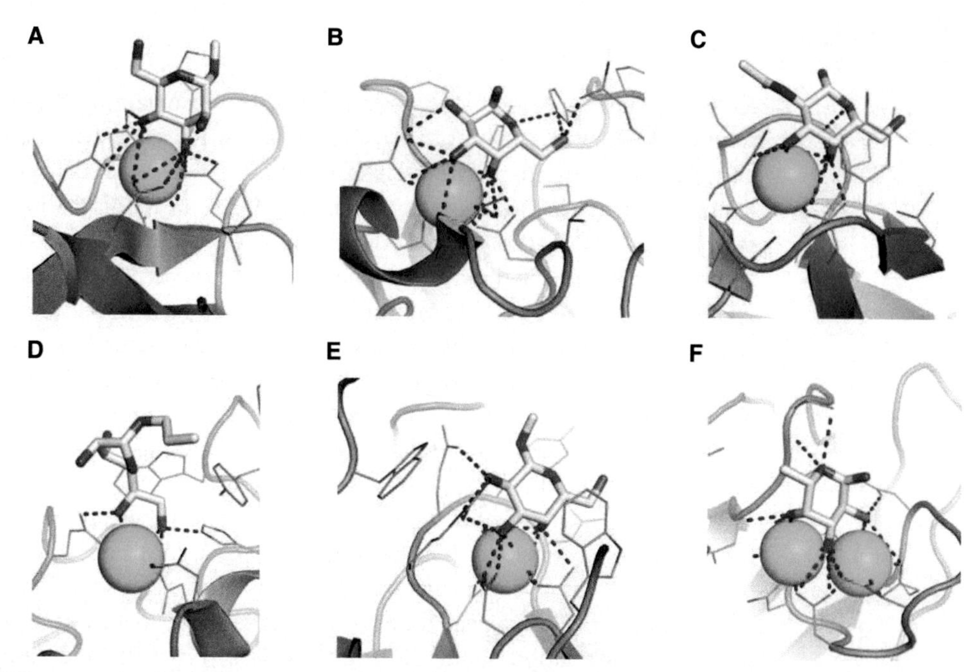

FIGURE 30.1. Graphical representation of six different calcium-dependent carbohydrate-binding sites found in crystal structures of lectins. (*A*) Human MBP-A complexed with mannoside (1KWU), (*B*) *Pseudomonas aeruginosa* LecA complexed with galactose (1OKO), (*C*) sea cucumber CEL-III complexed with GalNAc (2Z48), (*D*) human intelectin-1 complexed with galactofuranoside (4WMY), (*E*) *Candida glabrata* adhesin complexed with galactoside (4A3X), and (*F*) *P. aeruginosa* LecB complexed with fucose (1GZT).

side-chains of amino acids. Among the 350 different lectin families crystallized to date, more than 40 involve calcium ions in their binding sites. Most of them belong to the C-type lectin families (including selectins and DC-SIGN [dendritic cell–specific intercellular adhesion molecule-3-grabbing integrin]), but other types of lectins from different origins are also found to have one calcium ion in their binding site (Figure 30.1). LecB from *Pseudomonas aeruginosa* requires the presence of two closely located calcium ions. Calcium ions contribute to the specificity of lectins by selecting for precise stereochemistries of hydroxyl groups; the two calcium ions of LecB, for example, only coordinate monosaccharides bearing the specific sequence of two equatorial and one axial hydroxyl group present in "fuco" and "manno" configurations. The ions also play a role in enhanced affinity through delocalization of charge as evaluated by quantum chemical calculations and through compensation for binding entropy losses by releasing strongly coordinated water molecules.

Folding and Oligomerization Facilitate Binding to Cell Surfaces

Lectin structures adopt a limited number of folds (Figure 30.2). Among them, there is a strong predominance of β-sheet-containing domains, such as β-sandwich, β-prism, β-trefoil, or β-propeller. The β-sandwich fold, which is an assembly of two β-sheets, characterizes a large family with different structures that vary in size and localization of binding sites. For example, fimbrial adhesins are very different from galectins in that they use a site near the edge of a sheet as opposed to the concave surface of a sheet. Some structural convergence is nevertheless observed. Intracellular animal lectins, which are involved in the quality control of glycoprotein synthesis, share the same protein fold with legume lectins.

Convergence is also observed for the β-propeller fold, which is a circular arrangement of small β-sandwiches, called blades. Structures with five, six, or seven blades have been observed

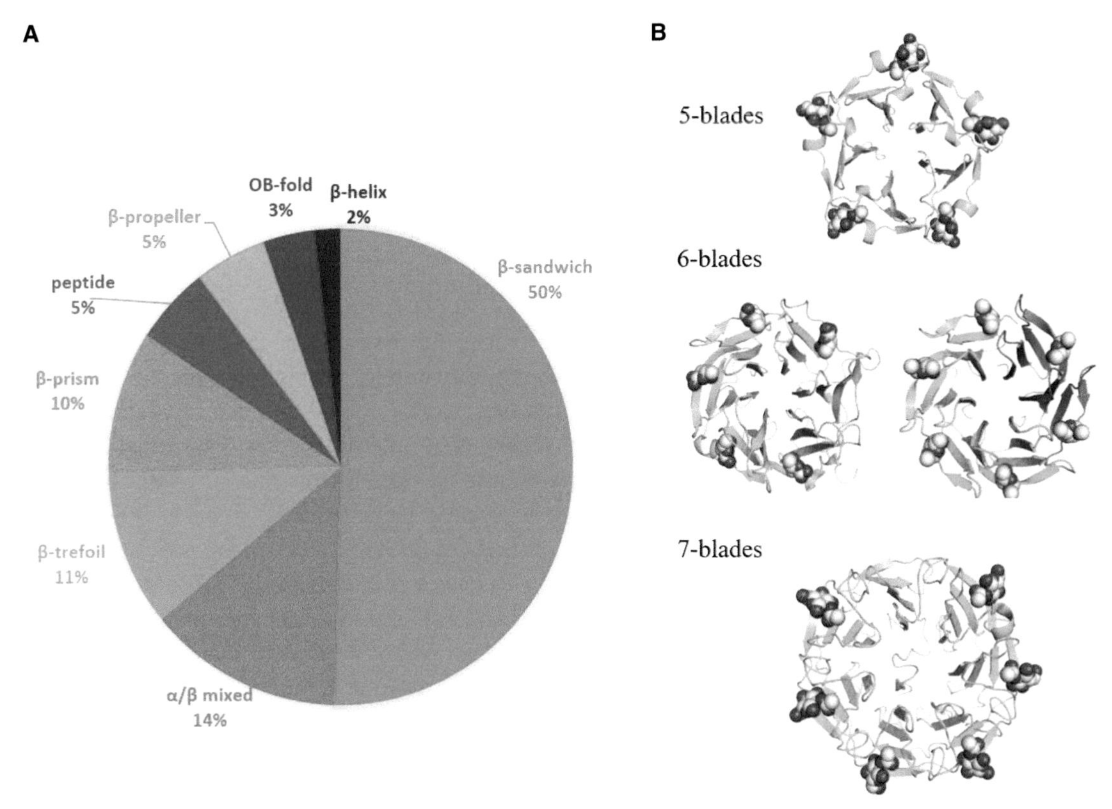

FIGURE 30.2. (*A*) Distribution of the lectins with structures available in the Unilectin3D database as a function of fold family. (*B*) Graphical representation of some convergent β-propeller folds for lectins. The polypeptide chains are represented as ribbons and the atoms in carbohydrates as spheres for the five-bladed β-propellers of tachylectin-2 from *Tachypleus tridentatus* complexed with GlcNAc (PDB entry 1TL2), the six-bladed β-propellers of (from *left* to *right*) *Aleuria aurantia* and *Ralstonia solanacearum* lectins complexed with fucose (1OFZ and 2BT9), and the seven-bladed β-propellers of *Psathyrella velutina* lectin complexed with GlcNAc (2C4D).

for lectins. With the exception of bacterial and fungal fucose-binding six-blade β-propellers, which are evolutionary related, these structures do not present sequence similarities. However, they share the same global shape allowing for the presentation of all binding sites on the same side of the "donut," providing for very efficient multivalent binding to glycoconjugates on cell surfaces. This multivalent effect results in high avidity: PVL from the fungus *Psathyrella velutina* has an affinity of only 100 μM for GlcNAc at each binding site but an apparent avidity of 10 nM for GlcNAc presented on chips. This high avidity makes PVL an excellent tool for identifying tumor cells presenting truncated glycans with exposed GlcNAc.

NUCLEAR MAGNETIC RESONANCE

NMR can provide de novo high-resolution structures of proteins and glycan–protein complexes. It can also provide dynamic information when parts of bound glycans retain some of the mobility displayed in solution. However, NMR-based structure determination usually requires uniform isotopic labeling with magnetic nuclei such as ^{13}C and ^{15}N, to complement data from the highly abundant nucleus, ^{1}H. Isotopic labeling can be accomplished when proteins can be expressed in bacterial hosts, but even then application is largely restricted to proteins of <20 kDa or of <40 kDa when perdeuteration can be used to improve resolution. The cost of uniform isotopic labeling often excludes application to many additional proteins of interest—in particular, glycoproteins—when expression in eukaryotic hosts proves essential. Hence, only a few complete structures of glycoproteins with native glycosylation have been produced by NMR methods. However, NMR has fewer restrictions when it builds on protein structures available from X-ray crystallography or computational modeling and capitalizes on its ability to focus on data involving actual glycan–protein interaction sites. We illustrate this potential in the following sections.

Chemical-Shift Mapping of Protein-Binding Sites for Glycans

The initial step on the route to produce a three-dimensional structure of a protein by NMR methods is usually the assignment of backbone resonances, including the proton and nitrogen resonances of all amide ^{1}H-^{15}N pairs. This step is quite robust and can be accomplished in much less time, and on much larger targets, than a complete structure determination. These assignments are based on a series of multidimensional experiments that correlate chemical shifts of a series of directly bonded, NMR-active nuclear pairs. Among these is the two-dimensional ^{1}H-^{15}N heteronuclear single quantum coherence (HSQC) experiment, which correlates an amide ^{1}H-^{15}N pair through the appearance of a crosspeak at the chemical shifts of the amide proton and nitrogen of a particular protein residue. Once crosspeaks in this experiment are assigned, changes in chemical shift on addition of a glycan ligand can be used to identify a binding site. These changes often arise from small perturbations in residue geometry rather than a direct effect of the ligand on chemical shift, but the effects are usually sufficiently localized to identify the binding site. Figure 30.3 shows an example of changes occurring on the interaction of a hexamer of chondroitin sulfate (CS), sulfated at the O4 position of each GalNAc residue (Chapters 3 and 17). There are actually two types of perturbations observed: gradual changes in chemical shift as ligand is added (arrows in Figure 30.3A) and the disappearance of one peak while another appears (ellipses in Figure 30.3A). These correspond to fast exchange on and off a weak binding site and slow exchange on and off a strong binding site, respectively. Perturbed residues can be mapped onto an existing structure of the protein, as shown in Figure 30.3B for the strong binding site. As with many complexes involving a sulfated GAG, positively charged residues are involved; in this case histidine residues and a lysine residue are among those showing chemical shift changes. The advantage of these experiments is that a range of ligands can be examined, even those that may fail to produce well-ordered crystals for crystallographic analyses. A limitation is that the backbone resonances of the protein need to be assigned first.

Identification of Bound Ligand Geometry and Ligand Interaction Surfaces

NMR also offers the potential for characterizing the geometry a ligand adopts on binding to a protein surface and the parts of the ligand that make contact with a protein. In both cases, the characterization stems from transfer of magnetization from one NMR active spin to another NMR active spin (usually protons) in a distance-dependent manner. In the case of bound ligand geometry, the experiment relies on a transferred nuclear Overhauser effect (trNOE). The basis is the same as for the NOE that is used in protein structure determination by NMR; however, as a large excess of ligand over protein is used ($>10:1$), only the ligand spectrum is observed. Measurements are usually made from crosspeaks in two-dimensional experiments similar to the HSQC experiment mentioned above, except that both dimensions are proton chemical shift, and crosspeaks have intensities dependent on the inverse sixth power of the distance between proton pairs ($1/r^6$) rather than direct bonding. An average over both bound and free ligands is observed, but contributions are heavily weighted by those coming from the ligand in a complex because of scaling in proportion to molecular weight. This makes it possible to conduct trNOE experiments with a large excess of ligand and very little protein. Also, there is no

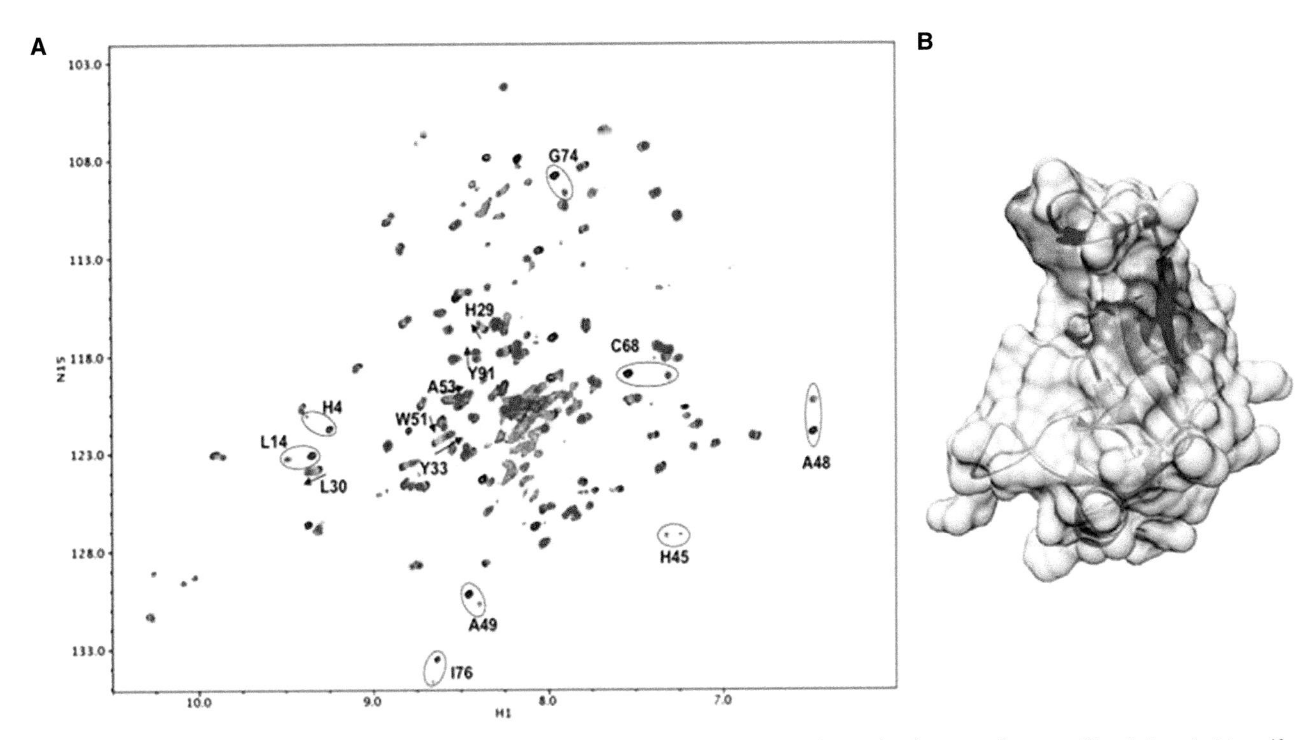

FIGURE 30.3. Chemical shift mapping of slow and fast exchange binding sites for a 4-sulfated chondroitin sulfate (CS) hexamer on the Link module of TSG6. (*A*) Crosspeaks from spectra with increasing amounts of hexamer are superimposed. Those from residues experiencing fast exchange show progressive shifts and are marked with arrows; those from residues experiencing slow exchange show a pair of peaks, one appearing while another disappears, and are enclosed in *ellipses*. (*B*) Residues showing slow exchange are mapped in *red* on a crystal structure (2PF5).

requirement for isotopic labeling of either ligand or protein, and having a high-molecular-weight complex is an advantage. The geometry of the bound ligand is derived primarily from distances measured between protons that fall on opposite sides of a glycosidic bond. This distance then restrains glycosidic torsion angles accessible to structural models. Although there are many cases in which the bound geometry is similar to that of the dominant conformer found in solution, there are cases in which the geometry differs. Here, trNOE experiments offer unique insight that can guide synthesis of competitive inhibitors.

Transfer of magnetization from protons on a protein to protons on a ligand in an intermolecular NOE-like fashion can also provide information on the parts of a ligand in contact with amino acids in a protein's binding pocket (the ligand's interaction surface or binding epitope). In some cases, NOEs between a ligand proton and a specific amino acid proton can be observed, but this requires work with near-equimolar concentrations of ligand and protein, as well as full resonance assignment for both the ligand and the protein. A far more widely applied experiment sacrifices knowledge about specific protons on the protein for an ability to work with very large unlabeled and unassigned proteins. This experiment is called a saturation transfer difference (STD) NMR experiment. In fact, STD NMR investigations have also been conducted on some very large and complex systems including receptors embedded in membrane fragments, whole cells, and viruses. The experiment, which can be conducted at ratios of ligand to protein approaching 100:1, involves selective perturbation (saturation) of the magnetization of a set of protein protons and relies on the fact that the magnetization transfer between protons in large proteins is so efficient that it makes little difference where the change in magnetization is initiated; it can be from saturation of a methyl proton having a resonance at one extreme of the spectrum (upfield) or an aromatic proton having a resonance at the other extreme (downfield). Ideally, the saturation effect diffuses all over the protein protons and eventually is transferred to ligand protons close to the protein surface, and the resonances of these protons are reduced in intensity in a way that inversely correlates with the distances of a proton from the protein surface. Data are collected as a difference between one-dimensional proton spectra with and without saturation in the extremes of the protein spectrum. The resulting difference spectrum is dominated by resonances from the ligand that have contact with the protein. Mapping the position of protons assigned to these resonances onto a ligand structure allows depiction of the ligand binding epitope.

Figure 30.4 shows an example that probes the interaction between a complex N-glycan (Chapters 3 and 9) and an HIV broadly neutralizing antibody. These antibodies specifically interact with surface glycans of HIV and are effective in inhibiting binding of the virus to target cells. Hence, there has been significant interest in exactly which glycans are recognized. Antibodies are large glycosylated proteins that are not usually amenable to NMR investigation by isotope-dependent methods, but STD NMR methods are applicable. The example uses a sample 20 μM in protein (Fab fragment) and 2 mM in glycan. Normal (reference) and STD NMR spectra are superimposed to show the saturated ligand resonances, which include some that come specifically from the Neu5Ac residues (Sia) on the termini of the glycan branches.

STD NMR applications involving long-chain multiantenna N-glycans, such as the one described above, are often hampered by the near-chemical equivalence of sites and degeneracy of resonances from the various branches of antenna. In these cases, the covalent attachment of lanthanide binding tags to the reducing end of the glycan has proven useful. The resulting

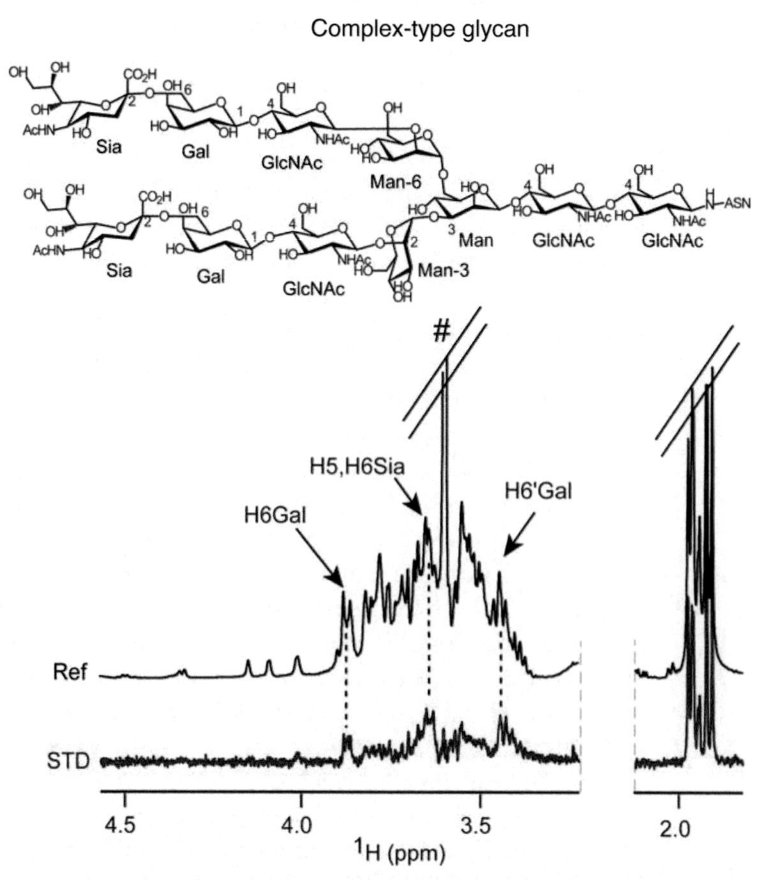

FIGURE 30.4. Binding epitope identification in a complex-type glycan bound to the HIV-1 neutralizing antibody PG16 using saturation transfer difference (STD) nuclear magnetic resonance (NMR) information. (Reproduced from Bewley CA, Shahzad-ul-Hussan S. 2013. *Biopolymers* **99:** 796–806, with permission from John Wiley and Sons.)

dispersion of glycan signals due to pseudo-contact shifts can allow unambiguous determination of binding epitopes (see Further Reading).

Although the rapid dispersion of saturation throughout the protein in standard STD NMR experiments suppresses information about what protein residues are involved in binding, there are ways to retrieve some of this information. In multifrequency STD NMR spectra, as in the case of the differential-epitope-mapping method (DEEP-STD NMR), two STD NMR spectra are obtained using two very different saturating frequencies in spectral regions devoid of glycan signals—for example, one in the aromatic region of protein protons and the other in the aliphatic region. In each STD NMR spectrum, the ligand protons close to directly saturated protein protons (e.g., aromatic protons) show a little increase in STD intensity, in comparison to protons further away from the saturated protein protons. Analyzing and mapping those differences along the structure of the ligand (so-called differential epitope mapping) allows retrieval of information about which areas of the glycan interact with those different types of amino acids in the binding pocket. If the geometry of the pocket is known, this allows elucidation of the orientation or polarity of the glycan in the binding pocket.

The above provides a glimpse of NMR experiments that can be used to investigate protein–glycan interactions. There are many others that take advantage of additional properties such as differences in translational diffusion constants and specific interactions with water molecules. Many of these have been adopted as screening methods used in fragment based drug discovery programs. Information about these is available in Further Reading.

CRYO-ELECTRON MICROSCOPY

With the development of direct electron detectors, cryo-electron microscopy (cryo-EM) has become one of the most powerful techniques to obtain high-resolution structural information on biological macromolecules. With reported resolutions exceeding 2 Å, many molecular details are revealed under native-like conditions. By eliminating the need for growing well-ordered three-dimensional (3D) crystals and operating at fairly low sample concentrations, cryo-EM can provide atomic level insights into many biological samples, from soluble and membrane-integrated protein complexes to filamentous polymers and entire viruses.

Cryo-EM can be roughly divided into two main directions: one working with single (usually purified) particles, single-particle analysis (SPA), and the other using a tomography approach to analyze species in larger assemblies, such as in vitro–assembled scaffolds or even native cells and tissues.

For SPA, samples are generally analyzed in a thin layer of vitreous ice containing the particles in random orientations. Data is collected in the form of movies, which allows correction for beam-induced drift, followed by estimation of the contrast transfer function for each micrograph; both are necessary to obtain near-atomic resolutions. The individual particles are then computationally extracted from the micrographs, sorted, and ultimately aligned in three dimensions to reconstruct the molecular structure.

For cryo-electron tomography of, for example, a vitrified cell or virus particle, a tilt series of images is acquired to obtain different "specimen views" necessary for 3D reconstruction. This technique continues to face technical challenges, in part because of limitations on the tilt angles that can be achieved. However, it is a powerful tool to image, for example, the glycocalyx of various tissues, plant and fungal cells walls, or microbial cell envelopes and capsules.

A major advantage of cryo-EM is that sample heterogeneity and/or conformational flexibility does not preclude analyses. To obtain well-diffracting crystals of a glycosylated protein, for example, the conformationally heterogenous glycans are often removed enzymatically to facilitate crystallization. For cryo-EM, these pretreatments are generally unnecessary, thereby providing molecular details of proteins in the context of posttranslational modifications. Analyses of

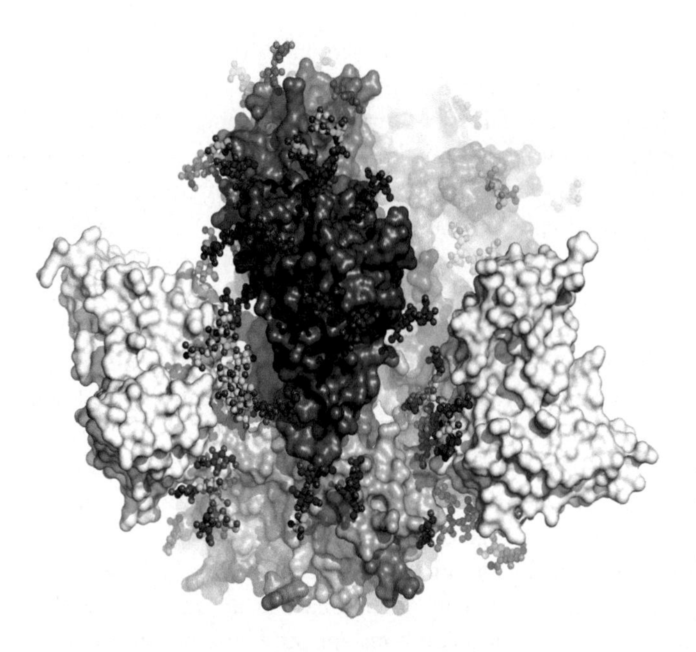

FIGURE 30.5. Cryo-electron microscopy structure of the native fully glycosylated HIV-1 envelope trimer. Protein subunits are shown as *gray* surfaces and glycans are shown as *ball and sticks* (PDB entry 5FUU). Carbohydrates are colored *green* and *pale green* (β- and α-D-mannopyranose, respectively), *blue* (N-acetylglucosamine), and magenta (α-L-fucopyranose).

fully glycosylated viral envelope proteins are fascinating examples that document the potential of cryo-EM for studying protein–carbohydrate interactions (Figure 30.5). Other examples include polysaccharide-synthesizing enzymes bound to their polymeric products as well as integral membrane transporters associated with lipopolysaccharide substrates.

We can thus look forward to correlating unprecedented structural insights on protein glycosylation and complex carbohydrate interactions with biochemical, functional, spectroscopic, and in silico approaches in the coming years.

COMPUTATIONAL MODELING

Experimental structural information obtained by crystallographic, NMR, and cryo-EM methods have clearly been of value in building an understanding of the molecular interactions that lead to glycan recognition by proteins. However, systems in which interactions are of interest far outnumber the cases in which these methods can be applied. Most crystal structures contain either small ligands or yield useful electron densities for only parts of larger ligands. The same is true to a certain extent for cryo-EM structures. NMR methods, although giving detailed information on bound ligand geometries, frequently give only qualitative information on parts of ligands or protein that are in intimate contact with each other. All three methods require substantial effort, particularly in preparing samples for investigation. A particular problem for glycans of interest is that they are often complex molecules that are difficult to prepare in highly pure forms or in the quantities needed for experimental investigation. There are also functionally important dynamic processes (e.g., enzymatic conversions of glycan substrates to products and transport of glycans) that are not well-represented by static, thermodynamically stable structures. Computational methods can extend analyses into these less accessible regions of structural investigation.

Computational Methods

Computational contributions to our understanding of glycan properties have a long history, beginning with a very fundamental understanding of factors influencing anomeric configuration and glycosidic torsion angles. These glycan-specific factors, such as the anomeric effect and the exo-anomeric effect, are described more thoroughly in Chapters 2, 3, and 50. When protein–glycan interactions are of interest, the situation becomes more complex with hydrogen bonding, van der Waals interactions, and electrostatic interactions between glycan and various amino acids becoming important. For very limited sets of atoms, it is possible to pursue an understanding of interactions using advanced quantum mechanical (QM) methods, but for larger systems other approaches based on semiempirical "force fields" are used, as in molecular mechanics (MM) and molecular dynamics (MD) simulations.

Empirical "force fields" used in MM and MD modules of packages such as Amber, CHARMM, and GROMOS are typically represented in terms of bond, bond angle, torsion angle, van der Waals, and electrostatic contributions to a molecular energy. Parameters in functions representing each of these terms have been optimized to reproduce QM as well as a selection of thermodynamic and spectroscopic data. Initially, these force fields were developed for proteins alone, so they did not include contributions such as the anomeric and exo-anomeric effects found in glycans. Subsequently, force fields explicitly designed to represent the energetics of glycans have been developed for use with these packages (e.g., the GLYCAM force field that is widely used with Amber). There still are challenges in simulating molecular interactions with these packages, among them perfecting models for solvent and accurately representing electrostatic interactions. These issues are very important for glycans, which are rich in hydroxyl groups that act as both hydrogen bond donors and acceptors in their interactions with water. Some glycans (e.g., GAGs) are highly charged, having both carboxylate groups and sulfate groups that interact strongly with positively charged amino acids in proteins and with water. Although early simulations were performed with implicit solvent models based on dielectric behavior, recent improvements in computational capabilities have allowed use of explicit solvent models, such as TIP3P and TIP5P.

MD, which uses the force fields directly in Newton's second law of motion, simulates movement of all atoms in addition to generating an ensemble of conformations and orientations that can be reached over times accessible to simulation (nsec to msec depending on the size of the system and efficiency of the computational platform). One advantage of MD is that certain important motional properties, such as the time for diffusion through a channel or the time needed for a conformational transition, can be modeled. One must remember, however, that force fields are meant to represent molecules near energy minima of a conformational surface and may not accurately represent the height of larger barriers separating different conformational states and certainly cannot represent changes in bonding that occur in a chemical reaction.

The actual characterization of how a ligand (a glycan in our case) interacts with a protein involves not just the conformational energetics of the free glycan, but also the conformational energetics of amino acid residues involved in the binding site and the energetics of the glycan–protein interaction. In some cases, there may be relatively little information on where the binding site on a protein is, so the characterization involves locating the best binding site, finding the best conformation for the ligand in the bound state, and finding the best conformations for the parts of the protein involved in binding. The whole process is referred to as "docking" a ligand onto a protein surface. Most docking programs (e.g., AutoDock, AutoDock Vina, and Glide) are designed to make the initial search for a site very efficient. To do this, they break the process into stages beginning with a rigid-body docking step that is designed to identify the best docking site and best initial "poses" for the ligand. Force fields are often simplified or interaction energies precalculated on a grid to speed up the process. Rigid-body docking generally works well for many small drug-like molecules. Also, in many situations, there is a crystal structure of the protein with a native ligand in the binding site, mitigating the problem of finding the binding pocket and optimizing side-chain conformations. For glycans, the situation is more complicated; the ligands are often flexible and protein structures with a native glycan in a binding site are often lacking.

In molecular docking, the objective is not to generate a single bound structure in the first stage but hundreds of "poses" that can be scored and ranked so that a subset can be selected for subsequent stages. Scoring functions are variable but usually include some sort of interaction energy as part of the score. Subsequent phases typically allow increased flexibility of side-chains and finally an MD refinement of poses, often in explicit water. Final scoring or ranking of poses by energy, even when performed with force fields used in MD programs, seldom leads to a single clear solution, and it has become common to filter poses with additional experimental information such as binding epitopes from STD NMR experiments or interactions with residues that have been identified as important in mutational studies.

Some docking programs are emerging (e.g., HADDOCK) that make use of experimental data in earlier stages to guide the selection of initial poses as well as maintain known preferences for glycan conformations or specific ligand–protein contacts. Some of the contributions to understanding of glycan–protein interactions that have come from docking exercises, as well as more advanced applications that merge QM with MD, are described in more detail in the following sections.

Docking of Heparan Sulfate Oligomers

Heparan sulfate (HS) chains, synthesized initially as a repeating disaccharide of glucuronic acid (GlcA) and *N*-acetylglucosamine (GlcNAc), and modified subsequently by sulfation and epimerization of some GlcA residues to iduronic acid (IdoA), are known to interact with a number of growth factors, receptors, and chemokines (Chapters 17 and 38). Despite the interest in the roles of these interactions in cell migration and differentiation, there are relatively few experimental structures depicting interactions with large HS fragments. The fact that suitable crystals are less apt to form in the presence of HS oligomers contributes to the lack of structures of complexes. Also, it is difficult to obtain homogeneous preparations of large oligomers, because of the variable sulfation patterns and variable conversion of GlcA to IdoA.

Computational modeling offers an alternative route to structures for many of these complexes. Specific patterns of sulfation and IdoA substitution are generated with ease. Yet, there are some challenges related to the flexibility of the HS chains and the ionic character of interactions that dominate their energetics; glycosidic angles in HS chains are variable, and IdoA rings sample several conformations, including a chair, $^1C_4$, and a skew-boat conformer, $^2S_0$. Moreover, orientations of the sulfate groups are variable, as are the side-chains of the lysine and arginine residues with which they tend to interact. Enhanced docking methodology combined with MD simulations overcomes some of these challenges.

Leukocyte common antigen-related protein (LAR) is a type IIa receptor protein tyrosine phosphatase (RPTP) important for signal transduction in biological processes, including axon growth and regeneration. Glycosaminoglycan chains, including HS, act as ligands that regulate LAR signal transduction. Knowing where HS binding sites are and what molecular interactions drive binding is an important step in the design of agents that could promote regeneration. Figure 30.6 shows a snapshot of an HS pentamer bound to LAR. The structure was generated using the docking program HADDOCK. It employed several types of NMR data (chemical shift perturbation, STDs, and trNOEs) to guide selection of an initial set of 20 docked structures. The top scoring structures were subjected to short (50-nsec) MD runs in explicit water using GLYCAM06 force-field parameters for the HS fragment. The snapshot is from a longer (1-μsec) run, now highlighted in a movie (see Online Appendix 38A).

The interactions are typical of many GAG–protein interactions in that charged sulfates and carboxylates of the HS fragment interact with lysine and arginine residues of the protein binding site. These interactions are further stabilized by hydrophobic and hydrogen-bonding interactions with neighboring groups (e.g., with the glutamine residue in Figure 30.6). The interactions with arginine are particularly important and, in addition to electrostatic contributions, often include those from bidentate hydrogen bonds between N-H groups on the arginine

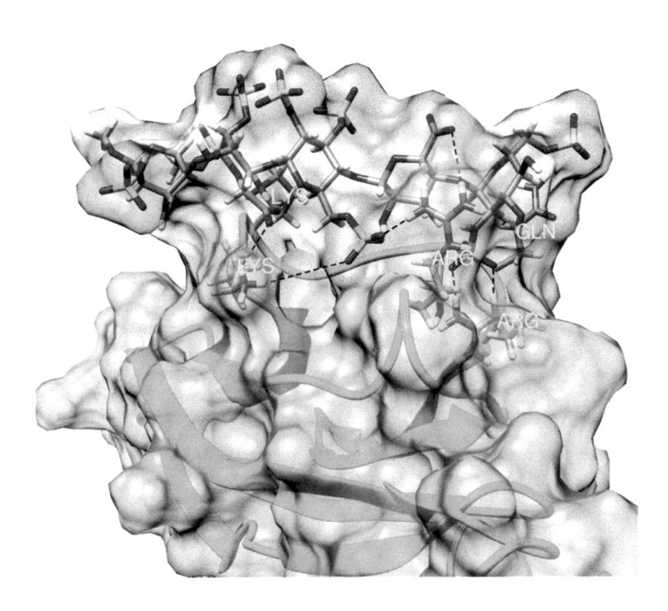

FIGURE 30.6. Docking of a heparan sulfate (HS) pentamer to the receptor protein tyrosine phosphatase, LAR (PDB entry 2YD5). (Docked structures as described in Gao N, et al. 2018. *Biochemistry* **57:** 2189 − 2199.)

side-chain terminus and oxygens of sulfate groups. An example can be seen in the lower right where arginine 77 of the protein interacts with the *N*-sulfate on the terminal GlcN of the HS fragment.

Docking of Enzyme Substrates

A large number of enzymes are involved in the synthesis and degradation of glycans (more than 300 human enzymes). Their relative activities, combined with cellular location, are essential to the proper balance of these processes and any alteration, including genetic mutation, can lead to disease in humans. Pathogens also depend on similar processes, and understanding such mechanisms can facilitate the design of selective inhibitors of pathogen enzymes. This is another area in which molecular docking can play a role. Structural studies of glycan–protein complexes usually require a stable system—not one that would continually convert substrates to products. Molecular docking can provide useful depictions of these reactive systems.

A good example involves the glycosyltranferase, ST6Gal1. This is the enzyme that adds a sialic acid (typically Neu5Ac) to the galactose terminated branches of N-glycans by transferring Neu5Ac from its nucleotide-sugar donor, CMP-Neu5Ac, to an acceptor terminated with a Galβ1-4GlcNAc moiety (Chapter 6). The production of crystal structures of ST6Gal1, from both human and rat, opened the possibility of modeling at least a pretransition complex with both donor and acceptor in place. For the study discussed here, the crystal structure of the rat enzyme that contained neither donor nor acceptor (4MPS) was used as a starting point. The CMP-Neu5Ac was modeled into the active site based on the inactive donor analog in the crystal structure of the CstII protein (1RO7), which has a $<20\%$ sequence identity overall, but a much higher identity in the part of the active site that contains the donor. An initial structure for the minimal acceptor, Galβ1-4GlcNAc, was generated using the GLYCAM WebTool, but glycosidic bonds and hydroxyl groups were allowed to rotate during docking. Docking used the program AutoDock Vina. As in the previous example, an additional MD step in explicit water was used to refine the top ranked docked structure containing protein, donor, and acceptor. Interaction energies were then generated by applying MM/GBSA routines from Amber12 to 100-nsec MD production runs.

Although the positions of donor and amino acid residues near the donor were modeled to be quite similar to those seen in other transferases, the docking/MD procedure provides a unique

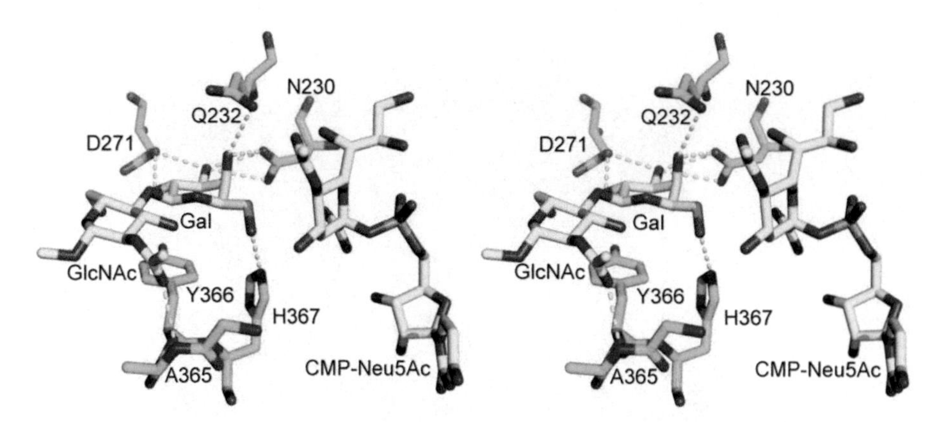

FIGURE 30.7. Stereo view of interactions between the donor (CMP-Neu5Ac), acceptor (GlcNAcβ1-4Gal), and protein residues in the active site of ST6Gal1. (Reproduced, with permission, from Meng L, et al. 2013. *J Biol Chem* **288:** 34680–34698.)

view of a possible acceptor position and its interactions. Most of the interaction energy holding the acceptor in place comes from interactions with the galactose ring, which is well-positioned to allow nucleophilic attack on the anomeric carbon of the nucleotide-activated Neu5Ac. This energy results from hydrophobic stacking of Tyr-366 with the nonpolar face of the pyranose ring and a network of hydrogen bonds between Asp-271, Asn-230, His-367, and Gln-232 of the protein and O2, O3, O4, and O6 hydroxyl groups of Gal. The position of the GlcNAc is more variable but does contribute to binding energy. The position and interactions among protein, donor, and acceptor are depicted in Figure 30.7.

FUTURE PROSPECTS

Structural biology is an evolving area of science both in terms of methodology and questions to be answered. The principle methodologies discussed here are each evolving: crystallographic methods using new X-ray sources (e.g., X-ray lasers) are allowing the analysis of microcrystals at room temperature and femtosecond timescales, thereby eliminating temperature- and beam-induced artifacts. Cryo-EM single-particle methods are approaching resolutions previously confined to X-ray crystallography. Single-particle and tomography EM approaches are continuing to undergo rapid development in terms of EM infrastructure, sample preparation, and data acquisition. Several user-friendly pipeline data processing packages exist, making this technology attractive to an increasing scientific audience. Hyperpolarization methods are reducing the sensitivity limitations of NMR, and solid-state NMR methods are allowing application to amorphous materials, including fibrils, cell-wall structures, and membrane fragments. Advances in computational technology are enabling simulation of ever larger systems and timescales. At the same time, structural targets are shifting from detailed characterization of single proteins and protein–glycan complexes to large-scale assemblies that cooperate to elicit a functional response. This is a promising situation for improved understanding of glycan function in biological systems.

ACKNOWLEDGMENTS

The authors appreciate helpful comments and suggestions from Barbara Mulloy, Dillon Chen, and Sean Stowell.

FURTHER READING

Bewley CA, Shahzad-ul-Hussan S. 2013. Characterizing carbohydrate–protein interactions by nuclear magnetic resonance spectroscopy. *Biopolymers* **99**: 796–806. doi:10.1002/bip.22329

Grant OC, Woods RJ. 2014. Recent advances in employing molecular modelling to determine the specificity of glycan-binding proteins. *Curr Opin Struct Biol* **28**: 47–55. doi:10.1016/j.sbi.2014.07.001

Pérez S, Tvaroska I. 2014. Carbohydrate–protein interactions: molecular modeling insights. In *Advances in carbohydrate chemistry and biochemistry* (ed. DA Baker, D Horton), Vol. 71, pp. 9–136. Elsevier, Amsterdam. doi:10.1016/b978-0-12-800128-8.00001-7

Bartesaghi A, Merk A, Banerjee S, Matthies D, Wu X, Milne JLS, Subramaniam S. 2015. 2.2 Å resolution cryo-EM structure of β-galactosidase in complex with a cell-permeant inhibitor. *Science* **348**: 1147–1151. doi:10.2210/pdb5a1a/pdb

Pomin VH, Mulloy B. 2015. Current structural biology of the heparin interactome. *Curr Opin Struct Biol* **34**: 17–25. doi:10.1016/j.sbi.2015.05.007

Canales A, Boos I, Perkams L, Karst L, Luber T, Karagiannis T, Domínguez G, Cañada FJ, Pérez-Castells J, Häusinger D, Unverzagt C, Jiménez-Barbero J. 2017. Breaking the limits in analyzing carbohydrate recognition by NMR spectroscopy: resolving branch-selective interaction of a tetra-antennary N-glycan with lectins. *Angew Chem Int Ed* **56**: 14987–14991. doi:10.1002/ange.201709130

Glaeser RM. 2017. How good can cryo-EM become? *Nat Methods* **13**: 28–32. doi:10.1038/nmeth.3695

Bonnardel F, Mariethoz J, Salentin S, Robin X, Schroeder M, Pérez S, Lisacek F, Imberty A. 2019. UniLectin3D, a database of carbohydrate binding proteins with curated information on 3D structures and interacting ligands. *Nucleic Acid Res* **47**: D1236–D1244. doi:10.1093/nar/gky832

31 R-Type Lectins

Richard D. Cummings, Ronald L. Schnaar, and Yasuhiro Ozeki

The R-type lectin superfamily is characterized by a carbohydrate-recognition domain (CRD) originally reported in a polypeptide of ricin, a plant toxin. Ricin was the first lectin discovered, and the R-type lectins are named after it. Through evolution, the R-type lectin domain has given rise to various protein forms, some containing this domain alone and others containing additional functional domains (e.g., enzyme or toxin). We will discuss the diversity of this superfamily in terms of molecular tinkering and "bricolage," based on structure–function relationships.

HISTORICAL BACKGROUND

In 1888, Peter Hermann Stillmark at the University of Dorpat (now University of Tartu, Estonia) reported that protein extracts from seeds of *Ricinus communis* (commonly called "castor oil plant"; seeds are called "castor beans") contained a factor (which he termed "ricin") capable of agglutinating erythrocytes. Ricin was already well known as a toxin, but its carbohydrate-binding specificity was not analyzed in detail until the mid-20th century, when it was classified as a lectin. Structural

published by Cold Spring Harbor Laboratory Press; doi: 10.1101/glycobiology.4e.31

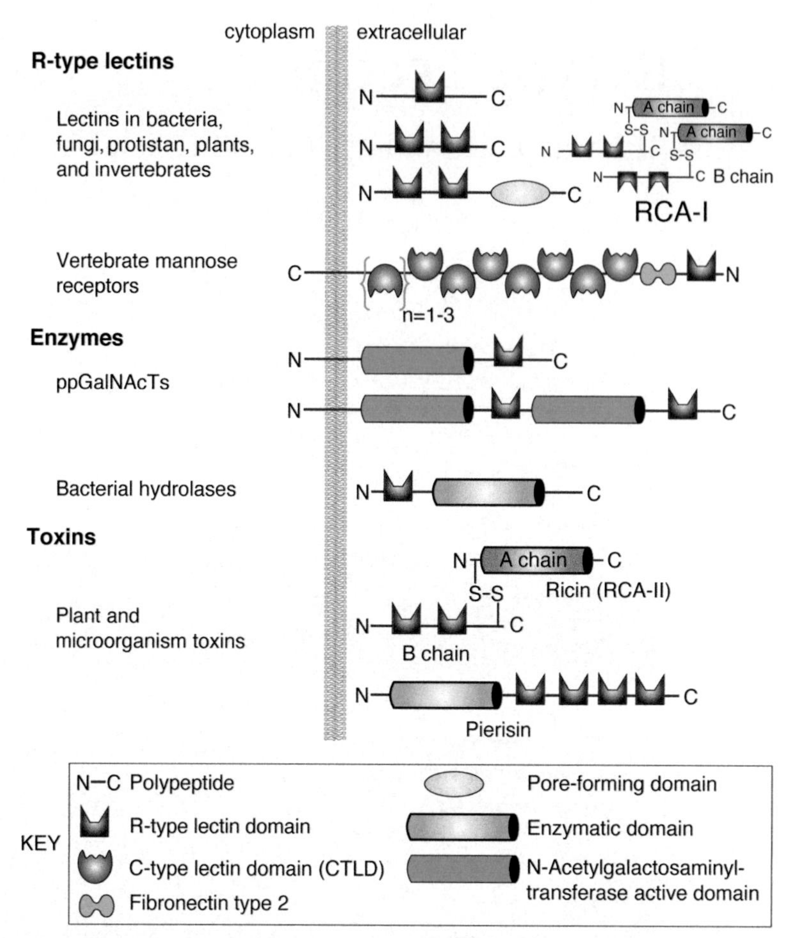

FIGURE 31.1. The R-type lectin superfamily. Different groups within the superfamily are characterized by the domain structures shown.

analysis of ricin, performed around a century after its discovery, showed that the CRD had a "β-trefoil" (three-leaved) fold with a common Q-x-W motif in each subdomain. This domain structure defines the R(icin)-type lectins, a superfamily also classified as carbohydrate-binding molecule (CBM) 13 (http://www.cazy.org/CBM13.html) in the Carbohydrate-Active Enzymes (CAZy) database (Chapter 6). These lectins are widely present in all three biological domains (and their viruses) as glycan-binding proteins, including enzymes and toxins (Figure 31.1).

R-TYPE LECTINS IN PLANTS

General Properties of Ricin

Ricinus communis is native to Africa and India and has been used for millennia, by many cultural groups, for numerous medical and industrial purposes. Two lectins, originally termed RCA-I and RCA-II, can be purified from the seeds. RCA-I (molecular mass ~120 kDa) is a hemagglutinin and a very weak toxin. RCA-II (~60 kDa), better known as ricin, is an agglutinin and a very potent toxin. Ricin is easily extractable from castor beans and can kill humans at very small doses.

RCA-II (ricin; Figure 31.2) is synthesized as a single prepropolypeptide of 576 amino acid (aa) residues containing a secretion signal peptide (residues 1–35), an A chain (residues 36–302), a 12-aa linker region (residues 303–314), and a B chain (residues 315–576). The A-chain toxin

A

B

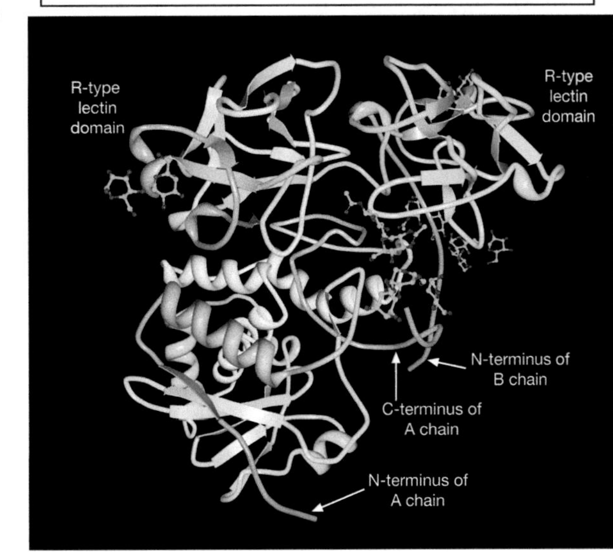

FIGURE 31.2. Structure of ricin. (*A*) *Ricinus communis* (beans and plants). (*B*) Crystal structure of ricin (RCA-II); resolution 2.5 Å. (Rutenber E, et al. 1991. *Proteins* **10:** 240–250. Image from PDB entry 2AAI.)

is an N-glycoside hydrolase (EC 3.2.2.22) that inactivates the 60S ribosome. The B-chain lectin domain binds to galactose (Gal) and β-galactosides. The amino-terminal signal sequence brings the propolypeptide to the endoplasmic reticulum (ER), where the signal is cleaved off, and the polypeptide is glycosylated. The A chain is linked to the B chain by a disulfide bond between cysteine residues 294 and 318 (Cys4 in the free B chain), and four intrachain disulfide bonds are formed. After glycosylation, ricin is transported to protein storage bodies in castor beans, where the mature ricin protein is generated by removal of the linker peptide by an endopeptidase.

RCA-I (agglutinin) is a tetramer consisting of two noncovalently associated ricin-like heterodimers. Each heterodimer contains an A chain disulfide-linked to a Gal-binding B chain. The A-chain sequences of RCA-I and RCA-II differ in 18 residues and are 93% similar, whereas the B-chain sequences differ in 41 of 262 residues and are 84% similar. The subunits are all N-glycosylated and usually display oligomannose-type N-glycans. Several other lectins having high homology with ricin are also encoded by the *R. communis* genome; some of these are termed ricin-A, -B, -C, -D (ricin), and -E.

Carbohydrate-Binding Specificity of Ricin

The B chain of both RCA-I (agglutinin) and RCA-II (ricin) contains two CRDs in the polypeptide. RCA-I binds preferentially to terminal β-linked Gal, whereas RCA-II binds to either terminal β-linked Gal or GalNAc. These lectins are often purified and separated by differential elution from Gal-based affinity resins; RCA-II is eluted with GalNAc, and RCA-I is then eluted with Gal. These lectins have very low binding affinities for monosaccharides (K_d in the range 10^{-3} to 10^{-4} M), but very high affinities (10^{-7} to 10^{-8} M) for cells, because of their multivalency and enhanced binding to multiple surface glycans having nonreducing terminal Galβ1-4GlcNAc-R (type-2 LacNAc) sequences. Both lectins also bind strongly to glycans with GalNAcβ1-4GlcNAc-R (LacdiNAc), and weakly to the isomer Galβ1-3GlcNAc-R (type-1 LacNAc). Neither lectin binds appreciably to glycoconjugates having nonreducing terminal α-linked Gal residues.

Toxicity of Ricin

Ricin is highly toxic and the effects of ingestion are severe, with symptoms appearing after a 2- to 24-h latent period. The lethal dose (LD_{50}) of ricin is extremely low (3–5 μg/kg body weight). Ricin is classified as a type II ribosome-inactivating protein (RIP-II). RCA-I (agglutinin) is less toxic than ricin, because of the weaker enzymatic activity of its A-chain. Type I ribosome-inactivating proteins (RIP-I), which lack a B chain with R-type lectin domains, are much less toxic than ricin because toxin entry into target cells is facilitated by B-chain carbohydrate-binding activity. RIP-I expressed in tissues of various types of plants affects disease resistance. Horizontal transmission of RIP-I gene to the genome of whitefly (Hemiptera, Insecta) feeding on such plants was recently demonstrated.

Ricin bound to a β-linked cell-surface Gal/GalNAc containing glycans is transported into endosomes (Figure 31.3), and then migrates by retrograde trafficking to ER via the *trans*-Golgi

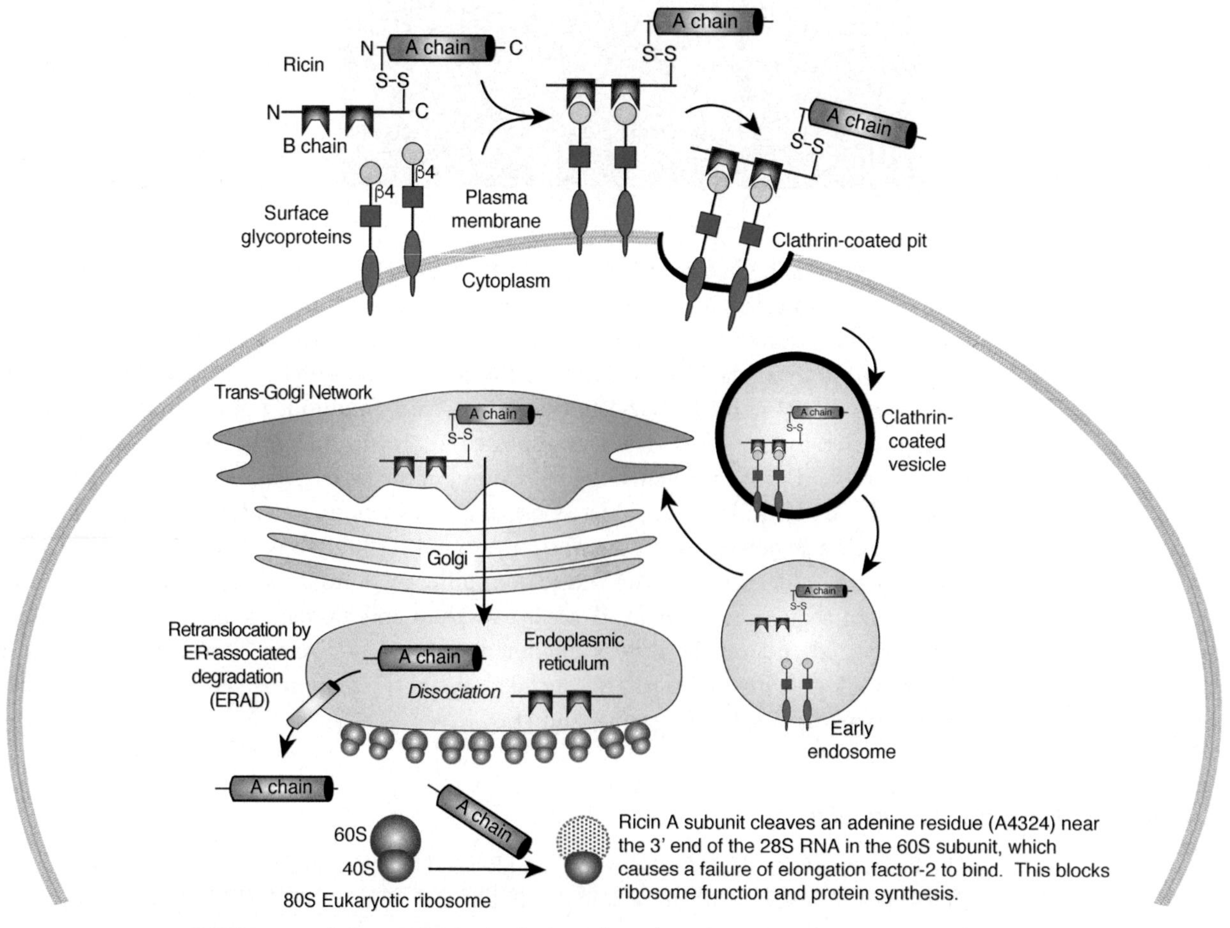

FIGURE 31.3. Pathway of ricin uptake by cells and mechanism whereby toxic activity of A chain in cytoplasm results in cell death.

network and Golgi apparatus. In the ER, A and B chains separate following reduction of the disulfide bond, perhaps catalyzed by the enzyme protein disulfide isomerase. A portion of the free A chain, which may be partially denatured in the ER, escapes by retrotranslocation through the Sec61 translocon and enters the cytoplasm.

Just a few molecules of ricin A-chain in the cytoplasm are sufficient to kill cells. The catalytic subunit of A-chain is an rRNA N-glycosylase having eight α-helices and eight β-strands, which cleaves one purine base (adenine 4324) in the exposed GAGA tetraloop of the 28S RNA of the eukaryotic 60S ribosomal subunit. This deletion results in loss of binding of elongation factor-2, inability of ribosomes to perform protein synthesis, and consequent cell death.

Sequence of R-Type Lectin Domain

Sequence analysis of RCA-I and RCA-II revealed that their B-chains contain two similar, characteristic regions in the polypeptide (i.e., the tandem-repeat R-type lectin). Each of the domains is approximately 120 aa long, with three tandem repeats of a 40-residue subdomain. Each subdomain contains a Q-x-W motif (x represents any aa). This tandem repeat is present in other toxins from plants of various phyla and is evidently characteristic of plant lectins. In the late 1990s, similar sequences were discovered in lectins from nonplant organisms, indicating that the ricin-B-like pattern is universal. In the early 2000s, these lectins were classified as the "R-type lectin superfamily," and this superfamily is now categorized as a single domain (Pfam: PF14200) or tandem-repeat domain (Pfam: PF00652).

β-Trefoil Fold

Crystal structure analysis of ricin (Figure 31.2) indicated that B-chain (PDB 2AAI) has a barbell-like shape with the two tandem R-type CRDs at the ends, ~35 Å apart. R-type lectins have a fold termed "β-trefoil fold" (from the Latin "trifolium" meaning "three-leaved," as seen for example in clover) (Figure 31.4). The β-trefoil fold presumably arose through triplication of an ancestral gene encoding the 40-residue subdomain. The three subdomains, termed α, β, and γ, each consist of 12 β-strands and loops arranged around a central axis, giving the protein internal pseudorotation symmetry. β-trefoil fold is maintained by the characteristic Q-x-W motif in each subdomain of ricin B-chain and common R-type lectins. More than 10% of lectins use this fold for glycan recognition (Chapter 30).

In view of the conserved structures consisting of variable loops and a short β-barrel, in combination with low sequence homology based on β-strands connected by loops, the ability of β-trefoil proteins to flexibly select a variety of carbohydrate partners is not surprising. β-Trefoil proteins have been divided into several groups on the basis of biological function; these include lectins, basic fibroblast growth factor (FGF) (PDB 2AXM; Chapter 38), interleukin-1 (PDB 1ILT), Kunitz-type protease inhibitors (PDB 4IHZ), and the actin cross-linking protein hisactophilin (PDB 1HCE), although they have no Q-x-W motif.

Each subdomain could conceivably have a carbohydrate-binding site; however, in most R-type lectins only one or two subdomains bind a carbohydrate ligand. Carbohydrate-binding of RCA-I and RCA-II arises from aromatic amino acids stacking against Gal/GalNAc residues and from hydrogen bonding between aa's and hydroxyl groups of carbohydrate ligands.

Diversification of R-Type Lectins in Plants

A wide variety of R-type lectins with diverse configurations, including RIP-IIs and heterotetramers such as RCA-II, have been purified from various plant tissues. Nontoxic lectins isolated from elderberry (genus *Sambucus*) bark include SSA (*Sambucus sieboldiana* agglutinin) and SNA

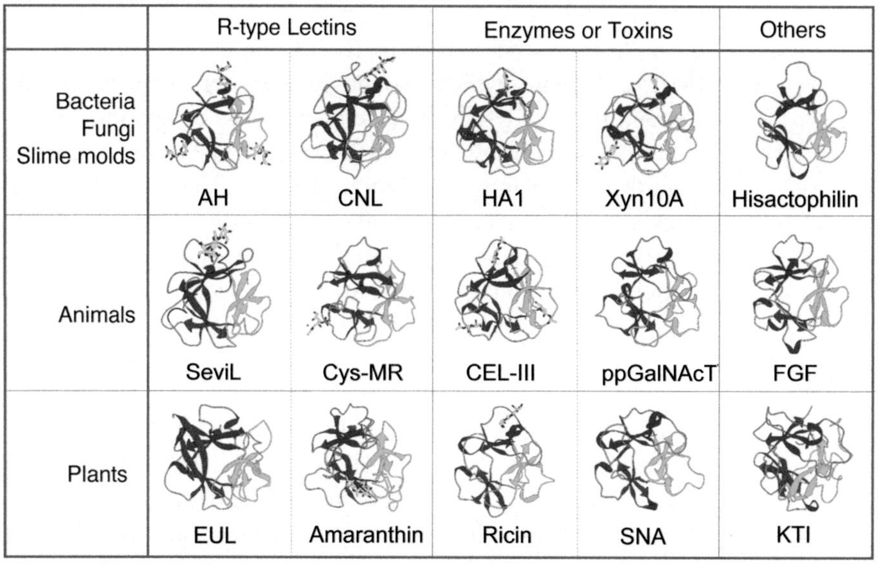

	R-type Lectins		Enzymes or Toxins		Others
Bacteria Fungi Slime molds	AH	CNL	HA1	Xyn10A	Hisactophilin
Animals	SeviL	Cys-MR	CEL-III	ppGalNAcT	FGF
Plants	EUL	Amaranthin	Ricin	SNA	KTI

FIGURE 31.4. Structures of β-trefoil folds of R-type lectin domains in various proteins. (*Top*) Structures derived from bacteria, fungi, and slime molds: AH (actinohivin in actinobacteria; PDB 4DEN), CNL (nematocidal lectin from conifer-loving mushroom *Clitocybe nebularis;* PDB 3NBE), HA1 (*Clostridium botulinum* neurotoxin; PDB 3AH2), Xyn10A (*Streptomyces lividans* endo-β1-4-xylanase 10A; PDB 1ISX), and hisactophilin (actin-binding protein; PDB 1HCE). (*Middle*) Structures derived from animals: SeviL (ganglioside-binding lectin from mussel; PDB 6LF2), Cys-MR (cysteine-rich mannose receptor in mouse; PDB 1DQO), CEL-III (hemolytic lectin from sea cucumber; PDB 2Z48), ppGalNAcT (polypeptide α-N-acetylgalactosaminyltransferase in freshwater snail *Biomphalaria glabrata;* Uniprot S5S833), and FGF (in human; PDB 2AXM). (*Bottom*) Structures derived from plants: EUL (lectin from European spindle-tree *Euonymus europaeus;* UniProt B3SV73), amaranthin (TF-antigen-binding lectin from amaranth seeds; PDB 1JLX), ricin (B-chain domain-1; PDB 2AAI), SNA (Siaα2-6-linked glycan-binding lectin from *Sambucus* [elderberry]; UniProt Q41358), and KTI (Kunitz-type protease inhibitor; PDB 4IHZ). Blue, green, and red ribbons, respectively, indicate α (amino-terminal side), β, and γ (carboxy-terminal side) subdomains in polypeptides.

(*Sambucus nigra* agglutinin UniProt Q41358). These are useful because it is the rare R-type lectin that binds strongly to α2-6-linked sialic acid–modified glycans, but not to α2-3-linked sialylated ligands. Both are heterotetramers (~140 kDa) composed of two heterodimers, each containing an A chain (resembling ricin A chain) disulfide-bonded to a B chain (R-type lectin).

Expression of R-type lectins in roots and shoots of rice (*Oryza sativa*) was shown to be up-regulated in response to changing environmental conditions, based on genome informatics of Man and Gal-binding β-trefoil fold group EUL (UniProt B3SV73) found in *Euonymus europaeus.* Analysis of the promoter region of rice EUL genes revealed diverse stress-responsive elements. Analysis of expression profiles revealed that the genes were regulated in response to various types of stress (drought, osmotic, hormonal). Thus, β-trefoil fold lectins are involved in plant stress signaling and defense.

Molecular structures of plant proteins having R-type lectin or β-trefoil fold domains vary extensively. Amaranth (*Amaranthus caudatus*) is an ancient South American crop plant with high nutritional value. Amaranthin (PDB 1JLX), purified from amaranth seeds, is a homodimeric lectin (Mr 66,000) consisting of two identical tandem-repeat β-trefoil fold domains. The primary structure of the lectin is ~20% similar to that of ricin B-chain. Amaranthin specifically recognizes Thomsen–Fredenreich (TF or T antigen) (Galβ1-3GalNAcα1-) and sialyl-T (Siaα2-3Galβ1-3GalNAcα1-) antigens. A GalNAc residue of the T-antigen disaccharide is captured by β- and γ-subdomains; thus, the mechanism of oligosaccharide binding by amaranthin differs from that of ricin B-chain and FGF (PDB 2AXM), in which a disaccharide is captured by a single subdomain.

Phylogenetic survey of plant genomes revealed that many families have amaranthin-like R-type lectin domain sequences in their polypeptides. The sequences consist either of solely lectin domains (prototype R-type lectin), and include tandem repeats, or of a combination of lectin with various functional domains (chimera-type R-type lectin). Plants use R-type lectin domain by combining it with functional domains such as "aerolysin," a pore-forming bacterial toxin. Tandem-repeat and chimera-type genes containing an amaranthin-like R-type lectin domain are expressed in many tissues and at different developmental stages, suggesting that R-type lectins respond to various external environmental stimuli.

R-TYPE LECTINS IN FUNGI, PROTISTA, AND ANIMALS

General Properties

R-type lectins are present in fungi, protists, invertebrates, and vertebrates. Almost all of them have the same Q-x-W sequence and bind to Gal, GalNAc, and Man. Some are dimeric or oligomeric forms of simple β-trefoil subunits (prototype R-type lectin); others are tandem-repeat R-type lectins having two CRDs in the polypeptide chain. Chimeric forms have functional domains (enzymes, toxins) associated with R-type lectin, similar to those in plants and bacteria.

Fungal R-Type Lectins

Diverse forms of R-type lectins have been found in fungi (Chapter 23). Some consist solely of β-trefoil subunits. A dimeric prototype R-type lectin CNL (PDB 3NBE), from the conifer-loving mushroom *C. nebularis*, exerts nematocidal activity by binding to GalNAcβ1-4GlcNAc glycan expressed in nematodes. CNL and another prototype R-type lectin, BEL (PDB 4I4X), from *Boletus edulis*, which binds to T-antigen, inhibit growth of cultured mammalian tumor cells.

Chimeric forms in fungi include R-type lectin domains connected to tyrosinase-domain (PPO3: PDB 2Y9W), peptide N-glycanase-like-domain (MOA: PDB 2IHO), and aerolysin-like domain (LSL: PDB 1W3A). LSL from *Laetiporus sulphureus* (a bracket fungus) has a β-trefoil fold but not a Q-x-W motif.

Protist R-Type Lectins

Studies of R-type lectins in protists are at an early stage. One was recently found in the anaerobic parasite *Entamoeba histolytica* (Chapter 43), a Gal/GalNAc-binding prototype R-type lectin (PDB 6IFB) consisting of two β-trefoil subunits. In each subunit, the α- and β-subdomain functioned, respectively, to capture carbohydrates and subunits connected by a disulfide bond.

Invertebrate R-Type Lectins

Prototype and tandem-repeat R-type lectins consisting solely of β-trefoil domains have been identified in various invertebrate groups, including sponges (phylum Porifera), earthworms (Annelida) (EW29: PDB 2ZQN), bivalves (Mollusca) (UniProt A0A646QVV9), and brachiopods (Brachiopoda) (UniProt A0A646QV64). SeviL (PDB 6LF2), in mussels, binds to the trisaccharide Galβ1-3GalNAcβ1-4Gal, a component of gangliosides GM1b (Siaα2-3 Galβ1-3GalNAcβ1-4Galβ1-4Glcβ1-1Cer) and GA1 (Galβ1-3GalNAcβ1-4Galβ1-4Glcβ1-1Cer). GM1b is a target antigen of Guillain-Barré syndrome (Chapter 11), which arises from *Campylobacter* infection because structures of some bacterial lipopolysaccharides mimic those of the ganglioside (Chapter 20). Interaction between prototype R-type lectin and ganglioside-induced cytotoxicity against mammalian neoplastic cells through activation of MAP kinases. Another family of mussel lectins (MytiLec: PDB 3WMV; CGL: PDB 5F90) has a β-trefoil fold but not Q-x-W motif. Similar to Shiga toxins (Chapter 37), these lectins captured three Gb3 (Galα1-4Galβ1-4Glc) glycans in three carbohydrate-binding pockets within the β-trefoil fold, and thereby induced signal transduction.

Chimera-type R-type lectins connect with various functional domains in invertebrates. Pierisin in cabbage butterfly (Arthropoda, order Lepidoptera) larvae expresses R-type lectin domain at the carboxyl terminus of an ADP-ribosyltransferase domain (Uniprot Q9U8Q4) and induces apoptosis of cancer cells. This lectin domain was found to bind to glycan ligands of Gb3 and Gb4 (GalNAcβ1-3Galα1-4Galβ1-4Glc) glycosphingolipids on cancer cells termed, respectively, P^k and P antigens (Chapter 14). Clotting factor G α-subunit (Uniprot Q27082) in horseshoe crab (Arthropoda, order Crustacea) blood contained two glycoside-hydrolase domains on either side of a central R-type lectin domain. Another R-type lectin, CEL-III (PDB 2Z48), in sea cucumber (Echinodermata, Deuterostomia; Chapter 27) has both a pore-forming domain and a tandem-repeat-type R-type lectin domain and displays hemolytic effect on mammalian erythrocytes.

Almost all the configurations seen in invertebrate R-type lectins are also found in bacterial R-type lectins. β-trefoil fold is a very effective protein scaffold for biological functions. Living systems have evidently improved their survival strategies by evolutionary tinkering of the R-type lectin domain per se, or by linking it with other functional domains.

UDP-GalNAc:Polypeptide α-N-Acetyl-Galactosaminyltransferases

UDP-GalNAc: polypeptide α-N-acetylgalactosaminyltransferases (ppGalNAcTs: EC 2.4.1.41) have chimeric-form R-type lectin structures consisting of a catalytic subdomain (amino-terminal) and an R-type lectin domain (carboxy-terminal) (PDB 1HXB). Mucin-type O-glycans have a common core structure GalNAcα1-Ser/Thr (Figure 31.5), which can be modified by addition of Gal or GalNAc residues (Chapter 10). Most proteins (~80%) that pass through the secretory apparatus have at least one Ser or Thr residue modified with α-linked GalNAc. A family of ppGalNAcTs functioning in the Golgi are responsible for modification of Ser/Thr residues.

These enzymes are widely distributed in vertebrates, invertebrates, and protists, but so far have not been found in bacteria, plants, or fungi (Chapter 26). In humans, 20 ppGalNAcT-encoding isoform genes have been studied with regard to structure, substrate specificity, and function (Chapter 6). A ppGalNAcT (Uniprot S5S833) was found in a snail that serves as an intermediate host for trematodes, the primary group of schistosomes that infect humans. The snail ppGalNAcT, in contrast to those in vertebrates, has affinity for a wide range of substrates and only one isoform.

Enzymes in this family of ppGalNAcTs have multidomain structure and are type II transmembrane proteins having the amino terminus in cytosol and carboxyl terminus on the other side of the membrane. They interact dynamically with their polypeptide substrates. In ppGalNAcT2, the R-type lectin domain interacts primarily with GalNAc on modified residues, whereas other regions of the polypeptide substrate lie in a groove next to the active site of the catalytic domain. The proximity of these two interactions positions the enzyme to bind special acceptor peptides. The parasitic protist *Toxoplasma gondii* (SAR supergroup, phylum Alveolata; Chapter 20) uses this enzyme to glycosylate cyst walls in the central nervous system to confer structural rigidity, thereby enabling the parasite to survive for the host's lifetime.

Vertebrate Mannose Receptors

The diversity of R-type lectin domains in vertebrates is not well-understood, relative to those in plants or invertebrates. Mannose receptors (MRs) are type I transmembrane glycoproteins and

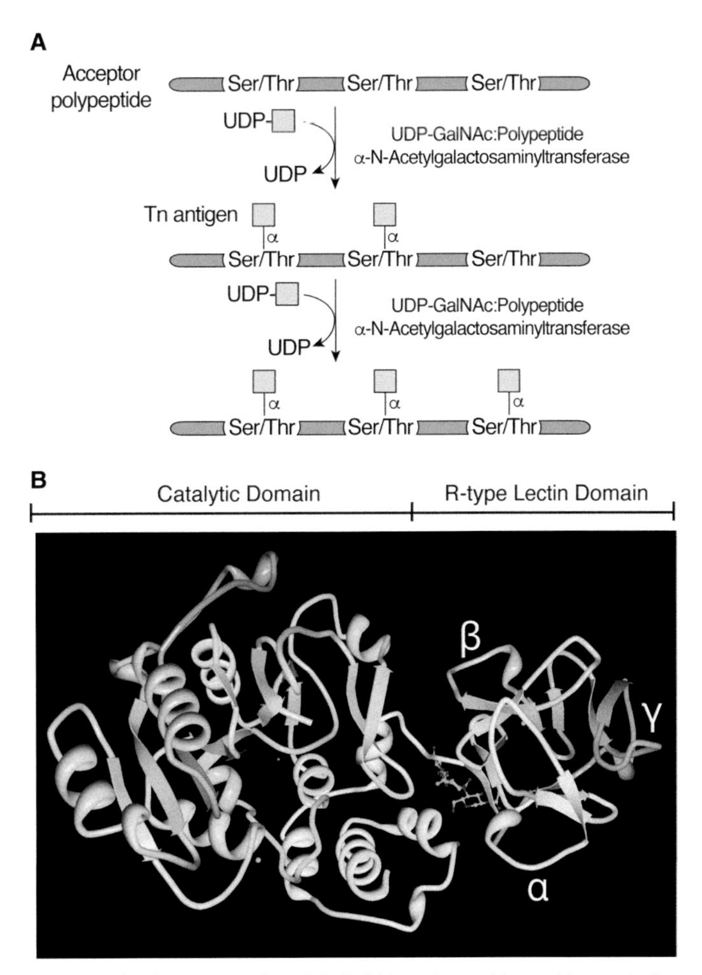

FIGURE 31.5. Structure and function of UDP-GalNAc:polypeptide α-N-acetylgalactosaminyltransferases (ppGalNAcTs). (*A*) ppGalNAcT activity with Ser/Thr-containing acceptor peptide and UDP-GalNAc as donor. Some ppGalNAcTs act preferentially on the product of this reaction, using peptides with attached *N*-acetylgalactosamine as acceptors. (*B*) Crystal structure of mouse ppGalNAcT1. *Left*: catalytic domain. (*Right*) R-type lectin domain. β-Trefoil structure and three-lobe repeating α, β, and γ loops are indicated. The carbohydrate chain attached to the R-type domain is partially visible. (Reprinted with permission from Fritz TA, et al. 2004. *Proc Natl Acad Sci* **101:** 15307–15312, © National Academy of Sciences, PDB 1XHB.)

unique multifunctional chimera-form lectins consisting of an amino-terminal cysteine-rich R-type lectin domain (PDB 1DQO), a single fibronectin type II domain, and multiple C-type lectin domains (CTLDs) (Chapter 34). MR CD206 (Uniprot Q9UBG0) is expressed on hepatic endothelial cells, Kupffer cells, other epithelial cells, macrophages, and dendritic cells and functions in innate and adaptive immune systems. The domain configurations of phospholipase A2 receptor, DEC-205, and Endo180 are the same as that of CD206 (Figure 31.1), although their ligands are not glycans.

When CD206 dimerizes on cell membrane, the R-type lectin domain binds to sulfated glycans. Pituitary hormones are captured by binding of this domain to 4-SO_4-GalNAcβ1-R residues, 3-O-sulfated Gal, 3-O-sulfated Le^x, or 3-O-sulfated Le^a for clearance from the bloodstream. CTLDs recognize surface mannose-rich glycoconjugates on *Candida albicans*, *Pneumocystis carinii*, *Mycobacterium tuberculosis*, *Klebsiella pneumoniae*, and *Leishmania donovani* (Chapter 43). CD206 is considered a pathogen-associated molecular pattern (PAMP) that functions in clathrin-dependent endocytosis and in phagocytosis of nonopsonized microbes. MRs also play roles in adaptive immunity through their ability to deliver antigens to major histocompatibility class II compartments. Their R-type lectin domains facilitate binding to dendritic cells that express sulfated glycan-containing ligands such as sialoadhesin and CD45.

MICROBIAL R-TYPE LECTINS

General Properties

Most of the known R-type lectin domains are found in bacterial lectins, enzymes, and toxins. Genes encoding R-type lectins have been observed in Archaea and viruses, evidently reflecting evolution from bacteria to eukaryotes. Microbial genes of the original R-type lectins may have been transmitted to eukaryotes; alternatively, lectins may have arisen through convergent evolution in the two groups.

Bacterial R-Type Lectins

Actinohivin (AH; PDB 4DEN), a prototype R-type lectin in bacteria, consists of 114-aa polypeptides isolated from actinobacteria (genus *Actinomyces*). It binds to high mannose-type oligosaccharides, commonly found on the viral envelope GP120 of human immunodeficiency virus 1 (HIV-1). AH displays strong anti-HIV activity against clinically isolated viral strains (Chapter 42).

R-type lectin domains are found in many bacterial enzymes and toxins. An example in *Streptomyces* is the enzyme endo-β1-4-xylanase (Xyn10A: EC 3.2.1.8; PDB 1ISX). Xyn10A acts as a glycoside hydrolase to catalyze cleavage of β1-4 xylose and can bind to xylan and a variety of soluble sugars, including Gal, lactose, and xylo- and arabino-oligosaccharides.

Pore-forming domains are found in exotoxins produced by pathogenic bacteria that hemolyze erythrocytes. A monomeric lysin domain comprised of β-sheets oligomerizes like a piece of fabric tube to create pores on the cell wall. Aerolysin (Chapter 12) and hemolysin are bacterial toxins having R-type lectin domains. Many eukaryotes (plants, fungi, invertebrates) have aerolysin-like domains. Hemolysin (UniProt P09545) of *Vibrio cholerae* has a β-prism domain with jacalin (jackfruit lectin) structure, as well as R-type lectin domain.

Botulinum neurotoxin (BoNT) from *Clostridium botulinum* has extremely high toxicity (LD50: 0.001 μg/kg) (Chapter 37). A component of BoNT termed hemagglutinin HA1 (PDB 3AH2) is a tandem-repeat-type dimeric R-type lectin subunit that binds to Gal at the carboxy-terminal subdomain. HA2, a subunit with β-trefoil fold, connects to the amino-terminal subdomains of two HA1s and to the jelly-roll-like β-sandwich fold subunit HA3 (PDB 2ZOE) (Chapter 32). HA1 helps orient BoNT to target intestinal cells by binding to Gal or Sia in gangliosides (e.g., GT1b) (Chapter 11) or cell-surface glycans.

MTX (UniProt Q03988), a mosquitocidal toxin found in *Bacillus sphaericus*, contains an ADP-ribosyltransferase domain and four R-type lectin domains, arranged in a structural configuration similar to that of pierisin in cabbage butterfly larvae. Subunit forms of MTX are the same as those of cholera toxin and ricin (i.e., they consist of A-subunit [toxic enzyme] and B-subunit [lectin]).

Archaeal and Viral R-Type Lectins

R-type β-trefoil lectins have been identified in numerous species in the families *Halobacteriaceae* and *Natrialbaceae* (class Haloarchaea), whose cell walls have mucin-containing glycoproteins (Chapter 22). Based on successful experimental culture of an archaeal strain, a model was proposed in 2020 for eukaryogenesis from Archaea. An intriguing possibility is that domain Archaea is the link for R-type lectins between domains Bacteria and Eukarya.

Type-C and type-D BoNTs have been produced from the genome of a *Clostridium* bacteriophage. Were R-type lectin genes somehow produced by viruses? Metagenomic analyses indicate the presence of R-type lectin genes (Uniprot Q5UPX2) in the viral family Mimiviridae (Chapters 8 and 20), and phylum Nucleocytoviricota. Recent surveys reveal that the genus *Mimivirus* is distributed worldwide in oceans as well as on land. Extensive future studies of comparative expression levels of viral R-type lectin genes will help clarify relationships between glycan-codes and specific environments in the world's ecosystems.

PERSPECTIVES

The term "bricolage" refers to creation of novel tools in primitive societies in which availability of materials was limited. Creation of such tools may be based on tinkering with frameworks comprising limited sets of products and techniques used previously for other purposes. Analogously, modification and/or combination of genes in different groups of organisms may generate molecules useful for novel environmental responses. The β-trefoil fold structure, characteristic of the R-type lectin superfamily, is evidently a core framework in living systems that arose via bricolage and has evolved to play key roles in a variety of biochemical processes, including carbohydrate binding.

Discovery of new R-type lectins is ongoing. The ancestor of this superfamily is not yet clear, but it is found in members of all biological domains. The origins and evolution of these lectins are therefore of interest to scientists in many fields. The presence of β-trefoil folding in proteins may reflect either convergent or divergent evolution. Elucidating the physiological functions of various R-type lectins will require identification of their endogenous ligands and the specific environmental stimuli that trigger their expression. Increasing knowledge of R-type lectin structure and function will enhance our understanding of lectin-glycan interactions in living systems and facilitate biotechnological applications of these lectins.

ACKNOWLEDGMENTS

The authors acknowledge contributions to previous versions of this chapter by Marilynn E. Etzler, helpful comments and suggestions by Sneha S. Komath, support in preparation of Figure 31.4 by Kenichi Kamata, and English editing of the manuscript by Stephen Anderson.

FURTHER READING

Murzin AG, Lesk AM, Chothia C. 1992. β-Trefoil fold: patterns of structure and sequence in the Kunitz inhibitors interleukins-1β and -1α and fibroblast growth factors. *J Mol Biol* **223:** 531–543. doi:10.1016/0022-2836(92)90668-a

Stirpe F, Barbieri L, Battelli MG, Soria M, Lappi DA. 1992. Ribosome-inactivating proteins from plants: present status and future prospects. *Biotechnology* **10:** 405–412. doi:10.1038/nbt0492-405

Lord JM, Roberts LM, Robertus JD. 1994. Ricin: structure, mode of action, and some current applications. *FASEB J* **8:** 201–208. doi:10.1096/fasebj.8.2.8119491

Hazes B. 1996. The $(QxW)_3$ domain: a flexible lectin scaffold. *Protein Sci* **5:** 1490–1501. doi:10.1002/pro.5560050805

Hirabayashi J, Dutta SK, Kasai K. 1998. Novel galactose-binding proteins in Annelida. Characterization of 29-kDa tandem repeat-type lectins from the earthworm *Lumbricus terrestris*. *J Biol Chem* **273:** 14450–14460. doi:10.1074/jbc.273.23.14450

Leteux C, Chai W, Loveless RW, Yuen CT, Uhlin-Hansen L, Combarnous Y, Jankovic M, Maric SC, Misulovin Z, Nussenzweig MC, Feizi T. 2000. The cysteine-rich domain of the macrophage mannose receptor is a multispecific lectin that recognizes chondroitin sulfates A and B and sulfated oligosaccharides of blood group Lewis(a) and Lewis(x) types in addition to the sulfated N-glycans of lutropin. *J Exp Med* **191:** 1117–1126. doi:10.1084/jem.191.7.1117

Matsushima-Hibiya Y, Watanabe M, Hidari KI, Miyamoto D, Suzuki Y, Kasama T, Kasama T, Koyama K, Sugimura T, Wakabayashi K. 2003. Identification of glycosphingolipid receptors for pierisin-1, a guanine-specific ADP-ribosylating toxin from the cabbage butterfly. *J Biol Chem* **278:** 9972–9978. doi:10.1074/jbc.m212114200

Uchida T, Yamasaki T, Eto S, Sugawara H, Kurisu G, Nakagawa A, Kusunoki M, Hatakeyama T. 2004. Crystal structure of the hemolytic lectin CEL-III isolated from the marine invertebrate *Cucumaria echinata*: implications of domain structure for its membrane pore-formation mechanism. *J Biol Chem* **279:** 37133–37141. doi:10.1074/jbc.m404065200

Nakamura T, Tonozuka T, Ide A, Yuzawa T, Oguma K, Nishikawa A. 2008. Sugar-binding sites of the HA1 subcomponent of *Clostridium botulinum* type C progenitor toxin. *J Mol Biol* **376:** 854–867. doi:10.2210/pdb2ehn/pdb

Broom A, Doxey AC, Lobsanov YD, Berthin LG, Rose DR, Howell PL, McConkey BJ, Meiering EM. 2012. Modular evolution and the origins of symmetry: reconstruction of a three-fold symmetric globular protein. *Structure* **20:** 161–171. doi:10.1016/j.str.2011.10.021

Fujii Y, Dohmae N, Takio K, Kawsar SMA, Matsumoto R, Hasan I, Koide Y, Kanaly RA, Yasumitsu H, Ogawa Y, et al. 2012. A lectin from the mussel *Mytilus galloprovincialis* has a highly novel primary structure and induces glycan-mediated cytotoxicity of globotriaosylceramide-expressing lymphoma cells. *J Biol Chem* **287:** 44772–44783. doi:10.1074/jbc.m112.418012

Abhinav KV, Samuel E, Vijayan M. 2016. Archeal lectins: an identification through a genomic search. *Proteins* **84:** 21–30. doi:10.1002/prot.24949

Hasan I, Gerdol M, Fujii Y, Rajia S, Koide Y, Yamamoto D, Kawsar SMA, Ozeki Y. 2016. cDNA and gene structure of MytiLec-1, a bacteriostatic R-type lectin from the Mediterranean mussel (*Mytilus galloprovincialis*). *Mar Drugs* **14:** 92. doi:10.3390/md14050092

Dang L, Rougé P, Van Damme EJM. 2017. Amaranthin-like proteins with aerolysin domains in plants. *Front Plant Sci* **8:** 1368. doi:10.3389/fpls.2017.01368

De Schutter K, Tsaneva M, Kulkarni SR, Rougé P, Vandepoele K, Van Damme EJM. 2017. Evolutionary relationships and expression analysis of EUL domain proteins in rice (*Oryza sativa*). *Rice (NY)* **10:** 26. doi:10.1186/s12284-017-0164-3

Liao J, Chien CH, Wu H, Huang K, Wang I, Ho M, Tu I, Lee I, Li W, Shih Y, et al. 2017. A multivalent marine lectin from *Crenomytilus grayanus* possesses anti-cancer activity through recognizing globotriose Gb3. *J Am Chem Soc* **138:** 4787–4795. doi:10.1021/jacs.6b00111.s001

Terada D, Voet ARD, Noguchi H, Kamata K, Ohki M, Addy C, Fujii Y, Yamamoto D, Ozeki Y, Tame JRH, Zhang KYJ. 2017. Computational design of a symmetrical β-trefoil lectin with cancer cell binding activity. *Sci Rep* **7:** 5943. doi:10.1038/s41598-017-06332-7

Franke H, Scholl R, Aigner A. 2019. Ricin and *Ricinus communis* in pharmacology and toxicology-from ancient use and "Papyrus Ebers" to modern perspectives and "poisonous plant of the year 2018". *Naunyn Schmiedebergs Arch Pharmacol* **392:** 1181–1208. doi:10.1007/s00210-019-01691-6

Hu Z, Wang Y, Cheng C, He Y. 2019. Structural basis of the pH-dependent conformational change of the N-terminal region of human mannose receptor/CD206. *J Struct Biol* **208:** 107384–107391. doi:10.1016/j.jsb.2019.09.001

Fujii Y, Gerdol M, Kawsar SMA, Hasan I, Spazzali F, Yoshida T, Ogawa Y, Rajia S, Kamata K, Koide Y, et al. 2020. A GM1b/asialo-GM1 oligosaccharide-binding R-type lectin from purplish bifurcate mussels *Mytilisepta virgata* and its effect on MAP kinases. *FEBS J* **287:** 2612–2630. doi:10.1111/febs.15154

Ismaya WT, Tjandrawinata RR, Dijkstra BW, Beintema JJ, Nabila N, Rachmawati H. 2020. Relationship of *Agaricus bisporus* mannose-binding protein to lectins with β-trefoil fold. *Biochem Biophys Res Commun* **527:** 1027–1032. doi:10.1016/j.bbrc.2020.05.030

Kamata K, Mizutani K, Takahashi K, Marchetti R, Silipo A, Addy C, Park S-Y, Fujii Y, Fujita H, Konuma T, et al. 2020. The structure of SeviL, a GM1b/asialo-GM1 binding R-type lectin from the mussel *Mytilisepta virgata*. *Sci Rep* **10:** 22102. doi:10.1038/s41598-020-78926-7

Khan F, Kurre D, Suguna K. 2020. Crystal structures of a β-trefoil lectin from *Entamoeba histolytica* in monomeric and a novel disulfide-bond-mediated dimeric forms. *Glycobiology* **30:** 474–488. doi:10.1093/glycob/cwaa001

32 L-Type Lectins

Richard D. Cummings, Marilynn E. Etzler, T.N.C. Ramya, Koichi Kato, Gabriel A. Rabinovich, and Avadhesha Surolia

The L-type lectins occur in the seeds of leguminous plants. They have structural motifs present in a variety of glycan-binding proteins (GBPs) from other eukaryotic organisms. The structures of many of these lectins have been characterized, and many L-type lectins are used in a wide range of biomedical and analytical procedures and implicated in biomedical processes. This chapter discusses the structure–function relationships of these lectins and the various biological roles they have in different organisms.

HISTORICAL BACKGROUND

The L-type lectins have a rich history going back to the end of the 19th century when it was found that extracts from the seeds of leguminous plants (Fabaceae family but also called the Leguminosae family) could agglutinate red blood cells. These agglutinins (later named lectins) were found to be soluble proteins abundant in the seeds of leguminous plants. Different species of legumes were found to differ in hemagglutination specificity. Much work on these proteins was performed in the early part of the 20th century, including the crystallization of concanavalin A (ConA; the hemagglutinin [HA] from jack beans), which was first isolated by James Sumner in 1919 and crystallized in 1936 and was the first commercially available lectin. The legume

published by Cold Spring Harbor Laboratory Press; doi: 10.1101/glycobiology.4e.32

lectins were also found to have hemagglutinating properties because of their ability to bind glycans on the cell surface.

The abundance of these proteins in the soluble extracts of legume seeds (up to ~5%–10% total protein) enabled many lectins to be isolated and characterized. Such seed lectins have considerable amino acid sequence homology, and the variety of carbohydrate-binding specificities found among these lectins made them useful tools in a wide variety of analytical and biomedical procedures.

The crystal structures of a number of legume seed lectins allowed the identification of the carbohydrate-binding sites. Structural similarities in the tertiary structures were identified among these lectins and several other lectins, including the galectins (see Chapter 36), as will be discussed in this chapter. For this reason, the term "L-type lectins" has recently been designated as a classification for all proteins with this legume seed lectin-like protein structure.

COMMON FEATURES OF L-TYPE LECTINS

The L-type lectins are distinguished from other lectins primarily based on tertiary structure and not the primary structure. In general, either the entire lectin monomer or the carbohydrate-recognition domains (CRDs) of the more complex lectins is composed of antiparallel β-sheets connected by short loops and β-bends, and they usually lack any α-helices. These sheets form a dome-like structure related to the "jelly-roll fold," which is often called a "lectin fold." The carbohydrate-binding site is generally localized toward the apex of this dome. The tertiary structure of the monomer of ConA, the lectin from the seeds of the legume *Canavalia ensiformis*, is shown in Figure 32.1. The crystal structures of at least 20 other legume L-type lectin monomers have been determined by high-resolution X-ray crystallography and are almost superimposable on this structure. Thus, it is not surprising that the amino acid sequences of

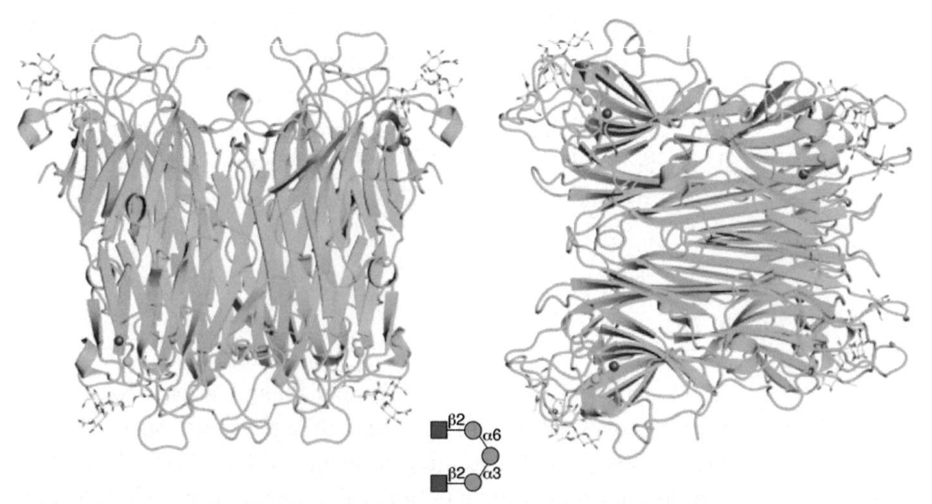

FIGURE 32.1. Structure of concanavalin A (ConA), a legume seed lectin in complex with a branched pentasaccharide GlcNAcβ1-2Manα1-3(GlcNAcαβ1-2Manα1-6)Man to 2.7 Å. The pentasaccharide is depicted in the *center*. The tetrameric Con A (monomer 26.5 kDa and 235 amino acids) is shown (*left*) each with the pentasaccharide and the orthogonal view is shown (*right*). The bound metal is indicated by the *purple* and *green balls* (Mn^{++} and Ca^{++}). The monomer is best described as a "jelly-roll fold," consisting of a flat six-stranded antiparallel "back" β-sheet, a curved seven-stranded "front" β-sheet, and a five-stranded "top" sheet linked by loops of various lengths. ConA dimers involve antiparallel side-by-side alignment of the flat six-stranded "back" sheets, giving rise to a contiguous 12-stranded sheet. The tetramerization of ConA occurs by a back-to-back association of two dimers. (Taken from the PDB ftp://ftp.wwpdb.org; Moothoo DN, Naismith JH. 1998. *Glycobiology* **8:** 173–181.)

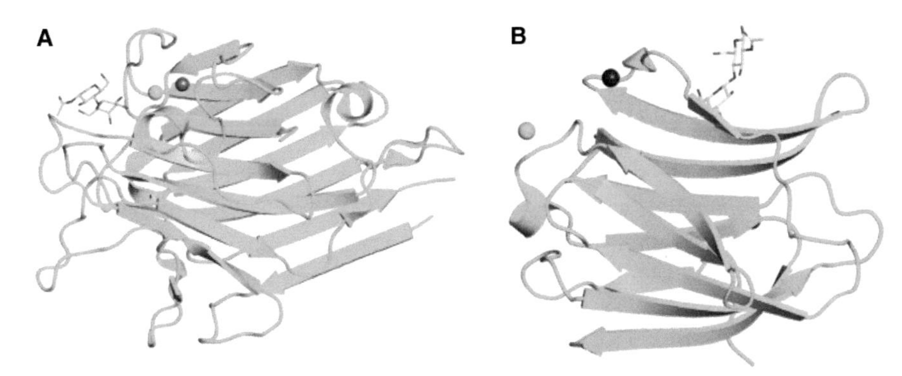

FIGURE 32.2. Comparison of the subunit structures of soybean agglutinin (*A*) complexed with a pentasaccharide containing Galβ1-4GlcNAc-R and human galectin-3 at 1.4 Å (*B*) complexed with Galβ1-4GlcNAc. Both lectins display a related β-barrel configuration. (Soybean agglutinin structure [1sbd] available at PDB and based on Olsen LR, Dessen A, Gupta D, Sebesan S, Sacchettini JC, Brewer CF. 1997. *Biochemistry* **36:** 15073–15080. Human galectin-3 structure [1kjl] available at PDB and based on Sorme P, Arnoux P, Kahl-Knutsson B, Leffler H, Rini JM, Nilsson UJ. 2005. *J Am Chem Soc* **127:** 1737–1743.) For PDB structures go to ftp://ftp.wwpdb.org/.

legume lectins show remarkable homology with one another and of the sequences of many other legume seed lectins sequenced but not yet crystallized. Fewer, but significant, homologies in primary structure have been found between legume L-type lectins compared with some L-type lectins from far distant sources, such as ERGIC-53 and VIP36, found in vertebrates. Yet, in other L-type lectins, no homology is found with the seed lectins, although they contain similar lectin folds. For example, a comparison of the tertiary structure of the legume soybean lectin with the structure of human galectin-3 shows that both proteins contain the typical L-type lectin fold, but no amino acid sequence homology exists between the two lectins (Figure 32.2).

Relationships between sequences of L-type lectins from legumes compared with the phylogeny of the various species within the Fabaceae family of plants suggest that these lectins most probably arose from divergent evolution. It remains an open possibility that the tertiary structures of some of the other members of the "L-type lectin family" arose by convergent evolution. It must also be noted that for a protein to be firmly placed in the L-type lectin category it must have the lectin fold and glycan-binding activity.

All soluble L-type lectins found to date are multimeric proteins, although all do not have the same quaternary structure. Thus, these lectins are multivalent with more than one glycan-binding site per lectin molecule. The same multivalent principle applies to the membrane bound L-type lectins because the presence of two or more molecules on a membrane surface essentially presents a multivalent situation. In addition to increasing the avidity of the lectins for branched and/or cell-surface glycans, this multivalence can have great biological significance. Binding of the lectins to the cell surface can lead to aggregation of specific glycan receptors, promoting a variety of biological responses such as mitogenesis and various signal transduction processes.

PLANT L-TYPE LECTINS

Distribution and Localization

Plant L-type lectins are primarily found in the seeds of leguminous plants and are synthesized during seed development several weeks after flowering; they are transported to the vacuole where they become condensed into specialized vesicles called protein bodies. They are stable during desiccation of the seeds and can remain in that state indefinitely until the seeds germinate. They represent one of several classes of proteins stored in high concentrations in the seeds and are often called storage proteins. During seed germination, the storage bodies become the vacuoles of the cotyledons, which appear as the first leafy appendages of the plant. During the first week of development, these cotyledons provide food for the plant and eventually shrivel up and disappear. L-type lectins have also been found in the bark of some leguminous trees, and very low amounts of these lectins are also found in other vegetative tissues of legumes. In some cases, these latter lectins are encoded by separate but very similar genes. More than 100 of the seed legume L-type lectins have been characterized and are the most extensively studied proteins of this class. It also should be noted that the L-type fold has been observed in noncarbohydrate binding proteins, such as lectin-like receptor kinases in plants that are important in development and stress responses.

Structure

A common feature of the legume L-type lectins is their monomeric structure. The structures of the monomers consist of three antiparallel β-sandwich: a flat six-stranded "back" sheet, a concave seven-stranded "front" sheet, and a short "top" sheet that keeps the two major sheets together (Figure 32.1A,B). All of these lectins require Ca^{++} and a transition metal ion (usually Mn^{++}) for their carbohydrate-binding activity. The glycan-binding and metal-binding sites are localized in close proximity to each other at the top of the "front" sheet.

The glycan-binding site is composed of four loops: A, B, C, and D (Figure 32.3, top). These loops contain four invariant amino acids that are essential for carbohydrate binding (Figure 32.3, bottom). Loop A contains an invariant aspartate, which forms hydrogen bonds between its side chain and the glycan ligand. This amino acid is linked to its preceding amino acid (usually alanine) by a rare *cis*-peptide bond, which is stabilized by the metal ions and is necessary for the proper orientation of the aspartate in the combining site. Loop B contains an invariant glycine, which also forms hydrogen bonds with the ligand. An exception to this case is found in two lectins (ConA and the closely related *Dioclea grandiflora* lectin) in which the glycine is replaced with an arginine. Both the glycine and arginine form hydrogen bonds with the ligand via their main-chain amides. Loop C contains an invariant asparagine, which forms a hydrogen bond with the ligand via its side chain, and an invariant hydrophobic amino acid. Besides these invariant amino acids, backbone atoms of the residues of loop D also contribute to monosaccharide recognition.

The legume L-type lectins are generally classified into groups based on their carbohydrate specificities as often identified by the ability of monosaccharides to inhibit their agglutinating activity. These differences in specificities are brought about by variability in the conformation and size of the D loop and to some extent by the C loop. Although the main specificity regions of the legume lectins are determined by the loops, there are sites other than these that contribute to lectin specificity. There are several additional modes of refining these specificities, such as interaction with water, posttranslational modifications, and state of oligomerization.

Legume L-type lectins are oligomeric—mostly dimeric or tetrameric in nature—and adopt a variety of quaternary structures. The back β-sheet in the monomeric unit is involved in oligomerization. Small differences in the monomeric structure of the different legume lectins result in different modes of association of the β-sheet during oligomerization. For instance, although both ConA and PNA are tetrameric—being dimers of dimers—the quaternary association in ConA involves the association of the two six-stranded back β-sheets to form an extended 12-stranded β-sheet in each dimer, whereas the quaternary association in PNA involves a back-to-back arrangement of the back β-sheets. The tetrameric structure of ConA is shown in Figure 32.1C,D. Although some of the other lectins occur as dimeric and tetrameric structures, several other different orientations of the β-sheets account for the variability in dimeric and tetrameric structures of other lectins in this class. Interestingly, some legume lectins have a

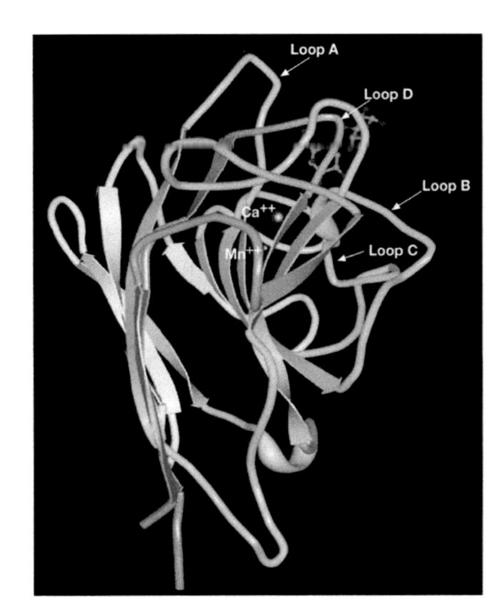

Lectin	Loop A	Loop B	Loop C	Loop D	Specificity/ Inhibitor
EcorL	GPPYT–RPLPADGLVF	AQ-GYGYLG	VEFDTFSN----PWDP	GLSGATG----AQRDAAETHDVYSW	GalNAc
DBL	APSK---ASFADGIAP	RR-NGGYLG	VEFDTLSNS---GWDP	GFSATTGLSDG----YIETHDVLSW	GalNAc
E-PHA	VPNN---EGPADGLAF	KD-KGGLLG	VEFDTLYNV---HWDP	GFTATTGITKG----NVETNDILSW	Complex
L-PHA	VPNN---AGPADGLAF	KD-KGGFLC	VEFDTLYNK---DWDP	GFSATTGINKG----NVETNDVLSW	Complex
SBA	APDT---KRLADGLAF	QT-HAGYLG	VEFDTFRN----SWDP	GFSAATGLDIP-----GESHDVLSW	GalNAc
PNA	KD--IKDYDPADGIIF	GSIGGGTLG	VEFDTYSNS–-EYNDP	GFSASGSL------GGRQIHLIRSW	Gal
LTA	IR--ELKYTPTDGLVF	GS-TGGFLG	VEFDSYHN----IWDP	GFSATTGN------PEREKHDIYSW	Fuc
UEA-I	SANP---KAATDGLTF	RRA-GGYFG	VEFDTI-GSPVNFDDP	GFSGGTYI------GRQATHEVLNW	Fuc
UEA-II	EPDE--KIDGVDGLAF	GS-SAGMFG	VEFDSYPGKTYNPWDP	GFSGGVGN------AAKFDHDVLSW	GlcNAc
LAA-I	PPIQSRKADGVDGLAF	GS-SAGMFG	VEFDTYFGKAYNPWDP	GFSAGVGN------AAKFNHDILSW	GlcNAc
LSL	RPNSDS-QVVADGFTF	RG-DGGLLG	VEFDTFHNQ---PWDP	GLSASTATYY-------SAHEVYSW	Man/Glc
Con A	SPDS----HPADGIAF	GS-TGRLLG	VELDTYPNT--DIGDP	GLSASTGL-------YKETNTILSW	Man/Glc
LenL	SPNG---YNVADGFTF	QT-GGGYLG	VEFDTFYNA---AWDP	GFSATTGAEF-------AAQEVHSW	Man/Glc
PSL	APNS---YNVADGFTF	QT-GGGYLG	VEFDTFYNA---AWDP	GPSATTGAEY-------AAHEVLSW	Man/Glc
Favin	APNG---YNVADGFTF	QT-GGGYLG	VEFDTFYNA---AWDP	GFSATTGAEY-------ATHEVLSW	Man/Glc
	** *	*	** **	* *	

FIGURE 32.3. (*Top*) Three-dimensional structure of a peanut agglutinin (PNA) monomer showing the four loops involved in sugar binding: loops A, B, C, and D. The bound sugar (lactose) is shown as a "*ball-and-stick*" model. Calcium and manganese ions are required for ligand binding. (*Bottom*) Sequence alignment of loops A–D in other legume lectins. The size of binding-site loop D and monosaccharide specificity show an explicit correlation. Monosaccharide specificity and number of gaps are indicated at the *right*. Key residues are highlighted in *blue* and highly conserved residues have been indicated with an *asterisk*. (Adapted, with permission of Elsevier, from Vijayan M, Chandra N. 1999. *Curr Opin Struct Biol* **9:** 707–714; and the PBD deposited structure 1V6I.) EcorL, *Erythrina corallodendron* lectin; DBL, *Dolichos biflorus* lectin; E-PHA, erythroagglutining phytohemagglutinin from *Phaseolus vulgaris*; L-PHA, leukoagglutinating phytohemagglutinin from *P. vulgaris*; SBA, soybean agglutinin *Glycine max*; PNA, peanut agglutinin *Arachis hypogaea*; LTA, *Lotus tetragonolobus* agglutinin; UEA-I, *Ulex europaeus* agglutinin-I; UEA-II, *U. europaeus* agglutinin-II; LAA-I, *Laburnum alpinum* agglutinin-I; LSL, *Lonchocarpus sericeus* lectin; ConA, concanavalin A *Canavalia ensiformis*; LenL, lentil lectin *Lens culinaris*; PSL, pea lectin *Pisum sativum*; Favin, *Vicia faba* lectin.

hydrophobic binding site that binds adenine and adenine-derived plant hormones with micromolar affinity; this is two to three orders of magnitude higher than their affinity for monosaccharides. Three of these lectins (the soybean agglutinin, phytohemagglutinin L [PHAL], and *Dolichos biflorus* lectin) have been crystallized and found to have a unique tetrameric structure, in that the dimer–dimer interface creates a channel running through the center of the tetramer. Two identical adenine-binding sites are found at opposite ends of this channel.

Another common feature of the legume lectins is that they are secretory proteins and undergo cotranslational signal peptide removal, which accompanies their entry into the secretory system. All but peanut agglutinin (PNA) are N-glycosylated as precursors; the N-glycans undergo normal posttranslational modification as they transit the Golgi apparatus. The lectins vary from one another as to whether the mature proteins contain oligomannose-type, complex-type, or a mixture of both types of N-glycans. The lectins may also undergo a variety of proteolytic modifications as they transit through the secretory system. Some of the lectins are cleaved to generate a β-chain, corresponding to the amino terminus and an α-chain corresponding to the carboxyl terminus. For example, the pea lectin and favin (the lectin from *Vicia faba*) are tetrameric glycoproteins that contain two types of subunits, α and β, which are ~5 kDa and ~21 kDa, respectively. These two lectins are each synthesized as single polypeptide precursors that contain the sequences of both chains in the following orientation: β-chain–α-chain. The chains associate to form dimers; they are then proteolytically processed in the protein bodies to form tetramers containing two separate α- and β-chains. Other lectins may undergo carboxy-terminal trimming of only some of their subunits. For example, the soybean agglutinin, phytohemagglutinin E (E-PHA), and *D. biflorus* lectins are tetramers of equimolar mixtures of intact and trimmed subunits.

The most intriguing proteolytic modification occurs in the case of ConA. A small segment is removed from the interior of the protein and the original amino terminus is ligated with the original carboxyl terminus. This forms what is termed a circularly permuted protein. The N-glycosylated segment of the protein is removed during this transpeptidation process; thus, the mature ConA is not a glycoprotein, in contrast to most seed lectins in the Fabaceae family, which have N-glycans. Thus, the protein sequence of isolated ConA aligns with other seed lectins, whereas the alignment of the DNA encoding the protein with other lectin genes indicates that the gene is circularly permuted.

Glycan-Binding Activities, Functions, and Applications

Despite years of research, the intrinsic biological roles of L-type lectins are poorly understood. The glycan-binding specificity can vary significantly; ConA binds mannose and glucose-containing glycans, whereas the lectins from the trees *Maackia amurensis* and *Sambucus nigra* bind to sialylated glycans (see Chapters 15, 29, and 31). There is evidence from many approaches that the legume lectins are insecticidal, antifungal, and antimicrobial and can be toxic to animals that eat raw seeds. Thus, the seed lectins may function in plant immunity as a type of pattern-recognition receptor (PRR) to protect the offspring of the plant. As highly abundant seed constituents, the lectins could also serve as a storage protein for the plant development. L-type lectins are involved in the symbiosis of plants with nitrogen-fixing bacteria, such as in Rhizobium-legume symbiosis, but the precise function of lectins in this regard is unclear. A recent study has shown that the *D. biflorus* seed lectin is also a lipoxygenase. It will be of interest to see how many other L-type lectins have this activity, which is necessary to initiate the wound-induced defense pathway in plants.

Regardless of their physiological function, legume L-type lectins might be useful in applications requiring agents with antitumor, antiviral, antibacterial, antifungal or antinociceptive properties. In addition, given their well-established glycan recognition preferences, these plant lectins are useful for glycophenotyping different cell types in physiological and pathological settings by means of immunofluorescence or immunohistochemical analysis.

L-TYPE LECTINS IN PROTEIN QUALITY CONTROL AND SORTING

Calnexin/Calreticulin

Calnexin (CNX) and calreticulin (CRT) are homologous molecular chaperones that mediate quality control of proteins in the endoplasmic reticulum (ER) (see Chapter 39). Although CRT is a soluble ER luminal component, CNX is membrane-bound and is perhaps closely associated with the protein-translocating channel that imports nascent proteins into the ER. Both CNX and CRT bind to monoglucosylated, high-mannose-type glycans and prevent their exit from the ER until they are properly folded and assembled into correct quaternary structures (Chapter 39). For example, generation of a functional major histocompatibility complex class I (MHC I) depends on its association with CRT or CNX as component of a peptide-loading complex and the antigen presentation machinery. During the binding and dissociation from CRT or CNX, if the glycoprotein folds correctly, then glucose removal by glucosidase-II allows its passage out of the ER. If a glycoprotein misfolds or aggregates, it is reglucosylated by UDP-Glc:glycoprotein glucosyltransferase (UGGT) containing four tandem thioredoxin-like domains; this enzyme only recognizes misfolded or aggregated glycoproteins. Following reglucosylation, the monoglucosylated protein binds again to CRT or CNX. Thus, there is a cycle of glucose removal and addition by the alternating actions of glucosidase-II and UGGT and interactions with CNX/CRT.

Both CRT and CNX are Ca^{++}-binding proteins, and their carbohydrate-binding activity is sensitive to changes in Ca^{++} concentration. CNX is a type I membrane protein with its carboxy-terminal end in the cytoplasm. The lumenal portion of the protein is divided into three domains: a Ca^{++}-binding domain (which is adjacent to the transmembrane domain), a proline-rich long hairpin loop called the P domain, and the amino-terminal L-type lectin domain. CRT has a similar structure, but it is missing the cytoplasmic and transmembrane regions; it is retained in the ER through its KDEL-retrieval signal at the carboxyl terminus (Figure 32.4).

ERGIC-53 and Its Related Proteins ERGL, VIP36, and VIPL

ERGIC-53 (human gene *LMAN1*) and its sequence-related proteins ERGL and VIP36 are type I membrane proteins that participate in vesicular protein transport in the secretory system (see Chapter 39). All share an L-type lectin folding motif. Orthologs of ERGIC-53 are found in plants and all animals and have been independently identified as being important in production of infectious viruses (e.g., coronavirus and filovirus). ERGIC-53 occurs in the ER–Golgi intermediate compartment (ERGIC), and its cytoplasmic carboxyl terminus contains the dilysine/diphenylalanine KKFF retention/retrieval motif. The dilysine is recognized by the COPI coatomer complex; this binding enables the coated vesicles to be recycled from the ERGIC back to the ER. The diphenylalanine helps to direct the COPII-coated vesicles to ER export sites by binding to the COPII coatomer. The location of VIP36 (vesicular integral membrane protein 36) is uncertain; overexpressed protein has been found in both the ER and ERGIC, as well as *cis*-Golgi. ERGIC 53 can associate with the soluble ER partner MCFD2 (multiple coagulation factor deficiency protein 2), which forms a cargo receptor complex important in factor V and VIII biosynthesis. ERGIC-53 binds to N-linked glycans and allosterically activates MCFD2, which thereby becomes able to capture specific polypeptide segments of cargo glycoproteins. Human mutations in ERGIC-53 are associated with deficiencies of circulating blood clotting factors V and VIII, which are glycoproteins with multiple N-glycans. ERGIC-53-deficient mice show FV and FVIII deficiencies and liver ER distension, with accumulation of α1-antitrypsin and GRP78. Partially penetrant, perinatal lethality occurs, dependent on inbred genetic background, suggesting potential roles for as yet unidentified ERGIC-53-dependent cargo proteins.

Both ERGIC-53 and VIP36 bind to oligomannose-type glycans and require Ca^{++} for their carbohydrate-binding activity. These two proteins were the first animal lectins found to share some sequence and structural homology with the legume seed lectins. Although the overall sequence identity of these proteins to the seed lectins is only ~19%–24%, those amino acids important for metal and carbohydrate binding in the seed lectins are conserved, including the invariant aspartate, glycine, and asparagine. The invariant aspartate also participates in a *cis*-peptide bond with its preceding amino acid; this is similar to the case of the legume seed lectins

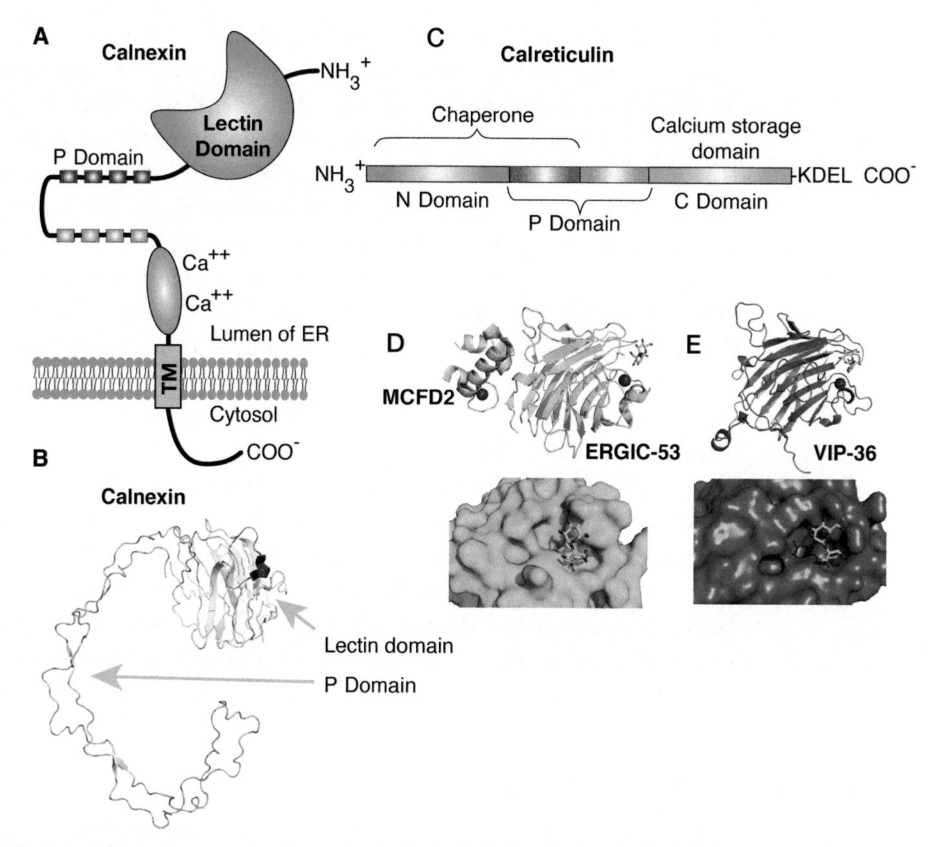

FIGURE 32.4. Schematic representations of calreticulin (CRT) and calnexin (CNX) showing the lectin domain, the P domain (containing the proline repeats), and the calcium-binding domain (*A,B*). Structure of CNX based on crystallographic data (*C*). (Adapted, with permission of Elsevier, from Schrag JD, et al. 2001. *Mol. Cell* **8:** 633–644.) Structures of the lectin domains of ERGIC-53 and VIP36 based on crystallographic data shown in cartoon and surface representation with the ligands indicated in *stick representation* (*D,E*).

discussed in the above section. The crystal structure of the lectin domain of ERGIC-53 has been determined and confirms the structural similarity of these lectins (Figure 32.4). In contrast to CRT/CNX, ERGIC-53 and VIP36 can bind the deglucosylated branch of high-mannose-type glycans and thereby contribute to vesicular transport of correctly folded glycoproteins in the early secretory pathway. During transport from the ER to the Golgi, release of cargo glycoproteins from ERGIC-53 is caused by the environmental acidification and a decrease in Ca^{++}. Like ERGIC-53, ERGL can associate with MCFD2 and is probably involved in FV and FVIII transport. VIPL (VIP36-like) is an ER-resident protein and functionally able to bind deglucosylated high-mannose-type glycans, and associate with ERGIC-53, which may be important in regulating ERGIC-53 localization.

OTHER L-TYPE LECTINS

A variety of other proteins have been described that have carbohydrate-binding domains with tertiary structures similar to the L-type lectin fold and may be considered as members of this family. Members of the galectin family of lectins also fit into this category and are the subject of a separate chapter in this volume (see Chapter 36). Other GBPs that may fit into this category are briefly discussed below.

Pentraxins and Related Proteins

The pentraxins are a superfamily of plasma proteins that are involved in innate immunity in invertebrates and vertebrates and are designated as PRRs. They contain L-type lectin folds and require Ca^{++} ions for ligand binding. Their name is based on the pentameric arrangement of their subunits. The short PTX C-reactive protein (CRP), which binds phosphocholine residues on polysaccharides and on phospholipids, and the serum amyloid P (SAP) component, which also binds carbohydrate derivatives on bacterial polysaccharides and to amyloid fibrils, are acute-phase proteins in humans and mice, respectively. This family also contains long PTX that have an unrelated long amino-terminal domain coupled to the PTX domain and includes neuronal PTX 1, neuronal PTX 2, neuronal PTX receptor, PTX 3, and PTX 4. A prototypical member, PTX3 is a soluble pattern recognition molecule involved in innate immunity and has been implicated in the pathogenesis of autoimmune diseases, cardiovascular inflammation, and cancer. In addition, PTX3 may also help in the assembly of a hyaluronan-rich extracellular matrix (ECM), which may involve binding to the hyaluronan–heavy chain-1 complex (inter-α-trypsin inhibitor or IαI or ITI) and regulation by TNF-stimulated gene-6 (TSG-6). Laminin G domain-like (LG) modules are made of 180–200 amino acid residues and were first identified in laminins. The carboxyl terminus of the laminin α-chain has five tandem laminin G domains, which are important in heparin and sulfatide binding and cell/basement membrane adhesion, such as to the novel glycans of α-dystroglycan (α-DG, see Chapter 13). Some LG modules share binding properties for cellular receptors and carbohydrate ligands, indicating that the LG fold may have evolved from the L-type lectin fold for participation in related functions.

VP4 and VP8 in Rotaviruses

VP4 is the surface spike protein in rotaviruses and is proteolytically cleaved to generate VP8, which is a GBP, and VP5, a hydrophobic region required for membrane entry. VP8 has an L-type lectin fold and structural features in common with galectins. This domain is required for infectivity of most animal rotaviruses. In some rotavirus strains, VP8 binds sialylated glycans, whereas most rotavirus VP8 recognizes type 1 and/or type 2 N- or O-glycans lacking sialic acid (Sia) and can especially bind to human milk oligosaccharides with appropriate sequences.

OTHER PROTEINS WITH JELLY-ROLL MOTIFS AND L-TYPE LECTIN DOMAINS

A number of proteins are known to contain the jelly-roll motif found in L-type lectins. These include the binding domain of *Clostridium* neurotoxins, which can bind to gangliosides (e.g., GT1b and GD1b, see Chapter 11); *Pseudomonas aeruginosa* exotoxin A in which one of its three domains, the amino-terminal domain Ia, displays an L-type lectin structure; *Vibrio cholerae* sialidase, a three-domain protein with a six-bladed β-propeller neuraminidase domain flanked by two L-type lectin domains; and the Leech intramolecular *trans*-sialidase (TS), which has a multidomain architecture with a lectin-like domain II and an irregular β-stranded domain III that is built around a canonical catalytic domain C. Domain II may be involved in carbohydrate recognition through sugar ring and aromatic side-chain interactions, as observed in many lectins.

Although the L-type lectin domain occurs in many proteins, in many cases evidence for carbohydrate binding and/or specificity is lacking. Examples include the L-type lectin receptor kinases (LecRKs) in plants, which is a large family of several thousand members comprising membrane receptors that are key to plant development, immunity, and adaptive responses to stimuli; and mammalian thrombospondins, a family of matricellular proteins implicated in angiogenesis and inflammation, including TSP-1, which has a carboxy-terminal domain that shows the typical β-sandwich of two curved antiparallel β-sheets, a feature of the jelly-roll topology, in which the L-type lectin domain is compactly assembled with three calcium-binding type 3 (T3) repeats

ACKNOWLEDGMENTS

The authors acknowledge helpful comments and suggestions from Yasuhiro Ozeki.

FURTHER READING

Lis H, Sharon N. 1998. Lectins: carbohydrate-specific proteins that mediate cellular recognition. *Chem Rev* **98:** 637–674. doi:10.1021/cr940413g

Moothoo DN, Naismith JH. 1998. Concanavalin A distorts the β-GlcNAc-(1→2)-Man linkage of β-GlcNAc-(1→2)-α-Man-(1→3)-[β-GlcNAc-)-(1→2)-α-Man-(1→6)]-Man upon binding. *Glycobiology* **8:** 173–181. doi:10.1093/glycob/8.2.173

Hamelryck TW, Loris R, Bouckaert J, Dao-Thi M-H, Strecker G, Imberty A, Fernandez E, Wyns L, Etzler ME. 1999. Carbohydrate binding, quaternary structure and a novel hydrophobic binding site in two legume lectin oligomers from *Dolichos biflorus*. *J Mol Biol* **286:** 1161–1177. doi:10.1006/jmbi.1998.2534

Loris R, Bouckaert J, Hamelryck T, Wynn L. 1999. Legume lectin structure. *Biochim Biophys Acta* **1383:** 9–36. doi:10.1016/s0167-4838(97)00182-9

Vijayan M, Chandra N. 1999. Lectins. *Curr Opin Struct Biol* **9:** 707–714. doi:10.1016/s0959-440x(99)00034-2

Schrag JD, Vergeron JJ, Li Y, Borisova S, Hahn M, Thomas DY, Cygler M. 2001. The structure of calnexin, and ER chaperone involved in quality control of protein folding. *Mol Cell* **8:** 633–644. doi:10.1016/s1097-2765(01)00318-5

Srinivas VR, Reddy GB, Ahmad N, Swaminathan CP, Mitra N, Surolia A. 2001. Legume lectin family, the 'natural mutants of the quaternary state', provide insights into the relationship between protein stability and oligomerization. *Biochim Biophys Acta* **1527:** 102–111. doi:10.1016/s0304-4165(01)00153-2

Sörme P, Arnoux P, Kahl-Knutsson B, Leffler H, Rini JM, Nilsson UJ 2005. Structural and thermodynamic studies on cation-Π interactions in lectin–ligand complexes: high-affinity galectin-3 inhibitors through fine-tuning of an arginine–arene interaction. *J Am Chem Soc* **127:** 1737–1743. doi:10.1021/ja043475p

Gouget A, Senchou V, Govers F, Sanson A, Barre A, Rouge P, Pont-Lezica R, Canut H. 2006. Lectin receptor kinases participate in protein–protein interactions to mediate plasma membrane–cell wall adhesions in *Arabidopsis*. *Plant Physiol* **140:** 81–90. doi:10.1104/pp.105.066464

Bouwmeester K, Govers F. 2009. *Arabidopsis* L-type lectin receptor kinases: phylogeny, classification, and expression profiles. *J Exp Bot* **60:** 4383–4396. doi:10.1093/jxb/erp277

Kouno T, Watanabe N, Sakai N, Nakamura T, Nabeshima Y, Morita M, Mizuguchi M, Aizawa T, Demura M, Imanaka T, et al. 2011. The structure of *Physarum polycephalum* hemagglutinin I suggests a minimal carbohydrate recognition domain of legume lectin fold. *J Mol Biol* **405:** 560–569. doi:10.1016/j.jmb.2010.11.024

Zhang B, Zheng C, Zhu M, Tao J, Vasievich MP, Baines A, Kim J, Schekman R, Kaufman RJ, Ginsburg D. 2011. Mice deficient in LMAN1 exhibit FV and FVIII deficiencies and liver accumulation of α1-antitryhpson. *Blood* **118:** 3384–3391. doi:10.1182/blood-2011-05-352815

Klaus JP, Eisenhauer P, Russo J, Mason AB, Do D, King B, Taatjes D, Cornillez-Ty C, Boyson JE, Thali M, et al. 2013. The intracellular cargo receptor ERGIC-53 is required for the production of infectious arenavirus, coronavirus, and filovirus particles. *Cell Host Microbe* **14:** 522–534. doi:10.1016/j.chom.2013.10.010

Croci DO, Cerliani JP, Dalotto-Moreno T, Méndez-Huergo SP, Mascanfroni ID, Dergan-Dylon S, Toscano MA, Carmelo JJ, García-Vallejo JJ, Ouyang J, et al. 2014. Glycosylation-dependent lectin–receptor interactions preserve angiogenesis in anti-VEGF refractory tumors. *Cell* **156:** 744–758. doi:10.1016/j.cell.2014.01.043

Satoh T, Suzuki K, Yamaguchi T, Kato K. 2014. Structural basis for disparate sugar-binding specificities in the homologous cargo receptors ERGIC-53 and VIP36. *PLoS ONE* **9:** e87963. doi:10.1371/journal.pone.0087963

Grandhi NJ, Mamidi AS, Surolia A. 2015. Pattern recognition in legume lectins to extrapolate amino acid variability to sugar specificity. *Adv Exp Med Biol* **842:** 199–215. doi:10.1007/978-3-319-11280-0_13

Kim DJ, Christofidou ED, Keene DR, Hassan Milde M, Adams JC. 2015. Intermolecular interactions of thrombospondins drive their accumulation in extracellular matrix. *Mol Biol Cell* **26:** 2640–2654. doi:10.1091/mbc.e14-05-0996

Wang Y, Weide R, Govers F, Bouwmeester K. 2015. L-type lectin receptor kinases in *Nicotiana benthamiana* and tomato and their role in *Phytophthora* resistance. *J Exp Bot* **66:** 6731–6743. doi:10.1093/jxb/erv379

Lamriben L, Graham JB, Adams BM, Hebert DN. 2016. N-glycan-based ER molecular chaperone and protein quality control system: the calnexin binding cycle. *Traffic* **17:** 308–326. doi:10.1111/tra.12358

Magrini E, Mantovani A, Garlanda C. 2016. The dual complexity of PTX3 in health and disease: a balancing act? *Trends Mol Med* **22:** 497–510. doi:10.1016/j.molmed.2016.04.007

Doni A, D'Amico G, Morone D, Mantovani A, Garlanda C. 2017. Humoral innate immunity at the crossroad between microbe and matrix recognition: the role of PTX3 in tissue damage. *Semin Cell Dev Biol* **61:** 31–40. doi:10.1016/j.semcdb.2016.07.026

Satoh T, Kato K. 2020. Recombinant expression and purification of animal intracellular L-type lectins. *Methods Mol Biol* **2132:** 21–28. doi:10.1007/978-1-0716-0430-4_3

33 P-Type Lectins

Nancy Dahms, Thomas Braulke, and Ajit Varki

Lysosomes are intracellular membrane-bound organelles that turn over and degrade many types of macromolecules, via the action of lysosomal enzymes (also called acid hydrolases because of the low-pH characteristic of lysosomes). These enzymes are synthesized in the endoplasmic reticulum (ER) on membrane-bound ribosomes and traverse the ER–Golgi pathway along with other newly synthesized proteins. At the terminal Golgi compartment, they are segregated from other glycoproteins and selectively delivered to lysosomes. In most "higher" animal cells, this specialized trafficking is achieved primarily through the recognition of N-glycans containing mannose 6-phosphate (M6P) by "P-type" lectins. As the first clear-cut example of a biological role for glycans on mammalian glycoproteins and the first shown link between glycoprotein biosynthesis and human disease, the interesting history of its discovery is described in some detail. More recent data on other proteins with "P-type" lectin domains are also mentioned.

HISTORICAL BACKGROUND

I-Cell Disease and the Common Recognition Marker of Lysosomal Enzymes

Early studies of human genetic "storage disorders" by Elizabeth Neufeld indicated a failure to degrade cellular components, which therefore accumulated in lysosomes (Chapter 44). Soluble

published by Cold Spring Harbor Laboratory Press; doi: 10.1101/glycobiology.4e.33

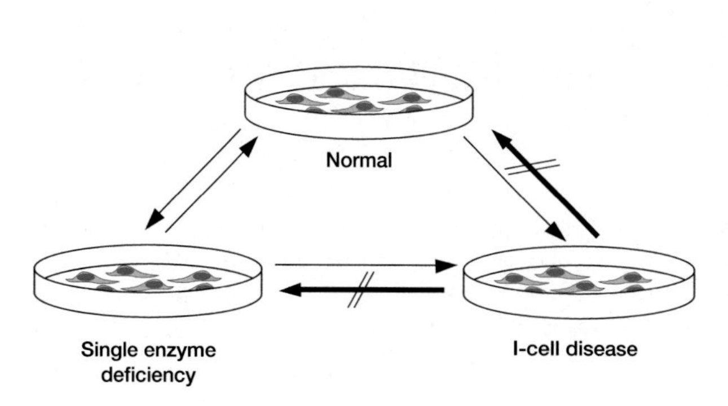

FIGURE 33.1. Cross-correction of lysosomal enzyme deficiencies in cultured cells. Small amounts of high-uptake lysosomal enzymes secreted by normal fibroblasts (*thin arrows*) were taken up by fibroblasts from a patient with a genetic lack of a single lysosomal enzyme, correcting that deficiency. In contrast, I-cell disease fibroblasts secrete large amounts of multiple lysosomal enzymes of the low-uptake variety (*thick arrows*). The latter forms cannot correct lysosomal enzyme deficiencies in other cells. However, I-cells retain the capability to accept high-uptake enzymes secreted by other cells (*thin arrows*). Uptake was blocked by the addition of mannose 6-phosphate (M6P) to the medium.

corrective factors from normal cells could reverse these defects when added to the culture media. These factors were later identified as lysosomal enzymes that were found to be deficient in patients with storage disorders. These enzymes, secreted in small quantities by normal cells in culture, or by cells from patients with a different "complementary" defect (Figure 33.1), existed in two forms: a "high-uptake" form, recognized by saturable, high-affinity receptors that could correct the defect in deficient cells, and an inactive "low-uptake" form that could not correct.

Fibroblasts from patients with a rare genetic disease showing prominent inclusion bodies in cultured cells (therefore termed "I-cell" disease) were found to be deficient in almost all lysosomal enzymes. Interestingly, all these enzymes are actually synthesized by I-cells, but are mostly secreted into the medium, instead of being retained in lysosomes. Neufeld showed that although I-cells incorporated the high-uptake enzymes secreted by normal cells, enzymes secreted by I-cells were not taken up by other cells (Figure 33.1). She therefore proposed that I-cell disease results from a failure to add a common recognition marker to lysosomal enzymes. This marker was assumed to be responsible for the proper trafficking of newly synthesized enzymes to lysosomes in normal cells. The "high-uptake" property was destroyed by strong periodate treatment, which oxidizes vicinal hydroxyl groups in glycans, predicting that the recognition marker was a glycan.

Discovery of the M6P Recognition Marker

High uptake of lysosomal enzymes was next found, by William Sly, to be blocked by M6P and its stereoisomer fructose 1-phosphate, but not by other sugar phosphates. Treatment of lysosomal enzymes with alkaline phosphatase also abolished high-uptake activity. By this time, the general pathway for N-glycan processing had been defined (see Chapter 9). Because oligomannosyl N-glycans are rich in mannose residues, these were predicted to be phosphorylated on lysosomal enzymes. Indeed, the M6P moiety detected on oligomannosyl N-glycans was released by endo-β-N-acetylglucosaminidase H (endo H) from high-uptake forms of lysosomal enzymes, confirming the hypothesis. The groups of Stuart Kornfeld and Kurt von Figura then found, surprisingly, that some of the M6P moieties were "blocked" by α-linked *N*-acetylglucosamine residues attached to the phosphate residue, creating a phosphodiester.

Enzymatic Mechanism for Generation of the M6P Recognition Marker

Comparison of glycans with phosphodiesters and phosphomonoesters predicted that the metabolic precursor of the M6P determinant was a phosphodiester and that phosphorylation was mediated not by an ATP-dependent kinase but by a UDP-GlcNAc-dependent GlcNAc-1-phosphotransferase. This was proven using a double-labeled donor and Golgi extracts:

Reactants: Uridine-P-^{32}P-[6-^{3}H]GlcNAc + Manα1-(N-glycan)-lysosomal enzyme,

Products: Uridine-P + [6-^{3}H]GlcNAcα1-^{32}P-6-Manα1-(N-glycan)-lysosomal enzyme.

Another Golgi enzyme phosphodiester glycosidase was shown to remove the outer *N*-acetylglucosamine residue and "uncover" the M6P moiety. Pulse-chase studies confirmed the order of events, documenting that multiple glycans on a given lysosomal enzyme could acquire one or two phosphate residues and that removal of some mannose residues on the N-glycan by processing Golgi mannosidases was also required (Figure 33.2). In most cell types, the phosphate is eventually lost after exposure to acid phosphatases 5 and 2 (ACP5, ACP2) in lysosomes. The overall biochemical pathway is shown in Figure 33.2.

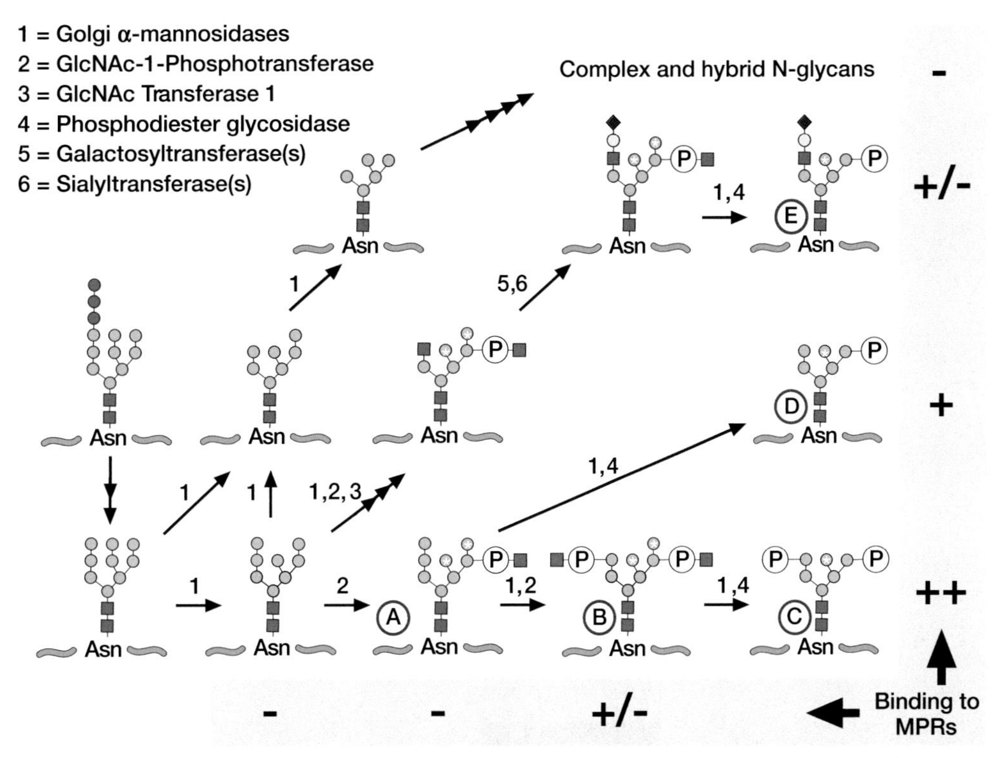

FIGURE 33.2. Pathways for biosynthesis of N-glycans bearing the mannose 6-phosphate (M6P) recognition marker. Following early N-glycan processing (Chapter 9), a single GlcNAc phosphodiester is added to the N-glycans of lysosomal enzymes on one of three mannose (Man) residues on the arm with the α1-6 Man linked to the core mannose (structure *A*). A second phosphodiester can then be added to the other side of the N-glycan (structure *B*) or onto other mannose residues (alternate sites for phosphorylation are denoted by *asterisks*). Removal of the outer *N*-acetylglucosamine residues and further processing of mannose residues give structures *C* and *D*. Further mannose removal is restricted by the phosphomonoesters. Thus, *C* and *D* represent only two of several possible structures bearing one or two phosphomonoesters. Glycans that are not phosphorylated become typical complex or hybrid N-glycans. Some hybrid N-glycans with M6P are also found (structure *E*). Binding studies of these N-glycans with purified mannose 6-phosphate receptors (MPRs) have shown the relative affinities indicated in the figure: ++, strong; +, moderate; +/−, weak; −, no binding.

The first two enzymes that mediate these reactions were then purified and characterized and the encoding genes cloned. UDP-GlcNAc:lysosomal enzyme GlcNAc-1-phosphotransferase (GlcNAc-P-T) is an $\alpha_2\beta_2\gamma_2$ complex encoded by two genes. The α- and β-subunits are encoded by the *GNPTAB* gene, whose 1256 amino acid type-III transmembrane product undergoes proteolysis in the Golgi by site-1-protease (S1P) to give rise to the two subunits, whereas the 305 amino acid γ-subunit is encoded by the *GNPTG* gene (Figure 33.3). The α/β subunits contain the catalytic function of the enzyme (Stealth domains) as well as elements (Notch1, Notch2, and DMAP [DNA-methylase-associated protein]) that mediate binding to protein determinants present on the lysosomal enzymes. The γ-subunit facilitates the phosphorylation of a subset of the lysosomal enzymes. It contains an M6P receptor homology (MRH) domain that is thought to bind oligomannosyl N-glycans on the lysosomal enzymes and present them to the catalytic site for phosphorylation, and a DMAP recognition domain. The second enzyme, α-N-acetylglucosamine-1-phosphodiester glycosidase encoded by the *NAGPA* gene, is a complex of four identical 68-kDa subunits, arranged as two disulfide-linked homodimers. Unlike other Golgi enzymes, this is a type-I membrane-spanning glycoprotein with its amino terminus in the lumen of the Golgi.

The oligomannosyl N-glycans of lysosomal enzymes are identical to those of many other glycoproteins passing through the ER–Golgi pathway. Thus, specific recognition by GlcNAc-P-T is crucial to achieve selective trafficking. This recognition is not explained by any similarities in the primary polypeptide sequences of lysosomal enzymes. Indeed, denatured lysosomal enzymes lose their specialized GlcNAc-P-T acceptor activity, indicating that features of secondary or tertiary structure are critical for recognition. Two complementary approaches were used to define elements of this recognition marker. In loss-of-function studies, various amino acids of the lysosomal enzyme were replaced with alanine, and the effect on phosphorylation determined. In gain-of-function experiments, residues of the lysosomal protease cathepsin D were substituted into the homologous secretory protease glycopepsinogen. These studies revealed that select lysine residues have a critical role in the interaction with GlcNAc-P-T. In fact, as few as two lysines in the correct orientation to each other and to an N-glycan can serve as minimal elements of the recognition domain. However, additional amino acid residues enhance the interaction with GlcNAc-P-T. In some instances (e.g., cathepsin D),

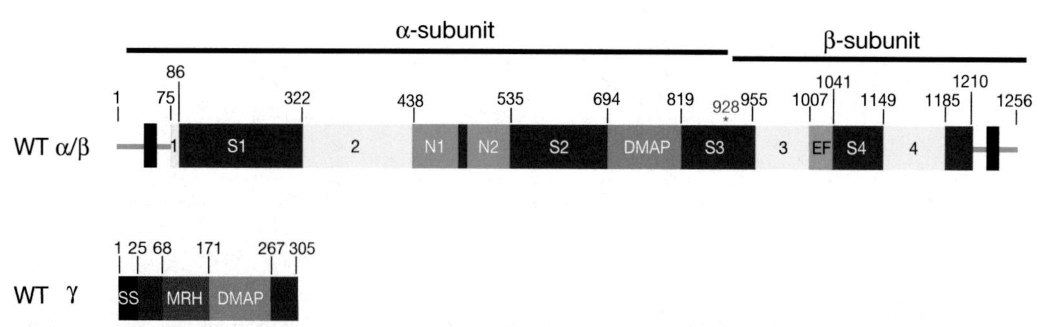

FIGURE 33.3. GlcNAc-P-T is an $\alpha_2\beta_2\gamma_2$ hexamer encoded by two genes. The *GNPTAB* gene encodes a catalytically inactive type 3 transmembrane precursor that undergoes a proteolytic cleavage between Lys-928 and Asp-929 in the Golgi by site-1-protease (S1P) to generate the active α- and β-subunits. These subunits mediate the catalytic function via the Stealth domains (*yellow*, numbered *1–4*). These domains are similar to bacterial genes that function as sugar-phosphate transferases in the biosynthesis of cell-wall polysaccharides. The Notch modules (N1, N2) and the DMAP (DNA-methylase-associated protein) domain participate in lysosomal enzyme recognition. The *GNPTG* gene encodes the γ-subunit that contains a DMAP lysosomal enzyme-binding domain and an MRH (M6P receptor homology) domain that is postulated to bind oligomannosyl N-glycans on lysosomal enzymes and present them to the catalytic site of the transferase. SS, signal sequence; EF, EF-hand calcium-binding motif; S1–S4, spacer-1 to spacer-4. The spacer S2 comprises the γ-subunit–binding domain.

the enzyme may contain a very extended determinant, or perhaps, more than one recognition domain.

The GlcNAc phosphodiester glycosidase that catalyzes the exposure of the M6P recognition marker is found primarily in the *trans*-Golgi network (TGN), and it cycles between this compartment and the plasma membrane. Thus, "uncovering" the recognition marker is a late event in the Golgi apparatus, occurring just before loading of the enzymes onto the M6P receptors (MPRs).

Enzymatic Basis for I-Cell Disease and Pseudo-Hurler Polydystrophy

Fibroblasts from patients with I-cell disease (also called mucolipidosis-II; ML-II) revealed a complete deficiency in GlcNAc-P-T enzyme activity. A milder variant called pseudo-Hurler polydystrophy (mucolipidosis-III; ML-III) subclassified by sequencing into MLIIIα/β or MLIIIγ showed a less severe deficiency of enzyme activity. Metabolic radiolabeling of fibroblasts corroborated the failure to phosphorylate mannose residues in these diseases, and asymptomatic obligate heterozygotes showed a partial deficiency, with slightly elevated levels of serum lysosomal enzymes. Mutations of various types in the two GlcNAc-P-T genes have since been detected in almost all patients with ML-II and -III examined, indicating that deficiency of this enzyme is the primary genetic cause of the disorder.

COMMON FEATURES OF P-TYPE LECTINS (M6P RECEPTORS)

The first candidate (~275-kDa) receptor for the M6P recognition marker was isolated by affinity chromatography and found to bind M6P in the absence of cations (CI-MPR [cation-independent MPR]). Certain cells deficient in this receptor still showed M6P-inhibitable binding of lysosomal enzymes, leading to the discovery of a second MPR of ~45 kDa, which required divalent cations for optimal binding (CD-MPR [cation-dependent MPR]). Both receptors bind with highest affinity to glycans carrying two M6P residues (Figure 33.2, structure C) whereas only the CI-MPR interacts with molecules bearing GlcNAc-P-Man phosphodiesters (Figure 33.2, structures A and B). Binding to molecules carrying one M6P (Figure 33.2, structure D) is intermediate in affinity. Removal of outer mannose residues by processing Golgi mannosidases enhances binding.

Genes encoding both MPRs have been cloned and extensively characterized. Both are type I transmembrane glycoproteins with extracytoplasmic, transmembrane, and carboxy-terminal cytoplasmic domains. The CI-MPR has 15 contiguous repetitive units of ~145 amino acids with partial identity to one another. The CD-MPR has a single extracellular domain, showing homology with the repeating domains of the CI-MPR (Figure 33.4A). Together with conservation of certain intron–exon boundaries, this homology suggests that the two genes evolved from a common ancestor. On the basis of their sequence relationships and unique binding properties to M6P, the two MPRs have been formally classified as P-type lectins. Structural homologs of the MPRs are present in yeast and Drosophila, but these proteins lack M6P-binding ability.

The CD-MPR exists mainly as a dimer, with each monomer binding one M6P residue. The CI-MPR also seems to be a dimer in the membrane, although it readily dissociates on solubilization. Two of its repeating units (3 and 9) bind M6P phosphomonoesters with high affinity, a third (5) binds M6P phosphodiesters, and a fourth (15) binds both M6P phosphomonoesters and diesters with equal affinity. Mutagenesis studies have identified the specific residues of these receptors involved in M6P binding, and crystal structures of several M6P-binding domains have been obtained in complex with M6P. Comparison of the binding pockets reveals four residues (Gln, Arg, Glu, Tyr) are positionally conserved and form the same contacts with the mannose

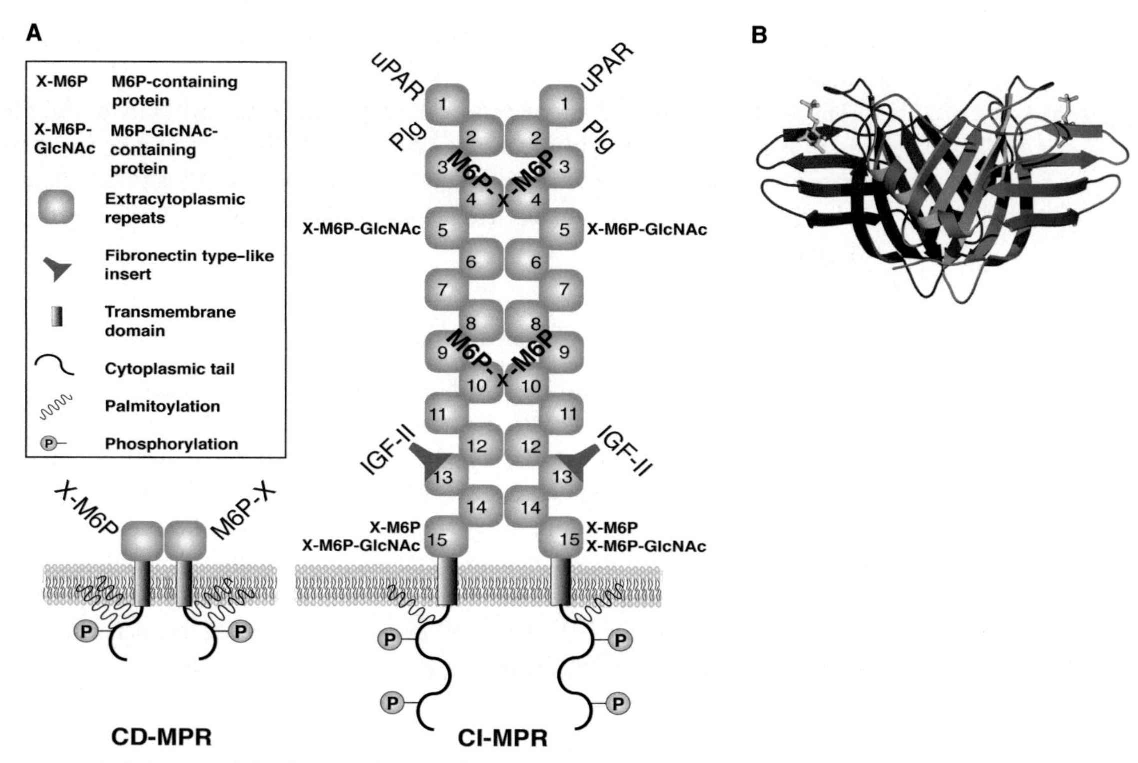

FIGURE 33.4. (*A*) Schematic diagram of cation-independent (CI-MPR) and cation-dependent mannose 6-phosphate receptors (CD-MPR). (*B*) Ribbon diagram of the bovine CD-MPR. Shown are the two monomers (*magenta* and *cyan ribbons*) of the dimer as well as the ligand M6P (*gold ball-and-stick model*). (*B*, Modified, with permission, from Roberts et al. 1998. *Cell* **93:** 639–648, © Cell Press.)

ring. The CD-MPR has been crystallized as a dimer, with each monomer folded into a nine-stranded flattened β-barrel that has a striking resemblance to the protein folds in avidin (Figure 33.4B). The distance between the two ligand-binding sites of the dimer provides a good explanation for the differences in binding affinity shown by the CD-MPR toward various lysosomal enzymes. Crystal structures of recombinant forms of the CI-MPR have also been solved that encompass all but two of the 15 contiguous domains in its extracellular region (i.e., domains 1–5, domains 7–11, and domains 11–14). Each domain shows a topology similar to that of the CD-MPR. Cryo-electron microscopy (cryo-EM) of the CI-MPR isolated from bovine liver has been carried out under conditions: (1) not conducive to ligand binding (pH 4.5), and (2) found at the cell surface (pH 7.4) and bound to the non-glycosylated ligand, insulin-like growth factor II (IGF-II) via CI-MPR's domain 11. Although the cryo-EM structure at pH 7.4 is unable to visualize the N- (domains 1–3) and C- (domain 15) termini with sufficient resolution, it does provide insight into the arrangement of the entire extracellular region of the CI-MPR. The proposed cryo-EM model does not position the two M6P-binding domains (3 and 9) sufficiently close to bind a single, diphosphorylated N-glycan. This suggests that the high-affinity binding of this N-glycan is due to the spanning of binding sites located on different monomers of the CI-MPR. An alternative possibility is that the receptor is dynamic, with the spacing between the two M6P-binding sites being flexible, to enhance interactions with lysosomal enzymes containing phosphorylated glycans at various positions on their protein backbones. The dynamic nature of the receptor is supported by structural and biophysical studies which show that the conformation of the CI-MPR is affected by pH. The CI-MPR adopts a more compact structure under acidic conditions that may facilitate unloading of lysosomal enzymes in the acidic environment of endosomal compartments.

INDUCED GENETIC DEFECTS IN THE MPRs

Targeted disruption of the CD-MPR gene in mice is associated with normal or only slightly elevated levels of lysosomal enzymes in the circulation and an otherwise grossly normal phenotype. However, thymocytes or primary cultured fibroblasts from such mice show an increase in the amount of phosphorylated lysosomal enzymes secreted into the medium. This indicates that mechanisms exist, which can compensate for the deficiency in vivo. Intravenous injection of inhibitors of other glycan-specific receptors capable of mediating endocytosis (e.g., the mannose receptor of macrophages and the asialoglycoprotein receptor of hepatocytes; see Chapter 34) gives rise to a marked increase in lysosomal enzymes in the serum of the deficient mice. Thus, such receptors are likely part of the compensatory mechanisms in vivo.

Like fibroblasts that lack only CD-MPR, fibroblasts that lack only CI-MPR have a partial impairment in sorting. Fibroblasts from embryos that lack both receptors show a massive missorting of multiple lysosomal enzymes. Thus, both receptors are required for efficient intracellular targeting of lysosomal enzymes. Comparison of lysosomal enzymes secreted by the different cell types indicates that the two receptors may interact preferentially with different subgroups of enzymes. Thus, the structural heterogeneity of the M6P recognition marker within a single lysosomal enzyme and between different enzymes is one explanation for the evolution of two MPRs with complementary binding properties: that is, to provide an efficient but varied targeting of lysosomal proteins in different cell types or tissues. What initially appeared to be a precise lock-and-key mechanism turned out to be a far more complex and flexible system, with functionally useful biological flexibility.

SUBCELLULAR TRAFFICKING OF THE MPRs

At steady state, MPRs are concentrated in the TGN and late endosomes, but they cycle constitutively between these organelles, early (sorting) endosomes, recycling endosomes, and the plasma membrane (Figure 33.5). MPRs avoid delivery to lysosomes, in which they would be degraded. This trafficking is governed by a number of short amino acid sorting signals in the cytoplasmic tails of the receptors. The TGN is the site in which newly synthesized lysosomal enzymes bind to MPRs that are then collected into clathrin-coated pits and packaged into clathrin-coated vesicles for delivery to the early endosome. This process involves interaction of the MPRs with two types of coat proteins: the GGAs (Golgi-localized, γ-ear–containing, ADP-ribosylation factor binding) and AP1 (adapter protein 1). In addition to binding MPRs, the coat proteins recruit clathrin for the assembly of clathrin-coated vesicles. Following delivery to early endosomes, lysosomal enzymes are released from MPRs as the endosomes mature to late endosomes and the pH decreases. Late endosomes then undergo dynamic fusion/fission with lysosomes, allowing selective transfer of lysosomal enzymes to the lysosomes and leaving the MPRs behind in subdomains of the late endosomes. These MPRs may then either return to the TGN mediated by a multiprotein retromer complex or move to the plasma membrane, in which internalization via clathrin-coated pits occurs, mediated by the coat protein AP2. There are several pathways for the MPRs to be returned to the TGN from the various endosomal compartments, although the relative importance of the different pathways is unclear.

The CI-MPR Mediates Uptake of Phosphorylated Lysosomal Enzymes—Implications for Enzyme Replacement Therapy

The original experiments of Neufeld showed that a portion of newly synthesized lysosomal enzymes are secreted into the medium, but may be recaptured by the same cell or by adjacent

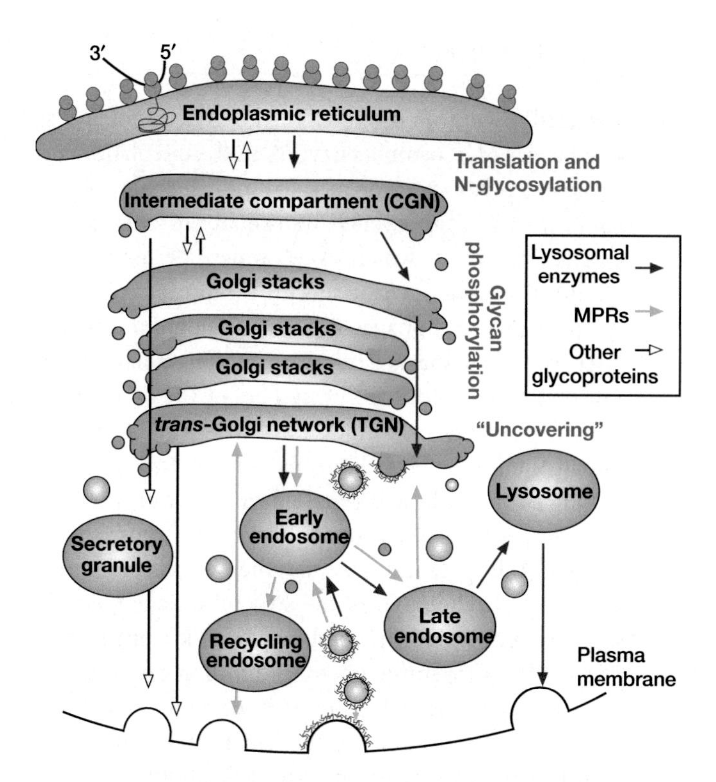

FIGURE 33.5. Subcellular trafficking pathways of glycoproteins, lysosomal enzymes, and M6P receptors (MPRs). Newly synthesized glycoproteins originating from the rough endoplasmic reticulum (ER) pass through the Golgi stacks and are then sorted to various destinations. Along this route, lysosomal enzymes are recognized by GlcNAc-P-T and phosphorylated in the intermediate compartment (*cis*-Golgi network) and then acted on by the uncovering enzyme (phosphodiester glycosidase) in the *trans*-Golgi network (TGN). Beyond the TGN, trafficking of lysosomal enzymes is primarily mediated by the MPRs through early endosomes to late endosomes, in which their lysosomal enzyme cargo is released. Smaller amounts of lysosomal enzymes can escape capture by MPRs and be secreted into the extracellular fluid. Lysosomal enzymes can also reenter the cell by binding to cell-surface MPRs and subsequent endocytosis. Once in the endocytic pathway, internalized lysosomal enzymes can intermingle with those following the biosynthetic route, as depicted. (*Open arrowheads*) Pathways general to many nonlysosomal glycoproteins; (*red arrows*) specific itineraries of lysosomal enzymes; (*green arrows*) pathways of MPRs. Additional pathways for MPRs include recycling from the early endosome to the cell surface and back to the TGN, and from the recycling endosome and the late endosome, following release of their lysosomal enzyme cargo.

cells expressing cell-surface CI-MPRs (Figure 33.1). Enzymes that bind to such cell-surface MPRs are endocytosed via clathrin-coated pits and vesicles, eventually reaching the same late endosomal compartments in which newly synthesized molecules arrive from the Golgi. This secretion–recapture pathway is a minor one in most cells, but plays a critical role in enzyme replacement therapy.

As described in Chapter 44, there are many genetic disorders in glycan degradation that result from decreased activity of a given lysosomal enzyme. Some of these enzymes that are targeted to lysosomes via the M6P pathway have been prepared in large quantities as recombinant soluble proteins and used in enzyme replacement therapy. To date, the benefits have been variable but less than optimal. There are a number of potential reasons for this. First, some of the preparations may not contain the physiologic complement of the phosphomannosyl recognition marker. It is reasonable to suggest that a greater M6P content may improve the efficacy of enzyme replacement in these patients. This has been shown to be the case in mouse and dog model systems. However, even with fully phosphorylated enzymes there may be obstacles that are difficult to overcome. For example, some cell types in the body may not express adequate levels of the CI-MPR on their surfaces to endocytose sufficient enzyme to restore normal lysosomal function. Also, the organ that is most seriously affected in many of these diseases (the brain) is inaccessible because of the blood–brain barrier. Although the first intraventricular M6P-dependent enzyme replacement therapy for children deficient for the lysosomal tripeptidyl peptidase 1 are very promising, further studies are needed.

EVOLUTIONARY ORIGINS OF THE M6P RECOGNITION SYSTEM

Although the MPR pathway has a major role in vertebrate lysosomal enzyme trafficking, its contribution in invertebrate systems is not prominent. Lysosomal enzymes are targeted in organisms such as *Saccharomyces*, *Trypanosoma*, and *Dictyostelium*, without the aid of identifiable MPRs. The slime mold *Dictyostelium discoideum* produces a novel methylphosphomannose structure on some of its lysosomal enzymes that can be recognized in vitro by the mammalian CI-MPR (not the CD-MPR). However, despite the presence of a GlcNAc-P-T that recognizes α1-2-linked mannose residues, the enzyme does not specifically recognize lysosomal enzymes, and no receptor for M6P has been found in this organism. Additionally, the *Dictyostelium* GlcNAc-P-T appears to differ from its mammalian hexameric counterpart because sequences related to the γ subunit have not been found in the *Dictyostelium* genome. The protozoan *Acanthamoeba* produces a GlcNAc-P-T that specifically recognizes lysosomal enzymes. However, this organism lacks a gene that could encode an "uncovering" enzyme, and so would not be expected to form M6P available to an MPR. Although some additional "lower" organisms do show evidence for an uncovering enzyme, no MPR activity has yet been found. The evolutionary divergence point at which the complete MPR system became established remains to be defined.

THE CI-MPR BINDS MANY OTHER LIGANDS

Although originally discovered as a receptor for lysosomal enzyme trafficking, the CI-MPR turns out to be a remarkably multifunctional molecule (Figure 33.4A). It binds IGF-II with high affinity even though this polypeptide lacks M6P residues. Many studies have explored potential interactions between these two disparate ligands. Although these two ligands bind to distinct sites on the CI-MPR receptor, there are conflicting reports regarding synergistic or antagonistic interactions between the two activities. Regardless, it appears that the CI-MPR acts primarily as a general "sink" for excess IGF-II in the extracellular fluid, carrying it to the lysosome for degradation, and reducing the amount available to bind to the IGF-I receptor. Several reports indicate that binding of IGF-II to the CI-MPR regulates motility and growth in some cell types. It has also been found that the CI-MPR binds retinoic acid with high affinity at a site that is distinct from those for M6P and IGF-II. This binding of retinoic acid seems to enhance the primary functions of the CI-MPR receptor, and the biological consequence appears to be the suppression of cell proliferation and/or induction of apoptosis. The significance of this unexpected observation is still being explored. Other ligands include urokinase-type plasminogen activator receptor (uPAR) and plasminogen, their functional significance is currently unknown.

There are also some unexplained changes in CI-MPR expression in relation to malignancy. Loss of heterozygosity at the CI-MPR locus occurs in dysplastic liver lesions and in hepatocellular carcinomas associated with the high-risk factors of hepatitis virus infection and liver cirrhosis. Mutations in the remaining allele were detected in ~50% of these tumors, which also seem to frequently develop from clonal expansions of phenotypically normal, CI-MPR-mutated hepatocytes. Thus, the CI-MPR fulfills many of the classic criteria to be classified as a liver "tumor-suppressor" gene.

SIGNIFICANCE OF M6P ON NONLYSOSOMAL PROTEINS

Interestingly, M6P-containing N-glycans have been found on a variety of nonlysosomal proteins. Some are hydrolytic enzymes that seem to have taken on a predominantly secretory route, for example, uteroferrin and DNase I. In the first case, the failure of removal of the blocking *N*-acetylglucosamine residues may be the cause for secretion. With DNase I, the native level of phosphorylation simply appears to be low. M6P has also been found on the transforming growth factor β (TGF-β) precursor and the phosphate is lost from the mature form. It appears that M6P may serve to target the precursor to CI-MPR for activation. However, like DNase I, the native level of phosphorylation of TGF-β appears to be low. Other nonlysosomal proteins reported to carry M6P include proliferin, LIF (leukemia inhibitory factor), and thyroglobulin. In the last case, M6P-containing N-glycans are suggested as a mechanism to target the protein for degradation and release of thyroid hormones. But one should not necessarily assume that M6P-containing N-glycans on all of these proteins are involved in intracellular trafficking. Just as phosphorylation of serine residues has diverse biological roles, M6P might be used for more than one purpose in a complex multicellular organism.

Herpes simplex virus (HSV) and varicella zoster virus (VZV) glycoproteins have also been shown to carry N-glycans with M6P. In these cases, M6P is on complex N-glycans, suggesting that it originates from a distinct biosynthetic pathway. Regardless of its mode of synthesis, interaction of cell-free VZV with CI-MPR at the cell surface is required for viral entry into endosomes. Interestingly, intracellular CI-MPR can also divert newly synthesized enveloped VZV to late endosomes in which the virions are inactivated before exocytosis. This is suggested as the mechanism by which this successful pathogen limits immediate excessive spread and avoids killing the host or, possibly, a form of host defense. Biopsies of VZV-infected human skin showed that CI-MPR expression is lost in maturing superficial epidermal cells, preventing diversion of VZV to endosomes and allowing constitutive secretion of infectious VZV. These data implicate CI-MPR in the complex biology of VZV infection.

OTHER PATHWAYS FOR TRAFFICKING LYSOSOMAL ENZYMES

Although the M6P recognition marker has a crucial role in trafficking newly synthesized lysosomal enzymes to vertebrate lysosomes, alternate mechanisms exist in some cell types. In patients with I-cell disease some cells and tissues, such as the liver and circulating granulocytes, have essentially normal levels of enzymes. B-lymphoblast lines derived from these patients also do not show the complete phenotype of enzyme deficiency seen in fibroblasts. One possible explanation is that the M6P pathway for trafficking of lysosomal enzymes is a specialized form of targeting, superimposed on some other basic mechanisms that more primitive organisms use, some of which remain undefined. In this regard, two lysosomal enzymes, acid phosphatase ACP2 and β-glucocerebrosidase, are not at all affected in their distribution in I-cell disease fibroblasts. With ACP2, the enzyme is synthesized initially as a membrane-bound protein, and once in the lysosome, it is proteolytically cleaved to generate the mature soluble form. Glucocerebrosidase is targeted to lysosomes independently of the MPR pathway by forming a complex with LIMP-2, a lysosomal membrane protein that contains a targeting signal in its cytoplasmic tail. Likewise, other integral membrane proteins of the lysosome use similar targeting motifs in their cytoplasmic tails to traffic to the lysosome. Furthermore, several M6P-independent cargo receptors for the transport of lysosomal enzymes have been identified such as the LDL receptor-related protein 1 (LRP1), LRP2 (megalin), sortilin, and mannose receptors.

Mannose 6-Phosphate Receptor Homology (MRH) Domain–Containing Proteins Function in the Secretory Pathway

A small number of proteins have now been identified that are related to the P-type lectins because of the presence of MRH domains. These MRH domains contain the residues that bind mannose, but lack the residues that bind phosphate. In several instances, the proteins have been shown to bind to oligomannosyl structures. Most of these proteins reside in the ER where they function either as lectins (e.g., OS-9 and XTP3-B) or as glycosidases (e.g., glucosidase II) in the ER quality control pathway. Interestingly, the γ-subunit of GlcNAc-P-T has an MRH domain that plays a critical role in the phosphorylation of a subset of lysosomal enzymes. Similar to GlcNAc-P-T, it has been proposed that the MRH domain in the β-subunit of glucosidase II acts to optimally present the glycan to the catalytic α-subunit of this heterodimeric enzyme for efficient removal of glucose residues. The precise biologic role of these MRH domains remains to be elucidated.

ACKNOWLEDGMENTS

The authors appreciate the contributions of Stuart Kornfeld to the earlier edition and helpful comments and suggestions from Sneha Komath and Anne Imberty.

FURTHER READING

Fratantoni JC, Hall CW, Neufeld EF. 1968. Hurler and Hunter syndromes: mutual correction of the defect in cultured fibroblasts. *Science* **162:** 570–572. doi:10.1126/science.162.3853.570

Hickman S, Neufeld EF. 1972. A hypothesis for I-cell disease: defective hydrolases that do not enter lysosomes. *Biochem Biophys Res Commun* **49:** 992–999. doi:10.1016/0006-291x(72)90310-5

Kaplan A, Achord DT, Sly WS. 1977. Phosphohexosyl components of a lysosomal enzyme are recognized by pinocytosis receptors on human fibroblasts. *Proc Natl Acad Sci* **74:** 2026–2030. doi:10.1073/pnas.74.5.2026

Hasilik A, Klein U, Waheed A, Strecker G, von Figura K. 1980. Phosphorylated oligosaccharides in lysosomal enzymes: identification of α-*N*-acetylglucosamine(1)phospho(6)mannose diester groups. *Proc Natl Acad Sci* **77:** 7074–7078. doi:10.1073/pnas.77.12.7074

Tabas I, Kornfeld S. 1980. Biosynthetic intermediates of β-glucuronidase contain high mannose oligosaccharides with blocked phosphate residues. *J Biol Chem* **255:** 6633–6639. doi:10.1016/s0021-9258(18)43616-2

Reitman ML, Varki A, Kornfeld S. 1981. Fibroblasts from patients with I-cell disease and pseudo-Hurler polydystrophy are deficient in uridine 5′-diphosphate-*N*-acetylglucosamine: glycoprotein N-acetylglucosaminylphosphotransferase activity. *J Clin Invest* **67:** 1574–1579. doi:10.1172/jci110189

Varki A, Kornfeld S. 1983. The spectrum of anionic oligosaccharides released by endo-β-N-acetylglucosaminidase H from glycoproteins. Structural studies and interactions with the phosphomannosyl receptor. *J Biol Chem* **258:** 2808–2818. doi:10.1016/s0021-9258(18)32790-x

Kornfeld S. 1986. Trafficking of lysosomal enzymes in normal and disease states. *J Clin Invest* **77:** 1–6. doi:10.1172/jci112262

von Figura K, Hasilik A. 1986. Lysosomal enzymes and their receptors. *Annu Rev Biochem* **55:** 167–193. doi:10.1146/annurev.bi.55.070186.001123

Kornfeld S, Mellman I. 1989. The biogenesis of lysosomes. *Annu Rev Cell Biol* **5:** 483–525. doi:10.1146/annurev.cb.05.110189.002411

Munier-Lehmann H, Mauxion F, Hoflack B. 1996. Function of the two mannose 6-phosphate receptors in lysosomal enzyme transport. *Biochem Soc Trans* **24:** 133–136. doi:10.1042/bst0240133

Bonifacino JS, Traub LM. 2003. Signals for sorting of transmembrane proteins to endosomes and lysosomes. *Annu Rev Biochem* **72:** 395–447. doi:10.1146/annurev.biochem.72.121801.161800

Ghosh P, Dahms NM, Kornfeld S. 2003. Mannose 6-phosphate receptors: new twists in the tale. *Nat Rev Mol Cell Biol* **4:** 202–212. doi:10.1038/nrm1050

Dahms NM, Olson LJ, Kim JJ. 2008. Strategies for carbohydrate recognition by the mannose 6-phosphate receptors. *Glycobiology* **18:** 664–678. doi:10.1093/glycob/cwn061

Vogel P, Payne BJ, Read R, Lee WS, Gelfman CM, Kornfeld S. 2009. Comparative pathology of murine mucolipidosis types II and IIIC. *Vet Pathol* **46:** 313–324. doi:10.1354/vp.46-2-313

Castonguay AC, Olson LJ, Dahms NM. 2011. Mannose 6-phosphate receptor homology (MRH) domain–containing lectins in the secretory pathway. *Biochim Biophys Acta* **1810:** 815–826. doi:10.1016/j.bbagen.2011.06.016

Kollmann K, Pestka JM, Schöne E, Schweizer M, Karkmann K, Kühn SC, Catala-Lehnen P, Failla AV, Marshall RP, Krause M, et al. 2013. Decreased bone formation and increased osteaclastogenesis cause bone loss in mucolipidosis II. *EMBO Mol Med* **5:** 1871 – 1886. doi:10.1002/emmm.201302979

Otomo T, Schweizer M, Kollmann K, Schumacher V, Muschol N, Tolosa E, Mittrücker H-W, Braulke T. 2015. Mannose 6 phosphorylation of lysosomal enzymes controls B cell functions. *J Cell Biol* **208:** 171 – 180. doi:10.1083/jcb.201407077

van Meel E, Lee WS, Liu L, Qian Y, Doray B, Kornfeld S. 2017. Multiple domains of GlcNAc-1-phosphotransferase mediate recognition of lysosomal enzymes. *J Biol Chem* **291:** 8295–8307. doi:10.1074/jbc.m116.714568

Bochel AJ, Williams C, McCoy AJ, Hoppe HJ, Winter AJ, Nicholls RD, Harlos K, Jones EY, Berger I, Hassan AB, Crump MP. 2020. Structure of the human cation-independent mannose 6-phosphate/IGF2 receptor domains 7-11 uncovers the mannose 6-phosphate binding site of domain 9. *Structure* **28:** 1300–1312.e5. doi:10.1016/j.str.2020.08.002

Khan SA, Tomatsu SC. 2020. Mucolipidoses overview: past, present, and future. *Int J Mol Sci* **21:** 6812. doi:10.3390/ijms21186812

Olson LJ, Misra SK, Ishihara M, Battaile KP, Grant OC, Sood A, Woods RJ, Kim JP, Tiemeyer M, Ren G, Sharp JS, Dahms NM. 2020. Allosteric regulation of lysosomal recognition by the cation-independent mannose 6-phosphate receptor. *Commun Biol* **3:** 498. doi:10.1038/s42003-020-01211-w

Wang R, Qi X, Schmiege P, Coutavas E, Li X. 2020. Marked structural rearrangement of mannose 6-phosphate/IGF2 receptor at different pH environments. *Sci Adv* **6:** eaaz1466. doi:10.1126/sciadv.aaz1466

34 C-Type Lectins

Richard D. Cummings, Elise Chiffoleau, Yvette van Kyook, and Rodger P. McEver

C-type lectins (CTLs) are Ca^{++}-dependent glycan-binding proteins (GBPs) that share primary and secondary structural homology in their carbohydrate-recognition domains (CRDs). The CRD of CTLs is more generally defined as the CTL domain (CTLD), because not all proteins with this domain bind either glycans or Ca^{++}. CTLs include collectins, selectins, endocytic receptors, and proteoglycans, some of which are secreted and others are transmembrane proteins. They often oligomerize, which increases their avidity for multivalent ligands and enhance recognition of pattern recognition receptor (PRRs). CTLs differ significantly in the types of ligands that they recognize with high affinity (e.g., glycans, proteins, lipids, and inorganic compounds). These proteins that recognize pathogens or self-expressed ligands function as adhesion, phagocytic, and signaling receptors in many pathways, including homeostasis and innate and adaptive immunity, and are crucial in inflammatory responses, leukocyte and platelet trafficking, and tissue remodeling.

DISCOVERY OF C-TYPE LECTINS AND COMMON STRUCTURAL MOTIFS

The CTL family is remarkably diverse and is the largest family of known GBPs. The first CTL identified in animals was the hepatic asialoglycoprotein receptor (ASGPR), also termed the hepatic Gal/GalNAc receptor or Ashwell–Morell receptor (AMR). The sequences of the

published by Cold Spring Harbor Laboratory Press; doi: 10.1101/glycobiology.4e.34

AMR and other CTLs revealed a CRD unique to this family of proteins. Glycan binding by CTLs is typically Ca^{++}-dependent because of specific amino acid residues that coordinate Ca^{++} and bind the hydroxyl groups of sugars, but some CTLDs bind glycans without coordinating Ca^{++}. The CTLD is defined by the sequence of amino acids and Cys positions, as well as the folded structure. Interestingly, comparisons of the folded structure of the CTLD to that in other proteins revealed a common folded structure, termed the CTL fold (CTLF), which is a structurally rigid scaffold that comprises a remarkable number of sequence variations, yet may have no sequence in common with CTLs, such as the major tropism determinant (Mtd), which is a receptor-binding protein of *Bordetella* bacteriophage. The evolutionarily ancient CTLF may occur in at least 10^{13} different sequences, a diversity that rivals the immunoglobulin fold in its conservation of structure using millions of different primary amino acid sequences.

The CRD of CTLs is a compact region of 110–130 amino acid residues with a double-looped, two-stranded antiparallel β-sheet formed by the amino- and carboxy-terminal residues connected by two α-helices and a three-stranded antiparallel β-sheet (Figure 34.1). The CRD has two conserved disulfide bonds and up to four sites for binding Ca^{++}, with site occupancy depending on the lectin. Amino acid residues with carbonyl side chains are often coordinated to Ca^{++} in the CRD, and these residues directly bind to sugars when Ca^{++} is bound in site 2. A ternary complex may form between a sugar, the Ca^{++} ion in site 2, and amino acids within the CRD, whereas the specific residues within the CRD determine sugar specificity. Key conserved residues that bind sugars include the "EPN" motif (which promotes binding to Man, GlcNAc, Fuc, and Glc) and "WND" motif (which promotes binding to Gal and GalNAc), as seen in mouse L-selectin and rat mannose-binding protein C (Figure 34.1). However, because the CTLD is relatively shallow with few contacts to sugars, it is not possible to predict the glycan structures that bind to a particular CTL. In several CTLs, such as P-selectin and the AMR, Ca^{++} binding induces structural changes in the CRD that stabilize the double-loop region. Loss of Ca^{++} can lead to destabilization of these loops and loss of ligand binding, even when Ca^{++} is not directly involved in complexing the ligand, as seen in the macrophage mannose receptor. This destabilization is also important in pH-induced changes that lead to loss of ligand-binding affinity, because of the pH-induced loss of Ca^{++}. In CTLD-containing proteins such as human tetranectin, which is not known to bind glycans, the CTLD can bind Ca^{++}, but in the absence of Ca^{++} the CTLD is important for interactions with kringle-domain-containing proteins, such as plasminogen.

CTLs occur as both monomers and oligomers, such as the trimeric rat mannose-binding protein (MBP)-A (Figure 34.2). The rat MBP was the first CTL structurally characterized with ligand by crystallography. The CRD of trimeric lectins is angled to the side of the stalk domain through which the protein associates to form the trimer. The CRDs are at the top of the trimer and enhance multivalent interactions with glycan ligands.

DIFFERENT SUBFAMILIES OF C-TYPE LECTINS

CTLDs have been categorized into 16 groups that are distinguished by their domain architecture. There are 86 proteins encoded in the human genome that contain a CTLD (mice have 123) (Figure 34.3). Most of these groups have a single CTLD, but the macrophage mannose receptor (group 6) is an example of a multi-CTLD protein and has eight of these domains. Several groups have CTLDs that lack critical Ca^{++} residues, but can bind glycans (e.g., dectin-1 and layilin in group 5); the REG group 7 lacks Ca^{++} binding but it is unclear whether they bind glycans, whereas tetranectin in group 9 binds Ca^{++} and again it is unclear if it binds glycans. From a functional perspective, we know most about collectins, endocytic receptors, myeloid lectins, and selectins, as discussed below.

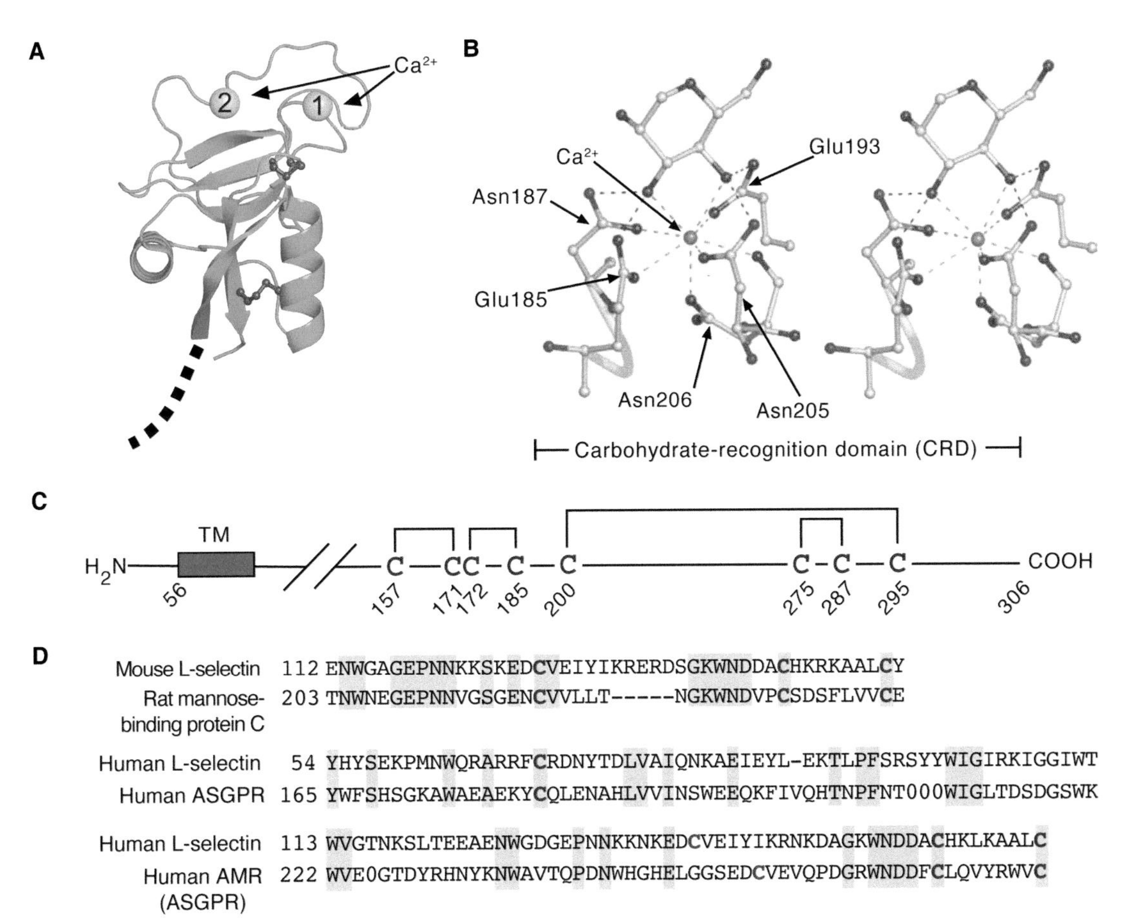

FIGURE 34.1. Structure of C-type lectins (CTLs). (*A*) Ribbon diagram of the carbohydrate-recognition domain (CRD) of rat mannose-binding protein A (MBP-A). (*Light green spheres*) Ca^{++}-binding sites, where 1 is the auxiliary binding site and 2 is the principal binding site; (*purple bars*) disulfide. The long-loop region that binds Ca^{++} ions is shown at the top of the CRD. A single disulfide bond helps to form this loop, and a second disulfide bond at the bottom of the CRD helps to form the whole loop domain. (Redrawn, with permission of the American Society for Biochemistry and Molecular Biology, from Feinberg H, et al. 2000. *J Biol Chem* **275:** 21539–21548.) (*B*) Stereo view of the complex between rat MBP-A and the terminal mannose residue in the N-glycan Man_6-$GlcNAc_2$-Asn. (*Orange*) Coordination bonds. Hydrogen bonds where sugar hydroxyl groups act as acceptor are *red* and those where they act as donor are *blue*. The interaction is through a ternary complex formed between the terminal mannose of the glycan, the Ca^{++} ion in binding site 2, and the protein. The complex is stabilized by a network of coordination and hydrogen bonds involving the 3- and 4-hydroxyl O atoms in the mannose, two coordination bonds with the Ca^{++} ion, and four hydrogen bonds with the carbonyl side chains that form the Ca^{++}-binding site. (Redrawn from PDB image 2msb in Weis WI, Drickamer K, and Hendrickson WA. 1992, with permission of Macmillan Publishers Ltd. *Nature* **360:** 127–134.) (*C*) The disulfide bonding in the CRD of the Ashwell–Morell receptor (AMR). The eight Cys residues in the extracellular domain are shown, with four disulfide bonds, including two in the CRD. The transmembrane domain (TM) is indicated. (Modified and with permission from the American Society for Biochemistry and Molecular Biology, from Yuk MH and Lodish HF. 1995. *J Biol Chem* **270:** 20169–20176.) (*D*) Primary sequence comparisons between different CTLs. Residues are numbered from the amino terminus. Cys residues are in *bold red* and homologous residues are *boxed in gray.*

CTLF-containing proteins are found in all metazoans and many nonmetazoans. This latter group includes bacterial toxins (e.g., pertussis toxin), outer-membrane adhesion proteins (e.g., invasin from *Yersinia pseudotuberculosis*), and viral proteins (e.g., envelope protein in Epstein–Barr virus). Interestingly, the viral proteins have more similarity to mammalian CTLD-containing proteins than the bacterial proteins. There are at least 278 genes encoding CTLDs within the genome of *Caenorhabditis elegans*, but only a small number have the key amino acid residues required to form the primary Ca^{++} binding site.

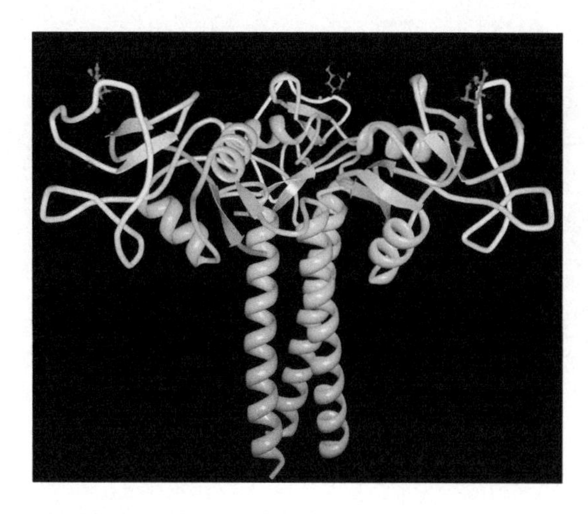

FIGURE 34.2. Crystal structure of trimeric rat mannose-binding protein-A complexed with α-methylmannoside. (Created from PDB deposited structure 1kwu and with permission of the American Society for Biochemistry and Biology from Ng KK, et al. 2002. *J Biol Chem* **277:** 16088–16095.)

THE ASHWELL–MORELL RECEPTOR

The AMR is expressed primarily on the sinusoidal surface of hepatocytes and was discovered serendipitously during studies on clearance of the serum glycoprotein ceruloplasmin, a major copper carrier in mammalian plasma. Glycosylation of ceruloplasmin was found to regulate its lifetime in the rabbit circulation; enzymatic desialylation of radiolabeled ceruloplasmin exposed penultimate β-linked Gal residues and caused rapid accumulation of asialoceruloplasmin into hepatocytes. Removal or modification of the exposed Gal residues lessened the rate of clearance. In 1974, the Ca^{++}-dependent AMR was purified from rabbit liver by affinity chromatography on asialoorosomucoid-Sepharose. The rabbit AMR is a heterooligomer containing two subunits of 48 kDa and 40 kDa, whereas the human AMR contains two subunits of 50 kDa (H1) and 46 kDa (H2) that occur as a heterooligomeric type I transmembrane protein (trimers and high-order oligomers) comprised of small and large subunits.

The purified rabbit AMR agglutinated desialylated human and rabbit erythrocytes and induced mitogenesis in desialylated peripheral lymphocytes, which represented the first demonstration that an animal GBP could profoundly affect cellular metabolism. Interestingly, hepatic AMR can also recognize some sialylated ligands (Siaα2-6Galβ1-4GlcNAc-R). A homologous GBP identified in chicken hepatocytes recognizes glycoproteins containing terminal GlcNAc, rather than Gal, residues. Interestingly, circulating glycoproteins in birds constitutively lack sialic acid (Sia) compared with their mammalian counterparts.

The AMR was one of the first proteins shown to participate in receptor-mediated endocytosis (RME) and, along with the low-density lipoprotein (LDL) receptor, transferrin receptor, and mannose 6-phosphate receptor (M6PR), represents some of the best characterized receptors. The AMR internalizes its ligands captured at the cell surface at physiological pH; via coated pits, the AMR in complex with its ligand is internalized into coated vesicles (Figure 34.4), and changes in pH in late endosomes lead to dissociation of Ca^{++} from the AMR and release of ligand. The uncomplexed AMR then recycles to the plasma membrane, whereas the ligand is delivered to lysosomes where it is degraded. For some nonrecycling receptors (e.g., dendritic cell [DC]-specific intercellular adhesion molecule-3-grabbing nonintegrin [DC-SIGN]), both the ligand and receptor are targeted to lysosomes and degraded. Internalization of ligands by the AMR in hepatocytes is rapid, occurring in 2–3 minutes, and the receptor recycles to the surface within 4–5 minutes. The efficiency of the AMR in endocytosis has led it to be a target for gene therapy and delivery of molecules to hepatocytes.

The precise orientation of the trimers within the AMR can dictate its affinity to specific glycans that have multivalent presentations. The AMR binds glycans with terminal β-linked Gal or

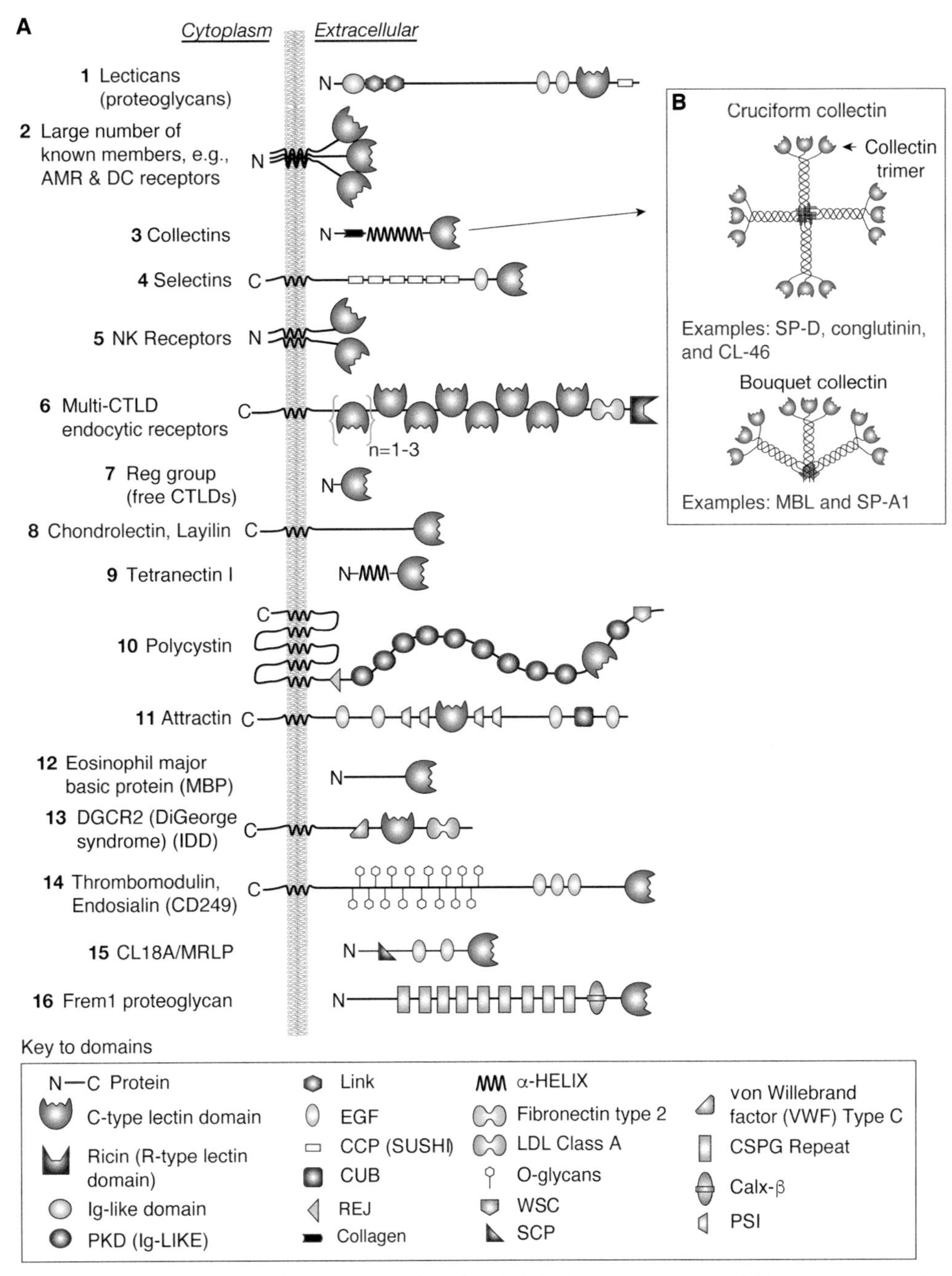

FIGURE 34.3. Different groups of C-type lectins (CTLs) and their domain structures. (*A*) Sixteen groups are shown, defined by their phylogenetic relationships and domain structures. Some of the groups are soluble proteins and others are transmembrane proteins. (*B*) The group 3 collectins form oligomeric structures, shown as cruciform and bouquet structures in the box. Each of the domains is named as indicated in the key. For abbreviations see the text.

GalNAc residues; however, tri- and tetra-antennary N-glycans, with appropriate branching and presentation of nonreducing terminal Gal/GalNAc, bind to the rat AMR with >100,000× higher affinity (~nM range) than ligands with a single terminal Gal/GalNAc residue.

Surprisingly, mice null for the AMR were found to be phenotypically normal without challenge. However, it is now known that the AMR is required to clear platelets (thrombocytes) during *Streptococcus pneumoniae*–related sepsis, which is accompanied by microbial-dependent

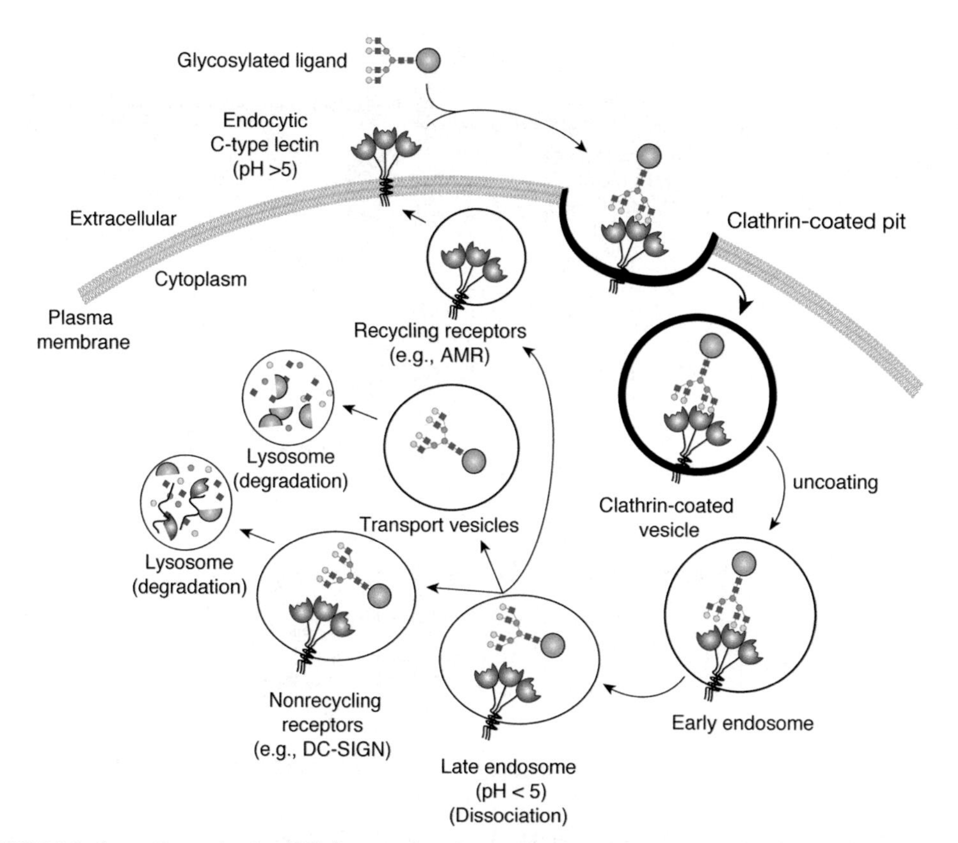

FIGURE 34.4. Some C-type lectins (CTLs) are endocytic receptors. Ligands are internalized by clathrin-dependent pathways and delivered to early and then late endosomes. Receptors may be recycled or degraded, depending on the receptor and the type of ligand it endocytoses. At a pH$<$5, CTLs typically lose Ca^{++}, which shifts the equilibrium to promote ligand dissociation. The cytoplasmic domains of these receptors dictate their fates in the endocytic pathway. The tyrosine-based motif in the cytoplasmic domains of the Ashwell–Morell receptor (AMR) and the mannose receptor promotes ligand delivery to early endosomes and receptor recycling to the cell surface. The triacidic motif (EEE or DDD) in the cytoplasmic domain of DEC-205 diverts it to late endosomes/lysosomes. Dendritic cell–specific intercellular adhesion molecule-3-grabbing nonintegrin (DC-SIGN) has a tyrosine-based, coated pit sequence-uptake motif similar to that in the macrophage mannose receptor, along with a dileucine motif essential for internalization and a triacidic cluster, which is key to targeting to endosomes/lysosomes.

desialylation of platelets and consequent thrombocytopenia. Clearance helps to limit disseminated intravascular coagulation (DIC) caused by the infection. Such clearance of platelets during sepsis or as a result of platelet aging is also linked to production of thrombopoietin that regulates platelet production.

OTHER ENDOCYTIC C-TYPE LECTINS

Many other CTLs, both type I and type II transmembrane proteins, also deliver bound ligands to lysosomes via RME through their cytoplasmic motifs (Figure 34.4). Endocytosis of ligands by CTLs in DCs and macrophages can lead to receptor accumulation and degradation in phagolysosomes or to recycling of the receptor to the cell surface. The pathway taken is dependent on the bound ligand. Dectin-1 is degraded when it internalizes zymosan, but it is recycled when it endocytoses soluble ligands. Stimulation of CTLs such as dectin-1 in myeloid cells activates mitogen-activated protein kinase (MAPK) and nuclear factor-κB (NF-κB), and enhances transcription of genes important in innate immune responses. Internalization of antigens via the

CTLs in DCs induces production of reactive oxygen species (ROS) and other responses. CLEC9A (DNGR-1) and the mannose receptor divert cargo away from lysosomal compartments to allow retrieval of antigens for cross-presentation.

Clustering and density of the CTLDs may determine both their specificity and their affinity for ligands because each individual CRD can act independently to bind sugar. Although the hepatic lectins subunits have a single CTLD, the macrophage mannose receptor has eight CTLDs in a single polypeptide. The adjacent CTLDs may promote binding to specific multivalent, mannose-containing glycans. The macrophage mannose receptor internalizes lysosomal enzymes containing oligomannose-type N-glycans and facilitates phagocytosis of several pathogens such as yeast, *Pneumocystis carinii*, and *Leishmania*. Interestingly, domains in some CTLs outside of the CTLD may also have receptor activity. For example, the cysteine-rich domain of the mannose receptor is an R-type domain (Chapter 31) that binds to glycans containing R-GalNAc-4-SO_4 on pituitary glycoprotein hormones and thus acts to clear these hormones from the circulation.

THE COLLECTINS

The collectins are a family of soluble and membrane-bound CTLs that contain a collagen-like domain amino-terminal to the CTLD and usually assemble in large oligomeric complexes containing 9–27 subunits (Figure 34.3). Bovine conglutinin was identified in the early 1900s by Bordet and Streng as a plasma protein that agglutinated erythrocytes following reactions with antibody and complement, now known to be through Ca^{++}-dependent interactions with the exposed glycans on iC3b after proteolytic conversion from C3b. Bovine conglutinin and human conglutinin bind yeast glycans, and GlcNAc can inhibit interactions. The term conglutinin was coined in 1993 based on evidence that conglutinin had a collagen-like sequence and exhibited lectin activity. To date, nine different collectins have been identified: conglutinin, mannose-binding lectin (MBL), surfactant proteins SP-A and SP-D (which were originally found in the lung but also expressed in the intestine), and collectins CL-43, CL-46, CL-P1, CL-L1, and CL-K1. MBL, conglutinin, CL-43, CL-46, CL-K1, SP-A, and SP-D are soluble, whereas CL-L1 and CL-P1 are membrane proteins.

Collectins are pattern-recognition receptors (PRRs) functioning in innate immunity that bind to surfaces of microbes or fungi expressing glycan ligands termed pathogen-associated molecular patterns (PAMPs). Thus, collectins and many other CTLs in myeloid cells, as discussed above, function as PRRs and cooperate with Toll-like receptors (TLRs) in recognition and responses of cells to pathogens. Examples of PAMPs include lipopolysaccharides (LPSs), β-glucans, lipoteichoic acids (LTAs), and glycoproteins of parasites. Collectins stimulate in vitro phagocytosis by recognizing PAMPs, and binding promotes leukocyte chemotaxis and stimulates the production of cytokines and ROS by immune cells. Lung surfactant lipids have the ability to suppress a number of immune cell functions such as proliferation, and this suppression of the immune response is further augmented by SP-A.

Binding of MBL and other collectins initiates the lectin pathway of complement activation. Such collectins are associated with proenzyme forms of MBL-associated serine proteases 1 and 2 (MASP-1 and -2), in which binding to ligand by the CRD activates MASP-1 to cleave MASP-2, which then activates complement via the classical pathway and generates opsonic C3b fragments that coat pathogens and lead to their destruction and phagocytosis. Some individuals with MBL deficiency syndrome have mutations in the Gly-X-Y repeat encoded within exon-1 of the MBL gene (*MBL2*). Mutations within exon-1, which are highly variable among human populations, inhibit assembly of the MBL subunit, leading to increased risk of microbial infections. Furthermore, several polymorphisms within the promoter region of MBL2 are associated with MBL deficiency and enhanced susceptibility to infections. Genetic variants of *MBL2* are also associated with Crohn's disease. In addition, polymorphisms in *MASP1* and *MASP2* are associated with altered serum levels and compromised activity of collectins. Activation of these proteases can also lead to prothrombin cleavage and promote clot formation.

THE MYELOID C-TYPE LECTINS

Myeloid cells, which include monocytes, macrophages, neutrophils, and DCs, have a large number of proteins with CTLDs, which belong mainly to groups 2, 5, and 6 (Figure 34.3). Myeloid cells also express many members of the galectin and Siglec families of lectins. All leukocytes express L-selectin (group 4), which is discussed below in a separate section on selectins. Group 2 CTLs are DC-SIGN (in humans, but there is no murine homolog), CD209a (also termed SIGN-R1 in mice, but there is no human homolog), macrophage CTL (MCL), dectin-2 (DC-associated C-type lectin-2), langerin, the macrophage galactose-binding lectin (MGL), DCIR and the macrophage inducible Ca^{++}-dependent lectin (MINCLE) (CLEC4E); in group 5, Ca^{++}-independent lectin dectin-1, LOX-1, CLEC12B, CLEC9A (DNGR-1), myeloid-DAP12-associating lectin (MDL-1), CLEC-1, CLEC-2, and dendritic cell–associated lectin-1 (DCAL-1); and in group 6, Ca^{++}-dependent macrophage mannose receptor (CD206) and DEC-205 (CD205).

There are approximately 20 members of the natural killer (NK) cell receptor group 5 of CTLD-containing proteins in humans, and additional ones in mice. Many of the genes encoding the NK cell receptor group 5 are present on human chromosome 12 and mouse chromosome 6. The dectin-1 cluster that is part of the NK gene complex of CTLDs (mouse chromosome 6 and human chromosome 12) includes dectin-1 (CLEC7A), CLEC-1 (CLEC1A), CLEC-2 (CLEC1B), LOX-1 (OLR), CLEC12b, and CLEC9A. These CTLD-containing receptors bind various ligands such as glycans, but also proteins, lipids, inorganic compounds, and even ice. The dectin-2 cluster of CTLDs (whose genes are clustered in the centromeric region of the NK gene cluster) and represent group 2 CTLs includes dectin-2 (CLEC6A), blood DC antigen 2 (BDCA-2) (CLEC4C), DC immunoactivating receptor (DCAR) Clec4b in mice, absent in humans), DC immunoreceptor (DCIR) (CLEC4A), CTL superfamily 8 (CLECSF8) (CLEC4D), and MINCLE (CLEC4E). Mice have the Ly49 family of receptors that contain CTLDs and are on murine chromosome 6. But the Ly49 members, also termed Ly49 NK cell receptors, may not bind sugar but primarily bind to major histocompatibility complex (MHC) class I ligands and MHC class I–like molecules expressed by viruses. These are most functionally related to the killer-cell immunoglobulin-like receptors (KIRs) in human NK cells and some T cells.

Numerous myeloid proteins with CTLDs can recognize not only pathogen moieties for host defense, but also modified self-antigens such as damage-associated molecular patterns (DAMPs) released from dead cells. The glycan and nonglycan ligands for the myeloid CTLs are diverse. MINCLE can recognize characteristic PAMPs such as α-mannose-containing glycans and trehalose-6,6-dimycolate, a key glycolipid virulence factor from *Mycobacterium tuberculosis* and *Mycobacterium bovis*, and thus induce immune responses against infection. MINCLE also maintains self-homeostasis and monitors the internal environment by sensing damaged cells by recognizing SAP130, a component of small nuclear ribonucleoprotein and the intracellular metabolite beta-glucosylceramide (β-GlcCer) both exposed by dying cells. BDCA-2 binds HIV-1 gp120 and galactose-containing glycans. Dectin-1, which binds ligands in the absence of Ca^{++}, is the functional receptor for β-glucans, which are polymers of a backbone of β1-3-linked glucose and side chains of β1-6-linked glucose. Dectin-2 binds in a Ca^{++}-dependent fashion to α-mannans, which are polymers of α1-6-linked mannose with α1-2-linked mannose side chains. Both dectin-1 and dectin-2 are important in fungal defense, and mice deficient in either one are more susceptible to specific yeast and fungal infections.

Interestingly, these receptors also play an important role in sterile inflammation; their dysregulations can lead to the development of diverse pathologies such as autoimmune diseases or cancers. For example, Dectin-1, by binding galectin-9, suppresses macrophage activation and promotes tumor oncogenesis. CLEC12A tempers inflammation by recognizing monosodium urate crystal formed by crystallization of soluble uric acid following contact with extracellular sodium ions. CLEC9A (DNGR-1) binds F-actin exposed by dying cells and regulates innate and adaptive immunity by tempering immunopathology associated with tissue damage; it also regulates adaptive immunity during acute tissue damage and by promoting phagosomal rupture and cross-presentation of internalized dead cell-associated antigens to $CD8^+$ T cells. It is not yet clear whether recognition of the endogenous and exogenous ligands involves similar binding sites in the CTLD. CLR signaling of both PAMPs released following viral infections and DAMP from collateral injured cells may ensure microbial control while preserving integrity of the infected organs.

The various signaling pathways by which many of these myeloid CTLDs function is illustrated in Figure 34.5. Most of these proteins express immunoreceptor tyrosine-based activation motif (ITAM) in their cytoplasmic domain. An example is dectin-1, which contains an activating ITAM-like motif, and like other activatory ITAM CTLs, also contains a tri-acidic motif that promotes downstream signaling. Some CTLs in myeloid cells, such as MICL (CLEC12A) and macrophage antigen h (MAH) (CLEC12B), contain an immunoreceptor tyrosine-based

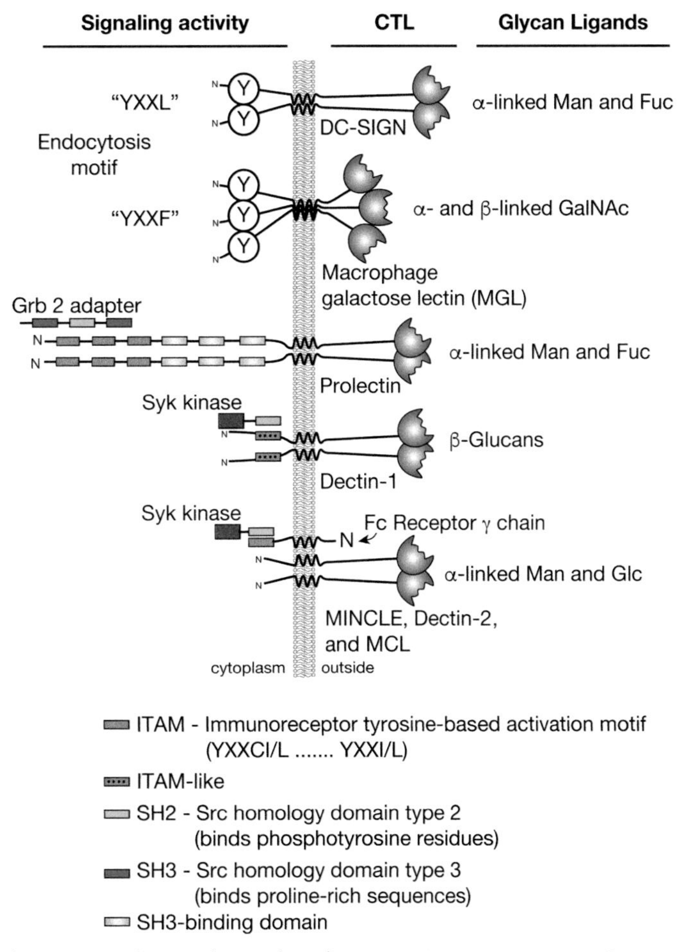

FIGURE 34.5. Signaling activity of C-type lectins (CTLs) in innate immune responses. CTLs expressed on dendritic cells (DCs) and macrophages interact with different glycans, often expressed on pathogen-derived glycans. Activation of CTLs occurs through glycan binding, and for each CTL indicated there are unique signaling pathways induced through the cytoplasmic domains.

inhibitory motif (ITIM). Such motifs on phosphorylation recruit tyrosine phosphatases to negatively regulate signaling pathways. Some CLRs, such as DC-SIGN, utilize alternative pathways (e.g., the serine/threonine kinase RAF-1). In fact, the same CLR can vary in function according to the ligand (physical nature, affinity, avidity) and thereby integrate distinct positive and negative signals, e.g., MINCLE and CLEC9A can trigger opposing signals through SYK or SHP-1.

Among the myeloid CTLDs, CLEC-2 is expressed by NK cells, DCs, megakaryocytes, and platelets, but it is also expressed on some tumor cells. CLEC-2 is a type II transmembrane glycoprotein whose signaling is initiated by tyrosine phosphorylation of a single tyrosine in the YXXL motif of its cytoplasmic domain. Signaling involves Src and Syk kinases and phospholipase Cγ2. CLEC-2 is also a receptor for the platelet-activating snake toxin rhodocytin, from the Malayan viper, *Calloselasma rhodostoma.*

The major endogenous glycoprotein ligand for CLEC-2 is podoplanin (also called Aggrus), expressed by lymphatic endothelial cells and lymph node fibroblastic reticular cells (FRCs). CLEC-2 binds to the sialylated O-glycans and peptide portions of podoplanin. This is physiologically important and serves both as an alternative pathway for in vivo platelet activation, as well as fundamentally serving to drive lymphangiogenesis through the signaling functions of podoplanin. In the absence of extended O-glycans on podoplanin, as engineered in mice lacking *C1GalT1* (*T-synthase*), due to loss of T-synthase or its chaperone *Cosmc* (*C1GalT1C1*), podoplanin is dysfunctional. This leads to defects in lymphangiogenesis, for lack of platelet involvement, a phenotype that closely phenocopies that seen in podoplanin-deficient mice. The CLEC-2-podoplanin pathway plays a critical role in dendritic cell motility and lymph node architecture and regulates inflammation notably during sepsis. Interestingly, different regions of the CTLD in CLEC-2 can bind the O-glycosylated podoplanin and the nonglycosylated rhodocytin.

Another potential adhesion molecule is prolectin (CLEC17A), which is expressed in dividing B cells in germinal centers, and interacts with glycans containing terminal α-linked mannose or fucose residues. It may function within lymph nodes as an adhesion or colonization factor for invading tumor cells and shows preference for binding to epithelial rather than mesenchymal cells, which is relevant for tumor cells undergoing epithelial–mesenchymal transition (EMT).

Both macrophages and DCs are antigen-presenting cells (APCs) that internalize antigens via specific endocytic receptors or by fluid-phase pinocytosis; they process antigens for presentation to $CD8^+$ cytotoxic T cells. DCs also contribute to the balance between tolerance and the induction of immunity and help to distinguish harmless "self-antigens" from pathogens. TLRs and CTLs (for myeloid cells, these have also been termed CTL receptors or CLRs) act to help DCs discriminate between pathogens and self-antigens. TLRs are PRRs that interact with PAMPs, but, unlike CLRs, TLRs cannot directly promote phagocytosis of bound ligands. A CLR can signal by itself, but can also interfere in signaling pathways induced by various other myeloid PRRs, such as TLRs, and hence can provide diverse immunological response outcomes. For example, mycobacteria interact with DCs via TLR-2 and TLR-4, resulting in strong T-helper 1 (Th1) responses by the activated DCs. However, some virulent strains of mycobacteria secrete glycosylated factors (e.g., mannosylated lipoarabinomannan or ManLAM) that are bound by CLRs, but apparently not by TLRs, which lead to down-regulation of TLR activation and limitation of DC maturation. DC-SIGN can tailor immune responses toward specific self and foreign structures in interaction with the signaling of other PRRs. The recognition of mannose-carrying PAMPs expressed by *M. tuberculosis* and HIV-1 activates the kinase Raf-1 to modulate TLR-induced NF-κB activation and enhance the production of pro-inflammatory cytokines, such as IL-6 and IL-12. In contrast, on recognition of fucose-carrying PAMPs being expressed by *Schistosoma mansoni* and *Helicobacter pylori*, DC-SIGN signaling leads to a suppression of the TLR-4-induced pro-inflammatory responses, abrogates Th1 and Th17 responses and favors a Th2 outcome. Recognition of pathogens through their glycans is important for protective immunity. In addition, glycans, when aberrantly expressed by cells in particular conditions, are now increasingly recognized as disease-driving factors in cancer, auto-immunity, and allergy. Therefore, myeloid CLRs represent important regulators in the balancing act between disease and homeostasis.

The myeloid C-type lectins have important innate immune interactions with viruses, through binding to the glycans on viral glycoproteins. DC-SIGN was initially identified through its interactions with gp120 on the HIV envelope. Examples of viruses known to bind to C-type lectins in dendritic cells and macrophages include hepatitis C virus, Ebola virus, West Nile virus, Nipah virus, Newcastle virus, and SARS coronaviruses. But almost all of the myeloid C-type lectins have been found to recognize viral glycoproteins. In particular, the SARS-CoV-2 spike glycoprotein interacts with DC-SIGN, LSECtin, CLEC10A, and others and may contribute to immune hyperactivation.

THE SELECTINS

A seminal experiment in the mid-1960s by Gesner and Ginsburg showed that ^{32}P-labeled rat lymphocytes injected back into rats homed to the spleen and lymph nodes, but when they were first treated with crude glycosidases homing was diminished. They hypothesized that "... sugars serve a physiological function by acting as sites of cellular interactions ... ." This exceptional insight eventually introduced the concept that a GPB might be involved in lymphocyte homing, which eventually led to the discovery of L-selectin. Interestingly, we now know that altering the glycans on the lymphocytes by such treatments blocks their homing by reducing chemokine receptor functions and extravasation events required for lymphocyte homing.

There are three members of the selectin family, P-, E-, and L-selectin, and they are among the best-characterized family of CTLs because of their extensively documented roles as cell adhesion molecules. They mediate (1) the earliest stages of leukocyte trafficking, (2) constitutive migration of lymphocytes to peripheral lymph nodes and to skin, and (3) hematopoietic stem cell trafficking to marrow. The three mammalian selectins are type-1 transmembrane glycoproteins that are expressed on platelets, endothelial cells, and leukocytes, hence the origin of their names (Figure 34.6A). Interactions between the selectins and cell-surface glycoconjugate ligands promote tethering and rolling of leukocytes and platelets in postcapillary venules and are important for leukocyte recruitment to sites of inflammation and injury. Rolling is a form of adhesion that requires rapid formation and dissociation of bonds between selectins and their ligands. Rolling adhesion enables leukocytes to encounter endothelium-bound chemokines. Signaling through chemokine receptors cooperates with signaling through selectin ligands to activate leukocyte integrins, which bind to immunoglobulin superfamily ligands on endothelial cells to slow rolling velocities and arrest leukocytes on vascular surfaces. The arrested leukocytes then crawl through or between endothelial cells into the underlying tissues (diapedesis or extravasation) (Figure 34.6D).

Each selectin has a C-type CRD at the amino terminus followed by a consensus epidermal growth factor (EGF)-like domain and a number of short consensus repeats composed of sushi domains (also called complement control protein [CCP] modules) (Figure 34.6A). The proteins have a single transmembrane domain (TM) and a cytoplasmic domain and are relatively rigid, extended molecules. The CRD of each selectin has modest affinity for the sialylated, fucosylated structure known as the sialyl-Lewis x antigen (SLe^{x}) and its isomer sialyl-Lewis a (SLe^{a}) (Chapter 14) and increased binding to some sulfated versions of these motifs. Each of the selectins binds with higher affinity to specific macromolecular ligands, and most contain sialylated,

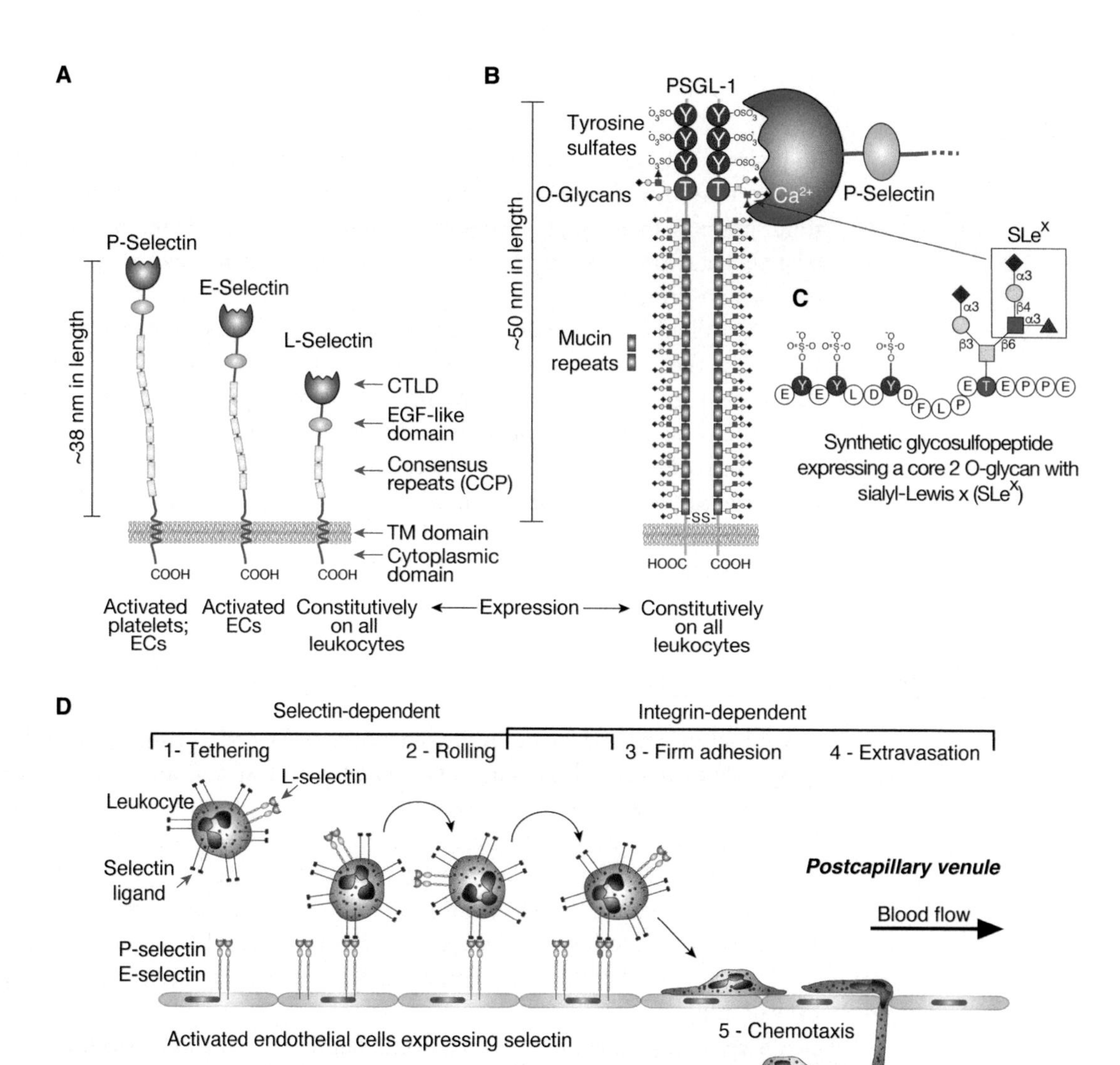

FIGURE 34.6. Structures and functions of selectins. (*A*) Overall domain structures of P-selectin, E-selectin, and L-selectin and their expression patterns are indicated. P-selectin also forms homodimers in the membrane. (The key for the domain structures is shown in Figure 34.3.) (*B*) The predicted disulfide-bonded dimeric form of PSGL-1 on leukocytes is shown. PSGL-1 has three potential sites for N-glycosylation and, as indicated, multiple sites for O-glycosylation. The sulfated tyrosine residues at the extreme amino terminus are indicated. The Ca^{++}-dependent interaction between P-selectin and PSGL-1 through their amino-terminal domains is indicated. L-selectin also binds to the same region of PSGL-1, although with different kinetics and affinity. (*C*) The major fucose-containing core-2 O-glycan identified in PSGL-1 is required for binding to selectins and is shown on a synthetic GSP generated based on the amino-terminal sequence of human PSGL-1. The sialyl-Lewis x determinant on a core 2 O-glycan is *boxed*. E-selectin may bind to more distal glycans expressing the SLe^x determinant. (*D*) Tethering of circulating leukocytes to activated endothelium via interactions between selectins and their ligands. In normal venules, leukocytes flow without adhesive interactions with the endothelium, but in inflamed vessels, selectins and integrin ligands are expressed on endothelial surfaces. This leads to tethering, rolling, and arrest of circulating leukocytes and their eventual extravasation from the circulation to the surrounding tissue. Selectin-dependent interactions also occur through E-selectin binding to glycans on leukocyte CD44 (HCELL). When adherent cells become activated by regionally presented chemokines or lipid autacoids, the activated leukocytes express integrins (e.g., LFA-1, CD11a/CD18 and Mac-1, CD11b/CD18) that interact with immunoglobulin-like counterreceptors on endothelial cells (ICAM-1, ICAM-2) to strengthen the adhesion and promote the transmigration of cells from the circulation into the underlying tissues. P-selectin is normally expressed in Weibel–Palade bodies of endothelial cells, but within minutes after endothelial cell activation by thrombin, histamine, hypoxia, or injury, these bodies fuse with the plasma membrane, promoting the expression of P-selectin on the endothelial cell surface. Similarly, P-selectin stored in the α-granules of platelets becomes expressed on the surfaces of platelets within minutes after platelet activation. E-selectin is expressed by activated endothelial cells through transcription-dependent pathways and is inducible by TNFα, IL1β, and LPS, mediated in part through NF-κB-dependent events. E-selectin cooperates with P- and L-selectin to recruit leukocytes to sites of inflammation.

fucosylated glycans. The major ligand for P-selectin, termed P-selectin glycoprotein ligand-1 (PSGL-1), has sulfated tyrosine residues adjacent to SLe^x expressed on a core-2-based O-glycan. This glycoform of PSGL-1 is also a high-affinity ligand for L-selectin, and additional PSGL-1 associated glycans also engage E-selectin. Although PSGL-1 is the major ligand for L- and P-selectin, E-selectin binds additional glycoconjugates also. Importantly, a major ligand for E-selectin is a glycoform of CD44 called "hematopoietic cell E/L-selectin ligand" (HCELL) that is constitutively expressed on human hematopoietic stem cells. This ligand carries SLe^x on its N-glycans. Murine leukocytes require core 1–derived O-glycans on CD44 and other glycoproteins to bind E-selectin. In addition to glycoproteins, P- and L-selectin, but not E-selectin, also binds to some forms of heparin/heparan sulfate (HS) in a divalent cation-independent manner. E-selectin is also reported to bind sialofucosylated glycolipids on human, but not murine, myeloid cells. Whereas the above glycoconjugates participate in myeloid leukocyte recruitment at sites of inflammation, the L-selectin ligands regulating lymphocytes homing to secondary lymphoid organs are distinct. These L-selectin ligands on endothelial cells of lymph node high endothelial venule (HEV) contain 6-sulfo-SLe^x determinants containing MECA-79 antigen on mucin-type O-glycans and on N-glycans.

P-Selectin

P-selectin (CD62P) was discovered as an antigen expressed on the surface of activated platelets. It is constitutively expressed in megakaryocytes, in which it is packaged into the membranes of α-granules of circulating platelets. It is also expressed in the Weibel–Palade bodies of vascular endothelial cells. Within minutes following activation of either platelets or endothelial cells by proinflammatory secretagogues such as histamine, thrombin, or complement components, P-selectin is expressed on the cell surface because of fusion of the intracellular storage membranes with the plasma membrane. Sequences within the cytoplasmic domain of P-selectin mediate its sorting to secretory granules as well as its rapid endocytosis from the plasma membrane and movement from endosomes to lysosomes, where it is degraded. Splice variants of human P-selectin transcripts yield forms of P-selectin that lack a TM, thus contributing to low-level soluble forms of P-selectin in the circulation. Leukocyte adhesion also stimulates proteolytic cleavage of the ectodomain of P-selectin from the plasma membrane, releasing it into the circulation. The inflammatory mediators tumor necrosis factor (TNF)-α, interleukin 1β (IL1β), and LPS augment transcription of mRNA for P-selectin in endothelial cells in mice and non-primate mammals but not in humans.

P-selectin contributes to leukocyte recruitment in both acute and chronic inflammation. Mice that lack P-selectin exhibit defective rolling of their leukocytes on endothelial cells of post-capillary venules, diminished recruitment of neutrophils or monocytes into tissues following injection of inflammatory mediators, and impaired recruitment of T cells into skin or other tissues following challenge with specific antigens. Mobilization of P-selectin to the surfaces of activated endothelial cells is important for all these responses. In addition, P-selectin expressed on the surfaces of activated platelets contributes to inflammation as well as to hemostasis and thrombosis. Activated platelets adhere through P-selectin to neutrophils, monocytes, NK cells, and some subsets of T lymphocytes. This adhesion augments the recruitment of leukocytes and platelets to sites of vascular injury. Expression of P-selectin on both platelets and endothelial cells contributes to experimental atherosclerosis in mice. Platelet-expressed P-selectin also stimulates monocytes to synthesize tissue factor, a key cofactor of blood coagulation that facilitates fibrin deposition during clot formation.

As discussed below, although the major leukocyte counter receptor for P-selectin is PSGL-1, P-selectin (and L-selectin) also binds less avidly to some forms of heparin/HS and to some other glycoproteins that express SLe^x. Although the physiological significance of this non-PSGL-1-dependent binding by P-selectin is unclear, clinically prescribed levels of heparin can block P-selectin functions. In addition, P-selectin can interact with mucins containing highly clustered O-glycans bearing SLe^x antigens and sulfate esters, which appear to be important in the metastasis of tumors bearing such ligands (Chapter 47). The therapeutic effect of heparin in blocking cancer metastases may thus be mediated at least partly through blocking P- and/or L-selectin-dependent adhesion of leukocytes and platelets to tumor cells expressing $SLe^{x/a}$ rich sulfated mucins. Results from a phase 2 trial suggest that a chemical selectin inhibitor (GMI-1070) may reduce the severity of vasoocclusive crisis in patients with sickle cell disease (SCD). Another therapeutic approach targeting P-selectin is based on the blockade of human P-selectin with an inhibitory humanized antibody. This approach has shown efficacy in the prophylactic treatment of SCD. The antibody (crizanlizumab) is marketed as the drug Adakveo and is approved in the United States as a monthly medication to treat pain crisis in patients with SCD.

PSGL-1

PSGL-1 (CD162) is a homodimeric, disulfide-bonded mucin with subunits of ~120 kDa. It is the major physiological ligand on leukocytes for P- and L-selectin and is also a ligand for E-selectin (Figure 34.6B). PSGL-1 contains 16 decapeptide repeating units (mucin repeats) with the consensus sequence spanning residues 118–277 in the ectodomain of the long form of the protein, which is the major form expressed in humans. Murine PSGL-1 has some sequence similarity to the human sequence, but the mouse protein has only 10 decameric repeats with the consensus sequence -E-T-S-Q/K-P-A-P-T/M-E-A-, which is different from human PSGL-1. The highest homology between human and murine PSGL-1 occurs in the transmembrane and cytoplasmic domains, and unexpectedly not in the P-selectin-binding domain.

The PSGL-1 polypeptide is expressed in most leukocytes (including neutrophils, monocytes, eosinophils, basophils, and subsets of T cells), and also on hematopoietic stem cells of mice and humans. Additionally, PSGL-1 is expressed in some activated endothelial cells, most notably the inflamed microvessels of the ileum in a spontaneous model of chronic ileitis in mice. In leukocytes, PSGL-1 is enriched in microvilli and is associated with lipid rafts.

In neutrophils, monocytes, and activated T cells, PSGL-1 undergoes the appropriate posttranslational modifications to express the SLe^x structure on core 2 O-glycans (Figure 34.6C). This glycoform of PSGL-1 interacts with each of the three selectins to support leukocyte rolling under flow. Engagement of PSGL-1 during rolling also transduces signals into leukocytes that activate leukocyte integrins to slow rolling velocities. Signaling through PSGL-1 cooperates with signaling through chemokine receptors to elicit other effector responses in leukocytes.

PSGL-1 has multiple functions in addition to being a selectin ligand. PSGL-1 is a check-point regulator in T cells, promotes T-cell exhaustion, and partly regulates the programmed cell death protein 1 (PD-1) expression. PSGL-1 via its sulfated tyrosine residues can interact with certain chemokines, including CCL21 and CCL19, which may promote T-cell entry into secondary lymphoid organs. PSGL-1 is an adhesion molecule for pathogens, including enterovirus 71, *Anaplasma phagocytophilum* (the tick-transmitted obligate intracellular bacterium that causes human granulocytic anaplasmosis), and *S. pneumonia*. In some cases, as for *A. phagocytophilum*, the SLe^x-containing O-glycan of PSGL-1 is a recognition factor, whereas for the other pathogens, sulfated tyrosines are recognized.

The key determinants for binding P-selectin are in the extreme amino terminus of PSGL-1. Antibodies to this region block binding of PSGL-1 to P- and L-selectin (but not to E-selectin, which may interact with other fucosylated sites on PSGL-1 and does not interact with sulfated tyrosines). Treatment of neutrophils with selective proteases that remove the first ten amino acids from the amino terminus of PSGL-1 abrogates its binding to P- and L-selectin. Site-directed mutagenesis of a specific O-glycosylated amino-terminal threonine and/or the three tyrosine residues in the amino terminus of PSGL-1 prevents binding to P- and L-selectin. Finally, a synthetic glycosulfopeptide (GSP) with the structure shown in Figure 34.6C binds to P-selectin with similar high affinity as that of native PSGL-1.

The data support the model in Figure 34.6B in which the combination of tyrosine sulfate residues and a glycan on PSGL-1 are required for high-affinity binding to P-selectin. The co-crystal structure of a PSGL-1-derived GSP with P-selectin confirms this model. The interactions between the GSP and P-selectin result from a combination of hydrophobic and electrostatic contacts. These include contacts of at least two of the three tyrosine sulfate residues as well as other PSGL-1 amino acids with multiple residues within the P-selectin CRD, and hydroxyl groups in the fucose residue of SLe^x that ligate the lectin domain-bound Ca^{++}, and there are additional binding interactions with the hydroxyl groups of galactose and the –COOH group of Neu5Ac.

Key enzymes mediating the formation of the selectin-ligand on human PSGL-1 include core-1 synthase (T-synthase/C1GalT1) and it chaperone Cosmc (C1GalT1C1) that facilitate initial O-glycan biosynthesis; β1-6-N-acetylglucosaminyltransferase-I (C2GnT-I) that facilitates core-2 structure formation; and the capping glycosyltransferases α2-3 sialyltransferases ST3Gal-4 and α1,3 fucosyltransferase FUT-VII that aid the formation of the functional SLe^x epitope. The enzymes mediating sulfation of the PSGL-1 peptide backbone on myeloid cells are likely PAPSS1 (3′-phosphoadenosine 5′-phosphosulfate synthase 1) and TPST2 (tyrosylprotein sulfotransferase).

E-Selectin

E-selectin (CD62E) was discovered as a leukocyte adhesion molecule expressed by activated vascular endothelial cells. In most tissues (the bone marrow and skin are exceptions), endothelial cells do not constitutively express E-selectin. Cytokine-dependent transcriptional processes lead to an inducible expression of E-selectin on the surface of the endothelium. Inducible transcription of the E-selectin locus by TNFα, IL1-β, and LPS is mediated at least in part through NF-κB-dependent events. In vitro, cytokine treatment increases E-selectin expression after two hours, with maximal expression at four hours. E-selectin expression then declines to basal levels within 12–24 hours in vitro, but may be expressed chronically at sites of inflammation in vivo. Decline of E-selectin expression is associated with decreased transcription of the E-selectin locus, degradation of E-selectin transcripts, and internalization and turnover of E-selectin protein. Acute and chronic inflammatory conditions associated with E-selectin expression include trauma, sepsis, rheumatoid arthritis (RA), and organ transplantation.

E-selectin cooperates with P- and L-selectin to recruit leukocytes to sites of inflammation. The physiological ligands for E-selectin contain the SLe^x antigen and occur on leukocytes (neutrophils, monocytes, eosinophils, memory/effector T cells, and NK cells) and on human hematopoietic stem cells. Each of the leukocyte subsets is found in acute and chronic inflammatory sites in association with the expression of E-selectin. PSGL-1 is one of the physiological ligands for E-selectin, but E-selectin can also interact with several other glycoproteins that express the SLe^x antigen on either N- or O-glycans, including HCELL (a human CD44 glycoform), E-selectin ligand-1 (in mice), and possibly long-chain glycosphingolipids expressing the SLe^x antigen. In this regard, whereas glycoprotein O- and N-glycans facilitate the initial recruitment and rapid rolling phase in the leukocyte adhesion cascade, sialofucosylated glycolipids may facilitate slow rolling and the transition to inflammatory leukocyte arrest. In human hematopoietic stem cells (HSCs), via adhesive interactions on bone medullary microvessels that constitutively express E-selectin, HCELL directs trafficking of HSCs into marrow. In addition to effects on leukocyte and hematopoietic stem cell migration, E-selectin receptor/ligand interactions may have a potential role in cancer metastasis, promoting malignant cell survival in bone marrow, and mesenchymal–epithelial transition in the bone marrow niche (Chapter 47).

L-Selectin

L-selectin (CD62L) is expressed on the microvilli of most leukocytes, including all myeloid cells, naïve T and B cells, and some memory/effector T cells. L-selectin was the first selectin discovered through efforts to define molecules that facilitate recirculation of lymphoid cells from the intravascular compartment to secondary lymphoid organs, including lymph nodes and Peyer's patches, from which the lymphoid cells then return to the circulation through the lymphatic system, as discussed above in work by Gesner and Ginsburg. This recirculation process provides lymphocytes with the opportunity to encounter foreign antigens displayed by APCs within secondary lymphoid organs. Early studies indicated that blood lymphocytes enter lymph nodes in specialized postcapillary HEVs. Endothelial cells in the HEV are cuboid in shape. Their surfaces are decorated with different types of small-sized mucins that mediate L-selectin-dependent adhesion of lymphocytes. These mucins are called peripheral node addressins. They include CD34, Sgp200, GlyCAM-1, MAdCAM-1, endoglycan, endomucin, and podocalyxin-like protein (PCLP). On cellular activation and migration L-selectin is proteolytically shed from the surface of activated leukocytes by a metalloproteinase TNF-α-converting enzyme (TACE/ADAM-17).

Rolling mediated by L-selectin exhibits a counterintuitive "shear threshold" requirement (e.g., a minimum flow rate is required for leukocytes to roll). As the flow rate drops below this threshold, leukocytes roll faster and more unstably and then detach. Flow-enhanced rolling operates through a force-dependent mechanism. As the flow rate increases, the force applied to adhesive bonds between L-selectin and its ligands increases. At threshold levels, the force actually strengthens the bonds, which prolongs their lifetimes. These are called "catch bonds." As the flow rate increases further, the applied force begins to weaken the bonds, which shortens their lifetimes. These are called "slip bonds." Catch bonds are regulated by force-dependent straightening of the angle between the lectin and EGF domains of L-selectin, which affects how ligand dissociates from the binding interface on the lectin domain. Transitions between catch and slip bonds are also seen for interactions of P-selectin and E-selectin with their ligands. However, these transitions occur at lower forces with less dramatic effects in the circulation.

A unique feature of the L-selectin ligands on HEV is the requirement for sulfated glycans, such as 6-sulfo-SLe^x on both core-2 O-glycans and on extended core-1 O-glycans. The 6-sulfo-SLe^x determinant is associated with the MECA-79 epitope on O-glycans (Figure 34.6), an antibody that binds to 6-sulfo-*N*-acetyllactosamine on extended core-1 O-glycans. The biosynthesis of the 6-sulfo-SLe^x determinant depends on two key α1-3 fucosyltransferases, Fuct-VII and Fuct-IV, along with at least four different sulfotransferases that may form the 6-sulfo-SLe^x determinant. Two of these sulfotransferases, GlcNAc6ST-1 and GlcNAc6ST-2, are expressed in HEV and appear to be most important. Mice lacking Fuct-VII or both Fuct-VII and Fuct-IV have dramatically reduced homing of lymphocytes to lymph nodes. Mice lacking both GlcNAc6ST-1 and GlcNAc6ST-2 do not express 6-sulfo-SLe^x or the MECA-79 epitope and exhibit markedly diminished lymphocyte homing to lymph nodes. A unique β1-3GlcNAcT generates the extended core-1 O-glycan; mice lacking both this β1-3GlcNAcT and the β1-6GlcNAcT branching enzyme for core-2 O-glycan biosynthesis do not express the MECA-79 antigen, but they have residual lymphocyte rolling on HEV and only a minimal decrease in lymphocyte numbers in peripheral and mesenteric lymph nodes. The residual L-selectin-dependent lymphocyte homing appears to result from 6-sulfo-SLe^x on N-glycans, suggesting that both N- and O-glycans on HEV glycoproteins contribute to L-selectin-dependent lymphocyte recirculation through lymph nodes. On the lymphocyte itself, the sialylated core-1-derived O-glycans on surface glycoproteins, including chemokine receptors, are crucial to migration of lymphocytes into lymph nodes. *Cosmc*-deficient murine

B cells or desialylated cells are unable to home to lymph nodes, thus providing a physiological mechanism for the observations of Gesner and Ginsburg on the block to homing by enzymatic deglycosylation of lymphocytes.

L-selectin also plays a role in adhesion of neutrophils, eosinophils, and monocytes to non-lymphoid vascular endothelium (Figure 34.6). The major ligand for L-selectin in these inflammatory settings is PSGL-1, which is expressed on adherent leukocytes and may also be deposited on inflamed endothelial cells as fragments left behind by previously rolling leukocytes. An alternate ligand on endothelial cells is HS. As with P-selectin, clinically relevant doses of the drug heparin can block L-selectin, and likely contributes to the inhibition of tumor metastasis and of mucinous-tumor associated hypercoagulability (Trousseau's syndrome) (Chapter 47). On human hematopoietic stem cells, HCELL is also a potent L-selectin ligand, yet unlike all other naturally-expressed L-selectin ligands, its binding to L-selectin is not sulfate-dependent. Mice lacking L-selectin have defects in neutrophil recruitment in the context of inflammation as well as defects in homing of naïve lymphocytes to secondary lymphoid organs.

PROTEOGLYCANS WITH C-TYPE LECTIN DOMAINS

The CTLD occurs in several proteoglycans (lecticans or hyalectins) that lack TMs and reside in the extracellular matrix (ECM) (group I; Figure 34.3). These include aggrecan, brevican, versican, and neurocan. Like the selectins, each of these core proteins contains a CTLD, an EGF-like domain, and a CCP domain, but their domain order is different and they are located in the carboxyl terminus of the protein. A large region containing attachment sites for chondroitin sulfate (CS) and keratan sulfate (KS) is proximal to the lectin domain. A more complete discussion of the proteoglycans is provided in Chapter 17. The exact functions of the CTLD in these proteins are still unknown. The CTLD of rat aggrecan is important for Ca^{++}-dependent binding to the fibronectin type II repeats 3–5 of rat tenascin-R. The CTLDs in other lecticans are probably also responsible for protein–protein interactions with other receptors, including tenascin-R, tenascin-C, and other tenascins, which are ECM glycoproteins highly expressed in the nervous system. Interestingly, the protein–protein interactions between rat fibronectin and the CTLD of rat aggrecan bear resemblance to the protein–protein interactions seen in P-selectin binding to PSGL-1. Tenascin-R is one of the main carriers of the unusual glycan antigen HNK-1, which was named after its identification on human NK cells. Brevican, a lectican found in the nervous system, binds to the HNK-1-containing glycosphingolipids. Thus, the CTLD in lecticans may represent a versatile structural feature that can be used for both protein–protein and protein–glycan interactions. Interestingly, several aggrecanopathies (inherited defects in aggrecan production/function) are associated with mutations in the exon encoding the CTLD domain of aggrecan.

OTHER PROTEINS WITH C-TYPE LECTIN DOMAINS

A number of proteins with CTLDs have been identified in the pancreas and kidney, but whether these proteins bind glycans is unclear. PKD1, one of two ADPKD gene products associated with autosomal-dominant polycystic kidney disease (ADPKD), has been implicated in cell–cell and cell–matrix interactions. The *PKD1* gene encodes polycystin-1 (PC1), a very large protein (4293 amino acids) with a single CTLD near its amino terminus. Carboxy-terminal proteolysis of PC1 during its secretion releases a cytoplasmic portion that becomes nuclear and regulates cell signaling pathways. Interestingly, the CTLD of PC1 appears to bind collagen in a Ca^{++}-dependent fashion that is inhibitable by sugars, including dextrans, but the precise nature of the interaction is still unclear.

Some small proteins with CTLDs include Reg3α and PSP (pancreatic stone protein), which are in the REG group 7 of CTLs, and are essentially isolated CTLDs that are preceded by a signal sequence. Reg3α (also known as HIP/PAP) is a secreted CTL that an amino-terminal signal sequence and a 16 kDa CTLD. Reg3α is an anti-inflammatory protein with scavenging activity toward ROS and can bind peptidoglycan and directly kill Gram-positive, but not Gram-negative, bacteria. Reg3α kills bacteria by first binding to peptidoglycan and then binding to membrane phospholipids and forms a hexameric membrane-permeabilizing oligomeric pore.

Lower vertebrates, invertebrates, and some viruses also contain CTLDs. The galactose-specific lectin from *Crotalus atrox* binds a variety of galactose-containing glycolipids in a Ca^{++}-dependent fashion. A number of related venom proteins inhibit platelet function and/or the coagulation cascade. Alboaggregin A from the white-lipped pit viper (*Trimeresurus albolabris*) binds to the platelet GPIb-V-IX complex and stimulates platelet agglutination, but the potential role of glycan recognition in this process is unclear. In contrast, the gp42 protein of Epstein–Barr virus (EBV) has a CTLD that is NK-receptor like, and ectodomain of the poxvirus (Poxviridae) protein A33, a target of subunit vaccine development, has two CTLDs in dimeric form, and has resemblance to the CTLDs of the NK lectins. But, neither of these domains in these viral proteins binds either Ca^{++} or glycans, and may have evolved by convergent evolution.

ACKNOWLEDGMENTS

The authors acknowledge the helpful comments of Robert Sackstein.

FURTHER READING

Gesner BM, Ginsburg V. 1964. Effect of glycosidases on the fate of transfused lymphocytes. *Proc Natl Acad Sci* **52:** 750–755. doi:10.1073/pnas.52.3.750

Ashwell G, Morell AG. 1974. The role of surface carbohydrates in the hepatic recognition and transport of circulating glycoproteins. *Adv Enzymol Relat Areas Mol Biol* **41:** 99–128. doi:10.1002/9780470122860.ch3

Rosen SD, Singer MS, Yednock TA, Stoolman LM. 1985. Involvement of sialic acid on endothelial cells in organ-specific lymphocyte recirculation. *Science* **228:** 1005–1007. doi:10.1126/science.4001928

Moore KL, Stults NL, Diaz S, Smith DF, Cummings RD, Varki A, McEver RP. 1992. Identification of a specific glycoprotein ligand for P-selectin (CD62) on myeloid cells. *J Cell Biol* **118:** 445–456. doi:10.1083/jcb.118.2.445

Norgard-Sumnicht KE, Varki NM, Varki A. 1993. Calcium-dependent heparin-like ligands for L-selectin in nonlymphoid endothelial cells. *Science* **261:** 480–483. doi:10.1126/science.7687382

McMahon SA, Miller JL, Lawton JA, Kerkow DE, Hodes A, Marti-Renom MA, Doulatov S, Narayanan E, Sali A, Miller JF, Ghosh P. 2005. The C-type lectin fold as an evolutionary solution for massive sequence variation. *Nat Struct Mol Biol* **12:** 886–892. doi:10.1038/nsmb992

Brown GD. 2006. Dectin-1: a signalling non-TLR pattern-recognition receptor. *Nat Rev Immunol* **6:** 33–43. doi:10.1038/nri1745

Kishore U, Greenhough TJ, Waters P, Shrive AK, Ghai R, Kamran MF, Bernal AL, Reid KB, Madan T, Chakraborty T. 2006. Surfactant proteins SP-A and SP-D: structure, function and receptors. *Mol Immunol* **43:** 1293–1315. doi:10.1016/j.molimm.2005.08.004

Ludwig IS, Geijtenbeek TBH, van Kooyk Y. 2006. Two way communication between neutrophils and dendritic cells. *Curr Opin Pharmacol* **6:** 408–413. doi:10.1016/j.coph.2006.03.009

Sperandio M. 2006. Selectins and glycosyltransferases in leukocyte rolling in vivo. *FEBS J* **273:** 4377–4389. doi:10.1111/j.1742-4658.2006.05437.x

Uchimura K, Rosen SD. 2006. Sulfated L-selectin ligands as a therapeutic target in chronic inflammation. *Trends Immunol* **27:** 559–565. doi:10.1016/j.it.2006.10.007

Zhou T, Chen Y, Hao L, Zhang Y. 2006. DC-SIGN and immunoregulation. *Cell Mol Immunol* **3:** 279–283.

Gupta G, Surolia A. 2007. Collectins: sentinels of innate immunity. *BioEssays* **29:** 452–464. doi:10.1002/bies.20573

Trinchieri G, Sher A. 2007. Cooperation of Toll-like receptor signals in innate immune defence. *Nat Rev Immunol* **7:** 179–190. doi:10.1038/nri2038

Gurr W. 2011. The role of Reg proteins, a family of secreted C-type lectins, in islet regeneration and as autoantigens in type 1 diabetes. In *Type 1 diabetes—pathogenesis, genetics and immunotherapy* (ed. Wagner D), pp. 161 – 182. InTech, Rijeka, Croatia. doi:10.5772/23887

Zarbock A, Ley K, McEver RP, Hidalgo A. 2011. Leukocyte ligands for endothelial selectins: specialized glycoconjugates that mediate rolling and signaling under flow. *Blood* **118:** 6743–6751. doi:10.1182/blood-2011-07-343566

Grewal PK, Aziz PV, Uchiyama S, Rubio GR, Lardone RD, Le D, Varki NM, Nizet V, Marth JD. 2013. Inducing host protection in pneumococcal sepsis by preactivation of the Ashwell–Morell receptor. *Proc Natl Acad Sci* **110:** 20218–20223. doi:10.1073/pnas.1313905110

Mukherjee S, Zheng H, Derebe MG, Callenberg KM, Partch CL, Rollins D, Propheter DC, Rizo J, Grabe M, Jiang QX, Hooper LV. 2014. Antibacterial membrane attack by a pore-forming intestinal C-type lectin. *Nature* **505:** 103–107. doi:10.1038/nature12729

Richardson MB, Williams SJ. 2014. MCL and Mincle: C-type lectin receptors that sense damaged self and pathogen-associated molecular patterns. *Front Immunol* **5:** 288. doi:10.3389/fimmu.2014.00288

Dambuza IM, Brown GD. 2015. C-type lectins in immunity: recent developments. *Curr Opin Immunol* **32:** 21–27. doi:10.1016/j.coi.2014.12.002

Drickamer K, Taylor ME. 2015. Recent insights into structures and functions of C-type lectins in the immune system. *Curr Opin Struct Biol* **34:** 26–34. doi:10.1016/j.sbi.2015.06.003

D'Souza AA, Devarajan PV. 2015. Asialoglycoprotein receptor mediated hepatocyte targeting—strategies and applications. *J Controlled Release* **203:** 126–139. doi:10.1016/j.jconrel.2015.02.022

Geijtenbeek TBH, Gringhuis SI. 2015. C-type lectin receptors in the control of T helper cell differentiation. *Nat Rev Immunol* **16:** 433–448. doi:10.1038/nri.2016.55

McEver RP. 2015. Selectins: initiators of leucocyte adhesion and signaling at the vascular wall. *Cardiovasc Res* **107:** 331–339. doi:10.1093/cvr/cvv154

Telen MJ, Wun T, McCavit TL, De Castro LM, Krishnamurti L, Lanzkron S, Hsu LL, Smith WR, Rhee S, Magnani JL, Thackray H. 2015. Randomized phase 2 study of GMI-1070 in SCD: reduction in time to resolution of vaso-occlusive events and decreased opiod use. *Blood* **125:** 1656–1664. doi:10.1182/blood-2014-06-583351

Hansen SWK, Ohtani K, Roy N, Wakamiya N. 2016. The collectins CL-L1, CL-K1 and CL-P1, and their roles in complement and innate immunity. *Immunobiology* **221:** 1058–1067. doi:10.1016/j.imbio.2016.05.012

Kedmi R, Peer D. 2016. Zooming in on selectins in cancer. *Sci Transl Med* **8:** p345fs11. doi:10.1126/scitranslmed.aag1802

Sackstein R. 2016. Fulfilling Koch's postulates in glycoscience: HCELL, GPS and translational glycobiology. *Glycobiology* **26:** 560–570. doi:10.1093/glycob/cww026

Barbier V, Erbani J, Fiveash C, Davies JM, Tay J, Tallack MR, Lowe J, Magnani JL, Pattabiraman DR, Perkins AC, et al. 2020. Endothelial E-selectin inhibition improves acute myeloid leukaemia therapy by disrupting vascular niche-mediated chemoresistance. *Nat Commun* **11:** 2042. doi:10.1038/s41467-020-15817-5

Busold S, Nagy NA, Tas SW, van Ree R, de Jong EC, Geijtenbeek TBH. 2020. Various tastes of sugar: the potential of glycosylation in targeting and modulating human immunity via C-type lectin receptors. *Front Immunol* **11:** 134. doi:10.3389/fimmu.2020.00134

Zeng J, Eljalby M, Aryal RP, Lehoux S, Stavenhagen K, Kudelka MR, Wang Y, Wang J, Ju T, von Andrian UH, Cummings RD. 2020. Cosmc controls B cell homing. *Nat Commun* **11:** 3990. doi:10.1038/s41467-020-17765-6

Lu Q, Liu J, Zhao S, Gomez Castro MF, Laurent-Rolle M, Dong J, Ran X, Damani-Yokota P, Tang H, Karakousi T, et al. 2021. SARS-CoV-2 exacerbates proinflammatory responses in myeloid cells through C-type lectin receptors and Tweety family member 2. *Immunity* **54:** 1304–1319. doi:10.1016/j.immuni.2021.05.006

Zhu Y, Groth T, Kelkar A, Zhou Y, Neelamegham S. 2021. A glycogene CRISPR-Cas9 lentiviral library to study lectin binding and human glycan biosynthesis pathways. *Glycobiology* **31:** 173–180. doi:10.1093/glycob/cwaa074

35 I-Type Lectins

Takashi Angata, Stephan von Gunten, Ronald L. Schnaar, and Ajit Varki

I-type lectins are defined as glycan-binding proteins (excluding antibodies and T-cell receptors) in which the binding domain is homologous to the large and varied immunoglobulin superfamily (IgSF) of proteins. Among I-type lectins, the Siglec family of sialic acid–recognizing lectins is the best characterized subgroup, both structurally and functionally, and is therefore the major focus of this chapter. Details of their discovery, characterization, binding properties, and biology are provided, along with discussions of their functional roles in vertebrate biology, with most currently available information being in mammals, and multiple unusual changes during human evolution.

published by Cold Spring Harbor Laboratory Press; doi: 10.1101/glycobiology.4e.35

HISTORICAL BACKGROUND AND OVERVIEW

The Ig fold is made up of antiparallel β-strands organized into a β-sandwich containing 100–120 amino acids and usually stabilized by an intersheet disulfide bond. Three types or "sets" of Ig domains are defined based on homologies in sequence and structure to domains of antibodies: the V-set variable-like domain, the C1- and C2-set constant-like domains, and the I-set domain that combines features of both V- and C-set domains.

Before the 1990s, it was thought that antibodies were the only IgSF members capable of recognizing glycans. The first direct evidence for nonantibody IgSF glycan-binding proteins came from independent studies on sialoadhesin (Sn), a sialic acid (Sia)–dependent binding receptor on mouse macrophage subsets, and on CD22, a molecule previously cloned as a B-cell marker. Various techniques showed that Sn functions as a lectin, including loss of binding following sialidase treatment of ligands, inhibition assays with sialylated compounds, and Sia-dependent binding of the purified receptor to glycoproteins and to red blood cells derivatized to carry Sias in different linkages. With recombinant CD22, abrogation of cell adhesive interactions by sialidase treatment led to the discovery that it was a Sia-binding lectin, with a high degree of specificity for α2-6-linked Sias. Cloning of Sn then showed that it was an IgSF member sharing homology with CD22 and with two other previously cloned proteins, CD33 and myelin-associated glycoprotein (MAG). Demonstration of Sia recognition by CD33 and MAG resulted in the definition of a new family of Sia-binding molecules, which were initially called "sialoadhesins." Meanwhile, preliminary evidence for glycan binding by additional IgSF members emerged, and a suggestion was made to classify all these molecules as "I-type" lectins. However, it became clear that these four Sia-binding molecules were a distinct subgroup sharing both sequence homology and Ig-domain organization, and that they were not all involved in adhesion. The term Siglec (sialic acid–binding immunoglobulin-like lectin) was therefore proposed in 1998. Subsequently, most of the CD33-related Siglecs (CD33rSiglecs) were discovered as a result of genomic and transcriptomic sequencing projects, which allowed in silico identification of novel Siglec-related genes and cDNAs.

Siglecs are divided into two major subgroups based on sequence similarity (Figure 35.1) and on conservation among mammalian species. The first group comprises Sn (Siglec-1), CD22 (Siglec-2), MAG (Siglec-4), and Siglec-15, for which there are clear-cut orthologs in all mammalian species examined and which share only ~25%–30% sequence identity among each other. The second group comprises the CD33rSiglecs, which share ~50%–80% sequence similarity but appear to be evolving rapidly and undergoing exon shuffling and gene conversions of Ig-domain-encoding exons, making it difficult to define orthologs, even between rodents and primates (see below). For this reason, Siglec nomenclature uses numbers and letters to differentiate between some nonhomologous human and mouse Siglecs. In retrospect, it turned out that many distantly related mammals have more clear-cut orthologs, and it is the rodent lineage that "discarded" these genes.

I-TYPE LECTINS OTHER THAN SIGLECS

Several IgSF members other than Siglecs have been proposed to bind glycans or glycoconjugates. The best evidence is for paired immunoglobulin-like type 2 receptors (PILR-α and PILR-β), both of which have a single V-set domain similar to that of Siglecs with a structurally defined Sia-binding site as well as a protein interaction site. PILR proteins bind a subset of mucin-like O-glycosylated membrane proteins involving simultaneous recognition of both the peptide backbone and Sia to mediate high-affinity interactions. Evidence suggests a role of PILRs in the control of innate immunity and host defense against microbial pathogens. PILRs are expressed on immune cells

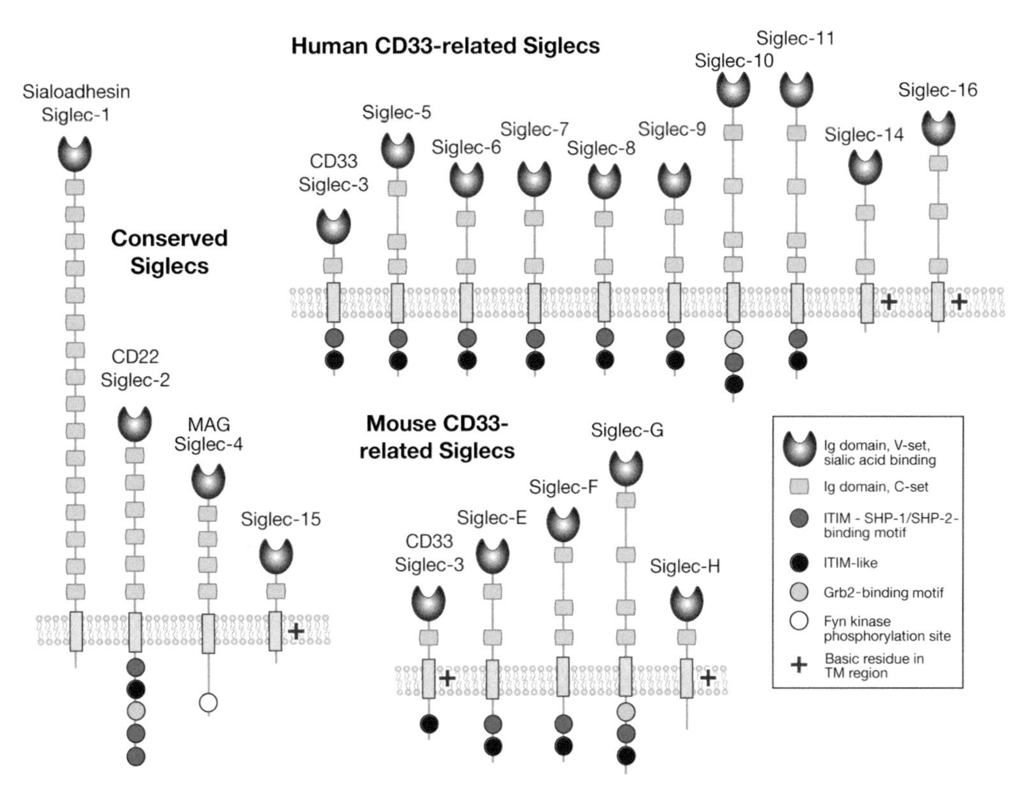

FIGURE 35.1. Domain structures of the known Siglecs in humans and mice. There are two subgroups of Siglecs: One group contains sialoadhesin (Siglec-1), CD22 (Siglec-2), MAG (Siglec-4), and Siglec-15, and the other group contains CD33-related Siglecs. In humans, Siglec-12 has lost its arginine residue required for sialic acid binding, Siglec-13 is absent, and Siglec-17 has been inactivated. The *red plus sign* indicates the presence of a positively charged residue in the transmembrane domain, which has been shown to interact with DNAX activation protein-12 (DAP12), which contains an immunoreceptor tyrosine-based activation motif (ITAM). ITIM, Immunoreceptor tyrosine-based inhibitory motif.

of the myeloid lineage, and PILR-β is also found on NK cells. PILR-α contains two immunoreceptor tyrosine-based inhibitory motifs (ITIMs) and delivers inhibitory signals by recruitment of inhibitory phosphatases. In contrast, PILR-β associates with DNAX activation protein-12 (DAP12), an adapter protein that has an immunoreceptor tyrosine-based activating motif (ITAM) and promotes innate immune cell activation. Platelet endothelial cell adhesion molecule (PECAM)-1 is composed of six extracellular Ig-like domains and selectively recognizes α2-6-linked Sia. PECAM-1 is broadly expressed on endothelial cells and leukocytes and plays multiple roles linked to leukocyte transendothelial migration, inflammation, and vascular biology. The neural cell adhesion molecule (NCAM) and basigin (CD147) have been claimed to recognize and bind oligomannose-type N-glycans on adjacent glycoproteins in the nervous system. NCAM contains five Ig-like domains followed by two fibronectin type III-like domains, and exerts essential functions in the development, plasticity, and regeneration of the nervous system. The different isoforms of basigin contain variable numbers of Ig-like domains and are widely expressed in different tissues where they have roles in reproduction, nervous system function, and immunity. Intercellular adhesion molecule (ICAM)-1 comprises five Ig-like domains and binds numerous ligands, including the integrins LFA-1 and Mac-1, hyaluronan and possibly certain mucin-type glycoproteins. Its expression on leukocytes and endothelial cells is enhanced on cellular activation. ICAM-1 plays crucial roles in cell–cell interactions, extravasation, inflammation, and host defense. Hemolin is an IgSF plasma protein from lepidopteran insects that binds lipopolysaccharide (LPS) from Gram-negative bacteria and lipoteichoic acid from Gram-positive bacteria. Hemolin appears to have two binding sites for LPS—one that interacts with the phosphate groups of lipid A and another that interacts with the O-specific glycan antigen and the outer-core glycans of LPS. There is indirect and less convincing evidence for interactions of other IgSF molecules with glycans, such as the peripheral myelin protein P0 with HNK-1, CD83 with Sias, and CD2 with Lewis x (Le^x). Overall, with the exception of the Siglecs and the PILRs, the direct assignment of an IgSF fold as the actual binding pocket for glycans has not been achieved. The rest of this chapter is devoted to Siglecs, the best-characterized I-type lectins.

COMMON FEATURES OF SIGLECS

The Amino-Terminal V-Set Sialic Acid–Binding Domain

All Siglecs are type-1 membrane proteins with a Sia-binding, amino-terminal V-set domain and varying numbers of C-set Ig domains that act as spacers, projecting the Sia-binding site away from the plasma membrane. The V-set domain and the adjacent C-set domain contain a small number of invariant amino acid residues, including an "essential" arginine on the F β-strand required for Sia binding, and an unusual organization of cysteine residues. Instead of the typical intersheet disulfide bond between the B and F β-strands, the V-set domain of Siglecs displays an intrasheet disulfide bond between the B and E β-strands, permitting increased separation between the β-sheets. The resulting exposure of hydrophobic residues allows specific interactions with constituents of Sia ligands. All Siglecs studied so far also appear to contain an additional unusual disulfide bond between the V-set domain and the adjacent C-set domain, which would be expected to promote tight packing at the interface between the first two Ig domains. Although the significance of this bond for ligand recognition is unclear, optimal Sia-binding activity of many Siglecs requires the adjacent C-set domain, probably for correct folding and stability.

Three-dimensional structures for the mouse Sn, human Siglec-7, and human CD33 V-set domains, the V-set and adjacent C-set domain of Siglec-5, first three Ig-like domains of human CD22, and all five Ig-like domains of mouse MAG have been determined by X-ray crystallography, in the presence and absence of Sia ligands (Figure 35.2). Along with the nuclear magnetic resonance (NMR) structure of Siglec-8, structural templates for Sia recognition by Siglecs appear to be shared among Siglecs. In all instances, an "essential" arginine residue is located in the middle of the F β-strand, making a bidentate salt bridge with the carboxylate of Sia. Siglecs also contain conserved hydrophobic amino acids on the A and G β-strands that interact with the N-acetyl group and glycerol side chain of Sia, respectively. Although all Siglecs appear to use a common template for recognizing Sias, their binding preferences for extended glycan chains vary greatly. The peptide loop between the C and C′ β-strands is highly variable among Siglecs and has a key role in determining their fine sugar specificity. For example, molecular grafting of the C–C′ loop between Siglec-7 and Siglec-9 resulted in switched sugar-binding specificities. Structural studies have shown that this loop appears to be highly flexible, being able to make specific and varied interactions with long glycan chains. In addition, V-set domain sequences are the most rapidly evolving regions in these molecules, likely explaining species differences in binding specificity.

Masking and Unmasking

The cell-surface glycocalyx of most vertebrate cells is richly decorated in glycoconjugates that contain Sias (Chapter 15). The high local concentration of Sias is likely to greatly exceed the K_d value of each Siglec, resulting in self-binding to sialoglycans on the same membrane (*cis*) that can be directly functional or that can "mask" the Sia-binding site from interacting with sialoglycans on

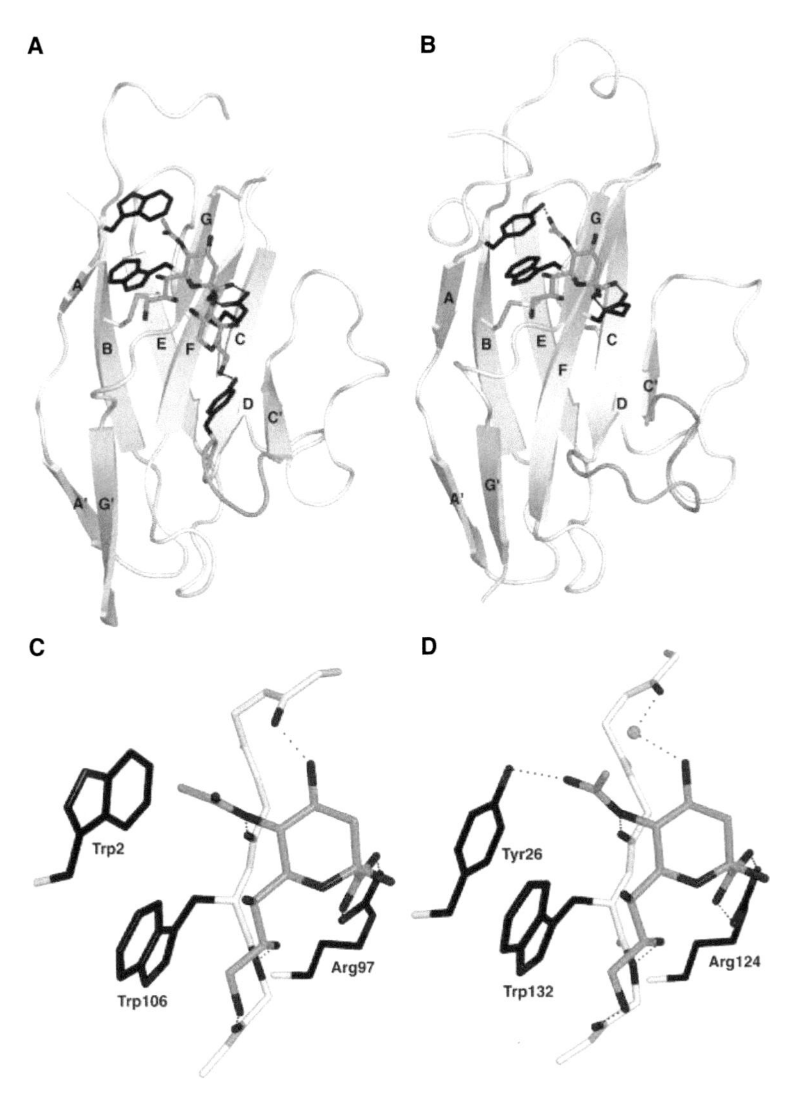

FIGURE 35.2. Structural basis of Siglec binding to ligands. X-ray crystal structures of the V-set domains of sialoadhesin (Sn) (*A*) and Siglec-7 (*B*) are shown complexed with sialic acid. (*C,D*) Molecular details of the interactions of sialic acid with Sn and Siglec-7. (Figure prepared by Dr. Helen Attrill.)

other cells (*trans*). Consequently, the Sia-dependent binding activity of most naturally expressed Siglecs is strongly enhanced following treatment of cells with sialidase to eliminate the *cis*-interacting sialylated glycans. A notable exception is Sn, which was discovered via its native property as a Sia-dependent cell-adhesion molecule on macrophages. The "masked" state of most Siglecs is a dynamic equilibrium with multiple *cis* ligands. Thus, an external probe, a cell surface or a pathogen bearing high-affinity ligands or very high densities of Sia residues, can effectively compete for binding to "masked" Siglecs. In addition, changes in expression of glycosyltransferases or sialidases of endogenous or exogenous origin could influence masking and unmasking of Siglecs at the cell surface, especially during immune and inflammatory responses.

Cell Type–Restricted Expression

Siglecs show restricted patterns of expression in single or related cell types (see below). This is most striking for the conserved Siglecs—Sn (macrophage), CD22 (B lymphocyte), MAG (myelin-forming cell), and Siglec-15 (osteoclast). This theme extends to some of the CD33rSiglecs, most notably in humans, including Siglec-6 on placental trophoblasts, Siglec-7 on NK (natural killer) cells, and Siglec-11 and -16 on tissue macrophages, including brain microglia in humans. In the mouse, Siglec-H and CD33 are excellent markers of plasmacytoid dendritic cells (DCs) and blood neutrophils, respectively, and Siglec-F is a useful marker of eosinophils and alveolar macrophages. These cell type–restricted expression patterns are thought to reflect discrete, cell-specific functions mediated by each of these Siglecs. However, certain key cells of the human immune system, such as monocytes and conventional DCs, express multiple CD33rSiglecs.

Tyrosine-Based Signaling Motifs

Most Siglecs have one or more tyrosine-based signaling motifs in their cytoplasmic tails or associate with membrane adaptor proteins containing cytosolic tyrosine motifs. The most prevalent motif is the ITIM with the consensus sequence (V/I/L)XYXX(L/V), where X is any amino acid. Up to 300 ITIM-containing membrane proteins are identified in the human genome, and many of these are established inhibitory receptors of the hematopoietic and immune systems. Following tyrosine phosphorylation by Src family kinases, they function by recruiting and activating SH2-domain-containing effectors, especially the protein tyrosine phosphatases SHP-1 and SHP-2. These counteract activating signals triggered by receptors containing ITAMs. Some Siglecs, notably Siglecs-14, -15, -16, and -H and mouse CD33, have a positively charged residue within the transmembrane region, which can associate with the DAP12 ITAM-containing adaptor, thereby mediating activating functions.

EXPRESSION PATTERNS AND FUNCTIONS OF THE CONSERVED SIGLECS

Sialoadhesin (Sn, Siglec-1, CD169)

Sn (CD169) was identified as a Sia-dependent sheep erythrocyte receptor expressed by mouse stromal macrophages. Sn has an unusually large number of Ig domains (seventeen), which are conserved among mammals and reptiles. These may be important for extending the Sia-binding site away from the plasma membrane to promote intercellular interactions. Sn prefers α2-3-linked Sias more than α2-6- and α2-8-linked Sias, and does not bind Sias modified by hydroxylation (Neu5Gc) or side-chain O-acetylation (e.g., $Neu5,9Ac_2$). Notably, this preference is very similar to the pattern of Sias expressed on commensal/pathogenic microbes (Neu5Ac>> Neu5Gc and α2-3>α2-8>>>α2-6).

In humans and mice, Sn expression appears specific for macrophage subsets, especially those resident in lymphoid tissues. These cells play important roles in antigen presentation to B cells and NKT cells, tolerance of self-reactive T cells but also as Trojan horses for viral infections to permit protective immune responses. Sn can also be strongly induced on monocytes, macrophages, and monocyte-derived DCs in vitro by type I interferons or agents such as viruses and Toll-like receptor (TLR) ligands that induce interferon production. Accordingly, Sn is up-regulated on circulating monocytes in human immunodeficiency virus (HIV)-infected individuals and on macrophages in patients with rheumatoid arthritis, primary biliary cirrhosis, and systemic lupus erythematosus (SLE). Sn expression on inflammatory macrophages has been associated with a favorable prognosis in colorectal cancer and with more severe disease in proliferative glomerulonephritis. Many of the above disease associations may reflect exposure of macrophages to interferons rather than being causally related. Indeed, in the BWF1 murine model of spontaneous SLE, there was no influence of Sn deficiency on disease severity. However, in mouse models of inherited neuropathy, autoimmune uveoretinitis, and experimental allergic encephalomyelitis (EAE), Sn-deficient mice showed reduced inflammation accompanied by reduced levels of T-cell and macrophage activation. These results suggest that Sn may suppress regulatory T-cell (Treg) expansion thereby promoting inflammation. Sn can also efficiently mediate the capture and uptake of exosomes released from B lymphocytes following apoptosis and therefore play a role in antigen presentation.

A role for Sn in phagocytic interactions of macrophages with various sialylated bacteria and protozoal pathogens has also been shown, including *Neisseria meningitidis*, *Campylobacter jejuni*, and *Trypanosoma cruzi*. A role in host protection was observed using an infection model with Group B streptococcus (GBS), in which Sn-deficient mice showed increased bacterial spread. Conversely, Sn expression on macrophages and monocyte-derived DCs can be exploited by enveloped viruses displaying host-derived Sias, leading to their capture, uptake, and dissemination. This mechanism was first seen with the porcine reproductive and respiratory syndrome virus, which targets lung alveolar macrophages of pigs, and more recently with HIV and other retroviruses. On HIV, Sn can recognize both sialylated gp120 glycoprotein and GM3, a monosialylated ganglioside terminating in Neu5Acα2-3Gal. GM3 is packaged into HIV during the budding from infected cells which occurs in lipid rafts. On monocyte-derived DCs, Sn interactions with GM3 on HIV lead to membrane invaginations containing viral particles that are very efficiently transferred to T cells in a process known as "*trans* infection."

CD22 (Siglec-2)

CD22 is a developmentally regulated cell-surface glycoprotein on B cells, expressed at approximately the time of Ig gene rearrangement and lost when mature B cells differentiate into plasma cells. CD22 has seven Ig-like domains and the intracellular region has six tyrosine-based signaling motifs, three of which could function as ITIMs. CD22 is a well-established negative regulator of B-cell activation, making an important contribution toward the threshold for signaling via the B-cell receptor (BCR) complex. Following BCR cross-linking, CD22 is rapidly tyrosine-phosphorylated on its ITIMs by the protein tyrosine kinase Lyn. This leads to recruitment and activation of the SHP-1 tyrosine phosphatase and subsequent inhibition of downstream signaling mediated via the BCR. Although some activating molecules are also recruited to the phosphorylated tyrosine motifs in CD22, the net phenotype of CD22-deficient mice is consistent with a primary role of CD22 in negative regulatory signaling, manifested by enhanced BCR-induced calcium signaling, enhanced B-cell turnover, reduced numbers of recirculating B cells in the bone marrow, and reduced numbers of marginal zone B cells.

Of all the Siglecs, CD22 has the most conserved specificity for sialylated ligands, binding primarily to α2-6-linked Sias of the type Neu5Ac(or Neu5Gc)α2-6Galβ1-4GlcNAc, which are common capping structures of many complex-type N-glycans. Additional specificity can be conferred by the nature of the Sia moiety as well as sulfation of the underlying glycan. Neither human nor mouse CD22 binds 9-O-acetylated Sias; mouse CD22 has a strong preference for Neu5Gc over Neu5Ac, whereas human CD22 binds both of the latter forms. Recombinant soluble CD22 can precipitate a subset of glycoproteins from B-cell lysates including CD45, a major sialoprotein of T and B cells. However, on B cells, CD22 appears to mainly be either *cis*-associated with other CD22 molecules in a glycan-dependent manner, or with the BCR, which it inhibits in a glycan-independent manner. These interactions are consistent with studies of mouse mutants that either lack CD22 glycan ligands or express mutated forms of CD22 unable to bind glycans. B cells from ST6Gal-I-deficient mice that lack CD22 ligands show an anergic phenotype, essentially the opposite of the phenotype observed in CD22-deficient mice. Mice expressing a lectin-inactive version of CD22 with a mutated binding site arginine also show reduced BCR signaling (anergy) and, similar to ST6Gal-I-deficient mice, they show increased CD22-BCR association and stronger CD22 phosphorylation (Figure 35.3). Likewise, *Cmah*-null mice deficient in Neu5Gc have reduced ligands for mouse CD22 and Siglec-G (see below) and show B cell hyperactivity.

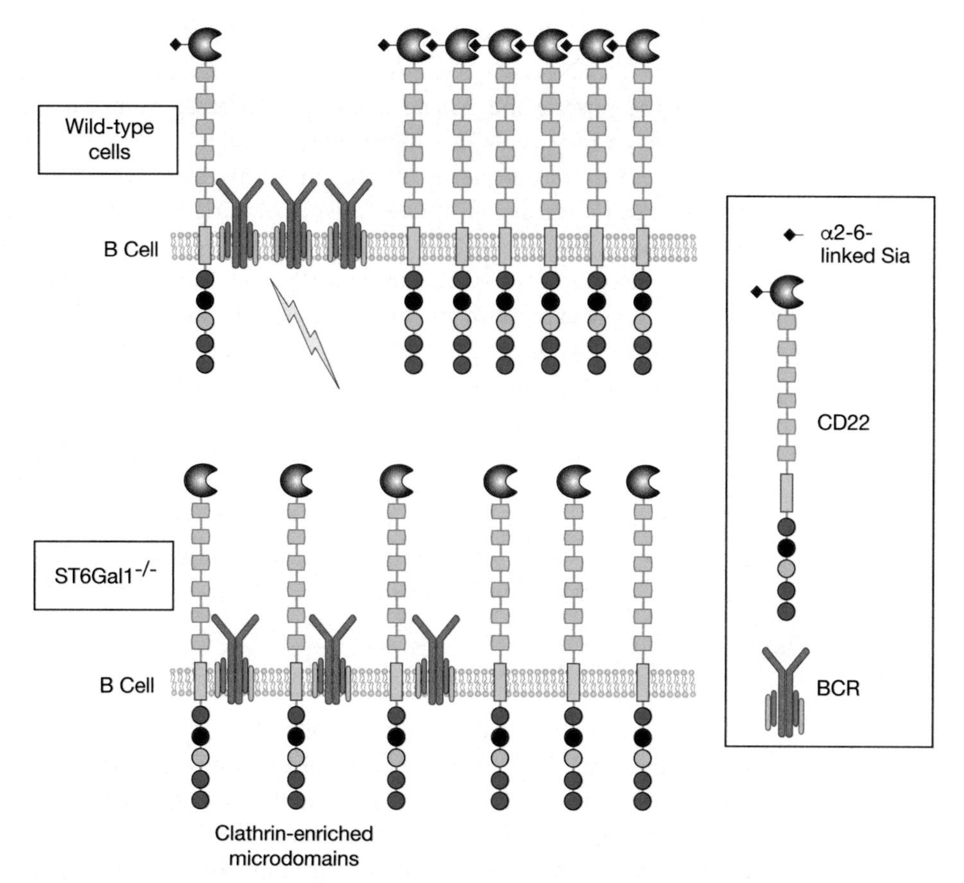

FIGURE 35.3. Proposed biological functions mediated by CD22. CD22 glycan-dependent homotypic interactions in equilibrium with CD22–BCR interactions. CD22 is clustered together by Sia-dependent homotypic interaction and kept away from BCR on wild-type B cells, allowing BCR-mediated cell activation. On the other hand, homotypic interaction among CD22 is lost on ST6Gal-I deficient B cells, tipping the balance in favor of Sia-independent interaction between CD22 and BCR, which suppresses BCR-mediated cell activation. The actual situation seems to vary between different cell types and analysis conditions. BCR, B-cell receptor; Sia, sialic acid.

Besides regulating B-cell functions via *cis*-interactions, CD22 can also mediate *trans*-interactions with sialylated ligands on other cells that sequester CD22 away from the BCR. This could be important for raising B-cell activation thresholds to "self"-antigens and may help to ensure that signaling through the BCR can only occur in lymphoid tissues in which CD22 α2-6-sialylated ligands are abundant. Although CD22-deficiency alone does not lead to extensive autoimmune reactions, mice deficient in both CD22 and the other major B-cell Siglec, Siglec-G, develop SLE-like symptoms, including production of IgG autoantibodies and glomerulonephritis.

The restricted expression and properties of CD22 make it an attractive therapeutic target. Antibody-based CD22-targeting agents, such as antibody–drug conjugate inotuzumab ozogamicin and recombinant antibody–toxin fusion protein moxetumomab pasudotox, have been approved for clinical use (for acute B-cell leukemia and hairy cell leukemia, respectively).

Myelin-Associated Glycoprotein (Siglec-4)

MAG, a minor constituent of central nervous system (CNS) and peripheral nervous system (PNS) myelin, has five Ig-like domains and is well conserved among vertebrates. It is expressed by myelin-forming cells: oligodendrocytes in the CNS and Schwann cells in the PNS. In mature myelinated axons it is found primarily on the innermost (periaxonal) myelin wrap, directly across from the axon, but not in the multilayers of compacted myelin. MAG-deficient mice develop normal myelin, but defects in myelin and axons increase as animals age, indicating a role for MAG in the maintenance of myelinated axons, rather than in the process of myelination. MAG-null mice display late-onset axonal degeneration leading to progressive loss of motor function. MAG-null mice fail to show characteristic myelin-induced increases in neurofilament phosphorylation and therefore reduced axon diameters, indicating that MAG signaling is required for optimal myelin–axon interactions. Mendelian disorders (autosomal recessive spastic paraplegia 75) caused by homozygous mutation in the *MAG* gene, including a point mutation in its Sia-binding arginine, have been reported, and clinical features of the disease are consistent with the roles of MAG revealed by the studies of *Mag*-deficient mice. MAG also directly inhibits neurite outgrowth from a wide variety of neuronal cell types in vitro. This contributes to the inhibitory activity of myelin on axon outgrowth after nervous system injury, hampering functional recovery.

Genetic and biochemical evidence indicate that gangliosides (sialylated glycolipids) are important physiological ligands for MAG, mediating both myelin–axon stability and inhibition of axon outgrowth. Recombinant forms of MAG bind selectively to the abundant axonal gangliosides GD1a and GT1b (Chapter 11). The phenotype of MAG-deficient mice is similar to that of mice lacking an N-acetylgalactosaminyltransferase (coded by the *B4galnt1* gene) required for synthesis of GD1a and GT1b, resulting in progressive motor dysfunction. Notably, human mutations in the same gene (*B4GALNT1*) result in hereditary spastic paraplegia, a progressive motor neuropathy reminiscent of that of *B4galnt1*-null mice. Binding of soluble MAG to some types of neurons, and subsequent MAG-mediated inhibition of neurite outgrowth, is Sia and ganglioside dependent. MAG-mediated inhibition of neurite outgrowth from *B4galnt1*-null mouse neurons is diminished, whereas MAG still inhibits neurite outgrowth from the neurons of mice lacking the “b-series” gangliosides (GD3 synthase-null; Chapter 11), which lack GT1b but express GD1a. These findings suggest that gangliosides GD1a or GT1b act as functional docking sites for MAG on neuronal cells. MAG binds to other axonal receptors including a family of GPI-anchored proteins (Nogo receptors NgR1 and NgR2) and paired immunoglobulin-like receptor B (PirB). Gangliosides, NgRs, and PirB may act independently or interactively as MAG receptors, linking MAG binding to axonal signaling in different neuronal cell types.

Siglec-15

Siglec-15 was first described as a highly conserved ancient Siglec in vertebrates. It lacks the typical arrangement of cysteines seen in the V-set Ig domain of other Siglecs and has an unusual intron–exon arrangement. Nevertheless, it can bind the Sialyl-Tn structure (Neu5Acα2-6 GalNAcα) and other structures containing a Neu5Acα2-6HexNAc determinant. It associates with DAP12 but also has an ITIM-like motif in its cytoplasmic tail. Although first reported on macrophages and DCs in human lymphoid tissues, Siglec-15 is most strongly expressed in osteoclasts and their precursors in which it plays an important role, together with receptor activator of nuclear factor (NF)-κB (RANK), in triggering osteoclast differentiation. Osteoclasts are key cells involved in bone degradation and share a common hemopoietic progenitor with macrophages. Mice lacking Siglec-15 show a mild osteopetrosis and impaired osteoclast differentiation. Specific antibodies against Siglec-15 phenocopy this state, because of antibody-induced internalization and degradation of Siglec-15. In addition, Siglec-15 is expressed in some tumor tissues (tumor-associated macrophages and cancer cells), and an antibody against Siglec-15 was shown to suppress the growth of a mouse melanoma in vivo. Siglec-15 therefore provides a novel target for diseases, such as menopause-related osteoporosis and cancer.

GENOMIC ORGANIZATION, EXPRESSION PATTERNS, AND FUNCTIONS OF THE CD33-RELATED SIGLECS

Genes encoding most of the CD33rSiglec subfamily are clustered on human chromosome 19q13.3-13.4 or the syntenic region of mouse chromosome 7. They include CD33 (Siglec-3), Siglecs-5 through -12, Siglec-14, and Siglec-16 in humans and CD33 and Siglecs-E, -F, -G, and -H in mice. Similar clusters are found in other mammals. It is difficult to assign all definitive orthologs between primates and rodents, resulting in different nomenclatures. One reason is that most IgSF domains are encoded by exons with phase-1 splice junctions, allowing exon shuffling without disrupting open reading frames, resulting in hybrid genes that are difficult to distinguish from similarly organized genes in other species. A second reason is that the Sia-binding V-set domains of the CD33rSiglecs are rapidly evolving, presumably to adjust their binding specificity to the rapid evolution of the endogenous host sialome, as well as evasion of binding by pathogens via molecular mimicry (Chapter 15) or specific protein-mediated interactions (see below). There are also multiple gene conversion events between adjacent genes and pseudogenes within this cluster. Of interest is the finding that humans show many CD33rSiglec differences compared with our closest evolutionary cousins (the chimpanzees), more than the differences between mice and rats, which shared a common ancestor much earlier (see below).

As mentioned earlier, an "essential" arginine residue in all known Siglecs is required for binding Sia-containing ligands. This residue is often mutated in nature, resulting in loss of binding ability. Examples include Siglec-12 in humans, Siglecs-5 and -14 in the chimpanzee, gorilla, and orangutan, Siglec-6 in the baboon, and Siglec-H in the rat. The common arginine codon (CGN, where N is any nucleotide) tends to be highly mutable because of the CpG sequence (which is prone to TpG or CpA transition via cytosine methylation–deamination). However, the frequency with which such events occur suggests that it might be a natural mechanism to eliminate Sia binding of a given Siglec, when such activity becomes inappropriate under changing evolutionary pressures, without causing a complete loss of the Siglec. Overall, it appears that this class of *SIGLEC* genes is subject to multiple "Red Queen" effects in evolution, in which evolutionary changes of sialyltransferases in response to the emergence of Sia-binding pathogens may lead to subsequent evolutionary changes of Siglec specificities (Figure 35.4).

Below, we provide a brief summary of the main features of the human CD33rSiglecs and, where relevant, their murine counterparts.

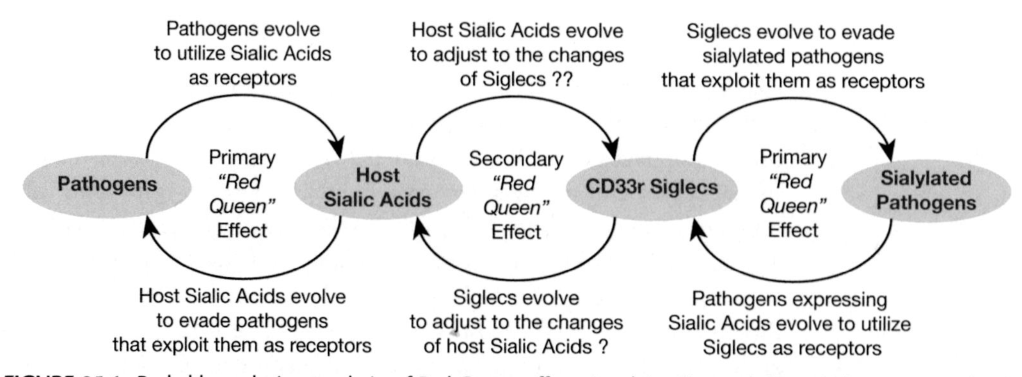

FIGURE 35.4. Probable evolutionary chain of Red Queen effects involving Sias and CD33rSiglecs. See text for discussion. (Redrawn, with permission, from Padler-Karavani V, et al. 2014. *FASEB J* **28:** 1280–1293.) To limit further complexity, two additional Red Queen forces driving rapid evolution are not shown. First, pathogens that encounter paired CD33rSiglecs need to evolve away from binding to such Siglecs. Second, some CD33rSiglecs and pathogens engage protein ligands via V-set domains, and subsequently evolve to adjust to such ligands.

CD33 (Siglec-3)

CD33 is a marker of early human myeloid progenitors and leukemic cells and is also expressed on monocytes and tissue macrophages, including brain microglia. It has two Ig domains and was the first of the CD33rSiglecs to be characterized as an inhibitory receptor, suppressing activation of FcγRI and recruiting SHP-1 and SHP-2 (Figure 35.5). CD33 has some preference for α2-6- rather than α2-3-sialylated glycans and binds strongly to sialylated ligands on myeloid leukemia cell lines. The restricted expression of CD33 has been exploited in the treatment of acute myeloid leukemia using gemtuzumab ozogamicin, a humanized anti-CD33 monoclonal antibody coupled to the toxic antibiotic calicheamicin. Binding of anti-CD33 mAbs to CD33 triggers endocytosis of the bound antibody. This depends on ITIM phosphorylation, recruitment of the E3 ligase Cbl, and ubiquitylation of the CD33 cytoplasmic tail. Selective expression of CD33 on leukemic progenitor cells also makes it an attractive target for therapy using chimeric antigen receptors expressed on cytotoxic T cells.

Recently, two coinherited single-nucleotide polymorphisms (SNPs) have been associated with protection of humans against late-onset Alzheimer's disease. These SNPs result in increased exon 2 skipping, leading to raised levels of CD33 lacking the V-set domain and reduced levels of full-length CD33. Because full-length CD33 can inhibit microglial cell uptake of Aβ protein in a Sia-dependent manner, it is thought that individuals lacking the protective SNPs may accumulate more toxic Aβ proteins, thus driving Alzheimer's disease pathology. Targeting CD33 using antibodies that either inhibit function or promote internalization and degradation may be a useful approach to Alzheimer's disease. A complexity arises because the truncated form of CD33 appears to be retained intracellularly in peroxisomes, and the biological consequence of this diversion is currently uncertain.

The murine ortholog of CD33 exists as two spliced forms that differ in the cytoplasmic region, neither containing the typical ITIM found in most other CD33rSiglecs. Furthermore, mouse CD33 has a lysine residue in the transmembrane sequence and couples to the DAP12 transmembrane adaptor, as shown for mouse Siglec-H and human Siglecs-14, -15, and -16. In contrast to human CD33, mouse CD33 in the blood is expressed mainly on neutrophils rather than monocytes, and at low levels in microglia, which also suggests a nonconserved function of this receptor.

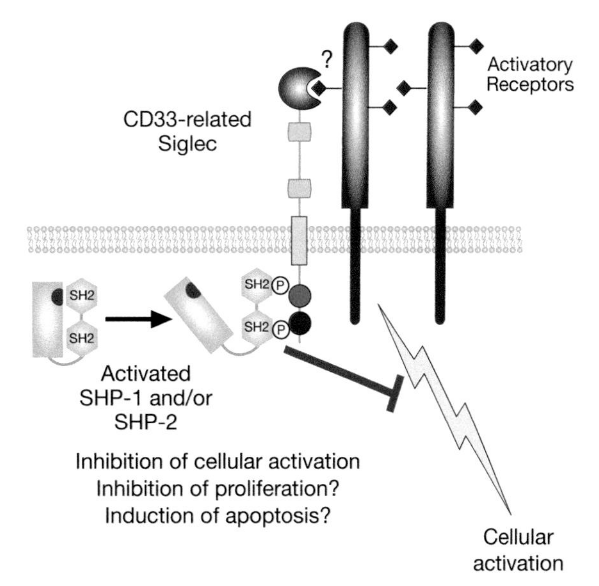

FIGURE 35.5. Proposed biological functions mediated by CD33-related Siglecs. A generic CD33-related Siglec is represented, showing the location of the immunoreceptor tyrosine-based inhibitory motif (ITIM) and the potential for inhibitory signaling.

Siglec-5 (CD170) and Siglec-14

The *SIGLEC5* and *SIGLEC14* genes are adjacent to each other on chromosome 19 and encode proteins containing four and three Ig-like domains, respectively. Because of ongoing gene conversions within most taxa, the first two Ig domains of Siglec-5 and Siglec-14 share >99% sequence identity but then diverge, with Siglec-5 being an inhibitory receptor with typical ITIMs, whereas Siglec-14 can complex with DAP12 and activate signaling. Both Siglec-5 and Siglec-14 bind similar ligands, with some preference for the sialyl-Tn structure (Neu5Acα2-6 GalNAcα) and α2-8-linked Sias. Although many antibodies to Siglec-5 cross-react with Siglec-14, specific antibodies revealed that although Siglec-5 is expressed on neutrophils and B cells, Siglec-14 is found on neutrophils and monocytes. A *SIGLEC14*-null allele is frequently present in Asian populations but is less common in Europeans. The null allele arose from a recombination event between the 5′ region of the *SIGLEC14* gene and the nearby 3′ region of the *SIGLEC5* gene, resulting in a fusion protein almost identical to Siglec-5 but expressed in a Siglec-14-like manner. Individuals with chronic obstructive pulmonary disease (COPD) who are also *SIGLEC14*-null showed reduced exacerbation attacks (sudden worsening of symptoms) compared with individuals expressing Siglec-14. Both Siglec-5 and Siglec-14 can bind sialylated strains of *Haemophilus influenzae* implicated in COPD exacerbations and trigger inhibition and activation responses, respectively. Thus, the absence of Siglec-14 on neutrophils would lead to reduced inflammatory responses in *SIGLEC14*-null individuals with the added impact of inhibitory Siglec-5 on monocytes. Besides expression on leukocytes, both Siglec-5 and Siglec-14 are found on human amniotic epithelium and may mediate dualistic responses to GBS infection and the frequency of preterm births in infected mothers. There are no obvious equivalents of Siglec-5 or Siglec-14 in mice, making it difficult to study this interesting pair of receptors in vivo.

Siglec-6

Siglec-6 was cloned from a human placental cDNA library and was also independently identified during a screen for proteins that bind leptin, a hormone that regulates body weight. It has three Ig-like domains and the typical arrangement of ITIM and ITIM-like motifs in its cytoplasmic tail. Low levels of Siglec-6 are expressed on B cells, but high expression is seen in the placental cytotrophoblasts and syncytiotrophoblasts in humans. Siglec-6 levels are increased in preterm deliveries associated with preeclampsia, but it is not yet known if there is a causal relationship. The heavily sialylated protein glycodelin is produced in the uterus and appears to bind Siglec-6 on cytotrophoblasts and to suppress their migration into decidua through inhibition of ERK and c-Jun signaling. A SNP in *SIGLEC6* is associated with SLE in Asians. Siglec-6 does not have an obvious ortholog in mice, but one is present in the chimpanzee and baboon. However, placental expression appears unique to humans.

Siglec-7, Siglec-9, and Siglec-E

Siglec-7 and Siglec-9 share a high degree of sequence similarity and appear to have evolved by gene duplication from an ancestral gene encoding a 3-Ig-domain inhibitory Siglec, represented in mice by Siglec-E. Siglec-7 is the major Siglec on human NK cells and is also seen at lower levels on monocytes, macrophages, DCs, and minor subsets of CD8 T cells. Disease-associated changes in expression patterns and levels of these Siglecs have been reported. Siglec-7 has also been detected in platelets, basophils, and mast cells in which it may modulate survival and activation. Siglec-9 is prominently expressed on neutrophils, monocytes, macrophages, and DCs, ~30% of NK cells, and minor subsets of CD4 and CD8 T cells. Based on glycan array binding experiments (Chapter 29), Siglec-7 binds strongly to α2-8-linked Sias present in “b-series” gangliosides (Chapter 11) and some glycoproteins, whereas Siglec-9 prefers α2-3-linked Sias. Sulfation of the sialyl-Le^x (SLe^x) structure can strongly influence recognition by both Siglecs, with Siglec-9 preferring 6-sulfo-SLe^x, and Siglec-7 binding well to both 6-sulfo-SLe^x and 6′-sulfo-SLe^x. Siglec-7 expression has also been reported in human pancreatic islet cell.

Siglec-E in mice shows a combination of some features of Siglec-7 and Siglec-9, being mainly expressed on neutrophils, monocytes, and macrophages, with Sia binding preferences that span those of both Siglec-7 and Siglec-9. Similar to T cells, NK cells in mice appear to lack expression of inhibitory Siglecs. Siglec-E is an important inhibitory receptor of neutrophils, as shown in multiple models such as LPS-induced lung inflammation in which Siglec-E-deficient mice show exaggerated CD11b-dependent neutrophil influx.

Tumor cells often up-regulate cell surface sialylated glycans and it appears that these may be important in Siglec-dependent dampening of antitumor immunity. Siglecs-7 and -9 can both suppress NK cell cytotoxicity against tumor cells expressing relevant glycan ligands. Siglec-9 and Siglec-E can also dampen neutrophil activation and tumor cell killing, whereas ligation of Siglec-9 or Siglec-E on macrophages by tumor-associated glycans seems to suppress formation of tumor-promoting M2 macrophages. In contrast to steady state, T cells in cancer patients express Siglec-9, and engagement of Siglec-9 by cancer-associated ligands may negatively impact tumor immunity, as do classic immune checkpoint receptor–ligand interactions. Overall Siglec effects on tumor cell biology may be dualistic, depending on the stage of tumor cell growth.

Studies with GBS have also shown that sialylated bacteria can subvert innate immune responses by targeting Siglec-9 or Siglec-E on neutrophils and macrophages, resulting in attenuation of phagocytosis, killing, and proinflammatory cytokine production. It has also been shown that myeloid Siglecs such as Siglec-9 and Siglec-E can modulate TLR signaling in response to pathogen ligands, leading to reduced secretion of proinflammatory mediators like TNF and increased production of anti-inflammatory cytokine, IL-10.

Siglec-8 and Siglec-F

Siglec-8 has three Ig domains and is expressed on eosinophils and mast cells, with weaker expression on basophils. It binds strongly to 6′-sulfo-SLe^x and to high-molecular-weight glycoproteins isolated from human airways. In mast cells, Siglec-8 ligation (with antibody) inhibits FcεRI-triggered degranulation responses, in line with its role as an inhibitory receptor. In eosinophils, Siglec-8 triggers apoptosis, which can occur following cross-linking with anti-Siglec-8 antibodies or sialoglycan polymers. Apoptosis depends on generation of reactive oxygen species and caspase activation, and is paradoxically enhanced in the presence of cytokine “survival” factors such as interleukin-5 (IL-5). The restricted expression of Siglec-8 on immune cells involved in allergy and the induction of cell death on antibody-mediated cross-linking make it an attractive therapeutic target for allergic diseases, and clinical trials of anti-Siglec-8 antibody in these diseases are ongoing.

Although there is no ortholog of Siglec-8 in mice, the four-Ig domain mouse Siglec-F is expressed in a similar way to Siglec-8 on eosinophils and has a similar glycan-binding preference for 6′-sulfo-SLe^x. It appears to have acquired similar functions through convergent evolution. There are some important differences, however. Siglec-F can recognize a broader range of α2-3-linked Sias, is also expressed on alveolar macrophages, and triggers weaker apoptosis using different signaling pathways.

Siglec-F-null mice show exaggerated eosinophilic responses in a lung allergy model, suggesting that its normal role is to dampen such responses. Interestingly, Siglec-F ligands in the airways and lung parenchyma were also up-regulated during allergic inflammation.

Siglec-10 and Siglec-G

Siglec-10 has five Ig-like domains, and in addition to the ITIM and ITIM-like motifs, displays an additional tyrosine-based motif (predicted to interact with Grb2) in its cytoplasmic tail. It is expressed at relatively low levels on several cells of the immune system, including B cells, monocytes, macrophages, and eosinophils. It can also be strongly up-regulated on tumor-infiltrating NK cells in hepatocellular carcinoma in which its expression was negatively associated with patient survival. It is the only CD33-related human Siglec that has a clear-cut ortholog in mice, designated Siglec-G. Both Siglec-10 and Siglec-G prefer Neu5Gc more than Neu5Ac in both α2-3 and α2-6 linkages. Similar to Siglec-10, Siglec-G is mainly expressed on B-cell subsets, DCs, and weakly on eosinophils. Mice deficient in Siglec-G show a 10-fold increase in numbers of a specialized subset of B lymphocytes, the B1a cells that make natural antibodies. These Siglec-G deficient B1a cells also show exaggerated Ca-fluxing following BCR cross-linking. Studies using "knock-in" mice carrying an inactivating mutation in the Sia-binding site of Siglec-G show a similar phenotype. This appears to be due to a requirement of Sia-dependent *cis*-interactions between Siglec-G and the BCR. On DCs, Siglec-G has been proposed to regulate cytokine responses to damage-associated molecular patterns (DAMPs) released by necrotic cells in sterile inflammation. This is thought to be due to a dampening effect of *cis*-interactions between Siglec-G and the heavily sialylated DAMP receptor, CD24. Disruption of this interaction through sialidases released by bacteria such as *Streptococcus pneumoniae* may be important in triggering inflammatory responses in sepsis. CD24 is overexpressed in some cancers and engages Siglec-10 on macrophages to suppress phagocytosis, in much the same way as does the classic "don't-eat-me" ligand–receptor pair, CD47 and SIRPα.

Siglec-11 and Siglec-16

Siglec-11 and Siglec-16 are paired inhibitory and activating receptors, with five and four Ig domains, respectively. In most humans, the *SIGLEC16* gene has a 4-base-pair deletion, and only ~35% of humans express one or two functional alleles. The extracellular regions of these proteins are >99% identical because of gene conversion events. Siglec-11 and Siglec-16 bind weakly to α2-8-linked Sias in vitro. These Siglecs appear to be absent from circulating leukocytes, but are expressed widely on populations of tissue macrophages, including resident microglia in the brain, where high levels of α2-8-linked Sias are present on gangliosides. Expression of Siglec-11 on microglia can impair their phagocytosis of apoptotic cells, yet alleviates neurotoxicity by suppressing neuroinflammation. Interestingly, microglial expression of Siglec-11 appears unique to humans. Siglec-16 is also present on microglia. These paired receptors exhibit dualistic responses to *Escherichia coli* K1, which expresses polysialic acid ligands.

HUMAN-SPECIFIC CHANGES IN SIGLEC BIOLOGY

The ancestral condition of some hominid Siglecs (e.g., Siglec-11) appears to have been preferential binding to Neu5Gc, a Sia specifically lost in human evolution ~2–3 million years ago (Chapter 15). This loss could have resulted in some Siglec unmasking, possibly leading to a state of heightened innate immune reactivity. Some human Siglecs have undergone an adjustment to allow increased Neu5Ac binding, and the question arises whether the adjustment is yet complete. Possibly, as a consequence of Neu5Gc loss, several Siglecs seem to have undergone human-specific changes in comparison to our great ape evolutionary cousins. For example, Sn is expressed on most human macrophages, whereas only subsets of chimpanzee macrophages are positive. This may be related to the fact that Sn has a strong binding preference for Neu5Ac rather than Neu5Gc, which is not synthesized by humans. Human Siglec-5 and Siglec-14 appear to have undergone a restoration of the "essential" arginine residue needed for Sia recognition, which is mutated in chimpanzees, gorillas, and orangutans. Siglec-7 is selectively expressed in human pancreatic islets. The gene encoding Siglec-11 has undergone a human-specific gene conversion, resulting in a new protein expression in brain microglia along with its paired receptor, Siglec-16. Siglec-12 has suffered a human-specific and human universal inactivation of the "essential" arginine residue with subsequent permanent pseudogenization by a frameshift in some humans. Siglec-13 underwent a human-specific gene deletion and Siglec-17 has a frameshift mutation, events possibly occurring close to the time of the common ancestor of modern humans. Expression patterns of some Siglecs have also undergone changes, with placental expression of Siglec-6 and amniotic epithelium expression of Siglec-5/14 being human-specific, and a general suppression of all CD33rSiglecs on human T cells compared with the chimpanzee. The functional implications of these human-specific changes in Siglec biology for physiology and disease are under exploration and may be related to human propensities for certain diseases (Chapter 15).

NONSIALYLATED LIGANDS FOR CD33-RELATED SIGLECS

Although CD33rSiglecs can modulate innate immune cell responses via recognition of endogenous sialoglycans as "self-associated molecular patterns" (SAMPs), many pathogens take advantage of this property by generating sialylated molecular mimics on their surfaces, using a variety of biochemical mechanisms. The naturally occurring "essential" arginine mutations mentioned above appear to be one way that the host can evade this subversion. However, pathogens in turn have evolved mechanisms to directly engage CD33rSiglecs via protein–protein interactions—for example, the β-protein of Type-Ia GBS engages human Siglec-5 and Siglec-14 independent of Sias. Early evidence indicates that more such examples exist, as well as alternate endogenous nonsialylated ligands within the host, such as the heat-shock protein HSP-70, cardiolipin, and another self-glycan, hyaluronan. As with many other classes of receptors whose original discovery was based on one canonical function, it is likely that evolutionary forces have generated many other functions for Siglecs, independent of their ability to recognize sialoglycans.

ACKNOWLEDGMENTS

The authors appreciate contributions of Paul Crocker to the earlier editions and helpful comments and suggestions from Pamela Stanley and Susan L. Bellis.

FURTHER READING

Powell LD, Varki A. 1995. I-type lectins. *J Biol Chem* **270**: 14243–14246. doi:10.1074/jbc.270.24.14243

van de Stolpe A, van der Saag PT. 1996. Intercellular adhesion molecule-1. *J Mol Med (Berl)* **74**: 13–33. doi:10.1007/bf00202069

Crocker PR, Clark EA, Filbin M, Gordon S, Jones Y, Kehrl JH, Kelm S, Le Douarin N, Powell L, Roder J, et al. 1998. Siglecs: a family of sialic-acid binding lectins. *Glycobiology* **8**: v–vi. doi:10.1093/oxfordjournals.glycob.a018832

Crocker PR, Varki A. 2001. Siglecs, sialic acids and innate immunity. *Trends Immunol* **22**: 337–342. doi:10.1016/s1471-4906(01)01930-5

Angata T, Brinkman-Van der Linden E. 2002. I-type lectins. *Biochim Biophys Acta* **1572**: 294–316. doi:10.1016/s0304-4165(02)00316-1

Kleene R, Schachner M. 2004. Glycans and neural cell interactions. *Nat Rev Neurosci* **5**: 195–208. doi:10.1038/nrn1349

Varki A, Angata T. 2006. Siglecs—the major subfamily of I-type lectins. *Glycobiology* **16**: p1R–27R. doi:10.1093/glycob/cwj008

Crocker PR, Paulson JC, Varki A. 2007. Siglecs and their roles in the immune system. *Nat Rev Immunol* **7**: 255–266. doi:10.1038/nri2056

Jandus C, Simon HU, von Gunten S. 2011. Targeting siglecs—a novel pharmacological strategy for immuno- and glycotherapy. *Biochem Pharmacol* **82**: 323–332. doi:10.1016/j.bcp.2011.05.018

Varki A. 2011. Since there are PAMPs and DAMPs, there must be SAMPs? Glycan "self-associated molecular patterns" dampen innate immunity, but pathogens can mimic them. *Glycobiology* **21**: 1121–1124. doi:10.1093/glycob/cwr087

Pillai S, Netravali IA, Cariappa A, Mattoo H. 2012. Siglecs and immune regulation. *Annu Rev Immunol* **30**: 357–392. doi:10.1146/annurev-immunol-020711-075018

Kitazume S, Imamaki R, Ogawa K, Taniguchi N. 2014. Sweet role of platelet endothelial cell adhesion molecule in understanding angiogenesis. *Glycobiology* **24**: 1260–1264. doi:10.1093/glycob/cwu094

Lu Q, Lu G, Qi J, Wang H, Xuan Y, Wang Q, Li Y, Zhang Y, Zheng C, Fan Z, Yan J, Gao GF. 2014. PILRα and PILRβ have a Siglec fold and provide the basis of binding to sialic acid. *Proc Natl Acad Sci* **111**: 8221–8226. doi:10.1073/pnas.1320716111

Macauley MS, Crocker PR, Paulson JC. 2014. Siglec-mediated regulation of immune cell function in disease. *Nat Rev Immunol* **14**: 653–666. doi:10.1038/nri3737

Müller J, Nitschke L. 2014. The role of CD22 and Siglec-G in B-cell tolerance and autoimmune disease. *Nat Rev Rheumatol* **10**: 422–428. doi:10.1038/nrrheum.2014.54

Angata T, Nycholat CM, Macauley MS. 2015. Therapeutic targeting of Siglecs using antibody- and glycan-based approaches. *Trends Pharmacol Sci* **36**: 645–660. doi:10.1016/j.tips.2015.06.008

Bochner BS, Zimmermann N. 2015. Role of siglecs and related glycan-binding proteins in immune responses and immunoregulation. *J Allergy Clin Immunol* **135**: 598–608. doi:10.1016/j.jaci.2014.11.031

Bochner BS. 2016. "Siglec"ting the allergic response for therapeutic targeting. *Glycobiology* **26**: 546–552. doi:10.1093/glycob/cww024

Bull C, Heise T, Adema GJ, Boltje TJ. 2016. Sialic acid mimetics to target the sialic acid–Siglec axis. *Trends Biochem Sci* **41**: 519–531. doi:10.1016/j.tibs.2016.03.007

Fraschilla I, Pillai S. 2017. Viewing Siglecs through the lens of tumor immunology. *Immunol Rev* **276**: 178–191. doi:10.1111/imr.12526

Siddiqui S, Schwarz F, Springer S, Khedri Z, Yu H, Deng L, Verhagen A, Naito-Matsui Y, Jiang W, Kim D, et al. 2017. Studies on the detection, expression, glycosylation, dimerization, and ligand binding properties of mouse Siglec-E. *J Biol Chem* **292**: 1029–1037. doi:10.1074/jbc.m116.738351

36 Galectins

Richard D. Cummings, Fu-Tong Liu, Gabriel A. Rabinovich, Sean R. Stowell, and Gerardo R. Vasta

Galectins are among the most widely expressed class of lectins in all organisms. They typically bind β-galactose-containing glycoconjugates and share primary structural homology in their carbohydrate-recognition domains (CRDs). Galectins have many biological functions, including roles in development, regulation of immune cell activities, and microbial recognition as part of the innate immune system. This chapter describes the diversity of the galectin family and presents an overview of what is known about their biosynthesis, secretion, and biological roles.

HISTORICAL BACKGROUND ON DISCOVERY OF GALECTINS

Following the discovery of the Ashwell–Morell asialoglycoprotein receptor (AMR) in the liver, many investigators sought other such receptors by affinity chromatography with immobilized asialoglycoproteins. In 1975, a novel lectin of ~15 kDa was isolated from electric organs of the electric eel, and named "electrolectin." This noncovalently linked homodimer displayed hemagglutinating activity toward trypsinized rabbit erythrocytes that was inhibitable by β-galactosides; inclusion of β-mercaptoethanol in the isolation buffers was required to maintain the binding activity. Soon thereafter, in 1976, similar β-galactoside-binding lectins were isolated from chicken muscle and from extracts of calf heart and lung (~15 kDa; now designated as galectin-1). These proteins were initially referred to as S-type lectins to denote their sulfhydryl

published by Cold Spring Harbor Laboratory Press; doi: 10.1101/glycobiology.4e.36

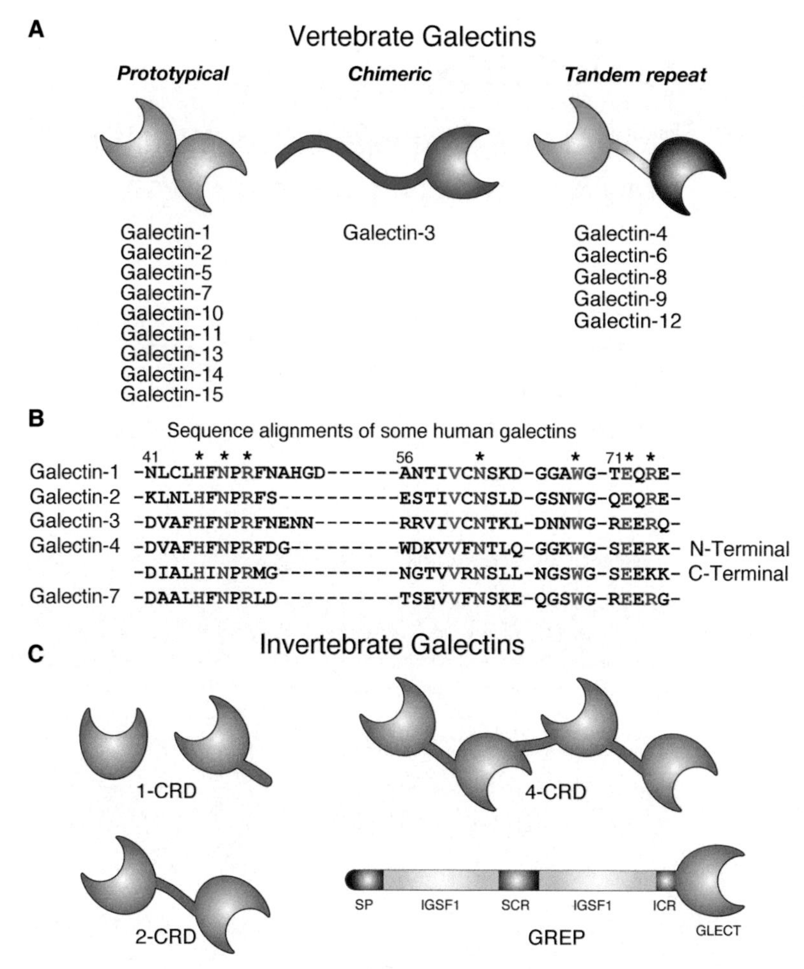

FIGURE 36.1. Different types of galectins in vertebrates and invertebrates and their organization and sequences. (*A*) Fifteen types of vertebrate galectins are known. (*B*) Sequence alignments of some representative human galectins are shown. The amino acid numbering is shown for galectin-1, which has 135 amino acids total, but the other galectins are aligned without showing their numbers. The highly conserved residues between galectins and those that are known to make contact with carbohydrate ligands are shown. (*C*) Examples of invertebrate galectins are shown. (GREP) Galectin-related protein, (N-terminal) amino-terminal, (C-terminal) carboxy-terminal.

dependency, presence of free cysteine residues, and solubility, along with shared primary sequence. In the early 1980s, a 35-kDa protein termed CBP35 that also bound to β-galactosides was identified in mouse fibroblasts. The same protein was studied by other groups under the names IgE-binding protein, L-29, and L-31 and is currently known as galectin-3.

The nomenclature for galectins was systematized in 1994 and the first galectin type found (~15 kDa, discussed above) retained the name galectin-1. All other members of this family were numbered consecutively by order of discovery. Multiple distinct galectins are usually expressed in a single species (up to 15 in the mouse). A list of the vertebrate galectins is shown in Figure 36.1A.

THE GALECTIN FAMILY

Galectins are identified based on a shared amino acid sequence motif in their CRDs. The CRD of galectins has ~130 amino acids, although only a small number of residues within the CRD directly contact glycan ligands. A comparison of more than 100 galectin sequences from many different sources reveals that eight residues, which have been shown to be involved in glycan binding by X-ray crystallographic analyses, are mostly invariant. In addition, another dozen residues appear to be highly conserved. Part of the conserved galectin sequence motif is shown in Figure 36.1B, along with a comparison of several human galectins. Many galectins contain variable numbers of free Cys residues, whose redox state is related to the stability of their binding activity. Interestingly, and unlike most galectins, electrolectin does not contain free Cys residues, but a key Trp in the binding site of its CRD can be oxidized, causing loss of activity.

Based on their CRD number and organization, galectins have been classified into three major groups: proto-type galectins (galectin-1, -2, -7, -10, -11, -13, -14, and -15), which contain a single CRD and may associate as homodimers; chimera-type galectins (galectin-3), which are characterized by having a single CRD and an amino-terminal polypeptide "tail" rich in proline, glycine, and tyrosine residues through which it can form oligomers; and tandem-repeat galectins (galectin-4, -8, -9, and -12), which comprise two protein domains that each carry a CRD connected by a peptide linker that can range from five to >50 amino acids in length. Galectin transcripts may be differentially spliced to generate multiple isoforms. Invertebrates have galectins with one, two, or four tandemly arrayed CRDs.

Some galectins appear to be species-specific, such as galectin-5 (prototype) and galectin-6 (tandem repeat) that are found in rodents but not humans, galectin-11 (ovagal11; prototype) that has only been reported in sheep, and galectin-15 that has only been found in sheep and goat. The amino-terminal domain of galectin-12 has significant homologies with other galectins, whereas the carboxy-terminal domain displays greater divergence from other galectins.

Several proteins with sequence homology to galectins, which do not bind typical β-galactosides, have been described as galectin-related proteins (GRPs). For example, galectin-10 is the Charcot–Leyden crystal protein, is highly abundant in eosinophil granules, and preferentially binds to β-mannosides. Thus, in retrospect, galectin-10 was actually the earliest galectin discovered. Several GRPs, including GRIFIN (galectin-related interfiber protein), PP-13, and PPL-13 that lack glycan-binding activity, are related to galectin-10. It is noteworthy that the DrGRIFIN homolog in the zebrafish (*Danio rerio*) is a functional β-galactoside-binding protein and, like its mammalian counterpart, is also highly expressed in the eye lens.

TAXONOMIC DISTRIBUTION AND EVOLUTION OF GALECTINS

Hallmarks of the galectin family are the extensive taxonomic distribution of its members, together with the evolutionary conservation of their primary structures, gene organization, and structure of the CRD. The early emergence and structural conservation of the galectin family in eukaryotic evolution is revealed in the fungus *Coprinopsis cinerea*, in the sponge *Geodia cydonium*, and in a protein that displays the galectin fold in the protozoan parasite *Toxoplasma gondii*. Galectin-like sequences have also been identified in plants (*Arabidopsis thaliana*) and some viruses that infect pigs and fish, such as the porcine adenovirus and lymphocystis disease virus. However, galectin-like proteins or sequence motifs in the latter should be interpreted with caution as they could result from horizontal gene transfer.

The evolution of the galectin family has been inferred from the rigorous analysis of primary structures, gene organization, and taxonomic distribution of family members identified in a variety of organisms, including fungi, invertebrates, and vertebrates. Among the arthropods and nematodes, galectins appear to be abundant in *Drosophila melanogaster* and *Caenorhabditis elegans* (6 and 26 candidate genes, respectively). Generally, the primary structures and domain organizations of galectins from invertebrates do not fit the canonical features of vertebrate galectins described above (Figure 36.1C).

Galectins are ubiquitous in vertebrates. In fish, galectins have been identified and characterized in numerous species. The zebrafish (*D. rerio*) genome has ortholog genes for all three galectin types (proto-, chimera-, and tandem-repeat) identified in mammals. Among amphibians, several galectin subtypes have been found in various tissues, mucus, and eggs from salamanders (the axolotl *Ambystoma mexicanum*), toads (*Rhinella arenarum*), and frogs (*Lithobates catesbeianus* and *Xenopus laevis*).

The evolution of galectins along the vertebrate taxa leading to mammals has been rationalized as the duplication of an early single-CRD galectin gene that would have led to a bi-CRD galectin gene. The amino- and carboxy-terminal CRDs subsequently diverged into two different subtypes, defined by an exon–intron structure (F4-CRD and F3-CRD). In this nomenclature, F refers to the β-sandwich strand structure discussed below. All vertebrate single-CRD galectins belong to either the F3- (e.g., galectin-1, -2, -3, -5) or F4- (e.g., galectin-7, -10, -13, -14) subtype, whereas the mammalian tandem-repeat galectins such as galectin-4, -6, -8, -9, and -12 contain both F4 and F3 subtypes. However, how the multiple CRD galectins from invertebrates relate to the vertebrate tandem-repeat galectins remains to be fully understood, but a preliminary phylogenetic analysis of the oyster galectin CvGal1 revealed that the individual CRDs cluster with the mammalian single-CRD galectins rather than with the tandem-repeat galectins, suggesting that the CvGal1 gene is the product of two consecutive gene duplications of a single-CRD galectin gene ancestral to the early invertebrate taxa that was conserved along the chordate lineages.

STRUCTURE OF GALECTINS

The crystal structures of several galectins in complexes with glycan ligands are known, including galectins-1, -2, and -7, the carboxy-terminal domain of galectin-3, and individual domains of galectins-4, -8, and -9. In all cases, the galectin CRD is composed of five- and six-stranded antiparallel β-sheets arranged in a β-sandwich or jelly-roll configuration that lacks an α-helix (Figure 36.2A). In the dimeric proteins, such as galectins-1, -2, and -7, the subunits are related by a twofold rotational axis perpendicular to the plane of the β-sheets. The glycan-binding sites in the CRD are located at opposite ends of the dimer, except the orientation of the subunits in the galectin-7 dimer differs from the other canonical galectin dimers. The compactly arranged structure of the CRD partly explains the protease resistance of the galectin CRD and the high degree of conservation and requirement for the 130 amino acids in the CRD.

The galectin-1 CRD displays highly specific interactions with Gal and GlcNAc residues in typical glycans. Interactions with glycans are generally through hydrogen bonding, electrostatic interactions, and van der Waals interactions through ring stacking with Gal and the highly conserved Trp residue (Figures 36.2B,C). In general, the open-ended structure of the glycan-binding site is predicted to allow access to extended Gal-containing glycans, such as the poly-*N*-acetyllactosamines and blood group–related structures (Chapter 14).

The binding preference of invertebrate galectins for blood group glycans can be illustrated by a unique structural feature of the binding site of the *C. elegans* and oyster galectins. They show a significantly shorter loop 4 that accommodates the 2′-fucosyl moiety that is a common feature of both the A and B blood groups (Figure 36.3A,B). The α(1-3)-linked GalNAc or Gal moieties of the A and B blood group tetrasaccharides, respectively, are recognized by the -NH at position 5 of the conserved Trp, and a hydrophobic pocket at the external side of the β3-strand (amino acids 30–38) recognizes the methyl of the -NAc group in α(1-3)GalNAc of the A1/2 antigens.

There are several known and predicted subsites on galectins near the carbohydrate-binding site that could serve to enhance affinity for more extended glycans. Some of these subsites have been identified by structural analysis of galectins binding to glycans and by glycan-binding

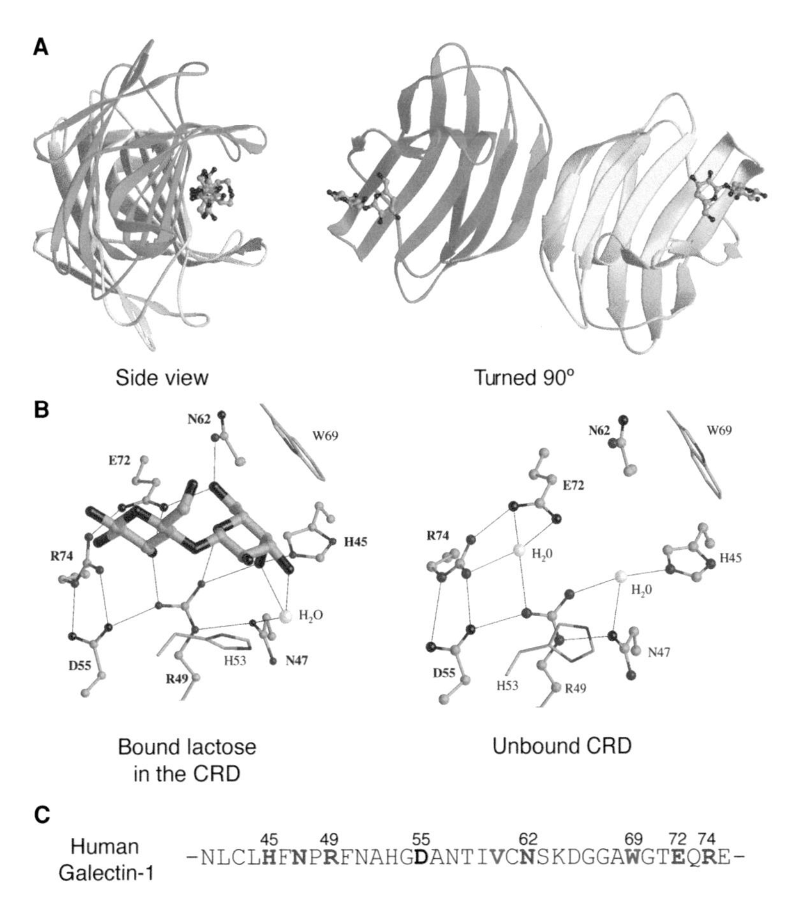

FIGURE 36.2. (*A*) Ribbon diagram of the crystal structure of human galectin-1 complexed with lactose. The homodimer is shown with each monomer colored differently and orthogonal views are presented. The subunit interface is based on interactions between the carboxy- and amino-terminal domains of each subunit. (*B*) A highlight of the interactions of key amino acid residues within the CRD with bound lactose, and the partial sequence of the CRD in human galectin-1. (*C*) Primary sequence of human galectin-1 with residues numbered and corresponding to those in the crystal structure. (Illustration kindly provided by Dr. Sean R. Stowell.)

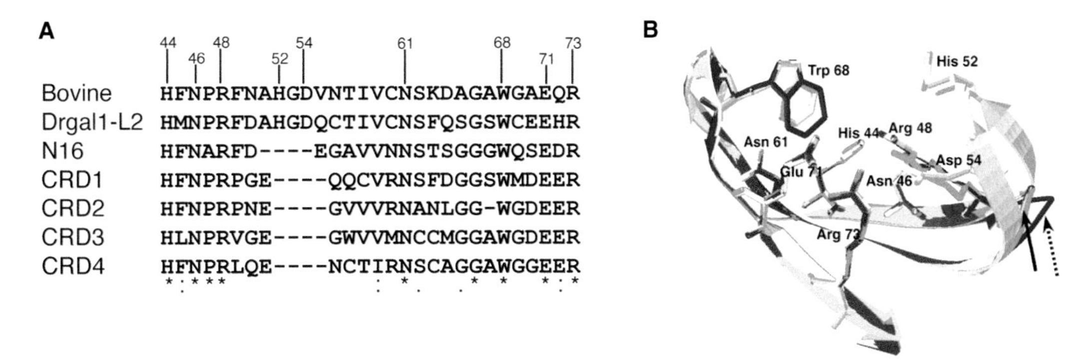

FIGURE 36.3. Structural aspects of galectins from mammals and invertebrates. (*A*) Sequence alignment of the CRD regions from bovine galectin-1, zebrafish Drgal1-L2, *Caenorhabditis elegans* N16, and the CRDs 1–4 of CvGal1. (*B*) Homology modeling of the CRDs from CvGal1: bovine galectin-1, CRD-1, -2, -3, and -4 are shown in *white, blue, yellow, red,* and *green*, respectively. Numbering of amino acid residues is based on bovine galectin-1. The *solid arrow* shows loop 4 of CvGal1 CRD-2, 3, 4, whereas a *dashed arrow* shows loop 4 of CRD-1. (*White ribbon*) Loop 4 in bovine galectin-1; (*solid arrow*) short loop 4 in *Crassotrea virginica* CvGal1 CRDs 1,2,3,4; (*dotted arrow*) short loop 4 in *C. elegans* N16.

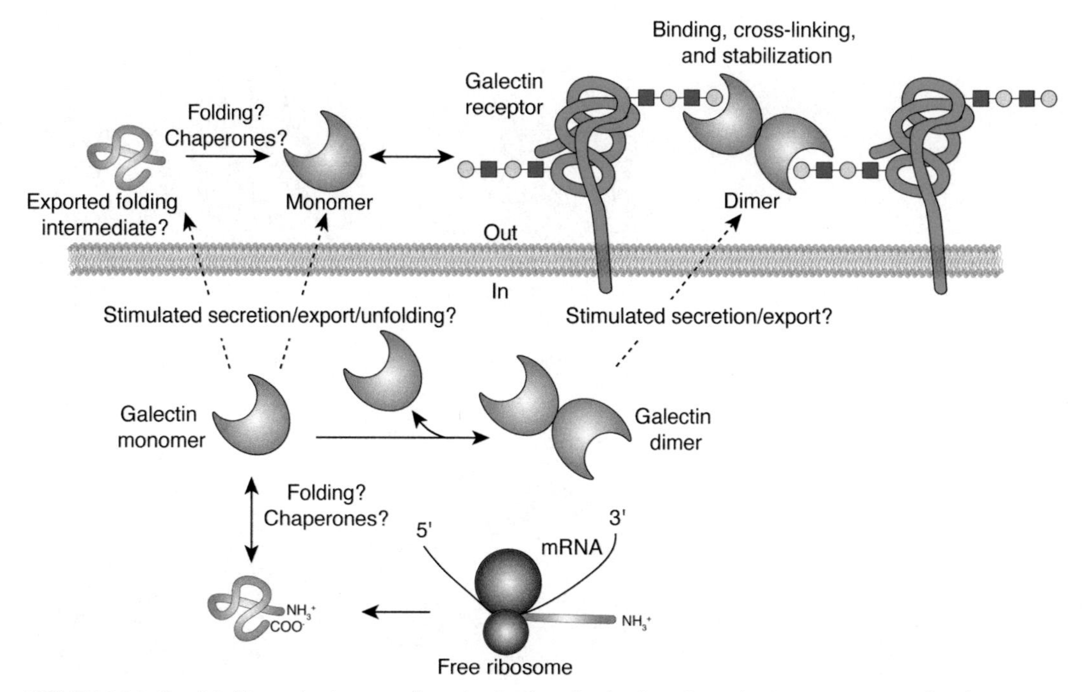

FIGURE 36.4. Possible biosynthetic routes for galectins in animal cells, using galectin-1 as an example. The mRNA for a galectin is translated on free polysomes in the cytoplasm and the newly synthesized protein is capable of binding glycan ligands or interacting with other proteins within the cell. After secretion or export from the cytoplasm by an undefined mechanism termed nonclassical export, the galectin appears to associate carbohydrate ligands that may stabilize its structure. The stable monomeric protein may form homodimers and other oligomers and interact with ligands on the cell surface and in the extracellular matrix, and directly interact with ligands on other cells.

studies. The crystal structure of bovine galectin-1 was derived in a complex with a biantennary N-glycan containing two terminal β-Gal residues. The N-glycan is bridged between two galectin-1 dimers, thus effectively creating a crystal latticework. This type of crystal latticework may be unique among vertebrate galectins and may be critical for their signaling and adhesive functions (Figures 36.4 and 36.5).

GLYCAN LIGANDS FOR GALECTINS

The interactions of galectins with glycans are complex and several factors contribute to high-affinity binding, including their natural multivalency and oligomeric state, as well as the multivalency of their natural ligands. Although most galectins bind weakly to simple β-galactosides, such as disaccharides or trisaccharides (K_d in the high micromolar to low millimolar range), they typically bind much stronger to natural glycoconjugate ligands (apparent K_d in the submicromolar range). Studies utilizing large libraries of glycans reveal that each galectin CRD recognizes different glycan ligands and shows highest affinity binding to different structures. For example, galectin-3 binds tightly to glycans with repeating [-3Galβ1-4GlcNAcβ-]$_n$ or poly-*N*-acetyllactosamine sequences containing 3-4 repeating units, regardless of the presence of terminal β-Gal residue, and binding is further enhanced if the penultimate β-Gal residues are substituted with Galα1-3, GalNAcα1-3, or Fucα1-2 residues. In contrast, human galectin-1 binds well only to glycans with terminal β-Gal residues, and does not bind blood group antigens. Galectin-8 has two CRDs within its single polypeptide and the two CRDs bind different glycans. Thus, the amino-terminal CRD of human galectin-8 binds α2-3-sialylated glycans with

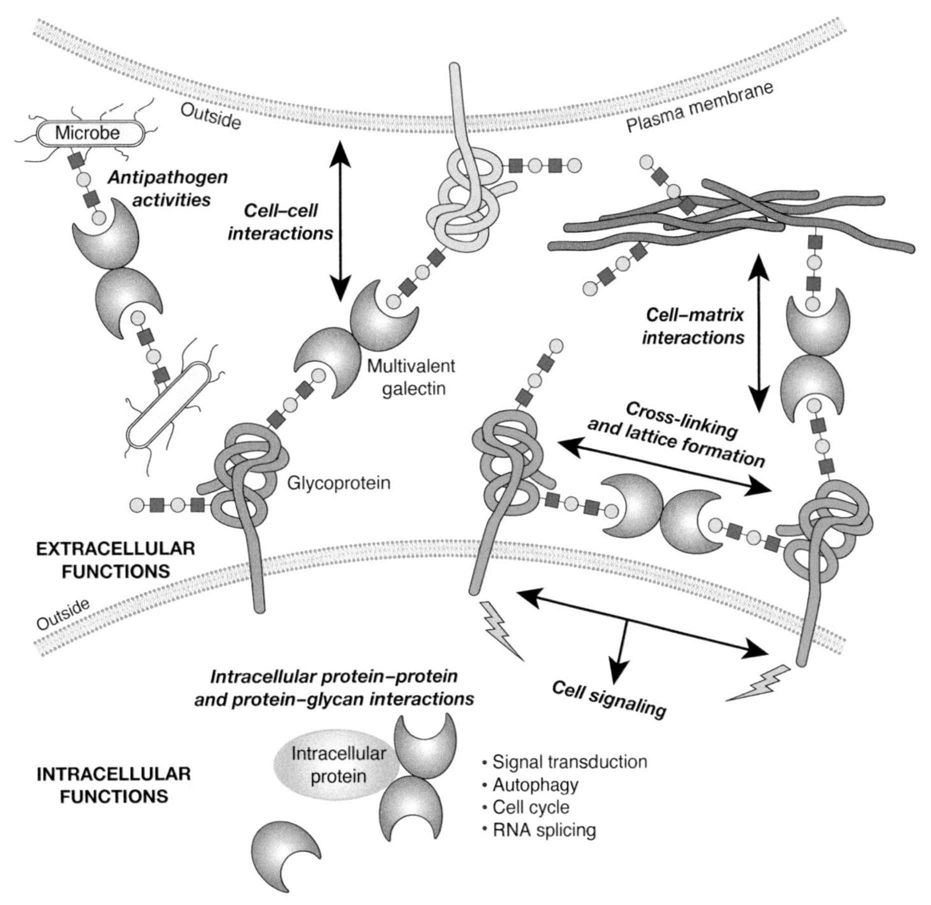

FIGURE 36.5. Functional interactions of galectins with cell-surface glycoconjugates and extracellular glycoconjugates leads to cell adhesion and cell signaling. Interactions of galectins with intracellular proteins may contribute to regulating intracellular pathways.

high affinity (K_d~50 nM), whereas the carboxy-terminal CRD binds to the blood group A determinant on a LacNAc core and does not bind sialylated glycans.

Cocrystallization of galectins with simple β-Gal-containing disaccharides has revealed that many galectins establish H-bonds to the C4 and C6 hydroxyl of Gal and the C3 hydroxyl of GlcNAc (Figure 36.3). The binding sites of galectins may be viewed as containing several subsites: one for Gal, another for GlcNAc, and other subsites that may be filled by other sugars and the aglycone moiety (e.g., protein or lipid). Thus, other sugar units linked to the disaccharide Gal-GlcNAc can enhance or decrease binding affinity, depending on the galectin. Some galectins from fungi, invertebrates, and mammals (such as galectins-4 and -8) preferentially bind selected ABO(H) blood group glycans as discussed above.

Galectins bind selectively to only some cell-surface and extracellular matrix ligands at low lectin concentrations but at higher concentrations they more often show broad binding to many different glycoproteins. However, the precise physiological ligands for each galectin and the physiological concentrations of galectins at sites of interactions are not well understood.

Galectins can also bind to exogenous ligands, such as exposed glycans on the surface of viruses, bacteria, fungi, and parasites, displaying a remarkable diversity of recognition. For example, galectin-1 can bind to complex type N-glycans on the HIV-1 gp120 envelope glycoprotein and to *Trichomonas vaginalis* lipophosphoglycan, whereas galectin-3 recognizes both terminal and internal *N*-acetyllactosamine units in lipopolysaccharides from meningococcus (*Neisseria meningitidis*), LacdiNAc from *Schistosoma mansoni*, and oligomannans from *Candida albicans*. The lipophosphoglycan from *Leishmania major* is recognized by both galectins-3 and -9, whereas galectin-4 recognizes *Escherichia coli* strains that display blood group B glycans.

BIOSYNTHESIS AND EXPORT OF GALECTINS

All members of the galectin family lack a classical signal sequence or membrane-anchoring domains and are synthesized on free polysomes in the cytoplasm before secretion. Galectins are unusual among all types of animal lectins in that they can be found in the nucleus, cytosol, outer plasma membrane, and extracellular matrix. One rather common modification of galectins in animal cells is blockage of the amino terminus, although galectin-3 has been shown to also have two serine residues in its unique amino-terminal region that can be phosphorylated. Newly synthesized galectins isolated directly from the cytosol of cells are functional in binding β-galactosides.

Interestingly, there is compelling evidence for galectins having nonglycan-binding partners in the cytosol. Galectin biosynthesis is unusual among all types of glycan-binding proteins as galectins are synthesized in the cytoplasm. The mechanism by which galectins can reach the outside of cells is poorly understood. A model of this complexity, as suggested for galectin-1, is illustrated schematically in Figure 36.4. Curiously, the export of galectins from cells does not involve direct movement through the secretory pathway. It appears the secretion and availability of glycan ligands in the cells generating the galectin may regulate export, activity, and stability of galectins.

BIOLOGICAL ROLES OF GALECTINS

Members of the galectin family display a striking functional diversification that ranges from key roles in early development, tissue regeneration, and cancer to immune homeostasis and recognition/effector functions against potential pathogens. By binding to endogenous glycans, galectins can contribute to cell–cell and cell–matrix interactions, and galectin signaling at the cell surface can also modulate cellular functions. In addition, intracellular galectins may, as in the case of galectin-3, regulate cellular activities and may contribute to some fundamental processes such as pre-mRNA splicing (Figure 36.5). The galectin-mediated recognition of glycans on the surface of potential pathogens and parasites discovered in the past few years has revealed their role(s) as pattern recognition receptors and effector factors in innate immune functions. Much information about the biological functions of galectins has been revealed largely through using galectin-deficient mouse models. Such studies, which have been performed primarily in mice deficient in galectins-1, -3, -7, and -9, have revealed roles of galectins in a wide range of physiological pathways (Chapter 41).

Roles of Galectins in Development and Tissue Regeneration

Studies in murine and zebrafish models revealed that during early embryonic stages, galectin-1 and -3 are expressed in the notochord and play subtle, albeit key roles in notochord development, somitogenesis, and development of muscle tissue and the central nervous system. Lack of galectin-3 in knockout mice is associated with several subtle and apparently unrelated phenotypic changes, whereas lack of galectin-1 in mice is associated with a different set of phenotypic changes, including decreased sensitivity to noxious thermal stimuli, altered primary afferent neural anatomy, and aberrant topography of olfactory axons. In addition, expression of galectin-3 at the onset of chondrification suggests its role in bone development. Recent studies have unraveled the critical role of galectin-1 in angiogenesis, as the basis for resistance of certain tumors to otherwise effective antivascular endothelial growth factor-targeted therapy. It should also be noted that some galectins are involved in regeneration of injured tissue. For example, on photoinduced retinal damage in the zebrafish, galectin-1 is one of the first proteins secreted by stem cells in the Müller glia, and is responsible for rod regeneration. Galectin-7 is important in the maintenance of epidermal homeostasis and skin repair.

Roles of Galectins in Apoptosis and Induction of Cell-Surface Phosphatidylserine Exposure

Several galectins induce apoptosis in some types of blood cells. For galectin-1, this activity has been studied mostly in human T cells, particularly Th1 and Th17 cells, in which apoptotic pathways may involve cell-surface glycoproteins, including CD7, CD29, CD45, and CD43. In contrast, for galectin-3 induction of apoptosis in T cells involves CD71 and CD45. In some cells, apoptotic signaling may function through down-regulation of Bcl-2 and activation of caspases. The interactions of galectin-9 with Tim-3 on Th1 cells may induce their apoptosis. In addition, some galectins, such as galectins-1, -2, and -4, also can induce exposure of cell-surface phosphatidylserine independently of apoptotic events. Moreover, overexpression of intracellular galectin-3 shows antiapoptotic activity, whereas overexpression of galectins-7 and -12 may promote apoptosis in cells. Some potential intracellular binding partners of galectins, especially for galectin-3, include several proteins involved in regulating apoptosis, such as Bcl-2, Fas receptor (CD95), synexin (a Ca^{++}- and phospholipid-binding protein), and Alix/Alg-2.

Roles of Galectins in Cancer

Many tumors overexpress various galectins and their heightened expression usually correlates with clinical aggressiveness of the tumor as well as progression to a metastatic phenotype. The immunosuppressive effects of galectin-1 can contribute to tumor survival, as revealed by knockdown studies, in which decreased galectin-1 expression is associated with decreased tumor survival, owing to increased survival of interferon- γ (IFN-γ)-producing Th1 cells and heightened T cell–mediated tumor rejection. Studies using galectin-1 knockout cell lines and mice have shown that expression of galectin-1 in tumor cell endothelium is essential for tumor angiogenesis, based on the binding of galectin-1 to complex N-glycans on VEGF receptor 2 (VEGFR2), and activation of VEGF-like signaling. Thus, galectins play important roles in tumor progression and metastasis through indirect effects in regulating tumor immune responses and direct effects in tumor angiogenesis. Although silencing of galectin-3 expression by methylation of its promoter is associated with early stages of prostate cancer, overexpression of galectin-3 correlates well with neoplastic transformation and tumor progression toward metastasis.

Roles of Galectins as Regulators of Immune Responses, Inflammation, and Adipogenesis

Galectins are important as regulators of immune and inflammatory responses. Galectins are expressed by a number of different immune and inflammatory cells. In addition, galectins can promote pro- or anti-inflammatory responses, depending on the inflammatory stimulus, microenvironment, and the target cell. Immune cell responses to galectins also depend on the specific glycosylation of surface glycoproteins to generate galectin ligands. However, many of these functions were shown by using recombinant galectins added to cultured cells. Whether these reflect the functions of endogenous galectins is not well understood.

Galectin-3, through its multivalent interactions, can form lattices with cell-surface glycoproteins—for example, with T-cell receptor (TCR) and the immune synapse (e.g., the interface between an antigen-presenting cell and a T/B cell or natural killer cell). Expression of such N-glycans is partly regulated by branching of N-glycans through β6 N-acetylglucosaminyltransferase V (MGAT5). *Mgat5* null T cells show enhanced clustering of TCRs suggesting that branched N-glycans interacting with galectin-3 may normally restrict TCR clustering and serve as a hindrance to the development of T-cell responses. The nature of glycan branching, interactions with galectins and perhaps other multivalent lectins, as well as metabolic regulation are intertwined. Interestingly, subcutaneous injections of recombinant human galectins-1 and -9 can reduce the severity of several autoimmune diseases in various animal models, but the underlying mechanisms for these in vivo immunosuppressive activities are unclear.

Studies of mice deficient in a certain galectin indicate that galectins have roles in immune responses and inflammation. Most studies suggest that endogenous galectin-1 functions as a homeostatic mechanism to counterbalance exuberant T-cell and dendritic-cell responses. Moreover, endogenous galectin-3 plays a role in phagocytosis by macrophages and mediator release as well as cytokine production by mast cells. These have been linked to this galectin functioning either extracellularly through binding to glycosylated receptors or intracellularly, in a fashion that is independent of glycan binding.

Current studies suggest a complex set of functions for galectin-3 in regulating many aspects of inflammatory responses and its role as a proinflammatory mediator. Although clearly complex, there have been some exciting developments linking galectins to inflammatory diseases in humans. Elevated serum galectin-3 levels are associated with a worse prognosis in patients with heart failure, and the plasma level of galectin-3 is a promising biomarker for acute coronary syndrome. Studies of galectin-3-knockout mice suggest a role for galectin-3 in cardiomyopathy and cardiac fibrosis. In other mouse disease models, endogenous galectin-3 contributes to fibrosis in various organs and tissues.

Recent studies have revealed critical and opposing roles of galectins-3 and -12 in adipogenesis and their association with obesity, inflammation, and type II diabetes. Galectin-12 is expressed in adipocytes and associated with lipid droplets, and functions as a negative regulator of lipolysis and insulin sensitivity. Galectin-12 knockout mice show increased lipolysis and reduced adiposity. In contrast, galectin-3 knockout mice display age-dependent increased adiposity, dysregulated glucose metabolism, and systemic inflammation. Furthermore, these mice show accelerated diabetes-associated kidney damage and diet-induced atherogenesis, with proinflammatory changes in visceral adipose tissue and pancreas.

Roles of Galectins as Recognition and Effector Factors on Infection

Galectins can also recognize exogenous glycans and related moieties on the surfaces of microbial pathogens and parasites, and function not only as pattern recognition receptors but also as effector factors, by directly inhibiting adhesion and/or entry into the host cell, or by directly killing or promoting phagocytosis of the potential pathogen. For example, galectin MjGal from the kuruma shrimp (*Penacus japonicus*) can recognize bacterial pathogens and promote their phagocytosis. Galectin-1 directly binds to dengue virus and inhibits viral adhesion and internalization into host cells. Similarly, the zebrafish galectins Drgal1-L2 and Drgal3-L1 interact directly with the glycosylated envelope of the infectious hematopoietic necrosis virus and with the glycans on the epithelial cell surface, significantly reducing viral adhesion. In mammals, the tandem-repeat galectins-4 and -8 can recognize and kill *E. coli* strains that display B-blood group oligosaccharides (BGB+*E. coli*). Mutation of key residues in either CRD revealed that the C-CRD mediates recognition the BGB+*E. coli* and promotes activity toward bacteria. Taken together, the results of these studies indicate that galectins can function not only in immune recognition but also as effector factors in innate immune responses against pathogens.

As lectins present in the cytosol with no commonly occurring glycan ligand, galectins are uniquely poised to detect abnormal cystosolic exposure of glycans during cell injury. Many intracellular bacteria are capable of inducing phagosomal lysis and escaping into the cytosol, after being phagocytosed into the host cells. This vacuolar lysis process exposes the host glycans that are initially present in the lumen of the phagosomes to the cytosolic milieu. Galectins-3 and -8 have been shown to bind to these exposed glycans, and galectin-8 has been shown to subsequently lead to autophagic activation for autophagosome degradation of the bacteria.

In contrast, some pathogens and parasites can "subvert" the roles of host galectins in immune defense, to attach to or gain entry into host cells. For example, galectin-1 promotes infection by HIV-1 by facilitating viral attachment to CD4 receptors, and increasing infection efficiency. *Leishmania* species, which are transmitted by sandflies, attach to the insect midgut epithelium via the sandfly galectin PpGalec that binds to Gal(β1-3) side chains on the *Leishmania* lipophosphoglycan (LPG), to prevent their excretion along with the digested bloodmeal, and allow them to differentiate into free-swimming infective metacyclics. Similarly, the protozoan parasite *Perkinsus marinus*, a facultative intracellular parasite of the eastern oyster *Crassostrea virginica* is recognized via the galectins CvGal1 and CvGal2, that are expressed by phagocytic hemocytes. The parasite is phagocytosed by the oyster hemocytes and proliferates, eventually causing systemic infection and death of the oyster host. Galectin-1 has been identified as the receptor for the protozoan parasite *T. vaginalis* and the bacterium *Chlamydia tracheomatis*, causative agents of the most prevalent nonviral sexually transmitted human infection in both women and men. Like *Leishmania* spp, *T. vaginalis* displays a surface LPG rich in Gal and GlcNAc, which is recognized in a glycan-dependent manner by galectin-1 expressed by the epithelial cells in the cervical linings (as well as placenta, prostate, endometrial, and decidual tissue), colonized by the parasite. Related types of studies also show the roles of galectins-1 and -3 in influenza/pneumococcal pneumonia infections of bronchoalveolar space, and that galectins might be involved in virus and bacterial binding. The expression of galectins in animal tissues is tightly regulated, but can be modulated by a variety of stimuli, including infectious challenge, and this may be especially important in innate immune responses.

ACKNOWLEDGMENTS

The authors acknowledge helpful comments and suggestions from Yasuhiro Ozeki.

FURTHER READING

Drickamer K. 1988. Two distinct classes of carbohydrate-recognition domains in animal lectins. *J Biol Chem* **263**: 9557–9560. doi:10.1016/s0021-9258(19)81549-1

Barondes SH, Castronovo V, Cooper DN, Cummings RD, Drickamer K, Feizi T, Gitt MA, Hirabayashi J, Hughes C, Kasai K, et al. 1994. Galectins: a family of animal β-galactoside-binding lectins. *Cell* **76**: 597–598. doi:10.1016/0092-8674(94)90498-7

Houzelstein D, Goncalves IR, Fadden AJ, Sidhu SS, Cooper DNW, Drickamer K, Leffler H, Poirier F. 2004. Phylogenetic analysis of the vertebrate galectin family. *Mol Biol Evol* **21**: 1177–1187. doi:10.1093/molbev/msh082

Liu FT, Rabinovich GA. 2005. Galectins as modulators of tumour progression. *Nat Rev Cancer* **5**: 29–41. doi:10.1038/nrc1527

van Die I, Cummings RD. 2006. Glycans modulate immune responses in helminth infections and allergy. *Chem Immunol Allergy* **90**: 91–112. doi:10.1159/000088883

Elola MT, Wolfenstein-Todel C, Troncosco MF, Vasta GR, Rabinovich GA. 2007. Galectins: matricellular glycan-binding proteins linking cell adhesion, migration, and survival. *Cell Mol Life Sci* **64**: 1679–1700. doi:10.1007/s00018-007-7044-8

Lau KS, Partridge EA, Grigorian A, Silvescu CI, Reinhold VN, Demetriou M, Dennis JW. 2007. Complex N-glycan number and degree of branching cooperate to regulate cell proliferation and differentiation. *Cell* **129**: 123–134. doi:10.1016/j.cell.2007.01.049

Mercier S, St-Pierre C, Pelletier I, Ouellet M, Tremblay MJ, Sato S. 2008. Galectin-1 promotes HIV-1 infectivity in macrophages through stabilization of viral adsorption. *Virology* **371**: 121–129. doi:10.1016/j.virol.2007.09.034

Stowell SR, Cummings RD. 2008. Interactions of galectins with leukocytes. In *Animal lectins: a functional view* (eds. GR Vasta, H Ahmed). CRC Press, Boca Raton, FL. doi:10.1201/9781420006971.ch28

Vasta GR. 2009. Roles of galectins in infection. *Nat Rev Microbiol* 7: 424–438. doi:10.1038/nrmicro2146

Dam TK, Brewer CF. 2010. Maintenance of cell surface glycan density by lectin-glycan interactions: a homeostatic and innate immune regulatory mechanism. *Glycobiology* **20**: 1061–1064. doi:10.1093/glycob/cwq084

Stowell SR, Arthur CM, Dias-Baruffi M, Rodrigues LC, Gourdine JP, Heimburg-Molinaro J, Ju T, Molinaro RJ, Rivera-Marrero C, Xia B, et al. 2010. Innate immune lectins kill bacteria expressing blood group antigen. *Nat Med* **16**: 295–301. doi:10.1038/nm.2103

Di Lella S, Sundblad V, Cerliani JP, Guardia CM, Estrin DA, Vasta GR, Rabinovich GA. 2011. When galectins recognize glycans: from biochemistry to physiology and back again. *Biochemistry* **50**: 7842–7857. doi:10.1021/bi201121m

Liu FT, Yang RY, Hsu DK. 2012. Galectins in acute and chronic inflammation. *Ann NY Acad Sci* **1253**: 80–91. doi:10.1111/j.1749-6632.2011.06386.x

Rabinovich GA, van Kooyk Y, Cobb BA. 2012. Glycobiology of immune responses. *Ann NY Acad Sci* **1253**: 1–15. doi:10.1111/j.1749-6632.2012.06492.x

Thurston TLM, Wandel MP, von Muhlinen N, Foeglein A, Randow F. 2012. Galectin 8 targets damaged vesicles for autophagy to defend cells against bacterial invasion. *Nature* **482**: 414–418. doi:10.1038/nature10744

Vasta GR, Ahmed H, Nita-Lazar M, Banerjee A, Pasek M, Shridhar S, Guha P, Fernández-Robledo JA. 2012. Galectins as self/non-self recognition receptors in innate and adaptive immunity: an unresolved paradox. *Front Immunol* 3: 199. doi:10.3389/fimmu.2012.00199

Yang RY, Havel PJ, Liu FT. 2012. Galectin-12: a protein associated with lipid droplets that regulates lipid metabolism and energy balance. *Adipocyte* **1**: 96–100. doi:10.4161/adip.19465

Chen HY, Weng IC, Hong MH, Liu FT. 2014. Galectins as bacterial sensors in the host innate response. *Curr Opin Microbiol* **17**: 75–81. doi:10.1016/j.mib.2013.11.006

Croci DO, Cerliani JP, Dalotto-Moreno T, Méndez-Huergo SP, Mascanfroni ID, Dergan-Dylon S, Toscano MA, Caramelo JJ, García-Vallejo JJ, Ouyang J, et al. 2014. Glycosylation-dependent lectin–receptor interactions preserve angiogenesis in anti-VEGF refractory tumors. *Cell* **156**: 744–758. doi:10.1016/j.cell.2014.01.043

Toledo KA, Fermino ML, Andrade CDC, Riul TB, Alves RT, Muller VDM, Russo RR, Stowell SR, Cummings RD, Aquino VH, Dias-Baruffi M. 2014. Galectin-1 exerts inhibitory effects during DENV-1 infection. *PLoS ONE* **9**: e112474. doi:10.1371/journal.pone.0112474

Vladoiu MC, Labrie M, St-Pierre Y. 2014. Intracellular galectins in cancer cells: potential new targets for therapy. *Int J Oncol* **44**: 1001–1014. doi:10.3892/ijo.2014.2267

Arthur CM, Baruffi MD, Cummings RD, Stowell SR. 2015. Evolving mechanistic insights into galectin functions. *Methods Mol Biol* **1207**: 1–35. doi:10.1007/978-1-4939-1396-1_1

Blidner AG, Méndez-Huergo SP, Cagnoni AJ, Rabinovich GA. 2015. Re-wiring regulatory cell networks in immunity by galectin–glycan interactions. *FEBS Lett* **589**: 3407–3418. doi:10.1016/j.febslet.2015.08.037

Than NG, Romero R, Balogh A, Karpati E, Mastrolia SA, Staretz-Chacham O, Hahn S, Erez O, Papp Z, Kim CJ. 2015. Galectins: double-edged swords in the cross-roads of pregnancy complications and female reproductive tract inflammation and neoplasia. *J Pathol Transl Med* **49**: 181–208. doi:10.4132/jptm.2015.02.25

Rabinovich GA, Conejo-García JR. 2016. Shaping the immune landscape in cancer by galectin-driven regulatory pathways. *J Mol Biol* **428**: 3266–3281. doi:10.1016/j.jmb.2016.03.021

Salvagno GL, Pavan C. 2016. Prognostic biomarkers in acute coronary syndrome. *Ann Transl Med* **4**: 258. doi:10.21037/atm.2016.06.36

Johannes L, Jacob R, Leffler H. 2018. Galectins at a glance. *J Cell Sci* **31**: jcs208884.

Popa SJ, Stewart SE, Moreau K. 2018. Unconventional secretion of annexins and galectins. *Sem Cell Dev Biol* **83**: 42–50. doi:10.1016/j.semcdb.2018.02.022

Mortales CL, Lee SU, Demetriou M. 2020. *N*-glycan branching is required for development of mature B cells. *J Immunol* **205**: 630–636. doi:10.4049/jimmunol.2000101

Vasta GR. 2020. Galectins in host–pathogen interactions: structural, functional and evolutionary aspects. *Adv Exp Med Biol* **1204**: 169–196. doi:10.1007/978-981-15-1580-4_7

Hong MH, Weng IC, Li FY, Lin WH, Liu FT. 2021. Intracellular galectins sense cytosolically exposed glycans as danger and mediate cellular responses. *J Biomed Sci* **28**: 16. doi:10.1186/s12929-021-00713-x

37 Microbial Lectins: Hemagglutinins, Adhesins, and Toxins

Amanda L. Lewis, Jennifer J. Kohler, and Markus Aebi

Symbiotic, commensal, and pathogenic microorganisms employ cell-surface glycans as targets for interaction with the host. Surface proteins (adhesins or agglutinins) mediate binding to such glycan "receptors," and many microbes depend on these interactions for success in the host. In addition, antagonistic interactions are mediated by secreted toxins that use surface glycan targets for internalization by various mechanisms. This chapter highlights examples of microbial lectins and their roles in pathogenicity. In addition, the modes of action of glycan-dependent toxins is discussed.

BACKGROUND

During the last 70 years, an enormous array of glycan-binding proteins (lectins) of viruses, bacteria, fungi, and protozoa have been discovered and characterized. Many of these microbial lectins were originally detected based on their ability to aggregate or induce the hemagglutination of red blood cells (erythrocytes). The influenza virus hemagglutinin was the first identified pathogen glycan-binding protein, shown by Alfred Gottschalk in the early 1950s to bind erythrocytes and other cells via the sialic acid component of host cell-surface glycoconjugates. Don Wiley and associates crystallized the influenza hemagglutinin and determined its structure in 1981. Later, they solved the structure of hemagglutinin cocrystals bound to sialyllactose, providing molecular insight into the affinity and specificity of the receptor-ligand binding sites.

Nathan Sharon and colleagues first described bacterial surface lectins in the 1970s. Their primary function is to facilitate the attachment or adherence of bacteria to host cells, a prerequisite for bacterial colonization and infection (Chapter 42). Thus, bacterial lectins are often called

published by Cold Spring Harbor Laboratory Press; doi: 10.1101/glycobiology.4e.37

adhesins, and these bind corresponding glycan receptors on the surface of the host cells via carbohydrate-recognition domains (CRDs) ("receptor" in this case is equivalent to "ligand" for animal cell lectins). Microbial adhesins can bind terminal sugar residues or internal sequences found in linear or branched oligosaccharide chains. These interactions are often an important determinant of host and tissue tropism. Detailed studies of the specificity of such microbial lectins have led to the identification and synthesis of powerful inhibitors of adhesion that may form the basis for novel therapeutic agents to combat infectious disease (Chapter 42). In fact, mammals often produce their own glycan decoys such as secreted mucins and milk oligosaccharides that help prevent binding of pathogen adhesins to their host cell targets.

Glycan-dependent interactions are important for both antagonistic and mutualistic interactions. For example, bacteria produce multimeric soluble toxins, whose glycan-binding subunits target and deliver the toxic cargo to the cell. In contrast, the normal flora of the lower gastrointestinal tract is determined by appropriate and desirable colonization by beneficial bacteria that engage in binding and digestion of host and dietary glycans. Likewise, the initial formation of nitrogen-fixing nodules in leguminous root tips by species of *Rhizobium* involves lectins on the root tip binding to glycan-containing Nod factors generated by the bacterium (Chapter 24).

VIRAL GLYCAN-BINDING PROTEINS

The first discovered and most well-studied example of a viral glycan-binding protein is the influenza virus hemagglutinin. Similar to most other lectin–glycan interactions, the affinity of this sialic acid–binding lectin is low. Avidity is increased by hemagglutinin trimerization and a high density of host cell surface glycan receptors. Binding is required for viral endocytosis and the subsequent pH-dependent fusion of the viral envelope with the endosomal membrane, ultimately triggering release of the viral RNA into the cytosol. The specificity of the host glycan–hemagglutinin interaction varies considerably for different subtypes of influenza. For example, human strains of influenza-A and -B viruses bind primarily to cells containing *N*-acetylneuraminic acid (Neu5Acα)2–6Gal-containing receptors. However, avian influenza viruses preferentially bind to receptors expressing Neu5Acα2–3Gal-, and porcine strains bind to both Neu5Acα2–6Gal- and -3Gal-containing receptors (Table 37.1). This linkage preference is a result of structural differences in the hemagglutinin (Figure 37.1). Viral adherence also depends on receptor abundance, such that tracheal epithelial cells in humans express glycans with a preponderance of Neu5Acα2–6Gal linkages, whereas other deeper airway surfaces contain many more Neu5Acα2–3Gal-terminated glycans. Thus, the specificity of the hemagglutinin determines the tropism of the virus with respect to species and target cells. As a result, avian-derived strains have a greater likelihood of producing lower respiratory tract infection (pneumonia) than common human strains because of the high abundance of Neu5Acα2-3Gal linkages deep in the lung. Conversely, avian-derived strains have a reduced likelihood of human-to-human transmission compared to human strains because of the paucity of Neu5Acα2-3Gal in the upper respiratory tract. The hemagglutinin is the major antigen against which neutralizing antibodies are produced and influenza strains continuously acquire genetic changes that affect both glycan binding and antigenicity. Such changes are responsible in part for new outbreaks and are considered during formulation of the annual influenza vaccine.

In addition to the hemagglutinin "H," influenza-A and -B virions express a sialidase (traditionally called neuraminidase "N", Chapter 15) that cleaves sialic acids from glycoconjugates. Its functions may include (1) prevention of viral aggregation by removal of sialic acid residues from virion envelope glycoproteins; (2) dissociation of newly synthesized virions inside the cell or as they bud from the cell surface; and (3) desialylation of soluble mucins at sites of infection to improve access to cell surface sialic acids. Inhibitors have been designed based on the crystal

TABLE 37.1. Examples of viral lectins and hemagglutinins

Virus	Lectin	Glycan-receptor preference	Site of infection
Myxoviruses			
Influenza A and B (human, ferret, and porcine)	hemagglutinin	Neu5Acα2-6Gal-	upper respiratory tract mucosa (tracheal epithelial cells)
Influenza A and B (avian and porcine)	hemagglutinin	Neu5Acα2-3Gal-	intestinal mucosa
Influenza C	hemagglutinin-esterase	9-O-acetyl-Siaα-	unknown
Newcastle disease	hemagglutinin-neuraminidase	Neu5Acα2-3Gal-	unknown
Sendai virus	hemagglutinin-neuraminidase	Neu5Acα2-8Neu5Ac-	upper respiratory tract mucosa
Polyomaviruses			
Polyoma	capsid protein VP1	Neu5Acα2-3Gal-, Neu5Acα2-3Galβ1-3 (Neu5Acα2-6)GalNAc- on gangliosides such as GM1 and GT1b/GD1a	kidney and brain glial cells
Herpesviruses			
Herpes simplex	glycoproteins gB, gC, and gD	3-O-sulfated heparan sulfate	mucosal surfaces of the mouth, eyes, genital, and respiratory tracts
Picornaviruses			
Foot-and-mouth disease (enterovirus)	caspid proteins	heparan sulfate	gastrointestinal and upper respiratory tracts
Retroviruses			
HIV	gp120 V3 loop	heparan sulfate	CD4 lymphocytes
Flaviviruses			
Dengue	envelope protein	heparan sulfate	macrophages?
Calciviruses			
Norovirus	capsid proteins	fucose, GalNAc, or Gal on A and B blood group antigens	secretory cells of the intestinal epithelium

structure of the sialidase from influenza-A virus. Some of these (e.g., oseltamivir) inhibit the enzyme activity at nanomolar concentrations and are used clinically as antiviral agents (Chapter 57). Influenza-C virions (and many coronaviruses) contain a single "hemagglutinin-esterase" that possesses both hemagglutinin and receptor-destroying activities, which in this case is an esterase that cleaves the 9-O-acetyl group from the target O-acetylated sialic acid receptors (Table 37.1). Some coronaviruses have evolved to adapt the spike "S" proteins to target O-acetylated sialic acid receptors. The severe acute respiratory syndrome coronavirus 2 (SARS-CoV-2), causative agent of the COVID-19 pandemic, has completely lost its hemagglutinin-esterase protein, but encodes an S protein that has evolved further, to bind more robustly to heparan sulfate (HS) proteoglycans, which can also act as a cofactor facilitating binding to a high-affinity protein receptor ACE2.

Rotaviruses, the major killer of children worldwide, can also bind to sialic acid residues. These viruses only bind to the intestinal epithelium of newborn infants during a period that appears to correlate with the expression of specific types and arrangements of sialic acids on glycoproteins. Claims regarding "sialic acid–independent" rotaviruses may be explained by internal sialic acids resistant to bacterial sialidases. Many other viruses (e.g., adenovirus, reovirus, Sendai virus, and polyomavirus) also use sialic acids for infection, and crystal structures are now available for several of their sialic acid–binding domains.

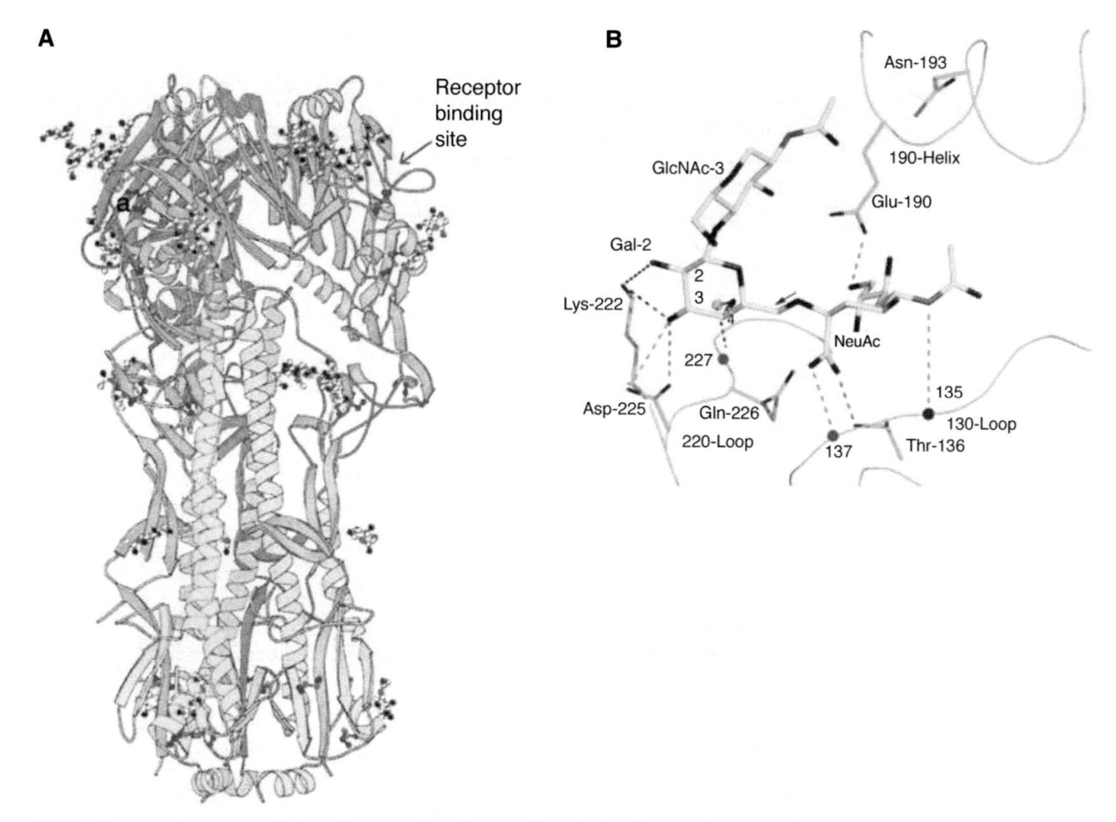

FIGURE 37.1. Structure of the influenza virus hemagglutinin (HA) ectodomain. (*A*) A schematic diagram of the trimeric ectodomain of the H3 avian HA from A/duck/Ukr/63 showing residues HA1 9–326 and HA2 1–172. Modeled carbohydrate side chains (*gray, red, blue*); disulfide bonds (*black, green*). The six polypeptide chains are shown in *light blue* (HA1), *magenta* (HA2), *dark blue* (HA1′), *light red* (HA2′), *green* (HA1″), and *yellow* (HA2″). (Redrawn, with permission of Elsevier, from Ha Y, et al. 2003. *Virology* **309:** 209–218.) (*B*) Receptor binding site of human influenza virus HA in complex with the human trisaccharide receptor NeuAcα2-6Galβ1-4GlcNAc. Hydrogen bonds (*dashed lines*); residues making interactions via main-chain carbonyl groups (*red spheres*) or nitrogens (*blue spheres*); trisaccharide carbon (*yellow*), nitrogen (*blue*), and oxygen (*red*) atoms; water molecules (*green spheres*). (Redrawn, with permission of AAAS, from Gamblin SJ, et al. 2004. *Science* **303:** 1838–1842.)

A number of viruses, including herpes simplex virus (HSV), foot-and-mouth disease virus, human immunodeficiency virus (HIV), and dengue flavivirus, use heparan sulfate proteoglycans as adhesion receptors (Table 37.1). In many cases, the proteoglycans may be part of a coreceptor system in which viruses make initial contact with a cell-surface proteoglycan and later with another receptor. For example, HSV infection is thought to initially involve the binding of viral glycoproteins gB and/or gC to cell-surface heparan sulfate proteoglycans. Glycoprotein gB promotes virus-cell fusion, syncytium formation (cell–cell fusion), and adherence, whereas gC binds to the C3b component of complement and blocks complement-mediated inhibition of the virus. These events are followed by HSV glycoprotein gD binding to one of several cell-surface receptors, including protein receptors and heparan sulfate, ultimately leading to fusion of the viral envelope with the host-cell plasma membrane. Interestingly, the interaction of gD with heparan sulfate shows specificity for a particular substructure in heparan sulfate containing a 3-O-sulfated glucosamine residue, the formation of which is catalyzed by specific isozymes of the glucosaminyl 3-O-sulfotransferase gene family. Thus, the heparan sulfate–binding adhesins appear to recognize carbohydrate units within the polysaccharide chains as opposed to binding terminal sugars.

Dengue flavivirus, the causative agent of dengue hemorrhagic fever, also binds to heparan sulfate. Computational modeling to compare the primary dengue viral envelope protein

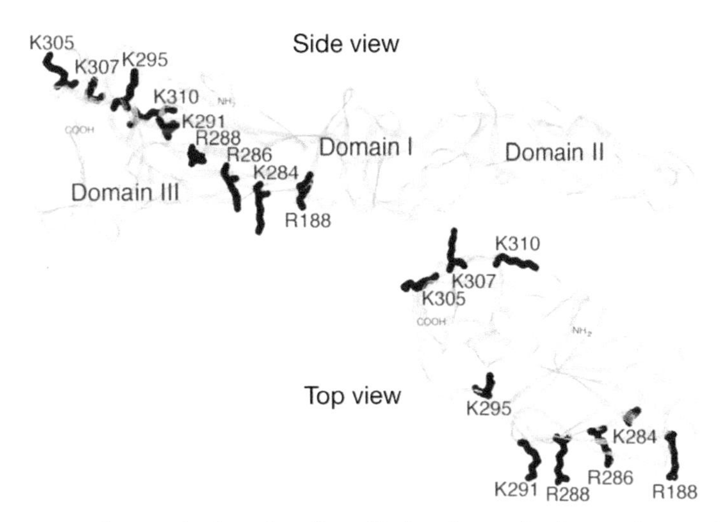

FIGURE 37.2. Two views of a putative heparin sulfate–binding site on the dengue virus envelope protein. The envelope protein monomer is shown in *ribbon* form, displayed along its longitudinal axis and as an external side view. Note the alignment of positively charged amino acids along an open face of the protein (*top*). (Redrawn, with permission, from Chen YP, et al. 1997. *Nat Med* **3:** 866–871, © Macmillan Publishers Ltd.)

sequence with the crystal structure of a related virion envelope protein suggests that the heparan sulfate–binding site may lie along a positively charged amino acid groove (Figure 37.2). The recently emerging Zika virus belongs to the same family. Furthermore, HIV can bind heparan sulfate and other sulfated polysaccharides by way of the V3 loop of its gp120 glycoprotein.

BACTERIAL ADHESION TO GLYCANS

Bacterial lectins occur commonly in the form of elongated multisubunit protein appendages, known as fimbriae (hairs) or pili (threads), which interact with glycoprotein and glycolipid receptors on host cells. Similar to viral glycan-binding proteins, adhesin-receptor binding is generally of low affinity. Because the adhesins and the receptors often cluster in the plane of the membrane, the resulting combinatorial avidity can be great. An analogy for adhesin–receptor binding is the interaction of the two faces of Velcro strips. The most well-characterized bacterial lectins include the mannose-specific type-1 fimbriae, the galabiose-specific P fimbriae, and the *N*-acetylglucosamine-binding F-17 fimbriae, which are produced by different strains of *Escherichia coli*. Fimbriated bacteria express 100 to 400 of these appendages, which typically have a diameter of 5–7 nm and can extend hundreds of nanometers in length (Figure 37.3)

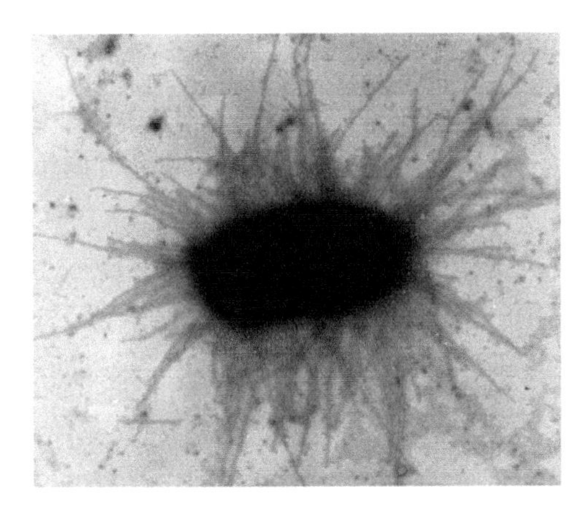

FIGURE 37.3. *Escherichia coli* express hundreds of pili, as indicated by the fine filaments extending from the bacterium. (Reprinted, with permission of Elsevier, from Sharon N. 2006. *Biochim Biophys Acta* **1760:** 527–537; courtesy of David L. Hasty, University of Tennessee, Memphis, TN.)

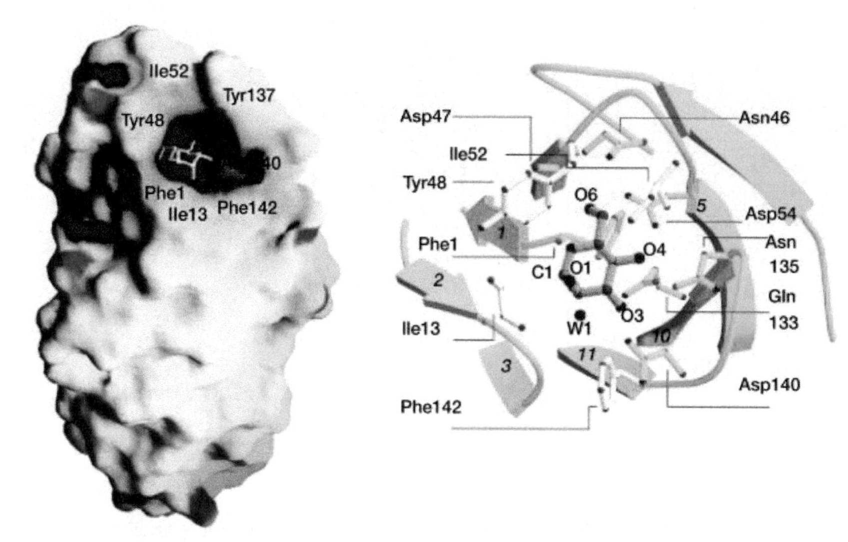

FIGURE 37.4. The α anomer of mannose in the binding site of FimH. The mannose residue is buried in a unique site at the tip of the carbohydrate-recognition domain (*left*) in a deep and negatively charged pocket (*right*). FimH prefers to bind D-mannose in the α-anomeric configuration. The hydroxyl groups of D-mannose interact with residues Phe1, Asn46, Asp47, Asp54, Gln133, Asn135, Asp140, and Phe142 by hydrogen-bonding and hydrophobic interactions. Residues in direct contact are shown as a *ball-and-stick* model. W1, Water. (Redrawn, with permission, from Hung CS, et al. 2002. *Mol Microbiol* **44:** 903–915, © John Wiley and Sons.)

Thus, pili extend well beyond the bacterial glycocalyx comprised of lipopolysaccharide and capsular polysaccharides (see Chapter 21).

Examination of a high-resolution, three-dimensional structure of FimH bound to mannose revealed that, although mannose exists as a mixture of α and β anomers in solution, only the former was found in the complex. Mannose binds FimH at a deep and negatively charged site of FimH (Figure 37.4). FimH has "catch-bond" properties, such that the binding strength of individual FimH adhesins is augmented by increasing the shear forces on surface-bound *E. coli* cells. This is relevant for urinary tract infections, in which type-1 fimbriae mediate binding of the bacteria to the glycoprotein uroplakin Ia on the surface of bladder epithelial cells. Uroplakin Ia presents high levels of terminally exposed mannose residues that are capable of specifically interacting with FimH. Alternatively, type-1 fimbriae can bind to high mannose glycans on the soluble urinary Tamm–Horsfall glycoprotein/uromodulin, that forms long fibers. The resulting bacterial aggregates are flushed out of the urinary tract. Mice lacking the gene encoding the Tamm–Horsfall glycoprotein are considerably more susceptible to bladder colonization by type-1-fimbriated *E. coli* than normal mice.

Most bacteria (and possibly other microorganisms) have multiple adhesins with diverse carbohydrate specificities. Many of these have been described (examples in Table 37.2), wherein the specificity of lectin binding can help define the range of susceptible tissues in the host (i.e., the microbe's ecological niche). The columnar epithelium that lines the large intestine expresses receptors with Galα1–4Gal-Cer residues, whereas cells lining the small intestine do not. Thus, *Bacterioides, Clostridium, E. coli,* and *Lactobacillus* only colonize the large intestine under normal conditions. P-fimbriated *E. coli* and some toxins bind specifically to galabiose (Galα1–4Gal) and galabiose-containing oligosaccharides, most commonly as constituents of glycolipids. Binding can occur to either internal (i.e., when the disaccharide is capped by other sugars) or terminal nonreducing galabiose units. P-fimbriated *E. coli* adhere mainly to the upper part of the kidney, where galabiose is abundant. The fine specificity of bacterial surface lectins and their relationship to the animal tropism of the bacteria can be further illustrated by *E. coli* K99. The K99 strain binds to glycolipids that contain *N*-glycolylneuraminic acid (Neu5Gc), in the form of Neu5Gcα2–3Galβ1–

TABLE 37.2. Examples of interactions of bacterial adhesins with glycans

Microorganism	Adhesin	Glycan-receptor specificity	Site of infection
Actinomyces naeslundii	fimbriae	Galβ1–3GalNAcβ-	oral
Bordetella pertussis	filamentous hemagglutinin (FHA)	sulfated glycolipids, heparin	ciliated epithelium in respiratory tract
Borrelia burgdorferi	ErpG protein	heparan sulfate	endothelium, epithelium, and extracellular matrix
Campylobacter jejuni	flagella, LPS	Fucα1-2Galβ1-4GlcNAcβ- (H-antigen)	intestinal cells
Escherichia coli	P fimbriae	Galα1-4Galβ-	urinary tract
	S fimbriae	gangliosides GM3, GM2	neural
	type-1 fimbriae	Manα1-3(Manα6Manα1-6) Man	urinary tract
	K99 fimbriae	gangliosides GM3, Neu5Gcα2-3Galβ1-4Glc	intestinal cells
Haemophilus influenzae	HMW1 adhesin	Neu5Acα2-3Galβ1-4 GlcNAcβ-, heparan sulfate	respiratory epithelium
Helicobacter pylori	BabA	sialyl Lewis x	stomach
H. pylori	SabA	Lewis b	stomach and stomach duodenum
Mycobacterium tuberculosis	heparin-binding hemagglutinin adhesin (HBHA)	heparan sulfate	respiratory epithelium
Neisseria gonorrhoeae	Opa proteins	LacCer; Neu5Acα2-3Galβ1-4 GlcNAcβ-, syndecans, heparan sulfate	genital tract
Pseudomonas aeruginosa	type IV pili	asialo GM1 and GM2	respiratory tract
Staphylococcus aureus	signal peptide of panton valentine leukocidin	heparan sulfate	connective tissues and endothelial cells
Streptococcus agalactiae	αC protein	heparan sulfate	brain endothelial cells
Streptococcus pneumoniae	carbohydrate-binding modules of β-galactosidase, BgaA	lactose or *N*-acetyl-lactosamine	respiratory tract

4Glc, but not to those that contain *N*-acetylneuraminic acid (Neu5Ac). These two sugars differ by a single hydroxyl group that is present only on Neu5Gc. Interestingly, Neu5Gc-containing receptors are expressed on the intestinal cells of newborn piglets, but disappear as the animals grow and develop. As Neu5Gc is not normally biosynthesized by humans, this may explain why *E. coli* K99 can cause often lethal diarrhea in piglets but not in adult pigs or humans.

SECRETED TOXINS THAT BIND GLYCANS

A number of secreted bacterial toxins also bind glycans (Table 37.3). The toxin from *Vibrio cholera* (cholera toxin), which consists of A and B subunits in the ratio AB_5, has been extensively studied. The crystal structure of cholera toxin shows that the carbohydrate-recognition domains are located at the base of the B subunits, which bind to the Galβ1–3GalNAc moiety of GM1 ganglioside (Chapter 11) receptors (Figure 37.5). On binding of the B subunits to membrane glycolipids, the AB_5 complex is endocytosed to the Golgi apparatus and then undergoes retrograde transport to the endoplasmic reticulum (ER). The A1 and A2 chains are proteolytically

TABLE 37.3. Examples of glycan receptors for bacterial toxins

Microorganism	Toxin	Glycan-receptor specificity	Site of infection
Bacillus thuringiensis	crystal toxins	Galβ1-3/6Galα/β1-3(±Glcβ1-6) GalNAcβ GlcNAcβ1-3Manβ1-4 GlcβCer	intestinal epithelium of insects/nematodes
Clostridium botulinum	botulinum toxins (A–E)	gangliosides GT1b and GQ1b	nerve membrane
Clostridium difficile	toxin A	GalNAcβ1-3Galβ1-4GlcNAcβ1-3 Galβ1-4 GlcβCer	large intestine
Clostridium tetani	tetanus toxin	ganglioside GT1b	nerve membrane
Escherichia coli	heat-labile toxin	GM1	intestine
Shigella dysenteriae	Shiga toxin	Galα1-4GalβCer, ;Galα1-4Galβ1-4 GlcβCer	large intestine
Vibrio cholerae	cholera toxin	GM1	small intestine

Gangliosides are defined using the Svennerholm nomenclature (see Chapter 11).
Cer, Ceramide.

cleaved on toxin secretion, but remain stably associated until arrival in the ER. There, the enzymatic A1 chain unfolds, dissociates from the A2-B_5 complex, and retrotranslocates to the cytosol where it rapidly refolds, thereby avoiding degradation by the proteasome. Catalytically active A1 then ADP-ribosylates a regulatory homotrimeric G-protein to activate adenylyl cyclase, severely altering ion homeostasis of the infected cell.

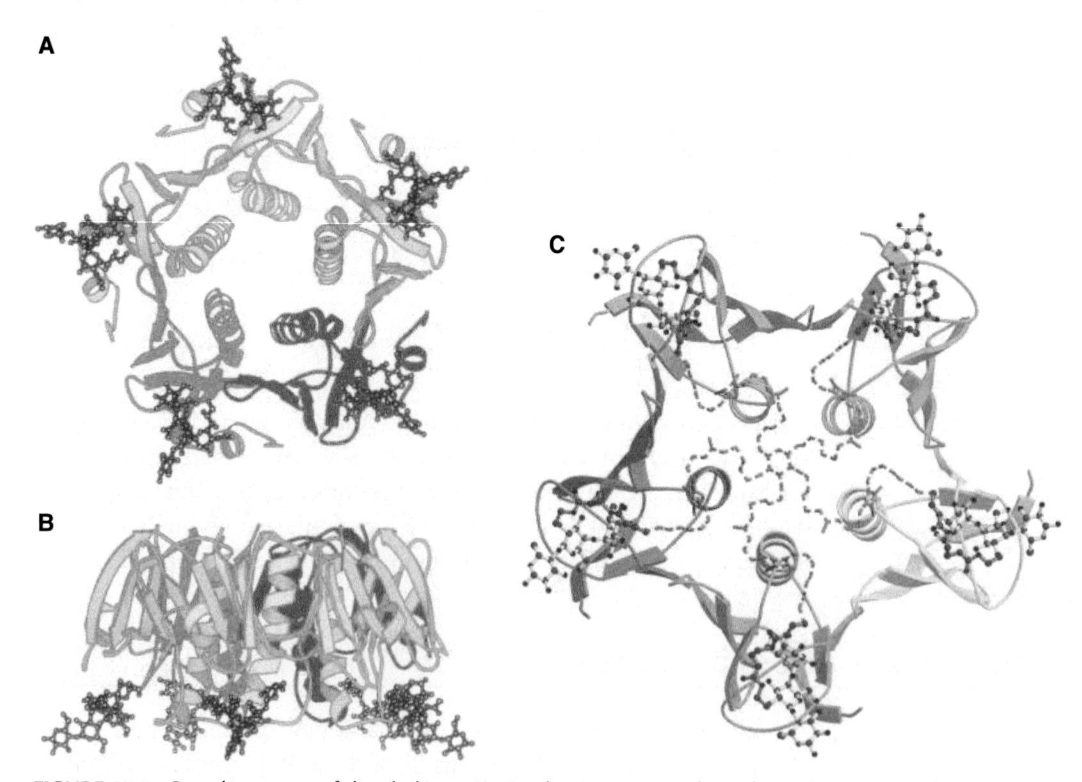

FIGURE 37.5. Crystal structure of the cholera toxin B-subunit pentamer bound to GM1 pentasaccharide, shown from the bottom (*A*) and the side (*B*). (Redrawn, with permission, from Merritt EA, et al. 1994. *Protein Sci* **3:** 166–175.) (*C*) Crystal structure of the Shiga toxin pentamer bound to an artificial pentavalent ligand, a powerful inhibitor of the toxin. The carbohydrate ligands are shown in a *ball-and-stick* representation. Possible conformation of the linker is indicated by the *dashed magenta lines.* (Redrawn, from Kitov PI, et al. 2000. *Nature* **403:** 669–672, © Macmillan Publisher Ltd.)

Shiga toxin, produced by *Shigella dysenteriae*, will bind to Galα1-4Gal determinants on both glycolipids and glycoproteins. However, only binding to the glycosphingolipid receptor Gb3 results in cell death. Similar to cholera toxin, the AB_5 complex is endocytosed and transported to the ER of the target cells, and then the A1 subunit is retrotranslocated to the cytosol. There, the catalytic A1 chain of Shiga toxin inactivates ribosomes, and thus the essential process of cytoplasmic protein synthesis, by an N-glycosidic cleavage event that depurinates 28S rRNA.

The most common class of bacterial toxins that are characterized by their pore-forming capability and a number of these toxins bind to glycans. For example, toxins produced by the soil-dwelling bacterium *Bacillus thuringiensis* (Bt) are used for crop protection by spraying plants or by genetically engineering crops to express the toxins. As such, these toxins function by binding glycolipids that line the insect gut and generating pores in the membrane. More specifically, the Bt toxin glycolipid receptors include in their structure the characteristic ceramide-linked, mannose-containing core tetrasaccharide GalNAcβ1-4GlcNAcβ1-3Manβ1-4GlcβCer. This structure is conserved between nematodes and insects (see Chapter 25) but is absent in vertebrates including humans. This explains why Bt can kill larval stages of insects but is harmless to humans.

Secondary glycan binding sites have been identified on some toxins. For example, cholera toxin can bind fucosylated glycans at a site distinct from the GM1 binding pocket, and *Clostridium difficile* toxin A (TcdA) binds sulfated glycosaminoglycans in addition to GalNAcβ1-3Galβ1-4GlcNAcβ1-3 Galβ1-4GlcβCer. Whereas these secondary interactions are lower affinity that those with the canonical receptors, the glycans recognized are abundant on the surface of host cells. Thus, glycan binding to secondary sites may serve as initial attachment events that capture and concentrate toxins on cell surfaces.

PARASITE LECTINS

In addition to viruses and bacteria, a number of parasites use glycans as receptors for adhesion (Table 37.4). *Entamoeba histolytica* expresses a 260-kDa heterodimeric lectin that binds to terminal Gal/GalNAc residues on glycoproteins and glycolipids via a cysteine-rich glycan-binding domain. This adhesin is essential for virulence, such that adhesin-glycan binding is required for

TABLE 37.4. Examples of glycan receptors for parasites

Parasite	Adhesin	Glycan-receptor specificity	Site of infection
Entamoeba histolytica	260-kDa surface-anchored lectin on trophozoites	terminal Gal/GalNAc residues	mucosa of human colon
Plasmodium falciparum	EBA-175; circumsporozoite (CS) protein	sialic acid–containing O-glycans (Neu5Acα2-3Galβ-) on glycophorins; heparan sulfate proteoglycans	erythrocytes (infected cells bind to placental vasculature) and hepatocytes
Trypanosoma cruzi	surface "mucins"	sialic acid–containing glycans and heparan sulfate	multiple cell types
Leishmania amazonensis	unknown	heparan sulfate	macrophages, fibroblasts, and epithelium
Cryptosporidium parum	lectin p30	terminal Gal-GalNAc	intestinal epithelium
Giardia lamblia	unknown	mannose-terminated oligosaccharides	duodenum and small intestine
Toxoplasma gondii	microneme protein 1 (TgMIC1)	α2-3-linked sialyl-*N*-acetyllactosamine sequences	intestinal epithelium

parasitic attachment, invasion, and cytolysis of the intestinal epithelium. Furthermore, it may function in binding *E. histolytica* to bacteria as a food source. This adhesin elicits protective immunity, and is a potential target to manage *E. histolytica* infection.

The initial interaction of *Plasmodium falciparum* (malaria) merozoites with red blood cells (erythrocytes) depends on sialic acid residues present on the host cell, in particular on the major erythrocyte membrane protein glycophorin. Parasite–host attachment is mediated by a family of sialic acid–binding adhesins on merozoites, the most prominent of which is called erythrocyte-binding antigen-175 (EBA-175). This adhesin preferentially binds Neu5Ac sialic acids, rather than 9-*O*-acetyl-Neu5Ac or Neu5Gc, and is sensitive to the linkage of the sialic acid to the underlying galactose. This is highlighted by the fact that Neu5Acα2–3Gal-containing oligosaccharides effectively inhibit the binding of EBA-175 to erythrocytes, whereas soluble Neu5Ac and Neu5Acα2–6Gal-containing oligosaccharides do not. Adhesin–glycan binding triggers invasion of the merozoites into red blood cells, where they develop into mature schizonts that rupture and release newly formed merozoites into the bloodstream. Many commonly used clinical antimalarial medications, such as chloroquine, target the parasite during this erythrocytic asexual reproduction stage.

THERAPEUTIC IMPLICATIONS

Various glycan-binding proteins mediate adhesion of microorganisms to host cells or tissues. These interactions are often a prerequisite for infection or symbiosis, with binding-deficient mutants being unable to initiate or maintain these relationships. Interestingly, glycans recognized by microbial surface lectins have been shown to block the adhesion of bacteria to animal cells in vitro and in vivo, and thus may protect animals against infection by such microorganisms. For example, coadministration of methyl α-mannoside with type-1-fimbriated *E. coli* into the bladder of mice can reduce microbial burden in a mouse model of urinary tract infection, whereas methyl α-glucoside, which does not bind to FimH, has no effect. Furthermore, lacto-N-neotetraose (LNnT) and its α2-3- and α2-6-sialylated derivatives block the adherence of *Streptococcus pneumoniae* to respiratory epithelial cells in vitro. In addition, these glycans prevent colonization of the nasopharynx and attenuate the course of pneumonia in rodent models of pneumococcal infection. Naturally occurring host glycans can also serve protective roles. For example, secreted mucins and human milk oligosaccharides (HMOs) act as natural receptor decoys, protecting the host from pathogens and shaping the microbiota composition.

Exogenous heparin and structurally related polysaccharides are known to inhibit viral replication, suggesting a potential approach for the development of polysaccharide-based antiviral pharmaceutical agents. For example, heparin octasaccharide decoy liposomes were recently shown to inhibit the replication of numerous viruses including HSV and respiratory syncytial virus. As more crystal structures are elucidated, the ability to design small molecule inhibitors that fit into the carbohydrate-recognition domains of adhesins should improve. Already, the structures of influenza hemagglutinin and sialidase have suggested numerous ways to modify sialic acid to fit better into the active sites. Some of these compounds are presently in clinical use to limit the spread of influenza.

ACKNOWLEDGMENTS

The authors acknowledge contributions to the previous version of this chapter from Jeffrey D. Esko, Victor Nizet, and the late Nathan Sharon and helpful comments and suggestions from Frederique Lisacek and Ajit Varki.

FURTHER READING

Rostand KS, Esko JD. 1997. Microbial adherence to and invasion through proteoglycans. *Infect Immun* **65**: 1–8. doi:10.1128/iai.65.1.1-8.1997

Kitov PI, Sadowska JM, Mulvey G, Armstrong GD, Ling H, Pannu NS, Read RJ, Bundle DR. 2000. Shiga-like toxins are neutralized by tailored multivalent carbohydrate ligands. *Nature* **403**: 669–672. doi:10.1038/35001095

Griffitts JS, Aroian RV. 2005. Many roads to resistance: how invertebrates adapt to Bt toxins. *BioEssays* **27**: 614–624. doi:10.1002/bies.20239

Olofsson S, Bergstrom T. 2005. Glycoconjugate glycans as viral receptors. *Ann Med* **37**: 154–172. doi:10.1080/07853890510007340

Mazmanian SK, Kasper DL. 2006. The love–hate relationship between bacterial polysaccharides and the host immune system. *Nat Rev Immunol* **6**: 849–858. doi:10.1038/nri1956

Sinnis P, Coppi A. 2007. A long and winding road: the *Plasmodium* sporozoite's journey in the mammalian host. *Parasitol Int* **56**: 171–178. doi:10.1016/j.parint.2007.04.002

Patsos G, Corfield A. 2009. Management of the human mucosal defensive barrier: evidence for glycan legislation. *Biol Chem* **390**: 581–590. doi:10.1515/bc.2009.052

Krachler AM, Orth K. 2013. Targeting the bacteria–host interface: strategies in anti-adhesion therapy. *Virulence* **4**: 284–294. doi:10.4161/viru.24606

Edinger TO, Pohl MO, Stertz S. 2014. Entry of influenza A virus: host factors and antiviral targets. *J Gen Virol* **95**: 263–277. doi:10.1099/vir.0.059477-0

Stencel-Baerenwald JE, Reiss K, Reiter DM, Stehle T, Dermody TS. 2014. The sweet spot: defining virus–sialic acid interactions. *Nat Rev Microbiol* **12**: 739–749. doi:10.1038/nrmicro3346

Bode L. 2015. The functional biology of human milk oligosaccharides. *Early Hum Dev* **91**: 619–622. doi:10.1016/j.earlhumdev.2015.09.001

Zajonc DM, Girardi E. 2015. Recognition of microbial glycolipids by natural killer T cells. *Front Immunol* **6**: 400. doi:10.3389/fimmu.2015.00400

Juge N, Tailford L, Owen CD. 2017. Sialidases from gut bacteria: a mini-review. *Biochem Soc Trans* **44**: 166–175. doi:10.1042/bst20150226

Moonens K, Remaut H. 2017. Evolution and structural dynamics of bacterial glycan binding adhesins. *Curr Opin Struct Biol* **44**: 48–58. doi:10.1016/j.sbi.2016.12.003

Raman R, Tharakaraman K, Sasisekharan V, Sasisekharan R. 2017. Glycan–protein interactions in viral pathogenesis. *Curr Opin Struct Biol* **40**: 153–162. doi:10.1016/j.sbi.2016.10.003

Ramani S, Hu L, Venkataram Prasad BV, Estes MK. 2017. Diversity in rotavirus–host glycan interactions: a "sweet" spectrum. *Cell Mol Gastroenterol Hepatol* **2**: 263–273. doi:10.1016/j.jcmgh.2016.03.002

Thomas GH. 2017. Sialic acid acquisition in bacteria—one substrate, many transporters. *Biochem Soc Trans* **44**: 760–765. doi:10.1042/bst20160056

Tytgat HL, de Vos WM. 2017. Sugar coating the envelope: glycoconjugates for microbe–host crosstalk. *Trends Microbiol* **24**: 853–861. doi:10.1016/j.tim.2016.06.004

Valguarnera E, Kinsella RL, Feldman MF. 2017. Sugar and spice make bacteria not nice: protein glycosylation and its influence in pathogenesis. *J Mol Biol* **428**: 3206–3220. doi:10.1016/j.jmb.2016.04.013

Wasik BR, Barnard KN, Parrish CR. 2017. Effects of sialic acid modifications on virus binding and infection. *Trends Microbiol* **24**: 991–1001. doi:10.1016/j.tim.2016.07.005

Poole J, Day CJ, von Itzstein M, Paton JC, Jennings MP. 2018. Glycointeractions in bacterial pathogenesis. *Nat Rev Microbiol* **16**: 440–452. doi:10.1038/s41579-018-0007-2

Thompson AJ, de Vries RP, Paulson JC. 2019. Virus recognition of glycan receptors. *Curr Opin Virol* **34**: 117–129. doi:10.1016/j.coviro.2019.01.004

38 Proteins That Bind Sulfated Glycosaminoglycans

Ding Xu, James H. Prestegard, Robert J. Linhardt, and Jeffrey D. Esko

Glycosaminoglycans bind to many different classes of proteins mostly through electrostatic interactions between negatively charged sulfate groups and uronic acids and positively charged amino acids in the protein. This chapter focuses on examples of glycosaminoglycan (GAG)-binding proteins, methods for measuring GAG–protein interaction, and information about three-dimensional structures of the complexes.

GAG-BINDING PROTEINS ARE COMMON

In contrast to lectins, which tend to fall into evolutionarily conserved families (Chapters 28–37), GAG-binding proteins do not have common folds and instead appear to have evolved by convergent evolution. Several hundred GAG-binding proteins have been discovered, which make up the GAG-interactome and fall into the broad classes presented in Table 38.1. To a large extent, studies of the GAG-interactome have focused on protein interactions with heparin, a more highly sulfated, iduronic acid (IdoA)-rich form of heparan sulfate (HS; Chapter 17). This bias reflects, in part, the commercial availability of heparin and heparin-Sepharose, which are frequently used for fractionation studies, and the partially incorrect assumption that binding to heparin mimics binding to HS present on cell surfaces and in the extracellular matrix. There are also a large number of proteins known to interact with chondroitin sulfate (CS) and dermatan sulfate (DS) with comparable avidity and affinity; there are fewer examples of specific

published by Cold Spring Harbor Laboratory Press; doi: 10.1101/glycobiology.4e.38

TABLE 38.1 Examples of various classes of glycosaminoglycan (GAG)-binding proteins

Class	Examples
Enzymes	Thrombin, cathepsin K, MMP-7, ADAM12, extracellular superoxide dismutase
Enzyme inhibitors	Antithrombin III, heparin cofactor II, cystatin C, tissue inhibitor of metalloproteinases
Cell adhesion proteins	P-selectin, L-selectin, some integrins
Extracellular matrix proteins	Laminin, fibronectin, collagens, thrombospondin, vitronectin, tenascin
Chemokines/cytokines	CXCLs, CCLs, γ- and β-interferons, midkine
Growth factors	FGFs, VEGFs, PDGF, HB-EGF, HGF, IGF–binding proteins
Morphogens	Hedgehogs, Wnts, BMPs, TGF-β family members
Guidance factors	Slits, ROBOs, semaphorins, neuropilins
Transmembrane receptors	FGF receptors, VEGF receptor, RAGE, RPTPs, Tie1
Lipid-binding proteins	Apolipoproteins B, E, and A–V, lipoprotein lipase, hepatic lipase, annexins
Plaque proteins	Prion proteins, amyloid proteins, Tau
Nuclear proteins	Histones, HMGB1
Pathogen surface proteins	Malaria circumsporozoite protein
Viral envelope proteins	Herpes simplex virus, dengue virus, SARS-CoV-2, Zika virus, human immunodeficiency virus, hepatitis C virus

(CXCL) CXC motif chemokine ligand, (CCL) CC motif chemokine ligand, (FGF) fibroblast growth factor, (VEGF) vascular endothelial growth factor, (PDGF) platelet-derived growth factor, (HB-EGF) heparin-binding epidermal growth factor, (HGF) hepatocyte growth factor, (IGF) insulin-like growth factor, (Wnt) Wnt ligands, (BMP) bone morphogenetic protein, (TGF-β) transforming growth factor-β, (ROBO) roundabout receptor, (RAGE) receptor for advanced glycation end products, (RPTP) receptor protein tyrosine phosphatase, (HMGB) high-mobility group box.

interactions with keratan sulfate (KS), but this may reflect fewer studies of KS. In some cases, CS may be the physiologically relevant ligand because CS predominates in many tissues. Determining the physiological relevance of these interactions is a major area of research.

Interactions between GAGs and proteins can have profound physiological effects on processes such as hemostasis, lipid transport and absorption, cell growth and migration, and development. Binding to GAGs can result in immobilization of proteins at their sites of production or in the extracellular matrix for future mobilization, regulation of enzyme activity, binding of proteins to their receptors, protein oligomerization, and protection of proteins against degradation. Several viruses and bacteria also exploit GAGs expressed in the extracellular matrix and on cell surfaces as attachment factors. For example, SARS-CoV-2, the causative agent of the COVID-19 pandemic, enters through the respiratory tract and infects epithelial cells lining the airways by interactions of the viral spike protein with heparan sulfate. The ability of virions to attach to cell-surface heparan sulfate facilitates capture of the virus and transfer to proteinaceous receptors (e.g., host cell receptor angiotensin converting enzyme 2 in the case of SARS-CoV-2) as well as subsequent viral glycoprotein processing and infection. The interactions of GAG-binding proteins are often driven by a general complementarity of charge (e.g., thrombin–heparin interactions). In some cases, the interaction has been shown to depend on rare but specific sequences of modified sugars in the GAG chain (e.g., antithrombin binding).

METHODS FOR MEASURING GAG–PROTEIN BINDING

Numerous methods are available for analyzing GAG–protein interactions, and some provide a direct measurement of K_d values. A common method involves affinity fractionation of proteins on Sepharose columns containing covalently linked GAG chains, usually heparin. The bound proteins are eluted with different concentrations of sodium chloride, and the concentration required for elution is generally proportional to the K_d. High-affinity interactions require 0.5–2 M NaCl to displace bound ligand, which usually translates into K_d values of 10^{-7}–10^{-9} M (determined under physiological salt concentrations by equilibrium binding). Proteins with lower affinity (10^{-5}–10^{-7} M) often require only 0.2–0.5 M NaCl to elute. This method

TABLE 38.2 Methods to measure glycosaminoglycan (GAG)–protein interaction

Affinity estimation	Affinity/kinetics determination	Stoichiometry determination	Structure determination
GAG affinity chromatography	Competition ELISA	Size exclusion chromatography	X-ray crystallography
FACS-based cell-surface GAG binding	SPR	SPR	Nuclear magnetic resonance
GAG oligosaccharide microarray	ITC	ITC	GAG oligosaccharide microarray
	Affinity electrophoresis	Ion mobility mass spectrometry	Computational docking
		Analytical ultracentrifugation	Fluorescence spectroscopy

FACS, Fluorescence-activated cell sorting; ELISA, enzyme-linked immunosorbent assay; SPR, surface plasmon resonance; ITC, isothermal titration calorimetry.

assumes that GAG–protein interactions are entirely ionic, which is not entirely correct. Nevertheless, it can provide an assessment of relative affinity, when comparing different GAG-binding proteins.

A number of more sophisticated methods are now in use that provide detailed thermodynamic data (ΔH [change in enthalpy], ΔS [change in entropy], ΔCp [change in molar heat capacity], etc.), kinetic data (association and dissociation rates), and high-resolution data on atomic contacts in GAG–protein interactions (Table 38.2). Regardless of the technique one uses, it must be kept in mind that in vitro binding measurements are not likely to be the same as those when the protein binds to proteoglycans on the cell surface or in the extracellular matrix, where the density and variety of GAG-binding proteins, proteoglycans, and other interacting factors varies greatly. To determine the physiological relevance of the interaction, one should consider measuring binding under conditions that can lead to a biological response. For example, one can measure binding to cells with altered GAG composition (Chapter 49) or after treatment with specific lyases to remove GAG chains from the cell surface (Chapter 17) and then determine whether the same response occurs as observed in the presence of GAG chains. The interaction can then be studied more intensively using the in vitro assays described above.

CONFORMATIONAL AND SEQUENCE CONSIDERATIONS

As mentioned above, most GAG-binding proteins interact with HS and/or heparin. The likely basis for this preference is greater sequence heterogeneity and more extensive and variable sulfation compared with other GAGs. The unusual conformational flexibility of iduronic acid, which is found in heparin, HS, and DS, also has a role in their ability to bind proteins. GAGs are linear helical structures, consisting of alternating residues of *N*-acetylglucosamine (GlcNAc) or *N*-acetylgalactosamine (GalNAc) with glucuronic acid (GlcA) or IdoA (with the exception of KS, which consists of alternating *N*-acetylglucosamine and galactose residues; Chapter 17). Inspection of heparin oligosaccharides containing highly modified domains ($[\text{GlcNS6S-IdoA2S}]_n$) shows that the *N*-sulfo and 2-*O*-sulfo groups of each disaccharide repeat lie on opposite sides of the helix from the 6-*O*-sulfo and carboxyl groups (Figure 38.1). Analysis of the conformation of individual sugars shows that *N*-acetylglucosamine and glucuronic acid residues assume a preferred conformation in solution, designated $^4C_1$ (indicating that carbon 4 is above the plane defined by carbons 2, 3, and 5 and the ring oxygen, and that carbon 1 is below the plane; Chapter 2). In contrast, IdoA2S assumes the $^1C_4$ or the $^2S_0$ conformation (Figure 38.1), which reorients the position of the sulfo substituents, thereby creating a different

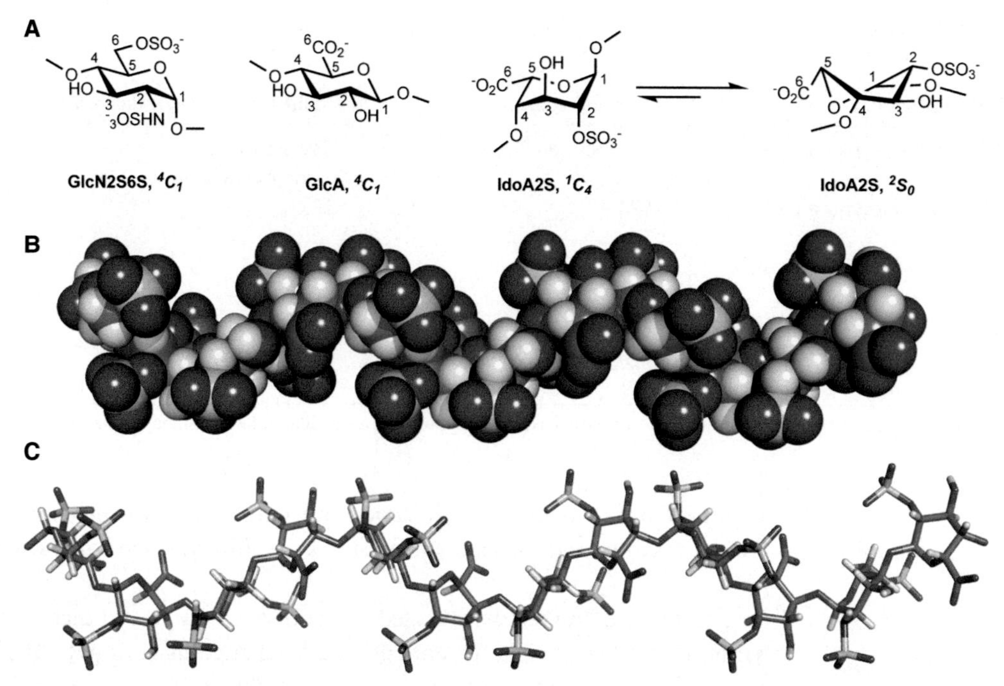

FIGURE 38.1 Conformation of heparin oligosaccharides. (*A*) Glucosamine (GlcN) and glucuronic acid (GlcA) exist in the $^4C_1$ conformation, whereas iduronic acid (IdoA) exists in equally energetic conformations designated $^1C_4$ and $^2S_0$. (*B*) Space-filling model of a heparin oligosaccharide (14-mer) deduced by nuclear magnetic resonance. (*C*) The same structure in stick representation. The renderings in *B* and *C* were made with RASMOL using data from the Molecular Modeling Database (MMDB Id: 3448) at the National Center for Biotechnology Information (NCBI).

orientation of charged groups. In many cases when a protein binds to an HS chain, it induces a change in conformation of the IdoA2S residue resulting in a better fit and enhanced binding. IdoA2S residues have always been found in domains rich in *N*-sulfo and *O*-sulfo groups (for biosynthetic reasons; Chapter 17), which is also where proteins usually bind. Thus, the greater degree of conformational flexibility in these modified regions may explain why so many more proteins bind with high affinity to heparin, HS, and DS than to other GAGs. The presence of an *N*-acetyl group in an *N*-acetylglucosamine residue changes the preferred conformation of the neighboring IdoA residue, showing that even minor modifications can influence conformation and chain flexibility. Binding to GAGs that have a low degree of sulfation may require larger domains in the protein to interact with longer stretches of an oligosaccharide. Molecular dynamic simulations on large heparin oligosaccharides, even in the presence of proteins, are possible with recent advances in computer performance (Online Appendix 38A). Such simulations can be used to predict the conformational flexibility of different domains within the chain and, when combined with recent advances in protein–GAG docking, can provide additional insights into GAG–protein interactions.

HOW SPECIFIC ARE GAG–PROTEIN INTERACTIONS?

The discovery of multiple GAG-binding proteins led a number of investigators to examine whether there is a consensus amino acid sequence for GAG binding. In retrospect, this strategy was overly simplistic because it assumed that all GAG-binding proteins have a common evolutionary origin and would recognize the same oligosaccharide sequence within heparin or, at least, sequences that would share many common features. It is now known that the convergently evolved GAG-binding proteins interact with different oligosaccharide sequences. The binding sites in the protein always contain basic amino acids (lysine and arginine), whose positive charges presumably interact with the negatively charged sulfates and carboxylates of the GAG chains. However, the arrangement of these basic amino acids can be quite variable, consistent with the variable positioning of sulfo groups in the GAG partner. Selectivity is also a function of H-bonding and van der Waal interactions of amino acid residues with the oligosaccharide.

Most proteins are formed from α-helices, β-strands, and loops. Therefore, to engage a linear GAG chain electrostatically, the positively charged amino acid residues must align along the same side of the protein segment. α-Helices have periodicities of 3.4 residues per turn, which would require the basic residues to occur every third or fourth position along the helix to align with an oligosaccharide. In β-strands, the amino acid side chains alternate sides every other residue. Thus, to bind a GAG chain, the positively charged residues in a β-strand would be located quite differently than in an α-helix.

On the basis of the structure of several heparin-binding proteins that were available in 1991, Alan Cardin and Herschel Weintraub proposed that typical heparin-binding sites had the sequence XBBXBX or XBBBXXBX, where B is lysine or arginine and X is any other amino acid. From the structural arguments provided above, it should be obvious that only some of the basic residues in these sequences could participate in GAG binding, the actual number being determined by whether the peptide sequence exists as an α-helix or a β-sheet. It is now known that the presence of these sequences in a protein merely suggests a possible interaction with heparin (or another GAG chain), but it does not prove that the interaction occurs under physiological conditions. In fact, the predicted binding sites for heparin in fibroblast growth factor 2 (FGF2) turned out to be incorrect once the crystal structure was determined. It is likely that binding involves multiple protein segments that juxtapose positively charged residues into a three-dimensional turn-rich recognition site. In many cases the binding involves loops which make the positioning more variable. An example of this phenomenon is observed in the chemokine CCL5, which contains a XBBXBX motif in a loop. The specific arrangement of residues should vary according to the type and fine structure of those oligosaccharides involved in binding.

In lectins, and in antibodies that recognize glycans, the glycan recognition domains are typically shallow pockets that engage the terminal sugars of the oligosaccharide chain (Chapters 29, 30, and 37). In GAG-binding proteins, the protein usually binds to sugar residues that lie within the chain or near the terminus. Therefore, the binding sites in GAG-binding proteins consist of clefts or sets of juxtaposed surface residues rather than pockets. These GAG-binding sites on the protein surface give rise to more rapid GAG–protein binding kinetics than are typically observed for protein–protein interactions. Given that GAG chains generally exist in a helical conformation, only those residues on the face toward the protein interact with amino acid residues; the ones on the other side of the helix are potentially free to interact with a second ligand (e.g., as observed in FGF dimers). Alternatively, residues in a binding cleft could interact with both sides of the helix (e.g., in dengue envelope protein). Finally, one should keep in mind that binding occurs to only a small segment of the GAG chain. Thus, a single GAG chain can potentially bind multiple protein ligands facilitating cooperative binding that can lead to protein oligomerization (e.g., some chemokines).

ANTITHROMBIN–HEPARIN: A PARADIGM FOR STUDYING GAG-BINDING PROTEINS

Perhaps the most studied example of a GAG–protein interaction is the binding of antithrombin to heparin and HS (Figure 38.2). This interaction is of great pharmacological importance because heparin is widely used clinically as an anticoagulant. Binding of antithrombin to heparin has a dual effect: first, it causes a conformational change in the protein and activation of the

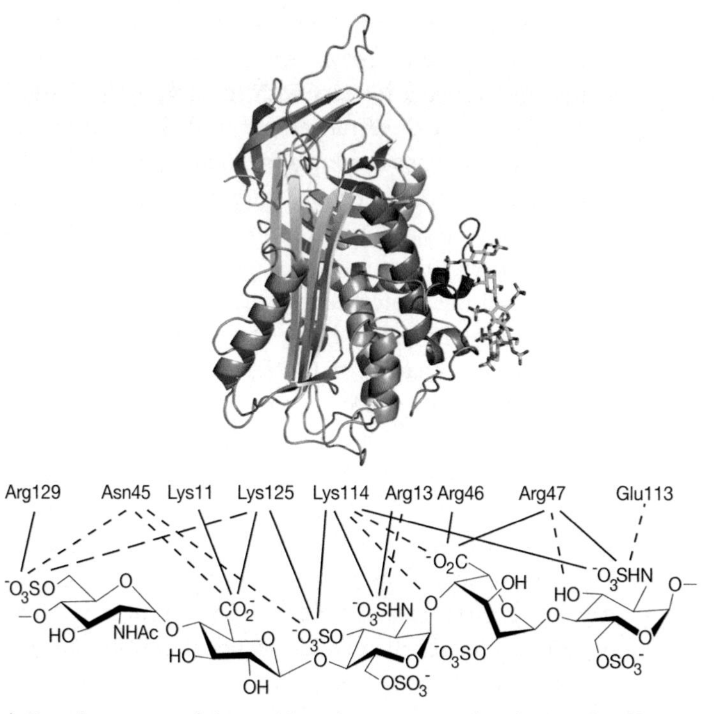

FIGURE 38.2 (*Top*) Crystal structure of the antithrombin–pentasaccharide complex (from Protein Data Bank). Three structural elements that make contact with heparin are shown, including a loop (*red*), and two α-helices (in *blue* and *magenta*). The pentasaccharide is shown in *stick representation*. The loop that interacts with thrombin is shown in *orange*. (*Bottom*) Interactions between key amino acid residues and individual elements in the pentasaccharide. (*Solid lines*) Electrostatic interactions between positively charged residues and carboxylates and sulfate group, (*short dashed lines*) hydrogen bonds, (*long dashed line*) water-mediated hydrogen bond. (Modified from Li et al. 2004. *Nat Struct Mol Biol* **11:** 857–862, with permission from Springer Nature.)

protease inhibiting action, resulting in a 1000-fold enhancement in the rate at which it inactivates thrombin and factor Xa. Second, the heparin chain acts as a template, enhancing the physical apposition of thrombin and antithrombin. Thus, both the protease (thrombin) and the inhibitor have GAG-binding sites. Heparin acts as a catalyst in these reactions by enhancing the rate of the reaction through apposition of substrates and conformational change. After the inactivation of thrombin or factor Xa by antithrombin occurs, the complexes lose affinity for heparin and dissociate. The heparin is then available to participate in another activation/inactivation cycle. Antithrombin is a member of the serpin family of protease inhibitors, many of which bind to heparin.

Early studies using affinity fractionation schemes showed that only about one-third of the chains in a heparin preparation actually bind with high affinity to antithrombin. Comparing the sequence of the bound chains with those that did not bind failed to reveal any substantial differences in composition, consistent with the later discovery that the binding site consists of only five sugar residues (Figure 38.2) (the average heparin chain is about 50 sugar residues). Within this pentasaccharide sequence, however, a centrally located 3-*O*-sulfated GlcNS6S unit plays an essential role in mediating antithrombin–heparin interaction. The binding sites in GAG chains in general represent a very small segment of the chains.

Crystals of antithrombin were prepared and analyzed by X-ray diffraction to 2.6-Å resolution. The docking site for the heparin pentasaccharide is formed by the apposition of two helices, which both contain critical arginine and lysine residues at the interface (Figure 38.2). The pentasaccharide is sufficient to activate antithrombin binding toward factor Xa, but it will not facilitate the inactivation of thrombin. For this to occur, a larger oligosaccharide of at least 18 residues is needed. As mentioned above, thrombin also contains a heparin-binding site, and the larger heparin oligosaccharide is thought to act as a template for the formation of a ternary complex with thrombin and antithrombin. In contrast to antithrombin, thrombin shows little oligosaccharide specificity. As might be expected, adding high concentrations of heparin actually inhibits the reaction, because the formation of binary complexes of heparin and thrombin or heparin and antithrombin predominate. This important principle of "activation at low concentrations and inhibition at high concentrations" also occurs in other systems in which ternary complexes form.

FGF–HEPARIN INTERACTIONS ENHANCE STIMULATION OF FGF RECEPTOR SIGNAL TRANSDUCTION

A large number of growth factors can be purified based on their affinity for heparin. The heparin-binding family of FGFs has grown to more than 22 members and includes the prototype FGF2, otherwise known as basic fibroblast growth factor. FGF2 has a very high affinity for heparin ($K_d \sim 10^{-9}$ M) and requires 1.5–2 M NaCl to elute from heparin-Sepharose. FGF2 has potent mitogenic activity in cells that express one of the FGF-signaling receptors (FGFRs; four FGFR genes are known and multiple splice variants exist). Cell-surface HS binds to both FGF2 and FGFR, facilitating the formation of a ternary complex. Both binding and the mitogenic response are greatly stimulated by heparin or HS, which promotes dimerization of the ligand–receptor complex.

The costimulatory role of HS (and heparin) in this system is reminiscent of the heparin/antithrombin/thrombin story. Indeed, the minimal binding sequence for FGF2 also consists of a pentasaccharide. However, this pentasaccharide is not sufficient to trigger a biological response (mitogenesis). For this to occur, a longer oligosaccharide (10-mer) containing the minimal sequence and additional 6-*O*-sulfo groups are needed to bind FGFR. The sequence that binds to both FGF2 and FGFR is prevalent in heparin but rare in HS. The requirement for this rare binding sequence reduces the probability of finding this particular arrangement in naturally occurring HS chains. Thus, some preparations of HS are inactive in mitogenesis, and those containing only one-half of the bipartite binding sequence are actually inhibitory.

The structure of FGF2 co-crystallized with a heparin hexasaccharide has since been obtained (Figure 38.3). The heparin fragment ($[\text{GlcNS6S}\alpha 1\text{-}4\text{IdoA2S}\alpha 1\text{-}4]_3$) was helical and bound to a turn-rich heparin-binding site on the surface of FGF2. Only one *N*-sulfo group and the 2-*O*-sulfo group from the adjacent iduronic acid are bound to the growth factor in the turn-rich binding domain, and the next GlcNS residue is bound to a second site, consistent with the minimal binding sequence determined with oligosaccharide fragments. Unlike antithrombin–heparin interaction, no significant conformational change in FGF2 occurs on heparin binding. The crystal structure of acidic FGF (FGF1) has also been solved and shows similar sequences on its surface. However, the oligosaccharide sequence that binds with high affinity to FGF1 contains 6-*O*-sulfo groups.

The cocrystal structure of the complex of $(\text{FGF2-FGFR})_2$, first solved in the absence of heparin/HS ligand, showed a canyon of positively charged amino acid residues, suggestive of an unoccupied heparin-binding site. Subsequently, the heparin–oligosaccharide-containing complex was solved after introduction of heparin oligosaccharides, suggesting a 2:2:2 complex of FGF2:FGFR1:HS (Figure 38.3). Another important feature of this complex is the orientation of the nonreducing ends of the HS chains that terminate in an *N*-sulfoglucosamine residue, which arises by endolytic cleavage of chains by the enzyme heparanase (Chapter 17). The structure of the FGF1–FGFR2–HS complex is not without controversy; structural analysis of complexes formed in solution and purified by gel filtration has suggested a very different structure consisting of a 2:2:1 complex (Figure 38.3).

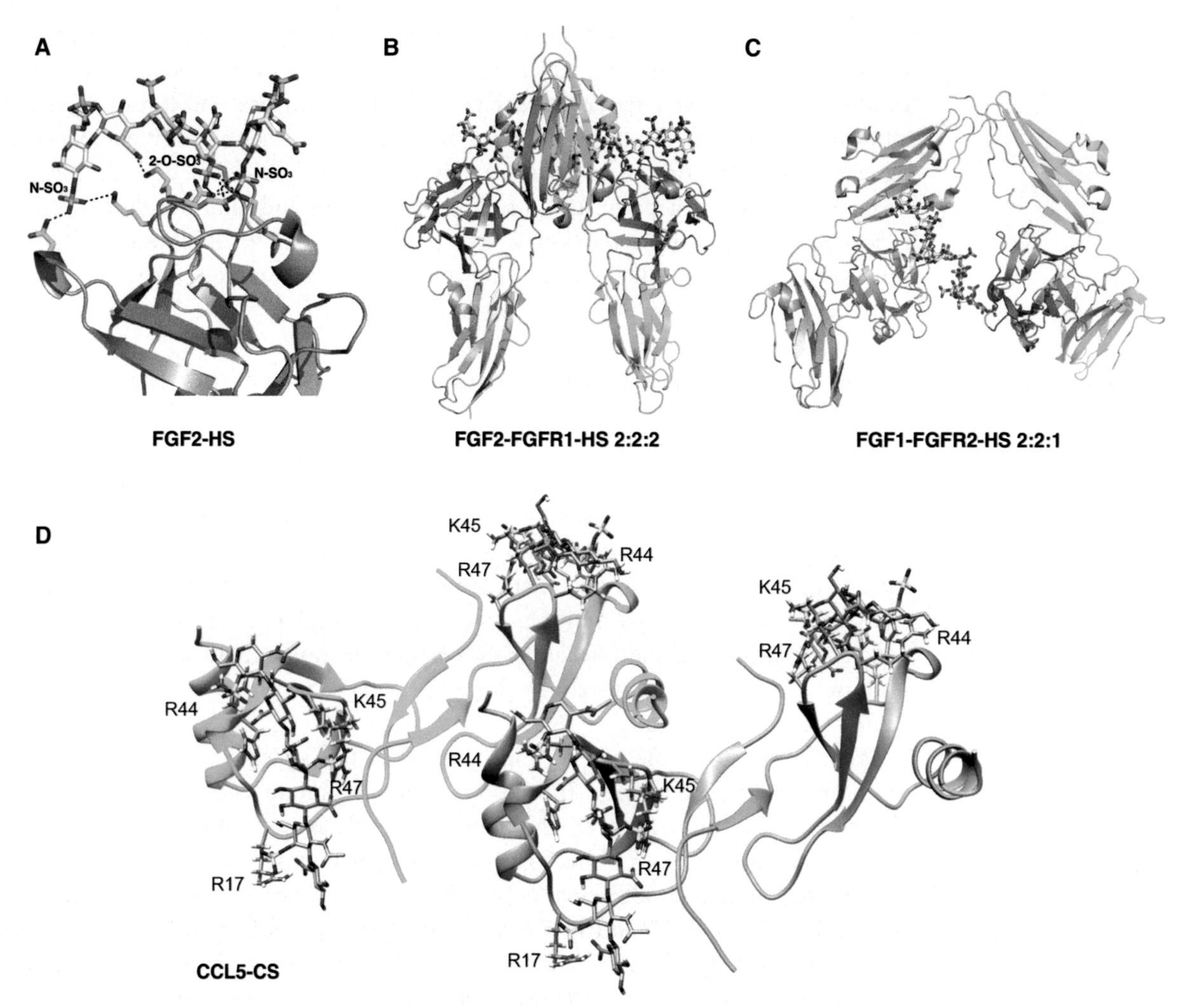

FIGURE 38.3 Crystal and solution structures of GAG–protein complexes. (*A*) Crystal structure of the FGF2–hexasaccharide complex (PDB: 1BFC). Key interactions between the hexasaccharide and FGF2 are shown in *black dotted lines.* The hexasaccharide is shown in *stick representation.* (*B*) Crystal structure of the 2:2:2 FGF2–FGFR1–heparin complex (PDB: 1FQ9). FGF2 is shown in *salmon*, and FGFR1 dimers are shown in *green* and *gold*, respectively. Heparin octasaccharides are shown in *stick representation.* (*C*) Crystal structure of the 2:2:1 FGF1–FGFR2–heparin complex (PDB: 1E0O). FGF1 dimers are shown in *salmon*, and FGFR2 dimers are shown in *green* and *gold*, respectively. Heparin decasaccharide is shown in *stick representation.* (*D*) Model of a CCL5-CS tetrameric complex assembled from the dimeric crystal structure using solution nuclear magnetic resonance (NMR) and small-angle X-ray scattering (SAXS) data. CS hexamers and interacting protein residues are shown as *stick figures.*

CCL5–CHONDROITIN SULFATE INTERACTIONS: STABILIZATION OF A CHEMOTACTIC GRADIENT

Chemokines are another class of proteins exhibiting strong interactions with GAGs. However, these interactions do not directly affect activities of inhibitors or signaling molecules as in previous examples. Instead, they stabilize chemotactic gradients of chemokines; these gradients, in turn, direct leukocytes to sites of injury or infection. They may also facilitate cell migration through the endothelial layer of the blood vessels in which they circulate. This is no easy task; blood is flowing rapidly in these vessels and would easily destroy gradients in the absence of these interactions. CCL5, also called RANTES, is a member of a chemokine subclass that has two adjacent cysteines (CC) near its amino terminus. It is among the earliest discovered and among the most studied of the chemokines. CCL5 binds both HS and CS; but given the abundance of CS chains in normal plasma and the association of CS proteoglycans with vascular development and disease, a specific discussion of CCL5–CS interactions seems appropriate.

Although CCL5 is a small protein (~8 kDa), it exists primarily as dimers ($K_d < 1$ μM) that assemble into larger filamentous structures, especially in the presence of GAGs. These characteristics have made it difficult to study CCL5 structurally. Most studies have involved a mutated form (E66S) and/or study at lower pH—conditions that limit oligomerization to a dimer. Nevertheless, complementary solution data, including nuclear magnetic resonance (NMR) and small angle X-ray scattering (SAXS) have allowed construction of filament models, both with and without GAGs present. Figure 38.3D shows a model of a tetramer built on the known dimer structure (PDB id, 1U4L) and complexed with CS hexamers 4-*O*-sulfated on all three *N*-acetylgalactosamine residues of the GlcAβ1-3GalNAc repeats. Individual dimers are shown as green and sea green ribbons. Additional dimers can be added to each end to build a filament structure.

The interactions between protein residues and charged groups on the CS involve a classic BBXB motif (R44, K45, N46, R47) plus a more remote arginine (R17). Each positively charged arginine is positioned to have strong electrostatic and hydrogen bond interactions with the negatively charged carboxylate of a glucuronic acid residue and 4-*O*-sulfo group of the following *N*-acetylgalactosamine residue. The positioning of these arginine residues may be unique to CS, and other GAG segments may bind quite differently. The model of the filament in Figure 38.3D is also not unique, and both positioning of dimeric subunits and flexibility in CS binding geometry may allow longer GAG chains to cross-link adjacent dimers and facilitate oligomerization. A final point of note is that the groups responsible for interaction with the CCR5 receptor on leukocytes are near the amino terminus, a segment deeply buried in the dimer interface. These groups may be transiently exposed on filament dissociation to release monomers or they may be more permanently exposed on filament ends. Much additional structural work will be required to fully understand the function of CCL5.

OTHER ATTRIBUTES OF GAG–PROTEIN INTERACTIONS

In some cases, the interaction of GAG chains with proteins may depend on metal cofactors. For example, L- and P-selectins have been shown to bind to a subfraction of HS chains and heparin in a divalent-cation-dependent manner. This observation raises the possibility that other examples of cation-dependent interactions with GAG chains may exist. Indeed, annexin A2 and annexin V have been shown to bind HS in a calcium-dependent manner. The co-crystal structures of annexin A2 and annexin V with heparin oligosaccharides revealed that calcium ions can either directly or indirectly (through sequestered water) participate in the interactions with HS. Another example is calcium-dependent interaction between cytokine S100A12 and HS. In this case, formation of the functional HS-binding site requires a calcium-induced conformational change of S100A12.

Although the vast majority of GAG–protein interactions were identified and studied under neutral pH, quite a few GAG-binding proteins have been shown to interact with GAG only under acidic pH. Such proteins include histidine-rich glycoproteins, cystatin-C, selenoprotein P, and serum amyloid A. There is good reason to believe that the interactions between these proteins and GAG are physiologically relevant because extracellular environments with acidic pH exist under many physiological and pathological conditions. In fact, for these proteins, pH change essentially functions as an on/off switch of their interactions with GAG, which might bear physiological significances. Not surprisingly, histidines are often found in the GAG-binding site of these proteins because of its pH-dependent protonation of the imidazole ring.

A main technical challenge to understand GAG–protein interactions is to dissect the essential structural elements of GAGs that contribute to the binding. Unlike researchers studying DNA-binding proteins who have access to all possible DNA sequences, researchers studying GAG-binding proteins traditionally do not have access to GAG oligosaccharides with defined structures. Fortunately, this situation has begun to change in the last 5 years. With the rapid progress of chemoenzymatic and chemical synthesis (or a combination of both) of HS/CS oligosaccharides, a growing number of structural defined oligosaccharides (dp4-dp20) have become available. Many of the oligosaccharides have been used to generate microarrays for rapid determination of structural preference of GAG-binding proteins. Although we have not yet reached a stage at which we can freely design any GAG structures that we wish, the current technology can already provide a surprisingly large array of structures for structure–function studies of GAG-binding proteins.

With more than 500 GAG-binding proteins already identified (and still counting), we cannot help contemplating how this huge GAG interactome really functions at the system level. It is obvious that in any cellular environment, if one GAG-binding protein is found, it is likely that many more GAG-binding proteins are present as well. Do these GAG-binding proteins live in harmony, or are they in constant conflict by competing for binding to GAGs? When one GAG-binding protein is down-regulated, how would other GAG-binding proteins respond to this sudden availability of the free GAG-binding sites? Addressing these questions requires a systems biology approach, which has become possible with the recent increase in information about individual GAG–protein interactions. Given that spatiotemporal changes of GAG structure have profound impacts on the physiological processes associated with GAG-binding proteins, it would be natural to study this question using a systems biology approach.

ACKNOWLEDGMENTS

The authors appreciate helpful comments and suggestions from Ryan Porell, Ulf Lindahl, and So Young Kim.

FURTHER READING

Li W, Johnson DJ, Esmon CT, Huntington JA. 2004. Structure of the antithrombin–thrombin–heparin ternary complex reveals the antithrombotic mechanism of heparin. *Nat Struct Mol Biol* **11:** 857–862. doi:10.1038/nsmb811

Mohammadi M, Olsen SK, Goetz R. 2005. A protein canyon in the FGF–FGF receptor dimer selects from an a la carte menu of heparan sulfate motifs. *Curr Opin Struct Biol* **15:** 506–516. doi:10.1016/j.sbi.2005.09.002

Thacker BE, Xu D, Lawrence R, Esko JD. 2014. Heparan sulfate 3-*O*-sulfation: a rare modification in search of a function. *Matrix Biol* **35:** 60–72. doi:10.1016/j.matbio.2013.12.001

Xu D, Esko JD. 2014. Demystifying heparan sulfate–protein interactions. *Annu Rev Biochem* **83:** 129–157. doi:10.1146/annurev-biochem-060713-035314

Deshauer C, Morgan AM, Ryan EO, Handel TM, Prestegard JH, Wang X. 2015. Interactions of the chemokine CCL5/RANTES with medium-sized chondroitin sulfate ligands. *Structure* **23:** 1066–1077. doi:10.1016/j.str.2015.03.024

Mizumoto S, Yamada S, Sugahara K. 2015. Molecular interactions between chondroitin-dermatan sulfate and growth factors/receptors/matrix proteins. *Curr Opin Struct Biol* **34:** 35–42. doi:10.1016/j.sbi.2015.06.004

Pomin VH, Mulloy B. 2015. Current structural biology of the heparin interactome. *Curr Opin Struct Biol* **34:** 17–25. doi:10.1016/j.sbi.2015.05.007

Xu D, Arnold K, Liu J. 2018. Using structurally defined oligosaccharides to understand the interactions between proteins and heparan sulfate. *Curr Opin Struct Biol* **50:** 155–161. doi:10.1016/j.sbi.2018.04.003

Chen YC, Chen SP, Li JY, Chen PC, Lee YZ, Li KM, Zarivach R, Sun Y J, Sue SC. 2020. Integrative model to coordinate the oligomerization and aggregation mechanisms of CCL5. *J Mol Biol* **432:** 1143–1157. doi:10.1016/j.jmb.2019.12.049

Toledo AG, Sorrentino JT, Sandoval DR, Malmstrom J, Lewis NE, Esko JD. 2021. A systems view of the heparan sulfate interactome. *J Histochem Cytochem* **69:** 105–119. doi:10.1369/0022155420988661

39 Glycans in Glycoprotein Quality Control

Tadashi Suzuki, Richard D. Cummings, Markus Aebi,
and Armando Parodi

N-Glycans affect glycoprotein folding because of their hydrophilic nature. In the endoplasmic reticulum (ER), the processing of N-glycans yields a series of truncated N-glycans that serve as checkpoints that dictate the life or death of many newly made membrane and secreted proteins. Other glycan modifications also may affect glycoprotein folding in the ER. This chapter describes glycan-mediated quality-control processes in the ER and Golgi apparatus and what happens to glycoproteins that fail their "final folding examination."

CHAPERONES FACILITATE PROTEIN FOLDING

During protein synthesis, nascent polypeptides begin to assume their final three-dimensional conformation by passing through folding intermediates dictated, to a large extent, by their primary sequence. Proper folding involves the formation of secondary structures (α-helices and β-strands), burying of hydrophobic residues in the interior of the protein, the formation of disulfide bonds, and quaternary associations via oligomerization or multimerization. Together, these processes prevent unwanted protein aggregation that would interfere with protein functions. To make folding more efficient, cells use various chaperones. Two major cytoplasmic chaperone families consist of heat-shock proteins in the *hsp60* and *hsp70* gene families that bind to exposed hydrophobic domains on misfolded proteins and maintain their solubility as they acquire their final conformation. They also participate in the repair of damaged proteins and proteins that have not properly multimerized. Improperly folded proteins that fail to mature are eventually tagged by ubiquitination, which then leads to degradation by the proteasome.

published by Cold Spring Harbor Laboratory Press; doi: 10.1101/glycobiology.4e.39

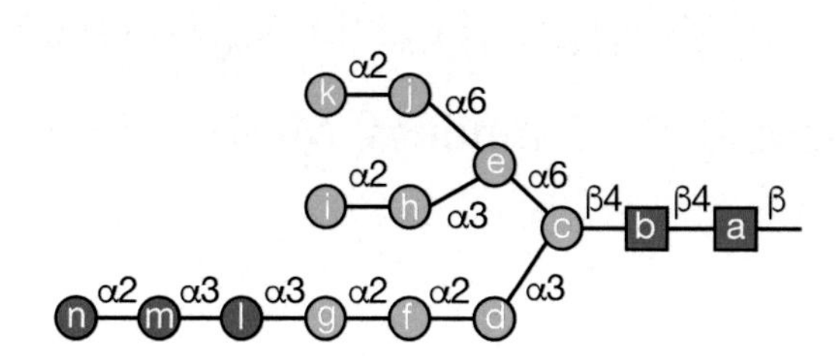

FIGURE 39.1. Mature N-glycan. Lettering (a–n) follows the order of addition of monosaccharides in the synthesis of $Glc_3Man_9GlcNAc_2$-P-P-dolichol (Chapter 9). α-Glucosidase I (GI) removes residue n, and α-glucosidase II (GII) removes residues m and l. UGGT adds residue l to residue g. $Man_8GlcNAc_2$ isomer A (M8A) lacks residues l–n and g. M8B lacks residues l–n and i, whereas M8C is devoid of residues l–n and k. The $Man_7GlcNAc_2$ isomer involved in driving misfolded glycoproteins to degradation (M7BC) lacks residues l–n, i, and k.

Membrane proteins and proteins destined for secretion also undergo quality control during folding. Membrane and secretory proteins originate on membrane-bound ribosomes; translation and translocation into the lumen of the ER occur simultaneously in most cases. However, certain proteins are fully synthesized in the cytosol and translocated into the ER lumen post-translationally. This latter process is more relevant in yeast than mammalian cells. The ER lumen is highly specialized for protein folding: Its oxidizing environment promotes disulfide bond formation and it is the main cellular reservoir of Ca^{++}. To aid proper folding, a battery of molecular chaperones is present, including BiP/Grp78 (a glucose-regulated protein and member of the hsp70 family of chaperones), Grp94, and Grp170. In addition, the ER contains enzymes that promote proline *cis–trans* isomerization and protein disulfide bond formation, such as protein disulfide isomerases (PDI and ERp57, ERp59, and ERp72). The expression of many of these proteins increases during stress responses.

Unlike cytoplasmic proteins, most membrane and secreted proteins undergo modification with N-glycans (Figure 39.1) as they enter the ER lumen. Oligosaccharyltransferase requires flexible domains of a polypeptide for glycosylation, and thus N-glycosylation evolved in eukaryotes to occur before protein folding. Accordingly, N-linked glycans affect the folding of glycoproteins directly by altering the biophysical properties of the protein. They can serve as signals to localize the folding machinery and to signal the folding status of the polypeptide. To achieve glycan-mediated folding, N-glycans are recognized by two lectin-like chaperones in the ER: calnexin (CNX) and calreticulin (CRT). These specialized chaperones require Ca^{++} for activity and bind to monoglucosylated forms of N-glycans, thus retaining the protein in the ER until proper folding occurs. The ER also contains three enzymes central to the process: α-glucosidases I and II (GI and GII), which trim glucose (Glc) from the $Glc_3Man_9GlcNAc_2$ initially transferred to the protein and release the glycoprotein from CNX/CRT. The enzyme UDP-glucose glycoprotein glucosyltransferase (UGGT) reglucosylates the N-glycan if proper folding has not occurred. Glycoproteins that fail to fold, or to oligomerize properly, are marked with specific truncated N-glycans, and eventually retrotranslocated into the cytoplasm in which they are destroyed by N-deglycosylation and proteasomal degradation, a process called ER-associated degradation (ERAD). There are several ERAD pathways, which are described in the following sections.

CNX/CRT AND UGGT DETERMINE WHEN GLYCOPROTEINS ARE PROPERLY FOLDED

N-Glycosylation in the ER is considerably different from other types of glycosylation. It begins with the "en bloc" transfer of a 14-monosaccharide glycan to the polypeptide as it exits the translocon complex (Chapter 9). In vertebrates, addition of the N-glycan occurs on incompletely folded polypeptides. In bacteria, N-glycosylation can occur both on nascent polypeptides and on fully mature proteins (Chapter 21). N-Glycans modify the physical properties of glycoproteins by providing noncharged, bulky, hydrophilic groups that keep the protein in solution during folding. They also modulate protein conformation by forcing amino acids near the N-glycan into a hydrophilic environment.

N-Glycan processing starts immediately after transfer of the precursor sugar chain to the protein in the lumen of the ER. GI removes the terminal α1-2Glc and GII then removes both α1-3Glc units (Figure 39.1). Although GII may remove the second Glc (designated m in Figure 39.1) from glycoproteins with a single N-glycan, the trimming of the second Glc occurs more efficiently when there is a second N-glycan on the same protein. Depending on cell type, a variable number of Man residues may also be removed while glycoproteins remain in the ER. GII is an ER-soluble, heterodimeric protein. Its 100- to 110-kDa α-subunit contains the catalytic activity, whereas the 50- to 60-kDa β-subunit displays a canonical ER retention/retrieval sequence (XDEL), which is partially responsible for the subcellular location of the whole enzyme. The carboxy-terminal portion of the β-subunit contains a mannose receptor homologous (MRH) domain that is homologous to the domain in the Golgi mannose 6-phosphate receptor responsible for recognizing lysosomal enzymes and for their delivery to lysosomes (Chapters 9 and 33). The affinity of the GII MRH domain is higher for the $Man_9GlcNAc_2$ (M9) N-glycan and decreases for truncated N-glycans. Consequently, the GII activity toward N-glycans diminishes for truncated N-glycans created by ER mannosidases.

CNX and CRT bind to glycoproteins containing a single α1–3Glc residue (Figure 39.2). CNX is a type I transmembrane protein and CRT is soluble, but both have ER retention/retrieval signals at their carboxyl terminus, retaining them in the ER. They have structurally similar lectin domains and comprise part of a large, weakly associated heterogeneous protein network that includes BiP/Grp78, ERp57, Grp94, and other ER-resident proteins that assist protein folding. Thereby, they act as interpreters of the N-glycan code to localize the folding machinery to a glycoprotein substrate. Both CNX and CRT are monovalent, low-affinity lectins for monoglucosylated, oligomannosyl N-glycans, but because of their different membrane-bound or soluble status within the ER lumen, their in vivo specificities are not identical. CNX mainly interacts with N-glycans close to the ER membrane, whereas CRT binds preferentially to glycoproteins in the ER lumen or to those that have large luminal domains. Binding of incompletely folded glycoproteins to the CNX/CRT complex prevents their exit from the ER and enhances folding efficiency by preventing aggregation and premature oligomerization/degradation and by facilitating the formation of native disulfide bonds. This latter task is performed by ERp57, a CNX/CRT-associated oxidoreductase. CNX and CRT are, therefore, unconventional chaperones that do not directly recognize protein moieties of folding intermediates like conventional chaperones, but rather they recognize particular N-glycans in the folding intermediates.

Removal of the final Glc residue from N-glycans by GII prevents binding of the protein to CNX/CRT. If properly folded, these glycoproteins are packaged into COPII-coated vesicles and transferred to the Golgi (Figure 39.2). However, if they remain unfolded, UGGT adds a single α1–3Glc to the glycan, thus recreating the identical N-glycan first recognized by CNX/CRT. Therefore, UGGT has the remarkable property of exclusively glucosylating glycoproteins that display a nonnative conformation. UGGT discriminates between different nonnative conformers, and preferentially glucosylates glycoproteins in their last stages of folding when they are displaying exposed hydrophobic amino acid patches (molten globules). In vitro, the enzyme also recognizes N-glycans linked to relatively short hydrophobic peptides or to hydrophobic aglycones, although less efficiently. Contrary to BiP/Grp78 that preferentially binds extended hydrophobic "sequences," UGGT recognizes hydrophobic "surfaces." This difference determines that glycoproteins are first recognized by chaperones like BiP/Grp78 and only later, at final folding stages, are epitopes created by UGGT recognized by CNX/CRT. UGGT modifies N-glycans located in the relative vicinity of amino acids that prevent rapid folding of a glycoprotein.

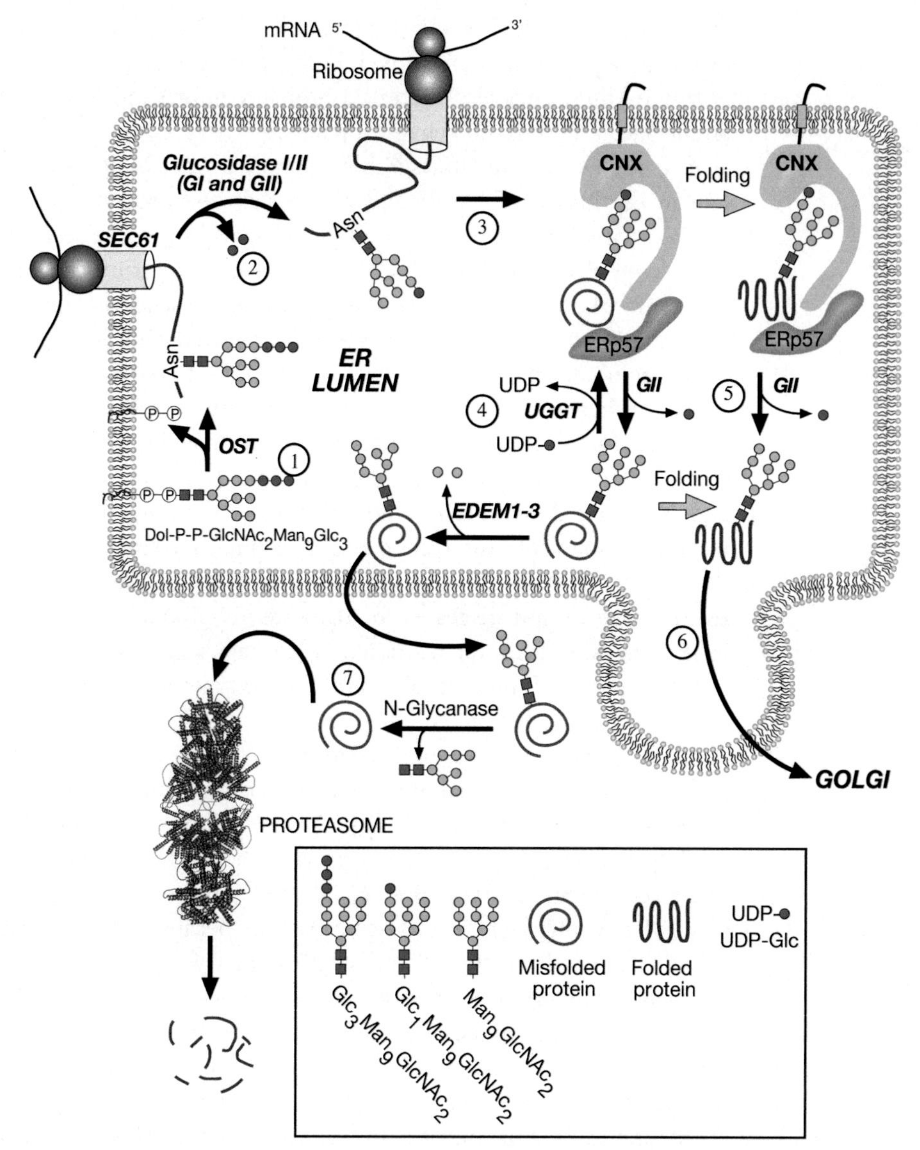

FIGURE 39.2. Model of quality control in glycoprotein folding. Proteins entering the endoplasmic reticulum (ER) are N-glycosylated by the oligosaccharyltransferase (OST) as they emerge from the translocon (SEC61) (*1*). Two glucose (Glc) residues are removed by the sequential action of α-glucosidases I and II (GI and GII) to generate monoglucosylated N-glycans (2) that are recognized by calnexin (CNX) and/or calreticulin (CRT) (only CNX is shown) that are associated with ERp57 (3). The complex between the lectins and folding intermediates dissociates on removal of the last Glc by GII and is reformed by UGGT (*4*). Once glycoproteins have acquired their native conformation, either free or bound to the lectins, GII hydrolyzes the remaining Glc and releases the glycoprotein from lectin anchors (5). Glycoproteins with native structures are not recognized by UGGT and are transported to the Golgi (6). Glycoproteins remaining in misfolded conformations are retrotranslocated to the cytoplasm, where they are deglycosylated and degraded by the proteasome (7). One or more Man residues may be removed during the whole folding process.

UGGT also glucosylates incompletely assembled multimeric complexes of fully mature monomers because it recognizes hydrophobic surfaces exposed in the absence of the appropriate subunits. UGGT is a relatively large, soluble, ER protein (150- to 170-kDa) that uses UDP-Glc as sugar donor and requires millimolar Ca^{++} concentrations, typically found within the ER lumen. UGGT has at least two domains. The amino-terminal one, comprising ~80% of the molecule, recognizes abnormal protein conformers. This domain contains three tandemly repeated, thioredoxin-like domains that are frequently found in ER proteins involved in protein folding and quality-control processes. A crystal structure of the third domain showed that it contains a large hydrophobic patch hidden by a flexible carboxy-terminal helix. The carboxy-terminal domain has homology with other glycosyltransferases and is responsible for the catalytic activity.

Together, UGGT and the CNX/CRT complex ensure that only properly folded and multimerized glycoproteins move from the ER to the Golgi. Unlike GII, the activity of UGGT does not decrease on reduction of N-glycan Man content. Therefore, ER α-mannosidase catalyzed removal of Man from N-glycans results in longer half-lives of UGGT reaction products that may be recognized by CNX/CRT. It is possible that the differential activities of UGGT and GII toward N-glycans with reduced Man content provides glycoproteins with a last opportunity to achieve proper folding before proteasomal degradation. The ER folding and quality-control process is not foolproof, however. Some misfolded proteins escape the ER and others may misfold after they exit. Thus, there are additional checkpoints to correct mistakes that occur later in the secretory pathway.

ER–Golgi intermediate compartment 53 (ERGIC53) is a type I membrane protein that binds to oligomannosyl N-glycans in a Ca^{++}-dependent manner. Its luminal, carbohydrate-binding domain is similar to those of soluble lectins from leguminous plants. ERGIC53 cycles between the ER and the ERGIC. The mammalian protein contains both ER-targeting (dilysine) and ER-exit (diphenylalanine) determinants at its carboxyl terminus, which bind, respectively, to COPI and COPII coatomers involved in trafficking between the ER and Golgi. ERGIC53 loads a subset of glycoproteins (coagulation factors V and VIII and cathepsins C and Z) into COPII-coated vesicles leaving the ER for the Golgi. Mutation of ERGIC53 causes combined factor V and VIII deficiency, indicating an important role in secretion of these, and probably other glycoproteins. Vesicular integral protein 36 (VIP36) is another Golgi lectin that also binds glycoproteins containing oligomannosyl N-glycans. VIP36 may facilitate transport of glycoproteins from the ERGIC to *cis*-Golgi cisternae, or retrieve glycoproteins bearing N-glycans that did not undergo conversion to $Man_5GlcNAc_2$ by *cis*-Golgi α-mannosidases IA, IB, and/or IC. Conceivably, this would provide an opportunity for new rounds of trimming and eventual formation of complex N-glycans. Both ERGIC53 and VIP36 are up-regulated as part of the unfolded protein response. Whether other lectins recognizing specific glycans are involved in post-ER quality control is still an open question.

REMOVING MISFOLDED GLYCOPROTEINS FROM THE CNX/CRT/UGGT CYCLE

Despite the presence of a battery of classical chaperones and the CNX/CRT/UGGT cycle in the ER, the process of glycoprotein folding is relatively inefficient, with as much as 80% of some newly made proteins never maturing. How do cells distinguish between misfolded glycoproteins and folding intermediates if both entities have almost identical structural features such as exposed hydrophobic patches? How do cells pull glycoproteins destined for degradation out from futile reglucosylation–deglucosylation cycles if those glycoproteins are indeed very efficiently glucosylated by UGGT? How are terminally misfolded glycoproteins driven to the proteasome for degradation? The key discriminating factor appears to be the structure of the truncated N-glycans that result from the relatively long stay of a misfolded glycoprotein in the ER lumen.

Mammalian cells have, in addition to Man1B1, three additional homologs called EDEMs (ER degradation-enhancing α-mannosidase-like) proteins, which initially were thought to be devoid of activity. However, when either Man1B1 or the EDEMs were individually knocked out in human or chicken cells, conversion of M9 to M8B (Figure 39.2) was mainly performed by

EDEM2, and the contribution of Man1B1 to that particular demannosylation step was minimal. On the other hand, conversion of M8B to M7 lacking residues i and k (M7BC; Figure 39.1) was performed by EDEMs 1 and 3. Recent evidence suggests that Man1B1, initially considered as ER mannosidase I, localizes to Golgi, and appears to work as a backup enzyme to cleave off residue i in the Golgi as this residue is rather resistant to the action of Golgi α-mannosidase I. The M7BC isomer exposes an α1–6Man (residue j; Figure 39.1), which is specifically recognized by an ER lectin (OS9 in mammals and Yos9p in *Saccharomyces cerevisiae*). Like GII, this lectin also contains an MRH domain that binds N-glycans smaller than M9, including M7BC. OS9/Yos9p forms part of the complex that translocates misfolded glycoproteins from the ER lumen to the cytosol (ERAD-L; see below). Both the conversion of M9 to M8B and the conversion of M8B to M7BC are slow processes in mammalian cells when compared with deglucosylation reactions. This affords two checkpoints ensuring that only misfolded glycoproteins, and not folding intermediates, are driven toward proteasomal degradation after a relatively long ER residence. In yeast, only an ER α-mannosidase and one EDEM-like protein called Htm1p are present. The former transforms M9 to M8B very rapidly in practically all glycoproteins, whereas the latter converts M8B to M7BC slowly, thus affording only one checkpoint for the exclusive degradation of terminally misfolded glycoproteins. Htm1p forms a complex with PDI. More recent evidence further suggests that mammalian EDEMs act as active α-mannosidases forming complexes with distinct oxidoreductases. In the case of yeast the oxidoreductase-Htm1p association enhances the α-mannosidase activity of the latter and the complex participates in the recognition of ERAD substrates. However, not all glycoproteins with N-glycans terminating in α1–6Man are driven to degradation. For instance, 3-hydroxy-3-methylglutaryl acetyl-coenzyme-A reductase, a key ER enzyme in sterol biosynthesis, carries $Man_5GlcNAc_2$ and $Man_6GlcNAc_2$ N-glycans that do not target it for degradation (see below).

RETROTRANSLOCATION OF MISFOLDED GLYCOPROTEINS TO THE CYTOSOL

Retrotranslocation of misfolded glycoproteins from the ER lumen to the cytosol for proteasomal degradation ultimately depends on various protein complexes, some of which contain integral ER membrane proteins. ER membrane proteins with folding defects in the cytosolic domain are extracted from the ER membrane by the Doa10 complex via ERAD-C, whereas those in which the folding defect is present in the luminal (ERAD-L) or membrane (ERAD-M) domains use the Hrd1 complex. Both Doa10p and Hrd1p are integral membrane proteins with E3 ligase activity in their cytosolic portion. Other proteins in these complexes include chaperones, proteins with E2 ubiquitin-conjugating activity or proteins that recognize misfolded proteins. For instance, glycoproteins with N-glycans containing a terminal α1–6Man are bound by Yos9p and undergo ERAD only if the N-glycans are in unstructured protein, a feature recognized by Hrd3p in the Hrd1 complex. Finally, proteins common to both complexes such as yeast Cdc48 (p97 in mammals) are responsible for membrane extraction of misfolded proteins in an ATP-dependent manner. Although the pore by which ERAD-L substrates are actually transported to the cytosol has not been identified, it is known that the substrates must be unfolded in the ER lumen. Although studies on ERAD complexes have been mainly performed in *S. cerevisiae*, mammalian cells have homologs of nearly all proteins described in the yeast complexes.

N-Glycans are removed from unfolded glycoproteins during proteasomal degradation. A cytoplasmic peptide:N-glycanase (PNGase, N-glycanase, NGLY1) plays an important role in both removing the glycan and constructing an efficient predegradation complex (Figure 39.3). NGLY1 recognizes only misfolded or denatured glycoproteins. The enzyme is bound to the proteasome by subunits S4 and HR23B as a complex with Cdc48, a component of

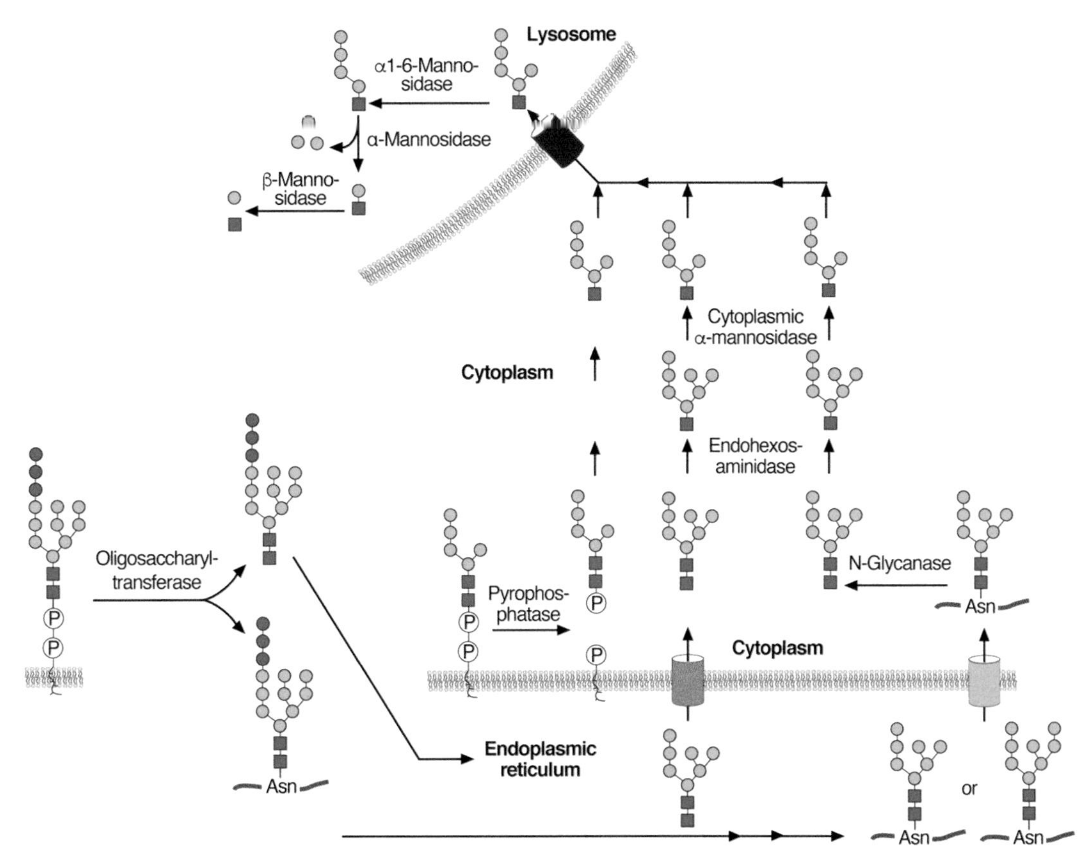

FIGURE 39.3. Degradation of oligomannosyl N-glycans in the endoplasmic reticulum (ER), cytoplasm, and lysosomes. Specific pathways exist for the degradation of free glycans released from misfolded glycoproteins and glycopeptides generated within the ER. Dolichol-linked oligosaccharides facing the cytoplasmic side of the ER can also be released by a pyrophosphatase and further degraded in lysosomes.

ERAD complexes. Degradation of the released N-glycan occurs in two stages. First, cleavage occurs between the two GlcNAcs (chitobiose) in the core of the N-glycan via a cytoplasmic endo-β-*N*-acetylglucosaminidase. A cytoplasmic α-mannosidase cleaves up to four Man residues to generate $Man_5GlcNAc$ (Man c, d, e, f, and g in Figure 39.1). This glycan is then taken into the lysosome via an ATP-dependent lysosomal membrane transporter for final degradation to monosaccharides. Whether these free glycans have any function in the cytosolic/nuclear compartment before they are degraded has not been investigated.

O-GLYCOSYLATION REACTIONS IN ER QUALITY CONTROL

Protein O-fucosyltransferases (Ofut1 in Drosophila; POFUT1 and POFUT2 in mammals) are soluble ER enzymes that O-fucosylate Ser/Thr residues. Ofut1 and POFUT1 add fucose to epidermal growth factor (EGF)–like repeats, whereas POFUT2 adds fucose to thrombospondin (TSR) type 1–like repeats (Chapter 13). EGF and TSR domains are present as tandem repeats in numerous proteins of multicellular organisms. Ofut1/POFUT1 and POFUT2 selectively fucosylate only properly folded protein domains, suggesting a role of this glycosylation in folding quality control. Indeed, several POFUT2 targets including ADAMTSL1 and ADAMTSL2 (scaffolding proteins in the extracellular matrix), ADAMTS9 (a matrix remodeling protease), and ADAMTS13 (the von Willebrand factor cleaving protease) strictly require O-fucosylation for efficient secretion. Indeed, O-fucosylation appears to stabilize folded TSRs. The case of

Ofut1/POFUT1 is a bit more ambiguous. Studies performed in Drosophila on the efficiency of cell-surface expression of the Ofut1 substrate Notch showed that Ofut1 is required as a chaperone for Notch folding and cell-surface expression, and that chaperone activity is largely independent of the O-fucosyltransferase activity. In contrast, POFUT1 does not appear to be absolutely required for cell-surface expression of Notch receptors in all mammalian cells, suggesting that cell-specific chaperones may be able to compensate for loss of POFUT1 in certain contexts. It has been reported recently that O-mannosylation in *S. cerevisiae* ER might be responsible for terminating futile cycles of association–dissociation between proteins unable to attain their native conformation and Kar2p (the budding yeast homolog of mammalian BiP/Grp78 chaperone).

THE ER QUALITY-CONTROL MACHINERY AND VIABILITY

The loss of CNX and CRT in most multicellular organisms causes a severe phenotype. Mice deficient in CNX reach full term, but half die within 2 days of birth, and very few survive beyond 3 months. The runts that survive have obvious motor disorders with loss of large myelinated nerve fibers. CRT-deficient mice and ERp57-null mice also show embryonic lethality. In contrast, CNX and CRT null *Caenorhabditis elegans* mutants are viable. Yeast has CNX or CNX-like lectins (but not CRT), whereas trypanosomatid protozoa have only CRT. Another important conserved mammalian ER lectin resident is calmegin, expressed only in the testes during spermatogenesis. Calmegin shares ~60% homology with CNX in mice. Mice deficient in calmegin are nearly sterile despite producing normal-looking sperm. The sperm are defective in migration into the oviducts and they do not adhere to the zona pellucida. UGGT is absent from *S. cerevisiae* but not from *Schizosaccharomyces pombe* cells. The enzyme is not required for the viability of the fission yeast or single mammalian cells grown under normal conditions. The enzyme is also dispensable for growth of some multicellular organisms, such as plants. However, loss of UGGT is embryonically lethal in mice. This fact, as well of the strict requirement of UGGT for the viability of *S. pombe* cells only when grown under severe ER stress conditions, point to a restricted set of glycoproteins absolutely requiring UGGT for proper folding.

In addition to combined factor V and VIII deficiency caused by ERGIC53 mutations (see above), other congenital diseases result in retention of defective proteins in the ER and their eventual proteasomal degradation. For instance, the most common mutation in cystic fibrosis patients, CFTR-ΔF508, leads to improper folding of this chloride channel and its retention in the ER. Importantly, the mutation affects CFTR transit to the plasma membrane but not its transport activity. Similarly, UGGT recognition of a brassinosteroid receptor mutant triggers retention in the ER of *Arabidopsis thaliana* cells. Thus, certain diseases caused by ER retention of defective glycoproteins might be ameliorated by inhibitors of UGGT activity. One beautiful example of how the fine sensitivity of UGGT allows modulation of basic physiological processes is its role during antigen presentation by the major histocompatibility complex I (MHC I). Here, MHC I complexes loaded with suboptimal peptides are recognized by UGGT, thus triggering their retention in the ER. On the contrary, complexes formed with high-affinity peptides are poor UGGT substrates and are able to be presented on the cell surface. Many congenital diseases such as lysosomal storage diseases also result in the accumulation of misfolded glycoproteins in the ER, thus overloading the folding machinery. An estimated 20%–30% of normal proteins misfold, so it is not surprising that small differences in the kinetics of conformational maturation and degradation can tip the balance to ERAD for lack of sufficient chaperones and other factors that facilitate folding.

Although most glycoproteins transiently associate with CNX/CRT during their maturation in the ER in vivo, other ER folding and quality-control mechanisms do the job quite efficiently in cultured cells. However, viral glycoproteins such as vesicular stomatitis virus G protein, HIV glycoprotein gp120, and M protein from hepatitis B virus, absolutely require CNX/CRT for proper folding and exit from the ER. Several viruses use components of the ERAD machinery to evade immune surveillance in the host. For instance, cytomegalovirus proteins bind to MHC I complexes and deliver them to E3 ligase complexes that result in their eventual proteasomal degradation. Similarly, expression of HIV ER protein Vpu results in the proteasomal degradation of CD4. SV40 and proteins such as cholera toxin are retrotranslocated to the ER through the secretory pathway where they use some ERAD components to reach the cytosol where they produce deleterious effects.

ACKNOWLEDGMENTS

The authors acknowledge contributions to the previous version of this chapter from Jeffrey D. Esko and Hudson H. Freeze and helpful comments and suggestions from Taroh Kinoshita and Kelley Moremen.

FURTHER READING

Helenius A, Aebi M. 2004. Roles of N-linked glycans in the endoplasmic reticulum. *Annu Rev Biochem* **73:** 1019–1049. doi:10.1146/annurev.biochem.73.011303.073752

Lehrman MA. 2006. Stimulation of N-linked glycosylation and lipid-linked oligosaccharide synthesis by stress responses in metazoan cells. *Crit Rev Biochem Mol Biol* **41:** 51–75. doi:10.1080/10409230500542575

Ruggiano A, Foresti O, Carvalho P. 2014. Quality control: ER-associated degradation: protein quality control and beyond. *J Cell Biol* **204:** 869–879. doi:10.1083/jcb.201312042

Vasudevan D, Haltiwanger RS. 2014. Novel roles for O-linked glycans in protein folding. *Glycoconj J* **31:** 417–426. doi:10.1007/s10719-014-9556-4

Benyair R, Ogen-Shtern N, Lederkremer GZ. 2015. Glycan regulation of ER-associated degradation through compartmentalization. *Semin Cell Dev Biol* **41:** 99–109. doi:10.1016/j.semcdb.2014.11.006

Caramelo JJ, Parodi AJ. 2015. A sweet code for glycoprotein folding. *FEBS Lett* **589:** 3379–3387. doi:10.1016/j.febslet.2015.07.021

Harada Y, Hirayama H, Suzuki T. 2015. Generation and degradation of free asparagine-linked glycans. *Cell Mol Life Sci* **72:** 2509–2533. doi:10.1007/s00018-015-1881-7

Lamriben L, Graham JB, Adams BM, Hebert DN. 2015. N-glycan based ER molecular chaperone and protein quality control system: the calnexin binding cycle. *Traffic* **17:** 308–326. doi:10.1111/tra.12358

Satoh T, Yamaguchi T, Kato K. 2015. Emerging structural insights into glycoprotein quality control coupled with N-glycan processing in the endoplasmic reticulum. *Molecules* **20:** 2475–2491. doi:10.3390/molecules20022475

Tannous A, Pisoni GB, Hebert DN, Molinari M. 2015. N-linked sugar-regulated protein folding and quality control in the ER. *Semin Cell Dev Biol* **41:** 79–89. doi:10.1016/j.semcdb.2014.12.001

40 Free Glycans as Bioactive Molecules

Antonio Molina, Malcolm A. O'Neill, Alan G. Darvill, Marilynn E. Etzler, Debra Mohnen, Michael G. Hahn, and Jeffrey D. Esko

There has been a growing recognition that free glycans are used as signals for the initiation of a wide variety of biological processes. Such signaling events are found in development and defense responses of plants and animals and in interactions of organisms with one another. This chapter covers the current information on this field of study.

NATURE AND SCOPE OF GLYCAN SIGNALING SYSTEMS

Glycan signaling systems are diverse. Sugars (glucose, fructose, and sucrose) may be used by sensing systems, which are typically linked with the metabolism of the sugar, and form complex webs of signaling events linked to hormones. Glycan-signaling systems also involve various glycoconjugates. The addition of O-linked *N*-acetylglucosamine (GlcNAc) to cytoplasmic and nuclear proteins results in changes in the cytoskeleton, gene transcription, and enzyme activation (Chapter 19). Glycosphingolipids may form lipid rafts, which act as a platform for sequestering signaling receptors or can associate with receptor tyrosine kinases and modulate their activity (Chapter 11). Membrane proteoglycans containing sulfated glycosaminoglycans—including syndecans, glypicans, and phosphacan—may act as signaling molecules by interacting with kinases or phosphatidylinositol-4,5-bisphosphate (Chapters 17 and 38). Most plasma membrane signaling receptors, including receptor tyrosine kinases and G-protein-coupled receptors, contain N-glycans and O-glycans that modulate their stability and activity (Chapters 9 and 10). Binding of galectins to these glycans (Chapter 36) or removal of sialic acids by cell-

published by Cold Spring Harbor Laboratory Press; doi: 10.1101/glycobiology.4e.40

surface sialidases (Chapter 15) is also thought to modulate signaling. These and other signaling processes are described elsewhere in this book and are not discussed further here.

There is increasing evidence that low concentrations of specific free glycans are signals that initiate numerous biological processes. The first evidence for such signals was obtained during studies of defense responses in plants. Subsequently, glycan perception and signaling have been shown to be important in plant and animal development, in innate immunity, and in the initiation of the nitrogen-fixing *Rhizobium*–legume symbiosis. The glycans that function as signals have been identified in many of these processes and have been given the generic term "oligosaccharins," regardless of their species of origin or the biological process(es) in which they participate. In contrast, only a few of the receptors and mechanisms of signal transduction have been identified and characterized.

GLYCAN SIGNALS TRIGGER THE INITIATION OF THE PLANT DEFENSE RESPONSE

Plants have several surveillance and defensive systems to control infection by pathogens. One of these systems shows similarities with the mammalian innate immune system and consists of plasma membrane-localized pattern-recognition receptors (PRRs) that trigger defense responses upon perception of specific molecules by their extracellular ectodomains (ECDs). These ECDs can bind nonself molecules derived from pathogens (pathogen-associated molecular patterns, PAMPs) or self-compounds that are released from plant cells or synthesized on infection by pathogens (named generically damage-associated molecular patterns or DAMPs; Figure 40.1). Among the molecules perceived by PRRs are glycans (oligosaccharides) released from microbial and plant cell wall polysaccharides by hydrolytic enzymes originating from the plant or the pathogen. On binding of a ligand by its specific PRR, a PRR coreceptor is often recruited to form a protein complex, which leads to the activation of the receptor's cytoplasmic kinase domain and the initiation of defense responses. Defense response signaling can involve changes in ion flux across the plasma membrane (including a cytoplasmic Ca^{++} burst), the formation of reactive oxygen species (ROS) by NADPH oxidases (RBOHs), the phosphorylation and activation of mitogen-activated and calcium-dependent protein kinases (MPKs and CDPKs, respectively) and the up-regulation of defense-associated genes (Figure 40.1). These defense responses can lead to changes in the plant cell wall, the production of glycanases that fragment the pathogen's cell wall, and the synthesis of metabolites (phytoalexins) and antimicrobial proteins/peptides that inhibit pathogen growth. These defense responses often lead to localized cell death in the plant tissue, which is visible as necrotic spots at the site of infection and limits the pathogen's spread. Of the vast number of ligands that ECD-PRRs can perceive, those composed of carbohydrate moieties are poorly studied. Only a limited number of PRR/glycan pairs have been identified and their triggered defense responses characterized. There is evidence for some similarities between the plant defense responses and animal innate immune system, most notably where specific pattern recognition occurs (Chapter 42).

Early studies showed that oligosaccharides derived from cell wall glycans of the plant or the pathogen elicit numerous plant defense responses at nanomolar/micromolar concentrations. Oligogalacturonides composed of 1-4-linked α-GalA residues are one example of oligosaccharins released from plant cell wall polysaccharides, in this specific case from homogalacturonans by endopolygalacturonases (EPGs) secreted by the pathogen. The activities of such pathogen-derived EPGs are frequently inhibited or modulated by polygalacturonase-inhibiting proteins (PGIP) produced by plants. Similarly, oligoglucosides composed of 1-6- and 1-3-linked β-Glc residues released from the mycelial walls of the soybean pathogen *Phytophthora sojae* by plant endo-glucanases were an early example of oligosaccharins generated from a pathogen's cell wall.

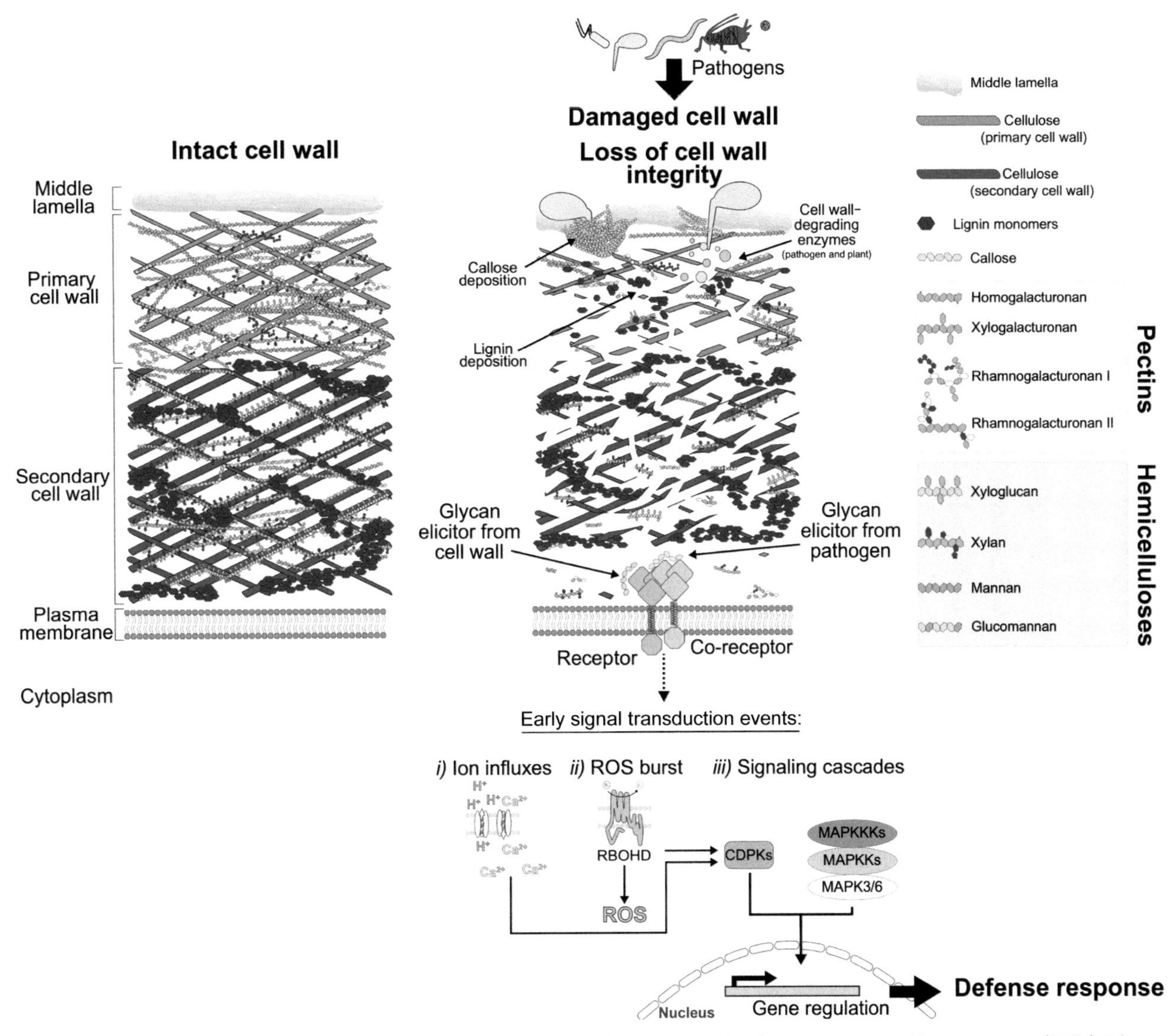

FIGURE 40.1. Plant defense responses upon glycan perception by pattern-recognition receptors (PRRs). These responses are initiated by a glycan generated by glycanases that fragment either the pathogen cell walls or the plant cell walls. The glycan interacts with specific PRR through its extracellular ectodomain (ECD), and a complex is then formed with a PRR coreceptor to initiate a phosphorylation cascade and/or numerous cellular responses that culminate in the expression of defense-related genes and killing of the pathogen or limiting its spread. The main components of the plant cell wall are indicated on the side bar.

Other pathogen-derived oligosaccharins are the PAMPs derived from linear homo-oligomers from fungal/oomycete cell walls that include chitin [1,4-β-D-$(GlcNAc)_n$], and its deacetylated form (chitosan), and β-1,3-glucan oligosaccharides (Figure 40.2). More recently, additional plant cell wall–derived oligosaccharins have been shown to trigger defense responses, including cellulose-derived oligomers (β-1,4-glucans), mixed-linkage glucans (MLGs: β-1,4/β-1,3 glucans), and oligosaccharides derived from xyloglucans, mannans, xylans, or callose, which trigger signaling cascades in *Arabidopsis* and other plant species, including crops (Figure 40.2). It has recently been shown that at least some "self" immune-active plant glycans are released in a regulated manner as part of normal plant developmental pathways and are not necessarily tied to cell-wall damage pathways (see below). It is also noteworthy that some immune-active oligosaccharides can be released from polymers present in the walls of both the pathogen and the plant. For

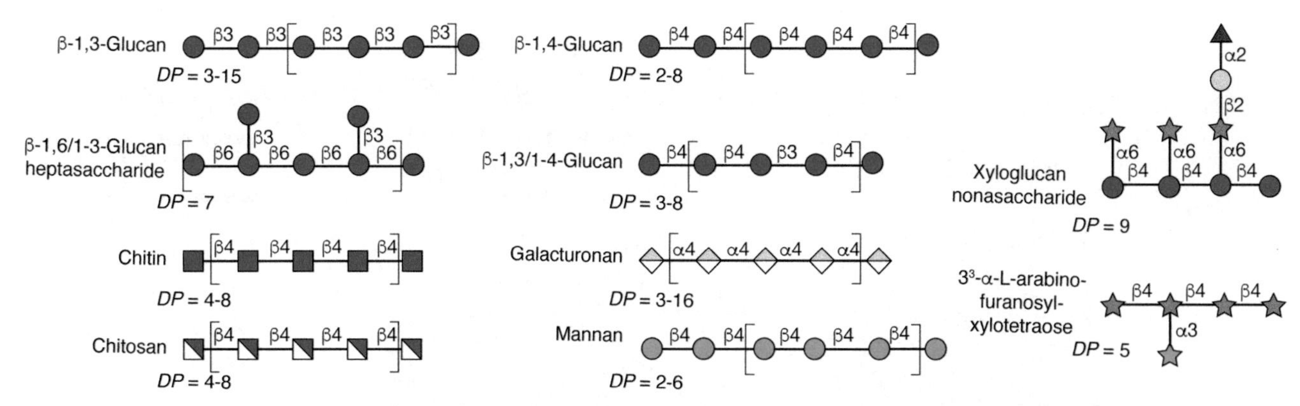

FIGURE 40.2. Oligosaccharins that are active in plants. Examples of oligosaccharides derived from fungal, oomycete, and plant cell walls are shown. The degree of polymerization (DP) of the minimally active oligosaccharide structures triggering some responses in plants is indicated between brackets.

example, β-1,3-glucan oligosaccharides are released from callose, which is produced by plants and is also a component of many fungal and oomycete cell walls. Similarly, β-1,4-glucan oligosaccharides are released from cellulose that is present in the cell walls of plants and some oomycetes.

The composition (monosaccharide units), degree of polymerization (DP), and branching of the oligosaccharins determine their biological activity in triggering plant defense. For example, oligogalacturonides that are biologically active typically require a DP of 10–14 for activity, whereas DPs of >7 and >4 are necessary for the oligochitosans and oligochitins, respectively (Figure 40.2). A single active hepta-glucoside from oomycete walls was isolated from a mixture of approximately 300 inactive structural isomers and found to trigger defense responses in soybean. Both the DP and location of the β-1-3 branches in this oligosaccharide are important for its biological activity (Figure 40.2). Elongating the oligosaccharide at the reducing end had no discernible effect on bioactivity and activity was also largely retained by removal of a single glucose from the reducing end. However, the hexa-glucoside was the minimal structure of this oligosaccharide that had appreciable activity in inducing defense responses in soybean.

The low quantities and different types of glycan signal molecules that elicit defense responses suggest that these glycans are recognized by specific plasma membrane-localized receptors. A 75-kDa plasma membrane protein with high-binding affinity for chitin elicitors was identified in rice cell plasma membranes and proposed to be involved in oligosaccharide perception and signal transduction. This rice chitin oligosaccharide elicitor-binding protein (CEBiP) contains a lysozyme motif (LysM) on its ECD and no appreciable portion of the protein on the cytoplasmic side of the membrane. Similarly, a plasma membrane-localized receptor-like kinase (CERK1) with a LysM-ECD is required for chitin-induced signaling in *Arabidopsis*. In both rice and *Arabidopsis*, the perception of chito-oligosaccharins involves an additional LysM-containing receptor, CERK1 and LYK5, respectively (Figure 40.2). A similar situation has been shown to exist in legumes (*Medicago*), which requires LYK9 (a AtCERK1 homolog) and LYR4 (a AtLYK5 homolog) for chito-oligosaccharide (DP=8) activation of defense signaling. Members of the LysM family of *Arabidopsis* are also involved, probably as coreceptors, in the perception of β-1,3-glucan and β-1,4/β-1,3-glucan oligosaccharides. Recognition of pectin-derived oligogalacturonides in *Arabidopsis* involves wall-associated receptor-like kinases (WAKs; WAK1) that have an ECD with similarities to the epidermal growth factor (EGF) domain of mammals. Reducing or impairing expression of genes encoding these receptors in plants by RNA interference (RNAi) or mutation results in suppression of defense responses.

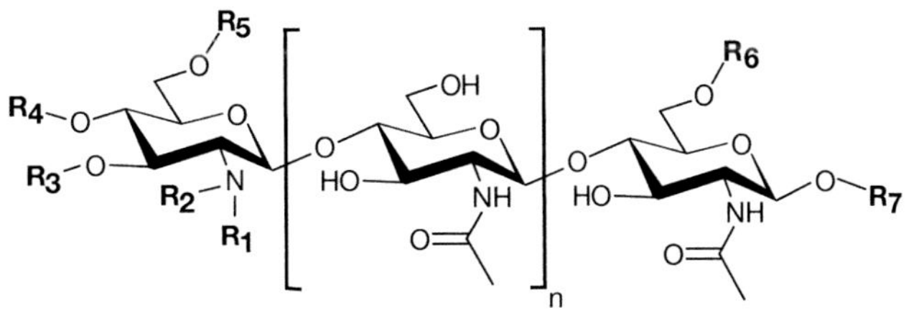

FIGURE 40.3. Generic structure of a Nod factor. Sites on the molecule where species-specific modifications can occur are designated by R1–R7. R1 = H or methyl; R2 = C16:2, C16:3, C18:1, C18:3, C18:4, C20:3, or C20:4; R3 = H, carbamate; R4 = H, carbamate; R5 = H, Ac; R6 = H, Ac, SO_3, Fuc, AcFuc, or MeFuc; R7 = H, glycerol. $n = 1-4$. (Adapted, with permission, from Dénarié J, et al. 1996. *Annu Rev Biochem* **65:** 503–535, © Annual Reviews.)

Nod FACTORS ARE SIGNALS FOR THE INITIATION OF THE NITROGEN-FIXING *RHIZOBIUM*–LEGUME SYMBIOSIS

The interaction between *Rhizobium* and the roots of legumes is an agriculturally and economically important symbiotic relationship because it enables the plant to fix atmospheric nitrogen. An early step in this process is the plant's recognition of lipooligosaccharide signals (Nod factors), which are produced by the bacteria (Figure 40.3). Nod factors have a chitin oligosaccharide backbone containing from three to five GlcNAc residues. However, the types of modifications of this backbone, which include methylation, acylation (typically with a C_{16} or C_{18} fatty acid), acetylation, carbamylation, sulfation, glycosylation, and the addition of glycerol, differ among *Rhizobium* strains.

Nod factors are effective at subnanomolar concentrations, are host-specific, and stimulate numerous changes in the plant's root hairs and roots that allow the bacteria to enter the root cortex and induce the formation of nodules where nitrogen fixation occurs. The initiation of nodule formation and *Rhizobium* entry into the root are host strain–specific; this specificity is determined by the structure of the Nod factor produced by a particular *Rhizobium* strain and the ability of a leguminous species to recognize that signal.

Genetic and biochemical approaches have been used to identify potential plant root Nod factor receptors and proteins involved in the signaling events. The putative receptors are transmembrane proteins with a serine/threonine receptor kinase motif on the cytoplasmic side of the membrane and LysM domains that may recognize Nod factors on the exterior of the membrane. Two receptors (NFR5 and NFR1) have been reported to bind Nod factor directly at high-affinity binding sites, although only limited carbohydrate-binding studies were conducted. A lectin nucleotide phosphohydrolase (LNP) has been identified in legume roots and reported to bind Nod factors from *Rhizobium* symbionts of the plant species from which it was obtained. LNP is a peripheral membrane protein that may function in a receptor complex with one or more LysM-type proteins or act downstream of the Nod factor receptors.

Rhizobium exopolysaccharides (EPS) also have important roles in the development of nitrogen-fixing root nodules. A root receptor–like kinase (EPR3) has been identified and shown to have a role in the recognition of the bacterial EPS. Thus, receptor-mediated recognition of Nod factors and EPS signals may be involved in plant-bacterial compatibility and bacterial access to legume roots.

OLIGOSACCHARIDE SIGNALS IN PLANT AND ANIMAL DEVELOPMENT

Several glycans have been shown to affect plant growth and plant organogenesis. Nanomolar concentrations of oligogalacturonides with a DP between 12 and 14 (Figure 40.2) induce flower formation but inhibit root formation. Oligogalacturonides also enhance cell expansion and thereby affect plant growth and development. Many of these effects may result from the ability of oligogalacturonides to alter the plant's responses to the hormone auxin. Oligogalacturonide receptor proteins (WAKs) have been reported to bind to cell wall pectin and thereby affect plant cell expansion. Auxin-induced elongation of pea stem segments is inhibited by nanomolar concentrations of a nonasaccharide-rich fragment of xyloglucan (Figure 40.2). Plants may also use endogenous Nod-factor-like signals to regulate their growth and development. Recent work on plants with altered lignin structure/composition suggests that plants have the ability to monitor changes in the structures of their walls and to trigger responses distinct from those involved in plant defense. In this case, the released active oligosaccharides appear to be fragments of rhamnogalacturonans, which are distinct from the oligogalacturonides involved in plant defense. These results suggest that plants are capable of using the substantial informational content of their wall glycans to release signals for diverse cellular pathways and responses.

Chitin oligosaccharides may have a role in animal embryogenesis. The *Xenopus* gene *DG42* encodes a protein with chitin synthase activity and is transiently expressed in endoderm cells during the mid-late gastrulation stage (Chapter 27). Homologs of *DG42* have also been identified in zebrafish and mice. The DG42 protein has sequence homology with the *Rhizobium* NodC chitin synthase. Transgenic expression of *DG42* results in the formation of glycans that are fragmented by chitinase. *DG42* is also homologous to a gene encoding a hyaluronan synthase, and studies suggest that the DG42 protein synthesizes chitin and hyaluronan, with the former perhaps as an initiation primer (Chapter 16). Injection of chitinases or expression of *NodZ* (which encodes a fucosyltransferase that can modify chitin) in animal cells has profound effects on development. Thus, chitin oligosaccharides are examples of free glycans that appear to act as intracellular signaling molecules in animals.

Human milk contains numerous compounds that affect newborn health including lactose, lipids, and the third most abundant component, human milk oligosaccharides (HMOs). HMOs are a set of more than 150 unique oligosaccharides synthesized from lactose in the mammary gland that are essentially not digested and function, among others, as prebiotics that selectively promote growth of mutualist intestinal microbes. HMOs contain lactose at their reducing end, may be fucosylated at O-2 of Gal (2′-fucosyllactose), at O-3 (3′-fucosyllactose), or may be sialylated at O-6 or O-3 of Gal. They may also be elongated by β1-3- or β1-6-linked lacto-N-biose or *N*-acetyllactosamine. HMOs can be linear or branched with α1-2, α1-3, or α1-4 fucosylation and/or α2-3 or α2-6 sialylation and may contain from three to more than 30 sugar units. The amount and composition of HMOs is genetically determined, varying among women and mirroring blood group characteristics. Increasing evidence suggests that HMOs protect breastfed infants from microbial infection through cell signaling and cell–cell recognition events resulting in enrichment of protective gut microbiota and inhibition of pathogenic microbe growth, adhesion, and invasion into the intestinal mucosa. For example, 2′-fucosyllactose, representing ~30% of HMOs in human milk, has been shown to inhibit binding and infection of distinct enteropathogens (Chapter 42) by competing for binding of microbes to mucosal surface human receptors terminating in α1-2-linked fucose, thereby inhibiting the first step of pathogenesis. Although more research is required to understand the mechanisms of HMO action, current data indicate that HMOs stimulate immunomodulatory activity at the neonatal intestinal surface and modulate cytokine production.

The presence of abnormal glycans or the accumulation of glycans in the wrong place may negatively impact signaling pathways in animal cells. Three prime repair exonuclease 1 (TREX1) is an ER-associated negative regulator of innate immunity. Mutations that affect TREX1 function are associated with numerous autoimmune and autoinflammatory diseases (Chapter 45). The ER-localized carboxyl terminus of TREX1 has been proposed to interact with, and stabilize, the catalytic activity of the ER oligosaccharyltransferase (OST) complex and thereby suppress immune activation. The OST complex becomes dysfunctional in the presence of carboxy-terminal truncated TREX1. This leads to the release of free glycans from dolichol-linked oligosaccharides, which has been hypothesized to lead to the activation of genes with immune system-related functions and the production of autoantibodies. Thus, TREX1 may safeguard the cell against free glycan buildup in the ER and thereby prevent glycan and glycosylation defects that can lead to immune disorders.

N-linked glycans have a role in the correct folding of glycoproteins in the ER. Misfolded glycoproteins are targeted for degradation by an ER-associated degradation (ERAD) process in which they are retrotranslocated into the cytosol (Chapter 39). The glycans are then released from the glycoprotein by the N-glycanase NGLY1. The protein is degraded by the proteasome, whereas the released glycans are likely partially de-mannosylated in the cytosol and then transported to the lysosomes by an as yet unidentified oligosaccharide transporter. It is not known if these free glycans have any signaling functions in the cytosolic/nuclear compartment. In the lysosome, glycosidases hydrolyze the glycans into monomeric sugars that can then be reused by the cell. Mutations that disrupt NGLY1 function may cause severe health problems in humans. Studies with *Ngly1* mutant mice cells suggest that in the absence of NGLY1, ERAD becomes dysfunctional because a cytosolic endo-β-N-acetylglucosaminidase generates proteins that contain only a single Asn-linked GlcNAc instead of completely deglycosylated proteins. The accumulation of these GlcNAc-proteins may result in the formation of aggregates that are harmful to the cell or they may interfere with intracellular signaling processes.

GLYCOSAMINOGLYCANS AND CELL SIGNALING

Glycosaminoglycans (GAGs) are signaling glycans because they interact with receptor tyrosine kinases and/or their ligands and facilitate changes in cell behavior (Chapters 16, 17, and 38). Hyaluronan oligosaccharides bind to specific membrane proteins, including CD44. In some cells, this binding leads to clustering of CD44, which activates kinases such as c-Src and focal adhesion kinase (FAK). Phosphorylation alters the interaction of the cytoplasmic tail of CD44 with regulatory and adaptor molecules that modulate cytoskeletal assembly/disassembly and cell survival and proliferation (Figure 16.6). Signaling by hyaluronan oligosaccharides depends on the DP of the glycans. Low molecular weight glycans are more active in triggering danger responses via binding to Toll-like receptors (TLRs).

In contrast to hyaluronan-dependent signal transduction, signaling via sulfated GAGs such as heparan sulfate (HS) and chondroitin/dermatan sulfate occurs by an indirect mechanism. Indeed, few membrane receptors have been described in which sulfated GAGs binding causes a specific downstream response, such as phosphorylation of the receptor or activation of a kinase. Instead, sulfated GAGs bind to many ligand/receptor pairs, thereby lowering the effective concentration of ligand required to engage the receptor or increasing the duration of the response. An example of this is the ability of exogenous heparin or endogenous HS proteoglycans to activate fibroblast growth factor (FGF) receptors by FGF (Chapter 38). No substantial conformational change in the ligand occurs on binding to sulfated GAG, consistent with the idea that the glycan primarily aids in the juxtaposition of components of the signal transduction pathway. Free HS oligosaccharides can be released by the action of secreted heparanase. These glycans may facilitate signaling through the mechanism described above or by the release of growth factors from stored depots in the extracellular matrix. Sulfated GAGs also facilitate the formation of morphogen gradients in tissues during early development. Because the gradient determines cell specification during development, the glycan indirectly affects signaling responses in receptive cells. These examples do not exclude the possibility that sulfated glycosaminoglycans may act as ligands and induce signaling directly (e.g., by ligating receptors).

GLYCANS AS MODULATORS OF INNATE IMMUNITY

In addition to mucins (Chapter 10), the innate immune system developed early in eukaryote evolution is a first line of defense against infection by microorganisms. A key feature of this system is its ability to distinguish self from infectious nonself. In more advanced eukaryotes, this is accomplished by receptors that recognize conserved molecular patterns specific to the pathogens. Many of these PAMPs are glycans located on the surfaces of the microorganism. The glycans include the lipopolysaccharides (LPS) of Gram-negative bacteria, the peptidoglycans and techoic acids of Gram-positive bacteria (Chapters 21 and 22) and the mannans and glucans of fungi, perceived by PRRs. Just as in plants (as described above), numerous PRRs are present in mammals that recognize diverse PAMPs and induce host-defense pathways, including TLRs and mannan-binding lectin (Chapter 42). Binding of PAMPs to TLRs activates various signaling pathways that induce inflammation and antimicrobial effector responses. Some TLRs are present on antigen-presenting cells and help to activate the adaptive immune response. TLRs also respond to tissue injury via binding of released hyaluronan fragments as DAMPs (Figure 16.6).

One of the best-studied models of mammal innate immunity involves the LPS of Gram-negative bacteria, which has a role in causing septic shock (see Chapter 42). Lipid A (endotoxin) is the glucosamine-based phospholipid anchor of LPS responsible for activating the innate immune system. Lipid A is an excellent PAMP as its structure is highly conserved among Gram-negative bacteria. Picomolar concentrations of lipid A are detected by TLR-4. The LPS is first opsonized and complexed with another host cell-surface protein, CD14. The binding of LPS leads to recruitment of the adaptor proteins MyD88 and IRAK. This complex initiates a signaling cascade of phosphorylation events that ultimately lead to the transcription of proinflammatory genes.

In contrast to PAMPs and DAMPs, inhibitory Siglec receptors (Chapter 35) on mammal innate immune cells recognize endogenous sialylated glycoconjugates as self-associated molecular patterns (SAMPs) and dampen unwanted immune reactions against the host. Details of the sialoglycan specificity involved require further investigation, but pathogens take advantage of the system via molecular mimicry (Chapters 7 and 42).

Plants also exhibit a type of innate immunity that on activation could confer resistance to pathogen attack to the entire plant. Preparations of β-1,3-glucans, MLGs and xyloglucan oligosaccharides (Figure 40.2) triggering defensive responses in plants are able to confer protection against different pathogens when applied exogenously to crops. Xyloglucan oligosaccharides effectively protected grapevine and *Arabidopsis* against the fungal/oomycete pathogens, whereas β-1,3-glucans improved, among others, tobacco, and grapevine protection against bacterial, fungal, and oomycete pathogens. Given the high abundance of β-1,3-glucans in brown seaweed, laminarin-based products have been successfully developed for use in agriculture as activators of plant natural defense against pathogens. Similarly, pretreatments with cellobiose reduced growth of some pathogens on *Arabidopsis* seedlings, although high doses were required to observe such effects. In all of these cases, PRRs involved in the perception of these oligosaccharide structures that are active in generating these resistance responses in plants have not yet been characterized in detail.

The examples given in this chapter indicate the diversity of free glycan structures that can function as signaling molecules. It is likely that further examples of glycan signals in both plant and animal cells, as well as in their interactions with microbes, will become apparent in the future.

ACKNOWLEDGMENTS

The authors appreciate the helpful comments, suggestions, and contributions from Xi Chen and Gabriel A. Rabinovich and by Laura Bacete.

FURTHER READING

Darvill A, Augur C, Bergmann C, Carlson RW, Cheong J-J, Eberhard S, Hahn M, Ló V-M, Marfa V, Meyer B. 1992. Oligosaccharins—oligosaccharides that regulate growth, development and defence responses in plants. *Glycobiology* **2:** 181–198. doi:10.1093/glycob/2.3.181

Cullimore JV, Ranjeva R, Bono J-J. 2001. Perception of lipo-chitooligosaccharidic Nod factors in legumes. *Trends Plant Sci* **6:** 24–30. doi:10.1016/s1360-1385(00)01810-0

Ronald PC, Beutler B. 2010. Plant and animal sensors of conserved microbial signatures. *Science* **330:** 1061–1064. doi:10.1126/science.1189468

Smeekens S, Ma J, Hanson J, Rolland F. 2010. Sugar signals and molecular networks controlling plant growth. *Curr Opin Plant Biol* **13:** 273–278. doi:10.1016/j.pbi.2009.12.002

Gough C, Cullimore J. 2011. Lipo-chitooligosaccharide signaling in endosymbiotic plant–microbe interactions. *Mol Plant Microbe Interact* **24:** 867–878. doi:10.1094/mpmi-01-11-0019

Broghammer A, Krusell L, Blaise M, Sauer J, Sullivan J, Maolanon N, Vinther M, Lorentzen A, Madsen EB, Jensen KJ. 2012. Legume receptors perceive the rhizobial lipochitin oligosaccharide signal molecules by direct binding. *Proc Nat Acad Sci* **109:** 13859–13864. doi:10.1073/pnas.1205171109

Ferrari S, Savatin D, Sicilia F, Gramegna G, Cervone F, Lorenzo GD. 2013. Oligogalacturonides: plant damage-associated molecular patterns and regulators of growth and development. *Front Plant Sci* **4:** 10.3389. doi:10.3389/fpls.2013.00049

Liang Y, Cao Y, Tanaka K, Thibivilliers S, Wan J, Choi J, ho Kang C, Qiu J, Stacey G. 2013. Nonlegumes respond to rhizobial Nod factors by suppressing the innate immune response. *Science* **341:** 1384–1387. doi:10.1126/science.1242736

Roberts NJ, Morieri G, Kalsi G, Rose A, Stiller J, Edwards A, Xie F, Gresshoff P, Oldroyd GE, Downie JA. 2013. Rhizobial and mycorrhizal symbioses in *Lotus japonicus* require lectin nucleotide phosphohydrolase, which acts upstream of calcium signaling. *Plant Physiol* **161:** 556–567. doi:10.1104/pp.112.206110

Kawaharada Y, Kelly S, Nielsen MW, Hjuler CT, Gysel K, Muszyński A, Carlson R, Thygesen MB, Sandal N, Asmussen M. 2015. Receptor-mediated exopolysaccharide perception controls bacterial infection. *Nature* **523:** 308–312. doi:10.1038/nature14611

He YY, Lawlor NT, Newburg DS. 2016. Human milk components modulate Toll-like receptor–mediated inflammation. *Adv Nutr* **7:** 102–111. doi:10.3945/an.115.010090

Kohorn BD. 2017. Cell wall–associated kinases and pectin perception. *J Exp Bot* **67:** 489–494. doi:10.1093/jxb/erv467

Bacete L, Mélida H, Miedes E, Molina A. 2018. Plant cell wall–mediated immunity: cell wall changes trigger disease resistance responses. *Plant J* **93:** 614–636. doi:10.1111/tpj.13807

Buhian WP, Bensmihen S. 2018. Mini-Review: Nod factor regulation of phytohormone signaling and homeostasis during Rhizobia–legume symbiosis. *Front Plant Sci* **9:** 1247. doi:10.3389/fpls.2018.01247

Plaza-Díaz J, Fontana L, Gil A. 2018. Human milk oligosaccharides and immune system development. *Nutrients* **10:** 1038. doi:10.3390/nu10081038

Plows JF, Berger PK, Jones RB, Alderete TL, Yonemitsu C, Najera JA, Khwajazada S, Bode L, Goran MI. 2021. Longitudinal changes in human milk oligosaccharides (HMOs) over the course of 24 months of lactation. *J Nutrition* **151:** 876–882. doi:10.1093/jn/nxaa427

41 Glycans in Systemic Physiology

Robert Sackstein, Sean R. Stowell, Karin M. Hoffmeister,
Hudson H. Freeze, and Ajit Varki

Glycans mediate or modulate numerous physiologic functions. This brief chapter focuses on vertebrate physiology (predominantly human), providing physiologists and physicians an overview of glycan impacts on organ system functions. Pathological aspects of glycan biosynthesis and degradation are discussed elsewhere. Given the breadth of physiologic functions of glycans, the individual sections highlight just a few representative examples, and listings are necessarily incomplete.

REPRODUCTIVE BIOLOGY

Glycans and glycan-binding proteins are important for both male and female reproduction. Studies in fish, frogs, and mammals indicate that glycans are involved in multiple steps in the process of fertilization (Chapter 27). In animals with internal fertilization, glycan-dependent recognition events occur during sperm interactions with reproductive tract mucins, in sperm–

published by Cold Spring Harbor Laboratory Press; doi: 10.1101/glycobiology.4e.41

fallopian tube interactions, in sperm–ova binding within the fallopian tubes, and during implantation of the early embryo. Glycosylation-deficient male mice are sometimes infertile or subfertile. Following birth, mammals produce milk containing an array of glycoproteins and a complex mixture of species-specific milk oligosaccharides (Chapter 14).

EMBRYOLOGY AND DEVELOPMENT

Genetic modifications that eliminate initial steps of major glycan synthetic pathways and of some monosaccharide biosynthetic pathways generally result in embryonic lethality, with one exception being deletion of individual genes needed for initiation of mucin O-GalNAc pathway, likely because of functional redundancy among the 20 different polypeptide:O-GalNAc transferases (Chapter 10). The causes of these lethal outcomes usually cannot be linked to specific glycoconjugates, but sometimes can be ascribed to a single mechanism such as the disruption of protein O-fucosylation, which impacts global Notch receptor signaling (Chapter 13). Conversely, loss of terminal glycan modifications are usually not embryonic-lethal, instead they have specific defects in some cell types. Elimination of glycosaminoglycans also causes developmental abnormalities, most likely because of their roles in modulating growth factor function and in setting up morphogen gradients (Chapter 17). Although the elimination of glycosaminoglycans causes systemic developmental abnormalities, elimination of some proteoglycan core proteins that carry these glycans can have tissue-specific consequences (Chapters 17, 25, and 26).

Given the significant role of glycans in development, many classic biomarkers of mammalian embryonic stem cells (ESCs), defined initially by monoclonal antibodies (MAbs), unsurprisingly turned out to be glycans. Many of these markers are species-specific. For example, stage-specific embryonic antigen-1 (SSEA-1), otherwise known as Lewis x (Le^x) or CD15, is a principal marker of mouse ESCs but is not present on human ESCs. Instead, human ESCs and induced pluripotent stem cells express globosides SSEA-3 and SSEA-4, and MAbs also react with TRA-1-60 and TRA-1-81, which detect keratan sulfate–related antigens and, also, the tetrasaccharide motif Galβ1-3GlcNAcβ1-3Galβ1-4GlcNAc.

HEMATOLOGY

Glycans affect functions of all classes of blood cells. ABO blood group antigens on human erythrocytes are glycans, and successful blood transfusion requires compatibility of these antigens between donor and recipient (Chapter 14). Glycans on glycoproteins (e.g., integrin GPIIb/IIIa [CD41/CD61]) on the surfaces of circulating platelets play key roles in hemostasis. The "platelet count" (i.e., number of platelets in blood) is tightly controlled; a low count (thrombocytopenia) causes bleeding and a high count (thrombocytosis) predisposes to pathologic clotting (thrombosis). Platelet count is affected by genetic alterations in enzymes responsible for synthesis of sialic acid (e.g., *GNE*) or of lactosamine units (e.g., *B4GALT1*) by increasing platelet destruction or inhibiting platelet production, respectively. Platelet aging is linked to loss of platelet surface sialic acid and increased clearance via the hepatic Ashwell–Morell receptor (AMR) (Chapter 34). Infection and inflammation can also cause thrombocytopenia because of human or microbial neuraminidases that render platelets available for AMR clearance. Selectins and their glycosylated ligands (Chapter 34) play critical roles in mediating the trafficking of hematopoietic stem cells into the marrow (a process critical for the success of hematopoietic stem cell transplantation) and of extravasation of leukocytes from the bloodstream into tissues. In each case, the ability of the circulating cells to engage target tissue endothelial beds depends on their expression of the tetrasaccharide, known as sialylated Lewis x (Sle^x) or CD15s, the canonical minimal binding determinant for selectins, found on N- and O-glycans and glycolipids. Leukocyte deficiency of SLe^x expression results in decreased extravasation with concomitant increased risk of infections. Alternatively, enforced expression of SLe^x via glycoengineering of stem cell surfaces, or of subsets of leukocytes, can promote their delivery to inflammatory sites and thereby enhance tissue regeneration or immunologic functions, respectively. Further information regarding the impact of selectin receptor/ligand interactions in hematologic disease is presented in Chapter 46.

Nearly all proteins in plasma are N-glycosylated for stability in circulation and their optimal function. Patients with N-glycosylation defects often have insufficient levels of coagulation factors such as antithrombin III and proteins C and S, because of instability and/or accelerated clearance (Chapter 45). The O-fucose glycans on Notch receptors regulate hematopoiesis and the hematopoietic stem cell niche (Chapter 13).

IMMUNOLOGY

In addition to SLe^x, other glycan moieties on N- and O-glycans of certain glycoproteins affect differentiation, adhesion, and survival of leukocytes (Chapters 36 and 45). Signaling in leukocytes is also regulated by Siglecs, which recognize sialic acid–containing ligands as "self-associated molecular patterns" (SAMPs; Chapter 35), and O-fucose glycans on Notch receptors regulate many cell differentiation processes, including development of T cells in the thymus (Chapter 13). Galectins (Chapter 36) play key roles in immune cell activation and function, as do C-type lectins such as dendritic cell–specific intercellular adhesion molecule-3-grabbing nonintegrin (DC-SIGN) on antigen-presenting cells (Chapter 34). Glycans are critical components of many antigens and may determine how epitopes are presented (e.g., presentation of glycolipid antigens by CD1a-positive lymphocytes and other antigen-presenting cells). Sialylated N-glycans in the IgG Fc domain play critical roles in antibody effector functions, and modulation of fucosylation and/or N-glycan display at Asn297 of human IgG regulates antibody-dependent cell-mediated cytotoxicity (ADCC) and complement activation.

CARDIOVASCULAR PHYSIOLOGY

Hyaluronan has a critical role in heart development (Chapter 16), and glycosaminoglycans modulate angiogenesis, in part because they bind growth factors such as vascular endothelial growth factor and fibroblast growth factor (Chapter 17). The structural integrity of the walls of blood vessels is thought to depend on glycans, including a high density of sialic acids at the luminal surface of endothelial cells, as well as glycosaminoglycans within the basement membrane underlying endothelial cells. Cardiac muscle integrity and optimal cardiovascular function depend on various glycans.

AIRWAY AND PULMONARY PHYSIOLOGY

The luminal surface of epithelial cells in the upper and lower airways are covered by a dense and complex array of mostly O-glycosylated mucins (Chapter 10) secreted by goblet cells and submucosal glands. Structural glycoproteins, glycolipids, and secreted mucin molecules form barriers that maintain epithelial surface hydration and protect against physical and microbial invasion. Embryonic stem cells lacking complex N-glycans cannot properly organize the bronchial epithelium. N-glycans are also important for healthy lung function, as mice lacking the core α1-6-linked fucose of N-glycans develop emphysema-like symptoms caused by overexpression of matrix metalloproteinases that degrade the lung tissue; this may result from aberrant transforming growth factor-β1 signaling through its misglycosylated receptor. Gene knockouts of individual mucin polypeptides show overlapping functions. Mice lacking O-fucose glycans in the lung do not generate secretory cells necessary for airway integrity.

ENDOCRINOLOGY

O-GlcNAc on proteins of the nucleus and cytoplasm modulate insulin action, and aberrant O-GlcNAcylation is linked to hyperglycemia (Chapter 19). N-glycans may also play a role in type II diabetes, as mice that cannot synthesize triantennary N-glycans develop diabetes when fed a high-fat diet. This deficiency alters the single N-glycan on the GLUT2 glucose transporter on pancreatic islet cells, accelerating its endocytosis, leading to loss from the cell surface and poor response to insulin. N-glycans may also be important for the production of functional thyroid hormones (e.g., targeting of endocytosed thyroglobulin to lysosomes for conversion to T3 and T4 in the thyroid gland; Chapter 33). The plasma half-life of several pituitary glycoprotein hormones is regulated by the presence of N-glycans that contain an unusual sulfated GalNAc (Chapters 14 and 31), which controls hormone clearance in the liver. O-GalNAc glycosylation can regulate peptide hormones by affecting proteolytic cleavage, as exemplified by O-glycan loss in the phosphate regulating hormone FGF23 causing familial tumoral calcinosis in patients deficient in GalNAc-T3 (Chapter 10).

ORAL BIOLOGY

Glycosaminoglycans (Chapters 16 and 17) are required for normal development, organization, and structure of both the gums and teeth. Oral commensal organisms interact with each other or the host epithelium using glycan recognition. Salivary gland–generated mucins help protect the oral cavity by preventing bacterial biofilm formation on teeth (Chapters 10 and 42). However, mucin sialoglycans also provide binding sites for tooth cavity–facilitating bacteria.

GASTROENTEROLOGY

Glycans physically separate the luminal contents of the gut from the cells by organizing the mucin barrier that helps block pathogens. However, the microbiome of the GI tract also contains microbial symbionts that maintain a complex physiologic equilibrium. Some organisms practice "glycan foraging" of host (Chapter 37) and help prevent invasion by pathogens such as *Helicobacter* species in the stomach and anaerobic bacteria in the colon (Chapters 37 and 42). *Helicobactor pylori* infection rarely occurs in the duodenum, in which unusual GlcNAcα(1-4)-terminated O-linked mucins are expressed. This "antimicrobial glycan" inhibits the synthesis of cholesteryl-glucoside, a major cell wall component. Heparan sulfate in the intestinal basement membrane acts as a permeability barrier, preventing protein loss from the plasma into the gut. O-fucose glycans in the small intestine regulate the balance of secretory and goblet cells necessary for intestinal development.

HEPATOLOGY

The liver synthesizes a large fraction of plasma proteins, and nearly all these proteins are heavily N-glycosylated, making hepatocytes a traditional cell type for studying the organization and function of the Golgi apparatus. Notably, acute phase reactants are glycoproteins, and variations in glycosylation patterns may reflect physiologic responses. Both hepatocytes and Kupffer cells in the liver have specific glycan-based recognition systems (e.g., the Ashwell–Morell receptor [AMR]) to clear unwanted circulating molecules (see Chapters 28, 31, 32, and 34 for more information on liver receptors). Heparan sulfate proteoglycans in the space of Disse between fenestrated endothelium and hepatocytes bind lipoproteins and aid in their clearance. Liver modification of bilirubin, hormones, and drugs with glucuronic acid ("glucuronidation") increases the water solubility improving clearance in the bile and/or the urine (Chapter 5).

NEPHROLOGY

Heparan sulfate glycosaminoglycans (Chapter 17) and sialic acid residues (Chapter 15) on podocalyxin are needed for the filtering function of the glomerular basement membrane. In addition, reduced branching of complex N-glycans causes kidney pathology that may result from an autoimmune response. As in the pulmonary and gastrointestinal tracts, mucins with O-GalNAc glycans (Chapter 10) and proteoglycans with glycosaminoglycans provide a barrier function at the luminal surfaces of the ureters and bladder.

SKIN BIOLOGY

Glucosylceramide and related glycosphingolipids help maintain the barrier function of the skin. Hyaluronan and dermatan sulfate each help maintain the structure of the dermis and participate in wound repair. The endothelial selectin "E-selectin" (CD62E) (Chapter 34) is induced by tumor necrosis factor (TNF), interleukin-1 (IL-1), lipopolysaccharide (LPS), or trauma on post-capillary venules. However, it is constitutively expressed on microvessels of skin, where it recruits SLe^x-bearing leukocytes and assures dermal immunosurveillance (Chapter 46).

MUSCULOSKELETAL BIOLOGY

Proper adhesion of skeletal muscle to extracellular matrix laminin requires unique O-mannose glycans on the sarcolemmal glycoprotein α-dystroglycan (Chapter 27). Various defects in this pathway cause mild to severe muscular dystrophies in both humans and mice (Chapter 45). Glycan-related interactions can promote clustering of acetylcholine receptors at neuromuscular junctions. Sialylated glycans on ion transport proteins are important, and their loss impairs control of calcium fluxes into skeletal muscle cells. Normal formation and ossification of cartilage into bone requires many glycosaminoglycans, including hyaluronan and heparan, chondroitin, and keratan sulfates all in the appropriate amounts (Chapters 16 and 17).

NEUROBIOLOGY

The central nervous system has the highest amount and concentration of sialic acid–containing glycolipids (gangliosides; Chapter 11), and alterations in these glycans affect neurological function. O-GlcNAcylation in specific cells of the brain sense nutrients and regulate satiety. The unusual polysialic acid chains on NCAM (neural cell adhesion molecule) differentially modulate the plasticity of the nervous system during embryogenesis (Chapter 15). The dystroglycanopathies mentioned above also typically have cognitive and/or neurologic defects in addition to muscle dysfunction (Chapter 45). There are additional instances wherein specific glycans appear to inhibit nerve regeneration after injury. Recognition of certain sialylated glycolipids by myelin-associated glycoprotein inhibits neuronal sprouting following injury (Chapter 35), and similar inhibitory effects may be mediated by the glycosaminoglycan chondroitin sulfate (Chapter 17). In both instances, targeted degradation of the glycan in vivo (by local injection of sialidase or chondroitinase, respectively) stimulates neuronal growth and repair, meaning that they normally prevent neuronal regeneration. Mutant mice lacking some complex N-glycans and glycosaminoglycans reveal the importance of these molecules in the development and organization of the nervous system (Chapters 9 and 17). Fucosylated N-glycans appear to play a role in modulating various aspects of neural development and function. The great majority of patients with inherited glycosylation disorders also have cognitive and/or neurological abnormalities, but specific mechanisms are mostly unknown (Chapter 45).

ACKNOWLEDGMENTS

The authors acknowledge contributions to previous versions of this chapter by Linda Baum and Victor Vacquier and appreciate helpful comments and suggestions from Ruth Siew and Hans Wandall.

FURTHER READING

Only a few references are given here. Please also see the citations at the ends of individual chapters referred to above.

Varki A. 2008. Sialic acids in human health and disease. *Trends Mol Med* **14**: 351–360. doi:10.1016/j.molmed.2008.06.002

Sackstein R. 2009. Glycosyltransferase-programmed stereosubstitution (GPS) to create HCELL: engineering a roadmap for cell migration. *Immunol Rev* **230**: 51–74. doi:10.1111/j.1600-065x.2009.00792.x

Stanley P, Okajima T. 2010. Roles of glycosylation in Notch signaling. *Curr Top Dev Biol* **92**: 131–164.

Natunen S, Satomaa T, Pitkänen V, Salo H, Mikkola M, Natunen J, Otonkoski T, Valmu L. 2011. The binding specificity of the marker antibodies Tra-1-60 and Tra-1-81 reveals a novel pluri-potency-associated type 1 lactosamine epitope. *Glycobiology* **21**: 1125–1130. doi:10.1093/glycob/cwq209

Hansson GC. 2012. Role of mucus layers in gut infection and inflammation. *Curr Opin Microbiol* **15**: 57–62. doi:10.1016/j.mib.2011.11.002

Marcobal A, Southwick AM, Earle KA, Sonnenburg JL. 2013. A refined palate: bacterial consumption of host glycans in the gut. *Glycobiology* **23**: 1038–1046. doi:10.1093/glycob/cwt040

Stanley P. 2017. What have we learned from glycosyltransferase knockouts in mice? *J Mol Biol* **428**: 3166–3182. doi:10.1016/j.jmb.2016.03.025

Lee-Sundlov MM, Stowell SR, Hoffmeister KM. 2020. Multifaceted role of glycosylation in transfusion medicine, platelets, and red blood cells. *J Thromb Haemost* **18**: 1535–1547. doi:10.1111/jth.14874

Smith BAH, Bertozzi CR. 2021.The clinical impact of glycobiology: targeting selectins, Siglecs and mammalian glycans. *Nat Rev Drug Discov* **20**: 217–243. doi:10.1038/s41573-020-00093-1

42 Bacterial and Viral Infections

Amanda L. Lewis, Christine M. Szymanski, Ronald L. Schnaar, and Markus Aebi

This chapter illustrates and discusses some key mechanisms by which glycans influence the pathogenesis and progression of bacterial and viral infections and describes examples of opportunities for therapeutic intervention.

BACKGROUND

Infectious diseases remain a major cause of death, disability, and social and economic disorder for millions of people throughout the world. Poverty, poor access to health care, human migration, emerging disease agents, and antibiotic resistance all contribute to the expanding impact of these illnesses. Prevention and treatment strategies for infectious diseases are derived from a thorough understanding of the complex interactions between specific viral or bacterial pathogens and the human (or animal) host.

Just as glycans are major components of the outermost surface of all animal and plant cells, so too are oligosaccharides and polysaccharides found on the surfaces of all bacteria and viruses of eukaryotes. Thus, most (if not all) interactions of microbial pathogens with their hosts are influenced to an important degree by the pattern of glycans and glycan-binding receptors that each expresses. This holds true at all stages of infection, from initial colonization of host

published by Cold Spring Harbor Laboratory Press; doi: 10.1101/glycobiology.4e.42

cell surfaces, to tissue spread, to the induction of inflammation or host cell injury that results in clinical symptoms. The microbial molecules most responsible for disease manifestations are known as virulence factors.

BACTERIAL SURFACE GLYCANS AS VIRULENCE FACTORS

Polysaccharide Capsules

A human in good health is colonized by as many as 10^{13} bacteria on the skin and mucosal surfaces, particularly in the gastrointestinal tract, a number that resembles the number of cells in our own body. Despite all these direct and continuous encounters, it is relatively rare that bacteria invade into the tissues to produce serious infections. Although not restricted to pathogenic species/strains, one feature shared by many of these disease-causing agents is the presence of diverse polysaccharide capsule structures, which encapsulate bacteria (Chapter 21). These structures can have many biological properties, both stimulating and thwarting immune detection in different settings. Capsules make key contributions to the virulence of multiple bacterial pathogens, reducing recognition by the innate and adaptive immune systems. However, capsular polysaccharides are also commonly the natural targets of the adaptive immune system. In fact, antibodies specific to different capsule structures often define the serology used to differentiate subtypes of particular bacterial species. *Streptococcus pneumoniae* (pneumococcus) encodes perhaps the most historically significant capsule in biology and medicine. Pioneering experiments by Frederick Griffith in 1928 showed the in vivo transfer of virulence from a dead encapsulated (smooth) disease-causing strain to an avirulent, nonencapsulated (rough) strain. Studies regarding immunogenicity of pneumococcal polysaccharides (Figure 42.1) provided the framework for the discoveries of Oswald Avery, Colin MacLeod, and Maclyn McCarty, which showed DNA to be the carrier of genetic information.

Killing of bacteria by phagocytes of the innate immune system, such as neutrophils or macrophages, is aided by opsonization, a process in which the bacterial surface is tagged with complement proteins or specific antibodies. Phagocytes express receptors for activated complement and antibody Fc domains, allowing them to bind, engulf, and kill bacteria. Together these processes are referred to as opsonophagocytosis. Genetic mutagenesis of capsule biosynthesis genes and infectious challenge in small animal models have illustrated the roles of bacterial capsules in resisting opsonophagocytosis. Compared with the wild-type parent bacterial strains, isogenic capsule-deficient mutants of group A *Streptococcus* (GAS), group B *Streptococcus* (GBS), pneumococcus, *Haemophilus influenzae*, *Neisseria meningitidis* (meningococcus), *Salmonella* serotype Typhi (typhoid fever), *Bacillus anthracis* (anthrax), and several other important human pathogens are rapidly cleared from the bloodstream by opsonophagocytosis and fail to establish systemic infection. Some bacteria mimic the anionic host molecule sialic acid in their capsules, including the neonatal pathogens GBS and *Escherichia coli* K1. These sialylated capsules bind the complement regulatory protein factor H and attenuate complement deposition. Independent of the complement system, GBS sialic acids also reduce neutrophil bactericidal activities by directly engaging the sialic acid–binding receptor Siglec-9.

Bacteria also use molecular mimicry to evade antibody generation by the adaptive immune system. Generally, humans can generate effective antibodies against bacterial polysaccharide capsules, but this ability is diminished early and late in life. Infants and the elderly are particularly prone to invasive infections with encapsulated pathogens. Molecular mimicry of common host glycan structures allows bacteria to masquerade as "self" to avoid being recognized by the adaptive immune system. For example, the GAS pathogen expresses a nonimmunogenic capsule of hyaluronan, identical to the nonsulfated glycosaminoglycan that is highly abundant in host skin

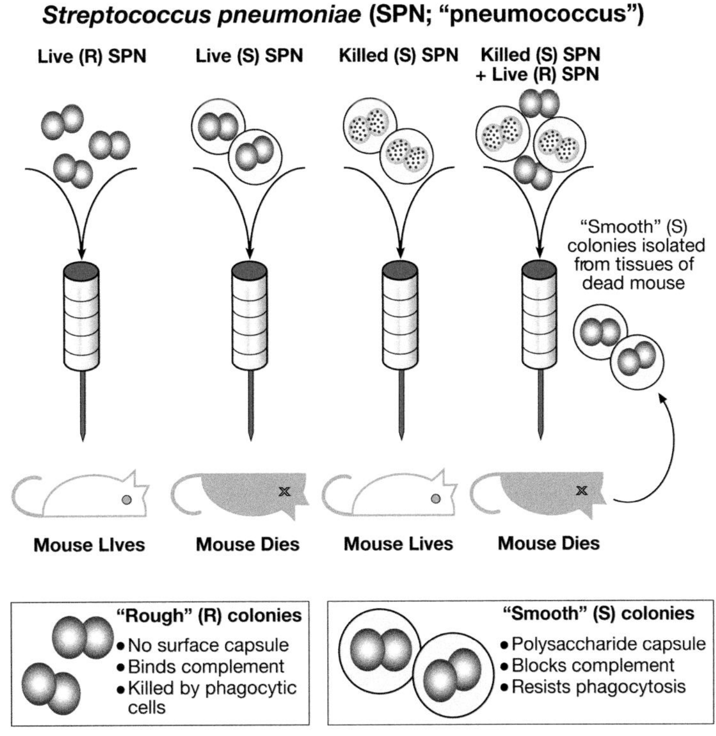

FIGURE 42.1. Classical experiments on the role of the pneumococcal polysaccharide capsule in virulence. *Streptococcus pneumoniae* (SPN) strains can be identified with either a "rough" (R) or a "smooth" (S) phenotype, the latter being due to expression of a thick polysaccharide capsule on their surface. In 1928, Frederick Griffith found that R SPN strains were avirulent for mice, whereas S strains were highly lethal. Heat-killed S strains did not cause disease, but when mixed with live R bacteria, the mice died, and the recovered bacteria expressed the S phenotype. Thus, the live R strains had been "transformed" to S strains by a factor present in the heat-killed preparation of S SPN. The factor later proved to be DNA carrying biosynthetic pathways for capsular polysaccharides, providing the first evidence that DNA was carrier of genetic information.

and cartilage (Chapter 16). The contribution of capsule-based host mimicry to bacterial immune evasion is also illustrated by the homopolymeric sialic acid capsules of *E. coli* and meningococcus, an important cause of sepsis and meningitis. Whereas the group C meningococcal capsule is composed of an α2-9-linked sialic acid polymer that is a unique bacterial structure, the group B meningococcal capsule is composed of an α2-8-linked sialic acid polymer that is identical to a motif present on neural cell adhesion molecules (NCAMs) found in human neural tissues (Chapter 15). The group C capsule has proven to be a successful vaccine antigen in human populations, whereas the group B capsule is essentially nonimmunogenic. Bacterial polysaccharide capsules can also cloak immunogenic surface proteins to which antibodies might be directed.

One of the challenges posed to the host is that different strains of the same bacterial species can display diverse compositions and linkages of repeating sugar units in their capsule structures. Often, these structures are immunologically distinct, allowing classification of different capsule "serotype" strains. For example, there are five major capsule serotypes of meningococcus (A, B, C, Y, and W-135), six different capsule serotypes of the respiratory pathogen *H. influenzae* (a–f), nine capsule serotypes of GBS (Ia, Ib, and II–VIII), and more than 90 different serotypes of pneumococcus, which is a leading cause of bacterial pneumonia, sepsis, and meningitis. Antibodies generated by the host against the capsule of one serotype strain typically do not provide cross-protective immunity. Thus, individuals can be repeatedly infected over their lifetime by different serotypes of the same bacterial pathogen. Figuratively, although the strategy of capsular molecular mimicry used for example by GAS renders the pathogen invisible to immune surveillance, the strategy of antigenic diversity of capsule types presents a moving target to the immune system. Genetic exchange of capsule biosynthetic genes among serotype strains of an individual species (e.g., the polysialyltransferase gene of meningococcus) can lead to capsule switching in vivo, which provides another means of pathogen escape from protective immunity.

Lipopolysaccharide

In addition to a polysaccharide capsule, Gram-negative bacteria have an outer membrane that is rich in lipopolysaccharide (LPS; Chapter 21), also known as endotoxin. As its name suggests, LPS consists of two parts, a lipid A moiety and glycan components. Lipid A is comprised of two glucosamines, acyl chains and phosphates, which are embedded in the outer membrane. The glycan components extend outward into the bacterial niche. A core oligosaccharide contains some sugars not found in vertebrates (such as ketodeoxyoctulonate [Kdo] and heptose) and a repeating polysaccharide known as the O-antigen that can vary widely among strains within an individual species. Many mucosal pathogens such as *H. influenzae*, *Campylobacter jejuni*, and *Neisseria gonorrhoeae* lack O-antigens; instead, they produce lipooligosaccharides (LOSs) that contain only lipid A and an extended core structure.

LPS is a pathogen-associated molecular pattern (PAMP) that is recognized by the innate immune system and stimulates inflammatory processes, including the classic fever response. Bacteria that have breached the barrier defenses of the skin or mucosa release soluble LPS, which is recognized by membrane receptors CD14 and Toll-like receptor 4 (TLR4), initiating downstream immune signaling processes (Figure 42.2). TLR4 belongs to an evolutionarily conserved family of receptors (TLRs) that can detect a wide array of microbe-derived ligands. For example, the TLR2 receptor can recognize peptidoglycan or lipoteichoic acid derived from the cell walls of Gram-positive bacteria that generally lack LPS. A signaling cascade ultimately leads to the activation of the transcription factor nuclear factor-κB (NF-κB). Translocation of NF-κB into the nucleus leads to the regulation of cellular processes ranging from immune cell differentiation, activation of the inflammasome, and up-regulation of proinflammatory chemokines and cytokines, including tumor necrosis factor-α (TNF-α) and interleukin-1 (IL-1).

Although TLR-mediated detection of LPS and other microbial molecules is critical for triggering host innate immunity, a dangerous condition known as sepsis can develop in the setting of overwhelming bacterial infections that lead to dysregulated systemic immune responses. Symptoms include fever, low blood pressure, rapid heart rate, abnormal white blood cell counts, and dysfunction of multiple organ systems that may lead to lung or kidney failure and death.

Processes of selection have led to many examples of Gram-negative bacteria that vary or modify the structure of LPS to interfere with antibiotic action and host defenses. For example, some LPS modifications reduce the overall negative charge of LPS (e.g., as phosphoethanolamine or 4-amino-4-deoxy-L-arabinose [L-Ara4N]) and in doing so can repel cationic antimicrobial peptides of host and microbial origin, such as host defensins or bacterial polymyxins. The plague bacillus *Yersinia pestis* changes the number and type of acyl groups on the lipid A of its LPS in response to temperature changes. At environmental temperatures (~21°C), *Y. pestis*, which lacks O-antigens, predominantly expresses a more immunostimulatory hexa-acylated lipid A, whereas at mammalian body temperature (37°C), the pathogen expresses a less immunostimulatory tetra-acylated lipid A. These data suggest that the production of a less immunostimulatory form of LPS on entry into a mammalian host might confer avoidance to immune detection.

Sialic acids and related nonulosonic acids are also synthesized de novo or scavenged from the host for incorporation into the LOS of several Gram-negative pathogens, including pathogenic members of the genera *Haemophilus*, *Neisseria*, *Campylobacter*, and *Vibrio*. This can be accomplished by several distinct mechanisms and often confers properties of immune evasive and

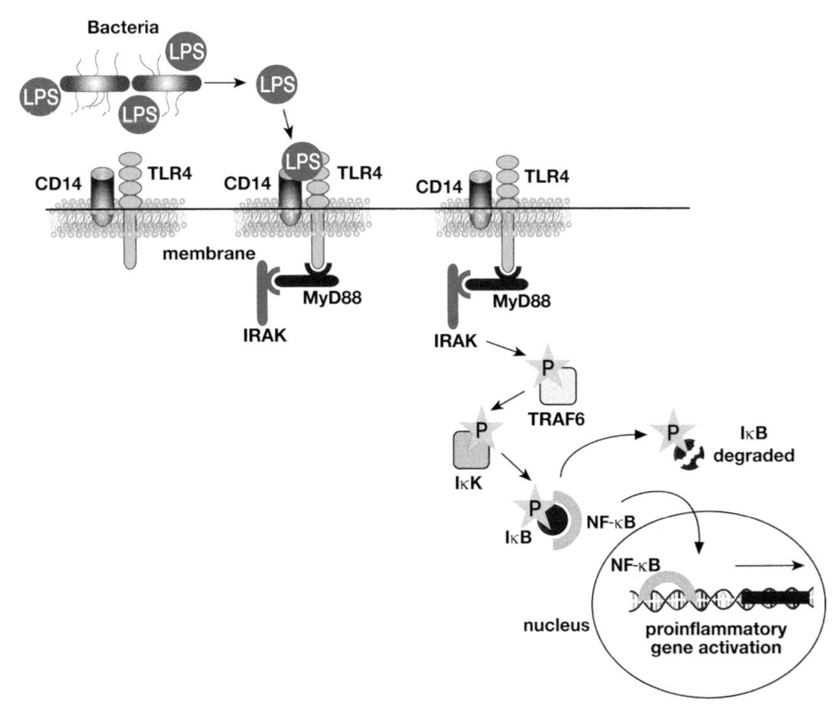

FIGURE 42.2. Activation of immune signaling by bacterial lipopolysaccharide (LPS). LPS from the cell wall of Gram-negative bacteria is bound by the pattern-recognition molecule Toll-like receptor 4 (TLR4) in conjunction with the cell-surface receptor CD14. The binding of LPS leads to recruitment of the adaptor proteins MyD88 and IRAK to the cytoplasmic domain of TLR4. This complex initiates a signaling cascade of phosphorylation events through TRAF6 and the kinase IκK. Finally, IκK phosphorylates IκB, an inhibitor bound to the transcription factor NF-κB. Phosphorylated IκB is degraded, releasing NF-κB, which migrates to the nucleus where it activates the transcription of proinflammatory genes. Similar signal transduction pathways are activated by Gram-positive cell wall constituents such as peptidoglycan and lipoteichoic acid via TLR2 or TLR6.

enhanced virulence. For example, the human-specific pathogen *N. gonorrhoeae* uses a surface sialyltransferase and scavenges the activated form of sialic acid (CMP-Neu5Ac) from the host. Such incorporation thwarts multiple arms of the host complement system and confers resistance to the cationic antimicrobial peptides (CAMPS). An Achilles' heel of this increasingly drug-resistant pathogen may be the promiscuity of its sialyltransferase toward related nonulosonic acids. Unnatural incorporation of legionaminic acid (CMP-Leg5,7Ac_2) in *N. gonorrhoeae* prevents the immune evasion benefits typically afforded by Neu5Ac; moreover, vaginal administration of CMP-Leg5,7Ac_2 leads to more rapid bacterial clearance of genital infections in mice.

Studies of another Gram-negative bacterium, *Vibrio vulnificus*, underscore that the presence of legionaminic acid in LOS may not always be a deterrence to virulence. *V. vulnificus* is normally present in environmental (aquatic) ecosystems and is pathogenic in humans only during accidental contact with susceptible individuals, often following consumption of raw or undercooked seafood. Although it is a rare human pathogen, the bacterium can cause bloodstream and disseminated infections that are commonly fatal when they strike. Legionaminic acid contributes to the ability of *V. vulnificus* to survive in bloodstream and cause disseminated infection following intravenous administration.

Occasionally, the host mounts an (auto)immune response to "self" antigens, including sialylated LOS. For example, the foodborne pathogen, *C. jejuni*, is capable of expressing a variety of sialylated LOS structures that mimic human gangliosides (Chapters 11 and 46). Sialylation of *C. jejuni* LOS results in CD14-dependent amplification of mucosal dendritic cell (DC) responses, which promote B-cell proliferation in a T-cell-independent manner. These events may help explain why in a small subset of infections, self-reacting (autoimmune) antibodies are inappropriately produced by B cells, leading to the attack of nerve fibers expressing the relevant gangliosides. These events can result in a life-threatening paralytic disorder known as Guillain–Barré syndrome (GBS). For most patients, GBS symptoms have to be managed by removing cross-reactive antibodies (plasmapheresis) or immunomodulatory therapies such as intravenous immunoglobulin (IVIg).

MECHANISMS OF COLONIZATION AND INVASION

Adhesins and Receptors

Adherence to skin or mucosal surfaces is a fundamental characteristic of the normal human microbiome and also an essential first step in the pathogenesis of many important infectious diseases (Chapter 34). Most microorganisms express more than one type of adherence factor or "adhesin" (Chapter 37). A large fraction of microbial adhesins are lectins that bind directly to cell-surface glycoproteins, glycosphingolipids, or glycosaminoglycans. Adhesion may be mediated through terminal sugars or internal carbohydrate motifs. In other cases, the bacteria express adhesins that bind matrix glycoproteins (e.g., fibronectin, collagen, or laminin) or mucin, mediating attachment to the mucosal surface. The specific carbohydrate ligands for bacterial attachment on the animal cell are often referred to as adhesin receptors, and they are quite diverse in nature. The tropism of individual bacteria for particular host tissues (e.g., skin vs. respiratory tract vs. gastrointestinal tract) is determined by specific combinations of adhesin–receptor pairs.

Pili or fimbriae are assemblies of protein subunits that project from the bacterial surface in hair-like threads whose tips often adhere to host glycans (Figure 42.3A). Such pili are usually composed of a repeating structural subunit providing extension and a different "tip adhesin" responsible for binding. The structural proteins for pilus assembly are often encoded in a bacterial operon. Lateral mobility of pili structures in the bacterial membrane provides a Velcro-like binding effect to epithelial surfaces. Certain strains of *E. coli* express pili that bind avidly to P-blood group–related glycosphingolipids in the bladder epithelium, leading to urinary tract infection. Pathogenic strains of *Salmonella* produce pili that facilitate adherence to human intestinal cell mucosa, thereby causing food poisoning and infectious diarrhea. In other cases, a surface-anchored protein (afimbrial adhesin) expressed by the bacteria represents a critical

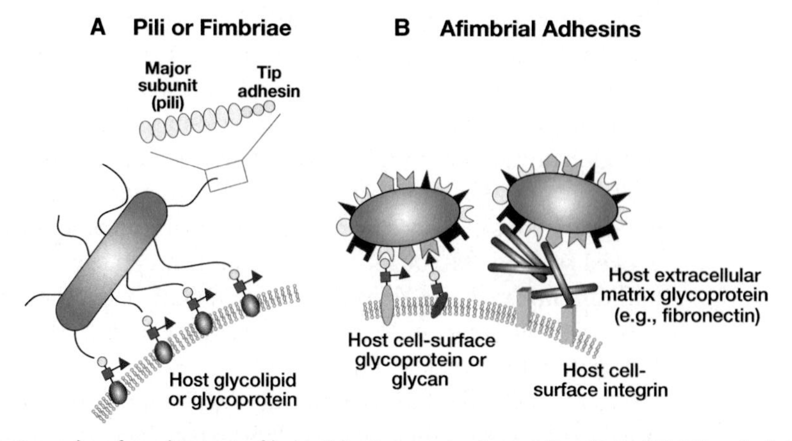

FIGURE 42.3. Examples of mechanisms of bacterial adherence to host cell surfaces. (*A*) Pili or fimbriae are organelles that project from the cell surface. They are made up of a repeating structural subunit and a protein at their tip that mediates recognition of a specific host cell glycan motif. (*B*) Afimbrial adhesins are integral bacterial cell wall proteins or glycoproteins that directly engage host cell receptors to promote colonization.

colonization factor (Figure 42.3B). For example, the filamentous hemagglutinin (FHA) of *Bordetella pertussis* promotes strong attachment of the bacteria to the ciliated epithelial cells of the bronchi and trachea, triggering local inflammation and tissue injury that results in the "whooping cough" disease. FHA is a component of modern pertussis vaccines given in infancy and early childhood to block infection.

Adhesins can be glycoproteins as well. In *Pasteurellaceae* and some *H. influenzae* strains adhesins are N-glucosylated by a cytoplasmic N-glucosylation system that is homologous to the cytoplasmic O-GlcNAc transferase of eukaryotes. Similarly, transfer of heptose residues by the dodecameric bacterial autotransporter heptosyltransferase (BAHT) family of enzymes to autotransporter adhesins in several different Gram-negative pathogens is essential for the adhesion process.

Invasion Factors

Glycan–lectin interactions play pivotal roles in enabling certain pathogens to penetrate or invade through epithelial barriers, whereupon they may disseminate into and through the bloodstream to produce deep-seated infections. *Samonella* Typhi causes typhoid fever in humans, a process that begins with intracellular invasion of intestinal epithelial cells. The outer core oligosaccharide structure of the LPS is required for internalization into epithelial cells. Removal of a key terminal sugar residue on the outer core markedly reduces the efficiency of bacterial uptake. Once invasion has occurred, the secreted A2B5 typhoid toxin mediates illness by first binding preferentially to the sialic acid Neu5Ac, which is enriched in humans. *Streptococcus pyogenes* (GAS), the cause of strep throat, as well as serious invasive infections, attaches to human pharyngeal and skin epithelial cells via interaction of its hyaluronan capsular polysaccharide with the host hyaluronan-binding protein CD44 (Chapter 16). This binding induces marked cytoskeletal rearrangements manifested by membrane ruffling and opening of intercellular junctions that allow tissue penetration by GAS through a paracellular route.

Bacterial glycosyltransferases are also involved in the intracellular manipulation of host responses. For example, injection of the NleB glycosyltransferase into host cells by enteropathogenic *E. coli* (EPEC) results in the N-GlcNAc modification of arginine residues of host proteins, leading to the suppression of NF-κB-driven responses. Another bacterial glycosyltransferase with an intracellular target is the GlcNAc transferase toxin of *Photorhabdus asymbiotica* PaTox. PaTox modifies the host Rho GTPase at tyrosine residues, resulting in reduced phagocytosis and disassembly of the actin cytoskeleton in insect and mammalian cells.

Biofilms

Biofilms are assemblies of bacteria affixed to environmental or host surfaces. Within the body, biofilms form on catheters, implanted devices, and other medical products in which their metabolic dormancy together with other physical, biochemical, and biological properties help resist the action of host defenses and antimicrobials. Layers of biofilm are often joined together by extracellular polysaccharides (EPS) that contribute architectural, adhesive, and protective functions. The EPS synthesized by bacteria in biofilms varies greatly in composition and in chemical and physical properties. Likewise, EPSs can have wide-ranging functions within biofilms, which depend on their exact compositions and the many facets of the niches in which they reside. Polysaccharides can also contribute to biofilm persistence by scavenging reactive oxygen species, trapping cationic antimicrobial peptides and antibiotics, and protecting against desiccation.

In the human mouth, polymicrobial biofilms contribute to dental plaque, caries, and periodontal (gum) disease. Dental plaque is a polymicrobial biofilm comprised of hundreds of species in which dense, mushroom-like clumps of bacteria pop up from the surface of the tooth enamel. Biofilm bacteria are interspersed with bacteria-free diffusion channels filled with

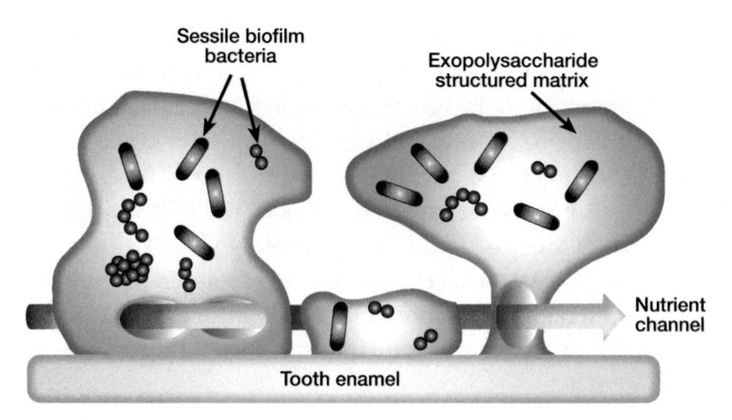

FIGURE 42.4. Structure of a polymicrobial biofilm. Dental plaque is an example of a polymicrobial biofilm in which *Streptococcus* species and other bacteria secrete a thick exopolysaccharide matrix and exist within this matrix in a dormant or sessile state of low metabolic activity. Biofilm bacteria have increased resistance to host immune clearance and antibiotic medicines.

nucleic acids, proteins, lipids, and EPS (Figure 42.4). *Streptococcus* species are often the initial colonizers of the tooth enamel and comprise 60%–90% of dental plaque. In periodontal (gum) disease, "early-" and "late-colonizers" of the biofilm are bridged by an impressive variety of lectin-glycan interactions involving the "middle colonizer" *Fusobacterium nucleatum.* This bacterium is found ubiquitously in the human mouth, but overgrows in the setting of periodontitis, binding to members of the outermost layers of the mature biofilm.

Many EPS types are polyanionic because of the presence of either uronic acids (D-glucuronic, D-galacturonic, or D-mannuronic acids) or ketal-linked pyruvate. Inorganic residues, such as phosphate or sulfate, also contribute to the negative charge and modifications such as O-acetylation or sugar epimerization further contribute to EPS complexity.

In several cases, EPS is a homoglycan composed of β-1,6-linked *N*-acetylglucosamine residues known as poly-*N*-acetylglucosamine (or PNAG), as exemplified by the adhesive polymer obtained from *Staphylococcus epidermidis* strains that produce biofilms on catheters. PNAG is common among many oral pathogens and multidrug-resistant bacteria resulting in concerted efforts to target this polymer, especially because antibodies raised against the deacetylated form of PNAG mediate opsonic killing. Interestingly, the periodontal pathogen *Aggregatibacter actinomycetemcomitans* secretes a PNAG hydrolase known as dispersin B (DspB) that has been shown to effectively disperse biofilms formed by PNAG-producing bacteria and is therefore also being developed as an adjuvant to antibiotic therapy.

VIRAL INFECTIONS

Viruses bind to host cells as a prerequisite for entry and intracellular replication. The enrichment of glycans on cell surfaces make them an attractive target for viral attachment and entry. Virus–glycan interactions are often responsible for species and tissue tropism (Table 37.1), as illustrated here with examples of three human pathogens: influenza virus, herpes simplex virus 1 (HSV-1), and human immunodeficiency virus (HIV), showing different modes of glycan-mediated viral interactions (Figure 42.5). Influenza viruses use a sialic acid–recognizing protein (hemagglutinin) for binding and entry (Figure 42.5A). HSV-1 has multiple envelope proteins that bind to heparan sulfate (HS) as the first step in recruitment of a multiprotein virus entry complex (Figure 42.5B). HIV expresses surface glycans that co-opt host lectins to enhance cell dissemination, but has also evolved to use specific protein receptors (Figure 42.5C). In contrast to these three examples,

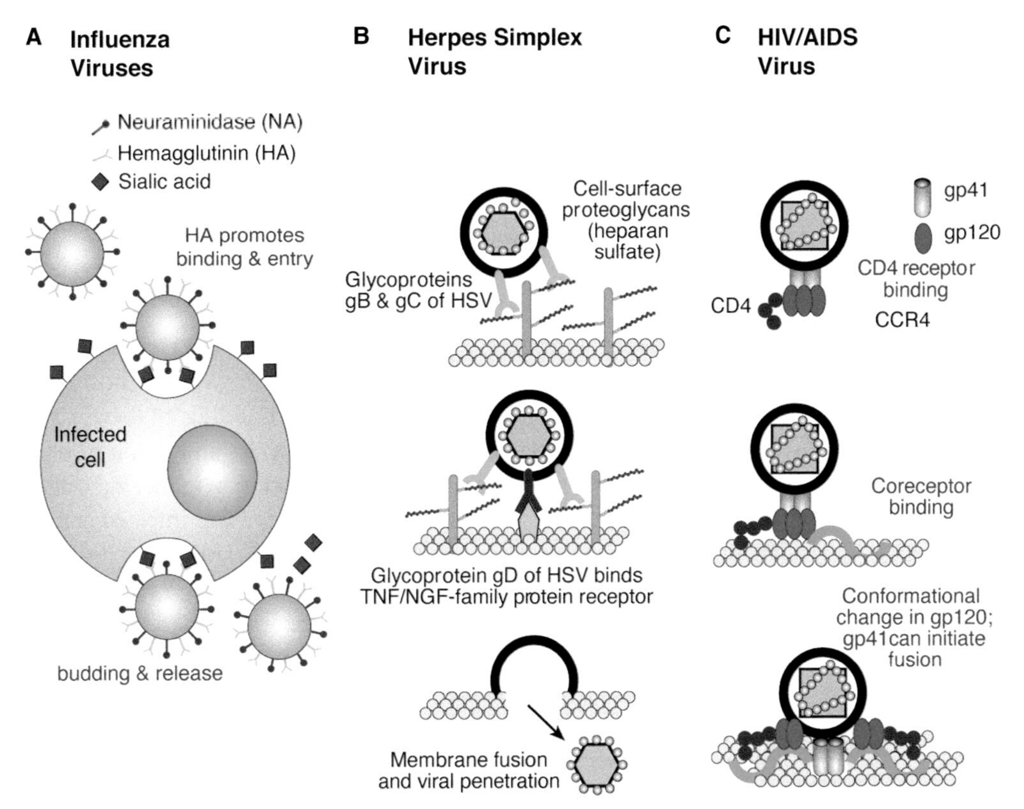

FIGURE 42.5. Mechanisms of viral entry into host cells. (*A*) Influenza virus initiates host cell contact and entry by binding to cell-surface sialic acid receptors through its surface glycoprotein hemagglutinin. After intracellular replication, a cell-surface neuraminidase cleaves sialic acid from the cell membrane allowing viral escape. (*B*) Herpes simplex virus (HSV) engages host cells first through a low-affinity engagement of heparan sulfate proteoglycans via its surface glycoproteins gB and gC. Subsequently, a higher-affinity binding of viral protein gD to a member of the tumor necrosis factor–nerve growth factor (TNF/NGF) receptor family promotes membrane fusion. (*C*) Human immunodeficiency virus (HIV) surface glycoprotein gp120 binds sequentially to the CD4 receptor on T cells and then to a coreceptor such as chemokine receptor CCR4. The latter interaction triggers a conformational change in gp120, which exposes gp41, the HIV factor initiating membrane fusion.

coronaviruses seem to be undergoing more rapid evolution, with regard to preferred ligands. Many of these viruses have a hemagglutinin-esterase (HE) that binds to specific O-acetylated sialic acids and they possess an esterase that cleaves the O-acetyl group. Some coronaviruses have eliminated the HE protein entirely and switched to sialic acid-binding via a spike protein. A few appear to have evolved further to preferentially bind to very specific host glycoproteins such as the angiotensin-converting enzyme 2 (ACE2, for the SARS viruses). SARS-CoV-2 (the cause of the COVID-19 pandemic) has also been shown to bind HS, an interaction that appears necessary for infecting ACE2-positive cells through stabilizing the up conformation of the spike receptor binding domain. Numerous reviews have been pulished on the glycosylation of SARS-CoV-2 proteins (including the spike glycoprotein shown on the cover image), the host glycoproteins involved in viral interaction, and the relevance of glycosylation in SARS-CoV-2 infectivity and immune avoidance, numerous reviews have been published.

Influenza

Influenza viruses are common human pathogens of the upper respiratory tract. Seasonal epidemics result in hundreds of thousands of deaths annually, with occasional, much more deadly pandemics. Influenza gains entry into cells of the human upper airway by binding to glycans terminated with α2-6-linked sialic acid (preferentially the human enriched one, N-acetylneuraminic acid or Neu5Ac) (Chapter 15). The entry process is mediated by influenza hemagglutinin, named for the method of its discovery. When first isolated in the 1930s, influenza was found to cause clumping (agglutination) of human red blood cells in vitro. Upon continued incubation, the red cells disaggregated and were not reagglutinated by fresh virus. It was hypothesized that a "receptor-destroying enzyme" was responsible. Isolation of the virus-released receptor revealed sialic acid that had been released by a sialidase, a term that is synonymous with neuraminidase. The familiar numbering system H1N1 represents hemagglutinin (H) and neuraminidase (N).

The life cycle of influenza starts with binding of the viral hemagglutinin to cell-surface sialic acids followed by hemagglutinin-mediated fusion with the host cell membrane, release of the virions intracellularly, replication, and then budding of newly assembled virions from the host cell surface. The hemagglutinin directs species and tissue tropism, whereas the neuraminidase is essential for propagation of the infection to neighboring cells. On completing its cellular replication cycle, the release of influenza viruses from infected cell surfaces relies on viral neuraminidase, which removes sialic acid from the surface of the host cells and the virion envelope. Without neuraminidase, newly formed virions stick together, form large rafts, and do not spread to other cells. Two rationally designed influenza virus neuraminidase inhibitor-based anti-influenza drugs, Relenza and Tamiflu, operate on this basis (Chapters 55 and 57).

Emergence of new human influenza strains occurs via transmission from other animal species, especially poultry. Sialic acid linkages in these different systems is key to understanding species tropism. Glycan array screening revealed that avian influenza binds to glycans terminated with α2-3 linked Neu5Ac, whereas human isolates bind to glycans terminated with α2-6 Neu5Ac. Neu5Ac in the α2-3 linkage is common in the intestinal tracts of birds but not found at substantial levels in the human upper airway. In contrast, α2-6-linked Neu5Ac predominates in the human upper airway. The switch from α2-3 to α2-6 binding is thought to underlie the emergence of new human influenza strains. This switch can occur by mutation of one or two amino acids in the hemagglutinin sialic acid–binding pocket. The emergence of animal influenza strains that bind α2-6 sialic acids is now monitored to detect potential new human influenza pathogens. Pigs are susceptible to both α2-3 (avian) and α2-6 (human) viruses, and can act as a "mixing vessel" to produce recombinant viruses capable of transmission from birds to humans. Direct avian–human transmission is often associated with enhanced morbidity, but human-to-human spread of avian influenza is uncommon. Notably, ferrets are the most effective animal model for human influenza studies as they have similar upper airway glycans terminated with α2-6 Neu5Ac.

Several nonenveloped viruses also have sialic acid–binding proteins on their icosahedral capsids including reovirus, adenovirus, parvovirus, and rotavirus, each of which binds to different sets of sialoglycans.

Herpes Simplex Virus

Heparan sulfate (HS) proteoglycans, which are widely distributed on vertebrate cells (Chapter 17), are implicated in the infective process of many pathogenic viruses including adeno-associated viruses, dengue viruses, hepatitis C virus, vaccinia virus, HIV, papillomavirus, and virtually all herpesviruses. In many cases, HS is a coreceptor, initiating attachment before recruitment of other host receptor proteins that support viral entry. A prominent example is HSV-1, also known as human herpesvirus 1 (HHV-1), one of the eight currently known human herpesviruses that cause widespread disease.

HSV-1 establishes latent, recurrent infections of mucous membranes, particularly lesions of the mouth and lips (cold sores, fever blisters) but also of genital tract and cornea, the latter of which may cause blindness. Unlike most other viruses, in which host cell binding and entry are mediated by one or two viral proteins, herpesvirus entry requires several viral entry glycoproteins, some of which are shared among all herpesvirus family members. One shared viral glycoprotein, gB, initiates virus attachment by binding to cell-surface HS, as does gC in HSV-1. Once bound, the virus "surfs" the cell surface until it encounters other receptors, including a specific HS structure, 3-O-sulfation on GlcNAc. This relatively rare HS modification induces binding of gD, another glycoprotein shared in the herpesvirus family. Once gD binds, it recruits additional proteins required for fusion and host cell entry.

Removal of HS from cell surfaces enzymatically or by selection of mutant cells defective in HS expression renders the cells resistant to HSV-1 infection by reducing virus attachment. Soluble heparin and HS mimetics inhibit viral infection by masking the HS-binding domain on the virus envelope. Immobilized HS columns bind to the HSV-1 viral entry proteins gB and gC, and HSV-1 deletion mutants lacking gB and gC exhibit impaired virus binding. Genetic evidence of a role of 3-O-sulfation of HS GlcNAc in HSV-1 infection was obtained by altering the expression of 3-O-sulfotransferase genes in cells and living organisms. Recent evidence suggests that up-regulation of the host's own heparanase helps newly budded HSV-1 to disseminate to neighboring cells, analogous to the role of influenza neuraminidase. Although the herpesviruses have evolved much more complex systems for host cell binding and entry, some of which remain to be established, it is clear that HS plays important roles in viral pathogenesis.

Human Immunodeficiency Virus

HIV is a retrovirus and the etiologic agent of the acquired immunodeficiency syndrome (AIDS), a pandemic disease affecting tens of millions of people worldwide. HIV is an enveloped virus with a surface dominated by spikes made of two proteins, gp120 and gp41, of which gp120 mediates viral attachment to host cells, primarily $CD4^+$ T cells, by binding to the host cell-surface receptor CD4 and a chemokine receptor coreceptor such as CCR5 or CXCR4. HIV gp120 is heavily glycosylated, with N-linked glycans comprising half of the spike mass and densely covering much of the spike surface. Dense glycosylation is thought to aid in immune evasion by masking the underlying polypeptide. However, gp120 glycans also actively support infection by co-opting host lectins, including C-type lectins on DCs (Chapter 34).

DCs are innate immune cells that capture and present antigens to T cells to initiate adaptive immunity. DCs capture antigens, in part, using C-type lectins that bind to glycan determinants common to pathogens but uncommon on host cells. Although DCs are not numerous, they are important in recognizing and presenting pathogen antigens to the adaptive immune system. DCs that reside in submucosal tissues of the vagina and rectum are early targets for HIV. Even with low levels of viral exposure, C-type lectins on DCs trap and concentrate HIV for subsequent presentation to T cells, the main site of HIV replication. DC-SIGN (dendritic cell–specific intercellular adhesion molecule-3-grabbing nonintegrin), mannose receptor, and Langerin are some of the C-type lectins that are important for this process. These lectins recognize dense arrays of mannose on pathogens including certain viruses (HIV, CMV [cytomegalovirus], hepatitis C virus, dengue virus), bacteria (*Helicobacter*, *Klebsiella*, *Mycobacteria*), fungi (*Candida*), and parasites (*Leishmania*, *Schistosoma*). Although T cells normally function to destroy pathogens and process their antigens for presentation, lectin-bound HIV evades destruction for extended periods. The natural role of DCs in presentation to T cells makes them ideal conduits for transmission of HIV to $CD4^+$ T cells, in which CD4 and cytokine receptors support binding, fusion, and viral replication. This process is termed *trans*-infection and facilitates early establishment of the HIV infection. DCs are not the only cells co-opted by HIV; macrophages express the same lectins and may also facilitate *trans*-infection. Other host lectins on DCs and macrophages, such as Siglec-1 (Chapter 35) play a similar role in facilitating uptake of viruses with heavily sialylated envelopes.

GLYCAN-BASED INTERACTIONS BETWEEN HOST AND GUT MICROBIOTA: COMMENSALS AND PATHOGENS

The nature of the relationship between microbes and the human host spans the spectrum from mutually beneficial (symbiotic), to benefiting the microbe without harming the host (commensal), to benefiting the microbe at the expense of the host (pathogenic). Glycans are key factors in the microbe–host interactions that occur along this continuum.

Bacteroides thetaiotaomicron is an anaerobic bacterium that is a common member of the normal colonic microbiota in mice and humans. This microbe has evolved mechanisms to maintain a mutually beneficial relationship with its mammalian host. A clue to this symbiosis came from examination of the gut epithelium of mice that are raised under germ-free conditions. Without bacterial exposure, the intestinal epithelium lacks expression of fucosylated glycoconjugates; when normal colonic bacteria are present, Fucα1-2Gal glycan expression is abundant on the surface of these host cells. *B. thetaiotaomicron* preferentially uses fucose both as an energy source and for incorporation into its own surface capsule and glycoproteins, phenotypes that are required for successful colonization and for proper immune development of the host. When dietary fucose is low, the bacterium induces the expression of host α1-2 fucosyltransferase, resulting in the expression of fucosylated (Fucα1-2Gal) glycoconjugates on the epithelium. *B. thetaiotaomicron* also expresses multiple fucosidases to cleave these terminal fucose residues and a fucose permease for uptake of the released sugar. Thus, the gut commensal has evolved a system for engineering the production of its own nutrient source from its host. This system is regulated for use only in times of need (during host fasting). *B. thetaiotaomicron* has evolved elaborate systems for regulating the expression of polysaccharide-binding proteins and glycosidases to forage and consume sugars from the host's diet or to switch over to glycans in the host mucus lining when sufficient polysaccharides are missing from the diet.

Helicobacter pylori colonizes nearly half the world's population, but it triggers chronic gastritis and stomach ulcers (conditions that are known to increase the risk of stomach cancer) in a subset of these individuals. Glycans of both the host and microbe contribute to whether *H. pylori* persists as a benign commensal or triggers disease pathology. *H. pylori* expresses an adhesin (BabA) that can interact with terminal Lewis b blood group antigen-containing glycans of the gastric epithelium. Lewis b expression in human intestines is limited to mucus-producing pit cells in the gastric epithelium. Transgenic mice engineered to express Lewis b show enhanced binding of *H. pylori* to their gastric epithelium, which triggers an enhanced cellular immune response and more severe gastritis. This microenvironment of immune activation appears to set the stage for a glycan-based process of molecular mimicry that can promote further host cell damage. *H. pylori* also expresses Lewis x–containing structures in the O-antigen of its LPS, which resemble Lewis x–modified glycans on the surface of parietal cells in the gastric lining. This Lewis antigen mimicry can be varied through the expression of two variable α1,3 fucosyltransferases, FutA and FutB. Similar to *B. thetaiotaomicron*, *H. pylori* has also developed mechanisms to obtain fucose from its host. The presence of *H. pylori* stimulates the host to secrete α-L-fucosidase 2 (FUCA2). This, in turn, increases the expression of the Lewis x–containing LPS O-antigen in *H. pylori*. Variation in both organism and host in the expression of Lewis x glycan structures and/or the adhesins may help explain the wide range of potential clinical outcomes following colonization by *H. pylori*.

This chapter provides only a glimpse of the diverse roles glycans play in the interactions between individual viruses and bacteria with their hosts. As scientists explore varying environments and their associated microbial communities (which also include parasites, fungi, and bacteriophages), it is becoming increasingly apparent that glycans influence every aspect of these interactions and there remains so much more to discover.

ACKNOWLEDGMENTS

The authors acknowledge contributions to the previous version of this chapter from Victor Nizet and Jeffrey D. Esko and helpful comments and suggestions from Frederique Lisacek.

FURTHER READING

Hooper LV, Gordon JI. 2001. Glycans as legislators of host–microbial interactions: spanning the spectrum from symbiosis to pathogenicity. *Glycobiology* **11**: 1R–10R. doi:10.1093/glycob/11.2.1r

Spear PG. 2004. Herpes simplex virus: receptors and ligands for cell entry. *Cell Microbiol* **6**: 401–410. doi:10.1111/j.1462-5822.2004.00389.x

Olofsson S, Bergström T. 2005. Glycoconjugate glycans as viral receptors. *Ann Med* **37**: 154–172. doi:10.1080/07853890510007340

Comstock LE, Kasper DL. 2006. Bacterial glycans: key mediators of diverse host immune responses. *Cell* **126**: 847–850. doi:10.1016/j.cell.2006.08.021

Ji X, Chen Y, Faro J, Gewurz H, Bremer J, Spear GT. 2006. Interaction of human immunodeficiency virus (HIV) glycans with lectins of the human immune system. *Curr Protein Pept Sci* **7**: 317–324. doi:10.2174/138920306778017990

Munford RS, Varley AW. 2006. Shield as signal: lipopolysaccharides and the evolution of immunity to Gram-negative bacteria. *PLoS Pathog* **2**: e67. doi:10.1371/journal.ppat.0020067

Wu L, KewalRamani VN. 2006. Dendritic-cell interactions with HIV: infection and viral dissemination. *Nat Rev Immunol* **6**: 859–868. doi:10.1038/nri1960

Akhtar J, Shukla D. 2009. Viral entry mechanisms: cellular and viral mediators of herpes simplex virus entry. *FEBS J* **276**: 7228–7236. doi:10.1111/j.1742-4658.2009.07402.x

Schwarz F, Fan YY, Schubert M, Aebi M. 2011. Cytoplasmic N-glycosyltransferase of *Actinobacillus pleuropneumoniae* is an inverting enzyme and recognizes the NX(S/T) consensus sequence. *J Biol Chem* **286**: 35267–35274. doi:10.1074/jbc.m111.277160

Koropatkin NM, Cameron EA, Martens EC. 2012. How glycan metabolism shapes the human gut microbiota. *Nat Rev Microbiol* **10**: 323–335. doi:10.1038/nrmicro2746

Needham BD, Trent MS. 2013. Fortifying the barrier: the impact of lipid A remodelling on bacterial pathogenesis. *Nat Rev Microbiol* **11**: 467–481. doi:10.1038/nrmicro3047

Stencel-Baerenwald JE, Reiss K, Reiter DM, Stehle T, Dermody TS. 2014. The sweet spot: defining virus–sialic acid interactions. *Nat Rev Microbiol* **12**: 739–749. doi:10.1038/nrmicro3346

Lu Q, Li S, Shao F. 2015. Sweet talk: protein glycosylation in bacterial interaction with the host. *Trends Microbiol* **23**: 630–641. doi:10.1016/j.tim.2015.07.003

Peng W, de Vries RP, Grant OC, Thompson AJ, McBride R, Tsogtbaatar B, Lee PS, Razi N, Wilson IA, Woods RJ, Paulson JC. 2017. Recent H3N2 viruses have evolved specificity for extended, branched human-type receptors, conferring potential for increased avidity. *Cell Host Microbe* **21**: 23–34. doi:10.1016/j.chom.2016.11.004

43 Parasitic Infections

Richard D. Cummings, Cornelis H. Hokke, and Stuart M. Haslam

Parasitic protozoans and helminths (worms) synthesize glycans with structures often different from those typically found in vertebrates, and are thus often antigenic. Parasites also express glycan-binding proteins (GBPs) involved in host invasion and parasitism. As part of the disease process, parasite glycans can trigger the host's innate immune system, which can lead to the induction of adaptive immune responses. This chapter discusses the major roles of glycoconjugates in parasitic infections.

BACKGROUND ON PARASITIC INFECTIONS

Parasitism is a condition in which one organism (the parasite) in some way lives at the expense of the host. Parasites infect millions of people worldwide and cause much suffering and death, especially in underdeveloped countries (Table 43.1). Like other types of infections (e.g., bacterial or viral) (Chapter 42), parasite GBPs interact with host glycomes and parasite glycans interact with host GBPs and antibodies. Thus, research on parasite glycobiology and biochemistry could provide therapeutics for reducing disease morbidity and mortality. Moreover, studies on the molecular pathology of organisms, such as parasites that have evolved to deceive and compromise the host immune system, provide novel insights into the regulation of human innate and adaptive immune responses.

The majority of parasitic diseases are divided into two categories: those caused by protists (single-celled organisms; Table 43.2) and those caused by helminths (worms/metazoans; Table 43.3). The major classes of protozoan parasites include *Plasmodium* (causing malaria), *Entamoeba histolytica* (causing amebiasis), *Leishmania* (causing leishmaniasis), and *Trypanosoma* (causing sleeping sickness and Chagas disease). Parasitic worms include trematodes (e.g.,

published by Cold Spring Harbor Laboratory Press; doi: 10.1101/glycobiology.4e.43

TABLE 43.1 Worldwide distributions of some major parasitic human diseases

Type of disease	Estimated human infections (need for treatment)	Deaths per year
Parasitic helminths (worms)		
Soil-transmitted helminths	880 million	~150,000
Roundworm (*Ascaris lumbricoides*)		
Whipworm (*Trichuris trichiura*)		
Hookworms (*Necator americanus, Ancylostoma duodenale*)		
Strongyloidiasis (*Strongyloides sterecoralis*)		
Schistosomiasis (*Schistosoma sp.*)	258 million	uncertain but ~20,000–200,000
Lymphatic filariasis (*Wuchereria bancrofti, Brugia malayi*, and *Brugia timori*)	120 million	~20,000–50,000 (plus millions disfigured or incapacitated)
Onchoceriasis (river blindness) (*Onchocerca volvulus*)	25 million	few deaths but 300,000 blinded
Enterobiasis (pinworms or threadworms) (*Enterobius vermicularis*)	200 million	rare
Parasitic protozoans		
Malaria (*Plasmodium vivax* and *Plasmodium ovale*)	214 million	438,000
Leishmaniasis (*Leishmania sp.*)	0.9–1.4 million new cases/year	20,000–30,000
African trypanosomiasis (sleeping sickness) (*Trypanosoma brucei gambiense* and *Trypanosoma brucei rhodesiense*)	60 million at risk and ~300,000 new cases/year	50,000
American trypanosomiasis (Chagas disease) *Trypanosoma cruzi*)	8 million	12,000

Information was condensed from the World Health Organization (www.who.int/en/) and Centers for Disease Control and Prevention (www.cdc.gov).

This table shows the total number of individuals with infections or in need of treatment worldwide, and the approximate number of infection-related deaths per year. Note that some individuals may have more than one infection.

TABLE 43.2 Some of the major parasitic protozoans of humans

Parasite	Effect
Amoeba infecting humans	
Entamoeba histolytica	causes amebic dysentery; can cause liver abscesses
Intestinal and genital flagellates	
Giardia lamblia	causes diarrhea; one of the most common parasites in North America
Trichomonas vaginalis	causes inflammation of reproductive organs; very common
Hemoflagellates	
Leishmania donovani	causes visceral leishmaniasis (kala-azar); hepatosplenomegaly
Leishmania mexicana	causes fulminating, cutaneous ulcers
Leishmania major	causes cutaneous ulcers
Trypanosoma brucei sp.	causes sleeping sickness in humans and nagana in cattle (African trypanosomiasis)
Trypanosoma cruzi	causes Chagas disease (South American trypanosomiasis)
Gregarines, coccidia, and related organisms	
Plasmodium falciparum	major cause of human malaria
Plasmodium vivax, Plasmodium ovale, Plasmodium malariae	also cause human malaria
Causes of opportunistic infections in immunodeficiency conditions	
Toxoplasma gondii	causes flu-like symptoms; latent cysts can reactivate to cause encephalitis or blindness
Pneumocystis carinii	causes interstitial cell pneumonia
Cryptosporidium parvum	intracellular parasite of intestinal cells that causes diarrhea

TABLE 43.3 Some of the major parasitic helminths of mammals

Parasite	Effect
Trematodes	
Blood flukes	
Schistosoma mansoni	causes human schistosomiasis (affects mesenteric veins draining large intestine)
Schistosoma haematobium	causes human schistosomiasis (affects urinary bladder plexus)
Schistosoma japonicum	causes human schistosomiasis (affects mesentery veins in the small intestine)
Liver flukes	
Fasciola hepatica	primarily infects ruminants and occasionally humans (worms live in biliary tract)
Clonorchis sinensis	most prevalent liver fluke in humans (can be acquired by eating raw fish)
Cestodes	
Taenia solium	long human tapeworm acquired by eating undercooked pork
Echinococcus granulosus	shorter human tapeworm acquired by eating undercooked lamb (parasitic cysts [hydatids] occur in liver and elsewhere)
Taeniarhynchus saginatus	long human tapeworm acquired by eating undercooked beef
Nematodes	
Ascaris lumbricoides	most common intestinal roundworm in humans
Trichuris trichiura	intestinal whipworm in humans
Enterobius vermicularis	tiny intestinal roundworm (causes perianal night itch in children)
Necator americanus	intestinal hookworm of humans (causes anemia)
Ancylostoma duodenale	intestinal hookworm of humans (causes anemia)
Strongyloides stercoralis	intestinal parasite (causes autoinfection)
Haemonchus contortus	intestinal parasite of sheep and goats
Trichinella spiralis	smallest nematode parasite of humans (trichinosis) residing in muscle fibers (acquired from eating undercooked pork)
Onchocerca volvulus	filarial parasite (causes river blindness)
Wuchereria bancrofti	filaria live in lymph nodes causing elephantiasis
Brugia malayi	filaria live in lymph nodes causing elephantiasis
Dirofilaria immitis	dog heartworm

Schistosoma mansoni, causing schistosomiasis), nematodes (e.g., *Ascaris lumbricoides*) and cestodes or tapeworms (e.g., *Taenia solium*, causing taeniiasis). Worms are very large relative to host cells. Thus, most worms live in the extracellular spaces of their hosts, and have evolved a variety of infective and protective strategies. Glycoconjugates are important in both the life cycles and pathology of most major parasites. For example, many parasitic protozoans and helminths have elaborated intriguing mechanisms to target GBPs (Chapter 29) or glycans in the host to promote parasitism, and to evade host immune responses.

PLASMODIA

Malaria is caused by *Plasmodium* species, prominently *Plasmodium falciparum* in humans. Malarial parasites lead a complicated life cycle, alternating between a sexual reproduction stage in the female *Anopheles* mosquito vector and an asexual reproduction stage in mammalian tissues (hepatocytes and erythrocytes) and the bloodstream (Figure 43.1). Cell–cell interactions between the parasite and host are critical for the successful completion of each stage.

Following inoculation into the bloodstream, the sporozoite's major circumsporozoite protein (CSP) interacts with heparan sulfate (HS) on the surface of hepatocytes, which allows invasion and the first replication in the mammalian host. Liver HS possesses an unusually high degree of sulfation compared to similar glycosaminoglycans from other organs, suggesting the basis for the selective targeting of *Plasmodium* to hepatocytes. A recombinant form of the pre-erythrocytic CSP made in yeast is a components of the RTS,S vaccine currently being

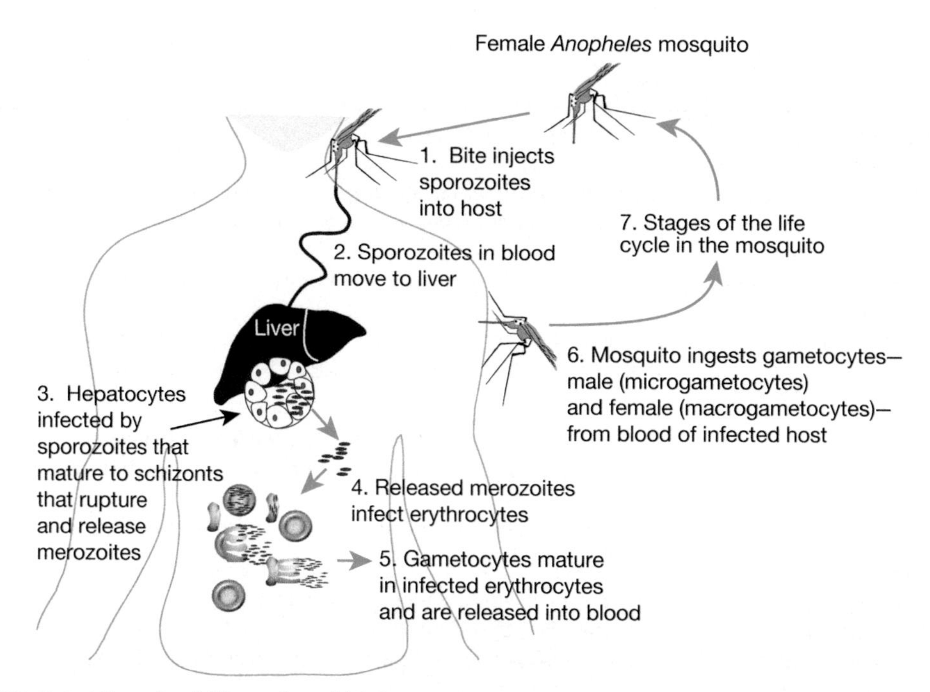

FIGURE 43.1. Life cycle of *Plasmodium falciparum*, a parasitic protozoan that causes the most severe form of malaria in humans. The bite of the female *Anopheles* mosquito introduces sporozoites into the human host, which mature as they travel to the liver and ultimately the bloodstream. After a blood meal from an infected person, the malarial gametocytes enter the midgut of the mosquito where they transform into male microgametes and female macrogametes. Their union leads to a zygote, which transforms into an ookinete, penetrates the intestinal wall of the mosquito, and is transformed into a circular oocyst. Inside the oocyst, the sporozoites develop from germinal cells known as sporoblasts. The sporozoites emerge from the oocysts and migrate to the salivary gland where they enter the human hosts during the blood meal of the mosquito.

implemented worldwide. Interestingly, the T-cell epitope of CSP is O-fucosylated, but the recombinant yeast-derived protein is not O-fucosylated.

On exit from the liver, *Plasmodium* merozoites use multiple ligand–receptor interactions to invade host erythrocytes. The merozoite proteins, such as the erythrocyte binding-like (EBL) proteins (Table 43.4), vary in their dependency on sialic acid residues on the erythrocyte surface and play an important role in erythrocyte invasion.

The glycophorins are the major sialic acid–containing glycoproteins on erythrocytes. The EBL protein EBA-175 (erythrocyte-binding antigen-175) recognizes clusters of sialylated O-glycans attached to glycophorin A, particularly within a 30-amino-acid region that carries 11 O-glycans. Desialylation of erythrocytes precludes interactions of some strains of *P. falciparum*, and individuals lacking glycophorins A or B are refractory to invasion. Some strains of *P. falciparum* can reversibly switch from sialic acid–dependent to sialic acid–independent invasion, and this has important implications for vaccine design against malaria parasites.

Pregnancy-associated malaria is a major cause of suffering and arises when *P. falciparum*–infected erythrocytes are sequestered in the placenta of pregnant women. The parasite expresses the VAR2CSA protein, which mediates adherence of the infected erythrocytes to placental chondroitin sulfate A (CSA) with very high specificity and affinity (K_d~15 nM), promoting parasite infection and invasion.

Eruption of the merozoites from infected erythrocytes releases free glycosylphosphatidylinositols (GPIs), which are prominent virulence factors contributing to malaria pathogenesis (Chapter 12). *P. falciparum* GPIs mimic host GPIs and activate host GPI-associated signaling pathways. The GPIs can activate host's macrophages, mainly through Toll-like receptor-2 signaling, leading to

TABLE 43.4 Some major parasites and their glycan-binding proteins

Parasite	Stage	Protein	Specificity
Plasmodium falciparum	merozoite	EBA-175	Neu5Acα2-3Gal/glycophorin=A
	merozoite	EBA-140	sialic acid/glycophorin B?
	merozoite	EBA-180	sialic acid
	sporozoite	circumsporozoite protein	heparan sulfate
	parasitized erythrocytes	the anchor protein VAR2CSA	chondroitin sulfate A
Trypanosoma cruzi	trypomastigote	*trans*-sialidase	Neu5Acα2-3Gal
	trypomastigote	penetrin	heparan sulfate
Entamoeba histolytica	trophozoite	Gal/GalNAc lectin	Gal/GalNAc
Entamoeba invadens (a reptilian pathogen)	cyst	cyst wall protein (Jacob lectin)	chitin
Giardia lamblia	trophozoite	taglin (α-1 giardin)	Man-6-PO_4-, heparan sulfate
Cryptosporidium parvum	sporozoite	Gal/GalNAc lectin	Gal/GalNAc
	sporozoite	Cpa135 protein	?
Acanthamoeba keratitis	trophozoite	136-kDa mannose-binding protein	mannose
Toxocara canis	larval	TES-32	?
Haemonchus contortus	gut-localized	galectin (Hco-gal)	β-galactosides

the production of inflammatory cytokines and up-regulation of cell-adhesion molecules, such as E-selectin in endothelial cells. Antibodies to these GPIs can neutralize their effects and mitigate the pathology of the disease independently of direct infection.

In addition, the *P. falciparum* genome encodes a number of enzymes involved in N-glycosylation and O-glycosylation (O-fucosylation) and C-mannosylation. Whereas N-glycans are truncated and typically contain only a chitobiosyl core, GlcNAcβ1-4GlcNAcβ-Asn, mannose is abundantly present in GPI anchors. There is also evidence suggesting that *Plasmodia* may express a number of unusual glycoconjugates, such as galactose- and glucose-containing glycoproteins and glucose-containing glycolipids.

TRYPANOSOMES

African trypanosomes (*Trypanosoma brucei*) transmitted by blood-sucking tsetse flies are the etiologic agents of nagana disease in cattle and sleeping sickness in humans. A remarkable feature of these organisms is their ability to survive extracellularly in the host bloodstream, where they are constantly exposed to the immune system. Evasion of the host immune response depends on "antigenic variation," a highly evolved survival strategy that relies on structural variance of GPI-anchored glycoproteins (VSGs) on the trypanosome surface (Figure 43.2). VSGs are dimeric proteins, consisting of two 55-kDa monomers that each carry N-linked oligomannose-type glycans, which make up a large component of the dense glycocalyx. As parasites multiply in the host bloodstream, the host mounts an immune response that is effective against only the population of trypanosomes expressing a particular VSG. Trypanosomes that have switched to an alternative VSG coat (encoded among 1000 distinct VSG genes) escape immunological destruction.

Within the gut of the tsetse fly, the trypanosome replaces its entire VSG coat with acidic glycoproteins called procyclins (Figure 43.2). These GPI-anchored proteins form a dense glycocalyx and are composed of polyanionic polypeptide repeat domains projecting from the membrane. Unusual features are the presence of a single type of N-glycan ($Man_5GlcNAc_2$) and GPI anchors that are modified with branched poly-*N*-acetyllactosamine $[Gal\beta1\text{-}4GlcNAc]_n$ glycans. The terminal β-galactose can be substituted with α2-3-linked sialic acid by a parasitic *trans*-sialidase that

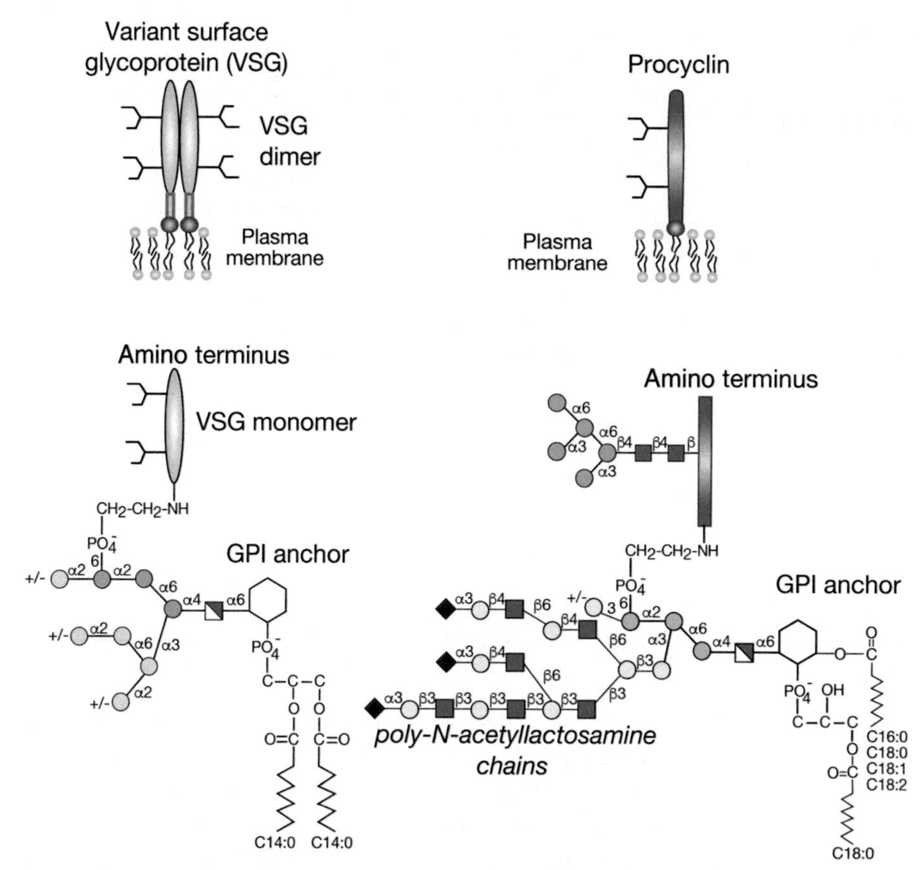

FIGURE 43.2. Schematic representation of the major surface glycoconjugates of procyclic and metacyclic *Trypanosoma brucei*. VSG (variant surface glycoprotein) is the major component of the metacyclic form, and each molecule consists of two GPI-anchored N-glycosylated monomers (*shaded ovals* are the protein component). The surface of the procyclic form is densely covered with procyclins, which are GPI-anchored polypeptides with polyanionic repeat domains (*shaded rectangle* is the protein component). The anchor structures are detailed below the schematic.

transfers sialic acid from host glycoconjugates to the parasite surface. This sialylation protects the parasite in the insect intestine and can compromise the human immune system. Among parasites, genes encoding *trans*-sialidases have only been found in the protozoan genus *Trypanosoma* and occur in both *T. brucei* and *Trypanosoma cruzi*. Interestingly, there is also evidence for *trans*-sialidase activities in some bacteria and in human serum, but these are far less well studied.

T. cruzi, transmitted by reduviid bugs, is the etiologic agent of Chagas disease, or South American trypanosomiasis. *T. cruzi* has a dense coat of glycosylinositolphospholipids (GIPLs) (Chapter 12) and mucins (Chapter 10) that project above the GIPL layer (Figure 43.3). The GIPLs contain the same basic structure as other GPI anchors, except that they are heavily substituted with Gal, GlcNAc, and host-derived sialic acid. The mucins contain large amounts of O-glycans composed of a serine- or threonine-linked αGlcNAc extended with one to five Gal residues (including Gal*f*) that can be substituted with sialic acid by a *trans*-sialidase. For *T. cruzi* sialylation is believed to reduce the susceptibility of the parasite to anti-α-Gal antibodies that are normally present in the mammalian bloodstream.

T. cruzi also express lipopeptidophosphoglycan (LPPG), the major surface glycan of the insect stage of the parasite (Figure 43.3). Depending on the life-cycle stage, LPPG is composed of an inositolphosphoceramide-anchored glycan or an alkylacylphosphatidylinositol-anchored glycan that includes nonacetylated glucosamine, mannose, galactofuranose, and 2-amino-

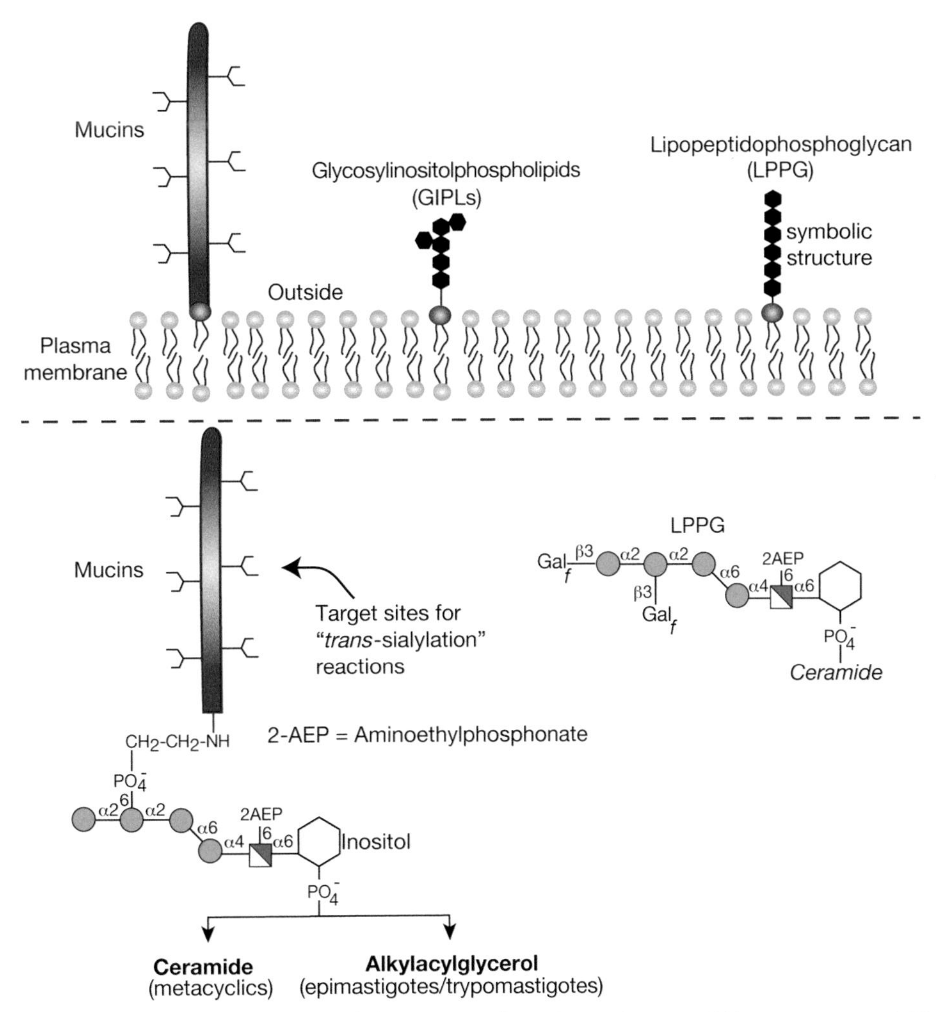

FIGURE 43.3. Schematic representation of the major surface glycoconjugates of *Trypanosoma cruzi*. The cell surface of *T. cruzi* is covered with a dense layer of mucins, with extensive O-glycosylation, glycosylinositolphospholipids (GIPLs), and lipopeptidophosphoglycan (LPPG). The structures of the mucin anchors and the predominant LPPG species are outlined. Aminoethylphosphonate residues are indicated by 2-AEP. For the monosaccharide symbol code, see Figure 1.5, which is also reproduced on the inside front cover.

ethylphosphonate (2-AEP). The lack of ceramide anchors and galactofuranose in mammalian cells suggests potential targets for the development of chemotherapeutic agents.

LEISHMANIA

Leishmania causes different forms of leishmaniasis, which manifest clinically in three forms: cutaneous, mucocutaneous, and visceral, the latter being fatal if untreated. *Leishmania* parasites have a remarkable capacity to avoid destruction in the hostile environments they encounter during their life cycle, alternating between intracellular macrophage parasitism and extracellular life in the gut of their sandfly vector (Figure 43.4).

Stage-specific adhesion is mediated by structural variation of the abundant cell-surface glycoconjugate lipophosphoglycan (LPG), which contributes to parasite survival in the hydrolytic midgut (Figure 43.5). The basic LPG structure in all *Leishmania* species consists of a 1-*O*-alkyl-2-*lyso*-phosphatidyl(*myo*)inositol lipid anchor, a heptasaccharide core, a long phospho-

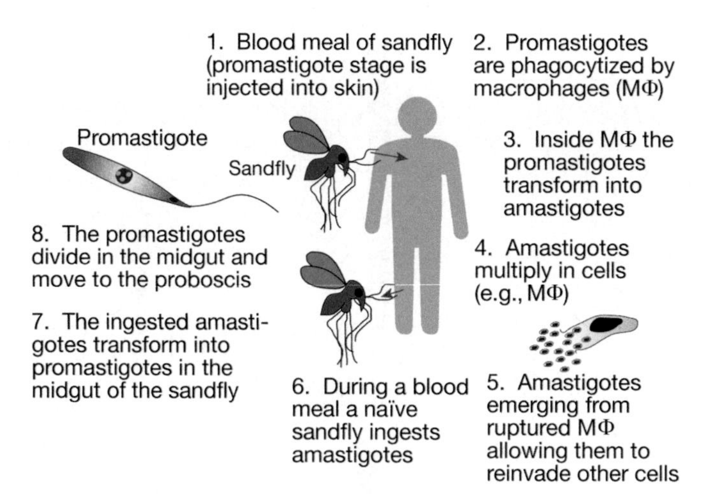

FIGURE 43.4. Life cycle of *Leishmania* species, the parasitic protozoan that causes leishmaniasis in humans.

glycan (PG) polymer composed of (-6Galβ1-4Manα 1-PO_4-) repeat units, and a small oligosaccharide cap. In many species, the PG repeats contain additional substitutions that mediate key roles in stage-specific adhesion. For example, in *Leishmania major*, the PG repeats bear β1-3-galactosyl side-chain modifications, which provide a binding site for the sandfly midgut galectin, PpGalec. As parasites differentiate in the sandfly from amastigotes into procyclic promastigotes, and finally to infective metacyclic stages, the β1-3 Gal-modified PGs are capped by α1-2 arabinosyl residues, giving rise to a structure that does not bind to PpGalec and thus facilitating detachment of the parasite from the sandfly midgut. LPG is highly abundant on the para-

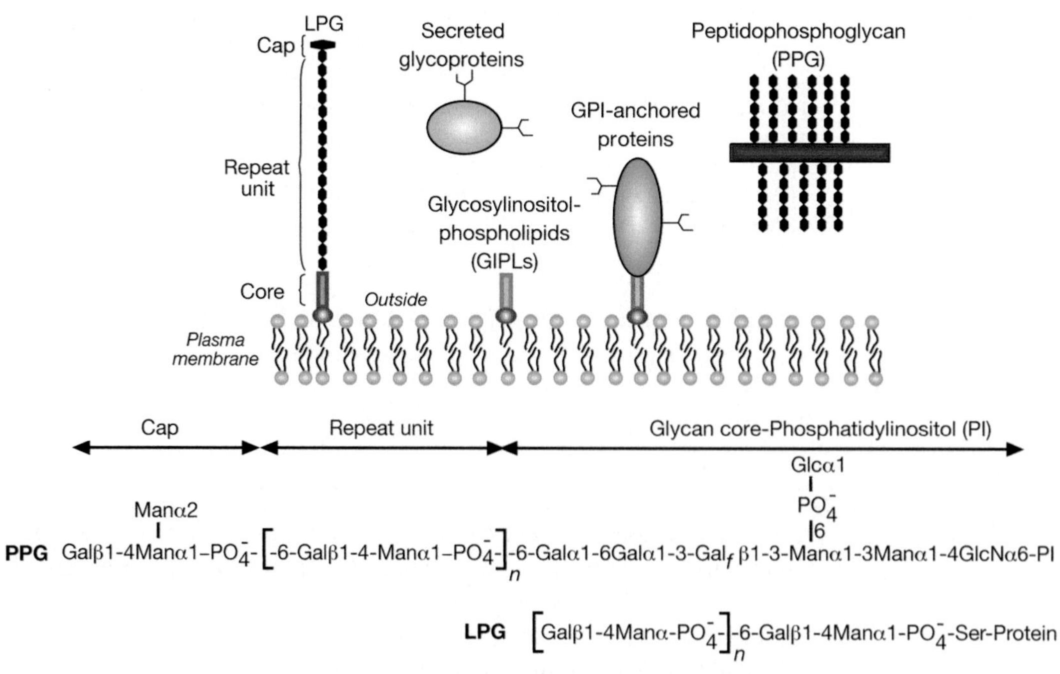

FIGURE 43.5. Schematic representation of the major cell-surface glycoconjugates of *Leishmania*. Lipophosphoglycan (LPG) consists of a Gal-Man-PO_4- repeat unit backbone attached to a lipid anchor via a glycan core. Secreted glycoproteins and mucin-like molecules called proteophosphoglycans (PPGs) are composed of serine/threonine-rich peptides that are heavily glycosylated on serine with phosphoglycan chains, similar to those found on LPG. The structures of *Leishmania donovani* LPG and PPG are shown.

site surface, suggesting a central role for the glycoconjugate in the parasite's infectious cycle, as well as allowing the parasite to establish successful infections in the mammalian host. In addition, galectin-3 in the vertebrate host may recognize some of the *L. major* glycans as pathogen-associated molecular patterns, facilitating leukocyte responses to infection.

Leishmania express also an abundance of other important glycoconjugates, such as GIPLs, GPI-anchored proteins, and secreted proteophosphoglycans (PPGs) (Figure 43.5) (also see Chapter 12). Most of the serine residues within PPGs are phosphoglycosylated with Gal-Man-PO_4- repeats via unique phosphodiester linkages. In *Leishmania mexicana*, these highly anionic polysaccharides form a gel-like matrix composed of interlocking filaments. These matrices enhance parasite development in the sandfly and contribute to the formation of a parasitophorous vacuole in macrophages of the mammalian host, where the parasite replicates.

Virtually every known glycoconjugate of *Leishmania* shows some intersection with the LPG biosynthetic pathway. The presence of molecules bearing similar modifications raises the possibility that they share biosynthetic steps. The early steps in LPG and GPI biosynthesis (up to Man-Man-GlcN-PI) occur in the endoplasmic reticulum, and galactosylation of the LPG glycan core and assembly of the PG repeats occurs in the Golgi apparatus (Chapter 12). The phosphoglycan portions of LPG and PPG are assembled by the sequential and alternating addition of mannose-PO_4- and galactose, forming the characteristic -Galβ1-4Manα1-PO_4- repeats. Depending on the species of *Leishmania*, additional branching sugars can then be added, creating a remarkable array of side chains that drive *Leishmania*–sandfly vectorial competence.

Leishmania spp. are also able to generate sialylated glycans, both $\alpha2\rightarrow3$ and $\alpha2\rightarrow6$ linkages with the latter being most prevalent. In addition, the parasite can generate O-acetylated versions of sialic acid. Interestingly, like for *Trypanosoma, Leishmania* does not synthesize sialic acid de novo, but instead acquires it from the host. Although Leishmania has several sialic acid transporters, it is not yet clear how the sialylated glycans are acquired and produced by *Leishmania*. The sialylated glycans they present may interact with multiple Siglecs, especially macrophage Siglecs, that contribute to infection and cell signaling.

ENTAMOEBA

The life cycle of *E. histolytica*, causing amebic dysentery and hepatic abscesses, includes the disease-inducing amebic or trophozoite stage, and the infectious cyst stage. Mature cysts are ingested by an animal, and once inside the host gut, excystation produces the trophozoite that causes pathology. The trophozoites can replicate or form cysts, which are passed via feces to complete the cycle. *E. histolytica* expresses several lectins, among them a Gal/GalNAc lectin, a GPI-anchored heterodimeric glycoprotein. This lectin mediates binding of the parasite to colonic mucins and plays a critical role in parasite viability, and is a promising vaccine candidate against the disease. The *Entamoeba* cyst wall is composed mostly of chitin, and chitin-binding lectins (including the Jacob and Jessie lectins) accompanied by remodeling via chitinases, help to create the cyst wall.

Entamoeba trophozoites synthesize cell-surface LPG and LPPG (Figure 43.6). The LPG consists of a lipid anchor and a phosphoglycan component that resembles the phosphoglycans of *Leishmania* LPG. The LPPG GPI anchors are unique in that they have a glycan backbone that contains the sequence Gal_1Man_2GlcN-(*myo*)inositol, in which α-Gal replaces the terminal α1-2 mannose residue found in typical protein anchors. *Leishmania* LPG and LPPG are important as virulence factors, because antibodies raised against these molecules inhibit the ability of the parasites to kill target cells. Thus, such glycans may be the target for vaccine development.

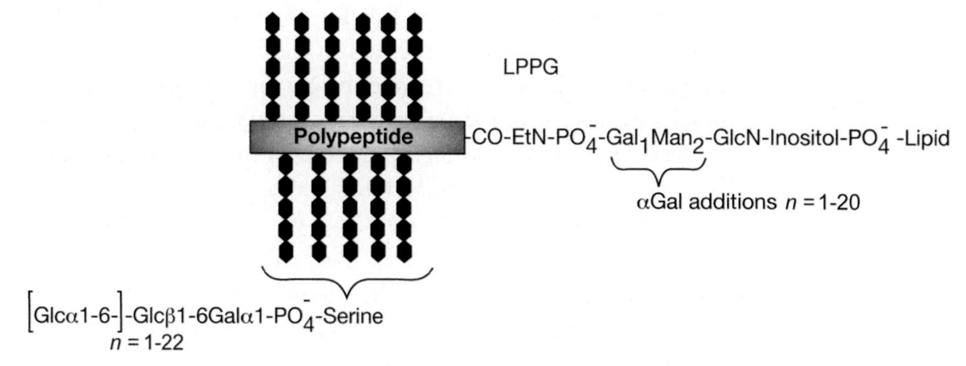

FIGURE 43.6. Structure of *Entamoeba histolytica* lipopeptidophosphoglycan (LPPG).

SCHISTOSOMA

Schistosomiasis, a major parasitic disease associated with a high morbidity worldwide is caused by parasitic trematodes. Three major species infect humans: *Schistosoma japonicum*, *Schistosoma mansoni*, and *Schistosoma haematobium* (Figure 43.7). Schistosoma is unique among helminths in that the male and female worms pair to live together in the blood vessels of their host. The female lays eggs ~4–6 weeks after infection, which adhere to the endothelium and migrate through vessels into tissues. The eggs may become lodged in the host tissues, where they sequester and induce granulomatous inflammatory responses. Eggs trapped in the peripheral circulation can cause portal hypertension and fibrosis, which are characteristic of chronic schistosomiasis and lead to morbidity and often death. However, many eggs eventually pass

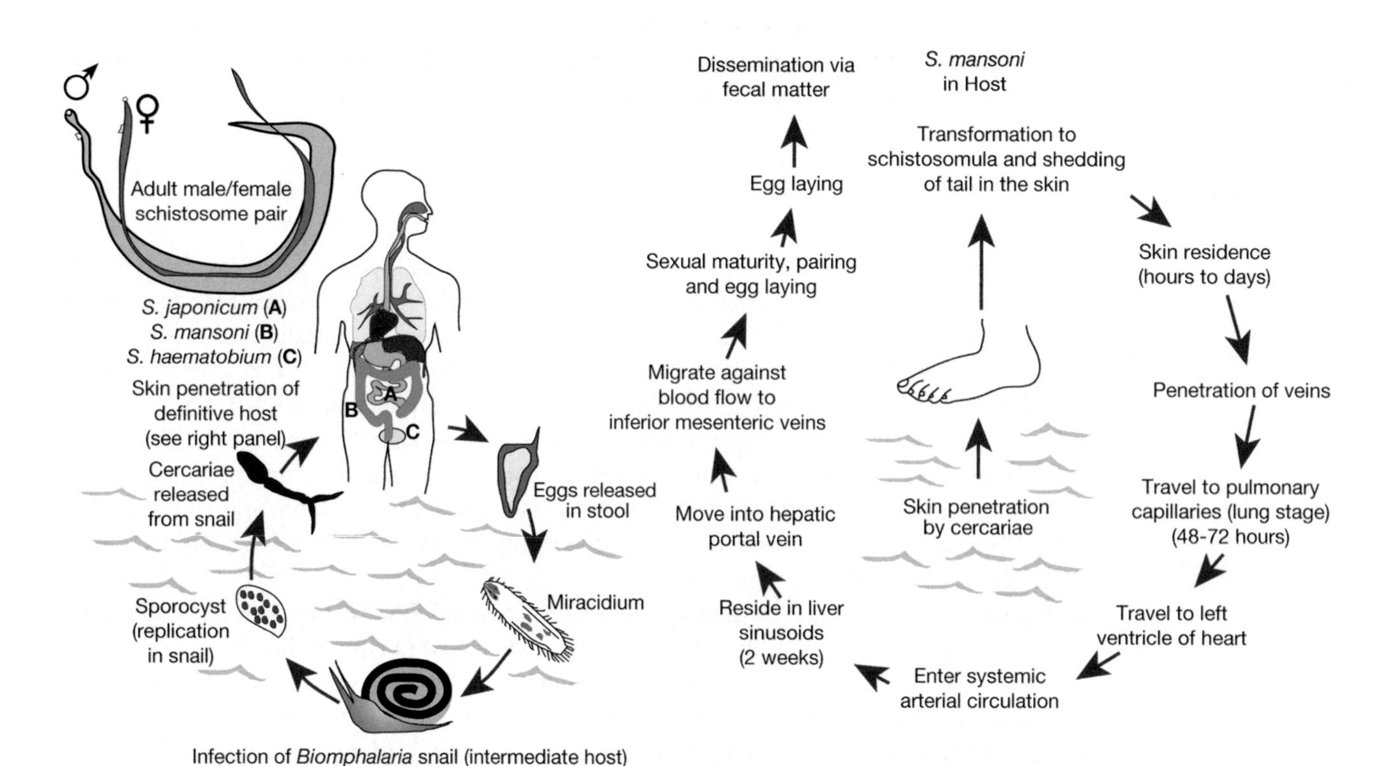

FIGURE 43.7. Life cycle of *Schistosoma* species, the parasitic helminth that causes schistosomiasis in humans. (*Left*) Alternating cycle between the invertebrate snail host and the human definitive host; (*right*) passage of the worms and their maturation to adult sexually active male/female pairs within the human host.

through the gut or bladder wall into the stool or urine and continue the cycle through intermediate snail hosts, which are unique for each *Schistosoma* species. Thus, the geographical limitation to the spread of this disease is the range of the intermediate host snails.

Schistosomes, especially the cercarial glycocalyx and eggs, generate huge quantities of membrane-bound glycoproteins and glycosphingolipids. Many glycoproteins derived from the tegument, gut, and eggs of the parasite are highly antigenic and occur in the circulation of the infected mammalian host. These proteins are generally rich in complex glycan structures, and contain an impressive array of O- and N-glycans. Fucosylated antigens are a common feature of most schistosome glycoconjugates, including Lewis^x (Le^x), LacdiNAc (LDN), and fucosylated LacdiNAc (LDNF) structures, and polyfucose branches (Figure 43.8). Like all helminths, glycoconjugates from schistosomes lack sialic acids, but in schistosomes and several other worms, glucuronic acid is present as a negatively charged monosaccharide.

The glycoconjugate structures expressed among schistosome species show some variation, but much less than between schistosomes and other helminths. For example, schistosomes have extended difucosylated oligosaccharides, but these structures are absent in other helminths. The LeX antigen, which is abundant in schistosomes also occurs in *Dictyocaulus viviparus*, a parasitic nematode in cattle, but overall is found in very few helminths. In addition to schistosomes, *Echinococcus granulosus*, *Dirofilaria immitis*, and *Haemonchus contortus* synthesize glycoproteins containing LDN and LDNF, in addition to other fucosylated and xylosylated glycans (Figure 43.8). The expression of many of these glycan structures is developmentally regulated and stage-specific, but their fundamental roles in parasite development and host pathogenesis are mostly unclear. In addition to the roles of schistosome glycans in the interaction with the mammalian host, essential interactions between schistosomes and the intermediate snail host are glycan-dependent.

There is mounting evidence that the glycan antigens expressed by schistosomes can modulate innate and adaptive immune responses. Individuals infected with *Schistosoma* species develop a

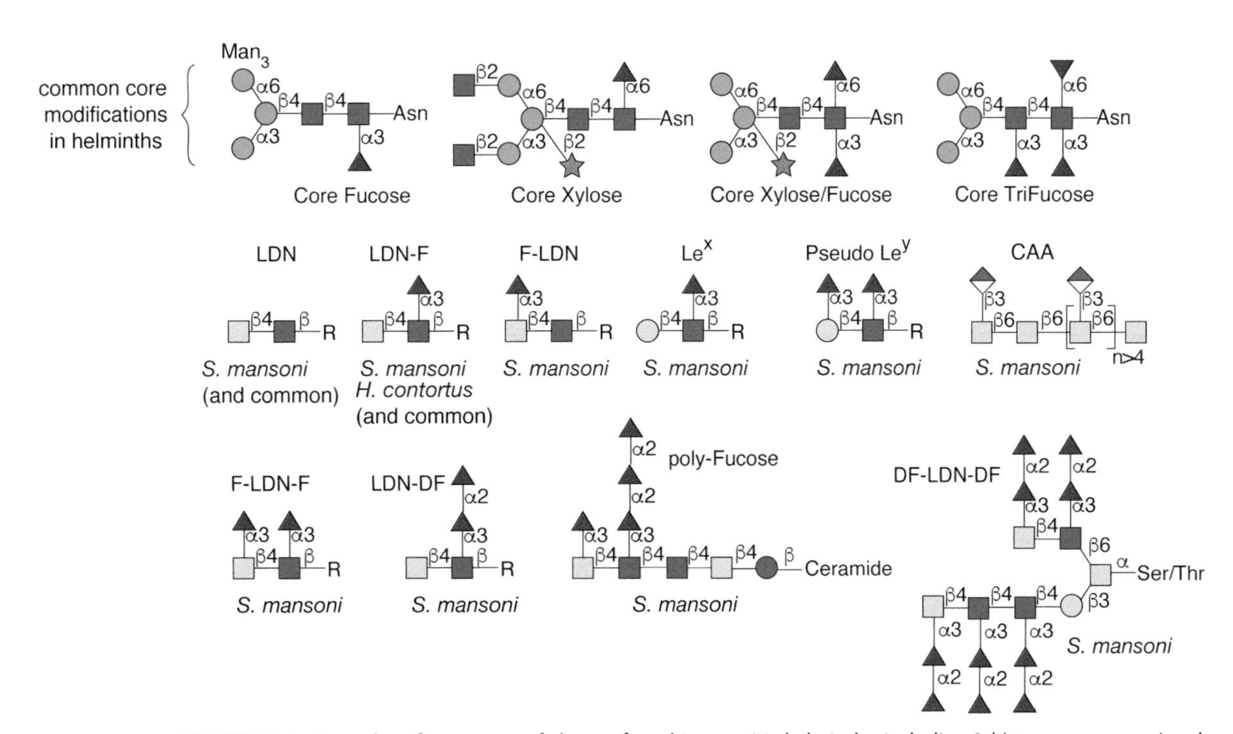

FIGURE 43.8. Examples of structures of glycans found in parasitic helminths, including *Schistosoma mansoni* and *Haemonchus contortus*. The antigens are indicated by the structures within the brackets. (LDN) LacdiNAc; (Le^x) Lewis x; (CAA) circulating anodic antigen; (F−) fucosylated; (DF−) difucosylated.

wide variety of antibodies to glycan antigens, which may provide partial protection against subsequent infections. A general feature of helminth infections, including chronic schistosome infections, is that T-helper 2 (Th2) immune responses (promoting humoral immunity) predominate over Th1 responses (promoting cellular immunity). Such Th2-driven responses also contribute to the generation of alternatively activated macrophages with wound-healing and anti-inflammatory properties. The induction of an anti-inflammatory immune response facilitates survival of the parasites in their hosts, but also benefits the host by reducing unrelated inflammatory responses (e.g., in autoimmune diseases). Worm-derived molecules from schistosomes and other helminths, including glycoconjugates, play a role in such modulation of the host immune response. Helminths, and their secreted products, are currently being evaluated as novel therapy for chronic inflammatory disorders such as Crohn's disease and multiple sclerosis.

Schistosome glycans are recognized by antigen-presenting cells such as dendritic cells and macrophages via GBPs, in particular C-type lectins (Chapter 34), including pattern-recognition receptors as dendritic cell–specific intercellular adhesion molecule-3-grabbing nonintegrin (DC-SIGN) and the mannose receptor. For immune responses to many helminths, collectins and surfactant protein D help to limit the pathology of infection. Internalization and processing of parasite glycoconjugates by dendritic cells can lead to polarization of adaptive immune responses and cellular activation (Chapter 34). The identification of many antigenic glycoconjugate structures from schistosomes is helping in the design of new diagnostic approaches for schistosomiasis. Because antibody responses to schistosomes are mainly directed against antigenic glycan motifs, and antibodies to such glycans can kill and inhibit parasite growth and egg deposition, such glycans may be targets of vaccine development. The characterization of schistosome glycosyltransferases responsible for antigenic glycan biosynthesis, together with enzymes from the intermediate snail hosts, may also help in the identification of new drug targets and the development of glycan-based vaccines and therapeutics.

GLYCOBIOLOGY OF OTHER PARASITES

Glycoconjugates and GBPs, as discussed above, play an impressive role in host infection by many different parasites. Several additional examples are shown in Table 43.4. Many protozoan parasites appear to use GBPs as a major mechanism for host-cell attachment and invasion. For example, *Acanthamoeba keratitis*, which causes severe eye infections involving the corneal epithelium, adheres to host cells via a lectin-glycan interaction and precludes amoeba-induced cytolysis of target cells. Adhesion is mediated by a mannose-binding protein, which is strongly inhibited by Manα1-3Man disaccharides. Furthermore, mannose and mannose-6-PO_4- can inhibit adhesion of *Giardia lamblia* trophozoites, and these glycans may be recognized by the parasite's antigenic protein taglin. Other antigenic proteins, including several galectins, have been cloned from parasitic nematodes including *Teladorsagia circumcincta*. Sporozoites from *Cryptosporidium parvum*, an opportunistic protozoan that infects individuals with compromised immunity, have hemagglutinating activity, and a lectin on the parasite surface may play a crucial role in host-cell attachment. Recently, a sialic acid–binding protein in *Toxoplasma gondii* has been identified (SABP1) that is important for infection by this protozoan.

In addition, the glycan antigens of many parasites are being characterized in an effort to develop vaccines and new diagnostics for the diseases they cause. For example, synthetic glycoconjugates based on glycan structural components of *P. falciparum* GPI to create a malaria vaccine and the *T. cruzi* cell surface mucins that are being utilized to develop serological assays to improve Chagas disease diagnostics, and the detection by lateral flow assay of secretory glycan antigens of *Schistosoma* in urine or blood allows sensitive diagnosis of active infections in the field.

ACKNOWLEDGMENTS

The authors acknowledge contributions to previous versions of this chapter by Salvatore Turco and Irma van Die.

FURTHER READING

Camus D, Hadley TJ. 1985. A *Plasmodium falciparum* antigen that binds to host erythrocytes and merozoites. *Science* **230:** 553–556. doi:10.1126/science.3901257

Farthing MJ, Pereira ME, Keusch GT. 1986. Description and characterization of a surface lectin from *Giardia lamblia*. *Infect Immun* **51:** 661–667. doi:10.1128/iai.51.2.661-667.1986

Parodi AJ. 1993. *N*-glycosylation in trypanosomatid protozoa. *Glycobiology* **3:** 193–199. doi:10.1093/glycob/3.3.193

Schenkman S, Eichinger D, Pereira ME, Nussenzweig V. 1994. Structural and functional properties of *Trypanosoma trans*-sialidase. *Annu Rev Microbiol* **48:** 499–523. doi:10.1146/annurev.mi.48.100194.002435

Hoessli DC, Davidson EA, Schwarz RT, Nasir-ud-Din. 1996. Glycobiology of *Plasmodium falciparum*: an emerging area of research. *Glycoconj J* **13:** 1–3. doi:10.1007/BF01049673

Mengeling BJ, Beverley SM, Turco SJ. 1997. Designing glycoconjugate biosynthesis for an insidious intent: phosphoglycan assembly in *Leishmania* parasites. *Glycobiology* **7:** 873–880. doi:10.1093/glycob/7.7.873-c

Cummings RD, Nyame AK. 1999. Schistosome glycoconjugates. *Biochim Biophys Acta* **1455:** 363–374. doi:10.1016/s0925-4439(99)00063-0

Dell A, Haslam SM, Morris HR, Khoo KH. 1999. Immunogenic glycoconjugates implicated in parasitic nematode diseases. *Biochim Biophys Acta* **1455:** 353–362. doi:10.1016/s0925-4439(99)00064-2

Ferguson MA. 1999. The structure, biosynthesis and functions of glycosylphosphatidylinositol anchors, and the contributions of trypanosome research. *J Cell Sci* **112:** 2799–2809. doi:10.1242/jcs.112.17.2799

Loukas A, Maizels RM. 2000. Helminth C-type lectins and host–parasite interactions. *Parasitol Today* **16:** 333–339. doi:10.1016/s0169-4758(00)01704-x

McConville MJ, Menon AK. 2000. Recent developments in the cell biology and biochemistry of glycosylphosphatidylinositol lipids. *Mol Membr Biol* **17:** 1–16. doi:10.1080/096876800294443

Guha-Niyogi A, Sullivan DR, Turco SJ. 2001. Glycoconjugate structures of parasitic protozoa. *Glycobiology* **11:** 45R–59R. doi:10.1093/glycob/11.4.45r

Petri WA Jr, Haque R, Mann BJ. 2002. The bittersweet interface of parasite and host: lectin–carbohydrate interactions during human invasion by the parasite *Entamoeba histolytica*. *Annu Rev Microbiol* **56:** 39–64. doi:10.1146/annurev.micro.56.012302.160959

Nyame AK, Lewis FA, Doughty BL, Correa-Oliveira R, Cummings RD. 2003. Immunity to schistosomiasis: glycans are potential antigenic targets for immune intervention. *Exp Parasitol* **104:** 1–13. doi:10.1016/s0014-4894(03)00110-3

Naderer T, Vince JE, McConville MJ. 2004. Surface determinants of *Leishmania* parasites and their role in infectivity in the mammalian host. *Curr Mol Med* **4:** 649–665. doi:10.2174/1566524043360069

Nyame AK, Kawar ZS, Cummings RD. 2004. Antigenic glycans in parasitic infections: implications for vaccines and diagnostics. *Arch Biochem Biophys* **426:** 182–200. doi:10.1016/j.abb.2004.04.004

Previato JO, Wait R, Jones C, DosReis GA, Todeschini AR, Heise N, Previato LM. 2004. Glycoinositolphospholipid from *Trypanosoma cruzi:* structure, biosynthesis and immunobiology. *Adv Parasitol* **56:** 1–41. doi:10.1016/s0065-308x(03)56001-8

Tolia NH, Enemark EJ, Sim BK, Joshua-Tor L. 2005. Structural basis for the EBA-175 erythrocyte invasion pathway of the malaria parasite *Plasmodium falciparum*. *Cell* **122:** 183–193. doi:10.1016/j.cell.2005.05.033

Petri WA Jr, Chaudhry O, Haque R, Houpt E. 2006. Adherence-blocking vaccine for amebiasis. *Arch Med Res* **37:** 288–291. doi:10.1016/j.arcmed.2005.09.012

Jang-Lee J, Curwen RS, Ashton PD, Tissot B, Mathieson W, Panico M, Dell A, Wilson RA, Haslam SM. 2007. Glycomics analysis of *Schistosoma mansoni* egg and cercarial secretions. *Mol Cell Proteomics* **6:** 1485–1499. doi:10.1074/mcp.m700004-mcp200

Debierre-Grockiego F, Schwarz RT. 2010. Immunological reactions in response to apicomplexan glycosylphosphatidylinositols. *Glycobiology* **20:** 801–811. doi:10.1093/glycob/cwq038

van Die I, Cummings RD. 2010. Glycan gimmickry by parasitic helminths: a strategy for modulating the host immune response? *Glycobiology* **20:** 2–12. doi:10.1093/glycob/cwp140

Schauer R, Kamerling JP. 2011. The chemistry and biology of trypanosomal *trans*-sialidases: virulence factors in Chagas disease and sleeping sickness. *Chembiochem* **12:** 2246–2264. doi:10.1002/cbic.201100421

Frank S, van Die I, Geyer R. 2012. Structural characterization of *Schistosoma mansoni* adult worm glycosphingolipids reveals pronounced differences with those of cercariae. *Glycobiology* **22:** 676–695. doi:10.1093/glycob/cws004

Tundup S, Srivastava L, Harn DA Jr. 2012. Polarization of host immune responses by helminth-expressed glycans. *Ann NY Acad Sci* **1253:** E1–E13. doi:10.1111/j.1749-6632.2012.06618.x

Van Diepen A, Van der Velden NS, Smit CH, Meevissen MH, Hokke CH. 2012. Parasite glycans and antibody-mediated immune responses in *Schistosoma* infection. *Parasitol* **139:** 1219–1230. doi:10.1017/s0031182012000273

Prasanphanich NS, Mickum ML, Heimburg-Molinaro J, Cummings RD. 2013. Glycoconjugates in host–helminth interactions. *Front Immunol* **4:** 24061. doi:10.3389/fimmu.2013.00240

Vázquez-Mendoza A, Carrero JC, Rodriguez-Sosa M. 2013. Parasitic infections: a role for C-type lectin receptors. *Biomed Res Intl* **2013:** 456352. doi:10.1155/2013/456352

Mickum ML, Prasanphanich NS, Heimburg-Molinaro J, Leon KE, Cummings RD. 2014. Deciphering the glycogenome of schistosomes. *Front Genet* **5:** 262. doi:10.3389/fgene.2014.00262

Sato S, Bhaumik P, St-Pierre G, Pelletier I. 2014. Role of galectin-3 in the initial control of *Leishmania* infection. *Crit Rev Immunol* **34:** 147–175. doi:10.1615/critrevimmunol.2014010154

Cabezas Y, Legentil L, Robert-Gangneux F, Daligault F, Belaz S, Nugier-Chauvin C, Tranchimand S, Tellier C, Gangneux J-P, Ferrières V. 2015. *Leishmania cell wall as a potent target for antiparasitic drugs.* A focus on the glycoconjugates. *Org Biomol Chem* **13:** 8393–8404. doi:10.1039/c5ob00563a

Cova M, Rodrigues JA, Smith TK, Izquierdo L. 2015. Sugar activation and glycosylation in Plasmodium. *Malaria J* **14:** 427. doi:10.1186/s12936-015-0949-z

Fleming JO, Weinstock JV. 2015. Clinical trials of helminth therapy in autoimmune diseases: rationale and findings. *Parasite Immunol* **37:** 277–292. doi:10.1111/pim.12175

Fried M, Duffy PE. 2015. Designing a VAR2CSA-based vaccine to prevent placental malaria. *Vaccine* **33:** 7483–7488. doi:10.1016/j.vaccine.2015.10.011

Singh RS, Walia AK, Kanwar JR. 2016. Protozoa lectins and their role in host–pathogen interactions. *Biotech Adv* **34:** 1018–1029. doi:10.1016/j.biotechadv.2016.06.002

Jaurigue JA, Seeberger PH. 2017. Parasite carbohydrate vaccines. *Front Cell Infect Microbiol* **7:** 248. doi:10.3389/fcimb.2017.00248

Kuipers ME, Nolte-'t Hoen ENM, van der Ham AJ, Ozir-Fazalalikhan A, Nguyen DL, de Korne CM, Konig RI, Tomes JJ, Hoffmann KF, Smits HH, Hokke CH. 2020. DC-SIGN mediated internalisation of glycosylated extracellular vesicles from *Schistosoma mansoni* increases activation of monocyte-derived dendritic cells. *J Extracell Vesicles* **9:** 1753420. doi:10.1080/20013078.2020.1753420

Malik A, Steinbeis F, Carillo MA, Seeberger PH, Lepenies B, Varón Silva D. 2020. Immunological evaluation of synthetic glycosylphosphatidylinositol glycoconjugates as vaccine candidates against malaria. *ACS Chem Biol* **15:** 171–178. doi:10.1021/acschembio.9b00739

Murphy N, Rooney B, Bhattacharyya T, Triana-Chavez O, Krueger A, Haslam SM, O'Rourke V, Pańczuk M, Tsang J, Bickford-Smith J, et al. 2020. Glycosylation of *Trypanosoma cruzi* TcI antigen reveals recognition by chagasic sera. *Sci Rep* **10:** 16395. doi:10.1038/s41598-020-73390-9

Ryan SM, Eichenberger RM, Ruscher R, Giacomin PR, Loukas A. 2020. Harnessing helminth-driven immunoregulation in the search for novel therapeutic modalities. *PLoS Pathog* **16:** e1008508. doi:10.1371/journal.ppat.1008508

Xing M, Yang N, Jiang N, Wang D, Sang X, Feng Y, Chen R, Wang X, Chen Q. 2020. A sialic acid-binding protein SABP1 of *Toxoplasma gondii* mediates host cell attachment and invasion. *J Infect Dis* **222:** 126–135. doi:10.1093/infdis/jiaa072

Cavalcante T, Medeiros MM, Mule SN, Palmisano G, Stolf BS. 2021. The role of sialic acids in the establishment of infections by pathogens, with special focus on *Leishmania. Front Cell Infect Microbiol* **11:** 671913. doi:10.3389/fcimb.2021.671913

West CM, Malzl D, Hykollari A, Wilson IBH. 2021. Glycomics, glycoproteomics, and glycogenomics: an inter-taxa evolutionary perspective. *Mol Cell Proteom* **20:** 100024. doi:10.1074/mcp.r120.002263

44 Genetic Disorders of Glycan Degradation

Hudson H. Freeze, Richard Steet, Tadashi Suzuki, Taroh Kinoshita, and Ronald L. Schnaar

This chapter explores degradation and turnover of glycans in lysosomes, especially with respect to human genetic disorders, with representative glycans, illustrating features unique to different pathways. Degradation of oligomannosyl N-glycans removed from misfolded, newly synthesized glycoproteins is covered in Chapters 39 and 45.

LYSOSOMAL ENZYMES

Most glycans are degraded in lysosomes by highly ordered pathways using endo- and exoglycosidases, sometimes aided by noncatalytic proteins. Insights that unraveled these complex

published by Cold Spring Harbor Laboratory Press; doi: 10.1101/glycobiology.4e.44

pathways emerged from studies of rare human genetic disorders called lysosomal storage diseases. In each disease, undigested molecules accumulate in lysosomes. Clever experiments combining enzymology with glycan structural analyses revealed the steps of the pathways and also unlocked the mechanisms of lysosomal enzyme targeting (Chapter 33).

Lysosomes contain approximately 50 to 60 soluble hydrolases that degrade various macromolecules. Most of the glycan-degrading enzymes (endo- and exoglycosidases and sulfatases) have pH optima between 4 and 5.5, but a few have higher pH optima, nearing neutral. Exoglycosidases cleave the glycosidic linkage of terminal sugars from the nonreducing end of glycans (the outermost left end of glycans for figures in this book, e.g., Figure 44.1). Exoglycosidases recognize only one monosaccharide (rarely two) in a specific anomeric linkage and are much less particular about the structure of the molecule beyond that glycosidic linkage. This lack of specificity allows these enzymes to act on a broad range of substrates. However, exoglycosidases do not usually work unless all of the hydroxyl groups of the terminal sugar are unmodified. Acetate, sulfate, or phosphate groups usually have to be removed before action of the glycosidases. Esterases cleave acetyl groups and specific sulfatases remove the sulfate groups on glycosaminoglycans (GAGs) and N- or O-linked glycans. Endoglycosidases cleave internal glycosidic linkages of larger chains. These enzymes are often more tolerant of modifications of the glycan; in some cases, they require a modified sugar for optimal cleavage.

Even though the lysosomal glycosidases perform similar reactions, their amino acid sequences are only ~15%–20% identical to each other. Thus, there are no highly conserved glycosidase catalytic domains. Lysosomal enzymes are all N-glycosylated, and most are targeted to the lysosome by the mannose 6-phosphate pathway (Chapter 33), share aspects of the recognition marker for assembly of mannose 6-phosphate on N-linked glycans, and have affinity for mannose 6-phosphate receptors. The concentration of enzymes within the lysosome is difficult to determine. Proteinases such as cathepsins are estimated to be ~1 mM; glycosidases are probably present at much lower concentrations.

GENETIC DEFECTS IN LYSOSOMAL DEGRADATION OF GLYCANS

About 50 known inherited diseases impair lysosomal degradation of macromolecules. Although each one is rare, together their occurrence is about one in every 5000 to 10,000 births. Loss of a single lysosomal hydrolase leads to the accumulation of its substrate as undegraded fragments in tissues and the appearance of related fragments in urine. Many of the human disorders have animal models. Tables 44.1, 44.2, and 44.3 show some of the major clinical symptoms of diseases associated with the degradation of three classes of glycans. Many of the diseases share overlapping symptoms, and yet each disease has unique clinical features that allow it to be diagnosed. Many of the diseases also present with a range of severities. Usually, an infantile onset is the most severe, and the juvenile or adult onsets have milder symptoms. The later-onset forms may even affect organ systems distinct from those affected by early-onset forms. Hundreds of mutations have been mapped in the different disorders. The severity usually depends on the combination of mutated alleles. Predicting the disease severity (prognosis) from the specific mutation is generally difficult. Complete absence of a lysosomal hydrolase is uniformly severe. Hypomorphic alleles have variable residual glycosidase activity making their prognosis difficult. If there is a defect on mannose 6-phosphate–mediated targeting of lysosomal hydrolases, it will also lead to a lysosomal storage disease such as I-cell disease (mucolipidosis II) or pseudo-Hurler polydystrophy (mucolipidosis III) (Chapter 33).

It is not clear whether accumulating different types of undegraded glycans leads to the different symptoms characteristic of each disease. There is no evidence that the stored material causes lysosomes to burst and spew their contents into the cytoplasm. However, increased lysosomal exocytosis or

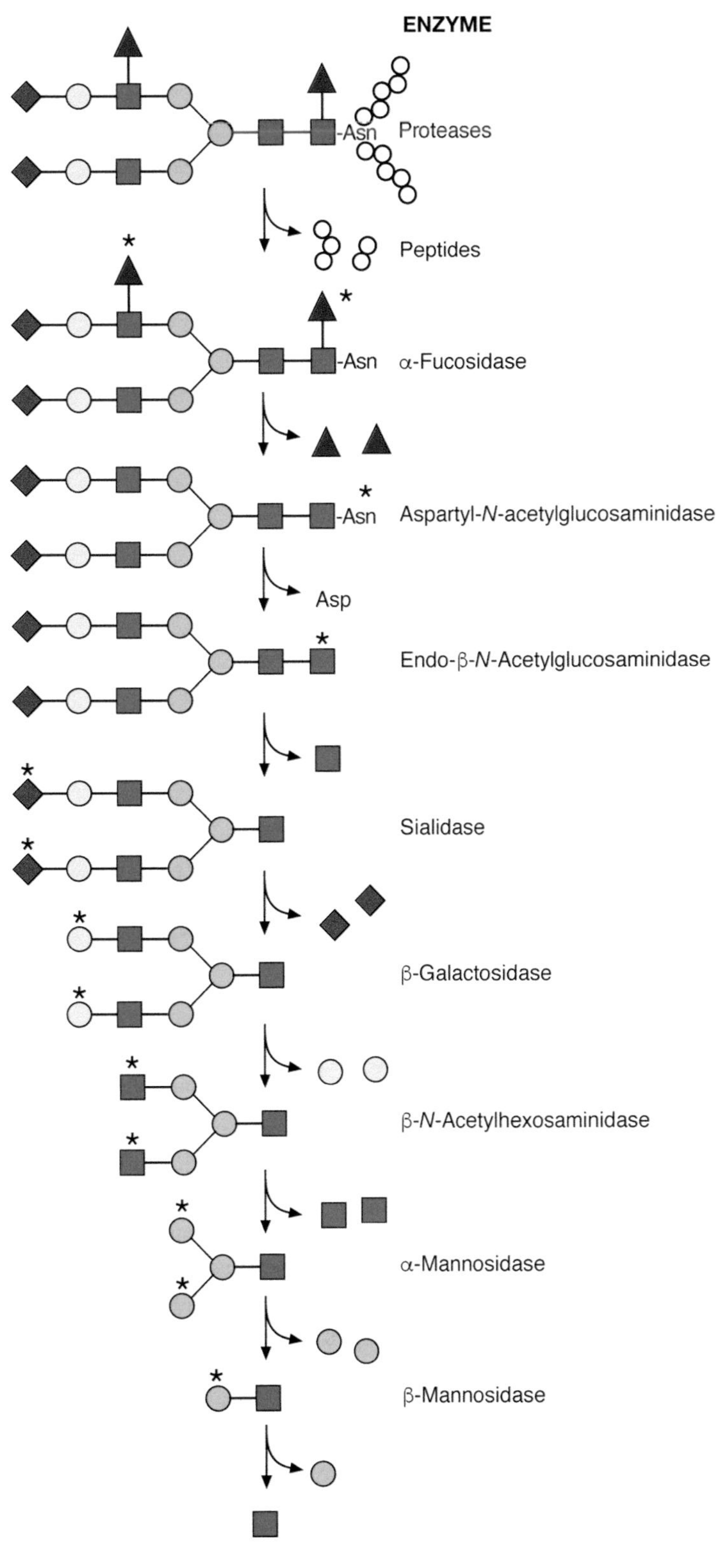

FIGURE 44.1. Degradation of complex-type N-glycans. The lysosomal degradation pathway of glycoproteins carrying complex-type glycans proceeds simultaneously on both the protein and glycan moieties. The N-glycans are sequentially degraded by the indicated exoglycosidases in a specific order as discussed in the text.

TABLE 44.1. Defects in glycoprotein degradation—the glycoproteinoses

Disorder	Defect	Effects on degradation of		Clinical symptoms
		Glycoprotein	Glycolipid	
α-Mannosidosis (types I and II)	α-mannosidase	major	none	type I: infantile onset, progressive intellectual disability, hepatomegaly, death between 3 and 12 yr type II: juvenile/adult onset, milder, slowly progressive
β-Mannosidosis	β-mannosidase	major	none	severe quadriplegia, death by 15 mo in most severe cases; mild cases have intellectual disability, angiokeratoma, facial dysmorphism
Aspartylglucosaminuria	aspartyl-glucosaminidase	major	none	progressive, coarse facies, intellectual disability
Sialidosis (mucolipidosis I)	sialidase	major	minor	progressive, severe mucopolysaccharidosis-like features, intellectual disability
Schindler (types I and II)	α-N-acetyl-galactosaminidase	yes	?	type I: infantile onset, neuroaxonal dystrophy, severe psychomotor delay and intellectual disability, cortical blindness, neurodegeneration type II: mild intellectual impairment, angiokeratoma, corpus diffusum
Galactosialidosis	protective protein/cathepsin A	major	minor	coarse facies, skeletal dysplasia, early death
Fucosidosis	α-fucosidase	major	minor	spectrum of severities includes psychomotor delay, coarse facies, growth retardation
GM1 gangliosidosis	β-galactosidase	minor	major	progressive neurological disease and skeletal dysplasia in severe infantile form
GM2 gangliosidosis	β-hexosaminidase	minor	major	severe form: neurodegeneration with death by 4 yr less severe form: slower onset of symptoms and variable symptoms, all relating to various parts of the central nervous system
MAN2B2 deficiency	α1,6-mannosidase	major	none	psychomotor developmental delay, microcephaly, growth retardation

fusion of lysosomes with the plasma membrane has been noted for some disorders. Many lysosomal disorders have defects in autophagy. This can result in impaired clearance of damaged organelles like mitochondria. The pathology likely depends on the cell type and the cellular balance of synthesis and turnover rates. For instance, dermatan sulfate (DS), a GAG (Chapter 17), predominates in connective tissue, which might explain the bone, joint, and skin problems in mucopolysaccharidosis (MPS) I, II, VI, and VII. Keratan sulfate (KS), another GAG, is present in cartilage; therefore, MPS IV is largely a skeletal disease. Gangliosides (Chapter 11) are most abundant in neurons; therefore, gangliosidoses are predominantly brain disorders. The importance of glycogen for muscle explains the impact of Pompe disease, a lysosomal disease caused by a mutation of *GAA* gene encoding α-glucosidase important for glycogen degradation, on the heart and diaphragm, leading to rapid lethality in that disease. Secondary storage of glycans is a common finding in many of these disorders, highlighted by the accumulation of glycolipids in the cholesterol storage disease, Niemann–Pick type C and in several disorders with GAG storage. Given the balance between synthesis and degradation of multiple glycans, reducing synthesis somewhat by the use of an inhibitor may help to retard the accumulation of primary or secondary storage and reduce the pathology of some diseases.

GLYCOPROTEIN DEGRADATION

The great majority of N- and O-glycans reaching the lysosome contain only six sugars linked in one or two anomeric configurations: β-*N*-acetylglucosamine (βGlcNAc), α/β-*N*-acetylgalactosamine (α/βGalNAc), α/β-galactose (α/βGal), α/β-mannose (α/βMan), α-fucose

TABLE 44.2. Defects in glycosaminoglycan (GAG) degradation—the mucopolysaccharidoses

Number	Common name	Enzyme (gene) deficiency	Glycosamino-glycan affected	Clinical symptoms
MPS I H	Hurler, Hurler–Scheie, Scheie	α-L-iduronidase (*IDUA*)	DS, HS	Hurler: corneal clouding, organomegaly, heart disease, intellectual disability, death in childhood Hurler–Scheie and Scheie: less severe, individuals survive longer
MPS II	Hunter	iduronic acid-2-sulfatase (*IDS*)	DS, HS	severe: organomegaly, no corneal clouding, intellectual disability, death before 15 yr less severe: normal intelligence, short stature, survival age 20–60 yr
MPS III A	Sanfilippo A	heparan N-sulfatase (*SGSH*)	HS	profound mental deterioration, hyperactivity, relatively mild somatic manifestations
MPS III B	Sanfilippo B	α-N-acetylglucosaminidase (*NAGLU*)	HS	similar to III A
MPS III C	Sanfilippo C	acetyl CoA: α-glucosaminide N-acetyltransferase (*HGSNAT*)	HS	similar to III A
MPS III D	Sanfilippo D	N-acetylglucosamine 6-sulfatase (*GNS*)	HS	similar to III A
MPS IV A	Morquio A	N-acetylgalactosamine-6-sulfatase (*GALNS*)	KS, CS	distinctive skeletal abnormalities, corneal clouding, odontoid hypoplasia, milder forms known to exist
MPS IV B	Morquio B	β-galactosidase (*GLB1*)	KS	same as IV A
MPS VI	Maroteaux–Lamy	N-acetylgalactosamine 4-sulfatase (*ARSB*)	DS	corneal clouding, normal intelligence, survival to teens in severe form; milder forms known to exist
MPS VII	Sly	β-glucuronidase (*GUSB*)	DS, HS, CS	wide spectrum of severity, including hydrops fetalis and neonatal form
MPS IX	Natowicz	hyaluronidase (*HYAL-1*)	hyaluronan	periarticular soft tissues masses, short stature, mild facial changes
	multiple sulfatase deficiency (MSD)	FGE converts cysteine→formyl glycine (*SUMF1*)	all sulfated glycans	hypotonia, developmental delay, neurodegeneration

The classification of mucopolysaccharidoses is based on that given in Neufeld E, Muenzer J. 2001. The mucopolysaccharidoses. In *The metabolic and molecular basis of inherited disease* (ed. Scriver CR, et al.), 8th ed, pp. 3421–3452. McGraw-Hill, New York.

MPS, Mucopolysaccharidoses; DS, dermatan sulfate; HS, heparan sulfate; KS, keratan sulfate; CS, chondroitin sulfate; FGE, formylglycine-generating enzyme.

TABLE 44.3. Defects in glycolipid degradation

Disease name	Enzyme or protein deficiency	Clinical symptoms
Tay–Sachs	β-hexosaminidase A	severe: neurodegeneration, death by 4 yr less severe: slower onset of symptoms, variable symptoms all relating to parts of the nervous system
Sandhoff	β-hexosaminidase A and B	same as Tay–Sachs
GM1 gangliosidosis	β-galactosidase	see Table 44.1
Sialidosis	sialidase	see Table 44.1
Fabry	α-galactosidase	severe pain, angiokeratoma, corneal opacities, death from renal or cerebrovascular disease
Gaucher's	β-glucoceramidase	severe: childhood or infancy onset, hepatosplenomegaly, neurodegeneration mild: child/adult onset, no neurodegenerative course
Krabbe	β-galactoceramidase	early onset with progression to severe mental and motor deterioration
Metachromatic leukodystrophy	arylsulfatase A (cerebroside sulfatase)	infantile, juvenile, and adult forms can include mental regression, peripheral neuropathy, seizures, dementia
Saposin deficiency	saposin precursor	similar to Tay–Sachs and Sandhoff

(αFuc), and α-sialic acid (αSia). Each linkage should theoretically require only one anomer-specific glycosidase, assuming that each glycosidase ignores the underlying glycan. This number is close to the known number of enzymes in most glycan degradation pathways. However, some linkages require a specific enzyme outside of this group. For example, β-N-acetylhexosaminidase cleaves both βGlcNAc and βGalNAc residues. Degradation of the GlcNAcβAsn and GalNAcαSer/Thr linkages also requires specific enzymes.

Lysosomal Degradation of Complex-Type N-Glycans

Much of what we know about this pathway comes from analysis of products that accumulate in patients' tissues or urine because of the absence of one of the degradative enzymes (Table 44.1). Structural analyses of monosaccharide-labeled glycoproteins during degradation in perfused rat liver aided elucidation of the pathway. By conducting the latter studies in the presence of inhibitors of different lysosomal enzymes, a picture of simultaneous and independent bidirectional degradation of the protein and carbohydrate chains emerged (Figure 44.1). The relative degradation rates vary depending on structural and steric factors of the protein and the sugar chains. The accumulation of GlcNAcβ1-4GlcNAcβAsn in cells that cannot cleave the GlcNAcβAsn linkage clearly shows that degradation of the sugar chain does not require advanced cleavage of the Asn. Much of the protein is probably degraded before N-glycan catabolism begins. Removal of core fucose (Fucα1-6GlcNAc) and probably any peripheral fucose residues linked to the outer branches of the chain (e.g., Fucα1-3GlcNAc) appears to be the first step in degradation, because patients lacking α-fucosidase still have intact N-glycans bound to asparagine. Glycosylasparaginase (aspartyl-N-acetyl-β-D-glucosaminidase) then cleaves the GlcNAcβAsn bond, producing a glycosylamine +Asp, not Asn. In rodents and primates, chitobiase (an endo-β-N-acetylglucosaminidase) removes the reducing *N*-acetylglucosamine, leaving the oligosaccharide with only one terminal GlcNAc. In many other species, splitting of the chitobiose linkage (GlcNAcβ1-4GlcNAc) uses the β-N-acetylhexosaminidase, mentioned below, as the last step in degradation. Either pathway appears to be effective, leaving the presence of chitobiase in some species unexplained. The oligosaccharide chain is then sequentially degraded by sialidases and/or α-galactosidase, followed by β-galactosidase, β-N-acetylhexosaminidase, and α-mannosidases. The remaining Manβ1-4GlcNAc (or Manβ1-4GlcNAcβ1-4GlcNAc in species lacking chitobiase) is cleaved by β-mannosidase to mannose and GlcNAc (or chitobiose, which is then cleaved to GlcNAc by β-N-acetylhexosaminidase).

Lysosomal sialidase (neuraminidase), β-galactosidase, and a serine carboxypeptidase called protective protein/cathepsin A form a complex in the lysosome that is required for efficient degradation of sialylated glycoconjugates. Cathepsin A protects β-galactosidase from rapid degradation and also activates the sialidase precursor, but the protection does not depend on the catalytic activity of cathepsin A. Mutations in this protective protein lead to galactosialidosis in which the simultaneous deficiencies in β-galactosidase and sialidase are secondary effects of defective cathepsin A.

Glycans that have GalNAcβ1-4GlcNAc, GlcAβ1-3Gal, or Galα1-3Gal on the outer branches must first have these residues removed by β-N-acetylhexosaminidase, β-glucuronidase, and α-galactosidase, respectively, before any further digestion of the underlying oligosaccharide chains.

Lysosomal Degradation of Oligomannosyl N-Glycans

Oligomannosyl N-glycans that enter the lysosome are hydrolyzed by an α-mannosidase to yield Manα1-6Manβ1-4GlcNAc, a common intermediate of hybrid and complex-type N-glycan. A second α1-6-specific mannosidase can cleave this linkage in humans and rats, but only on molecules that have a single core region *N*-acetylglucosamine (i.e., those generated by chitobiase cleavage). Finally, β-mannosidase completes the degradation. Oligomannosyl N-glycans derived from dolichol-linked precursors or misfolded glycoproteins are handled differently (Chapter 39).

Degradation of O-Glycans

Degradation of typical αGalNAc-initiated O-glycans has not been systematically studied. Many of the outer structures of N-glycans are also found on O-glycans (Chapter 14); therefore, degradation of these glycans probably uses the same group of exoglycosidases as discussed above. An exception to this is the linkage region, GalNAcα-O-Ser/Thr. Patients with Schindler disease lack the α-N-acetylgalactosaminidase specific for αGalNAc and will not cleave αGlcNAc. This same enzyme probably removes terminal αGalNAc from blood group A–containing glycans (GalNAcα1-3Gal) and some glycolipids such as the Forssmann antigen (GalNAcα1-3 GalNAcβ1-3Galα1-4Galβ1-4GlcβCer). Patients lacking α-N-acetylgalactosaminidase accumulate GalNAc-containing glycopeptides in their urine, but curiously they also accumulate more complex, extended glycopeptides containing *N*-acetylglucosamine, galactose, and sialic acid. The structures are the same as those found on some native glycoconjugates. Their production could result from a general slowdown of oligosaccharide degradation or may arise by reassembly of glycans on accumulated GalNAcα-O-Ser/Thr glycopeptides.

GLYCOSAMINOGLYCAN DEGRADATION

GAGs, including heparan sulfate (HS), chondroitin sulfate (CS), DS, KS, and hyaluronan, are degraded in a highly ordered fashion. The first three are O-xylose-linked to core proteins (Chapter 17), KS is N- or O-linked, and hyaluronan is a free glycan (Chapter 16). Some proteoglycans are internalized from the cell surface and the protein portion is degraded. The GAG chains are then partially cleaved by enzymes such as endo-β-glucuronidases or endohexosaminidases that clip at a few specific sites. Endoglycosidase cleavage creates multiple terminal residues that can be degraded by unique or overlapping sets of sulfatases and exoglycosidases. Structural analyses of partially degraded fragments in the lysosomes of cells from patients with MPS revealed both the degradation pathways and genetic defects responsible (Table 44.2). The clinical severities and manifestations vary even with mutations in the same gene. For instance, MPS I is clinically subdivided into Hurler (most severe), Hurler–Scheie, and Scheie syndromes, although the three disorders represent a continuum of the same disease (Table 44.2). Hurler–Scheie patients progress more slowly and die in early adulthood, whereas Scheie patients can survive to middle or old age. The milder forms of this disease do not cause intellectual disability.

Hyaluronan

Hyaluronan (Chapter 16) is the largest and most abundant GAG: A 70-kg person degrades 5 g of hyaluronan (molecular mass 10^7 Da) per day. Degradation of hyaluronan involves a series of two endo-β-N-acetylhexosaminidases called hyaluronidases (Hyal-1 and Hyal-2), β-glucuronidase, and finally β-N-acetylhexosaminidase. Hyal-2 is active at low pH and is GPI-anchored on the cell surface (Chapter 12). This enzyme associates both with hyaluronan in lipid rafts and with an Na^+/H^+ exchanger to create an acidic microenvironment. Cleavage generates fragments of ~20 kDa (approximately 50 disaccharides), which are internalized, delivered to endosomes, and then finally to lysosomes, in which the fragments are degraded by Hyal-1 into tetra- and disaccharides (Figure 44.2). The exoglycosidases are thought to participate in the degradation of the larger fragments as well as that of the di-and tetrasaccharide units. Mutations in *HYAL-1* cause MPS IX.

Heparan Sulfate

HS (Chapter 17) is first degraded by an endoglucuronidase, heparanase, followed by a well-ordered sequential degradation. An example is shown in Figure 44.2. A terminal iduronic acid-2-sulfate must be desulfated by iduronic acid-2-sulfatase to make the modified sugar a substrate for α-iduronidase. If glucuronic acid (GlcA)-2-sulfate is at this position, GlcA-2-sulfatase removes the sulfate before β-glucuronidase cleavage. The new terminal glucosamine sulfate ($GlcNSO_4$) is the next sugar for cleavage, requiring three steps. Sulfate is first removed by N-sulfatase forming glucosamine, which cannot be cleaved by α-N-acetylglucosaminidase. The free amino group is N-acetylated by an N-acetyltransferase embedded in the lysosomal membrane, converting glucosamine to *N*-acetylglucosamine, which is a substrate of α-N-acetylglucosaminidase. In the second step, acetyl CoA donates the acetyl group to a histidine residue in the cytoplasmic domain of the N-acetyltransferase. The acetyl group then becomes available on the luminal side of the lysosomal membrane and is transferred at low pH to the amino group of

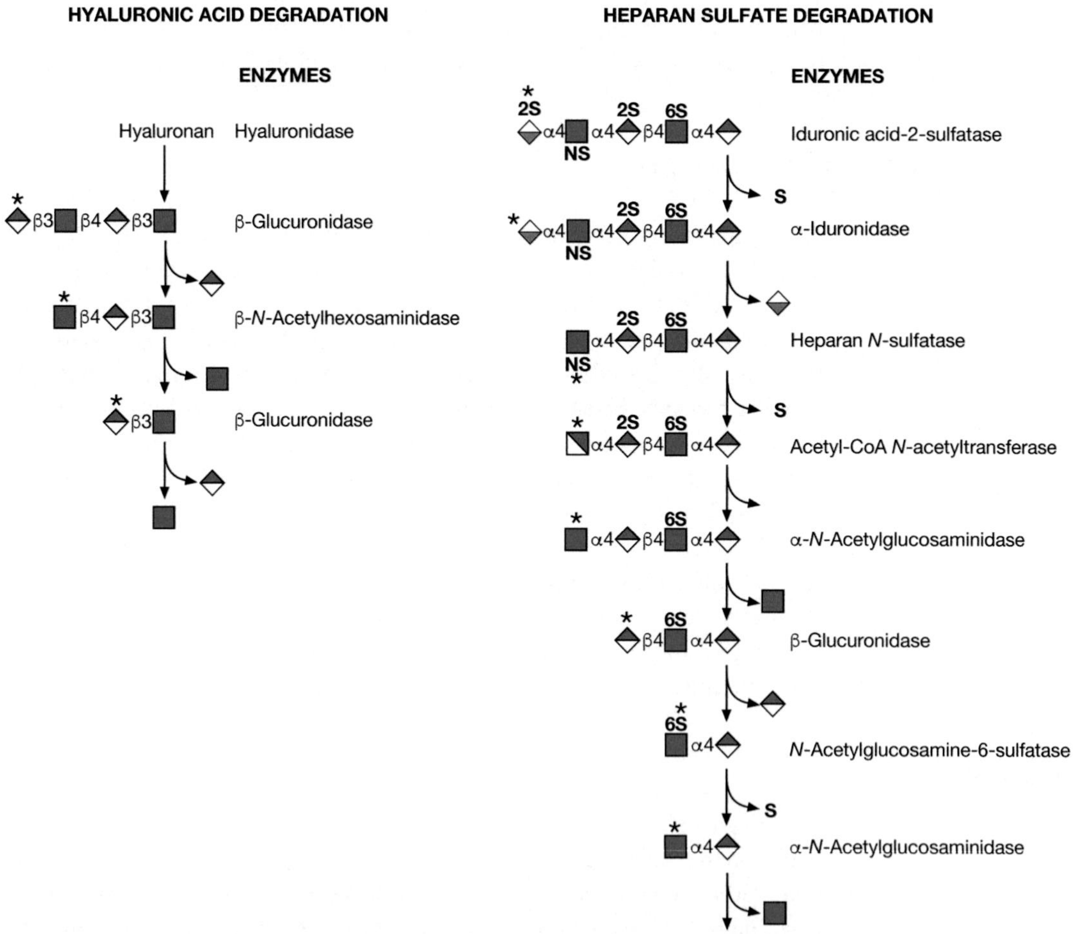

FIGURE 44.2. Degradation of hyaluronan and heparan sulfate (HS). (*Left*) Hyaluronidase (an endoglycosidase) cleaves large chains into smaller fragments, each of which is then sequentially degraded from the nonreducing end. (*Right*) Degradation of HS. An endo-β-glucuronidase first cleaves large chains into smaller fragments, and each monosaccharide is then removed from the nonreducing end as described in the text. N- and O-sulfate groups must first be removed before exoglycosidases can act. An unusual feature of HS degradation is that this process also involves a synthetic step. After removal of the N-sulfate residue on $GlcNSO_4$, the nonacetylated glucosamine must first be N-acetylated using acetyl-CoA before α-N-acetylglucosaminidase can cleave this residue.

glucosamine. If cleavage of glucuronic acid reveals a 6-O-sulfated αGlcNAc, the sulfate is removed by a specific GlcNAc-6-sulfatase before GlcNAc removal. Table 44.2 provides a list of the enzymatic defects in HS degradation and the disorders that result.

Dermatan Sulfate and Chondroitin Sulfate

DS and CS are related GAG chains based on a repeating polymer of βGlcA and βGalNAc (Chapter 17). Figure 44.3 shows that a combination of endoglycosidases, sulfatases, and exoglycosidases degrades DS in the lysosome. Iduronic acid-2-sulfatase is followed by α-iduronidase. The terminal GalNAc-4-SO_4 can be removed by either of two pathways. In the first pathway, a GalNAc-4-SO_4 sulfatase acts followed by β-N-acetylhexosaminidase A or B to remove *N*-acetylgalactosamine. In the second, β-N-acetylhexosaminidase A removes the entire GalNAc-4-SO_4 unit followed by sulfatase cleavage. β-Glucuronidase cleaves the β-glucuronic acid residue and the process is repeated on the rest of the molecule.

To degrade CS, GalNAc-6-SO_4 sulfatase and GalNAc-4-SO_4 sulfatase work in combination with β-N-acetylhexosaminidase A or B and β-glucuronidase. Hyaluronidases can also degrade CS, but no CS-specific endoglycosidases have been found.

Keratan Sulfate

KS is an N- or O-glycan with a heavily sulfated poly-*N*-acetyllactosamine chain. Mammalian cells do not have an endoglycosidase to break down KS (Figure 44.3). The sequential action of sulfatases and exoglycosidases is needed. Galactose-6-SO_4 sulfatase is the same enzyme that desulfates GalNAc-6-SO_4 in CS degradation. This hydrolysis is followed by β-galactosidase digestion, leaving a terminal GlcNAc-6-SO_4. Desulfation by the sulfatase followed by cleavage with β-N-acetylhexosaminidase A or B eliminates GlcNAc-6-SO_4. Alternatively, β-N-acetylhexosaminidase A can directly release GlcNAc-6-SO_4, followed by desulfation of the monosaccharide.

Linkage Region

Degradation of the core region of O-linked KS (skeletal type; type II) probably occurs by the same route as other O-glycans. The N-glycan corneal-type KS (type I) is probably degraded by the same set of enzymes used for N-glycan degradation. The O-xylose-linked GAG chains (DS, CS, and HS) all share a common core tetrasaccharide: GlcAβ1-3Galβ1-3Galβ1-4Xylβ-O-Ser. An endo-β-xylosidase has been detected in rabbit liver. How the linkage regions are degraded is not established.

Multiple Sulfatase Deficiency

A very rare human disorder is multiple sulfatase deficiency (MSD). All sulfatases are casualties of this defect because they all undergo a posttranslational conversion of an active site cysteine residue to a C_α-formylglycine (2-amino-3-oxopropionic acid) catalyzed by the C_α-formylglycine-generating enzyme (FGE), which is essential for activity. In essence, the SH group of cysteine is replaced by a double-bonded oxygen atom that probably acts as an acceptor for cleaved sulfate groups. Loss-of-function mutations in the *FGE* gene lead to inactive sulfatases. This deficit affects GAG degradation and any other sulfated glycan such as sulfatides.

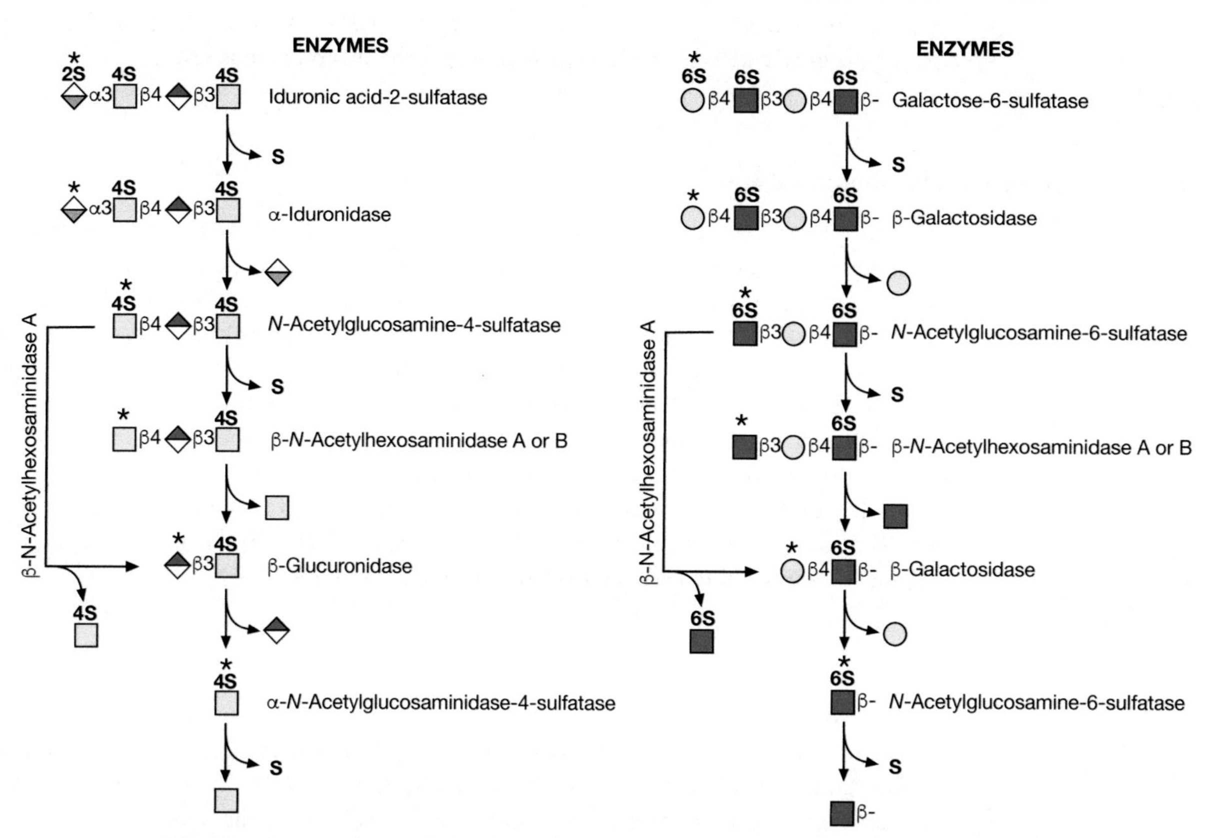

FIGURE 44.3. Degradation of chondroitin/dermatan sulfates (CS/DS) and keratan sulfate (KS). (*Left*) Only DS degradation is shown. After removal of sulfated iduronic acid in two steps, further sequential degradation can proceed by two different routes. One route (*straight down*) uses GalNAc-4-SO_4 sulfatase followed by cleavage with β-N-acetylhexosaminidase A or B. The other route uses only β-N-acetylhexosaminidase A, which is one of the few exoglycosidases that can cleave sulfated amino sugars at low pH. These two routes of sequential degradation also work in CS degradation. (*Right*) Degradation of KS. Sequential degradation of KS occurs from the nonreducing end like the degradation of DS and CS. The terminal GlcNAc-6-SO_4 can be cleaved sequentially by a sulfatase and then by β-N-acetylhexosaminidase A or B, or, alternatively, β-N-acetylhexosaminidase A can cleave GlcNAc-6-SO_4 directly at low pH.

GLYCOSPHINGOLIPID DEGRADATION

Glycosphingolipids (Chapter 11) are degraded from the nonreducing end by exoglycosidases while they are still bound to the lipid moiety ceramide. Because glycosphingolipids share some of the same outer sugar sequences found in N- and O-glycans (Chapter 14), many of the same glycosidases are used for their degradation (Figure 44.4). However, specific hydrolases cleave the glucose–ceramide and galactose–ceramide bonds. Besides specific enzymes, noncatalytic sphingolipid activator proteins (SAPs or saposins) help to present the lipid substrates to enzymes for cleavage. Ceramide glycanase (endoglycoceramidases) from leeches and earthworms release the entire glycan chain from the lipid, much like PNGase releases N-glycans from proteins.

Specialized Issues for Glycolipid Degradation

Some exoglycosidases are unique to glycolipid degradation, and their absence causes glycolipid storage diseases (Table 44.3). Glucocerebrosidase, also called β-glucoceramidase, is specific for the degradation of the GlcβCer bond. The loss of this enzyme causes Gaucher's disease.

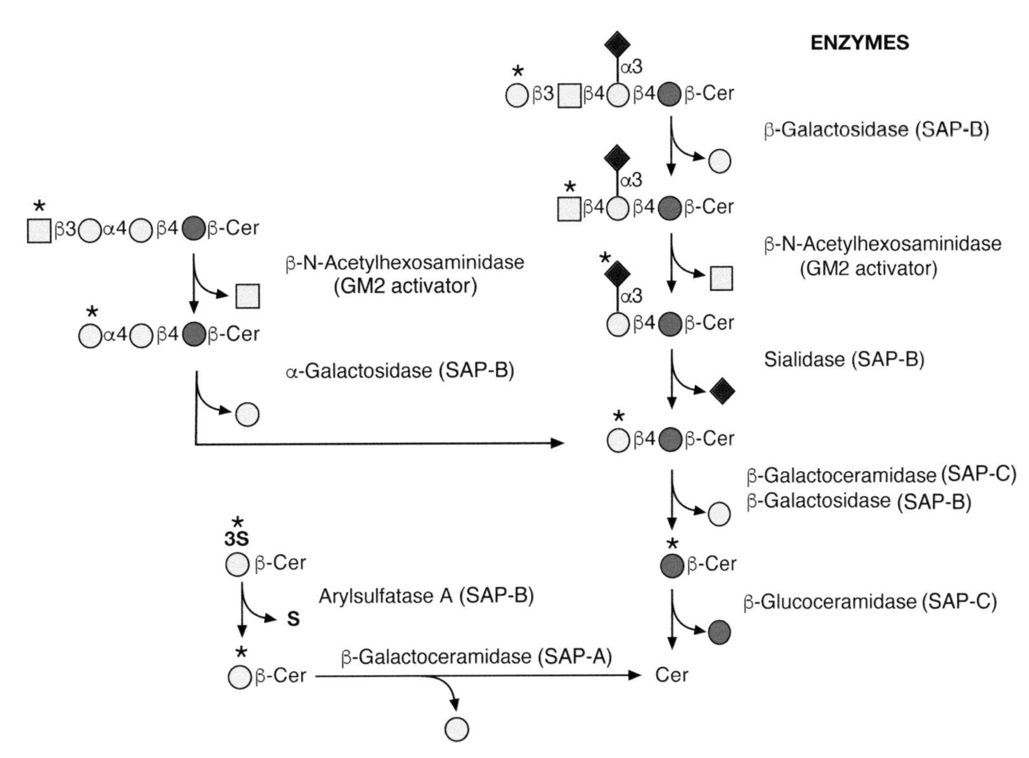

FIGURE 44.4. Degradation of glycosphingolipids. Required activator proteins are shown in *parentheses* and individual enzymes are as shown in previous figures in this chapter. (SAP) Saposin.

Heterozygous mutations in this enzyme have been shown to be among the most prevalent and robust genetic risk factors for Parkinson's disease, highlighting a role for glycolipid degradation in the context of more common neurological disorders. A specialized β-galactosidase, β-galactoceramidase, hydrolyzes the bond between galactose and ceramide and can also cleave the terminal galactose from lactosylceramide. Loss of this enzyme produces Krabbe disease. Galactosylceramide is often found with a 3-sulfate ester (sulfatide) and a specific sulfatase (arylsulfatase A) is needed for its removal before β-galactosylceramidase action. Loss of this sulfatase causes metachromatic leukodystrophy and the accumulation of sulfatide. Glycolipids terminated with α-galactose residues (globo-series; Chapter 11) are degraded by a specific α-galactosidase, the loss of which causes Fabry disease.

Activator Proteins

Sugars that lie too close to the lipid bilayer apparently give limited access to soluble lysosomal enzymes, and additional proteins called saposins are required to present the substrates to the enzymes. Saposins (SAPs), also called "liftases," form complexes with multiple degradative enzymes for more efficient hydrolysis of short glycolipids close to the membrane. SAPs are derived from a 524-amino-acid precursor called prosaposin, which is processed into four homologous activator proteins each comprising ~80 amino acids: saposins A, B, C, and D. Despite their homology, each saposin has different properties. SAP-A and SAP-C help β-galactosyl- and β-glucosylceramidase degradation, respectively. SAP-B assists arylsulfatase A, α-galactosidase, sialidase, and β-galactosidase. The activation mechanism of SAPs is thought to be similar to that of the GM2 activator described below. SAP-D and SAP-B also assist in sphingomyelin degradation by sphingomyelinase. Complete absence of prosaposin is lethal in humans, and deficiencies in SAP-B and SAP-C lead to defects that clinically resemble arylsulfatase A deficiency (metachromatic leukodystrophy) and Gaucher's disease. These symptoms might be predicted based on the enzymes that these saposins assist.

GM2 activator protein forms a complex with either GM2 or GA2 and presents them to β-N-acetylhexosaminidase A for cleavage of β-GalNAc. The activator protein binds a molecule of glycolipid, forming a soluble complex, which can be cleaved by the hexosaminidase. The resulting product is then inserted back into the membrane, and the activator presents the next GM2 molecule, and so on. Genetic loss of this activator protein causes the accumulation of GM2 and GA2, resulting in the AB variant of GM2 gangliosidosis.

Topology for Degradation and Roles of Lipids

Endocytic vesicles deliver membrane components to the lysosome for degradation and are often seen as intralysosomal multivesicular bodies (MVBs). They appear like vesicles within the lysosome and are especially prominent in patients with glycolipid storage disorders. How do "vesicles within vesicles" form and what is their function? The internal surface of the lysosomal membrane has a thick, degradation-resistant glycocalyx of integral and peripheral membrane proteins decorated with poly-*N*-acetyllactosamines, which protect the lysosomal membrane against destruction. However, these proteins would also shield incoming membranes from degradation if vesicle fusion simply occurred. By creating multiple internal membranes seen in typical MVBs, the target molecules are exposed to the soluble lysosomal enzymes, and digestion proceeds efficiently on the membrane surfaces of these internal vesicles.

MVB formation starts with inward budding of the limiting endosomal membrane. Lipids and proteins are sorted to either the internal or the limiting membrane. Ubiquitination of cargo proteins targets them to internal vesicles of MVBs, but ubiquitin-independent routes also exist. Intra-endosomal membranes and limiting endosomal membrane have different lipid and protein compositions. Membrane segregation and lipid sorting prepare the internal membranes for lysosomal degradation. During maturation of the internal membranes, cholesterol is continually stripped away (to $<1\%$) and a negatively charged lipid bis(monoacylglycero)phosphate (BMP) increases up to 45%. This molecule is highly resistant to phospholipases. BMP also stimulates sphingolipid degradation on the inner membranes of acidic compartments. Because BMP is not present in the lysosomal outer membrane, the unique lipid profile ensures that degradation by hydrolases and membrane-disrupting SAPs occurs without digesting the lysosomal outer membrane.

DEGRADATION AND RESYNTHESIS

Degradation of glycans is not always complete. Partially degraded or incomplete glycans on glycoproteins, glycopeptides, and glycosphingolipids can be internalized within a functional Golgi compartment containing sugar nucleotides and glycosyltransferases and then elongated. In the case of glycosphingolipids, this pathway makes a substantial contribution to total cellular synthesis, but in glycoproteins it probably makes a relatively small contribution. These processes can be salvage and repair mechanisms or may play an integral part in an unidentified physiological pathway.

Reglycosylation and Recycling of Glycoproteins

The half-life of certain membrane proteins is longer than the half-life of their sugar chains. Terminal monosaccharides turn over faster than those near the reducing end of the glycan, suggesting that terminal sugars are removed by exoglycosidases. Cleavage may occur at the cell surface or when proteins are endocytosed in the course of normal membrane recycling. Because the mildly acidic late endosomes contain lysosomal enzymes with a fairly broad pH range, terminal sugars such as sialic acid can be cleaved. If the endocytosed proteins are not degraded in lysosomes, the proteins may reencounter sialyltransferases in the Golgi, become resialylated, and appear again on the cell surface. A similar situation may occur if a protein has lost both sialic acid and galactose residues from its glycans. Some membrane proteins synthesized in the presence of oligosaccharide processing inhibitors can still reach the cell surface in an unprocessed form. Subsequent incubation in the absence of inhibitors leads to normal processing over time. The extent of processing and the kinetics depend on the protein and the cell type. Because Golgi enzymes are not distributed identically in all cells, the extent of reprocessing is variable. Most studies have monitored N-glycans, but membrane proteins with O-glycans probably behave similarly.

Glycosphingolipid Recycling

The majority of the newly synthesized glucosylceramide arrives at the cell surface by a Golgi-independent cytoplasmic pathway. Some of this material, as well as glucosylceramide generated by partial degradation of complex glycosphingolipids, can recycle to the Golgi. There, like glycoproteins, simple glycosphingolipids can serve as acceptors for the synthesis of longer sugar chains. In the lysosome, sphingosine is produced by complete degradation of glycosphingolipids and reused. Together, these pathways, especially the latter, may account for the majority of complex glycolipid synthesis in many cells. Depending on the physiological state of the cell and synthetic demands, the de novo pathway starting from serine and palmitoyl CoA may account for only 20%–30% of the total synthesis. Shuttling the recycled components through and among the various organelles appears to involve vimentin intermediate filaments. Mechanisms and details of this process are presently lacking.

SALVAGE OF MONOSACCHARIDES

Monosaccharide salvage is discussed in Chapter 5. Monosaccharides derived from glycan degradation are transported back into the cytoplasm using transporters specific for neutral sugars, N-acetylated hexoses, anionic sugars, sialic acid, and glucuronic acid. Turnover of glycans is high and the salvage of sugars can be quite substantial. There have been few studies comparing the actual contributions of exogenous monosaccharides, those salvaged from glycan turnover, and those generated from de novo synthesis. Cells probably differ in their preference for each source of monosaccharides. These differences may explain why monosaccharide therapy used to treat some glycosylation disorders is effective in certain cells and not others.

BLOCKING DEGRADATION

Lysosomal degradation of glycans can be blocked by raising the intralysosomal pH, by inactivating a particular glycosidase, by mutation, or by protein-specific inhibitors.

Lysosomal enzymes usually have acidic pH optima, and adding ammonium chloride or chloroquine slows degradation by increasing the intralysosomal pH. However, some lysosomal enzymes have relatively broad pH optima, higher than the lysosome. Some active "lysosomal" enzymes can be found in early and late endosomes, which have a higher pH than lysosomes.

Glycosylation inhibitors are covered in Chapter 55 but some of these agents also inhibit specific lysosomal enzymes. For example, swainsonine blocks lysosomal α-mannosidase as well as the α-mannosidase II involved in glycoprotein processing. Sheep and cattle become neurologically deranged by eating food rich in swainsonine, which is also called locoweed. These temporary symptoms are likely induced because lysosomal α-mannosidase is inhibited. Undegraded oligosaccharides probably accumulate in affected animals. Although many inhibitors block various enzymes in vitro, they may not be effective in cells or whole animals because they may not enter the lysosome at sufficient concentrations.

THERAPY FOR LYSOSOMAL ENZYME DEFICIENCIES

Study of lysosomal enzyme defects provided major insights into catabolic pathways and showed their importance to human health. Mannose 6-phosphate targeting of lysosomal enzymes was discovered by showing that cocultivation of fibroblasts from patients with different storage diseases led to the disappearance of stored material in both types of cells. Each supplied the corrective factors (i.e., lysosomal enzymes) that the other cell lacked by delivery of secreted enzymes through cell-surface mannose 6-phosphate receptors (Chapter 33). Cross-correction provides the mechanistic basis for an important mode of therapy and highlights the fact that only a relatively small amount of enzyme may be needed to prevent accumulation of storage products and the resulting pathology. Some estimates suggest that <5% of normal β-N-acetylhexosaminidase activity may be sufficient to prevent pathological symptoms of Tay–Sachs disease. In fact, the index Scheie patient with α-L-iduronidase deficiency had <1% normal activity in fibroblasts and lived 77 years.

Current therapeutic approaches include enzyme replacement therapy (ERT), substrate reduction therapy (SRT), enzyme enhancement therapy (EET), and hematopoietic stem cell transplantation (HSCT). Gene replacement therapy is a promising future approach because it would avoid the sustained delivery necessary for ERT. Accumulation of stored undegraded material occurs because the rate of glycan synthesis is greater than its degradation at steady state. SRT reduces the glycan's synthetic rate to counter low glycosidase activity. This hypothesis was tested in a mouse model of Sandhoff disease by using *N*-butyldeoxynojirimycin, an inhibitor of glucosylceramide synthase, which catalyzes the first step in glycosphingolipid biosynthesis. When wild-type mice were treated with this compound, the amount of glycosphingolipids fell 50%–70% in all tissues without obvious pathological effects. When this compound was given to Sandhoff disease mice, the accumulation of GM2 in the brain was blocked and the amount of stored ganglioside reduced. Thus, reducing the synthesis of the primary precursor reduced the load of GM2 to levels that were degradable by the glycosidase-deficient mice. Because this compound inhibits the first biosynthetic step, theoretically this drug could reduce the accumulation of storage products in any glycolipid storage disorder.

In clinical trials, *N*-butyldeoxynojirimycin (miglustat, Zavesca) was effective for many patients with mild-to-moderate Gaucher's disease, although improvement varied, as did side effects. Small trials in infantile Tay–Sachs disease patients did not arrest neurological deterioration but prevented macrocephaly. Further clinical studies are likely to reveal the potential and limitations of SRT.

EET or pharmacological chaperone therapy (PCT) is based on the concept that inhibitors of an enzyme can act as molecular chaperones to stabilize the mutated enzymes in the endoplasmic reticulum (ER). When the enzyme reaches the lysosome, the inhibitor dissociates at low pH and the abundant stored material is gradually digested. The overall increase in activity is small, but it can have significant clinical benefits. Clinical trials are in progress for treating Fabry disease, Gaucher type I, and Pompe disease, whereas preclinical trials are being performed for GM1 and GM2 gangliosidosis. A potential problem is that some inhibitors may not cross the blood–brain barrier, limiting their effectiveness in the central nervous system (CNS). The advent of high-throughput screening methods and large chemical libraries increase the potential of this approach. In addition, allosteric site binders can be selected that remain bound to the enzyme, and any successful compound could be used in combination with ERT to further enhance activity.

ERT has been quite successful for treating Gaucher's disease. Injection of glucocerebrosidase carrying mannose-terminated N-glycans targets the enzyme to the macrophage/monocytes, which are the primary sites of substrate accumulation. In use now for many years, the added enzyme consistently improves patients' clinical features with minimal side effects. The cost of treatment is high, and high doses are no more effective than lower doses for improving visceral and hematological pathologies. The proven effectiveness makes glucocerebrosidase the favorite therapy for these patients, but terminal mannose residues cannot be used to target the enzyme to other cells or organs or to enable treatment of other lysosomal storage disorders. Insufficient infused enzyme crosses the blood–brain barrier thus, the enzyme (Cerezyme) works very well for type I disease without neurological involvement but not for the less common Gaucher's with neurological involvement. In addition, patients sometimes develop antibodies against the injected human protein.

ERT trials with other recombinant lysosomal enzymes have progressed. Enzyme replacement therapy is available for Fabry disease, Pompe disease, and mucopolysaccharidosis types I, II, IVA, and VI and is under development for others. Combination therapy using ERT and a pharmacological chaperone to stabilize the therapeutic enzyme in the bloodstream is being considered. Recombinant α-galactosidase for Fabry disease works well and recombinant α-L-iduronidase for MPS I works very well with the intermediate severity form (Hurler–Scheie), but its effect is unclear for patients with the more common neurological form (Hurler syndrome). HSCT, presumably providing α-L-iduronidase in CNS from the donor-derived microglia, is effective to MPS I. Recombinant N-acetylgalactosamine-4-sulfatase (arylsulfatase B; Naglazyme) was approved for MPS VI, as was α-glucosidase (Myozyme) for treating Pompe disease patients. Iduronic acid-2-sulfatase (Elaprase) is approved for treating Hunter syndrome. These results clearly validate this approach. The efficacy of most of these enzymes relies on the engineering of Man-6-P-containing glycans. This is one of the foremost examples of how the clinically motivated exploration of a disorder (I-cell disease) led to the discovery of Man-6-P-based targeting of lysosomal enzymes (Chapter 33) and the development of an entire industry targeting therapy for multiple diseases. The availability of animals deficient in specific lysosomal enzymes offers appropriate systems to test the efficacy of corrective genes, compounds, and enzymes.

ACKNOWLEDGMENTS

The authors appreciate helpful comments and suggestions from Junko Matsuda and Nathan Lewis.

FURTHER READING

Neufeld EF, Lim TW, Shapiro LJ. 1975. Inherited disorders of lysosomal metabolism. *Annu Rev Biochem* **44**: 357–376. doi:10.1146/annurev.bi.44.070175.002041

Winchester B. 2005. Lysosomal metabolism of glycoproteins. *Glycobiology* **15**: 1R–15R. doi:10.1093/glycob/cwi041

Winchester B. 2014. Lysosomal diseases: diagnostic update. *J Inherit Metab Dis* **37**: 599–608. doi:10.1007/s10545-014-9710-y

Clarke LA, Hollak CE. 2015. The clinical spectrum and pathophysiology of skeletal complications in lysosomal storage disorders. *Best Pract Res Clin Endocrinol Metab* **29**: 219–235. doi:10.1016/j.beem.2014.08.010

Coutinho MF, Matos L, Alves S. 2015. From bedside to cell biology: a century of history on lysosomal dysfunction. *Gene* **555**: 50–58. doi:10.1016/j.gene.2014.09.054

Deng H, Xiu X, Jankovic J. 2015. Genetic convergence of Parkinson's disease and lysosomal storage disorders. *Mol Neurobiol* **51**: 1554–1568. doi:10.1007/s12035-014-8832-4

Espejo-Mojica ÁJ, Alméciga-Díaz CJ, Rodríguez A, Mosquera Á, Díaz D, Beltrán L, Díaz S, Pimentel N, Moreno J, Sánchez J, et al. 2015. Human recombinant lysosomal enzymes produced in microorganisms. *Mol Genet Metab* **116**: 13–23. doi:10.1016/j.ymgme.2015.06.001

Oh DB. 2015. Glyco-engineering strategies for development of therapeutic enzymes with improved efficacy for the treatment of lysosomal storage diseases. *BMB Rep* **48**: 438–444. doi:10.5483/bmbrep.2015.48.8.101

Parenti G, Andria G, Valenzano KJ. 2015. Pharmacological chaperone therapy: preclinical development, clinical translation, and prospects for the treatment of lysosomal storage disorders. *Mol Ther* **23**: 1138–1148. doi:10.1038/mt.2015.62

Rastall DP, Amalfitano A. 2015. Recent advances in gene therapy for lysosomal storage disorders. *Appl Clin Genet* **8**: 157–169. doi:10.2147/tacg.s57682

Breiden B, Sandhoff K. 2019. Lysosomal glycosphingolipid storage diseases. *Annu Rev Biochem* **88**: 461–485. doi:10.1146/annurev-biochem-013118-111518

Ryan E, Seehra G, Sharma P, Sidransky E. 2019. GBA1-associated parkinsonism: new insights and therapeutic opportunities. *Curr Opin Neurol* **32**: 589–596. doi:10.1097/wco.0000000000000715

Tancini B, Buratta S, Delo F, Sagini K, Chiaradia E, Pellegrino RM, Emiliani C, Urbanelli L. 2020. Lysosomal exocytosis: the extracellular role of an intracellular organelle. *Membranes (Basel)* **10**: 406. doi:10.3390/membranes10120406

45 Congenital Disorders of Glycosylation

Dirk J. Lefeber, Hudson H. Freeze, Richard Steet, and Taroh Kinoshita

This chapter discusses inherited human diseases that are caused by defects in glycan biosynthesis and metabolism (congenital disorders of glycosylation, CDGs). Representative examples are described of genetic defects in the major glycan families and what lessons we can learn from them about glycobiology. Among genetic disorders of glycosylation, those caused by somatic mutations are described in Chapter 46. Disorders affecting the lysosomal degradation of glycans are described in Chapter 44. Although the term "congenital disorders" by definition include those caused by nongenetic, unfavorable in utero conditions, the term "congenital disorders of glycosylation (CDG)" is now widely used as an equivalent of inherited disorders of glycosylation.

BACKGROUND AND DISCOVERY

Following the discovery of the genetic defect in human I-cell disease in 1980 (Chapter 33) it was anticipated that many more human defects in glycan biosynthesis would be found, but more than a decade went by before the next example was found. In retrospect, the difficulty arose from the

published by Cold Spring Harbor Laboratory Press; doi: 10.1101/glycobiology.4e.45

pleiotropic multisystem nature of the clinical presentations. Meanwhile Belgian pediatrician Jaak Jaeken noted unusual profiles of serum proteins in some children with multisystem disorders and decided to apply a previously established test to separate transferrin isoforms. The test was positive in these children, thereby, for the first time, indicating a generic defect in protein N-glycosylation. Lacking further details about glycan structures and the underlying genetic variants, Jaeken decided to call these cases "carbohydrate-deficient glycoprotein syndromes" (CDGS). Several years later, many people defined primary inherited genetic defects in glycosylation pathways in many such cases. To keep the acronym "CDG," those involved also decided to call them "congenital disorders of glycosylation," rather than inherited genetic defects in glycosylation.

INHERITED PATHOLOGICAL MUTATIONS OCCUR IN ALL MAJOR GLYCAN FAMILIES

Nearly all inherited disorders in glycan biosynthesis were discovered in the last 20 years. They are rare, biochemically and clinically heterogeneous, and usually affect multiple organ systems. CDGs are rare primarily because embryos with complete defects in a step of glycosylation do not usually survive to be born, documenting the critical biological roles of glycans in humans. CDG patients that survive are usually hypomorphic, retaining at least some activity of the pathways involved. Although rare, research on this group of genetically defined glycosylation disorders reveals important novel insights into the biology of the glycosylation process.

Some defects strike only a single glycosylation pathway, whereas others impact several. Defects occur in (1) the activation, presentation, and transport of sugar precursors; (2) glycosidases and glycosyltransferases; and (3) proteins that traffic glycosylation machinery or maintain Golgi homeostasis. A few disorders can be treated by the consumption of monosaccharides. The rapid growth in the number of discovered disorders, shown in Figure 45.1, has resulted in an evolution of disease nomenclature. Since 1999, CDGs were defined as genetic defects in N-glycosylation, but now the term is applied to any glycosylation defect. Nowadays, CDGs are categorized in four groups, comprising defects in N-glycosylation, O-glycosylation, and lipid and GPI-anchor glycosylation and defects that impact multiple glycosylation pathways. CDGs are named by the mutated gene followed by "-CDG" suffix (e.g., PMM2-CDG). Selected disorders are listed in Table 45.1 and all known disorders in Online Appendix 45A.

DEFECTS IN N-GLYCAN BIOSYNTHESIS

Clinical and Laboratory Features and Diagnosis

The broad clinical features of disorders in which N-glycan biosynthesis is defective involve many organ systems, but are especially common in the central and peripheral nervous systems and hepatic, visual, and immune systems. The generality and variability of clinical features makes it difficult for physicians to recognize CDG patients with defective N-glycosylation. The first were identified in the early 1980s based primarily on deficiencies in multiple plasma glycoproteins. The patients were also delayed in reaching growth and developmental milestones and had low muscle tone, incomplete brain development, visual problems, coagulation defects, and endocrine abnormalities. However, many of these symptoms are also seen in patients with other multisystemic genetic syndromes. CDG patients with defective N-glycosylation can be distinguished because they often have abnormal glycosylation of common liver-derived serum proteins containing disialylated, biantennary N-glycans. Serum transferrin is especially convenient because it has two N-glycosylation sites, each normally containing disialylated, biantennary N-glycans. Different glycoforms can be resolved by isoelectric focusing (IEF) or ion-

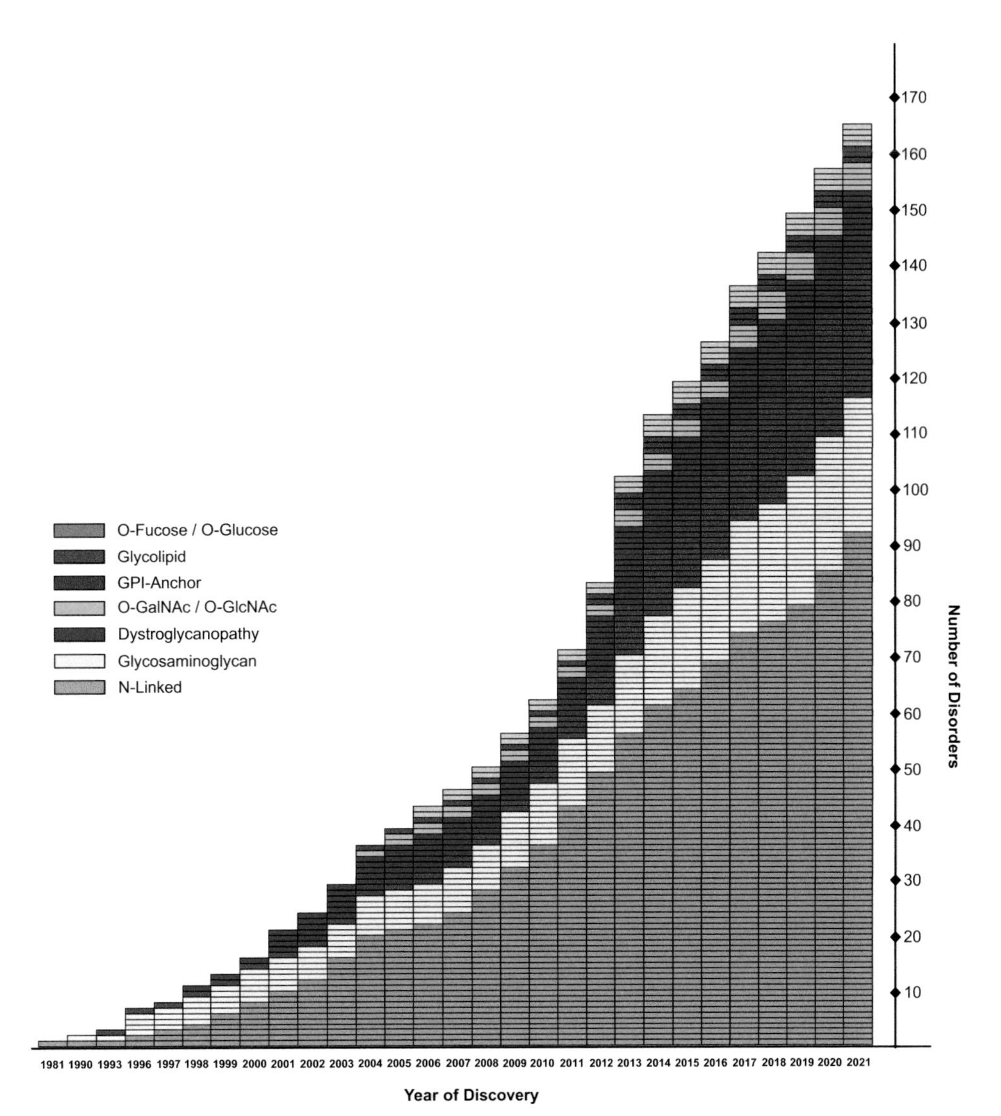

FIGURE 45.1. Glycosylation-related disorders. The graph shows the cumulative number of human glycosylation disorders in various biosynthetic pathways and the year of their identification. (For simplicity, disorders affecting multiple pathways are included in disorders of N-linked glycans.) In most cases, the year indicates the definitive proof of gene-specific mutations. In early years, discovery was based on compelling biochemical evidence. Introduction of next-generation sequencing, such as whole-exome sequencing, has spurred the identification of novel genetic defects since 2010.

exchange chromatography, but better accuracy and sensitivity is achieved by mass spectrometry (MS) of purified transferrin. This simple test alerts physicians to likely CDG patients without knowing the genetic or molecular basis of the disease.

N-glycosylation defects may be divided into two types based on transferrin glycoforms. Type I (CDG-I) patients lack one or both N-glycans because of defects in the biosynthesis of the lipid-linked oligosaccharide (LLO) or its transfer to proteins. Type II (CDG-II) patients have incomplete protein-bound glycans because of abnormal processing. The biosynthetic pathways and locations of N-glycan defects are shown in Figure 45.2.

Defects in ER N-Glycosylation

A complete absence of N-glycans is lethal. Therefore, known mutations mostly generate hypomorphic alleles encoding proteins with diminished activity. A deficiency in any of the steps required for

TABLE 45.1. Selected congenital disorders of glycosylation in humans

Disorder	Gene	Function	Disorder OMIM	Gene OMIM	Main Clinical Features	Pathway Affected
ALG1–CDG	*ALG1*	β1-4 mannosyltransferase	608540	605907	ID, Hy, Sz, M, infections, early death	N-linked pathway
NGLY1–CDG	*NGLY1*	N-glycanase-1	615273	610661	ID, DD, Sz, abnormal liver function, hypolacrima or alacrima	N-linked pathway
PMM2–CDG	*PMM2*	conversion of Man-6-P to Man-1-P	212065	601785	ID, Hy, Sz, strabismus, cerebellar hypoplasia, failure to thrive, cardiomyopathy, 20% lethality in the first 5 yr	potential to affect multiple pathways
MPI–CDG	*MPI*	interconversion of Fructose-6-P and Man-6-P	602579	154550	hepatic fibrosis, coagulopathy, hypoglycemia, protein-losing enteropathy, vomiting, no neurological symptoms	potential to affect multiple pathways
SRD5A3–CDG	*SRD5A3*	polyprenol reductase	612379	611715	ID, Hy, eye, and brain malformations, nystagmus, hepatic dysfunction, coagulopathy, ichthyosis	potential to affect multiple pathways
COG7–CDG	*COG7*	Golgi-to-ER retrograde transport	608779	606978	Hy, M, growth retardation, adducted thumbs, failure to thrive, cardiac anomalies, wrinkled skin, early death	potential to affect multiple pathways
CHIME syndrome Hyperphosphatasia mental retardation syndrome	*PIGL*	GlcNAc-PI de-N-acetylase	280000 239300	605947 605947	ID, colobomas, heart defect, early-onset ichthyosiform dermatosis, ear anomalies (conductive hearing loss), hyperphosphatasia with mental retardation syndrome	GPI-anchor pathway
Walker–Warburg syndrome (MDDGA1, B1, C1)	*POMT1*	O-mannosyl-transferase	236670 613155 609308	607423	Walker–Warburg syndrome, brain malformations, various eye malformations, elevated serum CK	O-mannose pathway
GNE myopathy	*GNE*	UDP-GlcNAc-2-epimerase/ MaNAc kinase	605820 269921	603824	proximal and distal muscle weakness, wasting of the upper and lower limbs, sparing of the quadriceps	potential to affect multiple pathways
Hereditary multiple exocytosis	*EXT1*	glucuronyltransferase/ GlcNAc transferase	133700	608177	multiple exocytosis of the bone	glycosaminoglycan

CDG, congenital disorder of glycosylation; OMIM, Online Mendelian Inheritance in Man database; ID, intellectual disability; Sz, seizures; Hy, hypotonia; M, microcephaly; DD, developmental delay; CK, creatine kinase.

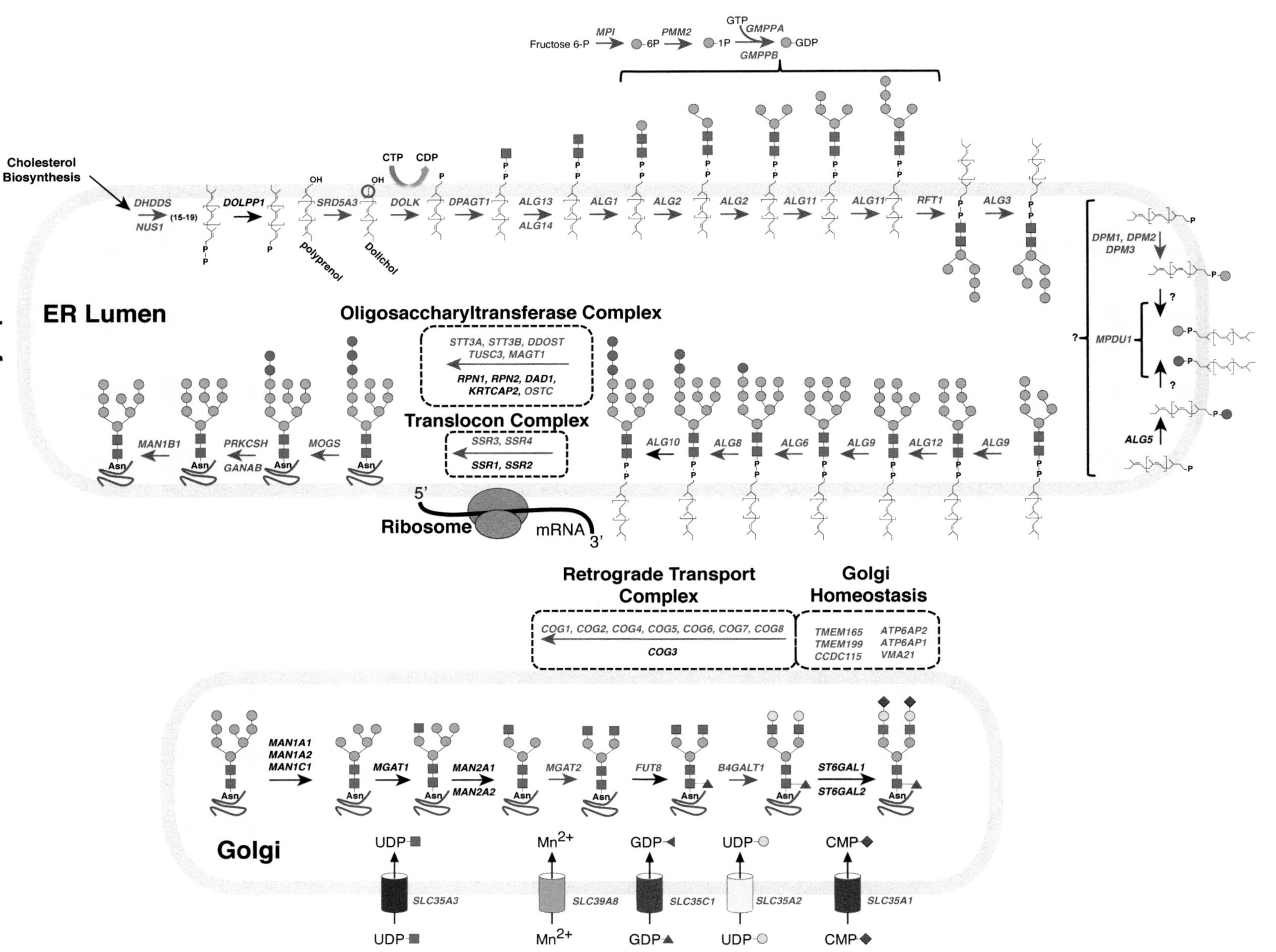

FIGURE 45.2. Congenital disorders of glycosylation in the N-glycosylation pathway. The figure shows individual steps in lipid-linked oligosaccharide (LLO) biosynthesis, glycan transfer to protein, and N-glycan processing similar to Figure 9.3 and Figure 9.4. The shuttling of the glycosylation machinery between the endoplasmic reticulum (ER) and Golgi and within the Golgi is organized and regulated by cytoplasmic complexes including the conserved oligomeric Golgi (COG) complex (see Figure 45.4 for Golgi homeostasis defects). *Red* gene names indicate CDGs. MAN1B1 shown in the ER is now known to act in the GOLGI as well.

the assembly of LLOs in the endoplasmic reticulum (ER) (e.g., nucleotide sugar synthesis or sugar addition catalyzed by a glycosyltransferase) (Chapter 9) produces a structurally incomplete LLO. Because the oligosaccharyltransferase (OST) prefers full-sized LLO glycans, this results in hypoglycosylation of multiple glycoproteins. This means that some N-glycan sites are not modified. In recent years, it has been shown that some intermediate LLO structures can also be transferred to proteins, although with much lower efficiency. For example, low amounts of an extended chitobiose glycan were identified in ALG1-CDG, which is deficient in the mannosyltransferase that adds mannose to the chitobiose core. Importantly, many deficiencies in LLO synthesis produce incomplete LLO intermediates. Most of the LLO assembly steps are not easy to assay, but LLO assembly is conserved from yeast to humans, and intermediates that accumulate in CDG patients often correspond to the intermediates seen in mutant *Saccharomyces cerevisiae* strains with known defects in LLO assembly. Some mutant mammalian cells (e.g., Chinese hamster ovary cells) have been shown to have similar defects (Chapter 49). The close homology between yeast and human genes for ER N-glycosylation enables the normal human orthologs to rescue defective glycosylation in mutant yeast strains, whereas mutant orthologs from patients do not. This provides substantial clues to the likely human defect, along with a system in which to perform functional assays.

PMM2-CDG in Table 45.1 is the most common CDG with more than 1000 cases identified worldwide. The patients have moderate to severe developmental and motor deficits, hypotonia, dysmorphic features, failure to thrive, liver dysfunction, coagulopathy, and abnormal endocrine functions. More than 100 mutations found in phosphomannomutase 2 (*PMM2*) impair conversion of mannose-6-P (Man-6-P) to Man-1-P, which is a precursor required for the synthesis of GDP-mannose (GDP-Man) and dolichol-P-mannose (Dol-P-Man). Both donors are substrates for the mannosyltransferases involved in the synthesis of $Glc_3Man_9GlcNAc_2$-P-P-Dol which is decreased in cells from PMM2-CDG patients. Patients have hypomorphic *PMM2* alleles because complete loss of PMM2 function is lethal. Mouse embryos lacking *Pmm2* die 2–4 days after fertilization, whereas some of those with hypomorphic alleles survive. There are currently no approved therapeutic options for PMM2-CDG patients.

MPI-CDG (CDG-Ib) in Table 45.1 is caused by mutations in *MPI* (mannose-6-phosphate isomerase). This enzyme interconverts fructose-6-P and Man-6-P. In contrast to PMM2-CDG, these patients do not have intellectual disability or developmental abnormalities. Instead, they have impaired growth, hypoglycemia, coagulopathy, severe vomiting and diarrhea, protein-losing enteropathy, and hepatic fibrosis. Several patients died of severe bleeding before the basis of this CDG was known. Mannose dietary supplements effectively treat these patients. Man-6-P can be generated directly by hexokinase-catalyzed phosphorylation of mannose (Chapter 5). This pathway is intact in MPI-CDG patients and its high activity in brain is thought to be the reason for the absence of neurological symptoms. Human plasma contains ~50 μM mannose because of export following glycan degradation and processing. Mannose supplements correct coagulopathy, hypoglycemia, protein-losing enteropathy, and intermittent gastrointestinal problems, as well as normalize the glycosylation of plasma transferrin and other serum glycoproteins. Because orally administered mannose is well-tolerated, this approach is clearly a satisfyingly effective therapy for this life-threatening condition, although not curative for long-term symptoms.

Patients with mutations in nearly all the remaining steps of LLO biogenesis have been found (Table 45.1; Online Appendix 45A; Figure 45.1) They have a broad range of clinical phenotypes including low low-density lipoprotein (LDL), low immunoglobulin G (IgG), kidney failure, genital hypoplasia, and cerebellar hypoplasia. The pathophysiology underlying these symptoms is still unknown. Defects in dolichol biosynthesis present with more specific symptoms as outlined below in the section Defects in Multiple Glycosylation Pathways. Mutations in six of the OST subunits also cause a CDG. The OST is a complex of multiple subunits of which STT3 is the catalytic subunit that contains OST activity. Two different genes encode STT3—namely, STT3A and STT3B. They are part of two different OST complexes, one with STT3A for cotranslational glycosylation and one with STT3B for glycosylation of proteins that escape glycosylation via the STT3A complex. Patients with defects in either subunit present with the typical multisystem phenotype. Interestingly, transferrin glycosylation is dependent on the STT3A-containing OST complex and thus abnormal in STT3A-CDG, but normal in STT3B-CDG. STT3B dependent protein substrates, such as sex hormone binding globulin, show reduced glycosylation in STT3B-CDG specifically. The phenotype of MAGT1-CDG is very different, presenting as a primary immunodeficiency characterized by chronic infections with the Epstein–Barr virus (EBV). Transferrin glycosylation is abnormal in MAGT1-CDG. MAGT1 is an oxidoreductase that is mostly associated with the STT3B-OST complex. It is required for glycosylation of protein sequences containing cysteine residues. Reduced expression on natural killer cells of the NKG2D protein that requires MAGT1 for glycosylation is proposed to contribute to the immunodeficiency against EBV infections.

Defects in Golgi N-Glycosylation

Golgi disorders (Figure 45.2) affect N-glycan processing and include defects in glycosyltransferases, nucleotide sugar transporters, vacuolar pH regulators, and multiple cytoplasmic proteins that traffic glycosylation machinery within the cell and maintain Golgi homeostasis. Most of these processes are required for multiple glycosylation pathways and are discussed below (see the section Defects in Multiple Glycosylation Pathways).

In B4GALT1-CDG, glycans show a dramatic loss of both galactose (Gal) and sialic acid (Sia) from transferrin because of deficient β1-4 galactosyltransferase activity. Patients present with a relatively mild clinical course without neurological symptoms, which can be explained by the fact that related genes are expressed in different tissues, such as B4GALT2 in the brain. A similar glycan pattern occurs in the X-linked SLC35A2-CDG, because of loss of UDP-Gal transporter activity. Surprisingly, within a few years after birth, abnormal glycosylation becomes normal. This is probably caused by somatic mosaicism of cells carrying the mutated *SLC35A2* gene along with unaffected cells and by selection against the affected cells. Patients suffer from a mainly neurological phenotype. Defects in the glycosyltransferase MGAT2 and glycosidase MAN1B1 also result in characteristic glycan abnormalities on transferrin and other proteins that are now used to diagnose patients, whereas FUT8-CDG shows normal glycosylation of transferrin but severe fucosylation defects on total serum proteins. These CDGs present with neurological symptoms and other multisystem features, whereas MAN1B1-CDG additionally gives rise to obesity.

Disorders of N-Glycoprotein Deglycosylation

Early on, it was assumed that glycosylation disorders would mainly result from defects in glycan biosynthetic enzymes, but that perspective has changed. Discovery of defects in Golgi organization and homeostasis, in ER chaperones such as COSMC or EDEM, and in ER quality control (such as EDEM-3-CDG) have broadened the perspective. A new defect in the ER-associated degradation (ERAD) continuum (Chapter 39) is caused by mutations in *NGLY1*, an enzyme that cleaves N-glycans from misfolded glycoproteins transported into the cytoplasm before their proteasomal degradation (Table 45.1). NGLY1 defects do not appear to induce the ERAD pathway, accumulate undegraded glycoproteins in vesicles, or trigger autophagy. It is unclear how these defects cause symptoms such as developmental delay, movement disorder, seizures, and a curious lack of tear production, but their clinical similarity to other CDGs emphasizes that glycosylation defects cannot simply be divided into "synthesis" or "degradation."

DEFECTS IN O-GLYCAN BIOSYNTHESIS

Defects in O-Man Synthesis (Congenital Muscular Dystrophies)

Mutations altering O-Man glycans (Chapter 13), primarily on α-dystroglycan (α-DG), cause a range of congenital muscular dystrophies (CMDs) termed dystroglycanopathies (Figure 45.3). α-DG at the sarcolemma links skeletal muscle cells to laminin in the extracellular matrix. A disruption of this linkage results in muscular dystrophy. Similarly, defective interaction of α-DG with other protein ligands in eye and brain results in a broad clinical spectrum of

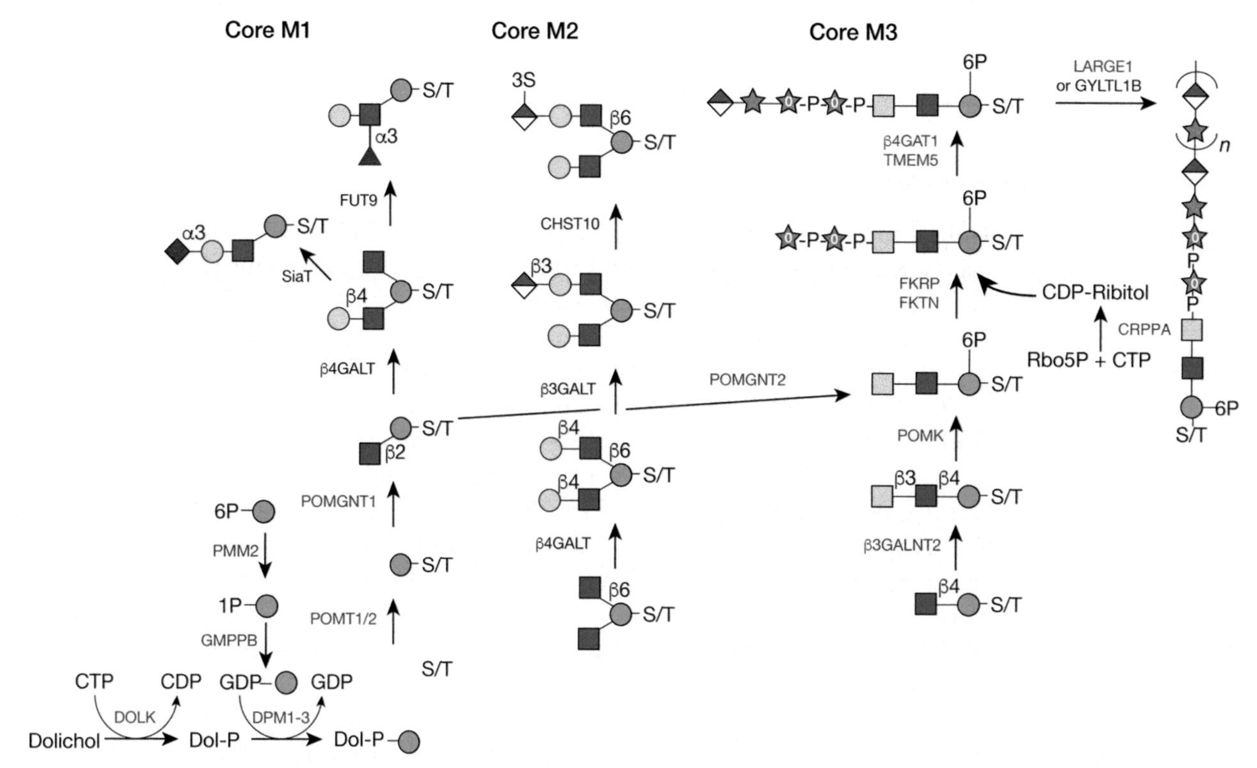

FIGURE 45.3. O-Man glycan biosynthetic pathway. The biosynthetic pathways of representative O-Man glycans are shown. Enzymes whose defects cause a disorder are indicated in *red*. Three main groups are identified: cores M1–M3. All O-Man glycans are initiated on either a serine (S) or a threonine (T) in the acceptor protein using two enzymes protein-O-mannosyltransferase 1 and 2 (POMT1/2) and Dol-P-Man donor. Several genes in the biosynthesis of Dol-P-Man (*GMPPB, DPM1–3*) are deficient in some patients with dystroglycanopathy, whereas others (*DOLK, DPM1,* and *PMM2*) cause a more generalized congenital disorder of glycosylation (CDG). Mannose receives either GlcNAcβ1-2- to form core M1 or β1-4GlcNAc to form core M3. Mutations in both genes (*POMGNT1* and *POMGNT2*) can cause a dystroglycanopathy. Core M2 can be formed if a branching β1-6GlcNAc is added. Core M1 and M2 are elongated by Gal and may terminate with Fuc, Sia, and GlcA with optional sulfation. None of the genes responsible for adding the terminal sugars have been associated with dystroglycanopathies. After the addition of the β1-4GlcNAc, core M3 is elongated with GalNAc, and then the Man is 6-O-phosphorylated via a specific kinase (POMK, also known as SGK196). With help of POMGNT1, FKTN and FKRP can act to add two ribitol-5-phosphate units to the GalNAc residue that is then extended with a single Xyl and GlcA by TMEM5 and B4GAT1, respectively. Ribitol-5-phosphate used by FKTN and FKRP is synthesized from CDP-ribitol by CRPAA, also known as ISPD. Mutations in these genes (*B3GALNT2, POMK, FKTN, FKRP, CRPPA, TMEM5,* and *B4Gat1*) can cause a dystroglycanopathy. The last defined step in core M3 biosynthesis is using this Xyl-GlcA primer for the stepwise addition of Xyl and GlcA to form a GAG-like repeating disaccharide termed matriglycan. This is catalyzed by the enzyme encoded by *LARGE1* (or its homolog *GYLTL1B*) with dual glycosyltransferase activities. Matriglycan binds laminin to α-DG and its reduction or loss is believed to be the cause of most of the dystroglycanopathies (*POMGNT1* mutations being the exception).

dystroglycanopathies, ranging from very severe, and often lethal, musculo-oculo-encephalopathies, such as Walker–Warburg syndrome (WWS), muscle–eye–brain disease (MEB), and Fukuyama congenital muscular dystrophy (FCMD), to milder forms of isolated limb-girdle muscular dystrophy in adults. Genetic analysis of these disorders has been indispensable for discovering α-DG's functional glycans and their biosynthesis. The complex pathway is presented in Figure 45.3 and Chapter 13. The pathway is initiated in the ER by the transfer of mannose to Ser/Thr of α-DG via the protein O-mannosyltransferase complex POMT1/POMT2 (Table 45.1). In the Golgi, a polymer glycan (matriglycan) is generated, recognized by several diagnostic monoclonal antibodies and necessary for the binding of laminin and other molecules to α-DG. An unusual feature on one subset of O-Man glycans is the existence of a Man-6-P generated by the kinase POMK. The Man-6-P-containing Core M3 glycan consists of Man-6-P, GlcNAc (transferred by POMGNT2), and GalNAc (transferred by B3GALNT2). The trisaccharide core is extended by two units of ribitol-5-phosphate sequentially transferred by FKTN (fukutin) and FKRP (fukutin-related protein) from CDP-ribitol, which is generated by CRPPA. This "core" glycan is elongated by an alternating disaccharide (β1-3Xylα1-3GlcA) to form matriglycan.

O-Man glycans can also play a traitorous role, because they are receptor molecules for Lassa virus entry into cells. This feature was cleverly exploited to screen libraries and identify genes required for virus entry. The method correctly identified all previously known WWS-causing genes and predicted new culprits. Dystroglycanopathies can be divided into primary (caused by mutations in the *DAG1* gene, encoding the α-DG core protein), secondary (caused by mutations in one of 11 glycosyltransferases [*POMT1, POMT2, POMGNT1, POMGNT2, FKTN, FKRP, LARGE1, RXYLT1, B3GALNT2, POMK,* or *B4GAT1*]) or tertiary (caused by defects in the synthesis of sugar building blocks [*CRPPA* for CDP-ribitol, *GMPPB, DOLK, DPM1, DPM2,* and *DPM3* for Dol-P-Man]).

WWS is the most severe CMD. Patients live about one year and have multiple brain abnormalities, and severe muscular dystrophy. About 20% of patients have mutations in *POMT1* and a few have mutations in *POMT2*. Others have defects in *FKTN* and *FKRP*, but mutations in both also cause milder forms of muscular dystrophy. *POMGNT1* is mutated in MEB, which is characterized by symptoms similar to, but milder than, WWS. The most severely affected MEB patients die during the first years of life, but the majority of mild cases survive to adulthood. FCMD is caused by a single 3-kb 3′-retrotransposon insertional event into the *FKTN* gene, which occurred 2000–2500 years ago. This partially reduces the stability of the mRNA, making it a relatively mild mutation. FCMD is one of the most common types of CMD in Japan with a carrier frequency of 1/188. *Fktn*-null mice die by E9.5 in embryogenesis and appear to have basement membrane defects. Congenital muscular dystrophy type 1C (MDC1C) is a relatively mild disorder that is caused by mutations in *FKRP*. Patients with MDC1D, a limb-girdle muscular dystrophy, contain mutations in *LARGE1*, originally described in myodystrophic mice (*myd;* now called $Large^{myd}$). The protein has two glycosyltransferase signatures (DXD) in different domains that account for xylosyl- and glucuronosyl-transferase activities, respectively (Chapter 13).

Defects in O-GalNAc Synthesis

Polypeptide GalNAc-transferases (ppGalNAcTs) are required for initiation of mucin type O-glycosylation (Chapter 10). Clinical phenotypes of defects in this family of enzymes depend on their tissue-specific expression and substrate specificity. A defect in O-GalNAc synthesis by a particular polypeptide GalNAc-transferase (GALNT3) causes familial tumoral calcinosis. This severe autosomal-recessive metabolic disorder causes phosphatemia and massive calcium deposits in the skin and subcutaneous tissues. Mutations in the O-glycosylated fibroblast growth factor 23 (FGF23) also cause phosphatemia, and further research has shown that ppGALNT3 glycosylates FGF23. Mutations in *GALNT2*, which encodes a ubiquitous ppGalNAcT, result in a multisystem neurological disease with remarkably low HDL cholesterol. Further research identified ANGPTL3 and APOCIII as substrates, the latter of which can be used for easy diagnostic screening by IEF. Tn syndrome, a rare autoimmune disease, is caused by somatic mutations in the X-linked gene *C1GALT1C1*, which encodes a highly specific chaperone COSMC required for the proper folding and normal activity of the β1-3galactosyltransferase C1GALT1, needed for synthesis of core 1 and 2 O-glycans (Chapters 10 and 46).

Defects in Other O-Glycosylation Families

The identification of patients with defects in other types of O-glycosylation (O-glucose, O-fucose, and O-GlcNAc) highlights the physiological importance of this class of glycans (Chapter 13). Each of these O-glycans is found on certain epidermal growth factor (EGF)-like repeats in the extracellular domain of Notch receptors, and mutations in the glycosyltransferases that modify NOTCH1 have been identified in different diseases, including Dowling–Degos disease (caused by haploinsufficiency of *POFUT1* or *POGLUT1*), a form of limb-girdle muscular dystrophy (caused by autosomal-recessive mutations in *POGLUT1*), Adams–Oliver syndrome (autosomal-recessive mutations in *EOGT*), and some cancers (amplification, gain or loss-of-function of *POFUT1*, Fringe GlcNAc-transferases, *POGLUT1*, *MGAT3*). Mutations in *OGT*, encoding the GlcNAc-transferase that acts on cytoplasmic and nuclear proteins (Chapter 18), cause intellectual disability. The mutations cluster in the region encoding TRP-repeats of the OGT enzyme, implicating altered substrate recognition and O-GlcNAc modification of specific cytosolic and nuclear proteins as a pathogenic mediator.

Defects in Glycosaminoglycan (GAG) Synthesis

Proteoglycans and their GAG chains are critical components in extracellular matrices. For a discussion of their biosynthesis, core proteins, and function, see Chapter 17. Genetic defects in GAG synthesis can be divided into so-called linkeropathies, because of defects in the synthesis of the tetrasaccharide linker, defects in the synthesis of the polysaccharides, and defects in GAG modification, mainly sulfation.

Genetic defects have been reported for all enzymes involved in biosynthesis of the linker tetrasaccharide (XYLT1, XYLT2, B4GALT7, B3GALT6, B3GAT3), linking the core proteins to the glycosaminoglycan. Skeletal and bone defects, joint laxity, and short stature are characteristic symptoms for this group of diseases. Bikunin has been used as a plasma proteoglycan marker for the biochemical confirmation of defects in this group of GAG biosynthesis defects. Ehlers–Danlos syndrome (progeroid type) is a connective tissue disorder characterized by failure to thrive, loose skin, skeletal abnormalities, hypotonia, and hypermobile joints, together with delayed motor development and delayed speech. The molecular basis of the disorder is reduced synthesis of the core region of GAGs initiated by Xyl. Galactosyltransferase I (B4GALT7) the enzyme that adds Gal to Xyl-Ser is mutated in this disease. A defect in the phosphorylation of the core xylose sugar by mutations in FAM20B results in Desbuquois dysplasia, a disease characterized by dwarfism, joint laxity, and skeletal abnormalities. This also shows the importance of Xyl phosphorylation in the regulation of GAG biosynthesis.

Linker elongation produces different polysaccharides, such as heparan sulfate, dermatan sulfate, chondroitin sulfate, and keratan sulfate. Defects in the formation of heparan sulfate (HS) cause hereditary multiple exocytosis (HME), an autosomal-dominant disease with a prevalence of about 1:50,000 births (Table 45.1). It is caused by mutations in one of two genes, *EXT1* or *EXT2*, that are both involved in HS synthesis. HME patients have bony outgrowths, usually at the growth plates of the long bones. Normally, the growth plate contains chondrocytes in various stages of development, which are enmeshed in an ordered matrix composed of collagen and chondroitin sulfate (CS). In HME, however, the outgrowths are often capped by disorganized, cartilaginous masses with chondrocytes in different stages of development. About 1%–2% of patients also develop osteosarcoma. HME mutations occur in *EXT1* (60%–70%) or *EXT2* (30%–40%). The encoded proteins probably form a complex in the Golgi as both are required for polymerizing *N*-acetylglucosamine (GlcNAc)α1-4 and glucuronic acid (GlcA)β1-3 into HS. However, the partial loss of one allele of either gene appears sufficient to cause HME. This means that haploinsufficiency decreases the amount of HS, and that EXT activity is rate-limiting for HS biosynthesis. This is unusual because most glycan biosynthetic enzymes are in substantial excess. Defects in GAG biosynthesis may also be caused by a lack of available nucleotide sugars. Defects in the UDP-GlcA/UDP-GalNAc Golgi transporter (SLC35D1) cause Schneckenbecken dysplasia. Patients have bone abnormalities similar to those seen in other chrondrodysplasias, and a mouse model of the disease shows similar features.

Modification of GAGs by sulfation is important for several functions. Three autosomal-recessive disorders—diastrophic dystrophy (DTD), atelosteogenesis type II (AOII), and achondrogenesis type IB (ACG-IB)—result from defective cartilage proteoglycan sulfation. These forms of osteochondrodysplasia have various outcomes. AOII and ACG-IB are perinatal lethal because of respiratory insufficiency, whereas DTD patients develop symptoms only in cartilage and bone, including cleft palate, club feet, and other skeletal abnormalities. Those DTD patients surviving infancy often live a nearly normal life span. All of these disorders result from different mutations in the DTD gene (*SLC26A2*), which encodes a plasma membrane sulfate transporter. Unlike monosaccharides, sulfate released from degraded macromolecules in the lysosome is not salvaged well. The heavy demand for sulfate in bone and cartilage proteoglycan synthesis probably explains why the symptoms are most evident in these locations.

DEFECTS IN LIPID AND GPI-ANCHOR BIOSYNTHESIS

Defects in GPI-Anchored Proteins

Complete deletion of the GPI pathway in mice causes embryonic lethality. This is not surprising because more than 150 membrane proteins require a GPI anchor for cell surface expression (Chapter 12). Hypomorphic mutations in multiple genes in the pathway lead to a partial reduction in GPI-anchored proteins. These include *PIGA, PIGH, PIGQ, PIGY, PIGC, PIGP, PIGL, PIGW, PIGM, PIGV, PIGB, PIGF,* and *PIGO* in GPI-anchor assembly (Table 45.1), and *PIGK, GPAA1, PIGS, PIGT,* and *PIGU* in the transfer of the GPI anchor to proteins. Defects in side-chain modifications (*PIGN* and *PIGG*) and maturation of the GPI glycan following attachment to proteins (*PGAP1, PGAP2,* and *PGAP3*) also cause inherited GPI deficiency, but not embryonic death. GPI deficiency has immense and variable consequences including neurologic symptoms, particularly developmental delay/intellectual disability and seizures, epileptic encephalopathy, progressive cerebral and/or cerebellar atrophy, hypotonia, cortical visual impairment, sensorineural deafness, and Hirschsprung disease. Nonneurologic phenotypes include brachytelephalangy, anorectal anomaly, renal abnormality, cleft palate, heart defect, and characteristic facial features such as hypertelorism, broad nasal bridge, and tented mouth. Other symptoms such as ichthyosis, iron deposition, hepatosplenomegaly, diaphragmatic hernia, and hepatic and/or portal vein thrombosis are reported in small fractions of the affected individuals. It is not easy to causally relate specific symptoms to deficiency of particular GPI-anchored proteins except in a few instances. Deficiency of tissue-nonspecific alkaline phosphatase (TNALP) accounts for seizures in some patients. Death within a year after birth due to aspiration or status epilepticus is not rare among severely affected individuals with GPI deficiency, whereas mildly affected individuals can live with a GPI deficiency.

Defects in Glycosphingolipid (GSL) Synthesis

Only three disorders in GSL synthesis (Chapter 11) are known in humans. Mutations in *ST3GAL5* cause autosomal-recessive, Amish infantile epilepsy syndrome and also "salt-and-pepper syndrome." This gene encodes a sialyltransferase required for the synthesis of the ganglioside GM3 (Siaα2-3Galβ1-4Glc-ceramide) from lactosylceramide (Galβ1-4Glc-ceramide). GM3 is also a precursor for some more complex gangliosides. The patients' plasma GSLs are nonsialylated. In contrast to the human form of the disease, mice that lack GM3 do not have seizures or a shortened life span. However, mouse strains that are null for the sialyltransferase and an N-acetylgalactosaminyltransferase that is required for making other complex gangliosides do develop seizures, suggesting that it is the absence of these more complex gangliosides that may be the underlying problem (Chapter 11). Mutations in *B4GALNT1* (also known as GM2/GD2 synthase) cause hereditary spastic paraplegia subtype 26 (SPG26). These patients have developmental delays and varying cognitive impairments with early-onset progressive spasticity owing to axonal degeneration. Cerebellar ataxia, peripheral neuropathy, cortical atrophy, and white-matter hyperintensities are also consistent across the disorder. A *B4galnt1*$^{-/-}$ mouse recapitulates several of the neurological characteristics of SPG26, most prominently the progressive gait disorder.

ST3GAL3 makes more complex gangliosides as well as N- and O-glycans. It is required for the development of high cognitive functions and is mutated in some individuals with West syndrome. An *St3gal3*$^{-/-}$ mouse model also exists, but these mice appear to have no overt neurological phenotype. GSL disorders are difficult to identify biochemically because no convenient biomarker tests have been developed. Therefore, novel genetic defects are likely to be found via next-generation sequencing.

DEFECTS IN MULTIPLE GLYCOSYLATION PATHWAYS

Defects in the Synthesis of Sugar Precursors

Sugar metabolism is essential to produce nucleotide sugars and dolichol-linked sugars for glycosylation reactions in the ER and Golgi. Because activated sugars are required to synthesize multiple glycan classes, genetic disorders affect multiple pathways. Below, some defects in the mannose, galactose, Sia, and fucose sugar activating pathways are discussed.

Galactose Pathway

Galactosemia refers to mutations in three genes involved in Gal metabolism. In "classical galactosemia," Gal-1-P uridyltransferase (GALT; Figure 45.4) is deficient. This results in excess Gal-1-P and decreased synthesis and availability of UDP-Gal. Defects in UDP-Gal-4′-epimerase (GALE; Figure 45.4) or galactokinase (GALK; Figure 45.4) also cause the disease, but they are rarer.

GALT-deficient infants fail to thrive and have an enlarged liver, jaundice, and cataracts. A lactose-free diet ameliorates most of the acute symptoms by reducing the amount of Gal entering the pathway and the accumulation of Gal and Gal-1-P. Reducing Gal decreases galactitol and galactonate, which are produced via reductive or oxidative metabolism of Gal, respectively. Galactitol is not metabolized further and has osmotic properties that contribute to cataract formation. Unfortunately, a Gal-free diet apparently does not prevent the appearance of cognitive disability, ataxia, growth retardation, and ovarian dysfunction that are characteristic of this disease. Abnormal glycosylation of glycoproteins and glycolipids has been observed in some GALT-deficient individuals; however, this has not been directly related to a lack of galactose. For example, some patients who mistakenly are given Gal or those before start of a Gal-free diet synthesize transferrin that is missing both N-glycans. In GALT deficiency, a Gal-free diet

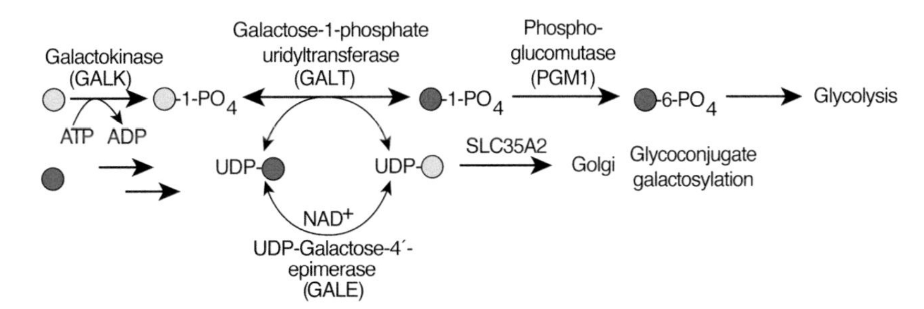

FIGURE 45.4. CDGs related with UDP-Gal metabolism. The most common form of galactosemia is caused by a deficiency of galactose-1-phosphate uridyltransferase (GALT). This enzyme uses Gal-1-P derived from dietary Gal. In the absence of GALT, Gal-1-P accumulates along with excessive Gal and its reductive and oxidative products, galactitol and galactonate (not shown). UDP-Gal synthesis may also be impaired in the absence of GALT, but not completely because UDP-Gal-4′-epimerase (GALE) can form UDP-Gal from UDP-Glc and can supply the donor to galactosyltransferases required for glycoconjugate biosynthesis. Phosphoglucomutase 1 interconverts Glc-1-P and Glc-6-P, which can be used for glycolysis or be metabolized to substrates for GDP-Man synthesis or the hexosamine biosynthesis pathway.

normalizes glycosylation. However, dietary supplementation with Gal was found to be an effective treatment for some of the symptoms in another CDG, PGM1-CDG. In PGM1-CDG, phosphoglucomutase activity results in a reduction of UDP-Gal and UDP-Glc, leading to hypogalactosylation and missing N-glycans. These biochemical abnormalities are restored by Gal feeding. Clinically, the hypoglycemia, liver and coagulation abnormalities, and gonadal hormones improved on treatment. Still, the exercise intolerance and dilated cardiomyopathy, linked to other functions of PGM1, did not improve.

Sialic Acid Pathway

Disorders in four of the Sia biosynthesis enzymes (Chapter 15) are known, resulting in rather different phenotypes. Recessive mutations in *GNE* cause adult-onset GNE myopathy (previously named hereditary inclusion body myopathy type 2 [HIBM2] or Nonaka myopathy) (Table 45.1). It occurs worldwide, but one mutation (p.Met745Thr) is especially common among Persian Jews (1:1500) and occurs in the kinase domain (Chapter 5). GNE mutations occur in various combinations in both GNE enzymatic domains and variably affect enzyme activity. The mutations moderately reduce enzymatic activity and reduce sialylation in mouse models. Oral ManNAc, as precursor for Sia synthesis, is being tested as a therapy for GNE myopathy patients, as well as for patients with primary glomerular diseases (focal segmental glomerulosclerosis, minimal change disease, and membranous nephropathy). Recessive mutations in other genes—*NANS*, *CMAS*, and *SLC35A1*—result in a neurological phenotype with intellectual disability and epilepsy. Additionally, varying degrees of thrombocytopenia may be seen. Interestingly, specific mutations in the *GNE* gene also give rise to a thrombocytopenia syndrome without the presence of neurological symptoms. The biological mechanism behind these differences is not yet understood. The catabolism of Sia is executed by neuraminate pyruvate lyase (NPL), the role of which was unknown in humans until the identification of a human genetic disease. Recessive mutations in *NPL* result in a phenotype of myopathy and cardiomyopathy, possibly indicating a role of Sia in (cardiac) muscle besides glycosylation.

Fucose Pathway

GDP-Fucose (Fuc) can be synthesized from GDP-Man or obtained via fucose salvage. In the salvage pathway, Fuc is converted into Fuc-1-P by fucose kinase and then to GDP-Fuc by GDP-Fuc pyrophosphorylase (Chapter 5). The contribution of the endogenous versus the salvage pathway in different cells and tissues is not yet known. A disease in the salvage pathway has been reported due to mutations in *FCSK* (fucose kinase), resulting in a severe neurological syndrome with encephalopathy, intractable seizures, and intellectual disability. Patients with leukocyte adhesion deficiency type II (LAD-II or SLC35C1-CDG) have mutations in *SLC35C1*, encoding the GDP-Fuc transporter in the Golgi. Here, transferrin sialylation was normal, so this defect was not detected by the usual test, but some serum proteins and O-linked glycans on leukocyte surface proteins were deficient in Fuc. One leukocyte protein carries a selectin ligand glycan, sialyl-Lewis x, that mediates leukocyte rolling before extravasation of leukocytes from capillaries into tissues (Chapter 34). This defect greatly elevates circulating leukocytes and decreases leukocyte extravasation, so patients have frequent infections. A few patients responded to dietary fucose therapy. Sialyl-Lewis x reappeared on their leukocytes, and circulating neutrophils promptly returned to normal levels. Fucose supplements must increase the amount of GDP-Fuc enough to correct the defect. A mouse model of Fuc deficiency lacks de novo biosynthesis of GDP-Fuc from GDP-Man (Chapter 5). The mice die without Fuc supplements, but providing Fuc in the drinking water rapidly normalizes their elevated neutrophils. The treatment also corrects abnormal hematopoiesis resulting from disrupted O-Fuc-dependent Notch signaling.

Defects in the Biosynthesis of Dolichol-Monosaccharides

Numerous defects in the biosynthesis of dolichol, the primary lipid carrier for monosaccharides and oligosaccharides in glycosylation, have been identified. These disorders share many of the features of other CDGs, although some of the mutations are associated with more specific phenotypes. As an example, variants in *DHDDS* have been identified in patients with retinitis pigmentosa, and heterozygous variants in *NUS1* have been found in patients with various neurological phenotypes. The NUS1 and DHDDS gene products are involved in the initial steps of dolichol synthesis. The two proteins form a complex with *cis*-prenyltransferase activity that catalyzes polyprenolpyrophosphate formation. SRD5A3 acts as a polyprenol reductase to form dolichol (Chapter 9). This dolichol is then phosphorylated by DOLK to make the activated lipid, dolichol-phosphate, on which both monosaccharide and oligosaccharide precursors are built. Although the primary functions of dolichol-linked sugars lie within the N-glycosylation pathway, Dol-P-Glc and Dol-P-Man are both integral for different O-glycosylation pathways, and Dol-P-Man is required for biosynthesis of GPI. Clinical symptoms of some of these defects, including muscular dystrophy and dilated cardiomyopathy in patients with DPM1-3 and DOLK defects, have been associated with abnormal O-mannosylation of dystroglycan.

Defects in Golgi Homeostasis

The discovery of CDG defects associated with trafficking proteins showed that the abnormal glycosylation linked to these disorders could arise from alterations in Golgi homeostasis and not only from impairment of the enzymes and transporters needed for glycosylation (Figure 45.5). Among the first of these trafficking proteins identified are subunits of the conserved oligomeric Golgi (COG) complex. This complex has multiple roles in trafficking within the Golgi, including the tethering of COPI-coated vesicles and recycling of Golgi-localized glycosyltransferases. COG7-CDG (Table 45.1) was discovered first. Trafficking of multiple glycosyltransferases and nucleotide sugar transporters were disrupted in COG7-CDG. The mutation affects the synthesis of both N- and O-glycans and glycosaminoglycan (GAG) chains. Mutations have now been found in all COG subunits except COG3. Mammalian cells deficient in the different COG subunits also show various degrees of altered glycosylation.

An unusual disorder called HEMPAS (hereditary erythroblastic multinuclearity with positive acidified-serum test) leads to abnormal red cell shape and instability (hemolysis) caused by

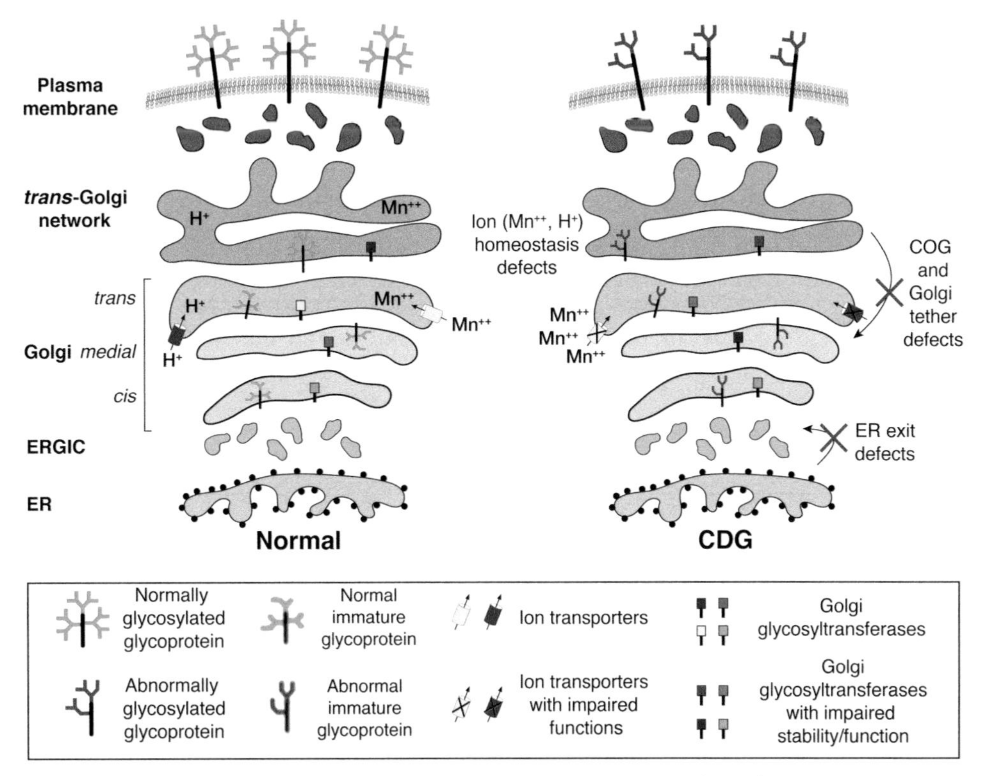

FIGURE 45.5. Golgi homeostasis defects. Congenital disorders of glycosylation (CDGs) are also caused by defects in proteins and transporters that control Golgi homeostasis. These CDGs include members of the vesicle tethering conserved oligomeric Golgi (COG) complex, which mediates the retrograde transport of glycosyltransferase-containing vesicles within the Golgi. COG defects cause mislocalization of glycosylation enzymes and transporters, resulting in altered processing of glycans. CDGs caused by proteins involved in the exit of cargo from the ER have also been identified. Ion transporters represent a growing class of CDGs. Defects in the transport of hydrogen ions and metals such as manganese alter the lumenal environment of the Golgi and can deplete cofactors necessary for metal-dependent glycosyltransferases.

mutations in *SEC23B,* another intracellular trafficking protein that produces abnormal red blood cell glycans in several pathways.

In addition to trafficking proteins, a number of mutations in Golgi-associated ion transporters and vacuolar H^+-ATPase subunits have also been shown to disrupt multiple glycosylation pathways, presumably because of an increase in Golgi pH, decrease in metal ion-dependent glycosyltransferase activities, or a more general disruption in Golgi recycling. Mutations in two Golgi manganese transporters, *TMEM165* and ZIP8 (*SLC39A8*), cause CDGs highlighting the role of manganese homeostasis in glycosylation.

FUTURE OUTLOOK—PATHOPHYSIOLOGY AND TREATMENT

Expression of the same CDG mutation can have highly variable impacts, even among affected siblings. Explanations based on residual activity from hypomorphic alleles of these "simple Mendelian disorders" are neither simple nor generally satisfying. The variability is often attributed to "genetic background." A knockout mutation may be lethal in one highly inbred mouse strain, but not in another because compensatory pathways exist in the latter. Dietary and environmental impacts are substantial as seen in MPI-CDG patients treated by oral mannose therapy. Multiple simultaneous or sequential environmental insults may impinge on borderline genetic insufficiencies to produce overt disease. A major question for future investigations is how CDG defects cause pathology. Addressing this challenge will require the implementation of analytical approaches like glycomics and glycoproteomics and of model systems that allow identification of sensitive pathways and glycoproteins in affected tissues.

In an interesting twist, it is possible to have diseases caused by "excessive" glycosylation. For example, Marfan syndrome results from mutations in *fibrillin1* (*FBN1*) and one of these creates an N-glycosylation site that disrupts multimeric assembly of FBN1. This may not be an isolated case. A survey of nearly 600 known pathological mutations in proteins traveling the ER–Golgi pathway showed that 13% of them create novel glycosylation sites. This is far greater than predicted by random missense mutations and may mean that hyperglycosylation leads to a new class of CDGs.

ACKNOWLEDGMENTS

The authors acknowledge contributions to previous versions of this chapter by Harry Schachter and Bobby G. Ng and appreciate helpful comments and suggestions from Pamela Stanley, Yan Wang, Jennifer Kohler, and Ajit Varki.

FURTHER READING

Dobson CM, Hempel SJ, Stalnaker SH, Stuart R, Wells L. 2013. O-Mannosylation and human disease. *Cell Mol Life Sci* **70**: 2849–2857. doi:10.1016/0968-0004(89)90175-8

Huegel J, Sgariglia F, Enomoto-Iwamoto M, Koyama E, Dormans JP, Pacifici M. 2013. Heparan sulfate in skeletal development, growth, and pathology: the case of hereditary multiple exostoses. *Dev Dyn* **242**: 1021–1032. doi:10.1002/1615-9861(200102)1:2<340::aid-prot340>3.0.co;2-b

Jaeken J. 2013. Congenital disorders of glycosylation. *Handb Clin Neurol* **113**: 1737–1743. doi:10.1093/glycob/cwj010

Rosnoblet C, Peanne R, Legrand D, Foulquier F. 2013. Glycosylation disorders of membrane trafficking. *Glycoconj J* **30**: 23–31. doi:10.1074/mcp.ra118.000799

Freeze HH, Eklund EA, Ng BG, Patterson MC. 2015. Neurological aspects of human glycosylation disorders. *Annu Rev Neurosci* **38**: 105–125. doi:10.1093/glycob/cwz078

Hennet T, Cabalzar J. 2015. Congenital disorders of glycosylation: a concise chart of glycocalyx dysfunction. *Trends Biochem Sci* **40**: 377–384. doi:10.1093/glycob/cwz045

Maeda N. 2015. Proteoglycans and neuronal migration in the cerebral cortex during development and disease. *Front Neurosci* **9**: 98. doi:10.1038/s41467-019-11131-x

Nishino I, Carrillo-Carrasco N, Argov Z. 2015. GNE myopathy: current update and future therapy. *J Neurol Neurosurg Psychiatry* **86**: 385–392. doi:10.1038/s41592-020-0879-8

Paganini C, Costantini R, Superti-Furga A, Rossi A. 2019. Bone and connective tissue disorders caused by defects in glycosaminoglycan biosynthesis: a panoramic view. *FEBS J* **286**: 3008–3032. doi:10.1093/glycob/cwz080

van Tol W, Wessels H, Lefeber DJ. 2019. *O*-glycosylation disorders pave the road for understanding the complex human *O*-glycosylation machinery. *Curr Opin Struct Biol* **56**: 107–118. doi:10.1093/nar/gkaa947

Kinoshita T. 2020. Biosynthesis and biology of mammalian GPI-anchored proteins. *Open Biol* **10**: 190290. doi:10.1098/rsob.190290

46 Glycans in Acquired Human Diseases

Robert Sackstein, Karin M. Hoffmeister, Sean R. Stowell, Taroh Kinoshita, Ajit Varki, and Hudson H. Freeze

published by Cold Spring Harbor Laboratory Press; doi: 10.1101/glycobiology.4e.46

Many acquired changes in glycan synthesis, turnover/degradation, or recognition are involved in human diseases. Knowing these changes can improve disease diagnosis and/or therapies. This chapter lists a few examples, proposes mechanisms, and suggests novel therapies. Glycosylation changes in cancer and in inherited human genetic disorders of glycosylation are discussed in more depth in Chapters 47 and 45, respectively.

CARDIOVASCULAR MEDICINE

Role of Selectins in Reperfusion Injury

Vascular thrombotic events such as stroke and myocardial infarction, plus others including acute ischemic injury (e.g., trauma, hypovolemic shock), temporarily interrupt of blood flow that must be fixed quickly. Natural or medical interventions that restore blood flow suddenly reintroduce leukocytes to the traumatized tissue and can cause tissue damage called "reperfusion injury." Selectins, cell adhesion molecules that are C-type lectins (Chapter 34), play major roles in this process. L-Selectin (CD62L) on leukocytes and P-selectin (CD62P) on the activated endothelium in the reperfused area initiate this cascade, and E-selectin (CD62E) participates later; however, among the selectins, the E-selectin-based leukocyte recruitment in humans (and primates, in general) is more prominent and sustained than in other mammals because the promoter region of the human E-selectin gene, but not that of P-selectin, contains elements responsive to inflammatory mediators tumor necrosis factor-α (TNF-α) and IL-1 (and to bacterial products like LPS). These increase transcription and expression of E-selectin (Chapter 34), whereas the P-selectin promoter in humans actually dampens gene transcription in response to such mediators. However, in mice, the promoters for both the P-selectin and E-selectin genes retain TNF/IL-1/LPS responsive elements, and mice thus display inflammation-driven inducible endothelial expression of both P-selectin and E-selectin. This regulatory distinction is critical in considering whether rodent models of P-selectin biology may reflect human pathophysiology. Nonetheless, animal models show that blocking initial selectin-based interactions ameliorates subsequent tissue damage. Pharmaceutical and biotechnology companies have made small-molecule inhibitors and biologics (e.g., monoclonal antibodies) to block these interactions in patients, and this approach has shown benefit and has achieved U.S. Federal Drug Administration (FDA) approval for clinical indications such as sickle cell vaso-occlusive crises (further described below; Chapters 55 and 57 also discuss selectin inhibitors). In addition, many preparations of the already-approved drug heparin can block P- and L-selectin interactions, but its dampening effects on reperfusion injury remain to be established.

Roles of Selectins, Glycosaminoglycans, and Sialic Acids in Atherosclerosis

Heart attacks, strokes, and other serious vascular diseases are associated with high levels of low-density lipoprotein (LDL) cholesterol and decreased high-density lipoprotein (HDL) cholesterol. These molecules increase the risk of atherosclerotic lesions in the large arteries. Atherosclerotic lesions begin with the development of the "fatty streak," in which monocytes enter into the subendothelial regions of the blood vessels. This process involves expression of P- and/or E-selectin on the endothelium, which recognizes glycosylated and sulfated P-selectin glycoprotein ligand-1 (PSGL-1) or sialyl Lewis x ("sLe^{x}"; CD15s) displayed on discrete glycoproteins and glycolipids of circulating monocytes. Indeed, P-selectin deficiency in mice delays progression of atherosclerotic lesions, which slows even more in combined P- and E-selectin-deficient mice. Oxidized lipids in LDL particles or inflammatory processes likely induces endothelial E-selectin in the early atheromatous plaque. Early intervention may be possible because these lesions develop slowly and relatively early in life. Retention of LDLs in the early plaque probably involves binding to proteoglycans and subsequent oxidation and uptake by macrophages and smooth-muscle cells. Clusters of basic amino acid residues in apolipoprotein B (the protein moiety of LDL) likely bind to the glycosaminoglycans. This binding also has a physiological function. Heparan sulfate (HS) found in the liver sinusoids regulates the turnover of lipoprotein particles. Reduced sialylation of LDL in patients also correlates with coronary artery disease. Although the mechanism(s) remain unclear, desialylated LDL may be more easily taken up and incorporated into atheromatous plaques. The nonhuman sialic acid Neu5Gc (enriched in red meats; see Chapter 15) can be metabolically incorporated into plaques, and recognition by circulating antibodies against Neu5Gc glycans may accelerate disease progression via chronic inflammation.

DENTAL DISEASE

The oral cavity contains numerous host and microbial glycans and glycan-binding proteins (Chapter 21). Dental caries and gingivitis involve certain viridans group streptococci that have serine-rich bacterial proteins whose stability requires O-glycosylation. They also have a carbohydrate recognition domain that recognizes O-linked sialoglycans or oral mucins to allow attachment to cells in the mucosa. If these bacteria reach the bloodstream, particularly during dental procedures, the same binding proteins recognize platelet specific glycoproteins such as GPIbα, delivering the bacteria to damaged heart valves, and leading to the serious disease of bacterial endocarditis.

DERMATOLOGY

Role of Selectins in Inflammatory Skin Diseases

In humans, E-selectin is constitutively expressed on microvessels of the skin, recruiting both adaptive and innate immune effector cells. E-selectin is up-regulated in all inflammatory skin diseases, further promoting extravasation of circulating leukocytes containing sLe^x. Dermal lymphocytes are recognized by the monoclonal antibody "HECA452" that binds sLe^x. This "cutaneous lymphocyte antigen" is displayed on glycoforms of PSGL-1, CD43, and CD44 molecules (sLe^x-decorated CD44 is called hematopoietic cell E-/L-selectin ligand (HCELL)) (Chapter 34). In mice, both E-selectin and P-selectin are constitutively expressed on dermal microvessels. In humans, T helper 1 (Th1) lymphocytes access the skin via binding to E-selectin, suggesting that blocking this interaction could treat dermal inflammatory diseases.

ENDOCRINOLOGY AND METABOLISM

Pathogenesis and Complications of Diabetes Mellitus

Diabetes mellitus causes long-term vascular complications, in part by increasing nonenzymatic glycation (not to be confused with glycosyltransferase-mediated glycosylation), wherein the open-chain (aldehyde) form of glucose reacts with lysine residues generating reversible Schiff bases that can rearrange to generate "browning" (Maillard) reactions and permanent cross-linked advanced glycation end products (AGEs). These adducts impair protein and cellular functions, disrupting the function of extracellular matrix proteins (e.g., collagen), contribute to neurodegenerative diseases (e.g., Alzheimer's disease), and bind to receptors, such as the receptor for advanced glycation end products (RAGE) and also the macrophage scavenger receptor that participates in atherogenesis. Notably, glycation of hemoglobin generates "Hemoglobin A1c," the biomarker for measuring long-term glucose control in diabetics.

Excess glucose increases UDP-GlcNAc through the glucosamine:fructose aminotransferase (GFAT) pathway, enhancing hyaluronan production (Chapter 16) and O-GlcNAcylation of multiple proteins that, in turn, alters their phosphorylation and functions (Chapter 19). In animal models, altered O-GlcNAcylation correlates with complications, such as diabetic cardiomyopathy (increased O-GlcNAcylation of various nuclear proteins) and erectile dysfunction (O-GlcNAcylation of endothelial nitric oxide synthase). Several cytoplasmic proteins involved in insulin receptor signaling and the resulting nuclear transcription changes are themselves O-GlcNAcylated and functionally altered in diabetes.

Kidney dysfunction is a very serious, potentially lethal, diabetic complication. Progressive albumin excretion eventually leads to nephrotic syndrome and to end-stage renal disease. This proteinuria correlates with reduced HS proteoglycan in the glomerular basement membrane. Reduced HS synthesis by glomerular epithelial cells may result from exposure to high glucose or increased porosity of the glomerular basement membrane. High glucose also increases plasminogen activator inhibitor-1 (PAI-1) gene expression in renal glomerular mesangial cells, via O-GlcNAc-mediated alterations in Sp1 transcriptional activity.

GASTROENTEROLOGY

Role of Gut Epithelial Glycans in Gastrointestinal Infections

Numerous gastrointestinal pathogens or the products they secrete recognize and bind to gut mucosal glycans (Chapter 37). For example, cholera toxin (CT) secreted by *Vibrio cholerae* can bind GM1 ganglioside, and *Helicobacter pylori*, the cause of peptic ulcers and gastritis, binds Lewis type glycans in the stomach mucosa. Soluble glycan inhibitors could block binding of these gut pathogens. This could explain that a time-honored treatment for peptic ulcers (before the identification of *H. pylori*) was a combination of antacids and milk (which contains large amounts of free sialyloligosaccharides). Cholera infection outcomes are correlated with ABO(H) blood group expression. Blood group O(H) individuals are not more frequently infected, but experience more severe disease. Although GM1 is the primary functional ligand for CT, recent studies have suggested that CT also has a binding site for the H antigen, possibly explaining this association.

Heparan Sulfate Proteoglycans in Protein-Losing Enteropathy

Protein-losing enteropathy (PLE) is the enteric loss of plasma proteins; however, its mechanisms are not well-understood. Some patients with congenital disorder of glycosylation type Ib (CDG-Ib or MPI-CDG) and CDG-Ic (ALG6-CDG) (Chapter 45) develop PLE, suggesting involvement of N-glycosylation. But other patients with normal N-glycosylation also develop PLE many years following (Fontan) surgery to correct congenital heart malformations. What causes this type of PLE? One concept is that environmental stress such as infections increase inflammatory cytokines, TNF-α and interferon-γ (IFN-γ), which, together with increased venous pressure resulting from the surgery, contribute to PLE. In both the N-glycosylation disorders and post-Fontan surgery patients, HS is lost from the basolateral surface of intestinal epithelial cells and returns when PLE subsides. HS binds cytokines, and its loss may increase the impact of inflammation at the cell surface and promote leakage. Combining increased venous pressure and cytokines along with localized HS depletion creates a downward spiral of disease. Traditional therapy for PLE includes albumin infusions, steroid hormones, or other anti-inflammatory drugs, but, interestingly, heparin injections can also reduce PLE, possibly by binding to and reducing circulating cytokines (Chapter 38).

Changes in Sialic Acid O-Acetylation in Ulcerative Colitis and Cancer

Ulcerative colitis is an inflammatory disease typically affecting the superficial epithelial layer of the rectum and the distal colon. The primary cause of the disease is unknown, but both genetic and environmental factors, such as changes in the microbiome, are involved and remissions and exacerbations are common. Normally, colonic mucosal proteins display heavily O-acetylated sialic acids, but these acetyl groups are diminished in ulcerative colitis. Whether this contributes to the pathology is unknown, but O-acetylated sialic acids are more resistant to bacterial sialidases (Chapter 15). Decreased sialic acid O-acetylation is also a feature of colon carcinomas.

HEMATOLOGY

Clinical Use of Heparin as an Anticoagulant

Heparin (a highly sulfated form of HS; Chapter 17) is a therapeutic agent extracted from porcine intestines or bovine lungs. It is a fast-acting and potent anticoagulant often used to avoid thrombosis, in procedures including dialysis and open-heart surgery. Its effectiveness relies on a specific sulfated heparin pentasaccharide that binds circulating antithrombin III and markedly enhances its ability to inactivate coagulation Factors Xa and IIa (thrombin) (Chapter 17). Animal-derived "unfractionated heparin" is now often replaced with low-molecular-weight heparins because they have fewer complications. One explanation is that the unfractionated heparin effects on thrombin require a long chain that interacts both with the antithrombin and with thrombin itself in a tripartite complex. In contrast, the shorter chains found in low-molecular-weight heparins only facilitate antithrombin inactivation of Factor Xa. Thus, low-molecular-weight heparins affect Xa but not thrombin activity. A synthetic pentasaccharide (fondaparinux) that binds and facilitates antithrombin inactivation of Factor Xa is an alternative to heparin. Although these improvements are valuable, unfractionated heparin has a variety of other biological effects besides anticoagulation. Thus, other beneficial effects of heparin, such as the blocking of P- and L-selectin, are reduced or even eliminated by the switch to low-molecular-weight heparins and the synthetic pentasaccharide.

An uncommon but feared complication of heparin treatment is heparin-induced thrombocytopenia (HIT). During HIT, complexes between heparin and platelet factor-4 form, and the de novo formed pathogenic antibodies against these complexes deposit on platelets, causing their aggregation and loss from circulation. Paradoxically, this process results in exaggerated thrombosis, rather than bleeding. The incidence of this complication is lower with low-molecular-weight heparins and significantly curtailed with the use of the pure pentasaccharide.

Selectin Inhibition to Obviate Sickle Cell Crises

Sickle cell anemia is an inherited disorder of hemoglobin that leads to various acute and chronic painful complications. Symptoms were thought to be caused by vaso-occlusion due to abnormally shaped and/or membrane-modified hypoxic erythrocytes. Now, it is clear that abnormal adhesion of multiple cell types is responsible, and that this is partially mediated by selectins. Blocking selectin action with analogs of natural ligands such as sLe^X or heparin can restore blood flow in a mouse model of sickle cell disease. A pan-selectin inhibitor (GMI-1070, "rivipansel") has undergone clinical trials for reduction of severity and duration of sickle cell crisis. Notably, glycopeptide mimetics of PSGL-1 and antibodies that block PSGL-1-P-selectin interactions (e.g., FDA-approved crizanlizumab) can mitigate sickle cell vaso-occlusive crises.

Hemolytic Transfusion Reactions

Blood transfusion medicine first identified the ABO blood group system, defined by differential expression of α-GalNAc and α-Gal transferases across populations (Chapter 14). These and other less prominent glycan antigens can cause hemolytic transfusion and rejection reactions, mostly because of errors in blood typing. Although anti-A and anti-B titers likely influence transfusion and transplantation outcomes, simple assessment of antibody levels fails to fully account for the clinical impact(s) of an ABO(H) incompatible transfusion. Efforts to generate "universal donor" red blood cells, via enzymatic conversion of blood group A and B antigens to the O state using bacterial enzymes, are underway.

Acquired Anticoagulation Due to Circulating Heparan Sulfate

Occasionally, patients with diseases such as cirrhosis and hepatocellular carcinoma spontaneously secrete a circulating anticoagulant and have an unusual coagulation test profile that makes it appear as if the patient has been treated with heparin. The anticoagulant activity can be purified from the plasma and has been identified as an HS glycosaminoglycan. Its source is unknown and therapy is difficult without correcting the underlying disease or transplanting the liver.

Abnormal Glycosylation of Plasma Fibrinogen in Liver Disorders

Plasma fibrinogen is heavily sialylated, and the sialic acids are involved in binding calcium. Certain genetic disorders of fibrinogen are associated with altered sialylation of its N-glycans, which causes altered function in clotting. Patients with hepatomas and other liver disorders can also sometimes manifest increased branching and/or number of N-glycans on fibrinogen, resulting in an overall increase in sialic acid content. This altered fibrinogen sialylation can present clinically as a bleeding disorder associated with a prolonged thrombin time. Patients with CDGs affecting N-glycan biosynthesis (Chapter 45) can also have thrombotic or bleeding disorders that are partly explained by altered glycosylation of plasma proteins and/or platelets involved in blood coagulation.

Paroxysmal Nocturnal Hemoglobinuria

Paroxysmal nocturnal hemoglobinuria (PNH) is an unusual form of acquired hemolytic anemia (excessive destruction of red blood cells) that usually appears in adults and arises through a somatic mutation in bone marrow stem cells generating one or more abnormal clones. The defect is an inactivation of the single active copy of the *PIGA* gene, an X-linked locus involved in the first stage of biosynthesis of glycosylphosphatidylinositol (GPI) anchors (for details on GPI anchor biosynthesis, see Chapter 12). Although several blood cell types show abnormalities, the red cell defect is the most prominent, being characterized by an abnormal susceptibility to the action of complement. This is now known to be caused by the lack of expression of certain GPI-anchored proteins, such as the decay accelerating factor and CD59, that normally downregulate complement activation on "self" surfaces. However, hypercoagulability also occurs, presumably because of loss of GPI-anchored proteins on other cells, such as monocytes and platelets. Interestingly, many of these patients had preceding, or later development of, bone marrow failure (aplastic anemia). Some of the latter also develop acute leukemia. Most normal humans already have a tiny fraction of circulating cells with a *PIGA* mutation, the PNH defect. These presumably represent the products of one or more bone marrow stem cells that develop this acquired defect because of a single hit on the active X chromosome but then do not become prominent contributors to the total pool of circulating red blood cells. In this scenario, the independent occurrence of a process damaging the proliferation of other (unaffected) stem cells allows the "unmasking" of the PNH defect.

Paroxysmal Cold Hemoglobinuria

Patients with the rare disorder paroxysmal cold hemoglobinuria have a cold-induced intravascular destruction of red cells (hemolysis). This hemolysis appears to be caused by a complement-fixing circulating IgG antibody directed against the red cell P blood group antigens that binds more efficiently at temperatures below the core body temperature of 37°C (such as within appendices) (Chapter 14). Biosynthesis of the P blood group antigens is dependent on the *B3GALNT1* (β1-3-N-acetylgalactosaminyltransferase 1) and *A4GALT* (α1-4-galactosyltransferase) genes. The pathogenesis of this disorder is unknown, but it tends to occur in the setting of some viral infections and in syphilis. The IgG antibody presence is detected by the so-called "Donath–Landsteiner test," in which the patient's serum is mixed either with the patient's own red cells or with those from a normal person and chilled to 4°C. Complement-mediated hemolysis occurs after warming to 37°C.

Cold Agglutinin Disease

Cold agglutinin disease is caused by autoimmune IgM antibodies directed against glycan epitopes on erythrocytes. High titers of IgM agglutinins are present in serum and are maximally active at 4°C. The IgMs are presumed to bind to erythrocytes that are circulating in the cooled blood of peripheral regions of the body. The antibody fixes complement, which then destroys the cells when they reach warmer areas of the body. There are several variants of the syndrome. One affects young adults and follows infection with *Mycoplasma pneumoniae* or Epstein–Barr virus (infectious mononucleosis). This antibody is characteristically polyclonal, and is generally short-lived, disappearing when the infection subsides. An idiopathic variant of cold agglutinin disease affects older individuals, involves a monoclonal IgM, and can be a precursor or an accompaniment to a lymphoproliferative disease such as Waldenström's macroglobulinemia, chronic lymphocytic leukemia, or other lymphomas. These antibodies are typically directed against the "I" antigen (β1-6-branched poly-*N*-acetyllactosamine) present on glycolipids and glycoproteins of erythrocytes (Chapter 14). Fewer less common variants of cold agglutinin disease involve antibodies directed against sialylated *N*-acetyllactosamines. In some patients on chronic hemodialysis, the syndrome occurs because of the formation of antibody directed against the sialylated blood group antigen N.

Tn Polyagglutinability Syndrome

In this condition a subset of bone marrow–derived blood cells express the Tn antigen (O-linked *N*-acetylgalactosamine, GalNAcα-O-Ser/Thr) and sialyl-Tn (NeuAcα2-6GalNAcα-O-Ser/Thr), thus becoming susceptible to hemagglutination by naturally occurring anti-Tn antibodies that exist in most normal humans. The underlying defect is a somatic stem cell–based loss of expression of the O-glycan core-1 β1-3 galactosyltransferase activity (also called the T synthase) (Chapter 10). This in turn is explained by acquired inactivation of Cosmc, a chaperone required for the biosynthesis of the T synthase. As with PNH, the existence of the *C1GALT1C1* gene encoding Cosmc on the X chromosome allows a single hit on the active X chromosome to cause a glycosylation defect in a single bone marrow stem cell. Some patients are picked up simply because polyagglutinability of their red blood cells is detected when blood typing is performed for a possible transfusion. Others have varying degrees of hemolytic anemia and/or decreases in other blood cell types. Some subsequently progress into frank leukemia. It is unclear how the primary syndrome predisposes to development of malignancy. As with PNH, the possibility exists that an underlying bone marrow disorder simply allows the "unmasking" of preexisting minor stem cell clones along with the defect. In keeping with this notion, leukemic clones that arise later may not have the same defect.

Altered Glycosylation Affects Platelet Count and Life Span

The sialylation of platelets regulates their count and production (thrombopoiesis). Aging platelets lose sialic acid moieties and are consequently cleared by multiple Gal- and GalNAc-recognizing lectins, including the Ashwell–Morell receptor (AMR) (Chapter 34) and the macrophage galactose lectin (MGL). Some patients with immune thrombocytopenia (ITP) express sialidase NEU1 on their platelet surface. Influenza antiviral drugs including anamivir (Relenza), oseltamivir (Tamiflu), and peramivir (Rapivab) mimic the transition state to block viral neuraminidase (Chapter 55). Administration of oseltamivir increases platelet counts in a cohort of patients with ITP and in healthy human subjects, suggesting that platelet sialic acid regulates platelet count. Loss of platelet sialic acid and thrombocytopenia can occur in patients with genetic mutations in sialylation, such as *GNE*, which makes a precursor of sialic acids. Mutations in the CMP-Sia transporter gene *SLC35A1* results in macrothrombocytopenia because of increased clearance. Mouse models such as sialyltransferase ST3GalIV knockout mice have profound thrombocytopenia. Crossing ST3GalIV-deficient mice with AMR-null mice results in platelet count recovery, implying that uncovering of galactose leads to AMR-mediated clearance. Multiple changes in O-glycans can also produce severe thrombocytopenia, including deficits in sialyltransferase ST3GalI, GalNT3, C1GalT1 or its chaperone, Cosmc.

IMMUNOLOGY AND RHEUMATOLOGY

Changes in IgG Glycosylation in Rheumatoid Arthritis

IgG immunoglobulins are N-glycosylated, and those in the constant (CH2 or Fc) region of human IgG have several unusual properties. First, they are buried between the folds of the two constant regions. Second, the crystal structure of the protein shows that they are immobilized by glycan–protein interactions. Third, the IgG complex-type biantennary N-glycans are rarely fully sialylated, and instead have one or two terminal β-linked galactose residues (termed G1 and G2, respectively). IgG from patients with rheumatoid arthritis (RA) have even less galactose or none at all (termed G0). The severity of this inflammatory disease tends to correlate inversely with the level of galactosylation. Spontaneous clinical improvement occurs during pregnancy and correlates with restored galactosylation. The basis of the decreased galactosylation in RA is unknown. Some evidence points to less β-galactosyltransferase activity in patients' lymphocytes, but whether altered glycosylation of IgG has a specific pathogenic role in rheumatoid arthritis is debatable, because G0 molecules are seen in other diseases including granulomatous diseases (e.g., tuberculosis) and Crohn's disease. The glycan changes are also seen to a lesser extent in osteoarthritis, a form of chronic degenerative arthritis with a different pathogenesis. One function attributed to the Fc N-glycans is to maintain the conformation of the Fc domains as well as the hinge regions. Other structural features are also necessary for effector functions such as complement and Fc receptor binding and Fc-dependent cytotoxicity. Nuclear magnetic resonance (NMR) studies show that the G0 N-glycans have an increased mobility resulting from the loss of interactions between the glycan and the Fc protein surface. Thus, it is thought that regions of the protein surface normally covered by the glycan are exposed in rheumatoid arthritis. Some studies suggest that the circulating mannose-binding protein recognizes the more mobile G0 N-glycan and activates complement directly. Rheumatoid arthritis patients also have increased circulating immune complexes consisting of antibody molecules (called rheumatoid factor) that recognize the Fc regions of other IgG molecules. However, the epitopes involved do not seem to be glycan-related. Another likely possibility is that the altered glycosylation changes interactions with Fc receptors.

INFECTIOUS DISEASE

Glycans and their binding proteins are key to the pathogenicity of essentially all infectious diseases, and this topic is covered in greater detail in Chapters 37, 42, and 43. Some key physiologic effects of glycan–host interactions are briefly covered here.

Urinary Tract Infections

Many urinary tract infections (UTIs) are caused by *Escherichia coli*, which adhere to bladder epithelial cells via a mannose-binding lectin, FimH, located on the F-pilus of the bacterium. A simple and effective antibiotic-independent treatment and prevention for this very common infection is drinking D-mannose because it competes bacterial glycan binding to the urinary tract when it is excreted in the urine. Another alternative is optimized synthetic α-mannosides, which could treat and prevent UTIs.

Recognition of Host Glycans by Bacterial Adhesins, Toxins, and Viral Proteins

Many pathogens initiate infection by binding to host cell-surface glycans (Chapter 37). Variable expression of these glycans may explain an individual's susceptibility to infection. For example, some pathogenic strains of *E. coli* infect the urinary tract using a lectin that binds the P blood group antigens (Chapter 14), with P negative individuals being immune. The *E. coli* P fimbriae lectins are also involved in spreading bacterial infections from the kidney to the bloodstream.

Binding of the spike glycoprotein of SARS-CoV-2, the virus causing COVID-19, to its ACE2 receptor on host epithelial cells first involves its initial binding to neighboring HS glycosaminoglycan chains, which induces a conformational change to enhance binding to ACE2. Importantly, mutations that alter spike glycosylation sites (e.g., variant B.1.1.7 (N501Y)) impact infectivity.

Desialylation of Blood Cells by Circulating Microbial Sialidases during Infections

Some pathogenic microorganisms secrete sialidases (neuraminidases), which usually remain at the site of infection. However, in some severe cases (e.g., *Clostridium perfringens*–mediated gas gangrene), sialidases reach the plasma where they desialylate red cells, resulting in clearance and anemia. Measuring plasma sialidase may aid diagnosis and prognosis. Similarly, the action of viral (e.g., influenza and dengue) or bacterial (e.g., *Streptococcus pneumoniae*) sialidases causes loss of platelet sialic acid and contributes to increased platelet clearance. Sialidase-producing *S. pneumoniae* can also cause hemolytic uremic syndrome, and selectively inhibiting sialidase could have therapeutic value. Paradoxically, the thrombocytopenia resulting from desialylation of platelets predisposes to bleeding but may serve to protect against sepsis-induced disseminated intravascular coagulation.

NEPHROLOGY

Loss of Glomerular Sialic Acids in Nephrotic Syndrome

Nephrotic syndrome occurs when the kidney glomerulus fails to retain serum proteins during initial filtration of plasma, allowing these proteins to leak into the urine. The epithelial/endothelial mucin molecule called podocalyxin on the foot processes (pedicles) of glomerular podocytes helps maintain pore integrity and excludes proteins from the glomerular filtrate. Sialic acids on podocalyxin are critical for this function. Loss of sialic acid is seen in spontaneous minimal-change renal disease in children and in the nephrotic syndrome that follows some bacterial infections. Animal models seem to mimic this situation with proteinuria and renal failure developing in a dose-dependent manner after a single injection of *V. cholerae* sialidase, which correlated with loss of sialic acids from the glomerulus. This was accompanied by effacement of foot processes and the alteration of tight junctions between podocytes. The anionic charge returned to endothelial and epithelial sites within two days, but the foot process loss remained. Another model is aminonucleoside nephrosis, induced in rats by injection of puromycin. Again, defective sialylation of podocalyxin and glomerular glycosphingolipids is seen in this model. A genetic mouse model with impaired sialic acid synthesis (UDP-*N*-acetylglucosamine-2-epimerase/*N*-acetylmannosamine kinase [GNE] deficiency) dies soon after birth because of kidney dysfunction and undersialylated podocalyxin.

Changes in O-Glycans in IgA Nephropathy and Chronic Kidney Disease

In humans, all immunoglobulin classes contain N-glycans in the Fc domain, but only IgA1 and IgD contain O-glycans (localized within the hinge regions). The IgA1 O-glycan chains are thought to stabilize the three-dimensional structure of the molecule. IgA nephropathy is a form of glomerulonephritis caused by the deposition of aggregated IgA1 molecules within the glomerulus, and patients with this condition harbor circulating IgA1 with O-glycan truncations. Underglycosylated IgA1 has the tendency to both self-aggregate as well as trigger immune reactions leading to IgG-IgA complexes, and both processes lead to glomerular IgA1 deposits. The primary mechanism of underglycosylation remains unknown. One possible scenario is a defect in the *C1GALT1C1* gene that encodes Cosmc, similar to that found in the Tn polyagglutinability syndrome (see above). Instead of affecting a bone marrow stem cell, the defect would presumably involve a clone of B cells that specifically expresses underglycosylated IgA1.

In chronic kidney disease, genome-wide association studies (GWASs) have indicated that deficiency in polypeptide GalNAc-transferase 11 (GALNT11) may be etiologic. GALNT11 adds O-GalNAc to LRP2 (megalin), a major endocytic receptor within renal proximal tubules. Studies with GALNT11 knockout mice have shown a critical role of LRP2 O-glycans in mediating its protein resorption function and, also, in preventing its loss with age.

Heparan Sulfate Changes in Systemic Lupus Erythematosus

Systemic lupus erythematosus (SLE) is an autoimmune disorder in which antigen–antibody complexes accumulate in various organs, especially the skin and kidney. How SLE begins is unknown, but the pathology may involve both cytokines and HS. HS is reduced on the glomerular basement membrane, and this was thought to result from masking of HS by complexes of nucleosomes and antinuclear antibodies, but the actual mechanism(s) is likely to be more complex. Even though anti-double-stranded DNA antibodies are the hallmark of SLE, circulating antibodies to HS strongly correlate with disease activity. In some studies, HS injections into dogs induce SLE symptoms within several weeks, and elevated HS is found in the urine of SLE patients, especially in severe cases. Some SLE patients also develop protein-losing enteropathy (PLE), perhaps as a consequence of misplaced or degraded HS and elevated cytokines, creating the appropriate environment for PLE (see above).

NEUROLOGY AND PSYCHIATRY

Pathogenic Autoimmune Antibodies Directed against Neuronal Glycans

A variety of diseases are associated with circulating antibodies directed against specific glycan molecules enriched in the nervous system, resulting in autoimmune neural damage. Such antibodies can arise via distinct pathogenic mechanisms. In the first situation, patients with benign or malignant B-cell neoplasms (e.g., benign monoclonal gammopathy of unknown significance [MGUS], Waldenström's macroglobulinemia, or plasma cell myeloma) secrete monoclonal IgM or IgA antibodies that are specific for either gangliosides or, more commonly, for sulfated glucuronosyl glycans (the so-called HNK-1 epitope). These antibodies react with glycolipids bearing the epitope 3-O-SO_3-GlcAβ1-4Galβ1-4GlcNAcβ1-3Galβ1-4Glc-Cer (3′sulfoglucuronosylparagloboside) and against the N-glycans on a variety of central nervous system (CNS) glycoproteins (MAG, P0, L1, N-CAM) that bear the same terminal sequence. The resulting peripheral demyelinating neuropathy can sometimes be more damaging than the primary disease itself. Therapy consists of attempts to treat the primary disease with chemotherapy or to remove the immunoglobulin by plasmapheresis. Both approaches are usually unsuccessful at lowering the immunoglobulin to a level sufficiently to diminish the symptoms. The second situation is an immune reaction to molecular mimicry of neural ganglioside structures by the lipo-oligosaccharides of bacteria such as *Campylobacter jejuni*. Following an intestinal infection with such organisms, circulating cross-reacting antibodies against gangliosides such as GM1 and GQ1b appear in the plasma. These are associated with the onset of symptoms of peripheral demyelinating neuropathy involving the peripheral or cranial nerves (the Guillain–Barré and Miller Fisher syndromes, respectively). The third situation is a human-induced disease arising from attempts to treat patients with stroke using intravenous injections of mixed bovine brain gangliosides. Although some evidence for benefits was seen, several cases of Guillain–Barré syndrome were reported as a likely side effect. One explanation is that the presence of small amounts of gangliosides with the nonhuman sialic acid, *N*-glycolylneuraminic acid (Neu5Gc), facilitates formation of antibodies that cross-react with gangliosides containing the human sialic acid Neu5Ac.

Role of Glycans in the Histopathology of Alzheimer's Disease

Alzheimer's disease is a common primary degenerative dementia of humans, with an insidious onset and a progressive course. The ultimate diagnosis is made by postmortem histological examination of brain tissue, which shows characteristic amyloid plaques with neurofibrillary tangles that are associated with neuronal death. Several types of glycans have been implicated in the histopathogenesis of the lesions: O-GlcNAc and HS glycosaminoglycans. Paired helical filaments are major component of the neurofibrillary tangle. These are primarily composed of the microtubule-associated protein Tau, which is present in an abnormally hyperphosphorylated state. This hyperphosphorylated Tau no longer binds microtubules and self-assembles to form the paired helical filaments that may contribute to neuronal death. Normal brain Tau is known to be heavily modified by Ser(Thr)-linked O-GlcNAc, the dynamic and abundant post-translational modification that is often reciprocal to Ser(Thr) phosphorylation (Chapter 19). The hypothesis currently being investigated is that site-specific or stoichiometric changes in O-GlcNAc addition may modulate Tau function and may also play a part in the formation of paired helical filaments by allowing excessive phosphorylation. Inhibitors of O-GlcNAcase that cross the blood–brain barrier are now in clinical trials.

HS proteoglycans may also have an important role in amyloid plaque deposition as investigators have shown high-affinity binding between HS proteoglycans and the amyloid precursor, as well as with the A4 peptide derived from the precursor. In addition, a specific vascular HS proteoglycan found in senile plaques bound with high affinity to two amyloid protein precursors. Further studies to determine the pathological roles of HS proteoglycans in Alzheimer's disease are needed. Recently, GWASs showed a strong correlation between Alzheimer's disease and the higher expression of a truncated form of CD33 (Siglec-3), which is expressed in brain microglia and may be suppressing clearance of amyloid.

ONCOLOGY: ALTERED GLYCOSYLATION IN CANCER

Altered glycosylation is a universal feature of cancer cells, but only certain specific glycan changes are frequently associated with tumors (for details, see Chapter 47). Findings include (1) increased β1-6GlcNAc branching of N-glycans; (2) changes in the amount, linkage, and acetylation of sialic acids; (3) truncation of O-glycans, leading to expression of Tn and sialyl Tn antigens, as well as N-glycan truncation yielding paucimannosidic glycans; (4) expression of the nonhuman sialic acid, Neu5Gc, incorporated from dietary sources; (5) expression of sialylated Lewis structures and selectin ligands; (6) altered expression and enhanced shedding of glycosphingolipids; (7) increased expression of galectins and poly-*N*-acetyllactosamines; (8) altered expression of ABH(O) blood-group-related structures; (9) alterations in sulfation of glycosaminoglycans; (10) increased expression of hyaluronan; (11) increased expression of the enzyme that attaches GPI anchors to proteins; and (12) increased O-GlcNAcylation on many proteins. Some of these changes have been shown to have pathophysiological significance in model tumor systems, and some are targets for diagnostic and therapeutic approaches to cancer. For example, the principal serum diagnostic/prognostic indicator of adenocarcinomas of the pancreas and gastrointestinal tract is the biomarker known as "CA19-9" which is the tetrasaccharide sialyl Lewis a (sLe^a), a binding determinant for E-selectin; its isomer, sLe^x, is displayed on bone marrow progenitors and by engagement with marrow microvessel E-selectin mediates hematopoiesis and leukemogenesis.

PULMONARY MEDICINE

Role of Selectins, Siglecs, and Mucins in Bronchial Asthma

Asthma is characterized by sporadic recurrent hyperresponsiveness of the tracheobronchial tree to various stimuli, resulting in widespread narrowing of the airways. The two dominant pathological features are airway wall inflammation and luminal obstruction of the airways by inflammatory exudates, consisting predominantly of mucins. Most cases are due to antigen-specific IgE antibodies, which bind to mast cells as well as to basophils and certain other cell types. Antigens can cross-link adjacent IgE molecules, triggering an explosive release of vasoactive, bronchoactive, and chemotactic agents from mast cell granules into the extracellular milieu. Eosinophils also contribute to the pathogenesis of asthma in several ways, by synthesizing leukotrienes, stimulating histamine release from mast cells and basophils, providing a positive feedback loop, and releasing major basic protein, a granule-derived protein that has toxic effects on the respiratory epithelium. Underlying all this, it appears that $CD4^+$ Th2 cells are responsible for orchestrating the responses of other cell types. Recent evidence indicates that the selectins are intimately involved in recruitment of eosinophils and basophils (and possibly T lymphocytes) into the lung, raising the hope that small-molecule inhibitors of selectin function and/or heparin can be used to treat the early stages of an asthmatic attack. Likewise, chemokine interactions with HS are important in leukocyte trafficking. Evidence from Siglec-F knockout mice also indicates that the functionally equivalent human paralog Siglec-8 is a good target for reducing the contributions of eosinophils to the pathology (Chapter 35). Finally, the large increase in mucus production is at least partly mediated by an up-regulation of synthesis of mucin polypeptides, under the influence of various cytokines that stimulate the goblet cells of the airway epithelium.

Role of Selectins in Acute Respiratory Distress Syndrome

Shock, trauma, or sepsis can all cause acute respiratory distress syndrome. Diffuse pulmonary endothelial injury causes pulmonary edema because of increased capillary permeability. Selectins and integrins help arrest neutrophils on the injured endothelium where they release injurious oxidants, proteolytic enzymes, and arachidonic acid metabolites, resulting in endothelial cell dysfunction and destruction. Bronchoalveolar lavage is rich in neutrophils and their secreted products documenting the inflammatory response. Here again, giving small molecule selectin inhibitors and/or heparin before serious lung damage and respiratory failure is a possible therapy.

Altered Glycosylation of Epithelial Glycoproteins in Cystic Fibrosis

Cystic fibrosis is a common genetic disorder caused by a mutation in the cystic fibrosis transmembrane conductance regulator (*CFTR*). This causes defective chloride conduction across the apical membrane of involved epithelial cells. Cystic fibrosis is associated with increased accumulation of viscous mucins in the pancreas, gut, and lungs, which leads to many symptoms of the disease. There are known to be widespread increases in sialylation, sulfation, and fucosylation of mucin glycoproteins. One possible explanation is that the primary CFTR defect allows a higher Golgi pH, resulting in abnormalities in glycosylation: however, there is controversy about this conclusion. Curiously, the CFTR is mainly expressed within nonciliated epithelial cells, duct cells, and serous cells of the tubular glands, but not highly expressed in the goblet cells and mucous glands of the acinar cells, which are the cells that synthesize respiratory mucins. Thus, the *CFTR* mutation may indirectly affect mucin glycosylation through the generation of inflammatory responses and/or changes in pH or chloride secretion. Another major cause of morbidity in the disease is the colonization of respiratory epithelium by an alginate-producing form of *Pseudomonas aeruginosa*. Certain glycolipids and mucin glycans have been suggested to be the *Pseudomonas* receptors that help to maintain the colonization. The changes in glycolipid and mucin glycosylation could enhance production of potential binding targets for organ colonization by the bacteria. The presence of bacterial products is also a proinflammatory condition, because the bacterial capsular polysaccharides may activate Toll-like receptors and lead eventually to neutrophil accumulation and organ damage.

Altered Glycosylation in Pulmonary Vascular Disease

Pulmonary vascular diseases include pulmonary embolism, pulmonary arterial hypertension (PAH), and arteriovenous malformations. These diseases increase pulmonary vascular resistance and pulmonary arterial pressure, ultimately leading to right ventricular hypertrophy and heart failure. PAH is the best studied—a progressive disease that shows nitric oxide deficiency, vasoconstriction, thrombosis, and enhanced vascular remodeling. Among other factors, dysregulated glucose metabolism may drive an increased flux into the hexosamine biosynthetic pathway. This results in PAH patients having increased hyaluronan in lung tissue, plasma, and pulmonary arterial smooth muscle cells. It also increases O-GlcNAc-modified proteins, a process shown to regulate pulmonary arterial smooth muscle cell proliferation associated with PAH progression, suggesting a potential therapeutic target.

ACKNOWLEDGMENTS

The authors appreciate helpful comments and suggestions from Morten Thaysen-Andersen and Priya Umapathi.

FURTHER READING

Because of the wide range of topics covered in this chapter, it is not feasible to provide literature citations for all of them. Some examples are provided but the reader should consult references at the end of the other cited chapters.

Varki NM, Varki A. 2007. Diversity in cell surface sialic acid presentations: implications for biology and disease. *Lab Invest* **87**: 851–857. doi:1010.1038/labinvest.3700656

Janssen MJ, Waanders E, Woudenberg J, Lefeber DJ, Drenth JP. 2010. Congenital disorders of glycosylation in hepatology: the example of polycystic liver disease. *J Hepatol* **52**: 432–440. doi:1010.1016/j.jhep.2009.12.011

Yuki N, Hartung H-P. 2012. Guillain–Barré syndrome. *N Engl J Med* **366**: 2294–2304. doi:1010.1056/nejmra1114525

Grewal PK, Aziz PV, Uchiyama S, Rubio GR, Lardone RD, Le D, Varki NM, Nizet V, Marth JD. 2013. Inducing host protection in pneumococcal sepsis by preactivation of the Ashwell–Morell receptor. *Proc Natl Acad Sci* **110**: 20218–20223. doi:1010.1073/pnas.1313905110

Ju T, Wang Y, Aryal RP, Lehoux SD, Ding X, Kudelka MR, Cutler C, Zeng J, Wang J, Sun X et al. 2013. Tn and sialyl-Tn antigens, aberrant O-glycomics as human disease markers. *Proteomics Clin Appl* **7**: 618–631. doi:1010.1002/prca.201300024

Lillehoj EP, Kato K, Lu W, Kim KC. 2013. Cellular and molecular biology of airway mucins. *Int Rev Cell Mol Biol* **303**: 139–202. doi:1010.1016/b978-0-12-407697-6.00004-0

Suh JH, Miner JH. 2013. The glomerular basement membrane as a barrier to albumin. *Nat Rev Nephrol* **9**: 470–477. doi:1010.1038/nrneph.2013.109

Swiecicki PL, Hegerova LT, Gertz MA. 2013. Cold agglutinin disease. *Blood* **122**: 1114–1121. doi:1010.1182/blood-2013-02-474437

Ehre C, Ridley C, Thornton DJ. 2014. Cystic fibrosis: an inherited disease affecting mucin-producing organs. *Int J Biochem Cell Biol* **52**: 136–145. doi:1010.1016/j.biocel.2014.03.011

Hayes JM, Cosgrave EF, Struwe WB, Wormald M, Davey GP, Jefferis R, Rudd PM. 2014. Glycosylation and Fc receptors. *Curr Top Microbiol Immunol* **382**: 165–199. doi:1010.1007/978-3-319-07911-0_8

Morawski M, Filippov M, Tzinia A, Tsilibary E, Vargova L. 2014. ECM in brain aging and dementia. *Prog Brain Res* **214**: 207–227. doi:1010.1016/b978-0-444-63486-3.00010-4

Reily C, Ueda H, Huang ZQ, Mestecky J, Julian BA, Willey CD, Novak J. 2014. Cellular signaling and production of galactose-deficient IgA1 in IgA nephropathy, an autoimmune disease. *J Immunol Res* **2014**: 197548. doi:1010.1155/2014/197548

Zhang GL, Zhang X, Wang XM, Li JP. 2014. Towards understanding the roles of heparan sulfate proteoglycans in Alzheimer's disease. *Biomed Res Int* **2014**: 516028. doi:1010.1155/2014/516028

Ghosh S, Hoselton SA, Dorsam GP, Schuh JM. 2015. Hyaluronan fragments as mediators of inflammation in allergic pulmonary disease. *Immunobiology* **220**: 575–588. doi:1010.1016/j.imbio.2014.12.005

Lauer ME, Dweik RA, Garantziotis S, Aronica MA. 2015. The rise and fall of hyaluronan in respiratory diseases. *Int J Cell Biol* **2015**: 712507. doi:1010.1155/2015/712507

McEver RP. 2015. Selectins: initiators of leucocyte adhesion and signalling at the vascular wall. *Cardiovasc Res* **107**: 331–339. doi:1010.1093/cvr/cvv154

Packman CH. 2015. The clinical pictures of autoimmune hemolytic anemia. *Transfus Med Hemother* **42**: 317–324. doi:1010.1159/000440656

Pilzweger C, Holdenrieder S. 2015. Circulating HMGB1 and RAGE as clinical biomarkers in malignant and autoimmune diseases. *Diagnostics* **5**: 219–253. doi:1010.3390/diagnostics5020219

Taniguchi N, Takahashi M, Kizuka Y, Kitazume S, Shuvaev VV, Ookawara T, Furuta A. 2016. Glycation vs. glycosylation: a tale of two different chemistries and biology in Alzheimer's disease. *Glycoconj J* **33**: 487–497. doi:1010.1007/s10719-016-9690-2

Bakchoul T. 2017. An update on heparin-induced thrombocytopenia: diagnosis and management. *Expert Opin Drug Saf* 7: 1–11. doi:10.1517/14740338.2016.1165667

Luzzatto L. 2016. Recent advances in the pathogenesis and treatment of paroxysmal nocturnal hemoglobinuria. *F1000Res* **5** (F1000 Faculty Rev): 209. doi:1010.12688/f1000research.7288.1

Pandolfi F, Altamura S, Frosali S, Conti P. 2017. Key role of DAMP in inflammation, cancer, and tissue repair. *Clin Ther* **38**: 1017–1028. doi:1010.1016/j.clinthera.2016.02.028

Schulz C, Schütte K, Malfertheiner P. 2017. *Helicobacter pylori* and other gastric microbiota in gastroduodenal pathologies. *Dig Dis* **34**: 210–216. doi:1010.1159/000443353

Silva M, Videira PA, Sackstein R. 2018. E-Selectin ligands in the human mononuclear phagocyte system: implications for infection, inflammation, and immunotherapy. *Front Immunol* **8**: 1878. doi:10.3389/fimmu.2017.01878 (eCollection).

Esposito M, Mondal N, Greco TM, Wei Y, Spadazzi C, Lin SC, Zheng H, Cheung C, Magnani JL, Lin SH et al. 2019. Bone vascular niche E-selectin induces mesenchymal–epithelial transition and Wnt activation in cancer cells to promote bone metastasis. *Nat Chem Biol* **21**: 627–639. doi:1010.1038/s41556-019-0309-2

Stowell SR, Stowell CP. 2019. Biologic roles of the ABH and Lewis histo-blood group antigens part II: thrombosis, cardiovascular disease and metabolism. *Vox Sang* **114**: 535–552. doi:1010.1111/vox.12786

Barbier V, Erbani J, Fiveash C, Davies JM, Tay J, Tallack MR, Lowe J, Magnani JL, Pattabiraman DR, Perkins AC et al. 2020. Endothelial E-selectin inhibition improves acute myeloid leukaemia therapy by disrupting vascular niche–mediated chemoresistance. *Nat Commun* **11**: 2042. doi:1010.1038/s41467-020-15817-5

Clausen TM, Sandoval DR, Spliid CB, Pihl J, Perrett HR, Painter CD, Narayanan A, Majowicz SA, Kwong EM, McVicar RN et al. 2020. SARS-CoV-2 infection depends on cellular heparan sulfate and ACE2. *Cell* **183**: 1043–1057. doi:1010.1016/j.cell.2020.09.033

Lee-Sundlov MM, Stowell SR, Hoffmeister KM. 2020. Multifaceted role of glycosylation in transfusion medicine, platelets, and red blood cells. *J Thromb Haemost* **18**: 1535–1547. doi:1010.1111/jth.14874

Smith BAH, Bertozzi CR. 2021. The clinical impact of glycobiology: targeting selectins, Siglecs and mammalian glycans. *Nat Rev Drug Discov* **20**: 217–243. doi:1010.1038/s41573-020-00093-1

47 Glycosylation Changes in Cancer

Susan L. Bellis, Celso A. Reis, Ajit Varki, Reiji Kannagi, and Pamela Stanley

Altered glycosylation is a universal feature of cancer cells, and certain glycans are well-known markers of tumor progression. This chapter discusses glycan biosynthetic pathways that are altered in cancer cells; correlations between altered glycosylation, diagnosis, and clinical prognosis; the genetic bases of some of these changes; and the functional role of glycans in cancer biology and pathogenesis.

HISTORICAL BACKGROUND

The earliest evidence of altered glycosylation as a hallmark of cancer was that some plant lectins show enhanced binding and agglutination of tumor cells. Next, it was found that transformation of cultured cells is frequently accompanied by a general increase in the size of glycopeptides of cell-surface glycoproteins. With the advent of monoclonal antibody technology, many "tumor-

published by Cold Spring Harbor Laboratory Press; doi: 10.1101/glycobiology.4e.47

specific" antibodies directed against glycan epitopes were found. In many instances, these epitopes represented "oncofetal antigens"—that is, glycan epitopes expressed on tumor cells and embryonic tissues. Like normal cells during embryogenesis, tumor cells also undergo rapid growth and invade tissues. Correlations between certain types of altered glycosylation and the prognosis of tumor-bearing animals or patients increased interest in glycan changes. In vitro cellular assays and in vivo animal studies have now supported the conclusion that glycan changes are critical to several aspects of tumor cell behavior.

GLYCOSYLATION CHANGES IN CANCER ARE NONRANDOM

Glycan changes in malignant cells take a variety of forms: loss of expression or excessive expression of certain glycans, increased expression of incomplete or truncated glycans, and, less commonly, the appearance of novel glycans. However, this is not simply the random consequence of disordered biosynthesis in tumor cells. It is striking that a very limited subset of changes are correlated with malignant transformation and tumor progression, highlighting a potential functional role in tumor biology. Given that cancer is a "microevolutionary" process in which only the fittest cells survive, and that tumors are under immune surveillance pressure, it is likely that these specific glycan changes are selected for during tumor progression.

ALTERED BRANCHING AND FUCOSYLATION OF N-GLYCANS

Classic reports of increased size of tumor cell glycopeptides have now been partly explained by an increase in β1-6 branching of N-glycans (Figure 47.1), resulting from enhanced expression of N-acetylglucosaminyltransferase V (GnT-V, MGAT5) (Chapter 9). Increased transcription of the *MGAT5* gene is induced by various oncogenic transcription factors, as well as viral- and chemical-induced carcinogenesis. Cells with up-regulated MGAT5 show an increased frequency of metastasis in mice, and spontaneous revertants lacking MGAT5 lose the metastatic phenotype. Overexpression of MGAT5 in cultured cells causes a transformed phenotype, whereas *Mgat5*-deficient mice show a striking reduction in the growth and metastasis of mammary tumors induced by a viral oncogene. Increased expression of MGAT4 to form a β1-4 branched tetra-antennary N-glycan also enhances tumor progression. Possible mechanisms by which increased N-glycan branching and extension enhances progression include lattice formation via galectin binding to poly-N-acetyllactosamines (LacNAc) leading to prolonged growth factor signaling, and the generation of sialyl-Lewis x (SLe^x) tetrasaccharides recognized by selectins because of increased outer-chain sialylation and fucosylation (Figure 47.1). Increased expression of FUT8 that transfers the core Fuc to N-glycans (Figure 47.1) is also observed in solid tumors and promotes tumor progression in lung cancer and melanoma. The expression of N-acetylglucosaminyltransferase III (GnT-III, MGAT3) that catalyzes the transfer of β1-4GlcNAc to Man on N-glycans to form the bisecting GlcNAc (Figure 47.1) is up-regulated in rat hepatomas and mouse mammary tumors. However, in this case, tumor progression is suppressed by the presence of N-glycans with the bisecting GlcNAc. Mice lacking MGAT3 show increased mammary tumors and lung metastases, whereas high MGAT3 expression correlates with better relapse-free survival in human breast cancer. However, liver tumors are reduced in mice lacking MGAT3, and MGAT3 activity endows ovarian cancer cells with cancer stem cell features. Thus, the effects of MGAT3 on cancer progression depend on tissue context.

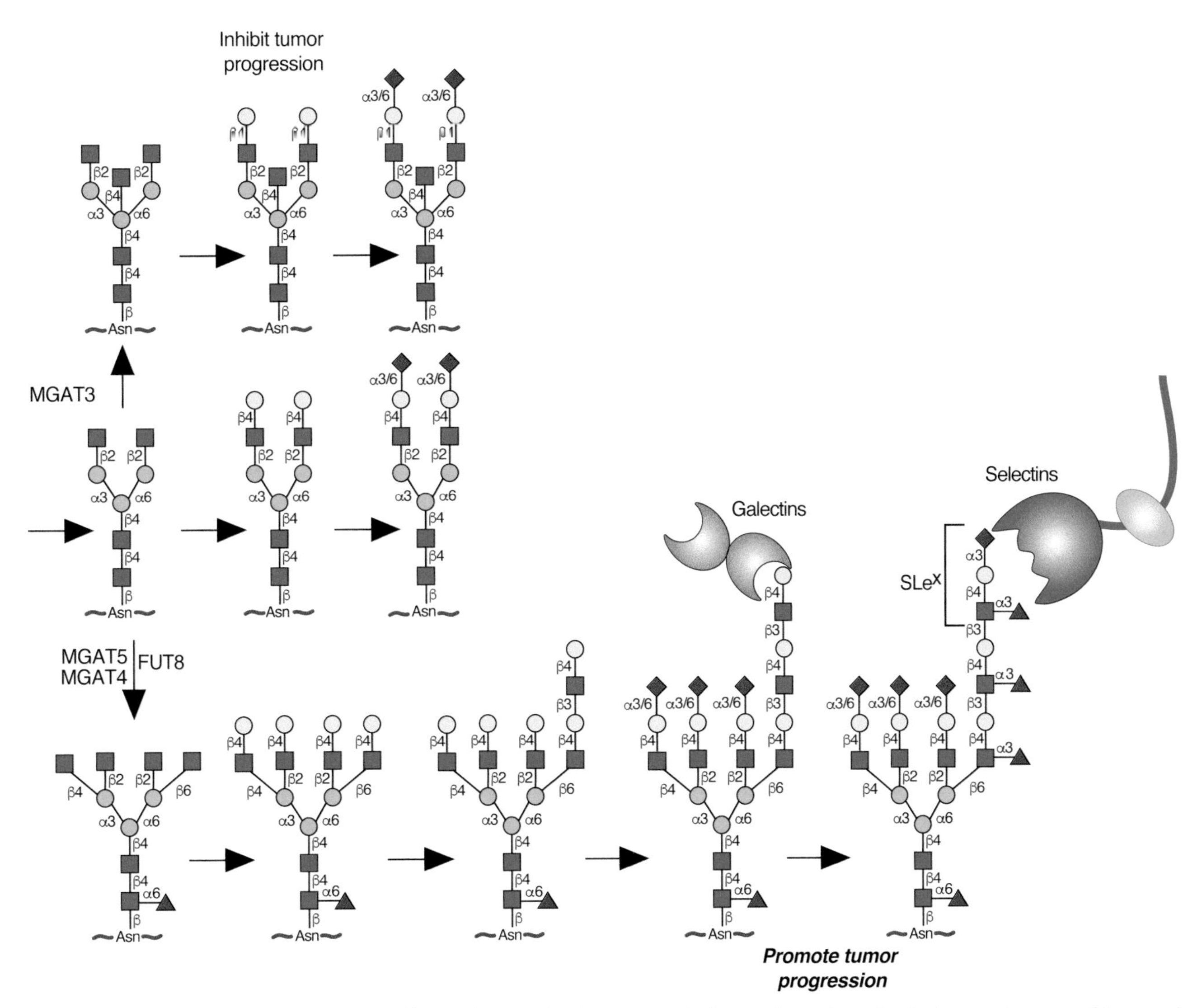

FIGURE 47.1. N-Glycans increase in size on neoplastic transformation of cells, in part because of increased MGAT4 and MGAT5 activity, which catalyzes GlcNAc branching of N-glycans. This may lead to increased numbers of LacNAc units, which can also be sialylated and fucosylated. The resulting glycans may be recognized by galectins and selectins. Increased FUT8, which transfers the core α1-6Fuc to N-glycans, also enhances tumor progression. In contrast, increased MGAT3, which catalyzes transfer of the bisecting GlcNAc to N-glycans, may suppress tumor progression, in part by inhibiting galectin binding.

ALTERED MUCIN EXPRESSION AND TRUNCATED O-GLYCANS

Mucins are large glycoproteins rich in O-GalNAc glycans on Ser or Thr in tandem repeat regions (Chapter 10). In normal polarized epithelium, mucins are expressed in the apical membrane, facing the lumen of a hollow organ, and soluble mucins are secreted exclusively into the lumen. Loss of adhesion junctions and topology in malignant epithelial cells destroys polarization, allowing mucins to enter the extracellular space and the blood. Loss of polarization and overexpression of mucins displaying aberrant glycosylation has made these glycoproteins a major source of biomarkers with clinical applications. The rod-like structure and negative charge of mucins are thought to repel intercellular interactions and sterically inhibit adhesion molecules such as cadherins and integrins from carrying out their functions. Thus, mucins may act as "antiadhesins" to promote displacement of a cell from the primary tumor during the initiation of metastasis. Tumor mucins bearing selectin ligands facilitate several aspects of cancer progression (see below). Mucins might also interfere with immune cell recognition and block or mask presentation of antigenic peptides by major histocompatibility complex (MHC) molecules.

A hallmark feature of carcinomas is the aberrant synthesis of incomplete O-glycans, present in mucins and other glycoproteins, typified by the Tn and T antigens and their sialylated glycoforms, sialyl-Tn (STn) and sialyl-T (ST) (Figure 47.2). Secreted glycoproteins expressing STn often appear in the bloodstream of patients with cancer. Because these glycans occur infrequently in normal tissues, they serve as prognostic biomarkers and therapeutic targets. Truncated O-glycans evoke immune responses in the patient, and accordingly, vaccines against these glycans have been developed. As an alternative, incomplete O-glycans can serve as targets for a variety of therapeutic antibodies and chimeric antigen receptor (CAR) T-cell therapies. The appearance of excessive Tn and STn on tumor cells correlates in some instances with silencing of the COSMC gene (*C1GALT1C1*). The β1-3 galactosyltransferase (C1GALT1) that synthesizes Galβ1-3GalNAcα1-Ser/Thr requires the chaperone C1GALT1C1 for activity. The gene for *C1GALT1C1* is on the X chromosome, thus a single mutation may be sufficient to eliminate expression. In other instances, STn accumulation results from overexpression of the sialyltransferase ST6GALNAC1 (Figure 47.2). The expression of Tn and STn can enhance tumorigenic and invasive properties, and promote immunosuppression.

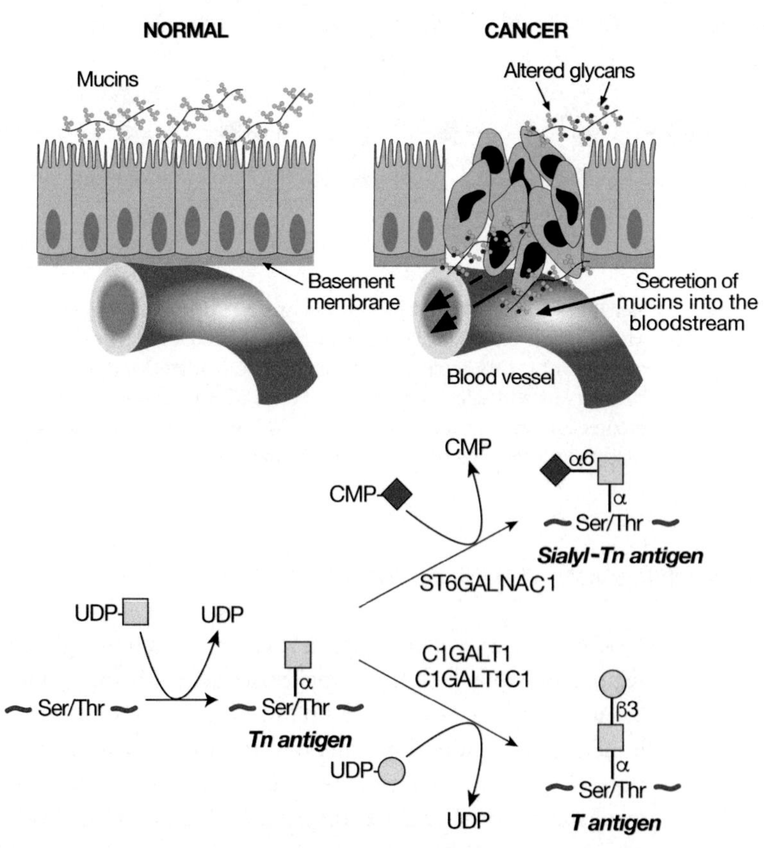

FIGURE 47.2. Loss of normal topology and polarization of epithelial cells in cancer results in secretion of mucins with truncated O-GalNAc glycans, such as sialyl-Tn (STn) and Tn, into the bloodstream. Tumor cells invading tissues and the bloodstream also present mucins on their cell surfaces.

ALTERED SIALIC ACID EXPRESSION

Sialylation is generally increased in tumor cells, both on N-glycans and O-glycans. N-glycan sialylation serves as a regulatory mechanism for multiple receptor tyrosine kinases, illustrated by activation of MET and RON receptors by α2-3Sia. Increased α2-6Sia on N-glycans, due to *ST6GAL1* up-regulation in cancer cells, enhances integrin-mediated cell motility and protects cells against apoptosis induced by galectins, death receptor ligands, and chemotherapeutic drugs. Additionally, some receptor glycoproteins and their downstream signaling pathways are dysregulated by addition of SLe^x and sialyl-Lewis a (SLe^a) structures on either N- or O-glycans. Other functions for tumor cell Sias include binding to factor H to limit complement activation and modulation of tumor cell attachment to matrix, facilitating invasion and metastasis. As well, tumor cell Sias play a prominent role in suppressing antitumor immunity through engagement of inhibitory Siglecs (Chapter 35), which are Sia-recognizing receptors expressed primarily on immune cells. There can also be changes in Sia modifications in tumor cells. Sia 9-O-acetylation can be increased (e.g., 9-O-acetylated GD3 [Figure 47.3] in melanoma cells) or decreased (e.g., on the O-glycans of colon carcinomas). Some tumor cells express small amounts of de-N-acetyl (deNAc) gangliosides (Figure 47.3). O-Acetylation of gangliosides appears to protect tumor cells from apoptosis, and de-N-acetylgangliosides may activate epidermal growth factor receptor (EGFR).

Many years ago, cancer patients were reported to express “Hanganutziu–Deicher” antibodies, which were later shown to recognize gangliosides carrying the nonhuman sialic acid N-glycolylneuraminic acid (Neu5Gc), which was also found in human tumors. This unusual phenomenon has now been explained by the metabolic incorporation of diet-derived Neu5Gc into

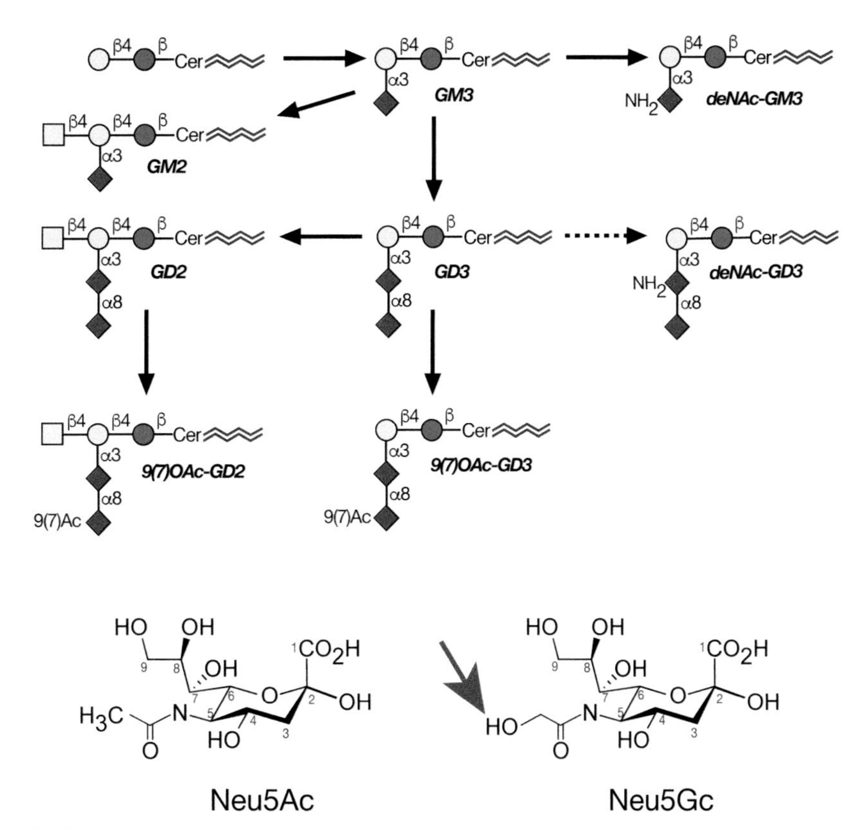

FIGURE 47.3. Gangliosides expressed in human neuroectodermal tumors. *Heavy arrows* indicate up-regulated reactions; *dashed arrow* indicates a potential reaction. O-Acetylation of Sia can occur at the 7- or 9-position. An O-acetyl group at position 7 will migrate to position 9 under physiological conditions. *Red arrow* identifies the difference between Neu5Ac and Neu5Gc.

human glycans (Chapter 15). This Sia differs from the common human Sia, Neu5Ac, by addition of a single oxygen atom (Figure 47.3). Humans show variable levels of circulating antibodies against Neu5Gc-containing epitopes. The resulting weak immune response (termed "xenosialitis") can promote tumor growth by enhancing chronic inflammation and angiogenesis. Given this selective advantage, it is not surprising that tumors are better at accumulating Neu5Gc in the face of an increased immune response. Because the primary dietary source of Neu5Gc is red meat, this may help explain the increased cancer risk associated with its consumption.

INCREASED SELECTIN LIGAND EXPRESSION

SLe^x and SLe^a epitopes (Chapter 14) were first identified as tumor antigens on glycosphingolipids. Expression of these antigens by epithelial carcinomas correlates with metastatic potential in mice and with tumor progression, metastatic spread, and poor prognosis in humans. SLe^x and SLe^a epitopes on glycoprotein ligands are key recognition determinants of the selectins (Chapter 34). Indeed, selectin ligands are expressed on carcinoma cells, and mucin-like tumor antigens carrying SLe^x and SLe^a are found in the blood of carcinoma patients (Figure 47.4). Transgenic overexpression of E-selectin in mouse liver causes carcinoma cells that would normally metastasize to the lung to be redirected toward colonization of the liver, supporting the concept that SLe/selectin interactions are important mediators of metastasis. Furthermore, metastasis is attenuated in mice lacking P-selectin or L-selectin or by administering heparin, which blocks binding by these selectins. Selectin interactions also help explain the classic observation that cancer cells entering the bloodstream form thromboemboli with platelets and leukocytes, which facilitate arrest in the vasculature, assist extravasation through the endothelium, and help in evasion of the immune system. Similar interactions involving soluble cancer mucins

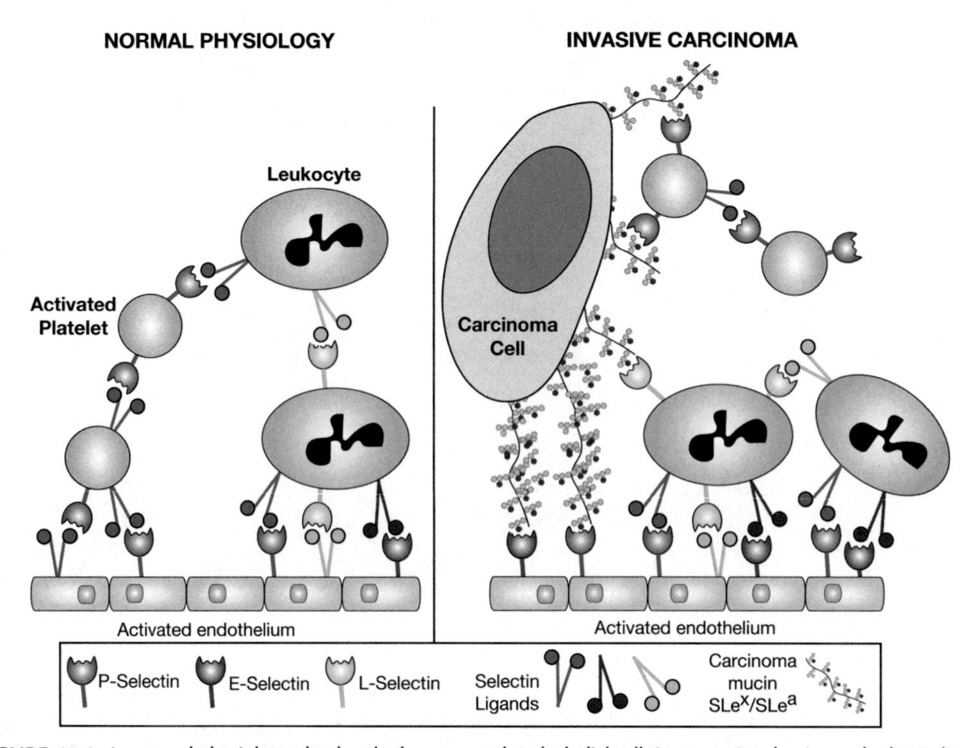

FIGURE 47.4. In normal physiology platelets, leukocytes, and endothelial cells interact via selectins and selectin ligands. In invasive carcinoma, interactions that promote invasion occur between tumor cells expressing selectin ligands and selectins on endothelial cells and platelets. (Modified from Stevenson JL, et al. 2005. *Clin Cancer Res* **11:** 7003–7011.)

may contribute to hypercoagulability (Trousseau's syndrome), a condition responsive to heparin treatment. Because of the prominent role of selectins in cancer progression, these receptors are major therapeutic targets (see below). Certain sialyl-Lewis-related structures may also influence carcinoma progression by interacting with Siglecs, which generally have immunosuppressive functions. For example, disialyl-Le^{a} and sialyl-6-sulfo-Le^{x} structures may protect against early carcinogenic events by binding to Siglec-7 on macrophages. This interaction suppresses macrophage production of the pro-oncogenic inflammatory mediator, Cox2, thereby exerting an anti-oncogenic function. However, during cancer development, many other sialylated glycan ligands for Siglecs are increased, and these bind to inhibitory Siglecs on various immune cell populations to induce immunosuppression and facilitate tumor progression.

ALTERED EXPRESSION OF BLOOD GROUPS

Loss of AB blood group expression in the context of carcinoma (accompanied by exposure of underlying H and Le^{y} epitopes) (Chapter 14) is associated with poor prognosis. The Sd^{a} (or Cad) antigen, a blood group glycan abundantly expressed in the colon, is lost in colon carcinoma. Sulfation of the C-3 position of terminal Gal residues is also reduced in cancers. Sialyl-6-sulfo-Lewis x and disialyl-Lewis a, expressed on colonic epithelial cells, are reduced in colon cancer cells. These changes may reflect the enhanced production of SLe^{x} and SLe^{a} in cancers. DNA methylation and histone deacetylation, epigenetic mechanisms for suppression of gene transcription commonly observed in cancers, are proposed to underlie these glycan alterations. There are also rare instances in which a tumor may present a "forbidden" blood group structure (i.e., expression of a B blood group antigen in an A-positive patient) (Chapter 14). Regardless of the underlying mechanism, tumor regression has been noted in a few such cases, presumably mediated by the naturally occurring endogenous antibodies directed against the foreign structure.

ALTERED EXPRESSION OF GLYCOSPHINGOLIPIDS

Many "tumor-specific" monoclonal antibodies raised against cancer cells recognize the glycan portion of glycosphingolipids. Some glycolipids are highly enriched in specific cancers (e.g., Gb3/CD77 in Burkitt's lymphoma and GM3, GD2, and GD3 in melanomas) (Figure 47.3). Several types of tumors (particularly melanoma and neuroblastoma) are characterized by the synthesis of very high levels of gangliosides (Chapter 11). Some of these (e.g., GD2) are not normally found at high levels in extraneural tissues and are therefore targets for both passive immunotherapy (monoclonal antibody infusion) and active immunotherapy (immunization with purified glycolipids). In some cases, gangliosides are also major carriers of modified Sias (Figure 47.3). Cell culture studies suggest that some gangliosides promote tumor cell growth and invasion. As a principal constituent of lipid raft membrane microdomains, gangliosides modulate cell signaling by numerous receptors. Additionally, gangliosides shed by some tumors appear to have immunosuppressive effects.

LOSS OF GPI-ANCHOR EXPRESSION

A complete loss of glycosylphosphatidylinositol (GPI)-anchored proteins is seen in some cases of malignant and premalignant states involving the hematopoietic system. This results from acquired mutations in hematopoietic stem cells in the *PIGA* gene (required for an early step in GPI-anchor biosynthesis [Chapter 12]). Conversely, some GPI-anchored proteins, such as members of the carcinoembryonic antigen family, like CEACAM5 (CEA), are overexpressed in cancers of the gastrointestinal tract, lung, breast, and female reproductive system, among others. Members of the CEACAM family have been implicated in tumor biology via interference with the integrin signaling pathway.

CHANGES IN HYALURONAN

Many classes of malignant tumors express high levels of hyaluronan, a large negatively charged polysaccharide composed of the repeating disaccharide unit [GlcAβ1-3GlcNAcβ1-4]$_n$ (Chapter 16). In carcinomas, hyaluronan is localized near the tumor cell surface, and is also enriched in tumor-associated stroma. In normal tissues, hyaluronan serves at least three functions, which may also contribute to tumor progression. First, it increases tissue hydration, facilitating cell movement through tissues. Second, it is intrinsic to the assembly of extracellular matrices through specific interactions with other macromolecules, and thus it participates in tumor cell–matrix interactions that facilitate or inhibit tumor cell survival and invasion. Finally, hyaluronan interacts with several types of cell-surface receptors, especially CD44, which mediate or modify cell signaling pathways. These interactions, notably with alternatively spliced isoforms of CD44 that are elevated in most cancer cells, are often crucial to tumor malignancy and are a current target for novel therapies.

In normal adult tissues, hyaluronan appears to be relatively inert with respect to cell signaling and behavior. However, during embryonic development, tissue regeneration, and in various pathologies, hyaluronan-CD44 signaling becomes activated, possibly because of limited cleavage of hyaluronan by hyaluronidases. The consequences of this signaling are dramatic because they promote cell proliferation, survival, epithelial–mesenchymal transition (EMT), and invasion, which are key elements of the malignant phenotype (Figure 47.5). Hyaluronan–CD44 interactions are required for the constitutive activation of some oncogenes, especially receptor tyrosine kinases such as *ERBB2*, which is amplified or mutated in numerous carcinomas. Accordingly, hyaluronan–CD44 interaction promotes downstream intracellular pathways that are also hallmarks of cancer, such as the phosphatidylinositol 3-kinase (PI3K)/AKT and mitogen-activated protein kinase pathways. Antagonists of hyaluronan–CD44 interaction cause inactivation of these pathways in malignant cells in culture and inhibit tumor growth and metastasis in animal models. Hyaluronan–CD44 interaction also stimulates multidrug resistance, and antagonists sensitize resistant cancer cells to chemotherapeutic drugs. In addition, interactions of hyaluronan with its receptors are important for the activities of several types of metabolic and multidrug transporters. These widespread effects may be due to stabilization of signaling platforms in the plasma membrane (e.g., lipid rafts) that are dependent on multivalent interactions of hyaluronan with CD44 or, in some cases, with another hyaluronan receptor (e.g., LYVE-1 or RHAMM/CD168) (Figure 47.5).

CHANGES IN SULFATED GLYCOSAMINOGLYCANS

Proteoglycans are comprised of core proteins decorated with negatively charged, sulfated glycosaminoglycan (GAG) side chains—namely, chondroitin sulfate (CS), dermatan sulfate (DS), keratan sulfate (KS), and heparan sulfate (HS) (Chapter 17). The content and distribution of many proteoglycans are altered during tumorigenesis, and the structurally and functionally diverse GAGs, particularly HS chains, have been implicated in tumor pathogenesis. HS chains are covalently linked to the core protein of proteoglycans (e.g., syndecans, glypicans, and perlecan). Other proteoglycans such as decorin, biglycan, versican, and lumican, carrying CS, DS, or KS chains, are also present in most tissues, including many tumor types.

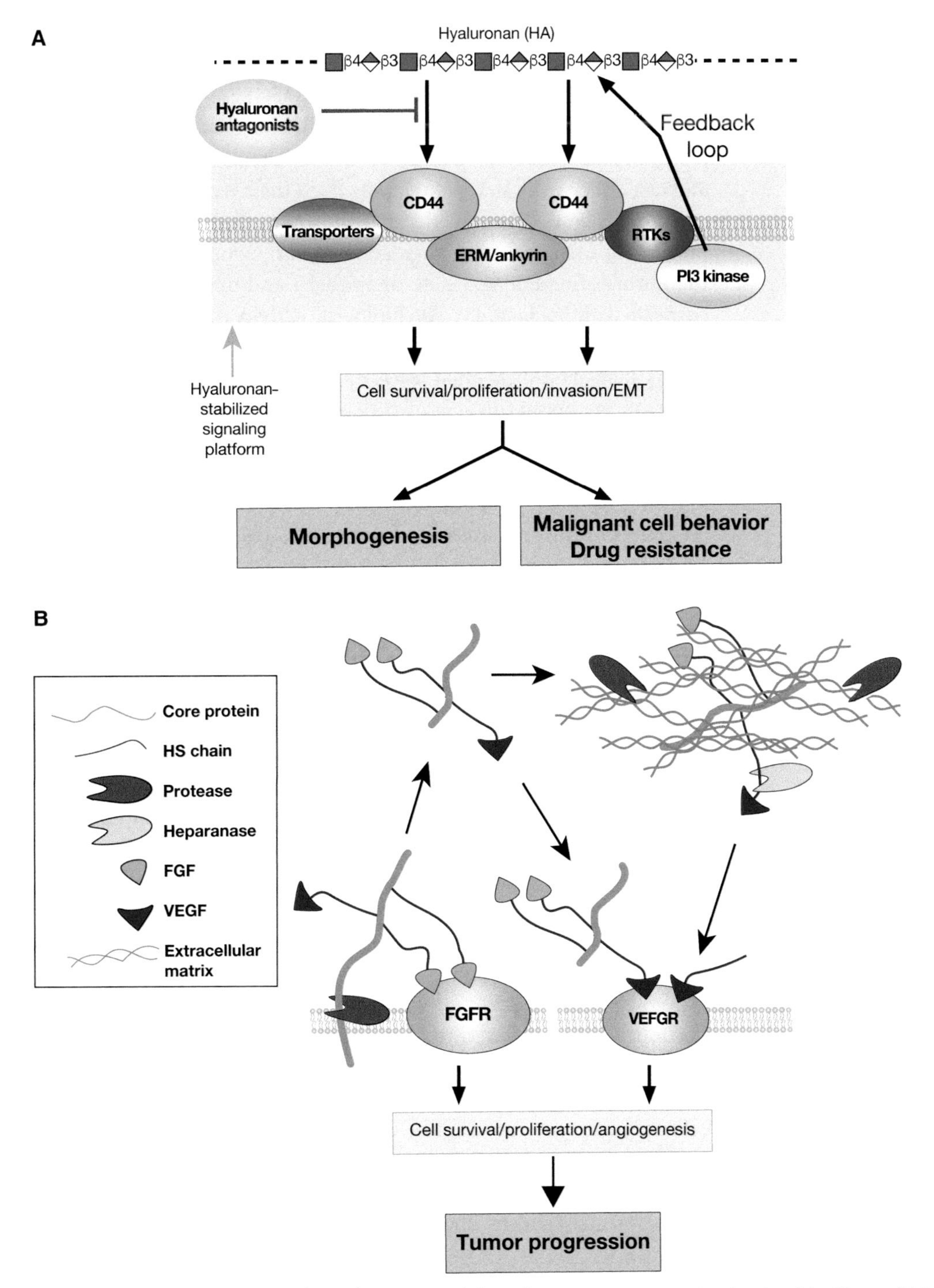

FIGURE 47.5. Glycosaminoglycans (GAGs) in cancer. (*A*) Hyaluronan interacts multivalently with CD44, which interacts with several types of signaling molecules in lipid rafts in the plasma membrane, thus promoting cell survival, proliferation, invasion, and epithelial–mesenchymal transition (EMT) in developing, regenerating, and malignant cells. In cancers, hyaluronan interactions can act cell-autonomously at the tumor cell surface or via hyaluronan produced by stromal cells. (*B*) Heparan sulfate (HS) proteoglycans present at the cell surface (e.g., syndecans and glypicans), or sequestered in the extracellular matrix (ECM) and basement membrane (e.g., perlecan), can be released by proteases in the microenvironment. Bioactive HS fragments can also be released by heparanase cleavage. The HS chains interact with several growth factors and amplify their interactions with their respective receptors on various cell types, thus resulting in increased angiogenesis or increased proliferation and invasion by tumor cells. However, HS proteoglycans can have inhibitory as well as stimulatory effects on tumor progression (e.g., via sequestration vs. release of HS-bearing growth factors from the ECM).

Major functions of HS proteoglycans relevant to tumor formation are to promote cell–cell and cell–matrix interactions important in tissue assembly and to bind a wide array of bioactive factors such as fibroblast growth factor-2 (FGF2), vascular endothelial growth factor (VEGF), hepatocyte growth factor, WNT, and numerous other cytokines and chemokines. Thus, HS proteoglycans can serve to build either inhibitory barriers or permissive pathways for cell invasion and can either sequester factors away from their receptors or present them efficaciously to those receptors (Figure 47.5). Syndecan-1 illustrates the multiple important roles of HS, especially in promoting metastasis and angiogenesis. Early work suggested that syndecan-1 helps maintain the normal differentiated state of epithelia and that low levels of tumor-associated syndecan-1 correlate with malignancy. The biological activity of syndecan-1 (and other HS proteoglycans) is influenced by enzymatic processing by proteases and heparanase, an enzyme that cleaves the HS chains at specific sites. High levels of proteolytically released HS-rich ectodomain fragments of syndecan are present in sera of patients with myeloma as well as some types of carcinoma and predict poor prognosis. Syndecan-1 ectodomains also accumulate in the tumor microenvironment, in which they play an important role in activation of both chemokine and growth factor signaling, and consequently tumor cell behavior. Especially significant is the finding that ectodomains promote metastasis to bone tissue, which is characteristic of myeloma and certain carcinomas (e.g., of breast or prostate) in human patients.

HS side chains of several HS proteoglycans are cleaved to fragments containing 10 to 20 sugar moieties by heparanase and endo-β-glucuronidase. Heparanase is elevated in numerous types of cancer, and increased heparanase activity can lead to induction of angiogenesis and metastasis, processes associated with a poor prognosis. HS cleavage promotes blood vessel remodeling required for angiogenesis and release of HS-bound angiogenic factors, growth factors, and chemokines. Furthermore, heparanase products are more bioactive compared with intact HS proteoglycan. Heparanase trimming of syndecan-1 HS chains and up-regulation of matrix metalloproteinase-9 (MMP-9) in myeloma cells results in enhanced shedding of syndecan-1, activating signaling pathways that promote cell migration and angiogenesis. Notably, in addition to their roles in regulating tumor angiogenesis and metastasis, heparanase and syndecans cooperate in regulating extracellular vesicle secretion by tumor cells. The therapeutic targeting of heparanase has potential to not only block tumor growth, but also interfere with the tumor microenvironment.

Along with HS cleavage, modifications in the sulfation pattern of HS, and other GAG chains, are important contributors to tumor biology. GAG sulfation patterns often determine their biological function and serve as specific recognition motifs for a variety of growth factors, cytokines, and chemokines. Many enzymes involved in GAG sulfation are altered in malignant cells. Changes in expression of CS 4-O-sulfotransferases are noted in both ovarian and breast cancers, whereas cancerous lung tissues display elevated 6-O-sulfated CS, when compared to nonmalignant tissue. Additionally, increased expression of 6-O-sulfotransferases in ovarian and colorectal cancers, and 3-O-sulfotransferases in breast and pancreatic cancers, have been reported.

CHANGES IN CYTOPLASMIC AND NUCLEAR O-GlcNAc

O-GlcNAc levels, regulated by the O-GlcNAc transferase OGT and the hexosaminidase OGA, are altered in many cancers. O-GlcNAcylation occurs outside of the secretory pathway and is prevalent on numerous cytosolic, nuclear, and mitochondrial proteins, as well as the intracellular domains of some transmembrane surface proteins. O-GlcNAc cycling serves as a nutrient sensor to regulate signaling, transcription, mitochondrial activity, and cytoskeletal functions. There is often cross-talk between O-GlcNAcylation and phosphorylation, not only at the same or proximal sites on polypeptides, but also through the regulation of enzymes that control cycles of these modifications. O-GlcNAcylation can influence cancer cell metabolism because of alterations in the stability or activity of transcription factors and kinases. O-GlcNAcylation affects phosphofructokinase 1 activity and redirects glucose flux through the pentose phosphate pathway. This metabolic switch can confer a growth advantage on cancer cells. Additionally, O-GlcNAcylation regulates glycolysis in cancer cells via the hypoxia-inducible factor-1α (HIF-1α) transcription factor. Many oncogene and tumor suppressor gene products, including c-Myc, cyclin D1, NF-κB/p65, PFK1, SNAIL, Rb, and p53, are O-GlcNAcylated.

MECHANISMS OF ALTERED GLYCAN EXPRESSION

The Cancer Genome Atlas (TCGA) has identified many alterations in expression of glycosylation genes that correlate with cancers. The mechanistic basis is known in some cases. For instance, the transcription of *MGAT5* (Figure 47.1) is induced by v-*src*, H-*ras*, and v-*fps* and the transcription factor *ets-1*. *ST6GAL1* is likewise up-regulated by oncogenic ras isoforms, as well as through gene amplification. The expression of *FUT7*, which generates SLe^x in leukocytes, is induced by the transcriptional activator protein Tax of HTLV-1, which causes leukemia. The consequent increase in SLe^x expression may underlie the strong tissue infiltrative capacity of adult T-cell leukemia cells, likely mediated by selectins. Under the hypoxic conditions found in advanced tumors, hypoxia-resistant cancer cells survive by up-regulating HIF-1α. HIF-1α induces transcription of several glycosylation genes, leading to altered tumor glycans, including enhanced SLe^x and SLe^a expression. Tumor hypoxia may also affect the incorporation of diet-derived nonhuman Neu5Gc in human tumor cells by up-regulating transcription of the lysosomal sialin Sia transporter. Down-regulation of glycosylation genes by DNA methylation or histone deacetylation/trimethylation may silence glycosylation genes that appear to suppress tumor progression in some types of cancers, like *MGAT3* or *ST6GAL1*. Posttranscriptional regulation of glycosylation pathways through microRNAs is also involved in tumor progression.

GLYCAN CHANGES IN CANCER STEM CELLS AND DURING EMT

Cancer stem cells, or tumor-initiating cells, constitute a small subpopulation of cancer cells that has tumor-initiating capability. Several glycans that are specific markers for embryonic stem cells (stage-specific embryonic antigen-3 [SSEA-3], SSEA-3 with fucose [Globo H], and SSEA-4) are also expressed by cancer stem cells. SSEA-1, an embryonic stem cell marker in mice, is found in cancer stem cells in human gliomas. Thus the expression of these glycans appears to be associated with the "stemness" of cells. Other cancer stem cell markers include the glycoproteins, CD133 (prominin-1), CD24, and CD44, all of which are regulated by their glycosylation status. Among these receptors, the activation of CD44 by its ligand, hyaluronan, is particularly important for maintaining cancer stem cell features.

Cancer stem cells are often investigated in the context of EMT. Cells that have undergone EMT are highly similar to cancer stem cells. EMT is a critical event in tumor progression that prepares cancer cells for metastasis. It is governed by several well-defined transcription factors such as SNAIL and ZEB. EMT induces the differential expression of a subset of glycosyltransferases. Cancer cells with a mesenchymal phenotype up-regulate ST6GAL1 and MGAT5, while down-regulating MGAT3. The corresponding changes in N-glycan branching and sialylation affect the stability and/or activity of many target molecules central to the process of EMT, such as cadherins and integrins. Other EMT-associated glycan modifications include the GD1 ganglioside, and decreased expression of Gg4 and GM2 glycolipids, along with increases in SLe^x

and SLe[a]. EMT is typified by alterations in cell adhesion and invasiveness; however, marked changes in cellular metabolism also occur, many of which are directed by O-GlcNAcylation. The addition of O-GlcNAc to E-cadherin, Snail, and other EMT-related proteins modulates protein stability or trafficking, leading ultimately to dysregulated expression of metabolic genes.

CLINICAL SIGNIFICANCE

Several glycan antigens are used for detecting and monitoring the growth status of tumors. A classical serological assay for CA19-9, which detects SLe[a], is used to monitor tumor burden, clinical response to therapy, and disease relapse in patients with gastric, colorectal, and pancreatic cancers. Other serological biomarkers include carcinoembryonic antigen (CEA), a highly glycosylated glycoprotein used in the early detection of recurrent or metastatic colorectal cancer, and CA125, which recognizes MUC16 and is used in monitoring ovarian cancer. In addition, the core-fucosylated α-fetoprotein (AFP-L3) can be a sensitive and specific circulating biomarker for the early diagnosis of hepatocellular carcinoma (HCC).

Truncated O-GalNAc glycans Tn, STn, and T (Figure 47.2) are good markers for the detection of cancer in tumor sections. Furthermore, the detection of specific glycoproteins bearing these truncated O-glycans, such as cancer-specific MUC16 glycoforms, enhances the specificity of diagnosis and monitoring of cancer. In addition, specific glycoproteins modified with truncated glycans, including Tn and STn, have been used as targets to generate glycopeptide-specific antibodies that have shown potential utility in clinical trials. Other important biomarkers include SLe[x]-related glycans, which are used in lung and breast cancer to monitor residual disease after surgery. In recent years, there has been a major interest in glycomic profiling of serum and other body fluid glycoproteins, in addition to glycoproteins in extracellular vesicles. Classic findings of glycan-specific antibodies associated with cancer, such as the antibodies against the nonhuman Sia, Neu5Gc, are also being reinvestigated. Another potential advance lies in improving the specificity of known cancer biomarkers by defining glycoprotein glycoforms uniquely expressed by cancer cells (e.g., prostate-specific antigen).

In tandem with their roles as biomarkers, glycans are important targets for cancer therapeutics. A monoclonal antibody against GD2 has been U.S. Federal Drug Administration (FDA)-approved for the treatment of pediatric neuroblastoma, and other GD2-targeting antibodies, as well as CAR-T cells, are in clinical trials for the treatment of neuroblastoma, glioma, melanoma, and other cancers. Vaccines against GD2 are also being explored. The Tn-bearing form of MUC1 is another promising target, with both therapeutic antibodies and CAR-T cells against Tn-MUC1 under active investigation. Additionally, several selectin inhibitors are showing efficacy in clinical trials, including E-selectin-targeting glycomimetic agents used in the treatment of multiple myeloma and acute myelogenous leukemia. Heparin has long been known to have potent antitumor effects, in part because of its ability to block P- and L-selectin binding to tumor and/or host ligands (Chapter 34). However, heparin therapy for cancer treatment can be limited by its anticoagulant properties, and thus low-molecular-weight heparins or heparinoids that lack such activity may be needed. Low-molecular-weight heparins may also inhibit tumor progression by blocking heparanase activity or by interfering with constitutive HS activities. Other heparanase inhibitors such as the sulfated phosphomannopentaose PI-88, which inhibits angiogenesis and metastasis, may work by blocking selectins or competitively inhibiting HS binding and function. Low-molecular-weight oligosaccharides of hyaluronan could also be useful therapeutically, because they inhibit the pro-oncogenic influences of constitutive polymeric hyaluronan, especially drug resistance and signaling events induced by hyaluronan–CD44 interaction. Disaccharides that can enter the cell and act as decoys to divert glycosylation pathways are also showing promise (Chapter 55).

Recently, there has been growing interest in targeting glycan-dependent molecular interactions to enhance immune checkpoint inhibition. N-glycosylation is important for stabilization of the PD-L1 checkpoint molecule, as well as recognition by PD-1. Thus, therapeutic interference with PD-L1 glycosylation could provide a mechanism for hindering the activation of PD-1 on immune cells. The sialoglycan–Siglec interaction represents another critical immune checkpoint, and function-blocking antibodies against select Siglecs have entered clinical trials. Additionally, methods are being developed to block or ablate tumor cell Sia ligands for Siglecs via administration of sialic acid mimetics or the targeted delivery of sialidases to tumor cell surfaces. Alternatively, Siglecs expressed by hematopoietic cancer cells (e.g., Siglec 2) can serve as recognition molecules for CAR-T cell immunotherapy or delivery of cytotoxic drugs. In general, the leveraging of glycans and glycan-binding proteins to modulate antitumor immunity holds much potential, given the well-known roles of these molecules in regulating the overall immune response.

ACKNOWLEDGMENTS

The authors appreciate contributions from Bryan Toole to the previous version and helpful comments and suggestions from Breeanna Urbanowicz, Melanie Simpson, and Koichi Furukawa to this update.

FURTHER READING

Feizi T. 1985. Demonstration by monoclonal antibodies that carbohydrate structures of glycoproteins and glycolipids are onco-developmental antigens. *Nature* **314:** 53–57. doi:10.1038/314053a0

Borsig L, Stevenson JL, Varki A. 2007. Heparin in cancer: role of selectin interactions. In *Cancer-associated thrombosis* (ed. AA Khorana, CW Francis), pp. 95–111. Taylor-Francis, Boca Raton, FL.

Kannagi R, Sakuma K, Miyazaki K, Lim KT, Yusa A, Yin J, Izawa M. 2010. Altered expression of glycan genes in cancers induced by epigenetic silencing and tumor hypoxia: clues in the ongoing search for new tumor markers. *Cancer Sci* **101:** 586–593. doi:10.1111/j.1349-7006.2009.01455.x

Schultz MJ, Swindall AF, Bellis SL. 2012. Regulation of the metastatic cell phenotype by sialylated glycans. *Cancer Metastasis Rev* **31:** 501–518. doi:10.1007/s10555-012-9359-7

Miwa HE, Koba WR, Fine EJ, Giricz O, Kenny PA, Stanley P. 2013. Bisected, complex N-glycans and galectins in mouse mammary tumor progression and human breast cancer. *Glycobiology* **23:** 1477–1490. doi:10.1093/glycob/cwt075

Knelson EH, Nee JC, Blobe GC. 2014. Heparan sulfate signaling in cancer. *Trends Biochem Sci* **39:** 277–288. doi:10.1016/j.tibs.2014.03.001

Ma Z, Vosseller K. 2014. Cancer metabolism and elevated *O*-GlcNAc in oncogenic signaling. *J Biol Chem* **289:** 34457–34465. doi:10.1074/jbc.r114.577718

Nabi IR, Shankar J, Dennis JW. 2015. The galectin lattice at a glance. *J Cell Sci* **128:** 2213–2219. doi:10.1242/jcs.151159

Pinho SS, Reis CA. 2015. Glycosylation in cancer: mechanisms and clinical implications. *Nat Rev Cancer* **15:** 540–555. doi:10.1038/nrc3982

Taniguchi N, Kizuka Y. 2015. Glycans and cancer: role of *N*-glycans in cancer biomarker, progression and metastasis, and therapeutics. *Adv Cancer Res* **126:** 11–51. doi:10.1016/bs.acr.2014.11.001

Alisson-Silva F, Kawanishi K, Varki A. 2017. Human risk of diseases associated with red meat intake: analysis of current theories and proposed role for metabolic incorporation of a non-human sialic acid. *Mol Aspects Med* **51:** 16–30. doi:10.1016/j.mam.2016.07.002

Groux-Degroote S, Guérardel Y, Delannoy P. 2017. Gangliosides: structures, biosynthesis, analysis, and roles in cancer. *ChemBioChem* **18:** 1146–1154. doi:10.1002/cbic.201600705

Pearce OM, Laübli H. 2017. Sialic acids in cancer biology and immunity. *Glycobiology* **26:** 111–128. doi:10.1093/glycob/cwv097

Steentoft C, Migliorini D, King TR, Mandel U, June CH, Posey AD Jr. 2018. Glycan-directed CAR-T cells. *Glycobiology* **28:** 656–669. doi:10.1093/glycob/cwy008

Furukawa K, Ohmi Y, Ohkawa Y, Bhuiyan RH, Zhang P, Tajima O, Hashimoto N, Hamamura K, Furukawa K. 2019. New era of research on cancer-associated glycosphingolipids. *Cancer Sci* **110:** 1544–1551. doi:10.1111/cas.14005

Mereiter S, Balmaña M, Campos D, Gomes J, Reis CA. 2019. Glycosylation in the era of cancer-targeted therapy: where are we heading? *Cancer Cell* **36:** 6–16. doi:10.1016/j.ccell.2019.06.006

48 Glycan-Recognizing Probes as Tools

Richard D. Cummings, Marilyn Etzler, Michael G. Hahn, Alan Darvill, Kamil Godula, Robert J. Woods, and Lara K. Mahal

Antibodies, lectins, microbial adhesins, viral agglutinins, and other proteins with carbohydrate-binding modules (CBMs), collectively termed glycan-recognizing probes (GRPs), are widely used in glycan analysis because their specificities enable them to discriminate among a diverse variety of glycan structures. The native multivalency of many of these molecules promotes high-affinity avidity binding to the glycans and cell surfaces containing those glycans. This chapter describes the variety of commonly used GRPs, the types of analyses to which they may be applied, and cautionary principles that affect their optimal use.

BACKGROUND

The first evidence that glycans were antigenic arose from the discovery of the human blood group ABO antigens (Chapter 14). A key tool in these studies were plant lectins that by the mid-1940s had found widespread use in typing blood. These lectins are relatively specific for blood types and they can be easily purified and are stable (Chapter 31). The discovery of the blood groups and the antibodies and lectins binding them indicated that such proteins could also be generally useful in identifying specific glycan sequences.

published by Cold Spring Harbor Laboratory Press; doi: 10.1101/glycobiology.4e.48

Hundreds of different plant and animal lectins and other proteins with CBMs have now been characterized. Thus, although monoclonal antibodies (mAbs) are often more specific for glycan determinants and bind with higher affinity, many plant and animal lectins and CBMs also have useful specificities for determinants beyond monosaccharides, have cloned sequences, are usually less expensive and commercially available, and have well-characterized binding specificities. GBPs are also found in many other organisms (Chapters 31, 36, and 37), and reagents from these organisms are currently in use in the field. The availability of GBPs and mAbs has helped to catapult the field of glycobiology into the modern era.

LECTINS MOST COMMONLY USED IN GLYCAN ANALYSIS

Many of the lectins currently used as tools in glycobiology originate from plants, but some also come from animals (e.g., snails) or mushrooms. Most were characterized initially by hapten inhibition assays, in which monosaccharides, their derivatives, or small oligosaccharides block binding to cells or other glycan-coated targets. Such small-sized molecules that compete with binding of a lectin or antibody to a larger ligand are termed haptens. Lectins are often grouped by specificity depending on the monosaccharide(s) that can inhibit their binding at millimolar concentrations and their distinct preference for α- or β-anomers of the sugar. However, lectins within a particular specificity group may also differ in their affinities for different glycans. The binding affinity (K_d) of lectins for complex glycans is often in the range of 1–10 μM, but for monosaccharides the affinity may be in the millimolar range. For complex glycoconjugates with multiple determinants or multivalency, the binding avidity of lectins may approach nanomolar values. For example, concanavalin A (ConA) is an α-mannose/α-glucose-binding lectin that recognizes N-glycans and is not known to bind common O-glycans on animal cell glycoproteins. However, it binds oligomannose-type N-glycans with much higher affinity than complex-type biantennary N-glycans, and it does not recognize more highly branched complex-type N-glycans (Figure 48.1). Other lectins, such as L-phytohemagglutinin (L-PHA) and E-PHA from *Phaseolus vulgaris*, as well as lentil lectin (LCA) from *Lens culinaris*, also recognize specific

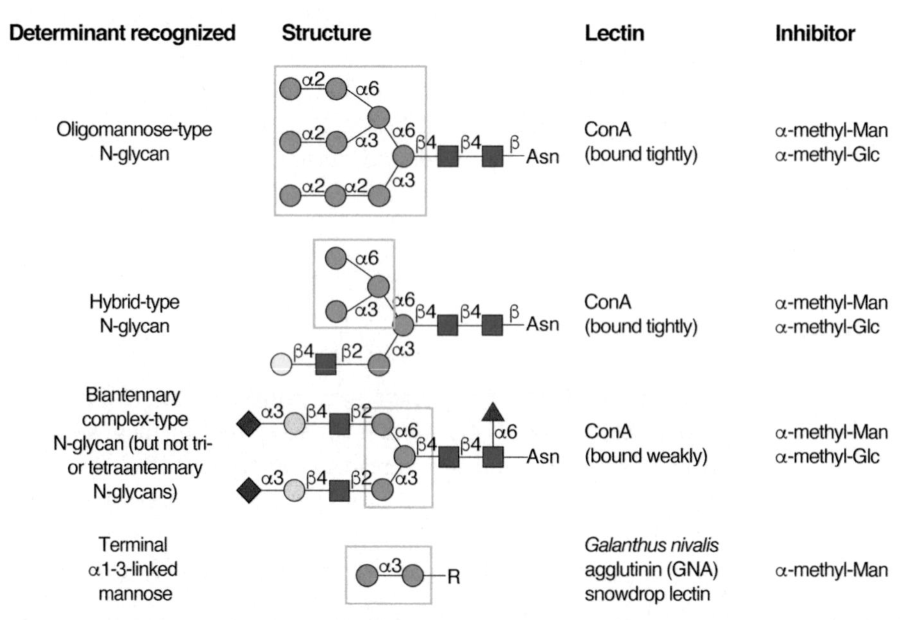

FIGURE 48.1. Examples of N-glycans recognized by concanavalin A (ConA) from *Canavalia ensiformis* and *Galanthus nivalis* agglutinin (GNA). The determinants required for binding are indicated in the *boxed areas*. Hapten sugars that can competitively inhibit binding of the lectin to the indicated glycans are shown on the *right*.

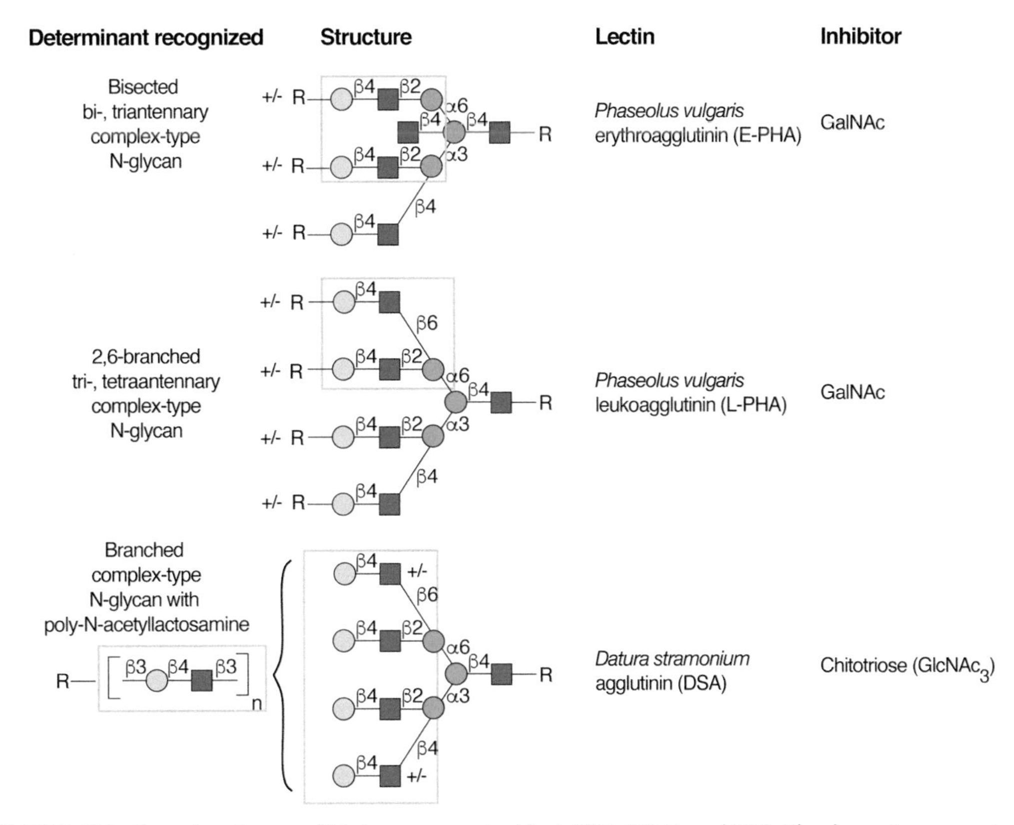

FIGURE 48.2. Examples of types of N-glycans recognized by L-PHA, E-PHA, and DSA. The determinants required for binding are indicated in the *boxed areas*. Hapten sugars that can competitively inhibit binding of the lectin to the indicated glycans are shown on the *right*.

determinants of N-glycans. Some animal lectins that are widely used include those from invertebrates, such as *Helix pomatia* agglutinin (HPA) from the snail. In fact, these lectins and others are frequently used to explore structural features of glycans on glycoproteins, glycolipids, and cells (Figures 48.2, 48.3, and 48.4 and Chapters 31 and 32). Our understanding of lectins and use of them as reagents has been improving because of the creation of recombinant lectins, including engineered varieties discussed below, and the increased definition of lectin specificity enabled by glycan microarrays and other techniques (Chapter 29).

GENERATION OF MONOCLONAL ANTIBODIES TO GLYCAN ANTIGENS

A number of mAbs have been generated against specific glycan determinants. Several approaches to obtain these antibodies have been described.

1. Whole cells and cell-derived membrane fractions—historically tumor cells—have been used to immunize mice to generate specific mAbs to various glycoprotein and glycolipid antigens, including the Tn antigen (GalNAcα1-Ser/Thr) and the stage-specific embryonic antigen-1 (SSEA-1), now known as the Lewis x (Le^x) antigen (Figure 48.5). In this approach, hybridomas are screened for mAbs that recognize the immunizing cell but no other types of cells.
2. Glycan-protein conjugates (glycans coupled to carrier proteins such as bovine serum albumin [BSA] or keyhole limpet hemocyanin [KLH]) have been used to generate both

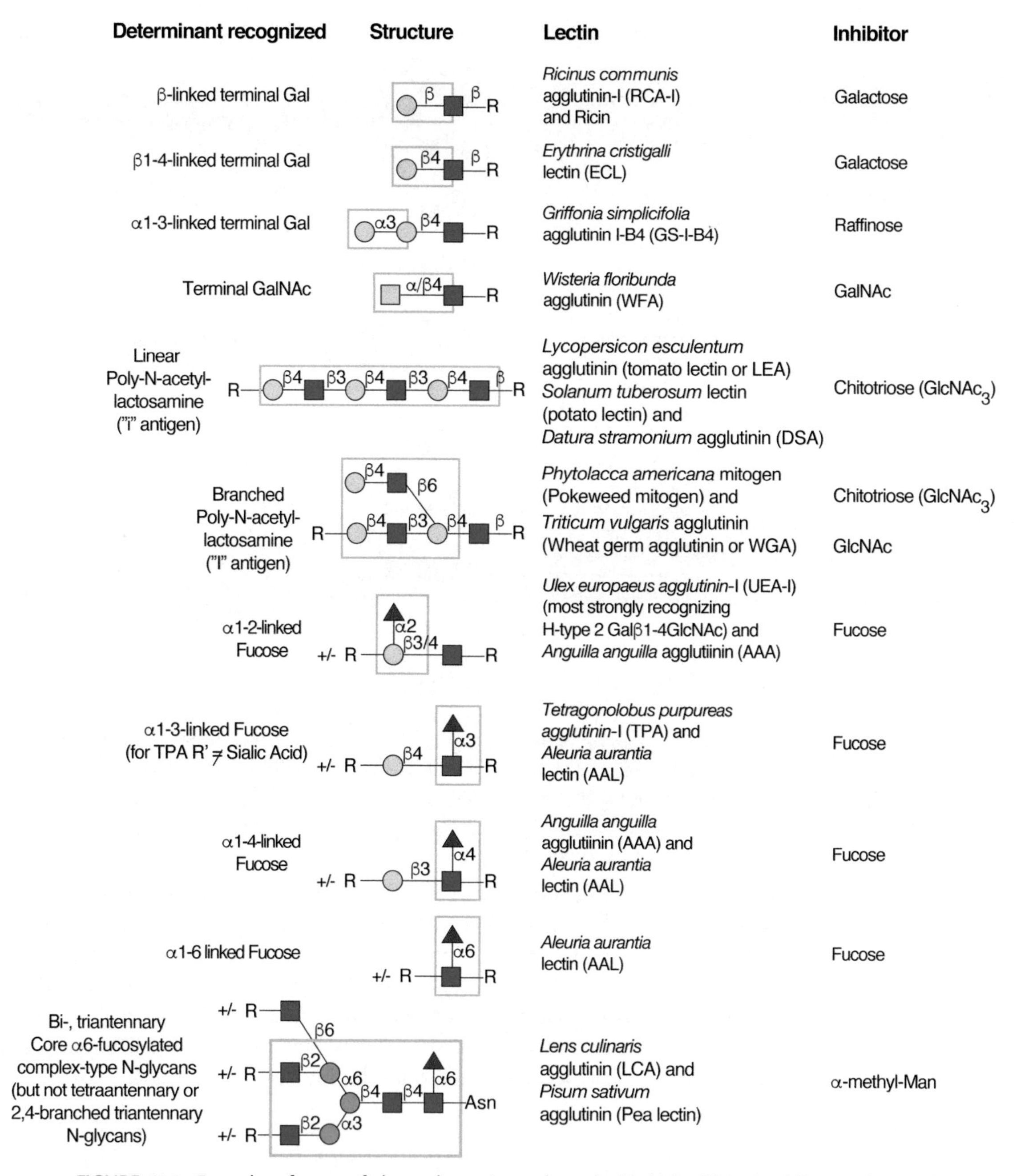

FIGURE 48.3. Examples of types of glycan determinants bound with high affinity by different plant and animal lectins. The determinants required for binding are indicated in the *boxed areas.* Hapten sugars that can competitively inhibit binding of the lectin to the indicated glycans are shown on the *right.*

polyclonal antibodies and mAbs to specific structures. Although BSA is not glycosylated, the unusual glycans on KLH may serve as adjuvants to enhance immune responses. Coupling of polysaccharides to protein carriers typically leads the generation of antibodies recognizing internal structural features, whereas coupling of small glycans typically yields antibodies that bind to terminal structural features. A common variation of this approach is to immunize mice directly with a glycoprotein, glycolipid, or glycosaminoglycan. For example, antibodies to the plant glycoprotein horseradish peroxidase are used to detect the presence of the unusual nonhuman modifications Xylβ1-2Man-R and the core fucose variation

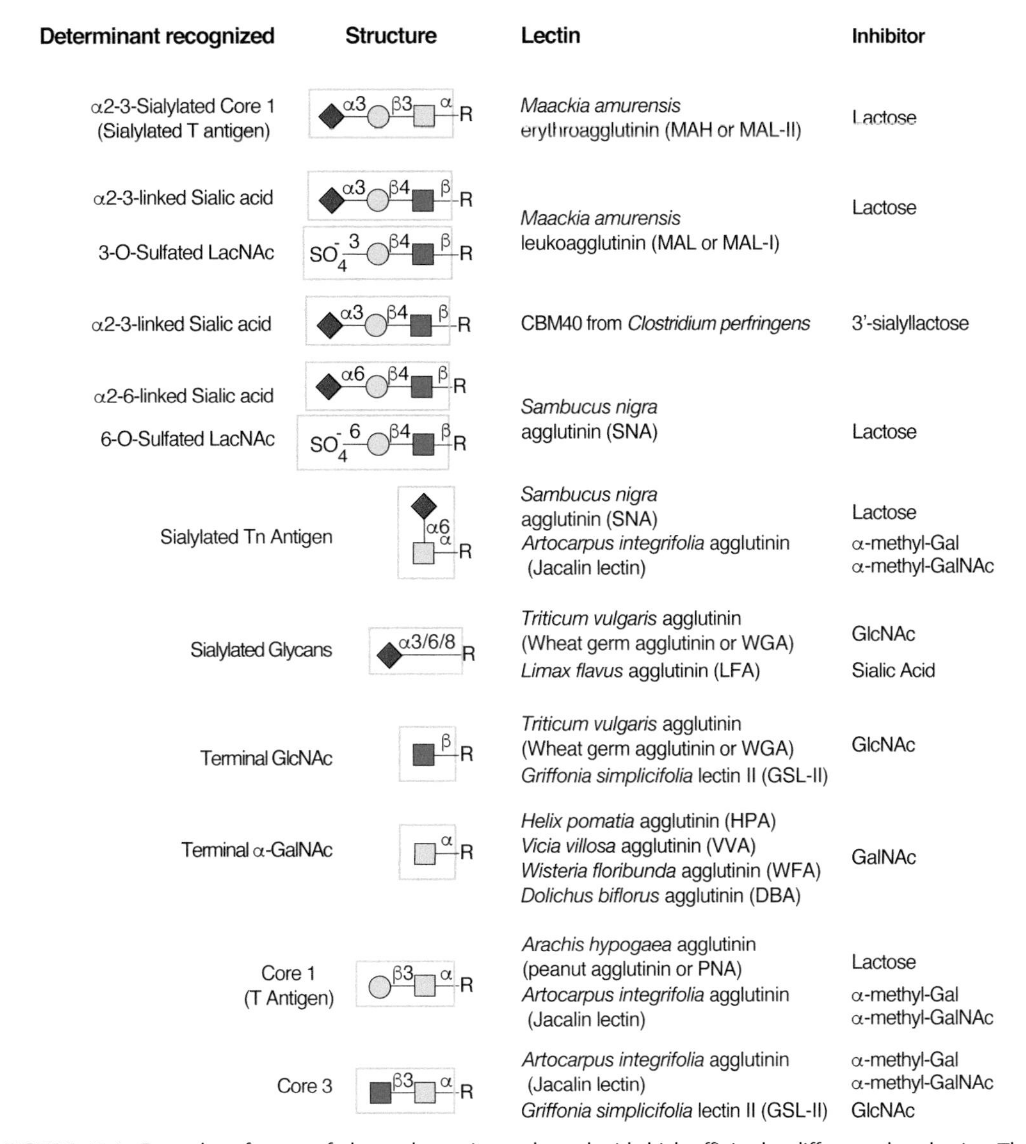

FIGURE 48.4. Examples of types of glycan determinants bound with high affinity by different plant lectins. The determinants required for binding are indicated in the *boxed areas.* Hapten sugars that can competitively inhibit binding of the lectin to the indicated glycans are shown on the *right.*

Fucα1-3GlcNAc-R in complex-type N-glycans. Immunization with glycan conjugates has also been used to generate polyclonal antisera in rabbits and chickens. Antibodies to specific glycan determinants can be purified from such antisera by affinity chromatography on immobilized glycans. Knockout mice lacking specific glycoconjugates are also useful for generating antibodies to common self-antigenic structures. For example, antibodies to the common glycolipid sulfatide (3-*O*-sulfate-Galβ1-ceramide) were obtained from mice that lacked the cerebroside sulfotransferase after immunization with sulfatide. If desired, recombinant single-chain antibodies to glycan determinants can be produced after cloning the V_H and V_L domains of antibodies from specific hybridomas.

3. A novel variation on the use of whole cells as immunogens to obtain antiglycan mAbs has been to use mice infected with specific parasites or bacteria followed by preparing hybridomas from splenocytes of the infected animals. This approach has been used to generate a variety of mAbs to pathogen-specific glycan antigens.

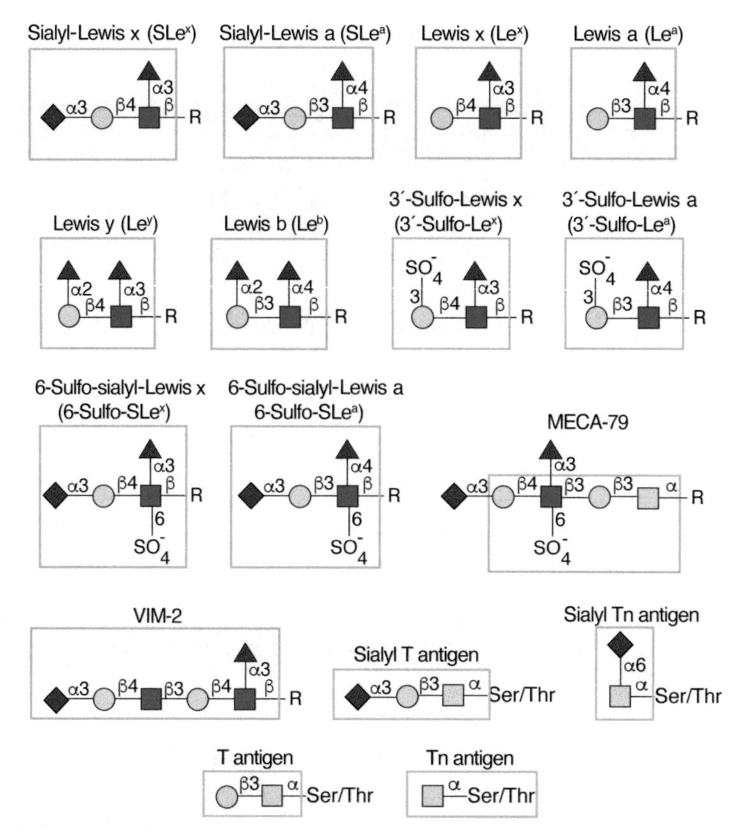

FIGURE 48.5. Examples of different mammalian glycan antigens recognized by specific monoclonal antibodies. The antigens have the structures shown within the *boxed area* and are named as indicated. Usually, the antigen shown in the box can be linked to almost any glycan and antibodies will still recognize the antigen.

4. A recent approach to generate specific antibodies to glycans involves immunization of the sea lamprey (*Petromyzon marinus*). When immunogenic material is introduced by intracoelomic injection into the larvae of the lamprey, the animal produces variable lymphocyte receptor (VLR)-based antibodies. An advantage of the sea lamprey for antiglycan reagent production is that there are tremendous antigenic differences in glycan antigens between sea lampreys and mammals, making the generation of these antibodies easier. Such VLRs contain leucine-rich repeats (LRRs), not immunoglobulin domains, within a single polypeptide, which comprises the antigen-binding domain. Yeast expression libraries can be prepared from the cDNA of the total lymphocytes and screening of such libraries with appropriate immobilized antigens (e.g., glycan microarrays or bead-based sorting) can enrich for yeast displaying specific VLRs. Insertion of these VLRs into an appropriate immunoglobulin framework produces chimeric mammalian-like antibodies. These antibodies have similar specificities and affinities to mammalian antibodies.

In addition, there is an emerging technology using phage display to identify single-chain variable fragments (scFvs) of antibodies that can also bind to glycan antigens. However, in many cases, it is difficult to define the specific epitopes recognized by these antibodies because isolated glycans are often characterized by microheterogeneities and related glycan antigens for comparison are often unavailable. Recently developed approaches, such as glycan microarrays and related techniques, combined with advances in the chemical synthesis of glycans, are helping to better define the specificities of mAbs (Chapter 29).

Antiglycan antibodies are widely used in glycobiology, and some common mammalian antigens they recognize are shown in Figures 48.5 and 48.6. Many of the murine mAbs to glycan antigens are

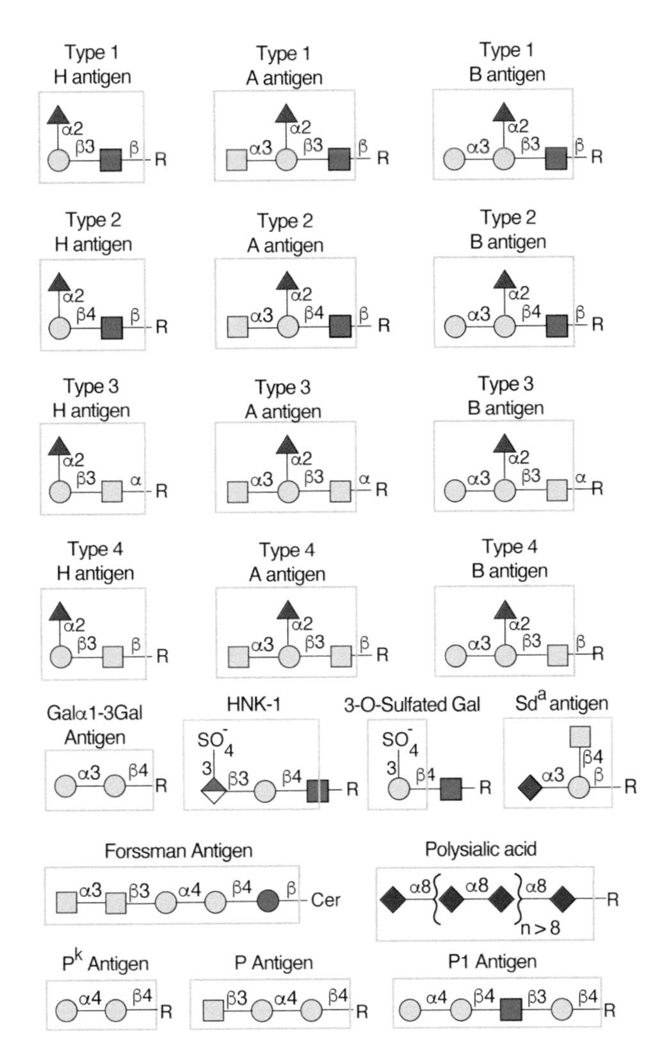

FIGURE 48.6. Additional examples of different mammalian glycan antigens recognized by specific monoclonal antibodies. The antigens have the structures shown within the *boxed area* and are named as indicated. In some cases, the importance of the reducing terminal end to antigen function is unclear, and the *right side* of the box may be ambiguous. Usually, the antigen shown in the box can be linked to almost any glycan and antibodies will still recognize the antigen. Antibodies are also available to glycosaminoglycans and to many glycolipid antigens, including globoside, GD3, GD2, GM2, GM1, asialo-GM1, GD1a, GD1b, GD3, and GQ1b.

of the IgM isotype. However, if the animal is well-immunized and serum titers are high, IgGs are frequently generated (e.g., ~35% of the plant-glycan-directed mAbs are IgG). The higher valency of IgM antibodies that form pentamers or hexamers can affect the binding specificity. Some antibodies to glycan antigens are commercially available, whereas others are obtainable only from individual laboratories or stock centers. Some antibodies against mammalian antigens are shown in Figures 48.5 and 48.6, and they recognize terminal glycan determinants, although subterminal sequences may also be required for binding and or in some cases may decrease binding. Many of the antibodies generated to date against plant glycans recognize internal structural features of these polysaccharides. Several antibodies against the plant cell wall homogalacturonans or xylans bind to internal backbone residues, frequently with varying sensitivities as to the level of backbone substitution by either glycan or nonglycan substituents. In some cases, the length of the unsubstituted backbone that is recognized by an antibody varies, thereby allowing the antibodies to be used to identify variations in backbone substitution density. In addition, the context of expression (free glycan vs. glycoprotein vs. glycolipid) or the class of glycan (N- vs. O-linked) may play a significant role in determining specificity and affinity. For example, antibodies to the 6-sulfo-SLe^x antigen require the fucose, sialic acid, and GlcNAc-6-*O*-sulfate residues. In contrast, the MECA-79 antibody recognizes extended core-1 O-glycans that contain internal GlcNAc-6-*O*-sulfate residues; it does not require either the fucose or sialic acid for recognition, but it does require the core-1 O-glycan. There are also many mAbs that recognize different glycolipids and glycosaminoglycans.

GENERATION OF RECOMBINANT AND ENGINEERED LECTINS AND INACTIVE ENZYMES THAT RECOGNIZE GLYCANS

With the growing understanding of the principles of glycan recognition (Chapter 29), there have been several novel glycan-detection reagents reported based on engineering mutant forms of endogenous lectins (engineered lectin) or carbohydrate-processing enzymes. Recombinant and engineered proteins can offer advantages over lectins or antibodies. They may be expressed, often in *Escherichia coli*, as opposed to being isolated from plant, fungal, or animal sources. Expression reduces batch variations that may result from isolation and offers a scalable solution to the demand for reagents. With protein engineering being based on the ever-increasing number of crystal structures of oligosaccharide-protein complexes, the structural origin of the specificity of an engineered protein may be well-understood. This stands in contrast to the case of antibodies that may be specific, but are rarely analyzed crystallographically, and thus the molecular origin of their specificity often remains unknown. For this reason, it is challenging to predict or interpret antibody specificity or cross-reactivity between glycans. In recent years, there has been increased interest in recombinant lectins derived from bacterial adhesins (e.g., "Siglec-like" binding regions [SLBRs] from *Streptococcus mutans*) and engineered lectins with altered specificities. Another avenue of new reagents has been the conversion of enzymes into reagents for detecting substrates. The concept was demonstrated more than 40 years ago, with anhydrotrypsin serving as an early example. In the specific case of glycan-binding reagents derived from inactivated glycosylhydrolases or other enzymes, the specificity is conferred by the endogenous enzyme, which is often well known. Inactivation—for example, by point mutation of the catalytic nucleophile—generally leads to a protein that retains the desired specificity and may still retain sufficient affinity to be useful as a detection reagent. As an example, an inactivated bacteriophage-derived endosialidase has been shown to have lectin-like properties and can be used as a probe in the specific detection of its substrate, polysialic acid. More often, however, the inactivated enzyme requires additional engineering to amplify its affinity and/or improve its selectivity/specificity for the glycan substrate of interest. Thus, the development of structure-based protein engineering promises a new avenue to generate glycan-detection reagents with complementary and sometimes preferable features to those offered by traditional lectins and antibodies.

USES OF GRPs IN GLYCAN IDENTIFICATION

Some of the important uses of lectins, CBMs, and antibodies are illustrated in Figure 48.7. Antibodies, lectins, and CBMs each have distinct advantages. Lectins are usually less expensive than antibodies and are often naturally available in seeds and plant materials. The genes encoding many CBMs have been cloned, enabling facile heterologous expression to obtain needed quantities of the protein. However, antibodies, lectins, and CBMs are often needed to bind a wide variety of glycan determinants. For example, no plant lectins specific for the SLe^a or the Le^x antigens have been identified, whereas mAbs to these antigens are available. Conversely, no mAbs have been identified that bind general determinants, such as α2-6-linked sialic acid and core fucose, whereas lectins have this discriminating ability (e.g., *Sambucus nigra* agglutinin [SNA] and *L. culinaris* [LcH], respectively). Of course, many GRPs recognize determinants independently of their presentation—that is, the same epitope may be presented on multiple types of glycans (e.g., O-glycans, N-glycans, or glycolipids). Although available antibodies thus far cannot distinguish common N-glycan structural motifs, such features are well-recognized by some plant lectins. For example, the plant lectin ConA does not bind mucin-type O-glycans in animal cells, but binds only to some specific classes of N-glycans (Figure 48.1). Additionally, E-PHA binds "bisected" complex-type N-glycans (Figure 48.2) and does not

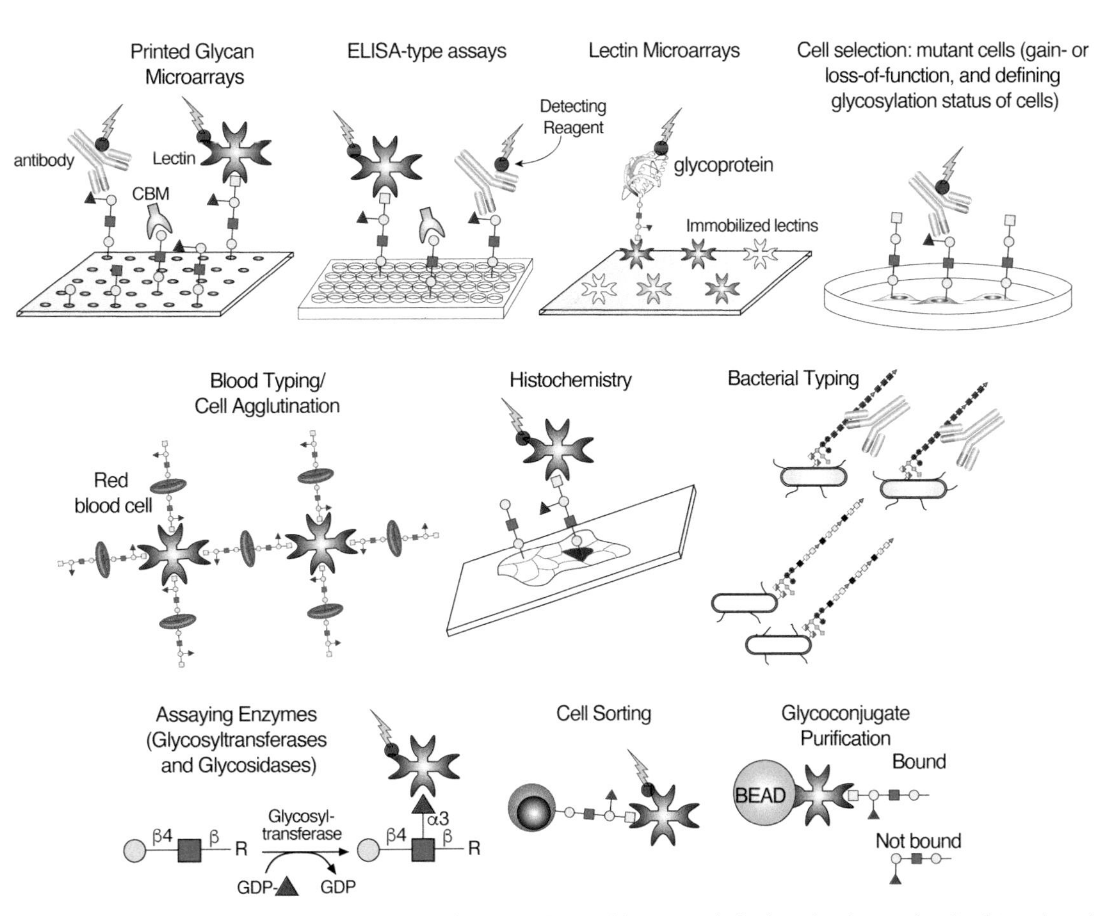

FIGURE 48.7. Examples of different uses of plant and animal lectins, carbohydrate-binding molecules (CBMs), and antibodies in glycobiology. Many plant and animal lectins are multivalent as shown, and antibodies are always multivalent, whereas CBMs are monovalent. They can be used to detect glycan structures in all of the formats shown.

bind known glycolipids or O-glycans. Great care should be taken, however, to use lectins and antibodies at appropriate concentrations in which their specificity can be exploited.

The glycan determinants bound with highest affinity by each of these probes have been identified by a combination of approaches, including affinity chromatography, glycan synthesis, and binding to specific glycoconjugates, cells, and glycan microarrays. A good example of this is the L-PHA, which is often used by immunologists as a mitogen to stimulate quiescent T cells to divide. L-PHA originates from the red kidney bean *Phaseolus vulgaris*, which also contains isolectins to L-PHA—notably E-PHA. L-PHA binds to certain branched, complex-type N-glycans containing the pentasaccharide sequence Galβ1-4GlcNAcβ1-2(Galβ1-4GlcNAcβ1-6)Manα1-R (the so-called "2-6-branch"), as shown in the boxed portion of the glycan in Figure 48.2. Curiously, the only monosaccharide that effectively inhibits either L-PHA or E-PHA is GalNAc, although this monosaccharide is not part of the N-glycan determinants recognized by these lectins (Figure 48.2). The binding of L-PHA is used to identify specific types of branched N-glycans in cells. L-PHA binding is dramatically decreased in mice genetically null for the branching β1-6 N-acetylglucosaminyltransferase (MGAT5 and -5B) (Chapter 9). The expression of L- or E-PHA-binding glycoproteins is increased in many tumor cells (Chapter 47). Similar studies show E-PHA binds bisected complex-type N-glycans containing the GlcNAcβ1-4Man-R in the core, and such structures are produced by MGAT3, and E-PHA-binding glycans are also elevated in some tumor cells.

Thus, using a variety of lectins and antibodies, it is possible to deduce many aspects of glycan structures. Microarrays in which a variety of lectins and antibodies are printed on a slide can also give valuable information about the glycosylation status of cells and glycoconjugates. This approach is especially sensitive in regard to defining whether biological samples differ in glycosylation. For example, such microarrays have been used to identify glycosylation differences in melanoma metastasis, leading to the identification of core fucose as a critical determinant of metastatic potential.

USES OF ANTIBODIES AND LECTINS IN GLYCAN AND GLYCOPROTEIN PURIFICATION

There are several approaches to using antibodies and lectins in glycan purification, including affinity chromatography or affinity binding and immunoprecipitation or lectin-induced precipitation. The proteins may be covalently coupled to a carrier such as Sepharose or biotinylated and captured on streptavidin-Sepharose. In addition, antibodies may be noncovalently captured on protein A (or G)-Sepharose. These bound antibodies and lectins can then be used to isolate glycoconjugates expressing specific glycan determinants. ConA-Sepharose is commonly used to isolate glycoproteins as it shows little binding to nonglycosylated proteins. However, it does not bind all glycoproteins because it recognizes specific N-glycan structures. ConA-Sepharose has also been used to isolate free oligo- and polymannose-, hybrid-, and complex-type biantennary N-glycans.

When combined in a serial format, multiple lectins can be used in affinity chromatography to isolate glycoconjugates containing most of the major glycan structures present in animal cells, with glycoconjugates being separated as classes that share common determinants. An example of serial lectin affinity chromatography is shown in Figure 48.8. Identification of glycoproteins through analysis of released glycopeptides gives important information on proteins carrying specific modifications. Release of glycans coupled with ion-exchange chromatography and high-performance liquid chromatography (HPLC) can yield highly pure glycans with predicted structures that can then be confirmed by mass spectrometry of native and permethylated derivatives (Chapters 50 and 51).

Mixed-bed lectin chromatography using a combination of different immobilized lectins is useful for simultaneously separating all types of glycoconjugates from nonglycosylated material (e.g., glycopeptides from peptides). Combinations of affinity chromatography and other chromatographic techniques can be very useful in identifying and separating glycopeptides. Thus, the ability of glycans to be recognized by lectins dependent on specific structural features in the glycans is a powerful tool for glycan identification and isolation. In some approaches, the glycans are tagged at the reducing end by fluorophores and radioisotopes or may be obtained by metabolic radiolabeling from cells or tissues grown in the presence of radiolabeled sugar precursors, such as [2-^{3}H]mannose or [6-^{3}H]glucosamine. Glycan fractionation shown on immobilized lectins in Figure 48.8 is currently not possible with antibodies because no antibodies are known that can distinguish such core structural features in glycans.

When intact glycoproteins or complex polysaccharides (e.g., those found in plant cell walls [Chapter 24]) are analyzed for their interactions with plant lectins or antibodies, the interpretation of data may be complicated by the multivalency of the glycoprotein/polysaccharide and the density of the immobilized lectin/antibody. For example, glycoproteins containing multiple high-mannose-type N-glycans bind so tightly to immobilized ConA that it is difficult to elute the bound glycoprotein, even with extremely high concentrations of hapten and under harsh conditions. Lower densities of ConA conjugation reduce its avidity for the glycoproteins and promote hapten dissociation of bound ligands with lower concentrations of sugars. When used in combination, multiple lectins, such as ConA, AAL (*Aleuria aurantia* lectin), LCA,

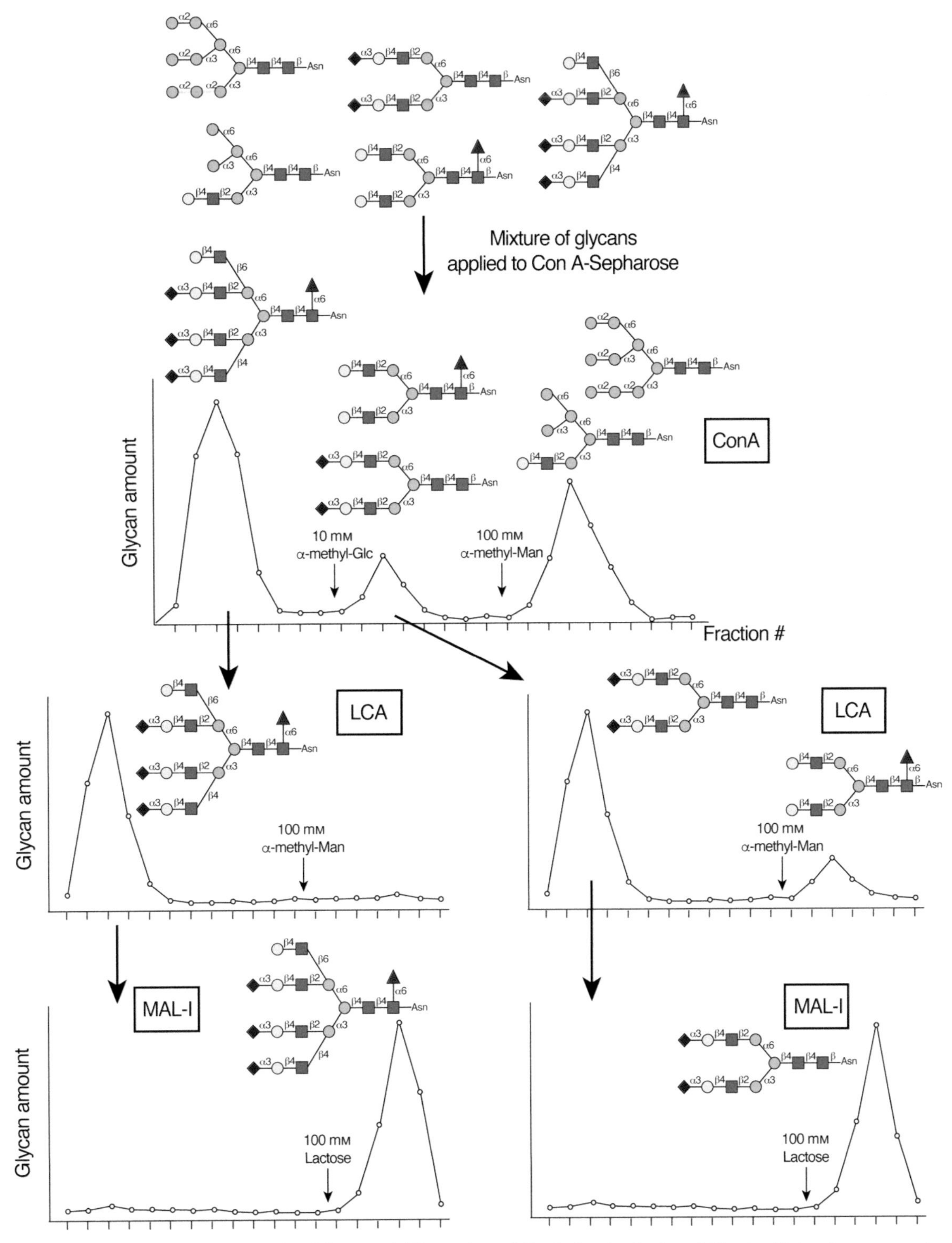

FIGURE 48.8. An example of the use of different immobilized plant lectins in serial lectin affinity chromatography of complex mixtures of glycopeptides. In this example, a mixture of glycopeptides is applied to a column of immobilized concanavalin A (ConA), and the bound glycans are eluted with increasing concentrations of hapten sugars (α-methylglucose and then α-methylmannose) (*arrows*). The recovered glycans are then applied to a second set of columns containing immobilized lentil lectin (LCA) and bound glycans are eluted with the hapten α-methylmannose. This process is repeated over other immobilized lectins, such as *Maackia amurensis* lectin (MAL), and bound glycans are eluted with the hapten lactose. The glycans bound at each step are shown in the *panels above* the chromatographic peaks.

and RCA (*Ricinus communis* agglutinin), can be used to isolate most glycoproteins containing N-and O-glycans from animal cells. This is a potentially powerful approach for glycoproteomics, or the identification of glycoproteins and their glycosylation status.

Another approach to enhance glycoproteomic analyses is to use multiple lectins in mixed-bed or multi-lectin affinity chromatography (M-LAC). In such an approach a number of different lectins are used that recognize different glycan features, such as those shown in earlier figures in this chapter. Glycoproteins (or glycopeptides) with these selected features can be isolated and enriched from complex mixtures, allowing for more focused analyses on the glycoconjugates of interest. In addition, in some types of analyses a genetic disruption can result in the collapse of classes of glycans to a single structure, allowing a lectin-enrichment strategy to be particularly useful in identifying the glycomic features of interest. For example, deletion of the gene *Cosmc* (*C1GalT1C1*) in cell lines leads to the collapse of GalNAc-type O-glycans, which can vary immensely in structure, to a single structure, the Tn antigen (GalNAcα1-O-Ser/Thr). These derived cell lines are termed SimpleCells. If glycopeptides are prepared from such cells, all glycopeptides that contain an O-glycan with the Tn antigen can be affinity purified using immobilized *Vicia villosa* agglutinin (VVA). The isolated glycopeptides can then be sequence-identified. In this type of approach the O-glycoproteome of cells regarding sites of O-GalNAc-type O-glycosylation can be more readily characterized, as there is no universal set of reagents that can generally recognize all the different types of O-glycans in normal cell-derived glycoproteins.

Lectins and antibodies can also be used in western blotting approaches to characterize protein and lipid glycosylation, in which biotinylated lectin or antibody probes are applied to material transferred to nitrocellulose or other supports after electrophoresis or chromatography. The bound lectins/antibodies are visualized by binding streptavidin-alkaline phosphatase and conversion of luminescent substrates. In these approaches, the concentrations of lectins and antibodies used must be low enough both to reduce false-positive, nonspecific binding and to allow for inhibition by appropriate haptens to confirm sugar binding. Removal of N-glycans by PNGaseF, or of sialic acid by neuraminidase, can eliminate binding to specific lectins and antibodies in western blotting, thus indicating the lectin/antibody-bound determinant.

Mention should also be made about the approaches to isolate novel lectins from natural sources, or even when made recombinantly or in the form of some type of GRP. Historically, such isolation methods have used affinity chromatography. For example, a high density of lactose linked to agarose in the form of an affinity column can be used to isolate galectins, which can be eluted with lactose as a hapten. The cation-dependent mannose 6-phosphate receptor (CD-MPR) was isolated by affinity chromatography on immobilized yeast mannans containing phosphorylated mannose. There are also other forms of isolation, including salt precipitation, ion exchange chromatography, gel filtration, and the capture of a tagged and recombinant chimeric-lectin by capture of the tag (e.g., His tag). A recent method exploits a "capture and release" approach, in which multivalent ligands recognized by a lectin are assembled, whereby they can capture the lectin of interest, and then filtration or centrifugation may be used to isolate the lectin complexed with ligand. The lectin can then be dissociated and is available for study. Any of these type of methods are also useful when combined with proteomic analyses in identifying novel lectins and protein complexes that bind to glycans.

USES OF GRPs IN CHARACTERIZING CELL-SURFACE GLYCOCONJUGATES

The classic approaches for using antibodies, CBMs, and lectins to characterize cell-surface glycoconjugates are histochemistry (lectins) and immunohistochemistry (antibodies), flow cytometry with cell sorting, and cell agglutination. In histochemistry and immunohistochemistry, tissues are prepared and fixed as usual for histological staining, and then incubated with appropriate biotinylated or peroxidase-labeled lectins or antibodies (glycolipids are extracted during standard paraffin embedding procedures and require frozen sections for sensitive detection). The bound lectins, CBMs, or antibodies are then visualized by means of secondary reagents, such as streptavidin-peroxidase or labeled secondary antibody. These approaches often yield information that is difficult to obtain by any other approach. For example, they can reveal the spatial orientation of different glycans, their relative abundance, and whether they are intracellular and/or extracellular. Three important controls in such studies are (1) use of lectins, CBMs, or antibodies at limiting concentrations to avoid nonspecific binding; (2) confirmation of the specificity of binding by appropriate inhibition by haptens or by destruction of the predicted target glycans with glycosidases; and (3) use of multiple lectins, CBMs, or antibodies to provide further confirmation of the conclusions.

Lectins and antibodies to glycans have also been widely used in flow cytometry and cell sorting. In such studies, cells are incubated with low, nonagglutinating levels of lectins or antibodies that are biotinylated and conjugated to fluorescently labeled streptavidin or directly fluorescently labeled. Cells with bound lectins or antibodies can then be identified by their fluorescence in the flow cytometer, and the degree of fluorescence can be correlated with the number of binding sites. A key consideration in flow cytometry is to avoid high concentrations of lectins or antibodies that cause agglutination of cells. Lectins and antibodies can also be used to identify specific membrane localization of different glycans, using confocal microscopy and electron microscopy.

Lectins and antibodies are useful for characterizing cell-surface glycans when limited numbers of cells are available. Studies on glycosylation of embryonic stem cells have been greatly aided by using panels of specific lectins to identify unique glycan determinants and changes in their expression during cellular differentiation. A recent variation of this approach is to use a microarray of immobilized lectins, which are probed with fluorescently labeled glycoproteins in extracts of cells. Such assays can reveal minor differences in protein glycosylation between different samples and give insight into the glycan structures that are present.

One of the oldest uses of lectins is in the agglutination and precipitation of glycoconjugates, cells, and membrane vesicle preparations. The easiest assay for many soluble lectins that are multivalent is agglutination of target cells, such as erythrocytes, leukocytes, or even bacteria or fungi. Agglutination can often be easily observed without a microscope, but it is also measurable in instruments such as aggregometers. In these assays, a lectin solution is serially diluted and the reciprocal of the final dilution that gives measurable cell agglutination is taken to define the activity. Bacterial agglutination by plant and animal lectins is often used to explore the surface glycocalyx and changes in glycocalyx on culture conditions and to define phenotypes of different strains or serotypes. Lectin precipitation and aggregation can be used to define the glycan composition and overall architecture of polysaccharides, as has been done for bacterial, algal, plant, and animal polysaccharides.

Combining many different GRP, including lectins and antibodies, into a lectin microarray is another way in which the glycome of cells may be explored. Such arrays have dozens of specific proteins immobilized in a microarray format and can provide valuable information about the glycosylation status of cells and glycoconjugates. Fluorescent-tagged glycoproteins from a source (e.g., cell or tissue extracts) can be applied to these arrays and the binding pattern compared to a control sample can reveal features of the glycome of the material, such as patterns of sialylation, fucosylation, and even linkages of residues with glycans using linkage-specific reagents. The advantages of this approach is the microscale in which it can be performed, the relatively high-throughput nature of the technology, and the precision of discrimination in terms of differential glycosylation between two samples. The system is particularly adaptable to include new reagents as they become available (e.g., new lectins or engineered proteins).

USES OF ANTIBODIES AND LECTINS FOR GENERATING ANIMAL CELL GLYCOSYLATION MUTANTS

An important use of lectins and antibodies has been in the selection of cell lines (e.g., Chinese hamster ovary [CHO] cells) that express altered cell-surface glycans. The common lectins that have been used are ConA, WGA (wheat germ agglutinin), L-PHA, LCA, PSA (*Pisum sativum* agglutinin), E-PHA, ricin, modeccin, and abrin. The latter three lectins are heterodimeric, disulfide-bonded proteins that are classified as type-II ribosome-inactivating proteins; they contain an A subunit that constitutes an enzyme called RNA-N-glycosidase, which inactivates the 28S ribosome, and a B subunit that is a galactose-binding lectin (Chapter 31). Other plant lectins such as ConA, WGA, or LCA (which lack an enzymatic or toxic A subunit) are still toxic to animal cells via poorly understood mechanisms, whereas yet others, such as soybean agglutinin (SBA) and peanut agglutinin (PNA), are not highly toxic to animal cells in culture. One mechanism of toxicity of plant lectins is the induction of apoptosis, perhaps by blocking receptor or nutrient transport functions in cells, or by cross-linking apoptotic receptors. Noncytotoxic lectins and antibodies to specific glycans can be rendered toxic by conjugation to ricin A subunit or other toxic proteins. Using such agents, both loss-of-function (e.g., loss of a glycosyl-transferase or glycosidase) and gain-of-function (e.g., activation of a latent transferase gene) mutants have been obtained. More details about glycosylation mutants of cultured cells are presented in Chapter 49.

USES OF ANTIBODIES AND LECTINS FOR CLONING GLYCOSYLTRANSFERASE GENES BY EXPRESSION

The specificity of lectins and antibodies to particular glycans has made them especially useful in cloning genes encoding glycosyltransferases or other proteins required for proper glycosylation, such as nucleotide sugar transporters. For example, CHO cells and the African green monkey kidney cell line COS lack glycans with terminal α-galactose residues. Consequently, the cells do not bind the plant lectin GSI-B4 (Figure 48.3). When transfected with a cDNA library prepared from cells that do express terminal α-galactose residues (such as murine teratocarcinoma cells F9), cells that have taken up a plasmid encoding the cognate α1-3 galactosyltransferase and express terminal α-residues bind to GSI-B4 and can be identified with plates coated with the lectin. Isolation of plasmids from the bound cells and repeated recloning and reexpression by this technique (called expression cloning) led to the identification of a specific gene encoding the murine α1-3 galactosyltransferase (Ggta1). Similar approaches using an antibody to the Lewis x antigen (SSEA-1) led to cloning of the Lewis blood group Fuct-III.

Related approaches using antibodies and lectins to other glycan structures and wild-type CHO cells and CHO mutants led to the identification of the genes encoding many other glycosyltransferases, including some involved in extending glycosphingolipids and nucleotide sugar transporters, such as the transporter for CMP-NeuAc. Such approaches based on lectin selection are also useful with yeast. For example, the gene encoding a yeast N-acetylglucosaminyltransferase (GlcNAcT) was identified by expression cloning using a GlcNAcT-deficient yeast. The mannan chains of the yeast *Kluyveromyces lactis* normally contain some terminal *N*-acetylglucosamine residues that are bound by the plant lectin GSL-II, which binds terminal *N*-acetylglucosamine residues (Figure 48.4). A mutant lacking the GlcNAcT and lacking terminal *N*-acetylglucosamine residues on mannoproteins was identified. Transformation of yeast with DNA containing the gene encoding the GlcNAcT led to yeast clones that were bound by the fluorescently labeled GSL-II. This strategy was used to identify the gene encoding the GlcNAcT. The transporters for UDP-Gal in *Leishmania* parasites and in the plant *Arabidopsis* were also identified by expression cloning in Lec8 CHO cells; these cells have a mutation in their endogenous UDP-Gal transporter, and consequently lack galactose-containing glycans on their surface. For the UDP-Gal transporter from *Arabidopsis*, Lec8 cells were cotransfected with a cDNA library encoding the putative *Arabidopsis* UDP-Gal transporter, along with a glucuronosyltransferase that can lead to synthesis of the unsulfated version of the HNK epitope (Figure 48.6), GlcAβ1-3Galβ-R that is bound by a specific antibody. This identification strategy led to the expression cloning of several UDP-Gal transporters from *Arabidopsis*.

CHO and COS cell lines have also been useful in characterizing the activities of novel glycosyltransferase genes that were identified by other approaches. For example, candidate fucosyltransferase genes encoding α1-2 or α1-3 fucosyltransferases have been expressed in CHO and COS cells, which lack these enzymes. Expression of these enzymes then leads to surface expression of antigens recognized by specific antibodies and lectins, such as the H-antigen (for the α1-2 FucT) or Le^x and SLe^x (for the α1-3 FucT).

USES OF ANTIBODIES AND LECTINS IN ASSAYING GLYCOSYLTRANSFERASES AND GLYCOSIDASES

Lectins and antibodies to glycan antigens have been useful in assaying specific glycosyltransferases and glycosidases in various formats (Figure 48.7). Immobilized lectins have been used to isolate products of glycosyltransferase assays, such as chitin polysaccharides on WGA or glycosylated peptides on mixed-bed lectin columns. α1-3 Fucosyltransferases that synthesize the Le^x and SLe^x antigens have been assayed based on capture of the product on immobilized antibodies to these antigens or binding of antibody to the immobilized fucosylated product in an ELISA-type format, or in flow cytometry using beads containing the acceptor glycan that is modified by an enzyme to a new structure recognized by an antibody or lectin. Likewise, α2-3 sialyltransferases and α2-6 sialyltransferases have been assayed using immobilized acceptors in an ELISA-type format and in BIAcore formats (Chapter 29), and their products have been measured with MAL (which binds to the α2-3-sialylated product) or SNA (which binds to the α2-6-sialylated product). Similarly, α1-3 galactosyltransferases have been assayed using immobilized acceptors in an ELISA-type format, and the product has been measured with GSI-B4 or *Viscum album* agglutinin, which binds to the α1-3 galactosylated product. The glycoprotein-specific β1-4 N-acetylgalactosaminyltransferase has been assayed using glycoprotein acceptors in solution, capture by a specific mAb in a microtiter plate, and measurement of product by an ELISA-type assay with a *Wisteria floribunda* agglutinin (WFA) that binds to terminal β1-4-linked GalNAc residues generated by the enzyme. Conversely, glycosidases can be assayed by measuring the gain or loss in binding of lectins or antibodies. For example, the lectin PNA, which binds to nonsialylated Galβ1-3GalNAcα1-Ser/Thr in O-glycans, can be used to measure bacterial sialidases by agglutination of treated erythrocytes. Many microbial hydrolases have evolved to attack their substrates in structurally complex contexts (e.g., plant cell walls [Chapter 24]). The number and diversity of glycan-directed probes now allows the detailed characterization of such enzymes acting on biologically relevant structures, yielding new insights into their activities. It is easy to envision other ways in which lectins, CBMs, and antibodies can be used to probe the products of specific glycosidases and glycosyltransferases, given the specificities of the glycan-directed probes described here.

FURTHER READING

Hakomori S. 1984. Tumor-associated carbohydrate antigens. *Annu Rev Immunol* **2:** 103–126. doi:10.1146/annurev.iy.02.040184.000535

Merkle RK, Cummings RD. 1987. Lectin affinity chromatography of glycopeptides. *Methods Enzymol* **138:** 232–259. doi:10.1016/0076-6879(87)38020-6

Osawa T, Tsuji T. 1987. Fractionation and structural assessment of oligosaccharides and glycopeptides by use of immobilized lectins. *Annu Rev Biochem* **56:** 21–42. doi:10.1146/annurev.bi.56.070187.000321

Osawa T. 1988. The separation of immunocyte subpopulations by use of various lectins. *Adv Exp Med Biol* **228:** 83–104. doi:10.1007/978-1-4613-1663-3_4

Esko JD. 1992. Animal cell mutants defective in heparan sulfate polymerization. *Adv Exp Med Biol* **313:** 97–106. doi:10.1007/978-1-4899-2444-5_10

Kobata A, Endo T. 1992. Immobilized lectin columns: useful tools for the fractionation and structural analysis of oligosaccharides. *J Chromatogr* **597:** 111–122. doi:10.1016/0021-9673(92) 80101-y

Cummings RD. 1994. Use of lectins in analysis of glycoconjugates. *Methods Enzymol* **230:** 66–86. doi:10.1016/0076-6879(94)30008-9

Knox JP. 1997. The use of antibodies to study the architecture and developmental regulation of plant cell walls. *Int Rev Cytol* **171:** 79–120. doi:10.1016/s0074-7696(08)62586-3

Lis H, Sharon N. 1998. Lectins: carbohydrate-specific proteins that mediate cellular recognition. *Chem Rev* **98:** 637–674. doi:10.1021/cr940413g

Bush CA, Martin-Pastor M, Imberty A. 1999. Structure and conformation of complex carbohydrates of glycoproteins, glycolipids, and bacterial polysaccharides. *Annu Rev Biophys Biomol Struct* **28:** 269–293. doi:10.1146/annurev.biophys.28.1.269

Morgan WT, Watkins WM. 2000. Unravelling the biochemical basis of blood group ABO and Lewis antigenic specificity. *Glycoconj J* **17:** 501–530. doi:10.1023/a:1011014307683

Rüdiger H, Gabius HJ. 2001. Plant lectins: occurrence, biochemistry, functions and applications. *Glycoconj J* **18:** 589–613. doi:10.1023/a:1020687518999

Goldstein IJ. 2002. Lectin structure-activity: the story is never over. *J Agric Food Chem* **50:** 6583–6585. doi:10.1021/jf0201879

Madera M, Mechref Y, Novotny MV. 2005. Combining lectin microcolumns with high-resolution separation techniques for enrichment of glycoproteins and glycopeptides. *Anal Chem* **77:** 4081–4090. doi:10.1021/ac050222l

Paschinger K, Fabini G, Schuster D, Rendic D, Wilson IB. 2005. Definition of immunogenic carbohydrate epitopes. *Acta Biochim Pol* **52:** 629–632. doi:10.18388/abp.2005_3422

Akama TO, Fukuda MN. 2006. N-Glycan structure analysis using lectins and an α-mannosidase activity assay. *Methods Enzymol* **416:** 304–314. doi:10.1016/s0076-6879(06)16020-6

Lehmann F, Tiralongo E, Tiralongo J. 2006. Sialic acid–specific lectins: occurrence, specificity and function. *Cell Mol Life Sci* **63:** 1331–1354. doi:10.1007/s00018-005-5589-y

Maeda Y, Ashida H, Kinoshita T. 2006. CHO glycosylation mutants: GPI anchor. *Methods Enzymol* **416:** 82–205. doi:10.1016/s0076-6879(06)16012-7

Patnaik SK, Stanley P. 2006. Lectin-resistant CHO glycosylation mutants. *Methods Enzymol* **416:** 159–182. doi:10.1016/s0076-6879(06)16011-5

Jokilammi A, Korja M, Jakobsson, Finne J. 2007. Generation of lectins from enzymes: use of inactive endosialidase for polysialic acid detection. In *Lectins: analytical technologies* (ed. Nilsson CL). Elsevier Science, Amsterdam. doi:10.1016/b978-044453077-6/50017-x

Varki NM, Varki A. 2007. Diversity in cell surface sialic acid presentations: implications for biology and disease. *Lab Invest* **87:** 851–857. doi:10.1038/labinvest.3700656

Cummings RD. 2009. The repertoire of glycan determinants in the human glycome. *Mol Biosyst* **5:** 2–12. doi:10.1039/B907931A

von Schantz L, Gullfot F, Scheer S, Filonova L, Gunnarsson LC, Flint JE, Daniel G, Nordberg-Karlsson E, Brumer H, Ohlin M. 2009. Affinity maturation generates greatly improved xyloglucan-specific carbohydrate binding modules. *BMC Biotechnology* **9:** 92. doi:10.1186/1472-6750-9-92

Pattathil S, Avci U, Baldwin D, Swennes AG, McGill JA, Popper Z, Bootten T, Albert A, Davis RH, Chennareddy C, et al. 2010. A comprehensive toolkit of plant cell wall glycan-directed monoclonal antibodies. *Plant Physiol* **153:** 514–525. doi:10.1104/pp.109.151985

Lam SK, Ng TB. 2011. Lectins: production and practical applications. *Appl Microbiol Biotechnol* **89:** 45–55. doi:10.1007/s00253-010-2892-9

Steentoft C, Vakhrushev SY, Vester-Christensen MB, Schjoldager KT, Kong Y, Bennett EP, Mandel U, Wandall H, Levery SB, Clausen H. 2011. Mining the O-glycoproteome using zinc-finger nuclease-glycoengineeered SimpleCell lines. *Nat Methods* **8:** 977–982. doi:10.1038/nmeth.1731

Pedersen HL, Fangel JU, McCleary B, Ruzanski C, Rydahl MG, Ralet M-C, Farkas V, von Schantz L, Marcus SE, Andersen MCF, et al. 2012. Versatile high resolution oligosaccharide microarrays for plant glycobiology and cell wall research. *J Biol Chem* **287:** 39429–39438. doi:10.1074/jbc .m112.396598

Gilbert HJ, Knox JP, Boraston AB. 2013. Advances in understanding the molecular basis of plant cell wall polysaccharide recognition by carbohydrate-binding modules. *Curr Opin Struct Biol* **23:** 669–677. doi:10.1016/j.sbi.2013.05.005

Hong X, Ma MZ, Gildersleeve JC, Chowdhury S, Barchi JJ Jr, Mariuzza RA, Murphy MB, Mao L, Pancer Z. 2013. Sugar-binding proteins from fish: selection of high affinity "lambodies" that recognize biomedically relevant glycans. *ACS Chem Biol* **8:** 152–160. doi:10.1021/cb300399s

Ribeiro JP, Mahal LK. 2013. Dot by dot: analyzing the glycome using lectin microarrays. *Curr Opin Chem Biol* **17:** 827–831. doi:10.1016/j.cbpa.2013.06.009

Smith DF, Cummings RD. 2013. Application of microarrays for deciphering the structure and function of the human glycome. *Mol Cell Proteomics* **12:** 902–912. doi:10.1074/mcp.r112.027110

Akkouh O, Ng TB, Singh SS, Yin C, Dan X, Chan YS, Pan W, Cheung RC. 2015. Lectins with anti-HIV activity: a review. *Molecules* **20:** 648–668. doi:10.3390/molecules20010648

Broecker F, Anish C, Seeberger PH. 2015. Generation of monoclonal antibodies against defined oligosaccharide antigens. *Methods Mol Biol* **1331:** 57–80. doi:10.1007/978-1-4939-2874-3_5

Oliveira C, Varvalho V, Domingues L, Gama FM. 2015. Recombinant CBM-fusion technology—applications overview. *Biotech Adv* **22:** 358–369. doi:10.1016/j.biotechadv.2015.02.006

Pattathil S, Avci U, Zhang T, Cardenas CL, Hahn MG. 2015. Immunological approaches to biomass characterization and utilization. *Front Bioeng Biotechnol* **3:** 173. doi:10.3389/fbioe.2015.00173

Blackler RJ, Evans DW, Smith DF, Cummings RD, Brooks CL, Baulked T, Liu X, Evans SV, Müller-Lennies S. 2016. Single-chain antibody-fragment M6P-1 possesses a mannose 6-phosphate monosaccharide-specific binding pocket that distinguishes *N*-glycan phosphorylation in a branch-specific manner. *Glycobiology* **26:** 181–192. doi:10.1093/glycob/cwv093

Ribeiro JP, Pau W, Pifferi C, Renaudet O, Varrot A, Mahal LK, Inberty A. 2016. Characterization of a high-affinity sialic acid-specific CBM40 from *Clostridium perfringens* and engineering of a divalent form. *Biochem J* **473:** 2109–2118. doi:10.1042/bcj20160340

Ruprecht C, Bartetzko MP, Senf D, Dallabernardina P, Boos I, Andersen MCF, Kotake T, Knox JP, Hahn MG, Clausen MH, Pfrengle F. 2017. A synthetic glycan microarray enables epitope mapping of plant cell wall glycan-directed antibodies. *Plant Physiol* **175:** 1094–1104. doi:10.1104/pp.17.00737

Walker JA, Pattathil S, Bergeman LF, Beebe E, Deng K, Mirzai M, Northen TR, Hahn MG, Fox BG. 2017. Glycome profiling of enzyme specificity during hydrolysis of plant cell walls. *Biotechnol Biofuels* **10:** 31. doi:10.1186/s13068-017-0703-6

Mahajan S, Ramya TNC. 2018. Nature-inspired engineering of an F-type lectin for increased binding strength. *Glycobiology* **28:** 933–948. doi:10.1093/glycob/cwy082

Hirabayashi J, Arai R. 2019. Lectin engineering: the possible and the actual. *Interface Focus* **9:** 20180068. doi:10.1098/rsfs.2018.0068

Narimatsu Y, Joshi HJ, Schjoldager KT, Hintze J, Halim A, Steentoft C, Nason R, Mandel U, Bennett EP, Clausen H, Vakhrushev SY. 2019. Exploring regulation of protein O-glycosylation in isogenic human HEK293 cells by differential O-glycoproteomics. *Mol Cell Proteomics* **18:** 1396–1409. doi:10.1074/mcp.ra118.001121

McKitrick TR, Eris D, Mondal N, Aryal RP, McCurley N, Heimburg-Molinaro J, Cummings RD. 2020. Antibodies from lampreys as smart anti-glycan reagents (SAGRs): perspectives on their specificity, structure, and glyco-genomics. *Biochemistry* **59:** 3111–3122. doi:10.1021/acs.biochem.9b01015

Riley NM, Bertozzi CR, Pitteri SJ. 2020. A pragmatic guide to enrichment strategies for mass spectrometry-based glycoproteomics. *Mol Cell Proteomics* **20:** 100029. doi:10.1074/mcp.r120.002277

Welch CJ, Talaga ML, Kadav PD, Edwards JL, Bandyopadhyay P, Dam TK. 2020. A capture and release method based on noncovalent ligand cross-linking and facile filtration for purification of lectins and glycoproteins. *J Biol Chem* **295:** 223–236. doi:10.1074/jbc.ra119.010625

Chen S, Qin R, Mahal LK. 2021. Sweet systems: technologies for glycomic analysis and their integration into systems biology. *Crit Rev Biochem Mol Biol* **56:** 301–320. doi:10.1080/10409238.2021.1908953

49 Glycosylation Mutants of Cultured Mammalian Cells

Jeffrey D. Esko, Hans H. Wandall, and Pamela Stanley

Rapid progress in understanding glycosylation pathways of eukaryotes came with the application of genetic strategies to isolate mutants of mammalian cells and yeast with alterations in glycan synthesis. This chapter reviews methods used to isolate mammalian cell glycosylation mutants and the diversity of mutants that may be obtained from selections and screens. The applications of glycosylation mutants to address functional roles of glycans and in glycosylation engineering are discussed briefly. Many of the cell lines described in this chapter are available through the American Type Culture Collection.

HISTORY

The success of bacterial and yeast genetics in isolating mutants and using them to define biochemical pathways led in the late 1960s to the development of somatic cell genetics using mammalian cells. Chinese hamster ovary (CHO) cells were selected by two independent groups for initial experiments to isolate stable mutants. Somatic cell genetic strategies were applied early to glycobiology, yielding numerous mutants in glycoprotein biosynthesis and later in proteoglycan, glycosylphosphatidylinositol (GPI) anchor, and glycolipid biosynthesis. The ability to isolate glycosylation mutants in mammalian cells made it possible to unravel pathways of glycan synthesis and degradation and to identify, isolate, and map structural and regulatory genes. CHO

published by Cold Spring Harbor Laboratory Press; doi: 10.1101/glycobiology.4e.49

cells thus became a focus for experiments to decipher glycosylation pathways and, importantly, provided mutant host cells for the production of viruses and glycoproteins with modified glycans. This proved to be extremely beneficial to the biotechnology industry because most recombinant therapeutics are glycoproteins. CHO cells and CHO glycosylation mutants are now the workhorse of the biotechnology industry. They are particularly useful because they produce only minor, if any, quantities of nonhuman glycans or glycan modifications that give rise to undesirable antibodies. Conserved glycosylation pathways in yeast were delineated by similar approaches (Chapter 23).

Mutants in any cell type often accumulate the precursor immediately upstream of the block in a pathway and thereby reveal the structure of their substrate(s). Sequencing of mutant alleles reveals specific mutations that may give rise to a glycosylation phenotype. In most cases, mutations are loss-of-function and they reduce or abrogate the activity of an enzyme in a pathway; but there are also gain-of-function mutations that activate a silent glycosylation gene, elevate the expression of an existing activity, or inactivate a negative regulatory factor (Figure 49.1). In nearly all cases, glycosylation mutations lead to the presence of altered glycans on cell-surface glycoconjugates and changes in cell properties that link glycan structure to function. Although gene editing techniques using CRISPR/Cas9 or transcription activator-like effector nucleases (TALENs) are now the method of choice for introducing a mutation that weakens or ablates a glycosylation gene (Chapters 27 and 56), initially such approaches did not allow for the serendipitous findings that often emerge from genetic screens. Subsequently, genetic screens were performed using HAP1 (haploid) human cells mutagenized by retroviral gene trap. Such unbiased screens led to the identification of multiple, previously unknown, glycosylation genes. Lately, however, the evolvement of CRISPR tools and the generation of genome-wide libraries have made it possible to conduct both loss-of-function and gain-of-function screens in nonhaploid cell lines, providing an unbiased strategy to discover new genes that influence cellular glycosylation (Figure 49.2).

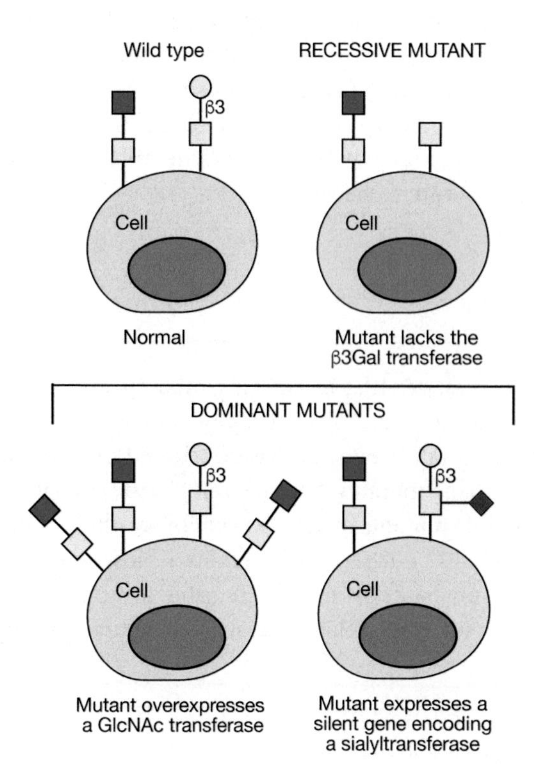

FIGURE 49.1. Alteration of cell-surface glycans by recessive and dominant glycosylation mutations.

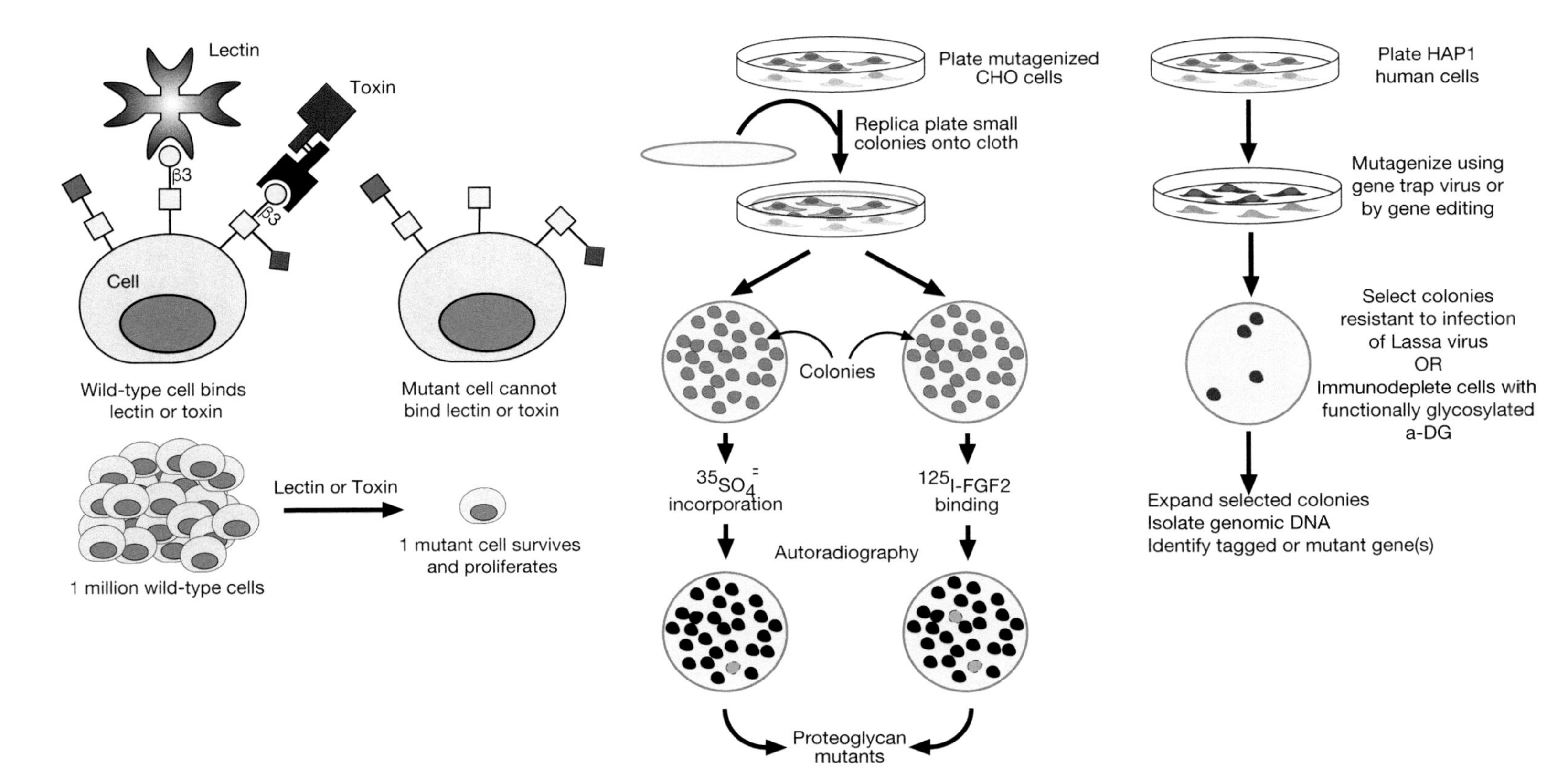

FIGURE 49.2. Selections for glycosylation mutants. Cytotoxic lectins or agents that bind to specific sugar residues select for resistant cells (*left*). Screen for mutants using replica plating. Colonies on plastic are transferred to discs and screened for defects in incorporation of radioactive precursors, binding to lectins and antibodies, or direct enzymatic assay. Mutants are colonies lacking a strong signal (*middle*). HAP1 human haploid cells mutagenized by infection with a gene trap retrovirus and selected for resistance to a Lassa pseudovirus (which needs glycosylated α-DG to infect cells) or immunodepleted for glycosylated α-DG (*right*).

ISOLATION OF GLYCOSYLATION MUTANTS

Cells in culture mutate at a low rate ($<10^{-6}$ mutations per locus per generation). In CHO cells, some loci are functionally haploid (single copy), and in HAP1 human cells, essentially all loci are haploid, which means that a single hit may generate a recessive mutant. However, typical mammalian cells are diploid and immortalized cells are often hyperploid, so the frequency of finding recessive mutants is low. To greatly increase the probability of finding desirable mutants, mutations may be induced by treating cells with chemical (e.g., alkylating agents), physical (e.g., ionizing radiation), or biological (e.g., a virus) mutagens or, perhaps more likely, by lentiviral transduction with CRIPSR/Cas9 genome-wide or focused libraries. Regardless of the method used to induce mutations, selection or enrichment is usually needed to find rare recessive or dominant mutants bearing a desired glycosylation phenotype (Figure 49.1). For example, direct selection for resistance to cytotoxic plant lectins (Chapters 31 and 32) that bind to cell-surface glycans gives a range of glycosylation mutants. Importantly, many mutants resistant to one or more lectins because of the loss of specific sugars become supersensitive to a different group of lectins that recognize sugar residues exposed by the mutation (Figure 49.2). The latter may be used to select for revertants in the original mutant population. Nontoxic lectins are also useful for enriching lectin-binding mutants (e.g., by flow cytometry). Mutations that affect all stages of glycosylation reactions, including the generation and transport of nucleotide sugars, have been identified using lectins as selective agents.

In principle, any glycan-binding protein (GBP), antibody, or other agent that recognizes cell-surface glycans or a glycoprotein can be used to isolate mutants with a glycosylation defect (Figure 49.2). Conjugation of a GBP or protein domain to a toxin that cannot enter the cell independently, but can kill the cell following entry, can be used to select mutants when cytotoxic lectins are not available. For example, basic fibroblast growth factor FGF2-saporin complexes have been used for the selection of mutants deficient in heparan sulfate (HS). Lectins, antibodies, or ligands that are fluorescently tagged may be used to enrich for mutants that are either deficient in binding or have acquired a novel binding ability because of altered glycosylation or reduced expression of an antigen at the cell surface. Panning or immunodepletion are related techniques. For example, coating a plate with FGF2 allows selection of mutant cells that fail to produce HS proteoglycans and consequently fail to adhere to a FGF2-coated plate. HAP1 cells that fail to glycosylate α-dystroglycan fail to bind a specific monoclonal antibody and may be enriched by immunodepletion (Figure 49.2). Radiation suicide is another direct selection method for obtaining glycosylation mutants. Incubation of cells with a radioactive sugar, sulfate, or other precursor of high radioactivity leads to labeled glycoproteins, glycolipids, or proteoglycans. After prolonged storage of the cells, radiation damage kills wild-type cells, whereas mutants with reduced incorporation of the label survive. Animal cells can also be replica-plated, much like microbial colonies, using porous cloth made of polyester or nylon as the replica (Figure 49.2). Colonies of cells on the disc can be used to identify mutants with reduced incorporation of radioactive precursors or to identify mutants that fail to bind to a lectin, an antibody, or a growth factor. An adaptation of this technique allows detection of mutants affecting a specific enzyme by direct assay for activity in colony lysates generated on a disc. Although this technique has great specificity, its limited capacity makes detection of rare mutants difficult, and mutagenesis before screening is usually a requirement.

The resulting strains must be cloned and carefully characterized for stability and the biochemical and molecular basis of mutation. Additional genetic analyses include somatic cell hybridization for dominance/recessive testing and assigning mutants to different genetic complementation groups. Specifically, when mutations are introduced by CRISPR/Cas9 and single-guide RNAs (sgRNAs), the affected genes can be identified through deep sequencing analysis of enriched sgRNA populations. Regardless of the technique used to isolate mutants, biochemical analysis involves the characterization of glycan structures produced by mutant cells (Chapter 50), the quantitation and analysis of intermediates, and assays for activities thought to be missing or acquired based on the properties of the mutant. Identifying the molecular basis of mutation requires isolation of a complementing cDNA that reverts the mutant phenotype and determining whether the mutation arose from defective transcription, translation, or stability of the gene product or from a missense or nonsense mutation in the coding region of the gene. Targeted gene mutation (Chapter 56) can also be used to validate a phenotype after a gene has been identified in a selected mutant.

CELL LINES FROM MICE OR HUMANS WITH A GLYCOSYLATION MUTATION

Transgenic mice that overexpress a glycosylation gene, or mutant mice that lack a glycosylation activity because of targeted gene inactivation (Chapters 27 and 56), are a source of mutant cells that may be used for glycobiology research. Fibroblasts or lymphoblasts can be obtained readily from humans with a disorder of glycosylation (Chapter 45). Cells may be grown as primary cultures or immortalized by viral transformation. By crossing mutant mice with the Immortomouse, which carries a temperature-sensitive SV40 T antigen in every cell, immortalized mutant cell lines can be derived from essentially any cell type. For mutations that cause embryos to die during gestation, mutant embryonic stem (ES) cells can be derived from blastocysts, provided the mutation does not result in cell-autonomous lethality. The resulting mutant ES cell lines can be used to investigate functions for specific glycans during differentiation in embryoid cell culture, or in vivo in mouse chimeras. A chimera is obtained by injecting wild-type or mutant ES cells into the inner cell mass of a mouse blastocyst. If the ES cells survive, the resulting mouse is a mixture of cells derived from the ES cells and cells derived from the blastocyst and is termed a chimera. Mutant ES cells may not contribute equally well to all tissues. For example, ES cells lacking MGAT1 are unable to make complex or hybrid N-glycans (Chapter 9), but they differentiate normally into many cell types in cultured embryoid bodies. However, following introduction into blastocysts, ES cells lacking MGAT1 do not contribute to the organized layer of bronchial epithelium in chimeric embryos.

Immortalized fibroblasts from patients with defects in glycosylation can be used to study the underlying defect (Chapter 45). Induced pluripotent stem cells derived from fibroblasts from patients with glycosylation disorders provide another approach for obtaining various differentiated cell lines for further study. The use of precise genetic engineering tools allows the correction of a genetic mutation in patient-derived fibroblasts, thereby creating pairs of isogenic cell lines for comparative studies. Alternatively, such isogenic pairs can be generated by introducing the patient-specific mutation in a cell or cell line of choice (Chapter 56).

RECESSIVE GLYCOSYLATION MUTANTS

Selection schemes based on isolating rare mutants resistant to cytotoxic plant lectins have yielded a large number of glycosylation mutants affected in diverse aspects of glycan synthesis (Table 49.1). Some mutations affect several types of glycans, such as mutants with reduced nucleotide sugar formation or transport into the Golgi. For example, the UDP-Gal transporter defect in Lec8 mutant cells affects transfer of galactose to O- and N-glycans on glycoproteins, as well as to glycosaminoglycans (GAGs) and glycolipids. The ldlD mutant is particularly interesting in this regard, because it lacks the epimerase responsible for converting UDP-Glc to UDP-Gal and UDP-GlcNAc to UDP-GalNAc (Figure 49.3). Because there are salvage pathways for importing Gal and GalNAc into cells (Chapter 5), the composition of different classes of glycans

TABLE 49.1. Examples of recessive glycosylation mutants

Mutant	Biochemical defect	Mutated gene	Glycosylation phenotype
Lec32 (CHO)	CMP-NeuAc synthetase	*Cmah*	reduced CMP-Neu5Ac synthesis; glycans lack terminal Sia; terminate in Gal
Lec2 (CHO)	CMP-NeuAc transporter	*Slc35a1*	reduced CMP-Neu5Ac transport into Golgi; glycans lack terminal Sia; terminate in Gal
Lec8 (CHO)	UDP-Gal transporter	*Slc35a2*	reduced UDP-Gal transport into Golgi; N-glycans terminate in GlcNAc; O-glycans terminate in GalNAc
Lec13 (CHO)	GDP-Man-4,6-dehydratase	*Gmds*	reduced synthesis GDP-Fuc; glycans lack fucose
ldlD (CHO)	UDP-Gal-4-epimerase	*Gale*	reduced UDP-Gal and UDP-GalNAc synthesis; N-glycans lack Gal; O-GalNAc glycans and chondroitin sulfate not synthesized
Lec1 (CHO)	GlcNAc-TI inactivated	*Mgat1*	no complex or hybrid N-glycans; replaced by $Man_5GlcNAc_2Asn$
Lec1A (CHO)	GlcNAc-TI defective	*Mgat1*	kinetic MGAT1 mutant; partial defect in complex and hybrid N-glycan synthesis
Lec4A (CHO)	GlcNAc-TV mislocalized	*Mgat5*	complex N-glycans lack the β1-6 GlcNAc branch
Lec4 (CHO)	GlcNAc-TV	*Mgat5*	MGAT5 inactivated; defect as above
Lec20 (CHO)	β1–4Gal-TI	*B4galt1*	many glycans low in β1-4 Gal
2A10 (CHO)	ST8SiaIV	*St8sia4*	reduced poly-Sia on N-glycans

Note on nomenclature: capital first letter and lowercase is used for loss-of-function recessive mutants (e.g., Lec32).
CHO, Chinese hamster ovary.

can be controlled in ldlD cells by nutritional supplementation with either of these two sugars. Mutations in glycosyltransferase genes may ablate activity or affect the kinetic properties of an enzyme (e.g., Lec1A; Table 49.1), or its subcellular localization (e.g., Lec4A; Table 49.1). Sequencing mutant alleles provides leads for further site-directed mutagenesis of the gene in order to define important functional domains of the protein required for catalysis or compartmentalization.

Some lectin-resistant mutants are defective in the formation of dolichol-P-oligosaccharides or in the processing reactions that remove Glc or Man after transfer of the glycan chain to a

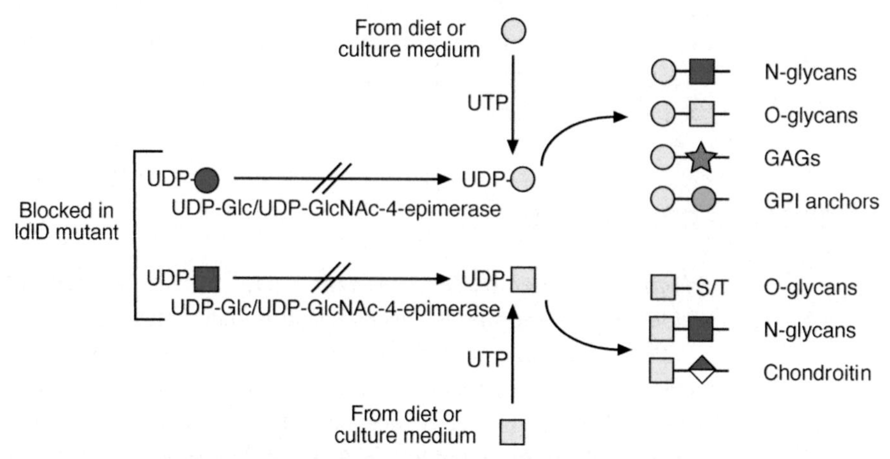

FIGURE 49.3. Mutation of UDP-Glc/UDP-GlcNAc-4-epimerase, also called UDP-Gal-4-epimerase or GALE, in ldlD mutant Chinese hamster ovary (CHO) cells prevents the generation of UDP-Gal and UDP-GalNAc preventing addition of Gal and GalNAc to all glycans. Salvage reactions that generate UDP-Gal or UDP-GalNAc by an alternate pathway may be used to differentially rescue the generation of UDP-Gal or UDP-GalNAc.

TABLE 49.2. Examples of dominant mutants expressing a new activity

Mutant	Biochemical change	Affected gene	Glycosylation phenotype
LEC10 (CHO)	GlcNAc-TIII expressed	*Mgat3*	complex N-glycans have the bisecting N-acetylglucosamine residue
LEC11 (CHO)	α3Fuc-TVI expressed	*Fut6A, Fut6B*	fucose on poly-N-acetyllactosamine generates Le^x, SLe^x, and VIM-2 determinants
LEC12 (CHO)	α3Fuc-TIX expressed	*Fut9*	fucose on poly-N-acetyllactosamine generates Le^x and VIM-2 determinants
LEC29 (CHO)	α3Fuc-TIX expressed	*Fut9*	fucose on poly-N-acetyllactosamine generates Le^x but not VIM-2
LEC30 (CHO)	α3Fuc-TIX expressed	*Fut4, Fut9*	fucose on poly-N-acetyllactosamine generates Le^x and VIM-2 determinants

Note on nomenclature: uppercase is used for gain-of-function dominant mutants (e.g., LEC10).
CHO, Chinese hamster ovary.

glycoprotein (Chapter 9). The latter mutants revealed the identity and importance of α-mannosidases in the formation of N-glycans. However, when the α-mannosidase II gene *Man2a1* was ablated in mice, no effect was seen in certain tissues because another previously unknown α-mannosidase gene (*Man2a2*) allowed N-glycans to be synthesized. This finding emphasizes a limitation of somatic cells in that they may not express glycosylation genes that are developmentally regulated in a tissue-specific manner, thereby precluding the isolation of mutants affected in those genes from that cell line.

DOMINANT GLYCOSYLATION MUTANTS

The recessive mutants in Table 49.1 lack a glycosylation activity or fail to make a precursor. Dominant mutations that activate a silent gene reveal activities that may normally be expressed only in a few, very specialized cells in the body. Therefore, dominant mutants are important in glycosylation gene discovery, for identifying mechanisms of glycosylation gene regulation, and for defining pathways of glycan biosynthesis. The mutants in Table 49.2 show a gain-of-function, dominant, lectin-resistant phenotype caused by the increased expression of a glycosyltransferase that is normally silent or expressed at very low levels. The activation of a glycosyltransferase gene may reflect a mutation in a regulatory region of the gene or in a *trans*-acting factor. The genetic bases of the mutants in Table 49.2 are not known, but their characterization may reveal novel genes or regulatory factors that may not have been previously known to exist.

MUTANTS IN GPI-ANCHOR BIOSYNTHESIS

Glycosylation defects in GPI-anchor biosynthesis reduce expression of GPI-anchored proteins at the cell surface (Chapter 12). Originally, many GPI-anchor mutants were isolated by strategies that took advantage of antibodies to a GPI-anchored glycoprotein. For example, lymphoma cells expressing Thy-1 on their surface were incubated with an antibody to Thy-1 and serum-containing complement, which lysed cells expressing the Thy-1 antigen. Loss of GPI-anchor biosynthesis reduced the expression of Thy-1 on the surface and conferred resistance to the cytolytic effect. Other mutants have been obtained by sorting cells that do not bind to a fluorescent antibody or with bacterial toxins that bind GPI glycans. The GPI-anchor mutants obtained to date fall into many genetic complementation groups, each having a different lesion in GPI-anchor biosynthesis (Chapter 12). These mutants reveal the complexity of GPI-anchor biosynthesis: Multiple gene products are involved in forming the N-acetylglucosamine linkage to phosphatidylinositol, the first committed intermediate in the pathway; dolichol-P-Man is used as the donor of Man; at least three enzymes are involved in the attachment of ethanolamine phosphate residues; and five genes are required for the transfer of the GPI anchor to protein. The available strains show the importance of genetic approaches for identifying genes that might not be obvious from measuring biosynthetic reactions in vitro.

MUTANTS IN PROTEOGLYCAN ASSEMBLY

A large collection of mutants defective in GAG/proteoglycan biosynthesis has been isolated (Table 49.3). Many of these mutants were obtained by replica-plating methods using sulfate incorporation to monitor GAG production in colonies (Figure 49.2). Mutants in the early steps of GAG biosynthesis (complementation groups A, B, and G) lack both CS and HS chains, and enzymatic assays showed that they lack enzymes responsible for the assembly of the core protein linkage tetrasaccharide shared by both these types of GAGs (Chapter 17). Another class of mutants (group D) is defective only in HS biosynthesis. This mutation defines a bifunctional enzyme (EXT1) that catalyzes the alternating addition of GlcNAc and glucuronic acid (GlcA) residues to growing HS chains. Some of the mutant alleles depress both enzyme activities, whereas others only affect the GlcA transfer activity. Thus, the mutants define different functional domains of the protein, which have been mapped by sequencing various mutant alleles. Mutants in another bifunctional enzyme, N-acetylglucosamine N-deacetylase/N-sulfotransferase (NDST1), have only a partial deficiency in N-sulfation of HS chains. Further analysis of the mutant showed that more than one isozyme is present in CHO cells and that the defect affects only one locus. Thus, the mutants revealed early on that the assembly of HS is much more complex than had been appreciated on the basis of known structures, enzyme reactions measured in cell extracts, or intermediates observed in pulse-labeling experiments. The many CHO mutants

TABLE 49.3. Examples of mutants defective in proteoglycan assembly

Strain	Biochemical defect	Mutated gene	Glycosylation phenotype
pgsA (CHO)	xylosyltransferase 2	*Xylt2*	lack of HS and CS
pgsI (CHO)	UDP-xylose synthase	*Uxs1*	lack of HS and CS
pgsB (CHO)	galactosyltransferase I	*B4galt7*	lack of HS and CS
pgsG (CHO)	glucuronyltransferase I	*B3gat1*	lack of HS and CS
pgsD (CHO)	GlcA and GlcNAc transferase	*Ext1*	HS deficient and accumulates CS
ldlD (CHO)	UDP-Gal/UDP-GalNAc-4-epimerase	*Gale*	lack of CS when starved for N-acetylgalactosamine and fed galactose; lack of all GAG chains when starved for galactose
Lec8 (CHO)	UDP-Gal transporter	*Slc35a2*	reduced KS
pgsC (CHO)	sulfate transporter	*Slc26a2*	normal GAG biosynthesis due to salvage of sulfate from oxidation of sulfur-containing amino acids
pgsE (CHO)	GlcNAc N-deacetylase/N-sulfotransferase	*Ndst1*	undersulfated HS
pgsF (CHO)	HS uronyl 2-O-sulfotransferase	*Hs2st*	defective 2-O-sulfation of uronic acids in HS; defective FGF2 binding
Mouse LTA cells	N-sulfoglucosamine 3-O-sulfotransferase	*Hs3st1*	defective 3-O-sulfation of N-sulfoglucosamine units; defective antithrombin binding

CHO, Chinese hamster ovary; HS, heparan sulfate; CS, chondroitin sulfate; KS, keratan sulfate.

have recently been complemented with several gene-edited cell libraries that produce and display distinct GAGs with a broad repertoire of modifications.

MUTANTS DEFECTIVE IN GLYCOLIPID OR O-GLYCAN SYNTHESIS

Glycolipids and glycans linked by O-GalNAc are often relatively simple in cultured cells. For example, CHO cells synthesize mainly gangliosides GM3 and lactosylceramide with a small amount of glucosylceramide. O-GalNAc glycans contain up to only four sugars in glycoproteins from CHO cells. O-Fuc, O-Glc, and O-Man glycans are expressed on only a small subset of glycoproteins and are generally not detected by glycomic profiling methods (Chapter 50). All of these glycans are affected in the mutants described in Table 49.1 in which CMP-Neu5Ac, UDP-Gal, UDP-GalNAc, or GDP-Fuc are reduced in the Golgi. Similarly, a defective sialyltransferase or galactosyltransferase may cause these glycans to be truncated. A mutant of B16 melanoma cells that is defective in ceramide glucosyltransferase (glucosylceramide synthase) lacks all glycolipids because this enzyme catalyzes the first step in the synthetic pathway (Chapter 11). However, cultured cell mutants defective in polypeptide O-GalNAc transferases (GALNTs) or protein O-fucosyltransferase 1 (POFUT1) have not been isolated. This may reflect the paucity of cytotoxic lectins or toxins that bind to O-glycans and glycolipids or, in some cases, due to redundancy of enzymes (Chapters 10 and 11). However, cell mutants defective in most of the GALNTs, POFUT1, POGLUT1, and the various O-mannosyl transferases, including POMT1/2 and TMTC1-3, are now made available through precise genetic engineering. The same is the case with mutants in the biosynthesis of glycosphingolipids. In addition, mice lacking specific glycolipid biosynthetic enzymes and glycosyltransferases that transfer GlcNAc or Fuc to protein have been generated and provide a source of mutant cells that may be studied in culture. Interestingly, cells lacking the O-GlcNAc transferase (OGT) that acts in the cytoplasm to transfer GlcNAc to protein have not been obtained, and mouse mutants defective in this transferase become arrested in development at the two-cell-stage embryo, showing that this O-GlcNAc addition is essential for cell viability.

USES OF MAMMALIAN GLYCOSYLATION MUTANTS

Fortunately, the vast majority of glycosylation mutations allow single-cell viability in vitro under ideal culture conditions. Glycosylation mutants of mammalian cells have thus been used to address many questions in glycobiology and for glycosylation engineering of recombinant glycoproteins (Chapter 56). Because mutant selections are broad and often not intentionally biased, they generate mutants defective in both known and novel reactions. Thus, glycosylation mutants play an important role in research to define the pathways and regulation of glycosylation in mammals. In this regard, they are more useful tools than mutant mice because cells in culture are viable in the absence of glycolipids, GPI anchors, proteoglycans, O-GalNAc, O-Fuc, O-Glc, O-Man glycans, and complex or hybrid N-glycans. Glycosylation mutants make truncated or altered glycans and thus provide an opportunity to study functional roles for cell-surface glycans in the context of a living cell. Important insights have been gained into specific sugars required for viral, bacterial, or parasite adhesion and infection, and for leukocyte cell adhesion and motility. In addition, functional roles for glycans in the intracellular sorting and secretion of glycoproteins, in growth factor binding and activation, and in receptor functions have been identified using glycosylation mutants. For example, a panel of CHO glycosylation mutants was used in a coculture assay to show that ligand-induced Notch signaling is reduced when GDP-Fuc levels are low, but is unaffected by reductions in Sia. Similarly, one of the first demonstrations for coreceptor functions for HS used mutant CHO cells defective in HS synthesis and engineered to express the FGF receptor. Most recently, genome-wide gene editing is being combined with screens for an altered phenotype to identify genes that contribute to that particular phenotype. For example, genes that facilitate infection by SARS-CoV-2 were uncovered in a genome-wide genetic screen.

Although glycosylation is in many cases dispensable for survival of isolated cells in a culture dish, it is often crucial in vivo. Gene ablation studies in mice have identified numerous cases in which an intact glycosylation pathway is essential for embryogenesis. Examples include mutants that lack complex and hybrid N-glycans and proteoglycan mutants defective in HS, whereas the corresponding mutants in CHO cells do not cause an obvious growth phenotype. To address the role of glycans in more complex functions, introduction of mutations in stem cells compatible with creation of organoid or even organotypic 3D tissue cultures is used. For example, a human 3D organotypic skin model was used to address the contributions of various glycans on distinct types of glycoconjugates (glycolipids, N-glycans, O-GalNAc, O-Fuc, O-Glc) by targeting of core extension steps. These studies demonstrated distinct tissue phenotypes for each of the mutant lines grown in 3D cultures, whereas no obvious growth phenotype was observed in conventional 2D cultures. Thus, glycosylation is often critical in the context of a multicellular organism but dispensable in isolated cells. This conclusion has been driven home by gene ablation studies in mice and further evidenced by the discovery of human genetic diseases which arise from mutations in genes involved in glycosylation (Chapter 45).

CHO cells have become the cells of choice for the biotechnology industry in the production of recombinant therapeutic glycoproteins and in glycosylation engineering (see Chapter 56). For example, CHO cells lacking FUT8, which adds fucose to the core GlcNAc of complex N-glycans, are used to produce cytotoxic therapeutic antibodies that have a greatly enhanced ability to kill their target cells. In another example, CHO cells with multiple mutations that simplify N- and O-glycans are being used by X-ray crystallographers to produce homogeneous preparations of membrane glycoproteins with highly truncated N- and O-glycans, greatly facilitating their crystallization.

Somatic cell genetics arose from the desire to manipulate the genome of cultured cells in vitro. Today, the availability of genomic sequences from multiple organisms has shifted the emphasis in genetics toward the generation of mutant organisms using the techniques of transgenesis, homologous recombination for gene replacement, conditional gene inactivation, and precise gene editing. However, the study of somatic cell mutants still plays an important role in glycobiology research because it provides a less-expensive and faster method for studying the effects of deleting or expressing particular glycosylation gene products in a cell. Gain-of-function mutants may of course be generated by transfection of cDNAs encoding glycosylation genes, and reduced expression of any gene can be achieved by the use of RNA interference (RNAi), antisense cDNA strategies, or genome-wide gene editing screens. The combination of genome-wide screening and specific agents to select for glycan changes make it possible to discover new genes by screening for phenotypic changes directly related to glycosylation changes. Additionally, cells and mutants with well-characterized glycosylation pathways are ideal hosts for investigating the activity encoded by a putative glycosylation gene identified in genome sequence databases. These mutant cells also provide a platform to test the severity of human mutations in a complementation test: The normal human gene rescues defective glycosylation when transfected into the mutant cell, but the same gene with a pathological mutation does not. Thus, somatic cell mutants provide access to novel genes involved in glycosylation, which in turn guide strategies for sophisticated gene-manipulation experiments in animals. By combining the two approaches, the biological function of a particular glycosyltransferase, sugar residue, or lectin can be defined. Coupled with powerful mass spectrometry techniques for determining glycan structures (Chapter 50) from small samples of tissue or cells, glycosylation mutants of cells and animals provide complementary material for structure/function analyses and identifying mechanistic bases of glycan functions in mammals.

ACKNOWLEDGMENTS

The authors acknowledge contributions to previous versions of this chapter by Carolyn R. Bertozzi and appreciate helpful comments and suggestions from Taroh Kinoshita and Nathan E. Lewis.

FURTHER READING

Stanley P. 1984. Glycosylation mutants of animal cells. *Annu Rev Genet* **18**: 525–552. doi:10.1146/annurev.ge.18.120184.002521

Esko JD. 1989. Replica plating of animal cells. *Methods Cell Biol* **32**: 387–422. doi:10.1016/s0091-679x(08)61183-8

Esko JD. 1991. Genetic analysis of proteoglycan structure, function and metabolism. *Curr Opin Cell Biol* **3**: 805–816. doi:10.1016/0955-0674(91)90054-3

Stanley P. 1992. Glycosylation engineering. *Glycobiology* **2**: 99–107. doi:10.1101/cshperspect.a005199

Stanley P, Raju TS, Bhaumik M. 1996. CHO cells provide access to novel N-glycans and developmentally regulated glycosyltransferases. *Glycobiology* **6**: 695–699. doi:10.1093/glycob/6.7.695

Esko JD, Selleck SB. 2002. Order out of chaos: assembly of ligand binding sites in heparan sulfate. *Annu Rev Biochem* **71**: 435–471. doi:10.1146/annurev.biochem.71.110601.135458

Maeda Y, Ashida H, Kinoshita T. 2006. CHO glycosylation mutants: GPI anchor. *Methods Enzymol* **416**: 182–205. doi:10.1016/s0076-6879(06)16012-7

Patnaik SK, Stanley P. 2006. Lectin-resistant CHO glycosylation mutants. *Methods Enzymol* **416**: 159–182. doi:10.1016/s0076-6879(06)16011-5

Zhang L, Lawrence R, Frazier BA, Esko JD. 2006. CHO glycosylation mutants: proteoglycans. *Methods Enzymol* **416**: 205–221. doi:10.1016/s0076-6879(06)16013-9

North SJ, Huang HH, Sundaram S, JangLee J, Etienne AT, Trollope A, Chalabi S, Dell A, Stanley P, Haslam SM. 2010. Glycomics profiling of Chinese hamster ovary cell glycosylation mutants reveals *N*-glycans of a novel size and complexity. *J Biol Chem* **285**: 5759–5775. doi:10.1074/jbc.m109.068353

Jae LT, Raaben M, Riemersma M, van Beusekom E, Blomen VA, Velds A, Kerkhoven RM, Carette JE, Topaloglu H, Meinecke P, et al. 2013. Deciphering the glycosylome of dystroglycanopathies using haploid screens for Lassa virus entry. *Science* **340**: 479–483. doi:10.1126/science.1233675

Steentoft C, Bennett EP, Schjoldager KT, Vakhrushev SY, Wandall HH, Clausen H. 2014. Precision genome editing: a small revolution for glycobiology. *Glycobiology* **24**: 663–680. doi:10.1093/glycob/cwu046

Narimatsu Y, Büll C, Chen YH, Wandall HH, Yang Z, Clausen H. 2021. Genetic glycoengineering in mammalian cells. *J Biol Chem* **296**: 100448. doi:10.1016/j.jbc.2021.100448

Zhu Y, Feng F, Hu G, Wang Y, Yu Y, Zhu Y, Segoe UI, Xu W, Cai X, Sun Z, Han W, Segoe UI et al. 2021. A genome-wide CRISPR screen identifies host factors that regulate SARS-CoV-2 entry. *Nat Commun* **12**: 961. doi:10.1038/s41467-021-21213-4

50 Structural Analysis of Glycans

Stuart M. Haslam, Darón I. Freedberg, Barbara Mulloy, Anne Dell, Pamela Stanley, and James H. Prestegard

This chapter surveys techniques for structural characterization of glycans, including composition, linkage, and attachment to aglycones. It covers detection of specific glycan sequences on glycoproteins and cellular surfaces and methods for characterizing structures in three dimensions. The techniques described range from classical chemical detection and characterization of isolated glycan products to sensitive fluorescence methods used in combination with glycan-binding proteins and microscopy on cells and tissues. In addition, nuclear magnetic resonance spectroscopy (NMR) and mass spectrometry (MS), approaches that allow more detailed structural characterization, are discussed.

BACKGROUND

The primary structure of a glycan is defined by the type and order of monosaccharide residues, by the configuration and position of glycosidic linkages, and by the nature and location of the nonglycan entity to which it is attached (the aglycone); see Chapters 2 and 3. For glycoproteins, different glycans can be attached to different sites in the protein, and these glycans can vary

published by Cold Spring Harbor Laboratory Press; doi: 10.1101/glycobiology.4e.50

when that glycoprotein is made in different cell types or at different stages of development. Moreover, it is often three-dimensional (3D) features, or particular surface distributions of glycans, that are recognized by glycan-binding proteins. Characterization of these diverse structural features requires an array of different methods, with the choice of methodology depending very much on the problem.

For a typical mammalian glycoprotein, the aim is often to identify the correct glycan structure from a range of known or predictable candidate structures, and a limited amount of structural data may suffice (Chapter 51). For glycans from bacteria or less well-characterized organisms, it is hard to make predictions, and therefore, a more complete data set may be required. The choice of methodology also depends on the amount and purity of material available, as well as the context in which data must be collected (e.g., tissue vs. isolated glycoprotein). If quantities are not limiting, the complete primary structure and even the tertiary (3D) structure may be determined. The need to respond to diverse circumstances and to understand the complexities of glycan structure has driven the development of many of the methods described in the following sections. The level of glycan structural characterization required to address the specific biological question that is being investigated can also direct experimental approaches.

DETECTION OF GLYCANS

Methods for glycan detection in glycoconjugates include direct chemical reactions with the constituent monosaccharides, metabolic labeling with either radioactive or chemically reactive monosaccharides, and detection with specific glycan-recognizing proteins (including lectins and antibodies) (Chapter 48). A general method for detecting the presence of glycans on proteins involves periodate oxidation of their hydroxyl groups followed by Schiff base formation with amine- or hydrazide-based probes (Chapter 2). This chemical modification, also known as the periodic acid–Schiff (PAS) reaction, can identify glycoproteins in gels. Commercially available kits allow detection of 5–10 ng of glycoprotein, using the periodate reaction with subsequent amplification by means of biotin-hydrazide/streptavidin-alkaline phosphatase or fluorescence-based detection. Lectin overlay of a blot of a sodium dodecyl sulfate polyacrylamide gel electrophoresis (SDS-PAGE) gel can detect the presence of specific glycans with comparable sensitivity and greater specificity. For example, the agglutinin from *Sambucus nigra* (SNA) binds to glycans that terminate with α2-6 sialic acid (Sia). Lectins (Chapters 31–36) recognizing terminal fucose (Fuc), galactose (Gal), N-acetylgalactosamine (GalNAc), and N-acetylglucosamine (GlcNAc) are also commercially available.

Metabolic labeling of glycoconjugates with radioactive sugars is another powerful tool for determining the composition of glycans. Cells incubated in media containing ^{3}H- or ^{14}C-labeled monosaccharides will incorporate the label into the glycans of glycoconjugates. Radiolabeled glycans can be detected following gel electrophoresis (SDS-PAGE) or thin-layer chromatography (TLC) by autoradiography or fluorography. They can also be released and studied in detail by various methods. The use of fluorescent probes and labels has enabled a reduction in the use of radioisotopes in applications in which the detection and quantitative determination of glycans, as opposed to glycosylation pathway information, is the primary objective. Fluorescent labels for sensitive detection of glycans after liquid chromatography (LC) include readily available 2-amino benzoic acid (2-AA) and 2-aminobenzamide (2-AB), which may be attached by reductive amination to the reducing sugar exposed on release of glycans from an aglycone.

Metabolic labeling can also be performed with synthetic monosaccharides that are modified with chemically reactive groups. For example, the azido monosaccharide N-azidoacetylmannosamine (ManNAz) is converted by cells to N-azidoacetyl Sia (SiaNAz), which is incorporated into sialylated glycans in place of a natural Sia. The azido group can then be selectively reacted with phosphine or alkyne reagents (Chapter 53) that introduce a fluorescent dye or an affinity probe such as biotin, thereby enabling detection of Sia in cells. Azido analogs of GalNAc and GlcNAc can be used to label O-GalNAc (Chapter 10) or O-GlcNAc glycans (Chapters 13 and 19), respectively. Use of fluorescent labels can also be coupled with confocal microscopy to give important insights into the location of glycans in cells and tissues. However, such chemical modifications may change the biosynthesis and/or biology of glycans, creating some uncertainty about the observed results, and no method has completely supplanted radioactive metabolic labeling for pulse-chase studies of naturally occurring glycans.

Glycoproteins

A glycosylated protein typically presents one or more diffuse bands during gel electrophoresis, resulting from heterogeneity of the attached glycans. Even when visualized by protein staining reagents, this phenomenon is often the first indication of the presence of glycans. Some high-molecular-weight glycoconjugates, such as mucins and proteoglycans, do not enter ordinary gels or, if they do, they migrate as heterogeneous smears. Agarose gels or combination polyacrylamide-agarose gels may be useful in this situation. Several analytical options are available to investigate the presence of glycans further (e.g., the PAS stain described above). Treatment of glycoproteins with endoglycosidases (e.g., peptide-N-glycosidase F [PNGase F], endoglycosidase F2 [Endo F2], endoglycosidase H [Endo H]) is another option; see Table 50.1 and Figure 50.1. If this results in a mobility change of one or more of the bands on the gel, the presence of N-glycans is indicated. O-glycanase (endo-α-N-acetylgalactosaminidase; Table 50.1) can be used for the specific identification of O-glycans. However, depending on the structure of the O-glycan, pretreatment with other enzymes to expose the disaccharide core is sometimes required. Removal of individual sugars by exoglycosidases such as sialidase or β-galactosidase (Figure 50.1) may also result in a mobility change if a sufficient number of residues is removed. Some glycans cannot be altered by these treatments as they are resistant to the enzymes used. Resistance can result from modifications to glycan hydroxyl groups (e.g., sulfation, acetylation, or phosphorylation; Chapter 2), glycosidic linkages that are not recognized by the enzymes, or steric inaccessibility of the glycan. Complete removal of N- and O-glycans can be achieved by chemical treatments (e.g., hydrazinolysis or β-elimination), but peptide damage usually precludes further analysis by gel electrophoresis. Partial degradation (e.g., by loss of O-acetylation) of the glycan may also occur.

Proteoglycans

Proteoglycans (Chapter 17) may be separated by agarose gel electrophoresis and by ion-exchange chromatography, which separates on the basis of charge conferred by sulfate groups. Treatment of proteoglycans with GAG lyases (Table 50.1) will produce a shift in mobility on a gel, condensing the proteoglycan smear into discrete bands. After removal of much of the glycan portion, antibodies that recognize the remaining "stub" glycans may be used in western analysis.

Glycolipids

Typically, the analysis of glycolipid glycans by NMR or MS is preceded by their chromatographic purification. Mixtures of glycolipids can be fractionated by TLC and TLC plates stained with glycan-reactive reagents, allowing detection of many individual glycolipids. Using different reagents, it is possible to recognize gangliosides (e.g., resorcinol-HCl detects Sia) or neutral monosaccharides (e.g., orcinol-sulfuric acid detects all monosaccharides) in TLC bands. Reagents are also available for detection of sulfate and phosphate groups on glycolipids. Some pre-purification of the crude extract is usually required (e.g., Folch partitioning and ion-exchange chromatography). These procedures separate nonpolar or nonionic lipids from polar lipids (e.g., glycosphingolipids; Chapter 11)

TABLE 50.1. Table of enzymes for glycan analysis

	EC number	Specificity (cleavage site)	Analytical Uses
Endoglycosidases			
Endoglycosidase H (Endo H)	3.2.1.96	between the two GlcNAc residues in the core of oligomannose or hybrid N-glycans	detection of N-glycosylation; release of glycans
Endoglycosidase F2 (Endo F)	3.2.1.96	between the two GlcNAc residues in the core of oligomannose or biantennary N-glycans	detection of N-glycosylation; release of glycans
Peptide-N-glycosidase F (PNGase F), or N-Glycanase	3.5.1.52	between Asn and GlcNAc in the core of oligomannose, complex, or hybrid N-glycans; requires at least one amino acid at both the amino and carboxy terminal of Asn	may not work if core is α1-3-fucosylated
Peptide-N-glycosidase A (PNGase A) or glycopeptidase A	3.5.1.52	between Asn and GlcNAc in the core of oligomannose, complex, or hybrid N-glycans; requires at least one amino acid at both the amino and carboxy terminal of Asn	will work if core is α1-3-fucosylated; efficacy for use on mammalian glycoproteins poorly defined
Endo-β-galactosidases	3.2.1.102	between Gal and GlcNAc in poly-N-acetyllactosmine units	detection of poly-*N*-acetyllactosamines and some keratan sulfates
Endoneuraminidases (endo-α-sialidase)	3.2.1.129	between Sia units of polysialic acids	detection of polysialic acids
Endo-α-N-acetylgalactosaminidase (O-glycanase)	3.2.1.97	between Ser or Thr and Galβ-GalNAcα- core of O-GalNAc glycans	detects O-glycosylation. May need prior treatment with other enzymes to expose the disaccharide core
Exoglycosidases			
Sialidases, neuraminidases	3.2.1.18	removes terminal α-Sia	
Fucosidase	3.2.1.51	removes terminal α-Fuc	some types and linkages can be resistant
α1-2 Fucosidase	3.2.1.63	removes terminal Fuc- linked α1-2	
β-Galactosidase	3.2.1.23	removes terminal β-Gal	
α-Mannosidase	3.2.1.24	removes terminal α-Man	
Glycosaminoglycan lyases			
Chondroitinase ABC	4.2.2.4	reduces chondroitin sulfates A and C and dermatan sulfate to disaccharides	detection and characterization of chondroitins, purification of other glycosaminoglycans
Chondroitinase AC	4.2.2.5	reduces chondroitin sulfates A and C to disaccharides	detection and characterization of chondroitins A and C
Chondroitinase B	4.2.2.19	reduces dermatan sulfate (chondroitin sulfate B) to disaccharides	detection and characterization of dermatan sulfate
Heparinase, heparin lyase, Heparinase I	4.2.2.7	between *N*-sulfated glucosamine and 2-*O*-sulfated iduronate residues.	detection and characterization of heparin and heparan sulfate
Heparan sulfate lyase, heparinase III (also called Heparitinase I), Heparinase II	4.2.2.8	between GlcNAc and GlcA residues. Heparinase I, II, and III often used in combination.	detection and characterization of heparin and heparan sulfate
Keratanase	3.2.1.103	between Gal and GlcA	detection and characterization of keratan sulfate
Keratanase II	3.2.1.103	between 6-O-sulfated GlcNAc and Gal (±6-O-sulfate)	detection and characterization of keratan sulfate

EC, Enzyme Commission.

and those that contain charged groups (i.e., gangliosides, phospholipids, and sulfatides). It is also common practice to deduce the presence of specific sugars by evaluating the shifts produced in the migration position of a band following a chemical or enzymatic treatment. Glycolipids on TLC plates can also be detected by monoclonal antibodies, lectins, or even intact microorganisms expressing glycan-specific receptors (Chapter 48). Detailed structural features may be identified by running the TLC in a second dimension following a specific treatment. On a larger scale, glycolipids are separated using column chromatography or by high-performance TLC on silica plates.

Glycosidases remove sugars

α-Mannosidase β-Mannosidase α-Fucosidase

N-Glycanase
Endoglycosidase F

Neuraminidase β-Galactosidase β-N-Acetylhexosaminidase

Exoglycosidases remove terminal sugars
Endoglycosidases remove glycans

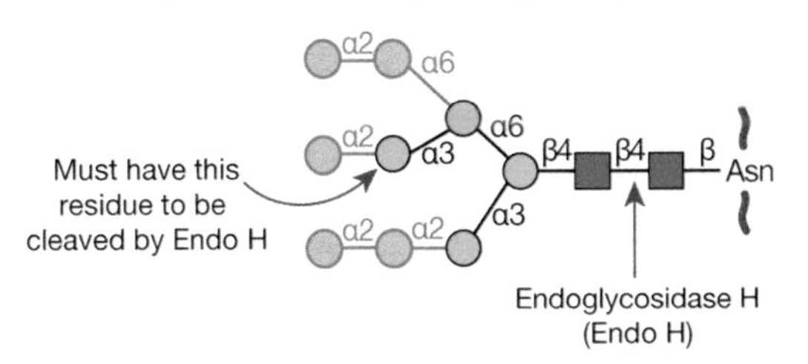

FIGURE 50.1. Glycosidases used for structural analysis. (*Left*) A biantennary N-glycan is shown with exoglycosidases that remove each monosaccharide sequentially. Exoglycosidases act only on terminal sugars. Also shown are endoglycosidases that digest in the core region of the N-glycan. N-Glycanase cleaves the GlcNAc-Asn bond, releasing the N-glycan and converting Asn into Asp, which is diagnostic of an N-glycan site. Endo F cleaves between the core GlcNAc residues and therefore leaves a GlcNAc, with or without an attached fucose, linked to the protein. (*Right*) Endo H cleaves between the core GlcNAc residues of oligomannose or hybrid N-glycans that have at least four Man residues as shown. Endo H does not act on complex N-glycans.

GPI Anchors

GPI-anchored proteins (Chapter 12), with their lipid, protein, and glycan constituents, have unique physicochemical properties that can be exploited for their detection. The nonionic detergent Triton X-114 at low temperature (4°C) extracts soluble and integral membrane proteins, as well as GPI-anchored proteins. When the solution is warmed, two phases separate, and GPI-anchored and other amphiphilic proteins remain associated with the detergent-enriched phase. GPI-specific phospholipases can be used to cleave GPI anchors for further characterization. Successful cleavage by GPI-specific phospholipases can be assessed using SDS-PAGE, because removal of the GPI anchor causes a shift in molecular mass. This is a common diagnostic method for identifying the presence of a GPI anchor on a protein of interest. Another method is to treat the GPI-anchored protein with nitrous acid, which cleaves the unsubstituted glucosamine residue that links the glycan to the phosphatidylinositol.

Plant and Bacterial Polysaccharides

This class of glycans contains many structures, including homopolysaccharides and heteropolysaccharides, neutral and ionic polysaccharides, and linear and branched structures, with widespread molecular sizes ranging from a few monosaccharide units to thousands (Chapters 3, 21, and 22). Most of these polymers are insoluble in aqueous solutions, occurring as complex and sometimes crystalline aggregates. But many can be extracted with water, salts, chaotropic agents, or detergents and isolated by precipitation with alcohols. Detection is based on refractive index (RI) or colorimetric reactions, because sample quantity is not usually a limitation.

RELEASE AND SEPARATION OF GLYCANS

Once the presence and general type of glycan has been established, the next challenge is to release specific types of glycans and separate different classes in sufficient quantities for structural characterization.

Release of Glycans from Glycoconjugates

When glycans are released before structural analysis, it is best to use a quantitative release procedure that neither destroys nor alters the glycan. Ideally, information regarding the type of linkage between the glycan and its liberated protein or lipid should be retained, although this is not always possible. Glycolipids can often be isolated and characterized by MS and/or NMR without the need for release of glycans, but, if necessary, enzymatic methods can be used, or for glycosphingolipids, ozonolysis will separate lipid from glycan. Complex, hybrid and oligomannose N-glycans can be released from glycoproteins with N-glycosidases such as PNGase F or PNGase A (Table 50.1). Endo H may be used for the selective release of high-mannose and hybrid N-glycans, but complex N-glycans are resistant (Table 50.1; Figure 50.1).

Chemical approaches suitable for release of glycans from a protein include hydrazinolysis, which releases N-glycans and/or O-glycans, depending on experimental conditions. Strong base treatment can, under carefully controlled conditions, release only O-glycans in a process called β-elimination; it is generally accompanied by reduction with borohydride to give an alditol. More recently methods have been developed in which base treatment is accompanied by derivatization with pyrazolone, which acts as a UV-absorbing label during chromatographic separation. However, all of the above chemical methods can result in partial or complete loss of labile modifications of the glycan, such as O-acetylation or sulfation, and degradation of the protein.

Separation of Glycans

Glycans released from a glycoconjugate usually form a complex mixture. Even when only one glycosylation site in the protein is occupied, individual molecules can bear different glycan species generating multiple glycoforms. The high-throughput analysis of these glycan mixtures using glycomics technologies is described in Chapter 51.

Preparation of glycan samples for structural analysis will rely on separation techniques. Chromatographic separations commonly used to isolate pure glycans, or at least glycan mixtures of reduced complexity, include size exclusion chromatography (SEC), strong or weak anion exchange chromatography (SAX or WAX), and some forms of reverse-phase high-pressure liquid chromatography (HPLC). Electrophoretic methods include capillary electrophoresis (CE) and fluorophore-assisted carbohydrate electrophoresis (FACE).

HPLC-based separation is feasible, even for large amounts (>10 mg), if the mixture contains fewer than approximately 50 glycan species. Individual fractions can be recovered for further structural analysis by MS or NMR. Once liberated from their glycoconjugates, glycans with free-reducing termini (Chapter 2) can be chemically labeled with fluorescent tags, such as 2-amino benzoic acid (2-AA) and 2-aminobenzamide (2-AB) or 8-aminonaphthalene-1,3,6-trisulfonic acid (ANTS), providing detection sensitivity that rivals the sensitivity achieved with radiolabels. Advantages of this method include more facile purification of the labeled glycans and a wider variety of options for chromatographic separations and structural analysis techniques. Nonlabeled glycans may be detected by (relatively insensitive) RI methods, or by more sensitive pulsed amperometric detection (PAD).

If a label is introduced at the reducing end, it is possible to use sequential exoglycosidase treatments (Table 50.1; Figure 50.1) and look for shifts of the oligosaccharide in a chromatographic system (e.g., paper chromatography, HPLC, TLC) as an indication of susceptibility to the enzyme. Comparison with known standards treated in the same manner allows tentative glycan identification. However, well-characterized standards of known structure are difficult to obtain in a pure form and there are nearly always "contaminating" peaks in a chromatogram that cannot be readily identified. It is very important to note that separation profiling should not be confused with actual structural analysis, because coelution with a standard could occur with a variety of different structures.

MONOSACCHARIDE COMPOSITION ANALYSIS

Some qualitative information on monosaccharide composition of a glycan may be derived from the procedures described above. However, it is often convenient and informative to determine monosaccharide composition without prior release of glycans. After total hydrolysis of a glycan into its monosaccharide constituents, colorimetric reactions can be used to determine the total amount of hexose, hexuronic acid, or hexosamine in a sample. These approaches only require common reagents and a spectrophotometer, but determination of total glycan content may not always be accurate because of variations in the sensitivities of different linkages to hydrolysis, variations in the degradation of individual saccharides, or a lack of specificity and/or sensitivity in the assays.

Quantitative monosaccharide analysis provides estimated molar ratios of individual monosaccharide components of glycans and may suggest the presence of specific oligosaccharide classes (e.g., N-glycans vs. O-glycans). The analysis involves the following steps: cleavage of all glycosidic linkages (typically by acid hydrolysis), fractionation of the resulting monosaccharides, detection, and quantification. Since the early 1960s, a variety of gas chromatography (GC) methods have been developed to quantify monosaccharides. The most useful involve coupling of GC and MS for linkage and composition information. GC requires volatile samples so the monosaccharides are first chemically modified at their hydroxyl, amide, carboxyl, and aldehyde groups. Reduction of the aldehyde of free monosaccharides followed by acetylation of their hydroxyl groups provides derivatives termed "alditol acetates." These modified monosaccharides can be readily analyzed by GC-MS and compared with authentic standards. The hydroxyl and amino groups of free monosaccharides generated by glycan hydrolysis can also be converted to trimethylsilyl ethers. These per-O-trimethylsilyl derivatives are widely used for monosaccharide compositional analysis by GC-MS. Incorporation of an optically pure chiral aglycone (e.g., [−]-2-butyl alcohol), in combination with trimethylsilylation, allows the GC separation of D and L isomers and thus determination of the absolute configuration of each monosaccharide.

An alternative to GC-MS is high-pH anion-exchange chromatography with pulsed amperometric detection (HPAEC-PAD), a special type of ion exchange chromatography which does not require monosaccharide derivatization. This technique is especially useful for analyzing acidic sugars such as Sia, which are refractory to GC-MS. A convenient method for quantitating Sia, that does not require expensive equipment, involves tagging with 1,2-diamino-4,5-methylene-dioxybenzene and measuring fluorescence. This method has a detection sensitivity in the femtomole range, and can also pick up many of the naturally occurring modifications of this diverse class of monosaccharides (Chapter 15). Other popular techniques for defining monosaccharide composition are HPLC and high-performance capillary electrophoresis (HPCE). The monosaccharides are usually tagged with a fluorescent derivative for high-sensitivity detection. Tagging with 8-amino-1,3,6-naphthalene trisulfonic acid yields anionic derivatives that can be conveniently analyzed by gel electrophoresis. This is referred to as fluorophore-assisted carbohydrate electrophoresis (FACE).

LINKAGE ANALYSIS

Determination of Linkage Positions

Linkage analysis is a well-established and ingenious approach for determining linkage positions. The principle of this method is to introduce a stable substituent (an ether-linked methyl group) onto each free hydroxyl group of the native glycan. Glycosidic linkages are then cleaved by acid hydrolysis, producing partially methylated monosaccharides with free hydroxyl groups at the positions that were previously involved in a linkage. The partially methylated monosaccharides are ring-opened with a reducing agent (normally borodeuteride) to introduce a new hydroxyl group and, more importantly, a deuterium atom at C-1, which helps identify the reducing end of each monosaccharide. All the free hydroxyl groups are then acetylated resulting in partially methylated alditol acetates (PMAAs) that can be identified by a combination of GC retention times and electron impact (EI)-MS (Figure 50.2). The masses of fragments produced by impact of high-energy electrons on PMAAs identify substitution sites in some cases, but fragmentation patterns of similarly substituted isomeric monosaccharides (e.g., Glc and Gal) can be nearly identical. Thus, definitive identification of monosaccharides requires, in addition to the analysis of the MS pattern, a comparison of GC retention times with those of known standards (e.g., all peracetylated 2,3,4-tri-O-methyl-hexoses produce the same EI-MS spectrum, but peracetylated 2,3,4-tri-O-methyl-galactitol elutes later than peracetylated 2,3,4-tri-O-methyl-glucitol). This type of analysis identifies terminal residues (they are methylated at every position except the hydroxyl group at C-1 and the ring oxygen), indicates how each monosaccharide is substituted, including the occurrence of linkage and branching points, and allows the determination of the ring size (pyranose [*p*] or furanose [*f*]) for each monosaccharide. However, linkage analysis provides no information on the nature of substituents or sequence information and cannot reveal α- or β-anomeric configuration.

Mass Spectrometry

EI-MS in monosaccharide composition and linkage analyses is covered above. In this section, the other main types of ionization used in MS analysis of glycans—matrix-assisted laser desorption/ionization (MALDI) and electrospray ionization (ESI)—are described. These technologies permit the direct ionization of nonvolatile substances, and are applicable to intact glycoconjugates, as well as fragments thereof. These techniques continue to exploit sample handling strategies and knowledge of glycan fragmentation pathways originally developed for the earlier fast atom bombardment (FAB)-MS. Among the structural features that can be defined by MS methods are (1) degree of heterogeneity and type of glycosylation (e.g., N-glycan vs. O-glycan; oligomannose, hybrid, or complex N-glycans); (2) sites of glycosylation and identity of the protein/lipid carrier; (3) glycan branching; (4) number and lengths of antennae, their composition, and substitution with Fuc, Sia, or other capping groups such as sulfate, phosphate, or acetyl esters; and (5) complete sequences of individual glycans.

In MALDI-MS experiments, the sample is dried on a metal target in the presence of a light-absorbing matrix until matrix crystals containing trapped sample molecules are formed. Ionization of the sample is affected by energy transfer from matrix molecules that have absorbed energy from laser pulses. MALDI-MS is the preeminent technique for screening for molecular ions in complex mixtures, and it is a powerful tool for glycomic profiling (Chapter 51). For ESI-MS, a stream of liquid containing the sample enters the source through a capillary interface, where the sample molecules are stripped of solvent, leaving them as multiply charged species. ESI-MS can be coupled to nano- or capillary-bore LC permitting online LC/ESI-MS analysis. This method is particularly useful when complex mixtures of peptides and glycopeptides are being analyzed (e.g., after proteolytic digestion of a glycoprotein).

FIGURE 50.2. (*See following page.*) An example of linkage analysis showing a bacterial O-linked branched hexasaccharide with a sequence of Rha1-3Glc1-(Glc1-3GlcNAc1-)2,6Glc1-6GlcNAc. The O-glycan is reductively eliminated from protein before successive steps of permethylation, acid hydrolysis, borodeuteride reduction, and acetylation result in a set of compounds in which the linkage positions can be identified based on their acetylated hydroxyl groups (selective ions are shown on structures of PMAAs). For the sequence illustrated, the terminal Glc and Rha give rise to alditols only acetylated at C-1 and C-5; the 3-linked Glc and GlcNAc units on either side of the branch point are acetylated at C-1, C-5, and C-3 and deuterated at C1 for the internal glycans; the branch-point Glc is acetylated at C-1, C-5, C-2, and C-6; and the 6-GlcNAcitol (it is at the reducing end of the glycan, therefore has been reduced during reductive elimination before linkage analysis, and it carries no deuterium) is acetylated only at C-6.

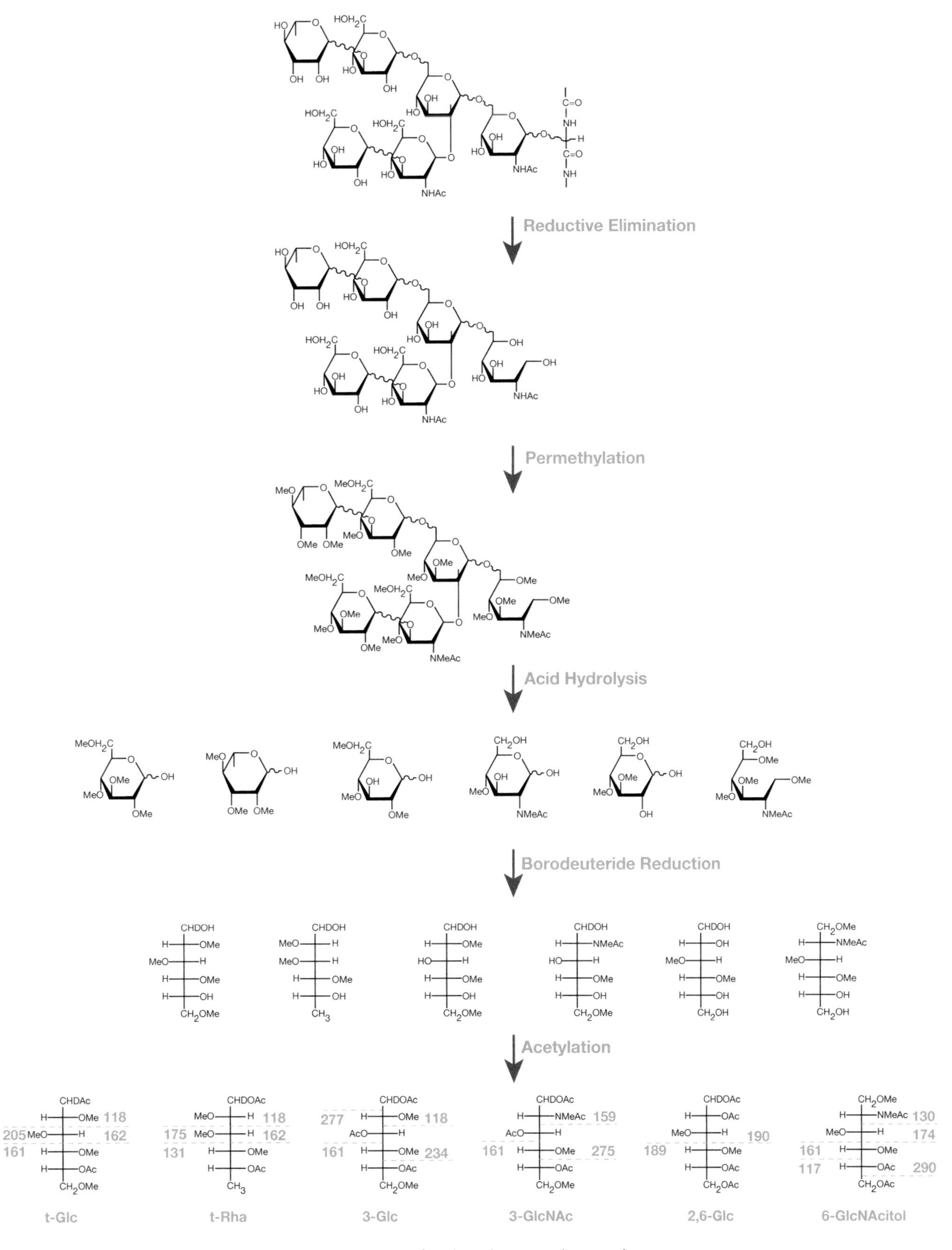

FIGURE 50.2. *(See legend on preceding page.)*

In principle, MS provides two types of structural information—the masses of intact molecules (the molecular ions) and the masses of fragment ions. However, in both ESI- and MALDI-MS, the ionization process is usually not sufficiently energetic for fragment ion formation. To overcome this, ESI and MALDI instruments normally have two analyzers in tandem, which allows detection by the second analyzer, of fragment ions produced after molecular ions selected by the first analyzer undergo collisions with an inert gas in a chamber placed between the two analyzers. This is referred to as collision-activated dissociation (CAD) or collision-induced dissociation (CID). MALDI-MS/MS is commonly performed on instruments that have two time-of-flight (TOF) analyzers in tandem (TOF-TOF), whereas the most popular instruments for ESI-MS/MS have either a quadrupole (Q) as the first analyzer and an orthogonal TOF as the second analyzer (QTOF) or a linear ion trap as the first analyzer and an Orbitrap as the second analyzer (LTQ-Orbitrap). Some MS/MS instruments have a single "ion trap" analyzer that traps ions and functions as the collision chamber. Ion traps are relatively inexpensive, but they lack the sensitivity, versatility, and mass range of most tandem instruments. The Fourier transform mass spectrometer (FTMS), a very expensive but very powerful instrument, has extremely high mass accuracy. An alternative to CAD for fragment ion production has been developed for FTMS. This is called electron capture dissociation (ECD) and involves the capture of an electron by the multiply charged molecular ion. A related technique, called electron transfer dissociation (ETD) was subsequently developed for tandem instruments. Both ECD and ETD yield radical cations generated by different fragmentation pathways than ions produced by CAD. ECD and ETD are especially useful for defining glycosylation sites in glycopeptides.

Although underivatized glycans can be analyzed by MALDI- or ESI-MS/MS, superior data can be obtained if the glycans are derivatized before MS analysis. Derivatization methods can be broadly divided into two categories: (1) "tagging" of reducing ends and (2) protection of all functional groups. Commonly used tagging reagents include p-aminobenzoic acid ethyl ester, 2-AP and 2-AB. This type of derivatization facilitates chromatographic purification and enhances useful fragmentation. Protection of hydroxyl groups by permethylation is by far the most important type of full derivatization used in glycan MS (although with accompanying destruction of acetyl esters and other base-labile functional groups during the derivatization process). In MS/MS experiments, permethylated derivatives form abundant fragment ions produced by cleavage at susceptible glycosidic linkages, notably on the reducing side of each HexNAc residue. In N- and O-glycans, this preferred fragmentation unambiguously establishes the antennae sequences. MALDI- and ESI-MS can equally be applied to the characterization of glycolipid and GAG derived glycans. However, in the latter case, because of the size and high degree of sulfation of GAGs, MS analysis is usually performed after hydrolysis of the polymeric chain.

As described above, mass spectrometry measures the mass to charge ratio of glycans during the analytical process. However, as multiple monosaccharides that make up a glycan can have the same mass (e.g., mannose, galactose and glucose) and monosaccharides can be linked together in different ways, a single peak in a glycan mass spectrum can actually be made up of multiple structural isomers. Ion mobility mass spectrometry additionally separates ions based on their size and shape which can resolve glycan structural isomers with the same mass to charge ratio (see Further Reading).

Broadly speaking, the unique strengths of MS can be exploited in two general ways in glycobiology. The first is to obtain detailed characterization of purified individual glycans or mixtures of glycans. In this type of study, it is essential to acquire sufficient rigorous data to define structure unambiguously; many different MS-based experiments will be required, often complemented by NMR, linkage analysis, and profiling of enzyme digests. The second is to pursue glycomics investigations in situations in which it may not be essential to define structures fully, and high-throughput glycomic profiling or mass-mapping procedures in combination with specific glycosylhydrolases can be exploited (Chapter 51).

NMR Spectroscopy

NMR spectroscopy is a powerful tool capable of full de novo structural characterization of isolated glycans and simple glycan mixtures. Among its several advantages are its broad applicability (glycans contain an abundance of hydrogen (^{1}H), the most easily detected magnetic nucleus), its nondestructive nature, the quantitative relationship between resonance intensity and residue concentration, the diverse set of experiments available, and the ability to return information on primary, secondary, and tertiary structure. Its principal limitation is sensitivity, typically requiring several nanomoles of glycan for even the simplest structure determination. Other limitations for both structural and metabolic applications are the relatively high cost of spectrometers and the level of expertise required for acquiring and interpreting NMR spectra. Nevertheless, there are emerging experiments using methods, such as dynamic nuclear polarization and looped, projected spectroscopy (L-PROSY), that provide several orders of magnitude improvement in sensitivity for applications that can monitor rapid in vivo conversions of metabolites such as glucose and fructose. Diffusion-based methods can also yield spectra of mixtures of components, separated by diffusion constants (effective molecular size).

When small amounts of sample are available and the source of the sample is known, the acquisition of a 1D ^{1}H NMR spectrum, combined with simple 2D total correlation spectroscopy (TOCSY), can frequently identify a glycan. A 1D ^{1}H NMR spectrum can be collected on ~30 µg of a decasaccharide in ~20 min on a state-of-the-art spectrometer (900 MHz with a cryoprobe). The anomeric ^{1}H resonances are usually well-resolved in a relatively noncrowded region of the spectrum and show characteristic scalar couplings that are small for most α anomers and significantly larger for most β anomers. Thus, a first glance at the ^{1}H -NMR spectrum gives a good estimate of how many residues there are (one anomeric ^{1}H per residue except for Sia and KDOs) in a given oligosaccharide and how many of them belong to each anomeric type. NMR also provides a good estimate of how many residues are in a repeating unit of a polysaccharide. Connections from the anomeric (H1) resonances to the H2 and subsequently other resonances of a scalar coupled spin system can usually be seen in 2D TOCSY data, which show chemical shifts (resonance positions) for coupled ^{1}Hs as crosspeaks on columns or rows emanating from an anomeric resonance (a ~12-h acquisition for a 30-µg sample). The chemical shifts of two or more connected ^{1}Hs are often sufficient to identify a residue. For example, the downfield (higher ppm) position and small couplings of three H2 resonances connected by crosspeaks to three α-anomeric resonances can identify these as belonging to core Man residues of an N-glycan. Historically, a collection of easily resolved resonances, often from 1D spectra, sensitive to both monosaccharide type and linkage type, were called reporter group resonances. These often prove sufficient to identify a glycan, particularly when options are restricted by knowledge of biosynthetic pathways in the system under study.

Today, it is possible to greatly extend glycan identification by including ^{13}C chemical shifts correlated with directly bonded ^{1}Hs using 2D heteronuclear single quantum coherence (HSQC) spectra (Figure. 50.3). ^{13}C is magnetically active, but ^{13}C spectroscopy is less sensitive than ^{1}H spectroscopy. To improve sensitivity, ^{13}C chemical shifts are indirectly detected through ^{1}Hs. Even with the improved detection, ^{13}C is only 1% naturally abundant, and unless the sample is isotopically enriched, approximately two orders of magnitude more material is required for collection of these data. Improvements in NMR hardware can reduce the quantities or time required and allow hybrid experiments, such as 2D HSQC-TOCSY, also illustrated in Figure 50.3. Furthermore, significant spectral overlap can be overcome with recent 2D NMR experiments optimized by using direct detection on ^{13}C combined with optimized data processing. Web-based computational tools that rely on empirical rules or databases that correlate chemical shifts with structural features now exist, and these allow facile prediction of possible structures from 1D and 2D NMR data (e.g., CASPER).

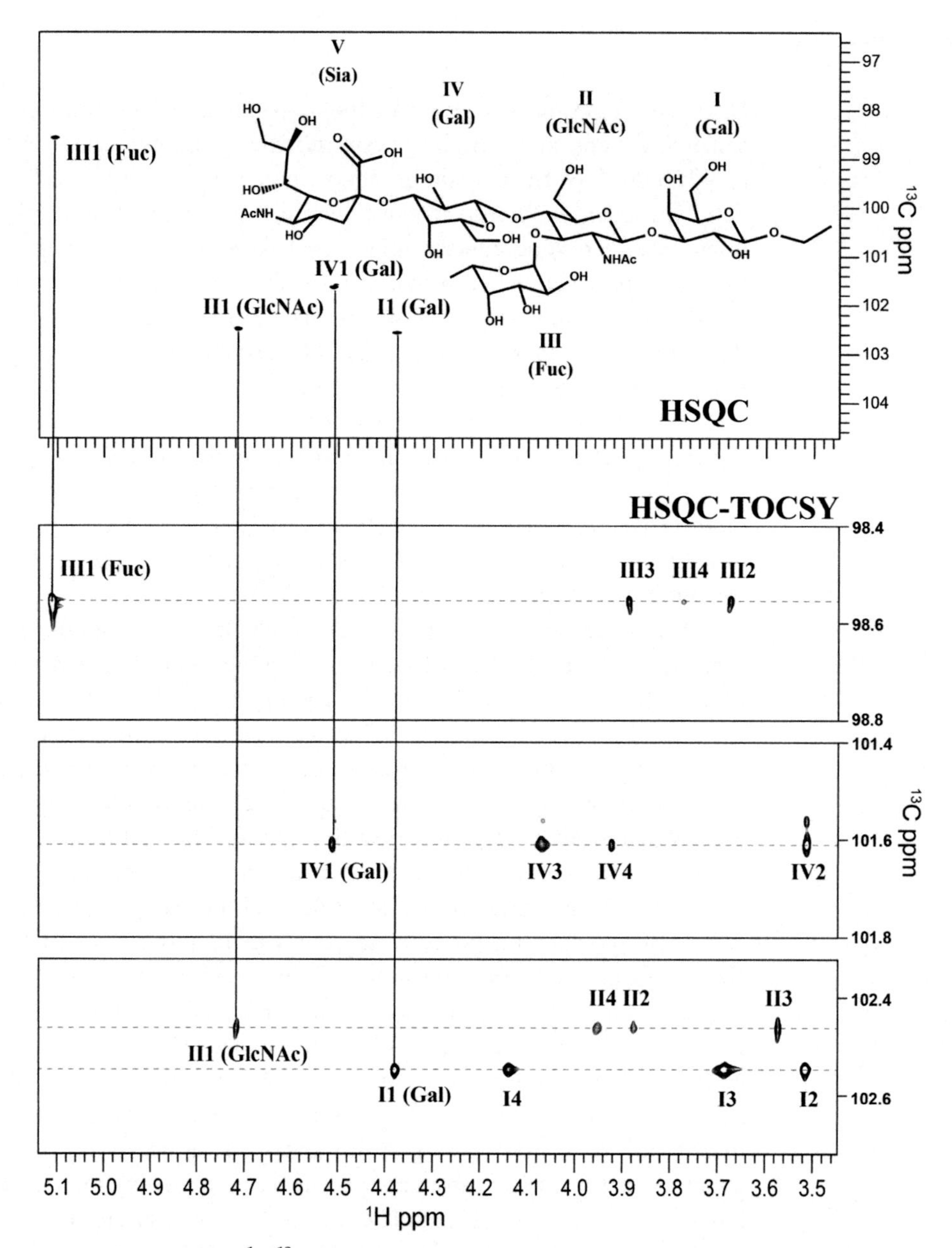

FIGURE 50.3. Sections of 2D ^{1}H-^{13}C 700 MHz NMR spectra of a sialyl Lewis x–capped glycan in D_2O. In this example, the combination of heteronuclear single quantum coherence (HSQC) and HSQC-total correlation spectroscopy (TOCSY) spectra is used to assign the chemical shift of each peak to an atom in the structure. Each residue in the structure is delineated by Roman numerals, and atom numbers in each residue are denoted using Arabic numerals. Assigning ^{1}H-^{13}C crosspeaks is accomplished by correlating data in the HSQC (*top panel*) and the HSQC-TOCSY (*bottom three panels*). The HSQC shows a peak for every ^{1}H atom covalently bonded to a ^{13}C atom. The HSQC-TOCSY shows crosspeaks for all ^{1}Hs in a ring (also called a spin system) connected to a particular ^{1}H-^{13}C crosspeak. For example, the *bottom* panel shows connections for the anomeric pairs of I-Gal (4.38 ^{1}H ppm, 102.6 ^{13}C ppm) and II-GlcNAc (4.72 ^{1}H ppm, 102.51 ^{13}C ppm). Along each of the horizontal dotted lines are the ^{1}H chemical shifts for the ^{1}Hs in a ring, annotated as I4, I3, and I2 for I-Gal and II4, II2, and II3 for II-GlcNAc. The positions are characteristic of a residue type and linkage. Once the residue is identified the peaks in the HSQC can be assigned. The resulting assignments can be used in subsequent three-dimensional structure determination.

When unprecedented structures emerge, it is also possible to perform a de novo structure determination using NMR data alone. This usually begins with a full assignment of the ^{1}H and ^{13}C resonances for each residue of a glycan using a combination of COSY, TOCSY, HSQC, and HSQC-TOCSY spectra. In addition, 2D heteronuclear multiple-bond correlation (HMBC or LR-HSQMBC) experiments are used to detect through-bond coupling between the anomeric ^{1}H and the carbon atom on the opposite side of the glycosidic linkage. These are, however, even less sensitive and more time-consuming experiments. It is possible to avoid both ^{13}C-based HSQC and HMBC experiments by substituting a 2D nuclear Overhauser effect (NOE) spectrum. In medium size glycans (1 kDa), NOE crosspeaks may go unobserved because of competition of coherence transfer pathways. Such crosspeaks can be observed using a rotating-frame Overhauser effect (ROE) spectrum. Both experiments correlate ^{1}H resonances based on inter-^{1}H distances, as opposed to through-bond couplings; they typically show crosspeaks for pairs of ^{1}Hs that are within ~4 Å of one another. Most connections are between ^{1}Hs in the same residue, but typically an extra connection will be between an anomeric ^{1}H and a ^{1}H in another residue. This ^{1}H is frequently across the glycosidic bond at the linkage site, providing a way to assemble residues into a complete structure. However, caution must be exercised. The identification of linkage sites is not as definitive as with ^{13}C-based experiments, because the distance for the trans-glycosidic pair is dependent on glycosidic torsion angles, and these NOE crosspeaks can be weak. Also, other non-linkage site protons can come within NOE distance and give false identification.

THREE-DIMENSIONAL GLYCAN STRUCTURE

The same NOE crosspeaks that assist in identifying linkages between residues can, in principle, be used to define the tertiary (3D) structure of a glycan. However, the interpretation is complicated by the presence of internal molecular motion, resulting in many different conformations in solution. This motion is nicely illustrated in $\phi-\psi$ energy plots much like the Ramachandran plots used to describe the energetics of torsional angles in the polypeptide backbone of a protein. For glycans, these plots are usually based on parameterized functions that represent a combination of quantum mechanical and experimental data. Such a plot is shown for a Glcβ1-4Glc linkage in Figure 50.4. The minima in these plots correlate well with angles found in structures

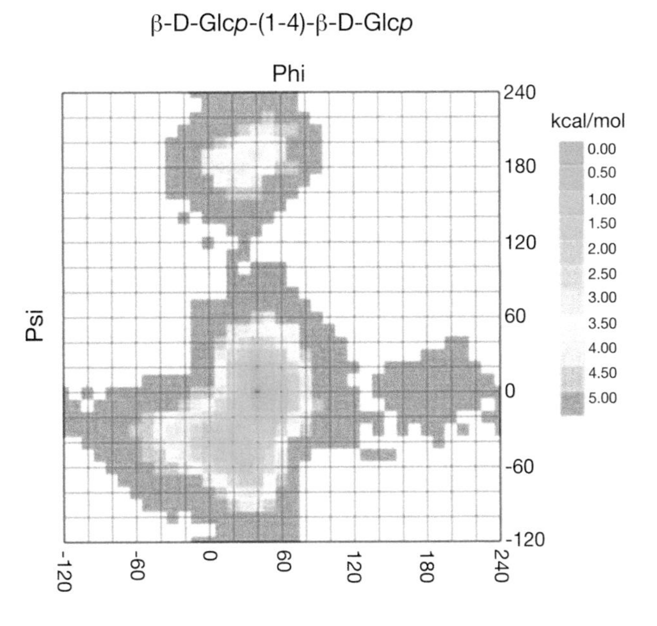

FIGURE 50.4. Energy plots showing likely $\phi-\psi$ values for the glycosidic torsion angles between Glc residues in Glcβ1-4Glc-OMe. Regions in *green* indicate a likely combination of ϕ and ψ, or conformation, whereas red regions represent less likely conformations. The *white* regions indicate that the molecule is not likely to be found in those regions of $\phi-\psi$ space. The plot was obtained with the Web-based tool CARP. ϕ, ψ angles are in the "NMR" convention (ϕ = H1′-C1′-O4-C4, ψ = C1′-O4-C4-H4).

of glycans deposited in databases such as the Cambridge Structural Database or the Protein Data Bank (PDB). Multiple minima indicate that multiple conformations are likely to be present in solution and therefore imply that there is conformational motion for the disaccharide. Thermal energies available to molecules at these temperatures are ~0.6 kcal, so in solution, states within ~1.4 kcal of the minimum have appreciable populations (>10%), representing a ±20° variation in torsion angles about those of the most stable form of the Glcβ1-4Glc linkage. In NMR experiments, intensities of NOE crosspeaks are roughly proportional to the average of the inverse sixth power of the distance sampled over these states, and the steep distance dependence results in skewing distances toward smaller numbers. Nevertheless, NOEs, particularly those representing close approach of remote parts of larger glycans, can guide construction of 3D structural models. Convenient tools for building structures are available, such as the carbohydrate builder of the GLYCAM package. Initial structures can be further refined with molecular mechanics or molecular dynamics software packages, or they can be manipulated in common molecular graphics packages to match solution data coming from NOEs and other NMR measurements, including residual dipolar couplings and paramagnetic effects, which provide angular and long range distance constraints respectively (Chapter 30).

ACKNOWLEDGMENTS

The authors acknowledge contributions to previous versions of this chapter by Carolyn R. Bertozzi and appreciate helpful comments and suggestions from Markus Pauly and Martina Delbianco.

FURTHER READING

Jiménez-Barbero J, Peters T. 2003. *NMR spectroscopy of glycoconjugates.* Wiley-VCH, Weinheim, Germany.

Lamari FN, Kuhn R, Karamanos NK. 2003. Derivatization of carbohydrates for chromatographic, electrophoretic and mass spectrometric structural analysis. *J Chromatogr B Analyt Technol Biomed Life Sci* **793**: 15–36. doi:10.1016/S1570-0232(03)00362-3

Haslam SM, North SJ, Dell A. 2006. Mass spectrometric analysis of N- and O-glycosylation of tissues and cells. *Curr Opin Struct Biol* **16**: 584–591. doi:10.1016/j.sbi.2006.08.006

Wuhrer M. 2013. Glycomics using mass spectrometry. *Glycoconj J* **30**: 11–22. doi:10.1007/s10719-012-9376-3

Zaia J. 2013. Glycosaminoglycan glycomics using mass spectrometry. *Mol Cell Proteomics* **12**: 885–892. doi:10.1074/mcp.R112.026294

Battistel MD, Azurmendi HF, Yu B, Freedberg DI. 2014. NMR of glycans: shedding new light on old problems. *Prog Nucl Magn Reson Spectrosc* **79**: 48–68. doi:10.1016/j.pnmrs.2014.01.001

Balagurunathan K, Nakato H, Desai UR. eds. 2015. Glycosaminoglycans: chemistry and biology. In *Methods in molecular biology*, Vol. 1229. Springer, New York.

Yan H, Yalagala RS, Yan F. 2015. Fluorescently labelled glycans and their applications. *Glycoconj J* **32**: 559–574. doi:10.1007/s10719-015-9611-9

Woods RJ. 2018. Predicting the structures of glycans, glycoproteins and their complexes. *Chem Rev* **118**: 8005–8024. doi:10.1021/acs.chemrev.8b00032

Mucha E, Stuckmann A, Marianski M, Struwe WB, Meijer G, Pagel K. 2019. In-depth structural analysis of glycans in the gas phase. *Chem Sci* **10**: 1272–1284. doi:10.1039/C8SC05426F

Gimeno A, Valverde P, Ardá A, Jiménez-Barbero J. 2020. Glycan structures and their interactions with proteins. A NMR view. *Curr Opin Struct Biol* **62**: 22–30. doi:10.1016/j.sbi.2019.11.004

51 Glycomics and Glycoproteomics

Pauline M. Rudd, Niclas G. Karlsson, Kay-Hooi Khoo,
Morten Thaysen-Andersen, Lance Wells, and Nicolle H. Packer

The term genomics arose from the availability of complete genome sequence data as well as computational methods for their analysis. However, <2% of genes in the human genome encode proteins. These genes are transcribed into messenger RNAs (mRNAs) that make up the "transcriptome," of which ~30% are assigned to protein coding. The total complement of proteins expressed by the cell is collectively termed the "proteome." Most eukaryotic proteins are post translationally modified (e.g., by phosphorylation, sulfation, oxidation, ubiquitination, acetylation, methylation, lipidation, or glycosylation). These modifications, combined with alternative RNA splicing in eukaryotes, render the proteome considerably more complex than the transcriptome. Although it has been estimated that approximately 120,000 different protein splice forms are expressed by human cells, the total number of modified proteoforms is likely to be at least an order of magnitude higher. The systems-level analysis of all proteins expressed by cells, tissues, or organisms is known as "proteomics." The proteome, like the transcriptome, but unlike the DNA sequence of the genome, is fundamentally dynamic. The repertoire of proteins expressed by a cell is highly dependent on tissue type, microenvironment, and stage within the life cycle. As cells receive external and internal cues in the form of growth factors, hormones, metabolites, or cell–cell interactions, the expression of various genes is modulated and may

published by Cold Spring Harbor Laboratory Press; doi: 10.1101/glycobiology.4e.51

be transcribed at levels ranging from silence to more than 10^4 mRNA copies per cell and 10^7 protein molecules per cell. Thus, proteomes and their modifications vary during cell differentiation, activation, trafficking, and malignant transformation.

However, no systems-level analysis of a biological process is complete without interrogating the glycome, defined as the entire complement of glycan structures (be it protein-/lipid-bound or -free) produced by cells, in addition to the genome, transcriptome, proteome, lipidome and metabolome.

THE GLYCOME

Vertebrates synthesize N-linked and O-linked glycoproteins, glycolipids (Chapters 11 and 12), proteoglycans, glycosaminoglycans (GAGs), and glycosylphosphatidylinositol (GPI) anchors covalently attached to proteins, as well as free oligosaccharides (Chapter 3). As with the proteome, each cell type has its own distinct glycome that is governed by local cues and the metabolic state of the cell. Other organisms have distinct glycomes; those of plants (Chapter 24) and prokaryotes (Chapter 21) are distinctly different from the vertebrate and invertebrate glycomes (Chapters 25–27).

The size of any particular cellular glycome has not yet been established, but the combinatorial possibilities that can occur with numerous glycan structures on multiple glycoconjugates means that determining a "complete" glycome is not straightforward. The notion that glycans should be studied as a totality (glycomics), as well as simply one glycan or glycoconjugate at a time, developed when it became apparent that glycans form patterns on cells that change during development (Chapter 41), cancer progression (Chapter 47), infection (Chapters 42 and 43), and many other diseases (Chapters 44–46). Many glycan-binding proteins, such as lectins, are oligomerized on the cell surface and interact with multivalent arrays of glycans on the same or opposing cells (Chapters 28–38). Sometimes, multiple discrete glycans and their matching glycan-binding proteins work together to engage two cells or to deliver signals between cells. Thus, the term "glycomics" was coined to describe the many aspects of glycobiology that can be understood only with a systems-level analysis of the glycome.

RELATIONSHIP OF THE GLYCOME TO THE GENOME AND PROTEOME

Clues regarding the composition and complexity of the glycome are found in the genome, transcriptome, and proteome of a cell. Thus, if a gene encoding a glycosyltransferase is not expressed (absent from the transcriptome), no glycans in that cell can carry the sugar transferred by that glycosyltransferase at that particular time. The action of many glycosyltransferases and glycoside hydrolases competing for the same substrates in the biosynthetic pathway renders the complete glycome impossible to predict with current tools and knowledge. As an example, the reduced expression of a single glycosyltransferase can perturb the biosynthesis of dozens of glycans. Furthermore, unlike the genome, the glycome is sensitive to exogenous nutrient levels and metabolic fluxes including salvage pathways. Thus, variations in dietary monosaccharides, such as glucose, galactose, glucosamine, fucose, mannose, and *N*-glycolylneuraminic acid (Chapter 15), may change the composition of the glycome. The numerous factors that influence the glycome (the transcriptome, the proteome, environmental nutrients, the secretory machinery, pH, and many other determinants) create a glycome that is highly diverse, adaptable, and dynamic. Thus, the glycome of a cell can change dramatically over time. It is this enormous structural plasticity in response to cellular and environmental states that underlies the essential roles of glycans in development, communication, and disease processes.

COMPARATIVE GLYCOMICS

Because the glycome is influenced by both genetic and environmental factors, the information contained therein sheds light on intra- and interspecies variations, including providing indicators of disease that can be used for diagnosis and for monitoring the efficacy of drugs. Comparative glycomics, the comparison of glycome profiles obtained from two or more individuals, tissues, or conditions of interest, is therefore an exciting frontier in biology and medicine.

For example, as discussed in detail in Chapter 47, numerous changes in the glycome have been associated with malignancy and metastasis, including altered N- and O- linked protein glycosylation, up-regulation of conjugated sialylated and fucosylated glycans, and altered GAGs. Regardless of functional consequence, a change in the glycome that is highly correlated with malignancy (or any disease) may serve as a diagnostic marker candidate. Notably, glycans altered in a disease may reflect downstream consequences of the disease on remote organs, changes in the patient's immune system, or other effects of the disease.

One major caveat is the currently unknown extent of natural variation among individual human glycomes. The glycome is known to respond to dietary and environmental changes and vary in elusive ways with age, gender, and acquired disease susceptibility (Chapter 46). Studies of evolutionary biology also have much to gain from comparative glycomics. Evolution of the vertebrate immune system, for example, was accompanied by the acquisition of new classes of glycan-binding proteins, including Siglecs (Chapter 35) and selectins (Chapter 34). Likewise, the glycomes and glycan-binding proteins of microbes and their vertebrate hosts appear to have coevolved in some instances (Chapter 42).

TOOLS FOR CHARACTERIZING THE GLYCOME

The glycome can be determined at different levels of granularity. First, "glycomics" constructs an inventory of glycans separated from their protein or lipid scaffolds from the cell, organ, or organism of interest. This is an important starting point for any comprehensive glycome analysis. The second level of analysis defines specific glycans associated with individual proteins or lipids. Analysis of the complete repertoire of a cell's glycoconjugates, including the microheterogeneity of glycans present at individual sites of attachment, lies, for example, at the intersection of glycomics and proteomics ("glycoproteomics") and of glycomics and lipidomics ("glycolipidomics"). A third level of complexity involves determining which glycans and/or glycoconjugates are expressed in specific cells, tissues, or secretions at specific times or conditions. This level of analysis is essential if the goal is to reveal new functions in cell–cell communication or to correlate particular glycomes with specific diseases. Of course, none of these approaches recapitulates the actual arrangement of the complex glycan forest present on the surface of cells or in extracellular matrices. The spatial organization of the glycome on cell surfaces is currently only amenable to microscopy-, array- and flow cytometry–based imaging using various glycan-recognizing probes (lectins/GRPs; Chapter 48). The spatial organization of glycans in tissue sections can also be explored by recent matrix-assisted laser desorption/ionization (MALDI) MS imaging methods.

GLYCOMICS AND GLYCOPROTEOMICS

The term "glycomics" thus describes studies designed to define the complete repertoire of glycans that a cell, tissue, or organism produces under specified conditions of time, location, and environment. "Glycoproteomics" describes this glycome as it appears on the cellular proteome. Glycoproteomics determines which sites on each glycoprotein of a cell are glycosylated and ideally includes the identification and quantitation of the heterogeneous glycan structures at each site. The molecular complexity of the glycoproteome makes glycomics and glycoproteomics both exciting and daunting. Because neither the genome, transcriptome, nor the proteome can accurately predict the dynamically expressed protein-linked glycans, the glycome and glycoproteome must be analyzed directly. Techniques used to characterize a single glycoprotein after isolation (Figure 51.1) or a complex mixture of glycoproteins (the glycoproteome) (Figure 51.2) are described in this chapter.

As described below, numerous techniques have been developed for interrogating the glycome and glycoproteome. Because no single technique can at present define all aspects of the glycome or glycoproteome, several approaches are typically used in parallel from assembling an individual cell type glycan repertoire to determining the global tissue expression for example. Different approaches and techniques are required to characterize, for example, the structures of glycoproteins versus glycolipids, N-glycans versus O-glycans, and sulfated GAGs versus neutral glycans

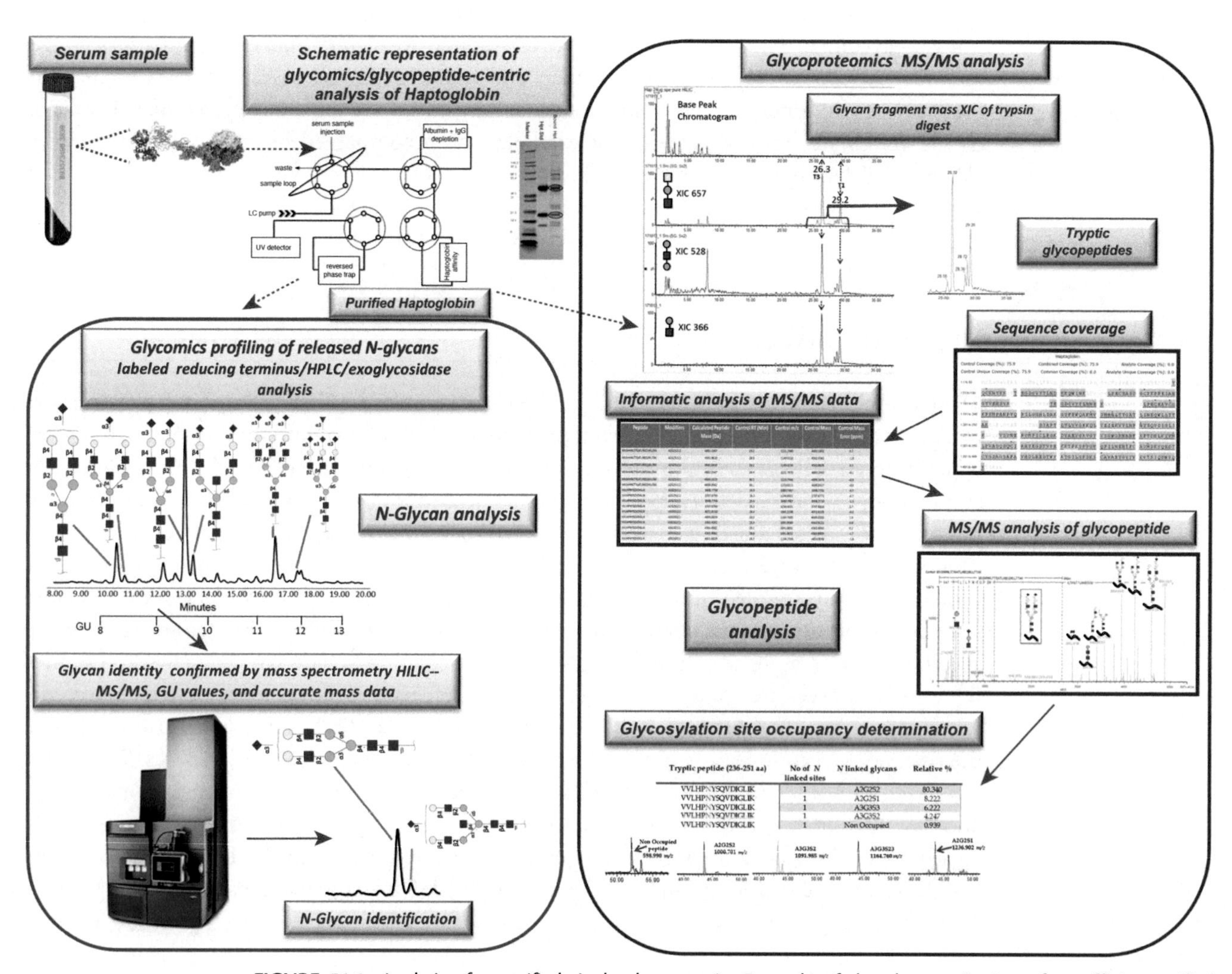

FIGURE 51.1. Analysis of a purified single glycoprotein. Example of the characterization of an affinity purified serum glycoprotein, haptoglobin, including (*left panel*) a total glycomics profile of N-glycans released from the protein, fluorescently derivatized with 2-aminobenzamide (2-AB) and separated by liquid chromatography (LC). Structures were determined by comparison with retention time standards (GU values) and validated by mass spectrometry (MS) and MS/MS; and (*right panel*) tryptic glycopeptides were analyzed by reversed-phase LC–MS/MS to give glycan composition and structural and site information, with good coverage of the amino acid sequence. (Figure courtesy of Mark Hilliard, NIBRT, Ireland.)

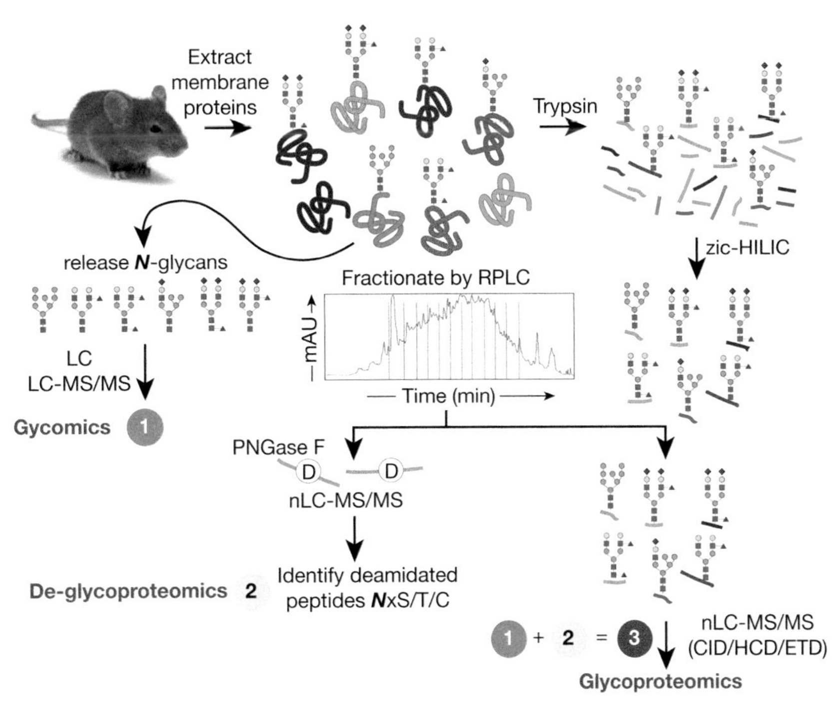

FIGURE 51.2. Glycomics-assisted glycoproteomics of a complex mixture of glycoproteins. In this workflow, the structures of the PNGase F released N-glycans from a total extract of rat brain membrane proteins are first analyzed in detail (*1*) and the sites of glycosylation are then determined by identification of the resulting deamidated peptides with the NxS/T motif (2). The information assists the assignment of the heterogeneous glycan structures on the separated tryptic glycopeptides (3). (Figure courtesy of Benjamin L. Parker, University of Melbourne, Australia.)

(Chapter 50). In contrast, a single technique such as RNA sequencing (RNA-seq) can be used to identify and quantitate all mRNA transcripts at once, a much easier task.

GLYCOMIC ANALYSIS

Glycomic methods and analyses have been mostly directed toward protein glycosylation because of the current proteomics focus, although considerable effort is being directed toward methods that encompass all glycan conjugates. As an example of a typical experiment, a glycoprotein-enriched sample is prepared from a cell lysate and their released glycans analyzed by liquid chromatography (LC) and/or mass spectrometry (MS). In the case of glycoproteins, the N-glycans can be selectively released enzymatically or chemically, separated by high-performance liquid chromatography (HPLC) methods, and are often sequenced online by tandem mass spectrometry (MS/MS) with or without exoglycosidase treatments. Separately, the O-glycans may be released chemically and sequenced in a similar manner. In contrast, glycolipids can often be directly sequenced with or without release from the lipid component. GAGs are more problematic because of their large size and negative charge, but disaccharide fragments can be sequenced by LC or MS in conjunction with enzymatic digestion (Chapter 17).

Depending on the level of detail desired, glycomic analyses may be divided into basic techniques: glycoprofiling, glycan class characterization, and full structural analysis of the glycans released from protein(s). The level of detail should ideally be tailored to the particular question at hand.

1. **Glycoprofiling** (fingerprinting, patterning) is the separation of a complex glycan mixture by a technique that provides a signature or fingerprint to give a simple overview or snapshot of the glycans in the sample. Technologies that provide different one-dimensional windows are

HPLC (separation by physiochemical parameters such as hydrophilicity, size or charge) with or without MS, capillary electrophoresis (separation of labeled glycans by mass/charge), and MALDI and/or electrospray ionization (ESI) MS (separation of unlabeled or labeled glycans by mass/charge).

2. **Glycan class characterization** uses technologies to separate glycan mixtures into types of glycans based on structural features. Examples include MS separations of di-, mono-, and non-galactosylated IgG glycans or the weak anion-exchange (WAX) LC charge-based separations that separate glycans into neutral, mono-, di, tri-, and tetrasialylated structure types. This approach is a convenient way to highlight defined critical features and provide relative quantitation of the different glycan classes. LC-MS analysis can also be used to separate released N-glycan structures into paucimannose, oligomannose, and hybrid and complex sialylated and nonsialylated glycan types.
3. **Detailed (full) structural analysis** requires the determination of the sequence and any modifications to the monosaccharides, branch points, anomericity, and glycosidic linkages of the glycans in a glycome. In this detailed analysis, orthogonal technologies are usually required, first to assign preliminary structures and then to confirm the assignments. For example, an anion-exchange separation into differently charged glycan classes can be complemented by hydrophilic interaction liquid chromatography (HILIC) separation of each class. Digestion by exoglycosidases can then be used to help determine the sequence, anomericity, and linkage of different glycans. On the other hand, MS assigns compositions that are consistent with the mass data, and structural details can be resolved, for example, by (LC) ESI MS/MS or MS^n fragmentation. Separate release and analysis of glycans with sialic acid residues that have labile modifications such as O-acetylation or polysialylation may be needed, if preparation for MS is likely to destroy them. Full structural analysis also can include absolute or relative quantitation of the assigned glycan structures such as, for example, the level of core fucose on antibodies designed to initiate antibody-dependent cellular cytotoxicity (ADCC), the levels of antigenic α-gal residues, sialyl-Lewis x epitopes that may be useful markers of inflammation and metastasis, or the specific structure binding to a bacterial protein.

Full structural analysis of glycans is challenging as different structures can have the same mass, often coelute on separation systems, and can require detailed manual annotation of MS/MS spectra. Usually glycans are given a preliminary structural assignment from one technology and are then confirmed by at least one orthogonal technology. Bioinformatic tools are being developed to try to mitigate this bottleneck (Chapter 52).

Release of Glycans from Proteins

The starting material in a glycomic analysis can be glycoproteins embedded in gels, protein extracts from whole cell lysates, homogenized tissues, enriched membrane fractions, or serum and other body fluids. For high-throughput analyses of the glycome, intact N-glycans are most often released from glycoproteins using an amidase (peptide N-glycosidase F [PNGase F]). PNGase F cleaves the linkage between the core GlcNAc and the asparagine residue in the NXT/S ($X \neq P$) sequon, of all classes of N-glycans, with the exception of specific N-glycans found in plant and invertebrate glycoproteins that contain fucose α(1,3) linked to the core GlcNAc residue attached to the protein. PNGase A, an enzyme extracted from almond emulsion, may be used to release all core fucosylated N-glycans from protease-generated glycopeptides. Treatment with other endoglycosidase enzymes that cleave between the two GlcNAc residues within the chitobiose core (e.g., endoglycosidase D, which releases paucimannosidic N-linked glycans, endoglycosidase H, which selectively cleaves oligomannose- and hybrid-type structures and various types of endoglycosidase F) is also possible. Before treatment, denaturation both with and without trypsin digestion can be used to relax the three-dimensional (3D) structure of the protein and improve enzyme accessibility. N-glycans can also be conveniently released from glycoproteins purified in SDS-PAGE gel bands. After the N-glycans have been cleaved, the protein remaining in the gel can be identified by traditional proteomics.

For O-linked glycan release from serine (S) and threonine (T) the method of choice is chemical reductive β-elimination. Drawbacks of the method include the fact that the resultant O-linked alditols cannot be further labeled at the reducing end, and that labile modifications can be destroyed during the release process. To date, there is no enzyme that can release all classes of O-linked glycans. The enzyme O-glycanase (in contrast to the PNGase enzymes) is restricted to releasing only simple core 1 (Galβ-1,4GalNAc-αS/T) O-glycans.

Analysis of Released Glycans

Derivatization of N- and O-Linked Glycans for LC, CE, and MS

Labeling of released glycans can optimize HPLC and capillary electrophoresis (CE) detectability and separability and may improve their MS properties. Many of these labeling approaches use reductive amination of the glycan at the reducing end or react the free amine at the reducing end that is left after PNGase F release. Fluorescent tags increase the sensitivity by lowering the limit of quantitation and detection. Permethylation and peracetylation, when all the mobile protons (present on hydroxyls, carboxyls, and amides) in a glycan are substituted by alkyls (e.g., -methyl) or esterified (e.g., -acetyl) convert the glycans from being hydrophilic to hydrophobic, making the glycans easier to purify and greatly improves the sensitivity and linkage determination by MS-based analyses.

Both N- and O-glycans can also be analyzed as alditols in which the reducing end monosaccharide ring is converted into a reduced linear alditol by reducing agents, such as sodium borohydride. This reduction removes the anomeric ambiguity of the carbohydrate, where the α- and β-isomers of the reducing end sugar may otherwise separate into two peaks chromatographically.

Tagging of the reducing ends of the released glycans (e.g., with 2-aminobenzamide [2-AB], aminobenzoic acid [2-AA], or aminopyridine [2-PA]) is often used for HILIC and reversed-phase LC separation of released glycans (Figure 51.1). Labeled dextran oligomer ladders are commonly used in LC as external standards to help define composition and size based on comparable retention times, for which an "incremental value" can be calculated for each monosaccharide present in the structure (Figure 51.1).

CE with laser-induced fluorescence (LIF) can also provide efficient, rapid, and quantitative separation of derivatized glycans. Glycans are mostly neutral structures so coupling a charged fluorescent label such as 1-aminopyrene-3,6,8-trisulfonic acid (APTS) is necessary to provide electrophoretic mobility and to enable sensitive fluorescence detection. Further details can be assigned after digestion of glycan mixtures by one or more exoglycosidases that specifically cleave glycosidic bonds of individual monosaccharide units from the terminal residue producing predictable shifts in the HPLC or CE (or MS) profiles of the digests.

Special derivatization protocols have also been developed to improve the quality of the MS fragmentation spectra of sialylated glycans—in particular, to target the charged carboxyl group of sialic acid residues. These can be converted into esters or amides to remove the acidic proton of the carboxylic acid that destabilizes the sialic acid in MALDI-MS and promotes undesirable in-source or post-source fragmentation in MS/MS. Specific derivatization of sialic acid residues can also help to determine by MS whether they are 2,3- or 2,6-linked to the glycan structure.

TABLE 51.1. Families of common monosaccharides found in mammalian N- and O-linked glycans

Name	Abbreviation	Symbol	Example	Monoisotopic mass (Da)
Hexose	Hex	○	glucose, Glc, ●	162.0528
			mannose, Man, ●	
			galactose, Gal, ○	
N-acetylhexosamine	HexNAc	□	*N*-acetylglucosamine, GlcNAc, ■	203.0794
			N-acetylgalactosamine, GalNAc, □	
Deoxyhexose	dHex	△	fucose, Fuc, ▲	146.0579
Sialic acid	Sia	◇	*N*-acetylneuraminic acid, Neu5Ac, ◆	291.0954
			N-glycolylneuraminic acid, Neu5Gc, ◇	307.0903

Symbols as per Symbol Nomenclature for Glycans (SNFG).

MS Profiling of N- and O-Linked Glycans

An advantage of mass spectrometric glycan profiling is that different glycans can be identified at once by their mass and diagnostic fragmentation ions, increasing the throughput of the glycomic analysis. However, mass spectrometry of glycans may miss potentially important labile modifications, such as sulfation and O-acetylation, depending on the sample preparation and MS techniques applied. MS has inherent challenges because of the isomeric and sometimes isobaric nature of the constituent monosaccharide units that form multiple glycan isomers exhibiting the same molecular mass.

Determining the molecular masses of glycans using MALDI- or ESI-MS gives a picture of the molecular distribution of glycans and allows a quantitative comparison of glycosylation between samples (Chapter 50). The limited number of masses of the monosaccharide units (Table 51.1) makes combinatorial translation of molecular ion masses to monosaccharide composition possible albeit often with some remaining ambiguity. There are available search engines that can provide a suggested list of glycan compositions based on an experimentally determined mass (e.g., GlycoMod; Chapter 52). MS alone, however, cannot distinguish between isomeric monosaccharides, so a nomenclature for generic monosaccharide compositions has been adopted. For instance, all of the isomeric 6-carbon-containing monosaccharides, such as glucose, mannose, and galactose are given the unifying name of hexose (Hex) (Table 51.1).

Elucidation of variable linkage configurations, and glycan branch points is another analytical challenge.

Generic glycomic workflows using all the derivatization approaches described above have been developed for MS analysis. Both the neutral and sialylated glycans in a sample can be analyzed after permethylation and MALDI-MS, in which glycan masses are detected as their singly charged alkali ion adducts in positive ion mode (e.g., as $[M+Na]^+$ ions or $[M+K]^+$ ions). Negative ion ESI-MS is widely adopted for intact glycan profiling in their alditol form, in which both neutral and sialylated glycans are detected as deprotonated $[M\text{-}nH]^{n-}$ ions. The number of charges (n) will increase with the size of the glycan and is also dependent on the number of acidic moieties present (e.g., sialic acids, sulfates, and phosphates). Positive ion MALDI-MS without permethylation usually requires that sialic acid residues are derivatized as described above to prevent their loss in-source.

To obtain orthogonal separations prior to the mass analysis, ESI-MS is often connected to HPLC. HILIC columns can be used for separation of reducing end derivatized glycans and provides separation based on hydrophilicity (which correlates with the glycan size), with some isomeric structure resolution (Figures 51.1 and 51.2). The retention of glycans on such columns is primarily based on hydrogen bonding to water surrounding the HILIC stationary phase (partitioning). Using an alternate stationary phase, porous graphitized carbon (PGC) has shown a unique ability to clearly separate isomers of released glycan alditols based on more complex retention mechanisms. If glycans are permethylated, the increased hydrophobicity allows separation using conventional C18 reversed-phase chromatography. Ion mobility has also been utilized to resolve glycan isomers entering the mass spectrometer. Data from two or more approaches can be used to give confidence in structural assignments and can be facilitated by software that allows automated data analysis (Chapter 52).

MS Fragmentation of N- and O-Linked Glycans

MS fragmentation has become the gold standard for glycan structural characterization. The goal is to generate information-rich fragment spectra that will allow unequivocal assignment of a glycan structure. However, in the current state of the art, fragments from colliding glycan molecular ions (collision-induced dissociation [CID]—either "beam-type" or "ion trap-type") can only partly determine the structure of interest. We can distinguish three types of glycan fragment ions depending on the type and amount of information they carry (Figure 51.3).

1. B- and C-type fragment ions comprise, by definition, the nonreducing end fragments that arise from a single glycosidic bond cleavage without or with the glycosidic oxygen, respectively. Reducing end fragments are assigned as Y- (with glycosidic oxygen) and Z-type (without glycosidic oxygen) fragment ions.
2. Cross-ring fragments can be assigned as nonreducing end fragments (A-type) and reducing end fragments (X-type) and require further annotation to specify which bonds in the carbon ring were dissociated to form the cross-ring fragment ion.
3. Internal fragments that occur from more than one fragmentation event from a combination of glycosidic and/or cross-ring fragmentation.

Ideally, a fragmentation spectrum containing all possible glycosidic fragments would allow the assignment of the primary sequence and branching of a glycan structure. In practice, however, most CID approaches provide glycosidic fragments, but several methods or a combination of methods (LC retention times, exoglycosidase digestions, ion mode, derivatization, multiple fragmentation steps, i.e., MS^n) are usually used to fully define a particular structure of interest. Biosynthetic rules can be and often are applied to aid the glycan characterization.

Glycan Modifications

A further challenge is that many key glycan modifications such as O-acetylation, pyruvylation, etc., are labile to, and/or missed with, current analytical methods. This problem can result in populating databases with misleading or biased information. As just one example, although many databases assume that a sialic acid at the terminus of the vertebrate glycan chain is *N*-acetylneuraminic acid, there are in fact dozens of kinds of modified sialic acids in nature, and the differences can have profound effects on biological functions (Chapter 15). The same is true of N- and O-sulfate esters on HexNAc and hexosamine residues (Chapters 14 and 17). These analytical challenges are still being pursued. Meanwhile, in many instances, it would make sense to not make definitive positional assignments of modifications on the glycan chains.

THE FUTURE OF GLYCOMIC ANALYSES

Releasing the glycans from the protein(s) is currently a prerequisite for glycomics. The described techniques above are, however, complementary to support the assignment of glycan structures still attached to peptides, the techniques for which are less well-developed and present greater

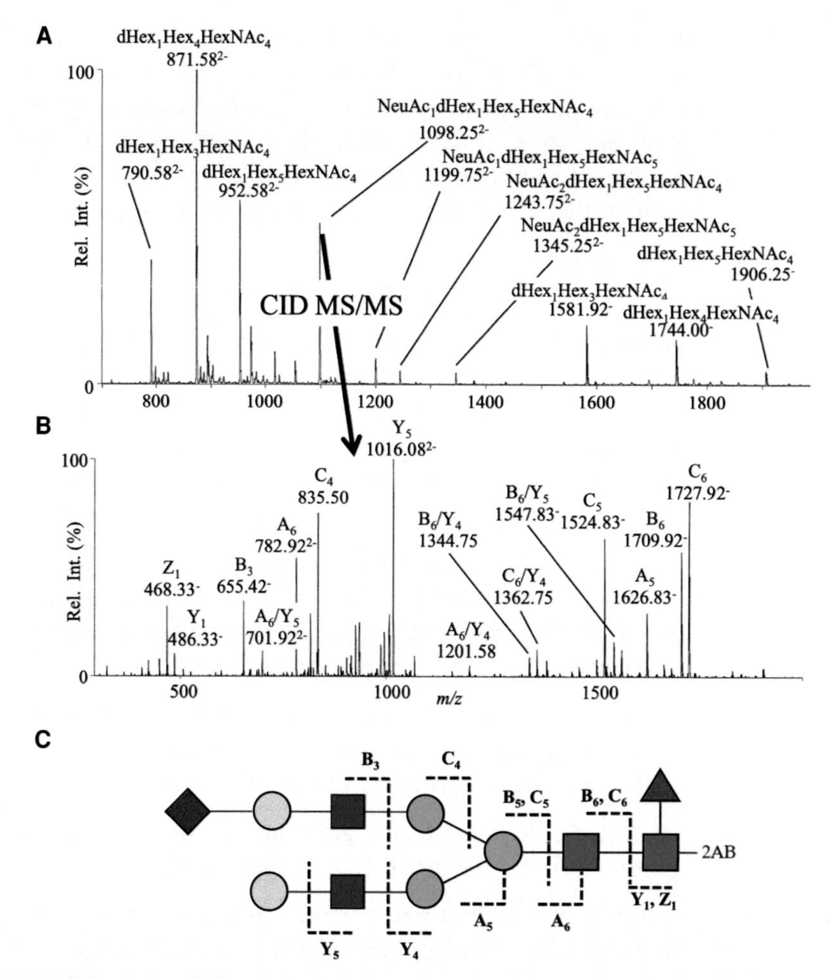

FIGURE 51.3. Collision-induced dissociation–tandem mass spectrometry (CID-MS/MS) of released N-glycans. (*A*) MS of released N-glycans: Masses are shown as doubly charged ions with corresponding monosaccharide compositions. (*B*) CID-MS/MS of m/z of $[1098.25]^{2-}$ depicting masses of the resultant fragments. (*C*) Illustration of Domon and Costello bond cleavages of the glycan structure that produced the fragments seen in *B*. Rel. Int., Relative intensity.

challenges (see below). N-linked and O-linked glycomics are still the only techniques that can detail the overall glycan structural landscape of the surface of a cell (if applied in combination with plasma membrane enrichment) so they will continue to be valid approaches in the foreseeable future to define cell–cell interactions and to discover disease biomarkers and new therapeutic targets.

However, there may not be a need for de novo analysis of every single glycan in the future. Matching of fragment spectra using fragmentation libraries containing many of the more ubiquitously expressed glycans (Chapter 52) is becoming a pathway to quickly assign a majority of structures fully or partially and allows researchers to focus on validation of only the structures that are important for addressing the biological question. Similar to proteomics, automated glycan identification should ideally be performed with stringent confidence thresholds by, for example, determining the false discovery rate of the reported identifications. Novel fragmentation techniques, the introduction of ion mobility MS, and in vivo or in vitro incorporation of heavy isotopic monosaccharides or tags are starting to provide other opportunities for the isomeric glycan structural characterization, quantitation, and visualization that are currently lacking in glycomic mass spectrometric techniques.

FROM GLYCOMIC TO GLYCOPROTEOMIC ANALYSIS

A key question that often follows the identification of important glycosylation features by glycomics is, which proteins carry the implicated glycan structures and at which sites? Given that protein glycosylation is dependent on the concerted action of multiple glycosyltransferases and glycoside hydrolases, one might expect all glycoproteins that traffic through the secretory pathway to be equally susceptible to similar modifications. However, protein- and site-specific glycosylation are well-recognized features of protein glycosylation. Although the underlying molecular basis for protein- and site-specific glycosylation is not entirely understood, contributing factors are likely to include common sequence motifs, protein structural conformation, unique physicochemical patches surrounding the glycosylation site, or some yet-unknown traits that collectively allow individual proteins to be sought out from among hundreds or thousands of other proteins to be acted on by the glycosylation machinery in a specific way.

MAPPING OF GLYCOSYLATION SITES

The mapping of N-glycosylation sites of formerly glycosylated (de-N-glycosylated) peptides of a complex protein sample is now routinely performed using conventional proteomic approaches. Notably, this approach does not give information on what glycan structure(s) are present on these defined glycosylation sites and as such does not fit well under glycoproteomics. By virtue of the action of PNGase F or Endo F/Endo H, the de-N-glycosylated peptides are mass-tagged by conversion of Asn to Asp (a +1 Da change in peptide mass) within the consensus sequon (Figure 51.2) or by retention of a GlcNAc(±Fuc) at the Asn site, respectively. These previously glycosylated proteins can then be subjected to standard trypsin digestion and LC-MS/MS analysis, and the MS/MS data searched against proteomic databases for rapid identification of the N-glycosylation sites on the tryptic peptides. This approach can be coupled to initial affinity or chemical capture of the glycoproteome subset, by use of lectins at the glycoprotein and/or digested glycopeptide level, or by hydrazine capture of the oxidized glycans on the glycopeptide followed by PNGase F release of the captured peptide containing the deamidation of Asn to Asp. In this manner, thousands of N-glycosylation sites can now be routinely identified and their relative occupancy quantified in a single experiment. The use of sequential Endo H followed by PNGase F, given that Endo H only acts on oligomannosidic glycans and PNGase F cannot remove the remaining single GlcNAc attached to an Asn, can provide additional information about N-glycan classes at an individual site. Apart from not being able to inform the site-specific glycoform structures, the most commonly criticized aspect of this approach is the potential of false positives being introduced by spontaneous deamidation of Asn to Asp, which is independent of the action of the PNGase F. Enzymatic deglycosylation in heavy water and/or the use of Endo F/H are ways to reduce the number of false positive identifications arising from this spontaneous deamidation.

Innovative approaches using the zinc finger nuclease gene targeting technique, or by using CRISPR/Cas9 knockout of the *C1GalT1* glycosyltransferase gene, which impairs O-glycosylation pathways by preventing extension of the O-GalNAc core, has led to the generation of so-called SimpleCells, in which all synthesized mucin-type O-glycoproteins carry only a single GalNAc or sialyl GalNAc (Chapter 56). The same approach applied to O-mannose initiated glycans resolved a long-standing question concerning the apparent abundance of O-mannose glycans in brain tissue although at the time only a few relatively low abundance proteins had been determined to possess this modification. The SimpleCell technology has greatly facilitated the experimental identification of hundreds and even thousands of O-glycosylation sites, although the actual sites and attached O-glycan structures under more natural and specific physiological states remain unknown.

GLYCOPROTEOMICS: DETERMINING THE HETEROGENEOUS SITE GLYCOSYLATION

Not every sequon (NXT/S) will carry an N-glycan, and in rare cases the T/S position of utilized sequons can be substituted with C. The actual glycosylated sites usually carry a heterogeneous collection of glycan structures. Glycoproteomics needs ideally to be able to identify all glycoproteins in a sample down to the level of which sites are occupied (macroheterogeneity) and to quantify and characterize their respective glycoforms at that site (microheterogeneity). The ultimate aim is to take snapshots of the distribution of disparate glycans in real-time on each glycoprotein in a cell to infer how site-specific glycosylation may promote or interfere with cellular interactions or signaling events. Ultimately, to understand the specific biological roles of protein glycoforms, we will need to define the population of each molecular species that arises from combinations of site-specific glycan diversity at multiple sites.

Although promising glycoproteomics methods have been developed in recent years, the unambiguous identification of intact glycopeptides in complex mixtures using LC-MS/MS remains challenging. The workflows are comparably less mature than the corresponding proteomics methods for the analysis of unmodified peptides or peptides carrying simple modifications. Appropriate false discovery rate (FDR)-based glycopeptide identification strategies are emerging but are still not sufficiently integrated into the workflows such that confident identifications are still dependent on manual interrogation of data by experienced users. Most, if not all, glycoproteomics methods rely on accurate mass analysis of the monoisotopic glycopeptide precursor ions (low ppm) to accurately specify the molecular mass of the intact glycopeptides. The fragment ions also benefit from detection at high resolution to aid the spectral assignment.

For N-glycopeptides, CID, either on an ion trap or quadrupole time-of-flight (QTOF) platform, or increasingly by higher-energy collision dissociation (HCD), available on a range of Orbitrap mass analyzers, induces mostly glycosidic cleavages, giving rise to abundant glycan oxonium ions in the low mass region, complemented by successive neutral losses of glycosyl residues from the glycopeptide precursors down to a single GlcNAc on the Asn (Figure 51.4A). The abundant diagnostic oxonium ions enable quick and accurate broad classification of all glycopeptide-containing spectra, with details of the monosaccharide composition and topology, and usually is sufficient to allow for confident glycopeptide identification. In favorable cases, the Y1 ion (peptide backbone+GlcNAc) can be assigned and its m/z can be used to define the molecular mass of the peptide carrying the glycan Thereafter, the precursor ion mass gives the composition of the attached glycan after the peptide mass has been subtracted. More recently the use of step-HCD (generating multiple fragmentation spectra of the same species with different collision energies) has been shown to generate rich spectra for assigning both the glycan and the peptide.

When submitting the more labile O-glycopeptides to CID/HCD fragmentation the glycosyl residues are usually detached from the peptide carrier and therefore produce b- and y-ions without the conjugated glycan. The limitation is also that the modification site(s) cannot be specified.

Electron transfer dissociation (ETD) performed with or without additional CID- or HCD-based activation, termed ETciD and EThcD, respectively, represents an alternative fragmentation strategy to achieve more information-rich glycopeptide spectra. In principle, ETD yields c- and z-type ions arising from peptide bond cleavages along the peptide backbone without inducing glycosidic cleavages (Figure 51.4B). The resulting c- and z-ions that still carry the intact glycan are therefore useful to identify both the peptide carrier and the modification site, a feature particularly useful for O-glycopeptides in which site allocation cannot reliably be predicted based on the peptide sequence. The practical problem is that the ETD efficiency is dependent on the charge density of the glycopeptide. Relative low dissociation efficiency is

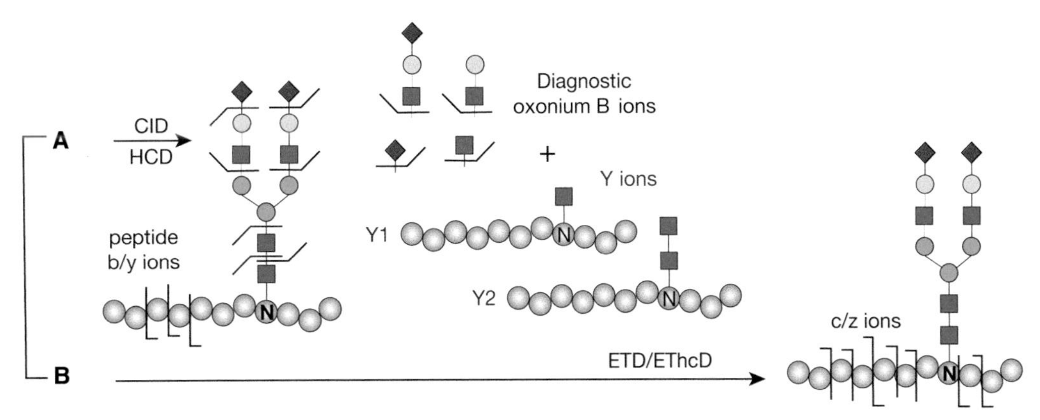

FIGURE 51.4. Complementary tandem mass spectrometry (MS/MS) fragmentation of N-glycopeptides. (*A*) Both collision-induced dissociation (CID)-MS/MS and higher-energy collision dissociation (HCD)-MS/MS of glycopeptides produce cleavages of the glycan to give diagnostic oxonium ions, along with a series of neutral losses of glycosyl residues, which often results in a single GlcNAc remaining on the peptide (Y1 ion). HCD can also produce sufficient peptide fragments (b/y ions) enabling identification of the peptide backbone and hence the source protein, but typically provides no direct evidence for the attachment site-often assigned to be the Asn of the sequon if a single site. (*B*) Electron transfer dissociation (ETD)-MS/MS results in cleavage of the peptide backbone while leaving the glycan intact on the site. It can be supplemented by HCD (EThcD) to produce a hybrid spectrum of both.

typically observed with doubly and triply charged N-glycopeptides, which constitutes a substantial portion of all tryptic peptides. One way to address this issue is to increase the charge state of the glycopeptides by either chemical derivatization, such as using a tandem mass tag (TMT), by using supercharging agents in the LC solvents and/or by using a different proteolytic enzyme such as LysC or GluC instead of trypsin to generate larger glycopeptides. Using EThcD (supplemental activation) and/or longer ETD activation times may also produce more information rich spectra by promoting dissociation of already charge-reduced nondissociated precursors. ETD works reasonably well with O-glycopeptides, particularly those decorated with only one or two glycosyl residues including O-GlcNAc, O-GalNAc (Tn), O-GalNAc-Gal (T), O-Fuc, and O-Man. By retaining these glycosyl substituents, ETD ideally allows identification of their distribution over several closely placed Ser/Thr residues which is not possible with CID or HCD.

Innovative LC-MS/MS acquisition strategies are increasingly being used to enhance the performance of glycoproteomics experiments. For example, abundant HexNAc oxonium ions at *m/z* 204 arising from HCD-MS/MS fragmentation are useful diagnostic ions for any type of N- and O-linked glycopeptides. Product ion-dependent ETD, or EThcD triggering of precursors giving rise to such HexNAc (or other) oxonium ions, are elegant strategies to tailor the instrument to acquire more glycopeptide specific informative data enabling a higher glycoproteome coverage without the need of prior glycopeptide enrichment.

Similar to the quantitation of peptides in proteomics, the quantitation of intact glycopeptides can be performed using either label-free or label-assisted (e.g., TMT) strategies. The type of quantitation depends on the question at hand but may involve relative quantitation of all glycans at each glycosylation site (glycoproteome mapping), comparison of the glycoprofiles of individual sites between conditions (comparative glycoproteomics), or, more simply, determination of the relative abundance of a single glycopeptide form between two or more conditions (biomarker discovery). Multiple quantitative approaches as well as information of the protein level and degree of site occupancy are often required to extract biologically relevant information from glycoproteomics data.

LIMITATIONS AND PROSPECTS OF GLYCOPROTEOMICS

A limitation of current glycoproteomics data is that the glycopeptide fragment data do not afford detailed glycan topology, linkage, and stereochemistry information of the attached glycans. Diagnostic fragment ions can sometimes be found confirming terminal epitopes such as a fucosylated Hex-HexNAc or sialylated fucosylated Hex-HexNAc (Lewis x and sialyl-Lewis x, respectively) but these cannot be distinguished from Lewis a and sialyl-Lewis a. Detailed glycan profiling achieved with site resolution (structure-focused glycoproteomics) is an important frontier that should be a goal of the next-generation of glycoproteomics. At this time, continued development of selective enrichment and/or preseparation of specific subsets of glycopeptides needs to be pursued, because no single LC-MS^n method or instrument is sufficient to handle the full dynamic range of the entire glycopeptide pool derived from a complete glycoproteome.

To limit the search space and enhance the detailed structural knowledge of the glycans encountered in a given glycoproteome, parallel profiling of the glycome can be performed. Such a "glycomics-assisted glycoproteomics" approach (Figure 51.2) can also be complemented with quantitative proteome and de-glycoproteome profiling (mapping of glycosylation sites), which reduce the search space even further and provide supporting evidence to pinpoint the mechanism(s) driving the observed glycoproteome alterations.

Although glycoproteomics is rapidly maturing and continues to reach a larger group of scientists, significant analytical challenges remain and the technology is still best performed in "specialist" laboratories because of the demand for manual expert interrogation and interpretation of data. Current glycoproteomics methods cannot identify the full complement of glycans on every site of a mixture of glycoproteins and cannot easily tackle multiply glycosylated peptides especially with both N- and O-linked glycan structures. Defining the relative proportions of different intact glycoform populations is important for understanding biological function. This has only been achieved in a few cases (e.g., by MS analysis of the intact IgG glycoprotein).

Several promising informatics initiatives aimed at automating the glycopeptide identification process, aiding the quantitation, and supporting the interpretation of the data output have appeared over recent years. A recent interlaboratory study conducted by the Human Glycoproteomics Initiative of HUPO highlights the importance of different software search parameters for the successful identification and characterization of glycopeptides in order to try to deal with false positives and the reporting of thousands of nonverified glycopeptides. At this stage, for most of the available glycopeptide identification software tools it is strongly advised that manual validation of the glycopeptide assignments still be performed to generate sufficient confidence in the reported identifications. It is therefore clear that the computational solutions require further improvements to advance the field of glycoproteomics and ultimately achieve full integration among other disciplines within the glycosciences (Chapter 52).

ACKNOWLEDGMENTS

The authors acknowledge contributions to previous versions of this chapter by Carolyn R. Bertozzi and Ram Sasisekharan and appreciate helpful comments and suggestions from Daron Freedberg and Chad Slawson.

FURTHER READING

Berger EG, Buddecke E, Kamerling JP, Kobata A, Paulson JC, Vliegenthart JF. 1982. Structure, biosynthesis and functions of glycoprotein glycans. *Experientia* **38:** 1129–1162. doi:10.1007/bf01959725

Domann PJ, Pardos AC, Fernandes DL, Spencer DI, Radcliffe CM, Royle L, Dwek RA, Rudd PM. 2007. Separation-based glycoprofiling approaches using fluorescent labels. *Proteomics* **1:** 70–76. doi:10.1002/pmic.200700640

Tissot B, North SJ, Ceroni A, Pang PC, Panico M, Rosati F, Capone A, Haslam SM, Dell A, Morris H. 2009. Glycoproteomics: past, present and future. *FEBS Letters* **583:** 1728–1735. doi:10.1016/j.febslet.2009.03.049

Zaia J. 2010. Mass spectrometry and glycomics. *OMICS* **14:** 401–418. doi:10.1089/omi.2009.0146

Jensen PH, Karlsson NG, Kolarich D, Packer NH. 2012. Structural analysis of N- and O-glycans released from glycoproteins. *Nat Protocols* **7:** 1299–1310. doi:10.1038/nprot.2012.063

Kolarich D, Jensen PH, Altmann F, Packer NH. 2012. Determination of site-specific glycan heterogeneity on glycoproteins. *Nat Protocols* **7:** 1285–1298. doi:10.1038/nprot.2012.062

Harvey DJ. 2015. Analysis of carbohydrates and glycoconjugates by matrix-assisted laser desorption/ionization mass spectrometry: an update for 2011–2012. *Mass Spectrom Rev* **34:** 268–422. doi:10.1002/mas.21411

Shajahan A, Heiss C, Ishihara M, Azadi P. 2017. Glycomic and glycoproteomic analysis of glycoproteins—a tutorial. *Anal Bioanal Chem* **409:** 4483–4505. doi:10.1007/s00216-017-0406-7

Chernykh A, Kawahara R, Thaysen-Andersen M. 2021. Towards structure-focused glycoproteomics. *Biochem Soc Trans* **49:** 161–186. doi:10.1042/bst20200222

Ross AB, Langer JD, Jovanovic MD. 2021. Proteome turnover in the spotlight: approaches, applications, and perspectives. *Mol Cell Proteomics* **20:** 100016 doi:10.1074/mcp.r120.002190

Thaysen-Andersen M, Kolarich D, Packer NH. 2021. Glycomics & glycoproteomics: from analytics to function. *Mol Omics* **17:** 8–10. doi:10.1039/d0mo90019b

52 Glycoinformatics

Kiyoko F. Aoki-Kinoshita, Matthew P. Campbell, Frederique Lisacek, Sriram Neelamegham, William S. York, and Nicolle H. Packer

Glycans are branched, biosynthetic metabolic products that are commonly encoded by multiple genes. Unique genes may be involved in the biosynthesis of specific glycan classes (glycoprotein, glycolipid, glycosaminoglycans etc.), and at the same time, many glycogenes participate in the biosynthesis of more than one glycan class. These intricacies relating to gene expression, enzyme specificity, endoplasmic reticulum (ER)–Golgi compartment–specific localization of enzymes, the branched nature of glycan structures, and species-specific variation in monosaccharide composition makes the analysis of glycosylation processes complicated. To aid this effort, a variety of analytical methods have been developed to identify and quantify the structure of glycans and their conjugates in biological samples. Glycoinformatics tools and software aim to use computers to integrate these experimental data, using our knowledge of glycan biosynthetic pathways as a backbone. Glycoinformatics databases ideally curate experimental data allowing glycan structures to be rigorously defined, archived, organized, searched, and annotated. When linked to other relational databases, glycoscience data may then be integrated with related genomic, transcriptomic, proteomic, lipidomic, and metabolomic information. This chapter describes the current status of glycoinformatics databases and software development, with focus on efforts to bridge the gap between glycan structure and function.

published by Cold Spring Harbor Laboratory Press; doi: 10.1101/glycobiology.4e.52

THE NEED FOR INFORMATICS IN GLYCOBIOLOGY

Informatics plays a critical role in virtually every aspect of modern biology. Our ability to compile the genomes of diverse organisms made it possible to predict the protein sequences. These sequences are widely used as the basis to predict protein function and to enable proteomic analyses. Proteomic studies in turn advance biological research by providing experimental evidence that supports the expression of specific proteins, in particular tissues or cell types (Figure 52.1). The informatics resources that support these endeavors are facilitated by the fact that genes and the translated polypeptides (i.e., proteins) are typically linear molecules whose sequences are readily specified as a series of characters. These representations are easy to digitize and store and many powerful informatics tools for comparing and classifying polypeptides have been developed. In contrast, the development of informatics tools for glycobiology is more difficult for several reasons. Notably, glycans are not directly encoded by genes. Rather, they are the metabolic products of multiple enzyme activities, both glycosyltransferases [GTs] and glycosidases (Chapter 6). These reactions are tightly controlled by the availability of the donor and acceptor substrates in the cellular compartment containing the GT as well as by modulation of enzyme activity and expression levels. In addition, the organization and function of the ER–Golgi biosynthetic pathway is sensitive to many factors such as the metabolic or developmental stage of the cell or its level of nutrients (Chapter 4), and this leads to structures that can be complex (Chapter 3). Furthermore, glycans are often highly branched and their structures cannot be described as a simple linear sequence (Chapter 3).

Because of this biosynthetic and structural complexity, it is not currently possible to accurately predict the structures of the glycans that an organism can produce under different

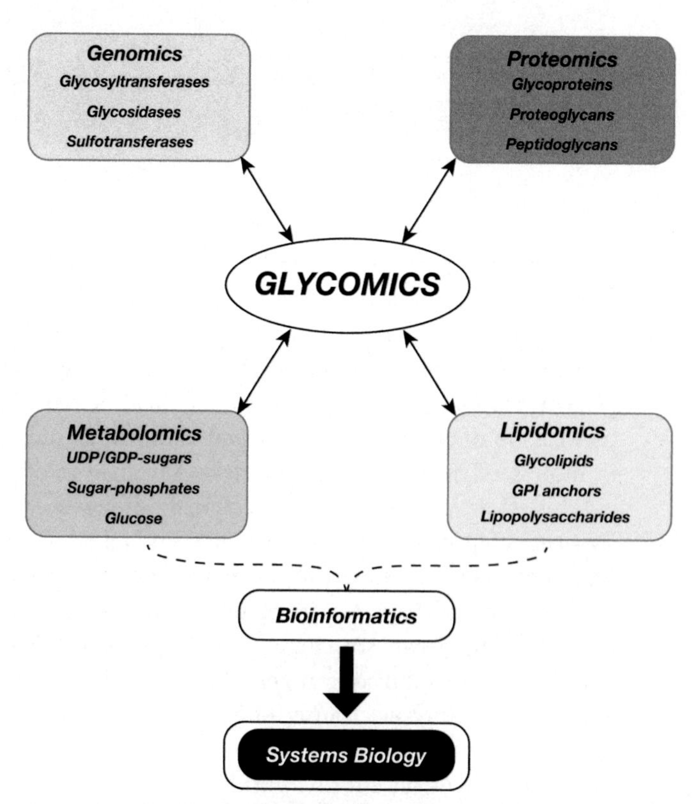

FIGURE 52.1. The critical role of glycomics in systems biology. Glycan structures have no template from which to be predicted, are regulated by cellular metabolism and glyco-enzyme expression, and modify both proteins and lipids. Glycomics thus requires the tools of genomics, transcriptomics, proteomics, lipidomics, and metabolomics to be integrated by bioinformatics.

environments or how these glycans are conjugated with other molecules, armed only with knowledge of the genome or proteome. Rather, the identity of each glycan in a biological sample must be identified using analytical methods (Chapter 50 and Chapter 51) that are sufficiently sophisticated to detect and discern the glycan's diverse structural features. Thus, research aimed at understanding the biological roles and consequences of glycan structures depends on the availability of integrated glycoinformatics databases. A consortium of international scientists under the GlySpace Alliance have made considerable progress in this area in recent years to streamline the annotation of glycans and related expression patterns such that they can be linked across diverse databases. Thus, a clearer path from glycan structure to biosynthetic pathways is beginning to emerge from these efforts.

Interpreting glycan structural information in the context of diverse types of biological and chemical information is a challenge. For example, most glycans in animals are covalently linked to proteins or lipids. The glycan moieties of a glycoprotein are linked to specific amino acids (usually asparagine, serine, or threonine) (Chapters 9 and 10). Which sites are glycosylated and which structures are present at a particular site often vary, depending on many factors, including the type, developmental stage, and disease state of the cell or tissue. Collection, storage, and retrieval of a description of each protein's glycosylation is time-, tissue-, organism-, interaction-, and disease-dependent and thus presents a major challenge to bioinformaticians working in the glycosciences (glycoinformaticians) as it requires integration of conceptually diverse information. Moreover, many different types of digital tools are necessary, ranging from basic visualization software to software that assists in the interpretation and structural annotation of glycoanalytical data (e.g., mass spectra), to algorithms that identify correlations between glycosylation and other biological phenomena (e.g., gene expression, cell differentiation, disease). A major challenge facing the glycoinformatician is the representation of the information that is processed and produced by these software tools in ways that are conceptually accessible to scientists who do not have an extensive background in glycobiology.

Glycoinformatics enables the development of streamlined data reporting and sharing standards. For glycoproteins, the structures of both the glycan and the protein must be represented along with the relationship between these two entities (e.g., the identity of the glycosylation site and the fraction of the protein molecules that bear the glycan in each physiological state). To make this information relevant, the scientist often requires explicit information about the biological context (e.g., tissue and disease state) corresponding to the specified glycosylation or describing how the glycosylation changes when the tissue or cell is perturbed. The glycoscience community is building on the efforts of the Human Proteome Organization-Proteomics Standards Initiative (HUPO-PSI) to develop similar resources that describe the information that should be included when reporting experimental data, with digital data exchange formats to facilitate communication of structural and biological information and controlled vocabularies that allow the data that is exchanged to be unambiguously interpreted. For example, the Minimum Information Required for A Glycomics Experiment (MIRAGE) initiative is modeled after the well-established Minimum Information for A Proteomics Experiment (MIAPE) initiative of HUPO-PSI. These standards are required for the emergence of glycobiology as a mature discipline that is accessible to the scientific community as a whole.

GLYCAN STRUCTURE DRAWING

A major component of databases is the standardized depiction of the glycan structures. The Symbol Nomenclature for Glycans (SNFG) (see Online Appendix 1B) universal symbol nomenclature for the graphical representation of glycan structures has been developed to facilitate such standardization and is used throughout this book. Online Appendix 52A lists current databases and journal publishers that have thus far accepted or strongly recommend the use of this nomenclature. A variety of drawing software has been developed to simplify the usage of the SNFG (Online Appendix 1B). Among them, GlycanBuilder is a tool that can be used as a stand-alone, embedded in webpages, and be integrated into other programs that allow for the drawing of glycan structures. Recently, an updated version of the tool was included in the GlyGen and GlyTouCan databases providing an intuitive method for searching structural content. Built-in features provide support for the conversion of various glycan display and formats allowing users to efficiently switch between the SNFG, Oxford, hybrid, and International Union of Pure and Applied Chemistry (IUPAC) symbol formats while supporting different text formats: LinearCode, KCF (KEGG Chemical Function), GlycoCT, GLYDE-II, Oxford, LINUCS, and WURCS.

RECOGNITION OF THE NEED FOR GLYCOINFORMATICS DATABASES

The Complex Carbohydrate Structure Database (CCSD) (commonly referred to as CarbBank) was established in the mid-1980s. It was developed and maintained by the Complex Carbohydrate Research Center of the University of Georgia (United States). The main design objective of CCSD was to allow researchers to find publications in which specific carbohydrate structures were reported. The need to develop CarbBank as an international effort was clearly recognized and resulted in worldwide curation teams responsible for specific classes of glycans that resulted in more than 30,000 entries into the database. During the 1990s, a Dutch group (led by Hans Vliegenthart) assigned nuclear magnetic resonance (NMR) spectra to CCSD entries (SugaBase). This was the first attempt to create a carbohydrate NMR database that complemented CCSD entries with proton and carbon chemical shift values.

Following the end of CCSD development in 1997, other large projects followed. Among these, the EUROCarbDB initiative (ceased in 2011) seeded integrated tools for streamlining European glycomics research through the development of databases, bioinformatics standards, analysis methods, and Web-based software components. The Australian company, Proteome Systems, provided commercial access to mammalian N- and O-glycan structures, and glycoprotein data curated from the literature, in GlycoSuiteDB. The final release of GlycoSuiteDB comprised more than 3000 glycoprotein-derived glycan structure entries and relevant metadata descriptors including taxonomy, disease, and methods of determination. This effort to provide curated information at the glycoprotein level now continues under the GlyGen and GlyConnect initiatives. In the United States, the Consortium for Functional Glycomics (CFG) was established in 2001. This project aimed to deepen understanding of the function of carbohydrate–protein interactions on the cell surface and in cell–cell communication. The CFG generated diverse data sets of (1) gene expression of glycosyltransferases and glycan-binding proteins (GBPs) from gene microarray experiments, (2) phenotypic analysis of transgenic mice, (3) mass spectrometric profiling of glycan structures isolated from selected cells and tissues, and (4) glycan affinity of proteins using glycan arrays.

CURRENT GLYCOINFORMATICS EFFORTS

Several large-scale initiatives to organize and integrate various glycan-related information and resources have been launched in recent years. Among these, GlyGen, GlyCosmos, and Glycomics @Expasy present integrated portals to query diverse databases related to glycomics, genomics, and proteomics. In this regard, GlyGen retrieves carbohydrate- and glycoconjugate-related data from several international data sources and integrates and harmonizes them. A user-friendly Web portal allows this information, including key associations of diverse data types, to be queried, browsed, displayed, and downloaded. GlyCosmos includes data previously collected by the Japan Consortium for Glycobiology and Glycotechnology Database (JCGGDB). This encompasses experimentally verified MS data, lectin affinity data, glycoprotein data, and glyco-gene information. For example, activity data about glycogenes such as glycosyltransferases and sugar nucleotide transporters from the GlycoGene Database (GGDB) have been integrated with KEGG Orthologs and are further linked with glycan structure (GlyTouCan) and disease information (OMIM). Pathways in which glycoproteins are involved are also integrated and can be cross-searched using the main GlyCosmos search form. Glycomics@Expasy aims to host and interconnect available resources to reflect the diversity of glyco-related interactions at the cell surface. The collection is organized around GlyConnect (glycoproteins) and UniLectin (glycan-binding proteins). It also provides tools to support data interpretation such as Compozitor that maps glycomes at any level (site, protein, cell, tissue) into interactive graphs.

Whereas the above resources are portals that query other data resources, additional unique, broadly useful data sets and webtools are available at Glycomics@Expasy, GLYCOSCIENCES.de, the Asian Community of Glycoscience and Glycotechnology database (ACGG), and the Consortium for Functional Glycomics (CFG) websites. Additional resources have curated non-mammalian glycan information (e.g., CSDB), lectin binding data (e.g., UniLectin, SugarBind), and enzyme data (e.g., CAZy, Kyoto Encyclopedia of Genes and Genomes [KEGG] GLYCAN). Data on the sites of addition of single O-GlcNAc to Ser/Thr on nuclear/cytoplasmic proteins is now being incorporated into the general glycoprotein databases.

Table 52.1 provides a more detailed description of the currently (in 2021) active individual database resources. Note that there is no certainty that all cited databases and tools in this chapter will be active over time. Maintenance and quality of databases and software are dependent on secure, funded, and curated input which has been historically not continuous.

DATA STANDARDIZATION AND ONTOLOGIES

A critical component of glycoinformatics is the availability of standardized approaches to share data and tools. To address this problem a number of international efforts are underway to establish standards for the presentation of glycomics data to facilitate data comparison, exchange, and verification. The Glycomics Ontology (GlycO) was the first ontology developed to provide standard terminologies for representing experimentally verified glycan structures as collections of chemically and contextually defined constituents, facilitating the association of these structural elements with biosynthetic and functional processes. Another effort, GlycoRDF, is a proposed standard ontology for glycan and related metadata using Resource Description Framework (RDF) to provide consistent terminologies for representing glycan sequences, related biological sources, publications, and experimental data on the Semantic Web. GlycoRDF is now being used by several glycomics database providers, to enable large-scale integration of diverse data collections in the glycosciences. For example, GlyTouCan uses the GlycoRDF ontology to represent the registered data such that other data resources also using this ontology can be integrated and queried by glycan structure. Different information resources (e.g., databases and publications) can reference these identifiers, thus facilitating identification and interpretation of diverse but complementary data sets that embody information about specific glycan structures.

The Semantic Web is a new technology that provides a framework for making data available directly on the Internet, provided with semantics, such that inferences can be made automatically based on the data. For example, a researcher often refers to various publications to derive a new hypothesis to test. Using the Semantic Web, the data in the publications would be formatted in such a way (using predefined vocabulary, or ontologies) that the meaning behind the data

TABLE 52.1. Glycoscience databases, repositories and web portals

	Description	URL
Glycomics-integrated portals		
GlyGen	enables search of multiple databases using glycan- and protein-based queries	https://www.glygen.org/
GlyCosmos	integration of glycan-related multi-omics data, and integrated queries across databases	https://glycosmos.org/
Glycomics@Expasy	Web tools and databases to relate glycans and proteins that carry or bind to them	https://www.expasy.org/
Specialized databases		
GLYCOSCIENCES.de	glycan structures, 3D structures, NMR data, software tools	http://www.glycosciences.de
ACGG-DB (includes previous Japanese JCCDB resource)	glycan structures, glyco-gene information, glycomics-related protocols, cross references	https://acgg.asia/db/
CFG	glycan structures, MS profile, GlycanArray data, glyco-gene expression data	http://www.functionalglycomics.org
MonosaccharideDB	repertoire of monosaccharides	http://www.monosaccharidedb.org
CSDB	bacterial, fungi, and plant glycan structures, NMR data	http://csdb.glycoscience.ru/
UniCarbKB	literature-based curated glycan structures, glycoprotein site/global information	http://www.unicarbkb.org
GlyConnect	literature-based curated glycan structures, glycoproteins, glycosylation sites, and glycome expression	https://glyconnect.expasy.org
Glycan repositories		
GlyTouCan	glycan structure registry	http://www.glytoucan.org
UniCarb-DR	glycomics-annotated mass spectra repository	https://unicarb-dr.glycosmos.org/
GlycoPOST	mass spectrometry data repository for glycomics and glycoproteomics data	https://glycopost.glycosmos.org/
Glycan-related enzyme and pathway databases		
CAZy	carbohydrate-active enzymes	http://www.cazy.org
KEGG GLYCAN	glycan structures, references to reactions and pathways, glyco-gene information	http://www.genome.jp/kegg/glycan
GlycoEnzDB	glycan enzyme definitions and use in pathway building	https://virtualglycome.org/GlycoEnzDB
CSDB_GT	glycosyltrasferase data for selected yeast, bacteria and plant species	http://csdb.glycoscience.ru/gt.html
Lectins and glycan-binding data and related tools		
UniLectin	classification curation and prediction of lectins	https://www.unilectin.eu/
SugarBind	curated information on pathogen–glycan binding	http://sugarbind.expasy.org/
MatrixDB	extracellular matrix interaction database	http://matrixdb.univ-lyon1.fr/
GLAD	dashboard to analyze glycan microarray data	https://glycotoolkit.com/Tools/GLAD/
Ontologies and guidelines		
GlycO ontology	canonical descriptions of the chemical structures of glycans for functional inference	http://bioportal.bioontology.org/ontologies/3169
GlycoRDF	ontology for glycan structures and related experimental data, biological source, and publication information	http://www.glycoinfo.org/GlycoRDF
GlycoCoO	glycoconjugate ontology, integrating GlycoRDF with proteins and lipids	https://github.com/glycoinfo/GlycoCoO
GNOME	glycan naming and subsumption ontology	https://gnome.glyomics.org/
MIRAGE	experimental reporting guidelines	http://www.beilstein-institut.de/en/projects/mirage

is preserved in a computable form, on the Web. Because a common vocabulary, or ontology, would be used across different publications in different websites (i.e., journals), the terminology used to encapsulate the semantics is preserved. Therefore, the Semantic Web becomes a virtual online database in which all linked data can be queried directly, without any need to transfer large amounts of data. Moreover, with such data available on the Semantic Web, machine learning technologies allow computers to make inferences based on the data available, just as a researcher would think of new hypotheses.

The essential accurate interpretation of glycoanalytical data for glycan structure determination requires well-documented metadata, including the parameters used to acquire and process the raw data along with supporting biological source information for the sample being analyzed. The MIRAGE initiative was formed to develop guidelines for researchers to report the qualitative and quantitative results obtained by diverse types of glycomics analyses (e.g., chromatography, mass spectrometry, and glycan/lectin arrays).

In an effort to allow glycan data to be shared seamlessly among glycan resources, GlyTouCan has been established as a stable repository and registry of chemically valid glycan structures. For each glycan, it provides unique accession numbers that are now used by diverse glycobiology data systems so that information on the glycan can be linked across databases. These unique GlyTouCan identifiers thus provide the foundation for linking glycan-related knowledge to the Semantic Web. The initial dataset of GlyTouCan structures comes from GlycomeDB, an undertaking that consolidated structures from a number of established glycan structure databases including CarbBank and provided links to the original sources. Registered users can submit any glycan structure whether they are fully defined, contain ambiguous linkages, or are simply monosaccharide compositions, independent of experimental evidence. All structures registered in GlyTouCan are checked only for representational and chemical consistency and not for biological relevance. Thus, databases that leverage GlyTouCan structural representations still require improved methods for establishing and validating the biological context of these structures. This registry facilitates the interpretation of results in the context of structural and biological information that is available from other sources. Nevertheless, the meaningful comparison of incompletely defined structures in terms of sequence, linkage, and anomericity remains an ongoing challenge for glycoinformaticians. It should be noted that, because of the reliance on analytical structure determination, a limitation for a majority of databases is the number of fully characterized entries (i.e., defined linkages and anomeric configurations between monosaccharides). For example, less than 15,000 out of the 40,000 structures in the amalgamated GlycomeDB are fully defined.

These unique identifiers, however, provide the semantic foundation required for individuals or databases to effectively communicate by recording the identifier of the structures they have characterized.

SOFTWARE TOOLS FOR EXPERIMENTAL DATA INTERPRETATION

Many advances in glyco-related databases and informatics tools development have focused on the interpretation and storage of analytical data, including liquid chromatography (LC), capillary electrophoresis (CE), interaction arrays, mass spectrometry (MS), and three-dimensional (3D)/modeling/nuclear magnetic resonance (NMR) (Table 52.2).

Mass Spectrometry

Most efforts so far have focused on tools that assist the interpretation of MS data. A number of commercial and publicly accessible software are now available. Introduced in 1999, the widely used GlycoMod was the first glycoinformatics Web-based tool to be released in this context, and its function is to suggest possible glycan compositions from experimental mass values of either free or derivatized glycans or glycopeptides. The Glycosciences.de portal started a toolset for analyzing and interpreting experimental data, and a decade later, the EUROCarbDB initiative launched GlycoWorkbench as a freely downloadable software tool to assist the interpretation of MS/MS data by matching a theoretical list of fragment masses against the experimental peak list derived from the mass spectrum. This tool has been integrated into several glycomics

TABLE 52.2. Glycoinformatics data analysis software tools

	Description	URL
Web tools and downloadable programs for glycomics and glycoproteomics		
GlycoMod	composition of glycan	http://web.expasy.org/glycomod
GlycoWorkBench	glycan MS/MS data analysis product from EuroCarb	http://code.google.com/archive/p/glycoworkbench/
GRITS Toolbox	integrated glycoanalytics	http://www.grits-toolbox.org/
SimGlycan	commercial; large-scale, rapid LC/MS/MS analysis	http://premierbiosoft.com/glycan/index.html
Glycoforest	partial de novo sequencing of glycan structures from MS/MS data	https://bitbucket.org/sib-pig/glycoforest-public/src/master/
UniCarb-DB	glycomics MS/MS spectra, LC/MS-MS, HPLC data	https://www.expasy.org/resources/unicarb-db
Byonic	commercial; supports high-throughput LC-MS/MS analysis using database search	https://www.proteinmetrics.com/products/byonic/
pGlyco	intact glycopeptide analysis performed as part of the pFind Studio suite	https://github.com/pFindStudio/pGlyco2/
GlycoPAT	LC-MS/MS analysis using scoring, false discovery rate, and decoys calculations	https://www.virtualglycome.org/glycopat
MSFragger-Glyco	implements "open glycan search" for the ultrafast identification of N- and O-glycopeptides within the MSFragger search engine	https://msfragger.nesvilab.org/
GlycReSoft	composition and MS/MS spectra based scoring, for both glycomics and glycoproteomics	http://www.bumc.bu.edu/msr/glycresoft/
GlycoProteomeAnalyzer (IQ-GPA)	database based search for N- and O-glycosylation, with scoring, false discovery rate estimation, and relative abundance quantitation	https://www.igpa.kr/
GPQuest	software tool for identification of intact N- and O-glycopeptides in HCD LC-MS/MS studies	https://www.biomarkercenter.org/gpquest
O-Pair	identify O-glycopeptides and localize O-glycosites on MetaMorpheus platform	https://github.com/smith-chem-wisc/MetaMorpheus
GlycopeptideGraphMS	MS1 intact glycopeptide identification	https://bitbucket.org/glycoaddict/glycopeptidegraphms/
glyXtoolMS	toolbox for targeted intact glycopeptide identification	https://github.com/glyXera/glyXtoolMS
SugarQb	automated intact glycopeptide identification from HCD-MS/MS data	https://ms.imp.ac.at/?action=sugarqb
MAGIC	MS/MS intact glycopeptide identification	http://ms.iis.sinica.edu.tw/comics/software_magic
SugarPy	database-independent identification of intact glycopeptides from in-source CID-MS data	https://github.com/SugarPy/SugarPy
3D/Modeling/NMR		
GlyCAM	collection of tools for 3D modeling and simulation of carbohydrate structures	http://www.glycam.org
Sweet-II	3D modeling	http://www.glycosciences.de/modeling/sweet2/doc/index.php
Glyco3D	3D structure database	http://glyco3d.cermav.cnrs.fr/home.php
PDB-Care	PDB carbohydrate validation	http://www.glycosciences.de/tools/pdb-care/
GlyVicinity	statistical analysis	www.glycosciences.de/tools/glyvicinity/
CASPER	determination of oligosaccharide and polysaccharide structures	http://www.casper.organ.su.se/casper
GAG-DB	3D structures of GAG binding proteins	https://gagdb.glycopedia.eu/
Glycosylation site analysis		
NetNGlyc & NetOGlc	prediction of N- and O-glycosylation sites	https://services.healthtech.dtu.dk/
GlycoMinesStruct	N- and O-glycosylation site predictor	http://glycomine.erc.monash.edu/Lab/GlycoMine_Struct/
SPRINT-Gly	N-/O-glycosylation site predictor for humans and mouse	https://sparks-lab.org/server/sprint-gly/
GlycoDomain	aligns amino acid sequences surrounding glycosites depending on glycan properties	https://glycodomain.glycomics.ku.dk/

(*Continued*)

TABLE 52.2. (*Continued*)

	Description	URL
ISOGlyP	Isoform-Specific O-Glycosylation Predictor: canonical descriptions of the chemical structures of glycans for functional inference	https://isoglyp.utep.edu/
GlycoSiteAlign	glycosylation site predictor	https://glycoproteome.expasy.org/glycositealign/
Other glycoanalytical tools		
GlycoPedia	information on glycobiology	http://www.glycopedia.eu/
NIST Glycan Reference	glycan MS/MS reference library	https://chemdata.nist.gov/glycan/about
GlycoEpitope	glycan epitopes	http://www.glycoepitope.jp
GlycoMob	ion mobility collisional cross sections	http://www.glycomob.org
GuCal	calculate GU values for separated N-glycan in an electropherogram with structure assignment	https://www.gucal.hu
HappyTools	stand-alone application for high-throughput targeted quantitation of U/HPLC data	https://github.com/Tarskin/HappyTools
GlycanAnalzyer	interpretation of *N*-glycan UPLC profiles with exoglycosidase digestion	http://glycananalyzer.neb.com
GlycoDigest	exoglycosidase digest prediction	http://glycoproteome.expasy.org/glycodigest/
RINGS	online data mining and algorithmic tools for glycan structure analysis	https://rings.glycoinfo.org/

MS, Mass spectrometry; 3D, three-dimensional; NMR, nuclear magnetic resonance; LC, liquid chromatography; GU, glucose unit; UPLC, ultra performance liquid chromatography; HPLC, high-performance liquid chromatography; PDB, Protein Data Bank.

resources as it provides an easy-to-use interface, a comprehensive collection of fragmentation types, and a list of annotation options. UniCarb-DB adopts the approach of storing annotated and curated experimental glycan MS/MS data against which spectral matching can be used to identify unknown structures. RINGS provides Web-based software (Glycan Miner Tool, ProfilePSTMM) to analyze distinguishable glycan fragments from glycan profiling (MS) data, and the GlycomeAtlas tool is a visualization tool for glycan profiling data in mouse and human, where the distribution of glycans across various tissues is visualized. The GRITS toolbox is also a freely available integrated environment for processing glycoanalytic data. It uses a plug-in approach to facilitate reuse and integration of data-processing modules. GRITS includes a module called GELATO for collecting, annotating, and comparing mass spectral data, manipulating the corresponding metadata, and generating reports. This also uses libraries from GlycoWorkBench.

With recent advances in high-accuracy, high-throughput mass spectrometers (Chapter 51), glycoinformatics tools have emerged (and continue to emerge and be improved) to analyze these data both for glycomics and glycoproteomics applications (Table 52.2). Despite the rapid speed of these glycoinformatics tools, the analysis may only reveal partial glycan structure information as it is challenging to determine glycosidic bond linkage type and position using MS/MS data alone, particularly in the absence of standards. To alleviate these problems in glycomics analysis, glycan spectral library data repositories have been established by various resources including NIST and UniCarb-DR, and their integration into computational programs is awaited. From the biological perspective, however, even partial information is helpful as it enables comparative glycomics analysis of different healthy and disease tissues.

Similar approaches have also been undertaken in the field of glycoproteomics, in which glycoproteins and glycopeptides are analyzed in their native form, with glycans attached to them, using MS techniques (Figure 52.1; Chapter 51). Glycoproteomics not only allows the identification of the glycoprotein and the sites of the attached glycans, but can also provide some specific microheterogeneity information on the glycan compositions. Different fragmentation methods (electron transfer dissociation [ETD], collision-induced dissociation [CID], and higher-energy collision dissociation [HCD]) each provide distinct information about some different structural features of glycoproteins. More recently both stepped-energy HCD, and ETD supplementation with HCD (EThcD) have emerged as powerful strategies for revealing both the composition of the attached glycans and the sites of glycosylation. The application of multiple fragmentation modes to a single candidate glycopeptide sets the stage for the development of glycoinformatics tools that may reveal some information on the attached glycan structure (in addition to composition) at specific sites. The mass spectrum matching algorithms used in these programs vary widely (e.g., database search, de novo sequencing/open-glycan searching, or spectral library matching) to identify specific spectral fragment masses and thereby assign glycan/glycopeptide structural features. Following the ability now to obtain large data sets on glycopeptides generated from complex mixtures of glycoproteins (Chapter 51), a bottleneck that has severely limited the field of glycoproteomics is the downstream glycopeptide structural identification. The identification process was, until recently, largely driven by manual expert annotation of the resulting MS/MS spectra. However, many glycoinformatics initiatives are under continuing development to automate this glycopeptide identification process using various strategies to identify intact glycopeptides by using characteristic fragment ions. Several software tools are freely available to address the challenge of analyzing glycoproteomics data, such as pGlyco, GlycoPAT, MSFragger-Glyco, GlycReSoft, GP Finder, IQ-GPA, O-Pair, and GPQuest. Licensed Byonic and open-source Protein Prospector and MASCOT, initially designed for proteomics studies, allow semi-automated identification of N- and O-glycopeptides from high-resolution MS/MS data. The recent HUPO-HGI (Human Glycoproteomics Initiative) interlaboratory glycoproteomics analysis provides a side-by-side comparison of some of these programs, with a focus on highlighting best practices in the different approaches. Quantitation analysis can either be performed using the generic SkyLine platform or dedicated tools in the Happy-Tools collection.

Liquid Chromatography

In comparison to MS, few software tools are available for supporting (ultra)high-performance liquid chromatography (U/HPLC) data analysis and storage. GlycoStore is a curated chromatographic and capillary electrophoretic composition database of labeled glycans (2-AB, RFMS, and 2-AA) of N-, O-, glycosphingolipid (GSL) glycans and free oligosaccharides. The database is built on publicly available experimental LC data sets from GlycoBase, which has now been commercialized. To assist analysis, GlycanAnalyzer is available for pattern-matching N-glycan LC peak shifts following exoglycosidase digestions, which can be used to assign structures to each peak. GlycoStore also provides access to CE migration data for a limited set of glycan structures.

Nuclear Magnetic Resonance Spectroscopy

NMR data was obtained on carbohydrate structures in the 1980s and 1990s and is still the best analytical technique available to obtain complete structural information on purified oligosaccharides, but is less used now because of the difficulty in obtaining sufficient material from biological sources. The CASPER (Computer-Assisted Spectrum Evaluation of Regular Polysaccharides) program predicts ^{1}H- and ^{13}C-NMR chemical shifts of glycans. As such, it is used for determining the glycan structures based on experimental NMR data.

Glycan-Binding Data and Interpretation

Another area of software analysis of experimental data has been in mining glycan array data sets to identify glycan sequence motifs recognized by various GBPs, such as plant and animal lectins, viral and bacterial pathogen proteins, and antibodies. Several data analysis tools for glycan array experiments have recently emerged, including MotifFinder, GLYMMR, CCARL, Glycan Miner, Glycan Microarray Database (GlyMDB), and GLAD. Such data analysis determines the relative binding strength/specificity of a GBP to a glycan motif or determinant on the array. RINGS also has analytical tools for predicting glycan-binding patterns from glycan array data.

A few databases collect information on glycan-binding proteins. The Lectin Frontier DataBase (LfDB) provides affinity data for a few hundred lectins. UniLectin, which includes the UniLectin3D collection of thousands of curated lectin 3D structures, suggests a classification based on protein folds and stores these predictions. These classes enable the definition of profiles that can be used to screen sequence databases and predict glycan-binding domains. SugarBind is a curated database of literature-derived knowledge of pathogen–glycan binding, and MatrixDB collects glycosaminoglycan-binding proteins.

3D Glycan Structure Modeling

Because of their inherent flexibility, oligosaccharides typically exist in solution or on proteins, as an ensemble of conformations, making it a challenge to describe their 3D structure (see Chapters 30 and 50 for a description of 3D structures). Computational chemistry is an essential tool in analyzing glycan experimental data, to make predictions that may be tested experimentally, and to unravel and explain chemical processes at the atomic level.

Web-based tools are available to generate a theoretical model of a carbohydrate 3D structure. A useful resource is GLYCAM-Web that provides tools for modeling oligosaccharides and glycoproteins in addition to providing downloadable structure files that can be used for molecular modeling. SWEET-II is also a carbohydrate 3D builder that is available on the GLYCOSCIENCES.de website.

The two major databases for storing experimentally determined 3D carbohydrate structures are the PDB and the Cambridge Structural Database. Crystal structures of oligosaccharides are also available at Glyco3D. A recent extension of the latter is GAG-DB, centered on the 3D description of glycosaminoglycan-binding proteins. Most of the carbohydrates in the PDB are either connected covalently to a glycoprotein or form a complex with a lectin, enzyme, or antibody. Recently, the PDB has undergone a carbohydrate remediation, ensuring that carbohydrates are accurately annotated. Therefore PDB entries now contain glycan annotations, which are available in LINUCS, GLYCAM (IUPAC-like), and WURCS formats.

Glycan Attachment Sites on Proteins

Despite the known sequon (NXT/S, X is not Pro) for N-linked glycosylation, many potential sites are not glycosylated in vivo, and there are no clear motif(s) for predicting O-linked glycosylation. Understanding the "rules" of attachment site specificity for the glycosylation of proteins is thus an ongoing challenge for glycoinformaticians. Over the past 20 years, neural networks, hidden Markov models (HMMs), and support vector machines (SVMs) have been implemented to predict N- or O-glycosylation and C-mannosylation. Although the original tools were hosted on the Danish CBS Prediction Servers, additional resources have emerged in the last few years.

Glycoprotein informatic resources of GlyGen, GlyCosmos, and Glycomics@Expasy now provide information on the glycan structures as attached to proteins in a complementary manner. The coverage and content depend on both automated and manual efforts to mine or curate current literature that contains characterized glycan structures and their sites of attachment to proteins and on supporting data from experimental conditions and biological sources. This collaborative, international bioinformatic integration of complex molecular data from all types of glycoanalytical techniques and interactions is in constant development and is essential for the continued progress of glycobiological research.

FUTURE PERSPECTIVES FOR GLYCOINFORMATICS DEVELOPMENT

Glycobiology as a Part of Systems Biology

Systems biology involves the development, simulation, and analysis of biological systems (including whole-body and environmental systems) at the molecular and cellular levels. As research on glycan biosynthetic pathway simulation progresses, its integration with genomics, transcriptomics, proteomics, lipidomics, and metabolomics data represents the next step. This will result in a holistic understanding of biological processes (Figure 52.1), such that glycomics data can be viewed in the context of complementary data. Such integrated knowledge will result in better elucidation of the glycosylation process, reveal new interactions with GBPs, and enhance our understanding of related functional consequences.

The current coordinated trend toward RDF-based data integration is already shaping future developments of glycoscience databases. It will help bridge the gap between glycomics and other -omics that have already adopted RDF ontologies but that are still very much DNA sequence–centered. Indeed, 50 years after the advent of sequencing technology, DNA sequences, along with their links to other data types (gene expression, protein structure, etc.), remain the most prominent entities in molecular biology databases and repositories. Scientists gather sequence-centered information in the course of elucidating a cellular process or a pathological behavior, simply because a gene/protein sequence is usually the common element shared across -omics domains. The problem here is that glycans are only linked to the gene via their biosynthetic enzymes and substrates. The advancement of glycoscience as a discipline depends on expanding the integration of data describing glycoproteins, glycolipids, glycosaminoglycans, lipopolysaccharides, and the genome-coded enzymatic machinery that generates or breaks down these glycans, together with the ever-increasing information about the interactions of these glycoconjugates with other components of the cell.

Linking Glycan Structures with Function

It is the ultimate goal of glycoscience research to be able to link glycan structures with their function. Although it is still difficult to completely identify fully defined glycan structures on their conjugates in a high-throughput manner, various hypotheses regarding the relationships between the specific structure of a glycan and its biological functions have been developed. One hypothesis maintains that the structural features characteristic of a group of glycans, rather than of a single structure, are required for biological function. This hypothesis is possible but is unlikely to be valid in all cases, as many discrete glycans are known to have quite specific functions, and changing a single monosaccharide or glycosidic conformation can greatly affect their capacity to realize their functions. Thus, additional work is required to accumulate as much of the experimental glycomics data as possible into standardized formats, such that comprehensive, integrated analyses can be performed using bioinformatics technologies.

Collaboration in Glycoinformatics

The future of current endeavors in glyco-related informatics lies in the consolidation of international consortia. The small size of the glycoscience community has prompted several cooperative initiatives across all continents for representing and collecting glycomics data (as described above). To favor interactions between these complementary initiatives, the international

Glycome Informatics Consortium (GLIC) was founded in 2015 to provide and maintain a centralized software resource for developers, thus enabling cooperative database and tool development. Also in 2015, a Glycomics section was established on the Swiss Institute For Bioinformatics (SIB) Expasy proteomics resource portal, and glycoprotein entries in UniProtKB have been also linked to glycan structural information, when known, in GlyGen and GlyConnect. In addition, the U.S. National Institutes of Health (NIH) as part of the Common Fund Glycoscience Program is now focused on creating new methodologies and resources to study glycans that include the development of data integration and analysis tools.

In 2018, the GlySpace Alliance (glyspace.org) was formed for further standardization and collaboration between glycan database portals. This alliance consists of GlyGen, funded by the U.S. NIH Common Fund, GlyCosmos funded by the Japan Science and Technology Agency–National Bioscience Database Center, and Glycomics@Expasy of the Swiss Institute of Bioinformatics. Ensuring that all data is available under a completely open license while providing the provenance of all data, sharing all data between resources, and quality-checking all data are the key goals of this alliance.

These international efforts are affirmation that the importance of bioinformatics resources for glycoscience is finally being recognized, such that the role of glycans may be more easily understood and accessed by the broader research community. However it is clear that there is a long way to go before the entire community can have routine access to what aficionados of nucleic acid and protein biology currently take for granted: reliable, well-curated, user-friendly, cross-referenced databases that are permanently and safely housed in major long-term government-funded central servers. Achievement of this goal will be critical to bringing the study of glycans into the mainstream of evolutionary, molecular, and cellular biology, and its applications to medicine, materials science, and other fields that benefit humankind.

ACKNOWLEDGMENTS

The authors acknowledge contributions to previous versions of this chapter by Ram Sasisekharan and appreciate helpful comments and suggestions from Manfred Wuhrer.

FURTHER READING

Doubet S, Bock K, Smith D, Darvill A, Albersheim P. 1989. The complex carbohydrate structure database. *Trends Biochem Sci* **14:** 475–477. doi:10.1016/0968-0004(89)90175-8

Cooper CA, Gasteiger E, Packer NH. 2001. GlycoMod—a software tool for determining glycosylation compositions from mass spectrometric data. *Proteomics* **1:** 340–349. doi:10.1002/1615-9861(200102)1:2<340::aid-prot340>3.0.co;2-b

Hashimoto K, Goto S, Kawano S, Aoki-Kinoshita KF, Ueda N, Hamajima M, Kawasaki T, Kanehisa M. 2006. KEGG as a glycome informatics resource. *Glycobiology* **16:** 63–70. doi:10.1093/glycob/cwj010

Mariethoz J, Alocci D, Gastaldello A, Horlacher O, Gasteiger E, Rojas-Macias M, Karlsson NG, Packer NH, Lisacek F. 2018. Glycomics@ExPASy: bridging the gap. *Mol Cell Proteomics* **17:** 2164–2176. doi:10.1074/mcp.ra118.000799

Aoki-Kinoshita KF, Lisacek F, Mazumder R, York WS, Packer NH. 2020. The GlySpace Alliance: toward a collaborative global glycoinformatics community. *Glycobiology* **30:** 70–71. doi:10.1093/glycob/cwz078

Neelamegham S, Aoki-Kinoshita K, Bolton E, Frank M, Lisacek F, Lütteke T, O'Boyle N, Packer NH, Stanley P, Toukach P, et al. 2020. SNFG discussion group updates to the symbol nomenclature for glycans guidelines *Glycobiology* **30:** 72–73. doi:10.1093/glycob/cwz045

Rojas-Macias MA, Mariethoz J, Andersson P, Jin C, Venkatakrishnan V, Aoki NP, Shinmachi D, Ashwood C, Madunic K, Zhang T, et al. 2020. Towards a standardized bioinformatics infrastructure for N- and O-glycomics. *Nat Commun* **10:** 3275. doi:10.1038/s41467-019-11131-x

Yamada I, Shiota M, Shinmachi D, Ono T, Tsuchiya S, Hosoda M, Fujita A, Aoki NP, Watanabe Y, Fujita N, et al. 2020. The GlyCosmos portal: a unified and comprehensive Web resource for the glycosciences. *Nat Methods* **17:** 649–650. doi:10.1038/s41592-020-0879-8

York WS, Mazumder R, Ranzinger R, Edwards N, Kahsay R, Aoki-Kinoshita KF, Campbell MP, Cummings RD, Feizi T, Martin M, et al. 2020. GlyGen: computational and informatics resources for glycoscience. *Glycobiology* **30:** 72–73. doi:10.1093/glycob/cwz080

Fujita A, Aoki NP, Shinmachi D, Matsubara M, Tsuchiya S, Shiota M, Ono T, Yamada I, Aoki-Kinoshita KF. 2021. The international glycan repository GlyTouCan version 3.0. *Nucleic Acids Res* **49:** D1529–D1533. doi:10.1093/nar/gkaa947

Kawahara R, Alagesan K, Bern M, Cao W, Chalkley RJ, Cheng K, Choo MS, Edward N, Goldman R, Hoffmann M, et al. 2021. Community evaluation of glycoproteomics informatics solutions reveals high-performance search strategies of glycopeptide data. bioRxiv doi:10.1101/2021.03.14.435332

53 Chemical Synthesis of Glycans and Glycoconjugates

Peter H. Seeberger and Hermen S. Overkleeft

Pure glycans of defined structure are essential research tools in glycobiology. Unlike proteins and nucleic acids, which can be obtained in homogeneous forms by recombinant expression and polymerase chain reaction (PCR), respectively, glycans produced in biological systems are heterogeneous. Furthermore, the quantities that can be obtained from biological systems are often small. Chemical synthesis can be used to obtain homogeneous glycans in larger quantities than are available from most cellular production systems. Chemical synthesis can be further employed to incorporate glycans into homogeneous glycoproteins. This chapter summarizes the current state of the art in chemical methods to produce glycans. Enzymes can be used together with chemical methods to prepare diverse glycans (Chapter 54).

CONTROLLING REGIOCHEMISTRY

The structure of oligosaccharides render their synthesis more complicated than the synthesis of the other major classes of biomolecules: oligonucleotides and oligopeptides. Fundamental challenges of glycan synthesis are the requirement for modifying one specific hydroxyl group in the presence of many others, and control over the stereochemical outcome in the creation of a glycosidic linkage. Commonly, glycan synthesis is characterized by the manipulation of various protecting groups, chemical moieties that mask the hydroxyl groups and prevent them from reacting with other chemical reagents. Hydroxyl-protecting groups are selectively added and removed from glycan structures, allowing for chemical alteration of the exposed hydroxyl

published by Cold Spring Harbor Laboratory Press; doi: 10.1101/glycobiology.4e.53

groups. Subsequently, in a typical glycan synthesis scheme, the exposed hydroxyl group serves as a point for further elaboration. The selective exposure of one hydroxyl group allows for regioselective addition of another saccharide unit. Synthetic schemes of this type are commonly applied to the generation of O- and N-linked glycans (see Chapters 9 and 10) as well as proteoglycans (see Chapter 17) and glycosphingolipids (see Chapter 11).

The choice of protecting groups and the order of protecting group installation are essential for a successful synthetic route. The most common protecting groups include benzyl ethers that stay in place through many synthetic steps (permanent protecting groups) or carbonates and esters that are removed during intermediate steps in a synthesis (temporary protecting groups). A wealth of information exists on the chemical generation of glycans and various protecting group manipulations used in the context of glycan synthesis.

CONTROLLING STEREOCHEMISTRY

A glycosidic linkage is generally formed through the activation of a glycosylating agent (glycosyl donor) to create an electrophilic species that reacts with the nucleophile, for example, a hydroxyl group present on the glycosyl acceptor. Other possible acceptors include serine/threonine in the case of glycopeptides or a sphingoid in the case of glycosphingolipids. The glycosylation reaction results in the formation of either an α- or β-glycosidic linkage (see Chapter 2). A major challenge in glycan synthesis is the stereoselective formation of glycosidic bonds (Figure 53.1A; Chapter 2). A variety of methods are available to stereoselectively generate glycosidic linkages. The yield and the stereochemical outcome of these reactions depend on the steric and electronic nature of the glycosylating agent (the glycosyl donor), the nature of the nucleophile, and the reaction conditions. One common method to control the stereochemistry at the anomeric center involves the use of certain protecting groups, such as esters or amides/carbamates, on the C2-hydroxyl or C2-amino group (Figure 53.1B). These "participating neighboring" protecting groups can form a cyclic oxonium ion intermediate during the glycosylation reaction that shields one face of the molecule,

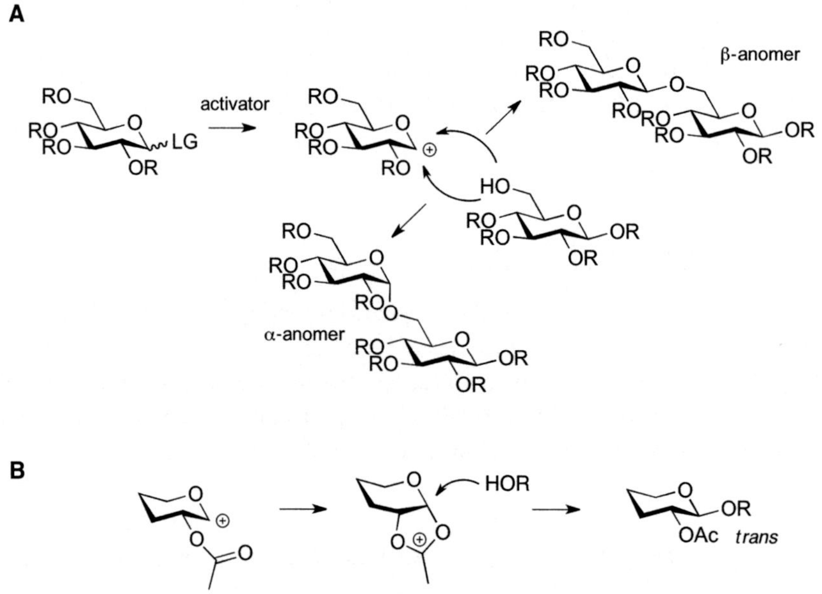

A

RO RO RO O LG OR activator RO RO RO O ⊕ RO HO RO RO O OR OR β-anomer α-anomer

B

O ⊕ O O → O O⊕O HOR → O OR OAc trans

FIGURE 53.1. (A) Stereospecific formation of glycosidic bonds as either an α- or β-linkage. (LG) Leaving group, (R) protecting group. (B) Formation of a cyclic oxonium ion intermediate leading to the formation of a β-glycosidic linkage.

leading exclusively to the formation of a "*trans*"-glycosidic linkage (i.e., where the anomeric substituent and the C2 group are on opposite sides of the ring, as in β-glucosides). The opposite anomeric stereochemistry, termed a "*cis*"-glycosidic bond, is more difficult to construct with high selectivity. Many different synthetic procedures have been developed for the stereoselective construction of *cis*-glycosidic linkages and linkages involving C2-deoxy sugars, though not as general as protecting group participation in the synthesis of *trans*-glycosidic linkages . The use of particular solvents/additives, remote protecting group participation, or preorganization of donor and acceptor and other methods have allowed for the stereocontrolled formation of an increasing number of *cis*-glycosidic linkages.

PROTECTIVE GROUP MANIPULATIONS

Carbohydrates are endowed with a wealth of functional groups. This complicates glycan synthesis, and considerable time and effort are devoted to the development of suitable protective group schemes that allow for the protection and deprotection of individual functional groups at will. Carbohydrate functional groups include hydroxyls, amines, and carboxylates, and distinguishing between these is relatively easy. Selective address of hydroxyls—the most abundant functional groups in carbohydrates—is somewhat more complicated. Primary hydroxyls and anomeric hydroxyls are relatively easy to manipulate—specifically, the former for steric reasons and the latter because they are part of a hemiacetal or hemiketal. Specific access to secondary hydroxyls requires chemistry tailored to capitalize on, for instance, their configuration (*cis*-hydroxyls vs. *trans*-hydroxyls, equatorial vs. axial) and their relative reactivity (i.e., exploiting steric and/or electronic differences). A representative protective group scheme starting from S-tolylglucopyranose (the S-tolyl group serves as both a masking group of the anomeric hydroxyl and as leaving group in glycosylation schemes) is depicted in Figure 53.2. In the first step, all hydroxyls are transformed into the corresponding trimethylsilyl ethers to render the polar carbohydrate soluble in organic solvents. Next and in a series of complementary functional group manipulations, either O-2 (A), O-4 (B), or O-6 (C) is selectively liberated. In all instances, the first step comprises creation of the 4,6-O-benzylidene species. Reductive benzylation of O-3 followed by benzoylation of O-2 and reductive opening of the benzylidene

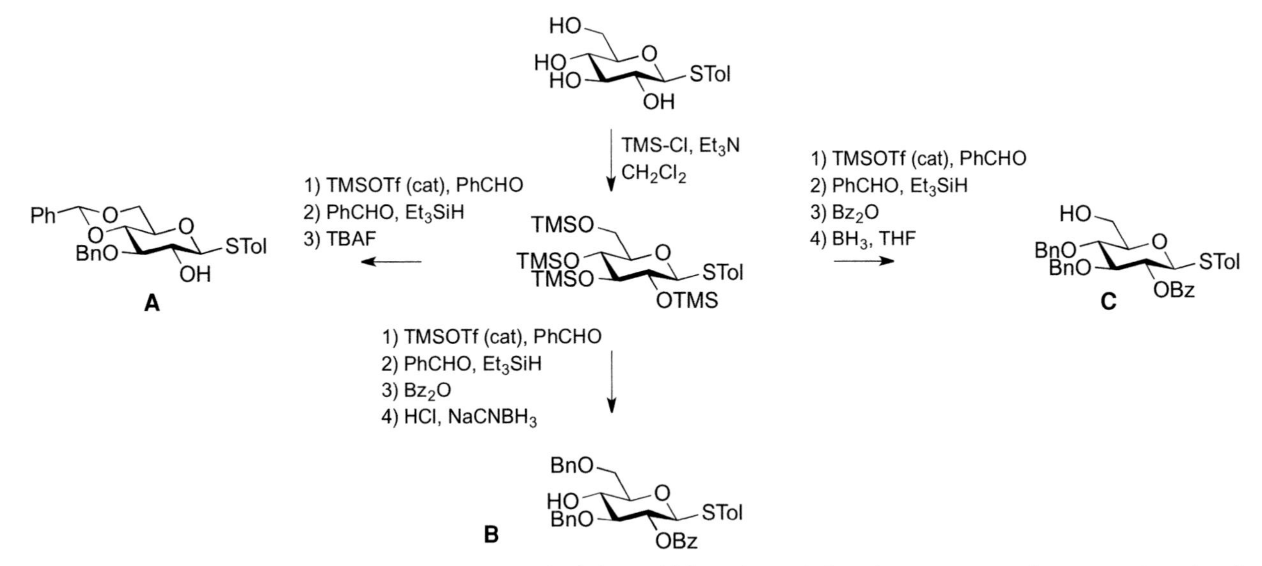

FIGURE 53.2. Protective group manipulations, which can be carried out in one-pot procedures, toward a series of glucopyranose-derived building blocks for glycan assembly.

toward the O-4-benzyl delivers glucose building block C. Alternative reductive opening of the benzylidene toward the O-6-benzyl delivers B, whereas leaving the benzylidene intact and removal of any remaining silyl protective groups delivers glucoside A.

A REPRESENTATIVE SOLUTION PHASE CHEMICAL GLYCAN SYNTHESIS

In the synthesis of a fragment of a putative repeating unit of the exopolysaccharide of *Pseudomonas aeruginosa* (Figure 53.3) the interplay between protective group pattern and glycosylation strategies becomes apparent. Glycosylation of 4,6-O-benzylidene protected thiomannoside **1** with orthogonally protected fucoside **2** promoted by benzenesulfinylpiperidine (BSP) and triflic anhydride (Tf_2O) affords disaccharide **3**. The stereoselective creation of the β-mannosidic linkage follows a S_N2 displacement of an intermediate anomeric α-triflate stabilized by the benzylidene moiety.

Oxidative removal of the naphthyl (Nap) protecting group and glycosylation with mannoside **4** gives trisaccharide **5**. The anomeric allyl (All) group on the reducing fucoside is removed and transformed into the glycosyl trichloroacetimidate, which is condensed with the 3-hydroxyl group of orthogonally protected glucoside **6** to give tetrasaccharide **7**. A 2-O-benzoyl (Bz) participating group in fucose building block **2** ensures the 1,2-*trans*-configuration. Elaboration of tetrasaccharide **7** led to fully protected decasaccharide **10** and illustrates the need for the two differently protected building blocks **1** and **4**. The benzyl (Bn) protective protecting group in **1** serves as a permanent protecting group destined for removal at the final stages of the synthesis. In contrast, The tertbutyldimethylsilyl (TBS) group in **4** can be removed selectively during the synthesis. The 2-hydroxyl groups in mannose residues are liberated and 1,2-*trans* (α) mannosylated with building block **8**. Global deprotection of **9** is effected in two steps: base treatment to remove acyl protective groups followed by catalytic hydrogenation to simultaneously remove benzyl, naphthyl, and benzylidene groups. The azide at the reducing end spacer is transformed into the amine for neoglycoprotein synthesis for immunization studies, or to produce glycan arrays.

REPRESENTATIVE AUTOMATED GLYCAN ASSEMBLY

Automated solid-phase synthesis has made great impact on peptide/protein and nucleic acid chemistry and biology. This development in synthetic chemistry made peptides and oligonucleotides available to biologists and facilitated studies of biopolymer structure and function. Likewise, the availability of a general automated (solid-phase) method for glycan synthesis vastly accelerates access to homogeneous material for biological studies. Because oligosaccharide synthesis is much more complex than the assembly of oligopeptides and nucleotides, automated procedures for glycan assembly were slower to evolve, but automated synthesizers are now commercially available. A prerequisite in automated glycan assembly (AGA) is full control over the stereochemistry of a newly formed glycosidic linkage. AGA is based on monosaccharide building blocks with orthogonal protecting groups and a linker. Important added value of AGA is the ability to rapidly generate libraries of related but distinct glycans, and the ability to prepare polysaccharides of a size not easily accessible with other methods.

The synthesis of a 100-mer polymannoside commences with placing a polystyrene Merrifield resin equipped with a photo-cleavable linker **11** in the reaction vessel of an automated synthesizer. Mannose thioglycoside building block **12** is employed for AGA using a four-step cycle (Figure 53.4A). The incorporation of each monomer relies on an acidic washing step to prevent quenching by any remaining base, followed by the coupling step using 5-6.5 equivalents of building block **12**. Next, a capping step prevents any unreacted nucleophiles from engaging

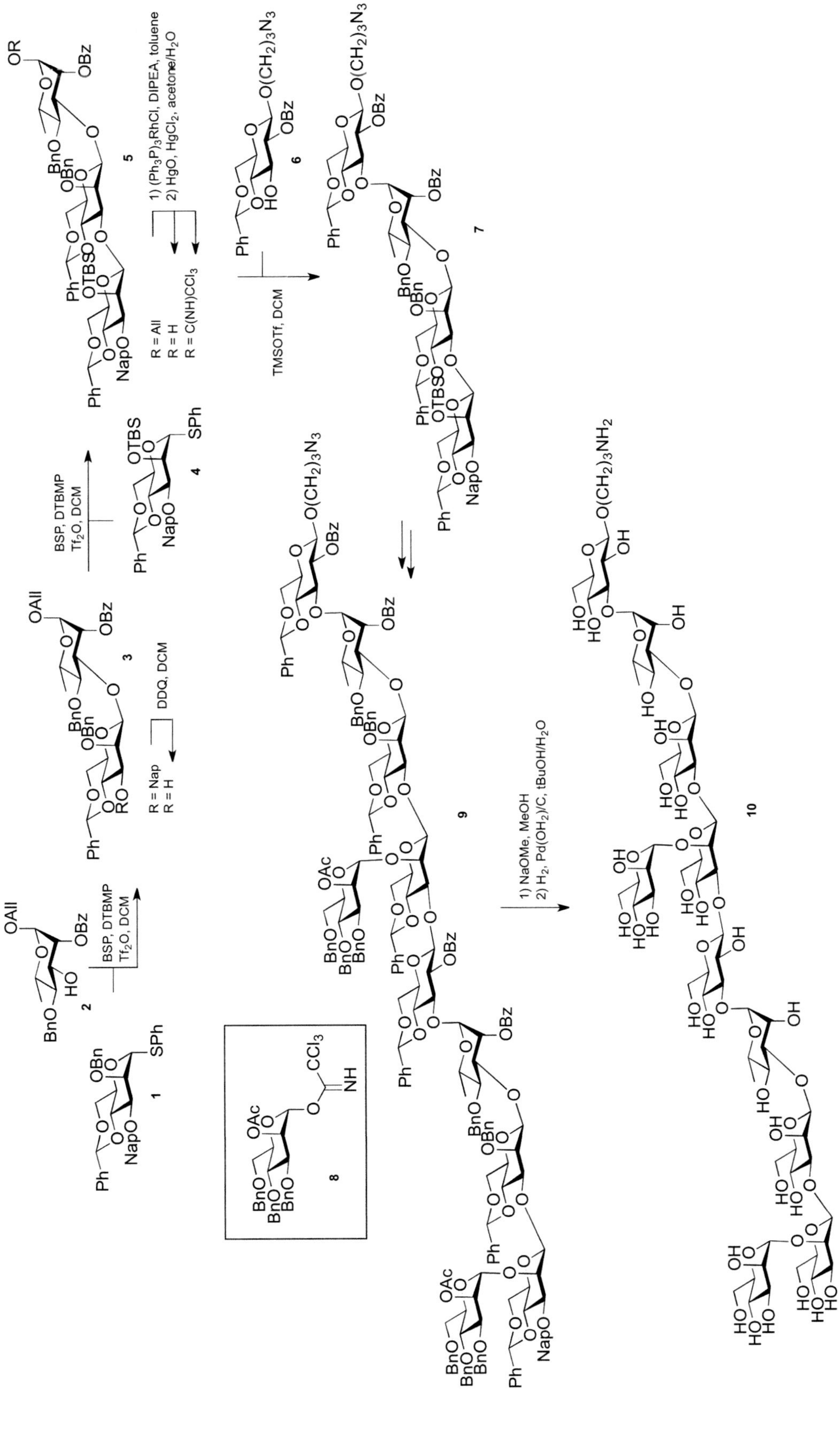

FIGURE 53.3. Solution phase synthesis of *Pseudomonas aeruginosa*–derived decasaccharide **10.**

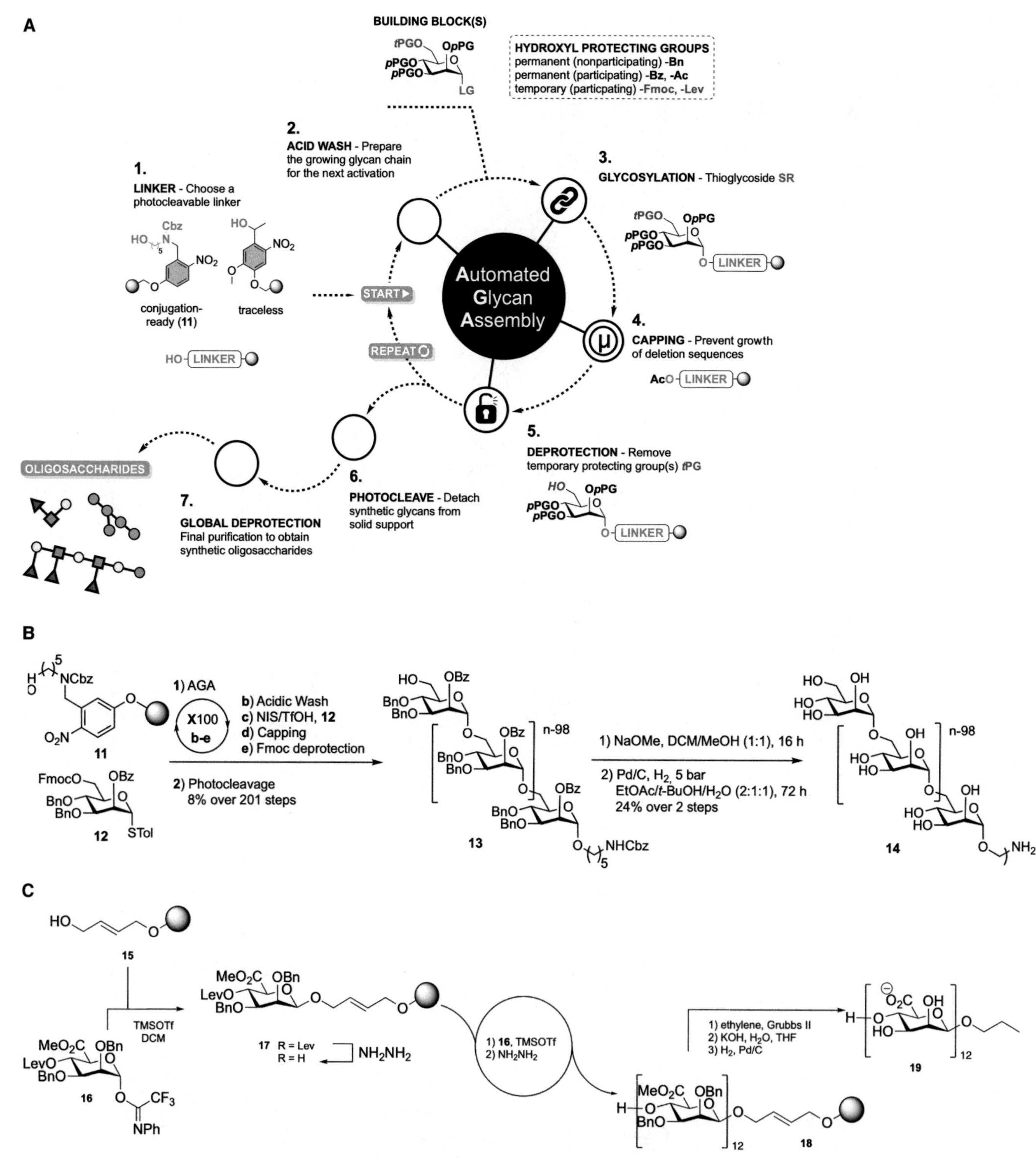

FIGURE 53.4. (*A*) Schematic overview of automated glycan assembly. (*B*) Automated glycan assembly (AGA) of 100-polymannoside **14**. (*C*) AGA of an alginate dodecasaccharide.

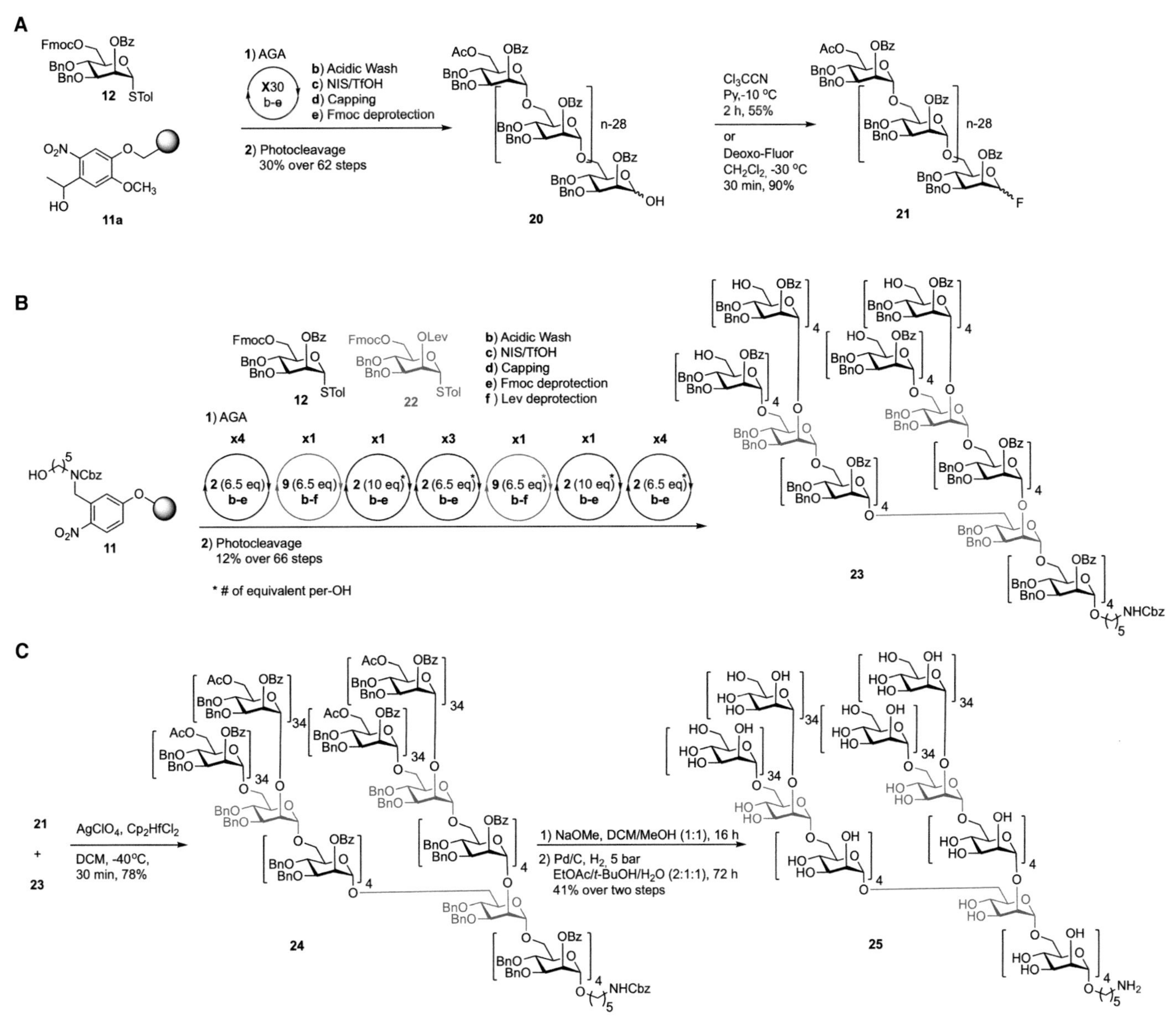

FIGURE 53.5. Synthesis of a 151-mer polymannoside by block coupling of polymannosides prepared by automated glycan assembly (AGA).

in subsequent couplings to produce (n-1)-polysaccharides that are difficult to separate from the desired product. Removal of the temporary Fmoc protecting group exposes the nucleophile for the subsequent coupling step. The 100 coupling cycles were executed automatically within 188 hours and yielded 100-mer α-(1-6) polymannoside **13** after photo-cleavage from the resin and normal-phase HPLC purification (8% overall yield). Cleavage of all ester protective groups by treatment with sodium methoxide, followed by palladium catalyzed hydrogenolysis under pressure to release all benzyl ether protective groups produced 100-mer polymannoside **14** after preparative reverse-phase HPLC.

The synthesis of an alginate dodecamer composed of the β-1,4-mannuronate moieties is of note because it required the repetitive installation of *cis*-glycosidic linkages. The synthesis makes use of immobilized allyl alcohol **15**. Reaction with donor mannuronic ester **16** under catalytic amounts of triflic acid (TfOH) yields immobilized monosaccharide **17**. Removal of the levulinoyl (Lev) group at O-4 is accomplished by treatment with hydrazine (H_2N-NH_2) monoacetate, after which the coupling-deprotection sequence is repeated for the desired number of times. Upon completion of the glycan assembly, the fully protected oligosaccharide is removed from the solid support by cross metathesis and subsequent removal of all protecting groups by treatment with strong base and hydrogenolysis.

More complex polysaccharides are accessible by block couplings of polysaccharides prepared by AGA. Two polysaccharide blocks, linear 30-mer α-(1-6) polymannoside **21** prepared by AGA from mannose building block **12** and branched 31-mer polymannoside **23** prepared by AGA using **12** and differentially protected mannose building block **22** for the branching positions were unified in a 31+30+30+30+30 coupling to create branched 151-mer **24** that was deprotected to furnish **25** (Figure 53.5).

COMPUTER-ASSISTED GLYCAN ASSEMBLY

A database of relative reactivity values (RRVs a number indicating how reactive a glycosyl donor is) of thioglycoside donors has been used to design one-pot glycosylation sequences. Several thioglycosides are combined and selective activation of one of the building blocks over the others can be achieved. The Optimer computer program aids in the selection of the combination of building blocks, equipped with the proper set of protecting groups that tune the reactivity. Figure 53.6 shows the one-pot assembly of the tumor associated N3 antigen octasaccharide using the Optimer approach. Three building blocks were designed and synthesized:

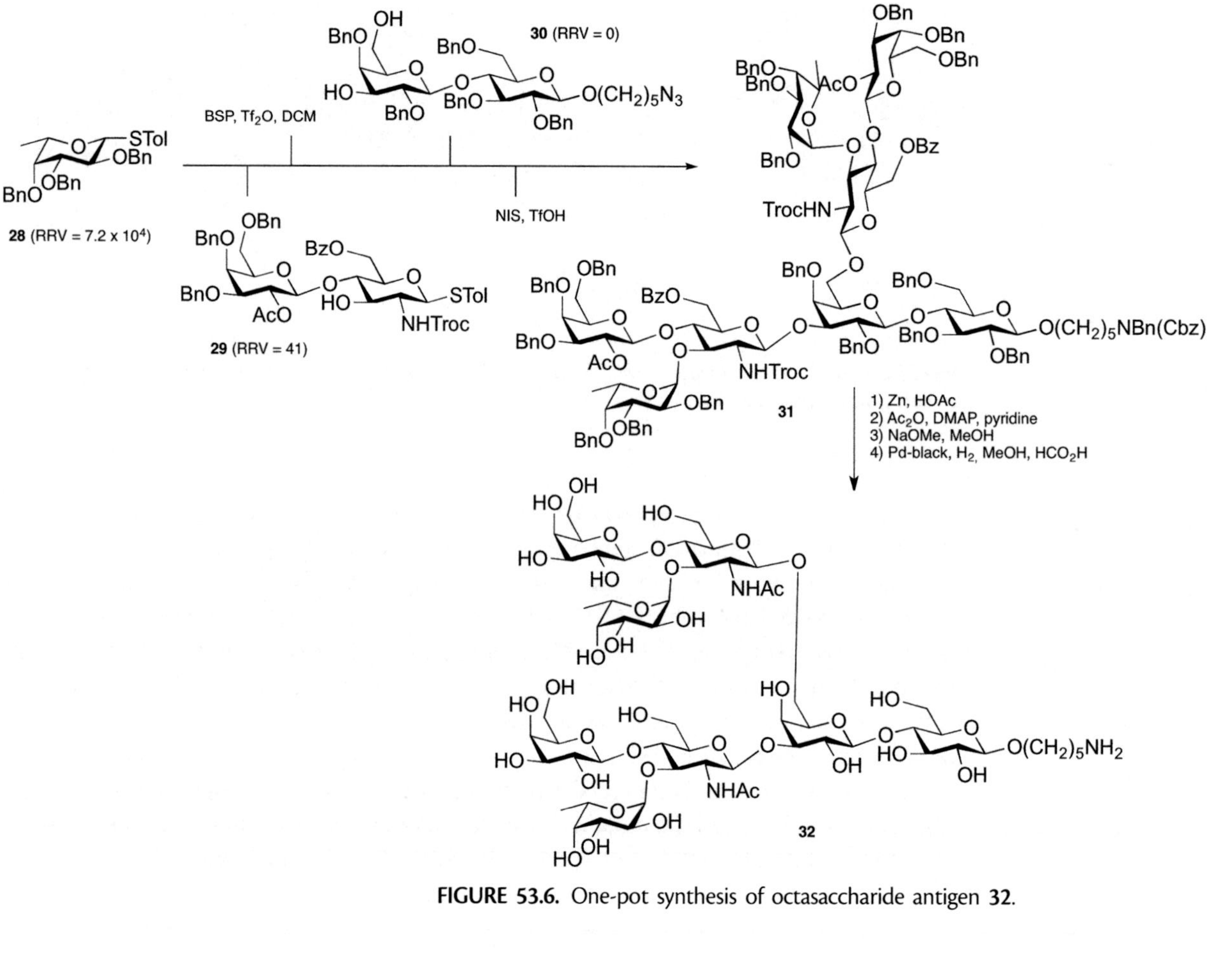

FIGURE 53.6. One-pot synthesis of octasaccharide antigen **32**.

fucosyl donor **26** (RRV = 7.2 × 10^4), lactosamine donor **27** (RRV = 41), and lactose building block **28** (RRV = 0). The latter disaccharides were prepared by chemoselective glycosylations exploiting RRVs. Combination of the first two building blocks in a BSP/Tf_2O mediated condensation gave the trisaccharide thioglycoside. Addition of lactose **28** to the reaction vessel and NIS/TfOH to promote the ensuing double glycosylation provided the octasaccharide **29**. After deprotection the spacer functionalized octasaccharide **30** was obtained in 11% yield.

OUTLOOK

Over the past quarter century, the chemical synthesis of glycans has advanced from an art practiced by synthetic chemists that engaged in heroic total synthesis efforts to a routine task for glycans that contain common monomeric constituents and *trans*-glycosidic linkages. With the commercialization of AGA in which a set of common building blocks is combined on a solid phase resin, the synthesis process can be greatly accelerated. AGA has become an important part of carbohydrate synthesis as polysaccharides as long as 100-mers as well as glycosaminoglycans are now rapidly accessible in multimilligram quantities. The production of highly branched structures will become routine in the coming years. Still, for challenging targets, containing rare monosaccharides and 2-deoxy and 1,2-*cis*-linkages, improved methods for solution phase synthesis will have to be developed. These methods will greatly benefit from the advances in fundamental knowledge on the glycosylation reaction and will be aided by knowledge gathered in databases. In addition, enzymatic synthesis and combinations of chemical and enzymatic synthesis are important assets to procure large quantities of glycans.

ACKNOWLEDGMENTS

The authors acknowledge contributions to previous versions of this chapter by Nathaniel Finney and David Rabuka and appreciate helpful comments and suggestions from Martina Delbianco.

FURTHER READING

Plante OJ, Palmacci ER, Seeberger PH. 2001. Automated solid-phase synthesis of oligosaccharides. *Science* **291:** 1523–1527. doi:10.1126/science.1057324

Wang CC, Lee JC, Luo SY, Kulkarni SS, Huang YW, Lee YW, Chang KL, Hung SC. 2007. Regioselective one-pot protection of carbohydrates. *Nature* **446:** 896–899. doi:10.1038/nature05730

Zhu X, Schmidt RR. 2009. New principles for glycoside-bond formation. *Angew Chem Int Ed* **48:** 1900–1934. doi:10.1002/anie.200802036

Crich D. 2010. Mechanism of a chemical glycosylation reaction. *Acc Chem Res* **43:** 1144–1153. doi:10.1021/ar100035r

Codée JDC, Ali A, Overkleeft HS, van der Marel GA. 2011. Novel protecting groups in carbohydrate chemistry. *C R Chimie* **14:** 178–193. doi:10.1016/j.crci.2010.05.010

Hsu CH, Hung SC, Wu CY, Wong CH. 2011. Toward automated oligosaccharide synthesis. *Angew Chem Int Ed* **50:** 11872–11923. doi:10.1002/anie.201100125

Walvoort MTC, van den Elst H, Plante OJ, Kröck L, Seeberger PH, Overkleeft HS, van der Marel GA, Codée JDC. 2012. Automated solid-phase synthesis of β-mannuronic acid alginates. *Angew Chem Int Ed* **51:** 4393–4396. doi:10.1002/anie.201108744

Li H, Mo KF, Wang Q, Stover CK, DiGiandomenico A, Boons GJ. 2013. Epitope mapping of monoclonal antibodies using synthetic oligosaccharides uncovers novel aspects of immune recognition of the Psl exopolysaccharide of *Pseudomonas aeruginosa*. *Chem Eur J* **19:** 17425–17431. doi:10.1002/chem.201302916

Hahm HS, Schlegel MK, Hurevich M, Eller S, Schuhmacher F, Hofmann J, Pagel K, Seeberger PH. 2017. Automated glycan assembly using the Glyconeer 2.1® synthesizer. *Proc Nat Acad Sci* **114:** E3385–E3389. doi:10.1073/pnas.1700141114

Guberman M, Seeberger PH. 2019. Automated glycan assembly: a perspective. *J Am Chem Soc* **141:** 5581–5592. doi:10.1021/jacs.9b00638

Joseph A, Pardo-Vargas A, Seeberger PH. 2020. Total synthesis of polysaccharides by automated glycan assembly. *J Am Chem Soc* **142:** 8561–8564. doi:10.1021/jacs.0c00751

54 Chemoenzymatic Synthesis of Glycans and Glycoconjugates

Hermen S. Overkleeft and Peter H. Seeberger

Glycosyltransferases are the biosynthetic enzymes responsible for the construction of interglycosidic linkages and glycosidases catalyze the opposite reaction, hydrolysis of interglycosidic linkages. The diversity of natural glycans is reflected by the numerous glycosyltransferases and glycosidases encountered in nature, each exhibiting a defined substrate specificity. Natural glycans are often encountered in heterogeneous form and often produced in minute amounts making their isolation and characterization from natural sources often cumbersome. Therefore, glycobiology relies heavily on synthetic glycans, and synthetic methodology to produce glycans has witnessed tremendous progress (see Chapter 53). Glycosyltransferases and glycosidases offer advantages in the construction of glycans as these biocatalysts are very powerful under controlled conditions. This chapter summarizes recent developments in the use of (mutant) glycosidases and glycosyltransferases for the synthesis of tailored glycans, including combinations of both enzyme classes and in conjunction with chemically synthesized intermediates are presented.

MECHANISM OF GLYCOSYLTRANSFERASES AND GLYCOSIDASES

The mechanism of action of glycosyltransferases resembles the way an interglycosidic linkage is installed by chemical synthesis (Chapter 53). An activated donor saccharide, represented by UDP-glucose (Figure 54.1) is condensed with an acceptor moiety (here, ceramide) to give, after expulsion of the leaving group (here, UDP), the glycoconjugate (glucosylceramide). In contrast to chemical synthesis, the regioselectivity is regulated by the enzyme active site that also selects for the donor and the acceptor. The glucosylceramide synthase (GCS)-catalyzed synthesis of glucosylceramide (Figure 54.1) proceeds with "inversion" of the anomeric configuration, with the α-glucosidic linkage in UDP-glucose transformed into the β-glucosidic linkage in glucosylceramide. Nature employs a limited set of donor glycosides—most prominently the sugar

published by Cold Spring Harbor Laboratory Press; doi: 10.1101/glycobiology.4e.54

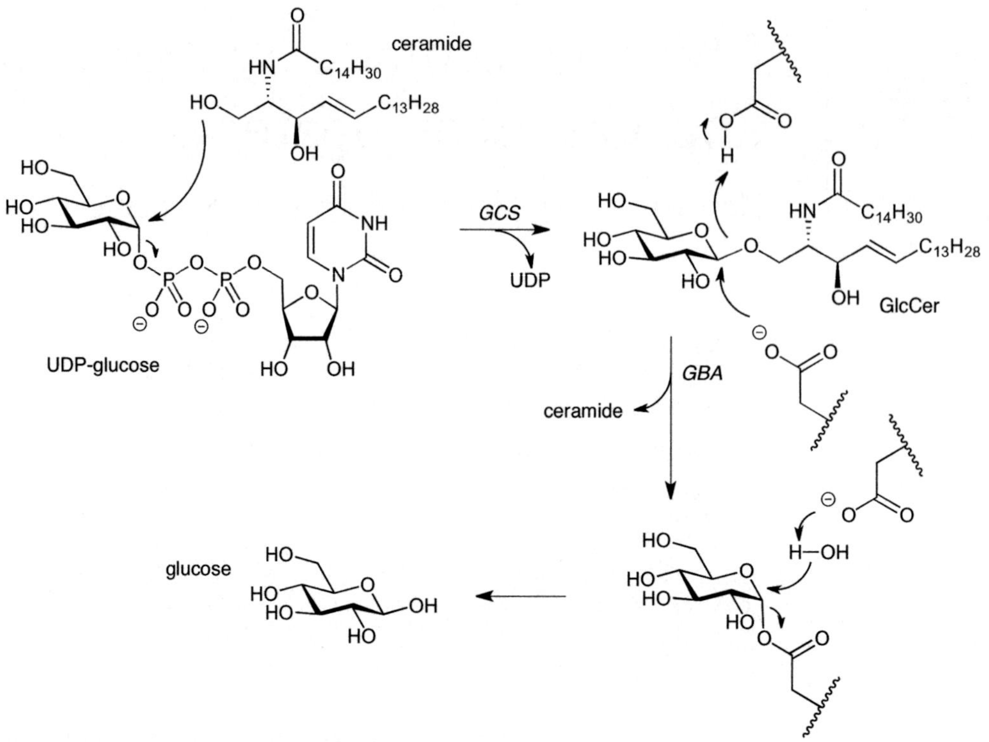

FIGURE 54.1. Formation and hydrolysis of the glycosphingolipid, glucosylceramide. (GCS) Glucosylceramide synthase, (UDP) uridine diphosphate glucose, (GBA) acid glucosylceramidase.

nucleotides for Leloir-type glycosyltransferases, next to UDP-glucose for instance CMP-sialic acid, GDP-mannose—and the nature of the glycosyltransferase determines whether glycosylation proceeds with "retention" or "inversion" of configuration at the anomeric center. The mechanisms employed by most Leloir glycosyltransferases are now resolved.

Glucosylceramide is hydrolyzed by the lysosomal exo-glucosidase, acid glucosylceramidase (GBA). Hydrolysis takes place with "retention of configuration" and is the result of a double displacement mechanism. Upon protonation of the aglycon, ceramide is displaced in an S_N2 substitution–like process to yield a covalent enzyme-glycoside adduct. Upon entry of water in the enzyme active site, the formed glycosyl linkage is hydrolyzed in another S_N2-like process to release glucose from the enzyme active site. Next to retention of configuration, the hydrolysis of interglycosidic bonds can also take place with "inversion of configuration" and is normally the result of an S_N2 displacement–like process of a protonated aglycon by water. Although not necessarily relevant for the product formation in nature (sugar hemiacetals being prone to mutarotation at physiological pH), the different mechanisms employed by "retaining glycosidases" (involvement of a covalent intermediate) and "inverting glycosidases" (no covalent intermediate involved) bears consequences for their use in glycan synthesis.

GLYCOSYLTRANSFERASE-MEDIATED SYNTHESIS OF GLYCANS

The use of Leloir-type glycosyltransferases in glycan synthesis requires access to the natural donor glycosides, which are sugar nucleotides. Thus, the intrinsic advantage of glycosyltransferase-mediated synthesis (excellent regio- and stereoselectivity) can be offset by the limitation on the access to the required donor glycosides. However, this challenge may be overcome by in situ biosynthesis/

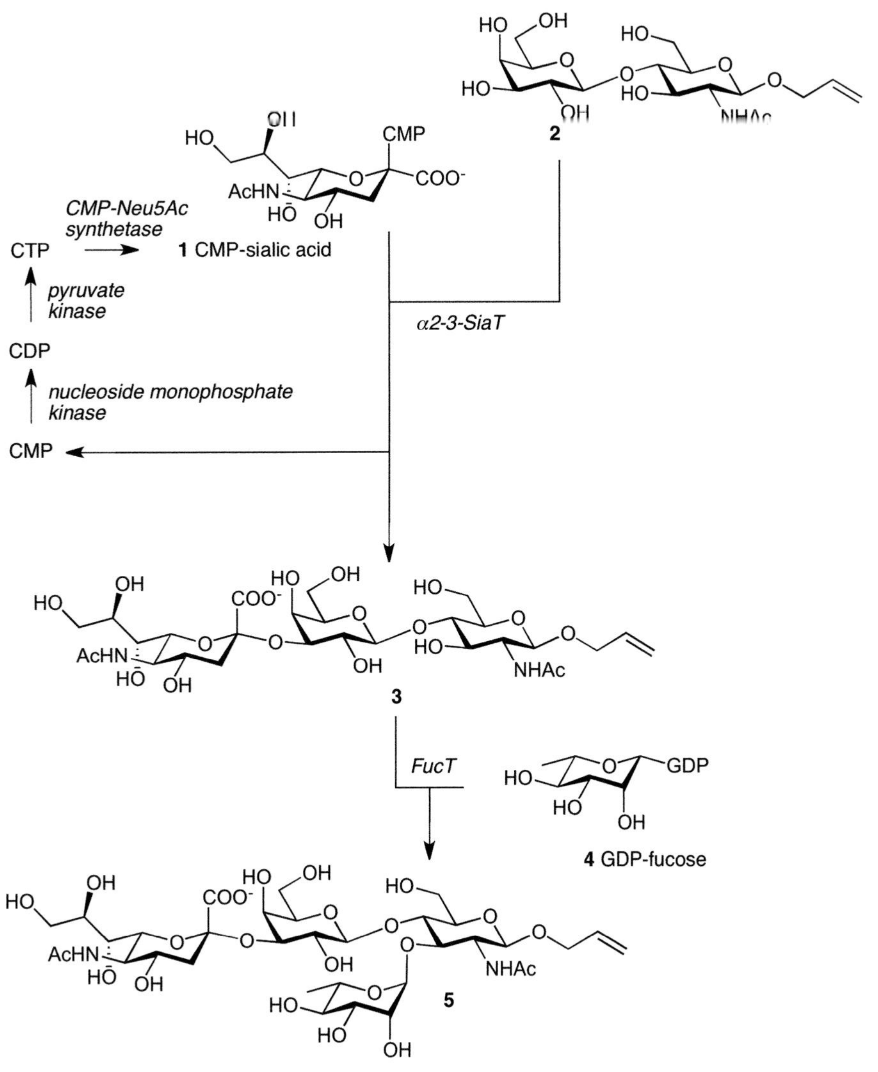

FIGURE 54.2. Glycosyltransferase-mediated synthesis of sialyl-Lewis x. (CMP) Cytidine monophosphate, (CDP) cytidine diphosphate, (CTP) cytidine triphosphate.

regeneration of the consumed sugar nucleotides. The power of glycosyltransferase-mediated glycan synthesis was demonstrated in 1992 by the synthesis of sialyl-Lewis x derivative **5** (Figure 54.2). Allyl lactoside **2**, derived by chemical synthesis (see Chapter 53), is reacted with CMP-sialic acid **1** using a recombinant α-2,3-sialyltransferase (α-2,3-SiaT) as the catalyst. The resulting trisaccharide **3** is further extended with GDP-fucose **4** as the donor and recombinant fucosyltransferase (FucT) as the catalyst to deliver allyl sialyl-Lewis x.

In the first step of the assembly line, the expensive sugar nucleotide, CMP-sialic acid, is consumed, and on transfer of the sialic acid, cytosine monophosphate (CMP) is produced. With the aid of two consecutive kinases (nucleoside monophosphate kinase and pyruvate kinase), CMP can be transformed in situ into the corresponding triphosphate (CTP), which is then condensed by the enzyme CMP-sialic acid synthetase with sialic acid to regenerate CMP-Neu5Ac **2**.

The synthesis of GalNAc-GD1a heptasaccharide **7** equipped with a biotin at the reducing end (replacing the sphingolipids present in the natural product) was accomplished by submitting synthetic lactoside **6** to the consecutive action of four glycosyltransferases, one of which (α-2,3-SiaT) was employed twice (Figure 54.3). By using this method with various donor sugar

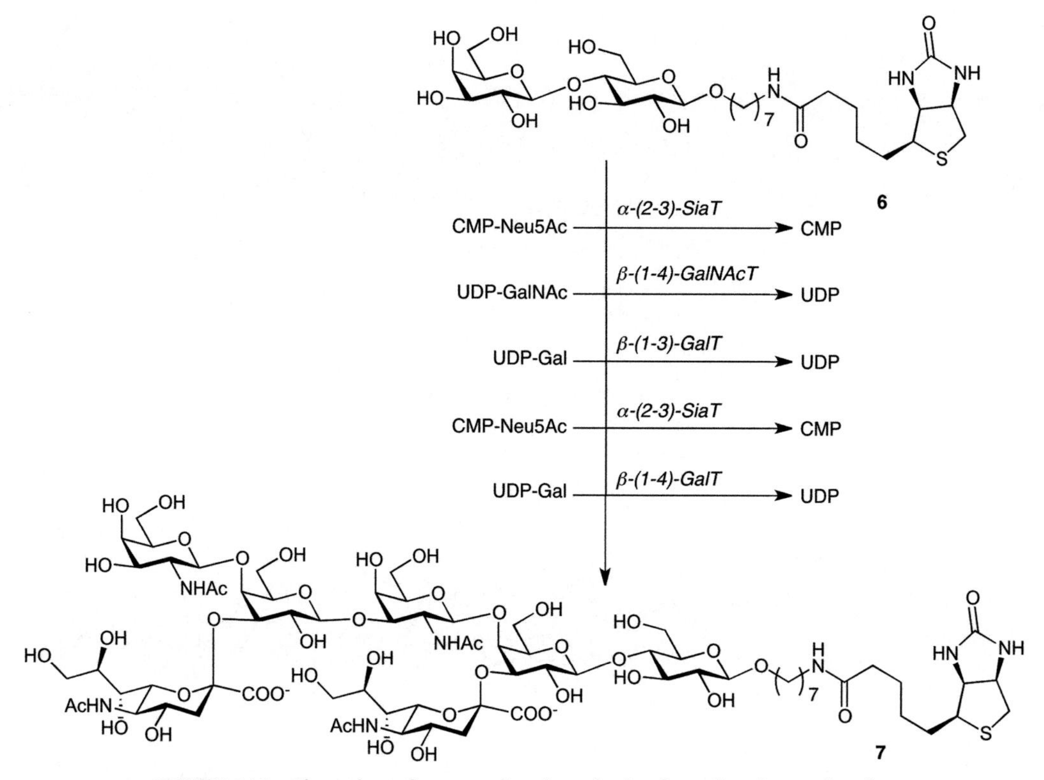

FIGURE 54.3. Glycosyltransferase-mediated synthesis of ganglio-oligosaccharides.

nucleotides and glycosyltransferases, a comprehensive series of glycosphingolipid glycans and their analogs have been obtained. The methodology, especially with respect to enzymatic sialic acid introduction, is competitive when compared to chemical gangliosides synthesis.

The synthesis of complex, asymmetrically branched mammalian N-glycans was accomplished using combined chemical and glycosyltransferase-mediated enzymatic synthesis. As an example, decasaccharide **8** was prepared via contemporary solution phase chemical oligosaccharide synthesis (see Chapter 53). The asymmetrically branched decasaccharide features two nonreducing galactopyranose moieties, one of which is introduced as the tetra-acetate (in bold) whereas the other is unprotected. Decasaccharide **8** is thus designed to allow for specific enzymatic sialylation of the unprotected galactose residue (Figure 54.4). After saponification, the intermediate undecasaccharide is expanded to well-defined oligosaccharide **9** making use of the substrate specificity of α-1,3-fucosyltransferase (α-1,3-FucT, introduction of two fucopyranoses), β-1,4-galactosyltransferase (β-1,4-GalT, twice), β-1,3-N-acetylglucosaminetransferase (β-1,3-GlcNAcT), and finally the sialyltransferase, ST6Gal-I. The chemoenzymatic strategy proved flexible and allowed for the generation of diverse N-glycans in which the reducing end is available for bioconjugation and for the preparation of glycan microarrays for protein binding studies.

FROM TRANSGLYCOSYLATION TO GLYCOSYNTHASES

In contrast to glycosyltransferase-mediated reactions, in which the equilibrium is shifted predominantly to (natural) product formation because of the intrinsic reactivity of donor glycosides, the equilibrium in a glycosidase-mediated reaction can be influenced such that the reaction proceeds in the opposite direction.

Under physiological conditions, with high water concentrations, glycosidases hydrolyze glycosidic linkages to produce the corresponding hemiacetal, either with retention (Figure 54.5A)

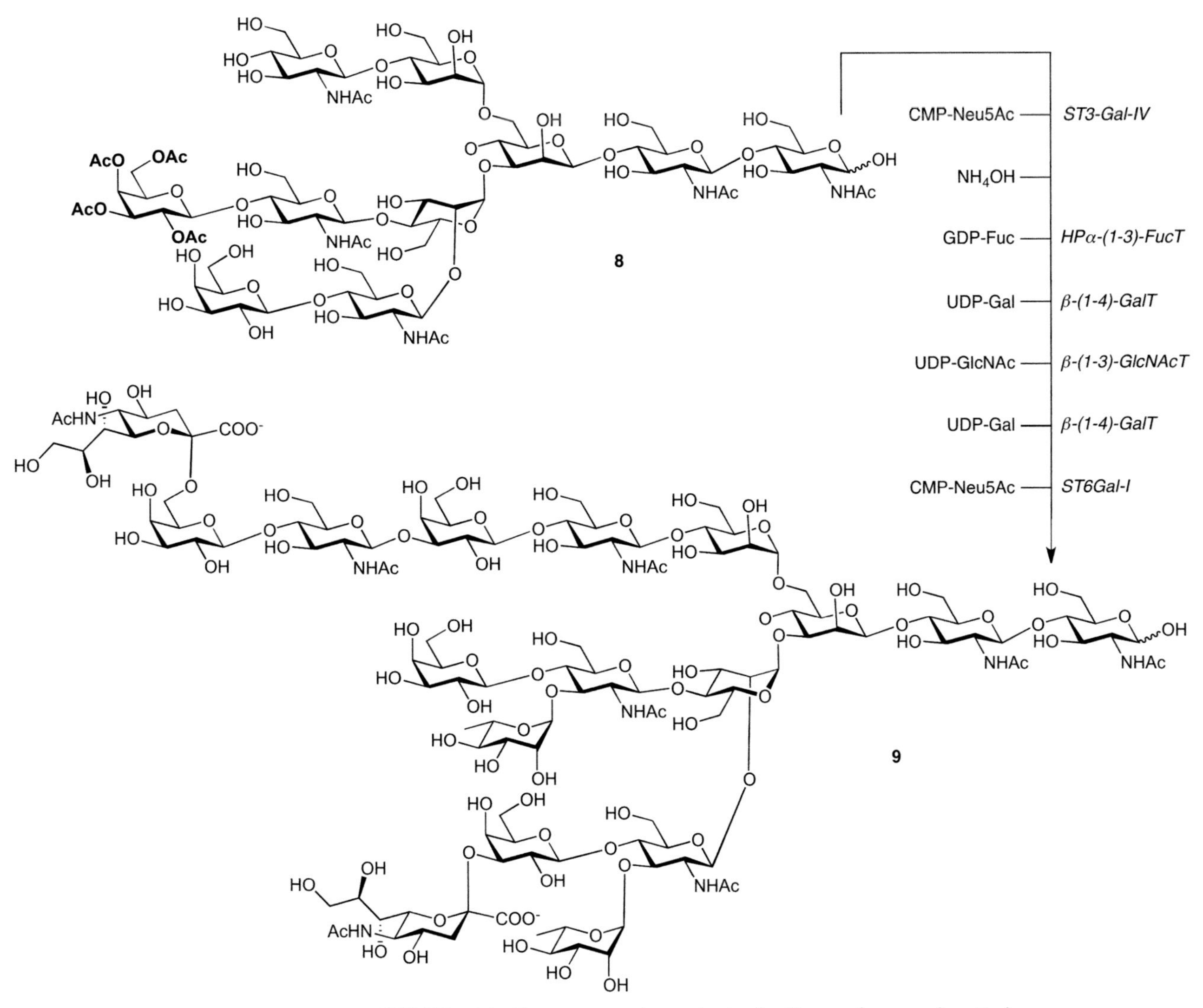

FIGURE 54.4. Chemoenzymatic synthesis of a library of mammalian N-glycans.

or inversion of configuration at the anomeric center. Performing a glycosidase reaction in partly nonaqueous conditions, by addition of large excess of aglycon, by inducing kinetic conditions or by a combination of these allows for partial reversal of the reaction equilibrium. By this means and in a "transglycosylation" event, glycans can be constructed. Disadvantages of this method are that reaction conditions may be adverse to enzyme reactivity and/or stability and, moreover, that the formed product is in essence a substrate for glycosidase-catalyzed hydrolysis. This caveat can be circumvented by making use of mutant glycosidases (Figure 54.5B) in which the catalytic nucleophile in case of retaining glycosidases is mutated to an innocent bystander (depicted is an Asp to Ala substitution). Such a "glycosynthase" can be used to react a synthetic donor glycoside that bears the anomeric configuration corresponding to the intermediate enzyme glycosyl covalent adduct (see Figure 54.1) with an appropriate nucleophile to construct a desired glycosidic linkage. The main advantage of this strategy is that, in principle, the mutant glycosidase has largely lost the ability to hydrolyze the formed product, because of the absence of the catalytic nucleophile. Many retaining glycosidases and, in recent years, inverting glycosidases have been mutated to glycosynthases, which have been used to produce a diverse array of glycans.

The strategy is exemplified by the synthesis of flavonoid glucoside **13** (Figure 54.6). Synthetic α-fluorolactoside **10** is treated with phenol **11** in the presence of the E197S mutant of the

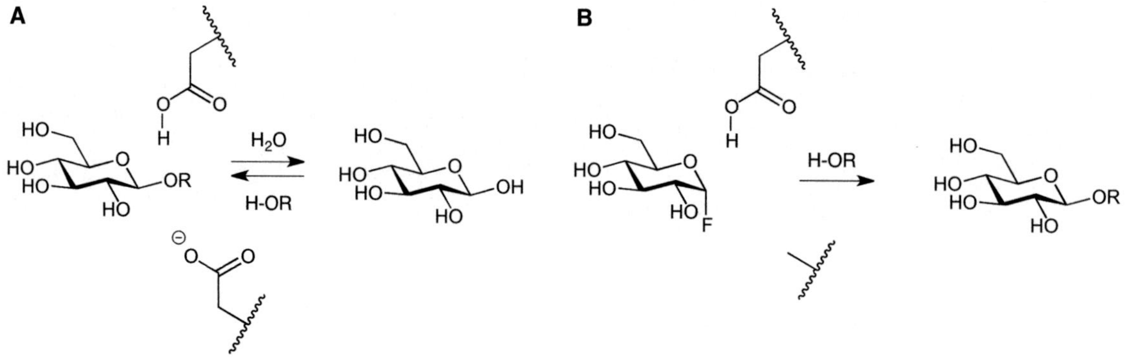

FIGURE 54.5. (*A*) Equilibrium in a retaining β-glucosidase. (*B*) Mutant-retaining β-glucosidase in which the catalytic nucleophile is substituted for a nonparticipating amino acid allows for the construction of β-glucosides.

Humicola insolens Cel7B endoglucosidase. This glycosynthase proved highly flexible with respect to the acceptor phenol, allowing for the construction of a small library of flavonoid glycosides represented by phenolic disaccharide **12**. Subsequent enzymatic removal of the nonreducing galactospyranoside provided flavonoid glucoside **13**.

By combining the strengths of glycosyltransferases, glycosynthases, and chemical synthesis, the total synthesis of ganglioside LLG-3 from the neurogenic starfish was accomplished (Figure 54.7). *N*-Carboxybenzyl(Cbz)-protected CMP-sialic acid **17** was prepared by first performing a Neu5Ac aldolase-catalyzed aldol reaction of mannosamine derivative **14** and pyruvate **15**. The resulting sialic acid derivative **16** was reacted with CTP under the agency of CMP-Neu5Ac synthetase to give donor sialoside **17**. In a sialyltransferase-catalyzed reaction, compound **17** was condensed with α-fluorolactoside **10** to give trisaccharide **18**, in which the amine was unmasked by palladium-catalyzed hydrogenolysis to give trisaccharide, equipped with an anomeric fluoride for glycosynthesis and an amine for chemical amide bond formation. The sequence commenced by condensation of synthetic sialic acid derivative **20** with the free amine in **19** under amide bond-forming conditions, to give **21**. Condensation of tetrasaccharidyl fluoride **21** with the double (E351S/D341Y) mutant of the bacterial endoglycosidase EGCase II gave lysolipid **23** in good yield. The free amine in **23** can be condensed with a fatty acid or alternatively with a fluorescent reporter group.

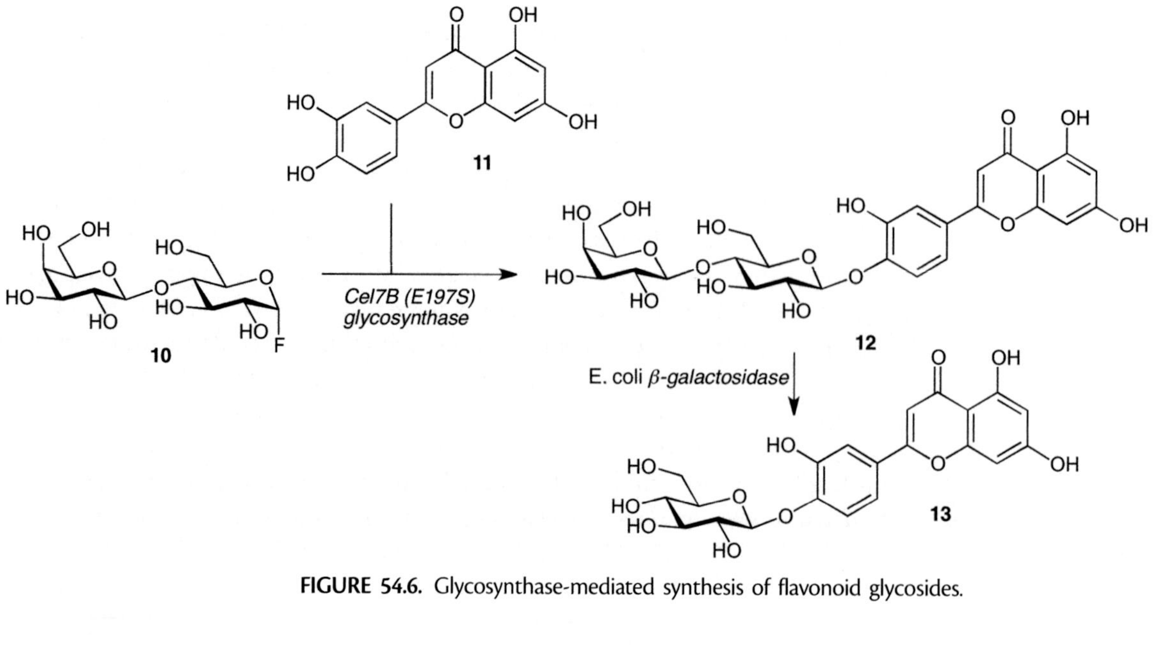

FIGURE 54.6. Glycosynthase-mediated synthesis of flavonoid glycosides.

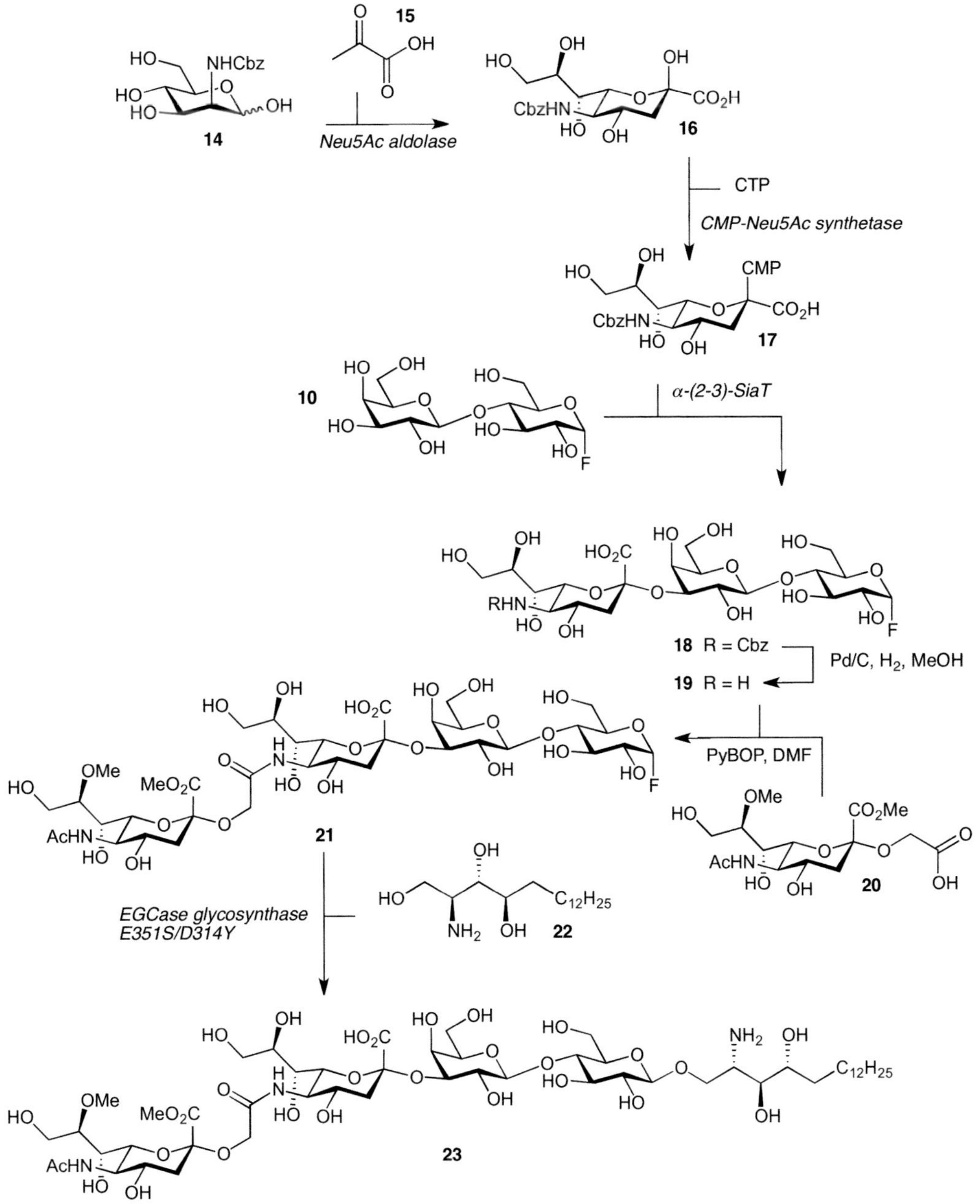

FIGURE 54.7. A combined glycosyltransferase/glycosynthase/chemical synthesis of a lysosphingolipid.

Glycosynthases derived from endoglycosidases have been used for the construction of structurally well-defined N-glycoproteins (Figure 54.8). Hexosaminidases, both exo- and endo-types, hydrolyze *N*-acetylglucosamine-containing glycosidic linkages with retention of the anomeric configuration. In contrast to most other retaining glycosidases, some retaining hexosaminidases do not employ an enzyme active site nucleophile in the nucleophilic displacement of the aglycon, but rather utilize the *N*-acetyl group in the substrate for this purpose (Figure 54.8, insert). As a result, an intermediate oxazolinium ion intermediate is produced that after nucleophilic attack of water yields the hemiacetal with retention of configuration. Endo-hexosaminidases can be employed in transglycosylation reactions and, in mutant form, as glycosynthases.

A demonstration of an endo-hexosaminidase-derived glycosynthase in action is given by the synthesis of $Man_9GlcNAc_2$-glycopeptide **28** from a glycoprotein from HIV-1. The sequence of

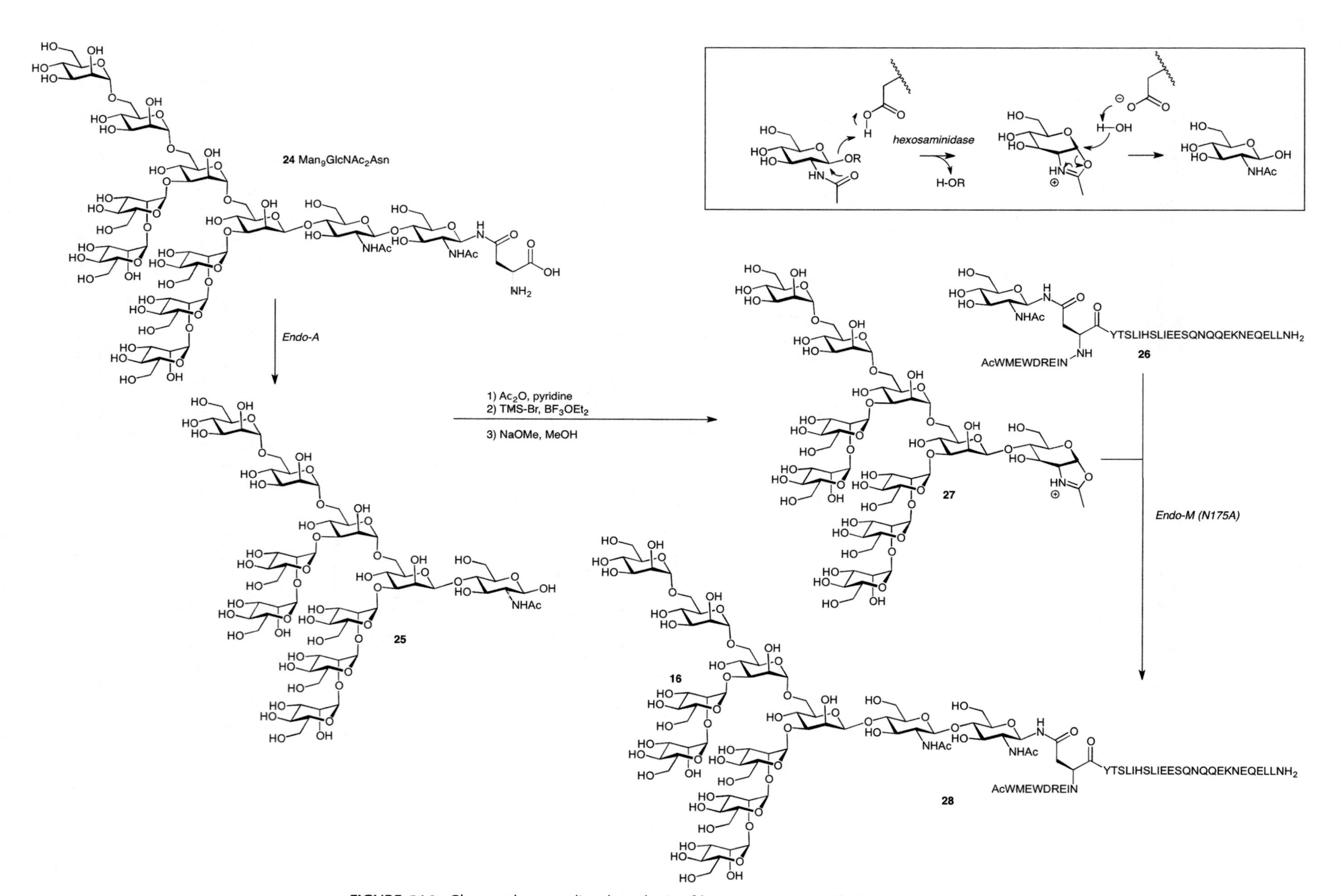

FIGURE 54.8. Glycosynthase mediated synthesis of homogeneous peptide N-glycans.

steps commences with homogeneous $Man_9GlcNAc_2Asn$ **24** that can be prepared by exhaustive proteolytic digestion of soybean glycoproteins by pronase. Wild-type endohexosaminidase, Endo-A, recognizes this high-mannose N-glycan and hydrolyses the GlcNAc-GlcNAc glycosidic linkage, to yield $Man_9GlcNAc$ **25**. After peracetylation, treatment with trimethylsilyl bromide and borontrifluoride etherate, and global deprotection provided oxazolidinium ion **27**. This compound is a good substrate for mutant (N175A) endohexosaminidase, endo-M, which was identified by site-directed mutagenesis as a putative glycosynthase activity. In the final step, the efficacy of the endo-M glycosynthase was demonstrated by the construction of high-mannose N-glycan **28** from **26** and **27**.

OUTLOOK

Enzymatic synthesis has proven to be a powerful methodology for the construction of complex glycans. Chemoenzymatic synthesis can be a suitable alternative for chemical synthesis, as demonstrated by the many applications of sialyltransferases for the construction of complicated or intractable sialic acid-containing glycans. The marriage of enzymatic synthesis and chemical synthesis is very promising. The elaboration by means of glycosyltransferases and/or glycosynthases has proven to be a highly effective strategy to prepare large and complex glycans and glycoconjugates. Advances in chemical and enzymatic syntheses provide the means to construct most natural or designed glycans.

ACKNOWLEDGMENTS

The authors acknowledge contributions to previous versions of this chapter by Nathaniel Finney and David Rabuka and appreciate helpful comments and suggestions from Xi Chen.

FURTHER READING

Ichikawa Y, Lin YC, Dumas DP, Shen GJ, Garcia-Junceda E, Williams MA, Bayer R, Ketcgam C, Walker LE, Paulson JC, Wong CH. 1992. Chemical-enzymatic synthesis and conformational analysis of sialyl Lewis X and derivatives. *J Am Chem Soc* **114:** 9283–9298. doi:10.1021/ja00050a007

Henrissat B, Davies G. 1997. Structural and sequence-based classification of glycoside hydrolases. *Curr Opin Struct Biol* **7:** 637–644. doi:10.1016/s0959-440x(97)80072-3

Williams SJ, Withers SG. 2002. Glycosynthases: mutant glycosidases for glycoside synthesis. *Aust J Chem* **55:** 3–12. doi:10.1071/ch02005

Coutinho PM, Deleury E, Davies GJ, Henrissat B. 2003. An evolving hierarchical family classification for glycosyltransferases. *J Mol Biol* **328:** 307–317. doi:10.1016/s0022-2836(03)00307-3

Blixt O, Vasiliu D, Allin K, Jacobsen N, Warnock D, Razi N, Paulson JC, Bernatchez, Gilbert M, Wakarchuk W. 2005. Chemoenzymatic synthesis of 2-azidoethyl-ganglio-oligosaccharides GD3, GT3, GM2, GD2, GT2, GM1, and GD1a. *Carbohydr Res* **340:** 1963–1972. doi:10.1016/j.carres.2005.06.008

Rising TWDF, Claridge TDW, Moir JWB, Fairbanks AJ. 2006. Endohexosaminidase M: exploring and exploiting enzyme substrate specificity. *ChemBioChem* **7:** 1177–1180. doi:10.1002/cbic.200600183

Bennet CS, Wong CH. 2007. Chemoenzymatic approaches to glycoprotein synthesis. *Chem Soc Rev* **36:** 1227–1238. doi:10.1039/B617709C

Yang M, Davies GJ, Davis BG. 2007. A glycosynthase catalyst for the synthesis of flavonoid glycosides. *Angew Chem Int Ed* **46:** 3885–3888. doi:10.1002/anie.200604177

Umekawa M, Huang W, Li B, Fujita K, Ashida H, Wang LX, Yamamoto K. 2008. Mutants of *Mucor hiemalis* endo-β-*N*-acetylglucosaminidase show enhanced transglycosylation and glycosynthase-like activities. *J Biol Chem* **283:** 4469–4479. doi:10.1074/jbc.m707137200

Pukin AV, Florack DEA, Brochu D, van Lagen B, Visser GM, Wennekes T, Gilbert M, Zuilhof H. 2011. Chemoenzymatic synthesis of biotin-appended analogues of gangliosides GM2, GM1, GD1a and GalNAc-GD1a for solid-phase applications and improved ELISA tests. *Org Biomol Chem* **9:** 5809–5815. doi:10.1039/c1ob00009h

Schmaltz RM, Hanson SR, Wong CH. 2011. Enzymes in the synthesis of glycoconjugates. *Chem Rev* **111:** 4259–4307. doi:10.1021/cr200113w

Rich JR, Withers SG. 2012. A chemoenzymatic total synthesis of the neurogenic starfish ganglioside LLG-3 using an engineered and evolved synthase. *Angew Chem Int Ed* **51:** 8640–8643. doi:10.1002/anie.201204578

Armstrong A, Withers SG. 2013. Synthesis of glycans and glycopolymers through engineered enzymes. *Biopolymers* **99:** 666–674. doi:10.1002/bip.22335

Wang Z, Chinoy ZS, Ambre SG, Peng W, McBride R, de Vries RP, Glushka J, Paulson JC, Boons GJ. 2013. A general strategy for the chemoenzymatic synthesis of asymmetrically branched *N*-glycans. *Science* **341:** 379–383. doi:10.1126/science.1236231

Zhang J, Chen C, Gadi MR, Gibbons C, Guo Y, Cao X, Edmunds G, Wang S, Liu D, Yu J, Wen L, Wang PG. 2018. Machine-driven enzymatic oligosaccharide synthesis by using a peptide synthesizer. *Angew Chem Int Ed Engl* **57:** 16638–16642. doi:10.1002/anie.201810661

Moremen KW, Haltiwanger RS. 2019. Emerging structural insights into glycosyltransferase-mediated synthesis of glycans. *Nat Chem Biol* **15:** 853–864. doi:10.1038/s41589-019-0350-2

55 Chemical Tools for Inhibiting Glycosylation

David J. Vocadlo, Todd L. Lowary, Carolyn R. Bertozzi, Ronald L. Schnaar, and Jeffrey D. Esko

The use of chemical tools to inhibit glycosylation is a powerful approach for studying glycan functions and can serve as a starting point for drug discovery. This chapter discusses various types of inhibitors including those present in nature, those identified through rational design followed by synthesis, and those found by screening of chemical libraries.

ADVANTAGES OF INHIBITORS

Chapters 44, 45, and 49 describe natural and induced mutants with defects in glycosylation. These mutants have helped to define genes that encode various transferases and glycosidases, and in some cases alternate biosynthetic pathways have been uncovered. Mutants also provide insights into the function of glycosylation in cells and tissues and models for human inborn errors in metabolism and disease. However, one limitation of studying mutants is that the analyses are usually restricted to the cell or organism from which the mutant strain was isolated, and mutations in essential genes require conditional alleles.

Inhibitors of carbohydrate-processing enzymes—in particular, glycosyltransferases and glycosidases—provide another approach for studying glycosylation in cells, tissues, and whole organisms that avoids some of the problems associated with genetic models. Many of these

published by Cold Spring Harbor Laboratory Press; doi: 10.1101/glycobiology.4e.55

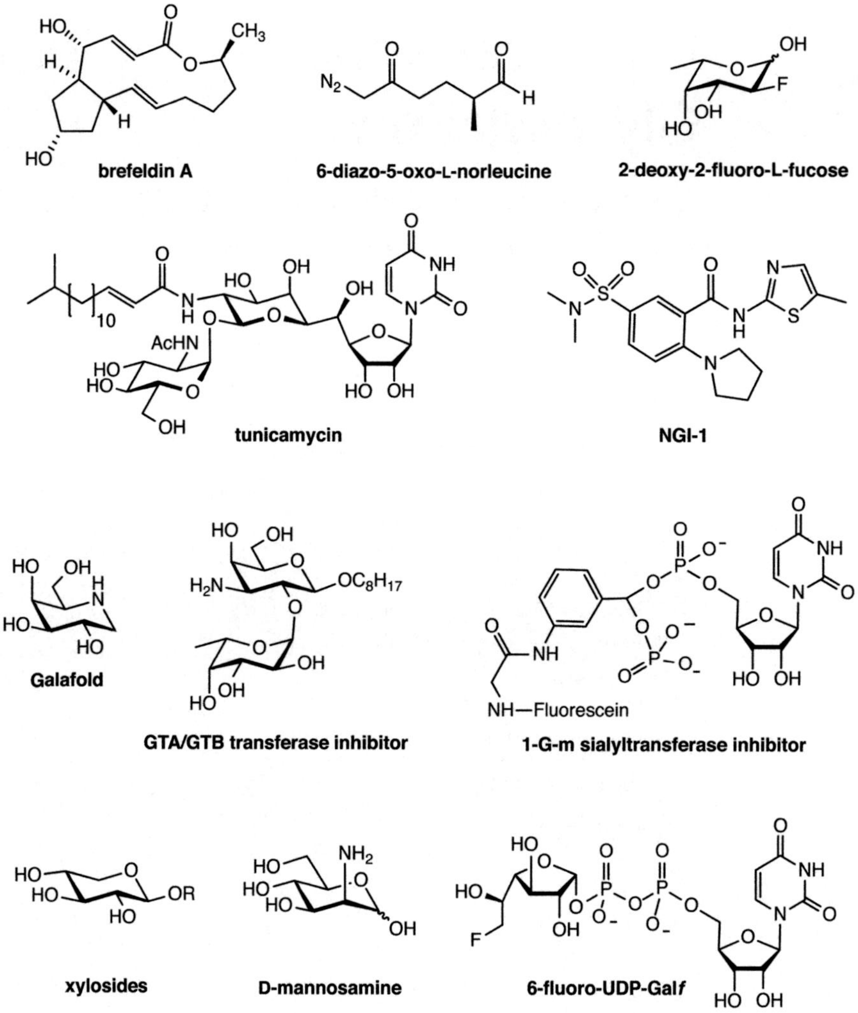

FIGURE 55.1. Different classes of compounds for inhibiting glycosylation including those that prevent the formation of biosynthetic precursors, those that directly act on glycosidases and glycosyltransferases, and those that serve as primers/decoys and chain terminators.

compounds are small molecules that are taken up readily by cells and the effects can be reversed, enabling experimental designs that are difficult to achieve using genetic methods. Some compounds can also be absorbed through the gut, providing an opportunity for designing drugs to treat human diseases and disorders correlated with altered glycosylation (Chapter 57). Because the field is broad, only a selection of inhibitors that act on specific enzymes or metabolic pathways and that illustrate certain basic concepts are discussed here (Figure 55.1).

INHIBITORS OF BIOSYNTHETIC PRECURSORS

A number of inhibitors have been described that block glycosylation by interfering with the metabolism of common precursors or intracellular transport activities. Some of these compounds act indirectly by impeding the transit of proteins between the endoplasmic reticulum (ER), Golgi, and *trans*-Golgi network. For example, brefeldin A (Figure 55.1) causes retrograde transport of Golgi components located proximal to the *trans*-Golgi network back to the ER.

Thus, treating cells with brefeldin A separates enzymes located in the *trans*-Golgi network from those found in the ER and Golgi and uncouples the assembly of the core structures of some glycans from later reactions, such as sialylation or sulfation. The drug can be used to examine if two pathways reside in the same compartment or share enzymes. Because the localization and array of the enzymes vary considerably in different cell types, extrapolating the effects of brefeldin A from one system to another is often difficult.

Some inhibitors act at key steps in intermediary metabolism in which glycosylation precursors are formed. For example, a glutamine analog, 6-diazo-5-oxo-L-norleucine (DON; Figure 55.1) blocks many glutamine-dependent amidotransferases including glutamine:fructose-6-phosphate amidotransferase, the enzyme of the hexosamine biosynthetic pathway that forms glucosamine from fructose and glutamine (Chapter 5). Depressing glucosamine production in this way has a pleiotropic effect on glycan assembly because all of the major families contain *N*-acetylglucosamine or *N*-acetylgalactosamine. Given the nonspecific activity of DON, care should be taken to understand and limit nonspecific side effects.

An array of sugar analogs have been made with the hope that they might show selective inhibition of glycosylation. Some examples include 2-deoxy-D-glucose and fluorosugar analogs (3-deoxy-3-fluoro-D-glucosamine, 4-deoxy-4-fluoro-D-glucosamine, 6-deoxy-6-fluoro-D-*N*-acetylglucosamine, 2-deoxy-2-fluoro-D-glucose, 2-deoxy-2-fluoro-D-mannose, 2-deoxy-2-fluoro-L-fucose, and 3-fluorosialic acid), which all inhibit glycoprotein biosynthesis (Figure 55.1). Early studies with 2-deoxyglucose showed that the analog was converted to UDP-2-deoxyglucose, as well as to GDP-2-deoxyglucose and dolichol-P-2-deoxyglucose. Inhibition of glycoprotein formation apparently occurs as a result of accumulation of various dolichol oligosaccharides containing 2-deoxyglucose, which cannot be elongated or transferred to glycoproteins normally. Although the mechanism of action of many of these molecules remains poorly understood, one better studied case is 2-deoxy-2-fluoro-L-fucose (Figure 55.1). This compound acts as a biosynthetic precursor of the fucose salvage pathway (Chapter 5) to form GDP-2-deoxy-2-fluoro-fucose within cells, but the analog is a poor substrate for mammalian fucosyltransferases because the highly electronegative fluorine atom close to the anomeric center inductively stabilizes the oxocarbenium ion-like transition states used by the superfamily of glycosyltransferases (Chapter 6). This compound also causes feedback inhibition of the GDP-fucose biosynthetic enzyme GDP-mannose 4,6-dehydratase (GMD) leading to a loss of natural GDP-fucose within cells and a consequent decrease in the levels of fucosylation on all types of glycoconjugates. Notably, 2-deoxy-2-fluoro-L-fucose is orally available and is active both in vitro on cells as well as in vivo, leading to decreases in the levels of the carbohydrate epitope sialyl-Lewis x (sLe^x) (Chapter 14). Consistent with these effects, 2-deoxy-2-fluoro-L-fucose has shown potential, in preclinical models, at blocking tumor growth and metastasis as well as vaso-occlusive crisis caused by sLe^x-mediated adhesion of blood cells to selectin-expressing endothelial surfaces (Chapter 34) in transgenic sickle-cell mice. Similarly, 3-fluorosialic acid is converted by the NeuAc-salvage pathway (Chapter 5) within cells into CMP-3F-NeuAc and blocks formation of CMP-NeuAc, leading to profound decreases in sialylation of glycoconjugates both in cells and in vivo, suggesting potential use in cancers. Other sugar analogs, including 5-thio-*N*-acetylglucosamine and 4-deoxy-4-fluoro-*N*-acetylglucosamine, also lead to formation of unnatural donor sugars formed within cells that reduce pools of the natural nucleotide sugars and lead to consequent decreases in protein glycosylation. These compounds are proving useful research tools; however, care must be taken in interpreting the results of experiments using these compounds because they may have pleiotropic effects on glycan assembly caused by overlap of nucleotide precursors in different pathways.

A number of natural products have been found to alter glycosylation. Tunicamycin belongs to a class of nucleoside antibiotics composed of uridine, an 11-carbon disaccharide called 2-amino-2,6,-dideoxyundecodialdose (tunicamine), and a fatty acid of variable length (13 to 17 carbons), branching, and unsaturation (Figure 55.1). Tunicamycin derives its name from its antiviral activity, which occurs by inhibiting viral coat (or "tunica") formation. The biosynthesis of tunicamycin has been defined, which may lead to analogs that are selective for different species.

Tunicamycin inhibits N-glycosylation in eukaryotes by blocking the transfer of *N*-acetylglucosamine-1-phosphate (GlcNAc-1-P) from UDP-GlcNAc to dolichol-P (catalyzed by GlcNAc phosphotransferase; GPT), thereby decreasing the formation of dolichol-PP-GlcNAc (Chapter 9). Other GlcNAc transferase reactions are not inhibited (e.g., GlcNAcTI–V), but the transfer of GlcNAc-1-P to undecaprenyl-P and the formation of undecaprenyl-PP-MurNAc pentapeptide (which is involved in bacterial peptidoglycan biosynthesis) are sensitive to tunicamycin (Chapter 21). The compound is a tight-binding competitive inhibitor, presumably because it resembles the donor nucleotide sugar. The K_i value for tunicamycin is $\sim 5 \times 10^{-8}$ M, whereas the K_m value for UDP-GlcNAc is $\sim 3 \times 10^{-6}$ M. Given the key role of N-glycosylation in protein folding and quality control in the ER (Chapter 39), tunicamycin is cytotoxic to cells, and resistant mutants overproduce GPT. Similarly, transfection of cells with the cloned GPT confers resistance, suggesting that the variable dose of inhibitor required in different cells may reflect variation in enzyme levels. A more recently identified compound, NGI-1, offers an approach to blocking N-glycosylation that is complementary to use of tunicamycin. This compound acts as a direct inhibitor of oligosaccharyltransferase, the enzyme responsible for transferring the dolichol-PP-loaded glycan to asparagine.

Tunicamycin has been used extensively for studying the role of N-glycans in glycoprotein maturation, secretion, and function. The drug induces apoptosis preferentially in cancer cells, presumably because of alterations in glycosylation of various cell-surface receptors and signaling molecules and by inducing ER stress (Chapter 39). Thus, inhibition of N-glycan formation could be useful for treating cancer patients. Other potential applications include substrate reduction therapy for treatment of lysosomal storage disorders (Chapter 44), congenital disorders of glycosylation (Chapter 45), or naturally occurring mutations that create N-glycosylation sites in cell-surface receptors (gain-of-glycosylation mutants; Chapter 45). In a related vein, amphomycin, a lipopeptide, inhibits dolichol-P-mannose synthesis by apparently forming complexes with the carrier lipid dolichol-P. Other lipophilic compounds that bind lipid intermediates in bacterial cell wall synthesis also have been studied (Chapter 21).

INHIBITORS OF GLYCOSIDE FORMATION OR CLEAVAGE

Many different inhibitors of enzymes involved in glycoside formation or cleavage have been described, spanning a wide range of compound classes. These include natural products and synthetic derivatives, as well as acceptor and donor analogs obtained by chemical synthesis. The high-throughput screening of compound libraries and subsequent optimization of hits using medicinal chemistry approaches has also led to compounds that are both useful chemical tools and drugs. In this section, we highlight a few examples.

Example 1: Natural Product Inhibitors of Glycosidases

Plant alkaloids block N-linked glycosylation by inhibiting the processing glycosidases (α-glucosidases and α-mannosidases) involved in trimming nascent chains (Figure 55.2). Unlike tunicamycin, which blocks glycosylation of glycoproteins entirely, these alkaloids inhibit the trimming reactions that occur after the $Glc_3Man_9GlcNAc_2$ oligosaccharide is attached to a glycoprotein (Chapter 9). Treatment of cells with alkaloids results in the display of glycoproteins on the cell surface lacking the characteristic termini found on mature N-glycans (Chapter 14).

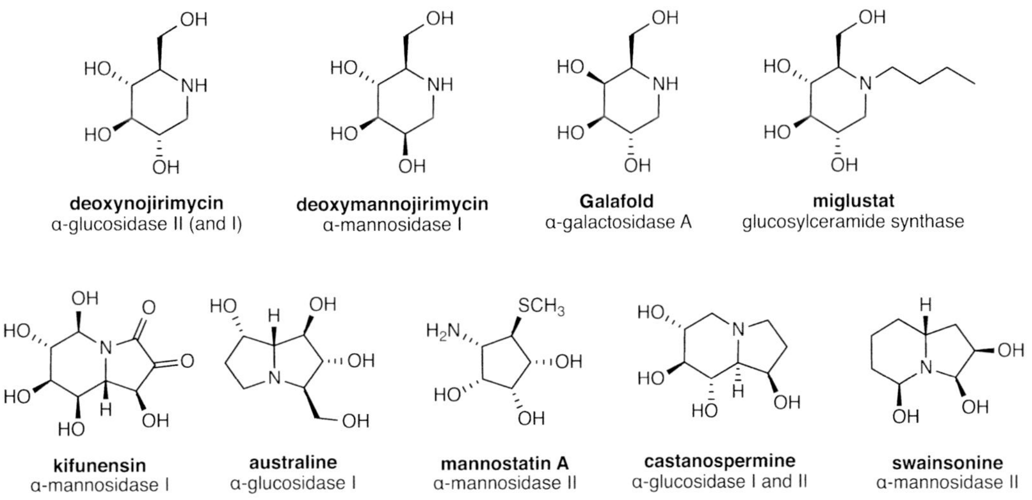

deoxynojirimycin α-glucosidase II (and I)

deoxymannojirimycin α-mannosidase I

Galafold α-galactosidase A

miglustat glucosylceramide synthase

kifunensin α-mannosidase I

australine α-glucosidase I

mannostatin A α-mannosidase II

castanospermine α-glucosidase I and II

swainsonine α-mannosidase II

FIGURE 55.2. Examples of alkaloids that inhibit glycosidases involved in N-linked glycan biosynthesis.

α-Glucosidase inhibitors involved in the initial processing of N-glycans and in quality control of protein folding (Chapter 39) include castanospermine (from the seed of the Australian chestnut tree, *Castanosperum australe*), which inhibits α-glucosidases I and II, australine (also from *C. australe*), which preferentially inhibits α-glucosidase I, and deoxynojirimycin (from *Streptomyces* species), which preferentially inhibits α-glucosidase II (Figure 55.2). Castanospermine and australine cause accumulation of fully glucosylated chains, whereas deoxynojirimycin results in chains containing one to two glucose residues. Unexpectedly, treating cells with these inhibitors revealed that some trimming of the mannose residues occurs independently of removal of the glucose residues through the action of the mechanistically intriguing GH99 *endo*-mannosidase (Chapter 9).

Swainsonine was first discovered in plants from the western United States (*Astragalus* species, also known as locoweed) and Australia (*Swainsona canescens*). Consumption of these plants by animals causes a severe abnormality called locoism and accumulation of glycoproteins in the lymph nodes. Swainsonine inhibits α-mannosidase II, causing accumulation of paucimannose oligosaccharides ($Man_4GlcNAc_2$ and $Man_5GlcNAc_2$) and hybrid-type chains at the expense of complex oligosaccharides. In addition, swainsonine inhibits the lysosomal α-mannosidase. Mannostatin A works in a similar way, but differs significantly in structure from swainsonine (Figure 55.2). Other mannosidase inhibitors include deoxymannojirimycin and kifunensin, which selectively inhibit α-mannosidase I. These agents cause the accumulation of $Man_{7-9}GlcNAc_2$ oligosaccharides on glycoproteins.

All of the above listed inhibitors have in common polyhydroxylated ring systems that mimic the orientation of hydroxyl groups in the natural substrates, but a strict correlation between stereochemistry and enzyme target (α-glucosidase vs. α-mannosidase) does not exist. The compounds contain nitrogen, usually in place of the ring oxygen. To explain their activity, it is proposed that the nitrogen, which is protonated at physiological pH, mimics the transition state during the hydrolysis reaction, which has substantial positive character. Crystal structures for the α-mannosidase are available with a range of different bound inhibitors.

Diastereomers as well as alkylated and acylated analogs of the alkaloids have interesting and useful properties. Notably, the galactose-configured analog of deoxynojirimycin (Galafold; Figure 55.1) has been approved as a pharmacological chaperone of the lysosomal α-galactosidase for the treatment of the lysosomal storage disease known as Fabry disease (Chapter 44). Mutations in the gene encoding the enzyme generally impair its ability to fold and traffic from the ER to lysosomes. Galafold can bind to the mutant α-galactosidase in the ER and help it fold and avoid degradation leading to its improved trafficking and activity within lysosomes. *N*-Butylation of deoxynojirimycin makes this glucosidase inhibitor an inhibitor of glycolipid biosynthesis, which is discussed in more detail further below. In other cases, alkylation of the amino group or acylation of the hydroxyl groups can improve the potency of the compound, presumably by facilitating uptake across the plasma and Golgi membranes. Some of these compounds have shown positive effects for treating diabetes, cancer, HIV infection, and lysosomal storage diseases (see Chapter 57), but some also induce male sterility. A major challenge with all of these enzymes is their limited specificity; thus while convenient and widely used, some caution is needed when interpreting resulting observations.

Example 2: Inhibitors of O-GlcNAc-Processing Enzymes

The importance of O-GlcNAc addition to many cytoplasmic and nuclear proteins (Chapter 19) has stimulated interest in developing agents to inhibit its addition by O-GlcNAc transferase (OGT) or its removal by O-GlcNAc-specific β-glucosaminidase (O-GlcNAcase [OGA]). These enzymes provide excellent examples of successful rational design and medicinal chemistry efforts directed to the development of glycoside hydrolase and glycosyltransferase inhibitors. Alloxan and streptozotocin affect O-GlcNAc addition, but these compounds lack specificity. The first potentially useful OGT inhibitor (OSMI-1; Figure 55.3) was obtained by screening chemical libraries for compounds that displaced a fluorescent derivative of the donor sugar, UDP-GlcNAc. Structure-guided medicinal chemistry led to quinoline-containing ester (OSMI-4), which is a high-quality OGT inhibitor that is effective in cells. The active compounds do not block other *N*-acetylglucosamine addition reactions—for example, one involved in formation of the polysaccharide backbone of bacterial peptidoglycan (Chapter 21). In addition, 5SGlcNAc is another inhibitor of OGT that acts as a metabolic precursor leading to formation of UDP-5SGlcNAc. Its peracetylated form (Ac-5SGlcNAc) crosses cell membranes and undergoes deacetylation by nonspecific esterases to generate cell-active inhibitors. Modification of the N-acetyl group to give S5GlcNHex

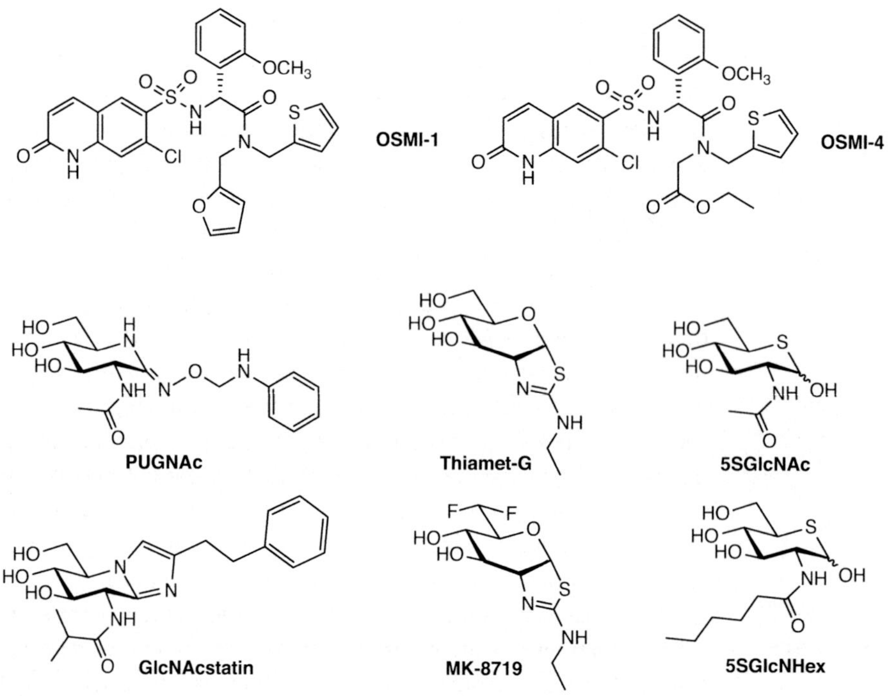

FIGURE 55.3. Inhibitors of O-GlcNAc-specific β-glucosaminidase (OGA) and O-GlcNAc transferase (OGT).

eliminated the need to have O-acetylation and provided a compound that could be used in vivo, although, like other metabolic precursor inhibitors, their use requires care because of possible off-target effects on other enzymes (e.g., GlcNAc transferases [Chapter 9]).

Several O-GlcNAcase inhibitors are based on *N*-acetylglucosamine. The first described, PUGNAc (Figure 55.3) inhibits O-GlcNAcase at nanomolar concentrations, but also inhibits lysosomal β-hexosaminidases (HexA and HexB; Chapter 44). 1,2-Dideoxy-2′-propyl-α-D-glucopyranoso-[2,1-d]-Δ2′-thiazoline (NButGT) and the more potent aminothiazoline Thiamet-G are more specific and potent in cells than PUGNAc (Figure 55.3). A rationally designed glucoimidazole, GlcNAcstatin, also potently inhibits O-GlcNAcase with good selectivity over HexA and HexB. These compounds inhibit the enzyme in cells and tissues, and Thiamet-G, in particular, has been used in animal models, providing new tools to study the function of O-GlcNAc. Use of these compounds has uncovered the potential of O-GlcNAcase inhibitors in various neurodegenerative diseases, which has spurred industrial pharmaceutical interest that has led to OGA inhibitors, such as the Thiamet-G analog MK-8719 (Figure 55.3), advancing into Phase I clinical trials. Noncarbohydrate O-GlcNAcase inhibitors have also been uncovered, and highly potent brain permeable analogs have been used as positron emission tomography agents to examine the inhibition and distribution of O-GlcNAcase in human brain.

Example 3: Rational Design of Acceptor and Donor Analogs

A number of specific glycosyltransferase inhibitors have been developed based on the concept that donor and acceptor substrate analogs might serve as inhibitors. For acceptor substrate analogs, the general strategy is to modify the hydroxyl group that acts as the nucleophile during formation of the glycosidic bond or groups in its immediate vicinity (Table 55.1). Many designer compounds lack inhibitory activity, because modification of the targeted hydroxyl group prevents binding of the analog to the enzyme by interfering with hydrogen bonding networks that position the substrate; that is, these groups serve as "key polar groups." In a smaller number of cases, the analogs show K_i values in the approximate range of the K_m values for the unmodified substrate. As one might expect, the analogs usually act competitively with respect to the unmodified substrate, but in a few cases the inhibition pattern is more complex, suggesting possible binding outside the active site.

Nucleotide sugar analogs provide opportunities for blocking classes of enzymes that use a common donor (e.g., all fucosyltransferases use GDP-fucose). Many nucleotide sugar derivatives have been synthesized (e.g., N- and O-substituted analogs of UDP-GalNAc), and several inhibit the enzymes in vitro. These have proven less useful in living cells because of poor uptake,

TABLE 55.1. Synthetic substrate-based inhibitors of glycosyltransferases

Enzyme	Substrate	Inhibitor	Substrate K_m (μM)	Inhibitor K_i (μM)
α2FucT	β3GlcNAcβ-*O*-R	2-deoxyGalβ3GlcNAcβ-*O*-R	200	800
β4GalT	GlcNAcβ3Galβ-*O*-R	6-thioGlcNAcβ3Galβ-*O*-Me	1000	1000
α3GalT	Galβ4GlcNAcβ-*O*-R	3-aminoGalβ4GlcNAcβ-*O*-R	190	104[a]
β6GlcNAcT	Galβ3GalNAcα-*O*-R	Galβ3(6-deoxy) GalNAcα-*O*-R	80	560
β6GlcNAcT-V	GlcNAcβ2Manα6Glcβ-*O*-R	GlcNAcβ2(6-deoxy) Manα6Glc β-*O*-R	23	30
β6GlcNAcT-V	GlcNAcβ2Manα6Glcβ-*O*-R	GlcNAcβ2(4-*O*-methyl) Manα6 Glcβ-*O*-R	23	14
β6GlcNAcT-V	GlcNAcβ2Manα6Glcβ-*O*-R	GlcNAcβ2(6-deoxy,4-*O*-Me)Manα6Glcβ-*O*-R	23	3
α6SialylT	Galβ4GlcNAcβ-*O*-R	6-deoxyGalβ2GlcNAcβ-*O*-R	900	760[a]
α3GalNAcT-A	Fucα2Galβ-*O*-R	Fucα2(3-deoxy)Galβ-*O*-R	2	68
α3GalNAcT-A	Fucα2Galβ-*O*-R	Fucα2(3-amino)Galβ-*O*-R	2	0.2[a]

The aglycone (R) varies in the different compounds.

[a] Inhibition mixed or noncompetitive.

but there are some notable exceptions, including the fluorescently tagged CMP-NeuAc analog in which the carbohydrate is replaced with an aryl group (1-G-m, Figure 55.1), which has a nanomolar K_i values for a range of sialyltransferases. This compound has also been reported to block sialyltransferases within cells, although this compound has not been more widely adopted by the community. "Bisubstrate" transition-state analogs consist of the nucleoside sugar donor covalently linked to the acceptor substrate by way of a bridging group. In principle, this approach could lead to high-affinity inhibitors, in turn justifying the usually complex routes needed for their synthesis. However, the compounds reported to date have been only modest inhibitors with K_i values, at best, in the range of the K_m values for the donors. Although such transition state mimics remain attractive targets, the results to date suggest new features need to be incorporated into their design.

Example 4: Inhibitors of Glycolipids and GPI Anchors

Reagents that alter the assembly of glycolipids in cells have been described. Xylosides (Figure 55.1) have a mild effect on glycolipid formation, possibly because of the similarity between xylose and glucose and the assembly of a GM3-like compound (Neu5Acα2-3Galβ1-4Xylβ-*O*-R) on the primer. Because cells take up intermediates in glycolipid biosynthesis, they behave like synthetic glycoside primers. For example, glucosylceramide produces complex glycolipids when fed to cells. More direct competitive inhibitors of glucosylceramide synthase (GCS) have been generated with a view to their potential benefit in Gaucher disease (Chapter 44), which stems from accumulation of glucosylceramide caused by loss-of-function mutations in β-glucocerebrosidase (GCase). D-Threo-1-phenyl-2-decanoylamino-3-morpholino-1-propanol (D-PDMP; Figure 55.4) has been a widely used compound; however, it also inhibits the activity of purified lactosylceramide synthase. Medicinal chemistry efforts have led to the close analog eliglustat, which although it does not gain access to the central nervous system (CNS), is approved for treating type I Gaucher disease, a lysosomal storage disorder in which glucocerebrosidase is missing (Chapter 44). Its beneficial activity occurs through "substrate deprivation" by blocking synthesis of glycosphingolipids, thereby "depriving" the lysosome of substrate. Subsequent efforts have led to CNS permeable GCS inhibitors including venglustat, which is advancing through late stage clinical trials. The α-glucosidase inhibitor, *N*-butyldeoxynojirimycin, known commercially as miglustat, also inhibits GCS and is approved for treating Niemann–Pick disease type C and Gaucher disease.

The search for active compounds often benefits from serendipity and the synthesis of disaccharide (or larger) acceptor analogs with appropriate modifications is labor intensive, as is the

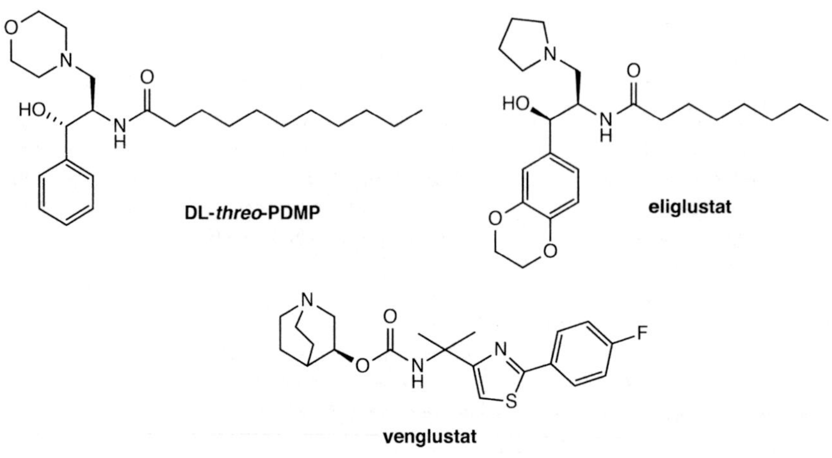

DL-*threo*-PDMP

eliglustat

venglustat

FIGURE 55.4. Inhibitors of glycosphingolipid metabolism.

preparation of complicated nucleotide sugar analogs. Nevertheless, the approach has yielded insights into the binding and reactivity of the glycan-processing enzymes, and substrate analogs with selectivity for particular enzymes have been developed in this way. Improvements in recombinant expression systems have led to increasing numbers of X-ray crystal structures, which provides clues for deriving mechanism-based inhibitors in the future (Chapter 6).

GLYCOSIDE PRIMERS AND CHAIN TERMINATORS

Glycoside Primers

The utility of any glycosyltransferase inhibitor ultimately depends on its ability to cross the plasma membrane and enter the Golgi where the glycosyltransferases reside. Unfortunately, many of the compounds described above lack activity in cells, presumably because their polarity and charge prevents their uptake. More than 40 years ago, Okayama and colleagues found that D-xylose in β-linkage to a hydrophobic aglycone (the noncarbohydrate portion of a glycoside) was taken up efficiently and inhibited the assembly of glycosaminoglycans on proteoglycans. Xylosides mimic the natural substrate, xylosylated serine residues in proteoglycan core proteins, and thus act as a substrate. "Priming" of chains occurs on the added xyloside, which diverts the assembly process from the endogenous core proteins and causes inhibition of proteoglycan formation. In general, cells incubated with xylosides secrete large amounts of individual glycosaminoglycan chains and accumulate proteoglycans containing truncated chains. The success of β-D-xylosides in altering proteoglycan biosynthesis suggested that other glycosides might function similarly (Figure 55.5). Subsequent studies showed that β-*N*-acetylgalactosaminides prime oligosaccharides found on mucins and inhibit O-glycosylation of glycoproteins. Other active glycosides include β-glucosides, β-galactosides, β-*N*-acetylglucosaminides, and even disaccharides and trisaccharides. These latter compounds require conjugation to appropriate aglycones and acetylation to mask the polar carbohydrate hydroxyl groups. Cells contain several carboxyesterases that remove the acetyl groups and render the compounds available to the transferases in the Golgi.

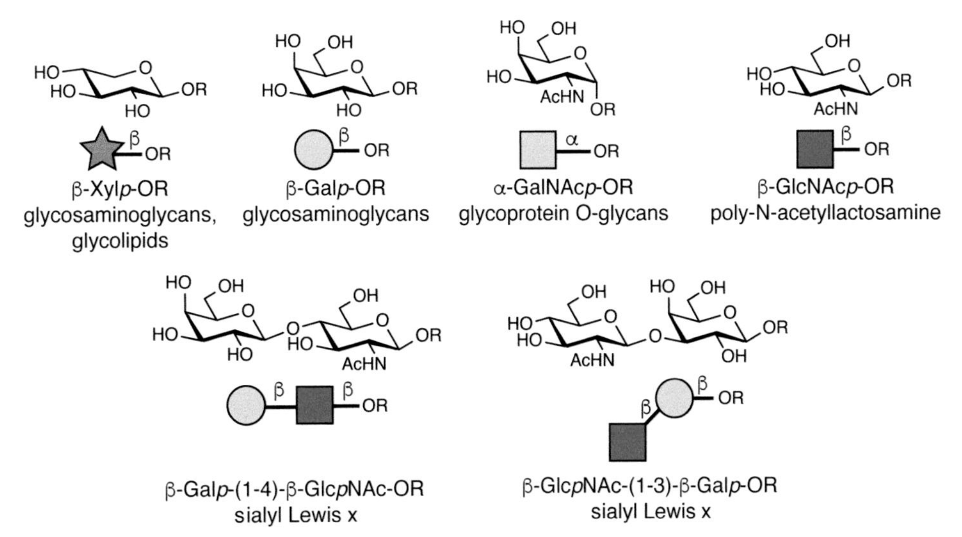

FIGURE 55.5. Examples of glycoside primers. Structures shown are the compounds administered to cells; the depictions of the disaccharides do not show their acetylation. Below the names of the compounds are the glycan classes that are impacted.

Priming by glycosides occurs in a concentration-dependent manner, but the efficiency varies widely among different compounds and cell types. These variations may relate to the relative abundance of endogenous substrates, enzyme concentration and composition, the solubility of different glycosides, their susceptibility to hydrolysis, their uptake across the plasma membrane and into the Golgi, and their relative affinity for the glycosyltransferases. The type of chain made on a given primer also depends on concentration and aglycone structure, which may reflect selective partitioning of primers into different intracellular compartments or into different branches of biosynthetic pathways. Like priming, inhibition of glycoprotein, glycolipid, or proteoglycan formation occurs in a dose-dependent fashion, but the blockade is rarely complete, probably because of the inability of glycosides to mimic the entire endogenous substrates.

Primers represent starting points for tight-binding inhibitors with the properties described above. The compounds described in Figure 55.5 could be converted to permeable acylated glycosides and tested in live cells for inhibitory activity. Active compounds could potentially become lead compounds for drugs to treat glycosylation-dependent diseases. Oligosaccharide priming may have beneficial effects as well. Xylosides, for example, can be absorbed through the gut, and when consumed at sufficient concentration, show antithrombotic activity. Many glycosides occur naturally because various organisms (especially plants) produce hydrophobic compounds as part of chemical defense and conjugate them to sugars to render them soluble. Thus, the human diet may contain various types of glycosides with interesting (and unknown) biological activities.

Care must be taken in interpreting the results of experiments using glycoside primers. For example, β-D-xylosides also prime glycans related in structure to glycosphingolipids and HNK-1. In some cases priming per se is not the mechanism responsible for inhibition of glycosylation, but rather inhibition occurs because of competitive binding of the primer to a target enzyme. Finally, primers could deplete cells of nucleotide sugars and have multiple effects on glycosylation. For example, 4-methylumbelliferone can be used to block hyaluronan biosynthesis. The precise mechanism of action is unknown but is thought to involve depletion of UDP-GlcA due to glucuronidation of the glycoside primer. Reduction in cellular UDP-GlcA levels in turn affects formation of sulfated glycosaminoglycans and other glucuronic acid–containing glycans and alter the pools of other nucleotide sugars, such as UDP-Xyl, which is produced in one step from UDP-GlcA (Chapter 5).

Chain Terminators

Chain terminators are compounds that are introduced into a growing glycan by a glycosyltransferase, but in so doing introduces a functionality that prevents further elongation. Mannosamine acts as a metabolic inhibitor that inhibits GPI anchor formation both in *Trypanosoma brucei* and in mammalian cells by the formation of ManNH$_2$-Man-GlcNH$_2$-PI. Apparently, mannosamine in its activated form (GDP-ManNH$_2$) is used as a substrate in the second mannosyltransferase reaction, but the ManNH$_2$-Man-GlcNH$_2$-PI intermediate will not act as a substrate for the next α2-mannosyl-transferase (Chapter 12). GlcNR-phosphatidylinositols with different substituents (R) act as substrate analogs and some act as suicide inhibitors in vitro. Another class of inhibitors is based on fatty acid analogs that only trypanosomes incorporate into GPI anchors. Trypanosomes, unlike their mammalian hosts, incorporate myristic acid into GPI anchors by exchanging myristic acid for other fatty acids in the phosphatidylinositol moiety. By making a series of analogs, an inhibitor was found that is highly toxic to trypanosomes in culture and nontoxic to mammalian cells (10-(propoxy)decanoic acid). Such reagents are drug candidates for treating trypanosomiasis, which is endemic in sub-Saharan regions of Africa. Additional examples of chain terminators have been reported. For example, fluorinated sugar nucleotides (Figure 55.1) are incorporated into growing oligosaccharides by carbohydrate polymerases, but the resulting products lack the hydroxyl group to enable further polymerization.

STRUCTURE-BASED RATIONAL DESIGN

The increasing number of X-ray crystallographic and cryo-EM investigations of carbohydrate processing enzymes has provided structural information that has facilitated the rational design of new inhibitors, often with very high potency. These compounds have been used as research tools and some have also advanced to clinically used drugs.

Studies of influenza neuraminidase exemplify the power of rationally designed drugs that have successfully been marketed as drugs. The crystal structure for influenza neuraminidase was obtained in 1983, and many other enzymes have since been characterized from other sources. Even before the crystal structure had been obtained, a neuraminidase inhibitor was designed by assuming that the hydrolysis reaction involved an oxocarbenium ion–like transition state with significant positive charge accumulation at the anomeric center. This would result in C-2 and C-3 adopting a planar configuration, and therefore compounds that mimicked this geometry were hoped to have inhibitory activity. Indeed, Neu5Ac-2-ene (DANA; Figure 55.6) has a micromolar K_i value. Interestingly, this compound inhibits most sialidases, but does not inhibit the trypanosome *trans*-sialidase and only weakly bacterial sialidases.

A visual inspection of the X-ray structure of influenza neuraminidase with DANA bound showed that two glutamate residues lined a pocket near carbon 4 of the sialic acid analog. The pocket is fairly open, suggesting that a bulkier substituent at this position might be tolerated. A substrate analog containing a positively charged guanidinium group instead of the hydroxyl at carbon 4 (4-guanidino-DANA; Figure 55.6) is a remarkably potent influenza neuraminidase inhibitor (K_i=10^{-11} M). The higher affinity is presumably due to a salt bridge

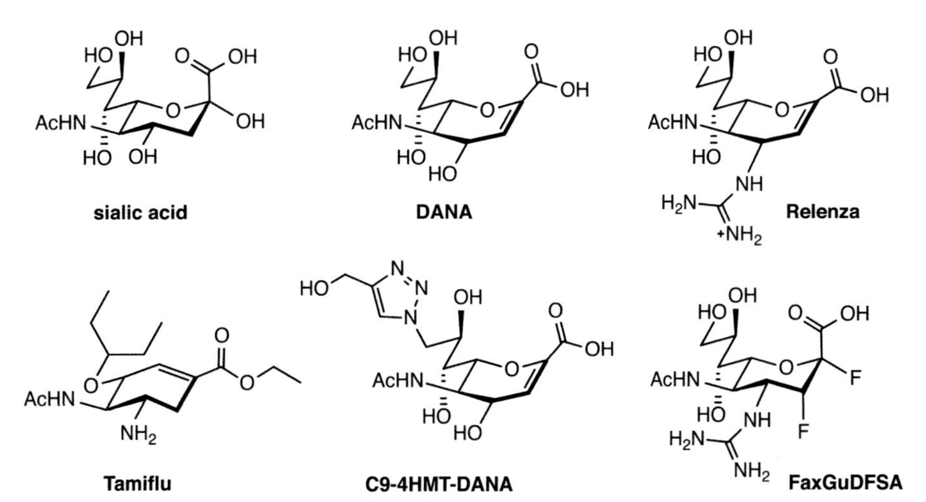

FIGURE 55.6. Structure of neuraminidase inhibitors. Chemical structure of sialic acid (Neu5Ac, 2-deoxy-2,3-dehydro-*N*-acetyl neuraminic acid), DANA; 4-amino-DANA; 4-guanidino-DANA (Relenza, zanamivir); (3*R*, 4*R*, 5*S*)-4-acetamido-5-amino-3-(1-ethylpropoxyl)-1-cyclohexane-1-carboxylic acid ethyl ester (Tamiflu, oseltamivir), C9-4HMT-DANA and FaxGuDFSA. DANA is thought to resemble the transition state in hydrolysis, and addition of the guanidinium group in Relenza provides higher affinity binding to the active site. The ethyl ester in Tamiflu enhances oral availability and then is quickly removed in the body by nonspecific esterases. C9-4HMT-DANA is a selective inhibitor of human neuraminidase 4, and FaxGuDFSA is an example of a non-DANA-like scaffold that inhibits viral neuraminidases.

formed between the charged guanidinium group and the carboxylates lining the pocket. The analog is nearly a million times less potent on human sialidases, leading to its approval and use as the anti-influenza drug Relenza (Chapter 57). It does not work on bacterial sialidases, however, because the equivalent pocket is filled with an arginine group and is only a modest (low-micromolar) inhibitor of the human neuraminidases. Next-generation analogs of Relenza including the close derivative laninamivir as well as the guanidinium-containing peramivir have also been developed.

The presence of the guanidinium group requires these drugs to be either inhaled or injected. Subsequent studies focused on dispensing this functionality. Replacement of the pyranose ring with a cyclohexene to mimic the planar ring of the proposed intermediate in hydrolysis, protecting the carboxylate as an ester that is hydrolyzed after ingestion, and replacing the guanidinium with an amine led to an orally active, widely used analog, the anti-influenza drug Tamiflu (Figure 55.6; Chapter 57). Crystal structures for other sialidases have allowed the design of species-specific analogs. This rational approach to inhibitor design holds great potential, not only for neuraminidase inhibitors, but also, as described above, for O-GlcNAcase and OGT inhibitors. These examples illustrate how structure can guide design of carbohydrate-based inhibitors as well as heterocyclic inhibitors that have been more commonly advanced as drugs.

These advances have stimulated greater interest into influenza neuraminidase as well as the four human neuraminidases, which have poorly understood physiological roles. A major challenge associated with viral neuraminidase inhibitors—most strikingly, Tamiflu—is the rapid development of mutations that confer resistance. An alternative approach to circumvent this problem has been to develop mechanism-based inhibitors that covalently inhibit the enzyme. One example, FaxGuDFSA, has a fluoride leaving group, the displacement of which by the neuraminidase forms a glycosyl–enzyme intermediate. This intermediate has an appreciable half-life because the electronegative fluorine destabilizes the oxocarbenium ion-like transition state that leads to its breakdown. Accordingly, this compound can protect mice against infection. Moreover, the emergence of resistant influenza strains is slower than against Tamiflu.

In addition to driving increased effort against influenza neuraminidase, efforts to create inhibitors of the human neuraminidases have benefited from the fact that DANA and Relenza are modest and nonselective inhibitors of these enzymes. Medicinal chemistry efforts using DANA as a starting point has yielded useful inhibitors for these enzymes by exploiting structural differences in their active sites. One example, C9-4HMT-DANA (Figure 55.6), is the most potent and selective inhibitor of any of these enzymes having an ~100 nM K_i value and exhibiting 500-fold selectivity over any of the other family members. Doubtless, structures of these human enzymes are likely to accelerate the development of new selective inhibitors.

ACKNOWLEDGMENTS

The authors appreciate helpful comments and suggestions from Manfred Wulhrer.

FURTHER READING

Brown JR, Crawford BE, Esko JD. 2007. Glycan antagonists and inhibitors: a fount for drug discovery. *Crit Rev Biochem Mol Biol* **42**: 481–515. doi:10.1080/10409230701751611

Gloster TM, Vocadlo DJ. 2012. Developing glycan processing enzyme inhibitors as enabling tools for glycobiology. *Nat Chem Biol* **8**: 683–694. doi:10.1038/nchembio.1029

Tu Z, Lin YN, Lin CH. 2013. Development of fucosyltransferase and fucosidase inhibitors. *Chem Soc Rev* **42**: 4459–4475. doi:10.1039/c3cs60056d

Galley NF, O'Reilly AM, Roper DI. 2014. Prospects for novel inhibitors of peptidoglycan transglycosylases. *Bioorg Chem* **55**: 16–26. doi:10.1016/j.bioorg.2014.05.007

Kallemeijn WW, Witte MD, Wennekes T, Aerts JM. 2014. Mechanism-based inhibitors of glycosidases: design and applications. *Adv Carbohydr Chem Biochem* **71**: 297–338. doi:10.1016/b978-0-12-800128-8.00004-2

Shayman JA, Larsen SD. 2014. The development and use of small molecule inhibitors of glycosphingolipid metabolism for lysosomal storage diseases. *J Lipid Res* **55**: 1215–1225. doi:10.1194/jlr.r047167

Selnick HG, Hess JF, Tang C, Liu K, Schachter JB, Ballard JE, Marcus J, Klein DJ, Wang X, Pearson M, et al. 2019. Discovery of MK-8719, a potent O-GlcNAcase inhibitor as a potential treatment for tauopathies. *J Med Chem* **62**: 10062–10097. doi:10.1021/acs.jmedchem.9b01090

Howlader MA, Guo T, Chakraberty R, Cairo CW. 2020. Isoenzyme-selective inhibitors of human neuraminidases reveal distinct effects on cell migration. *ACS Chem Biol* **15**: 1328–1339. doi:10.1021/acschembio.9b00975

56 Glycosylation Engineering

Henrik Clausen, Hans H. Wandall, Matthew P. DeLisa, Pamela Stanley, and Ronald L. Schnaar

Knowledge of the cellular pathways of glycosylation across phylogeny provides opportunities for designing glycans via genetic engineering in a wide variety of cell types including bacteria, fungi, plant cells, and mammalian cells. The commercial demand for glycosylation engineering is broad, including production of biological therapeutics with defined glycosylation (Chapter 57). This chapter describes how knowledge of glycan structures and their metabolism (Chapters 2–27) has led to the current state of glycosylation engineering in different cell types. Perspectives for rapid advances in this area using precise gene editing technologies are also described.

published by Cold Spring Harbor Laboratory Press; doi: 10.1101/glycobiology.4e.56

GOALS FOR GLYCOENGINEERING OF CELLS

There is a long history of engineering glycosylation in mammalian cells, plants, fungi (yeast), and bacteria using genetic strategies, and many well-characterized glycosylation mutants are available (Chapters 20–27 and 49). This chapter focuses on approaches and methods for designing glycosylation in cells, whereas other active areas of engineering glycans to produce structural bioproducts, foods, and fuels are not covered. Today, cellular glycoengineering is often used to produce recombinant therapeutic glycoproteins that require glycosylation for their efficacy and at the same time must have human-compatible glycosylation to avoid immune responses to nonhuman glycans. Glycosylation can alter the size, charge, and solubility of therapeutic glycoproteins to prevent rapid clearance from the circulation. In addition, glycoengineering has been used to improve or develop new therapeutic modalities (Chapter 57). Glycans can also serve as ligands for lectin receptors that target therapeutics to certain cells. Of particular importance is the role of N-glycosylation for effector functions of IgG antibodies; therapeutic IgG antibodies with N-glycosylation designed to improve their cytotoxic properties are in clinical use. In the past decade, new methods have emerged to precisely engineer glycosylation by gene editing, and with increased knowledge the field seems to be limited only by imagination.

Cell lines are widely used as factories to produce recombinant glycoproteins from introduced gene constructs. The most common factories for glycoproteins include yeast, plants, insect cells, nonhuman mammalian cells, and, more rarely, human cells. More recently, bacteria are also being engineered to accommodate production of glycoproteins. The glycosylation capabilities of different species vary substantially in terms of both the sites of glycan attachment and the glycans attached (Figure 56.1 and Chapters 9–27). A first step in glycoengineering strategies is therefore to consider which cell type to use. This decision requires detailed knowledge of glycosylation pathways and genes. Historically, the mammalian Chinese hamster ovary (CHO) cell line has played a dominant role, and today most biologics are produced in CHO cells (Chapter 49). The CHO line was selected for human therapeutic production because its glycosylation capacities are relatively simple and resemble those of humans. The CHO cell line produces a comparatively narrow repertoire of glycans that are not immunogenic in humans; glycoengineering can expand their native glycosylation capabilities and provide optimization of glycoforms. Alternate host species in which native glycosylation (or lack thereof) provides a simpler starting point for engineering can also be selected. For example, glycoproteins for enzyme replacement therapies have been produced in yeast and glycan vaccines in bacteria.

There have been major achievements in glycoengineering of cells from bacteria to yeast and "higher" eukaryotes (Table 56.1 and Chapter 49). New precise gene editing technologies described below enable glycoengineering in a wide variety of species and open opportunities for selection of host cells based on optimal production efficiency and production of human-like glycans. Common principles in glycoengineering via gene editing are described below, followed by more detailed descriptions of progress in cells from various species.

KNOWLEDGE OF GLYCOSYLATION PATHWAYS ENABLES GLYCOENGINEERING

Although the glycomes of different species have distinct features (Figure 56.1), the basic biosynthetic machinery and pathways are remarkably conserved in eukaryotes, and there are even similarities with glycosylation pathways in some bacteria and Archaea. Most enzymes involved in glycosylation in eukaryotes are highly conserved in fungi, plants, and animals, facilitating the design and execution of glycoengineering strategies in these organisms. Nevertheless, current knowledge is far from complete, and glycoengineering across species is still in its infancy. Whereas expression of a particular protein in a heterologous host may require only introduction

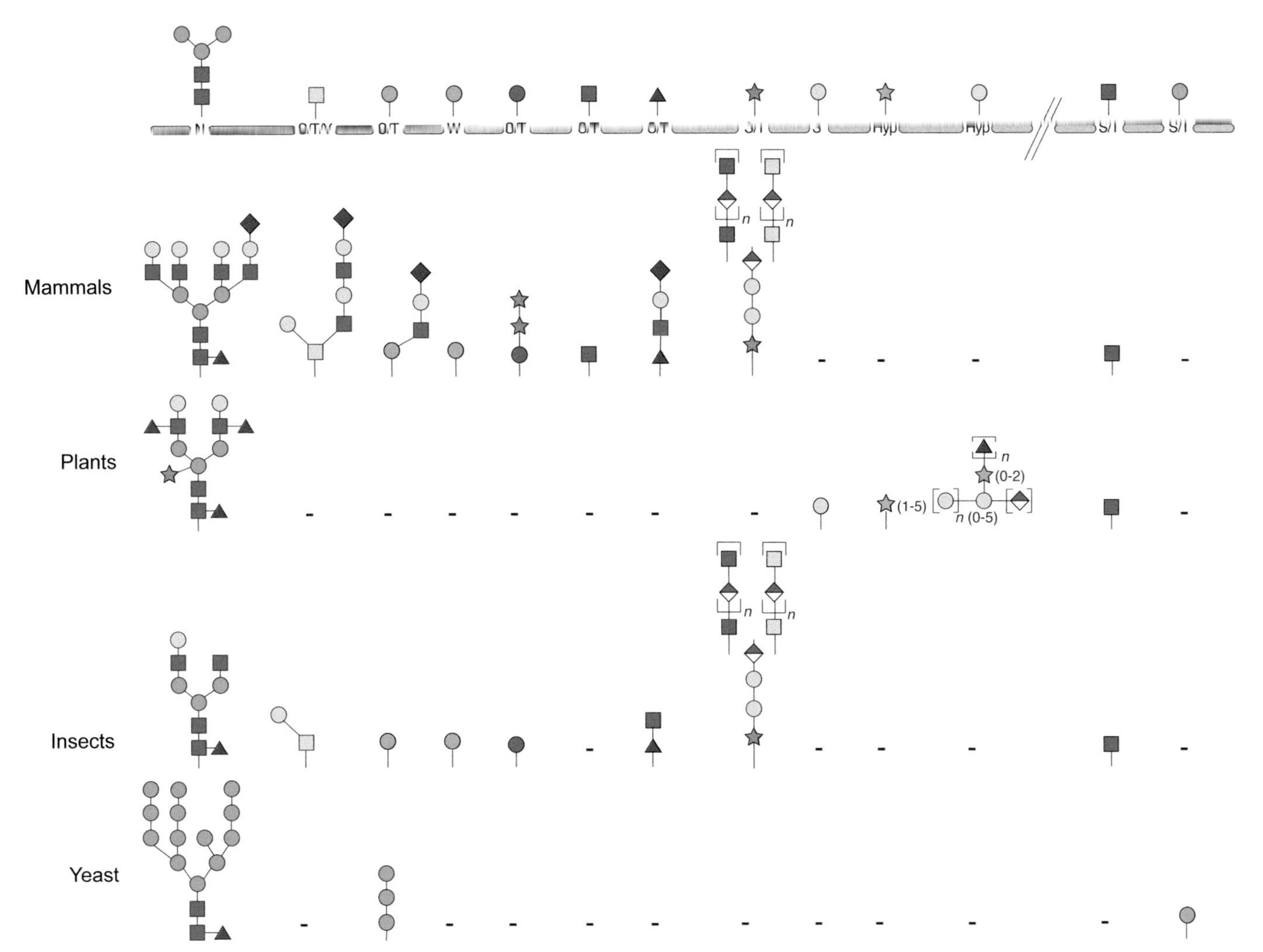

FIGURE 56.1. Overview of species-specific glycosylation features. The figure presents the different classes of glycoconjugates present in mammalian, plant, insect, and yeast cells with a representative glycan from each class. Structures to the *right* of the break are found in the cytoplasm and nucleus of the indicated organisms. Glycans of bacteria and Archaea (not shown) are more varied and often contain nonhuman monosaccharides (Chapters 20 and 21).

of the single gene for that protein, precise glycosylation engineering of that protein may require introduction of a suite of genes, including those required for the biosynthesis and transport of appropriate activated nucleotide sugar donors, as well as multiple glycosyltransferases.

Successful glycoengineering requires knowledge of the glycosyltransferase genes and substrates required to direct synthesis of a particular glycan. Certain genes may need to be removed and others inserted to create biosynthetic pathways that produce the glycans of interest. Four decades of glycogene discoveries have resulted in the identification of many genes encoding glycosyltransferases, hydrolases, and other enzymes involved in synthesizing and metabolizing the glycans of eukaryotic cells and the biosynthetic pathways involved (Chapters 8–19). Different glycosylation pathways may function independently using different sets of enzymes or, in some cases, may share enzymes. Enzymes working in consecutive order to assemble mature glycans generally work independently, although there may also be cooperative effects. In principle, there is sufficient knowledge to predict the role of individual enzymes and assign them to specific pathways, allowing prediction of the enzyme repertoire required to generate a particular glycan on a particular glycoconjugate. An excellent resource in this regard is the classification

TABLE 56.1. Examples of major achievements in glycoengineering of cells

Cell	Gene modification	Effect	Purpose	Structure
Mammalian				
CHO	KO of *Fut8*	Elimination of core fucose	IgG1 with increased binding affinity ti FcγRIIIa and ADCC	
CHO	OE of *MGAT3*	Reduction of core fucose by addition of bisecting GlcNAc	IgG1 with increased binding affinity ti FcγRIIIa and ADCC	
CHO	KO combinations of 19 glycosyltransferses KI *ST6GAL1*	N-glycan deconstruction and construction controlling branching, polyLacNAc, and sialylation	Design matrix to create homogeneous N-glycans of choice	
CHO	OE of GDP-6-deoxy-D-lyxo-4-hexulose reductase (*Pseudomonas aeruginosa rmd* gene)	Depletes endogenous GDP-Fuc and provides control of fucosylation by exogenous addition of fucose	IgG1 with increased binding affinity to FcγRIIIa and ADCC	
CHO	KO of *Fut8* and *B4galt1*, KI of integrated circuits expressing synthetic glycosyltransferase genes under constitutive or inducible promoters	Small molecule control of antibody fucosylation and galactosylation levels	IgG1 with increased binding affinity to FcγRIIIa and ADCC	
HEK293	KO of *MGAT1* combined with Golgi-targeted expression of endo-β-N-acetylglucosaminidases	Single GlcNAc N-glycan "stumps" that can be recognized and modified by galactosyltransferases and sialyltransferases	Simple and homogenous N-glycosylation	
Insect				
S2 *Drosophila melanogaster* cells	KI of sialic acid synthase, CMP-sialic acid synthase, CMP-sialic acid transporter, and a N-acetylglucosamine-6-phosphate 2′-epimerase	Biantennary N-glycans with galactosylation and sialic acid	Human-like N-glycans in insect cells	
Yeast				
Yarrowia lipolytica	KO of *och1*, KI of *mnn4*	N-linked glycans with mannose-6-phosphates; uncapping can be performed in vitro with glycosidases and α-mannosidases	Lysosomal targeting of enzymes used for enzyme replacement in lysosomal storage diseases	PO_4^- PO_4^-

Pichia pastorius	KO of *ochl, pnol, mnn4B, bmt2*; KI of 14 genes: UDP-GlcNAc transporters, *MnsI, MGAT1, MGAT2, MnsII*, Gal epimerase, UDP-Gal transporter, *B4GALT1*, and sialic acid biosynthetic pathway, and transporters	Biantennary N-linked glycans with sialylation	Human-like N-glycans in yeast	
P. pastorius	KO of *ochl*; OE of *MGAT1, B4GALT1* targeted to Golgi and α2- mannosidase retained in the endoplasmic reticulum	Biantennary N-linked glycans	Human-like N-glycan core in yeast	
Saccharomyces cerevisiae	KO of *ALG3* and *ALG11* and OE of artificial flippase and protozoan OST	Lipid-linked $Man_3GlcNAc_2$ assembled on cytoplasmic side of endoplasmic reticulum, flipped and transferred to protein	Mammalian core $Man_3GlcNAc_2$	
Bacteria				
Escherichia coli	KI of OST *pglB* from *Campylobacter jejuni* together with four glycosyltransferases from *S. cerevisiae* encoded by *ALG1, ALG2, ALG13*, and *ALG14*	Introduce N-glycosylation into major prokaryotic production organism	Mammalian core $Man_3GlcNAc_2$	
E. coli	OE/KI of OST *pglB* and NeuBCA enzymes from *C. jejuni*, LsgCDEF glycosyltransferases from *Haemophilus influenza*, and α2-6 sialyltransferase from *Photobacterium leiognathi*	Introduce Neu5Acα2-6Galβ1-4GlcNAc termini to N-linked glycans	Simplified sialylated N-linked glycans	
E. coli	OE/KI of OST *pglB* and select *pgl* cluster genes from *Campylobacter jejuni*	Introduce distinctive N-linked glycan that can be trimmed to GlcNAc-Asn for chemoenzymatic elongation	Scaffold for human-like N-glycans	
E. coli	OE/KI of OST *PglO* from *Neisseria gonorrhoeae* or *PglL* from *Neisseria meningitidis*	Introduce GalNAc O-glycosylation into major prokaryotic production organism	Mammalian-type Tn, T, sialyl-Tn, and sialyl-T glycans	
Plants				
Nicotiana tabacum	KO of β-hexosaminidase, α3-fucosyltransferase, and β2-xylosyltransferase; KI of *B4GALT1, ST6GAL1*, CMP-sialic acid synthase, and transporters	Production of biantennary N-glycans with α2-6NeuAc and without core fucose	Human-like N-glycans in plants	
N. tabacum	KO of β-hexosaminidase, α3-fucosyltransferase, and β2-xylosyltransferase; OE of *GALNT2, C1GALT1, ST3GAL1, SLC35A1, e1/2, ST6GALNAC3/4*	Introduce N-glycosylation into major prokaryotic production organism	Mammalian-type Tn, T, sialyl-Tn, and sialyl-T glycans	

(KO) knockout, (OE) overexpression, (KI) knock-in, (CMP) cytidine monophosphate, (OST) oligosaccharyl transferase, (ADCC) antigen-dependent cellular cytotoxicity.

of homologous gene families from diverse species in the "Carbohydrate-Active enZYmes" (CAZy) database (Chapter 8).

Among the prerequisites for glycoengineering a desired glycan in a chosen host is that the appropriate repertoire of activated sugar donors and their transporters are present (Chapter 5). This is especially important when engineering glycosylation in prokaryotes or nonmammalian eukaryotes in which the nucleotide sugar donors required to synthesize therapeutics with human glycosylation may not be present. For example, yeast does not produce UDP-GalNAc, and many organisms do not produce CMP-sialic acids.

THE IN/OUT STRATEGIES OF GENETIC GLYCOENGINEERING IN EUKARYOTES

Different genetic strategies may be used to alter the glycosylation capabilities of cells. Knockdown and nontargeted overexpression in eukaryotes have been used for many years, and precisely targeted gene editing strategies are now well-established.

Knockdown

Reducing undesirable glycosyltransferase activities in cells has been achieved by gene silencing strategies. Whereas this has been particularly successful in plants and Drosophila, silencing has not gained wide use in glycoengineering mammalian cell lines because the low efficiency of knockdown often leaves undesirable levels of target glycosyltransferase activity remaining.

Overexpression

Adding desirable glycosyltransferase activities to eukaryotic cells is achieved by transfection of glycogenes from any organism, random integration of plasmid DNA, and antibiotic selection of stable clones. Although this strategy is successful, it provides no control over site(s) of genomic integration (unless specific strategies are used), gene copy number, or gene expression levels. Overexpression of enzymes can lead to disruption of normal glycosylation patterns and unpredictable glycosylation. Instability of the introduced glycosylation genes and the use of antibiotics for selection have also been problematic for the long-term use of such engineered cells for production of therapeutic glycoproteins.

Knockout and Knock-In by Precision Genome Editing

Knockout of glycosylation genes to eliminate unwanted glycans has long been a simple task in bacteria and yeast. Although powerful, knockout or knock-in strategies have been time-consuming and difficult to use in "higher" eukaryotic cells (see *Fut8* knockout below). However, these difficulties were substantially reduced with the introduction of nuclease-based precise gene editing tools including zinc-finger nucleases, transcription activator–like effector nucleases (TALENs), and clustered regularly interspaced short palindromic repeat/targeted Cas9 endonucleases (CRISPR/Cas9), which enable highly specific gene manipulation in all cell types (Figure 56.2; Chapters 27 and 49). These tools can also be used to activate endogenous silent genes, edit gene sequences to mimic hypomorphic disease mutations, and insert foreign genes at specific genomic sites.

Precise gene editing can insert foreign genes at "safe harbor" sites in the genome to ensure stable expression and avoid interfering with endogenous gene expression. One such safe harbor in human cells is the AAVS1 site on chromosome 19 known to enable stable expression of transgenes without adverse effects. However, precise genetic engineering allows insertion of one or more

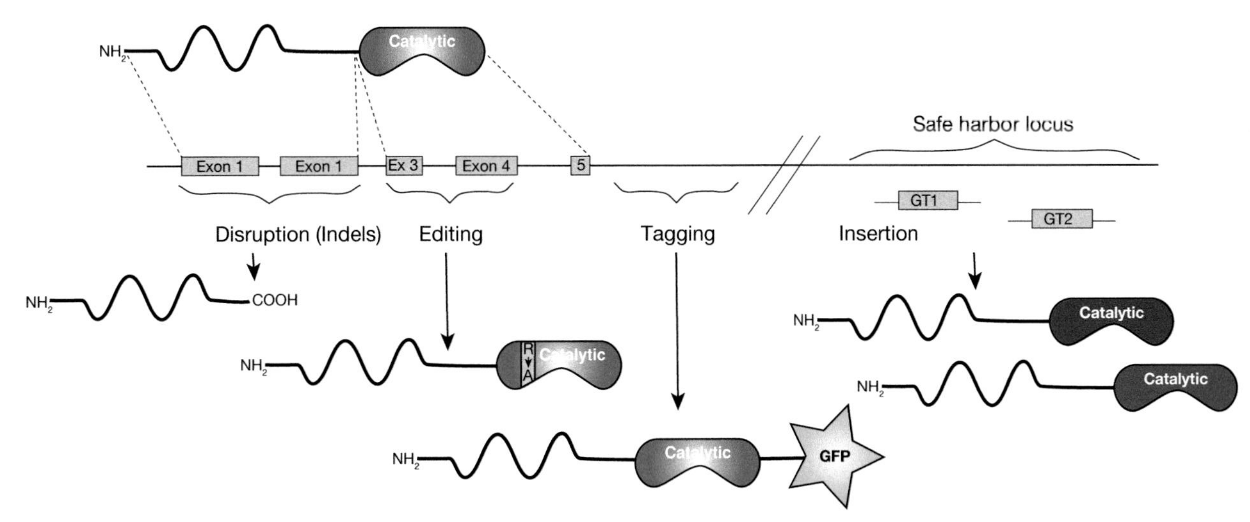

FIGURE 56.2. Precise gene editing modalities. Glycosyltransferases are represented with their catalytic domains and amino-terminal membrane spanning domains. Examples of gene editing for gene disruption (knockout), mutation, tagging, and insertion of heterologous glycosyltransferases at a "safe harbor" locus are shown. Precise gene editing uses targeted nucleases to introduce double-stranded DNA breaks at specific locations. In mammalian cells, these breaks are repaired by nonhomologous end joining (NHEJ) or homologous recombination (HR). NHEJ is an error-prone mechanism that often introduces small insertions or deletions (indels) resulting in frameshift mutations. Editing (to introduce targeted mutations) or insertion of glycosyltransferase genes mainly relies on HR following cotransfection of appropriate donor DNA constructs. For efficient knockout of type 2 transmembrane glycosyltransferases, it is often sufficient to target one of the first coding exons, which leads to nonsense-mediated decay of mRNA or causes alternative splicing that usually gives rise to nonfunctional proteins, but more generally it is advisable to target exons encoding functional domains.

foreign genes at any position in the genome. For example, precise knock-in of glycogenes can be combined with knockout by inserting an exogenous gene in place of an unwanted endogenous gene. Precise knock-in strategies also enable control of the number of copies inserted and can be used to insert entire landing platforms of multiple genes inserted consecutively.

Successful expression of an enzyme or transporter for glycoengineering requires that the expressed protein finds its way to the correct subcellular compartment. Heterologous expression of type 2 Golgi transmembrane glycosyltransferases, for example, often requires testing different Golgi retention sequences. Although design guidance is available, this is often a trial and error exercise, and in some cases a combinatorial screening is required to identify the optimal construct.

GLYCOENGINEERING IN BACTERIA

The most common bacterium used for heterologous protein production, *Escherichia coli*, does not have the native capacity to glycosylate proteins. However, research in the last two decades identified N- and O-glycoproteins, and the glycosylation pathways responsible for their biosynthesis, in the pathogenic proteobacterium *Campylobacter jejuni* and other bacterial species. Moreover, several bacterial toxins have glycosyltransferase domains that exert their pathogenicity by glycosylating highly specific amino acids in key host proteins to interfere with essential cellular functions. Chapter 21 discusses glycosylation in Eubacteria. Another unique feature of bacteria is that sugar nucleotide donors are synthesized and remain in the cytosol, so that engineered glycan assembly on a lipid carrier (for systems based on en bloc glycosylation) or on proteins directly (for systems based on processive glycosylation) must take place in the cytosol unless nucleotide sugar transporters are introduced to the periplasmic membrane.

Engineering Free and Lipid-Linked Oligosaccharides

Bacteria efficiently produce free and lipid-linked oligo- and polysaccharides including capsular polysaccharides (CPSs) and lipopolysaccharides (LPSs) (Chapter 21), and these pathways have been engineered to produce a variety of complex human-like glycans. In particular, the LPS pathway has been used to engineer and display glycans in *E. coli*. LPS consists of a base lipid (lipid A) linked to a core oligosaccharide followed by highly diverse O-polysaccharides (O-antigens). Genetically interrupting biosynthesis of the lipid A core oligosaccharide prevents coupling of the O-antigen, thereby allowing engineering of novel glycans on lipid A for display on the cell surface. This strategy has been used to engineer the synthesis of a variety of human glycan epitopes, including blood group antigens (Chapter 14) and cancer-associated glycolipid glycans (Chapter 11).

In a complex example of engineering large free complex glycans in bacteria, 4-O-sulfated chondroitin sulfate, a sulfated glycosaminoglycan (Chapter 17), was produced in *E. coli* by deleting a fructosyltransferase (*kfoE*), introducing bacterial UDP-Glc/-GlcNAc 4-epimerase and chondroitin synthase genes, and introducing a mutated human chondroitin-4-O-sulfotransferase gene, opening the way for glycosaminoglycan production in bacteria.

Engineering N-Glycosylation

Two types of native protein N-glycosylation occur in some bacteria, although not in *E. coli*. One type is similar to eukaryotic cell N-glycosylation with production of a lipid-linked oligosaccharide in the cytosol, which is then transferred en bloc to Asn by an oligosaccharyltransferase (OST) in the periplasm. The OST of *C. jejuni*, PglB, is a single polypeptide related to the catalytic STT3 subunits of the multiprotein OST complex of eukaryotes (Chapter 9). PglB shows a more restricted acceptor sequence motif than the eukaryotic N-X-S/T, with requirement of an acidic residue (D/EXNXS/T, in which X cannot be P) (Chapter 21). This places some restrictions on the usefulness of engineering human-like N-glycans, because most N-glycan sites in mammalian proteins do not conform to this extended consensus sequence. PglBs from other species, or mutants evolved by adaptive evolution, have been identified to address this problem, but further improvements are needed. Importantly, PglB has rather relaxed donor substrate specificity. Although bacterial lipid-linked oligosaccharides are distinct from those in eukaryotes, PglB can use mammalian-type lipid-linked oligosaccharides as donors.

An important feature for using prokaryotes for glycoengineering is the arrangement of entire glycosylation machineries in multigene operons. This enables the transfer of large genetic elements of 10–20 kb between species. A major achievement was the successful transfer of the entire N-glycosylation operon from *C. jejuni* to *E. coli*, producing N-linked glycoproteins in *E. coli*. Production of glycoproteins carrying a $Man_3GlcNAc_2$ core N-glycan has been achieved by introduction of eukaryotic enzymes (Table 56.1). Bacterial N-linked glycosylation is being exploited as an alternative method for glycoconjugate vaccine production, and vaccines against both Gram-negative and Gram-positive bacteria have been developed.

Another type of N-glycosylation found in γ-proteobacteria involves a cytosolic N-glycosyltransferase (NGT) that targets the N-X-S/T acceptor sequence motif recognized by mammalian OST (Chapter 9). NGT transfers a single monosaccharide (e.g., Glc) from an activated sugar nucleotide donor, with loose donor substrate specificity that includes both UDP and GDP sugar nucleotides. This provides an entirely new approach to engineering alternate types of N-glycosylation. Engineering this pathway in bacteria resulted in the assembly of glycan motifs including α1-3-galactose epitopes as well as fucosylated and sialylated lactose or poly-*N*-acetyllactosamine (LacNAc) units primed by Glc residues on glycoproteins (Table 56.1).

Although most bacteria do not have the capacity for sialylation, there are exceptions (Chapters 15 and 21). Bacterial genes for CMP-sialic acid synthesis and for sialyltransferases with specificities similar to those in mammals have been introduced with plasmids or integrated into the genome of host bacteria cells, enabling the production of sialylated N- and O-glycoproteins.

Engineering O-Glycosylation

Some bacteria have processive O-glycosylation pathways controlled by glycosyltransferases using activated sugar donors. These pathways inspired the engineering of human O-glycosylation reactions in *E. coli*. By introducing mammalian polypeptide GalNAc-transferase genes and a UDP-Glc/GlcNAc 4-epimerase, O-GalNAc protein glycosylation (Chapter 10) has been achieved. Further introduction of a β1-3-galactosyltransferase enabled biosynthesis of core 1 O-glycans (T antigen) on cytoplasmic acceptor proteins. Introduction of GalNAc residues has been used for postexpression enzymatic addition of polyethylene glycol (PEG)-derivatized sialic acids to enhance the therapeutic properties of protein drugs.

Other bacteria possess a protein O-glycosylation mechanism that is unlike the stepwise biosynthesis of O-glycans in eukaryotes (Chapter 10), in that preassembled undecaprenol-PP-linked oligosaccharides are transferred en bloc to proteins by several OSTs with relaxed donor substrate specificities and poorly understood acceptor substrate specificities. Engineering of this endogenous glycosylation machinery has been used for en bloc transfer of human O-GalNAc glycans (Tn, T, sialyl-Tn, and sialyl-T antigens) onto acceptor proteins (Table 56.1).

GLYCOENGINEERING IN YEAST

Yeast natively produce N-glycans and O-mannosyl glycans on diverse glycoproteins. The general features of the biosynthetic pathways for initial glycan transfer are common in eukaryotes from yeast to human, and the enzymes involved are highly homologous. However, subsequent glycan processing in yeast generally results in polymannosylated glycans instead of the complex N- and O-glycans found in "higher" eukaryotes (Figure 56.1). Yeast have similar systems for protein folding, quality control, and posttranslational modifications to other eukaryotic cells, in contrast to bacteria. Because genetic engineering in yeast has long been rapid and easy, experience with glycoengineering is more advanced in this organism compared with most others. Several commercial ventures have been based on engineering "humanized" N-glycosylation in yeast; the Pichia GlycoSwitch platform uses engineered yeast to add simple human N-glycans to expressed proteins.

Engineering N-Glycosylation

N-glycans on yeast glycoproteins differ from those in vertebrates (Chapter 9), comprising large polymannosyl glycans on a poly(Manα1-6)$_n$ backbone, which are highly immunogenic in mammals (Chapter 23). A key α1-6-mannosyltransferase, Och1p largely initiates polymannosylation. Knockout of the *och1* gene, however, does not completely abrogate polymannosylation, and additional knockouts of mannosyl- and phosphomannosyltransferases, depending on yeast strain, are needed to achieve a homogenous $Man_8GlcNAc_2$ N-glycan suitable for further engineering. Reducing $Man_8GlcNAc_2$ to $Man_5GlcNAc_2$ is achieved by expressing an α1-2-mannosidase in the endoplasmic reticulum (ER), creating a convenient platform for generating complex N-glycans. Introduction of GlcNAcT-I (*MGAT1*) in the Golgi initiates complex N-glycan synthesis, and further addition of α3/6-mannosidase II (*MAN2A1*) and GlcNAcT-II (*MGAT2*) results in the biantennary $GlcNAc_2Man_3GlcNAc_2$ N-glycan suitable for appending galactose and sialic acid by further engineering. Some yeast species, including *Pichia pastoris*, do not contain UDP-Gal, and all yeast lack the native ability to synthesize CMP-Neu5Ac, so considerable engineering with introduction of multiple genes is required to obtain mature complex N-glycans. Although the engineering appears simple in silico, considerable efforts have been devoted to identifying optimal chimeric gene constructs with respect to both catalytic efficiency and ER/Golgi targeting.

Engineering O-Glycosylation

Yeast perform extensive co- and posttranslational ER protein O-mannosylation (Chapter 23) using several polypeptide mannosyltransferases (PMTs). *Saccharomyces cerevisiae* has six PMTs, and only a subset can be knocked out without reducing viability. Protein O-Man residues undergo polymannosylation in the Golgi. Muticellular eukaryotes also perform O-mannosylation and express two PMT orthologs, *POMT1* and *POMT2* (Chapter 13), but these have narrower acceptor substrate specificities. However, multicellular eukaryotes perform several other types of O-glycosylation (Figure 56.1) (Chapters 10, 13, and 14), and their O-GalNAc glycans tend to be located in similar regions and protein sites as O-Man glycans in yeast. This means that expression of human O-glycoproteins in yeast may result in O-mannosylation at sites that carry O-GalNAc in mammals. Examples of this include the hinge region of IgA and mucin sequences. Because it is still difficult to predict types of O-glycosylation, human proteins expressed in yeast must be tested to determine if they are O-mannosylated.

Human O-GalNAc glycans have been successfully engineered into yeast by introducing human polypeptide GalNAc-transferases (Chapter 10) along with UDP-Glc/GlcNAc C4-epimerase and a UDP-Gal/GalNAc Golgi transporter. The entire biosynthetic machinery for CMP-Neu5Ac synthesis and transport has also been introduced together with a human sialyltransferase, and sialylated O-glycans have been produced in yeast. The problem with competing endogenous O-mannosylation can be partly eliminated by including a mannosyltransferase inhibitor (rhodanine-3-acetic acid). A deeper understanding of the yeast O-Man and human O-GalNAc glycosylation pathways is needed to provide new strategies to circumvent competition between the two systems and enhance O-glycan engineering in yeast.

GLYCOENGINEERING IN PLANT CELLS

Plants offer a simpler starting point than yeast for N-glycan humanization because the predominant native N-glycans of plants are paucimannose ($Man_3GlcNAc_2$) and biantennary terminating in GlcNAc ($GlcNAc_2Man_3GlcNAc_2$). The abundance of paucimannose N-glycans appears to be due to a β-hexosaminidase that removes attached GlcNAc residues in competition with GlcNAc-transferases, a feature also found in insect cells. Two plant-specific N-glycan modifications include core α1-3-Fuc (instead of mammalian core α1-6-Fuc) and β1-2-Xyl linked to the β-Man in the N-glycan core. Both modifications are potentially immunogenic in humans. Plants also produce unique types of O-glycosylation not found in other species that pose potential problems for the generation of therapeutic glycoproteins.

Engineering N-Glycosylation

Great advances in engineering plants for human-like N-glycosylation have been achieved. Knockdown or knockout of the β-hexosaminidase that inhibits complex N-glycan formation, as well as the α1-3 fucosyltransferase and β1-2 xylosyltransferase, have been achieved in different plants, including *Arabidopsis thaliana* and *Nicotiana benthamiana*. Nearly homogeneous biantennary $GlcNAc_2Man_3GlcNAc_2$ N-glycans were produced. These were further engineered by the introduction of Gal (using *B4GALT1*) and sialic acid (using *ST6GAL1* along with the enzymes needed to synthesize and transport CMP-Neu5Ac) in an engineering design using up to six gene constructs. Such humanized plants produced α2-6-Neu5Ac capped biantennary N-glycans without core fucose on a variety of recombinant glycoproteins. These achievements depended on combinatorial screening strategies to identify appropriate chimeric constructs of exogenous enzymes to drive the engineered glycosylation toward homogeneity.

Glycoproteins produced in plants carrying native paucimannose N-glycosylation are in use as approved drugs. For enzyme replacement therapy, the terminal mannose N-glycans of glucocerebrosidase (taliglucerase alfa) produced in carrots is beneficial for targeting to endogenous human mannose receptors, despite α1-3-Fuc and β1-2-Xyl modifications, and is in clinical use. Moreover, glycoengineered *N. benthamiana* cells without α1-3-Fuc and β1-2-Xyl have been used to produce a triple-antibody cocktail used to treat Ebola virus infections.

Engineering O-Glycosylation

Plants do not have the types of O-glycosylation found in other eukaryotes but produce extensins and arabinogalactan proteins with two unique O-glycans. A family of prolyl-4-hydroxylases (P4H) converts selected Pro residues to hydroxyproline that may be arabinosylated by a series of enzymes. In addition, Ser residues may be O-glycosylated by the addition of Gal residues. Although a number of the P4Hs and glycosyltransferases have been knocked out in different plants, it is unclear whether these modifications can be completely eliminated without affecting viability. Nonetheless, the human machinery for O-GalNAc glycosylation has been engineered into plants by introducing the necessary polypeptide GalNAc-transferases and elongation enzymes, whereas UDP-Glc/-GlcNAc 4-epimerase and a UDP-GalNAc transporter may not be required. Human core 1 O-glycan biosynthesis and sialylation machinery including ST3GalI sialyltransferase have also been successfully introduced into plants. If issues related to hydroxyproline can be resolved, plants offer a valuable system in which different types of mammalian O-glycosylation could be engineered and exploited. A clear highlight of glycoengineering in plants was the combined introduction of 14 genes for production of the major human therapeutic glycoprotein erythropoietin with human sialylated biantennary N-glycans and core 1 O-glycans in tobacco cells (Table 56.1).

GLYCOENGINEERING IN INSECT CELLS

Engineering in insect cells involves multiple strategies. Two different platforms are generally used for recombinant expression of proteins—transient expression in the baculovirus-insect cell system and constitutive expression in Sf9 *Spodoptera frugiperda* or S2 *Drosophila melanogaster* cells. The baculovirus-insect cell platform can be glycoengineered by including glycosylation genes in either the recombinant baculovirus vector genome or the insect cell line host genome. Engineering host insect cell lines has been the more common strategy, but remarkable success has been achieved by incorporating up to nine glycogenes in a baculovirus vector (Table 56.1). CRISPR/Cas gene targeting of Sf9 insect cells has been established, and their use for glycoengineered baculovirus protein expression is feasible.

Engineering N-Glycosylation

Insect cells produce mostly high-mannose and paucimannose N-glycans despite having the genetic capacity to produce complex sialylated N-glycans (Figure 56.1). This is due in part to the action of a processing β-hexosaminidase, FDL, which removes attached GlcNAc residues from the α1-3-Man branch, and in part to low levels of GlcNAcT-II (MGAT2) activity. Like plants, some insect cells may add a potentially immunogenic core α1-3-Fuc and do not typically add terminal sialic acids. However, sialylation has been engineered by introducing genes encoding a CMP-sialic acid synthase and an *N*-acetylglucosamine-6-phosphate 2′-epimerase into insect cells. For efficient sialylation, a dedicated CMP-sialic acid transporter appears to be needed as well. Using different strategies, production of glycoproteins carrying biantennary N-glycans with galactosylation and sialic acid capping has been achieved. Precision gene editing was used to knock out *fdl* in Sf9 and S2 cells to greatly improve complex N-glycan formation.

Engineering O-Glycosylation

Insect cells perform the same range of O-glycosylation reactions as mammalian cells (Figure 56.1), although the extent to which O-GalNAc glycans are attached at the same sites as in mammals is unexplored. Moreover, processing of O-glycans is limited to mainly truncated core 1 structures (Tn and T). Although insect cells offer a straightforward host for production of glycoproteins with human O-glycans, little has been investigated in this regard.

GLYCOENGINEERING IN MAMMALIAN CELLS

The cores of all types of glycoprotein glycans (Figure 56.1) are highly conserved among mammals, although there are terminal glycan variations (Chapter 14). At least 16 different glycosylation pathways have been delineated in mammalian cells; maps of the predicted genetic regulation of biosynthetic steps by more than 170 distinct glycosyltransferases have been generated. The most popular mammalian cell line used for glycoengineering is the CHO cell line established more than 60 years ago. The success of the CHO cell line is partly due to the ease with which glycosylation mutants could be isolated (Chapter 49), and it was the first cell used to manufacture a recombinant therapeutic with relatively simple human-type terminal glycans without expression of antigenic nonhuman glycans or unusual modifications of the glycans. As discussed in Chapter 49, the CHO cell line has an important place in glycoengineering history, exemplified by the Lec mutant lines generated by lectin selection. These cell lines with distinct mutations in glycosylation genes have provided tools for the scientific community for more than three decades and illustrate the importance of access to recombinant proteins with particular glycoforms for discovery of biological functions of glycans.

CHO cells can be considered as Glycobiology's gift to Biopharma. Major successes have been achieved in engineering CHO and other mammalian cell lines for production of human therapeutics (Table 56.1 and Chapters 49 and 57). The field is, however, undergoing a revolution with the new methods for facile, targeted, precise gene editing that allow the design of almost any conceivable glycosylation capacity in any mammalian cell by combining knockout and knock-in events.

Engineering N-Glycosylation

The first major feat in gene editing of mammalian cells was elimination of the core α1-6-Fuc for production of recombinant IgG antibodies with enhanced binding to the Fcγ-IIIa receptor (Table 56.1). Overexpression of bisecting GlcNAcT-III (*MGAT3*) resulted in stable CHO cells with highly limited capacity for core fucosylation (commercialized by Roche). A second strategy involved a tour-de-force approach using homologous recombination (HR) to knock out the two *Fut8* alleles in CHO cells. More than 10,000 CHO clones were screened to identify the final knockout cell. Although this was an impressive achievement, such laborious random selection limits options for selecting cell clones that retain the attributes needed for optimal bioprocessing. Using precise gene editing, the same engineering was rapidly replicated, providing ample clones for selection of those with optimal properties. Glycoengineered CHO lines optimized for antibody production are now commercially available (Potelligent CHOK1SV, Lonza/Kyowa Kirin BioWa). Another elegant strategy introduced GDP-6-deoxy-D-lyxo-4-hexulose reductase to deflect the endogenous production of GDP-Fuc and enable fine-tuning of fucosylation by exogenous addition of fucose.

Engineering N-glycan sialylation has been another focus in the field. CHO cells produce only α2-3-linked sialic acids on N-glycans, whereas human HEK293-T cells (for example) produce a mixture of α2-3- and α2-6-linked sialic acids. Most soluble glycoproteins in human blood (including IgG) have α2-6-linked sialic acids on N-glycans, and reports have suggested that the sialic acid linkage may influence immunomodulatory functions as well as circulatory half-life. It has therefore been of interest to engineer more homogeneous α2-6-sialylation in cells. These efforts have mainly been limited to the overexpression of α2-6-sialyltransferases to override endogenous α2-3-sialylation with variable results, illustrating the complexity of engineering glycosylation in cells with competing pathways.

An innovative glycoengineering strategy (GlycoDelete) reduced the inherent heterogeneity of mammalian N-glycan structures. Human HEK293-T cells lacking *MGAT1* were stably transfected to express a fungal endo-N-acetylglucosaminidase (EndoT) that efficiently truncated N-glycans to a single GlcNAc, which was an acceptor for galactosylation and sialylation. Recombinant antibodies with truncated N-glycans had lower affinity for Fcγ receptors, suggesting that this glycoengineering strategy may be suitable for use with neutralizing antibodies.

Deconstruction of the N-glycosylation pathway in CHO cells was performed by precise gene editing to knock out 19 glycosyltransferases, including all four α2-3-sialyltransferases that function on N-glycans (Figure 56.3). Combining knockout of *St3gal4* and *St3gal6* with site-specific knock-in of *St6gal1* resulted in homogeneous α2-6-sialylation. Combinatorial knockout of all isoenzymes

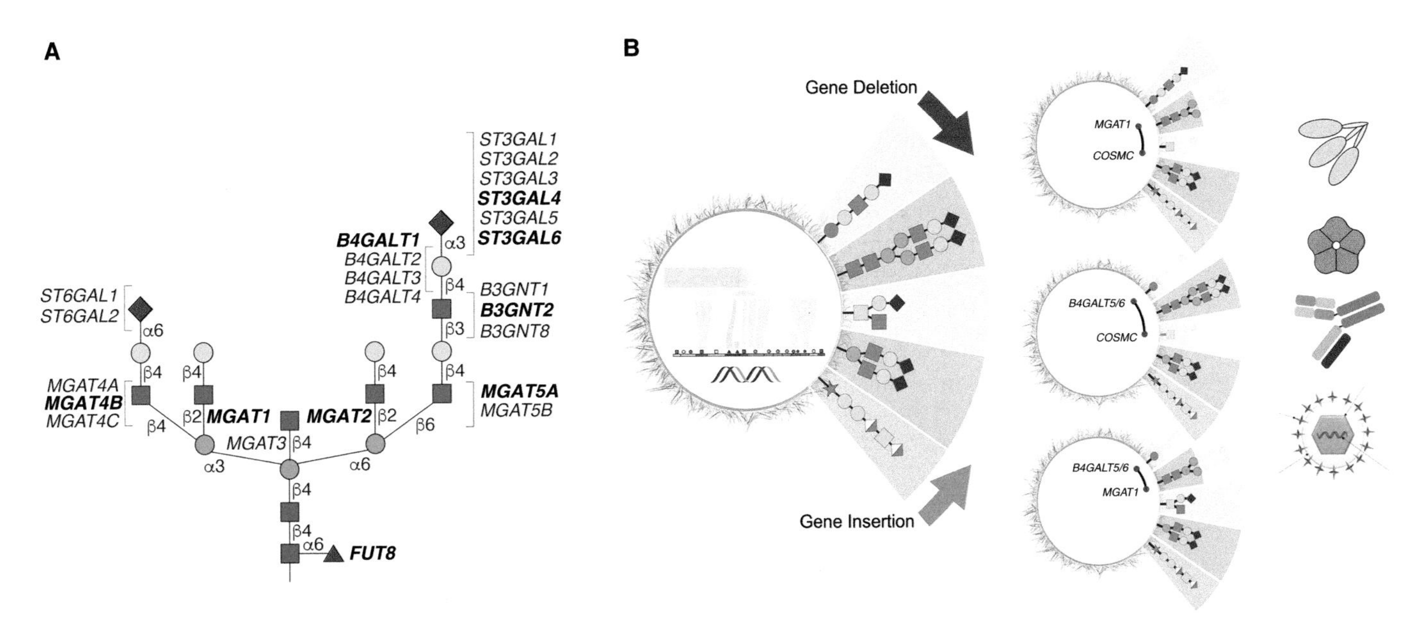

FIGURE 56.3. (*A*) A complex N-glycan with glycosyltransferases responsible for each reaction. Combinatorial knockout of the glycosyltransferase isoenzyme genes indicated led to the identification of the primary genes (highlighted in *bold*) controlling N-glycan branching (*Mgats*), elongation (*B3gnts* and *B4galts*), and sialylation (*St3gals*) in Chinese hamster ovary (CHO) cells. The methods discussed are equally applicable to the glycosyltransferases responsible for other types of glycosylation. (*B*) A generalized scheme for genetically altering the expression of different classes of glycans on cells. Gene deletions, insertions, or activation of glycosyltransferases can be used to generate isogenic cell lines that display different glycan features on endogenous surface glycoconjugates. Libraries of such engineered isogenic cells can be used to determine binding specificities of lectins, toxins, antibodies, or viruses. For illustration, isogenic cells with selective loss of different types of glycans are shown (loss of elaborated N-glycans [KO *MGAT1*], O-glycans [KO *C1GALT1*], and glycosphingolipids [KO *B4GALT5/6*]).

involved in N-glycan sialylation, galactosylation/LacNAc formation, branching, and core fucosylation has provided a design matrix for improving the homogeneity of N-glycans in CHO cells. A combination of five gene knockouts and the knock-in of *St6gal1* created the glycoprotein therapeutic erythropoietin having homogeneous biantennary N-glycans with terminal α2-6-Neu5Ac. Wider engineering of almost all genes involved in N-glycosylation in CHO cells has shown that there are few limitations for engineering of glycosylation. For example, the GlcNAc-1-phosphate transferase (*Gnptab*) that tags select oligomannose N-glycans on glycoproteins destined for lysosomal targeting was knocked out to produce lysosomal enzymes bearing complex-type sialylated glycans with extended blood circulation and improved biodistribution.

Therapeutic glycoprotein production still suffers from heterogeneity, including variations in which Asn residues are glycosylated (site occupancy, macroheterogeneity) and/or the diversity of mature glycan structures at any one site (microheterogeneity). This is currently addressed by ensuring reproducibility in batch-to-batch production through the use of highly standardized bioprocessing protocols, but this strategy is far from optimal. For example, incompletely sialylated therapeutic glycoproteins may be cleared by the hepatic asialoglycoprotein receptor (Ashwell–Morell receptor), resulting in inconsistent circulatory half-lives of therapeutic glycoproteins (Chapter 34). Considerable efforts have been devoted to improving sialylation by overexpressing relevant sialyltransferases as well as inhibiting or knocking out endogenous sialidases in host cells. Protein-specific glycosylation patterns and heterogeneity are more difficult to control.

Nonhuman mammalian cell lines can produce two immunogenic nonhuman glycans: α1-3-Gal added to *N*-acetyllactosamine and Neu5Gc added to Gal or GalNAc (Chapters 14 and 15). The α1-3-galactosyltransferase and CMP-*N*-acetylneuraminic acid hydrolase genes responsible are inactive in humans. Although α1-3-Gal and Neu5Gc are not produced in CHO cells, both genes have been knocked out as a precaution. Even so, Neu5Gc scavenged from animal glycoproteins used in cell culture can appear in expressed glycoproteins, so use of defined media lacking nonhuman glycoproteins is also necessary. In engineering mammalian cell lines, it is important to consider that the glycosylation capacity is driven by the expression of a subset of available enzyme genes, but unexpressed genes can become activated. Thus, cell-specific glycosylation features are generally controlled by transcriptional regulation rather than mutations or gene aberrations. Analysis of all known glycosylation genes in five distinct CHO production cell lines derived from the original CHO-K1 cell line found no apparent deleterious mutations or loss of genes, despite severe chromosomal alterations. This suggests that one must consider all the known glycogenes in a mammalian cell line for glycoengineering strategies.

Engineering the glycosylation capacity of CHO cells has also enabled more homogeneous bioconjugation of therapeutic drugs. For example, therapeutic drugs may be chemically conjugated with PEG chains to enhance circulatory half-life, but chemical conjugation is difficult to direct to specific sites in glycoproteins. A strategy for enzymatic modification of glycans post-production has been developed that involves desialylation of recombinant glycoproteins, followed by in vitro transfer of a modified (PEGylated) Neu5Ac (as its CMP analog) to exposed Gal/GalNAc residues by a sialyltransferase. The process is in use with approved drugs, although heterogeneous modifications occur when multiantennary N-glycans are targeted. CHO cells have now been engineered to produce monoantennary, unsialylated N-glycans, which circumvent heterogeneity while retaining multiple exposed Gal acceptor sites for sialo-PEGylation.

Engineering O-Glycosylation

Mammalian cells perform many different types of O-glycosylation (Figure 56.1), and although these exert diverse and important biological functions, the interest in O-glycans for recombinant therapeutics has been limited. Nevertheless, recombinant coagulation factors in clinical use carry O-GalNAc, O-Fuc, and/or O-Glc glycans, and many other approved drugs including erythropoietin and Enbrel have O-GalNAc glycans. O-GalNAc glycans are also used for site-specific bioconjugation.

Engineering O-GalNAc glycans involves a new level of complexity because up to 20 isoenzymes (polypeptide GalNAc-transferases) direct the initiation of O-GalNAc glycans. It may therefore be important to consider the repertoire of these enzymes in a cell line. In theory, a protein that is naturally found with an O-glycan may not be O-glycosylated when expressed in a specific production cell line, and vice versa. An illuminating case is the important phosphaturic factor FGF23, a potential drug for patients with a congenital deficiency associated with hyperphosphatemia, which requires an O-GalNAc glycan for activity. The repertoire of polypeptide GalNAc-transferases in CHO and HEK293 cells has been extensively engineered by knockout and also knock-in of *GALNT* genes, revealing adaptation of mammalian cells to loss of O-glycosylation capacities.

GLYCOENGINEERING IN GLYCOSCIENCE

Glycoengineering of cell lines has vast potential to address a number of unmet needs in the glycosciences. As already mentioned, the CHO Lec, Ldl, Pgs, and Pig mutant cell lines (Chapters 12 and 49) served the research community for decades by providing defined alterations in glycosylation that enable studies of the functional roles of glycans. For example, CHO and HEK293-T mutant cells with MGAT1 deficiency have been widely used to produce recombinant proteins with homogeneous N-glycans suitable for crystallization studies.

Moreover, targeting glycosylation genes in whole organisms has provided immense insight into the importance of glycosylation and has revealed biological functions of specific glycosylation genes (Chapter 41). However, discovery of distinct biological functions of specific glycans, and the molecular mechanism(s) involved in multicellular organisms, is complicated by cell-type regulation of glycosylation and the cellular heterogeneity of tissues. Cell lines help to answer certain specific questions and complement whole organism studies.

Precise gene editing provides vast opportunities for glycoengineering cell lines and designing new strategies to probe glycan functions. Truncation of O-glycan elongation was used to produce homogeneous simple O-glycoproteomes (SimpleCell strategy), which enabled enrichment and sensitive mapping of the O-GalNAc and O-Man glycoproteomes of human cell lines. The development of isogenic cell lines differing in only one glycosyltransferase gene allows comparative studies to explore the function of a particular glycan or glycosylation pathway. For example, truncation of O-GalNAc glycans by targeting of the COSMC chaperone (Chapter 10) induced oncogenic features (proliferation, growth, and invasive behavior) of human nontransformed keratinocytes, an interesting finding in light of the frequent overexpression of truncated O-glycans (Tn, sialyl-Tn) in cancer. Large libraries of isogenic cells with comprehensive engineered glycosylation are used for cell-based glycan arrays and are useful for studying glycan binding in the natural context of the cell surface. The strategy has been expanded to organotypic tissue models used to address the more complex functions of distinct types of glycoconjugates (glycolipids, N-glycans, O-GalNAc, O-Fuc, O-Glc glycans) in human tissue formation.

A related approach is to use glycoengineering for discovery of host glycans required for microbial and viral infectivity. In one remarkable study using binding of Lassa virus to a haploid cell line, the large number of glycogenes required for synthesis of the extended O-Man glycan termed matriglycan (Chapter 13) that are bound by Lassa virus were identified and validated using a combination of selection for virus resistance and TALEN-mediated gene knockout.

FUTURE PERSPECTIVES

Glycoengineering of cells has entered a phase that may be described as "LEGO toying" because of the efficiency of precise gene editing. Entire glycosylation machineries can be deconstructed and rebuilt in various cell types. Cells show remarkable plasticity for engineering glycosylation pathways, with only a few glycosylation enzymes essential for cell growth in vitro. The essential functions for viability of mammalian cells are the initial steps of N-glycosylation (Chapter 9) and nuclear/cytosolic O-GlcNAc modifications (Chapter 19). These aside, essentially all glycosylation pathways in cells can be genetically deconstructed. Combining engineering of glycosylation pathways in creative ways by introducing completely foreign enzymes can be used to produce novel glycans for study.

Large-scale glycosylation screening and discovery strategies are possible. The CRISPR/Cas9 editing tool is particularly suited for multiplexed screening strategies, and whole-genome lentiviral-based knockout libraries have already been used for screening mutations that result in altered biological function. Although knockdown strategies have been successful for screening biological functions of glycogenes in multiple organisms (worms, flies, frogs, and zebrafish), these have generally not been effective in mammalian cells because of the low efficiency of knockdown. It is now possible to apply whole-glycogenome screening strategies in cell lines to probe and dissect the roles of glycosylation. These tools are dramatically improving options for dissection of structure–function relationships in the field.

Glycoengineered cell lines and validated targeting constructs (knockout, knock-in, and mutated) including libraries for screening will become important community resources that will advance the glycoscience field and help disseminate and integrate glycosciences more broadly in biology. Large libraries of engineered isogenic cells with subtle differences of all types of glycosylation are available for dissection of glycan functions using different assays not limited to binding. Expanding these to organotypic tissue models is providing deeper insights. The ability to produce glycoproteins with a large variety of glycans opens up for unbiased testing of different glycoforms and determines the optimal design for therapeutic uses. Glycans can be custom designed, for example, to improve homogeneity, circulation time, targeting to select organs, or stimulating immunity.

A word of caution, however: Whereas glycoengineering by knockout is fairly straightforward, there is still considerable work needed to establish methods to build robust complex glycosylation capabilities requiring multiple gene insertions in cells. Here, activation of endogenous genes in mammalian cells may provide a solution. Gene editing technologies caused a revolution in the glycoscience field, and we have only begun to see the new possibilities for manipulating glycosylation in cells and organisms and for exploiting glycoengineering in therapeutic glycoproteins and biologics.

Finally, the well-established power and cost efficiency of using bacteria as reactors for production of proteins is now being harnessed to generate designer glycoproteins. There remains much to be explored in this rapidly developing area, but the potential is enormous.

ACKNOWLEDGMENTS

The authors acknowledge contributions to previous versions of this chapter by Catherina Steenhoft and appreciate helpful comments and suggestions from Jenny Mortimer.

FURTHER READING

Wacker M, Linton D, Hitchen PG, Nita Lazar M, Haslam SM, North SJ, Panico M, Morris HR, Dell A, Wren BW, et al. 2002. N-linked glycosylation in *Campylobacter jejuni* and its functional transfer into *E. coli*. *Science* **298:** 1790–1793. doi:10.1126/science.298.5599.1790

Hamilton SR, Gerngross TU. 2007. Glycosylation engineering in yeast: the advent of fully humanized yeast. *Curr Opin Biotechnol* **18**: 387–392. doi:10.1016/j.copbio.2007.09.001

Malphettes L, Freyvert Y, Chang J, Liu PQ, Chan E, Miller JC, Zhou Z, Nguyen T, Tsai C, Snowden AW, et al. 2010. Highly efficient deletion of *FUT8* in CHO cell lines using zinc-finger nucleases yields cells that produce completely nonfucosylated antibodies. *Biotechnol Bioeng* **106**: 774–783. doi:10.1002/bit.22751

North SJ, Huang HH, Sundaram S, Jang-Lee J, Etienne AT, Trollope A, Chalabi S, Dell A, Stanley P, Haslam SM. 2010. Glycomics profiling of Chinese hamster ovary cell glycosylation mutants reveals N-glycans of a novel size and complexity. *J Biol Chem* **285**: 5759–5775. doi:10.1074/jbc.m109.068353

Baker JL, Celik E, DeLisa MP. 2013. Expanding the glycoengineering toolbox: the rise of bacterial N-linked protein glycosylation. *Trends Biotechnol* **31**: 313–323. doi:10.1016/j.tibtech.2013.03.003

Bosch D, Castilho A, Loos A, Schots A, Steinkellner H. 2013. N-glycosylation of plant-produced recombinant proteins. *Curr Pharm Des* **19**: 5503–5512. doi:10.2174/1381612811319310006

Merritt JH, Ollis AA, Fisher AC, DeLisa MP. 2013. Glycans-by-design: engineering bacteria for the biosynthesis of complex glycans and glycoconjugates. *Biotechnol Bioeng* **110**: 1550–1564. doi:10.1002/bit.24885

Strasser R, Altmann F, Steinkellner H. 2014. Controlled glycosylation of plant-produced recombinant proteins. *Curr Opin Biotechnol* **30**: 95–100. doi:10.1016/j.copbio.2014.06.008

Castilho A. 2015. Glyco-engineering. Preface. *Methods Mol Biol* **1321**: v–vii. doi:10.1007/978-1-4939-2760-9

Laukens B, De Visscher C, Callewaert N. 2015. Engineering yeast for producing human glycoproteins: where are we now? *Future Microbiol* **10**: 21–34. doi:10.2217/fmb.14.104

Laukens B, De Wachter C, Callewaert N. 2015. Engineering the *Pichia pastoris* N-glycosylation pathway using the GlycoSwitch technology. *Methods Mol Biol* **1321**: 103–122. doi:10.1007/978-1-4939-2760-9_8

Yang Z, Wang S, Halim A, Schulz MA, Frodin M, Rahman SH, Vester-Christensen MB, Behrens C, Kristensen C, Vakhrushev SY, et al. 2015. Engineered CHO cells for production of diverse, homogeneous glycoproteins. *Nat Biotechnol* **33**: 842–844. doi:10.1038/nbt.3280

Dabelsteen S, Pallesen EMH, Marinova IN, Nielsen MI, Adamopoulou M, Rømer TB, Levann A, Andersen MM, Ye Z, Thein D, et al. 2020. Essential functions of glycans in human epithelia dissected by a CRISPR-Cas9-engineered human organotypic skin model. *Dev Cell* **54**: 669–684. doi:10.1016/j.devcel.2020.06.039

Schjoldager KT, Narimatsu Y, Joshi HJ, Clausen H. 2020. Global view of human protein glycosylation pathways and functions. *Nat Rev Mol Cell Biol* **21**: 729–749. doi:10.1038/s41580-020-00294-x

Narimatsu Y, Büll C, Chen YH, Wandall HH, Yang Z, Clausen H. 2021. Genetic glycoengineering in mammalian cells. *J Biol Chem* **296**: 100448. doi:10.1016/j.jbc.2021.100448

57 Glycans in Biotechnology and the Pharmaceutical Industry

Peter H. Seeberger, Darón I. Freedberg, and Richard D. Cummings

Several classes of successful commercial products are based on isolated or synthetic glycans or agents that alter their expression and recognition. This chapter summarizes the use of glycans as vaccines and therapeutics. Applications of glycan mimics as drugs are also discussed.

GLYCANS AS COMPONENTS OF SMALL-MOLECULE DRUGS

Many well-known small-molecule drugs, such as antibiotics and anticancer therapeutic agents, are natural products that contain glycans as part of their core structure and/or as a sugar side chain (i.e., a glycoside). Some modern examples of natural products that bear glycan side chains

published by Cold Spring Harbor Laboratory Press; doi: 10.1101/glycobiology.4e.57

are shown in Figure 57.1. The well-established area of natural product chemistry will not be reviewed in detail here, though many natural products are glycosylated such as the well-known digitalis. Modified glycans gave rise to synthetic drugs such as small-molecule inhibitors of influenza virus neuraminidase (see also Chapter 55). Recent advances in the functional understanding of carbohydrate–protein interactions have enabled the development of glycomimetics, a new class of small-molecule drugs that are briefly described.

Rational Drug Design—Small-Molecule Inhibitors of Influenza Virus Neuraminidase

Influenza virus has two major surface proteins, hemagglutinin and neuraminidase (see Chapter 34). The hemagglutinin initiates infection by binding to cell-surface sialic acids. The neuraminidase assists virus release by cleaving sialic acids to prevent unwanted retention of newly synthesized virus on the cell surface. Neuraminidase may also function during the invasion phase by removing sialic acids on soluble mucins that would otherwise inhibit cell-surface binding. Because neuraminidase is essential to the viral life cycle, based on the crystal structure of the enzyme a rational drug design program yielded zanamivir (Relenza). The addition of a bulky guanidino side chain at C-4 of a previously known neuraminidase inhibitor, 2-deoxy-2,3-dehydro-*N*-acetyl-neuraminic acid, markedly increased the affinity for influenza neuraminidase, without affecting host-cell neuraminidases (Figure 57.2; see also Chapter 55). Relenza blocks the influenza virus life cycle by interrupting the spread of the virus during the early phase of an infection and subsequently preventing infection. Because of poor oral availability, Relenza has to be inhaled to work at the mucosal sites of infection in the upper airway. Because of the ease of use the orally available drug oseltamivir (Tamiflu) achieves the same effects and has taken over most of the market.

The fear of avian influenza virus ("bird flu") spreading into human populations has prompted stockpiling of Tamiflu. Fortunately, widespread use of the drug has not become necessary to date. The development of Tamiflu is a textbook example for rational glycan-based drug design resulting in a powerful drug against a devastating disease.

THERAPEUTIC GLYCOPROTEINS

Most biotherapeutic products are glycoproteins, including erythropoietin, various other cytokines, antibodies, glycosyltransferases, and glycosidases. This class of molecules sells in the tens of billions of U.S. dollars annually worldwide. Therapeutic glycoproteins or glycan-processing/recognizing proteins are typically produced recombinantly in cell culture systems or, less commonly, in the milk of transgenic animals. Control of glycosylation is of major importance during the development of these drugs, because their glycan chains have marked effects on stability, activity, antigenicity, and pharmacodynamics in intact organisms. Frequently, glycosylation must be optimized to ensure prolonged circulatory half-life in the blood. Manipulation of glycans to promote targeting to specific tissues and cell types has also been a useful element of drug design.

Optimizing Glycans of Therapeutic Glycoproteins for Prolonged Serum Half-Life

Erythropoietin (EPO) is the most successful biotechnology product to date. It is a circulating cytokine that binds to the erythropoietin receptor, inducing proliferation and differentiation of erythroid progenitors in the bone marrow. EPO was developed to treat anemias caused by bone marrow suppression after chemotherapy or lack of erythropoietin (e.g., renal failure). Natural and recombinant forms of erythropoietin carry three sialylated complex N-glycans and one

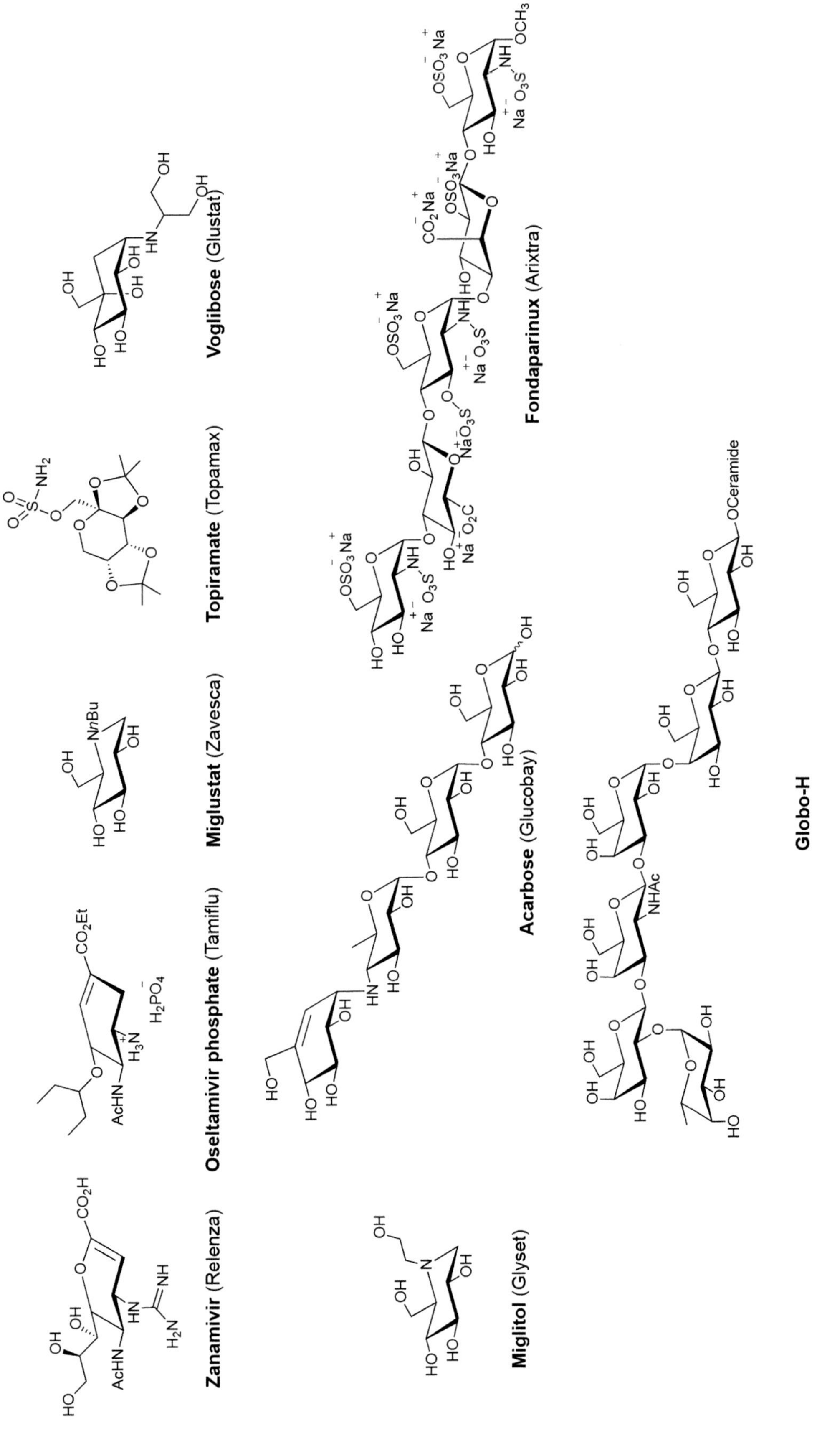

Zanamivir (Relenza)

Oseltamivir phosphate (Tamiflu)

Miglustat (Zavesca)

Topiramate (Topamax)

Voglibose (Glustat)

Miglitol (Glyset)

Acarbose (Glucobay)

Fondaparinux (Arixtra)

Globo-H

FIGURE 57.1. Examples of natural products that contain glycan components. Streptomycin and erythromycin A are antibiotics, doxorubicin is chemotherapeutic drug, and digoxin is used to treat cardiovascular disease.

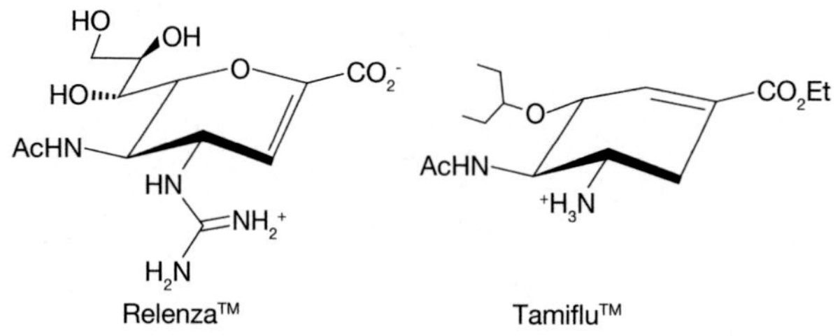

FIGURE 57.2. The synthetic influenza neuraminidase inhibitors Relenza and Tamiflu.

sialylated O-glycan. Although in vitro the activity of deglycosylated erythropoietin is comparable to that of the fully glycosylated molecule, its activity in vivo is reduced by ~90%, because poorly glycosylated EPO is rapidly cleared by filtration in the kidney. Undersialylated EPO is also rapidly cleared by galactose receptors in hepatocytes and macrophages (see Chapter 31). Fully sialylated chains and increased tetra-antennary branching reduce these problems and increase EPO activity in vivo nearly tenfold. Addition of an N-glycosylation site also increases half-life and activity in vivo. Covalently linking polyethylene glycol to the protein also reduces clearance by the kidney.

Erythropoietin is unusual because it is small enough to be cleared by the kidney if it is underglycosylated. For most glycoprotein therapeutics, a more important consideration is minimizing clearance by galactose-binding hepatic receptors by ensuring full sialylation of glycans. Glycans greatly influence the efficacy of these drugs; therefore, controlling glycosylation during production is crucial considering regulatory requirements for batch-to-batch product consistency. Changes in culture pH, the availability of precursors and nutrients, and the presence or absence of various growth factors and hormones can each affect the extent of glycosylation, the degree of branching, and the completeness of sialylation. Sialidases and other glycosidases that are either secreted or released by dead cells can also degrade the previously intact product in the culture medium. These issues were hotly debated with the advent of "biosimilars" or generic versions of glycoproteins. The need to prove composition has fueled efforts devoted to glycan analysis and sequencing.

Impact of Glycosylation on Licensing and Patentability of Biotherapeutic Agents

Patenting new therapeutics is typically based on the composition of matter in the claimed molecule. Small molecules of defined structure and nonglycosylated proteins are easily captured in this manner. However, glycoproteins, especially those with multiple glycosylation sites, render it virtually impossible to obtain preparations that contain only a single glycoform. For examples of glycosylated therapeutics, please see Table 57.1. Thus, most biotherapeutic glycoproteins consist of a mixture of glycoforms. Licensing bodies allow for a certain range of variation in glycoforms and complexity of the mixture. However, the manufacturer and the agency must agree on the extent such variation is acceptable for a given drug formulation. Biopharmaceutical companies therefore spend considerable effort in assuring that their products fall within these defined ranges, once these are approved by licensing bodies. The inherent difficulty in reproducing complex glycoform mixtures also complicates efforts to make generic forms of recombinant glycoprotein drugs. Given the complexities of producing glycotherapeutic agents in mammalian cells, licensing agencies use consistency in glycoform composition as an indirect measure of the quality of process control in production. Glycosylation differences can have implications for the patentability of agents in which the polypeptide remains constant. Marked glycosylation differences have been used to define agents as being unique. However, it is usually necessary to show that the differences in glycosylation being claimed have a correspondingly significant effect in changing the functionality of the drug in question. The associated pharmaceutical licensing and legal issues are rapidly evolving to keep pace with scientific advances in this area.

TABLE 57.1 Examples of glycan-based drugs, their target diseases, and modes of action

Drug	Source	Target disease	Mode of action
Targeting sialic acids			
zanamivir (Relenza)	Biota/ GlaxoSmithKline	influenza type A and B	inhibits neuraminidase
oseltamivir (GS 4104, Tamiflu)	Gilead/Roche	chemoprophylaxis	inhibits neuraminidase
Targeting glycosaminoglycans			
heparin	multiple brands	anticoagulant; possible value in cancer metastasis prevention	activates antithrombin; inhibits heparanase and selectins and blocks interactions between growth factors and heparan sulfate
hyaluronan (HA)	multiple brands	ocular surgery; osteoarthritis; plastic surgery	tissue space filler; anti-inflammatory agent
laronidase (Aldurazyme)	Genzyme	mucopolysaccharidosis type I (MPSI); α-idu-ronidase deficiency	ERT
galsulfase (Naglazyme)	Biomarin	mucopolysaccharidosis type VI; arylsulfatase B deficiency	ERT
hyaluronidase (Cumulase)	Halozyme	in vitro fertilization; in development as an adjuvant for cancer chemotherapy	degrades HA around oocytes improving fertilization; degrades HA in tumors to decrease intratumor pressure
ALZ-801	phase III trials (Alzheon, Inc.)	amyloid diseases, Alzheimer's disease, and possibly other amyloidoses	binds to amyloid plaque, blocks its formation
Targeting glycosphingolipids			
N-butyl-deoxynojirimycin (DNJ) (Miglustat, Zavesca)	Acetelion	type 1 Gaucher's disease; Niemann–Pick's disease type C; late-onset Tay–Sach's disease; type 3 Gaucher's disease	substrate reduction therapy; inhibits glucosylceramide synthase
imiglucerase (Cerezyme)	Genzyme	type 1 Gaucher's disease	ERT
β-agalsidase (Fabrazyme)	Genzyme	Fabry disease; α-galactosidase A deficiency	ERT
Others			
acarbose (Glucobay)	Bayer	type 2 diabetes	blocks intestinal α-glucosidases involved in digestion of dietary glycans
alglucosidase alfa (Myozyme)	Genzyme	Pompe's disease (glycogen storage disease); α-glucosidase A deficiency	ERT
allosamidin	Industrial Research	insecticide	chitinase inhibitor

Modified and updated from Brown JR, Crawford BE, Esko JD. 2007. *Crit Rev Biochem Mol Biol* **24:** 481–515.
Compounds targeted at microbial glycans, such as the aminoglycoside antibiotics or other inhibitors of cell wall assembly, have not been included.
(ERT) Enzyme replacement therapy.

GLYCOSYLATION ENGINEERING

There are limits as to how much of a biotherapeutic glycoprotein an animal cell line can produce. Production becomes an issue in cases where very large amounts of a particular glycoprotein are needed. Glycoprotein production in plants or yeast is attractive but makes it necessary to eliminate risks arising from the nonhuman glycans of plant and fungal cells that could cause excessively rapid clearance and/or antigenic reactions. Many plant and yeast glycans are immunogenic and elicit glycan-specific IgE and IgG antibodies in humans when delivered parenterally. A variety of mammalian genes have been introduced into yeast and/or genes that are producing nonhuman glycosylation have been eliminated. Extensively engineered yeast strains are capable of producing biantennary N-linked glycans with the human sialic acid *N*-acetylneuraminic acid (Neu5Ac) but the productivity of such yeast strains is often low. Efforts to engineer yeast to make human-like O-glycans are underway.

Plants and algae as well as insect cell lines have also been used to engineer recombinant glycoproteins, but, as in yeast, the glycans produced by plants differ from those found in vertebrates. The antigenic differences that arise in recombinant glycoproteins produced in plants become less problematic if used for topical or oral administration, because humans are normally exposed to plant glycans in the diet. The cost of production is much lower than in animal cell culture systems and animal sera are not needed. As in yeast, "humanizing plants" with respect to glycosylation may allow the production of nonimmunogenic glycoproteins. Chemical methods for synthesizing entire glycoproteins from scratch have been developed and single glycoforms of EPO have been prepared by total synthesis. Given the complexity of glycoprotein synthesis, scale-up of these processes is challenging and an area of intense research activity (see Chapter 49).

GLYCAN THERAPEUTIC APPROACHES TO METABOLIC DISEASES

Salvage versus De Novo Synthesis

All monosaccharides needed for cellular glycan synthesis can be obtained from glucose through metabolic interconversions (see Chapter 4). Alternatively, monosaccharides can be derived from the diet or salvaged from degraded glycans. The relative contributions of different sources can vary with the cell type. For instance, even though all mammalian cells use sialic acid, only some contain high amounts of UDP-GlcNAc epimerase/*N*-acetylmannosamine kinase (GNE), which is required for the de novo synthesis of CMP-sialic acid. But sialic acid salvage from degraded glycans is quite efficient, decreasing the demand on the de novo pathway. Similarly, galactose, fucose, mannose, *N*-acetylglucosamine, and *N*-acetylgalactosamine can come from diet or be salvaged for glycan synthesis, whereas glucuronic acid, iduronic acid, and xylose cannot. All monosaccharides derived from diet or degraded glycans can be catabolized for energy, and, again, cells vary in their reliance on the different pathways.

The variable contributions of these pathways are important for therapy of some diseases. Patients with congenital disorder of glycosylation type Ib (CDG-Ib), who are deficient in phosphomannose isomerase, benefit greatly from oral mannose supplementation to bypass the insufficient supply of glucose-derived mannose 6-phosphate. A few CDG-IIc patients defective in fucose transporters have been treated with fucose to restore synthesis of sialyl-Lewis x on leukocytes (see Chapter 42). Some patients with Crohn's disease show clinical improvement with oral *N*-acetylglucosamine supplementation, but the mechanism is unknown. Mice deficient in GNE activity have kidney failure, but providing *N*-acetylmannosamine in the diet prevents this outcome. Clinical trials using *N*-acetylmannosamine to treat GNE-deficient patients with hereditary inclusion body myopathy type II (HIBM-II) have been conducted but have yielded inconclusive results.

Special Diets

Some monosaccharides and disaccharides can be toxic to humans who lack specific catabolizing enzymes. For example, people who lack fructoaldolase (aldolase B) accumulate fructose-1-phosphate, which ultimately causes ATP depletion and disrupts glycogen metabolism. Prolonged fructose exposure in these people can be fatal, and fructose-limited diets are critical. Deficiencies in the ability to metabolize galactose (see Chapter 4) are mostly due to a severe reduction in galactose-1-phosphate uridyl transferase activity and cause galactosemia. Although these patients are asymptomatic at birth, ingesting milk leads to vomiting and diarrhea, cataracts, hepatomegaly, and even neonatal death. Low-galactose or galactose-free diets can prevent these life-threatening symptoms. However, even these diets do not prevent unexplained long-term complications, which include speech and learning disabilities and ovarian failure in females with galactosemia.

Infants hydrolyze lactose (Galβ1-4Glc) quite well, but the level of intestinal lactase catalyzing the breakdown of lactose can be much lower or absent in adults because of down-regulation of lactase gene expression after childhood. About two-thirds of the human population has lactase nonpersistence, making milk products a dietary annoyance. Unabsorbed lactose provides an osmotic load and is metabolized by colonic bacteria, causing diarrhea, abdominal bloating, flatulence, and nausea. Lactase persistence has evolved in certain pastoral populations from northwestern Europe, India, and Africa, allowing milk consumption in adult life. However, many adults either avoid lactose-containing foods or use lactase tablets to improve lactose digestion.

Substrate Reduction Therapy

The failure to turn over glycans by lysosomal degradation driven by a plethora of glycoside hydrolases causes serious problems for patients with lysosomal storage disorders. Deficiencies in individual lysosomal enzymes or their trafficking lead to pathological accumulation of their substrates in inclusion bodies inside the cells (see Chapter 41). One approach to treating these disorders is to inhibit initial glycan synthesis, a strategy termed substrate reduction therapy (SRT). Reduced synthesis of the initial compound decreases the load on the impaired enzyme, and some patients show significant clinical improvement. A small-molecule drug used for SRT is *N*-butyldeoxynojirimycin (or *N*-butyl-DNJ) (miglustat, Zavesca), which was approved in 2002 to treat Gaucher's disease (glucocerebrosidase deficiency).

Lysosomal Enzyme Replacement Therapy

Another approach for treating lysosomal storage disorders is enzyme replacement therapy. Unlike most therapeutic glycoproteins that interact with target receptors on the surface of cells, lysosomal enzymes developed for replacement therapy must be delivered intracellularly to lysosomes, their site of action. During the normal biosynthesis of lysosomal enzymes, their N-glycans become modified with mannose 6-phosphate (Man-6-P) residues, which target them to lysosomes using Man-6-P receptors (see Chapter 30). The challenge for enzyme replacement therapy is to get the enzymes targeted properly to lysosomes, where they can degrade accumulated substrate. Enzyme replacement therapy for Gaucher's disease targets the lysosomes of macrophages via the cell-surface mannose receptor (see Chapter 31). The four recombinant enzyme products imiglucerase (approved in 1995), velaglucerase (approved in 2010), taliglucerase alfa (Elelyso, approved in 2012), and eliglustat (Cerdelga, approved in 2014) are marketed to treat Gaucher's disease.

The success of glucocerebrosidase treatment stimulated the development of lysosomal enzymes for treatment of other lysosomal storage diseases such as Fabry's disease, mucopolysaccharidoses type I, II, and VI, and Pompe's disease. The replacement therapies clearly have beneficial effects and prolong life but are extremely expensive.

Chaperone Therapy

A third approach for treating lysosomal storage disorders takes advantage of the fact that some genetic defects lead to misfolding of the encoded enzyme in the endoplasmic reticulum (ER). Low-molecular-weight competitive inhibitors of some of these enzymes can act as "chaperones," which stabilize the folded enzyme in the ER and effectively rescue the mutation and increase the steady state concentration of active enzyme in the lysosome. The dose of the inhibitor must be carefully adjusted to ensure that the inhibitory effects on enzyme function do not overshadow beneficial effects on folding. Only a low level of enzyme restoration is needed to significantly reduce the accumulation of undigested glycan substrates, indicating that lysosomal hydrolases are normally present in large catalytic excess.

THERAPEUTIC APPLICATIONS OF GLYCOSAMINOGLYCANS

The use of purified glycans as therapeutics has received less attention than the development of glycoprotein-based treatments. Difficulties in establishing structure–activity relationships because of the large number of stereocenters and functional groups, undesirable pharmacokinetics of available formulations, poor oral absorption of the compounds, and low-affinity interactions with drug targets have limited their development. Some successful glycan drugs, such as the anticoagulant heparin, are given by injection, although efforts are under way to convert heparin into an orally absorbable form by complexing it with positively charged molecules. It may be possible to deliver other hydrophilic and/or negatively charged glycan drugs in this way to allow penetration of the intestinal barrier. Glycans are also sometimes attached to hydrophobic drugs to improve the drugs' solubility and ability to cross biological membranes to enter cells and alter their pharmacokinetics.

The anticoagulant heparin is, as discussed in Chapters 16, 43, and 46, one of the most widely prescribed drugs today. Heparin binds and activates antithrombin, a protease inhibitor of the coagulation cascade. Antithrombin activation leads to rapid inhibition of thrombin and factor Xa, shutting down the production of fibrin clots. Billions of doses of heparin (several metric tons) are produced by autodigestion of pig intestines, followed by graded fractionation of the products. Unfractionated heparin produces a variable anticoagulant response as it also binds to several plasma, platelet, and endothelial proteins. Low-molecular-weight (LMW) heparins are derived from chemical or enzymatic cleavage of heparin to form smaller fragments. The pharmacological properties and the relative efficacy of the various LMW heparins are superior to those of unfractionated heparin and fewer secondary complications are reported. LMW heparins have replaced unfractionated heparins as therapeutic of choice in virtually all developed countries. In price-sensitive markets the unfractionated products are still heavily used. The preparation of recombinant heparin based on heparin biosynthesis enzymes is still under development. Arixtra, a synthetic heparin pentasaccharide that binds antithrombin exactly as isolated heparin is used to prevent deep-vein thrombosis and pulmonary embolism and has gained market share in recent years even though it is more costly.

To prevent excessive bleeding, rapid neutralization of heparin is desirable. Administration of the basic protein protamine, which binds to heparin, neutralizes its activity, and results in clearance of the complex by the kidney and liver. Heparin is also used to treat protein-losing enteropathy (PLE), likely working by competing for proinflammatory heparin-binding cytokines that trigger PLE in susceptible patients (see Chapter 43).

Hyaluronan (see Chapter 15) is a naturally occurring glycosaminoglycan that is extensively used in surgical applications. Because of its viscoelastic properties, hyaluronan has lubricating and cushioning properties that have made it useful for protecting the corneal endothelium during ocular surgery. Hyaluronan is also useful in postsurgical wound healing. The mechanism of action is not well-understood, but may involve hyaluronan-binding proteins that mediate cell adhesion (see Chapter 15). Intra-articular injections of hyaluronan are used to treat knee and hip osteoarthritis. Modest improvement in patients treated with hyaluronan, may be the result of a mechanical (as a viscosupplement) and/or a biological (via signaling pathways) effect. Hyaluronan is used in very large quantities as a tissue filler in cosmetic medicine.

GLYCONUTRIENTS

"Glyconutrient" is a term used by the nutritional supplement industry to describe some of their products with wide-ranging claims concerning potential benefits. In most cases, these claims have not been substantiated through placebo-controlled, double-blind trials with defined, quantifiable outcomes. Much work is needed in this area to obtain insight into the potential role of dietary glycans on human health and to help consumers make informed decisions regarding their use.

Mixtures of plant polysaccharides such as larch bark arabinogalactan and glucomannan are often termed "glyconutrients" that are claimed to contain "essential monosaccharides" needed for "cell communication." Because all monosaccharides can be made in vivo from glucose (except in patients with rare genetic deficiencies; see Chapter 42), none of the other monosaccharides are actually known to be "essential." Moreover, arabinogalactan and glucomannan are not degraded to available monosaccharides in the stomach or small intestine. Instead, anaerobic bacteria in the colon metabolize them and produce short-chain fatty acids. No peer-reviewed clinical studies support the efficacy of such "glyconutrients" for any disease or condition. Nevertheless, the following examples demonstrate how dietary glycans might have beneficial effects.

Glucosamine and Chondroitin Sulfate

Glucosamine (often mixed with chondroitin sulfate) has been promoted to relieve symptoms of osteoarthritis, which involves the age-dependent erosion of articular cartilage. Cartilage provides a cushion between the bones to minimize mechanical damage, and a net loss of cartilage occurs when the degradation rate exceeds the synthetic rate. Claims that glucosamine improves osteoarthritis symptoms and restores partially the structure of the eroded cushion in the knees are controversial. Superficially, this would seem to make sense, because primary glycans of cartilage include hyaluronan (see Chapter 15) and chondroitin sulfate, both of which contain hexosamines within their structure (see Chapter 16). Nevertheless, veterinarians report positive results after treating animals with glucosamine for more than two decades.

Positive effects of chondroitin sulfate on osteoarthritis are not well-documented. It remains unclear how the acidic chondroitin sulfate polymer can be absorbed and delivered to its proposed site of action.

Xylitol and Sorbitol in Chewing Gum

Some studies suggest that chewing gum containing sugar alditols, such as xylitol and sorbitol, can help control the development of dental caries. The benefit of these reduced sugars seems to be based on stimulation of salivary flow and an antimicrobial effect by inhibiting a glucosyltransferase that blocks glucose utilization by *Streptococcus mutans*. Xylitol also inhibits the expression and secretion of proinflammatory cytokines from macrophages and inhibits the growth of *Porphyromonas gingivalis*, one of the suspected causes of periodontal disease.

Milk Oligosaccharides

Human milk contains ~70 g/L of lactose and 5–10 g/L of free oligosaccharides. More than 130 different glycan species have been identified with lactose at the reducing end, including poly-*N*-acetyllactosamine units. Some glycans are α2-3- and/or α2-6-sialylated and/or fucosylated in α1-2, α1-3, and/or α1-4 linkages. In contrast, bovine milk, the typical mainstay in human infant formulas, contains much smaller amounts of fucose oligosaccharides. These differences may account for some of the physiological advantages seen for breastfed versus formula-fed infants. The glycans may also favor growth of a nonpathogenic bifidogenic microflora and/or block pathogen adhesion that causes infections and diarrhea. Surprisingly, a substantial number of human milk oligosaccharides remain almost undigested in the infant's intestine and are excreted intact into the urine. Whether supplementing infant formula with specific, biologically active free glycans enhances infant health is unknown.

GLYCANS AS VACCINE COMPONENTS

Microbial Vaccines

Polysaccharide vaccines consisting solely of glycan components typically elicit poor immunity, especially in infants. Since glycans are T-cell-independent antigens they do not effectively stimulate T-helper-dependent activation and class switching of B-cell-mediated immunity. Conjugate vaccines consisting of glycans coupled to carrier proteins have proven to be highly effective. Three major conjugate vaccines are marketed today: *Haemophilus influenzae* type b (Hib) causes an acute lower respiratory infection among young children that constitute a high-risk group for *H. influenzae* type B infections. Consequently, the Hib polysaccharide vaccine introduced in 1985 was withdrawn from the market in 1988 and replaced by capsular polysaccharide (CPS-)protein conjugate vaccine formulations. A conjugated form of a Hib-derived oligosaccharide coupled to a protein carrier is part of routine vaccination schedules and has been so successful that infectious diseases caused by this bacterium are nearly eradicated in vaccinated populations. The first semisynthetic glycoconjugate vaccine QuimiHib marketed in Cuba contains glycan chains with an average length of 16 monosaccharide residues. Pneumococcal conjugate vaccines have been developed to cover an increasing number of serotypes, and current formulations are 10- (Synflorix, GlaxoSmithKline) and 13-valent (Prevnar13, Pfizer). Prevnar 13 provides protection against serotypes that account for >70% of cases of invasive pneumococcal disease worldwide and is the best-selling vaccine with revenues of six billion U.S. dollars in 2019. Conjugate vaccines to protect from *Neisseria meningitides* are also very successful on the market. Because of the fast onset and rapid progression of meningococcal infections, vaccination is required to protect against this disease. Several conjugated CPS vaccines are licensed in different parts of the world: The tetravalent serogroup A, C, W, and Y (Menactra, Menveo, and Nimenrix) and a few monovalent vaccines based on serogroup C CPS (Meningitec, Menjugate, NeisVac-C). Two combination vaccines for *N. meningitidis* and *H. influenzae* type b (Hib) are available against meningococcal serogroups C/Y (MenHibrix) and against meningococcal serogroup C (Menitorix). A monovalent serogroup A vaccine (MenAfriVac) is widely used in the sub-Saharan meningitis belt of Africa.

Currently, several new vaccines based on synthetic oligosaccharide antigens are being developed to protect children and the elderly from a variety of bacterial infections. Vaccines to protect from hospital acquired infections caused by *Clostridium difficile* and *Klebsiella pneumoniae* that are increasingly antibiotic resistant are in preclinical evaluation.

Cancer Vaccines

Several carbohydrate-based cancer vaccines are at different stages of development to treat cancer. Ganglioside immunogens present on certain types of cancer cells such as gangliosides GM2 and GD2 in melanomas and Globo H in breast cancer are being explored. Treatment strategies targeting the shorter glycan sequences such as sialyl-Tn (sialylα2-6GalNAcα-) found on cancer mucins (see Chapter 44) have seen little progress in 20 years. The synthetic Globo H hexasaccharide (see Figure 57.1) resembling the breast and prostate cancer antigen reached Phase 3 clinical trials but failed to gain marketing approval. Further human clinical trials are ongoing. Instead of active immunization of immunocompromised patients, the use of humanized anti-glycan antibodies (passive immunization) is now explored in preclinical evaluations.

BLOCKING GLYCAN RECOGNITION IN DISEASE

Blocking Infection

As discussed in Chapter 34, many microbes and toxins bind to mammalian tissues by recognizing specific glycan ligands. Thus, small soluble glycans or glycan mimetics can be used to block the initial attachment of microbes and toxins to cell surfaces (or block their release), and thus prevent or suppress infection. Because many of these organisms naturally gain access through the airways or gut, the glycan-based drugs can be delivered directly without being distributed systemically. Milk oligosaccharides are natural antagonists of intestinal infection in infants (see above). Glycosylated polymers block the binding of viruses such as influenza. Although backed by a strong scientific rationale and robust in vitro studies, such "antiadhesive" or "mimicry" therapies have not yet found much practical application.

Inhibition of Selectin-Mediated Leukocyte Trafficking

When specific glycan-protein interactions are responsible for selective cell–cell interactions and a resulting pathology, then administration of small-molecule glycomimetics of the natural ligand is a useful means of intervention. Selectin-mediated recruitment of neutrophils and other leukocytes into sites of inflammation or ischemia/reperfusion injury involves specific selectin–glycan interactions in the vascular system (see Chapters 31 and 46). The use of sialyl-Lewis x tetrasaccharide derivatives failed because of poor oral availability and a short serum half-life. Glycomimetics that preserve the essential functionality of the parent tetrasaccharide but eliminate unwanted polar functional groups and synthetically cumbersome glycan components have been successful. The design of a monosaccharide glycomimetic starting from sialyl-Lewis x is shown in Figure 57.3. First, the sialic acid residue was replaced with a charged glycolic acid group, the *N*-acetylglucosamine residue was then replaced with an ethylene glycol linker, and finally the galactose residue was replaced with a linker moiety. The resulting glycomimetic had E-selectin binding affinity comparable to sialyl-Lewis x.

GMI-1070, targeting E-, P-, and L-selectin, yielded unconvincing results in Phase 3 clinical trials as a treatment for sickle cell crisis and is under further investigation.

TRANSFUSION AND TRANSPLANTATION REJECTION BY ANTIGLYCAN ANTIBODIES

A variety of glycans, including the classical A and B blood group determinants, can act as barriers to blood transfusion and transplantation of organs (Chapters 13 and 46). Rejection of mismatched blood or organs occurs because hosts have a high titer of preexisting antibodies against the glycan epitopes, presumably as a prior reaction to related structures found on bacteria or other microbes. In the case of the ABO blood groups, incompatibility is routinely managed by blood and tissue typing and finding an appropriate donor for the recipient. Bacterial enzymes can be used in vitro to remove the A and B blood group determinants from A and B red cells, converting them into "universal donor" O red cells.

A related problem is found in xenotransplantation (i.e., the transplantation of organs between species), which is actively being pursued as a solution for the shortage of human organs for patients. The animal donors of preference are pigs, because many porcine organs resemble

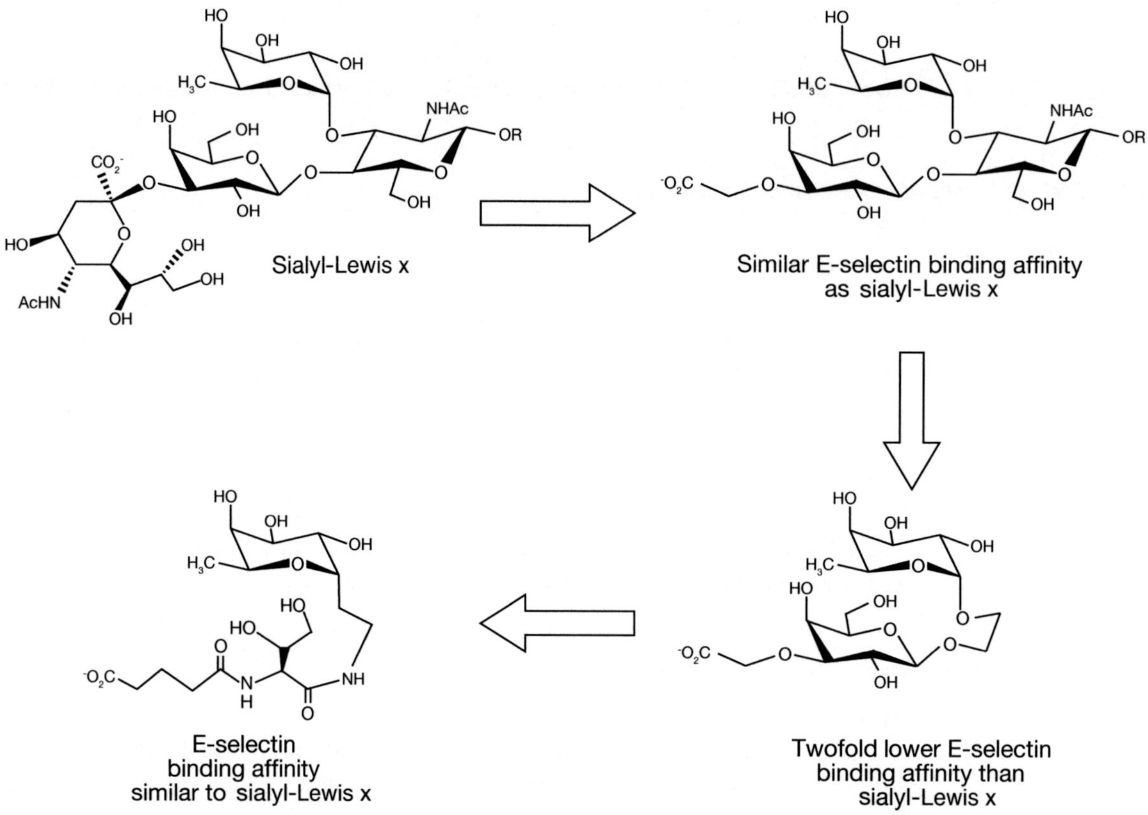

Sialyl-Lewis x

Similar E-selectin binding affinity as sialyl-Lewis x

Twofold lower E-selectin binding affinity than sialyl-Lewis x

E-selectin binding affinity similar to sialyl-Lewis x

FIGURE 57.3. Glycomimetic E-selectin inhibitors based on sialyl-Lewis x.

those of humans in size, physiology, and anatomy. However, unlike humans and certain other primates, pigs and most other mammals produce the terminal "α-Gal" epitope on glycoproteins and glycolipids. Because humans have naturally occurring high-titer antibodies in blood directed toward this epitope, this results in hyperacute rejection of porcine organ transplants, via reaction of the antibodies with endothelial cells of blood vessels. Attempts to prevent this reaction, include blood filtration over glycan affinity columns to remove xenoreactive antibodies and blockade of the interaction by infusing soluble competing oligosaccharides. Transgenic pigs lacking the reactive epitope have also been produced, as have animals with an excess of complement-controlling proteins on their cell surfaces. Pig organs also have high levels of the nonhuman sialic acid (Neu5Gc), against which most humans have antibodies. Even if this problem is solved, there are other glycan and protein structural differences between humans and pigs that cause later stages of graft rejection, thus necessitating immunosuppression.

ACKNOWLEDGMENTS

The authors acknowledge helpful comments and suggestions from Morten Thaysen-Andersen.

FURTHER READING

Kunz C, Rudloff S, Baier W, Klein N, Strobel S. 2000. Oligosaccharides in human milk: structural, functional, and metabolic aspects. *Annu Rev Nutr* **20:** 699–722. doi:10.1146/annurev.nutr.20.1.699

Gomord V, Chamberlain P, Jefferis R, Faye L. 2005. Biopharmaceutical production in plants: problems, solutions and opportunities. *Trends Biotechnol* **23:** 559–565. doi:10.1016/j.tibtech.2005.09.003

Joshi L, Lopez LC. 2005. Bioprospecting in plants for engineered proteins. *Curr Opin Plant Biol* **8:** 223–226. doi:10.1016/j.pbi.2005.01.003

Pastores GM, Barnett NL. 2005. Current and emerging therapies for the lysosomal storage disorders. *Expert Opin Emerg Drugs* **10:** 891–902. doi:10.1517/14728214.10.4.891

Beck M. 2007. New therapeutic options for lysosomal storage disorders: enzyme replacement, small molecules and gene therapy. *Hum Genet* **121:** 1–22. doi:10.1007/s00439-006-0280-4

Brown JR, Crawford BE, Esko JD. 2007. Glycan antagonists and inhibitors: a fount for drug discovery. *Crit Rev Biochem Mol Biol* **42:** 481–515. doi:10.1080/10409230701751611

Eklund EA, Bode L, Freeze HH. 2007. Diseases associated with carbohydrates/glycoconjugates. In *Comprehensive glycoscience* (ed. JP Kamerling, et al.), Vol. 4, pp. 339–372. Elsevier, New York. doi:10.1016/b978-044451967-2/00098-2

Hamilton SR, Gerngross TU. 2007. Glycosylation engineering in yeast: the advent of fully humanized yeast. *Curr Opin Biotechnol* **18:** 387–392. doi:10.1016/j.copbio.2007.09.001

Schultz BL, Laroy W, Callewaert N. 2007. Clinical laboratory testing in human medicine based on the detection of glycoconjugates. *Curr Mol Med* **7:** 397–416. doi:10.2174/156652407780831629

von Itzstein M. 2007. The war against influenza: discovery and development of sialidase inhibitors. *Nat Rev Drug Discov* **6:** 967–974. doi:10.1038/nrd2400

Schnaar RL, Freeze HH. 2008. A "glyconutrient sham". *Glycobiology* **18:** 652–657. doi:10.1093/glycob/cwm098

Ernst B, Magnani JL. 2009. From carbohydrate leads to glycomimetic drugs. *Nat Rev Drug Discov* **8:** 661–677. doi:10.1038/nrd2852

Wilson RM, Dong S, Wang P, Danishefsky SJ. 2013. The winding pathway to erythropoietin along the chemistry–biology frontier: a success at last. *Angew Chem* Int Ed Engl **52:** 7646–7665. doi:10.1002/anie.201301666

Bonam SR, Wang F, Muller S. 2019. Lysosomes as a therapeutic target. *Nat Rev Drug Discov* **18:** 923–948. doi:10.1038/s41573-019-0036-1

Seeberger PH. 2021. Discovery of semi- and fully-synthetic carbohydrate vaccines against bacterial infections using a medicinal chemistry approach. *Chem Rev* **121:** 3598–3626. doi:10.1021/acs.chemrev.0c01210

Smith BAH, Bertozzi CR. 2021.The clinical impact of glycobiology: targeting selectins, Siglecs and mammalian glycans. *Nat Rev Drug Discov* **20:** 217–243; 244. doi:10.1038/s41573-020-00093-1

58 Glycans in Nanotechnology

Martina Delbianco, Benjamin G. Davis, and Peter H. Seeberger

Nanomaterials offer tunable chemical and physical properties, such as electronic, photonic, and magnetic properties. Decoration of nanomaterials with glycans increases solubility and biocompatibility and lowers cytotoxicity, while allowing for multivalent glycan presentation. Given the central role of multivalency in glycobiology, glycosylated nanomaterials are interesting probes to study cellular, tissue, and organismal interactions. Nanomaterials purely composed of glycans, such as polysaccharide nanoparticles or nanocrystals, are interesting imaging agents, drug delivery systems, and tissue scaffolds that illustrate the potential of glycans in nanotechnology.

INTRODUCTION

Glycoproteins and glycolipids are natural glycoconjugates that take part in cellular communication, inflammation, and immune response using carbohydrate–protein or carbohydrate–carbohydrate interactions. Certain glycan sequences are characteristic markers of diseases such as cancer, asthma, and diabetes. The elucidation of molecular mechanisms requires tools that mimic the presentation of glycans on the cell surface.

Individual protein–carbohydrate interactions are often of low affinity and broad specificity, complicating the description of glycan function. Nature enhances specificity by utilizing

published by Cold Spring Harbor Laboratory Press; doi: 10.1101/glycobiology.4e.58

multivalent interactions. The number and the presentation of carbohydrate residues on a biomolecule are major determinants of binding avidity of ligands to cell-surface receptors. The transition from monovalent to multivalent is often associated with a larger variation in affinity/avidity, suggesting a "thresholding" effect, and in some cases cooperativity.

To elucidate those mechanisms, glycans have to be displayed in a scenario closer to that found on the cellular scale. "Nanotechnology" moves from the angstrom to the nanometer range (from $\sim 10^{-10}$ to $\sim 10^{-7}$ m), offering the tools to create, manipulate, and characterize structures on those scales.

Large glycoconjugates bearing multiple copies of a carbohydrate on various scaffolds, such as glycodendrimers or glycopolymers, have been generated to probe carbohydrate–protein interactions. Several glyconanomaterials with inherent high surface/volume ratios have been developed to allow for a greater contact surface area and explore multivalency effects. The integration of nanomaterials in the glycosciences has already enabled biomedical applications such as drug delivery systems, imaging agents, diagnostic platforms, or precise sensing tools that operate through biological mimicry. Further cooperation between the glycosciences and nanotechnology will improve our understanding of glycobiology.

TYPES AND APPLICATIONS OF GLYCONANOMATERIALS

Glyconanomaterials based on metal particles, semiconductors, or carbon-based scaffolds take advantage of the unique physical properties of the nanoscale, such as catalytic, photonic, electronic, or magnetic properties that are not seen in the bulk. The glycan portion ensures water solubility, exceptional stability in water and biological buffers, biocompatibility, structural diversity, and passive and active targeting properties. As a result, biocompatible nanomaterials offering a multivalent glycan presentation are generated for sensing and drug delivery applications.

Nanoformulations, exclusively based on polysaccharides, have been developed as well. Polysaccharides such as chitosan, dextran, hyaluronic acid, and heparin have given rise to polysaccharide-based nanoparticles (NPs) for pharmaceutical use with superior biocompatibility and biodegradability. Low toxicity, low cost, and easy chemical modifications are additional advantages of polysaccharide-based NPs compared to synthetic polymers. These glyconanomaterials are currently used for drug delivery and tissue engineering, and electronics and device applications are emerging.

INORGANIC NANOPARTICLES

Hybrid materials based on inorganic nanostructures and biomolecules are a major focus of nanotechnology. Iron oxide, noble metal, and semiconductor nanoparticles served as synthetic scaffolds to multimerize glycans and enhance the affinity for receptors (Figure 58.1). The physical properties, such as magnetism or fluorescence, of these hybrid materials have given rise to applications in sensing, delivery, and imaging.

Gold Nanoparticles

The unique optoelectronic properties and facile chemical modification make gold nanoparticles (AuNPs) an important tool to monitor biological binding events. The functionalization of AuNPs with glycans generates materials with high aqueous solubility/dispersibility and biocompatibility. The resonance between collective oscillations of electrons in AuNPs (plasmons) and the incident electromagnetic radiation, gives rise to localized surface plasmon resonance (LSPR).

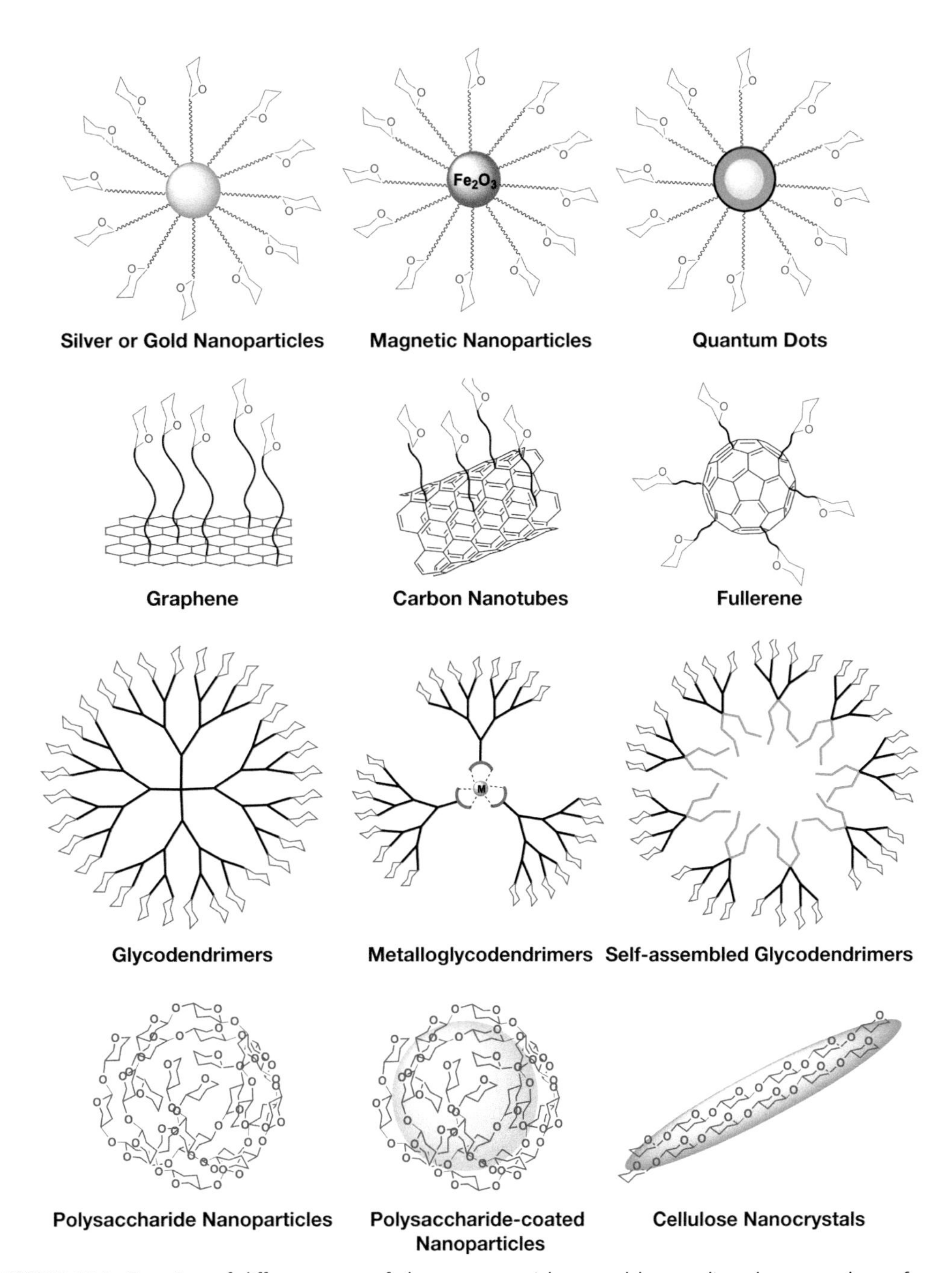

FIGURE 58.1. Overview of different types of glyconanomaterials created by coupling glycans to the surface of diverse nanomaterials.

Resonance frequencies of gold surface plasmon bands lie in the visible region (400–750 nm), giving rise to color effects. The high surface area/volume ratio results in high LSPR sensitivity and colorimetric changes, making AuNPs extremely valuable analytical reporters.

AuNPs are produced by reduction of gold salts with sodium citrate and surface modification with "capping" agents. Size, shape, and morphology are tuned by adjusting reaction conditions, allowing access to the near-infrared (NIR) spectrum. Colorimetric carbohydrate–lectin analyses exploiting LSPR of AuNPs have typically used 10-nm particles capped with thiol-poly(ethylene glycol) (thiol-PEG) aldehydes decorated with simple mono- or disaccharides. AuNPs *Ricinus communis* agglutinin (RCA_{120}) or cholera toxin induced a reversible color change.

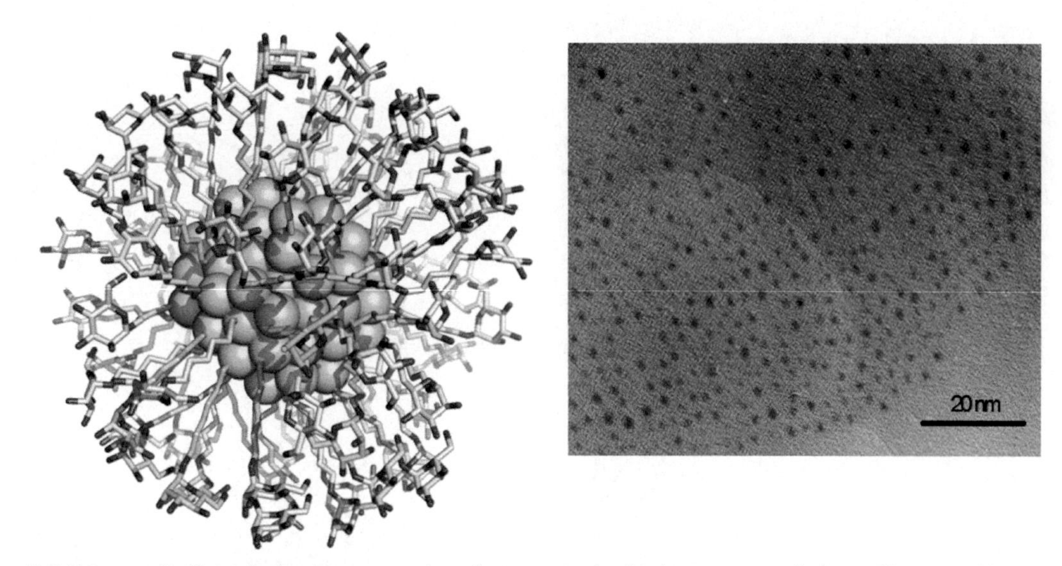

FIGURE 58.2. (*Left*) A calculated representation of a 2-nm-sized gold glyconanoparticle formed by 102 gold atoms and coated with 44 molecules of 5-mercaptopentyl α-D-mannopyranoside and (*right*) the corresponding transmission electron microscopy (TEM) image.

Direct visualization is a particularly attractive feature in biology. Stable colloidal gold was first described by Faraday in 1857, but applied in biology only in the 1970s when immunogold-staining procedures were used to observe microorganisms by transmission electron microscopy (TEM). Immunogold staining with mannosylated AuNPs permitted users to probe complement activation and opsonization processes in macrophage-mediated endocytosis. Carbohydrate–protein interactions were visualized by TEM thanks to the mannosylated AuNPs ability to target *Escherichia coli* type 1 pili mannose-specific receptors.

Very small gold glyconanoparticles can be prepared by reducing a gold salt in the presence of glycosylated thiol ligands (Figure 58.2). Ligand density and composition can be adjusted precisely. These glyconanoparticles conserve the chemical properties of the ligands and can be advantageously characterized by ultraviolet-visible (UV-Vis) spectroscopy, infrared (IR) spectroscopy, elemental analysis, nuclear magnetic resonance (NMR), TEM, and X-ray photoelectron spectroscopy (XPS).

The smaller glyco-clusters lack a LSPR band, but can be observed by TEM. Visualization of AuNPs helped to unambiguously demonstrate some weak, previously controversial, effects. For example, Le^x-decorated AuNPs provided visual evidence for the existence of Ca-mediated sugar–sugar interactions and were used to explore potential mechanisms of sugar-mediated self-assembly of sponge cells.

Small gold glyconanoparticles were helpful in clarifying mechanistic aspects of multivalent carbohydrate recognition and have been exploited as antiadhesion agents to prevent melanoma metastasis, as vaccine candidates, and for cellular and molecular imaging.

Magnetic Nanoparticles

Magnetic nanoparticles (MNPs), including iron oxide and manganese oxide nanoparticles, are interesting contrast agents for magnetic resonance imaging (MRI). MRI can generate internal tomographic tissue images by using a radio frequency (RF)-induced electromagnetic field; modulation of that field's signal by MNPs (so-called "contrast") helps to detect their location. In clinical practice, gadolinium complexes are commonly used as MRI "contrast" agents. MNP–biomolecule hybrids are typically more sensitive, as they can be loaded with multiple

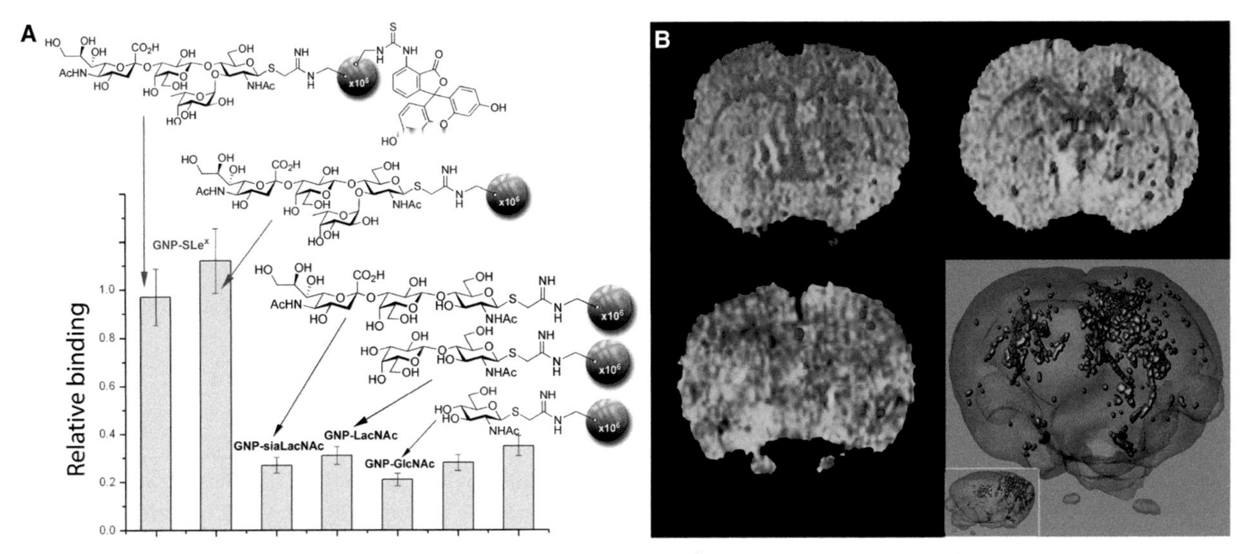

FIGURE 58.3. (*A*) In vitro binding studies using SLe^x-MNPs to rat E-selectin; (*B*) magnetic resonance images (MRIs) and their 3D reconstruction of SLe^x magnetic nanoparticles. (Adapted from van Kasteren SI, et al. 2009. *Proc Natl Acad Sci* **106:** 18–23.)

copies of the same ligand for better receptor targeting. Multimodal imaging can be achieved by attaching labels (i.e., a fluorescent dye) that allow for additional detection modes.

The surface functionalization of MNPs is crucial for molecule-specific binding and "molecular MRI." Antibodies have been widely used as targeting ligands because of their superb specificity, but suffer from high cost, short lifetime due to thermal instability, and potential immunogenicity. Structurally defined ligands, such as glycans, provide an attractive alternative.

Glyco-MNPs allow for detection of early stage disease by successfully mimicking leukocyte recruitment during inflammation (see Figure 58.2). By taking advantage of their high surface/volume ratio, glyco-MNPs can display multiple copies of oligosaccharides, thus increasing multivalency of binding interactions. Tetrasaccharide sialyl-Lewis x (SLe^x)-functionalized MNPs successfully targeted E-/P-selectins. Notably, SLe^x-MNPs detected inflammation events, both in vitro and in vivo, without any significant signs of associated cytotoxicity. Specific binding to the activated endothelium of blood vessels allowed for the detection of lesions in clinically relevant brain mouse models of stroke (Figure 58.3). Cross-species efficacy is easier to achieve with sugars than with antibody ligands. Thus, glyco-MNPs may be translated more readily from mammalian models to the clinic.

Quantum Dots

Quantum dots (QDs) are luminescent semiconducting nanomaterials, typically made of binary cadmium or zinc selenides or sulfides. QDs can emit light in the entire spectrum, and their optical properties are tunable depending on their size. Compared with organic dyes, QDs also have a broader excitation spectrum and sharper emission bands, allowing for multicomponent analysis with a single excitation source. Glyco-QDs functionalized with carboxymethyldextran and polylysine were used to study carbohydrate–protein interactions. The addition of maltose-modified dendrimers rendered the QDs water-soluble and biocompatible, while enhancing stability.

Using a host–guest strategy, β-cyclodextrin-quantum dots (β-CD-QDs) were prepared. Synthetic β-CD-QDs behaved much like PEGylated QDs and agglutinated lectins such as ConA, *Galanthus nivalis* agglutinin (GNA), and peanut agglutinin (PNA).

CARBON-BASED GLYCONANOMATERIALS

Elemental carbon has several allotropes, including tetravalent diamond and trivalent graphitic structures, which provide potential scaffolds for glycan presentation. Graphene is a two-dimensional carbon allotrope and the basic structure of important carbon-based materials. Buckminsterfullerene (C_{60}) is a discrete spherical construct that, although larger than many small molecules, can be manipulated using techniques common to small molecules. Carbon nanotubes (CNTs) can be considered cylindrical, elongated fullerenes. As a consequence of their curvature, hybridization, and boundary/inner atom ratios, both fullerenes and CNTs possess different reactivity that other carbon allotropes.

Fullerenes

Glycosylated-fullerenes such as α-D-mannosyl fullerenes and fullerenols inhibit erythrocyte aggregation. These "sugar balls" are generated by introduction of a reactive group, such as a terminal alkyne, followed by attachment of azido sugars. Thereby, a near-spherical display of glycans is possible. The glycosylated C_{60}s with 0.7-nm diameter can be considered the smallest "nanoparticles" in glyconanotechnology.

Carbon Nanotubes

CNTs are classified based on the number of graphene-like sheets that make up the sidewalls of the cylinder. Single-walled CNTs (SWCNTs) have a typical diameter of 1–2 nm and multiwalled CNTs (MWCNTs) have diameters of ~2–25 nm. The CNT length can range from 10s of nanometers to 10s of micrometers or even longer. Both the inner hollow space and the outer surface may be exploited to create functionalized CNTs (f-CNTs) to serve as delivery systems.

Broader use of CNTs has been hampered by their cytotoxicity and poor solubility. CNTs have been coated with glycopolymers. A C_{18}-lipid tail "wrapped" the CNT surface through hydrophobic interaction, to expose α-GalNAc residues. Glycopolymer-coated CNTs were nontoxic in vitro, whereas noncoated CNTs induced death in certain cells. Noncovalent functionalized CNTs have the risk of losing their coating materials once introduced into a biological milieu, and the ultimate fate of such products is unclear.

Covalent surface glycosylation or glycoconjugation of CNTs creates more stable probes for in vivo studies. Oxidation of the CNT surface introduces carboxylic acids that can be used for the covalent attachment of amino-functionalized sugars. Galactosylated SWCNTs can "capture" pathogenic *E. coli.* Scanning electron microscope (SEM) images show a strongly bound matrix formed by cells binding to the glycosylated nanotubes.

Direct attachment of β-GlcNAc residues via "one-pot" Staudinger reduction and amidation allows for good control of the anomeric configuration. Upon monosaccharide attachment, glycosyltransferases were used for regio- and stereoselective glycan elaboration. The sugar hydroxyl groups were used as "tagging" sites for heavy element-bearing labels to visualize the glycans via TEM.

1,3-Dipolar cycloaddition of reactive azomethine ylides, generated in situ from α-amino acids, created pyrrolidine derivatives of fullerenes and CNTs. This covalent approach avoided oxidative "cutting" and provided filled-and-functionalized glycosylated CNTs for in vivo applications.

CNTs can be considered "1D hollow pores" with an associated capillarity. Through capillary action, molten salts or their solutions can be encapsulated inside CNTs. Glyco-CNTs were used for encapsulation of the radioemitter $Na^{125}I$ and in vivo localization of high levels of this

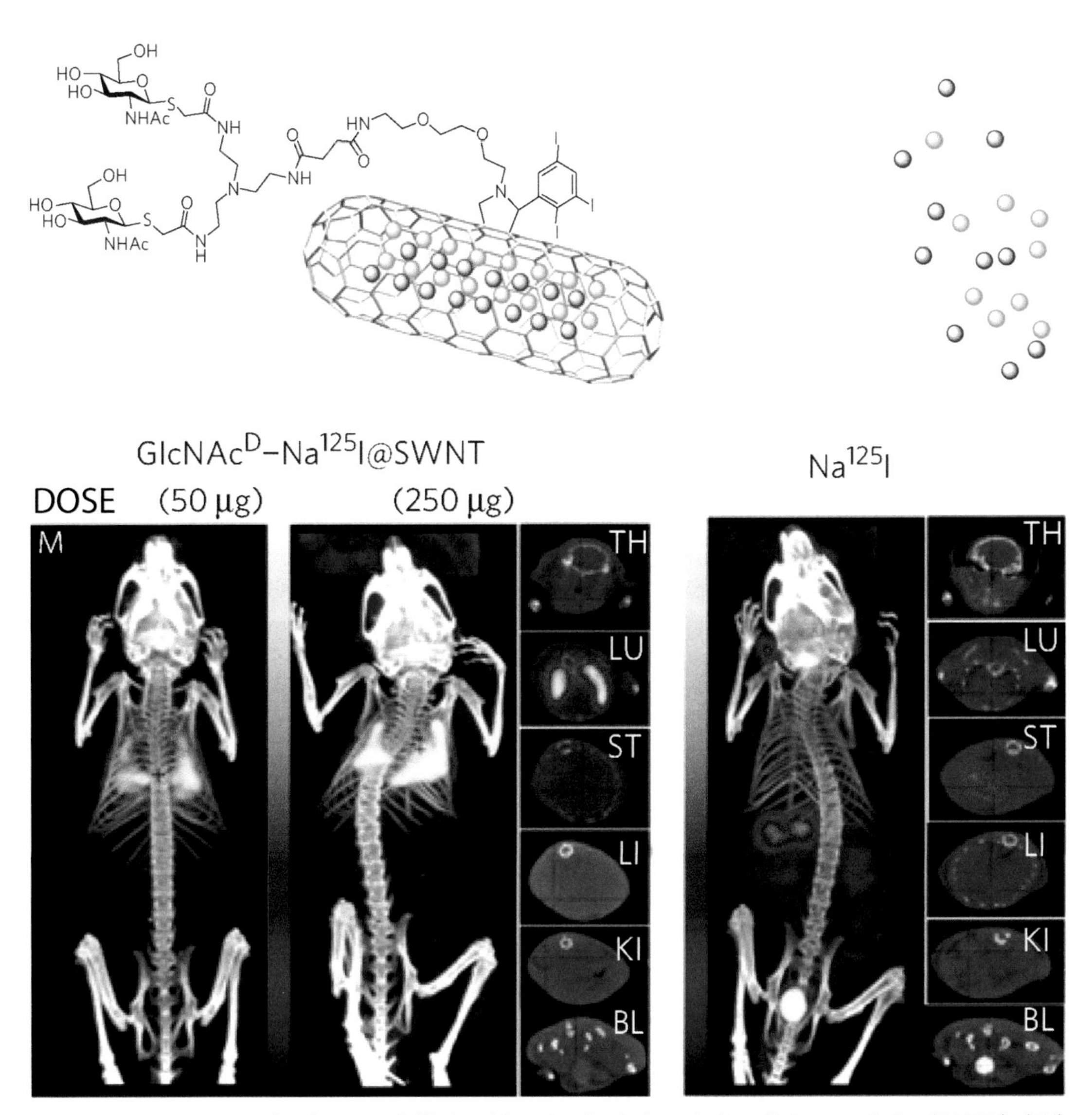

FIGURE 58.4. In vivo localization of filled-and-functionalized glyco-single-walled nanotubules (SWNTs). (TH) thyroid, (LU) lungs, (ST) stomach, (LI) liver, (KI) kidney, (BL) bladder. (Adapted from Hong SY, et al. 2010. *Nat Mater* **9:** 485–490.)

radionuclide. Multiple copies of GlcNAc improved both water dispersibility and biocompatibility. Thanks to the high aspect ratio and surface area to volume ratio of such CNTs, sugars can be efficiently displayed in a multivalent format (Figure 58.4). These glyco-CNTs are alternative radiotracers for in vivo imaging or radiation-delivery systems with high radioisotope-loading capacity and high sensitivity. The rapid uptake of iodide by the mammalian thyroid served as a test of any potential leakage of radioactive iodide "cargo." Although "free" iodide ^{125}I rapidly entered the thyroid, iodide encapsulated in the glyco-CNT remained at its target site even after a month.

Graphene

The chemical flexibility of graphene-based materials allowed for the formation of dynamic supramolecular constructs. A carbohydrate-functionalized two-dimensional (2D) surface was prepared by decorating thermally reduced graphene sheets with multivalent sugar ligands.

Host–guest inclusion provided a versatile strategy to present biofunctional ligands on the carbon surface. The multivalent sugar-functionalized graphene sheet agglutinates bacteria and inhibits their motility. Taking advantage of the supramolecular design, the captured bacteria can be partially released by adding a competitive guest. The unique thermal IR-absorption properties of graphene allow for the killing of the captured bacteria by IR-laser irradiation of the captured graphene–sugar–*E. coli* complex.

GLYCODENDRIMERS

Dendrimers are nanosized branched compounds that can be decorated with ligands, permitting control over their number and orientation. Several scaffolds based on organic molecules, metal complexes, and supramolecular assemblies have been exploited for the formation of multivalent glycodendrimers to study lectin-binding properties. "Click" chemistry and amide bond formation have been used to tether sugar molecules to the dendrimer scaffold. Mannose-conjugated glycodendrimers bind specifically to ConA. Galactose functionalized dendrimeric arms have been conjugated to β-cyclodextrin (βCD). The βCD can incorporate either a drug, such as doxorubicin, or a fluorescent dye, to monitor the uptake of the dendrimeric structure. Specific delivery of doxorubicin to hepatocytes was achieved with the help of targeting galactose units.

Metal complexes like $Ru(bpy)_3$ were also employed as scaffold for the assembly of three-dimensional supramolecular glycodendrimers. βCDs decorated with seven mannose functional groups were bound, via hydrophobic interactions, to the ruthenium core appended with adamantyl groups. The resulting supramolecular assemblies bind *E. coli* that express mannose receptors in the bacterial pili.

Supramolecular glycodendrimers have been prepared to display carbohydrates in a defined manner, offering controlled multivalency. Supramolecular particles, vesicles, or fibers have been created, offering a dynamic representation of a biological system. Synthetic molecules that mimic glycolipids self-assembled into nanovesicles displaying oligomannosides on their outer surface, resembling the glycocalyx coating of eukaryotic cells, bacteria, and viruses. The supramolecular approach permitted the formation of raft-like nanomorphologies on the synthetic vesicle that could shine light on important binding mechanisms.

Glycodendrimers can also be combined with particles to create multiple levels of multivalency. In one example, glycodendrimers displayed on protein-derived particles (virus-like particles) allowed for picomolar inhibition of Ebola virus–related adhesion events.

POLYSACCHARIDE-BASED NANOMATERIALS

Natural polysaccharides are very useful for the preparation of nanosized carriers. The low toxicity, biocompatibility, stability, low cost, hydrophilic nature, and availability of reactive sites for chemical modification render polysaccharides attractive building blocks for pharmaceutical applications. Polysaccharides can be used as NP backbone or coating. Polysaccharide-based nanoparticles are prepared by covalent or ionic cross-linking, polyelectrolyte complexation, or self-assembly of hydrophobically modified polysaccharides.

Chitosan-based NPs are common drug delivery systems. Positively charged chitosan gives rise to ionic cross-linked particles with polyanions to deliver proteins, oligonucleotides, and plasmid DNA. Multifunctional chitosan NPs, incorporating a NIR fluorophore for fluorescence imaging, can encapsulate anticancer drugs or complex small interfering RNA (siRNA) for sequential drug delivery. Hyaluronic acid and heparin-based NPs are promising platforms in cancer therapy; unlike chitosan NPs, they display inherent targeting properties. Hyaluronic acid binds to CD44, a transmembrane glycoprotein that is overexpressed in many types of cancer. This targeting property has promoted the application of hyaluronic acid-based NPs as theranostic agents.

Polysaccharides have also been used to coat polymeric or metallic nanoparticles, improving their water solubility, stability, and long-term circulation. Chitosan-PEG-coated iron oxide NPs improve intracellular delivery of a DNA repair inhibitor (O^6-benzylguanine) to glioblastoma multiform cells and enable treatment monitoring by MRI. Dextran was also employed to increase the water solubility and stability of iron oxide MNPs. Sulfated dextran electrostatically interacts with positively charged polycations. Functionalization of dextran-coated iron oxide NPs with sLe^x tetrasaccharide helped to monitor inflammation events in mouse brains in vivo. Hyaluronic acid–coated superparamagnetic iron oxide NPs have been used for imaging and for drug delivery to cancer cells.

Nanocarriers based on heparin and heparin derivatives have been applied to combat cancer via targeted, magnetic, photodynamic, and gene therapy. Gold and magnetic NPs have been coated or modified with heparin to improve their biocompatibility for applications in heparin-mediated events. Polysaccharide-functionalized gold NPs have given rise to multifunctional NPs with a wide range of applications including imaging, photodynamic therapy, and induction of apoptosis of metastatic cells.

Because of their natural abundance, nanomaterials based on cellulose have become extremely popular, offering biocompatibility, easy surface functionalization, and outstanding mechanical properties. Cellulose nanocrystals (CNCs) are rod-like particles obtained on acid hydrolysis of cellulose fibers. Because of their stiffness, twist, and aspect ratio, CNCs can assemble into chiral nematic phases, resulting in iridescent colors. Applications include functional paper, optoelectronics, engineered tissues, drug delivery, and biosensors.

GLYCONANOMATERIALS IN DIAGNOSIS AND THERAPY

Glycans are suitable biomarkers for medical diagnostics. Glyconanotechnology aids the development of biosensors and methods for the detection of glycans, lectins, or cancer cells and pathogens. Nanoengineered glycan sensors may help glycoprotein profiling, avoiding labeling or glycan liberation steps. A variety of nanomaterials are being explored as specific probes for label-free lectin or glycan detection. AuNPs and CNTs are the most widely used nanomaterials. Nanoengineered materials can be detected by mass changes with quartz crystal microbalance (QCM) and cantilever sensors, by field-effect transistor (FET) sensors based on carbon nanotubes, or optical sensors based on surface plasmon resonance (SPR) in combination with self-assembled monolayers (SAMs) of glycans or lectins.

Cancer cells and pathogens can be detected using glyconanomaterials. Glycan-functionalized nanodiamonds functioned as cross-linkers for bacteria and permitted water decontamination by filtration of the resulting aggregates through a 10-mm membrane. Mannan-coated gold glyconanoparticles incubated with a human gastric cell line in the presence of the mannose-binding lectin ConA detect and quantify cell-surface mannose glycans and silver signal amplification. The specific detection of cancer cells requires further refinement of the system, such as the substitution of ConA by a panel of lectins that recognize aberrant glycosylation specific to cancer.

The translation of biomolecular binding events into nanomechanics using a nanosized cantilever array of mannosides was applied to detect and discriminate different strains of *E. coli.* Methods exploiting intrinsic optical properties can be explored for biomedical "nonlabeled" imaging applications. QDs and AuNPs allow for spectral tuning, whereas SWCNTs show characteristic Raman peaks as well as photoluminescence in the NIR range. These intrinsic properties have yet to be systematically explored in vivo, but hold significant potential.

CONCLUSIONS

Glyconanomaterials based on different glycan-coated scaffolds have been created. Given the specific roles of endogenous glycoconjugates, nanomaterials that display glycans on similar length scales provide chemical platforms to advance our understanding of carbohydrate-mediated biological events.

Tailored glycomaterials that resemble more closely natural systems are under development to uncover hidden biological mechanisms. The challenge is now to create functional structures based on the appropriate scaffold and combine it with the proper glycan. Systems with increased control on the glycan presentation (i.e., spacing and conformation) is the next goal. At the same time, more complex glycans need to be incorporated into glyconanomaterials. Given the many functions carbohydrates in Nature, the glyconanotechnology field should evolve from simple binding studies involving the prototypical mannose-concanavalin A interaction to the exploration of new and more relevant avenues.

By taking advantage of the physical properties and inducible sizes, translation of fundamental studies to biomedical glyconanotechnology appears to be imminent. Promising examples include the use of glycomaterials for pathogen detection and as diagnostic tools. Biomedical imaging, using multimodal imaging techniques, such as PET/MRI or fluorescence/MRI, is an attractive vision for using glyco-NPs. Still, much effort has to be put to improve glycan stability in biological samples. Glycomimetics offer an interesting approach to avoid enzymatic degradation.

Polysaccharide-based materials are a green alternative to petroleum-based chemicals. A better molecular level understanding of these materials will fuel their application in biotechnology, as tissue scaffolds or nanoformulation for drug delivery. Polysaccharides are major component of the extracellular matrix, playing important roles in cellular communication and evolution. Being able to reproduce these complex natural environments will be game changing for tissue engineering.

ACKNOWLEDGMENTS

The authors acknowledge contributions to previous versions of this chapter by Soledad Penades and appreciate helpful comments and suggestions from Sriram Neelamegham.

FURTHER READING

Bertozzi CR, Bednarski MD. 1992. Antibody targeting to bacterial cells using receptor-specific ligands. *J Am Chem Soc* **114:** 2242–2245. doi:10.1021/ja00032a046

Mammen M, Choi SK, Whitesides GM. 1998. Polyvalent interactions in biological systems: implications for design and use of multivalent ligands and inhibitors. *Angew Chem Int Ed* **37:** 2755–2794. doi:10.1002/chin.199909293

de la Fuente JM, Barrientos AG, Rojas TC, Rojo J, Cañada J, Fernández A, Penadés S. 2001. Gold glyconanoparticles as water-soluble polyvalent models to study carbohydrate interactions. *Angew Chem Int Ed* **113:** 2317–2321. doi:10.1002/1521-3757(20010618)113:12<2317::aid-ange2317>3.0.co;2-u

Chen X, Lee GS, Zettl A, Bertozzi CR. 2004. Biomimetic engineering of carbon nanotubes by using cell surface mucin mimics. *Angew Chem Int Ed* **43:** 6111–6116. doi:10.1002/anie.200460620

Gu L, Elkin T, Jiang X, Li H, Lin Y, Qu L, Tzeng T-RJ, Joseph R, Sun Y-P. 2005. Single-walled carbon nanotubes displaying multivalent ligands for capturing pathogens. *Chem Commun* **2005:** 874–876. doi:10.1039/b415015e

Chen X, Tam UC, Czlapinski JL, Lee GS, Rabuka D, Zettl A, Bertozzi CR. 2006. Interfacing carbon nanotubes with living cells. *J Am Chem Soc* **128:** 6292–6293. doi:10.1021/ja060276s

Kiessling LL, Gestwicki JE, Strong LE. 2006. Synthetic multivalent ligands as probes of signal transduction. *Angew Chem Int Ed* **45:** 2348–2368. doi:10.1002/anie.200502794

Hong SY, Tobias G, Ballesteros B, El Oualid F, Errey JC, Doores KJ, Kirkland AI, Nellist PD, Green MLH, Davis BG. 2007. Atomic-scale detection of organic molecules coupled to single-walled carbon nanotubes. *J Am Chem Soc* **129:** 10966–10967. doi:10.1021/ja069080i

Wu P, Chen X, Hu N, Tam UC, Blixt O, Zettle A, Bertozzi CR. 2008. Biocompatible carbon nanotubes generated by functionalization with glycodendrimers. *Angew Chem Int Ed* **47:** 5022–5025. doi:10.1002/anie.200705363

Chen X, Wu P, Rousseas M, Okawa D, Gartner Z, Zettl A, Bertozzi CR. 2009. Boron nitride nanotubes are noncytotoxic and can be functionalized for interaction with proteins and cells. *J Am Chem Soc* **131:** 890–891. doi:10.1021/ja807334b

Csaba N, Köping-Höggård M, Alonso MJ. 2009. Ionically crosslinked chitosan/tripolyphosphate nanoparticles for oligonucleotide and plasmid DNA delivery. *Int J Pharm* **382:** 205–214. doi:10.1016/j.ijpharm.2009.07.028

van Kasteren SI, Campbell SJ, Serres S, Anthony DC, Sibson NR, Davis BG. 2009. Glyconanoparticles allow pre-symptomatic in vivo imaging of brain disease. *Proc Natl Acad Sci* **106:** 18–23. doi:10.1073/pnas.0806787106

Hong SY, Tobias G, Al-Jamal KT, Ballesteros B, Ali-Boucetta H, Lozano-Perez S, Nellist PD, Sim RB, Finucane C, Mather SJ, et al. 2010. Filled and glycosylated carbon nanotubes for in vivo radioemitter localization and imaging. *Nat Mater* **9:** 485–490. doi:10.1038/nmat2766

Kikkeri R, Grünstein D, Seeberger PH. 2010. Lectin biosensing using digital analysis of Ru(II)-glycodendrimers. *J Am Chem Soc* **132:** 10230–10232. doi:10.1021/ja103688s

Grünstein D, Maglinao M, Kikkeri R, Collot M, Barylyuk K, Lepenies B, Kamena F, Zenobi R, Seeberger PH. 2011. Hexameric supramolecular scaffold orients carbohydrates to sense bacteria. *J Am Chem Soc* **133:** 13957–13966. doi:10.1021/ja2036767

El-Dakdouki MH, Zhu DC, El-Boubbou K, Kamat M, Chen J, Li W, Huang X. 2012. Development of multifunctional hyaluronan-coated nanoparticles for imaging and drug delivery to cancer cells. *Biomacromolecules* **13:** 1144–1151. doi:10.1021/bm300046h

Mizrahy S, Peer D. 2012. Polysaccharides as building blocks for nanotherapeutics. *Chem Soc Rev* **41:** 2623–2640. doi:10.1039/c1cs15239d

Reuel NF, Mu B, Zhang J, Hinckley A, Strano MS. 2012. Nanoengineered glycan sensors enabling native glycoprofiling for medicinal applications: towards profiling glycoproteins without labeling or liberation steps. *Chem Soc Rev* **41:** 5744–5779. doi:10.1039/c2cs35142k

Ribeiro-Viana R, Sánchez-Navarro M, Luczkowiak J, Koeppe JR, Delgado R, Rojo J, Davis BG. 2012. Virus-like glycodendrinanoparticles displaying quasi-equivalent nested polyvalency upon glycoprotein platforms potently block viral infection. *Nat Commun* **3:** 1303. doi:10.1038/ncomms2302

Marradi M, Chiodo F, García I, Penadés S. 2013. Glyconanoparticles as multifunctional and multimodal carbohydrate systems. *Chem Soc Rev* **42:** 4728–4745. doi:10.1039/c2cs35420a

Delbianco M, Bharate P, Varela-Aramburu S, Seeberger PH. 2016. Carbohydrates in supramolecular chemistry. *Chem Rev* **116:** 1693–1752. doi:10.1021/acs.chemrev.5b00516

59 Glycans in Bioenergy and Materials Science

Malcolm A. O'Neill, Robert J. Moon, William S. York, Alan G. Darvill, Kamil Godula, Breeanna Urbanowicz, and Debra Mohnen

Plants provide large amounts of glycans that are used by humans for many purposes. Wood, which is composed predominantly of lignified secondary walls, is used as an energy source, as a building material, and for papermaking. Pectins isolated from the primary cell walls of fruits and polysaccharides isolated from seeds are used as thickeners, stabilizers, and gelling agents in many foods and beverages. Plant cell walls are the major component of forage used as animal feed. These walls, as dietary fiber, also contribute to human health. Recent concerns about the environmental costs of fossil fuel extraction and consumption have led to renewed interest in using plant glycans as feedstocks for energy production, for the generation of polymers with improved or new functionalities, and for the generation of high-value chemical precursors. In this chapter, we briefly describe four broad categories—bioenergy, fine chemicals and chemical feedstocks, polymeric materials, and nanomaterials—in which plant glycans have the potential to replace or to provide alternatives to petroleum-based products.

INTRODUCTION

Plant glycans are used by humans as an energy source, as a building material, and for making numerous bioproducts including paper. Cellulose from diverse plant sources is the primary component of many valuable materials, including textiles and plastics. Pectins are used as thickeners, stabilizers, and gelling agents in many foods and beverages. Plant cell walls are used as animal feed and, as dietary fiber, also contribute to human health. The well-established adverse

published by Cold Spring Harbor Laboratory Press; doi: 10.1101/glycobiology.4e.59

effects of the extraction and use of fossil fuels on the Earth's climate have led to worldwide efforts to develop plant-derived glycans as a renewable raw material to displace or supplement fossil fuels for energy production, for the generation of polymers with improved or new functionalities, and for the generation of high-value chemical precursors.

GLYCANS AND BIOENERGY

The process of photosynthesis by terrestrial plants has been estimated to assimilate at least 100 billion metric tons of CO_2 annually. The chemical energy generated in this manner is stored predominantly in the form of carbohydrates. Some of these carbohydrates are used directly for plant growth and development, whereas others are converted to storage polysaccharides (starch and fructans) that provide plants with a readily available form of energy. A considerable portion of the carbohydrate formed via photosynthesis is used to produce the polysaccharide-rich walls that surround plant cells (Chapter 24). Thus, plant cell walls account for a substantial amount of biological carbon sequestration and are a potentially sustainable and economical source of nonpetroleum-based energy and high-value chemicals.

First-generation bioethanol produced by fermenting the starch present in corn grains currently accounts for virtually all of the liquid transportation fuel generated from plant materials in the United States. The starch is first treated with enzymes that convert it to glucose, which is then fermented to ethanol and carbon dioxide by adding yeast. Yeast can convert 1 kg of glucose to 0.33 gallons (1.25 L) of ethanol and an equivalent amount of carbon dioxide. The United States and Brazil together account for as much as 84% of world ethanol production (https://afdc.energy.gov/). In 2019, 16.9 billion gallons (64 billion liters) of ethanol were produced in the United States according to the U.S. Energy Information Administration (www.eia.gov). Brazil produces approximately 8.57 billion gallons (32 billion liters) of ethanol annually by fermenting the sucrose extracted from sugarcane (https://afdc.energy.gov/). Corn and cane bioethanol are then blended with gasoline in varying amounts or used directly as a transportation fuel.

Concerns about the negative impacts of the large-scale production of corn-based ethanol on food production and the environment has led to renewed interest in generating ethanol and other liquid transportation fuels from sustainable plant lignocellulosic biomass that can be grown on marginal land. This biomass is comprised predominantly of lignified secondary walls (Chapter 24) that are composed of cellulose (40%–50% w/w), hemicellulose (25%–30% w/w), and lignin (15%–25% w/w) and lesser amounts of pectin and protein. Several different plants, including poplar, switchgrass, sorghum, miscanthus, eucalyptus, and sugarcane, are being considered for use as bioenergy crops.

The biomass from energy crops can be converted to liquid fuel by fermentation or gasification. In gasification, the biomass is heated in a low-oxygen environment to generate syngas (hydrogen, carbon monoxide, and carbon dioxide) and heat. The syngas can then be reacted to produce diverse chemicals including alcohols or alkanes via Fischer–Tropsch synthesis that can be further transformed into fuels (mainly diesel oil and jet fuel). Most of the technical challenges to commercial biomass gasification are understood. However, the process has not been widely adopted because of the high capital costs involved.

The generation of liquid fuel from lignocellulose by fermentation currently involves using a cocktail of enzymes to convert the biomass to sugar, which is then fermented to give the desired product. This approach may be superseded by the development of consolidated bioprocessing (CBP) technologies in which the microorganism that deconstructs plant biomass into sugars also converts those same sugars into products such as fuels and chemicals. The fermentation approach is simple in concept but there are many technical challenges that must be solved before it becomes commercially viable. One major obstacle is that the cellulose and hemicellulose in lignocellulosic biomass are not readily accessible to hydrolytic enzymes and thus are not efficiently converted to fermentable sugars. The biomass must be pretreated with dilute acid, ammonia, or steam to decrease this recalcitrance. Cost-effective and environmentally sound pretreatment technologies need to be developed if the commercial production of bioproducts by fermentation is to become a reality. The efficiency of the enzymes used to convert the cellulose and hemicellulose to sugar must also be improved. To this end there is extensive ongoing research to engineer thermophilic microorganisms to more efficiently deconstruct biomass and to convert the released sugars to the desired product, avoiding the necessity of releasing sugars from the biomass with enzyme cocktails before fermentation.

Increased understanding of cell wall structure, together with knowledge of polysaccharide and lignin biosynthesis, is expected to facilitate the engineering of plants to produce biomass that is more amenable to bioprocessing and an improved resource for biofuel, value-added chemicals, and bioproducts. However, the susceptibility of such modified plants to biotic and abiotic stress in the field will need to be addressed. Such concerns, together with those regarding the introduction of genetically modified plants into the environment, may be lessened by identifying natural plant variants that produce lignocellulosic biomass with the desired properties including reduced recalcitrance to saccharification.

Lignin is poised to become a major by-product of commercial biorefineries, as it is not converted to a liquid fuel during fermentation. Early concepts of biorefineries envisioned burning the lignin to generate power. However, there is now a greater emphasis on valorization of the recovered lignin to produce value-added compounds for the chemical industry. Considerable resources are being allocated to research and development worldwide to create a viable and sustainable lignocellulosic advanced biofuels and bioproducts industry. Nevertheless, many technical, environmental, and societal challenges must be solved if this industry is to develop and contribute to a biobased economy and to reduce the demand for fossil fuels.

FINE CHEMICALS AND FEEDSTOCKS

Several of the sugars released from lignocellulosic biomass, including glucose and xylose, are being investigated for use in the production of functional chemical precursors that can be used to make industrially relevant compounds and polymers including plastics. Some examples of functional chemical precursors are alcohols (ethanol, propanol, and butanol), sugar alcohols (xylitol, and sorbitol), furans (furfural, hydroxymethylfurfural), biobased hydrocarbons (isoprene and long-chain hydrocarbons), organic acids (lactic acid, succinic acid, and levulinic acid), and biobased polyurethanes. Current research is focused on optimizing the bioconversion of polysaccharides (yield, rate, separation, titer, and product specificity) by identifying and engineering improved fermentation organisms and fermentation processes and developing enhanced chemical catalysts.

POLYMERIC MATERIALS

Plant-derived cell wall polysaccharides (Chapter 24) including cellulose, xyloglucan, mannan, and xylan (Figure 24.1) are used to produce diverse polymeric materials used by industry. They are both biorenewable and biocompatible, making them advantageous over their petroleum-based counterparts. Cellulose has been extensively modified to develop synthetic cellulose-based polymers. Cellulose films (cellophane) and fibers (rayon) are produced using regenerated cellulose that is itself formed by dissolving natural cellulose (predominantly from wood pulp) in alkali and carbon disulfide and then precipitating the polymer in a process that has been used for at least 125 years (the viscose process).

With society's ongoing need for polymers with new properties and functions, there is increased effort to develop chemical or biocatalytic reaction pathways to modify the structure of a polysaccharide backbone or side chains to enable the production of polysaccharide derivatives with enhanced or new properties. Cellulose is one example of a plant polysaccharide that has been extensively modified to develop new biosourced polymers. Reaction pathways have been developed to generate specific cellulose derivatives by substituting accessible hydroxyl groups with other chemical groups. Such derivatives include cellulose acetate, cellulose acetate propionate, cellulose acetate butyrates, carboxymethyl cellulose, and cellulose butyrate succinate. These products are used in many industrial applications as coatings, inks, binders, and thickening/gelling agents. They are also used in the pharmaceutical industries to produce controlled-release drug tablets and in the cosmetics and food industries as thickening and gelling agents.

Chitin is the second most abundant natural polysaccharide after cellulose. It is present in crustacean shells and insect cuticles and may also be produced by fungi and algae (Chapter 23). Chitin is composed of 1-4-linked β-D-GlcNAc residues. It can be enzymatically or chemically deacetylated to produce chitosan, the cationic and more water-soluble form of the glycan. Large amounts (~5 million metric tons) of chitin are produced as waste by the seafood processing industry, and thus, there is considerable interest in developing biobased processes to convert this waste into value-added products. Chitosan has reactive amino and hydroxyl groups that can be modified to generate materials with diverse properties and applications.

Hemicellulosic polysaccharides, including xylan and mannan, have a backbone structure similar to cellulose and are abundant in agricultural and forestry sidestreams, including the pulping and the viscose processes. With the complexity and variability of polysaccharide structures there is considerable potential for the development of unique synthetic polysaccharides with new or enhanced functionality. Noncellulosic matrix polysaccharides present an attractive target for enzymatic synthesis and functionalization. They are easily extracted from biomass and, unlike cellulose, are typically soluble in aqueous solutions and are often substituted with both glycosyl and nonglycosyl substituents that can be modified to influence their material properties. To this end, current research aims to further understand and use new reaction pathways that target chemical or enzymatic modifications to functionalize and/or alter specific locations on the polysaccharide and thereby generate regioselective functionalization.

Synthetic or naturally derived oligosaccharides can also be covalently appended to polymer chains built from petroleum-based monomers. This gives rise to glycopolymers with architectures resembling those of glycoproteins or proteoglycans. Such materials have found increasing use as research tools to study the biological functions of glycans and are currently explored as biomaterials for drug delivery or as antifouling and antifreeze agents.

NANOMATERIALS

Nanomaterials from plants and crustacean shells offer new materials for the development of biorenewable and biocompatible products. These nano-sized particles, consisting of bundled polymer chains, have properties and function that are different from the isolated polymer chains from which they are made. Such nanomaterials can be produced from cellulose, hemicellulose, pectin, chitin and chitosan. A more focused description is given below on nanomaterials from cellulose.

Polysaccharides that have little or no branching of their backbone can self-assemble to form ordered structures in which the individual polymer chains stack along the chain axis, thus forming a crystalline structure. Cellulose is one example of a plant polysaccharide that has this type of crystalline structure. During cellulose biosynthesis individual glucan chains assemble to form microfibril structures that contain both crystalline and disordered arrangements (see Figure

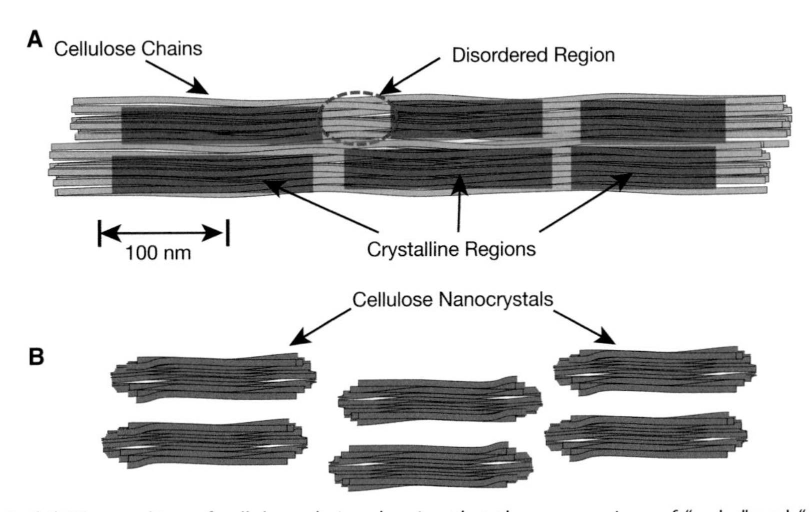

FIGURE 59.1. (*A*) The stacking of cellulose chains showing that there are regions of "order" and "disorder." (*B*) During one type of cellulose nanomaterial extraction process that uses acid hydrolysis, the disordered regions are preferentially dissolved and only the crystalline regions are left.

59.1A). The high mechanical stiffness and tensile strength along the length of the cellulose microfibrils provide high mechanical strength, high strength-to-weight ratio, and toughness to plant tissues and organs.

The cellulose fibril structures and the crystalline regions can be isolated using specialized chemical–mechanical extraction methods. The resulting nano-sized particles, typically referred to as cellulose nanomaterials (CNs), have properties and functions that are considerably different than individual cellulose chains (CNs; Chapter 58). The CNs' morphology, properties, and surface chemistry vary depending on the plant source and the conditions used to extract the cellulose. Plant CNs are typically classified as cellulose nanocrystals or cellulose nanofibrils. Cellulose nanocrystals are rod-like particles, which are between 3- and 20-nm-wide and between 50 and 500 nm in length (Figure 59.2A). Cellulose nanofibrils are fibril-like particles, which are between 5- and 100-nm-wide and between 500 nm to several microns in length (Figure 59.2B). Both types of CNs have high stiffness and tensile strength, high surface-area-to-volume ratio, and surfaces that can be readily chemically modified to alter their physicochemical and materials properties. CNs are biorenewable and biocompatible and have minimal environmental, health, and safety risks. Thus, CNs are being used to develop new products including barrier films, separation membranes, antimicrobial films, food coatings, cement/concrete modifiers, rheology modifiers, biomedical applications, template scaffolds for catalytic supports, batteries, supercapacitors, and many others. Research, development, and commercialization in CNs is accelerating and covering an ever-broadening scope, all of which would benefit through advancements in glycoscience and the new characterization and synthesis tools being developed.

PERSPECTIVES AND FUTURE CHALLENGES

Considerable resources are being allocated to research and development to create economically viable and sustainable biofuels and bioproducts industries that use plant glycans as feedstocks. In the future, there is likely to be a greater emphasis on using both the glycan and the lignin components of plant biomass to produce liquid transportation fuels and value-added chemicals. Such technologies will require the development of new chemical catalysts and robust enzymes in addition to plants engineered to produce biomass with enhanced processing and value characteristics. Many technical, economic, environmental, and societal challenges must be met if these

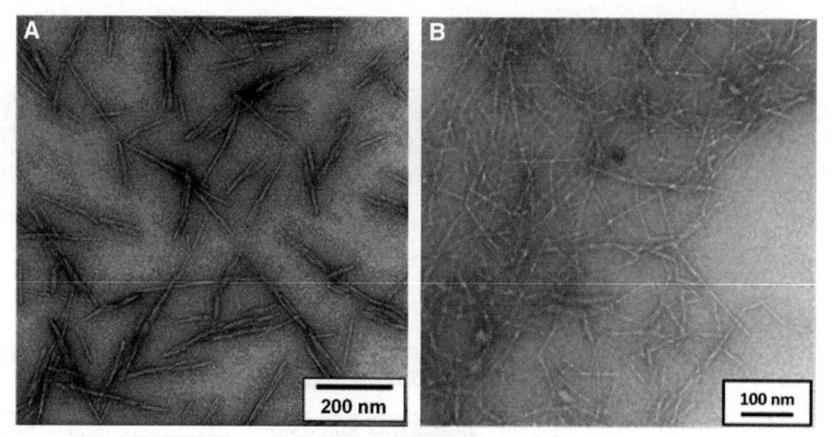

FIGURE 59.2. Transmission electron microscopy images showing two types of cellulose nanomaterials. (*A*) Cellulose nanocrystals produced by acid hydrolysis. (*B*) Cellulose nanofibrils produced by mechanical defibrillation of wood pulp.

industries are to develop and contribute to a biobased economy and to reduce the demand for fossil fuels. Despite the commercial and technological success of natural cellulose-based materials, glycans are yet to be systematically explored as building blocks for new materials. The development of controlled polymerization methods for making well-defined glycans, matching those used for producing polymers from petroleum-based building blocks, will need to be developed to fully realize the potential of glycans to produce materials with new properties and functions.

ACKNOWLEDGMENTS

The authors acknowledge helpful comments and suggestions from Martina Delbianco and Markus Pauly.

FURTHER READING

Klemm D, Heublein B, Fink H-P, Bohn A. 2005. Cellulose: fascinating biopolymer and sustainable raw material. *Agnew Chem Int Ed* **44**: 3358–3393. doi:10.1002/anie.200460587

Ragauskas AJ, Williams CK, Davison BH, Britovsek G, Cairney J, Eckert CA, Frederick WJ, Hallett JP, Leak DJ, Liotta CL, et al. 2006. The path forward for biofuels and biomaterials. *Science* **311**: 484–489. doi:10.4324/9781315793245-98

Hansen N, Plackett D. 2008. Sustainable films and coatings from hemicelluloses: a review. *Biomacromolecules* **9**: 1493–1505. doi:10.1021/bm800053z

Carroll A, Somerville C. 2009. Cellulosic biofuels. *Ann Rev Plant Biol* **60**: 165–182. doi:10.1146/annurev.arplant.043008.092125

Mishra A, Malhotra AV. 2009. Tamarind xyloglucan: a polysaccharide with versatile application potential. *J Mater Chem* **19**: 8528–8536. doi:10.1039/b911150f

Tilman D, Socolow R, Foley JA, Hill J, Larson E, Lynd L, Pacala S, Reilly J, Searchinger T, Somerville C, et al. 2009. Beneficial biofuels—the food, energy, and environment trilemma. *Science* **325**: 270. doi:10.1126/science.1177970

Chung D, Cha M, Guss AM, Westpheling J. 2014. Direct conversion of plant biomass to ethanol by engineered *Caldicellulosiruptor bescii*. *Proc Natl Acad Sci* **111**: 8931–8936. doi:10.1073/pnas.1402210111

Doblin MS, Johnson KL, Humphries J, Newbigin EJ, Bacic A. 2014. Are designer plant cell walls a realistic aspiration or will the plasticity of the plant's metabolism win out? *Curr Opin Biotechnol* **26**: 108–114. doi:10.1016/j.copbio.2013.11.012

Habibi Y. 2014. Key advances in the chemical modification of nanocelluloses. *Chem Soc Rev* **43**: 1519–1542. doi:10.1039/c3cs60204d

Ragauskas AJ, Beckham GT, Biddy MJ, Chandra R, Chen F, Davis MF, Davison BH, Dixon RA, Gilna P, Keller M, Langan P. 2014. Lignin valorization: improving lignin processing in the biorefinery. *Science* **344**: 1246843. doi:10.1126/science.1246843

Zhao Q, Dixon RA. 2014. Altering the cell wall and its impact on plant disease: from forage to bioenergy. *Annu Rev Phytopathol* **52**: 69–91. doi:10.1146/annurev-phyto-082712-102237

Liao C, Seo S-O, Celik V, Liu H, Kong W, Wang Y, Blaschek H, Jin Y-S, Lu T. 2015. Integrated, systems metabolic picture of acetone–butanol–ethanol fermentation by *Clostridium acetobutylicum*. *Proc Natl Acad Sci* **112**: 8505–8510. doi:10.1073/pnas.1423143112

Smith PJ, Wang HT, York WS, Peña MJ, Urbanowicz BR. 2017. Designer biomass for next-generation biorefineries: leveraging recent insights into xylan structure and biosynthesis. *Biotechnol Biofuels* **10**: 1–14. doi:10.1186/s13068-017-0973-z

Chen C, Kuang Y, Zhu S, Burgert I, Keplinger T, Gong A, Li T, Berglund L, Eichhorn SJ, Hu L. 2020. Structure–property–function relationships of natural and engineered wood. *Nat Rev Mater* **5**: 642–666. doi:10.1038/s41578-020-0195-z

Smith PJ, Ortiz-Soto ME, Roth C, Barnes WJ, Seibel J, Urbanowicz BR, Pfrengle F. 2020. Enzymatic synthesis of artificial polysaccharides. *ACS Sustainable Chem Eng* **8**: 11853–11871. doi:10.1021/acssuschemeng.0c03622

60 Future Directions in Glycosciences

Gerald W. Hart and Ajit Varki

This closing chapter discusses the future potential of glycosciences for impacting basic and applied research, human health, material science, and renewable energy. Technological advances predicted to occur in the coming years are mentioned. Finally, a sampling of glycoscience-related questions that remain to be addressed in the future is presented.

PERVASIVE IMPACT OF GLYCOSCIENCES ON SCIENCE AND SOCIETY WILL INCREASE IN THE FUTURE

Basic and Applied Research

Every living cell in nature generates a complex and diverse array of glycans that is critical for the evolution, development, functioning, and survival of all natural biological systems (Chapter 1). A strong basic knowledge base regarding the genomics, chemistry, biochemistry, biosynthesis, and biological roles of these ubiquitous and diverse molecules is now well-established (Chapters 2–19). The broad outlines of their natural occurrence and evolution (Chapters 20–27) and their recognition by glycan-binding proteins (Chapters 28–38) are becoming clear, as is their important role in normal and abnormal physiology and disease (Chapters 39–47). Facile methods for their analysis, manipulation, and synthesis of glycans have also been established (Chapters 48–56), and their significance in the biotechnology and pharmaceutical industries, nanotechnology, and bioenergy and materials science is obvious (Chapters 57–59). Given all these major advances in basic and applied research on glycans, there is no reason why this major class of biomolecules should continue to be the "dark matter of the biological universe." However, since the 1980s an entire

published by Cold Spring Harbor Laboratory Press; doi: 10.1101/glycobiology.4e.60

generation of scientists have been largely trained without much exposure to, or knowledge of, glycans. Thus, it will still be awhile before these molecules return to their rightful place in the mainstream of thinking in conventional molecular and cellular biology and medicine. Continued advances in basic and applied research on glycans will occur, but this needs to be coupled with the training of a new generation of scientists, engineers, and physicians for whom these molecules will be an obvious aspect of their understanding of living systems in health and disease. The National Institutes of Health (NIH) National Heart, Lung, and Blood Institute (NHLBI) of the United States has recognized the importance of training scientists and clinicians in glycosciences by establishing a National Career Development Consortium for Excellence in Glycosciences (K12), which focuses on immersive training of medical and research professionals in all aspects of glycosciences. In addition, the National Institute of General Medical Sciences (NIGMS) has funded the first graduate student training grant (T32) in glycosciences. Clearly, the teaching of glycosciences in all training programs for graduate and medical professionals will be essential if we are to continue to make medical advances in nearly all areas.

Health and Development

As emphasized in a report from the National Research Council of the National Academy of Sciences, nearly every disease process that affects humans and other animals involves glycans (Chapters 39–47). In past decades, it has been realized that most functions of complex glycans are required at the multicellular (organismic) level. In contrast, cycling monosaccharides (e.g., O-GlcNAc in the nucleus and cytoplasm) serve regulatory functions at the single-cell level. The significance of nucleocytoplasmic O-glycosylation was broadened by the discoveries that an evolutionary branch of O-GlcNAc transferase is in fact an O-fucosyltransferase (OFT) that modifies many proteins in plants, protists, and protist pathogens like *Toxoplasma* and *Cryptosporidium*, with O-Fuc instead of O-GlcNAc (Chapters 18 and 19).

The critical roles of complex glycans in the biology of intact organisms have been dramatically illustrated by the contrast between the viability of glycosyltransferase mutant cell lines in culture (Chapter 49) with the often lethal outcome caused by inactivation of the same enzymes in living organisms (Chapter 41). Transgenic mouse studies and the severe phenotypes of human congenital disorders of glycosylation (Chapter 45) have dramatically revealed the critical importance of glycans in development, physiology, and disease.

Most major diseases also involve disordered inflammation and immunity, in which the glycan-binding selectins, Siglecs, galectins, and other glycan-binding proteins (Chapters 34–36) play critical roles. Most pathogens, viruses, bacteria, and parasites gain entry to cells by binding to glycans on the cell surface (Chapters 42 and 43). Recently, many studies have shown an essential role for complex glycans in SARS-CoV-2 infections and a role for O-GlcNAcylation in viral-induced cytokine storms (Chapter 19). In addition, many vaccines against infectious agents are directed against microbial glycans. Proteoglycans play critical roles in development, tissue morphogenesis, and cardiovascular disease and in regulating the actions of cytokines and growth factors. The glycosaminoglycan heparin is one of the oldest and most commonly used "drugs" in the clinic. Notch signaling, which plays a major role in controlling morphogenesis in development and cell fate decisions, is controlled by glycans (Chapter 13), and glycans on the surface of tumor cells play critical roles in tumor progression and metastasis (Chapter 47). Many of the current therapeutics in use or under development are glycoproteins, like monoclonal antibodies, which often require particular types of N-glycans for functional efficacy. Defects in the synthesis of glycan chains on dystroglycan underlie many types of congenital muscular dystrophy (Chapter 45). Dysregulated O-GlcNAcylation contributes to the etiologies of diabetes, neurodegeneration, cardiovascular disease, and cancer (Chapter 19). Although the few scientists and physicians well-educated in glycobiology are acutely aware of the importance of glycans in disease, most others have not learned much about this major class of molecules. However, it is now clear that studies of glycans will be essential for understanding the pathophysiology of most diseases and the development of effective therapies. For the great potential of glycobiology to be realized, much greater efforts need to be directed toward education of all undergraduate, graduate and postgraduate students, clinicians, and basic scientists and toward advancing technologies that will allow ease of experimentation on glycans.

Renewable Energy

Clearly, in the future, we will run out of fossil fuels, which are not renewable resources. Plants are by far the most efficient source of renewable energy because they efficiently use photosynthesis to trap the sun's energy, mostly in the form of glycans. The challenge in using plants as a source of fuel is the difficulty in degrading plant cell walls into smaller glycans that can be converted into usable fuels at a low cost (Chapter 59). In the future, glycoscience will play a critical role in the evolution of our society from one based on burning fossil fuels to one based on the use of sustainable energy, in part derived from rapidly growing plants and algae.

Industry and Materials

Carbon sequestration by photosynthesis to produce glycans is the major process preventing carbon dioxide from building up to levels that are already causing severe "greenhouse" warming of the planet, with unpredictable climate disruption and increased frequency of extreme weather events as major consequences. Wood, which is mostly made up of complex glycans, is already a major building material. Paper, textiles, cellophane, and rayon are other examples of everyday materials made from glycans. As our supply of petroleum runs out, glycans will increasingly be used to provide materials for the manufacture of plastics and a myriad of polymers (Chapter 59). Various forms of modified cellulose, for example, will be critical for the manufacturing of exotic materials and as carbon sources for various chemicals.

TECHNOLOGICAL ADVANCES TO ELUCIDATE STRUCTURE/FUNCTIONS OF GLYCANS

Analytical Methods

Advances in mass spectrometry instrumentation and methods in recent years have been remarkable (Chapters 50 and 51). The ability to use mass spectrometry to map glycan attachment sites, to profile glycan structural variations, and to determine detailed fine structures of glycans has progressed rapidly because of technological advances in instrumentation, including electron capture and electron transfer dissociation fragmentation methods, and the development of ion traps with very high mass accuracy and sensitivity. In the future, ion-mobility separations in the gas phase will allow the identification of glycan isomers that have identical molecular masses in mixtures. In addition, there have been major advances in the chromatographic isolation and separation of complex glycans. New ultra-high-pressure pumps and very small particulate high-pressure liquid chromatography (HPLC) resins that can withstand high pressures will also greatly increase the resolving power of chromatography. Advances in ion mobility and high-resolution mass spectrometric methods are, for the first time, allowing us to begin to decode the information content of glycosaminoglycans (Chapter 17). The development of induced hyperpolarization methods shows great promise in greatly increasing the sensitivity of nuclear magnetic resonance (NMR) analyses of glycans (Chapter 50), which has been a major limitation in the analyses of biological samples. Recent advances in solution NMR spectroscopy that allows measurements of NOEs between OH groups have increased the number of distance restraints and improve the quality of glycan three-dimensional structures (Chapter 50). Bioorthogonal labeling methods continue to be developed that will allow the study of changes in glycans in living cells in real time. Currently, for atomic structural analyses of glycoproteins, including X-ray crystallography, it is usually necessary to remove glycans to allow for crystallization. As electron imaging methods, such as cryo-electron microscopy (cryo-EM), continue to reach higher resolution approaching that achieved by crystallographic analyses, we will be able to obtain images of glycoconjugates at atomic resolution, without the need to remove the glycans. In fact, recent cryo-EM structures of the oligosaccharyltransferases have significantly impacted our molecular understanding of N-linked glycosylation. As NMR becomes more sensitive and can handle larger molecules, one will also be able to visualize glycoconjugates in physiologically relevant solvents and at appropriate temperatures. These developments will allow a much more accurate view of the 3D structures of glycoconjugates. Eventually, one will need to define the population of molecular species (glycoforms of glycoproteins and glycolipids) at an organismal and cellular level to fully understand structure/function relationships of glycoconjugates in response to extracellular stimuli. Existing technologies that show great promise include top-down glycoproteomics and high-throughput, high-resolution imaging methods. In the past few years, there have also been major advances in glycoproteomics. In contrast to glycomics, glycoproteomic approaches do not lose the context of where on a polypeptide glycans are attached (Chapter 51). In theory, if these methods live up to their promise, one day it may be possible to define the complete set of molecular species of glycoconjugates in a population and thus elucidate how they collectively contribute to the many individual functions of a gene product. However, all these methods still result in loss of information about labile modifications of glycans and also involve breaking apart the intact glycome into pieces before analysis. Ultimately, one will need to understand the actual structure of the intact glycome in situ in living systems.

In parallel with advances in such sophisticated methods, it is also important to "democratize" the practical approach to glycobiology by developing simple methods that can be used by the average biologist working at the bench without sophisticated instrumentation. The NIH Common Fund has recently supported research designed to make the study of glycosciences more accessible to biologists and biochemists not in the field. In this regard, one should be able to take advantage of the fact that millions of years of pathogen and symbiont interactions with hosts have already generated a large number of highly specific glycan-binding proteins, which could be harnessed to interrogate glycosylation at multiple levels of resolution.

The international glycoscience community has recently established a guideline called MIRAGE (minimum information required about a glycomics experiment) to help nonexperts to ensure that publications are understandable and reproducible. These guidelines address sample preparation, mass spectrometry, glycan arrays, and liquid chromatographic methods (Chapters 50–52).

Chemical and Enzymatic Synthesis

Because of their stereochemistry and water solubility, organic synthesis of glycans has proven to be one of the most challenging areas of synthetic organic chemistry (Chapter 53). The combined use of purified glycosyltransferases with chemically synthesized glycan precursors has proven to be invaluable in the stereoselective synthesis of complex glycans (Chapter 54). Automated synthesis of glycans is rapidly becoming a reality, and even the synthesis of the most complex glycans, such as glycosaminoglycans, is becoming possible. In the past five years, major advances have been reported in both the automated chemical and the automated chemico-enzymatic synthesis of complex glycans. In the future, it seems likely that the synthesis of glycans will become nearly as facile and widely available as the synthesis of nucleic acids and proteins. It should be possible for biologists to easily obtain glycoconjugates that have homogeneously uniform glycans to further explore the structural and functional roles of the glycans. The availability of nearly all of the major glycosyltransferases and glycosidases in recombinant form should also greatly increase our ability to modify glycoconjugates for investigations of structure/function relationships. The National Institute of General Medical Sciences (NIGMS) of the NIH has established a Resource for Integrated Glycotechnology that makes widely available cDNAs and proteins for nearly all of the major mammalian glycoenzymes, including glycosyltransferases, glycosidases, and other enzymes involved in glycan synthesis and degradation (Chapters 5 and 6). This resource is greatly facilitating the study of glycans by nonexperts in various areas of biology (http://glycoenzymes.ccrc.uga.edu). In addition, glycoengineering of proteins and lipids will make it possible to generate molecules with novel properties for a wide variety of applications.

Genomics and Enzymology of Glycans

The complete mapping of multiple genomes continues to have a huge impact on the advancement of glycoscience. The identification, sequencing, and classification of genes important to glycosciences have already allowed us to understand the evolution and functions of genes that affect glycosylation ("glycogenes") at an unprecedented rate (Chapters 8 and 52). The wealth of information in these databases has only just begun to be mined. These studies will lead to the identification of novel glycosylation enzymes and should shed light on how various species have evolved to produce their glycoforms and glycotypes. Current projects to make cDNAs, mRNA, and proteins for most of the common glycoenzymes available to the wider research community will greatly facilitate the study of glycoconjugates by all investigators.

An international Computational and Informatics Resource for Glycosciences (GlyGen) supported by NIH has been established, which is a powerful tool to advance the field. Two annotated and searchable databases for O-GlcNAcylated (OGN) proteins have been established, which currently list nearly 8,000 OGN proteins in human cells. The CAZy database (www.cazy.org) of carbohydrate-active enzymes has also played an important role in advancing the field. The overall goal is to coordinate all of these efforts into bioinformatic databases that a nonexpert can explore in a manner similar to current resources for DNA, RNA, and proteins.

The combination of organic synthesis and chemoenzymatic synthesis will allow the production of well-defined glycan arrays that contain the various binding epitopes for glycan-binding proteins. Recombinant glycosyltransferases will become common tools to generate specific glycans on glycoproteins and living cells, and these enzymes will greatly enhance the many studies that now rely on the use of lectins (Chapter 48), whose specificity is often not well-defined. Finally, the rapidly expanding analytical toolkit will be instrumental to decipher the regulatory mechanisms that mediate homeostasis of glycoconjugates at a cellular level. As with analytical methods, it is also important to "democratize" the approaches by developing simple methods for the average biologist working at the bench.

High-Throughput Analyses

In recent years, great advancements in the glycomic profiling of glycans from cells, tissues, or organisms have occurred. Matrix-assisted laser desorption/ionization mass spectrometry (MALDI-MS) has allowed the rapid profiling of the majority of N- and O-glycans in a sample (Chapter 51). As matrix-assisted laser desorption/ionization time-of-flight (MALDI-TOF) instruments also gain the ability to perform tandem mass spectrometry, confidence in the structure of profiled glycans will continue to improve. However, detailed structural analysis of individual glycans still remains a low-throughput method that requires considerable expertise and expensive instrumentation. The development of glycan arrays presenting hundreds of different glycans of defined structure is perhaps one of the most impactful recent developments in glycobiology (Chapter 56). These arrays have allowed for rapid, high-throughput analysis of the binding specificities of numerous biologically important glycan-binding proteins, including lectins, antibodies, viruses, and bacteria. It has been estimated that these arrays will need to contain between 10,000 and 20,000 different glycans to cover most possible glycotopes (this highly conservative estimate does not take into account various common and labile modifications of glycans). In addition, more specialized arrays will be required to study glycan-binding proteins made by prokaryotes or to study proteins that bind to glycosaminoglycans, for example. Fortunately, the ability to make such large and diverse glycan arrays is on the horizon, and the remaining challenge will be to better understand the parameters of the glycan arrays, besides the glycans themselves, that influence binding, such as linkers, spacing, and density of glycans on solid supports. Mixtures of glycans need to be studied next, to mimic "clustered saccharide patches" that more closely mimic natural states.

Significant progress has also been made in Systems Glycobiology by better integration of glycomics with genomics, transcriptomics, and proteomics. Our understanding as to how miRNAs and transcription factors regulate glycan expression has also improved in the last few years (Chapters 51 and 52).

Need for More Facile Methods to Explore Glycobiology

At the present time, the study of glycan structure/function relationships requires highly specialized expertise in synthetic and analytical chemistry. The inherent difficulty in studying glycans remains a major challenge preventing glycobiology—the study of glycan function—from entering mainstream conceptual frameworks of biology. In response to this challenge, strong emphasis has recently been placed on the development of facile technologies that will allow non-glycobiologists without specialized analytical chemistry skills to study glycans in a meaningful way. In fact, the use of glycosyltransferase and glycosidase probes to manipulate glycans is not any more difficult than the use of restriction enzymes and endonucleases to analyze DNA. The use of antibodies and lectins specific for glycans also exemplifies methods that can be readily applied by biologists lacking expensive analytical equipment or extensive glycobiology expertise (Chapter 49). In fact, several kits for the study of glycans have emerged, as companies recognize the future market potential of glycobiology. In addition, companies are offering highly purified enzymes, antibodies, and chemistries, making it much easier to study glycans. In contrast, few glycans are available for experimental purposes and most enzymes that act on glycans are not commercially available. Equally important, most investigators are not trained to know what questions to ask regarding the roles of glycans attached to their favorite glycoconjugate. In addition to new and facile methods to study glycans, education of the next generation of cell biologists and biochemists with respect to the importance of glycans will be key to advancing glycoscience to the next level.

SOME MAJOR FUNDAMENTAL QUESTIONS FOR THE FUTURE

It is clear from the chapters in this book that glycans affect all aspects of life on this planet. Yet, there is still a great deal to learn about the functions of glycans in biology. Below is a sampling of some "big" questions in glycobiology that remain to be answered in the future.

How Common Is the Occurrence, and What Are the Functions of "glycoRNAs" on the Surfaces of Mammalian Cells?

Recently, using chemical and biochemical approaches, it was reported that small noncoding RNAs bear N-glycan-type (Chapter 9) sialylated (Chapter 15) glycans. These so-called "glycoRNAs" were found on the surfaces of multiple cell types and mammalian species, in cultured cells, and in vivo. This remarkable finding, if confirmed by other researchers, may bridge RNA biology and glycobiology and expand the roles of RNA at the cell surface.

What Are the Roles of Glycans in the Organization of the Components in the Plasma Membrane, Glycocalyx, and Extracellular Matrix?

Glycans affect the stability and turnover of cell-surface resident molecules, and many can self-associate in the plane of the membrane. Do they play a role in the organization and regional concentration of molecules at the cell surface? How do the various glycoconjugates on cell surfaces collaborate among themselves and with glycan-binding proteins to organize the glycocalyx of a given cell or its extracellular matrix? After all, most living cells in a multicellular organism exist within a glycan-enriched gel, not in the artificially fluid medium of the tissue culture dish.

How Do Glycans Regulate Cellular Signaling from the Cell Surface to the Nucleus?

Complex glycans clearly regulate the function, stability, and residence time of receptors at the cell surface, and the addition of O-GlcNAc to nuclear and cytoplasmic proteins regulates many of the signaling pathways downstream from these receptors. Yet, still very little is known about the mechanistic roles of glycans in these pathways.

Is There a Biological Significance for Site-Specific Glycan Diversity on Glycoproteins?

Glycoproteins typically contain an array of many glycans with different structures at individual sites in the polypeptide. Evidence suggests that these arrays of glycans are site-specific and cell type–specific, and not random in distribution. How does site-specific diversity arise and how is it controlled? Does each glycoform have different biological functions or variable strengths and/or does the site-specific diversity control molecular interactions? Or is the complexity the natural outcome of evolutionary arms races between hosts and their pathogens, which exploit glycans in various ways?

Can Specific Glycoforms of Glycoproteins Be Used to Improve Biomarkers for Disease?

Many investigators are trying to discover biomarkers to help diagnose disease, such as cancer. It has been proposed that identification of specific glycoforms of a glycoprotein might improve the specificity and sensitivity of these biomarkers.

How Can the Marked Mobility of Glycans in Biological Systems Be Taken into Account?

Unlike most other biomolecules, glycans have a substantial degree of mobility in aqueous solution. This further increases the potential information content of these molecules but also raises further challenges for their analysis and biological exploration.

How Do Cell-Surface Glycans and Matrix Glycans Contribute to Cancer Progression and Metastasis?

Data from glycan analysis, animal models, and correlations from the clinic strongly indicate that cell-surface glycans are critical to tumor progression and metastasis. Yet mechanistically, relatively little is known with respect to the roles glycans play in these processes.

The Glycosciences Should Have a Huge Impact on Antiviral, Antibacterial, and Antifungal Diagnostics and Therapeutics—How Does One Make This Happen?

It is known that glycans are critical to infectious disease and are key antigens in the generation of several vaccines. As conventional antibiotics become less effective, how does one exploit glycans to fight infectious disease?

Can Glycan Analytical Methods Be Improved to Avoid Loss of Labile Modifications?

Most current analytical methods for glycans result in a marked loss of labile modifications, particularly ester groups of various kinds. Given their location and potential biological importance, there is likely a large unexplored area of glycobiology that can only be approached with new techniques.

What Is the True Extent of Diversity of the Nonulosonic Acids and What Are the Functions of This Diversity?

The discovery that the highly diverse sialic acids are just the proverbial tip of the iceberg of the ancestral family of nonulosonic acids has raised many new questions. It appears that investigators are just beginning to scratch the surface of this enormous diversity and its potential biological significance.

How Can One Decode the Information Content in Glycosaminoglycans and Proteoglycans at the Molecular Level?

The potential information content of glycosaminoglycans and proteoglycans is enormous. Yet, investigators are only just beginning to be able to understand how the arrangement of sulfate esters and uronic acids affects affinity and protein-binding selectivity. Elucidating how such diversity is regulated temporally and spatially is also a major challenge going forward.

What Are the Molecular Mechanisms by Which O-GlcNAcylation Regulates Transcription and Signaling in Response to Nutrients?

Current data indicate that O-GlcNAc serves as a nutrient sensor to regulate signaling and transcription. Yet, mechanistically, almost nothing is known with respect to how the cycling of this sugar on and off proteins regulates these processes at the molecular level. More than 40% of proteins in the brain are O-GlcNAcylated. What are the roles of O-GlcNAc cycling at the synapse, in brain development, and in learning and memory?

Has Nature Only Made Use of a Small Subset of All the Possible Monosaccharides?

There are hundreds of known monosaccharides, and many can be present in different ring and/or modified forms. So far, it appears that natural systems have only incorporated a fraction of this diversity. Is this because of rate limits on genomic and enzymatic evolution in the opportunity to incorporate new monosaccharides or is this an ascertainment bias, owing to the fact that only a small fraction of species on the planet has been sampled. Is it possible that every possible monosaccharide has been used somewhere in nature?

What Are the Complete Structures of All the Glycans within the Human Glycome or Any Animal or Plant Glycome?

Every day witnesses new glycan structures being identified, with previously unknown and novel sugar linkages, along with unique and unanticipated modifications, such as acetylation, methylation, phosphodiesters, and sulfation. How many such glycan structures exist within a glycome, to what glycoproteins or lipids are they attached, and where are they attached? Much of our knowledge of the human glycome comes from studying serum glycoproteins and several cell lines, but there is little information about the overall complexity of the human glycome either at the glycan level or the glycoprotein level.

What Is the True Extent of Diversity of Glycans, Glycan-Binding Proteins, and Glycan-Degrading Enzymes in the Microbiome?

It is clear that investigators are just beginning to scratch the surface of the organismal diversity within the complex diverse microbiomes that occupy various niches in nature. Each time a new species within a given microbiome is sequenced, the genomes are found to predict hundreds of "carbohydrate-active" enzymes. Understanding the mechanisms and functions of all these molecules is a major challenge for the future, as is the deciphering of the structures of the complex glycans they generate, bind, or degrade.

Do More Complex Biological Systems Have Simpler Glycosylation?

Of the hundreds of known monosaccharides in nature, less than a dozen are found in vertebrates, and many more tend to be found in simpler, earlier evolved systems. Is this an ascertainment bias or does increasing biological complexity place limits on the fraction of the genome that can be committed to controlling glycosylation?

What Is the Significance of Lineage-Specific Gains and Losses of Specific Glycans during Evolution?

Current data indicate that glycans are the most rapidly diverging components of life-forms, with distinct gains and losses that sometimes become (or stay) polymorphic in various lineages and clades. It seems likely that this is a consequence of the intimate involvement of glycans in the most rapidly evolving aspects of biological systems (i.e., host–pathogen interactions, immunity, and reproduction). Further studies are needed to ascertain if glycans are indeed involved in the process of speciation.

When Will the Integration of Glyosciences into the Mainstream of Bioinformatics Be Achieved?

Currently, searches of the main genome and protein databases used by essentially all biologists yield almost no information on the glycosylation state of glycoproteins. Indeed, there is as yet no central comprehensive database of glycan structures that an investigator can mine to determine the occurrence, provenance, or phylogeny of a specific glycan of interest. An international effort led by nearly every major country involved in the glycosciences has come together to agree on uniform standards for describing glycans in computer databases and to develop methods to curate and maintain glycan structural data. Together with the expanding use of uniform symbol nomenclature (Chapter 1), this is the first step toward integrating glycomic data into mainstream protein databases (Chapter 52).

When Will Glycobiology Completely Merge into a Holistic Approach to Biology?

As with other fields like protein sciences, there will always be a small cadre of investigators whose primary interest is in the structure, chemistry, biochemistry, and biology of glycans. But a combination of advances in methodology and education will eventually integrate glycosciences into the general awareness of biologists who are currently not trained to understand the significance of glycans. As our understanding of the functions of glycans increases, biochemists and biologists will come to recognize that glycan modifications are no less important than the amino acids that make up the polypeptide backbone or the nucleotides and nucleotide derivatives that make up DNA and RNA. In the long run, glycobiology will merge into a holistic approach to biological systems. When that intellectual singularity is eventually achieved, further editions of this book may no longer be necessary.

FURTHER READING

National Research Council Committee (US) on Assessing the Importance and Impact of Glycomics and Glycosciences. 2012. *Transforming glycoscience: a roadmap for the future.* National Academies Press, Washington, D.C.

Agre P, Bertozzi C, Bissell M, Campbell KP, Cummings RD, Desai UR, Estes M, Flotte T, Fogleman G, Gage F. 2017. Training the next generation of biomedical investigators in glycosciences. *J Clin Invest* **126:** 405–408. doi:10.1172/jci85905

Glossary

ABO: Gene locus comprising three major allelic glycosyltransferases that generate the A, B, and O blood groups.

Acetal: An organic compound derived from a hemiacetal by reaction with an alcohol. If the hemiacetal is a sugar, the acetal is a glycoside.

Adhesin: A protein on the surface of bacteria, viruses, or parasites that binds to a ligand present on the surface of a host cell.

Affinity: A measure of the strength of interaction between a receptor and its ligand.

Agglutination: The clumping of cells in the presence of a protein (e.g., antibody or lectin). The related term *hemagglutination* denotes the specific case wherein the cells are red blood cells.

Aglycone: Non-carbohydrate portion of a glycoconjugate or glycoside that is glycosidically linked to the glycan through the reducing terminal sugar.

Aldose: A monosaccharide with an aldehyde group or potential aldehydic carbonyl group (by definition, this is the C-1 position).

Amino sugar: A monosaccharide in which a hydroxyl group is replaced by an amino group.

Anomeric carbon: The carbon atom of a monosaccharide that bears the hemiacetal functionality (C-1 for most sugars; C-2 for sialic acids).

Anomers: Stereoisomers of a monosaccharide that differ only in configuration at the anomeric carbon of the ring structure.

Antenna: A branch of an oligosaccharide emanating from a "core" structure.

Arixtra: *See* **Fondaparinux**.

Asparagine-linked oligosaccharide: *See* **N-glycan**.

Automated glycan assembly: Rapid method for the chemical synthesis of oligo- and polysaccharides on a solid support.

Avidity: A measure of the combined strength of interaction from the multiple affinities of a multivalent complex.

Azide: A functional group comprising three nitrogen atoms bound in a linear arrangement (N_3).

Azido sugar: A monosaccharide to which an azido group has been introduced synthetically.

Bactoprenol: *See* **Undecaprenol**.

Beta elimination: The cleavage of a C-O or C-N bond positioned on the beta carbon with respect to a carbonyl group. The process is used to cleave O-glycans from Ser or Thr residues.

Biofilm: Community of bacteria that adheres to a moist surface (e.g., surface of ponds or teeth).

C-type lectins: A class of Ca^{++}-dependent lectins recognizable by a characteristic sequence comprising their carbohydrate recognition domain.

Calnexin: Membrane-bound protein chaperone that mediates quality control of protein folding in the endoplasmic reticulum.

Calreticulin: Soluble protein chaperone that recognizes N-glycans and mediates quality control of glycoprotein folding in the endoplasmic reticulum.

Capillary electrophoresis (CE): An analytical technology using high voltage across the span of a small-diameter capillary to accomplish separation. It is applicable to small quantities of carbohydrates and can interface with a mass spectrometer.

Capsule: A protective extracellular polysaccharide coat surrounding certain bacteria. Presence of a capsular polysaccharide is often associated with virulence.

Carbohydrate: A generic term used interchangeably in this book with sugar, saccharide, or glycan. Includes monosaccharides, oligosaccharides, and polysaccharides, as well as derivatives of these compounds.

Carbohydrate-recognition domain (CRD): The domain of a polypeptide that is specifically involved in binding to carbohydrate; in lectins, often a highly evolutionarily conserved region of the polypeptide.

CAZy database: Denoting "Carbohydrate Active enZYmes," this database describes the families of structurally related catalytic and carbohydrate-binding modules (or functional domains) of enzymes that degrade, modify, or create glycosidic bonds.

Cellulose: A repeating homopolymer of β1-4-linked glucose residues, which is the main constituent of plant cell wall of green plants and many forms of algae and oomycetes and is produced by some bacteria.

Ceramide: The common lipid component of glycosphingolipids, composed of a long-chain amino alcohol (sphingosine) and an amide-linked fatty acid.

Cerebroside: A glycolipid composed of ceramide with an attached galactose (galactosylceramide) or glucose (glucosylceramide).

Chemical shift: A term referring to the position of a resonance in an NMR spectrum.

Chemoenzymatic synthesis: Glycan synthesis that uses both chemical and enzymatic reactions to obtain the desired product.

Chitin: A repeating homopolymer of β1-4-linked *N*-acetylglucosamine residues; the main component of the cell walls of fungi and the exoskeletons of arthropods, among other functions.

Chondroitin sulfate: A type of glycosaminoglycan defined by the disaccharide unit $(\text{GalNAc}\beta\text{1-4GlcA}\beta\text{1-3})_n$, modified with ester-linked sulfate at certain positions and typically found covalently linked to a proteoglycan core protein.

Congenital disorder of glycosylation: An inheritable genetic disorder in which mutations have led to improper assembly of glycans.

Conjugate vaccine: A vaccine consisting of an antigen (frequently a glycan) coupled to a carrier protein.

Coupling correlated spectroscopy (COSY): An NMR technique producing a two-dimensional map of connections, usually between vicinal protons. Useful in assignment of spectra and identification of carbohydrate residues.

Cryo-electron microscopy (cryo-EM): An imaging technique capable of providing near-atomic level resolution for three-dimensional structures of proteins and glycoproteins in frozen noncrystalline preparations.

Deoxy sugar: A monosaccharide in which a hydroxyl group is replaced by a hydrogen atom.

Dermatan sulfate: A modified form of chondroitin sulfate in which a portion of the D-glucuronate residues are epimerized to L-iduronates.

Dolichol: A terminally saturated polyisoprenoid lipid carrier utilized during the assembly of N-glycans and GPI anchors and the O- and C-mannosylation of proteins in the endoplasmic reticulum.

Electron transfer dissociation (ETD): A mass spectrometry fragmentation and ionization technique useful in determining sites of glycosylation on peptides and proteins.

Electrospray ionization (ESI): A commonly used method for producing charged species for mass spectrometry analysis.

Endoglycosidase: An enzyme that catalyzes the cleavage of an internal glycosidic linkage in an oligosaccharide or polysaccharide.

Endotoxin: *See* **Lipid A.**

Epidermal growth factor (EGF)-like repeats: Small protein motifs (~40 amino acids) with six conserved cysteine residues that form three disulfide bonds. Certain EGF repeats may contain sites for glycan modification.

Epimerase: An enzyme that catalyzes racemization of a chiral center in a sugar.

Epimers: Two isomeric monosaccharides differing only in the configuration of a single chiral carbon. For example, mannose is the C-2 epimer of glucose.

Epitope: The part of a molecule that is recognized by a specific antibody or receptor.

Erythropoietin: A circulating glycosylated cytokine used to treat anemias.

Exoglycosidase: An enzyme that cleaves a monosaccharide from the outer (nonreducing) end of an oligosaccharide, polysaccharide, or glycoconjugate.

Exotoxins: Heat-labile, proteinaceous toxins secreted by bacteria that cause illness.

Expressed protein ligation (EPL): A method for generating semisynthetic proteins by the condensation of a synthetic peptide and a recombinant protein. Glycoproteins can be generated by condensation of a synthetic glycopeptide and a recombinant protein.

Extracellular matrix: A complex array of secreted molecules including glycoproteins, proteoglycans, and/or polysaccharides and structural proteins. In plants, the extracellular matrix is also referred to as the cell wall.

Extrinsic glycan-binding proteins: Receptors that recognize glycans from a different organism and consist mostly of pathogenic microbial adhesins, agglutinins, or toxins.

Fimbrae: Proteinaceous fiber-like appendages found in many Gram-negative bacteria.

Fischer projection: A two-dimensional representation of a three-dimensional organic molecule devised by Hermann Emil Fischer.

Fluorophore-assisted carbohydrate electrophoresis (FACE): A technology that combines glycan derivatization and gel electrophoresis for the analysis of small quantities of carbohydrates.

Fondaparinux: A synthetic heparin (also called Arixtra) used as an anticoagulant.

Fringe: Family of proteins that modify Notch activity by catalyzing the transfer of *N*-acetylglucosamine from UDP-GlcNAc to fucose on an EGF-like repeat.

Furanose: Five-membered (four carbons and one oxygen, i.e., an oxygen heterocycle) ring form of a monosaccharide named after the structurally similar compound furan.

Galectins: S-type (sulfhydryl-dependent) β-galactoside-binding lectins, usually occurring in a soluble form, expressed by a wide variety of animal cell types and distinguishable by the amino acid sequence of their carbohydrate recognition domains.

Ganglioside: Anionic glycosphingolipid containing one or more residues of sialic acid.

Gene chip: A DNA microarray used to quantify transcript levels in high-throughput format.

Genome: The complete genetic sequence of one set of chromosomes.

Glycan: A generic term for any sugar or assembly of sugars, in free form or attached to another molecule, used interchangeably in this book with saccharide or carbohydrate.

Glycan array: A collection of glycans attached to a surface in a spatially determined manner.

Glycan-binding protein: Protein that recognizes and binds to specific glycans. *See* **Lectin** and **Glycosaminoglycan-binding protein.**

Glycation: The nonenzymatic, chemical modification of proteins by addition of carbohydrate, usually through a Schiff-base reaction with the amino group of the side chain of lysine and subsequent Amadori rearrangement to give a stable conjugate. Not to be confused with (enzymatic) glycosylation.

Glycobiology: Study of the structure, chemistry, biosynthesis, and biological functions of glycans and their derivatives.

Glycoengineering: Altering the biosynthetic machinery for glycosylation in a given cell for the production of defined glycans on glycoconjugates.

Glycocalyx: The cell coat consisting of glycans and glycoconjugates surrounding animal cells that is seen as an electron-dense layer by electron microscopy.

Glycoconjugate: A molecule in which one or more glycan units are covalently linked to a noncarbohydrate entity.

Glycoforms: Different molecular forms of a glycoprotein, resulting from variable glycan structures at one or more sites of glycosylation and/or differences in the occupancy of glycosylation sites.

Glycogen: A polysaccharide comprising α1-4- and α1-6-linked glucose residues that functions in short-term energy storage in animals; sometimes referred to as animal starch.

Glycogenin: A protein that acts as a primer for glycogen synthesis.

Glycolipid: General term denoting a molecule containing a saccharide linked to a lipid aglycone. In "higher" organisms, most glycolipids are glycosphingolipids, but glycoglycerolipids and other types exist.

Glycome: The total collection including a glycon of glycans synthesized by a cell, tissue, or organism under specified conditions of time, space, and environment.

Glycomics: Systematic analysis of the glycome.

Glycomimetics: Noncarbohydrate compounds that mimic the properties of saccharides.

Glycone: Carbohydrate component of a glycoconjugate.

Glycopeptide: A peptide having one or more covalently attached glycan.

Glycoprotein: A protein with one or more covalently bound glycans.

Glycoproteomics: The systems-level analysis of glycoproteins, including their protein identities, sites of glycosylation, and glycan structures.

Glycosaminoglycans: Polysaccharide side-chains of proteoglycans or free complex polysaccharides composed of linear disaccharide repeating units, each composed of a hexosamine and a hexose or a hexuronic acid (*see* **Heparin, Heparan sulfate, Chondroitin sulfate, Dermatan sulfate, Keratan sulfate,** and **Hyaluronan**).

Glycosaminoglycan-binding protein: Protein that recognizes and binds to specific glycosaminoglycans.

Glycosidase: An enzyme that catalyzes the hydrolysis of glycosidic bonds in a glycan. *See* **Exoglycosidase** and **Endoglycosidase.**

Glycoside: A glycan containing at least one glycosidic linkage to another glycan or an aglycone.

Glycosidic linkage: Linkage of a monosaccharide to another residue. The linkage generally results from the reaction of a hemiacetal with an alcohol (e.g., a hydroxyl group on another monosaccharide or amino acid) to form an acetal. Glycosidic linkages between two monosaccharides have defined regiochemistry and stereochemistry.

Glycosphingolipid: Glycolipid containing a glycan glycosidically attached to the primary hydroxyl group of ceramide.

Glycosyl acceptor: The nucleophile in a glycosylation reaction, usually containing a free hydroxyl group.

Glycosyl donor: The electrophile in a glycosylation reaction; the nucleotide sugar in an enzymatic glycosylation reaction.

Glycosylation: The enzyme-catalyzed covalent attachment of a carbohydrate to a polypeptide, lipid, polynucleotide, carbohydrate, or other organic compound, generally catalyzed by a glycosyltransferase, utilizing a specific nucleotide sugar donor.

Glycosylphosphatidylinositol (GPI) anchor: A membrane anchor that consists of a glycan bridge between phosphatidylinositol and a phosphoethanolamine in amide linkage to the carboxyl terminus of a protein.

Glycosyltransferase: Enzyme that catalyzes transfer of a sugar from a nucleotide sugar or sugar-phosphate lipid donor to a substrate.

Hapten: Any small molecule, including a glycan, that is recognized by a receptor or antibody.

Haworth projection: A representation of monosaccharides wherein the cyclic structures are depicted as planar rings with the hydroxyl groups orientated above or below the plane of the ring.

Hemagglutination: The clumping of red blood cells in the presence of a protein (e.g., antibody or lectin).

Hemagglutinin: A lectin that recognizes carbohydrates on the surface of red blood cells and causes hemagglutination.

Hemiacetal: A compound formed by reaction of an aldehyde with an alcohol group, as in ring closure of an aldose.

Hemiketal: A compound formed by reaction of a ketone with an alcohol group, as in ring closure of a ketose.

Heparan sulfates: Glycosaminoglycans defined by the disaccharide unit (GlcNAcα1- 4GlcAβ1-4/IdoAα1-4)$_n$, containing N- and O-sulfate esters at various positions, and typically found covalently linked to a proteoglycan core protein.

Heparin: A type of heparan sulfate made by mast cells that has the highest amount of iduronic acid and of N- and O-sulfate residues. Heparin binds and activates antithrombin.

Heteronuclear single quantum coherence (HSQC): An NMR technique producing a two-dimensional map correlating chemical shifts of heteronuclei (usually ^{13}C for carbohydrates) and chemical shifts of directly bonded protons.

Heteropolysaccharide: A polysaccharide containing more than one type of monosaccharide.

Hexosamine: Hexose with an amino group in place of the hydroxyl group at the C-2 position. Common examples found in vertebrate glycans are the N-acetylated sugars, N-acetylglucosamine and N-acetylgalactosamine.

Hexose: A 6-carbon monosaccharide typically with an aldehyde (or potential aldehyde) at the C-1 position (aldohexose) and hydroxyl groups at all other positions. Common examples in vertebrate glycans are mannose, glucose, and galactose.

High pressure liquid chromatography (HPLC): A separation technique frequently used to isolate glycans for analysis or subsequent study.

Homopolysaccharide: A polysaccharide composed of only one type of monosaccharide.

Hyaluronan: A glycosaminoglycan defined by the disaccharide unit $(GlcNAc\beta 1\text{-}4GlcA\beta 1\text{-}3)_n$ that is neither sulfated nor covalently linked to protein; referred to in older literature as hyaluronic acid.

Hydrazinolysis: A chemical method that uses hydrazine to cleave amide bonds (e.g., the glycosylamine linkage between a sugar residue and asparagine or the acetamide bond in N-acetylhexosamines).

I-type lectins: A class of lectins belonging to the immunoglobulin superfamily.

Intrinsic glycan-binding proteins: Receptors that recognize glycans from the same organism. Typically they mediate cell–cell interactions or recognize extracellular molecules, but they can also recognize glycans on the same cell.

Jelly-roll fold: Description of tertiary structure common to L-type lectins.

Keratan sulfate: A polylactosamine $[Gal\beta 1\text{-}4GlcNAc\beta 1\text{-}3]_n$ with sulfate esters at C-6 of N-acetylglucosamine and galactose residues, found in a keratan sulfate proteoglycan.

Ketal: An organic compound derived from a hemiketal by reaction with an alcohol. If the hemiketal is a sugar, the ketal is a glycoside.

Ketose: A monosaccharide with a ketone group or a potential ketonic carbonyl group (typically at the C-2 position in natural compounds).

L-type lectins: Superfamily of glycan-binding proteins with a common feature of tertiary structure called a "jelly-roll" fold.

Lactose: The disaccharide Galβ1-4Glc; an abundant milk sugar.

Lectin: A protein (other than an anticarbohydrate antibody) that specifically recognizes and binds to glycans without catalyzing a modification of the glycan.

Lewis blood group antigens (e.g., Le^x, Le^y, and Le^a): A related set of glycans that carry α1-3/1-4 fucose residues covalently linked to galactose or N-acetylglucosamine.

Ligand: A molecule that is recognized by a specific receptor. In the case of lectins, the ligands are partly or completely glycan-based.

Link module: A protein fold that interacts specifically with hyaluronan.

Linkage analysis: A technology employing a combination of derivatization of hydroxyl groups, gas chromatography (GC) separation, and mass spectrometry to identify types between monosaccharides in a glycan.

Lipid A (also known as endotoxin): Lipid that contains fatty acids linked to glucosamine with a variable number of phosphate groups and 1–4 units of ketodeoxyoctulosonic acid (Kdo). *See* **Lipopolysaccharide.**

Lipid-linked oligosaccharide (LLO): An oligosaccharide linked to dolichol or lipid.

Lipid rafts: Small lateral microdomains of self-associating membrane molecules.

Lipooligosaccharide (LOS): Similar to lipopolysaccharide but lacking the O-antigen polysaccharide side chain repeats.

Lipopolysaccharide (LPS): A bacterial glycolipid composed of a polysaccharide (O-antigen), connected via a core oligosaccharide to lipid A that makes up the major portion of the outer leaflet of the outer membrane of Gram-negative bacteria. A major determinant of antigenic specificity, also known as heat-stable toxin or endotoxin.

Lysosomal storage disorder: Human genetic disorder in which a defect in a lysosomal enzyme results in the accumulation of undigested glycoconjugates in the lysosomes (e.g., Tay–Sachs disease).

Lysozyme: An endo-β-N-acetylhexosaminidase that cleaves the polysaccharide backbone of bacterial peptidoglycan.

Mannan: Mannose-rich polysaccharide found in certain bacteria, fungi, and plants.

Mannose 6-phosphate receptors: *See* **P-type lectins.**

Mass spectrometry (MS): An analysis technique providing the mass of ionizable glycans and their fragments in the gas phase. The small sample amounts required make it particularly useful for glycan analysis.

Matrix-assisted laser desorption/ionization (MALDI): A procedure commonly used to produce charged species for mass spectrometric analysis.

Membrane-derived oligosaccharides (MDOs): Highly charged β-glucans that create an osmotic buffer in the periplasmic space of Gram-negative bacteria.

Metabolic labeling: A procedure dependent on metabolic processes in cells to incorporate isotopic or derivatized monosaccharides (or other moieties) into glycans for subsequent analysis.

Methylation analysis: A method for carbohydrate structure analysis based on the acid stability of methyl ethers and the acid lability of glycosidic linkages; used to determine the linkage positions of monosaccharide residues in an oligosaccharide chain.

Michael addition: The chemical reaction in which a nucleophile attacks the β carbon of an α,β-unsaturated carbonyl compound. The reaction is used after O-glycan β elimination in order attach probes to those sites.

Microarray: A collection of molecules (e.g., DNA, proteins, or glycans) spatially arrayed on a surface with micrometer dimensions.

Microheterogeneity: Structural variations in a glycan at any given glycosylation site on a protein (microheterogeneity generates glycoforms).

Molecular mimicry: Strategy some microbial pathogens use to evade immune reactions by decorating themselves with glycans similar to their hosts.

Molecular dynamics (MD): A computational technique based on Newton's laws of motion useful in simulating motions and structures of glycans or glycan–protein complexes.

Molecular docking: A computational technique useful in predicting binding sites and geometry for glycan–protein complexes.

Monosaccharide: Carbohydrate that cannot be hydrolyzed into a simpler carbohydrate. The building block of oligosaccharides and polysaccharides. Simple monosaccharides are polyhydroxyaldehydes or polyhydroxyketones with three or more carbon atoms.

Mucin: Large glycoprotein with a high content of serine, threonine, and proline residues and numerous O-GalNAc glycans, often occurring in clusters on the polypeptide.

Mucopolysaccharide: An out-of-date term replaced by the term, glycosaminoglycan. Still used as a group name for human disorders ("mucopolysaccharidoses") involving glycosaminoglycan accumulation due to genetic deficiency of certain lysosomal enzymes.

Mutarotation: The interconversion of stereoisomers at the anomeric center of a monosaccharide.

Multivalent: Having multiple points of interaction. Often seen in oligomeric lectins in which low affinity individual interactions combine to provide high avidity.

N-acetyllactosamine: A disaccharide with the sequence Galβ1-4GlcNAc.

N-glycan (N-linked oligosaccharide, N-linked glycan): Glycan covalently linked to the side-chain amide of asparagine residue of a polypeptide chain in the consensus sequence: -Asn-X-Ser/Thr. Unless otherwise stated, the term N-glycan is used generically in this book to denote the most common linkage region, Manβ1-4GlcNAcβ1-4GlcNAcβ1-N-Asn.

Native chemical ligation: (NCL) A technique used to generate large polypeptides by condensation of smaller peptide fragments.

Neuraminidase: *See* **Sialidase.**

Nod factor: Lipooligosaccharide produced by *Rhizobium* bacteria that stimulates nodule formation and initiates nitrogen fixation in leguminous plants.

Nonreducing terminus (nonreducing end): Outermost end of an oligosaccharide or polysaccharide chain, opposite to that of the reducing end.

Notch: Family of cell-surface receptors that are glycosylated on EGF-like repeats. Ligands include Delta and Serrate/Jagged.

Nuclear magnetic resonance (NMR): A spectroscopic technique based on detecting precession of magnetically active nuclei in strong magnetic fields. Useful in structure determination of both glycans and protein–glycan complexes as they exist in solution.

Nuclear Overhauser effect (NOE): An effect on intensities of resonances in NMR spectra that is often related to internuclear distances in the case of proton–proton experiments.

Nucleotide sugar: Activated form of monosaccharide, such as UDP-Gal, GDP-Fuc, and CMP-Sia, typically used as a donor substrate by glycosyltransferases.

Nucleotide sugar transporter: Membrane-bound protein that specifically transports nucleotide sugars from the cytosol into the lumen of intracellular organelles (e.g., the Golgi).

O-GalNAc glycan: *See* **O-glycan.**

O-GlcNAcylation: Dynamic modification of proteins by β-linked N-acetylglucosamine (a posttranslational modification often reciprocal with protein phosphorylation).

O-glycan (O-linked oligosaccharide, O-linked glycan): A glycan glycosidically linked to the hydroxyl group of the amino acid serine, threonine, tyrosine, or hydroxylysine. Unless otherwise stated, the term O-glycan is used in this book to denote the common linkage GalNAcα1-O-Ser/Thr.

Oligosaccharide: Linear or branched chain of monosaccharides attached to one another via glycosidic linkages. The number of monosaccharide units can vary; the term polysaccharide is usually reserved for large glycans with repeating units.

Peptidoglycan: A bacterial polysaccharide consisting of MurNAcβ1-4GlcNAcβ1-4 repeat units, covalently cross-linked to short peptides. Also known as murein, peptidoglycan represents the major structural component of the periplasm.

Periodate oxidation: A reaction using periodate to cleave C–C bonds with vicinal hydroxyl groups (e.g., within carbohydrates) to form the two corresponding aldehydes.

Pili (Fimbriae): Hair-like appendages on the surface of some bacteria that often contain adhesins.

Polyisoprenoid: A lipid polymer composed of repeating units of the unsaturated 5-carbon isoprene unit. *See* **Dolichol** and **Undecaprenol.**

Poly-N-acetyllactosamine: Repeating units of N-acetyllactosamines [Galβ1-4GlcNAcβ1-3]$_n$, of variable length (sometimes called PolyLacNAc).

Polymerase chain reaction (PCR): The process used to amplify DNA starting from a template DNA strand and complementary oligonucleotide primers.

Polysaccharide: Glycan composed of repeating monosaccharides, generally greater than ten monosaccharide units in length.

Polysialic acid: A homopolymer of sialic acids abundant in the brain and fish eggs and found on certain pathogenic bacteria.

Protecting group: A chemical moiety commonly used in glycan synthesis that masks hydroxyl groups in order to prevent them from reacting with other chemical reagents.

Protein data bank (PDB): A repository containing atomic coordinates primarily for proteins, but also nucleic acids and protein-carbohydrate complexes. Each entry is given a four-element alpha-numeric code.

Proteoglycan: Any protein with one or more covalently attached glycosaminoglycan chains.

Proteome: The total collection of proteins in a cell, tissue, or organism, under specific conditions of time, space, and environment.

P-type lectins: Class of lectins that recognize mannose-6-phosphate (also called M6P receptors).

Pulsed amperometric detection: A detection procedure based on conduction at high pH and used in HPLC analysis of monosaccharide composition.

Pyranose: Six-membered (five carbons and one oxygen, i.e., an oxygen heterocycle) ring form of a monosaccharide; the most common form found for hexoses and pentoses. The name is based on the structural similarity to the compound "pyran."

R-type lectins: Superfamily of glycan-binding proteins that contain a carbohydrate-recognition domain similar to that in ricin.

Receptor: A protein that binds to a ligand and initiates signal transmission or other cellular activity. In this book, most receptors are lectins (i.e., they recognize glycans). In microbiology, adhesins or agglutinins on the microbes bind to receptors which are glycans on the host cell.

Reducing terminus (reducing end): End of a glycan that has reducing power because it is unattached to an aglycone and is thus a hemiacetal. In a glycoconjugate, reducing terminus is also used as a synonym for a potential reducing terminus, referring to the end of a glycan covalently attached to the aglycone by a glycosidic bond (i.e., it would have reducing power if it were released).

Regiochemistry: The region among many possible regions of a molecule that is involved in a chemical reaction. For glycosidic linkages, regiochemistry denotes the hydroxyl group of one monosaccharide that is bound to the anomeric position of the other (i.e., 1-3 vs. 1-4 linkage).

S-layer (surface-layer): A protein monolayer coating often containing covalently linked glycans and found in the cell envelope of many bacteria and archea.

Saccharide: A generic term for any carbohydrate or assembly of carbohydrates, in free form or attached to another molecule, used interchangeably in this book with carbohydrate and glycan.

Saccharolipid: A glycoconjugate comprising fatty acyl chains covalently attached directly to a sugar backbone (e.g., Lipid A).

Saturation transfer difference (STD): An NMR technique useful in detecting protein–glycan binding and identifying parts of glycans in close approach to protons in the binding sites of protein receptors.

Selectin: A C-type (Ca^{++}-dependent) lectin expressed by cells in the vasculature and bloodstream. The three known selectins are L-selectin/CD62L (expressed by most leukocytes), E-selectin/CD62E (expressed by cytokine-activated endothelial cells), and P-selectin/CD62P (expressed by activated endothelial cells and platelets).

Ser/Thr-linked oligosaccharide: *See* **O-glycan.**

Sialic acids: Family of acidic sugars with a nine-carbon backbone, of which the most common is *N*-acetylneuraminic acid, in vertebrates.

Sialidase: Enzyme that releases sialic acid residues from a glycoconjugate. Older name was neuraminidase, now used only to refer to the influenza sialidase.

Sialome: Total array of sialic acid types and linkages expressed by a particular cell tissue or organism, under specified conditions of time, space, and environment.

Siglec: Sialic acid–binding protein that is a member of the I-type lectin family and has an amino-terminal V-set domain with typical conserved residues.

Single-particle analysis (SPA): An analysis technique used in some cryo-EM studies that extracts 3D structures from 2D images of randomly oriented objects.

Small-angle X-ray scattering (SAXS): An X-ray scattering technique useful in defining the shape of large molecular complexes as they exist in solution.

Sphingolipids: Lipids with ceramide as their core structure.

Sphingosine: Long-chain amino alcohol (forms ceramide when in amide linkage with a fatty acid).

Sugar: A generic term often used to refer to any carbohydrate, but most frequently to low-molecular-weight carbohydrates that are sweet in taste. Table sugar, sucrose, is a nonreducing disaccharide (Fruβ2-1αGlc). Oligosaccharides are sometimes called "sugar chains" and individual monosaccharides in a sugar chain are sometimes referred to as "sugar residues."

Surface plasmon resonance (SPR): An optical technique for the measurement of adsorption of materials onto a surface. Frequently used to estimate affinities of proteins for glycans on coated chips.

Teichoic acid: A complex polymer consisting of either phosphoglycerol- or phosphoribitol-carrying carbohydrates or amino acids, found on the surface of Gram-positive bacteria.

Time of flight (TOF): An analysis technique used in mass spectrometry to separate species of different mass based on velocity differences for particles with equal kinetic energy.

Transfer nuclear Overhauser effect (trNOE): An NMR technique particularly useful in determining the conformation of glycans noncovalently bound to proteins.

Thrombospondin repeat (TSR): Small protein motif (50–60 amino acids) with six conserved cysteine residues that form three disulfide bonds. Serves as a site for glycan modification.

Transcriptome: The total collection of RNA transcripts in a cell, tissue, or organism, under specific conditions of time, space, and environment.

Undecaprenol (bactoprenol, C55 isoprenoid): A polyisoprenoid lipid carrier for membrane-bound glycan synthesis in bacteria.

X-ray crystallography: A structure determination technique dependent on scattering of X-rays from molecules ordered in crystals. Useful in the determination of three-dimensional structures of protein–glycan complexes.

Study Guide

CHAPTER 1: HISTORICAL BACKGROUND AND OVERVIEW

1. What factors have deterred the integration of studies of the biology of glycans ("glycobiology") into conventional molecular and cellular biology?
2. Why has evolution repeatedly selected for glycans to be the dominant molecules on all cell surfaces?
3. Explain how extracellular and nuclear/cytosolic glycans differ from one another.
4. What are the various factors that can affect glycan composition and structure on cell-surface and secreted molecules?
5. Discuss the various ways in which glycans participate in critical intracellular functions.

CHAPTER 2: MONOSACCHARIDE DIVERSITY

1. What are the nine most common monosaccharides found in vertebrate glycans?
2. Define the following terms: D- and L-stereochemistry, epimer and anomer, axial and equatorial, reducing end and nonreducing end, and α- and β-linkages.
3. α-Glycosides of glucose position the aglycone group in the axial orientation, whereas α-glycosides of sialic acid position this group in the equatorial orientation. Explain this apparent discrepancy by applying the definitions of α- and β-anomeric stereochemistry to these two monosaccharides.
4. In nature, D-galactose can be converted to L-galactose in just two enzymatic steps. Using Fischer projections, show the chemical transformations that can accomplish this two-step interconversion.
5. Based on the atoms and functional groups within monosaccharides, describe the ways in which they might interact with proteins (e.g., electrostatic interactions, hydrogen bonding, van der Waals forces, and hydrophobic interactions).

CHAPTER 3: OLIGOSACCHARIDES AND POLYSACCHARIDES

1. If there are 21 amino acids and only 10 major monosaccharides in eukaryotes, why are there so many more possible combinations of monosaccharides in a hexasaccharide than amino acids in a hexapeptide?
2. What amino acid serves as the aglycone of an N-linked oligosaccharide?
3. What is the repeating unit of chondroitin sulfate?
4. What unique structural characteristic contributes to the flexibility of heparan sulfate but not chondroitin sulfate?
5. Name three types of bacterial polysaccharides that serve as antigens.

CHAPTER 4: CELLULAR ORGANIZATION OF GLYCOSYLATION

1. Consider the advantages and disadvantages of topologically restraining glycosylation to the ER/Golgi compartments.
2. What are the differences between physical and functional localization of glycan-modifying enzymes?
3. Describe mechanisms that determine Golgi localization of transferases.
4. Explain how localization of transferases can affect glycan composition of cell-surface and secreted molecules.
5. Propose functions for secreted soluble glycosyltransferases or sulfotransferases generated from membrane-bound enzymes.

CHAPTER 5: GLYCOSYLATION PRECURSORS

1. "Essential" monosaccharides are defined as those that an organism cannot make de novo. Are there essential monosaccharides in mammals?
2. Why do animals usually not require mannose, fucose, or galactose in the diet? In what situations would an individual require dietary supplementation of any of these sugars?
3. Why are nucleotide sugar transporters required in ER and Golgi membranes? What might be the outcome of congenital mutations of nucleotide sugar transporters?

4. Why would humans fail to metabolize cellulose as a source of energy? How do cows and other ruminants metabolize cellulose?
5. Multiglycosyltransferase complexes may exist in the Golgi apparatus. How might such multienzyme complexes affect glycan synthesis?

CHAPTER 6: GLYCOSYLTRANSFERASES AND GLYCAN-PROCESSING ENZYMES

1. Explain how glycosyltransferases achieve a strict donor substrate specificity.
2. Glycosyltransferases and glycosidases have been assigned to families in the CAZy database. What is the basis for assignment to a particular family?
3. Give an example of a glycosyltransferase that recognizes acceptor substrates possessing a specific peptide sequence motif or protein domain. Explain why this glycosyltransferase may have evolved to possess such acceptor specificity.
4. What is meant by the terms "inverting" and "retaining" when used to describe a glycosidase or glycosyltransferase? In mechanistic terms, what do we know about how inverting glycosyltransferases catalyze their reactions?
5. Why is the K_m of a glycosyltransferase, for both its donor and acceptor substrates, an important parameter in establishing what glycan structures are produced by a cell?

CHAPTER 7: BIOLOGICAL FUNCTIONS OF GLYCANS

1. What are the different ways in which glycans can mediate or modulate biological functions?
2. Explain the difference between intrinsic and extrinsic functions of glycans.
3. Why are the biological consequences of altering glycosylation in cultured cells and intact animals so variable?
4. Given intra- and interspecies variations in glycosylation, how can one narrow down critical functions?
5. Why does it appear that some glycans may not have specific functions when their assembly is genetically determined?

CHAPTER 8: A GENOMIC VIEW OF GLYCOBIOLOGY

1. What is a sequence-based classification of glycosyltransferases?
2. Describe the ways in which gene sequences predict or fail to predict functionality in transferases, hydrolases, and glycan-binding proteins.
3. Give examples of bifunctional enzymes involved in glycosylation. Suggest the driving force for the evolution of bifunctional transferases.
4. What can you learn about the way of life of an organism ("ecology") based on the relative number of glycosidase and glycosyltransferase genes in its genome?
5. How could an organism effectively augment the number of glycosidases and glycosyltransferases at its disposal?

CHAPTER 9: N-GLYCANS

1. What are some advantages for a glycoprotein in having a large number of N-glycosylation sites?
2. Consider the topology of N-glycosylation and provide possible explanations for segregating the formation of $Man_5GlcNAc_2$-Dol from the formation of $Glc_3Man_9GlcNAc_2$-Dol.
3. How is N-glycan biosynthesis different in yeast, invertebrates, plants, and mammals?
4. What is N-glycan microheterogeneity? What might be some advantages of N-glycan microheterogeneity?
5. Describe how the branching of N-glycans can regulate growth factor signaling.

CHAPTER 10: O-GalNAc GLYCANS

1. What are the factors that determine the O-GalNAc glycan composition of a cell?
2. What characteristics make a polypeptide a good acceptor for O-GalNAc glycosylation? Can you predict sites of O-GalNAc glycosylation based on these characteristics?
3. How does the assembly of O-GalNAc glycans differ from the assembly of N-glycans?
4. Explain the most important functional features of a typical secreted mucin.
5. What are the advantages of having so many polypeptide-N-acetylgalactosaminyltransferases?

CHAPTER 11: GLYCOSPHINGOLIPIDS

1. The lipid moiety of glycosphingolipids, ceramide, endows them with self-associating properties in the plane of the membrane. Explain why.

2. Glycosphingolipids function via *cis* regulation and *trans* recognition. Explain these terms and provide examples of each.
3. Some humans and experimental animals with mutations in the enzyme responsible for glucosylceramide breakdown suffer from dehydration. Explain why.
4. Several human lysosomal storage diseases result from mutations in the enzymes responsible for breakdown of glycosphingolipids leading to toxic buildup of the uncleaved substrate. Similar buildup of glycosphingolipids sometimes occurs even when there is ample enzyme present. Explain why.
5. The animal brain is enriched in glycosphingolipids compared to other glycan classes. Describe two distinct structural classes of brain glycosphingolipids and two examples of their physiological functions.

CHAPTER 12: GLYCOSYLPHOSPHATIDYLINOSITOL ANCHORS

1. What do GSLs and GPI anchors have in common? How do they differ?
2. Describe differences in the behavior of proteins that have transmembrane domains from those with GPI anchors.
3. Explain how GPI-anchored proteins might facilitate signal transduction across the plasma membrane.
4. Devise an assay to measure the distribution of GPI anchor intermediates across the ER membrane and the mechanism for flipping intermediates across the ER.
5. Why are clinical symptoms of diseases caused by glycosylphosphatidylinositol biosynthetic defects so variable?
6. How is the GPI biosynthetic pathway in *Trypanosoma brucei* or yeast different from that in humans? What implications does elucidation of these differences have for our understanding and manipulation of the GPI pathway?

CHAPTER 13: OTHER CLASSES OF EUKARYOTIC GLYCANS

1. Propose a mechanism that could explain how altering the glycosylation of Notch affects the binding of different Notch ligands.
2. Why do you think a protein like ADAMTS13, which has eight tandem TSRs, requires POFUT2 for proper folding?
3. What advantage is there in the O-mannose matriglycan on α-dystroglycan compared to simpler O-mannose glycans?
4. What effects could O-glycosylation of collagens have on their folding and/or structure?
5. C-mannosylated tryptophan was first detected in human urine. Provide an explanation for why this amino acid glycoside was excreted into the urine rather than as free tryptophan and mannose.

CHAPTER 14: STRUCTURES COMMON TO DIFFERENT GLYCANS

1. Propose a function for the allelic variation observed in the ABO blood group system. Nonprimates do not express the ABO locus—how does this affect your answer?
2. Hyperacute (graft) rejection (HAR) occurs after transplantation of organs from nonhuman donors into humans and results from an immediate reaction of circulating anti-Galα1-3Gal antibodies with the transplanted tissue. Suggest ways to modify the donor or recipient to prevent HAR.
3. Compare and contrast "LacNAc" and "LacdiNAc" units. How does the presence of these terminal disaccharides affect the addition of sialic acid and fucose?
4. Based on what you know about terminal structures on follicle-stimulating hormone and lutropin, propose several glycan-based mechanisms that could account for infertility in humans.
5. Certain strains of *Escherichia coli* bind to P blood group antigens and cause urinary tract infections. What evolutionary advantage might exist for retaining the transferases that make a deleterious glycan?

CHAPTER 15: SIALIC ACIDS AND OTHER NONULOSONIC ACIDS

1. Compare and contrast the structure of sialic acids with other monosaccharides.
2. What advantages does sialic acid diversity provide in vertebrate systems?
3. What are the unique features of the sialic acid biosynthetic pathways in comparison to those of other monosaccharides?
4. How would you determine if a previously unstudied organism contains sialic acids?
5. Contrast the addition of α2-6-linked sialic acids to O-GalNAc glycans and N-glycans and their recognition by sialic acid–binding lectins.

CHAPTER 16: HYALURONAN

1. Why do small molecules diffuse readily through a high-molecular-weight hyaluronan (HA) solution such as the vitreous of the eye, whereas larger macromolecules (e.g., certain proteins) do not?
2. What are some of the dominant physical and molecular factors that influence diffusion rate through HA solution, and how might the rate be different in a purely HA matrix or in a heterogeneous extracellular matrix consisting only partially of HA?
3. HA solutions have unusual viscoelastic properties; for example, HA acts like a gel, yet it can function as a lubricant. How do you explain these properties in terms of the molecular structure of the chains?
4. Why are HA-binding proteins considered lectins, but proteins that bind to sulfated glycosaminoglycans are not? How do these two classes of glycan-binding proteins differ?
5. How could a cell-surface HA receptor (e.g., CD44) respond differently to HA oligosaccharides with 6–10 sugar units compared to high-molecular-weight HA? Why might different responses need to be elicited through the same receptor?
6. How would you demonstrate whether an HA chain assembles from the reducing end versus the nonreducing end?
7. How would you demonstrate hyaluronidase activity of a new putative hyaluronidase family member in vivo or in a cellular context? Specifically, how could you distinguish HA degradation from HA clearance if it occurred intracellularly?
8. High-molecular-weight HA has been shown to have tissue-protective effects in lung pathologies, but the presence of high-molecular-weight HA is also obstructive to lung function. How can these observations be reconciled? Is there a way to overcome these contrasting impacts of HA so it can be used effectively as a therapeutic?

CHAPTER 17: PROTEOGLYCANS AND SULFATED GLYCOSAMINOGLYCANS

1. What factors can affect the fine structure of sulfated glycosaminoglycans in cells?
2. Overexpression of Ext2 (which is part of the heparan sulfate copolymerase complex) increases the extent of sulfation of the chain. Provide an explanation for this finding.
3. Compare and contrast the biological functions of GPI-anchored proteoglycans from those that contain transmembrane domains.
4. Give examples of ways to modify the metabolism of glycosaminoglycans in cells and animals.
5. What are the options for generating drugs based on glycosaminoglycan–protein interactions?

CHAPTER 18: NUCLEOCYTOPLASMIC GLYCOSYLATION

1. What biochemical criteria would you require to demonstrate the attachment of a glycan to a specific nuclear or cytoplasmic protein?
2. What conventional glycosylation pathways have steps that occur on the cytoplasmic side of membranes that could be a source of nucleocytoplasmic glycans?
3. Compare and contrast the initiating glycosylation reactions on mucins, proteoglycans, Notch, glycogenin, and Skp1.
4. How would you demonstrate the presence of glycosaminoglycans in the nucleus?
5. Give examples of glycoconjugates that are initially formed in the cytoplasm but later transit to and function at the cell surface or in the extracellular space.

CHAPTER 19: THE O-GlcNAc MODIFICATION

1. O-GlcNAc is now known to be the most common form of glycosylation in the cell. Why did it take so long for this fact to be appreciated? What was the serendipity involved in its discovery?
2. O-GlcNAc is thought to compete with phosphorylation for the same or similar sites on nuclear or cytoplasmic glycoproteins. What are the similarities and differences between O-GlcNAcylation and phosphorylation?
3. What are the mechanistic differences between O-GlcNAc glycosylation and cell-surface glycosylation?
4. How does O-GlcNAc act as a "metabolic sensor"?
5. Speculate as to how O-GlcNAc might contribute to "glucose toxicity" in diabetes.

CHAPTER 20: EVOLUTION OF GLYCAN DIVERSITY

1. What processes could maintain glycan gene polymorphisms (i.e., structural heterogeneity) within populations?
2. What changes in sialic acid biology occurred during human evolution?
3. Is it possible to predict glycan function by examining glycan composition across phylogeny?

4. What are the problems in using "comparative glycobiology" for determining evolutionary relationships (phylogeny)?
5. Which glycosylation pathways support a common origin of eukaryotes?

CHAPTER 21: EUBACTERIA

1. Plants, bacteria, and yeast all have cell walls that provide resistance to osmotic pressure. Compare the composition and architecture of these barriers.
2. Both bacteria and animal cells utilize polyisoprenoids for the assembly of glycans. Compare and contrast these lipid intermediates.
3. Compare the structure of lipopolysaccharide to glycerolipids and gangliosides.
4. Cell wall biosynthesis in Gram-negative bacteria requires a coordinated synthesis of peptidoglycan and LPS. Propose potential regulatory mechanisms that ensure homeostasis.
5. Compare the architecture of the mycobacterial and Gram-negative cell wall.

CHAPTER 22: ARCHAEA

1. Compare and contrast the pathways of glycoprotein N-glycosylation in Archaea, Bacteria, and eukaryotes.
2. All cells produce acidic glycans, but the source of the negative charge varies. What are the acidic groups on the glycans present in *Escherichia coli*, Archaea, yeast, and animal cells?
3. Compare the S-layer in Archaea with surface glycoproteins in eukaryotic cells.
4. Search for molecular similarities in the extracellular matrix of eukaryotes and the archaeal cell wall.
5. Compare the bacterial murein and the archaeal pseudomurein.

CHAPTER 23: FUNGI

1. Compare the composition and structure of yeast cell walls and the envelope of Gram-negative bacteria.
2. What changes in the yeast cell wall might occur in a mutant that produces less β-glucan? What effects might an abnormal cell wall have on the shape, growth, or viability of this mutant?
3. Compare and contrast N-glycan synthesis in yeast and mammals. What is the functional significance of the differences?
4. Describe a unique feature of GPI-linked proteins in fungi. How does this process change protein localization in these organisms?
5. A pharmaceutical company has hired you to assess glycan synthesis as a target for drug development to combat a newly described and highly virulent pathogenic fungus. Describe a set of reasonable targets and some important issues you need to consider.

CHAPTER 24: VIRIDIPLANTAE AND ALGAE

1. Why do plants that do not express sugars present in animal cells (e.g., sialic acids) have lectins that bind to glycans containing these sugars?
2. Pectins in plants are sometimes compared to glycosaminoglycans in animals. How do they differ? How are they similar?
3. Why are recombinant mammalian glycoproteins generated in plants immunogenic?
4. Compare the structures of glycoglycerolipids in plants, lipid A in bacteria, and glycosphingolipids in animals.
5. Elicitors and Nod factors are active at very low concentration, and therefore one might predict that their affinity for their signal-transducing receptors would be very high (in the pM range). Based on what you know about other glycan-binding proteins, how would such high affinity be achieved?

CHAPTER 25: NEMATODA

1. Propose some evolutionary forces driving the large expansion of some glycosyltransferase families in *Caenorhabditis elegans* (e.g., fucosyltransferases) compared with others (e.g., mannosyltransferases).
2. Compare and contrast chondroitin proteoglycan synthesis in *C. elegans* and in vertebrates.
3. How would you go about selecting mutants of *C. elegans* defective in N-glycan formation?
4. In contrast to vertebrate systems, O-GlcNAc addition to nuclear and cytoplasmic proteins is dispensable in *C. elegans*. How do you explain this finding?
5. Given the absence of sialic acids in *C. elegans*, what might you predict about the types and specificity of glycan-binding proteins in *C. elegans*?

CHAPTER 26: ARTHROPODA

1. Compare and contrast what happens to the first *N*-acetylglucosamine residue attached to the mannosyl core of an N-glycan in Drosophila, *Caenorhabditis elegans*, and vertebrates.
2. Compare the structural differences in the O-glycan modifications of Drosophila Notch with those of vertebrate Notch EGF repeats. Why is Notch glycosylation in Drosophila less complex than in vertebrates?
3. Compare the core structure of glycosphingolipids in Drosophila with those present in *C. elegans* and vertebrates. How do the outer chains differ?
4. Transgenic expression of a β1-4 galactosyltransferase substitutes for Egghead (*egh*), which is a mannosyltransferase. What does this tell you about the function of the glycans present in Drosophila glycosphingolipids?
5. Explain how the overexpression or deletion of *dally*, a glypican homolog, can reduce the diffusion of a morphogen, such as dpp.

CHAPTER 27: DEUTEROSTOMES

1. In studying the glycoproteins that mediate sperm–egg interactions during fertilization, why is it important to use several model animals?
2. If you were an enzymologist, how would you study the synthesis of fucose sulfate polymers?
3. Sulfated fucans are also extremely potent inhibitors of coagulation and inflammation in mammalian systems. Propose a mechanism for this action based on the similarity of their structure to other bioactive glycans.
4. Why do some glycan-related gene knockouts in laboratory mice exhibit no obvious phenotype?
5. If you were to discover a new glycan in humans, which model organism(s) would you pick for further studies and how would you manipulate it genetically?

CHAPTER 28: DISCOVERY AND CLASSIFICATION OF GLYCAN-BINDING PROTEINS

1. How are sulfated glycosaminoglycan-binding proteins distinguished from lectins?
2. Suppose you discovered a new glycan-binding protein. How would you determine its classification?
3. Compare and contrast the functions of soluble and membrane-bound lectins.
4. Contrast the functions of animal lectins that recognize self and non-self glycans.
5. Compare the methods for characterizing glycan-binding proteins in organisms with well-annotated whole genomes with those from organisms in which whole-genome sequences are unavailable.

CHAPTER 29: PRINCIPLES OF GLYCAN RECOGNITION

1. What determines the affinity of a glycan for a GBP?
2. Many types of protein–glycan interactions are low affinity, and in some cases high avidity is achieved by clustering receptors and ligands. What are the advantages and disadvantages of achieving high-affinity interactions through multivalency?
3. How does the density of glycan ligands affect binding of a GBP? Is this relevant in vivo?
4. Provide examples of GBPs that bind with relatively low affinity to highly abundant glycans and other GBPs that bind with relatively high affinity to glycans that are scarce.
5. To measure the binding kinetics and/or affinity of a GBP to a glycan, there are several techniques, including isothermal titration calorimetry and surface plasmon resonance. Choose one of these or other techniques and design an experiment to measure the K_a of binding, assuming the glycan is easy to derivatize, if needed, at its reducing end.

CHAPTER 30: STRUCTURAL BIOLOGY OF GLYCAN RECOGNITION

1. Cholera toxin binds to the ganglioside GM1 with high affinity ($K_d \sim 0.1$ nM) relative to the binding of many other GBPs to their ligands (which exhibit K_d values in the range of 0.1 μm to 0.1 mm). How do you explain this observation?
2. Name four types of molecular interactions important in carbohydrate recognition.
3. What amino acid residues are likely to play important roles in binding highly sulfated glycosaminoglycans?
4. What NMR experiment can return information on glycan geometry as it exists in the protein-bound state?
5. Name a database in which you can find structures of GBPs.

CHAPTER 31: R-TYPE LECTINS

1. Describe the differences and similarities between *Ricinus communis* agglutinin-I and ricin.

2. For ricin and other ribosome-inactivating toxins to kill cells, they must first gain access to the cytoplasm. How does this occur? How would you exploit this mechanism to deliver cargo to different sites in a cell?
3. Explain how a cell that becomes resistant to one type of toxic lectin could become sensitive to another.
4. What are the functions of R-type lectin domains found in enzymes such as glycosyltransferases and glycosidases?
5. Describe examples of animal lectins in cells that engage glycan ligands in both *cis* and *trans* topologies.

CHAPTER 32: L-TYPE LECTINS

1. Describe possible functions for L-type plant lectins present in the seeds of leguminous plants.
2. If L-type lectins are involved in defense, why does each plant produce only a very limited number of lectins?
3. Why are both plant seed lectins and GBPs involved in protein quality control classified as L-type lectins?
4. Compare and contrast the "jelly-roll" fold in L-type lectins, the C-type lectin fold, and the link module.
5. Plant lectins are typically glycoproteins and therefore mature through the ER/Golgi secretory pathway. Propose a mechanism to prevent their interaction with other Golgi glycoproteins during their assembly and secretion.

CHAPTER 33: P-TYPE LECTINS

1. Why was it important to use a double-labeled substrate donor [β-^{32}P]UDP[^{3}H]GlcNAc in studies of Man-6-P recognition marker biosynthesis?
2. Compare and contrast the process of assembling the Man-6-P recognition marker on lysosomal enzymes via formation of GlcNAc-P-Man and subsequent removal of the *N*-acetylglucosamine moiety versus a mannose-specific ATP-dependent kinase.
3. The Man-6-P recognition marker assembles mainly on lysosomal enzymes by selective recognition of peptide determinants in the substrate proteins by GlcNAc-P-transferase. Describe other examples of selective modification of glycans on subsets of glycoproteins. How do the recognition determinants differ?
4. How would the number of N-glycans on a lysosomal enzyme affect its affinity for one of the Man-6-P receptors?
5. Compare and contrast the packaging of the Man-6-P receptors into clathrin-coated vesicular carriers at the *trans*-Golgi network versus the cell surface.

CHAPTER 34: C-TYPE LECTINS

1. Many proteins that contain C-type lectin domains do not bind glycans, and the ones that do are called C-type lectins. What is the difference in structure that distinguishes these two classes of proteins?
2. Why is it difficult to predict the type of glycan to which a C-type lectin will bind?
3. Some C-type lectins can form oligomers, which greatly increase the avidity of interactions with glycan ligands. Explain how oligomerization can also affect the specificity of the interaction.
4. Some C-type lectins, notably the selectins, bind with higher affinity to some glycoproteins than to others on the same cell, even though several glycoproteins may display similar glycan structures. Consider mechanisms that confer such preferential binding.
5. Compare the interaction of P-selectin with PSGL-1 to the binding of a plant lectin to PSGL-1.

CHAPTER 35: I-TYPE LECTINS

1. There are now more than a dozen human Siglecs known. Why were these and other sialic acid–binding proteins not discovered until relatively recently?
2. Compare the potential function of Siglecs with inhibitory motifs in their cytosolic tails with those that can recruit activating motifs.
3. Why are Siglec homologs found primarily in "higher" animals?
4. Explain the likely mechanisms and driving forces for the rapid evolution of some Siglecs.
5. Why do plants and invertebrates that do not express sialic acids have sialic acid–binding proteins?

CHAPTER 36: GALECTINS

1. How do you explain the finding that galectins are not routinely found in large amounts in body fluids, even though most of them are soluble proteins and are often found extracellularly?
2. Why do changes in glycan branching pathways and sialylation have the potential to impact galectin function?

3. How do galectins achieve high-affinity binding to cell-surface glycans? How do galectins form lattices with cell-surface glycans?
4. Explain how a galectin, as an innate immune effector, might act as a receptor to fight microbial infection.
5. Galectins bind to a variety of cells and trigger various responses in different cell types. How do galectins send signals through cell-surface receptors?

CHAPTER 37: MICROBIAL LECTINS: HEMAGGLUTININS, ADHESINS, AND TOXINS

1. What kinds of cytoplasmic glycosylation events are associated with infection and pathology?
2. Compare the carbohydrate-recognition domains of bacterial and viral adhesins to those of animals and plant lectins.
3. What agents other than simple sugars could be used for anti-adhesion therapy of microbial diseases?
4. A serious problem limiting the use of antibiotics is the rapid emergence of resistant bacteria. To what extent could this also become a problem with anti-adhesion therapy?
5. Multivalent and polyvalent sugars are more powerful inhibitors of microbial lectins than simple monomeric ones. Explain the reasons for this phenomenon and discuss its applications.

CHAPTER 38: PROTEINS THAT BIND SULFATED GLYCOSAMINOGLYCANS

1. Proteins that bind to sulfated glycosaminoglycans (GAGs) are not considered lectins. Why?
2. The extent of modification of heparin is much greater than that of heparan sulfate. How would this affect conformation and the interaction of GAG-binding proteins?
3. What are the main types of bonding forces that contribute to GAG-protein interactions?
4. Interactions between proteins and sulfated glycosaminoglycans are important in various physiological and pathophysiological settings. Are they specific?
5. Explain how HS plays similar roles in promoting antithrombin–thrombin interactions and FGF–FGFR interactions?

CHAPTER 39: GLYCANS IN GLYCOPROTEIN QUALITY CONTROL

1. What are the prerequisites for protein-bound glycans to function as signaling molecules in protein folding and quality control?
2. Describe the types of chaperones present in the ER.
3. The addition and removal of glucose residues constitutes part of the quality control system for monitoring protein folding. What is the role of mannose trimming?
4. How is the ER stress response (ERAD) coordinated with N-glycan synthesis?
5. Compare the processing of N-glycans in the ER and Golgi with degradative pathways for N-glycans in lysosomes and in the cytoplasm.

CHAPTER 40: FREE GLYCANS AS BIOACTIVE MOLECULES

1. What are the advantages of using glycans derived from host organisms as signals of danger?
2. How can glycans mediate the interaction between nonglycan signals and their receptors?
3. Can you think of disadvantages of the use of glycans as pathogen-associated molecular patterns (PAMPs) by host immune systems?
4. What structural modifications of generic chitin oligosaccharide Nod factors provide host specificity in the bacterial–plant interactions that lead to root nodule formation and nitrogen fixation in legumes?
5. How can you explain the observation that different β-glucans trigger plant immune systems through different PRRs?

CHAPTER 41: GLYCANS IN SYSTEMIC PHYSIOLOGY

1. If glycans have roles in almost every aspect of systemic physiology, why can loss of a glycosyltransferase and subsequent alteration of glycan structure in some cases have no obvious effect on development or physiology?
2. Explain the evidence supporting the idea that glycans and GBPs are involved in immune responses.
3. How does differential glycosylation regulate the plasma half-life of glycoproteins, and where are these glycoproteins cleared?
4. What are the different functions of glycans on mucins on different epithelial surfaces?
5. How do glycans regulate neuronal growth and repair?

CHAPTER 42: BACTERIAL AND VIRAL INFECTIONS

1. How can bacteria benefit by coating their surface with a polysaccharide capsule?
2. How do pathogenic bacteria initially colonize tissues?
3. Would a mouse lacking Toll-like receptor 4 be more or be less susceptible to bacterial infection? What about susceptibility to lipopolysaccharide-induced sepsis?
4. How do influenza and herpes simplex virus engage the host cell surface to initiate infection?
5. Can one manipulate glycans to prevent or treat microbial infection?

CHAPTER 43: PARASITIC INFECTIONS

1. Explain the role of glycoconjugates in the high fever typically associated with the pathogenesis of malaria.
2. How do African trypanosomes avoid destruction by the immune system after inoculation by the bite of the tsetse fly?
3. What is the mechanism by which the protozoan parasite *Leishmania* attaches and eventually detaches from its sandfly vector midgut during transmission?
4. Many glycans made by the parasitic worm *Schistosoma mansoni* are highly antigenic in the infected hosts. What property of these glycans makes them so antigenic, and would this offer a possibility to make a vaccine?
5. What glycosyltransferases and sugar/nucleotide sugar transporters may be unique to parasites and therefore potential targets for chemotherapeutic intervention?

CHAPTER 44: GENETIC DISORDERS OF GLYCAN DEGRADATION

1. Predict which glycans and tissues/organs would be affected most if β-galactosidase was altered.
2. In lysosomal storage disorders, undegraded or partially degraded glycans and glycopeptides are often excreted in the urine. Propose a mechanism for how these partial degradation products escape from lysosomes and cells.
3. Provide possible explanations for the accumulation of glycopeptides with O-glycans in the urine of patients deficient in α-N-acetylgalactosaminidase.
4. How do multivesicular bodies arise and what purpose do they serve?
5. It would seem counterintuitive to use an enzyme inhibitor as a molecular chaperone to restore enzyme activity in a lysosomal storage disorder. Explain the rationale behind this therapeutic approach.

CHAPTER 45: CONGENITAL DISORDERS OF GLYCOSYLATION

1. How do you define a "glycosylation" disorder? Describe the methods used today to identify a glycosylation disorder.
2. Serum transferrin has two N-glycosylation sites, and each glycan consists of biantennary sugar chains with sialic acid. What kinds of glycan patterns would you predict in patients with congenital disorders of glycosylation (CDGs)?
3. What types of cells might be especially susceptible to loss of heterozygosity or spontaneous mutations that cause glycosylation disorders?
4. Explain how "gain-of-function" mutations can cause a glycosylation disorder.
5. How would you assess the genetic and environmental contributions to a glycosylation disorder?

CHAPTER 46: GLYCANS IN ACQUIRED HUMAN DISEASES

1. What are the common underlying mechanisms for the roles of selectins in various diseases?
2. Although heparin is primarily used as an anticoagulant, its use has been proposed in connection with several other diseases. How can one drug have relevance to so many different mechanisms?
3. Give two examples in which altered glycosylation has resulted in acquired blood cell diseases involving the hematopoietic stem cell. Why is it possible for somatic mutations to give rise to a phenotype?
4. Describe the common underlying molecular mechanism that causes changes in O-glycans in blood cell diseases, in IgA nephropathy, and in the altered glycosylation of cancer.
5. Describe how pathogens exploit host glycans in establishing gastrointestinal and urinary tract infections.

CHAPTER 47: GLYCOSYLATION CHANGES IN CANCER

1. Explain why many cancer-specific markers detected by monoclonal antibodies turn out to be directed against glycan epitopes.
2. Many cancer cell types exhibit altered branching of N-glycans, excessive expression of mucins, changes in

hyaluronan production and turnover, and decreased expression and sulfation of heparan sulfate. Discuss how these changes come about and how they would affect cancer growth and metastasis.

3. Sialyl-Tn expression is a prominent feature of many carcinomas. What explains the high frequency of this expression despite the fact that the enzyme responsible for its synthesis is not always up-regulated?
4. Consider the potential roles of selectins and selectin ligands in cancer progression and metastasis.
5. What are the potential ways in which alterations in glycan structure could be used advantageously for diagnosing or treating cancer?

CHAPTER 48: GLYCAN-RECOGNIZING PROBES AS TOOLS

1. What are the advantages and disadvantages of using monoclonal antibodies versus plant lectins for determining the presence or absence of glycans in a preparation?
2. What are important controls when using lectins or anti-glycan antibodies to determine the presence or absence of a glycan in a tissue, on a cell, or in a mixture of glycans?
3. Select from the large number of available lectins a subset that would allow you to determine the relative amounts of oligomannosyl, hybrid, and complex type N-glycans in a preparation.
4. Propose methods for using a monoclonal antibody to a glycan determinant for the isolation of a mutant cell line deficient in the expression of the glycan.
5. By observing gene homology, you suspect that insects produce a novel β-glucuronidase that acts on terminal glucuronic acid residues present in insect glycans. Propose a nonradioactive method to measure the activity of this enzyme in cell extracts.

CHAPTER 49: GLYCOSYLATION MUTANTS OF CULTURED MAMMALIAN CELLS

1. What are the advantages and disadvantages of isolating mutants in cultured cell lines compared to deriving cell lines from mutant animals or humans afflicted with glycosylation disorders?
2. Discuss the advantages and disadvantages of different schemes used to isolate mutants (i.e., selection with lectins or toxins, using gene editing strategies, selection by complement-mediated lysis, screening by replica plating, and sorting by flow cytometry).
3. How might you use ldlD cells, in the presence and absence of galactose and *N*-acetylgalactosamine, to test the role(s) of glycans in biological processes?
4. Describe various types of gain-of-function glycosylation mutations. Consider mutations that create protein glycosylation sites as well as those that change the expression of glycosylation genes.
5. Propose a method to identify animal cell mutants blocked in the synthesis of O-mannose glycans.

CHAPTER 50: STRUCTURAL ANALYSIS OF GLYCANS

1. Describe a selective way of removing N-glycans and a selective way of removing O-glycans from a glycoprotein.
2. An oligosaccharide has a molecular weight of 972, and yet its NMR spectrum is that of a single monosaccharide of α-glucose. Methylation analysis yields a single product, methylated at the C-2, C-3, and C-6 positions. What is the glycan structure?
3. What characteristic in an NMR spectrum allows distinction of anomeric configuration in a linkage?
4. Describe a nonradioactive tagging procedure that allows sensitive detection of glycans in HPLC and TLC applications.
5. Name two types of ionization methods and two types of mass separation methods used in mass spectrometers.

CHAPTER 51: GLYCOMICS AND GLYCOPROTEOMICS

1. What is the "glycome" of an organism? Does it differ for individual cells in that organism?
2. What information from the genome and the proteome might be useful in predicting a cell's glycome?
3. What is the difference between glycomics and glycoproteomics?
4. Propose an experimental strategy to characterize the sequence and linkages of different glycan subtypes that comprise the glycome. For example, how might protein-associated N- and O-glycans be structurally characterized?
5. How can mass spectrometry help to characterize the sites of glycosylation on a protein?

CHAPTER 52: GLYCOBIOINFORMATICS

1. What are the limitations of obtaining a complete database of glycan structures?

2. Why are standards initiatives such as MIRAGE (Minimum Information Required for A Glycomics Experiment) required for glycobioinformatics resources?
3. Is it possible to have different unique identifiers for the same monosaccharide composition?
4. What aspects of genomics and proteomics databases could be linked to glycomics databases?
5. What types of software tools are required for glycomics experimental data analysis?

CHAPTER 53: CHEMICAL SYNTHESIS OF GLYCANS AND GLYCOCONJUGATES

1. β-Glucosides are readily synthesized by exploiting protecting groups at C-2 capable of neighboring group participation. Without such protecting groups, the preferred product in most chemical glucosylation reactions is the α-glucoside. Explain this finding.
2. Why are β-mannosides so difficult to generate chemically?
3. In solid phase synthesis of glycans, glycosidic bonds are most often constructed with the glycosyl acceptor bound to the solid support and the activated glycosyl donor in solution. Why is this situation preferred to the alternative approach in which the glycosyl donor is bound to the solid support?
4. Benzyl groups are often used as protective groups to mask those alcohol functions that are unmodified in the final oligosaccharide/glycoconjugate synthesized. Explain why.
5. Using solid phase glycan synthesis, repeating oligosaccharides can be synthesized of a size exceeding the size that can be attained in solution. Explain why, also considering the intrinsic advantages solid phase peptide synthesis offers in comparison with solution phase peptide synthesis.

CHAPTER 54: CHEMOENZYMATIC SYNTHESIS OF GLYCANS AND GLYCOCONJUGATES

1. We think of glycosidases as enzymes that cleave rather than synthesize glycosidic bonds. How are the substrates and reaction conditions of glycosidases manipulated to convert them from degrading enzymes to synthetic enzymes?
2. Enzymatic synthesis of glycans can be far more efficient than chemical synthesis of the same structures, but production of large quantities of a glycan requires significant amounts of the required glycosyltransferases or glycosidases. Pick a source of enzymes and explain why you think it is more promising with respect to production of specific glycan products in large quantity.
3. In a transglycosylation event, the equilibrium needs to be shifted from hydrolysis to glycosidic bond formation. Explain how this can be done, taking into consideration the reaction conditions (substrates, solvents).
4. Transglycosylation also happens in nature. Which of the two classes of glycosidases—inverting glycosidases or retaining glycosidases—would be more prone to give transglycosylation?
5. Solution phase glycan synthesis can be merged with enzymatic synthesis to arrive at complex glycan structures. What would be the requirements to combine solid phase glycan synthesis with enzymatic glycan synthesis?

CHAPTER 55: CHEMICAL TOOLS FOR INHIBITING GLYCOSYLATION

1. Explain how an inhibitor of glutamine:fructose aminotransferase (GFAT) would affect glycosylation?
2. From a mechanistic point of view, how can an alkaloid that inhibits a glycosidase also block a glycosyltransferase?
3. How would you go about obtaining an inhibitor of glycans that are initiated by the addition of O-fucose to EGF repeats in Notch?
4. Propose chemical modifications to galactose to create an inhibitor of sialyltransferases.
5. How can an enzyme inhibitor also act as a chemical chaperone?

CHAPTER 56: GLYCOSYLATION ENGINEERING

1. Why are Chinese hamster ovary (CHO) cells the preferred cell line used by industry to produce recombinant glycoprotein drugs for human use?
2. List important design elements required for engineering cellular glycosylation.
3. Describe differences between N-glycosylation in bacteria, yeast, insect, and mammalian cells that are important for glycosylation engineering.
4. Describe examples of glycosylation engineering in yeast and plant cells for enhanced delivery of lysosomal enzyme replacement.
5. Describe an example of glycoengineering of a recombinant antibody currently in clinical use that affects its function.
6. Describe the main principles of precise genetic editing technologies.

CHAPTER 57: GLYCANS IN BIOTECHNOLOGY AND THE PHARMACEUTICAL INDUSTRY

1. Explain the mechanism of action of influenza neuraminidase inhibitors.
2. Design a glycan-based therapeutic that acts by blocking the interaction of a naturally occurring glycan with GBPs on an intact (or live) microbe.
3. A portion of erythropoietin (EPO) produced by CHO cells is not fully sialylated (i.e., some glycoforms have exposed galactose residues on their N-glycans). What sugars might be added to the cell culture media to increase the overall level of EPO sialylation?
4. Explain how increasing the extent of glycosylation of recombinant glycoproteins can increase their half-life in vivo.
5. Describe the potential deleterious effects of producing recombinant therapeutic proteins in cultured animal cells of nonhuman origin.

CHAPTER 58: GLYCANS IN NANOTECHNOLOGY

1. What is the difference between avidity and affinity? Why is this particularly relevant to glycoconjugates? Design a glycan-based therapeutic that acts by blocking the interaction of a naturally occurring glycan with GBPs on an intact (or live) microbe.
2. Draw schematic representations for all the modes of multivalency that you can envisage between a protein and a glycan or glycoconjugate. Use these representations to show the different modes of interaction that glycodendrimers, glycopolymers, and glyconanoparticles might have with a protein.
3. For each of these glycoconjugate types explain the acronym (where applicable) and then place in typical order of size: glycoQDs, glycoAuNPs, glycoMNPs, glycofullerenes, glycoCNTs, glyco-dendrimers. How might size affect the application of any of these in vivo or in vitro?
4. Give one example each (including key linker structures or bonding patterns) of how platforms in glyconanotechnology can be decorated with glycans either covalently or noncovalently.
5. Name some potential applications of glyconanotechnology to clinical problems.

CHAPTER 59: GLYCANS IN BIOENERGY AND MATERIAL SCIENCE

1. What are the major uses by humans of plant glycans?
2. Discuss the positive and negative impacts of fermenting the starch present in corn grains into bioethanol.
3. What enzymes are likely required for the breakdown of biomass into sugars?
4. What chemical modifications of cellulose would lead to polysaccharide derivatives with enhanced or new properties?
5. Describe some uses for cellulose nanomaterials.

CHAPTER 60: FUTURE DIRECTIONS IN GLYCOSCIENCES

1. Explain to a nonexpert why an understanding of glycosciences is critical to advances in the understanding and treatment of nearly every disease affecting humans.
2. Generalize about what the congenital diseases of glycosylation (CDGs) tell us about the roles of glycans in human development and biology.
3. Describe the possible future value of large-scale glycan arrays in the diagnosis of infectious diseases and in the study of other diseases.
4. Describe how advances in genomics and proteomics have accelerated progress in glycobiology.
5. At the end of this chapter there is a list of some major fundamental questions for the future of glycobiology. Can you think of at least one additional question that is not listed?

Index

A

Page references followed by f denote figures; page references followed by t denote tables.

B

D

E

F

H

M

N

O

P

Q

R

S